NOTICE TO THE READER

Publisher does not warrant or guarantee any of the products described herein or perform any independent analysis in connection with any of the product information contained herein. Publisher does not assume, and expressly disclaims, any obligation to obtain and include information other than that provided to it by the manufacturer.

The reader is expressly warned to consider and adopt all safety precautions that might be indicated by the activities herein and to avoid all potential hazards. By following the instructions contained herein, the reader willingly assumes all risks in connection with such instructions.

The publisher makes no representation or warranties of any kind, including but not limited to, the warranties of fitness for particular purpose or merchantability, nor are any such representations implied with respect to the material set forth herein, and the publisher takes no responsibility with respect to such material. The publisher shall not be liable for any special, consequential, or exemplary damages resulting, in whole or part, from the readers' use of, or reliance upon, this material. Autodesk does not guarantee the performance of the software and Autodesk assumes no responsibility or liability for the performance of the software or for errors in this manual.

Trademarks

Autodesk, the Autodesk logo, and AutoCAD are registered trademarks of Autodesk, Inc. Microsoft and Windows 95 are registered trademarks of the Microsoft Corporation. Windows NT is a trademark of Microsoft Corp. All other product names are acknowledged as trademarks of their respective owners.

COPYRIGHT © 1998
Delmar Publishers Inc.
Autodesk Press imprint
an International Thomson Publishing company
The ITP logo is a trademark under license.
Printed in the United States of America

For more information, contact:

Delmar Publishers/Autodesk Press
3 Columbia Circle, Box 15-015
Albany, New York USA 12212-5015

International Thomson Publishing Europe
Berkshire House 168-173
High Holborn
London, WC1V 7AA
United Kingdom

Thomas Nelson Australia
102 Dodds Street
South Melbourne, Victoria 3205
Australia

Nelson Canada
1120 Birchmont Road
Scarborough, Ontario
Canada, M1K 5G4

International Thomson Publishing Southern Africa
Building 18, Constantia Park
240 Old Pretoria Road
P.O. Box 2459
Halfway House, 1685 South Africa

International Thomson Editores
Campos Eliseos 385, Piso 7
Colonia Polanco
11560 Mexico D. F. Mexico

International Thomson Publishing GmbH
Konigswinterer Strasse 418
53227 Bonn Germany

International Thomson Publishing France
Tour Maine-Montparnasse
33, Avenue du Maine
75755 Paris Cedex 15, France

International Thomson Publishing -Japan
Hirakawacho Kyowa Building, 3F
2-2-1 Hirakawa-cho Chiyoda-ku
Tokyo 102 Japan

International Thomson Publishing Asia
221 Henderson Road
#05-10 Henderson Building
Singapore 0315

All rights reserved. No part of this work covered by the copyright hereon may be reproduced or used in any form or by any means – graphic, electronic, or mechanical, including photocopying, recording, taping, or information storage and retrieval systems without the written permission of the publisher.

4 5 6 7 8 9 10 XXX 03 02

Library of Congress Cataloging-in-Publication Data

Fundamentals of Drafting with AutoCAD® LT
ISBN: 0-538-65982-3

AF778702

Fundamentals of DRAFTING with AutoCAD® LT

2nd Edition

PAUL ROSS WALLACH
DEAN CHOWENHILL
JAMES CULLEN

Autodesk® Press

ITP® An International Thomson Publishing Company

Albany • Bonn • Boston • Cincinnati • Detroit • London • Madrid
Melbourne • Mexico City • New York • Pacific Grove • Paris • San Franc
Singapore • Tokyo • Toronto • Washington

Preface

A Word From the Authors

We would like to welcome you to this new concept for an engineering drawing textbook. With the increasing development of our industry and constant new breakthroughs in technology, there will always be a large demand for qualified engineers and drafters. With this demand, there must be qualified schools and instructional materials to give students the skills they need to compete in today's workplace. We strove to fulfill these needs when we revised *Fundamentals of Drafting with AutoCAD LT*.

Each chapter of the text is designed to teach basic drafting concepts and drafting skills. These fundamental skills and concepts should be taught first with freehand sketching and/or with drawing instruments before you begin working on a computer-aided drafting (CAD) system. Otherwise, the CAD system could become a very expensive and complicated teaching aid.

Previous Edition

Previously titled *Drafting in a Computer Age*, we renamed this text *Fundamentals of Drafting with AutoCAD LT* for several reasons. First, we wanted to reflect the changes in our engineering and drafting industry by emphasizing the importance of drafting in conjunction with using computer-aided design systems. Second, we wanted to give you and your teachers a new tool that would enable you to learn the fundamental skills of drafting first and the transference to a well-known, user-friendly CAD system second. We believe that the elements of this instructional program will help you learn and retain these skills more easily and more assertively.

Second Edition

The second edition of *Fundamentals of Drafting with AutoCAD LT* fulfills the need for an instructional drafting text that will teach the fundamentals of engineering drawing through sketching, instrument drafting, and introductory CAD skills. Some of the special features of this second edition include the following:

- Each chapter opens with objectives and sets the stage for clear and concise learning.
- Over 1,300 illustrations and photographs help clarify the content and aid students in reading two-dimensional and three-dimensional working drawings.
- Step-by-step procedural illustrations take the students through the concepts of drafting, design, and layouts.
- Each chapter ends with step-by-step AutoCAD Instructions, Drafting Exercises, and Design Exercises, allowing students to practice their new skills both manually and using AutoCAD LT.
- Key terms are highlighted within the text and listed at the end of each chapter to reinforce important concepts and terminology.
- Explanations, tutorials, and graphics introducing new CAD concepts are kept simple to allow students to get a clear overview of how a CAD system operates.
- Each chapter will develop and strengthen specific technical concepts, allowing the student to develop proficiency in solving drafting problems.
- All chapters are organized in a logical sequence; however, each chapter may be used as a stand-alone unit of instruction.
- No previous drafting knowledge is required to use this textbook.
- Exercises at the end of each chapter start with simple concepts and become progressively more complex.
- Students use the following methods to solve end-of-chapter exercises:
  - freehand sketching
  - instrument drafting
  - CAD system drawings
  - special design exercises
  - engineering change orders (ECOs)
  - inch-decimal, inch-fraction, and metric units of measure

Chapter Overview

Chapter 1 presents an introduction to modern industry. Specific careers related to drafting also give the students an insight into occupational options.

Chapters 2 through 7 give the student the background needed to learn and draw the basic drafting concepts with: sketching, instruments, drafting supplies, lettering, formats, conventions and CAD. The latest ASME standards are used throughout the text.

Chapters 8 through 15 teach the students the concepts of mechanical drafting required to design and draw finished multi-view drawings, dimensioning, tolerancing, sectional drawings, auxiliary drawings, revolutions, descriptive geometry, development drawings, and pictorial drawings.

Chapters 16 through 24 teach students how to prepare finished working drawings that are required for production: fasteners, drafting systems, working drawings, welding drawings, gear drawings, cam drawings, piping drawings, electronics drawings, jig and fixture drawings, and architectural drawings. CAD procedures and tutorials for each of the subject areas are covered.

Student Workbook

The Student Workbook has many additional exercises and learning aids to supplement the end-of-chapter exercises in the textbook.

Teacher's Resource Guide

The Teacher's Resource Guide is a valuable tool for teaching a drafting class. It contains all of the tools and resources necessary to conduct a drafting course. The TRG contains information about: drafting software and hardware, classroom procedures, classroom organization, evaluating and managing CAD files, performance goals for every chapter, course curriculums, tests and answers for each chapter, transparency masters, additional AutoCAD LT information, and selected solutions to the textbook and workbook exercises.

Template Diskette

New with this edition is an optional 3.5-inch, IBM template diskette for student use. The template diskette contains a symbols library, prototype drawings, and drawing practice for the chapters of the text. Students can open drawing templates on the disk and complete them using AutoCAD LT.

Standards

The language of drafting is a uniform and standardized system that is used throughout the world. The standards for the U.S. Customary system are developed by the American Society of Mechanical Engineers (ASME). The standards for the metric system are developed by Systems International (ISO). Careful attention was given to the dimensioning and tolerancing chapters (ASME Y14.5M).

Acknowledgments

The authors would like to express their appreciation and gratitude to the following individuals who took the time to offer their expertise and wisdom in the development of this textbook.

Viviene Fredrick, Technology Teacher
Saratoga Springs City School District
Saratoga Springs, NY

John P. Bartlett, Technology Teacher
Telfair County High School
McRae, GA

Daniel Engstrom, Technology Teacher
Indiana High School
Indiana, PA

Dr. Lewis Denunzio
Council Rock High School
Newtown, PA

Bruce Bainbridge
Clinton Community College
Clinton, Iowa

Dr. Ron Phillips, Department of Technology
South West Texas State University
San Marcos, TX

David Brodersen, Professor
Community College of Southern Nevada
Henderson, NV

Contents

Chapter 1

Introduction to Contemporary Drafting

Objectives

The student will be able to:

- relate the historical development of drafting
- state the importance and need for drafting as a technical communication skill
- identify the role and responsibilities of various drafting specialists
- state how drafting is used in different fields as the major source of communication
- recognize the levels of education, training, and experience required for the various drafting positions

History

Drawings have been used throughout history as an art form and a method of communication. Early humans drew crude pictures of animals on cave walls **(Fig. 1-1)**, and by 4000 B.C., stone tablet drawings **(Fig. 1-2)** were used to show the outline of structures. By 2000 B.C., drawings were prepared on parchment. Around 500 to 1000 B.C., detailed construction drawings were widely used. The Egyptian pyramids, Greek Parthenon, and the Roman Coliseum are examples of structures designed and constructed with detailed drawings on parchment **(Fig. 1-3)**.

Figure 1-1 Early humans drew crude pictures on cave walls.

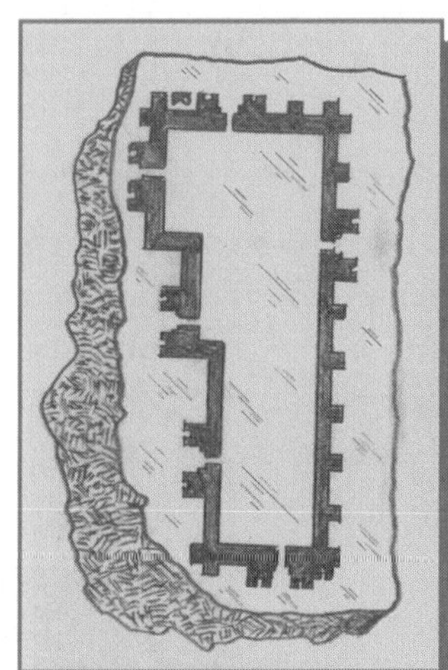

Figure 1-2 By 4000 B.C. stone tablets were being used to define the outline of fortifications.

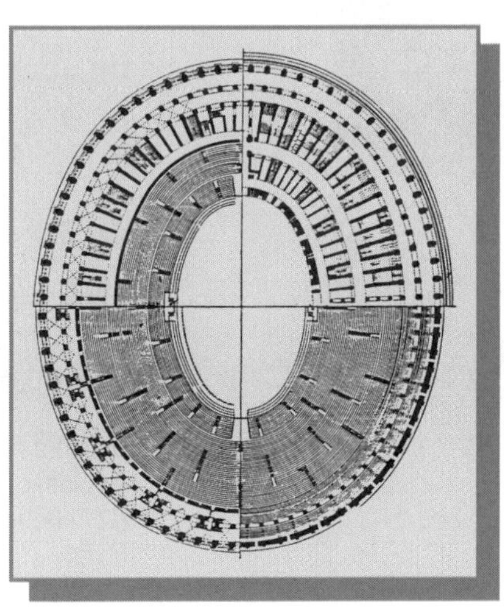

Figure 1-3 Detailed construction drawings were used to build the Egyptian pyramids, the Greek Parthenon, and the Roman Coliseum (shown left).

By the first century A.D., Romans were using detailed and dimensioned drawings as guides in building their structures, roadways, and aqueducts. It was not until the fifteenth century that two-dimensional working drawings were used for product drawings. An example is Leonardo da Vinci's sketch in **Figure 1-4**. Can you tell what his invention will do from his five hundred-year-old sketch?

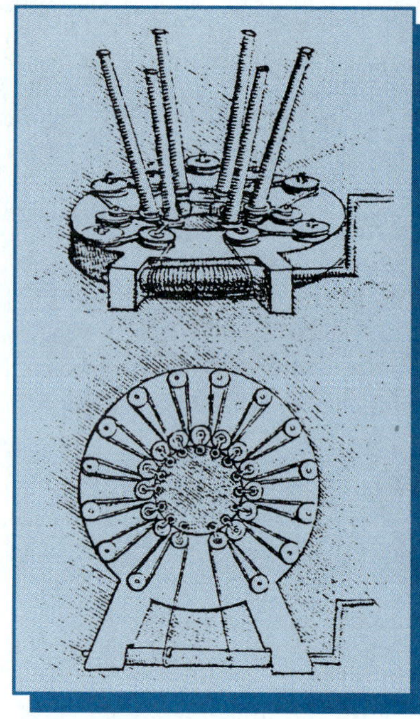

Figure 1-4
Leonardo da Vinci used pictorial and two-dimensional working drawings to develop his inventions. Can you figure out what this one will do?

Today's Drafters

Today's drafters must possess a broad understanding of the drafting knowledge and skills covered in this text. In addition, drafters must gain specific knowledge of the manufacturing methods and standards practiced in the industry in which they work. Regardless of the level of responsibility or the drafting specialization, all drafters must:

- understand the basics of drafting and design.
- communicate ideas with freehand sketches.
- use conventional drafting instruments quickly and clearly.
- use computer-aided drafting and design systems.

There are now nearly one million women and men working in drafting, design, and related positions. As more products are developed and manufactured, and more buildings are designed and constructed, the need for personnel with high levels of drafting skills and knowledge will continue to increase.

Today's Industries

In today's advancing technologies we are continually improving existing industries and creating new ones. All industries will require some degree of graphic representation for manufacturing, assembly, maintenance, and sales. A few of the major industries are: aerospace **(Fig. 1-5)**, passenger airlines **(Fig. 1-6)**, architectural and structural

Figure 1-5
A flight with the space shuttle
(National Aeronautics and Space Adminstration)

Figure 1-6
A well-designed passenger airliner
(Lockheed Martin)

engineering **(Fig. 1-7)**, construction **(Fig. 1-8)**, defense **(Fig. 1-9)**, nuclear **(Fig. 1-10)** petroleum **(Fig. 1-11)**, robotics **(Fig. 1-12)**, sporting goods **(Fig. 1-13)**, and transportation **(Fig. 1-14)**.

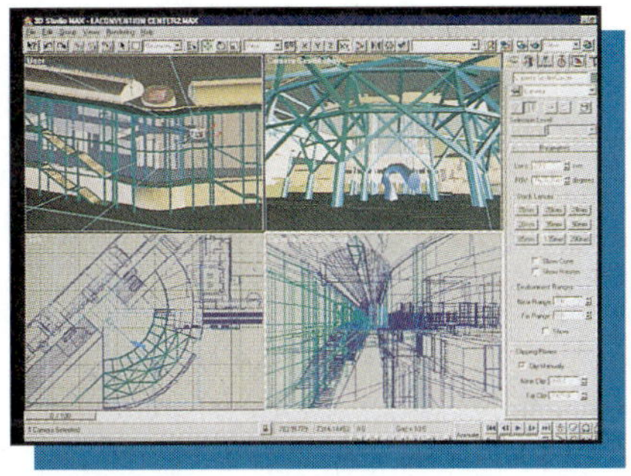

Figure 1-7
Computer-generated architectural drawings *(©Autodesk, Inc.)*

Figure 1-8
Construction industry

The Language of Drawing

Humans now routinely communicate with each other through the use of verbal sounds, written words, body movements, and many types of drawings. Written, verbal, or body language is very effective in communicating personal and social ideas or emotions. Only drawings are effective in describing the precise size and shape of objects. To illustrate this point, verbally describe the helicopters in **Figure 1-15** to a friend. Now show the picture to your friend. How good was your verbal description? This exercise should clearly illustrate the old Chinese proverb, "A picture is worth a thousand words."

Figure 1-9
Product of the defense industry *(Courtesy of McDonnell Douglas Corp.)*

Figure 1-10
Nuclear industry

Figure 1-11
Petroleum industry

Figure 1-12
Robotics industry

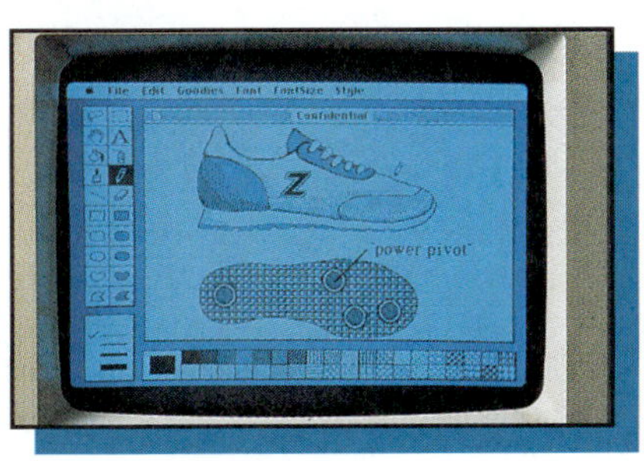

Figure 1-13
Sporting goods industry *(Courtesy of Apple Computer)*

Figure 1-14
Transportation industry

The Graphic Creators

As each industry is highly specialized, the drafters in each industry must have the knowledge to correctly design and draw each part for manufacture. The personnel involved with designing and producing the working drawings may be placed into three general categories depending on their formal education, knowledge, creativity, experience, and work ethics. These categories are top-level professionals, mid-level drafters, and intern-level drafters.

Figure 1-15
"A picture is worth a thousand words."

Top-Level Professional

This position requires a college degree (B.A., M.A., or Ph.D.) and often a state license in their specialty. They have the responsibility for successful operations. That is why they draw the largest salaries. Following is a brief description of several professional positions:

The **aeronautical engineer** performs a variety of work with research, planning, designing, and testing of airplanes, satellites, and rockets.

The **architect** does the planning, designing, structural engineering, and supervising of construction for all types of structures.

The **cartographer** designs and produces all types of maps.

The **civil engineer** does the planning, designing, and supervising for roads, airports, harbors, dams, tunnels, and most construction systems that are not inhabited buildings.

The **chemical engineer** plans the research to develop new and improved chemical manufacturing processes and production procedures.

The **developmental engineer** manager does the data research for the development of new ideas and new products.

The **electrical engineer** does the planning, designing, and supervising of the manufacturing of electrical and electronics components, equipment, and systems.

The **industrial designer** is a creative person who will use new ideas and materials to design a functional and attractive external package for all industrial items.

The **industrial engineer** or **mechanical engineer** works in all areas of industry, applying math and physics to a design so it will function properly and safely.

The **instructor of drafting** is a rewarding job for people who enjoy working with and helping others. Positions in secondary schools, technical schools, colleges, and universities are open to people with degrees, teaching credentials, and industrial drafting experience.

The **project coordinator** is a very talented person who is versed in most of the engineering, designing, and manufacturing processes. This person will coordinate all the engineers, designers, drafters, manufacturing, and assembly for a smooth process and troubleshoot any problems that may arise.

The **drafting/engineering supervisor** coordinates all the workers involved with the production of all the working drawings involved with specific projects. It is their responsibility to get well-designed and error-free working drawings finished on schedule.

The **tool designer** plans and designs the tools used to produce the manufacturing systems, machinery, and tools for all areas of industrial production.

Mid-Level Drafters

These are the technicians. They have a very broad and diversified job classification depending on their education, knowledge, creativity, drafting skills, and attitudes. They are classified as semi-professionals. Most will be supervised by the top-level professionals. Some technicians, even though they do not have a college degree, are highly capable and may perform the top-level job requirements, but they will not reach the top salary levels. It is recommended that they have an A.A.S degree from a 2-year college.

The **checker** is an experienced drafter whose responsibility is to see that all the drafter's working drawings are properly drawn and are error-free. This is critical, because correcting an error during production is very expensive.

The **chief drafter/senior drafter** will supervise the drafting personnel and set the parameters for the standards, practices, schedules, and drawing procedures for the project's set of working drawings.

The **commercial artist** prepares attractive illustrations for magazines, books, posters, and so on, to help promote recognition and sales of the manufactured item.

The **design technician** combines design skills and drafting ability, usually working from the top-level designer's sketches.

The **senior detailer** is skilled in understanding the engineer's concepts and can produce the complex working drawings needed for manufacture.

The **technical illustrator** creates three-dimensional drawings from the working drawings. The drawings are used to view a complicated part for better understanding **(Figure 1-16)** or to show an exploded drawing of a part to simplify its assembly.

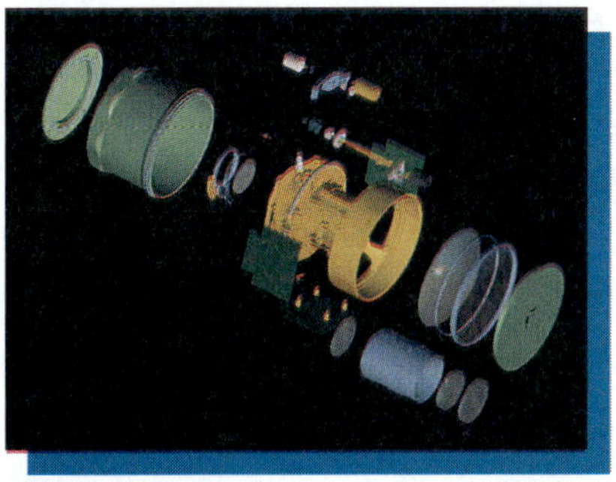

Figure 1-16
Example of technical illustration of a turbofan engine *(© Autodesk, Inc.)*

Intern-Level Drafters

This position usually does not have a formal education requirement, but receives on-the-job training. Interns may advance to technician levels depending on their abilities. Many engineers and technicians will have interns working with them.

The **computer-aided drafting operator** has a skill that is important for all designers and drafters. It is an asset in gaining employment and advancement in the workplace.

The **junior detailer** prepares working drawings from the sketches of the senior detailer and corrects drawing errors marked by the checker.

The **junior drafter** must have good manual drafting and lettering skills and be trainable on a CAD system. He or she starts with simple working drawings and is closely supervised.

The **printing operator** makes reproductions of working drawings on various equipment such as copy machines, cameras, diazo printers, and printing presses. Copies are then stored, filed, and/or distributed to assigned places.

The **tracer** is the starting entry level. This job requires good drafting and lettering skills because it involves copying or recopying sketches or quick drawings from engineers and designers who did not take the time to do a neat drawing.

The Design Process

The design process and the personnel involved in each industry will vary considerably, but the goal of an efficient product remain the same. A typical process may be:

1. Recognition of Needs
2. Scientific Investigation
3. Proposal of Concept
4. Layout and Development Drawings
5. Conceptual Design Reviews and Analysis
6. Component Testing of Prototype
7. Final Design Review
8. Detail Design
9. Quality Assurance
10. Manufacturing
11. Assembly
12. Installation
13. Final Testing
14. Operation
15. Planning for the Second-Generation Upgrade

An example of the design process for an automobile is shown in **Figures 1-17a** through **1-17d**.

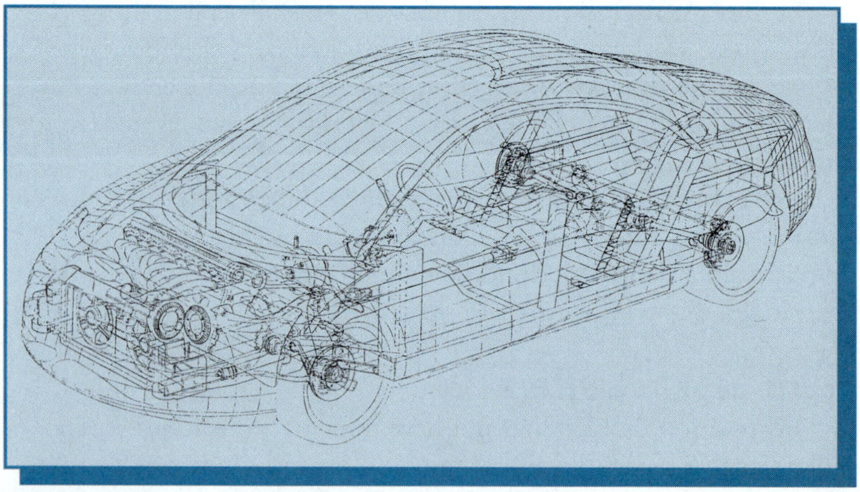

Figure 1-17a
CAD-generated design

Drafting Today

Drafting is the universal language of industry and construction. Through the use of standardized symbols and engineering drawing conventions, the drawings help overcome international language barriers by describing manufactured products, buildings, and process without verbal communication. Drawings are now the primary method of communication between designers and manufacturers, architects and builders, engineers and production personnel, and advertisers and their customers.

When used to show the material, size, and shape of a product, a drawing is known as a *technical working drawing*. All manufactured products of structures, regardless of size, must start with a working drawing. The design and manufacture of any product, no matter how simple or small, still requires a working drawing **(Figure 1-18)**. Larger and more complex products may require hundreds or thousands of working drawings if they are to be properly manufactured. Each step of the manufacturing process must be specified. There must be no guesswork or redesigning during the manufacturing process.

Figure 1-17b Mechanical design phase *(Courtesy of Ford Motor Co.)*

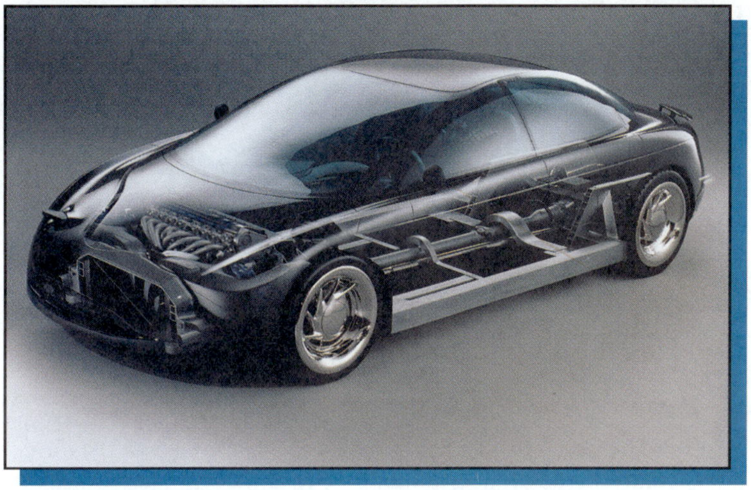

Figure 1-17c
Body design around the power drive *(Courtesy of Ford Motor Co.)*

Figure 1-17d Design process completed *(Courtesy of Ford Motor Co.)*

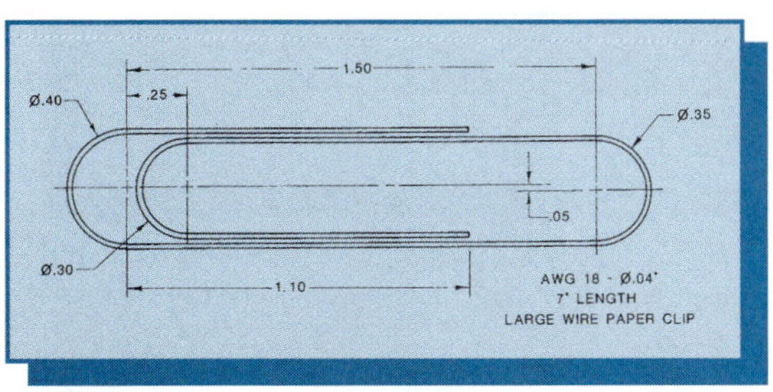

Figure 1-18
The design and manufacture of a simple paper clip requires technical working drawings.

Types of Basic Engineering Graphics

Most of the aspects of engineering graphics (drawings) are common to all the industrial areas of drafting. The types of drawings used are:

1. *Multiview drawings* are several views of the object drawn using orthographic projection. The complete shape of the item with all its details and dimensions must be shown in the drawing **(Fig. 1-19)**.

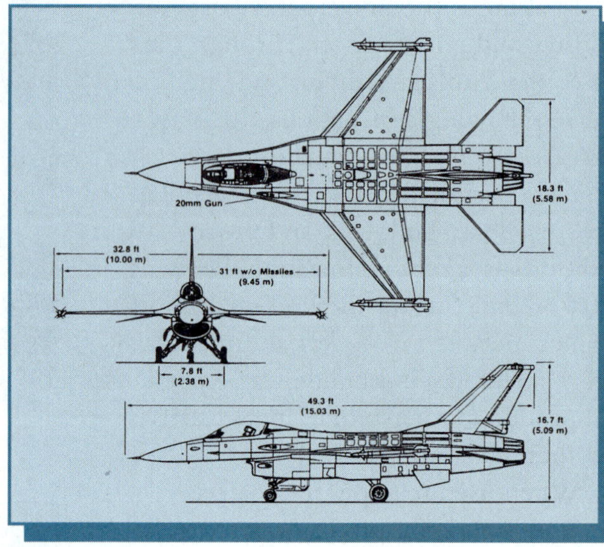

Figure 1-19
Multiview drawing for the F-16A aircraft *(Courtesy of General Dynamics)*

Figure 1-20
Pictorial cutaway of the F-16B aircraft *(Courtesy of General Dynamics)*

2. *Pictorial drawings* show an object as in a photograph. Usually three adjacent surfaces are shown in one drawing **(Fig. 1-20)**.
3. *Schematic drawings and block diagrams* are used to show the flow of energy or work such as electricity or fluids **(Fig. 1-21)**.

Nearly every design process during the development of the ideas through the working drawings will use some degree of freehand sketching, manual drafting, and a CAD system for the creation of the multiview drawings, pictorial drawings, and schematics.

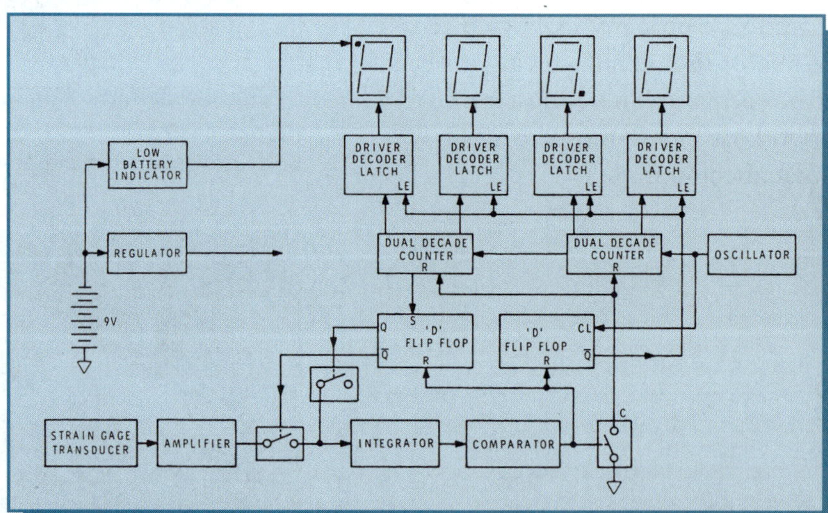

Figure 1-21
A block diagram schematic showing the electrical flow for a digital scale

Exercises

1. Pick one of the professional positions that you find the most interesting and write a short paper on the education, training, and type of work it involves.

2. Select one of the mid-level drafting positions that you find the most interesting and write a short paper on the education, training, and type of work it involves.

3. If you had to go to work directly from high school, what type of intern position would you prefer? Write a short paper on how you would apply for the internship.

4. Take a field trip to an industry with a CAD system. Watch a demonstration and interview the CAD operator. Take notes and write a short report on her or his background, education, training, and work performed on the CAD system.

5. Take a good look at the picture of the space shuttle **(Fig. 1-5)**. Try to list the personnel involved at each step of the design process described in this chapter.

6. Interview a college counselor and list the high school prerequisites needed for acceptance into an engineering school.

7. Interview a vocational or community college counselor and list the prerequisites needed for acceptance into a vocational training program.

8. Talk to your high school counselor about applying for a scholarship or an educational grant.

9. Practice writing a resumé to use to apply for an internship drafting position.

10. Note the advanced design of the automobile driver's area in **Figure 1-22.** Make a list of the innovations that you can recognize. Make another list or sketch with innovations of your own ideas.

Figure 1-22
Innovative design for driver

Chapter 2

Drafting Equipment and Supplies

Objectives

The student will be able to:

- understand the function of manual drafting instruments and supplies
- complete a pencil drawing using drafting instruments
- measure with a civil engineer's scale
- measure with a mechanical engineer's scale
- measure with a metric scale
- measure with an architect's scale

Introduction

Because of the many advantages of CAD systems **(Figure 2-1)** over manual drafting systems, the number of computer-generated technical drawings in manufacturing and construction is growing at a rapid rate. However, a successful drafter, designer, engineer, or architect must also be able to create drawings using conventional drafting instruments and supplies the following reasons:

Figure 2-1
Computer-aided drafting systems are gradually replacing manual drafting. *(Courtesy of Hewlett Packard)*

1. A CAD system may not always be available.
2. Not all companies can or will supply CAD systems to every drafter.
3. Field drawings and instructional sketches must be completed in environments hostile to computers.
4. Creative design in many areas is still done in pencil.
5. There may be times when the CAD system cannot create a specific type of detail. It would be expedient to make a print of your drawing and complete the detail with manual drafting instruments.

Whether a drawing is prepared as a pencil sketch, inked instrument drawing, or computer-generated hard copy **(Figure 2-2)**, the final output must be orthographically correct, dimensionally accurate, symbologically true, and easily readable.

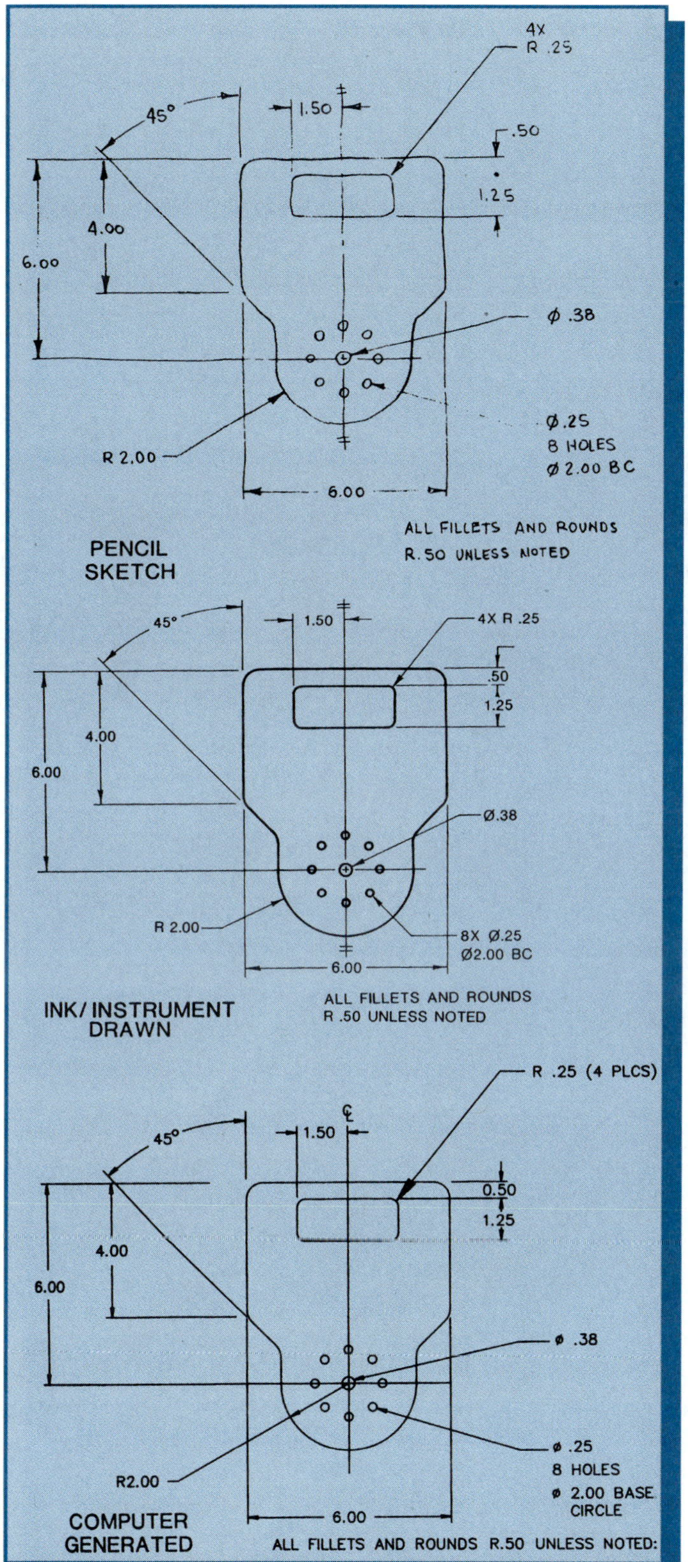

Figure 2-2
All finished working drawings must be easy to read.

Conventional Drafting Supplies

Conventional drafting supplies consist of surface media and marking devices. **Surface media** upon which pencil, ink, or CAD-plotted drawings are drawn include a variety of materials and surfaces such as **tracing paper, opaque drawing paper, vellum, film (Figure 2-3)**, and preprinted **grid papers (Figure 2-4)**. All drafting surfaces are held in place with drafting tape.

Marking devices used to make lines on surface media include pencils, pens, transfer film, and tapes. **Wood pencils** are the oldest type of drafting marking device **(Figure 2-5)**.

Pencils and Lead Holders

Drafting pencils are classified by degree of lead hardness **(Figure 2-6)**. Hard leads are used for fine layout work, medium leads for most object lines, and soft leads for lettering and rendering. To insure line weight consistency, pencils must be sharpened evenly **(Figure 2-7)**. Sharp points produce fine lines, while round or dull points produce broad lines. Sharpening devices for drafting pencil points include hand or electric rotating sharpeners, cylindrical lead pointers, and sandpaper for hand-pointing leads.

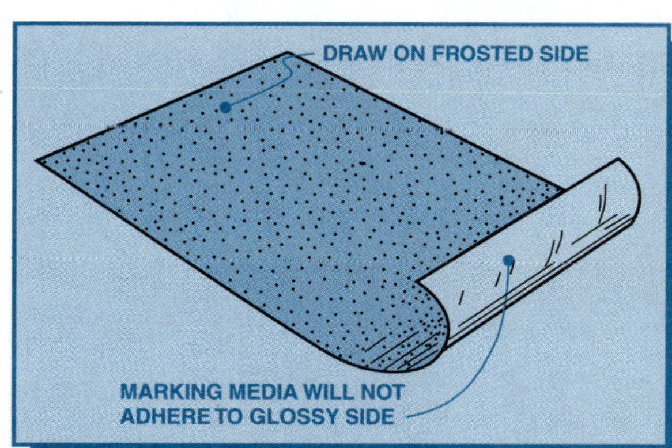

Figure 2-3
Drawing on polyester drafting film

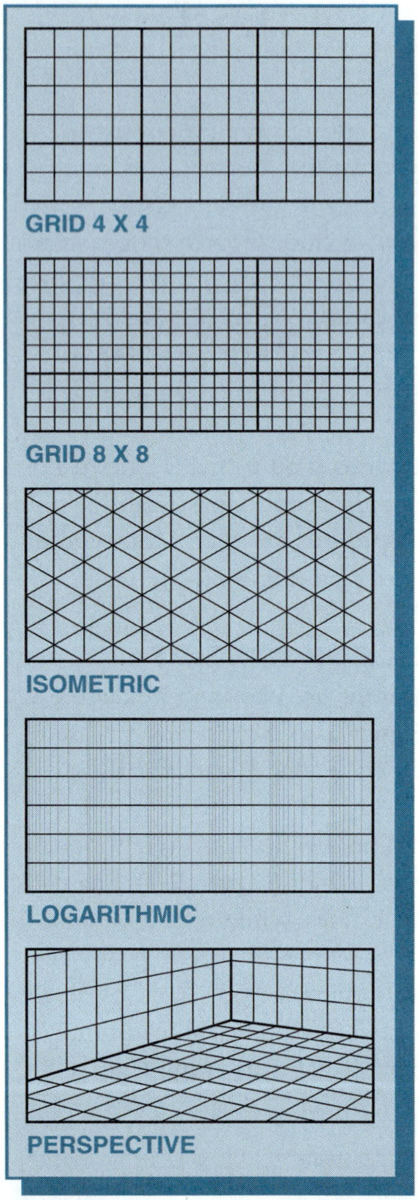

Figure 2-4
Examples of grid paper

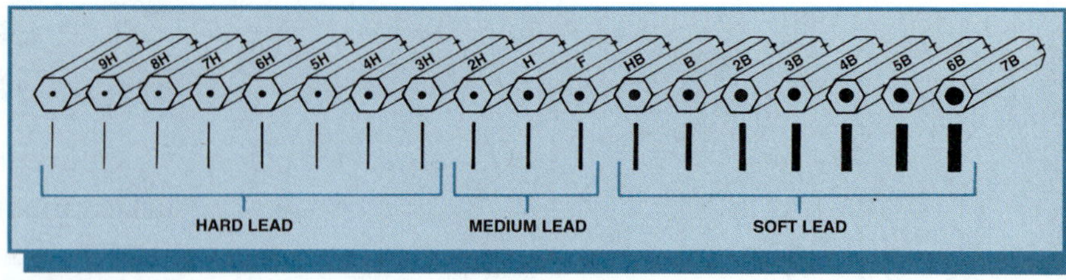

Figure 2-6
The various degrees of drafting pencils and their graphite lead widths

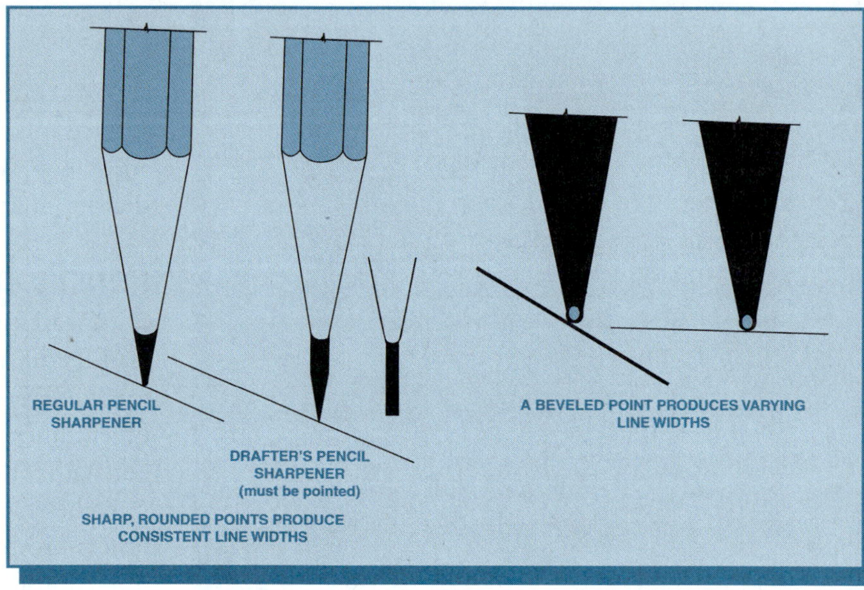

Figure 2-7
Pointing the drafting pencil

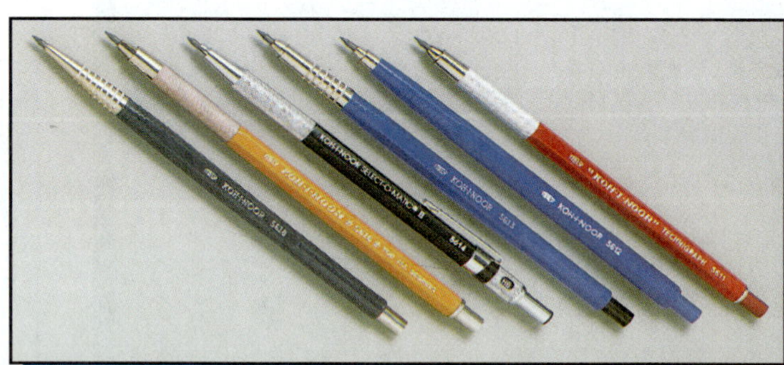

Figure 2-8
The lead holder grips the drafting lead. The lead can be sharpened on sandpaper, a file, or a mechanical sharpener. Leads are available in all drawing grades. *(Courtesy of Koh-I-Noor)*

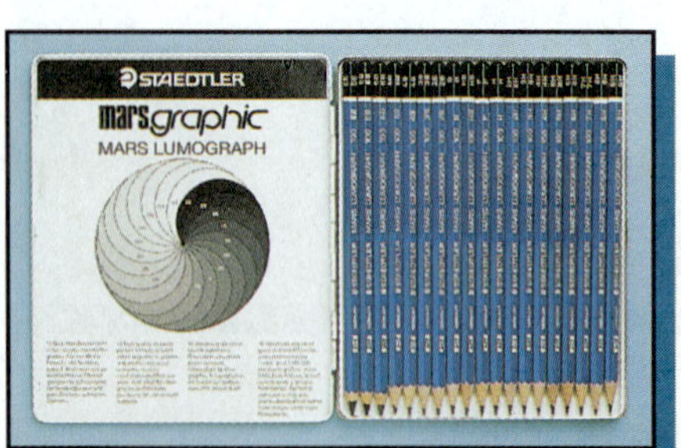

Figure 2-5
Typical wood-encased drafting pencil

Many conventional drafters prefer mechanical **lead holders (Figure 2-8).** These leads are not sharpened with a sharpener but pointed with a mechanical pointer or on a sandpaper block. Thin-lead mechanical pencils **(Figure 2-9)** do not require either sharpening or pointing. Leads for mechanical holders are available in all grades; fine-line leads are available in thicknesses of 0.3, 0.5, 0.7, and 0.9 mm. Drafting leads are made of graphite, not lead. However, some drafting leads are made of plastic specially designed for use on drafting film.

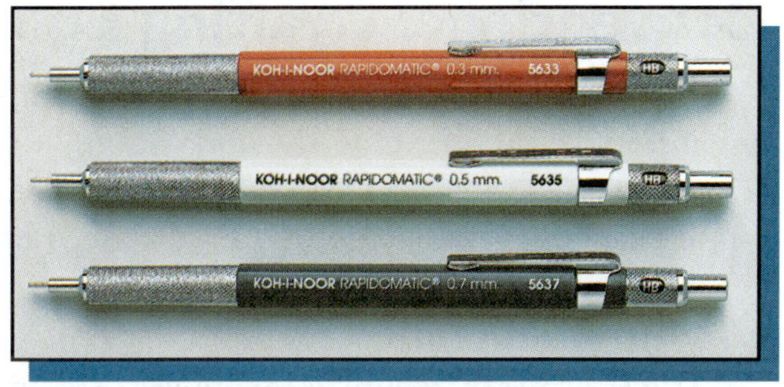

Figure 2-9
Fine-line mechanical pencils do not require sharpening. Leads are available in all drawing grades. *(Courtesy of Koh-I-Noor)*

After drawing with any graphite lead, remember to brush the surface periodically with a dusting brush to remove the accumulation of foreign matter and eraser leavings.

Technical Pens

A **technical pen (Figure 2-10)** is an instrument used to ink lines and lettering on a technical drawing. Inking procedures are similar to pencil drawing except that care must be taken not to smear ink lines before they dry.

Technical pens are available in sizes from .13 to 2.0 mm. The line widths are coded in several ways, depending on the manufacturer. **Figure 2-11** shows typical line widths and two coding systems.

Figure 2-10
Technical inking pen *(Courtesy of Koh-I-Noor)*

Erasers

Layout and construction lines often have to be removed, drafting mistakes corrected, and graphite smudges erased from drawing surfaces. An electric eraser **(Figure 2-12)** is the most efficient line remover, since a gentle touch removes graphite or ink without damaging the surface. A medium-hard eraser with no abrasives is best for most drafting work. To erase ink from drafting film, a drop of water or commercial ink remover can be used. Very small ink areas can be removed by gently scraping with a sharp knife.

Softer graphite lines can be removed with a rubber or vinyl eraser, and smudges can be reduced by sprinkling the drawing surface with a powder cleaner. When using any eraser, always use an erasing shield **(Figure 2-13)** to insure that only the desired line area is removed.

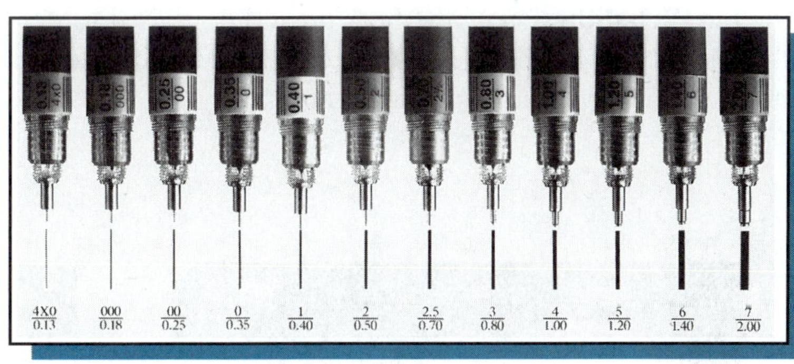

Figure 2-11
Technical pen identification designating line widths *(Courtesy of Faber-Castell)*

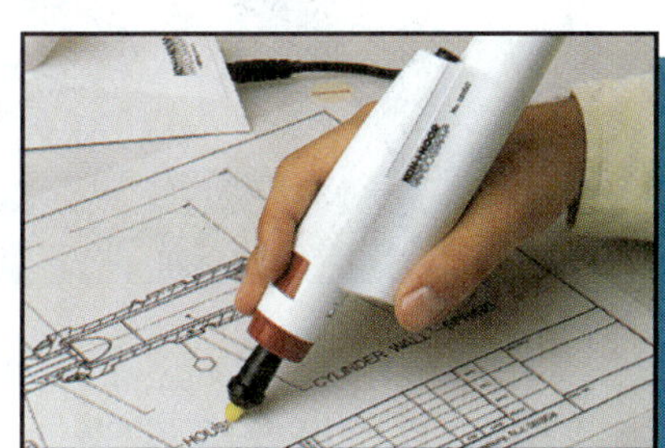

Figure 2-12
The electric eraser is the most efficient cleaner. *(Courtesy of Koh-I-Noor)*

Conventional Drafting Equipment

Regardless of the amount of work being done with CAD systems, or the number of CAD workstations, it would be convenient at times to have at least one manual drafting station for minor changes or detailing on CAD-generated drawings.

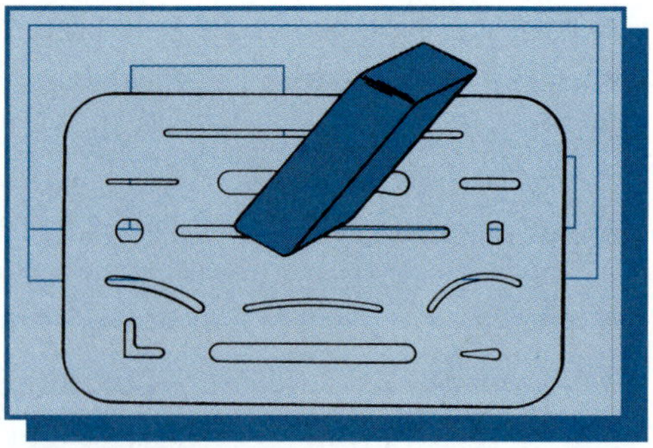

Figure 2-13
The erasing shield protects lines not to be erased.

Drafting Machines

Track drafters (Figure 2-14) are very efficient manual drafting machines. A track drafting machine includes a vertical track that slides left and right on a horizontal track. A protractor head is attached to the vertical track and allows the vertical scale (arm) and horizontal scale (arm) to be rotated to any angle. **Figure 2-15** shows the parts of a protractor head. Although the protractor head can be moved to any angle, an indexing button allows the arms to be automatically stopped every 15°. These are the angles commonly used in drawings.

Elbow drafting machines include a protractor head similar to the track drafter. The elbow machine **(Figure 2-16)** is connected at the top of a drawing board. It cannot cover an area as large as that covered by a track drafting machine, but it is just as effective.

Large-scale drawings with continuous lines longer than the length of track or elbow machine arms are cumbersome to draw with an elbow or track drafter. For this reason, many large architectural drawings are made with a parallel slide drafting system. In a **parallel slide** system, the horizontal slide is used to draw all horizontal lines. Triangles are placed on the horizontal slide and used for all vertical and angular lines.

Figure 2-14
The track drafter is a very efficient manual drafting machine. *(Courtesy of Vemco Corporation)*

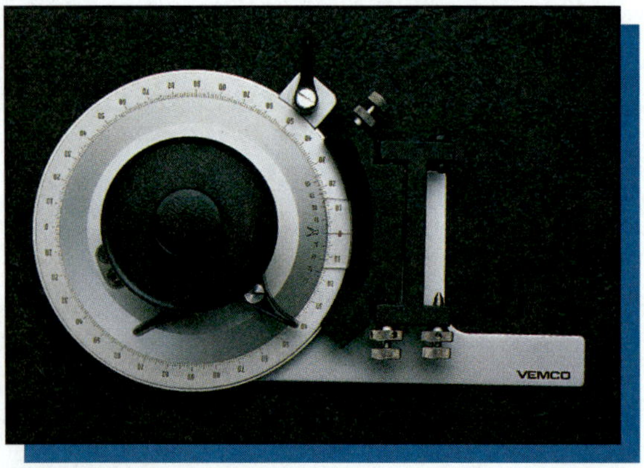

Figure 2-15
Parts of a protractor head for a drafting machine *(Courtesy of Vemco Corporation)*

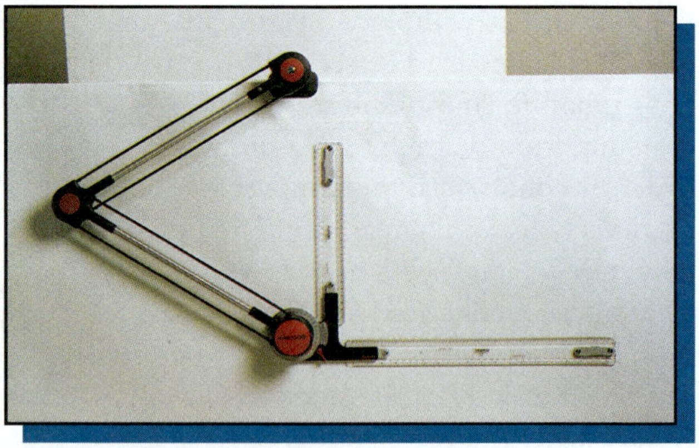

Figure 2-16
An elbow drafting machine and its parts *(Courtesy of Vemco Corporation)*

T Square and Triangles

The use of the **T square (Figure 2-17)** and **triangles** is one of the oldest drafting systems but tends to produce inaccurate lines if the T square is not rigidly held perpendicular to an edge at all times. **Figures 2-18** and **2-19** show the recommended method of mounting drawing paper onto a drawing board. When used with a T square or parallel slide, triangles **(Figure 2-20)** are used to draw 15°, 30°, 45°, 60°, 75°, and 90° (vertical) lines. A 45° triangle is used to draw 90° (vertical) and 45° lines **(Figure 2-21)**. A 30°– 60° triangle is used to draw 90°, 30°, and 60° lines **(Figure 2-22)**. By combining triangle arrangements 15° and 75° lines can be drawn **(Figure 2-23)**. For drawing angles other than at 15° increments, the protractor or the adjustable triangle can be used **(Figure 2-24)**.

Figure 2-17
Example of a T square *(Courtesy of Teledyne Post)*

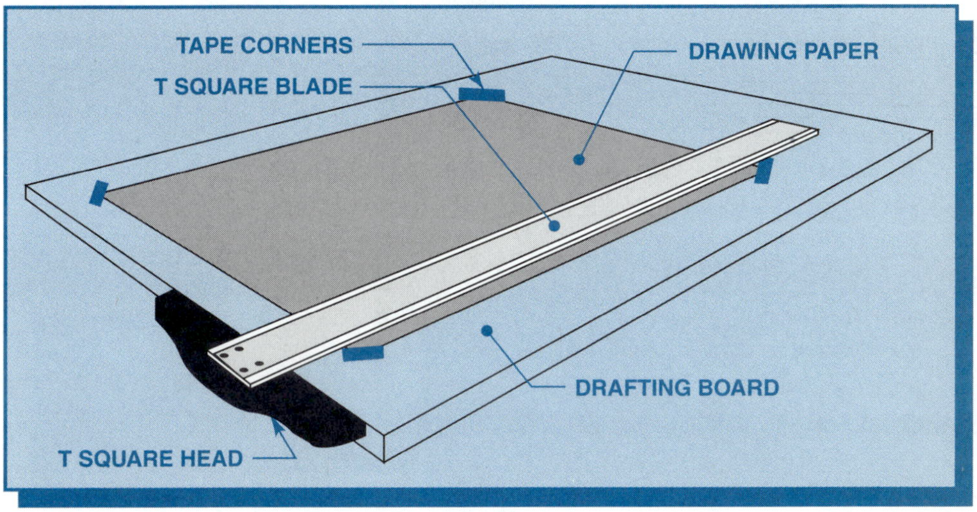

Figure 2-18
Line up drawing paper with T square on the drawing board as shown. Tape down corners of drawing media.

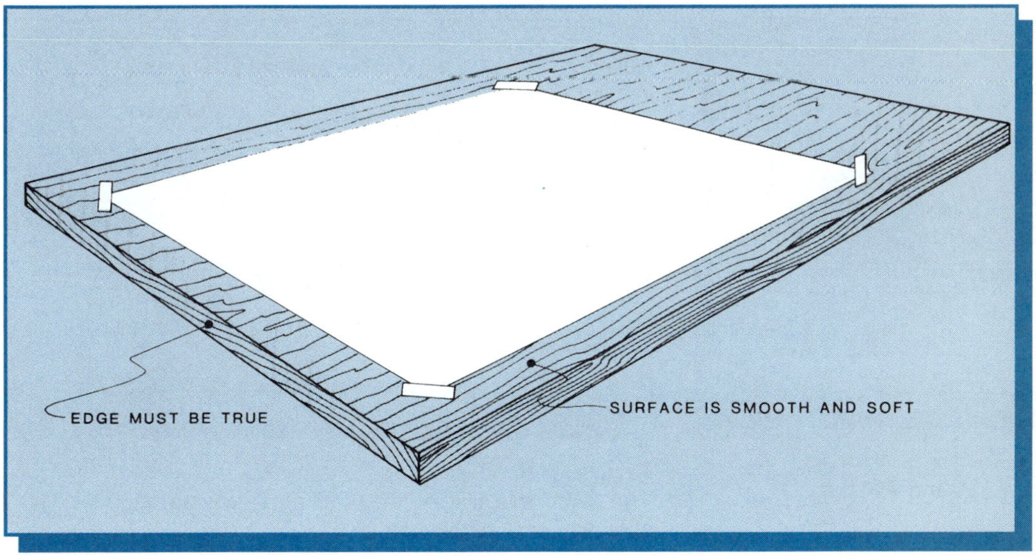

Figure 2-19
Drawing board qualities

Figure 2-20
The 45° and 30°– 60° triangles
(Courtesy of Koh-I-Noor)

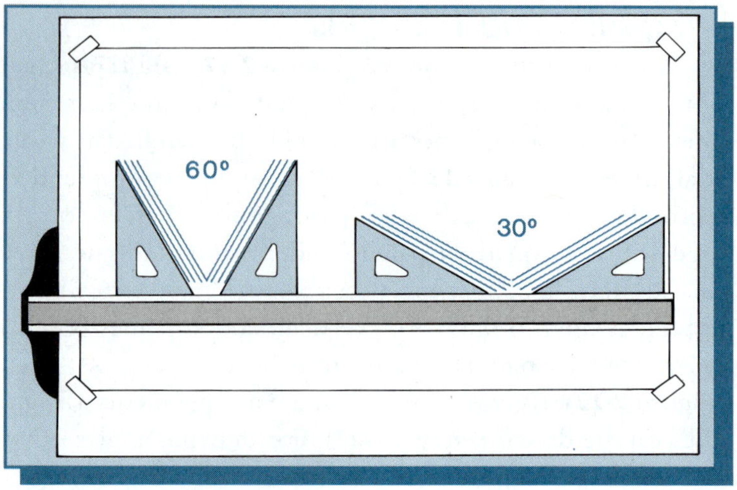

Figure 2-22
Positioning triangles to draw 30° and 60° lines

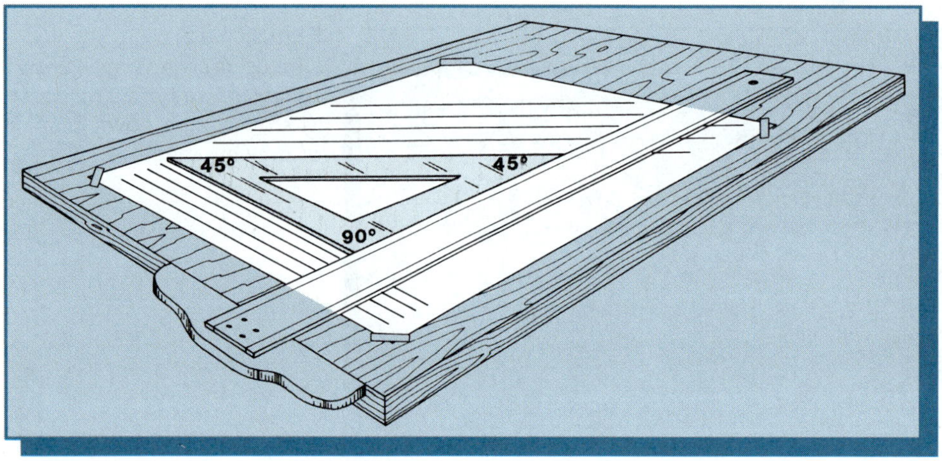

Figure 2-21
Drawing 45° and vertical lines

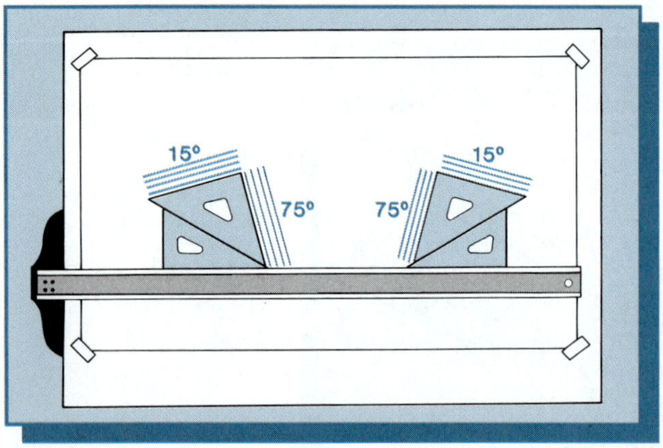

Figure 2-23
Positioning triangles to draw 15° and 75° lines

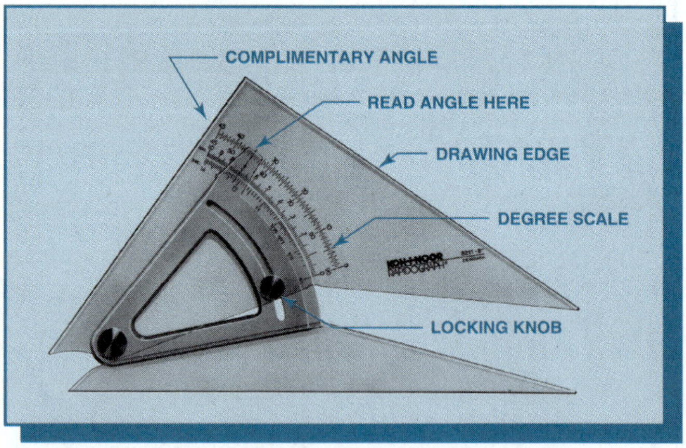

Figure 2-24
The adjustable triangle may be set to any angle.
(Courtesy of Koh-I-Noor)

Arc and Circle Instruments

All drawings are comprised of three kinds of lines: straight lines, arcs and circles, and irregular curved lines. Drafting machines can only be used to draw straight lines. Regular curved lines that radiate from a center point are drawn with a compass or circle template. There are three types of compasses used in drafting: bow, beam, and friction. **Bow compasses (Figure 2-25)** are used for circles up to 12" in diameter, **friction compasses** are not recommended because they may slip while drawing circles, and **beam compasses** are used for large circles over 12" in diameter.

The bow compass is the most common of the three. It should be tilted and revolved between the thumb and index finger for best results **(Figure 2-26)**. **Figure 2-27** shows the correct method of setting and sharpening a compass point and lead to produce the clearest and most consistent line quality.

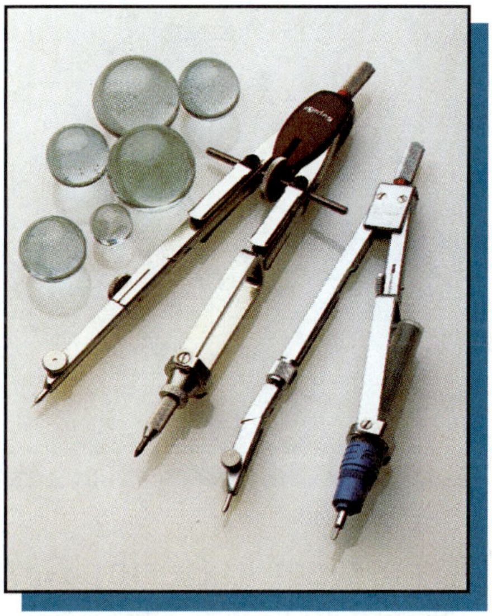

Figure 2-25
Bow compasses and inking compasses are used for drawing circles and arcs. *(Courtesy of Vemco Corporation)*

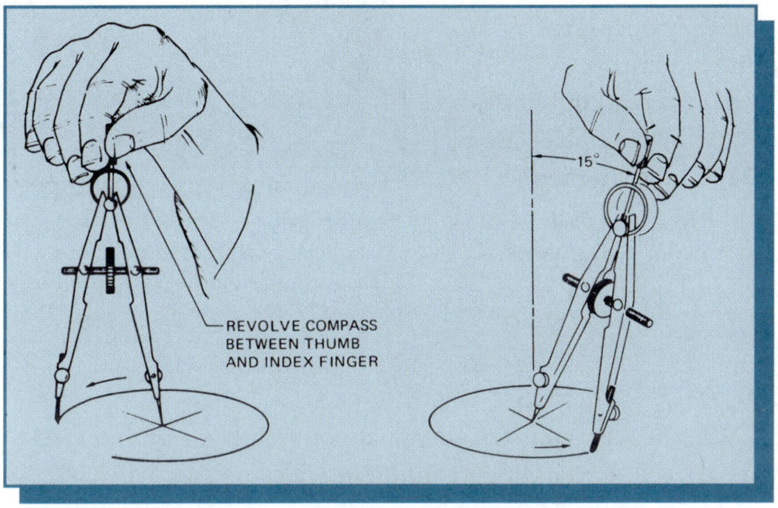

Figure 2-26 Drawing procedure for handling a bow compass. *(Reprinted from* Technical Drawing and Design *by Goetsch and Nelson, ©1986 Delmar Publishers Inc.)*

Dividers are similar to compasses except both tips are points. Dividers are used to step off or transfer distances on a drawing.

Irregular curves on technical drawings are drawn with the aid of an irregular curve template, sometimes called a French curve **(Figure 2-28)**. When using a French curve, follow preplotted points in gradual overlapping progression **(Figure 2-29)**.

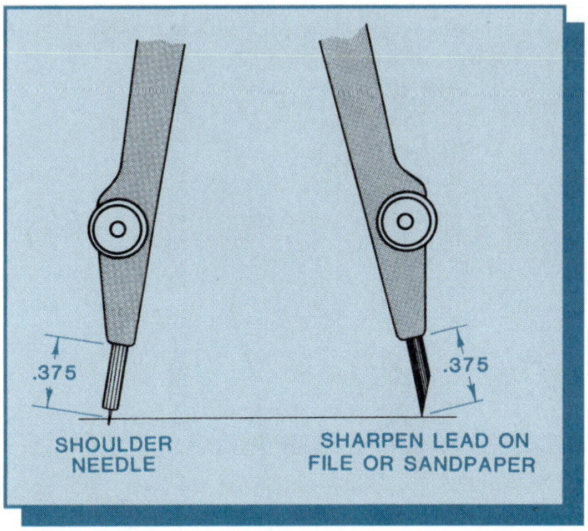

Figure 2-27
The compass point and the compass lead

Figure 2-28
A French curve is used to draw irregular curves. *(Courtesy of Teledyne Post)*

Protractor

When doing manual drafting, a **protractor** is used to measure and lay out various angles **(Figure 2-30)**.

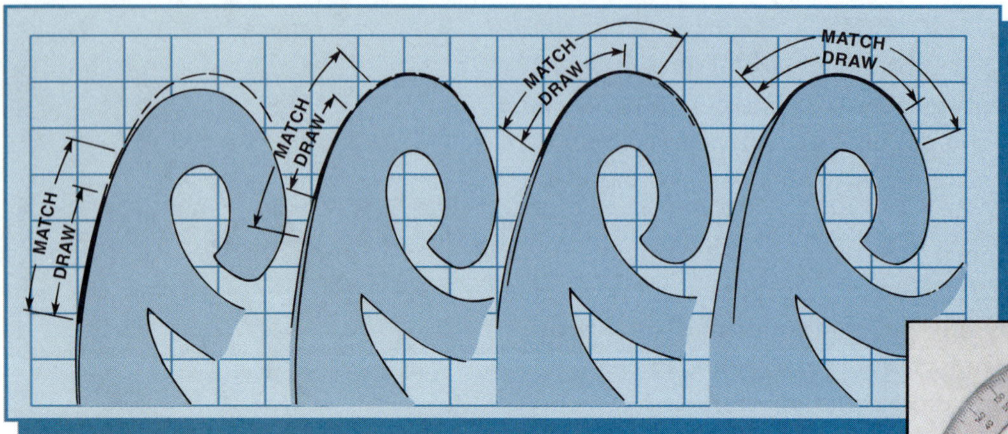

Figure 2-29
The procedural steps to draw an irregular curve with a French curve *(Reprinted from* Mechanical Drafting *by Madsen, Shumaker, and Stewart, © 1986 Delmar Publishers Inc.)*

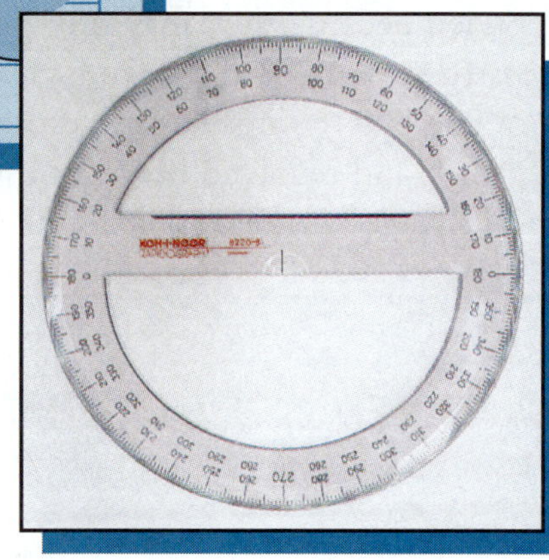

Figure 2-30
A semicircular protractor is used to measure angles in degrees. *(Courtesy of Koh-I-Noor)*

Mechanical Lettering Aids and Templates

For standard circles, ellipses, and arcs, one of the **templates** shown in **Figure 2-31** may be preferable to using a compass.

Templates also have the additional advantage of providing guides for drawing complete symbols, letters, and numerals. A template used for lettering guide lines is shown in **Figure 2-32**. Guide line templates are used by sliding the template along the top of a T square, slide, or blade with a sharp pencil inserted into a hole. The guides can be positioned to draw guide lines for different heights of lettering.

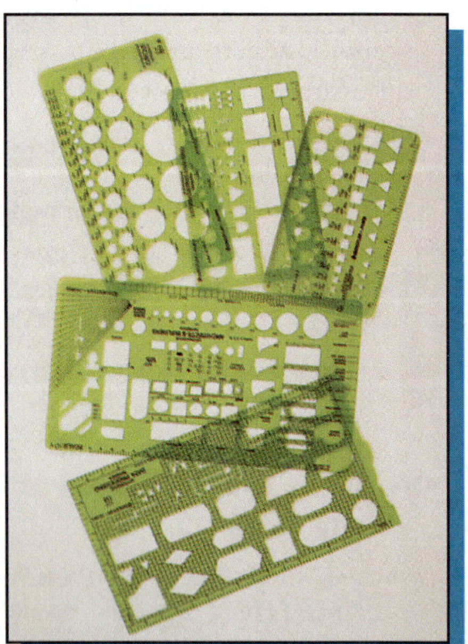

Figure 2-31
Sampling of some typical drawing templates

Drafting Scales

Whether drawing with manual equipment or with a CAD system, an understanding of different scales is essential. Scales are available in several different shapes **(Figure 2-33)**. Scales allow drafters to create accurate drawings that are proportioned to the actual size of the object being drawn. Scales are based on either

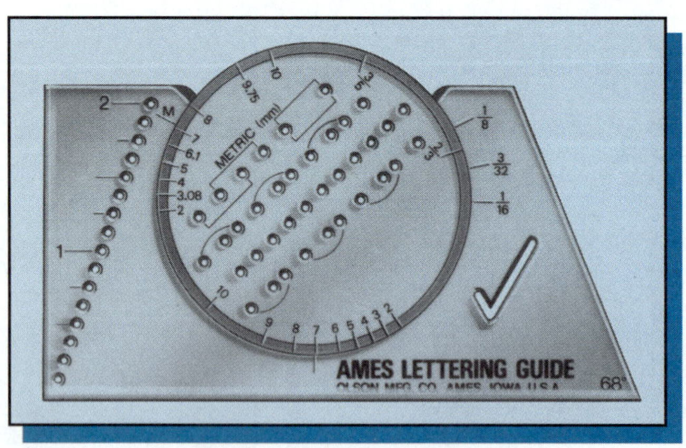

Figure 2-32
A guide line template is used by sliding the template along the top of the T square blade with a sharp pencil inserted into a hole. Various spaced guide lines may be drawn.
(Courtesy Olson Manufacturing & Distribution, Inc.)

the **U. S. customary system** or the **metric system**. Scales used in drafting include the mechanical engineer's scale, the civil engineer's scale, the architect's scale, and the metric scale.

The **mechanical engineer's scale** uses a U. S. customary inch-fraction unit of measure **(Figure 2-34)**. Scale subdivisions include 1/32", 1/16", 1/8", 1/4", 3/8", 3/4", 1/2", and 1" units.

The **civil engineer's scale** uses a U.S. customary inch-decimal unit of measure **(Figure 2-35)**. The inch-decimal unit is used by most industries using the U.S. customary system. Civil engineer's scale subdivisions include 10, 20, 30, 40, 50, and 60 parts per inch.

The **architect's scale** is used to prepare plans for structures. It is divided into twelve parts, so inches on a drawing can equal the actual inch or feet dimensions of a building **(Figure 2-36)**. For example, most architectural drawings are prepared to a scale 1/4" = 1'-0". This means that a 1/4" line on a drawing represents one foot on a building. Thus, at 1/4" = 1'-0" scale, an 8' wall would appear 2" long on the drawing. The various architect's scales are 3/32", 3/16", 1/8", 1/4", 3/8", 3/4", 1/2", 1", 1 1/2", and 3".

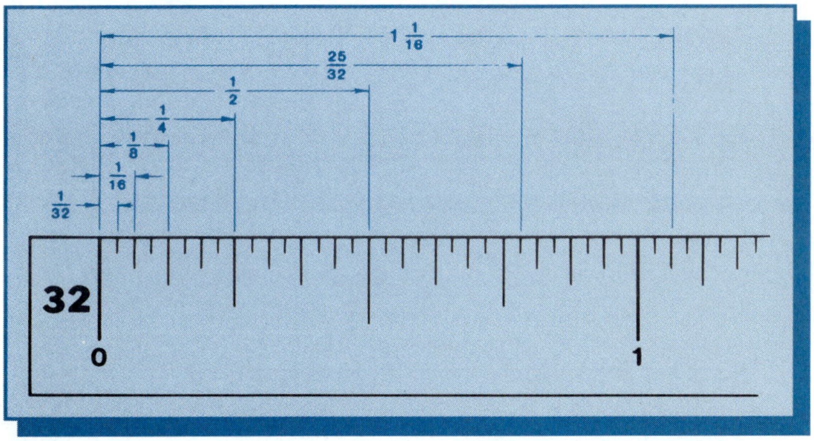

Figure 2-34
The mechanical engineer's scale uses inch-fraction units of measure.

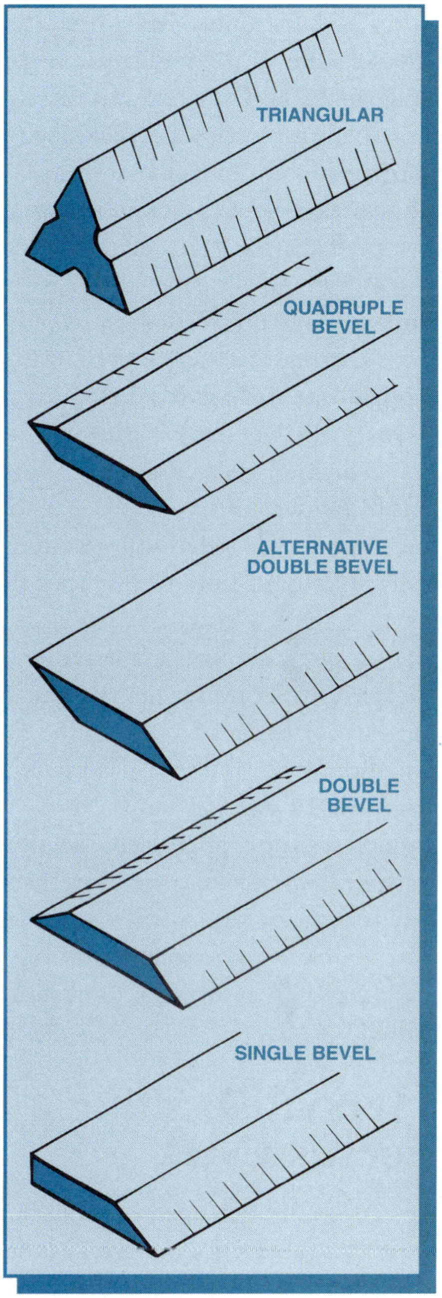

Figure 2-33
Various scale shapes

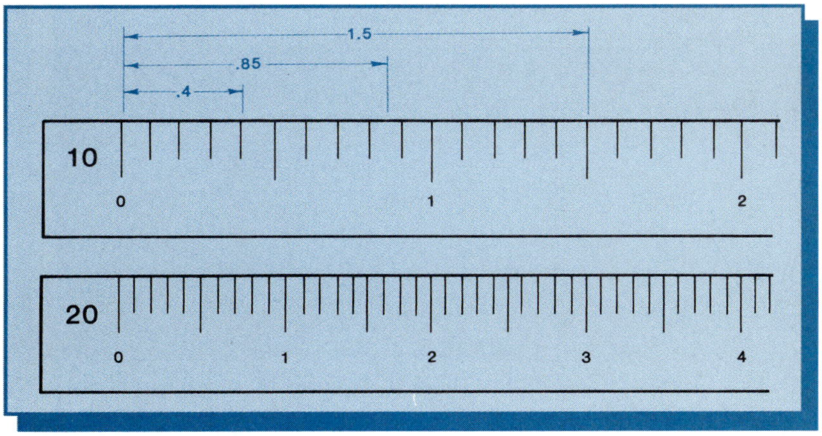

Figure 2-35
The civil engineer's scale uses inch-decimal units of measure.

The basic unit of measure for metric scales is the **millimeter (mm) (Figure 2-37)**. The abbreviation mm is not used in metric drawings since all dimensions are in millimeters. All countries except the United States use the metric system for technical drawing and manufacturing, although we are gradually converting.

Drafting scales are either **open-divided** or **full-divided (Figure 2-38)**. Only one major unit of open-divided scales is graduated with a full-divided unit. It is adjacent to the zero. Full-divided scales contain full subdivision lines throughout the entire length of the scale. In selecting the proper scale for each drawing, the drafter must consider the amount of space available, readability of the finished drawing, and ease of drawing. **Figure 2-39** provides some basic guidelines for proper scale selection. In selecting a scale, remember that a decimal or fractional part of an inch can be made equal to any unit of measure such as an inch, foot, yard, or mile.

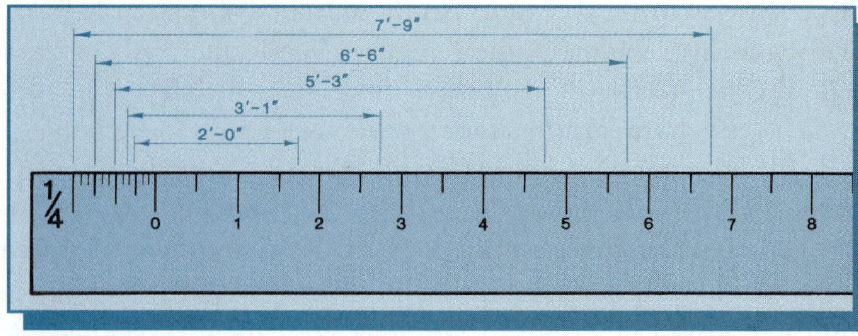

Figure 2-36
The architect's scale uses foot-inch units of measure. The most often used architect's scale is 1/4" = 1'-0".

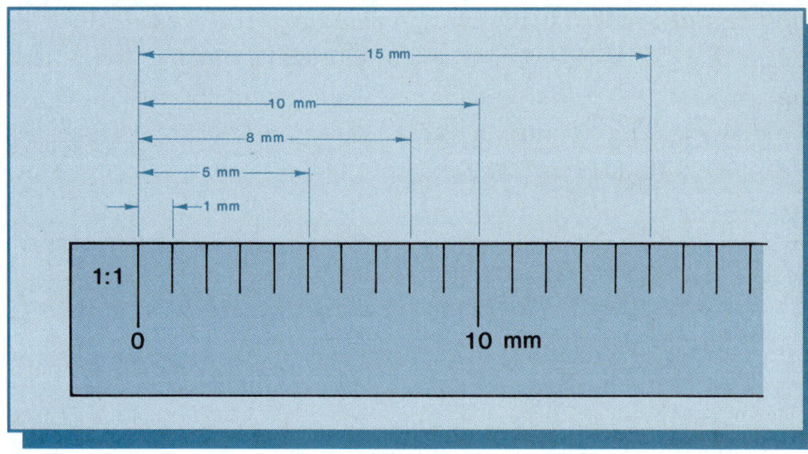

Figure 2-37
The basic unit of measure for metric scales is the millimeter (mm).

Manual Drafting versus CAD

As computer-aided drafting becomes more prevalent in schools and industry, a decision must be made as to how many CAD stations and how many manual drafting stations should be installed. The decision will depend on the needs, goals, and budgets of each school, industry, and business.

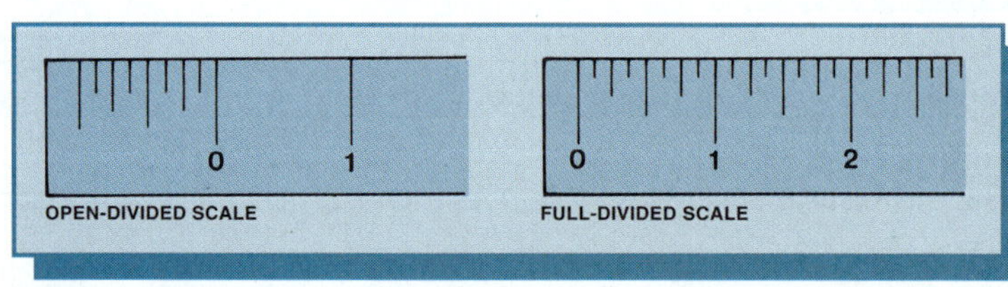

Figure 2-38
Drafting scales are either open-divided or full-divided.

SCALE SELECTIONS FOR ENGINEERING DRAWING			
Types of Drawings	**Mechanical Engineer's Scale**	**Civil Engineer's Scale**	**Metric Scale**
The object to be drawn must be smaller than the drawing format. All the drawings, dimensions, and notations will fit at actual size.	Full Size (1:1)	1:1	1:1
The object to be drawn is larger than the drawing format and must be reduced in half to fit the format.	1/2" = 1"	1:2	1:2
The object to be drawn is much larger than the drawing format and must be reduced eight to ten times in size to fit the drawing format.	1/8" = 1"	1" = 10" 1:10	1:10
The object to be drawn is very large, such as a building, and must be reduced at least fifty times in size to fit the drawing format.	1/4" = 1'–10" (1:48)	1:50	1:50
The object to be drawn is small and cannot easily be drawn full size. Doubling the drawing size makes the drawing easier to draw and interpret.	1" = 1/2" (2:1)	2:1	2:1
The object to be drawn is very small. For ease of drawing and interpreting, thr original size must be increased eight or ten times.	1" = 1/8" (8:1)	10:1	10:1
When an industrial product is extremely small, such as circuitry chips, the drawings must be drawn approximately 100 to 500 times larger.	Not used	100:1 500:1	100:1 500:1

Figure 2-39
Typical scale selection for engineering drawings

Exercises

1. Divide a "B" (11" x 17" or 12" x 18") drawing format into 11 parts as shown in **Figure 2-40**. Practice sketching the line work with drafting pencils.

2. Divide a "B" drawing format into 11 parts as shown in **Figure 2-40**. Practice drawing with instruments and drafting pencils.

Measure each line in **Figure 2-41** with these scales:

full size—inch-decimal

full size—inch-fraction

1/2" = 1"

1" =10'

full size—metric (millimeters)

1/4" = 1"-0"

3. Practice drawing the line work in **Figure 2-42**.

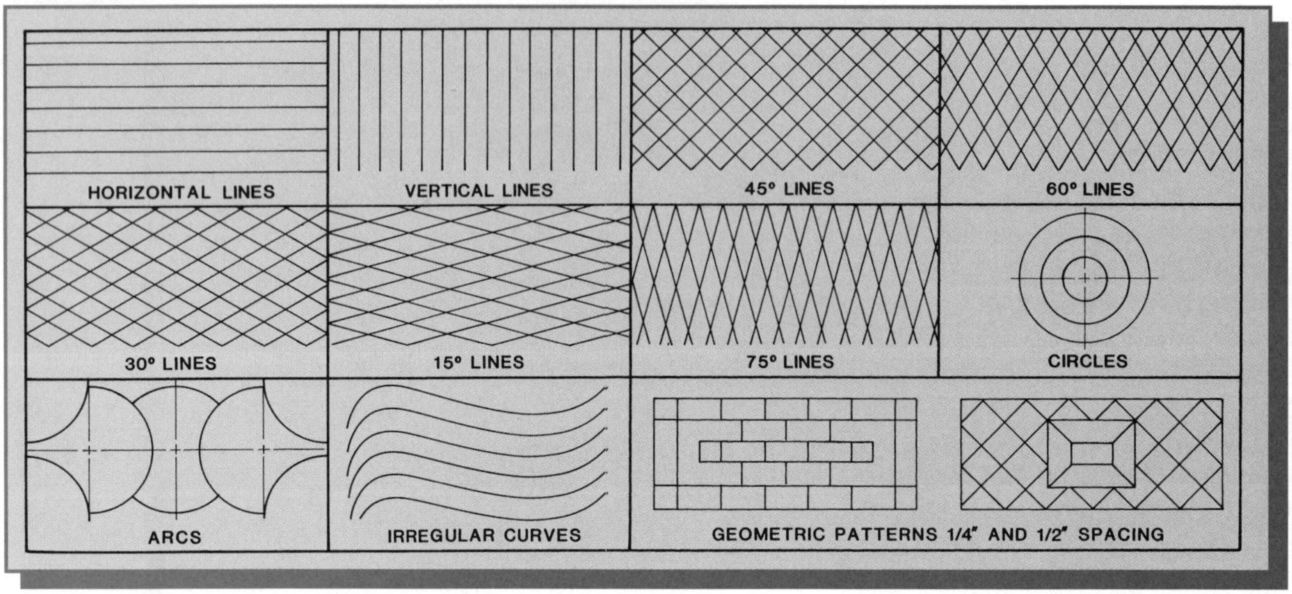

Figure 2-40
Divide four B-size drawing formats into eleven parts as shown. Practice the line work with pencil sketching, drawing instruments and pencil, drawing instruments and a CAD drawing.

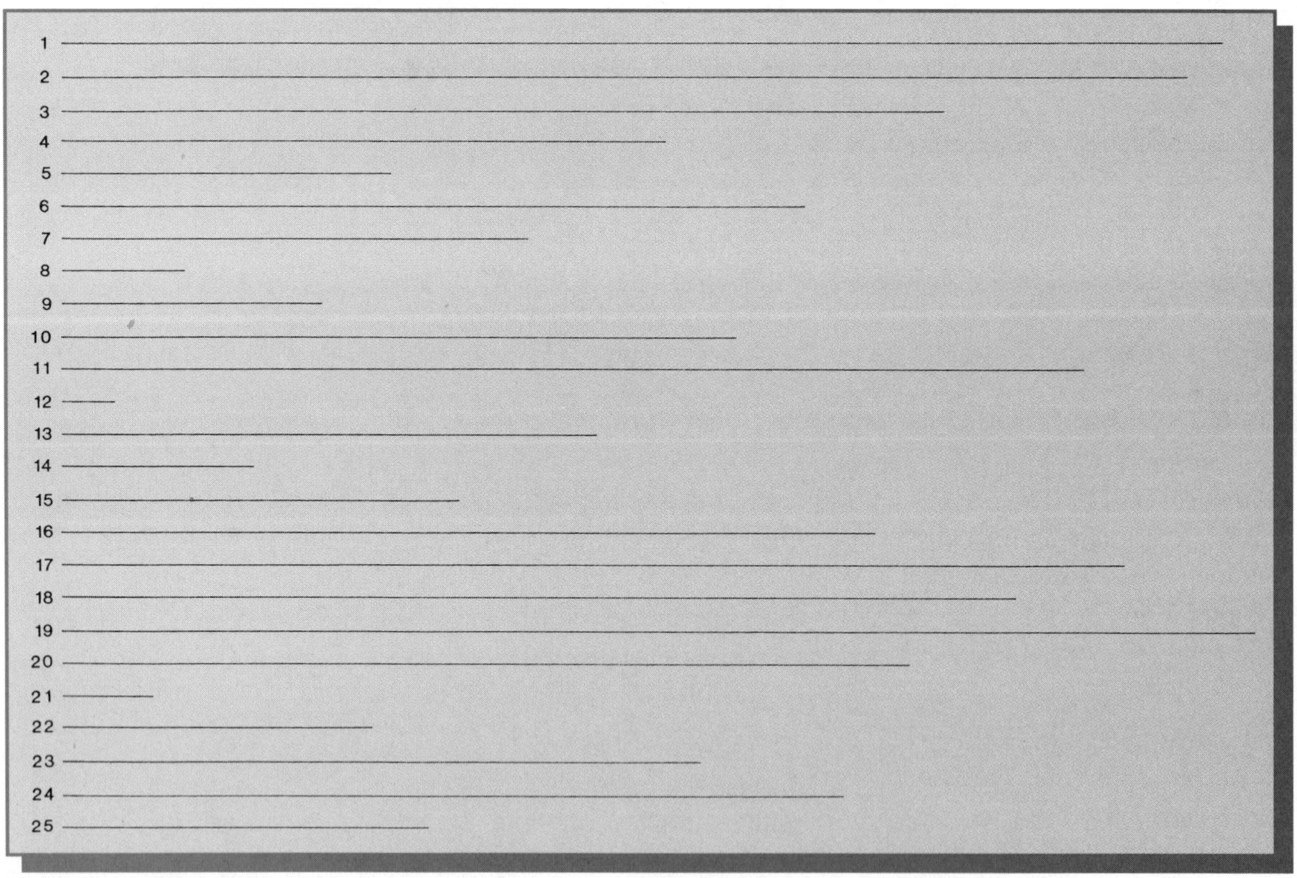

Figure 2-41
Measure each line with as many different scales as are available.

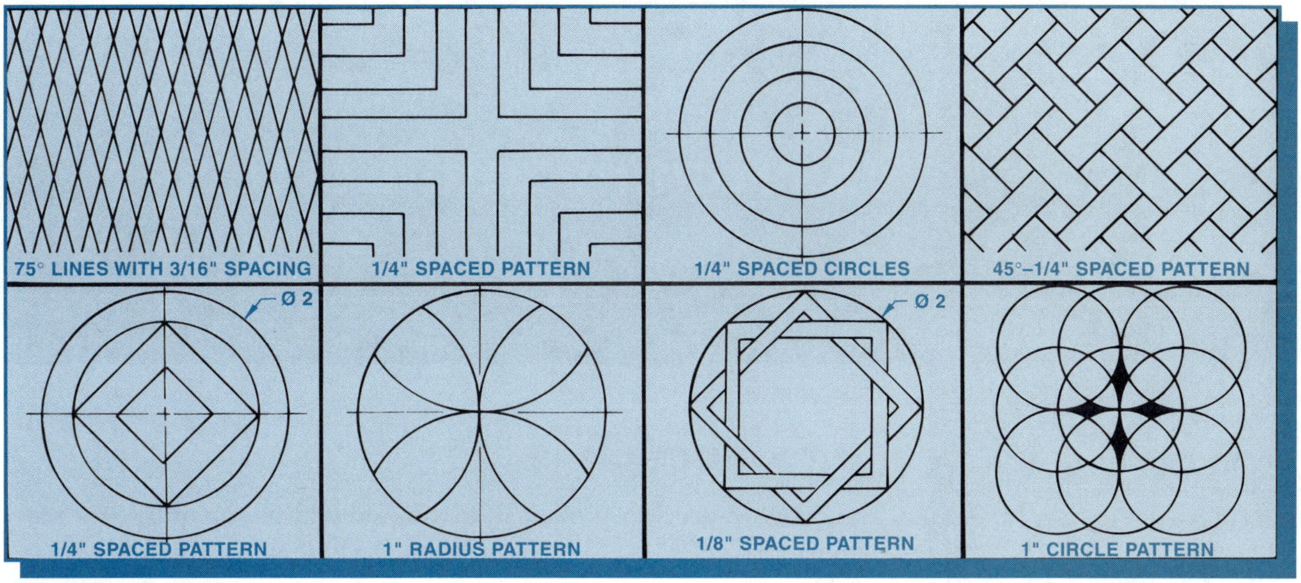

Figure 2-42
Practice drawing the line work.

Design Exercises

1. With your drafting instruments, make a geometric design of your initials about 3" in height.

2. Design a dart board target with the scoring values.

3. Design a rifle range target.

4. With your drafting instruments, design and cut out a puzzle on stiff paper.

Key Terms

Architect's scale
Beam compass
Bow compass
Civil engineer's scale
Dividers
Drafting pencils
Elbow drafter
Film
Friction compass
Full-divided scale
Grid paper
Irregular curve
Lead holder
Marking devices
Mechanical engineer's scale
Metric system
Millimeter (mm)
Opaque drawing paper
Open-divided scale
Parallel slide
Protractor
Surface Media
Technical pen
Template
Tracing paper
Track drafter
Triangles
T square
U.S. customary system
Vellum
Wood pencils

Chapter 3

Sketching and Lettering

Objectives

The student will be able to:

- use proper line weights to sketch and letter
- complete a two-dimensional sketch of simple objects
- complete a three-dimensional sketch of simple objects
- file a record of sketched ideas
- select and sharpen sketching and lettering pencils
- keep sketches in proportion
- dimension a sketch
- shade a sketch
- letter clearly on a sketch

Introduction

Highly-developed manual drafting skills are not required for the preparation of CAD drawings **(Figure 3-1)**; however, a competent CAD operator must possess a thorough understanding of the principles of drafting and be skilled in technical sketching.

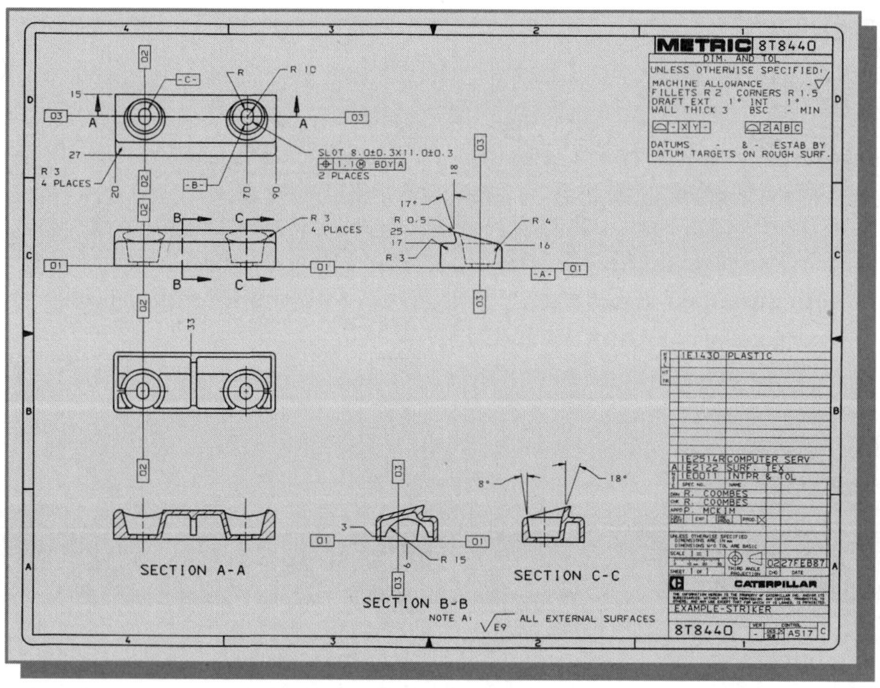

Figure 3-1
No manual drafting skills were required for this CAD-generated working drawing. *(Courtesy of Caterpillar Inc.)*

Observe how the sketch shown in **Figure 3-2** is used to describe the actual object shown in **Figure 3-3**. This is another example of the proverb, "A picture is worth a thousand words."

Supplies

Sketching supplies consist of a wide variety of pencils, erasers, and drawing media. Soft pencils in the range shown in **Figure 3-4** are used for sketching. Pencil points **(Figure 3-5)** are rounded for sketching compared to the sharp points used for instrument drawing. However, sharp points on soft pencils are used for layout and lettering guide line sketching. Chisel points are used for shading. Sketching pencils are sharpened with a mechanical sharpener, or with a knife and file, or sandpaper to shape the point.

Although many types of drawing media are used for technical sketching, **grid paper** is the most practical and easy to use **(Figure 3-6)**. Grids provide **guide lines** for rough scaling and proper alignment of perpendicular and parallel lines. Use light-colored surfaces if sketches are to be photocopied, and use translucent surfaces if diazo prints are to be made. Translucent surfaces are also helpful if progressive design sketches are to be traced.

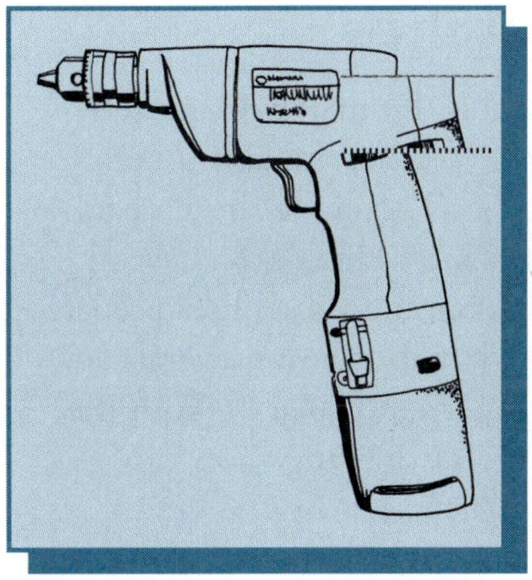

Figure 3-2
A freehand sketch is worth a thousand words.

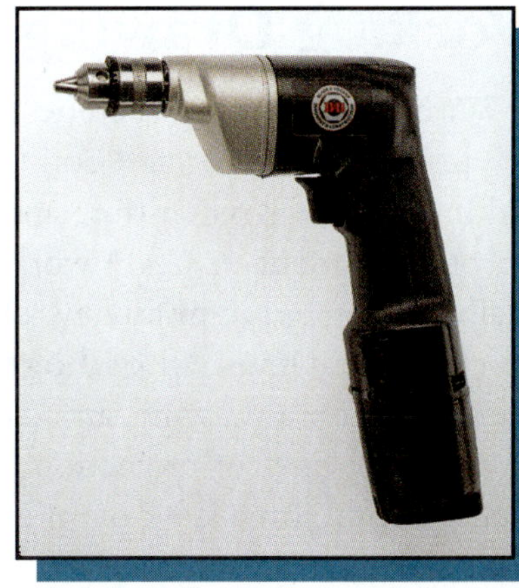

Figure 3-3
An industrial product *(Courtesy of Black and Decker U.S. Power Tools Group)*

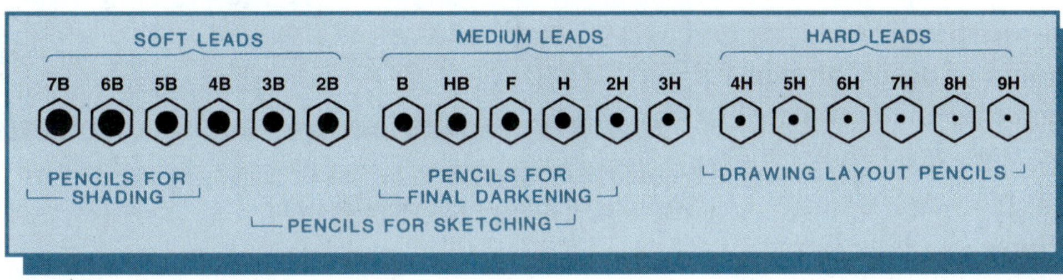

Figure 3-4
Recommended pencil grades for sketching

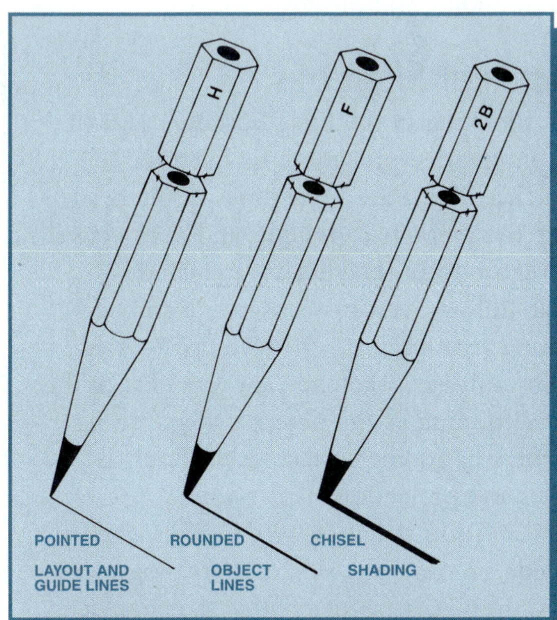

Figure 3-5
Recommended pencil points for sketching

Figure 3-6
Drawing on grid paper is faster and neater. *(Photo by Ann Ross Wallach)*

Working Drawing Sketches

Working drawing sketches are used as the major design reference in manufacturing and construction. Sketches are also used as working drawings when time and conditions preclude the preparation of instrument or CAD working drawings. These sketches are usually orthographic multiview drawings, but are sometimes prepared as pictorial drawings. An **orthographic** drawing shows several views of an object on a drawing surface that is perpendicular to both the view and the lines of projection. A **pictorial** drawing shows an object's depth; three sides of an object can be seen in one view.

Orthographic Multiview Sketches

Multiview sketches provide the greatest amount of detail for manufacturing and construction. **Figure 3-7** shows the steps recommended to complete a multiview sketch. As with instrument drawing (see Chapter 18), blocking in the overall outline before completing the internal details is the key to maintaining correct scales, angles, and proportions.

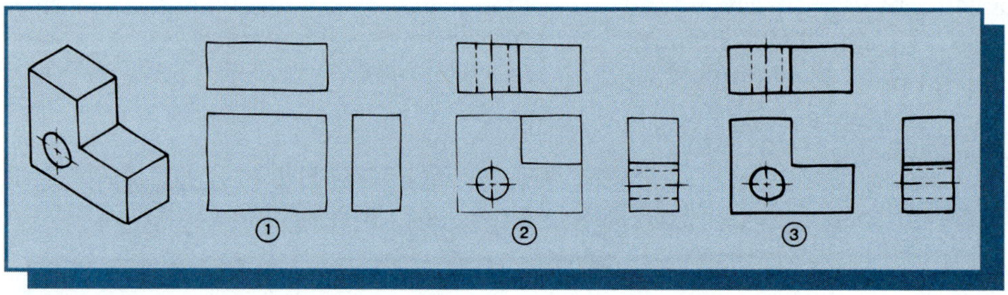

Figure 3-7
Steps to sketch a multiview drawing

Pictorial Sketches

Blocking in the basic outline is also the recommended procedure for pictorial sketches. There are three types of pictorial sketches: isometric, oblique, and perspective.

Isometric sketches are prepared by establishing a vertical corner line and projecting receding lines 30° **(Figure 3-8)**. As with multiview sketches, always block in the entire outline of the object before cutting corners or adding surface details such as holes and projections.

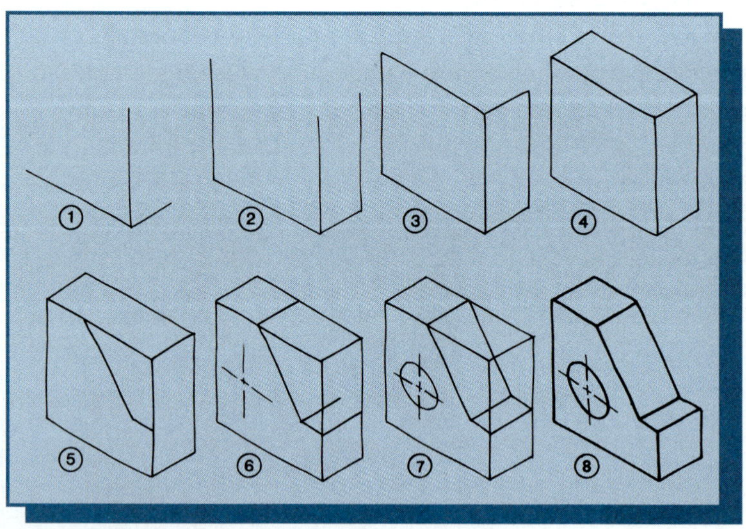

Figure 3-8
Steps to sketch an isometric drawing

Oblique sketches are pictorials that recede on only one side of an orthographic view. In preparing an oblique sketch, first sketch a front view of the object **(Figure 3-9)**. Then extend lines from each corner upward at 45° or 30°. Connect the ends of these lines with lines that are parallel to the lines of the front view. Oblique sketches are easy to dimension since the width and length of the front view are sketched to actual scale. Only the depth dimension is foreshortened by the receding lines.

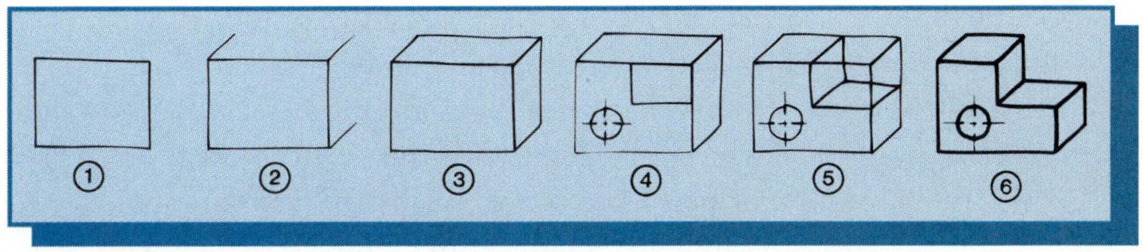

Figure 3-9
Steps to sketch an oblique drawing

Perspective sketches are pictorials that contain receding sides designed to approximate the actual appearance of an object. There are three types of perspectives: one-point, two-point, and three-point.

1. **One-point perspective** sketches are similar to oblique drawings, except the receding lines do not follow a consistent angle. Receding lines are connected to a **vanishing point**. A vanishing point represents the point at which all receding lines appear to come together. It is similar to road or train tracks disappearing on the horizon. In sketching a one-point perspective, first outline the front view, as in oblique sketching. Establish a vanishing point and sketch lines from each corner to this point **(Figure 3-10)**. Sketch lines representing the back of the object parallel to the lines on the front view. This blocks in the sketch. Now any angular cuts and details can be added.

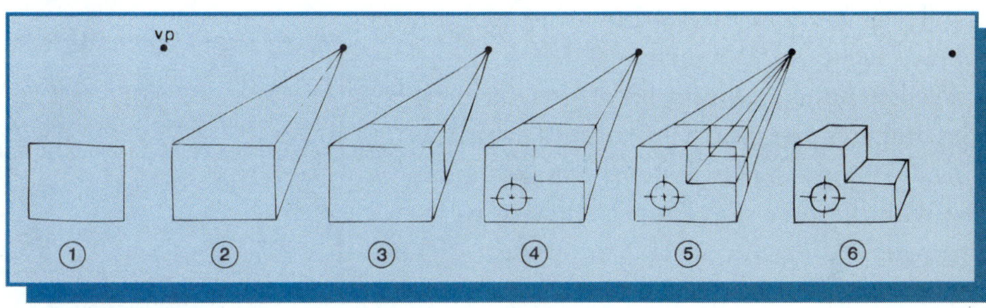

Figure 3-10
Steps to sketch a one-point perspective drawing

2. **Two-point perspective** sketches are similar to isometric sketches, except side lines recede to two vanishing points. First sketch the front vertical corner **(Figure 3-11)**. Establish two vanishing points above or below the front corner line, and connect the top and bottom of the front corner line to each of these points. Vanishing points on two-point perspective sketches must always be located on a horizontal line representing the horizon. Next, establish the depth of the sides and

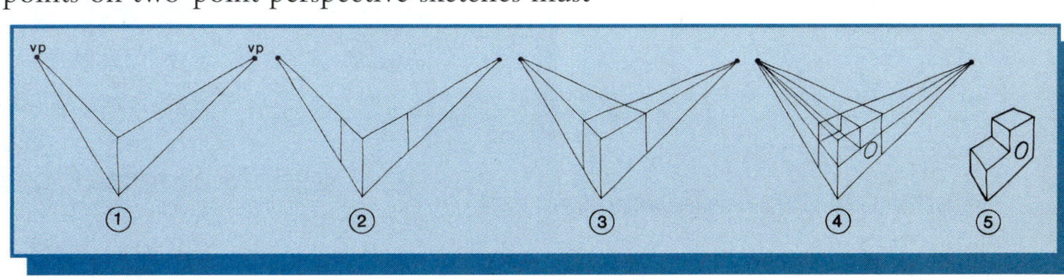

Figure 3-11
Steps to sketch a two-point perspective drawing

connect the back corner lines also to the vanishing points. In two-point perspectives, vertical lines are all parallel.

3. In **three-point perspectives**, vertical lines are projected to a third vanishing point which is aligned with the vertical front corner line. Three-point perspectives are usually used for architectural pictorials, and rarely for technical drawings.

Sketching Guidelines and Procedures

Line conventions and lettering for technical sketches are similar to those used for instrument drawings **(Figure 3-12)**. Sketching standards differ only in the degree of line raggedness. Lines are dark and wide for object lines, dark and thin for dimension and center lines, and thin and light for layout and guide lines. **Figure 3-13** shows the application of line conventions to a technical sketch.

All sketches are comprised of straight lines, circles and arcs, irregular curves, and letters and numerals **(Figure 3-14)**. Sketches cannot be prepared with the precision and accuracy of an instrument or CAD drawing. Care must be taken to insure that dimensions are relatively proportional. If dimensional proportions are grossly inaccurate, the sketch will misrepresent the actual appearance of the object **(Figure 3-15)**.

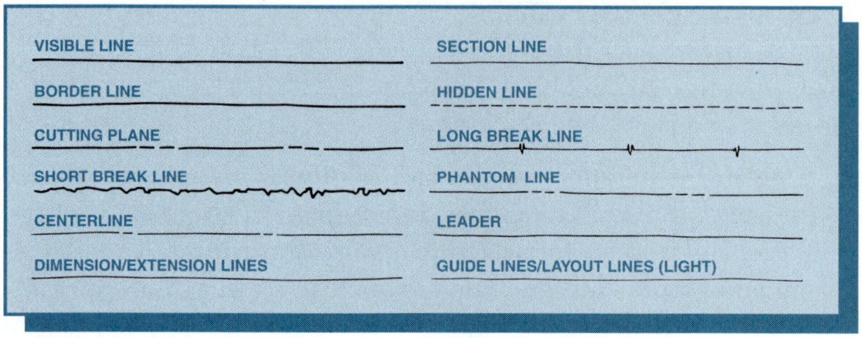

Figure 3-12
Line conventions for sketching working drawings

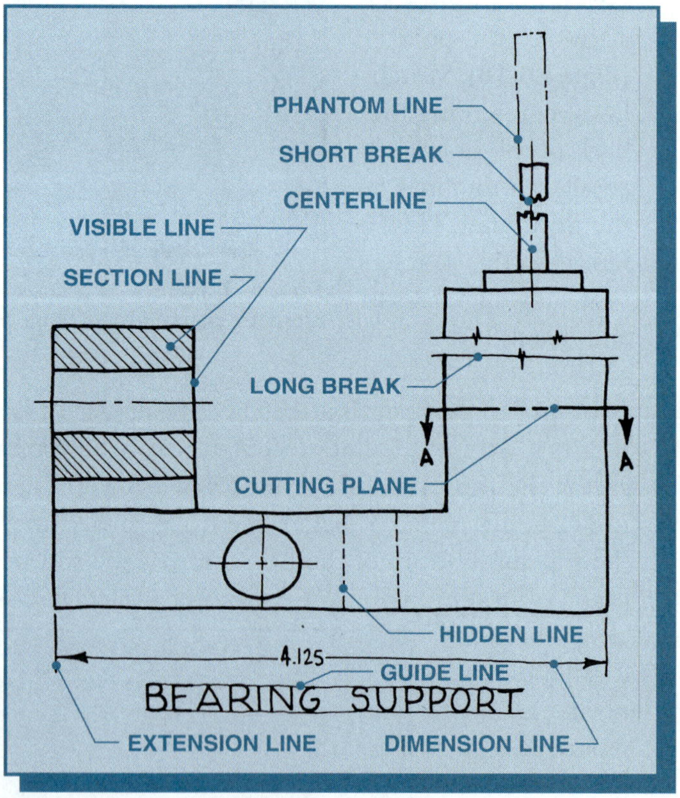

Figure 3-13
Using line conventions on a working drawing

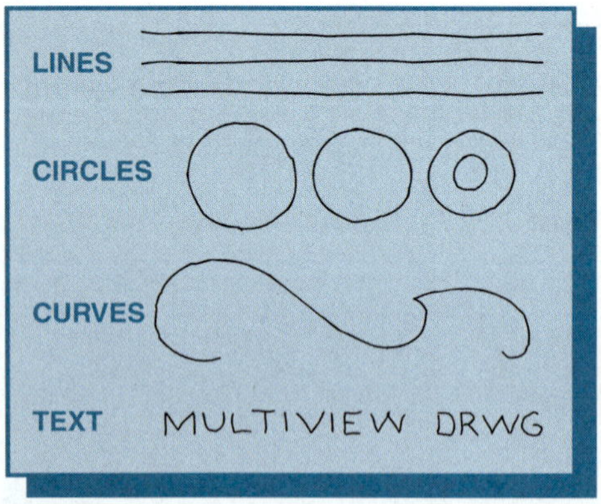

Figure 3-14
Four basic drawing forms make up all drawings.

When sketching straight lines, squares, and rectangles, use short strokes. Do not attempt to draw continuous lines. Right-angle lines, unless sketched on grid paper, should be laid out and sketched **(Figure 3-16)**.

Circles and arcs can be accurately and symmetrically sketched by following the sequence shown in **Figure 3-17a**. Just as the circle was derived from the square in **Figure 3-17a**, all fillets and rounds should first be blocked in square, as shown in **Figure 3-17b**. By following this procedure, proper proportions and symmetry can be maintained.

The procedure and sequence for sketching ellipses is similar to circle sketching **(Figure 3-18)**. Sketching accurate angles, other than right angles (90°), can be difficult without using a triangle or protractor; however, by estimating and dividing a right angle into even angles, an acceptable level of accuracy can be achieved **(Figure 3-19)**.

When sketching, hold the pencil comfortably **(Figure 3-20)** and pull it **(Figure 3-21)**, never push it. To maintain a consistently rounded point and avoid wearing a flat chisel point, rotate the pencil frequently **(Figure 3-22)**. When erasing soft pencil sketches, use a good quality medium-soft eraser.

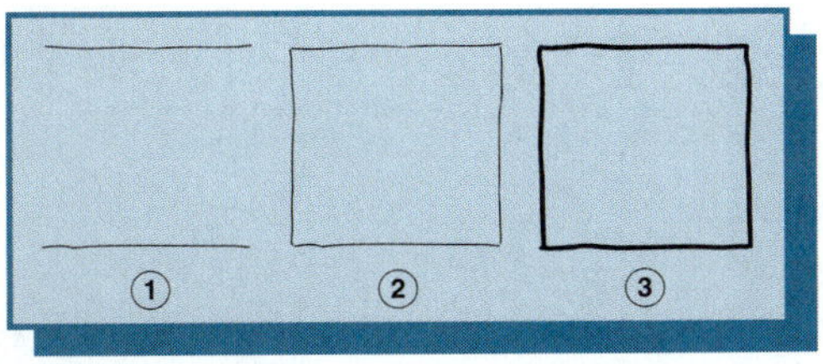

Figure 3-16
Sketching straight lines, squares, and rectangles

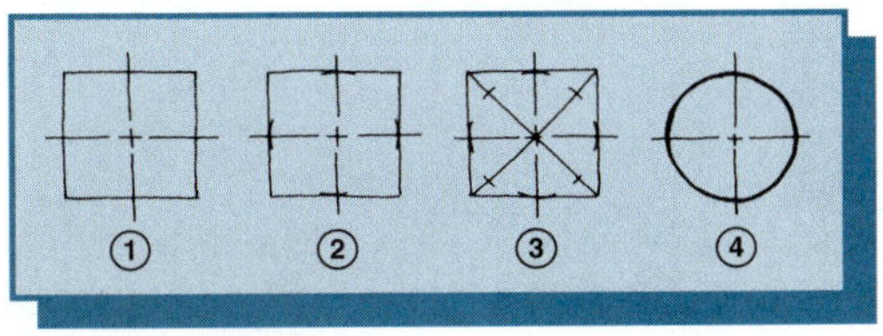

Figure 3-17a
Sketching circles

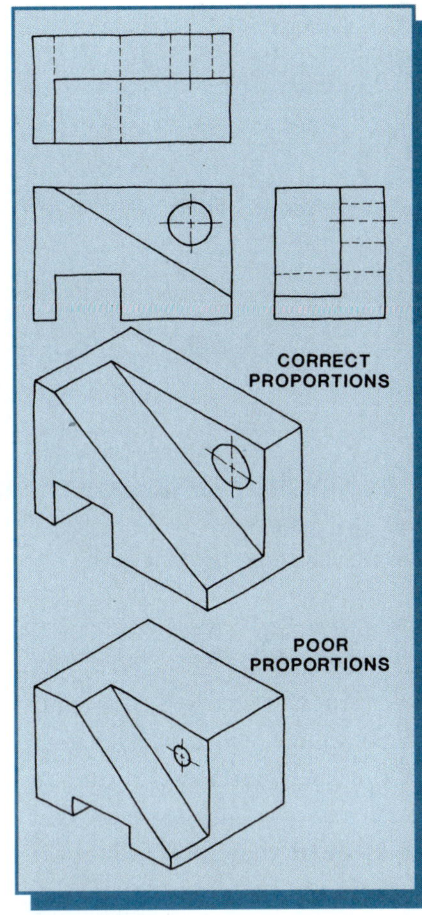

Figure 3-15
Well-proportioned drawings have superior communication.

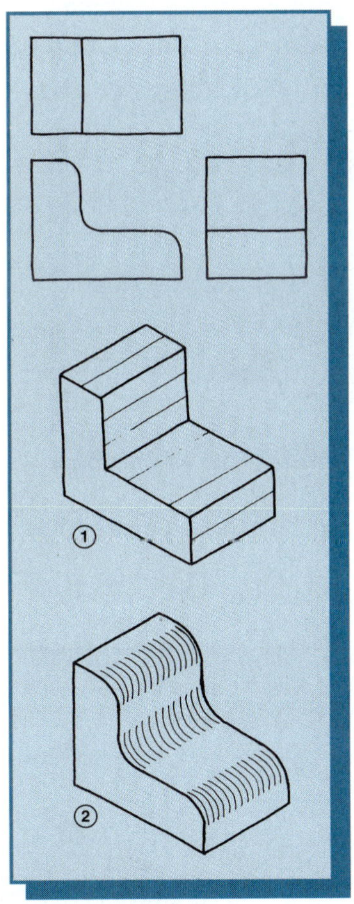

Figure 3-17b
Sketching fillets and rounds

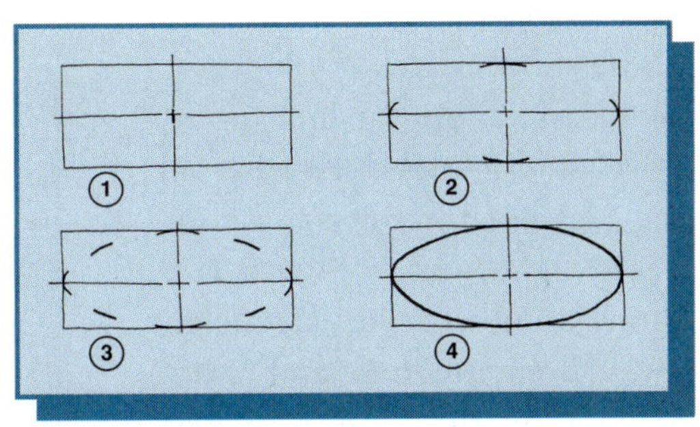

Figure 3-18
Sketching ellipses

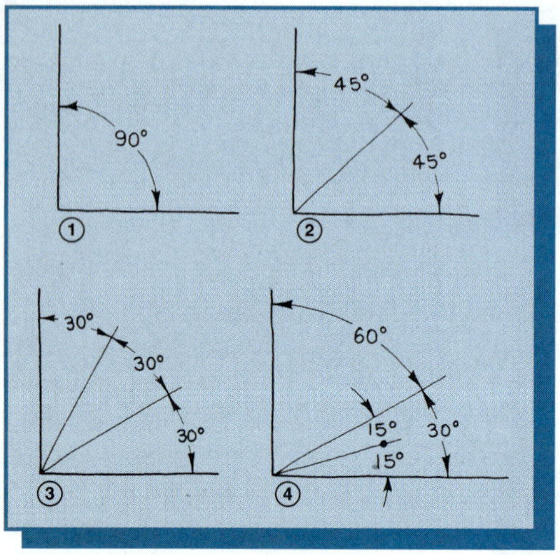

Figure 3-19
Sketching and estimating angles

Figure 3-20
Use a comfortable pencil grip for sketching. *(Photo by Ann Ross Wallach)*

Shading

Surfaces exposed to a major light source will appear bright. Conversely, surfaces not directly exposed to a light source will be shaded; therefore, adding shading to sketches creates realism **(Figure 3-23)**. In **Figure 3-23**, the light source is located above and to the left of the object. The opposite areas are shaded because direct light is blocked from these surfaces. **Figure 3-24** shows techniques for shadowing (**shading**) these areas.

Surfaces are rarely totally hidden from a light source. Some surfaces are very light, very dark, or appear in a variety of shadow grades depending on the position, intensity, and number of light sources. **Figure 3-25** shows an object with three light levels: light, medium, and dark. In addition, this figure shows a separate shadow cast by the object.

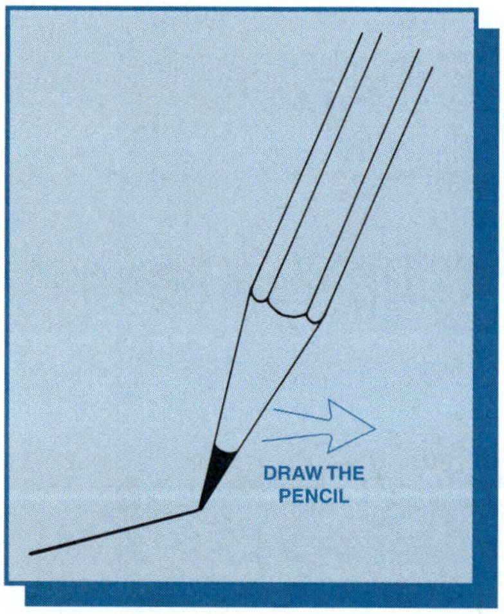

Figure 3-21
Always pull the pencil—never push it.

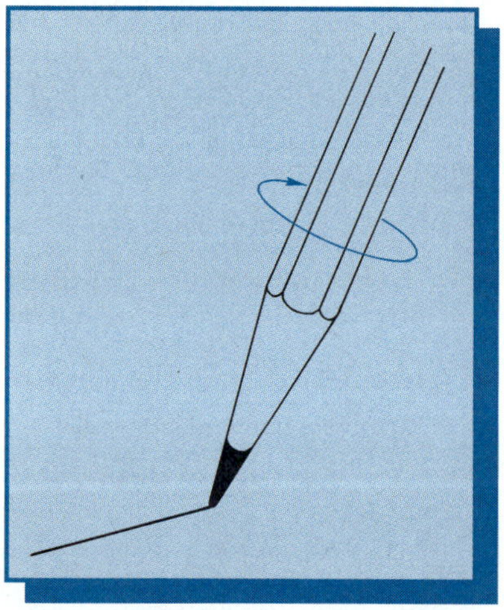

Figure 3-22
Rotate the pencil frequently for rounded points.

Figure 3-23
Shade (darken) the surfaces opposite the light source.

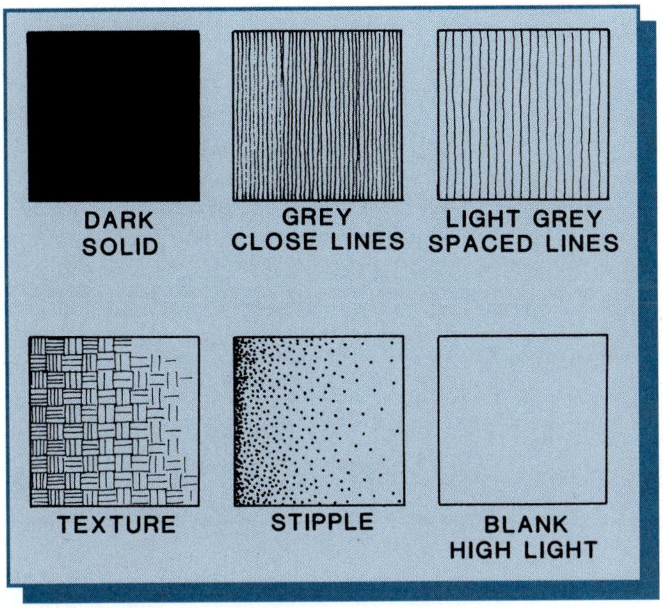

Figure 3-24
Shading techniques

Figure 3-25
The light source dictates the position of the shadows.

Light travels in a straight line and cannot bend around corners unless reflected; therefore, light intensity on surfaces always changes at the corner of objects. **Figure 3-26** shows several methods of sketching these differences to add realism to a sketch. On objects without corners, such as spheres and cylinders, light and dark areas change gradually. **Figure 3-27** shows several methods of shadowing a cylinder sketch to add realism.

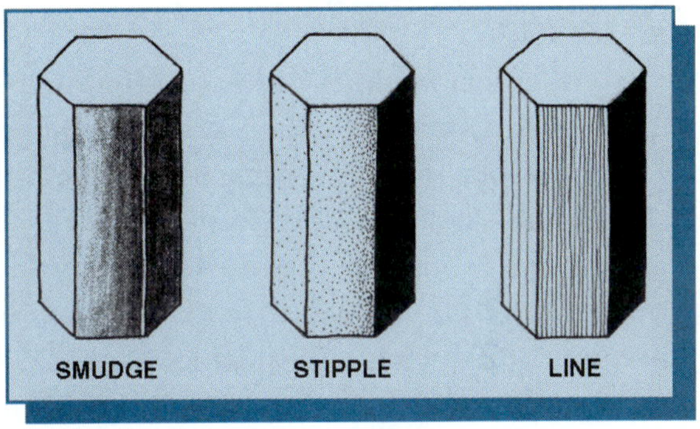

Figure 3-26
Freehand shading of sharp corners

Dimensioning

When sketches are used for instruction, manufacturing, or construction, dimensions are usually necessary to adequately describe the object. When sketches are dimensioned, the overall width, depth, and height dimensions are placed on the outside of the **location dimensions (Figure 3-28).** These are known as **overall dimensions.** Dimension lines are sketched parallel to object lines and connected to the object with extension lines and arrows. Location dimensions show the location of parts of an object. **Size dimensions** show the size of any hole or projection on the object and are placed between the object and the overall dimension lines. Dimensions are usually shown on multiview sketches. Pictorial sketches **(Figure 3-29)** are normally used only to show the general appearance of products, and do not require dimensions.

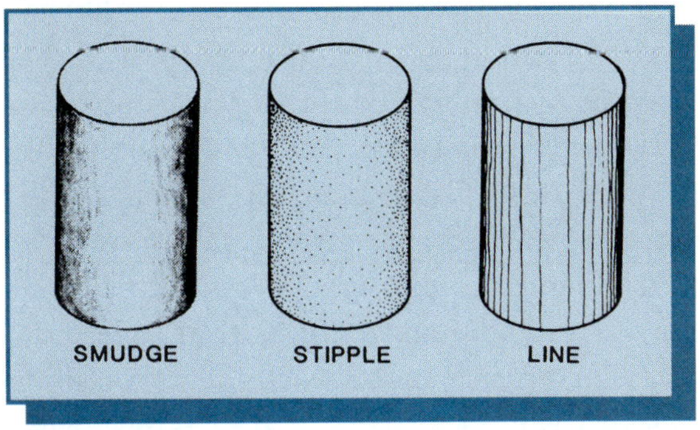

Figure 3-27
Freehand shading of rounded surfaces

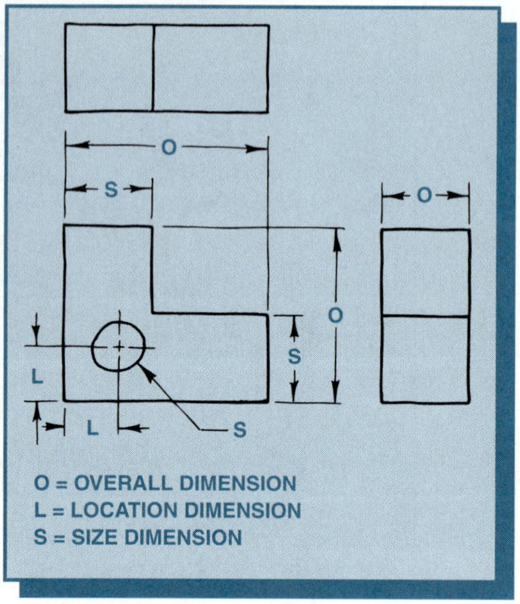

Figure 3-28
Types of dimensions

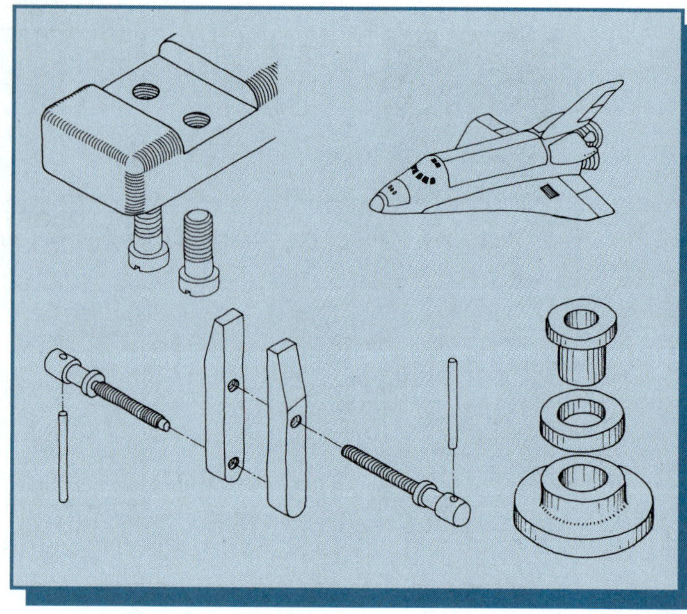

Figure 3-29
Examples of pictorial sketches

Lettering

Some information is best communicated with words or numerals. Almost every sketch or drawing contains notes, labels, and dimensions that must be legible and consistent. In **Figure 3-30**, the title, material, and hole size are described with a lettered notation. Each numeral used for dimensioning must be distinct from all other numerals. The letter *D* which looks like an *O*, or the numeral *3*, which looks like an *8*, can be easily misread and create costly manufacturing or construction mistakes. For this reason, the **American Society of Mechanical Engineers (ASME)** style of letters and numerals **(Figure 3-31)** is used by most American industries. **Figure 3-32** shows the most efficient and quickest method of making these letters and numerals. A slanted form of this style **(Figure 3-33)** is acceptable, but is rarely used or recommended for engineering drawings.

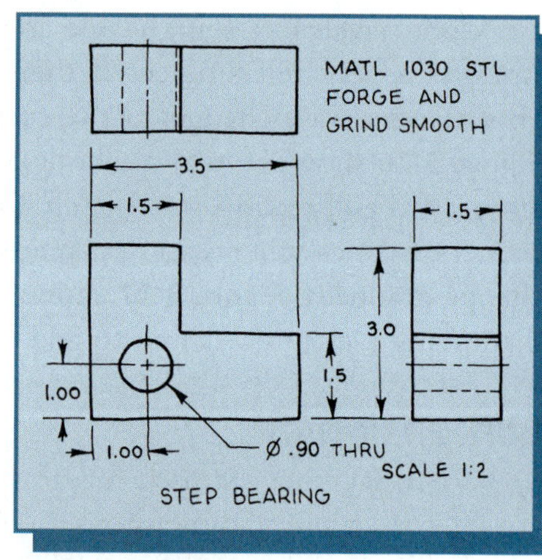

Figure 3-30
Working drawing sketch with dimensions and notations

To letter most efficiently and legibly, always use guide lines **(Figure 3-34)**. Guide lines keep letters consistent and contained in the area selected. As a general rule, most lettering is $^{1}/_{8}$" high with $^{3}/_{16}$" high title headings. Drawings to be microfilmed should be $^{3}/_{16}$" or $^{1}/_{4}$" high, using the **microfont** style shown in **Figure 3-35**.

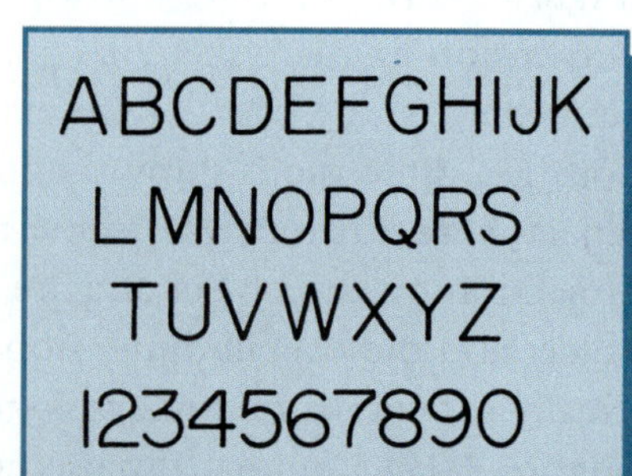

Figure 3-31
Single-stroke vertical Gothic lettering (ASME standards)

Correct spacing between letters is also needed to produce lettering that is most readable. **Figure 3-36** shows lettering that has the area between the letters approximately equal. By contrast, **Figure 3-37** shows wide and inconsistent spacing of letters. Notice how difficult it is to quickly read **Figure 3-37** compared to **Figure 3-36**. Spacing between words and sentences is also important for effective

Figure 3-32
Recommended lettering strokes

Figure 3-33
Single-stroke slant (68°) Gothic lettering

Figure 3-34
Guide lines will aid freehand lettering.

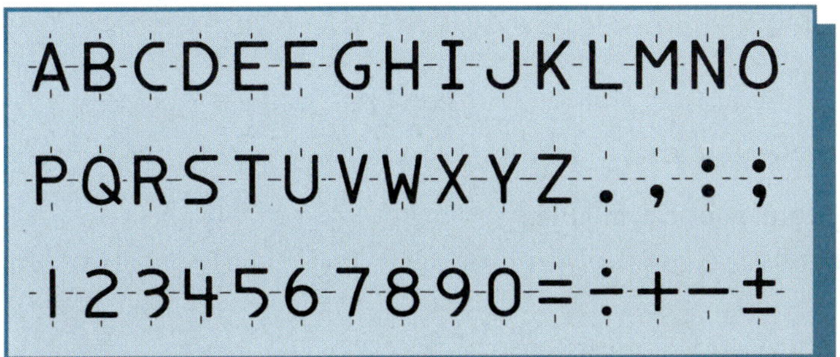

Figure 3-35
Microfont lettering is recommended if drawing is to be microfilmed.

reading. The space between words should be approximately equal to the height of the letters. The space between sentences should be twice the height of the letters **(Figure 3-38)**.

Fractions are one of the most frequent causes of dimensional errors. If a numerator or denominator is misread as a whole number, the result can be disastrous. For this reason, the ASME spacing for fractions **(Figure 3-39)** should be used to avoid confusion.

When all of the procedures and guidelines for sketching are combined, the result can be a readable and functional technical sketch as shown at the left in **Figure 3-40**. If these guidelines are ignored, the confusing series of lines and letters shown at right in **Figure 3-40** can result.

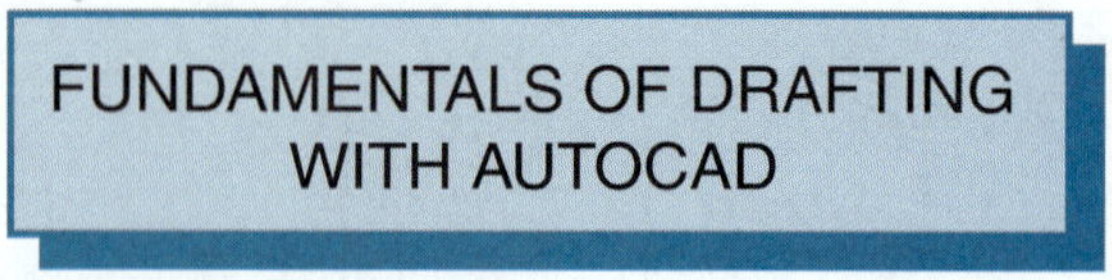

Figure 3-36
The area between letters should be as equal as possible.

Figure 3-37
Inconsistent areas between letters create reading difficulties.

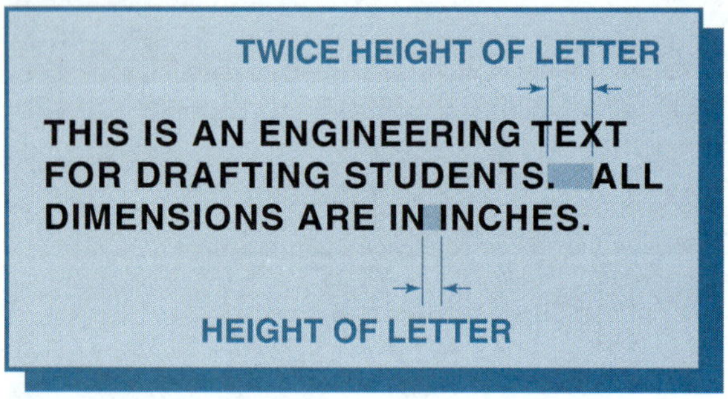

Figure 3-38
Spacing between words and sentences

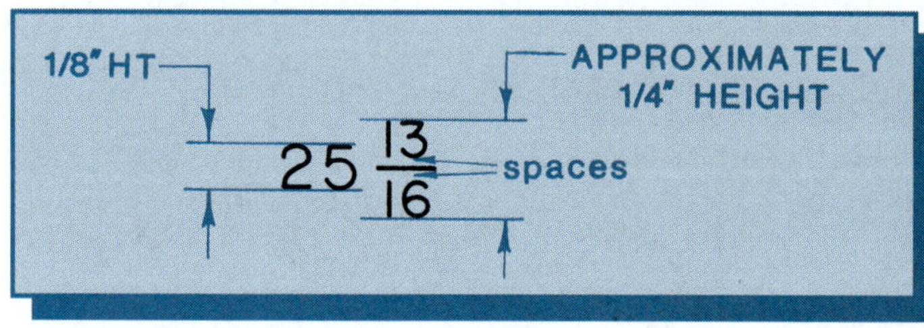

Figure 3-39
Fractions on the working drawing

Computer-Aided Drafting and Design

Sketching and the Computer

Before the CAD operator starts designing with a CAD system it is important to have the critical design features and ideas made with a freehand sketch. Examples of a designer's sketches and the finished CAD drawing are shown in **Figures 3-41**, **3-42**, and **3-43**. Creative thinking is usually easier with a sketch. The ability to quickly and legiblly sketch ideas and products is often a prerequisite for the CAD operator.

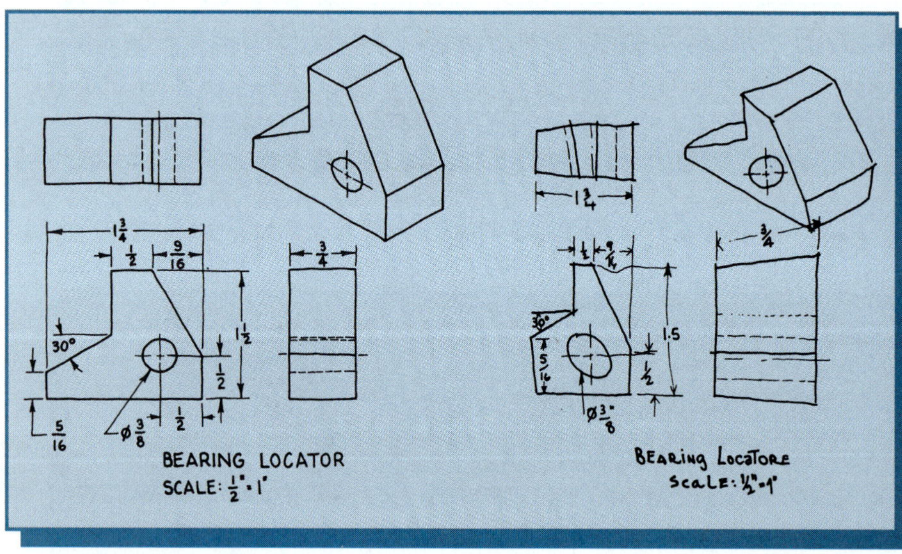

Figure 3-40
Which sketch gives the best communication?

The CAD operator's sketch may have many uses, such as:

1. to record ideas immediately when a CAD system is not available.
2. to help organize new designs with many versions.
3. to help formulate new ideas quickly.
4. to make certain that others understand your ideas and requests by reading the sketch.

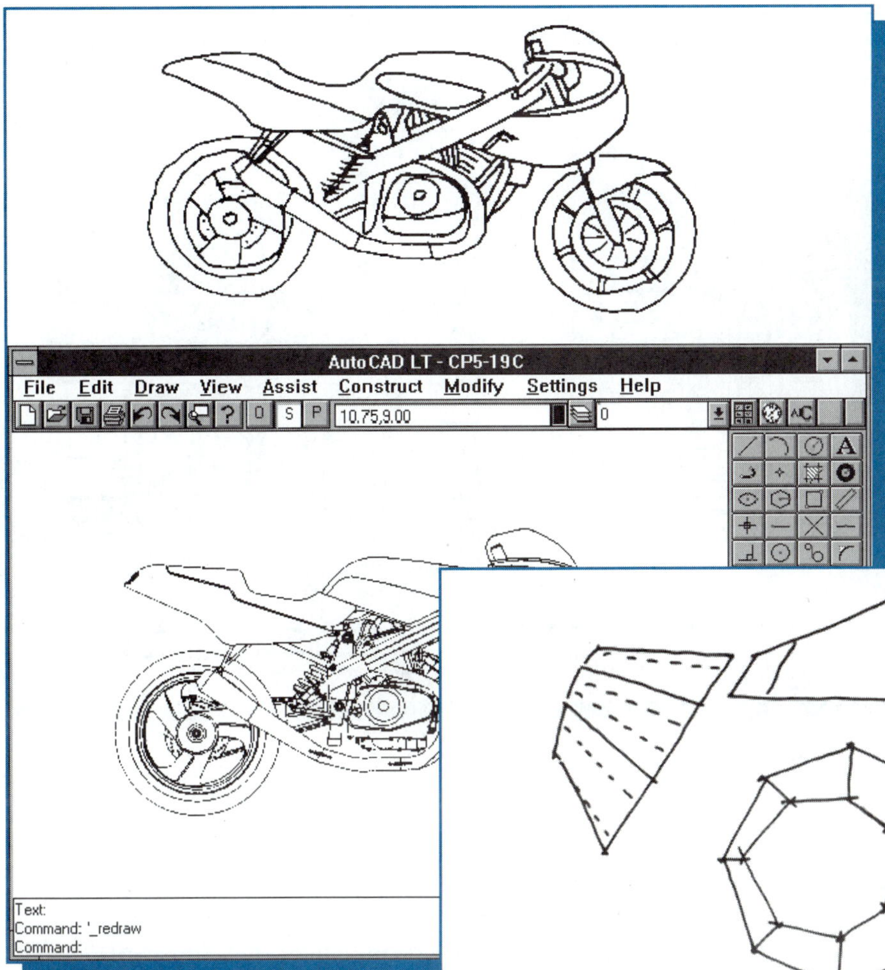

Figure 3-41
A sketch of a motorcycle and the CAD drawing

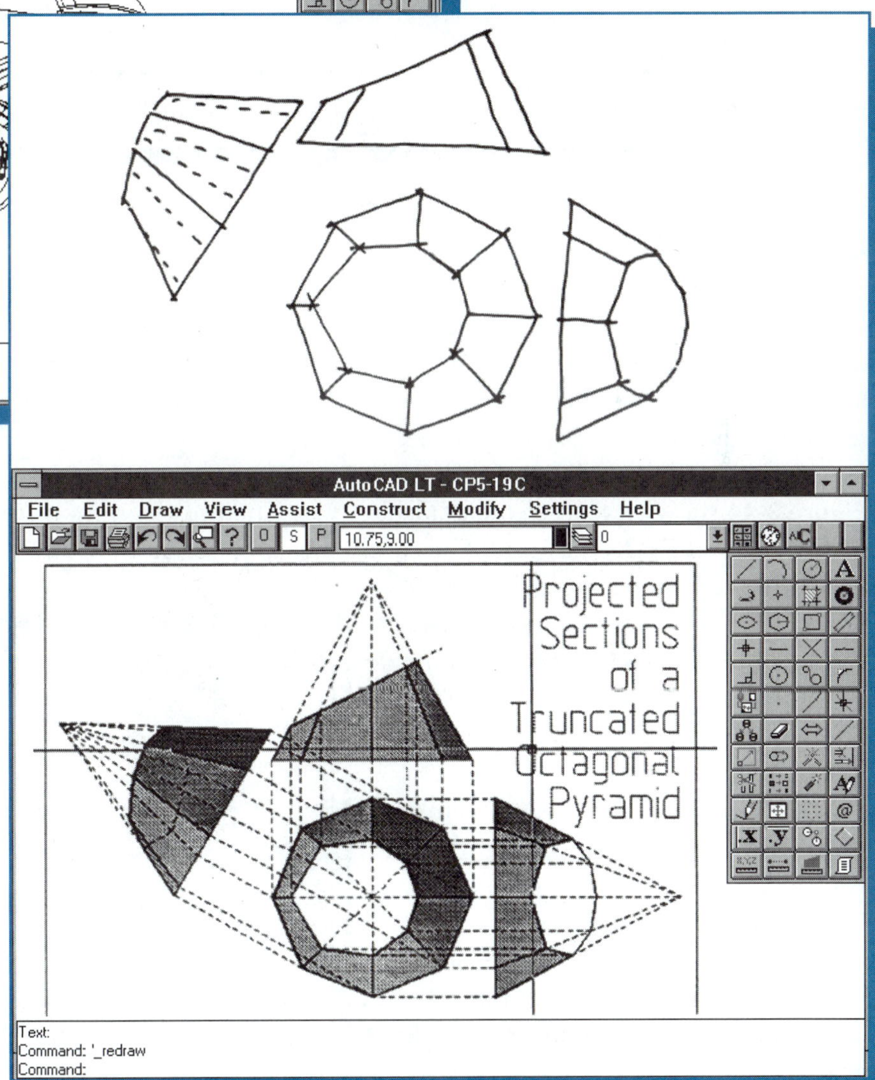

Figure 3-42
A sketch of a truncated cone and the CAD drawing

5. to explain complex details prior to developing the CAD working drawings.
6. to make design changes and updates prior to developing the CAD working drawings.
7. to discuss with others the creative design.
8. to help plan the sequences to be used in preparing the CAD working drawings.
9. to file the various alternative design sketches for future reference.
10. to obtain approval from supervisors before starting the CAD working drawings.

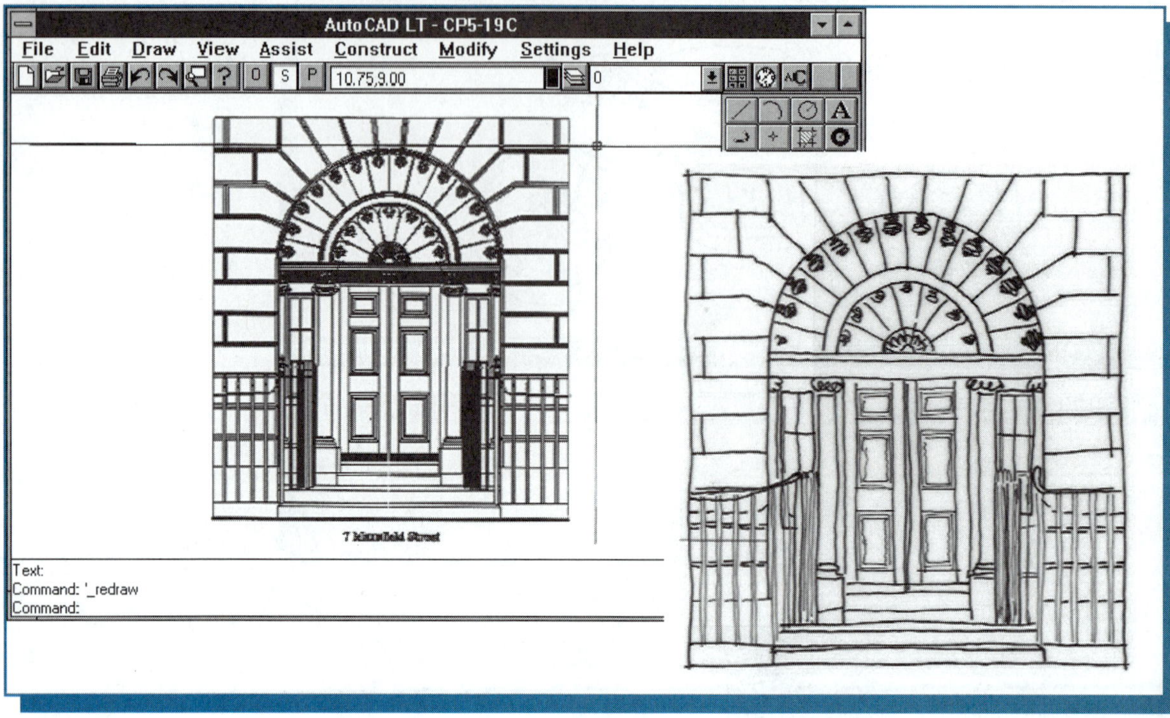

Figure 3-43
Example of a sketched entry and the CAD drawing

Computer-Aided Lettering

CAD lettering represents one of the greatest advantages of CAD drafting over manual drafting. When using a CAD system, all notes, dimensions, labels and material and parts lists are keyboarded rather than hand lettered. Lettering on a CAD system not only increases the speed of the lettering process but also produces dense, consistent letters and numerals in a variety of styles and sizes.

Lettering using CAD is called TEXT. There are many STYLES of text available to the CAD drafter. **Figure 3-44** shows several of the available styles. The most common style used for mechanical drafting is ROMANS. CITY BLUEPRINT is commonly used for architectural drafting.

In addition to having different styles, you can modify each style as shown in **Figure 3-45**. There are many combinations and possibilities that may be useful on drawings, charts, graphs, and graphic designs.

Another advantage to CAD lettering (TEXT) is that it is easily modified. After text has been placed on a drawing, it is simple to correct its location, size, and style, as well as the spelling.

Chapter 5 shows how to put text on a drawing, change styles, and modify it.

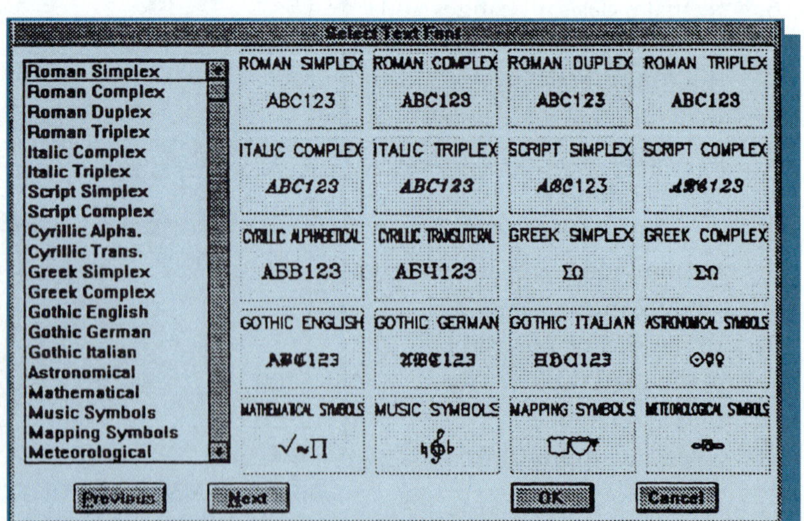

Figure 3-44
Some of the available text styles

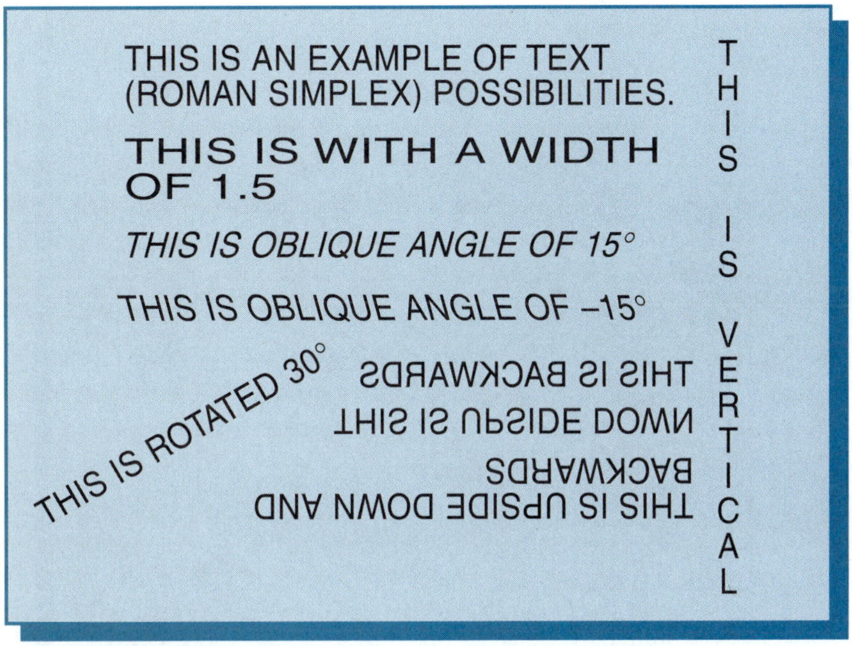

Figure 3-45
Text possibilities

Exercises

1. Sketch the isometric drawings in **Figures 3-46** through **3-56** on isometric grid paper.
2. Sketch the multiview drawings in **Figure 3-57**.
3. Sketch the pickup truck in **Figure 3-58**.
4. Sketch the objects in **Figure 3-59**.
5. Sketch the tool-bit holder in **Figure 3-60**.
6. Practice lettering single-stroke Gothic lettering on $^{1}/_{8}$" grid paper.
7. Practice placing various types of fonts in different sizes and angular positions with CAD system.

Figure 3-46
Sketch the approximate geometric forms on isometric grid paper.

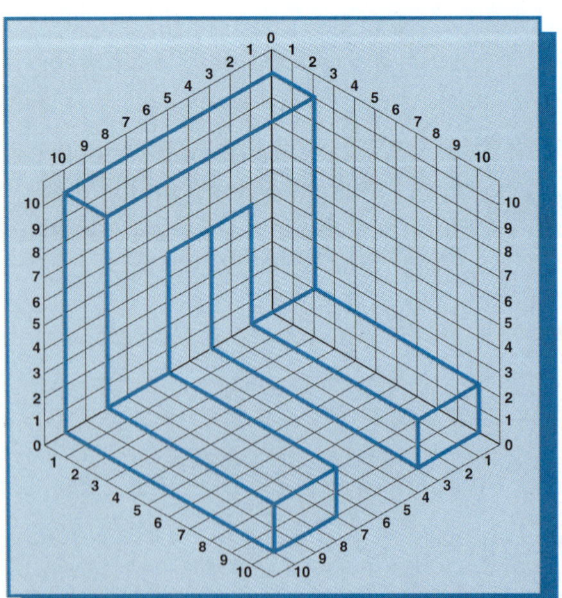

Figure 3-47

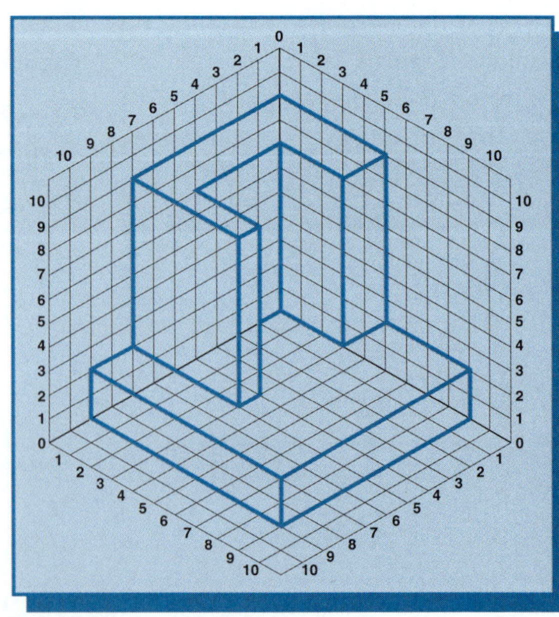

Figure 3-48

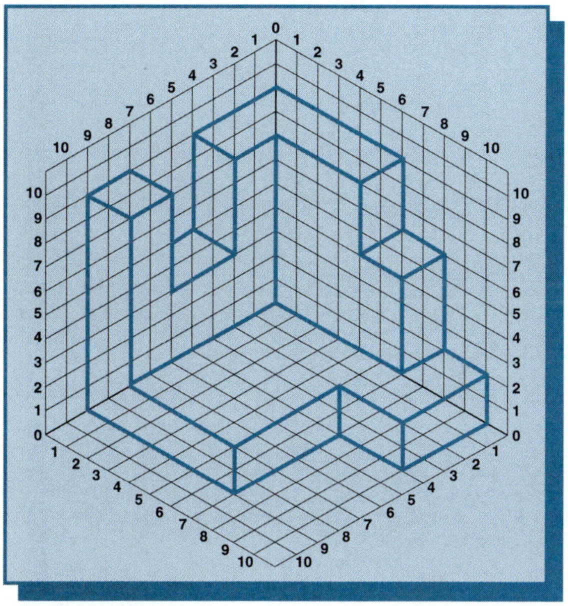

Figure 3-49

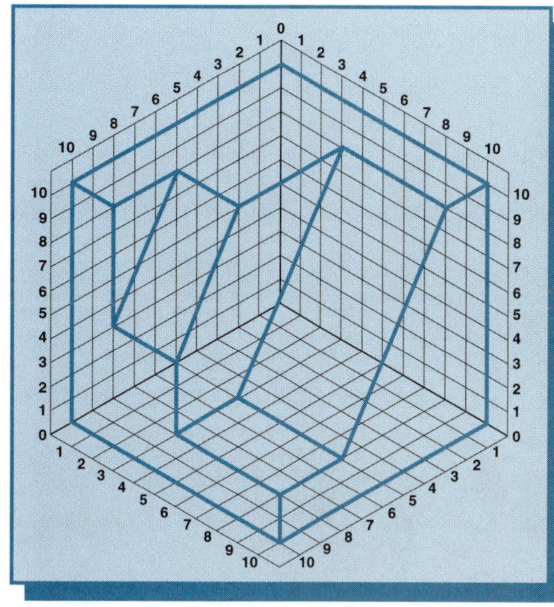

Figure 3-50

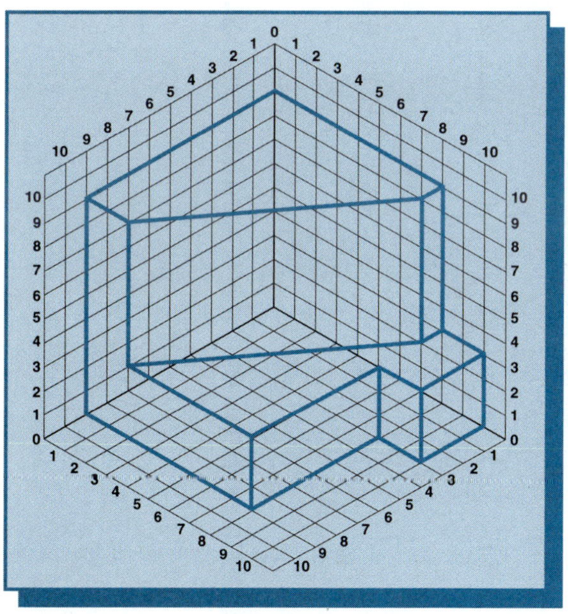

Figure 3-51

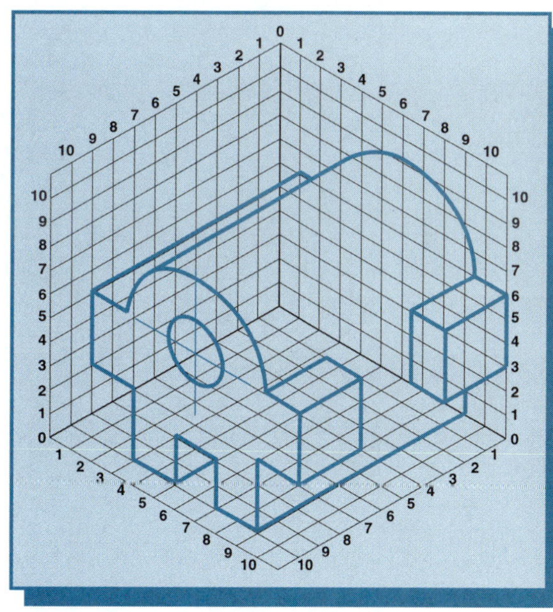

Figure 3-52

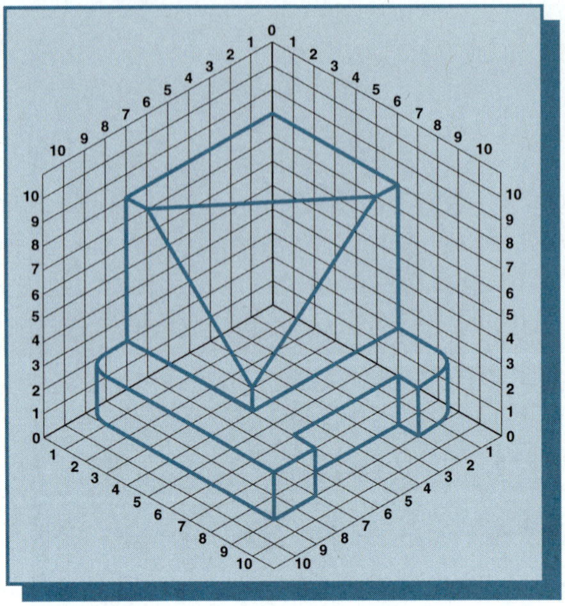

Figure 3-53

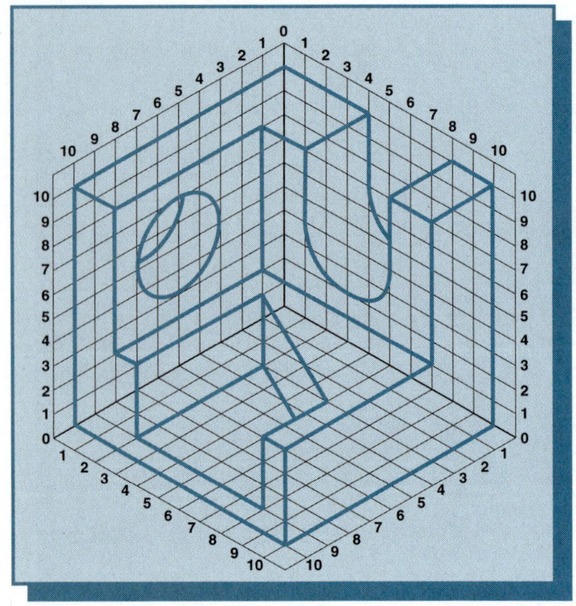

Figure 3-54

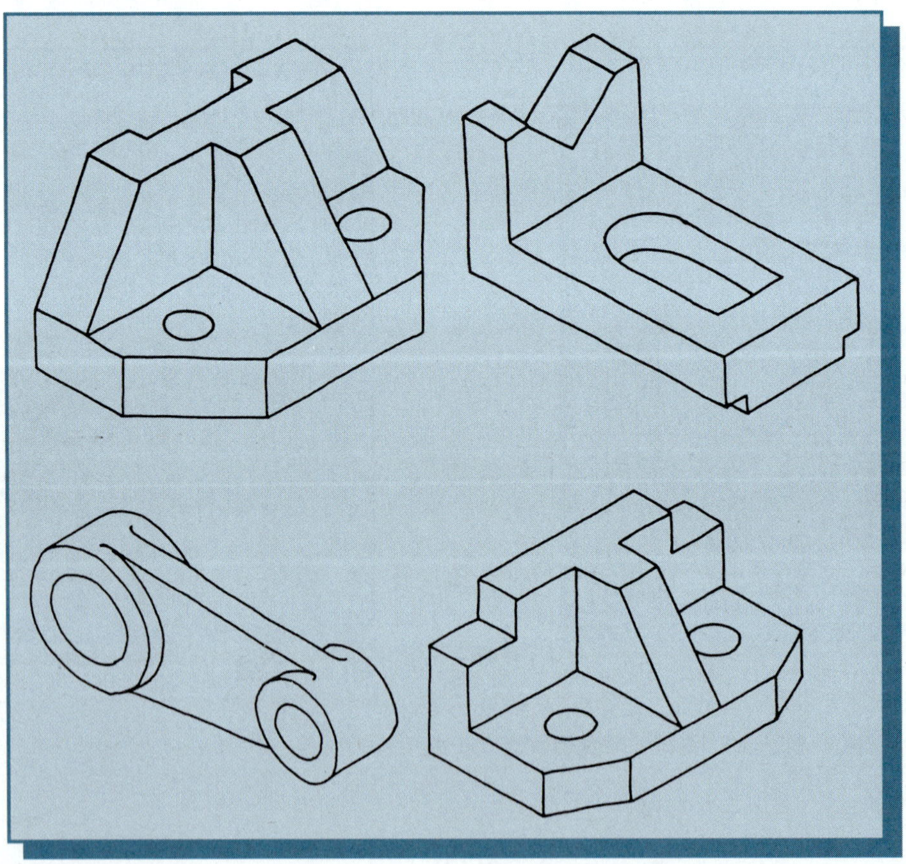

Figure 3-55
Practice drawing the isometric blocks freehand without isometric grid paper.

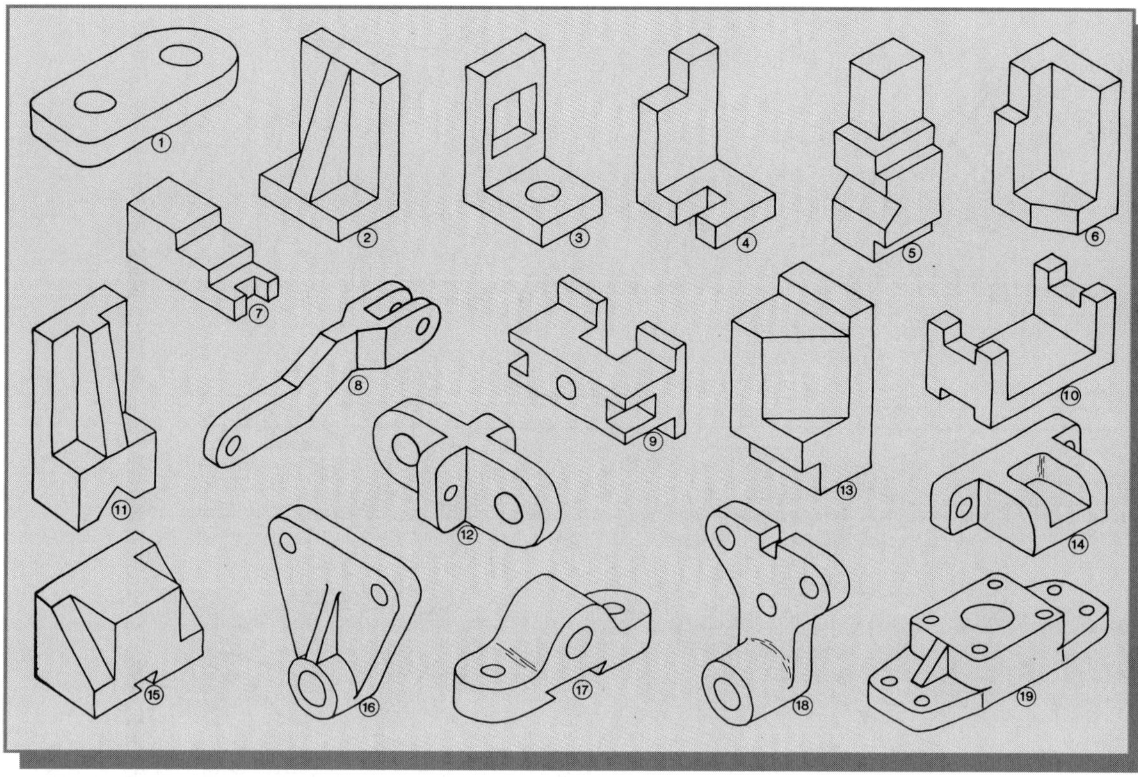

Figure 3-56
Isometric sketching practice

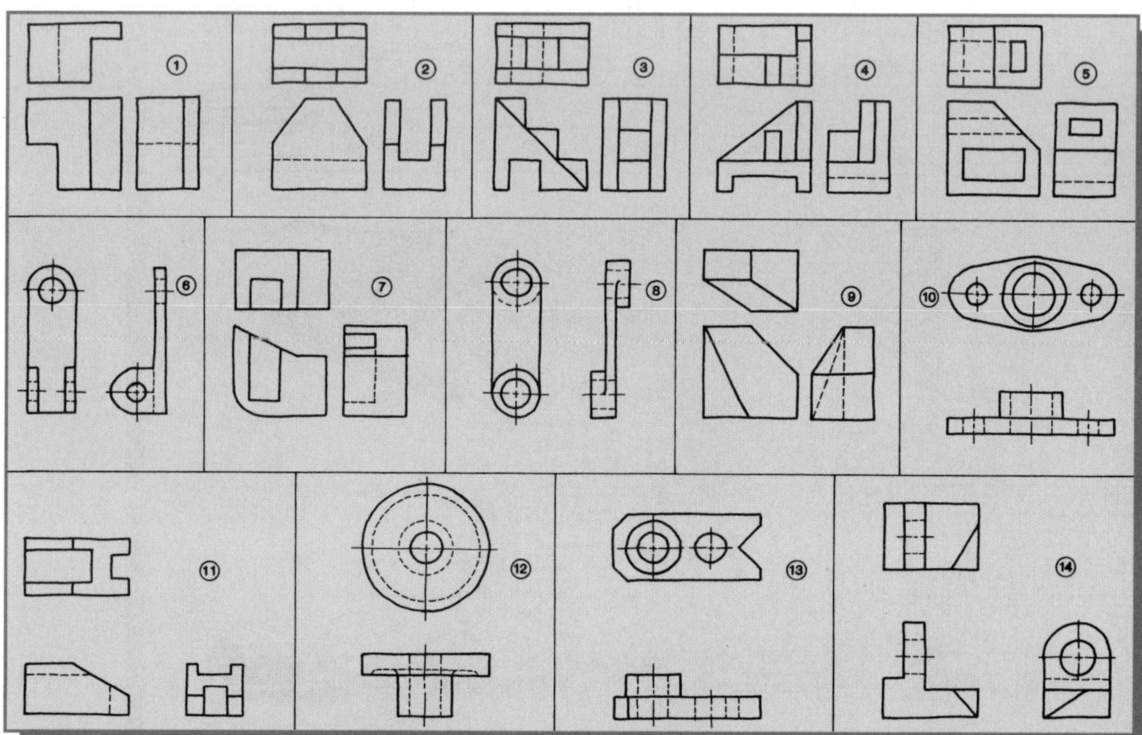

Figure 3-57
Orthographic multiview sketching practice

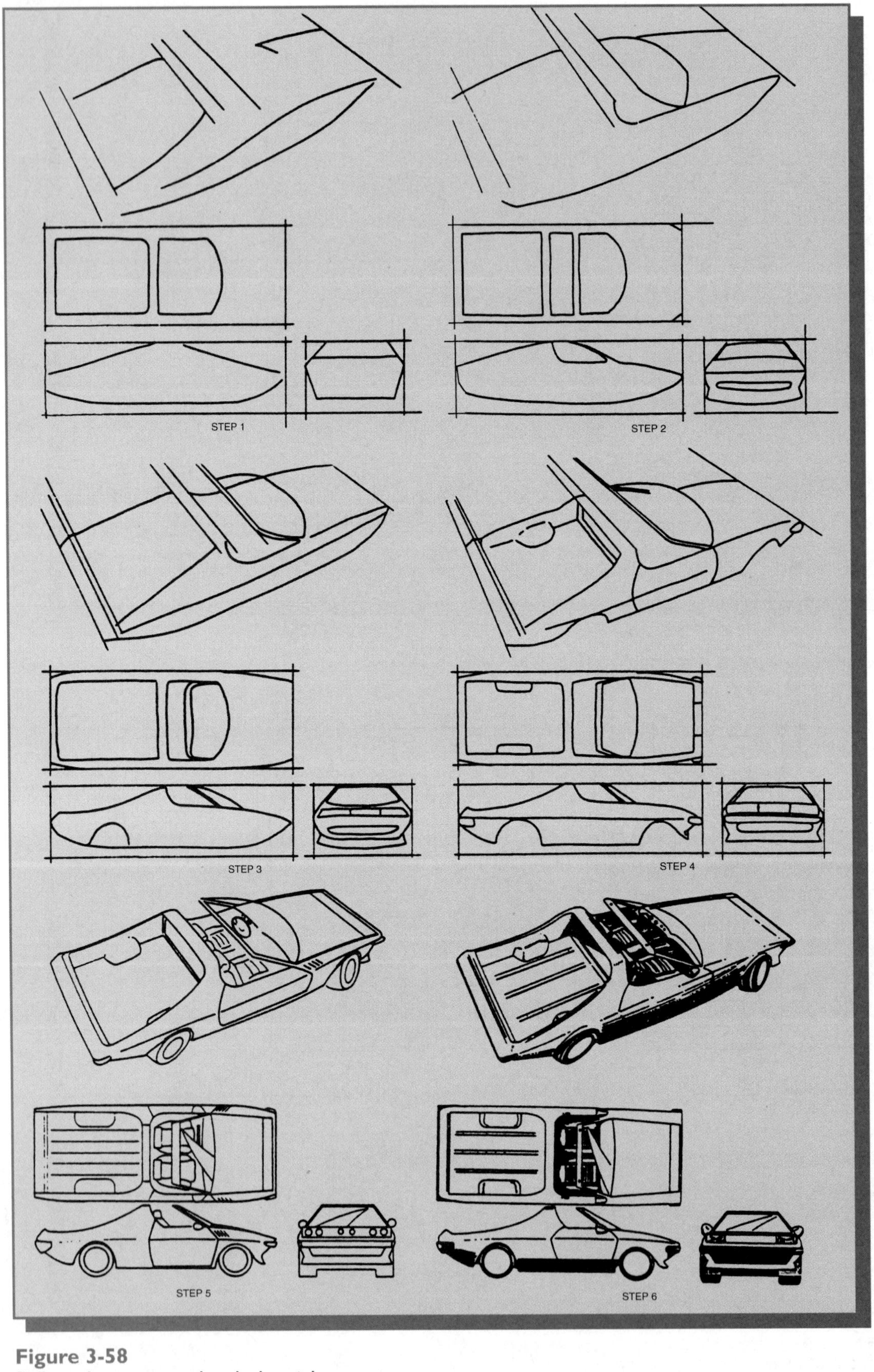

Figure 3-58
Follow the steps to sketch the pickup.

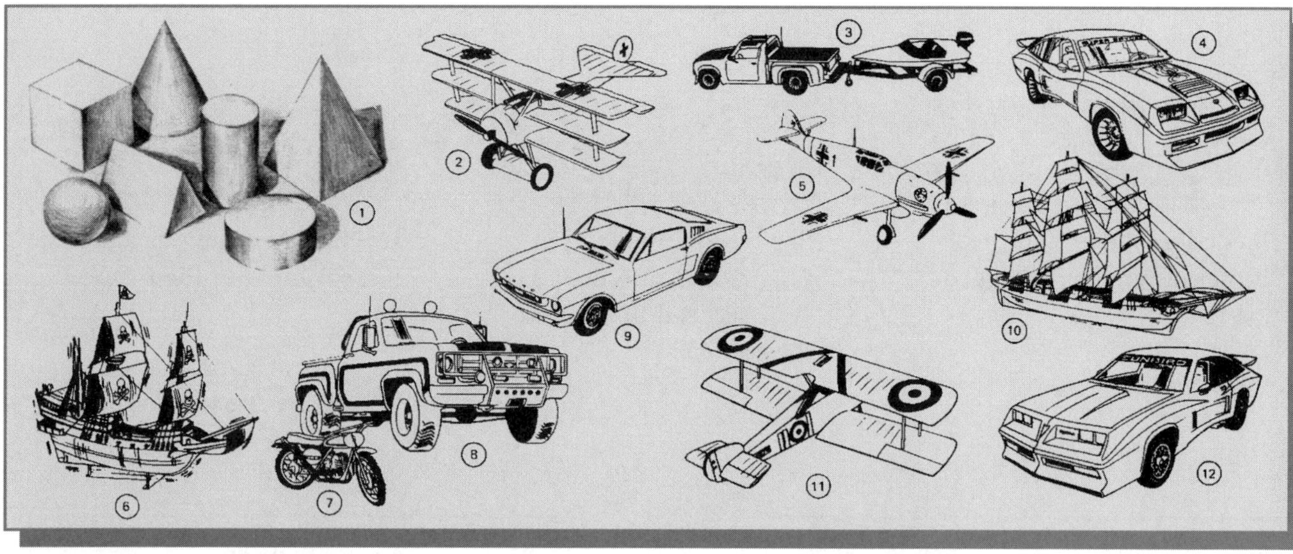

Figure 3-59
Sketching practice with real objects

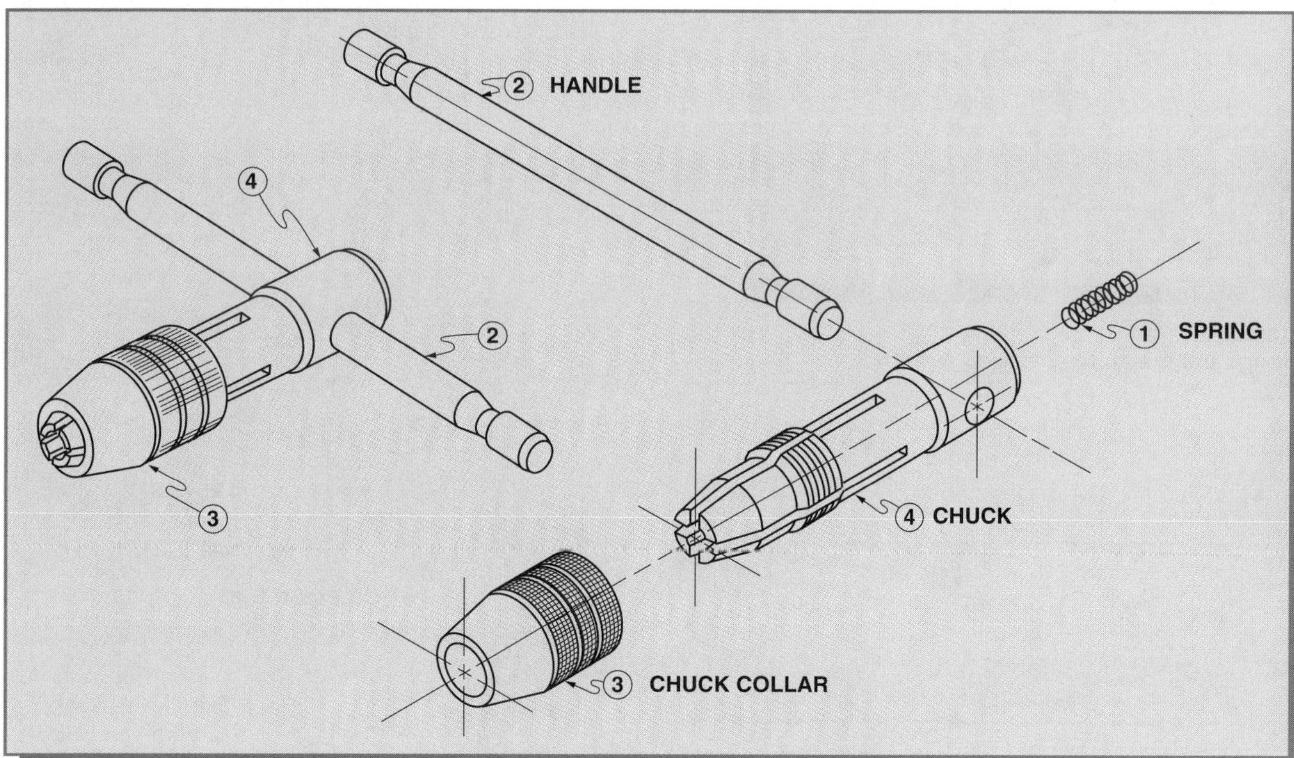

Figure 3-60
Sketch the tool-bit holder.

Design Exercises

1. Redesign and sketch the hexagon wrench in **Figure 3-61**.
2. Sketch the bookend in **Figure 3-62** and design your stylized initials into the surface.
3. Design a new tent stake as specified in **Figure 3-63**.
4. Select one or more of the industrial products in **Figure 3-64** and sketch a new design.

Figure 3-61
Redesign and sketch the hexagon wrench.

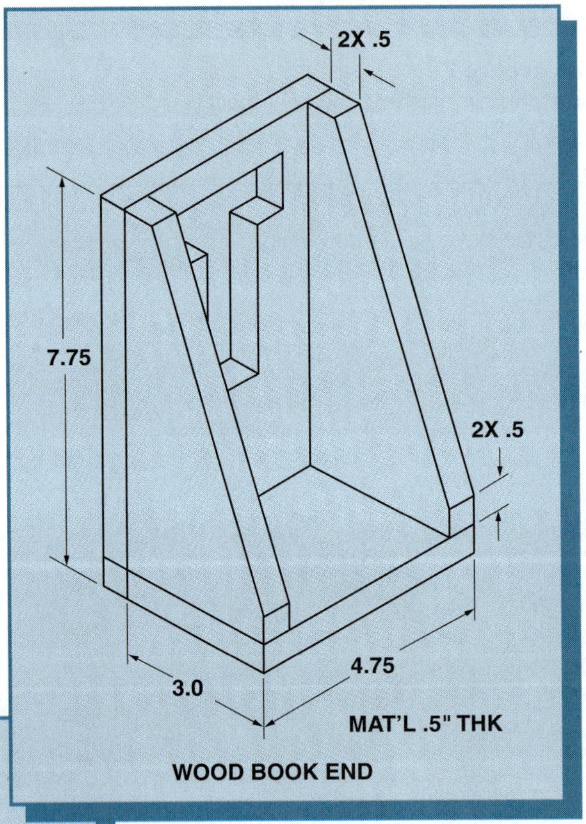

Figure 3-62
Sketch the bookend with your initials.

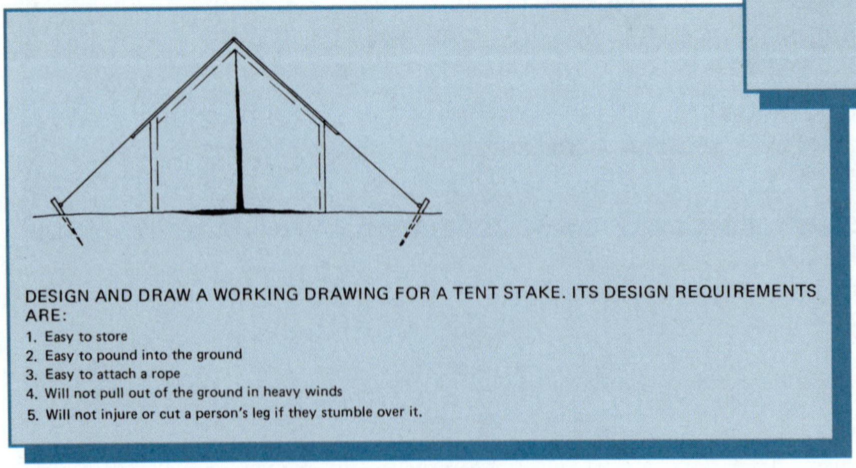

Figure 3-63

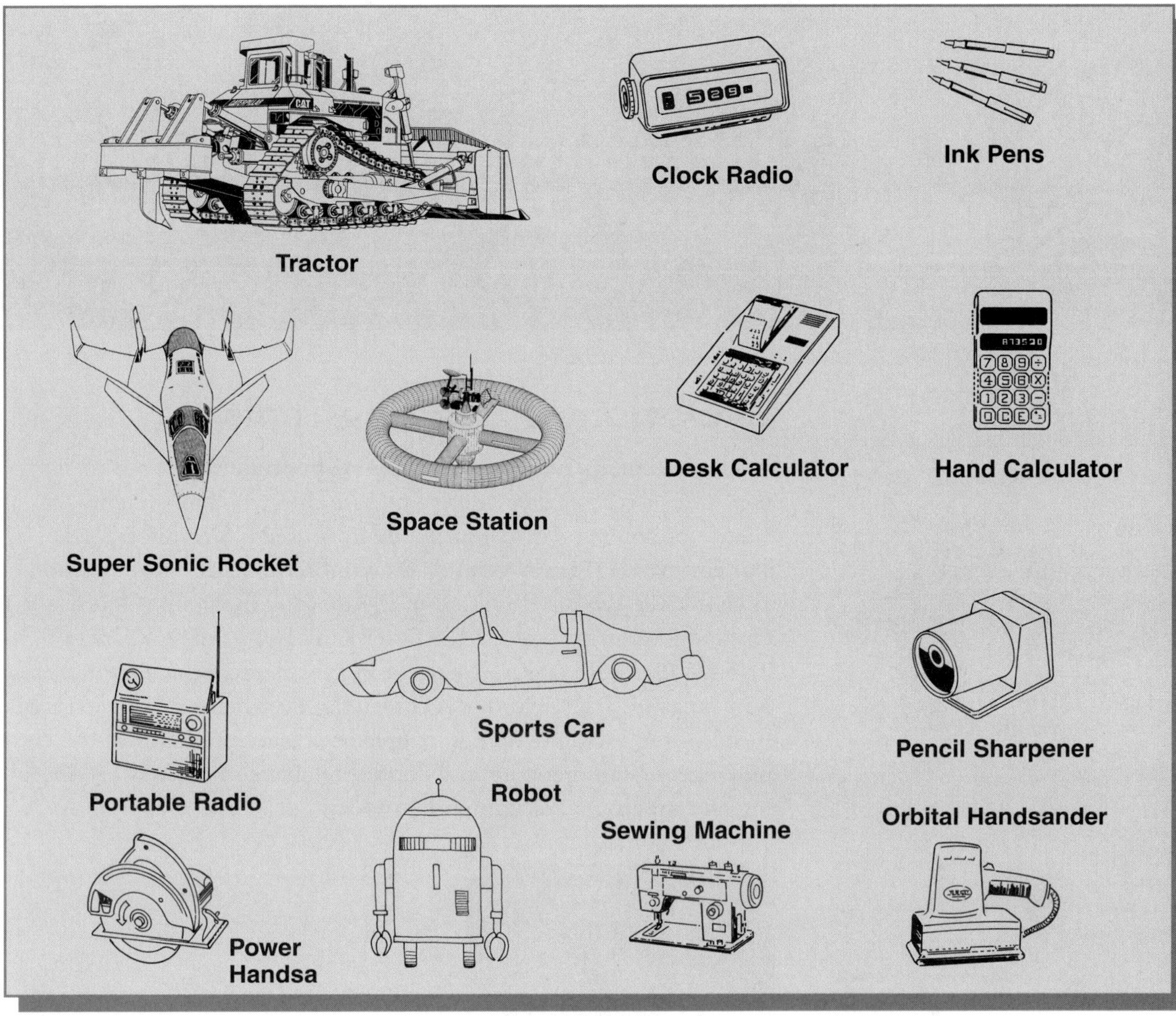

Figure 3-64
Sketch a new design for one or more of these industrial products.

Key Terms

American Society of Mechanical Engineers (ASME)
Grid paper
Guide lines
Isometric
Line conventions
Location dimensions
Microfont
Multiview
Oblique
One-point perspective
Orthographic
Overall dimensions
Perspective
Pictorial
Shading
Size dimensions
Three-point perspective
Two-point perspective
Vanishing point
Working drawing

Chapter 4

Introduction to Computer-Aided Drafting Systems

Objectives

The student will be able to:

- differentiate between different computer-aided drafting (CAD) systems
- describe the role of CAD in industry today
- identify the major components of a microcomputer-based CAD system
- describe the function of major CAD components
- list the equipment needed to run AutoCAD LT

Evolution of Drafting Tools

Early drafters measured distances with crude scales and made freehand sketches. Later drafters used simple wooden straightedges as an aid to drawing. For centuries, drafters used T squares, triangles, and compasses **(Figure 4-1a)** to draw more accurate lines, arcs, and circles. Later, the development of the drafting machine **(Figure 4-1b)** enabled drafters to draw much faster and with greater accuracy. Today, computer systems **(Figure 4-2)** give drafters the ability to draw with total accuracy and much faster than ever before. The increased speed of producing drawings, coupled with the precision of the computer-generated drawings, has drastically changed the way engineering and drafting are accomplished today.

Figure 4-1a
Hand held drawing instruments *(Courtesy of Teledyne Post)*

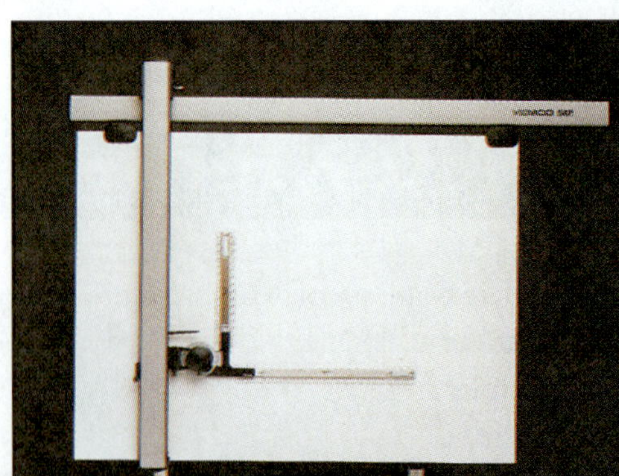

Figure 4-1b
Track drafting machine
(Courtesy of Teledyne Post)

Development and Growth of CAD

Computer systems designed to perform drafting functions are known as **computer-aided drafting** systems, or CAD. Drafters interact with CAD systems to rapidly produce drawings that are extremely accurate, consistent, and easily changeable.

Early CAD systems of the 1960s, like all early computer systems, were extremely large, expensive, complicated, and difficult to use. Some systems filled entire rooms and cost millions of dollars. Today's **personal computer systems (PCS)** are smaller, faster, less expensive, easier to use, and can perform more complex tasks than their predecessors. These improvements and changes were made possible by the development and continual refinement of the **integrated circuit (IC)** chip. Computer advances have led to an enormous expansion in the use of CAD systems in all phases of engineering, design, drafting, manufacturing, and construction as shown in **Figure 4-2**. CAD drawings, and the power of the computer, have actually added many new manufacturing systems. Computer systems have been designed to integrate technical drawings with manufacturing processes. These systems are called **computer-integrated manufacturing (CIM)** or **computer-aided manufacturing (CAM)** systems. **Figure 4-3** shows the common acronyms used to describe these different computer systems.

Figure 4-2
Computer-aided drafting (CAD) system

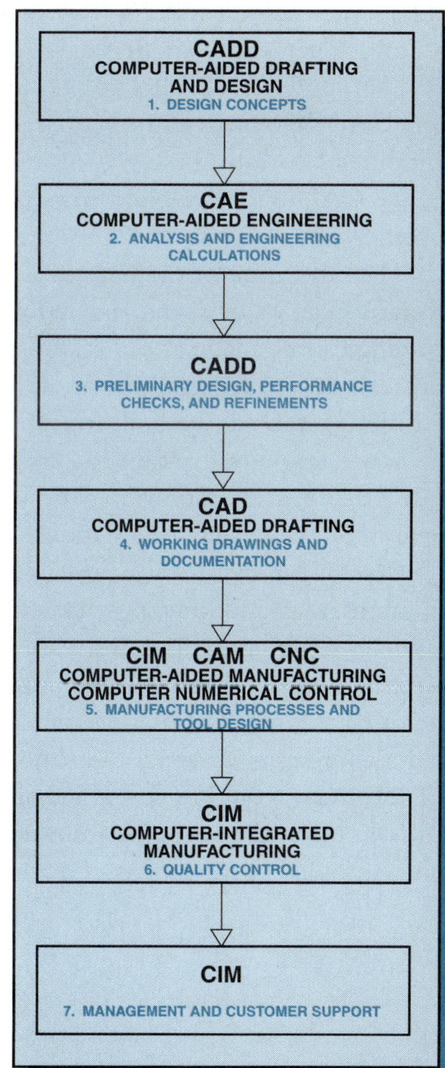

Figure 4-3
Flow chart for CAD-related processes

CAD, the Drafting Tool

The most advanced and sophisticated CAD system cannot draw or design without human input. Although faster, more accurate, and more consistent than the T square, triangle, and the drafting machine, a CAD system is simply an electronic drafting tool. It is important to remember the middle word in CAD is *Aided*. CAD systems aid the drafter in designing, drawing, and calculating faster, more easily, and more accurately.

Just as knowledge of mathematics is necessary to effectively use a calculator, knowledge of drafting fundamentals such as orthographic projection, dimensioning, sectioning, symbology, and descriptive geometry is necessary to effectively use a CAD system. Otherwise, it is not a tool, only a toy.

Computer Systems

Two general classifications of computer systems are mainframe and microcomputer systems. **Mainframe** systems consist of a large main computer with many terminals (workstations) that access the main computer memory **(Figure 4-4)**. These remote terminals are often referred to as dumb terminals because they are totally dependent on the mainframe computer.

Mainframe computer systems are large and very expensive to purchase or lease, and to maintain. One advantage of mainframe computers is that the computer programs only have to be loaded onto one machine (the mainframe), thereby simplifying the control of files and drawings. Your school administration most likely uses a mainframe computer system for keeping student and school records. The mainframe computer is at the school district office, and each school has terminals they can use to access the information. A disadvantage is that if the mainframe computer goes down (has a computer problem due to hardware or software), it affects the entire system, and work is halted until the problem is resolved.

Figure 4-4
Multiple work stations (terminals) of a mainframe CAD system *(Courtesy of Hewlett Packard)*

Microcomputers are the newest and fastest-growing CAD systems. The microcomputer is often referred to as a **personal computer (PC)**. Microcomputers are relatively inexpensive and are stand-alone systems **(Figure 4-5)**. It is the improvements in computer technology that have made the PC the most common computer for CAD. The two most common ways of running CAD are **stand-alone** computers and networked computers. Both of these systems use PCs as the workstation for the CAD operator. **Networks (Figure 4-6)** tie the PCs together into one system.

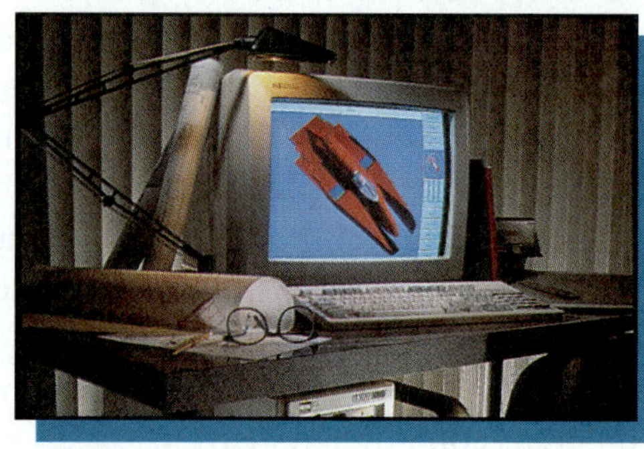

Figure 4-5
Stand-alone PC CAD system *(Courtesy of Hewlett Packard)*

The advantage of the networked computers is that different operators can easily share data and programs with each other. A drawing can be made at one station and all the other stations can view and print the drawing, and if given permission, modify it. Safeguards can be built into the network system to control the access and modification capabilities. In training or conference situations, all stations may be viewing the same drawing at the same time, while only one station is modifying the drawing.

One of the main advantages of a network is the ease of sharing programs, files, and drawings and still having each workstation independent. Keeping each workstation independent means that if one workstation goes down (has a computer problem), it does not affect the other stations. This is a distinct advantage over the large mainframe computers where, if the main computer went down, none of the terminals would function and all work would stop.

The PC is the type of system used throughout this text for running AutoCAD programs.

PC CAD System Components

The components of a PC CAD system are divided into hardware and software items. **Hardware** includes the computer, monitor, and the **peripherals**, which include the input, output, and storage devices. **Figure 4-7** shows the relationship of peripherals and software to a computer. CAD software is a set of computer instructions (computer programs) that cause the computer to act as a drafting system. This text will be using a CAD program called AutoCAD LT.

The CPU

The **central processing unit (CPU)** is the brain of the system. Here, all operator input is translated into electrical pulses, the computer's language, and stored in the memory storage system for immediate or later use. The CPU is the integrated circuit (IC), whose advances have made the PC such a valuable CAD tool. The IC is referred to as "the chip." There has been a continual evolution of ICs:

8088 → 286 → 386 → 486 → Pentium → 586 → 686 → ?

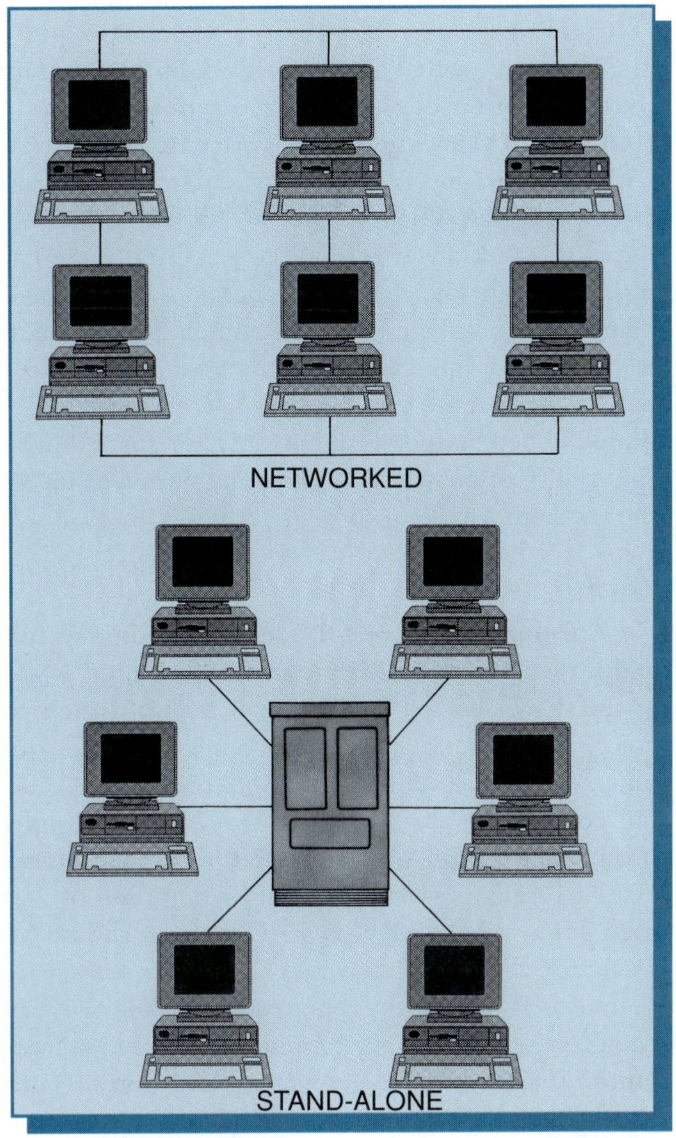

Figure 4-6
Networked stand-alone CAD stations

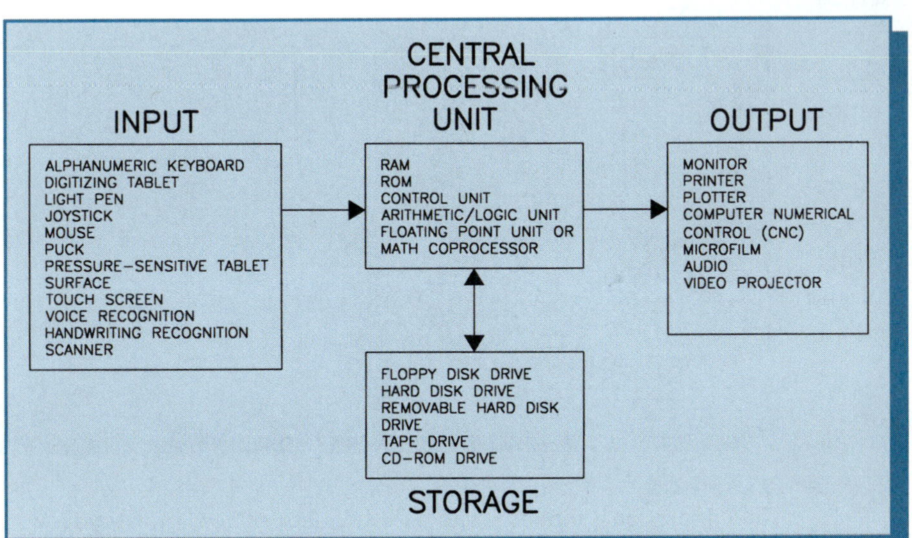

Figure 4-7
A block diagram of a CAD system and peripherals

Memory

There are many types of memory, but the two most basic are **read-only memory (ROM)** and **random-access memory (RAM)**. RAM is extremely important for rapid use of programs with large amounts of calculating and graphics capability such as CAD programs. When an operator starts a computer program, the most critical data is transferred to RAM from the hard drive. It is much faster for the computer program to access data from RAM than from a disk. The larger the computer program, the more RAM is needed. Word processing programs work well with 4 MB of RAM, but CAD programs work best with 8, 16 or 32 MB of RAM; the more the better. Keep in mind, however, that RAM also drives up the cost of a system. RAM chips are located inside the computer on the motherboard.

Monitor

A **monitor (Figure 4-8)** uses a **cathode ray tube (CRT)** and is used with all CAD programs. Basically it is what you see as you draw. The first CAD systems used **monochrome** (single-color) monitors, usually green or amber. However, just as computers have improved, so have monitors. Today, CAD systems use full-color monitors, commonly with 256 colors and very sharp images. The sharpness of the monitor is called the **resolution**. The picture on a monitor screen is comprised of a series of dots, each dot is called a **pixel**. A CAD monitor should have a higher resolution than a monitor used for word processing because of the precision and clarity required for reading the drawing lines. Monitors with fewer pixels produce very jagged lines (low resolution) **(Figure 4-9)**. An acceptable resolution for a 13" color monitor used for CAD is 680 x 520. This means that there are 680 pixels horizontally across the screen and 520 pixels vertically on the screen. This display will have 353,600 pixels on the monitor's screen. Monitors commonly used are 15" to 17" SVGA (Super Video Graphics Array), which have a resolution up to 1280 x 1024 and are capable of displaying over 16 million colors. On these monitors, colors are bright and crisp, lines do not appear jagged, and circles and arcs are very smooth.

Figure 4-8
Monitor, high resolution for CAD

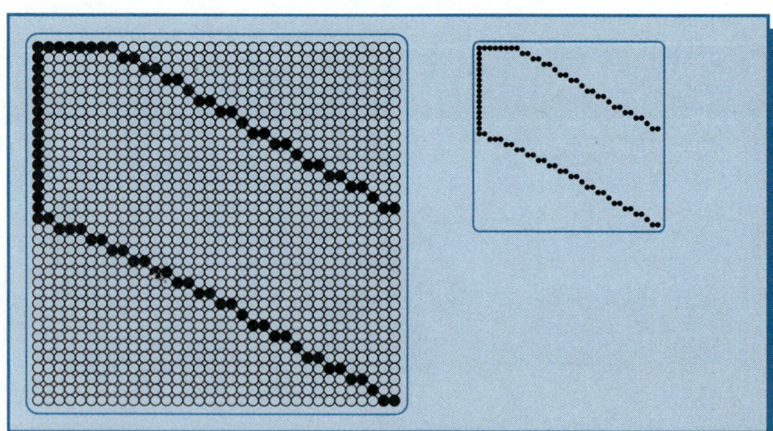

Figure 4-9
The display on a monitor with a low number of pixels will have low resolution.

Input Devices

Input is all the information fed into a CAD system by an operator. **Input devices** are extremely important because they are the link between the computer and the operator. The selection of the proper input devices is critical for efficient use of the computer and software.

Keyboards (Figure 4-10) are similar to standard typewriter keyboards. In addition to normal typing functions, CAD keyboards are used to select drafting tasks. Tasks such as line, arc, circle, erase, and so on, are assigned to keys. To draw a line, all the operator has to do is type **L** and the line command appears. Function keys on the keyboard may be assigned specific CAD functions such as Grid and Snap. Keyboards can also be used to draw lines by entering the grid coordinates of lines (see Chapter 5, CAD Procedures).

A **mouse (Figure 4-11)** is a handheld input device. It contains a ball that rolls on a flat surface. As the ball rotates it moves a **cursor**, a crosshair or an arrow on the monitor that the operator uses to select CAD commands and create drawings **(Figure 4-12)**.

Figure 4-10
Typical 101 keyboard

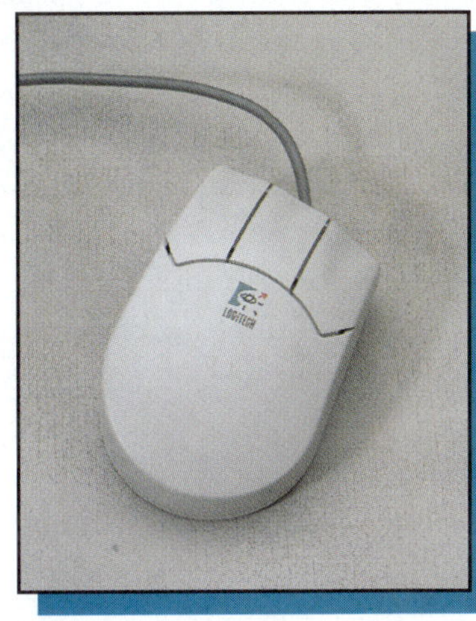

Figure 4-11
The mouse comes in various shapes.

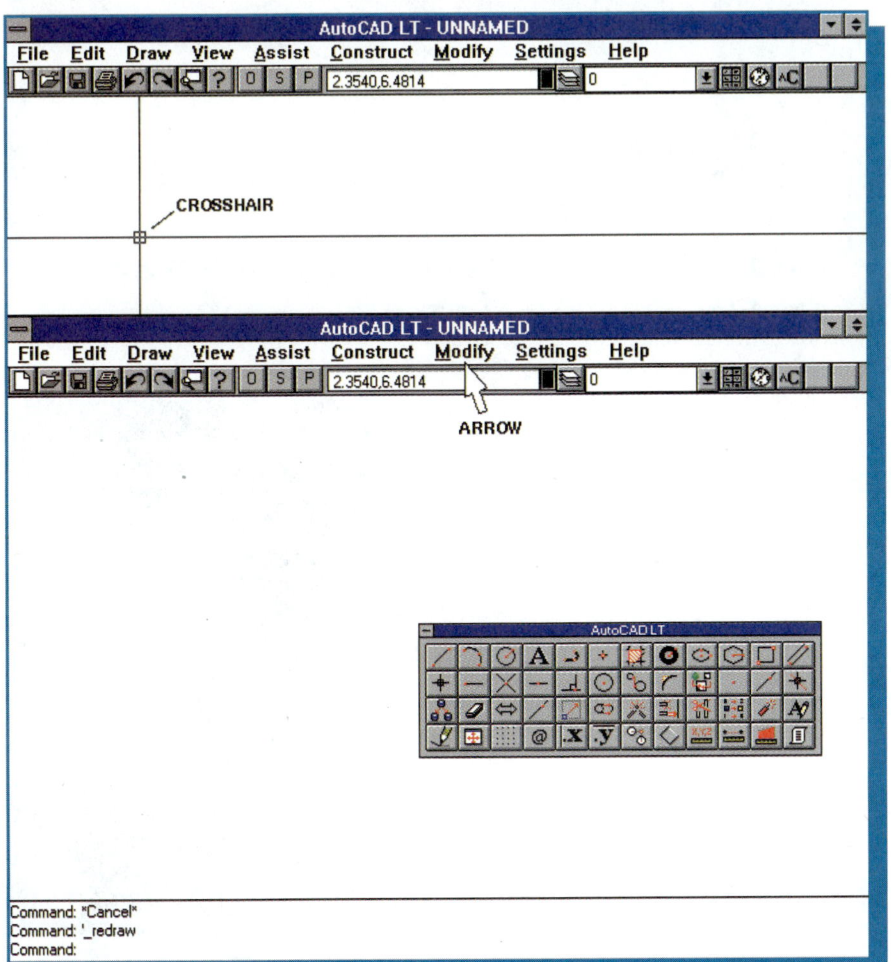

Figure 4-12
Arrow and crosshair are located by using the mouse.

A **puck (Figure 4-13)** looks like a mouse but functions differently. A puck will have between three and sixteen buttons, each of which may be programmed for the selection of a CAD command. The commands most frequently used such as RETURN, CANCEL, ERASE, and LINE are often programmed to buttons on the puck. The puck is used on a graphics tablet, also called a digitizer, to select commands with the crosshair on the puck **(Figure 4-14)**.

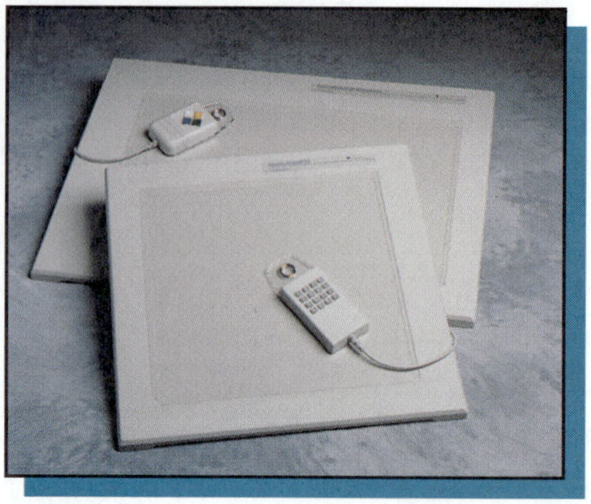

Figure 4-13
Pucks: sixteen and four-button types on a graphics tablet. Both have crosshair sites.

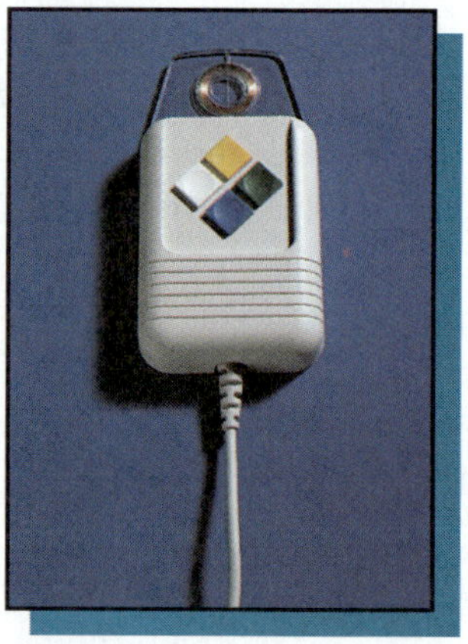

Figure 4-14
Closeup of a four-button puck

Stylus pens (Figure 4-15) are shaped like a pen or pencil and operate like a mouse or puck.

A **graphics tablet (Figure 4-16)** is a flat surface upon which the operator draws by moving the mouse, puck, or stylus across the tablet and selecting CAD commands, which are programmed into the tablet.

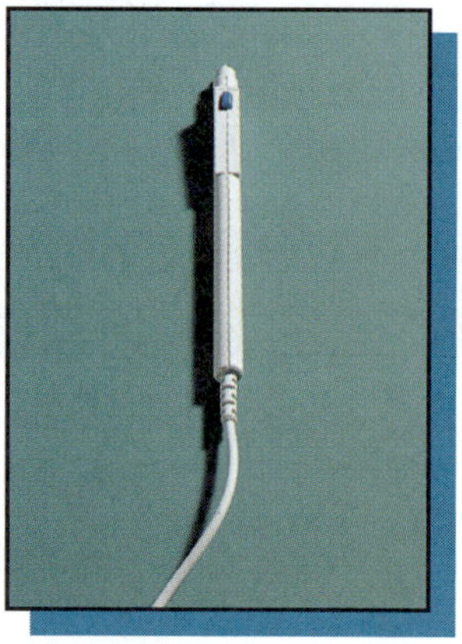

Figure 4-15
Stylus used for selecting drawing commands *(Credit: Versacad)*

Figure 4-16
Graphics tablet

Light pens (Figure 4-17) are electronic pointing devices that sense light from a monitor. With a light pen, a CAD operator can point directly to the screen to pick menu items or to draw.

On **pressure-sensitive screens** or tables, the cursor is controlled by slight pressure from a finger or a small pointed object.

Voice-activation controls the computer through commands spoken into a microphone interface.

Handwriting activation controls the computer through written commands on a small electronic pad that recognizes handwriting.

Optical scanners are the quickest way to enter existing drawings and text into the computer's memory. Scanning can convert conventional drawings into digitized CAD drawings.

Figure 4-17
Light pen

Output Devices

In conventional drafting, the output was an ink or pencil drawing prepared on a paper or vellum surface. The output from a CAD-produced drawing is an electronic file, used to create a print or plot.

The primary **output devices** for CAD systems are plotters and printers.

Plotters convert images shown on a CAD monitor to ink drawings on paper or vellum. When CAD drawings are printed or plotted onto paper, the result is called a **hard copy**.

Plotters use pens. Plotters with multiple pens can produce multicolor drawings depending on how the input was entered into the computer. Different line weights can also be drawn with different pens. There are several types of plotters.

Flatbed plotters (Figure 4-18) have a flat surface and pen carriage. Drawing paper is placed on the flat surface and the pen carriage moves over the surface, drawing lines similar to handheld pens, but much faster, more accurately, and with more consistency in line weight.

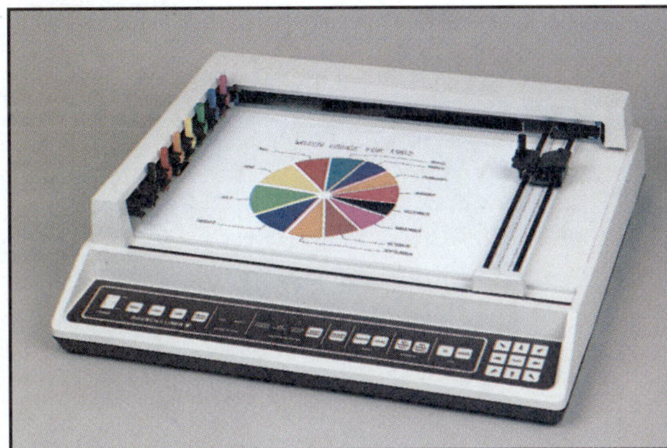

Figure 4-18
Flatbed pen plotter

Drum plotters are the most common plotters used. They include a cylinder and pen carriage. Drawing paper is placed on the cylinder, which moves back and forth to produce lines on the vertical axis. The pen carriage moves along the horizontal axis of the cylinder to produce lines on the opposite axis **(Figure 4-19)**.

Figure 4-19
Drum pen plotter

Plotter sizes vary and can plot drawings from A ($8\frac{1}{2}$" x 11") size to E (26" x 48") size. The scale of the finished drawing can also be adjusted on the plotter. An architect can plot the floor plan at $^{1}/_{4}$" = 1'-0" to produce a standard size floor plan, or plot the same drawing or a portion of the drawing at a larger scale to show more detail. This flexibility is a big advantage.

Printers (Figure 4-20) are output devices that transform monitor images into hard copy in a few seconds. Printers do not use pens. The first models, **dot matrix printers**, were slow, noisy, and of mediocre quality, only useful for rough copies and check prints. However, printers have made great advances. **Laser printers** are faster than pen plotters, very precise, and may print in color with very high resolution. **Figure 4-21** is an example of a laser print of a CAD-generated drawing.

Other computer output devices include audio output—yes drawings can actually talk. In fact with the use of multimedia CAD, drawings can be coupled with videos and sound.

The CAD drawing is an electronic file that can be sent over the telephone lines via a modem from person to person or via shared networks such as the Internet.

Figure 4-20
Laser printer

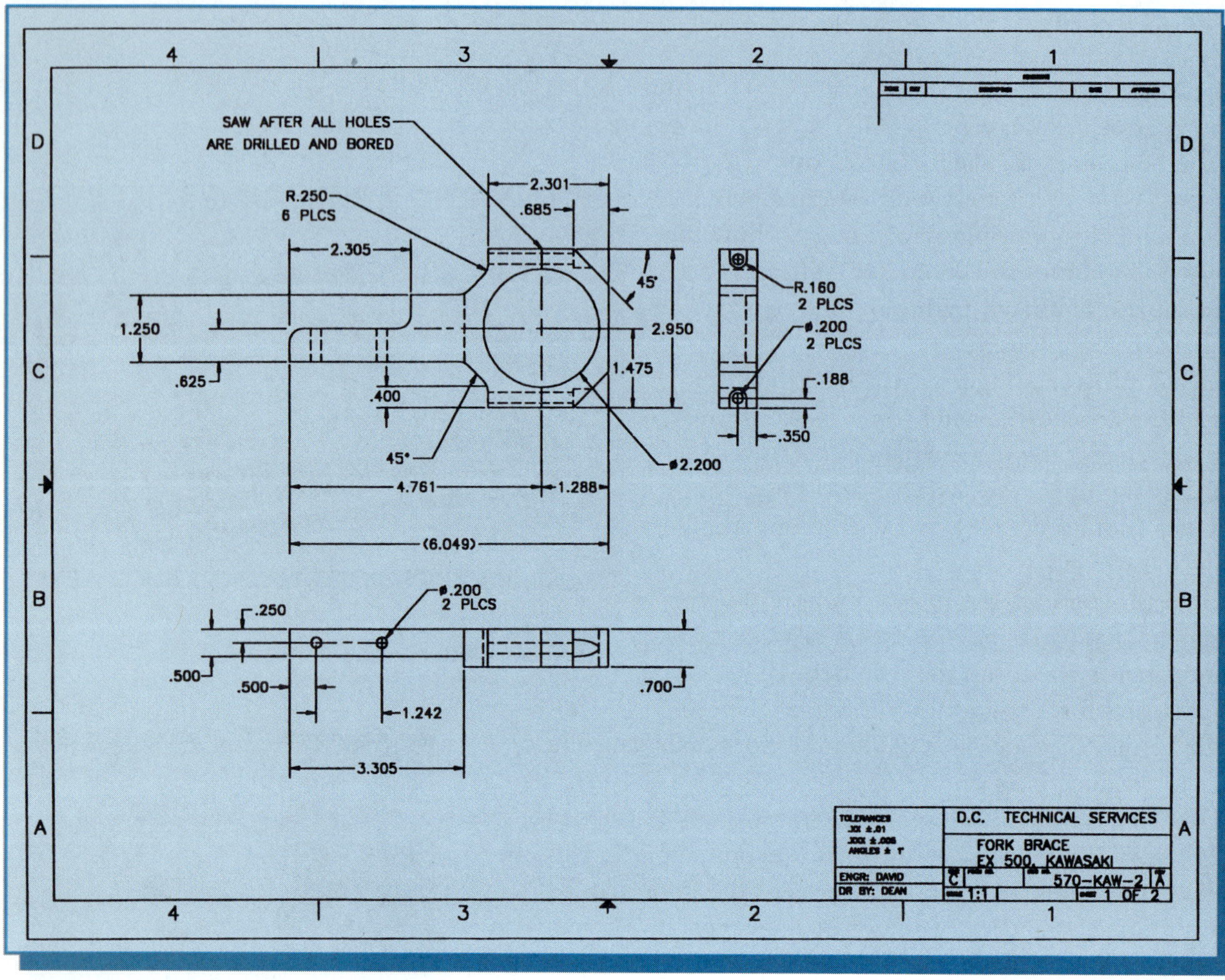

Figure 4-21
Laser printed drawing of a motorcycle fork brace

Data Storage

One of the greatest benefits of CAD over manual drafting is not in the preparation of original single drawings, but in the use of stored data. CAD storage capacity enables drafters to draw a detail or component only once, store it in a computer library, and then retrieve it to be used or altered any number of times. This elimination of repetitive tasks saves thousands of drafting hours on many large projects.

Storage capacity continues to increase as IC chips get more powerful and the PC hard drives have more capacity.

There are two kinds of **disk drives** that act as storage devices for the PC: hard drives and floppy disk drives. **Hard disk drives** are the primary storage device in the PC. The hard drive holds all the **software** (CAD programs) as well as all the documents, symbols, and drawings created by the CAD operator. Just a few years ago, a 40 MB (40,000,000 bytes) hard drive was considered sufficient. The typical hard drive capacity for a PC CAD workstation rapidly increased to 125 MB, 230 MB, and then 500 MB. Now it is not uncommon for a PC to have a 1 GB (1,000,000,000 bytes) hard drive or larger. This increase is necessary because software programs such as Windows and AutoCAD have gotten bigger and the demands drafters and designers put on CAD have increased.

Floppy disks, which move data in and out of the computer through a **floppy disk drive**, are commonly used for storing drawing files, for transferring data from one machine to another, and to back up files. There are two common types of floppies: 5.25" and 3.5" **(Figure 4-22)**. A high-density 3.5" floppy has a capacity of 1.44 MB of data.

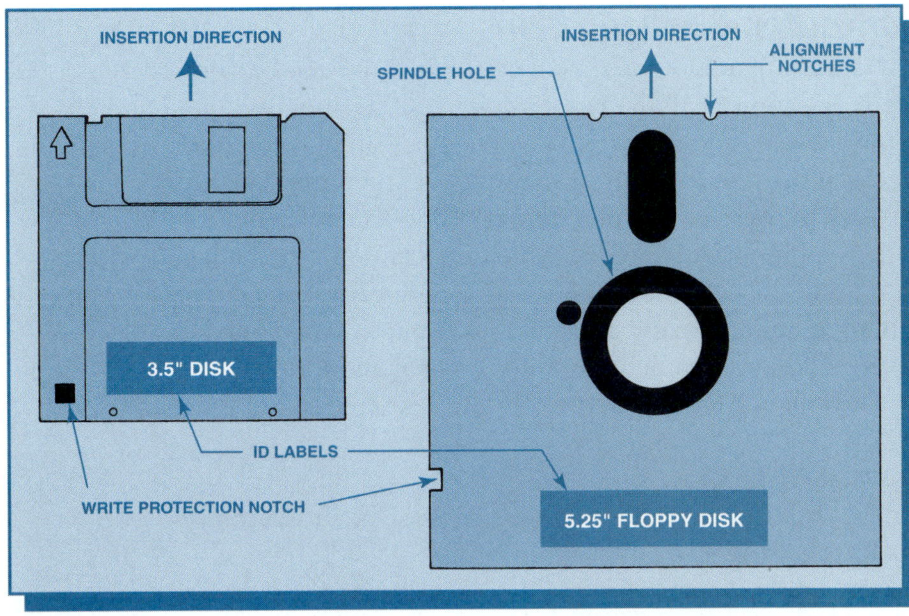

Figure 4-22
Floppy disks

Magnetic tape drives are used to store data. The magnetic tape drive may be internal (inside the computer case) or external. The main function of a tape drive is for data backup.

Software

The most sophisticated computer hardware is ineffective without quality software and a knowledgeable operator.

Computer **software** for CAD is the set of instructions given to a computer. The instructions are written in a computer language such as Basic, Assembler, C, or C++ by computer programmers. The instructions comprise a *computer program*. Drafters do not write or create computer programs, but use existing programs for operating CAD systems. Quality CAD computer programs include a wide variety of drafting tasks and options and are relatively easy to use. As CAD software is improved, the ease of use (user-friendliness) is increased. Good CAD software will include many drawing and editing commands as well as automatic functions such as dimensioning, sectioning, color features, and standard symbols. The AutoCAD program used in this text has all of these features plus many more and is still user-friendly **(Figure 4-23)**.

Figure 4-23
Computer software

Computer Configurations

The configuration of a computer means the combination and interconnection of all the bits and pieces: CPU, monitor, input and output devices, and software. The configuration of a computer needs to match the requirements of the software to be run. The more complex the software is, the more powerful the computer must be. Today's multimedia programs, which use extensive graphics and sound, and the CAD software used by engineers, designers, and drafters, require similarly configured computers. The computer requirements for running the AutoCAD LT and AutoCAD R13, which is the CAD software used in this text are:

AutoCAD LT:
CPU: 386 with Math Coprocessor, 486DX, or Pentium
RAM: 8 MB
Hard disk: Program needs 16MB, recommended 120MB
Monitor: SVGA 14", 28 dpi., resolution 1280 x 1024
Windows 3.1 or higher

AutoCAD R13:
CPU: 486 or Pentium
RAM: 16 MB
Hard drive: 500 MB or bigger
Monitor: SVGA 15", .28 dpi., 1280 x 1024 resolution
DOS, Windows, NT Windows, or UNIX
3-button mouse or graphic tablet

Exercises

1. Take a field trip to a vendor selling CAD systems. Make a list of various systems and the manufacturers' addresses. Write for additional information.
2. Take a field trip to an industry or office with a CAD system. Observe its operation. Make a list of the software observed.
3. Compare prices of CAD hardware and software.
4. Make a sketch of a CAD system and label all the hardware and peripherals.
5. Refer back to the listed computer requirements for running AutoCAD LT and AutoCAD R13. Make a sketch of each item with a label.
6. Check the prices of each item in exercise #5 with CAD vendors and/or computer magazines. Put the best price by each label and provide a total.
7. Note the B size drawing format in **Figure 4-25**, open the AutoCAD LT drawing files and Examine the other drawing formats.

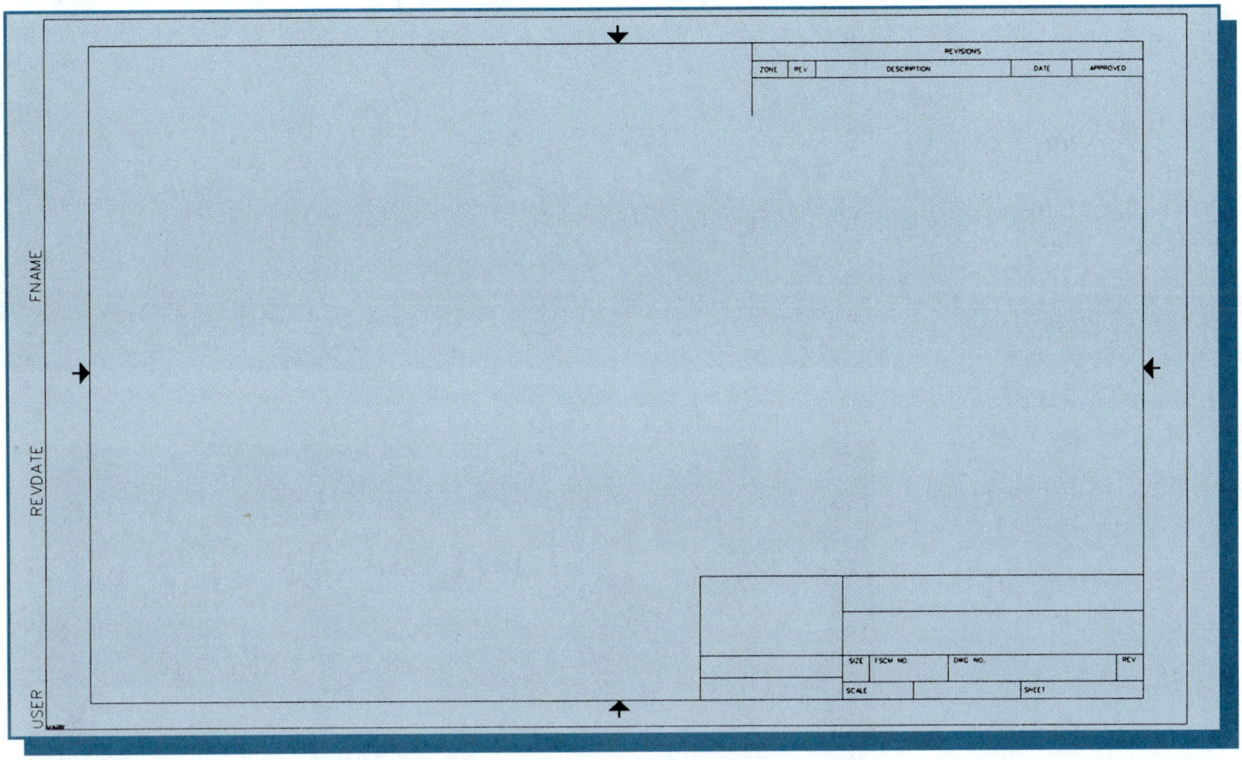

Figure 4-24
AutoCAD LT B size drawing format. The file name is ANSI_B.

Key Terms

Cathode ray tube (CRT)
Central processing unit (CPU)
Computer-aided drafting (CAD)
Computer-aided drafting and design (CADD)
Computer-aided engineering (CAE)
Computer-aided manufacturing (CAM)
Computer-integrated manufacturing (CIM)
Cursor
Disk operating system (DOS)
Disk drives
Dot matrix printer
Drum plotter
Flatbed plotter
Floppy disk
Floppy disk drive
Graphics tablet
Hard copy
Hard disk drive
Hardware
Input devices
Integrated circuit (IC)
Keyboard
Laser printer
Light pen
Magnetic tapes
Mainframe
Microcomputer
Monitor
Monochrome
Mouse
Networks
Optical scanners
Output devices
Peripherals
Personal computer (PC)
Pixel
Plotters
Pressure-sensitive screens
Puck
Random-access memory (RAM)
Read-only memory (ROM)
Resolution
Software
Stand-alone
Stylus pens
Voice-activation

Chapter 5

AutoCAD LT Procedures

Objectives

The student will be able to:

- understand the AutoCAD conventions used in this text
- start AutoCAD LT
- explain the use of the mouse and keyboard to input information
- set up drawing area
- identify and use the different areas of the AutoCAD LT drawing editor
- draw lines that are accurately located and are the correct size
- draw circles that are accurately located and are the correct size
- use selected edit commands
- correct mistakes
- move around the drawing
- save drawings

Figure 5-1
AutoCAD LT workstation

Introduction

AutoCAD LT is the best-selling 2D CAD program worldwide. Of the many CAD programs on the market, AutoCAD LT Release 2 for Windows was selected for use in this text because of its many features that are similar to AutoCAD R13, its compatibility with other Autodesk products, its learnability, and its price.

The format of AutoCAD LT is so similar to the other AutoCAD products for Windows that the skills learned transfer directly. Most of the **commands** and their locations in AutoCAD LT are the same as in other AutoCAD for Windows applications.

Mouse

The **mouse** is used more than any other method to input information into the computer. It may have two or three **buttons (Figure 5-2)**. One of the buttons, usually the left, is called a **Pick button**. The right button is usually the **Enter button** (EB) and has the same function as the Enter key on the keyboard (sometimes you must use the Enter key).

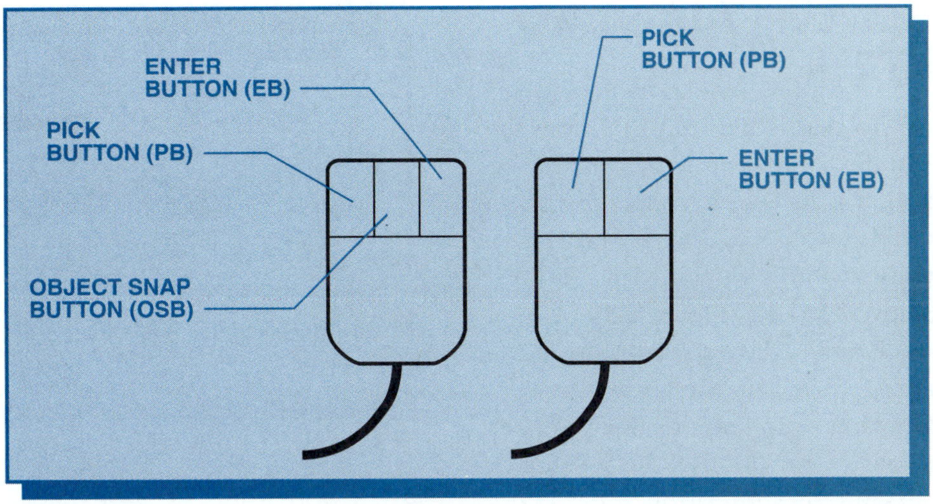

Figure 5-2
Mouse

The Pick button is used to select the command or item we choose. When the Pick button is used to select something, it is pushed and released. This is called *clicking* or *picking*. There are times when we push and release it two times in rapid succession. This is referred to as a double click. Other times we will push the Pick button down, and while holding it down, move the mouse. This is called *click and drag*.

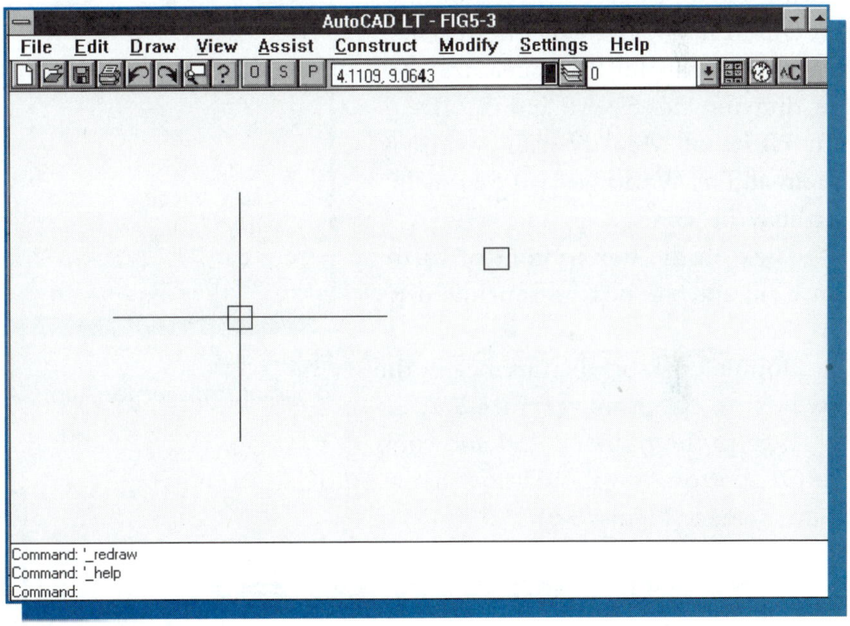

Figure 5-3
Crosshair, pick box, and arrow

The Enter button has several uses. It is used to end a command. When the cursor is over certain command buttons, the Enter button calls up a dialog box. In most cases, the Enter button is like hitting the ↵ key or space bar on the keyboard.

The mouse controls the cursor on the screen. The **cursor** will appear as a crosshair or as an arrow, depending on where it is located on the screen **(Figure 5-3)**. The cursor may appear as a box, depending on the mode that has been selected in AutoCAD LT.

Starting AutoCAD LT2

To start AutoCAD LT2 from the Windows Program Manager **(Figure 5-4)**, it is necessary to double-click on the AutoCAD LT2 icon. This will take you into the Program Manager for AutoCAD LT2 **(Figure 5-5)**.

Double-click on the AutoCAD LT2 icon to bring up the Create New Drawing dialog box for the program **(Figure 5-6)**. Use the Quick setup by clicking the OK button.

The Quick Drawing Setup dialog box **(Figure 5-7)** is now on the screen. In this box we can set the type of units to be used in the drawing, the full size **drawing area** (not the paper size), and our drawing aids SNAP and GRID. The **Units of Measurement** will stay Decimal. The World (actual) Size will also stay the same.

Click on the box in front of Turn On Grid and the box in front of Turn On Snap.

Double click on the number in the box next to X Spacing **(Figure 5-8)**.

Type in the number **.250** and click the OK button. You should now be in AutoCAD's Drawing Editor screen **(Figure 5-9)**.

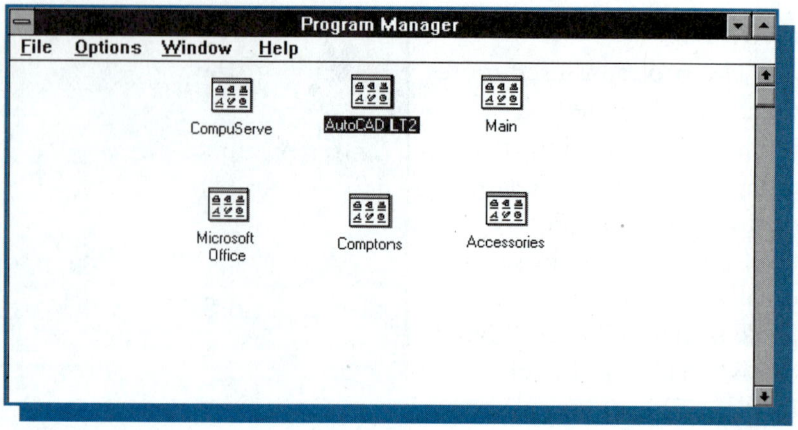

Figure 5-4
Windows Program Manager

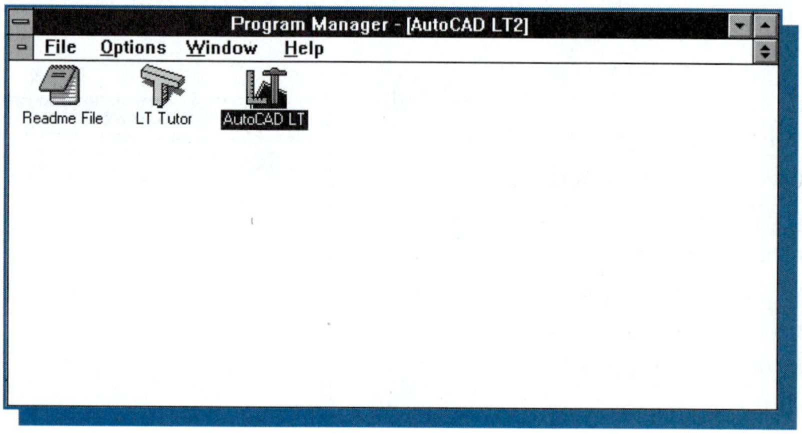

Figure 5-5
Program Manager for AutoCAD LT

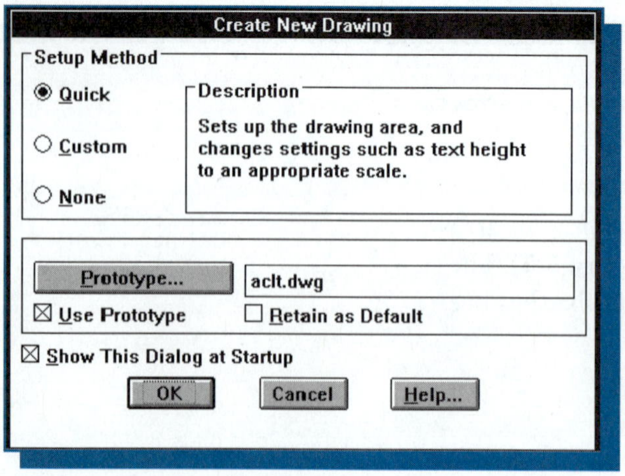

Figure 5-6
Create New Drawing dialog box

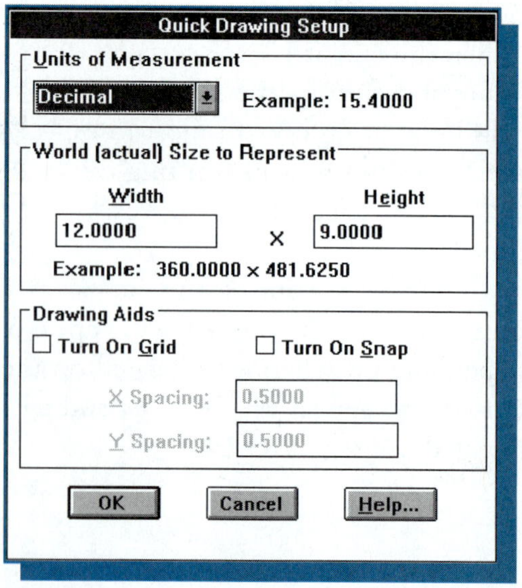

Figure 5-7
Quick Drawing Setup dialog box

Drawing Editor

The **Drawing Editor** is the working screen for creating drawings. The important areas are labeled in **Figure 5-9**. Use this diagram as a reference for this and the following chapters.

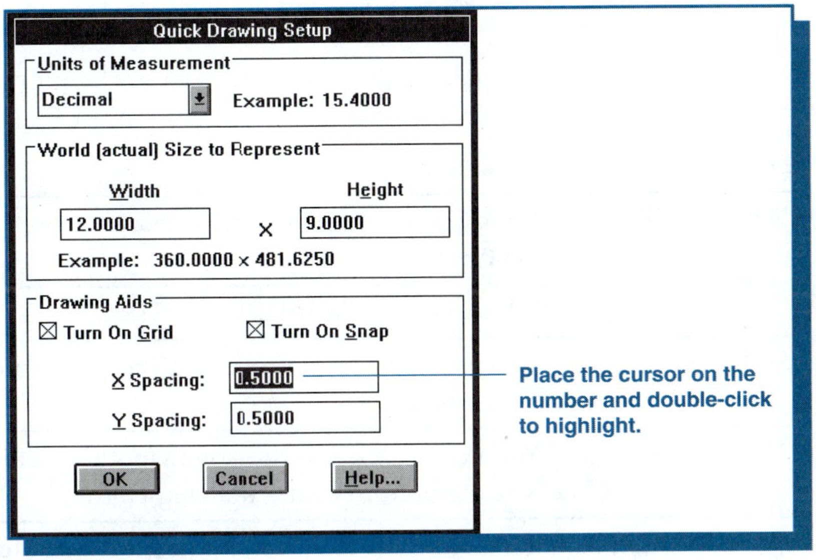

Figure 5-8
Setting Snap and Grid

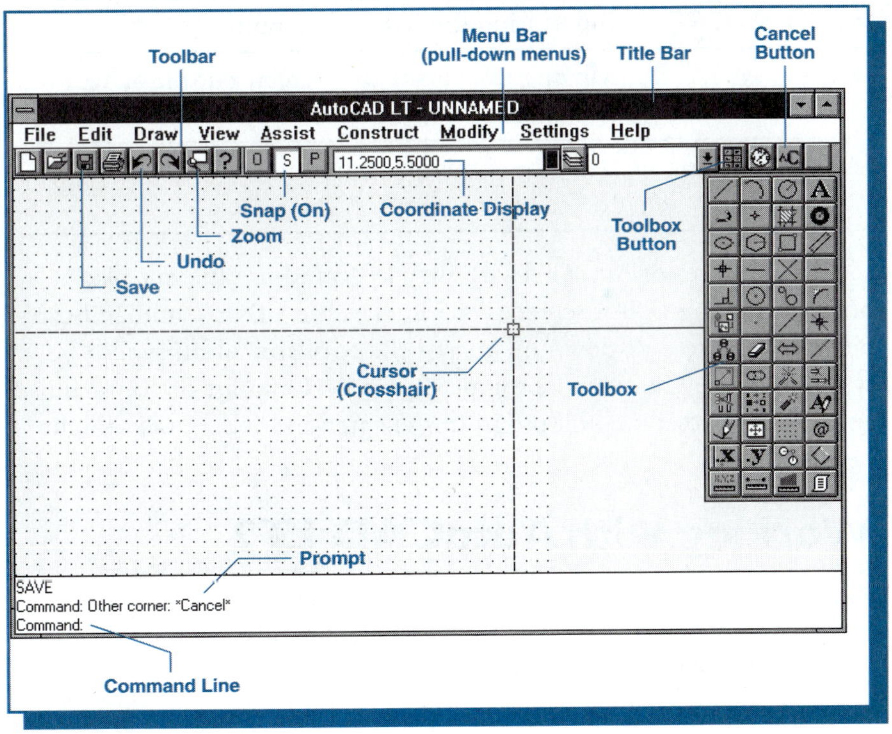

Figure 5-9
Drawing Editor

Conventions Used in This Text

To be consistent throughout this text, here is an explanation of the directions to be followed for drawing with AutoCAD LT.

EXAMPLES	MEANINGS
DELETE, ESC	Words in boxes are keys that you press on the **keyboard**.
	↵ is the same as ENTER or RETURN
CTRL + C	When you press keys together, they are joined with a plus sign.
Command:	You need to look at the command line, which is at the bottom of the drawing editor.
Command: **LINE**	Text you enter from the keyboard is in **bold**.
Pull-down menus	Menus and **submenus** are in **bold** **Draw-circle**
click	Input using the left mouse button.
Select	Moving the cursor to a screen button or word and clicking.
Move the cursor	Slide the mouse over the mouse pad, causing the cursor to move on the screen.

The features of AutoCAD LT may be found in different places on the Drawing Editor screen. For instance, the LINE command can be found in the pull-down menu and the **toolbox** and may also be typed in at the Command: prompt. **Figure 5-10** is a list of commands and functions that we will use in this chapter and where they might be found.

Working with AutoCAD LT2

One of the best ways to learn a drawing program is to use that program to create a specific drawing. While creating the following drawing **(Figure 5-11)**, keep in mind that understanding the concepts is the goal. The drawing will be started in this chapter and finished in the following chapters. At each step during the process, you will be given explanations and tips to help speed the learning process. The skills learned in this chapter are the foundation for the CAD work in the rest of the text.

Draw Commands			
From the command line	**From the menu bar (pull-downs)**	**From the toolbox**	**From the toolbar**
Command: line or l	Draw – line		
Command: circle or c	Draw – circle –		
Command: arc or a	Draw – arc –		
Edit Commands			
Command: erase or e	Modify – erase		
Command: undo or u	Edit – Undo		
Command: redo or re	Edit – Redo		
Command: move or m	Modify – move	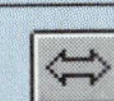	
Help Commands			
Command: toolbox	Settings – toolbox style		
Command: Ctrl+C or Cancel			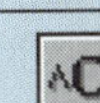
Command: snap or sn	Settings – Drawing aids		
Command: zoom or Z	View – zoom		
Command: redraw or rr	View – redraw		
File Commands			
Command: save or sa	File – save		

Figure 5-10
AutoCAD LT2 commands introduced in this chapter

First Drawing Procedures

The dots you see on the screen and in **Figure 5-9** represent the grid that was set during Quick Drawing Setup. Each dot is .250" apart both horizontally and vertically. The dot in the lower-left corner represents the 0,0 location on the screen. The dot in the upper-right corner represents the 12,9 location. That point is the upper limit we set in setup.

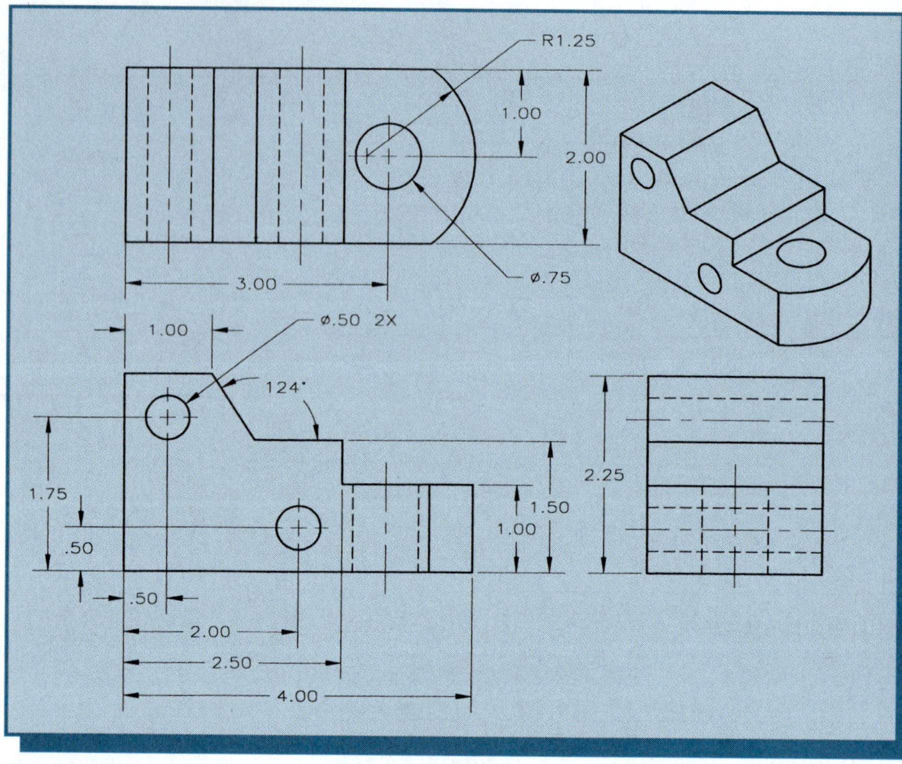

Figure 5-11
AutoCAD drawing

1. Use the mouse to move the cursor to the dot in the lower-left corner. **Coordinate** display should read 0.0000,0.0000.
2. Move the cursor to the dot in the upper right corner. Coordinate display should read 12.0000,9.0000.
3. Click the toolbox button in the toolbar. **(Figure 5-12)** (The cursor changes to an arrow when on the **toolbar**.) The **toolbox** should appear on the right side of your screen.
4. Click the toolbox button again.
5. Click the toolbox button again.

Each time you click the button, the toolbox moves to a different location. In this text it will be placed on the bottom of the Drawing Editor screen as shown. Place it there now. This is known as a *floating toolbox*.

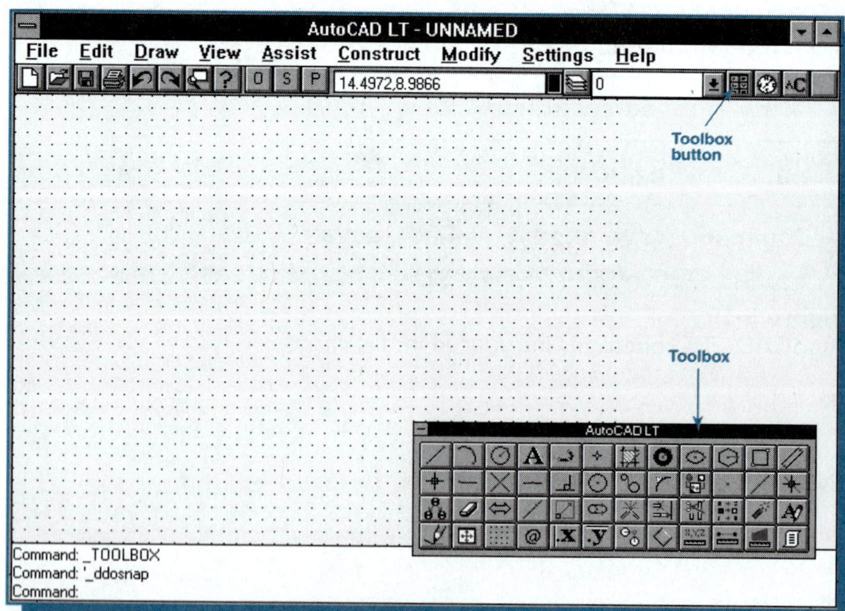

Figure 5-12
Toolbox location

Drawing the Front View of the Object

During setup we turned on the Snap and Grid options and set both of them to .250. The Grid is the display of the dots on the screen. Snap is what causes our cursor to jump .250 as you move it across the screen. Make sure the Snap button in the toolbar is highlighted or depressed **(Figure 5-12)**.

tip: *It is usually a good idea to draw the front view first and use it to create the other views.*

1. Click the Line icon from the toolbox.
2. Move the cursor to point 2,2 on the grid and click that point. The coordinate display should show 2.0000, 2.0000. The first number in the coordinate display is the X distance or horizontal distance. The second number is the Y distance or the vertical distance.
3. Move the cursor to point 6,2 on the grid and click that point. This creates the first line of the part.
4. Move the cursor to point 6,3 and click.
5. Move the cursor to point 4.5,3 and click.
6. Move the cursor to point 4.5,3.5 and click.
7. Move the cursor to point 3.5,3.5 and click.
8. Move the cursor to point 3.0,4.25 and click. This move created an angle.
9. Move the cursor to point 2,4.25 and click.
10. Type the prompt **Close** or **c** at the command line. Now at the command line. It should look like this: To point: **c**
11. Use the right button (EB) to enter this information.

This completes the outline of the front view **(Figure 5-13)**.

Drawing Holes in the Front View

1. Click the Circle icon from the toolbox. Look at the command line. What is it asking?

tip: *Did you notice that if the cursor is paused on a toolbox icon, it will tell you what it does?*

2. Move cursor to 4.5,2.5 and click. Look at the command line.
3. Move cursor over one grid dot and pick. This is called *drag* in AutoCAD. The circle changes size as you move the cursor.

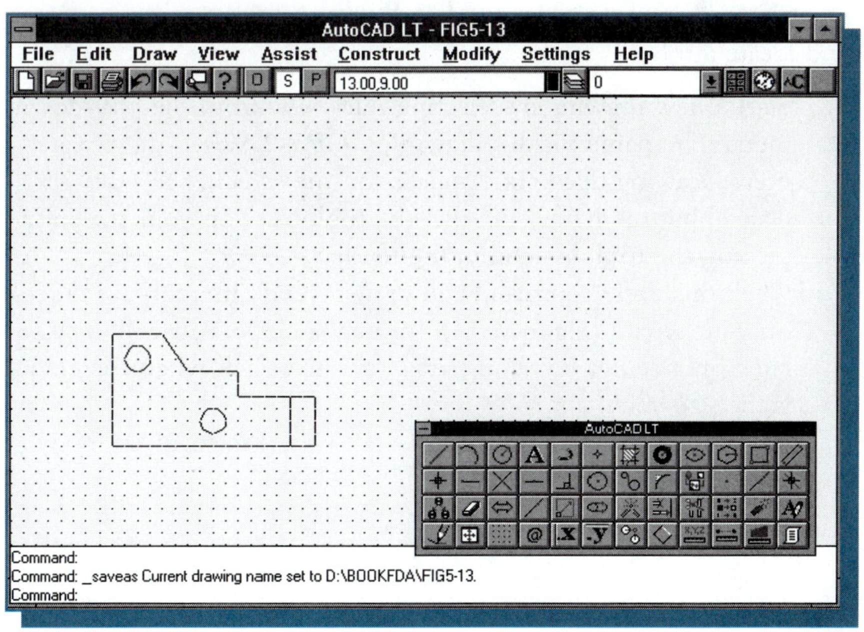

Figure 5-13
Front view of object

The **prompt** that is seen in the command line that is in < > is the prompt that is active at that time. This is called the *default*. If it is necessary to use one of the other prompts, the alternate prompt would need to be typed.

4. Draw the other hole in its proper place (center is at 2.5,3.75) and make it the correct size. The drawing should look like **Figure 5-13**.

This is a good start in learning to draw accurate drawings with AutoCAD. Let's save this drawing so we can use it later and go on to having a little fun with AutoCAD.

Saving a Drawing

It is important that the drawing be saved at regular intervals. If the computer loses power, locks up, or has one of many other mishaps, the unsaved drawing could be lost. Saving the drawing stores the drawing's environment as well as the drawing.

To Save a Drawing

1. Select Save from the toolbar or File-Save from the pull-down menus.
2. In the File Name box, type **CP5-1** and choose OK.

You have saved a drawing named cp5-1.dwg to the hard disk of your computer.

Exploring AutoCAD LT2

This time you are going to do a little exploring on your own and create a robot.

1. Start a new drawing by selecting File-New from the pull-down menu. The name for this drawing is **CP5-2**. When the Create New Drawing dialog box appears on the screen, click the Cancel button. Make sure that the toolbox is on your screen by picking the toolbox icon in the toolbar.
2. Click the circle command and draw several different-size circles on your screen **(Figure 5-14)**. To draw a circle, you will need to click on a center point and drag a radius. Click when the circle is the size you want.
3. Using the move command move the circles to form the body of the robot **(Figure 5-15)**.

tip: *To undo what you just did, select the Undo button in the toolbar.*

tip: *To refresh or clean up the screen, select the Redraw button in the toolbox.*

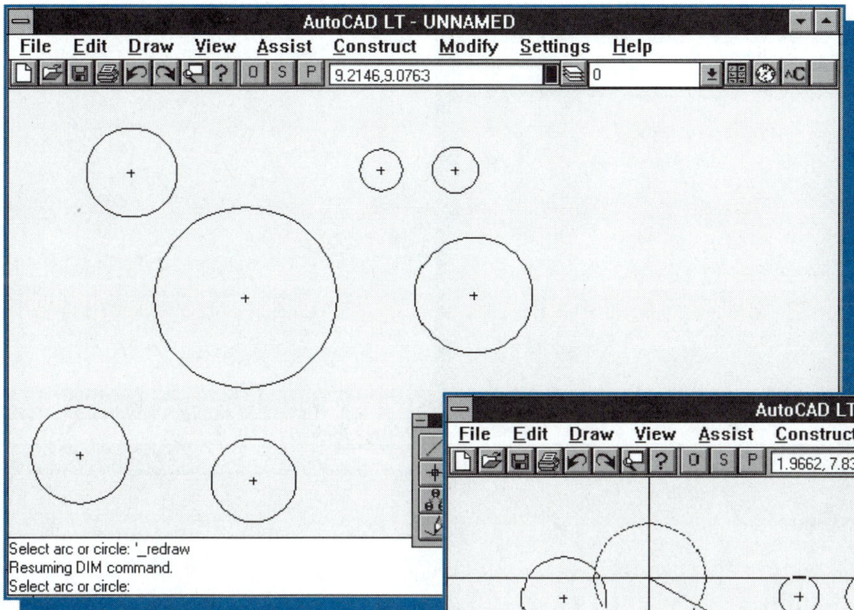

Figure 5-14
Different-size circles

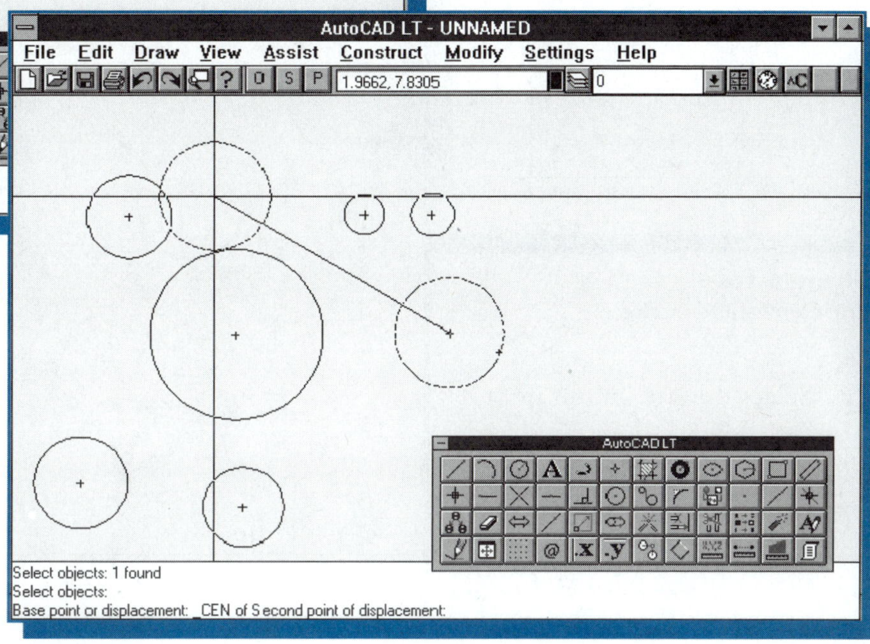

Figure 5-15
Moving circles

Select the object or objects to be moved by touching them with the pickbox and clicking. Notice that the object's lines turn to short dashes, indicating that it has been picked. Use the Enter button (EB) when you have finished selecting objects.

Now select the base point (can be any point, usually some identifiable point on the object.) Next, move the cursor to place the object in the new location, click to complete the move.

4. Erase the circles that are not needed. Select the Erase button from the toolbox **(Figure 5-16)**. Select the objects to erase and hit the Enter button (EB) on the mouse. More than one object may be selected before hitting the Enter button.
5. Select the Line button from the toolbox to add details to the robot **(Figure 5-17)**. Remember to hit the Enter button (EB) when you want the line to end.
6. Select the Zoom button in the toolbar to get close for the detail work. To zoom up on the drawing, select the Zoom button. Click to pick two diagonal corners to form a **window**. This area in the window will be the new drawing screen. To return to the full-size drawing screen, select the Zoom button and type an *A* (for *zoom all*) from the keyboard.

Figure 5-16
Erase unwanted circles

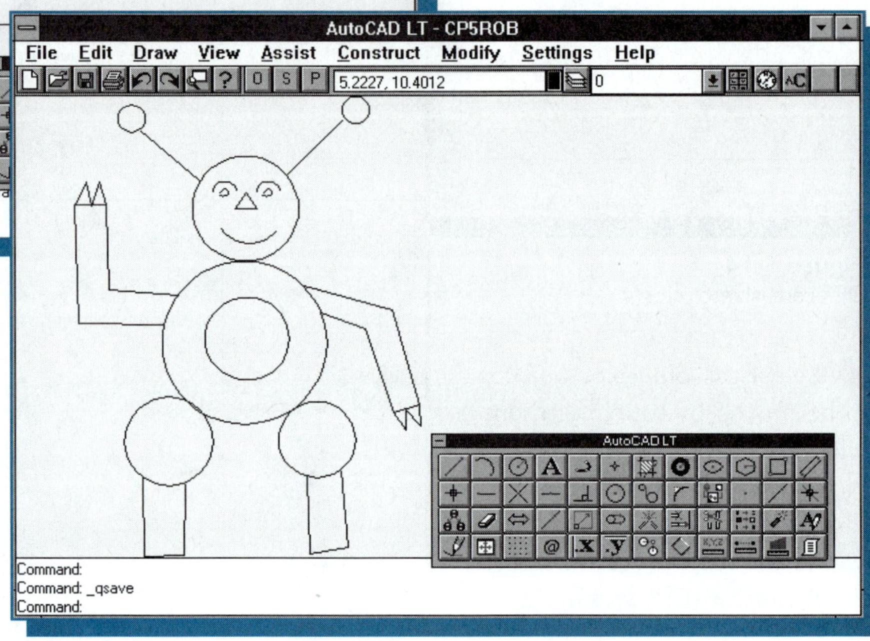

Figure 5-17
Complete robot

7. Select the Arc button from the toolbox to put finishing touches on your robot. Click two points on the screen and drag the arc until you get what you want, and then pick the third point to end your arc. Make several arcs and erase the ones not needed.
8. Use lines, circles, arcs, move, zoom, and erase to complete a simple robot.

Congratulations. You have finished your robot and it looks great. You have also found your way around the AutoCAD LT2 Drawing Editor. More commands will be explained as you progress through the text. By the end of the book you should have a good understanding of drafting and AutoCAD LT2.

Exercises

1. Try some of the commands from **Figures 5-18a-c**. See how many you can make work.
2. Using all the commands you know, draw a robot that is doing some specific task.
3. On 1/4" grid paper, sketch the drawings from **Figures 5-19a-f**.
4. Using AutoCAD LT make a drawing of each of the shapes in **Figures 5-19a-f**. Set Grid to .250 and Snap to .250. Make each shape the size shown to get an accurate drawing.
5. Customize and add detail to the mini-truck **(Figure 5-19e)**.
6. Visit a CAD vendor for a demonstration.
7. Visit an industry with a CAD system and ask for a demonstration.
8. Visit a school with a CAD system.
9. Read the manual for AutoCAD LT and become familiar with its vocabulary.
10. With a CAD system draw figures **5-19a** through **5-19g**. Use the zoom command for small details.

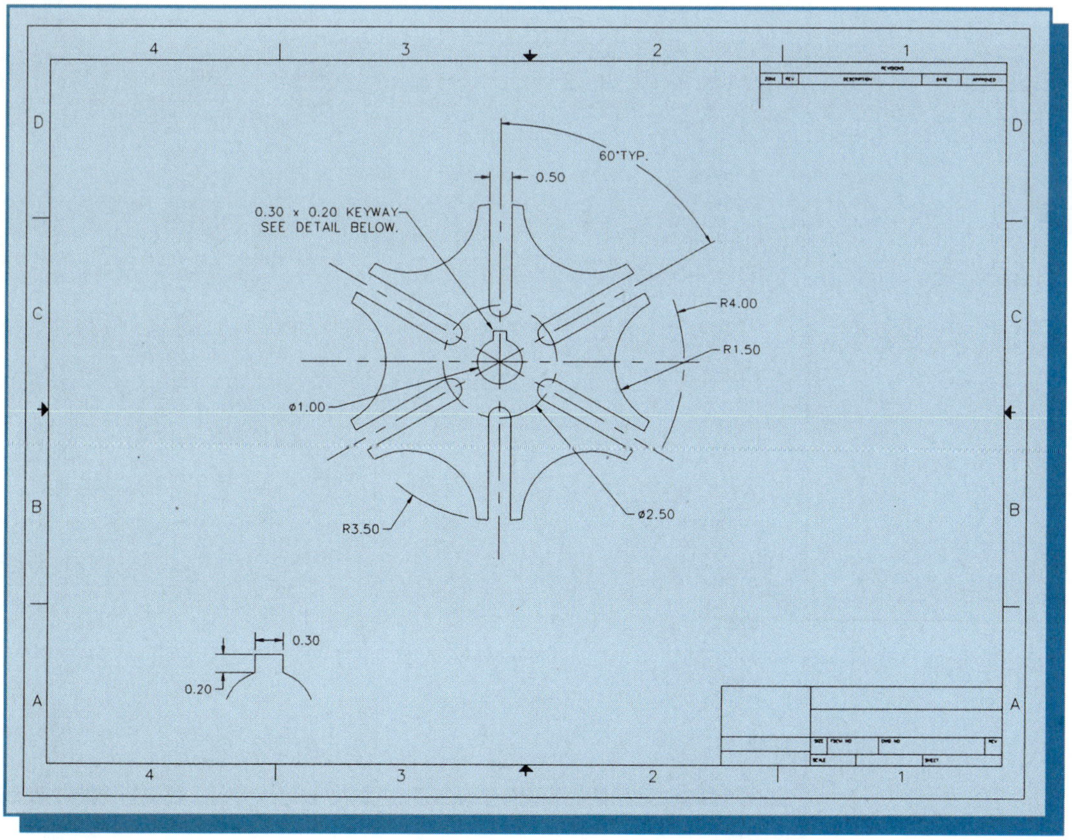

An Example of a CAM Drawing. Located in AutoCAD Drawing File CAM.

Figures 5-18a, b, and c
These AutoCAD commands are standard on most CAD systems. In the following explanations, the terms **selected** and **specified** are used. A point may be **selected** with a mouse, a puck, or by typing the coordinates on the keyboard. To **specify** is to give a measurement via the keyboard or show the distance with the puck. The command name is to the left and an explanation to the right of the diagrams.

COMMAND	ICON DIAGRAMS	DESCRIPTION
LINE	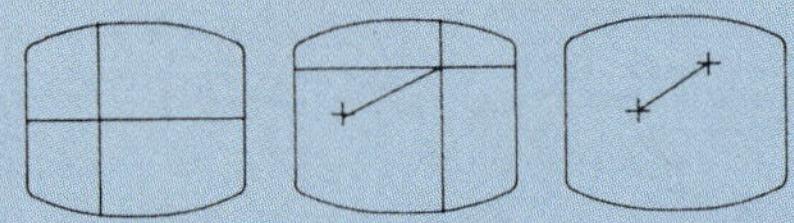	DRAWS A LINE FROM ONE SELECTED POINT TO ANOTHER
ARC	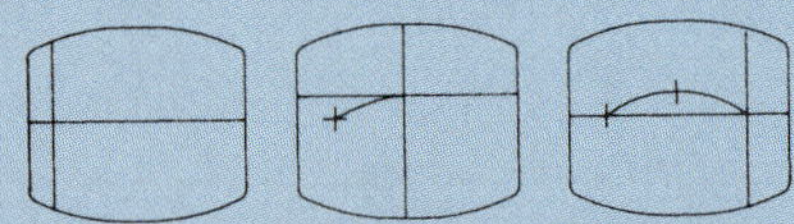	DRAWS AN ARC THROUGH THREE SELECTED POINTS
CIRCLE	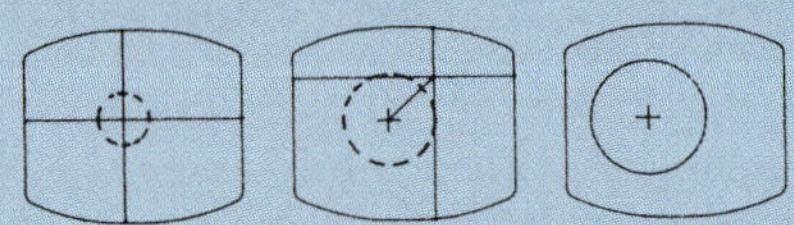	DRAWS A CIRCLE. THE CENTER IS SELECTED, THEN THE RADIUS. THE RADIUS MAY BE TYPED IN OR SPECIFIED WITH THE PUCK.
ELLIPSE	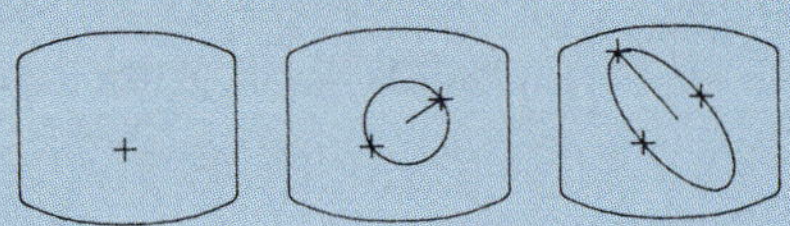	DRAWS AN ELLIPSE. THE MINOR AND MAJOR DIAMETERS ARE SPECIFIED AND THE ELLIPSE IS DRAWN. IT MAY THEN BE ROTATED AND MOVED.
POLYGON	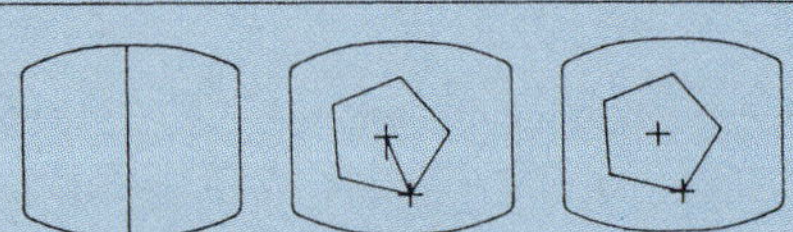	DRAWS A POLYGON OF A SPECIFIED NUMBER OF SIDES AND SIZE. THE NUMBER OF SIDES IS TYPED IN, AND THE SIZE MAY BE SPECIFIED BY ITS INSIDE OR OUTSIDE DIAMETER.
POINT	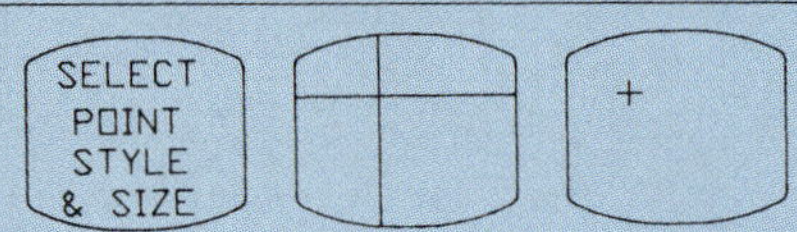	PLACES A POINT ON THE DRAWING. THE SIZE AND SHAPE OF THE POINT MAY BE SELECTED BY THE OPERATOR. THIS ONE IS A SIMPLE CROSS.
DONUT	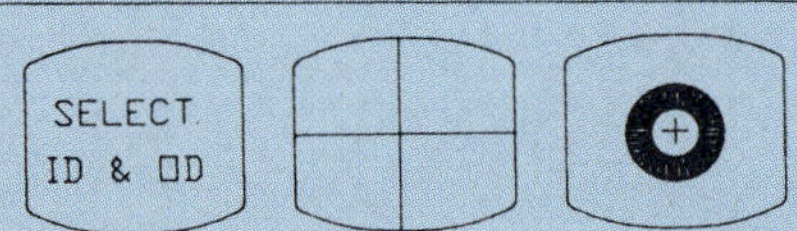	DRAWS 2 CONCENTRIC CIRCLES WITH A SPECIFIED INSIDE AND OUTSIDE DIAMETER.
DIMENSION, DIAMETER	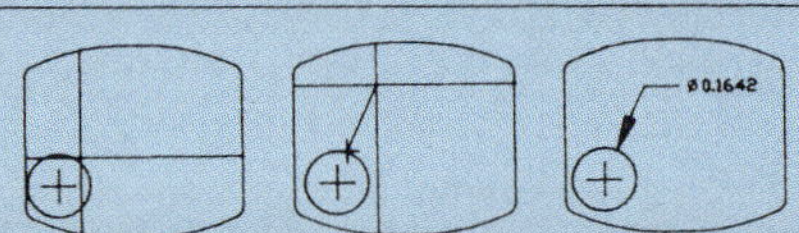	IF THE DIAMETER DIMENSION IS TOO SMALL TO APPEAR INSIDE THE CIRCLE, IT IS DISPLAYED ON THE OUTSIDE WITH A LEADER.
DIMENSION, DIAMETER	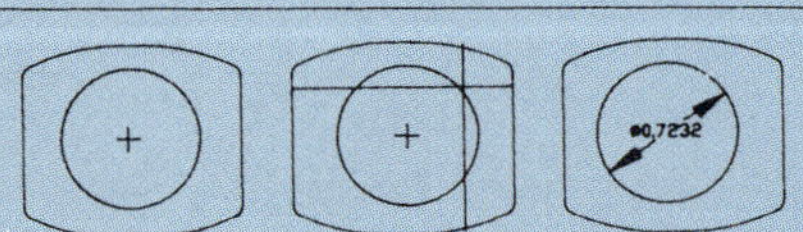	DIMENSION THE DIAMETER OF A CIRCLE. THE CIRCLE IS SELECTED, THE COMPUTER MEASURES IT AND DISPLAYS THE DIMENSION PRECEDED WITH THE O SYMBOL FOR DIAMETER.
DIMENSION, RADIUS	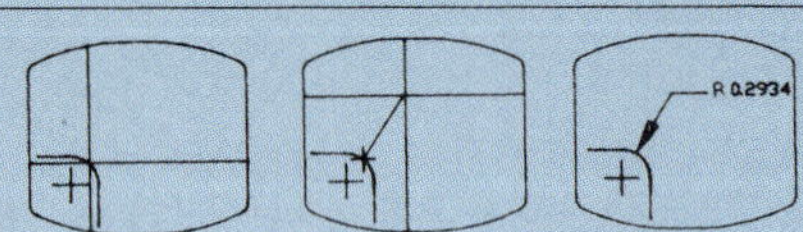	DIMENSION AN ARC. THE ARC IS SELECTED, THE COMPUTER MEASURES IT AND DISPLAYS THE DIMENSION PRECEDED BY AN R.
DIMENSION, ARC	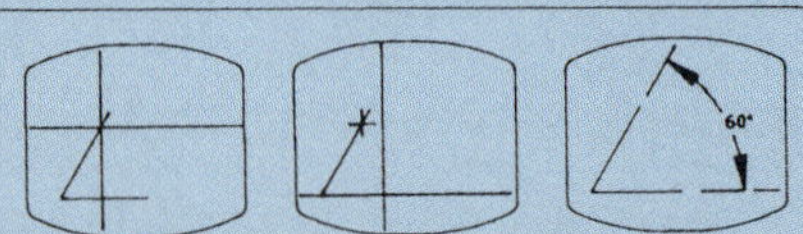	DIMENSION AN ARC. THE TWO LINES OF THE ARC ARE SELECTED AND THE DIMENSION IS DISPLAYED WITH THE ANGLE SYMBOL.
DIMENSION, VERTICAL	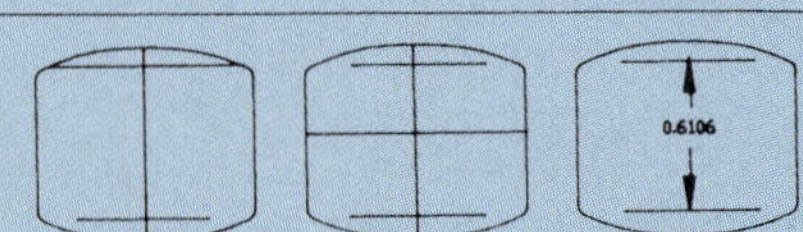	A VERTICAL DIMENSION IS DISPLAYED BY SELECTING THE BEGINNING OF THE TWO EXTENSION LINES AND THE LOCATION OF THE DIMENSION.

Figure 5-18a

COMMAND	ICON DIAGRAM	DESCRIPTION
DIMENSION, HORIZONTAL	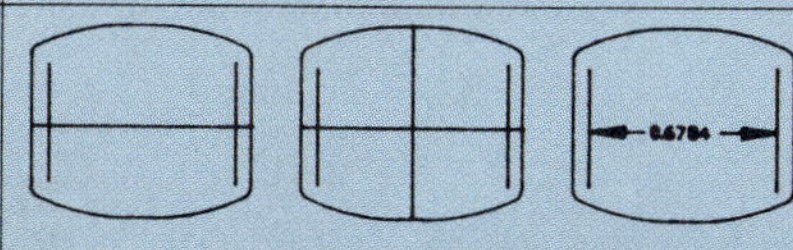	A HORIZONTAL DIMENSION IS DISPLAYED BY SELECTING THE BEGINNING OF THE TWO EXTENSION LINES AND THE LOCATION OF THE DIMENSION. NOTE THAT THE PROCESS IS THE SAME FOR VERTICAL AND HORIZONTAL DIMENSIONS. THE COMMAND SELECTION MAKES THE DIFFERENCE.
TEXT, LEFT	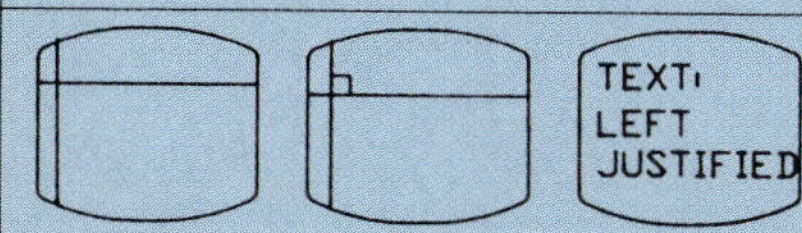	TEXT, LEFT JUSTIFIED ALIGNS THE TEXT TO ITS LEFT ALONG THE VERTICAL CURSOR.
TEXT, CENTER	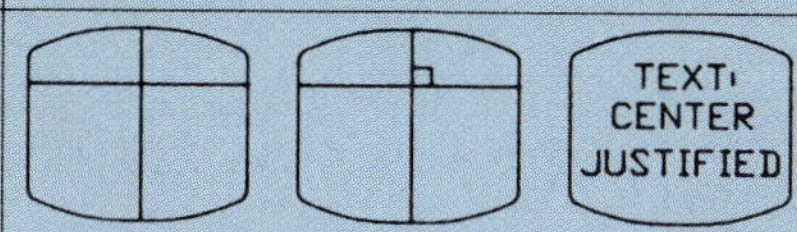	TEXT, CENTER JUSTIFIED CENTERS THE TEXT ON THE VERTICAL CURSOR.
TEXT, RIGHT	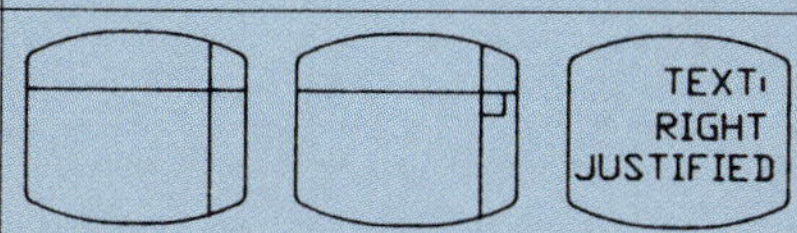	TEXT, RIGHT JUSTIFIED ALIGNS THE TEXT TO ITS RIGHT ALONG THE VERTICAL CURSOR.
SCALE	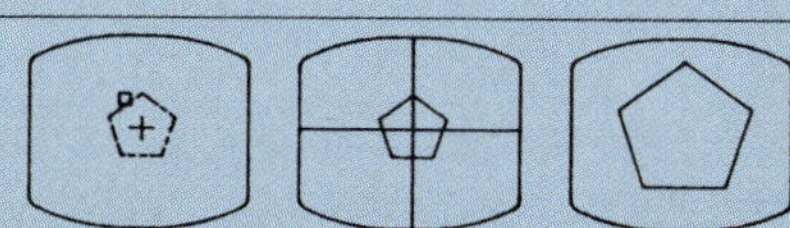	SCALE CHANGES THE SIZE OF SELECTED SHAPES. HERE A POLYGON HAS BEEN SCALED UP – ENLARGED. THE OPERATOR SPECIFIES THE AMOUNT OF SIZE CHANGE (2 = DOUBLE SIZE, .5 = ONE HALF SIZE, ETC.).
ROTATE	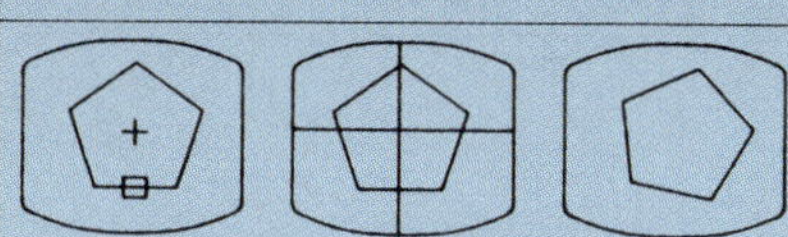	ROTATE REVOLVES A SELECTED OBJECT ABOUT A SELECTED POINT. THIS POLYGON HAS BEEN ROTATED 10 DEGREES ABOUT ITS CENTER.
OFFSET	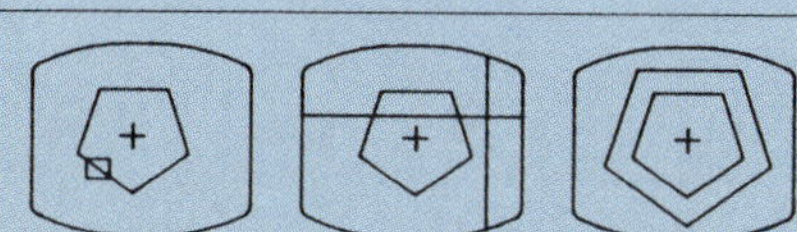	OFFSET DRAWS A LINE OR LINES PARALLEL TO SELECTED OBJECTS. THE OPERATOR CHOOSES THE AMOUNT OF OFFSET AND ON WHICH SIDE IT WILL APPEAR.
MOVE	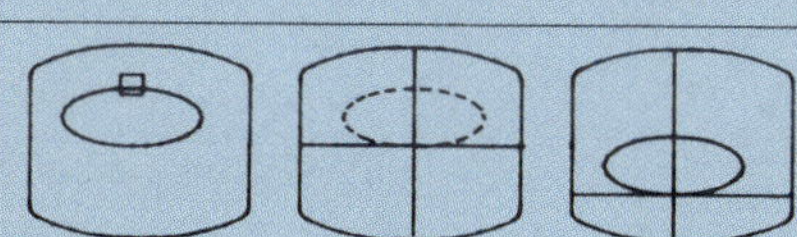	MOVE CHANGES THE LOCATION OF SELECTED OBJECTS. THIS ELLIPSE HAS BEEN MOVED DOWN.
ERASE, WINDOW	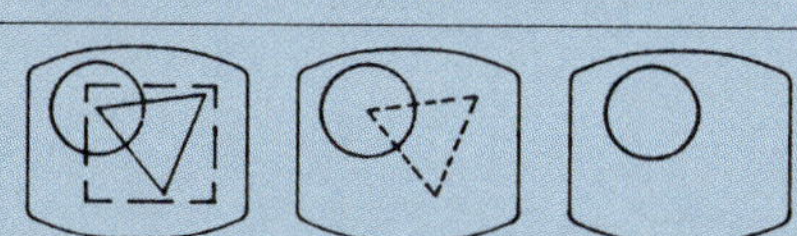	ERASE, WINDOW ELIMINATES THE SELECTED FEATURES THE OPERATOR INCLUDES IN THE WINDOW.
ERASE, OBJECT	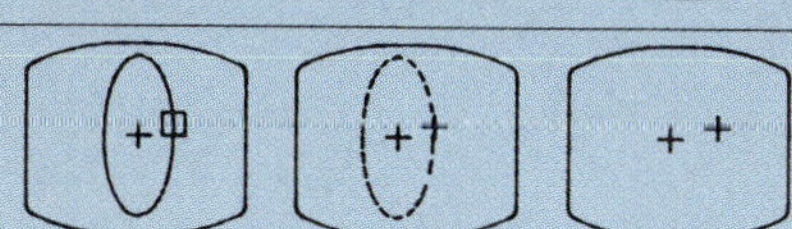	ERASE, OBJECT ELIMINATES THE FEATURES THE OPERATOR PICKS WITH THE PUCK.
ERASE, LAST	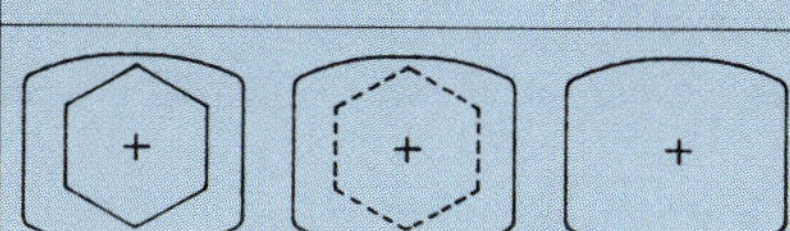	ERASE, LAST ELIMINATES THE LAST FEATURE DRAWN BY THE OPERATOR.

Figure 5-18b

COMMAND	ICON DIAGRAMS	DESCRIPTION
ARRAY, POLAR	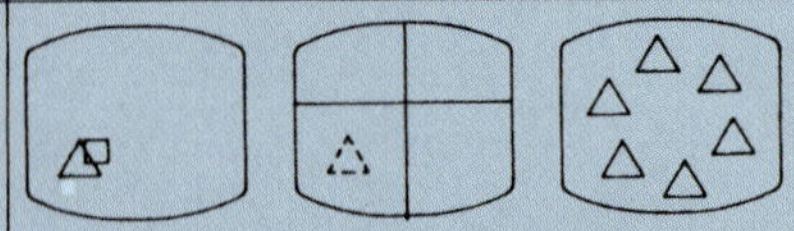	ARRAY, POLAR (CIRCULAR) COPIES A SELECTED FEATURE IN A CIRCULAR PATTERN ABOUT A SPECIFIED CENTER POINT. THE NUMBER OF TIMES THE FEATURE APPEARS IS SPECIFIED AND THEY ARE EQUALLY SPACED.
ARRAY, RECTANGULAR	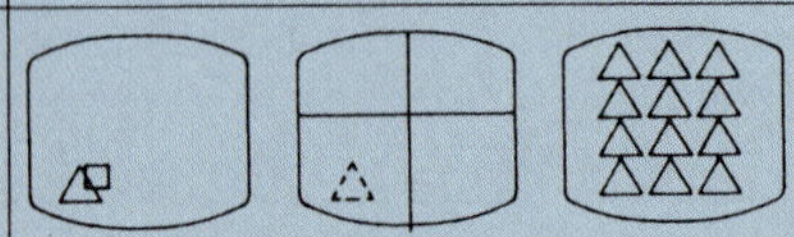	ARRAY, RECTANGULAR COPIES FEATURES HORIZONTALLY AND VERTICALLY. THE SPACING AND THE NUMBER OF COLUMNS AND ROWS ARE SPECIFIED. THIS EXAMPLE HAS 3 COLUMNS AND 4 ROWS.
DIVIDE	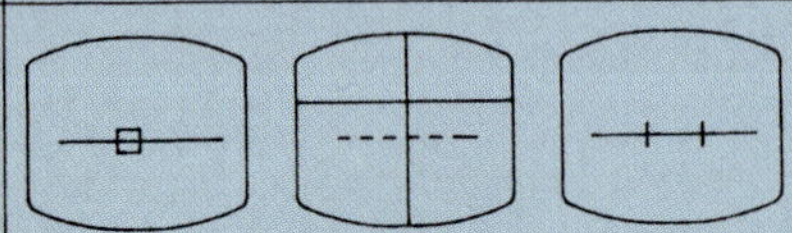	DIVIDES A LINE INTO A SPECIFIED NUMBER OF EQUAL SPACES. THIS LINE HAS BEEN DIVIDED INTO 3 EQUAL PARTS.
MIRROR	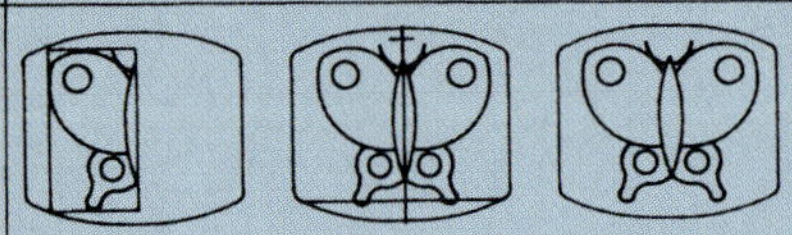	MIRROR DRAWS A MIRRORED IMAGE OF THE SELECTED FEATURE. THE MIRROR LINE MAY BE AT ANY ANGLE. THIS EXAMPLE HAS BEEN MIRRORED VERTICALLY, THAT IS, THE MIRROR LINE IS A VERTICAL LINE.
BREAK	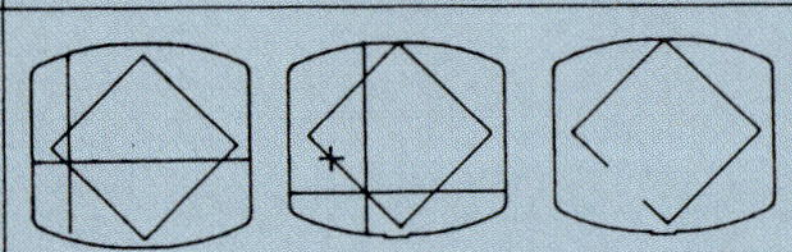	BREAK PUTS A GAP IN A LINE, CIRCLE OR ARC. THE LINE AND THE SIZE OF THE GAP IS SPECIFIED.
COPY	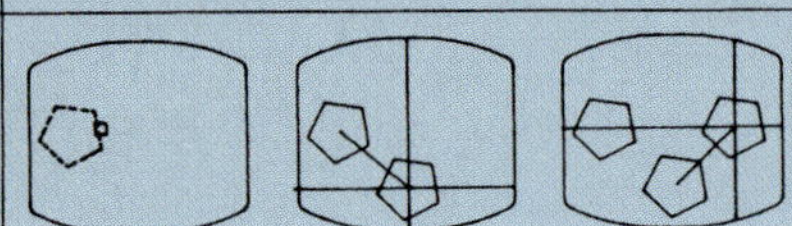	COPY REPEATS A SELECTED FEATURE ON THE DRAWING AS MANY TIMES AS DESIRED TO SELECTED LOCATIONS.
STRETCH	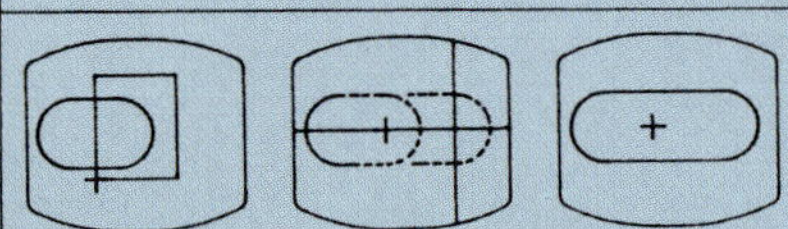	STRETCH ELONGATES OR SHORTENS FEATURES. THIS SLOT HAS BEEN ELONGATED HORIZONTALLY.
JOIN	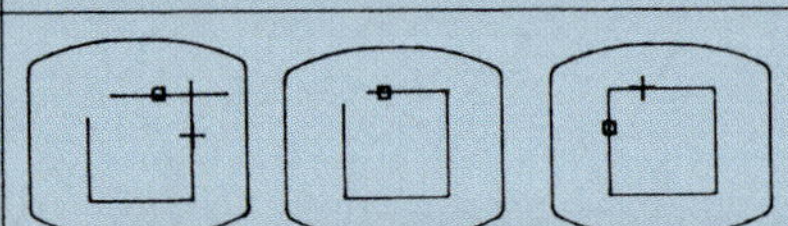	JOIN MAKES A CLEAN CORNER AT THE JUNCTION OF TWO LINES. THE LINES ARE SELECTED AND AUTOMATICALLY JOIN AT THEIR INTERSECTION.
FILLET	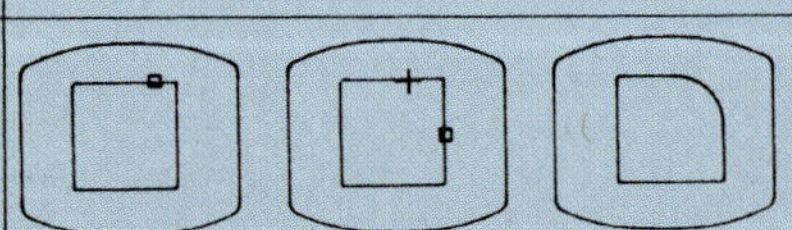	FILLET MAKES A RADIUS AT THE JUNCTION OF TWO LINES. THE SIZE OF THE RADIUS IS SPECIFIED AND THE LINES ARE SELECTED.
CHAMFER	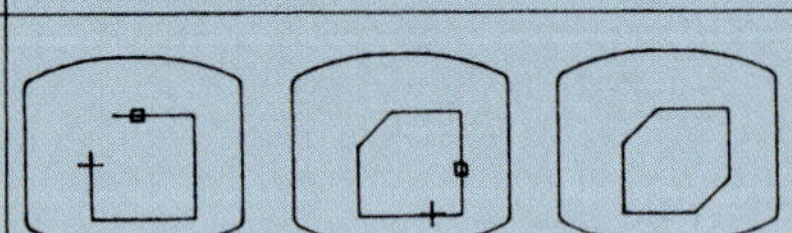	CHAMFER MAKES A SPECIFIED ANGLE BETWEEN TWO LINES AND TRIMS THE LINES.

Figure 5-18c

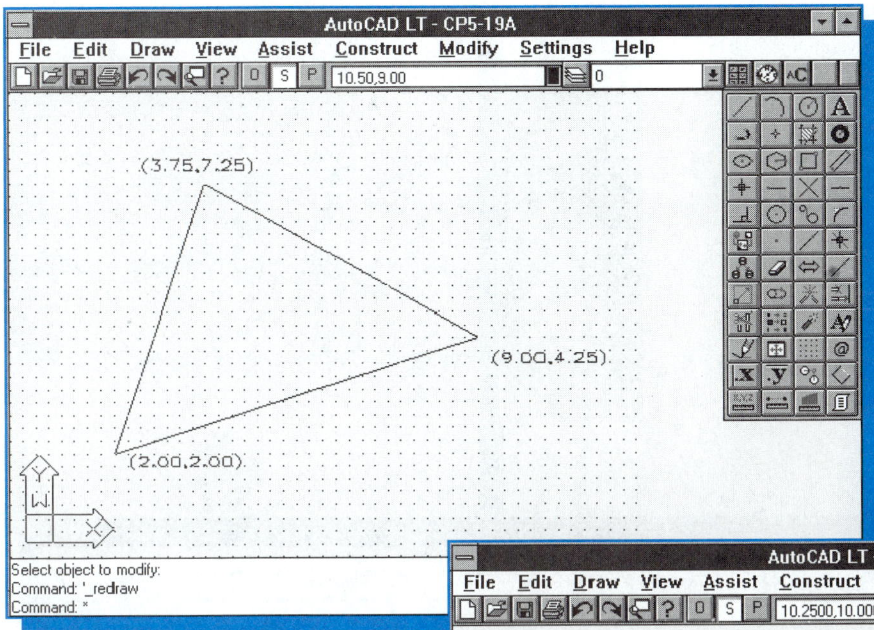

Figure 5-19a

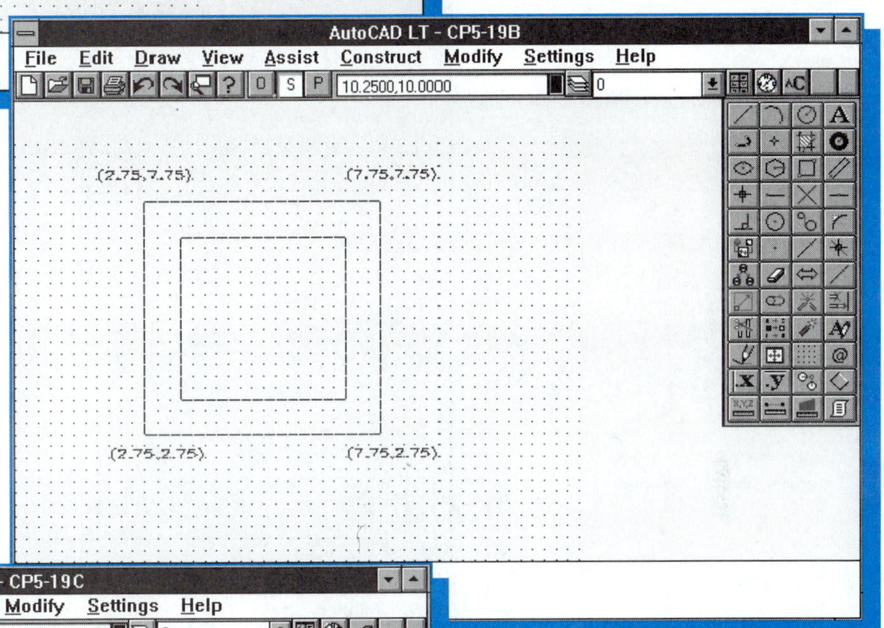

Figure 5-19b

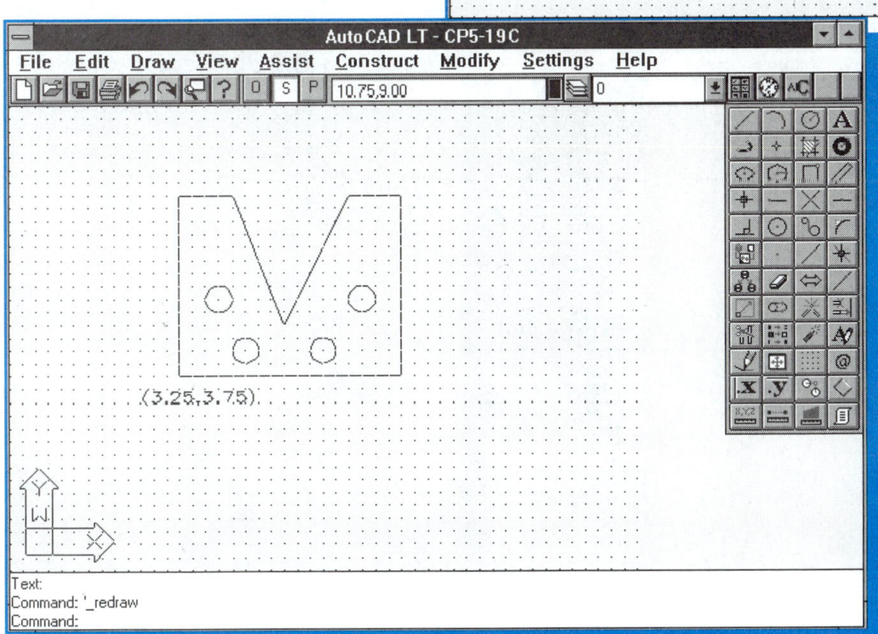

Figure 5-19c

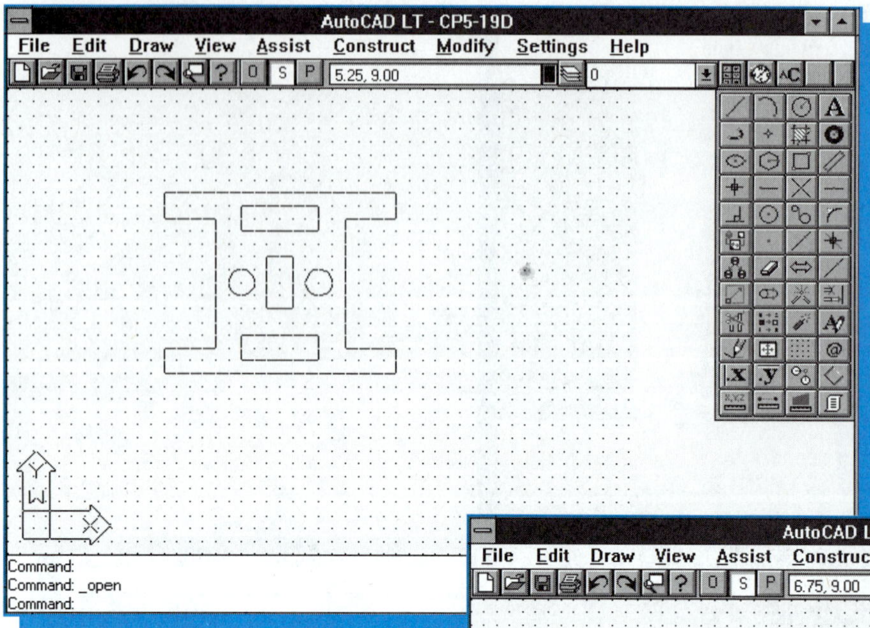

Figure 5-19d

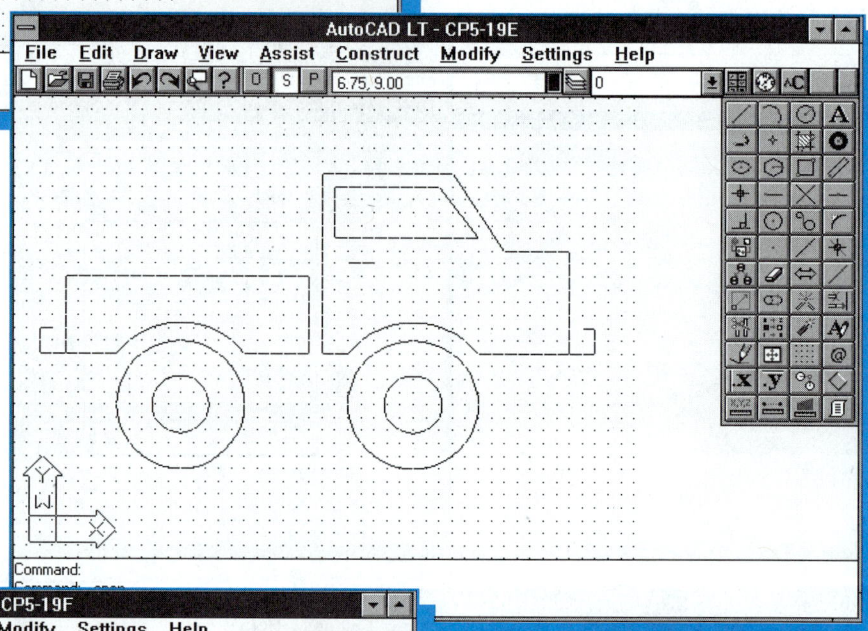

Figure 5-19e

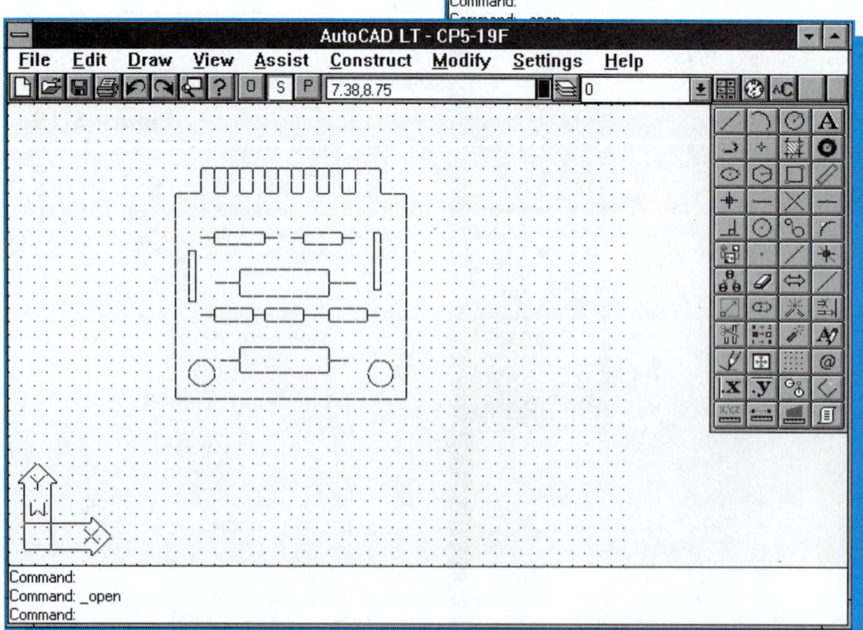

Figure 5-19f

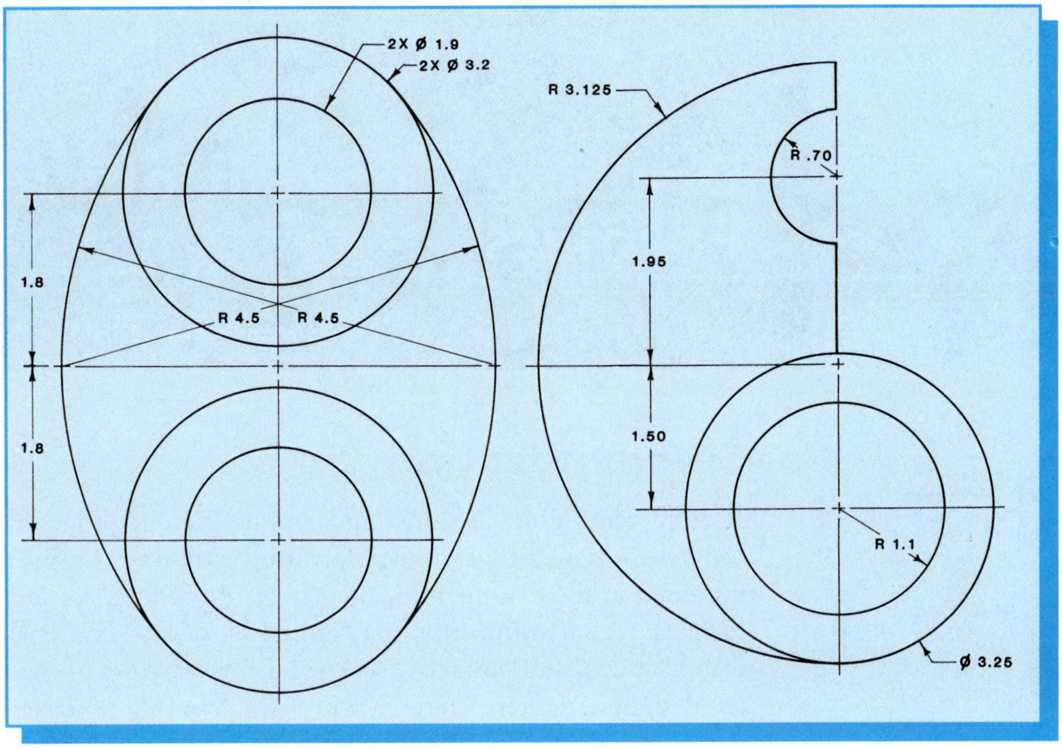

Figure 5-19g
Draw the geometric forms with your CAD system.

Key Terms

Commands
Coordinates
Cursor
Drawing area
Enter button
Entity
Drawing Editor
Window
Buttons
Pick button
Keyboard
Prompt
Mouse
Toolbox
Toolbar
Prompt area
Selected
Specify
Units of Measurement
Submenus

Chapter 6

Drafting Conventions and Formats

Objectives

The student will be able to:

- draw a basic working drawing using ASME line conventions
- describe the types of working drawing conventions
- lay out and draw a school drafting format sheet
- lay out and draw an industrial company format sheet

Standardization

Different industries and companies use unique processes and procedures for designing and manufacturing products. However, drawings used in the design and manufacture of products are prepared to identical sets of **standards**. Standardization of drawings enables all manufacturers and builders to interpret the drawings of many different designers, drafters, engineers, or architects. Before standardization, working drawings could only be interpreted by the designer or drafter who prepared the drawing **(Fig. 6-1)**.

Figure 6-1
Early working drawings usually required interpretation because only the drafter or designer could read his or her drawings. *(Courtesy of The Celotex Corporation)*

National and World Standards

Dictionaries are used to standardize words in a written or spoken language. Likewise, the **American Society of Mechanical Engineers (ASME)** publications are used as the universal reference work for drafting standards in the United States. The **International Drafting Standards (ISO)** publication R-1000 describes metric standards used throughout the world. The consistent use of these two international standards insures that a product can be designed anywhere in the world, and built or manufactured in any other part of the world. Each specialized standard for drafting is found in ASME's Y-14 publication with the following identification numbers:

Y14.1	Drawing Sheet Size and Format
Y14.2	Line Conventions and Lettering
Y14.3	Multiview and Sectional-View Drawings
Y14.4	Pictorial Drawings
Y14.5	Dimensioning and Tolerancing
Y14.6	Screw-Thread Representation
Y14.7	Gears, Splines, and Serrations
Y14.7.1	Gear-Drawing Standards—Part 1
Y14.9	Forgings
Y14.10	Metal Stampings
Y14.11	Plastics
Y14.14	Mechanical Assemblies
Y14.15	Electrical and Electronics Diagrams, Including Supplements Y14.15a and Y14.15b
Y14.17	Fluid Power Diagrams
Y14.26	Dictionary of Terms for Computer-Aided Preparation of Product Definition Data

Line Conventions

Technical drawings contain many types of lines, each with a special meaning. Through the consistent use of the ASME line conventions, technical drawings become a language that can be read by anyone who understands the standard. The ASME **line conventions** are shown in **Figure 6-2**.

Figure 6-2
ASME line conventions

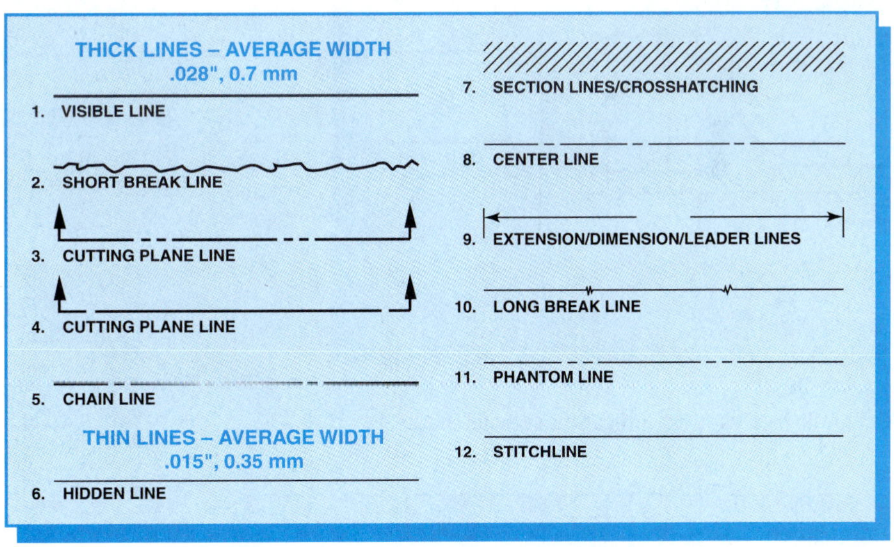

Line Widths

All lines are either thick (.032" or 0.7 mm average width) or thin (0.16" or 0.35 mm average width). Both thick and thin lines are always dark (**opaque**) to insure high reproduction quality. Lines used for layout and lettering guide lines **(Fig. 6-3)** are drawn extremely light and should be invisible on the

Figure 6-3
Construction lines and guide lines are drawn very lightly so they will not be seen in the finished working drawing or print.

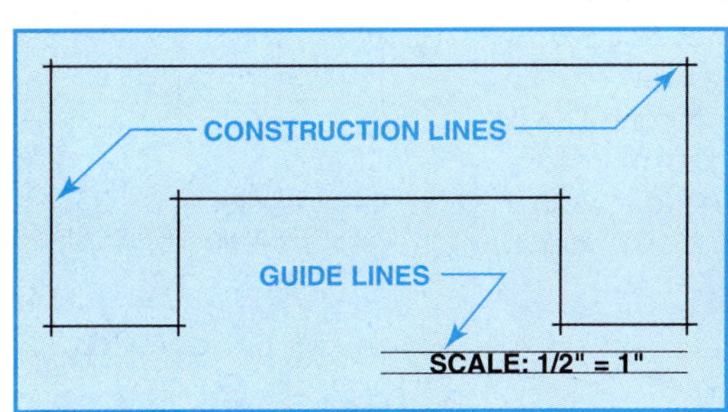

finished drawing reproduction. Some drafters use nonreproducible blue pencils to insure invisibility.

Line Types

Visible **object lines (Fig. 6-4)** are solid, thick lines used to define the outline of objects. **Break lines** are used to interrupt the drawing if the object will not fit on a drawing sheet. Short break lines are thick, wavy lines, and long break lines are thin lines with zigzags spaced at approximately 1.5" intervals **(Fig. 6-5)**. When objects are visually cut (sectioned) to show an interior view, a thick **cutting plane line** is used to show the location of the cut. Thin **section lines** are used to represent the material **(Fig. 6-6)** through which the cut is made. **Chain lines (Fig. 6-7)** are used to indicate specific areas. Object lines that fall behind the front plane of an object are drawn with dashed lines representing **hidden lines (Figs. 6-8a** and

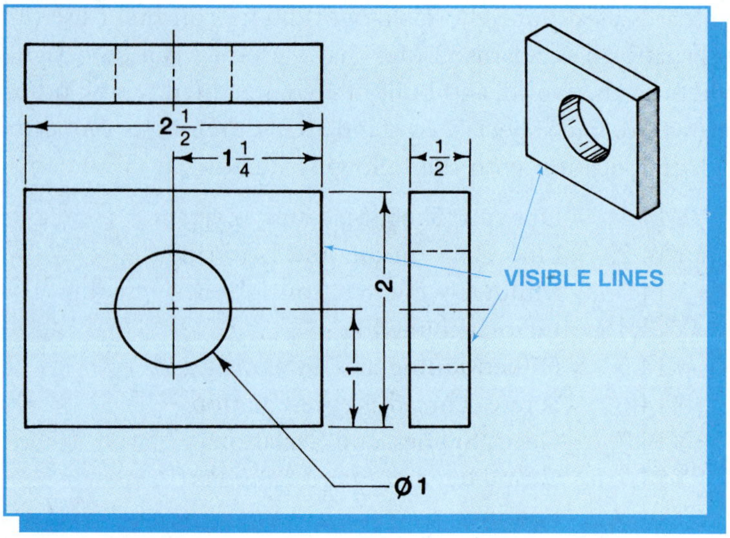

Figure 6-4
Note the width of the object lines and compare them to the widths of the other lines.

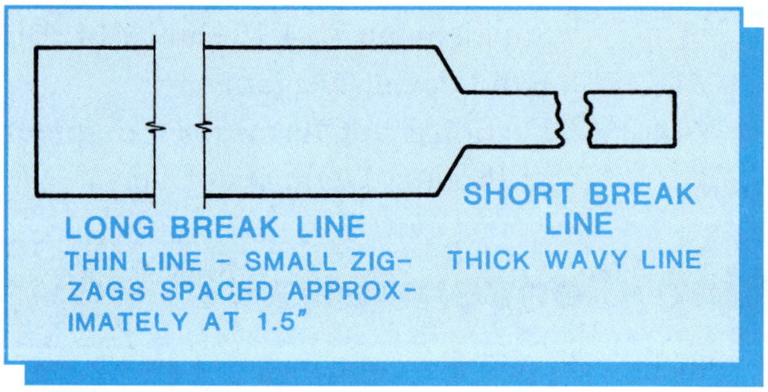

Figure 6-5
Long and short break lines

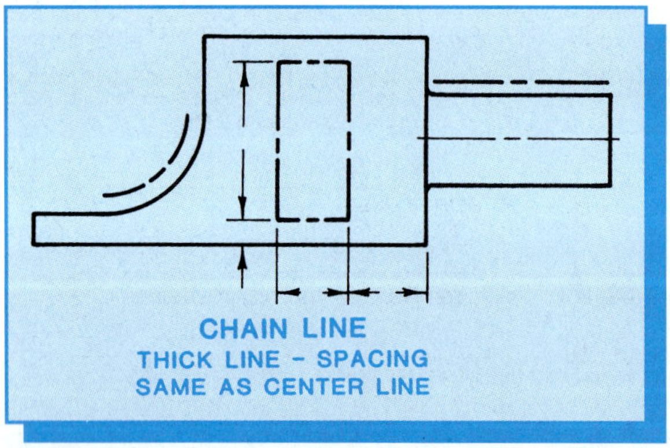

Figure 6-7
A chain line is used to indicate a specific area.

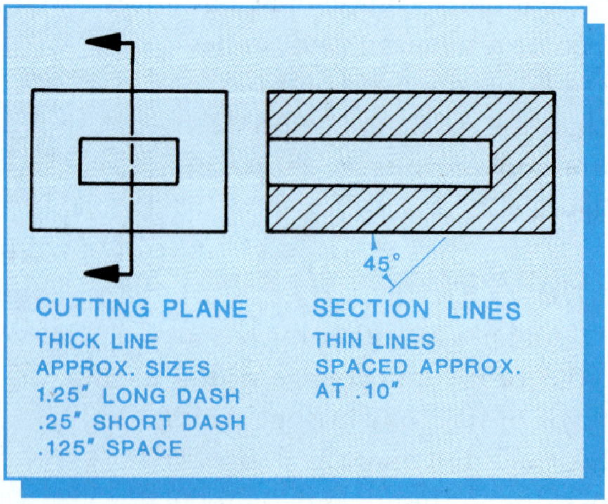

Figure 6-6
The cutting plane and the section lines (cross-hatching)

Figure 6-8a
Drafting conventions for hidden lines in a working drawing. The hidden line is approximately 1/8" long with a 1/32" space.

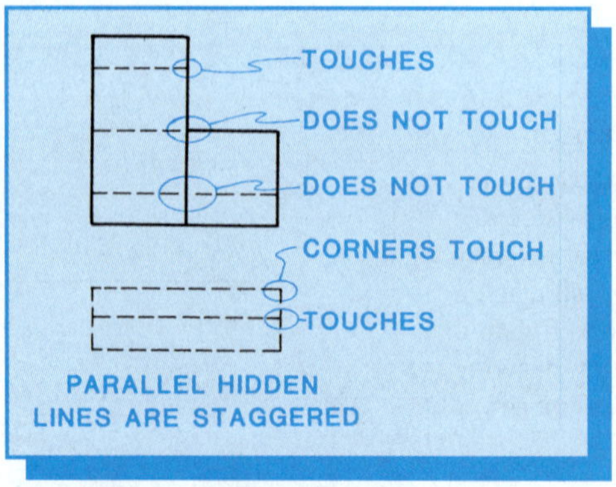

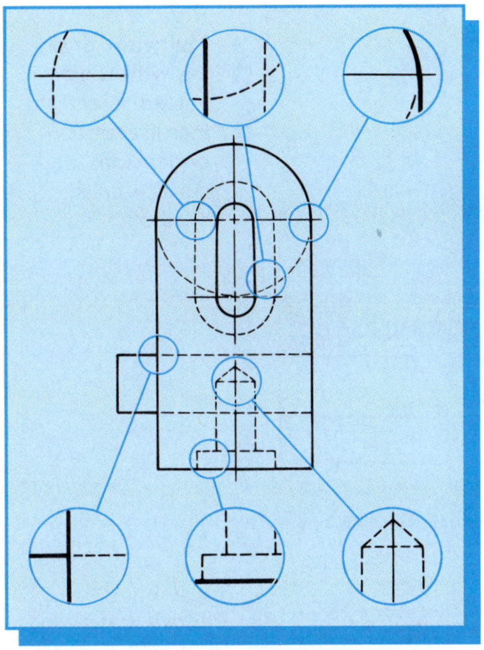

Figure 6-8b
General rules for hidden lines

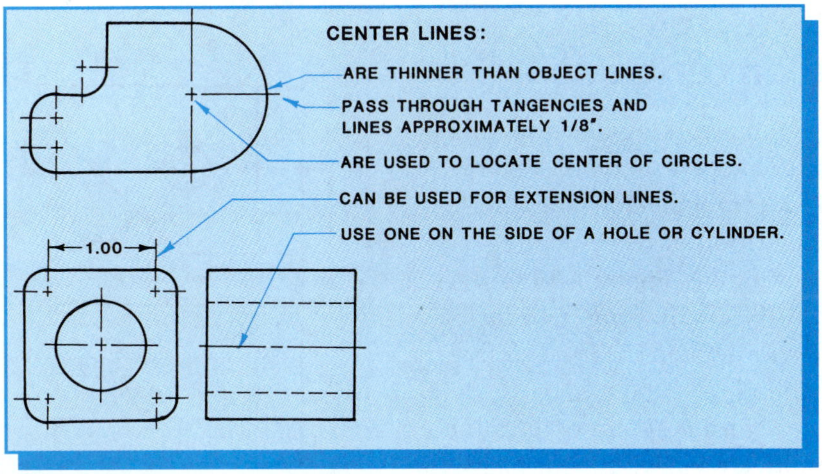

Figure 6-9
Drafting conventions for center lines on a working drawing. The approximate dimensions are: long dash 1 1/2", short dash 1/8", space 1/16".

6-8b). (Note that when a hidden line appears as the continuation of a solid object line, the drafter must leave a space.) When objects are symmetrical with a common center, the center is defined with a **centerline** constructed with thin long and short dash lines **(Fig. 6-9)**. The size of objects is described with the use of thin **leaders**, **extension lines**, and **dimension lines (Fig. 6-10)**. When alternate positions of an object or matching part must be shown without interfering with the main drawing, thin **phantom lines** are used **(Fig. 6-11)**. To show the location of stitches in a material, **stitch lines** are used **(Fig. 6-12)**.

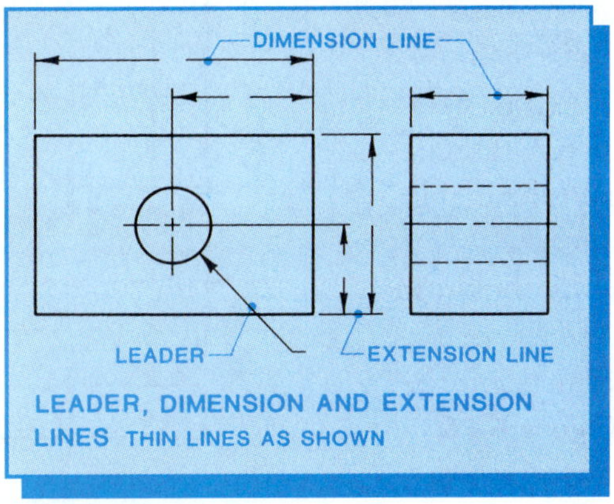

Figure 6-10
Leader, dimension, and extension lines.

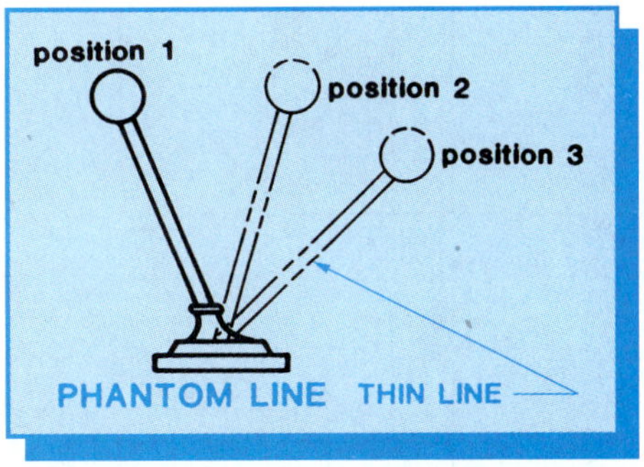

Figure 6-11
Phantom lines show alternate positions. Line sizes are the same as center lines.

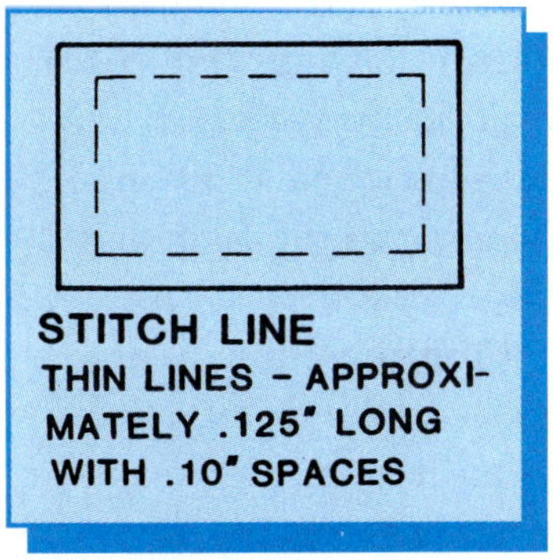

Figure 6-12
Stitch lines show various stitches sewn into materials.

Drafting Conventions

The working drawings of a machine part shown in **Figures 6-13 through 6-24** show the application of many drafting conventions. **Figure 6-13** is a **multiview** drawing with inch-fraction dimensions and surface symbols.

Figure 6-14 shows the same multiview drawing with inch-decimal dimensions. In this figure, a cutting plane line is positioned on the top view, while the corresponding full section is shown in the right-side view. Note the standard weld symbol and dual inch-millimeter chart in **Figure 6-14**.

Metric (millimeter) dimensions and positional tolerancing symbols are added to the machine part drawing shown in **Figure 6-15**. Tolerance dimensions are used to dimension the same drawing in **Figure 6-16**.

A pictorial drawing is usually used to show the finished shape of the assembled object.

Isometric drawings **(Fig. 6-17)** and isometric **section** drawings **(Fig. 6-18)** are all drawn with receding sides positioned at 30°, and extending to the right and left.

Oblique pictorial drawings show the front view of an object in true scale and form, with receding angles projected to only one side at 30° or 45°.

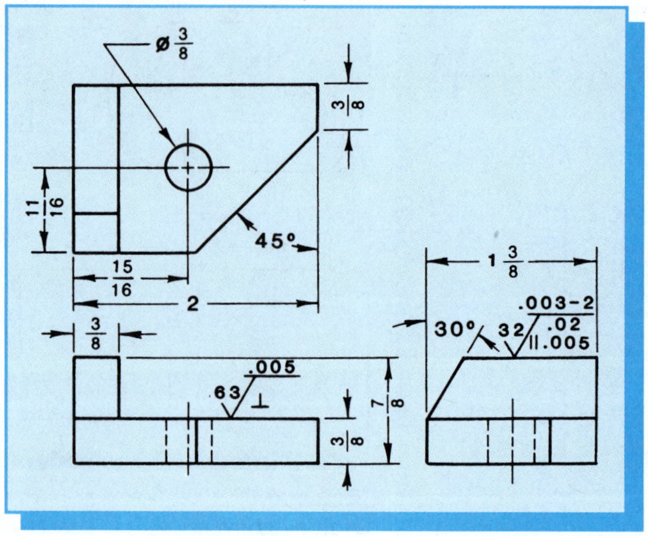

Figure 6-13
Multiview drawing with third angle projection, inch-fraction dimensions, and surface finish symbols.

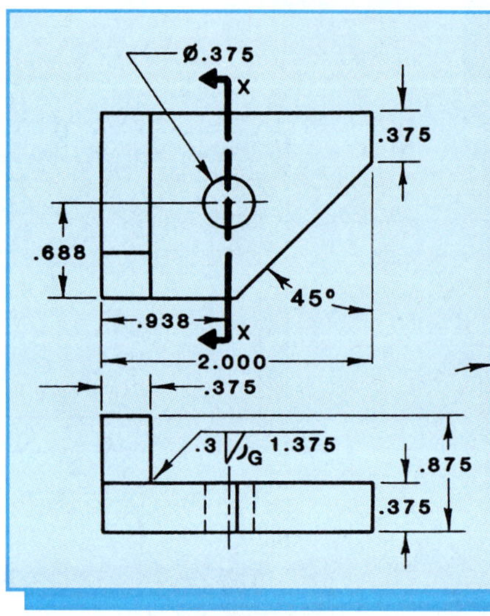

IN.	mm
.375	9.53
.688	17.46
.875	22.23
.938	23.81
1.375	34.93
2.000	50.8

Figure 6-14
Multiview drawing with inch-decimal dimensions, full section, weld symbol, and dual dimension table

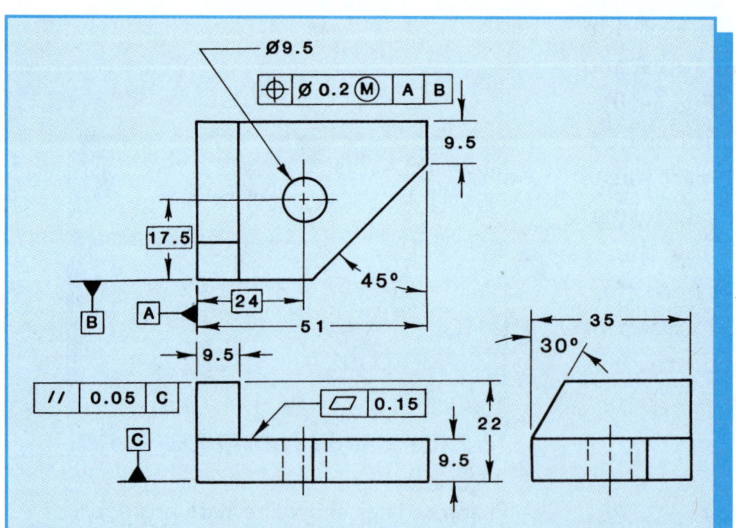

Figure 6-15
Multiview drawing with metric (millimeters) dimensioning and positional tolerancing symbols

When receding sides are drawn at one-half the actual depth, the drawing is known as a **cabinet** drawing **(Fig. 6-19)**. When sides are drawn at full-scale depth, the drawing is known as a **cavalier** drawing **(Fig. 6-20)**. Since cavalier oblique drawings appear more distorted than cabinet drawings, using half-depth dimensions on the receding angle helps reduce the appearance of distortion.

Perspective drawings **(Fig. 6-21)** are classified into one-, two-, and three-point perspective drawings (see Chapter 15). **Dimetric** pictorial drawings **(Fig. 6-22)** are similar to isometric drawings, except that two axes have equal angles other than 30°. In **trimetric** pictorial drawings **(Fig. 6-23)**, all three axes' angles are different.

Pattern development drawings show the outline of all connected surfaces of an object when laid flat. Pattern development drawings are used to define the outline of an object when a continuous material is used for all surfaces. The pattern development drawing represents all the surfaces of an object **(Fig. 6-24)**.

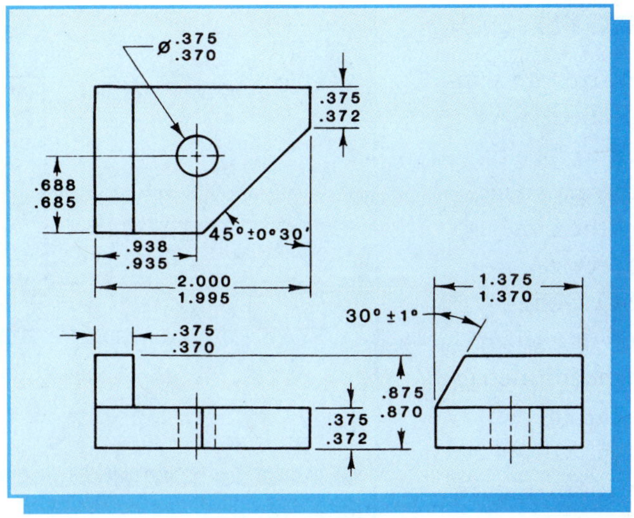

Figure 6-16
Multiview drawing with tolerance dimensions

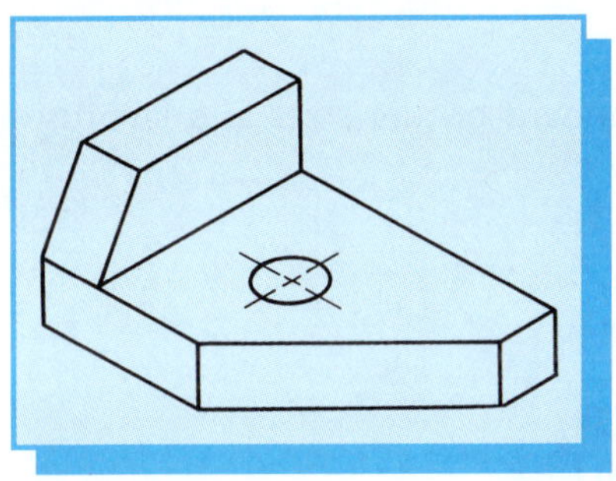

Figure 6-17
Isometric drawing

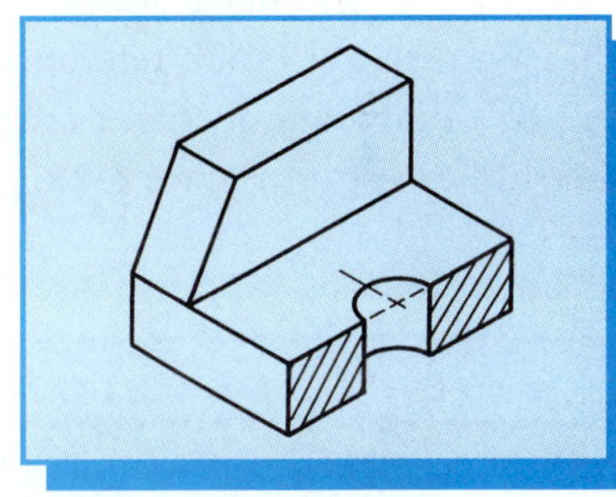

Figure 6-18
Isometric section drawing

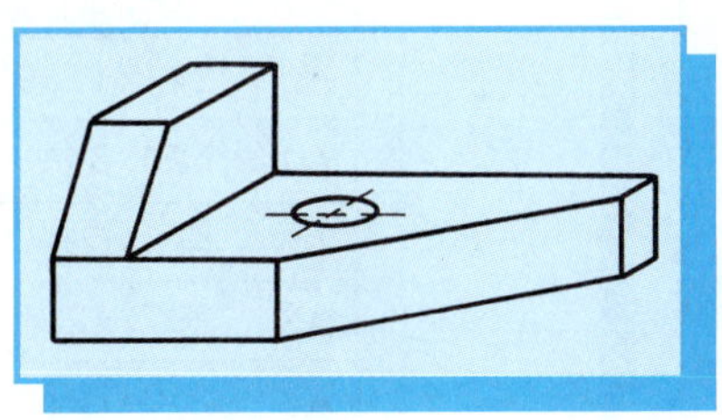

Figure 6-19
Cabinet drawing

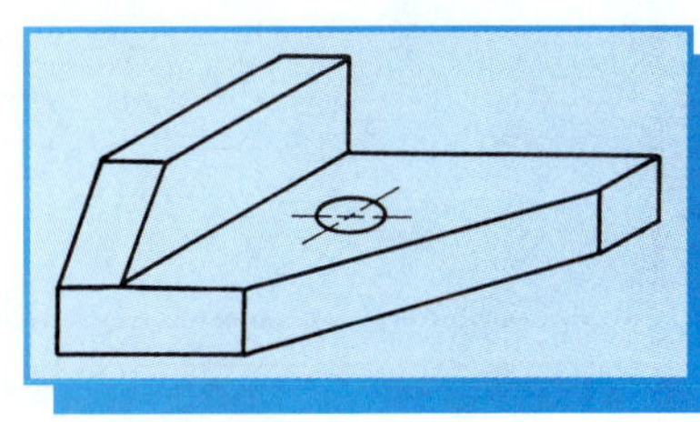

Figure 6-20
Cavalier drawing

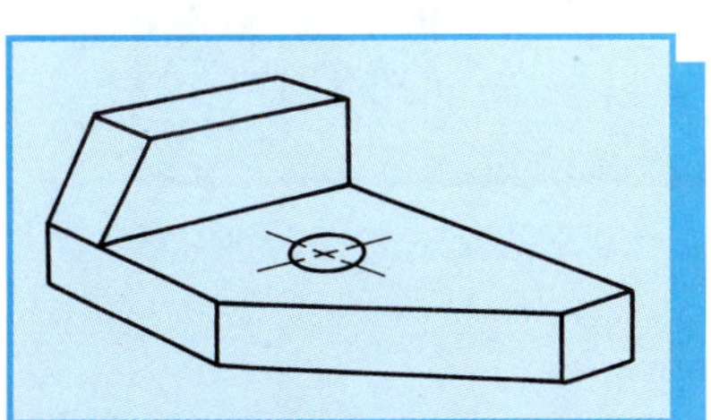

Figure 6-21
Perspective drawing

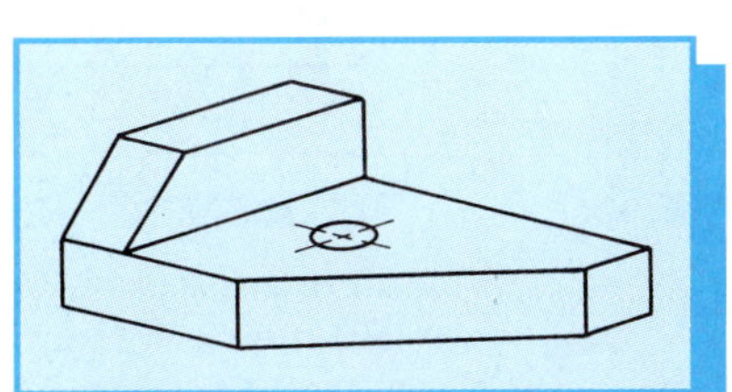

Figure 6-22
Dimetric drawing

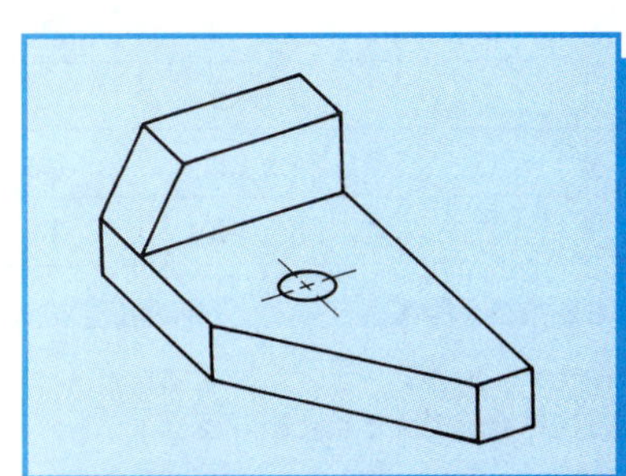

Figure 6-23
Trimetric drawing

Drawing Formats

Drawing sheet sizes are standardized by ASME. Letters A through E are used to specify standard U.S. customary sheet sizes **(Fig. 6-25)**. In the same manner, metric sheet sizes are standardized by ISO. Designations of A0 through A10 are used to identify standard metric sheet sizes **(Fig. 6-26)**. In addition to sheet sizes, ASME specifies **border** designations. **Figure 6-27** shows drawing sheets with standard alphanumeric zones that are used for quick referencing of a drawing's quadrant areas. Standard border sizes and layouts are shown also in **Figure 6-28**.

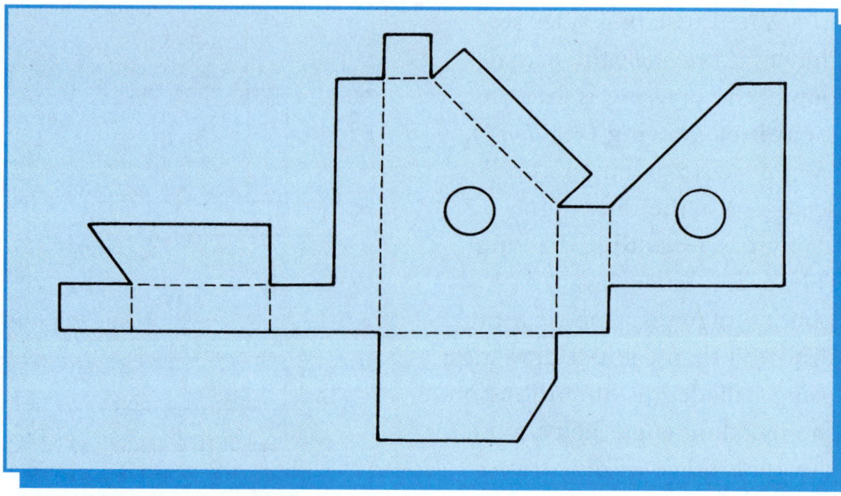

Figure 6-24
Pattern development drawing

E = TOTAL SHEET SIZE

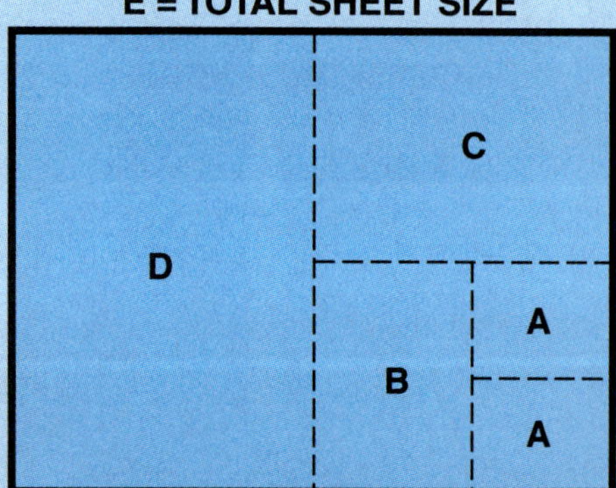

AMERICAN STANDARD SHEET SIZES (INCHES)		
DESG	WIDTH x LEN	WIDTH x LEN
A	9 x 12	8.5 x 11
B	12 x 18	11 x 17
C	18 x 24	17 x 22
D	24 x 36	22 x 34
E	36 x 48	34 x 44

Figure 6-25
ASME standard paper sizes

DES	WIDTHxLEN (mm)
A0	841x1189
A1	594x841
A2	420x594
A3	297x420
A4	210x297
A5	148x210
A6	105x148
A7	74x105
A8	52x74
A9	37x52
A10	26x37

AO = TOTAL SHEET SIZE

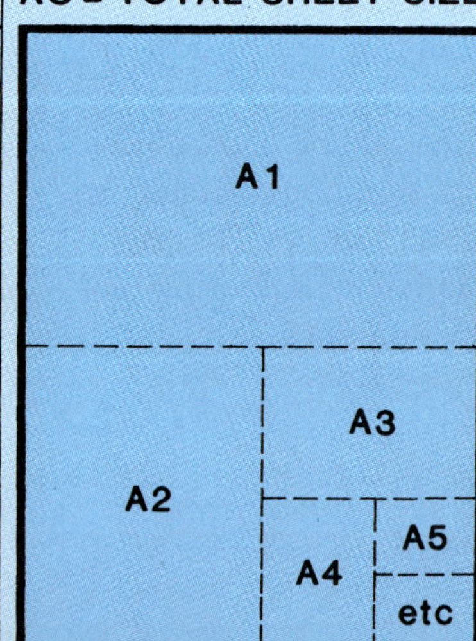

Figure 6-26
ISO metric A-size paper series

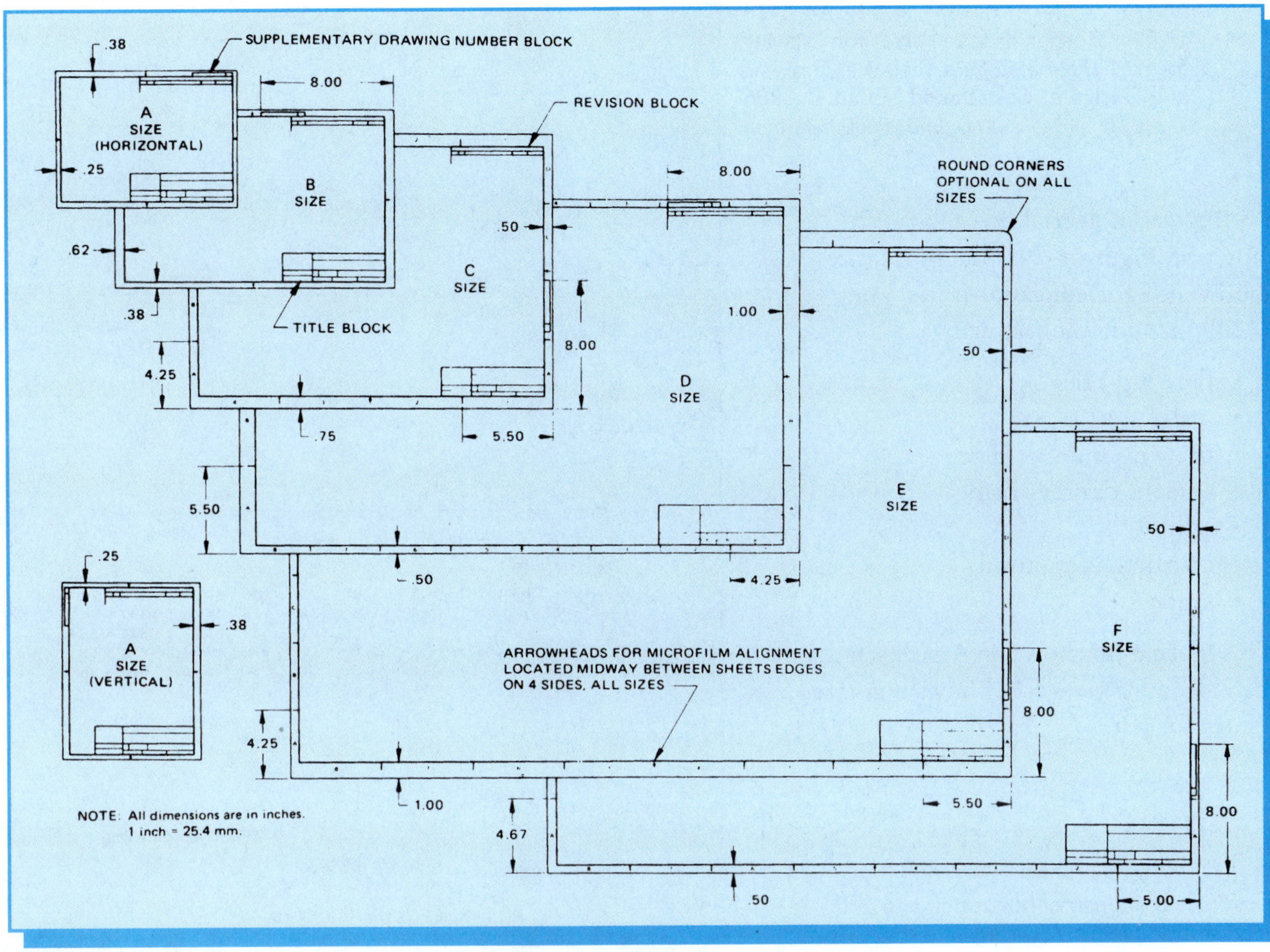

Figure 6-27
ASME standard drawing formats Y14.1 © 1980 *(Reprinted with permission of the American Society of Mechanical Engineers)*

Although the layout of drawing **title blocks** varies greatly among companies, title block information usually contains space for the following:

- company name and division
- drawing number
- date
- surface quality
- drafter's name
- angle of projection
- scale
- microfilming marks
- revision record
- drawing title
- part identification number
- tolerance limits
- designer's name
- checker's name
- material type
- next assembly
- photoreduction marks
- material lists

Figure 6-28
Border data and sizes for drawing formats *(Reprinted from* Technical Drawing and Design *by Goetsch and Nelson, © 1986 Delmar Publishers Inc.)*

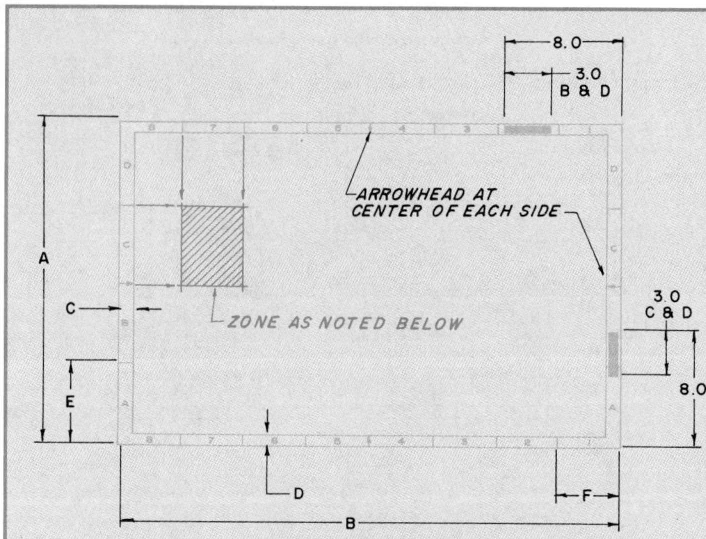

STANDARD BORDER SIZES

DRAWING SIZE	A	B	C	D	E	F
A HORIZONTAL	8.5	11.0	.25	.38	2AT 4.25	2AT 5.50
A VERTICAL	11.0	8.5	.38	.25	2AT 5.50	2AT 4.25
B	11.0	17.0	.62	.38	4AT 2.75	4AT 4.25
C	17.0	22.0	.50	.75	4AT 4.25	4AT 5.50
D	22.0	34.0	1.00	.50	4AT 5.50	8AT 4.25

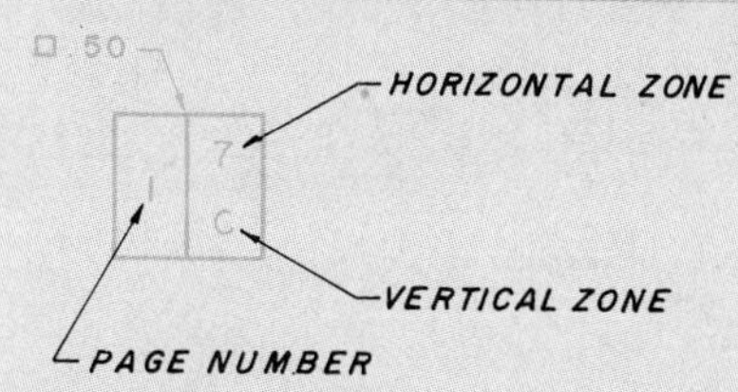

ZONE IDENTIFICATION

Typical industrial title blocks are shown in **Figure 6-29**. Title blocks are abbreviated for educational use. They usually include the following:

- drawing title
- scale
- drawing number date
- student drafter's name
- school
- instructor's approval
- grade

Figure 6-30 shows an A-size school format, and **Figure 6-31** shows a B-size school format.

Figure 6-29
Examples of industrial horizontal and vertical title blocks. *(Courtesy of Bishop Graphics, Inc., Westlake Village, CA)*

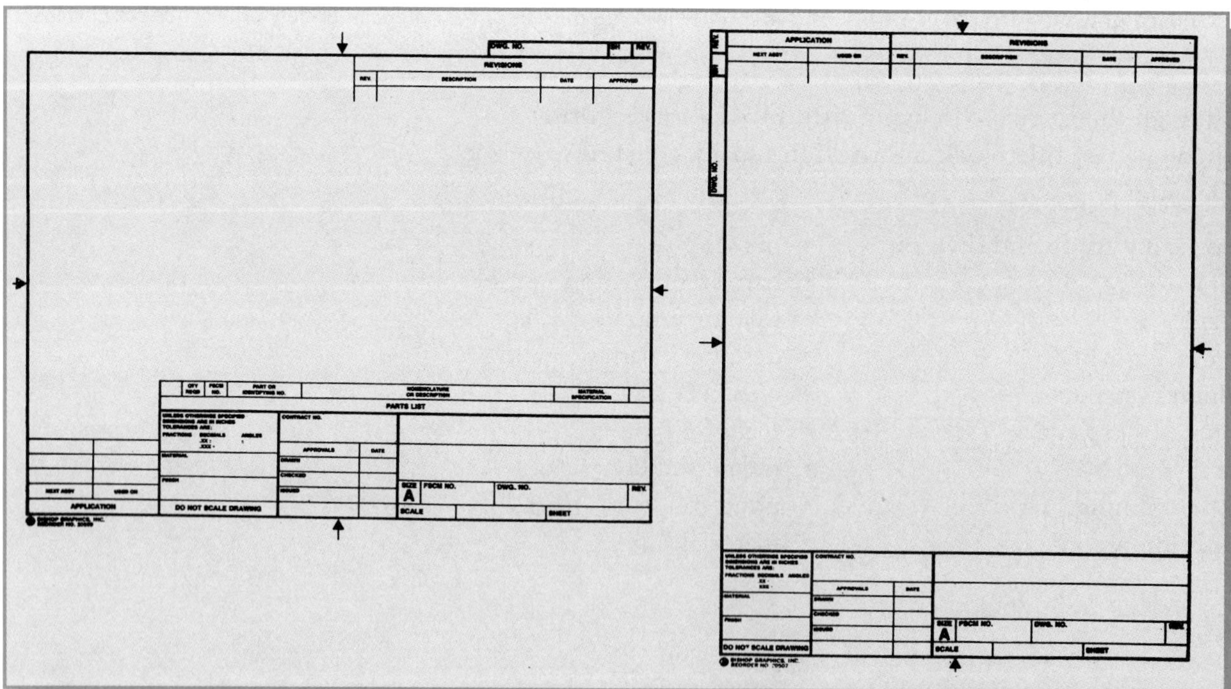

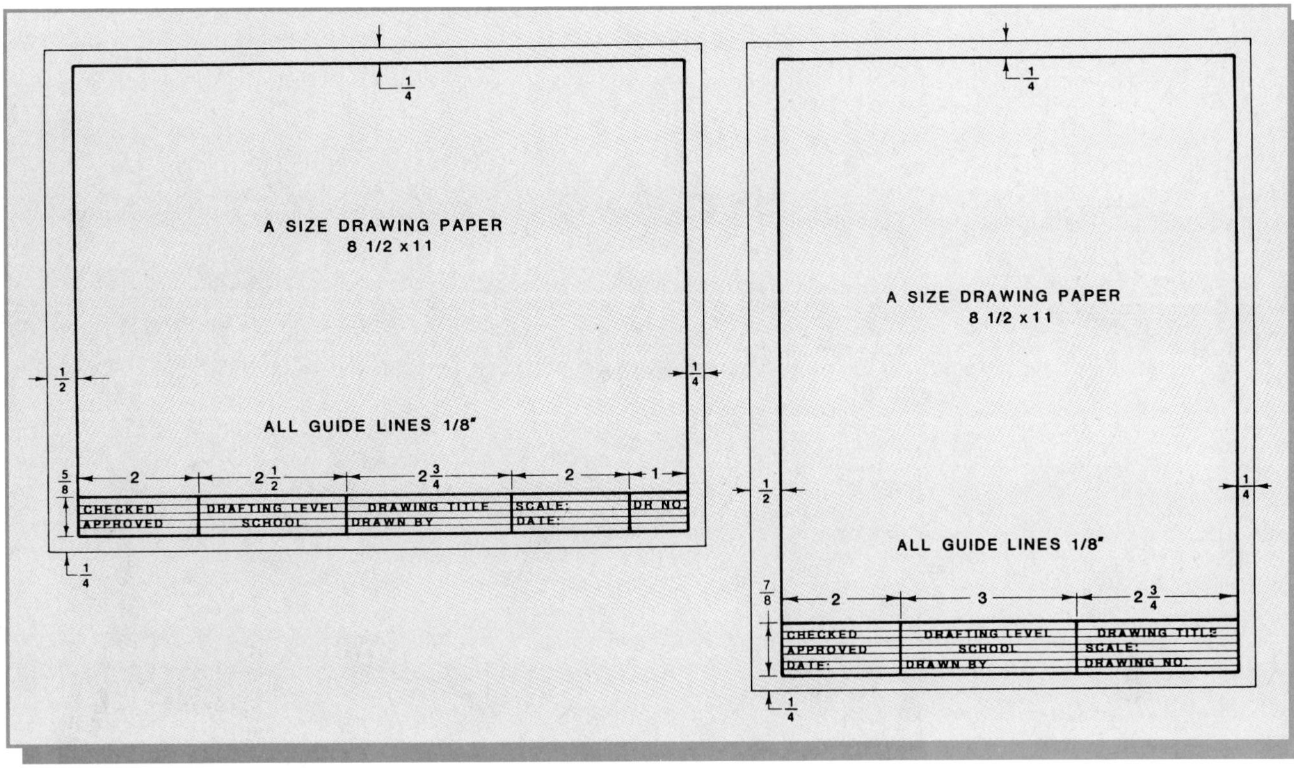

Figure 6-30
An example of an A-size horizontal format used in drafting classes

Folding methods for prints are standardized for consistency and ease of use. **Figure 6-32** shows how standard sheet sizes fold to 8.5" x 11" or 9" x 12". Prints are always folded with the title block and drawing identification number on top for convenient recognition and filing.

Figure 6-33 is an example of an A-size industrial format of a small metric valve check.

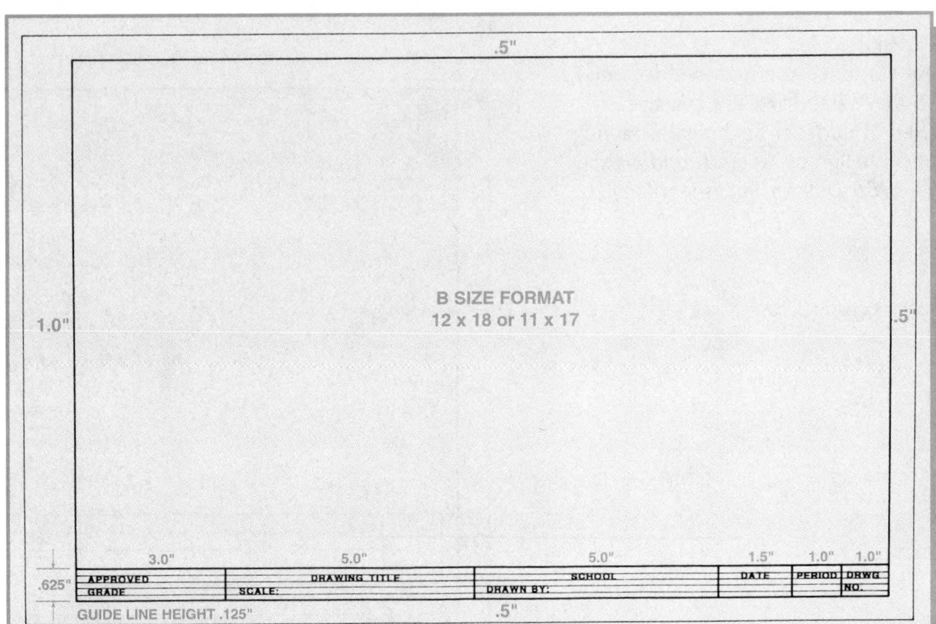

Figure 6-31
An example of a B-size format used in drafting classes

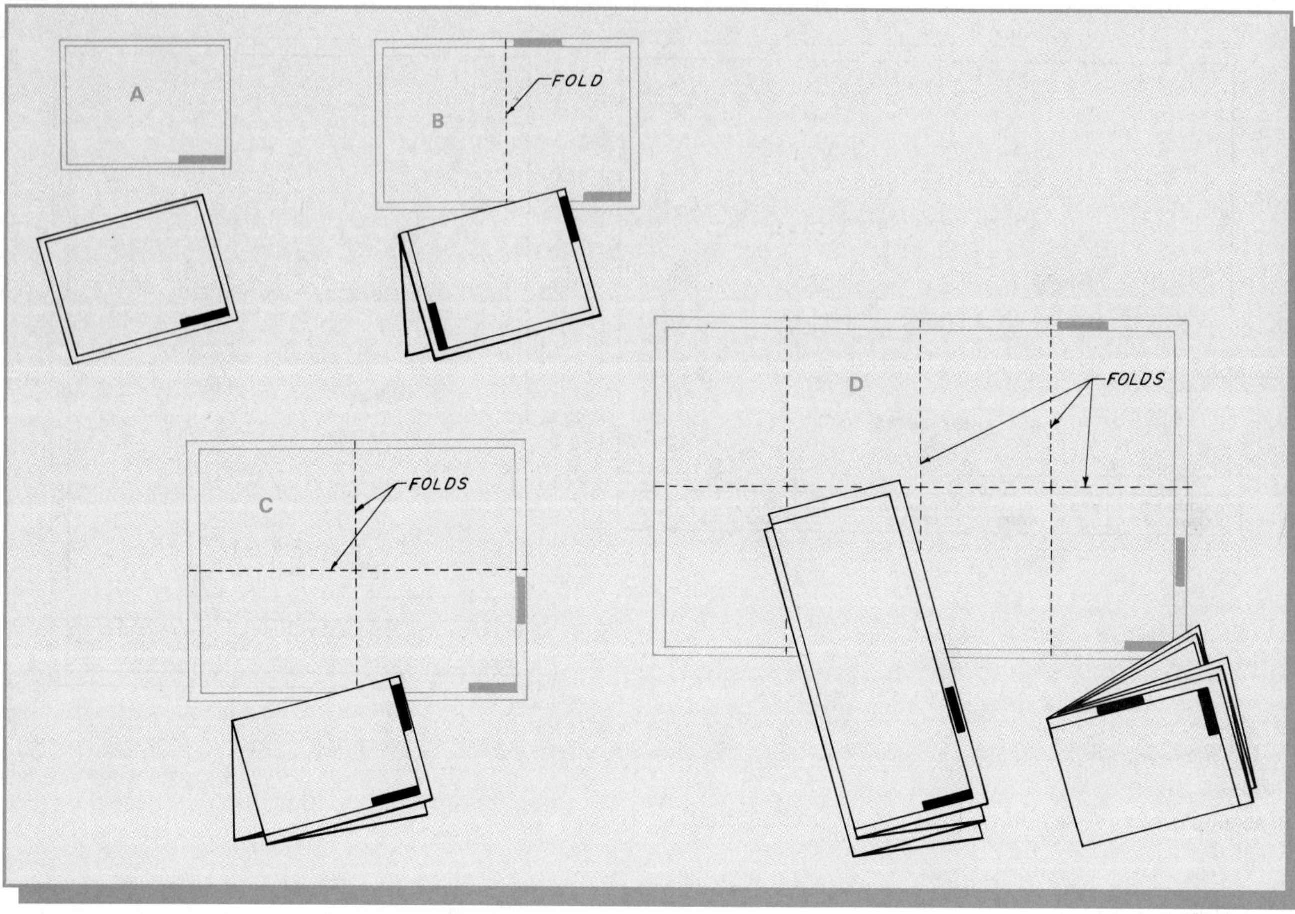

Figure 6-32
All standard formats will fold down to an A size for easy storage. *(Reprinted from* Technical Drawing and Design *by Goetsch and Nelson, © 1986 Delmar Publishers Inc.)*

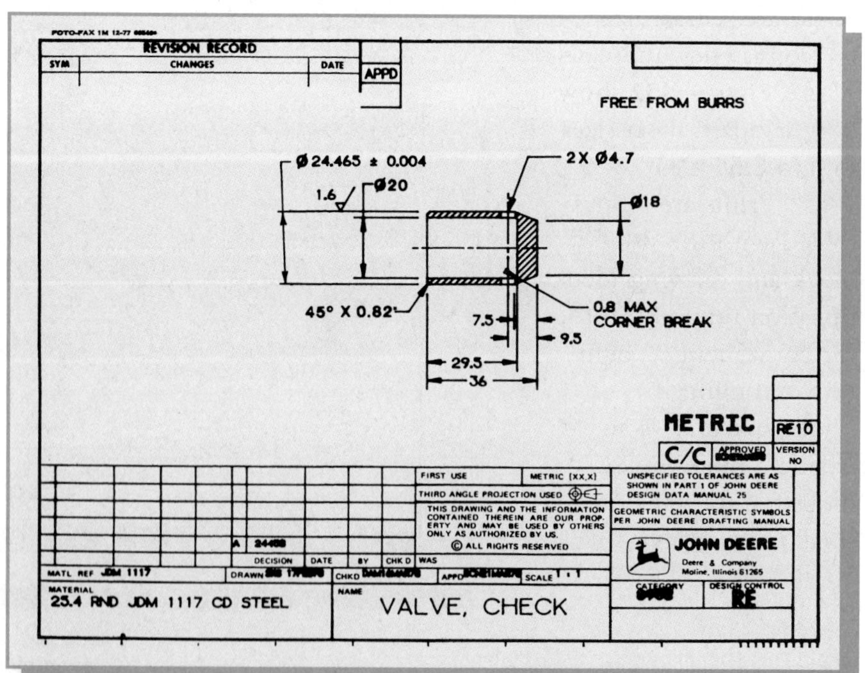

Figure 6-33
An example of an A-size industrial drawing and format
Credit: John Deere & Co.

AutoCAD LT Exercises

CAD Line Conventions

CAD systems with multipen plotters or printers are able to produce drawings with constant and near-perfect line thickness. These drawings are usually referred to as hard copies. In order to create a professionally correct drawing, CAD operators must understand the correct use of ASME standards for line widths and types.

Line Widths

When viewed on a monitor, the line widths are identically thin and dense **(Fig. 6-34)**. Plotters and printers create the difference in line widths **(Fig. 6-35)**. In multiple pen plotters, this is accomplished by using different-size pen points. In printers, the different line width is determined while setting the perimeters to plot during the print/plot command. On the CAD drawing, the color of the line determines which pen or line widths are chosen. A white line and a red line may have the same width on the monitor, but may be plotted at different widths on the finished drawing.

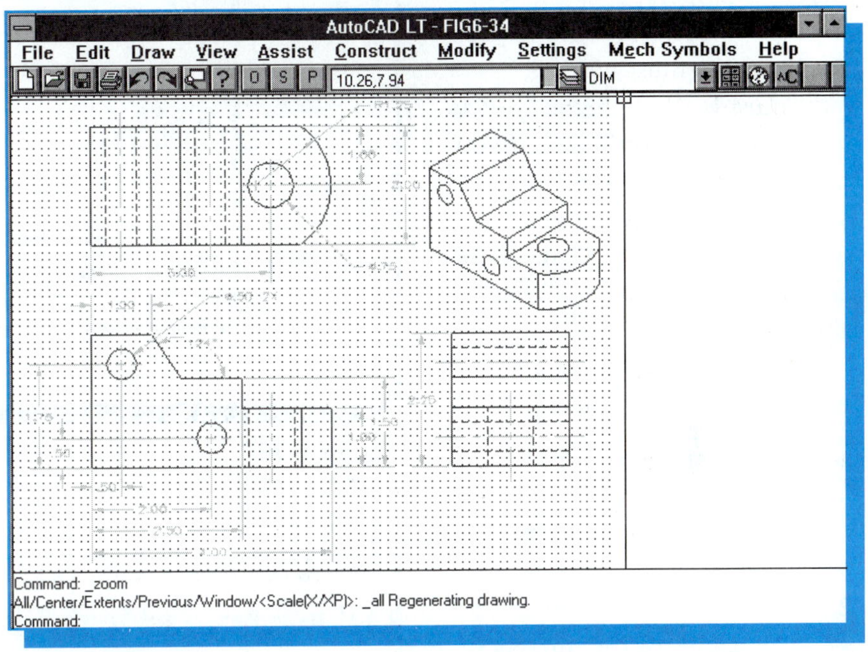

Figure 6-34
Lines as viewed on monitor

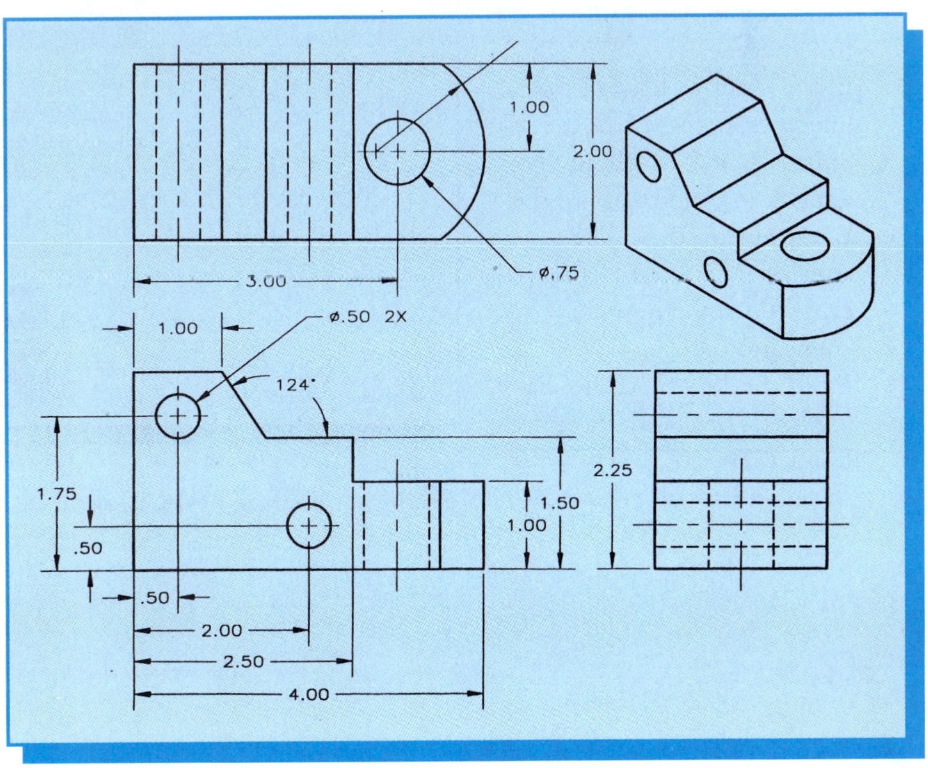

Figure 6-35
Printed or plotted drawing showing line contrast

Line Types

To produce proper linetypes, the layer command is used. Each layer is assigned a name, color, and linetype by the operator. Before the operator can select a linetype in the layer command, the linetypes must be loaded. This must be done before you select the layer command. If you select layer and linetype before you load the linetypes, your screen will look like **Figure 6-36** and only one linetype is available (continuous).

In order to load linetypes, you must do the following.

1. From the menu bar select **Settings**, then **Linetype Style**, and then **Load (Fig. 6-37)**.
2. On the command line, type the names of the linetypes you want to load from the list **(Fig. 6-38)**. It is recommended that you load all of the linetypes.
3. If you want to load all the linetypes, type *. Use the Shift key + 8 to get an asterisk (*) and then use the Enter button (EB). The asterisk (*) is used as a wild card in many computer applications.
4. From the Linetype File dialog box, select OK and press Enter (EB).

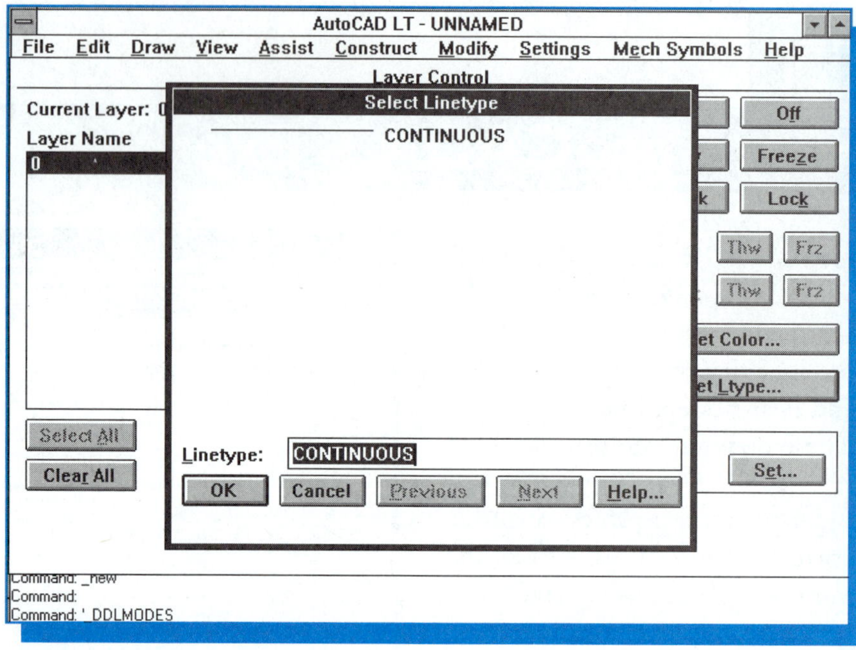

Figure 6-36
Linetypes not loaded yet

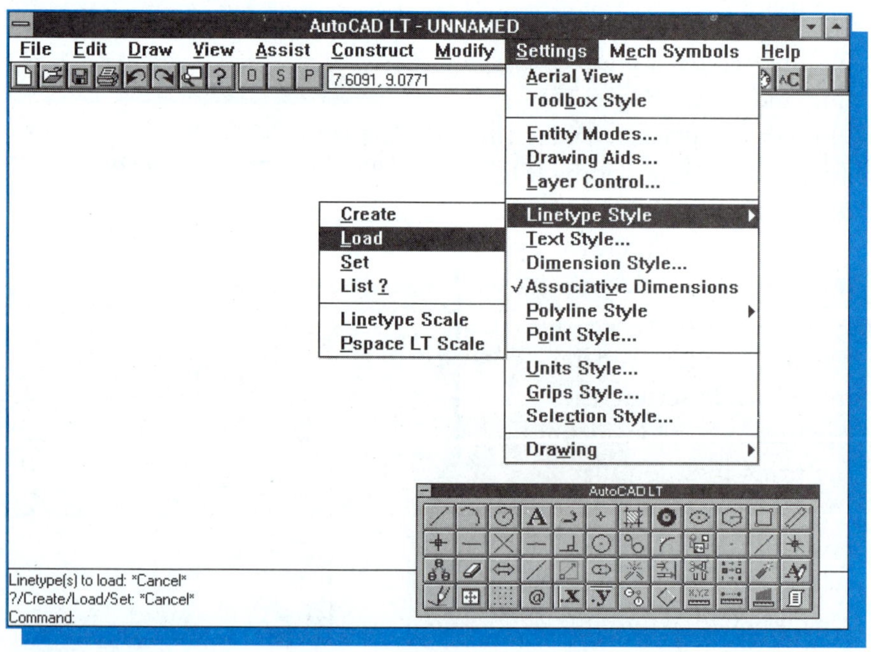

Figure 6-37
Settings, Linetype Style, Load to set linetypes

Now the linetypes are loaded and when you select the layer command, a layer, and the Linetype button, your screen should look like **Figure 6-39**.

Layers

Layers are like transparent pieces of paper laid on top of each other with only the lines on each one showing. In AutoCAD LT, layers are the primary means by which your drawing is organized. You use layers to assign colors and linetypes. Layers are used to organize different parts of a drawing, for instance, hot water lines for a floor plan. Layers can hold as many objects as you can create in that layer. There is no limit on the number of layers a drawing can have. Most companies standardize the layers used in their company for convenience. In this text we will use seven standard layers.

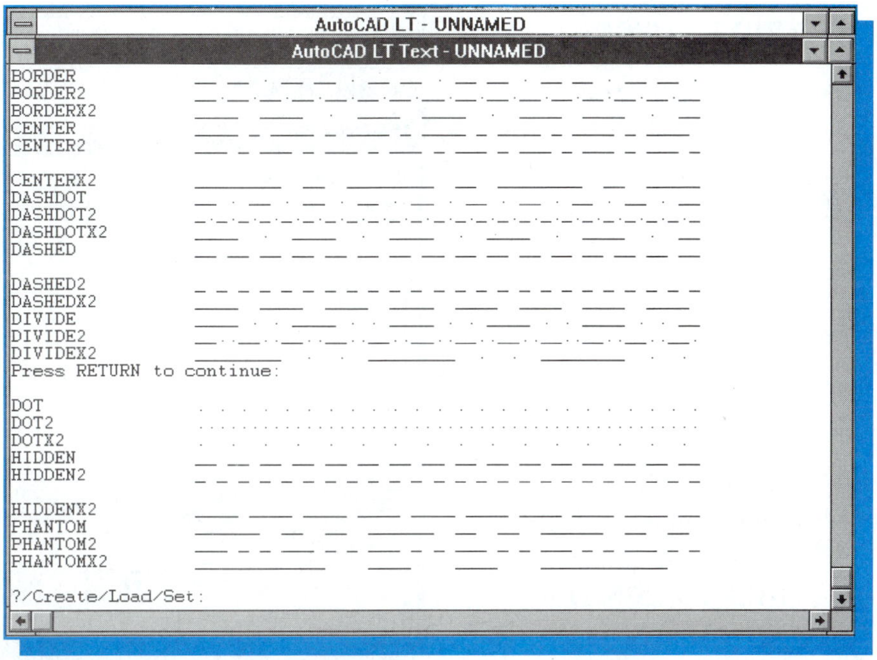

Figure 6-38
List of AutoCAD linetypes

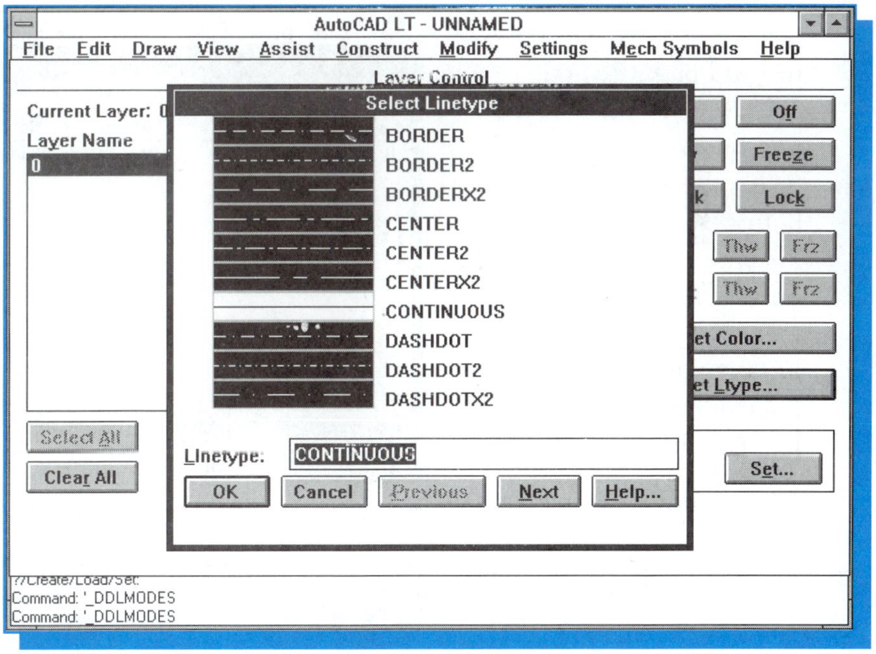

Figure 6-39
Layer Control, Linetypes

Creating Layers

Starting a New Drawing

Use the same settings as in Chapter 5, Starting AutoCAD LT.

1. From the toolbar, select the Layer button, The screen should look like **Figure 6-40**.
2. Use the (PB) to select the Text box (if it is not already selected), and type in the names of the layers separated by commas: **OBJ,HID,CEN,DIM, CON,PHAN (Fig. 6-41)**.
3. Select the New Button **(Fig. 6-42)**.
4. Select the CEN layer using (PB) **(Fig. 6-43)**.
5. Select the Set Color button and pick (PB) the color green. Click on the OK button.
6. Select the Set Linetype button and pick (PB) the linetype center by clicking on the Icon. Click on OK.
7. Do the same for all the layers until the layers look like **Figure 6-44**.

When using a white graphics screen, selecting a white color in layers appears black on the screen.

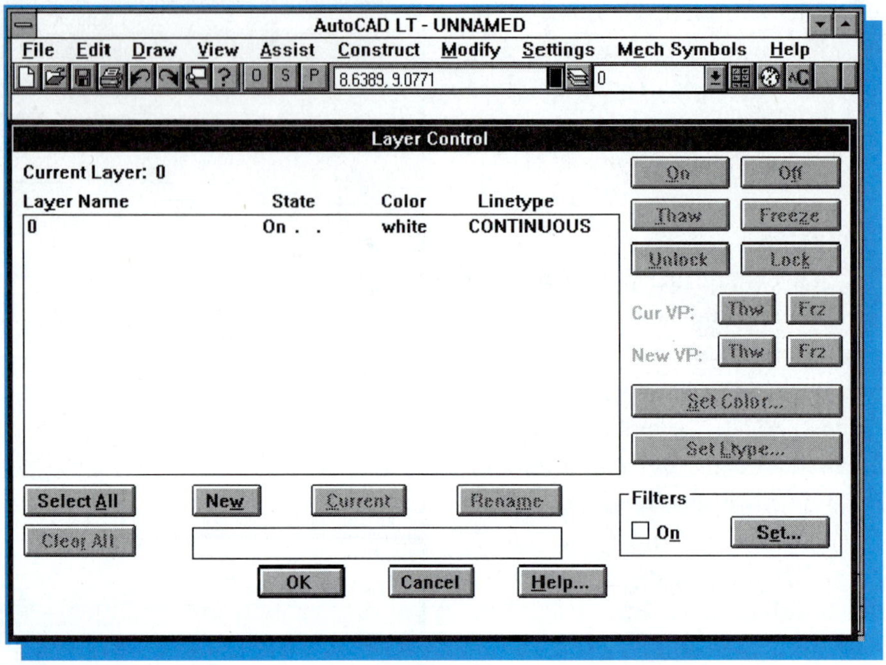

Figure 6-40
Layer Control

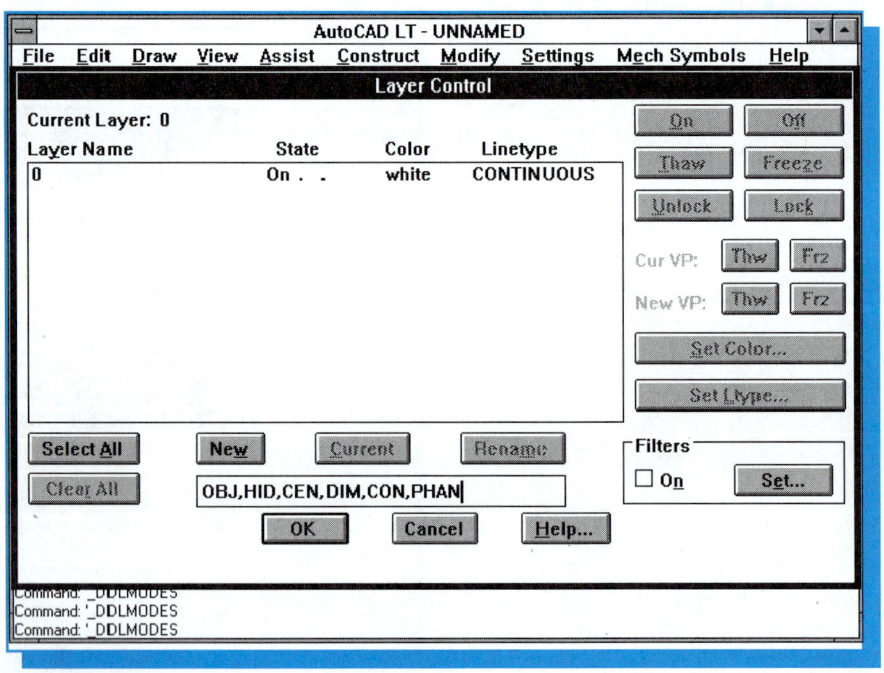

Figure 6-41
Typing in new layers

tip: *When selecting linetypes, it may be necessary to use the Next button to find the appropriate linetype.*

tip: *Be sure that there is only one layer selected when setting color and linetype. Select and unselect are both done with the Pick button (PB). If layers have a common color or linetype, it is a good idea to select more than one layer at a time.*

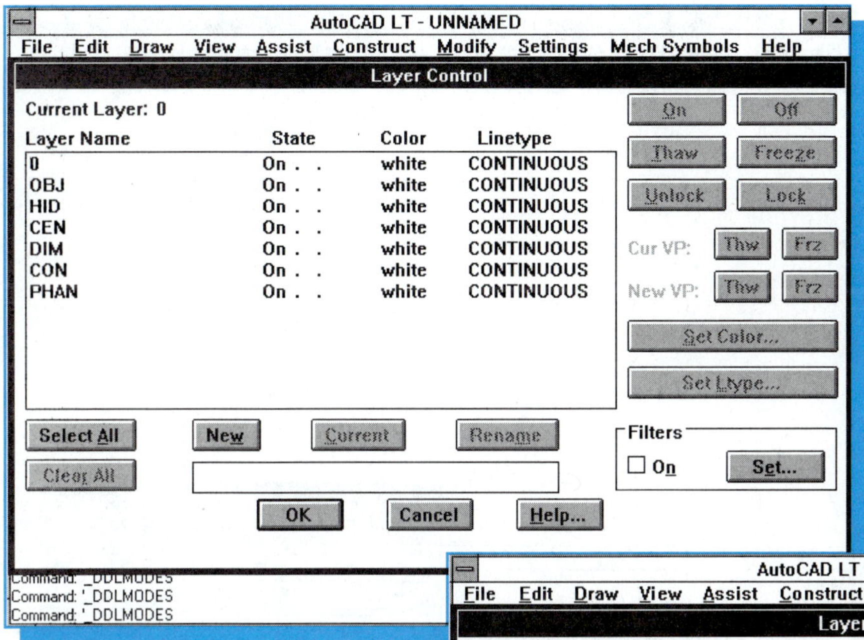

Figure 6-42
New displays the newly created layers.

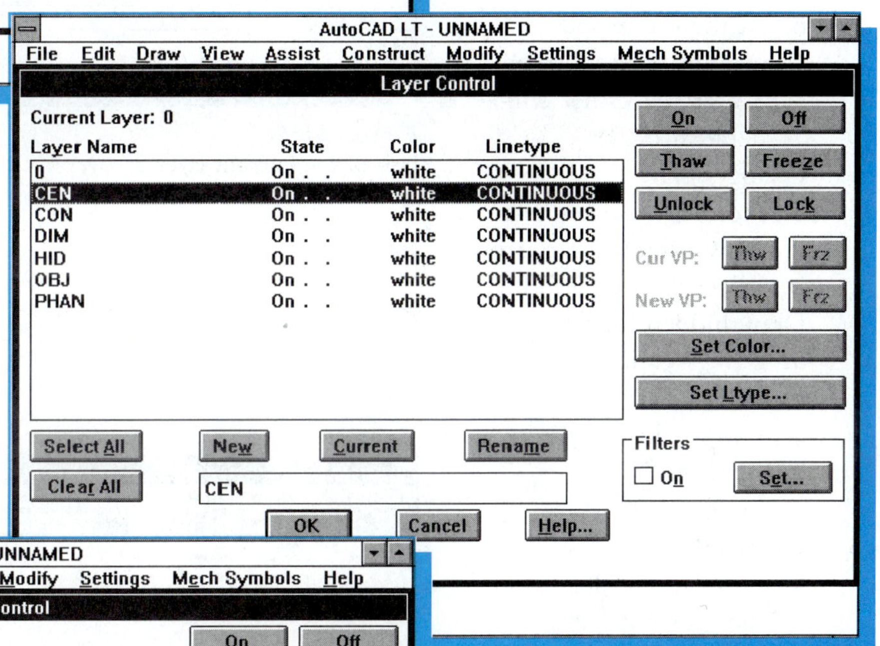

Figure 6-43
Selecting a layer

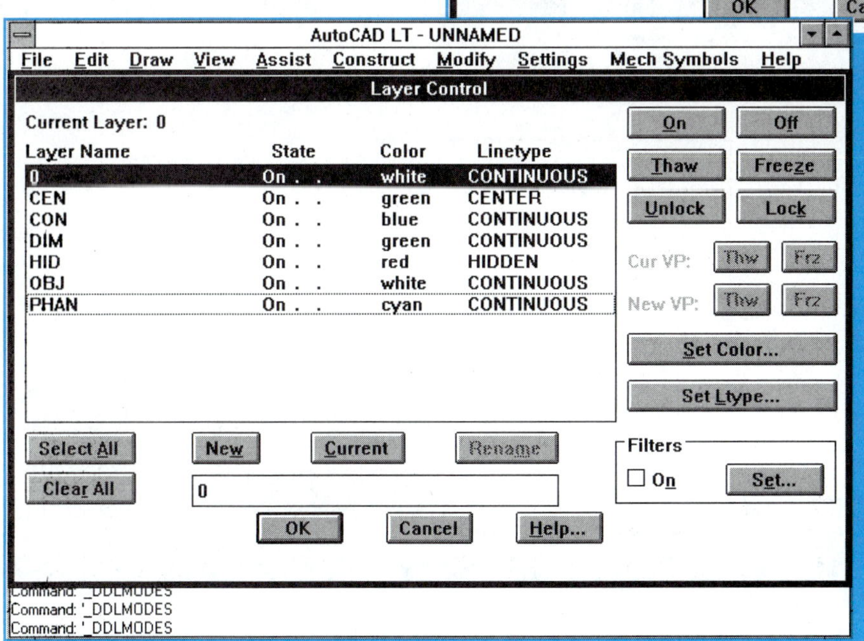

Figure 6-44
Layers completed, including color and linetype

Layer Practice Drawing

1. Select a layer:
   a. From the toolbar, select the Current Layer arrow. The screen should look like **Figure 6-45**.
   b. Select layer OBJ using the (PB). The things that are drawn now are on layer OBJ.
2. Draw a rectangle:
   a. From the toolbox, select the Rectangle button
   b. At the command prompt: First Corner: **2,2 (EB);** Other corner: **9,6 (EB)**
3. Change layer.
   a. From the toolbar, select the Current Layer arrow
   b. Select layer HID using the (PB).
4. Draw hidden lines:
   a. Line — line from: 2,5 (EB) to: 9,5 (EB)
   b. Line — line from: 2,3 (EB) to: 9,3 (EB)
5. Change layer.
   a. From the toolbar, select the Current Layer arrow
   b. Select layer CEN.
6. Draw center line:
   Line — line from: 1.5,4 to: 9.5,4

The drawing should look like **Figure 6-46**.

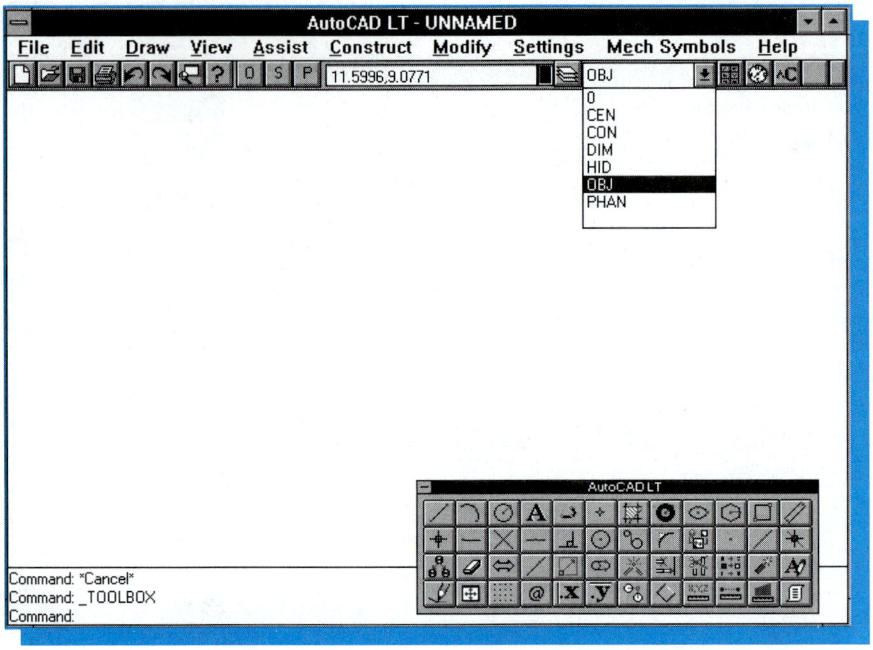

Figure 6-45
Use of the current layer arrow

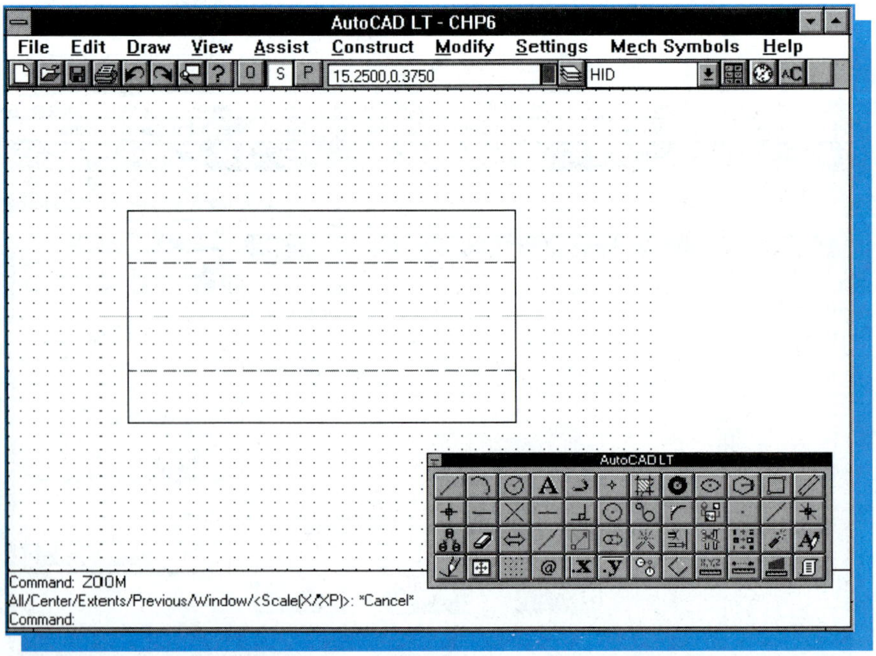

Figure 6-46
Drawing using layers to get proper colors and linetypes

Adding Text to Your Drawing

Text is how we add notation to a drawing. Many of the important things drawings communicate are communicated through the use of lettering or text.

Start a new drawing: (**File, New, OK, OK**)

1. From the toolbox or draw menu, select text. or Text.
2. Pick a point any place within the drawing area for the beginning of the text.
3. Type in height for the text if the default is not what you want (use .125).
4. Select the (ES) to accept the default of 0 for the angle of rotation.
5. Enter the text you want from the keyboard. At the end of each line, select (EB). When you have completed all the text, select (EB) twice. (EB)(EB)

   Standard practice on drawings is to use all uppercase letters (all capitals).
6. Practice by adding this text to the drawing:

   LEARNING AUTOCAD LT IS FUN AND EASY.
7. There is no need to save the practice text drawing.

Prototype Drawings

A prototype drawing is one that has been set up to have all the things we need in a drawing. New drawings in AutoCAD LT are based on a prototype drawing. We can use a drawing that we created, or AutoCAD LT will default to the ACAD prototype drawing. A prototype drawing can have the layers you want to use. It can have the border and title block you want to use. Now, instead of starting with the AutoCAD prototype drawing, start with the prototype drawing that you set up.

Creating a Prototype Drawing

Start a new drawing

1. Create the six layers we talked about earlier in the chapter.
2. Using layer 0, draw the border and title block shown in **Figure 6-30**.
3. Use the text command to add text to your title block.
4. Save the drawing as drawing PRO6.
   - **a.** Select save from the toolbar or File, Save from the pull-down menus.
   - **b.** In the File Name Box, type PRO6 and choose OK.

Using the Prototype Drawing

1. Start a new drawing.
   a. **File - New**
   b. Select Prototype button
   c. Select scroll arrows to move the files up or down **(Fig. 6-47)**.
   d. Select the PRO6.dwg.
   e. Select OK.
2. Continue drawing setup as before.
3. The new unnamed drawing, based on the prototype drawing (PRO6), should look similar to **Figure 6-48**.
4. Select the current layer arrow to see if all the standard layers are there.
5. Do not save this drawing.

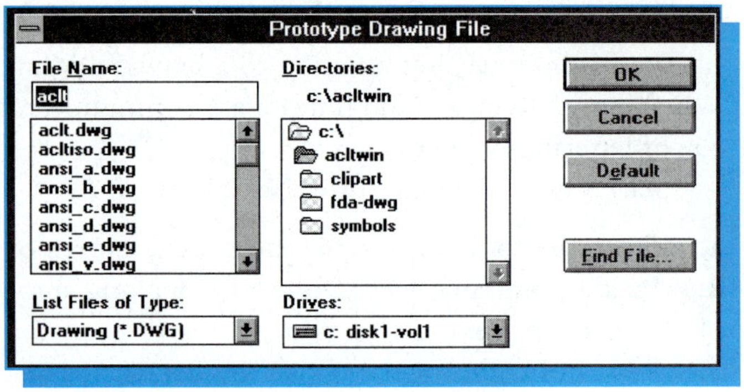

Figure 6-47
Scroll to find PROG6 prototype dwg

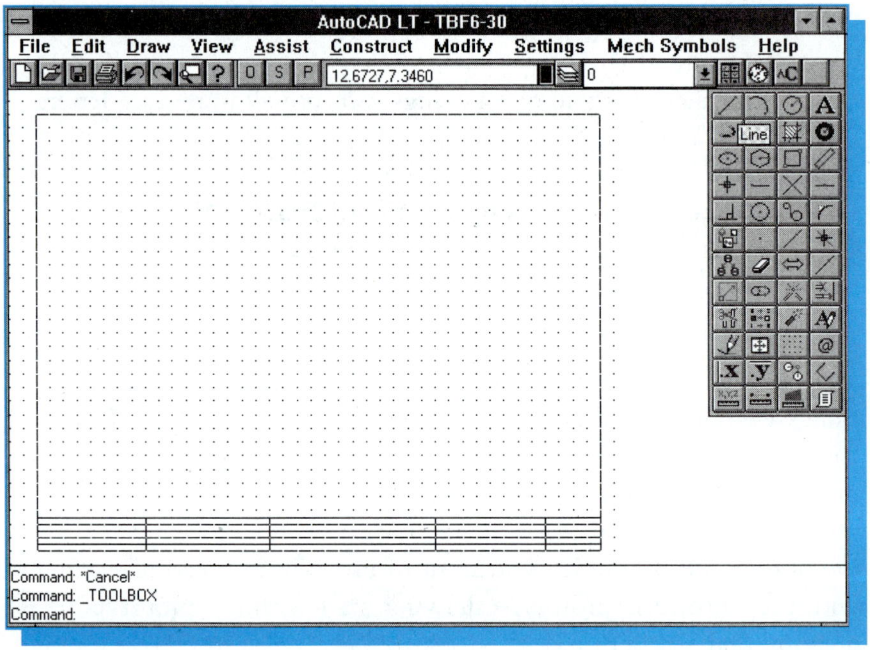

Figure 6-48
New drawing based on prototype dwg

Drafting Exercises

1. Draw **Figures 6-49a** and **6-49b** on a CAD system and/or with instruments using the proper line weights or line types.

2. Design a C-size drawing format for school use. Make it a prototype drawing.

3. Design a C-size drawing format for the Ace Tool Design Company.

4. List each ASME drafting standard used in this chapter with the ASME reference page numbers.

5. Set up a standard drawing format for A-, B-, and C-size paper on a CAD system, and create prototype drawings.

6. With a CAD system, draw the figures shown in **Figure 6-50**. Adjust the line widths as needed for the final hard copy plot. Double all dimensions.

7. Name each line convention in **Figure 6-51**.

8. Practice drawing **Figure 6-52** with a CAD system.

Figure 6-49a
Draw these patterns with instruments or CAD.

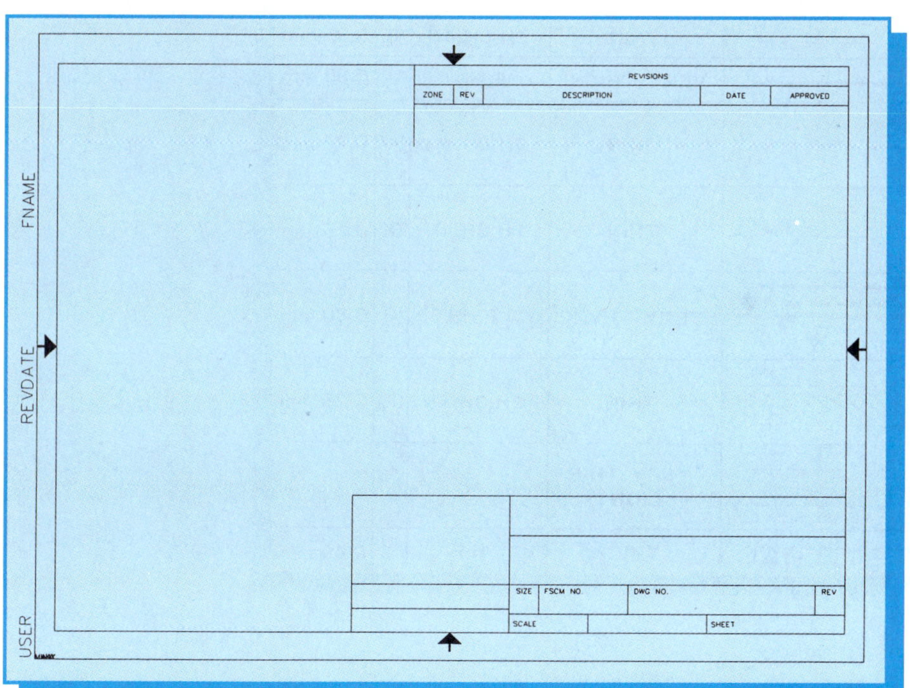

Figure 6-49b
Bring up the A size drawing format in the AutoCAD LT drawing file and create your own design.

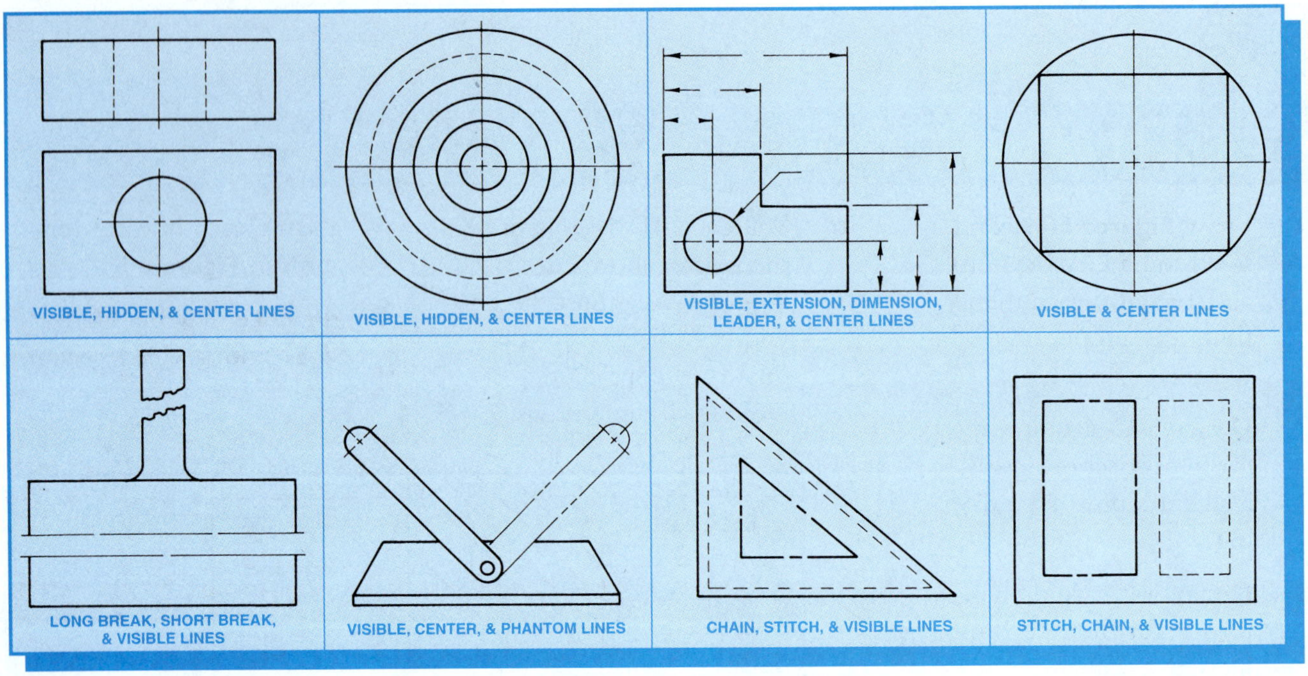

Figure 6-50
Redraw each figure two times as large using correct line widths

LINE SYMBOLS	LINE WIDTH	PENCIL	PEN SIZES	
	THICK	H,F	2	0.50 mm
	MEDIUM	2H,H	0	0.35 mm
	THIN	2H,3H,4H	0	0.35 mm
	THIN	2H,3H,4H	0	0.35 mm
	THICK	H,F	2	0.50 mm
	THIN	2H,3H,4H	0	0.35 mm
	THIN	2H,3H,4H	0	0.35 mm
	VERY THICK	F,HB	3	0.80 mm
	THIN	2H,3H,4H	00	0.25 mm
	THIN	2H,3H,4H	00	0.25 mm
	VERY THICK	F,HB	3	0.80 mm
	THIN	2H,3H,4H	00	0.25 mm
	VERY THIN LIGHT	4H		
ARCHITECTURAL	THICK	H,F	1	0.40 mm

Figure 6-51
Name each line convention.

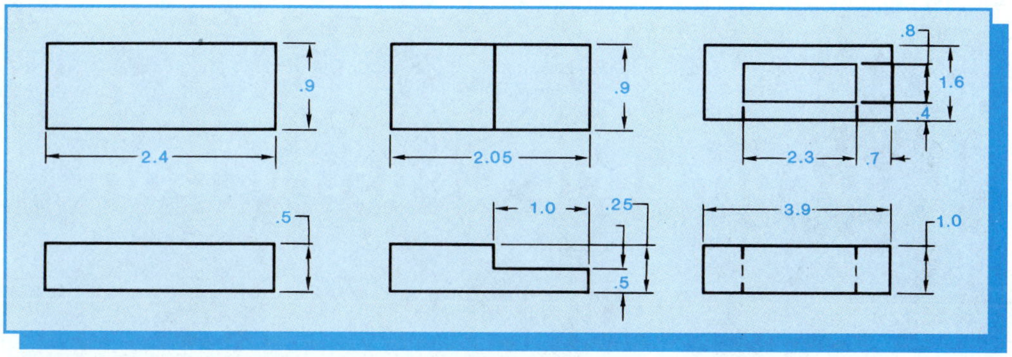

Figure 6-52
Practice drawing these figures with your CAD system.

Design Exercises

1. Design a B-size format for a small mechanical engineering company named D&P Engineering Inc. Include a full title block, borders, zones, and company logo.
2. Design an A size format for your classwork.
3. Turn off your monitor then create a 3" x 5" rectangle with a 1" hole in the center.

Key Terms

American Society of Mechanical Engineers (ASME)
Border
Break line
Cabinet drawing
Cavalier drawing
Centerline
Chain line
Cutting plane line
Dimension line
Dimetric drawing
Extension line
Hidden line
International Standards Organization (ISO)
Isometric drawing
Leader
Line conventions
Multiview drawing
Object line
Oblique drawing
Opaque
Pattern development drawing
Perspective drawing
Phantom line
Section drawing
Section line
Standards
Stitch line
Title block
Trimetric

Chapter 7

Geometric Construction

Objectives

The student will be able to:

- understand geometric form terminology
- recognize two-dimensional geometric forms
- recognize three-dimensional geometric forms
- construct basic geometric forms using drafting instruments
- construct basic geometric forms using a CAD system

CAD versus Drawing Instruments

Construction of geometric forms with drawing instruments has always been a tedious but necessary drafting task **(Fig. 7-1)**. The use of geometric drafting templates **(Fig. 7-2)** is a significant timesaver for the contemporary drafter; however, drawing geometric forms on a CAD system is simpler and faster. A CAD drafter must be familiar with the basics of **geometric construction** in order to fully utilize the geometric drawing capacity of a CAD system.

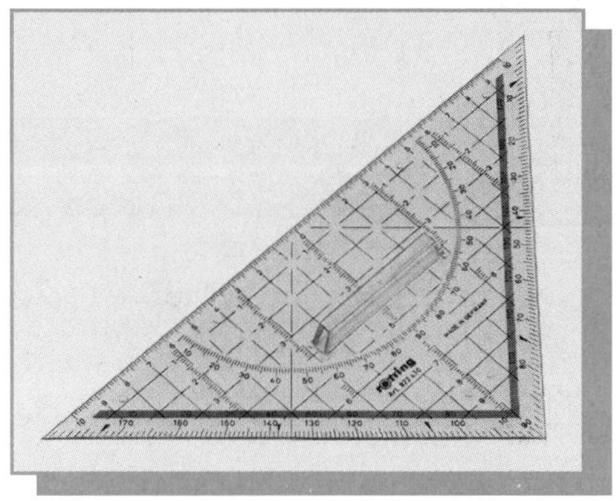

Figure 7-2
Drawing templates help with the drawing of geometric constructions. *(Courtesy of Koh-I-Noor)*

Figure 7-1
Early Greek architecture refined the use of geometric construction. *(Courtesy of The Celotex Corporation)*

Geometric Forms

To communicate effectively, a CAD drafter must be able to use geometric terminology correctly. Geometric terminology commonly used in most fields of drafting and design is shown in **Figure 7-3**. Geometric forms are either **two-dimensional** or **three-dimensional**.

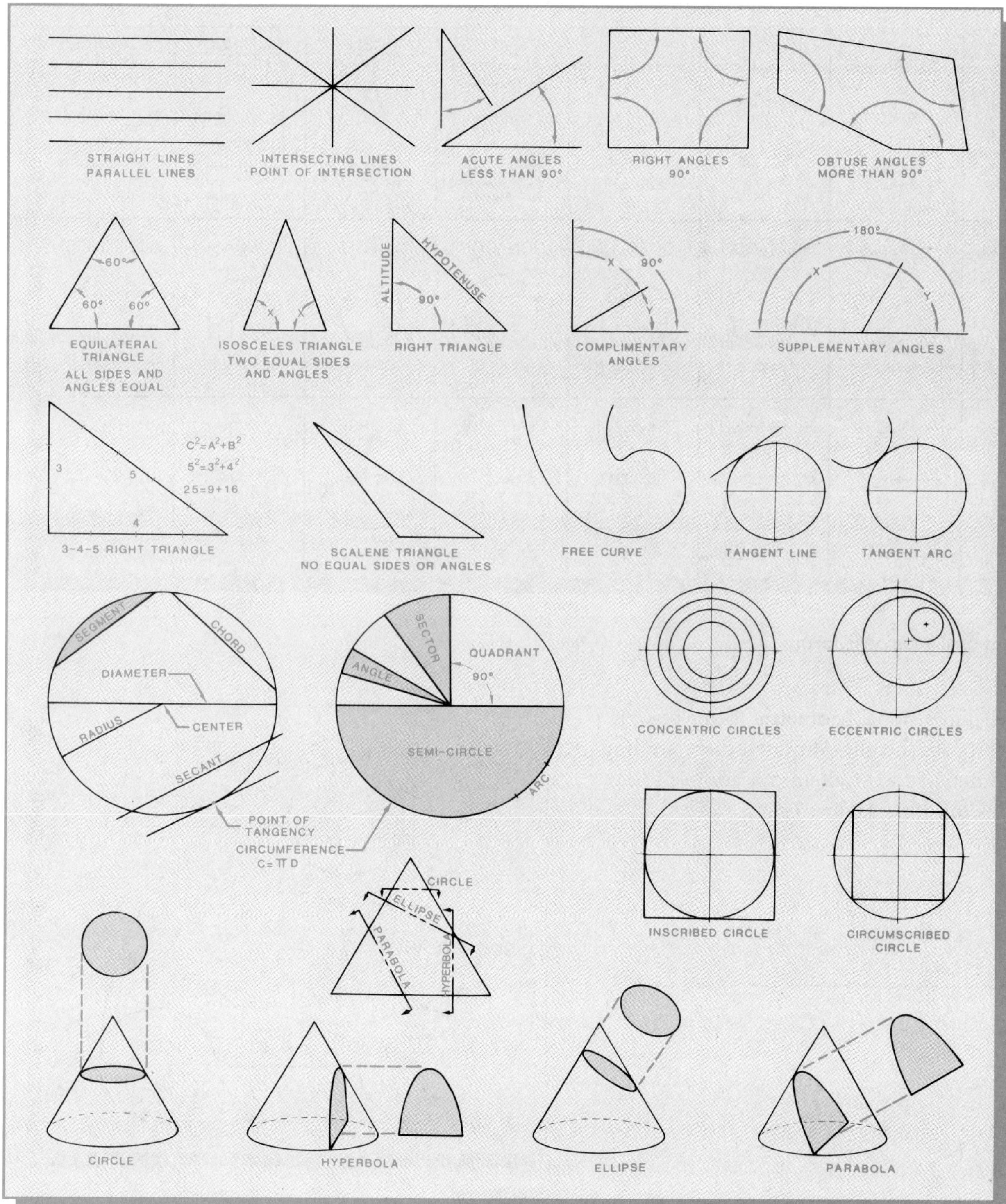

Figure 7-3
Geometric terminology

Figure 7-4 shows basic two-dimensional forms that are variations of circles, triangles, squares, polygons, ellipses, and free forms. **Figure 7-5** shows a two-dimensional engineering drawing constructed with a combination of straight lines, arcs, circles, and curves. Study **Figures 7-4** and **7-5**, and identify the different types of geometric forms.

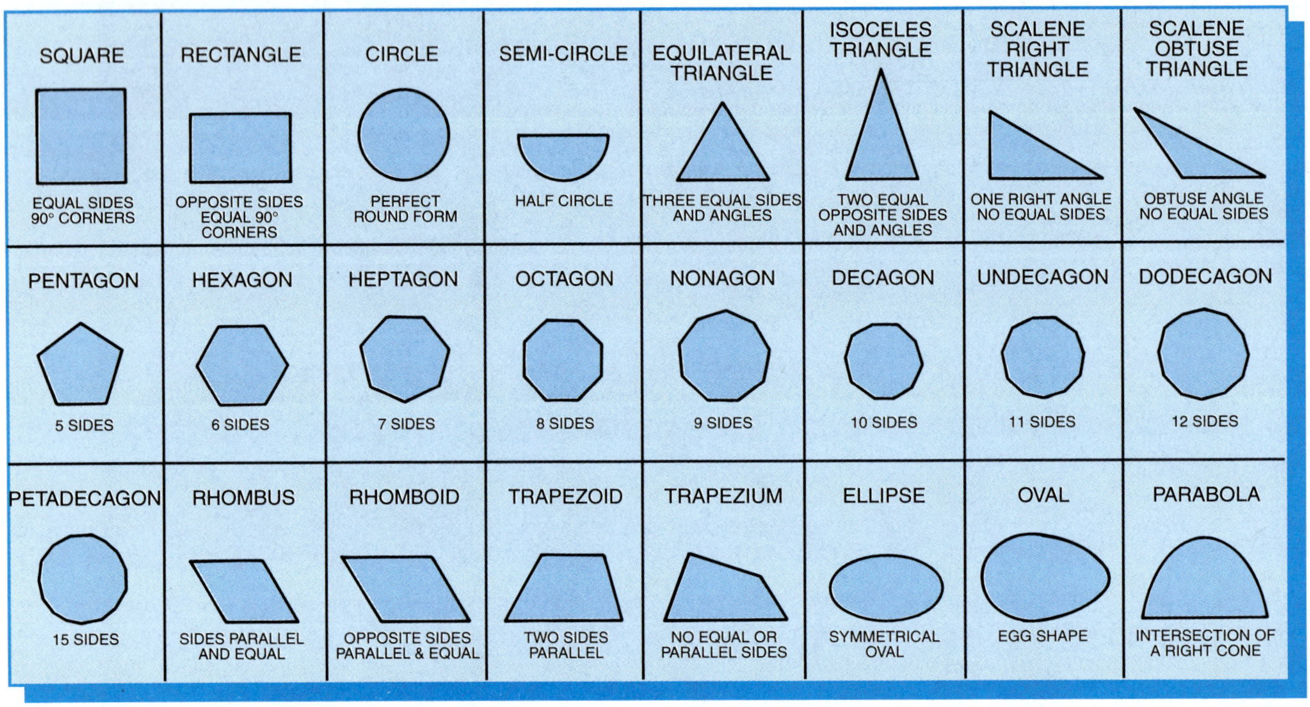

Figure 7-4
Two-dimensional geometric forms

Three-dimensional geometric forms possess width, depth, and height. Although there are hundreds of different forms, all are variations of the basic forms found in **Figure 7-6**.

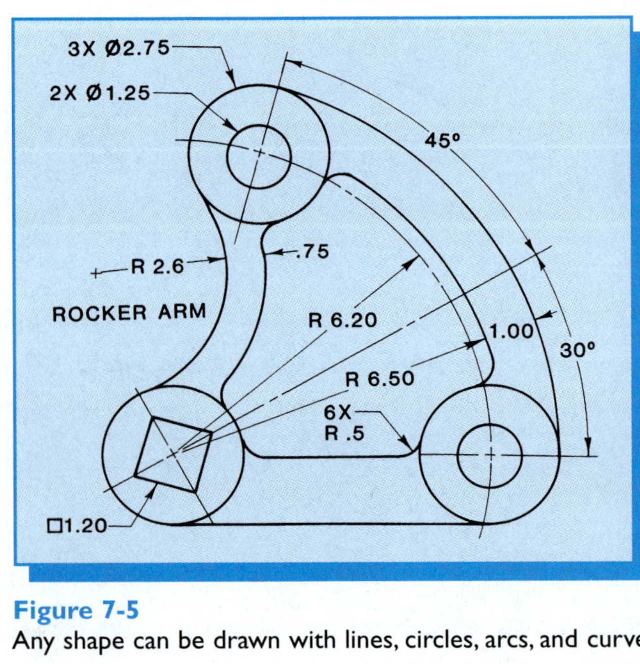

Figure 7-5
Any shape can be drawn with lines, circles, arcs, and curves.

X080100010497K

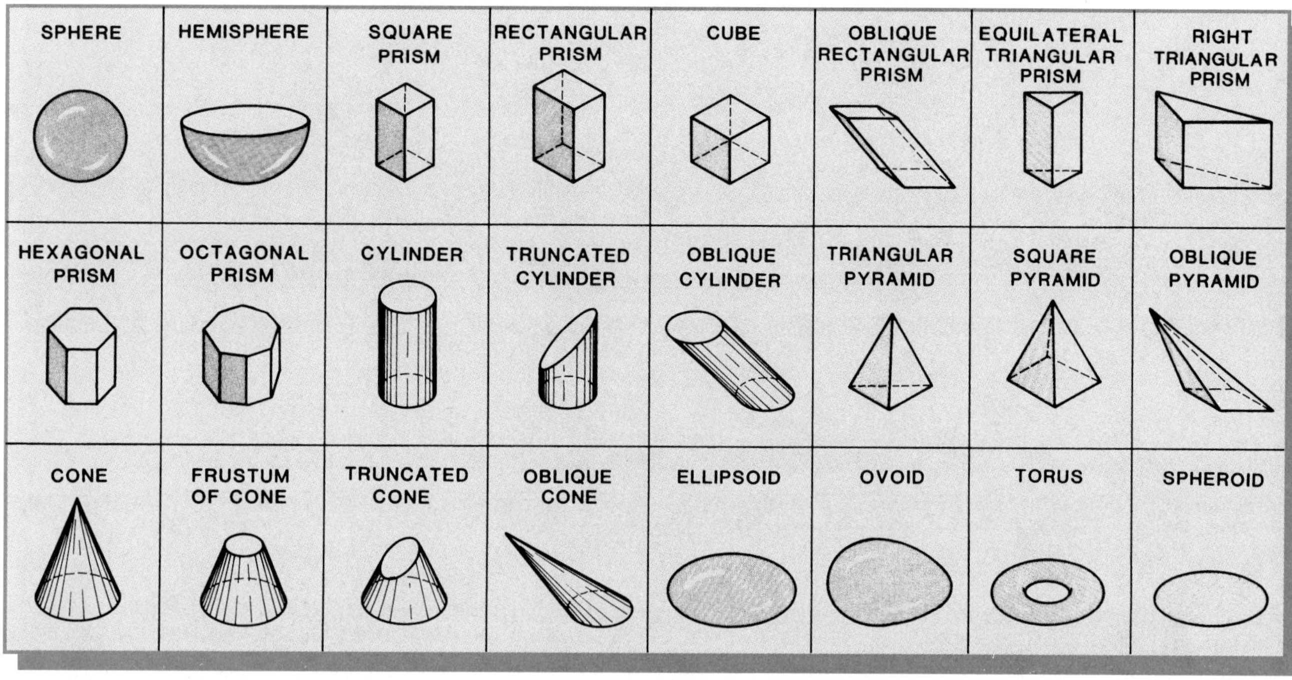

Figure 7-6
Three-dimensional geometric forms

Conventional Geometric Construction

The procedures for manually constructing the most common geometric forms are shown in **Figures 7-7** through **7-26**.

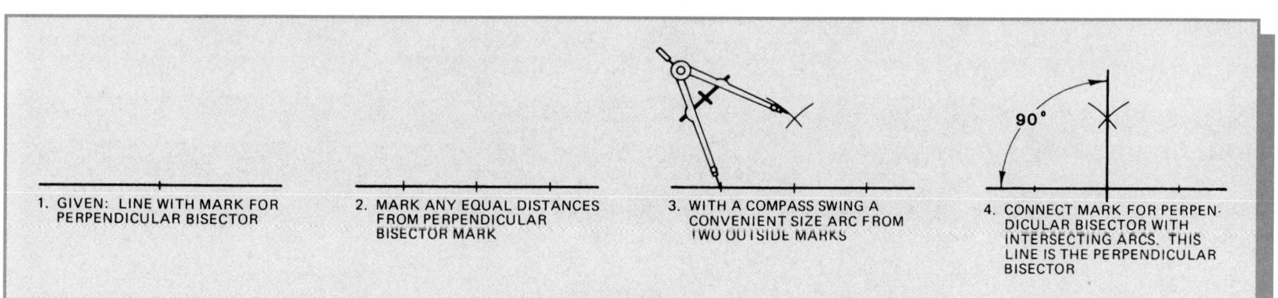

Figure 7-7
Perpendicular bisector

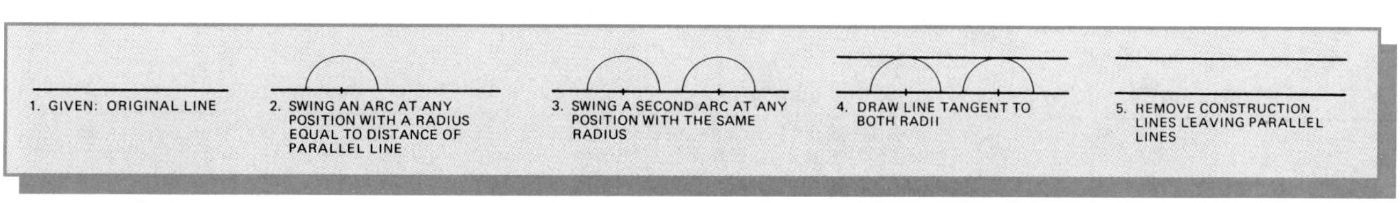

Figure 7-8
Parallel lines

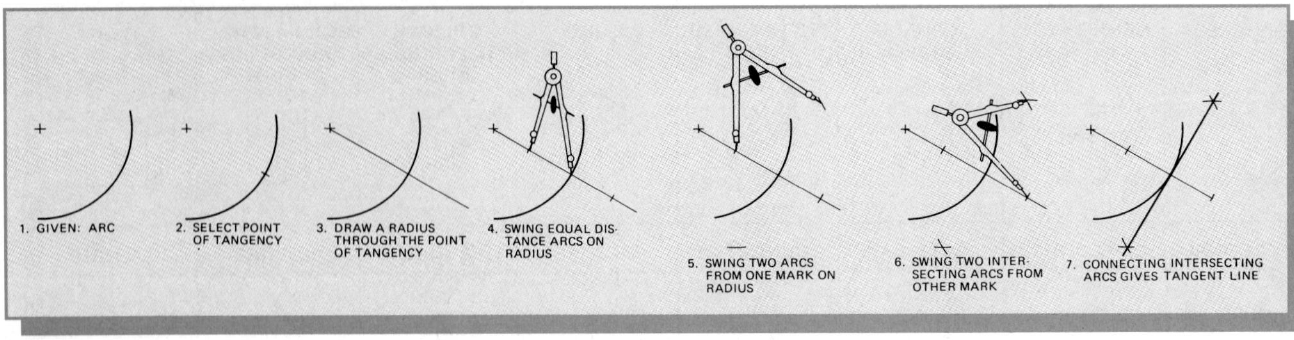

Figure 7-9
Tangent line

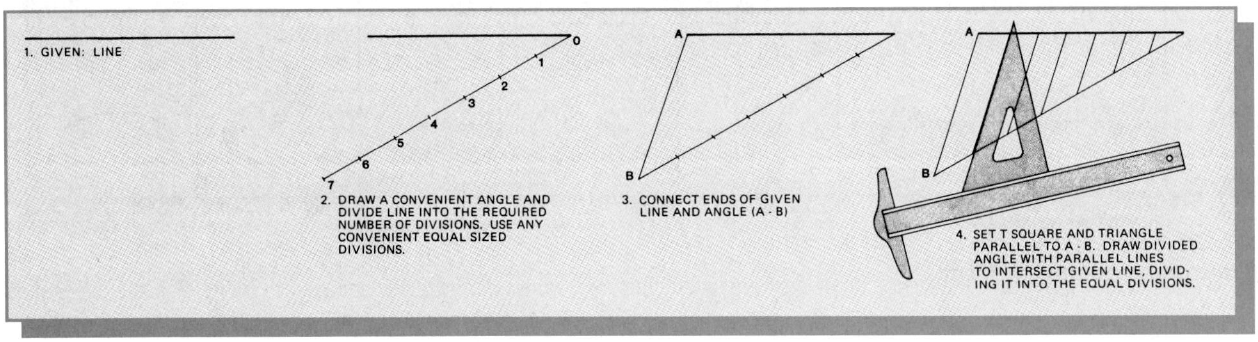

Figure 7-10
Divide a line into equal parts.

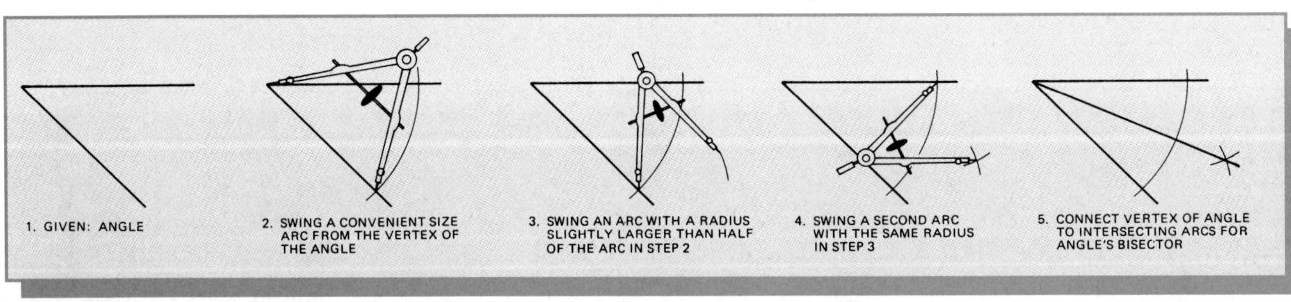

Figure 7-11
Bisect an angle.

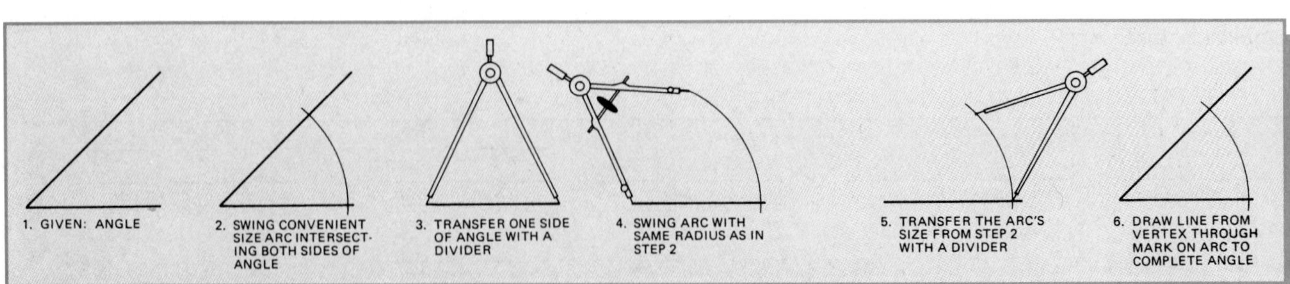

Figure 7-12
Duplicate an angle.

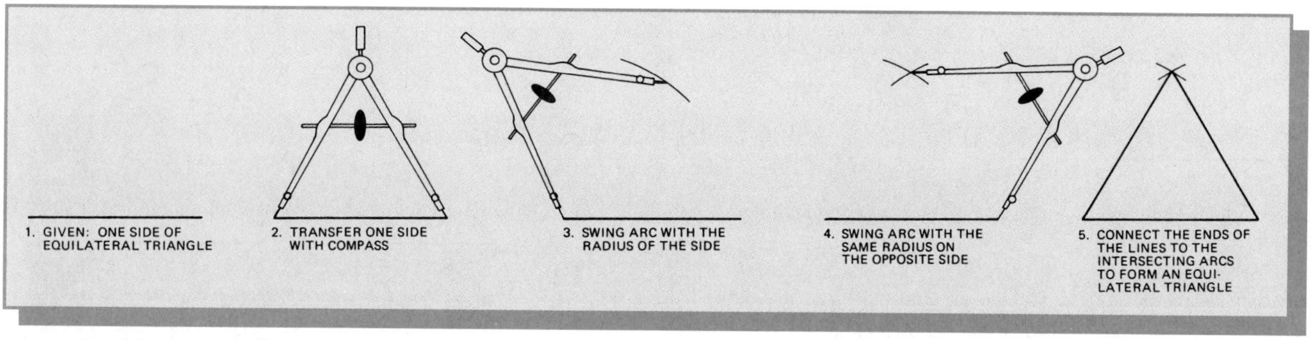

Figure 7-13
Constructing an equilateral triangle

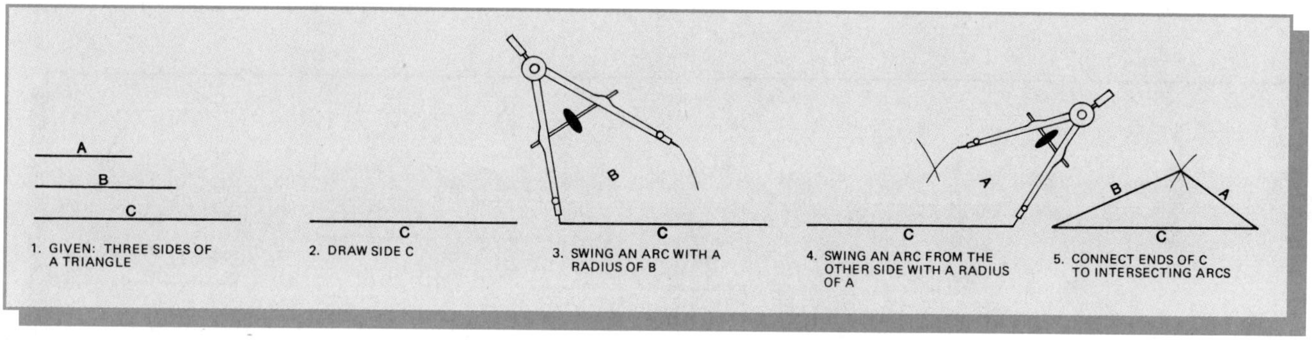

Figure 7-14
Constructing a triangle

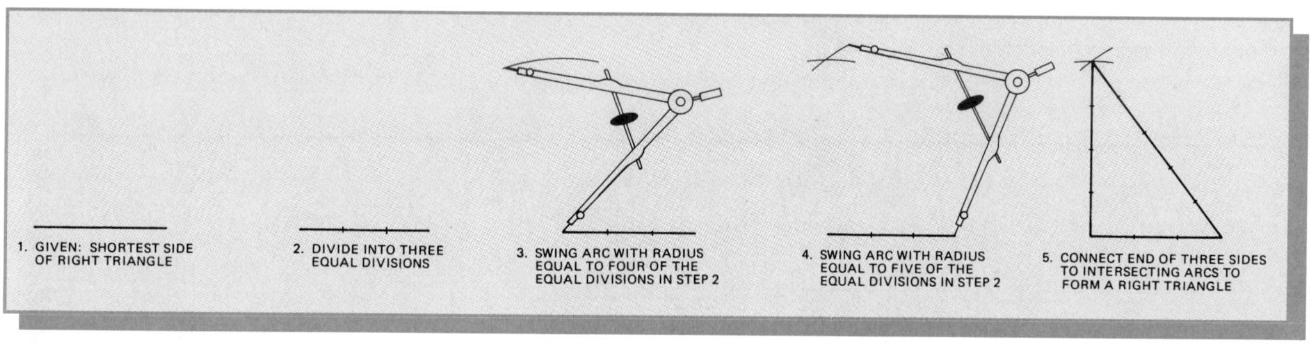

Figure 7-15
Constructing a right triangle

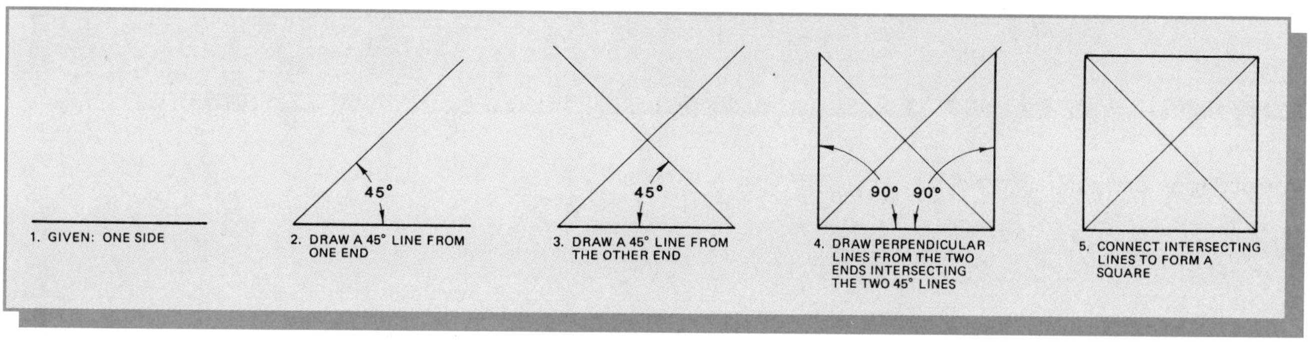

Figure 7-16
Constructing a square

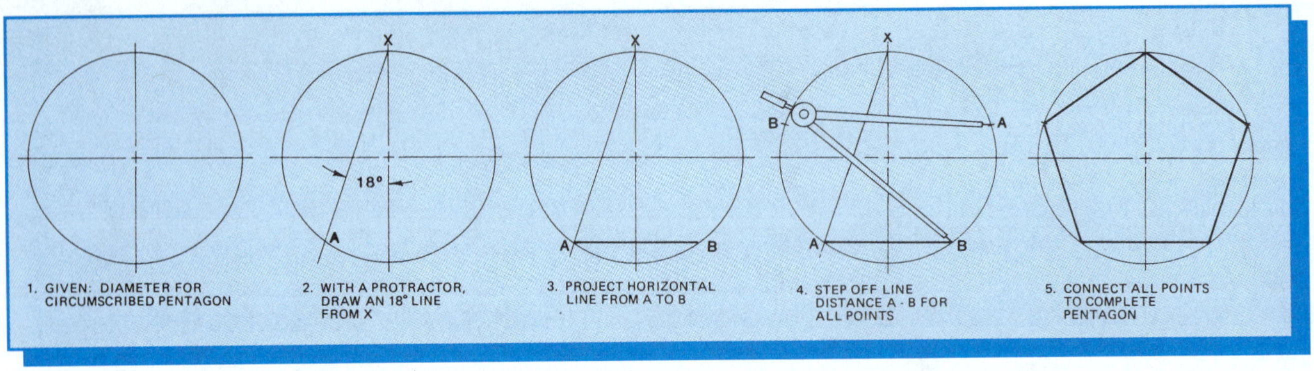

Figure 7-17
Constructing a pentagon (five sides)

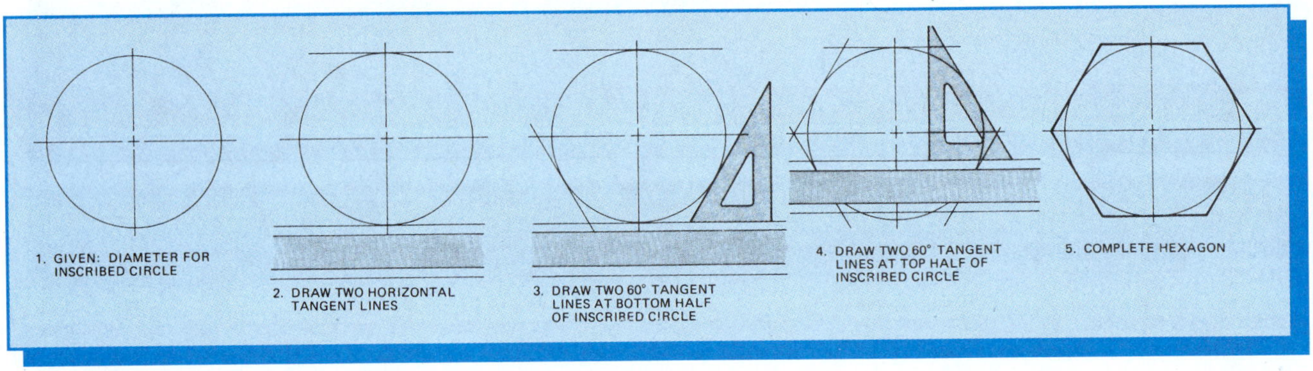

Figure 7-18
Constructing a hexagon (six sides)

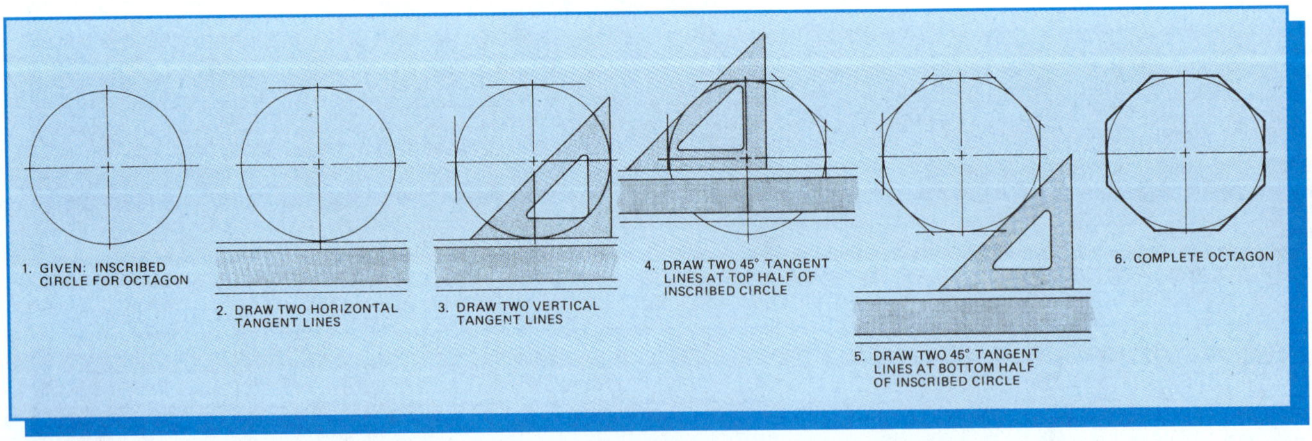

Figure 7-19
Constructing an octagon (eight sides)

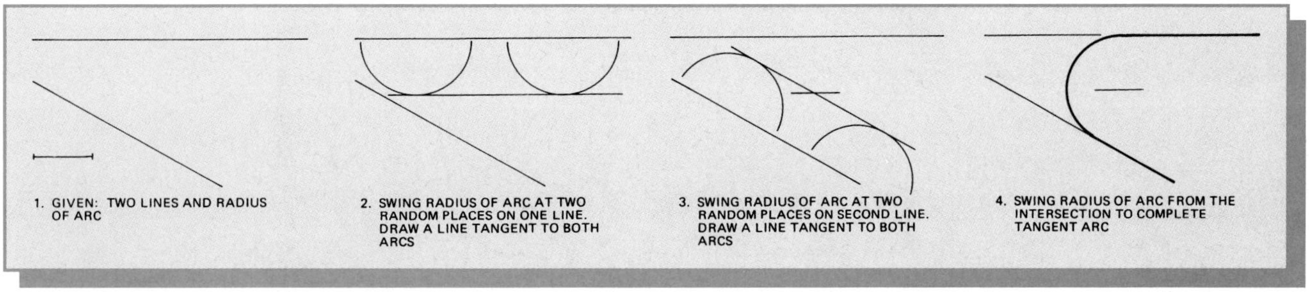

Figure 7-20
Drawing an arc tangent to two lines

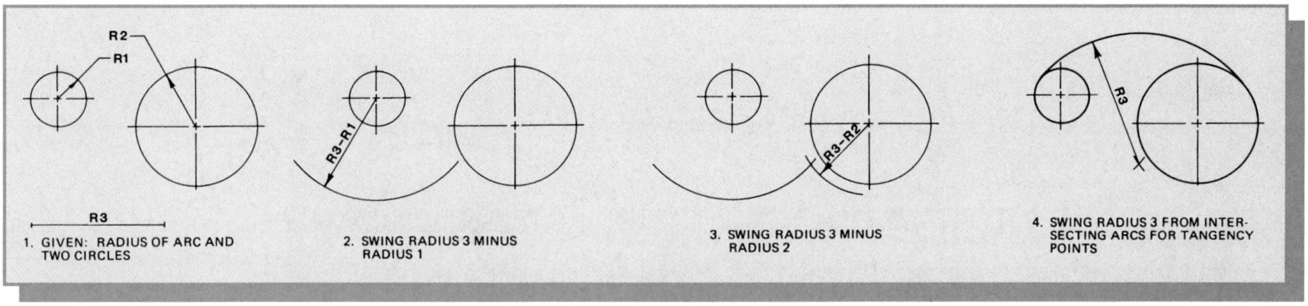

Figure 7-21
Drawing an arc tangent to two circles

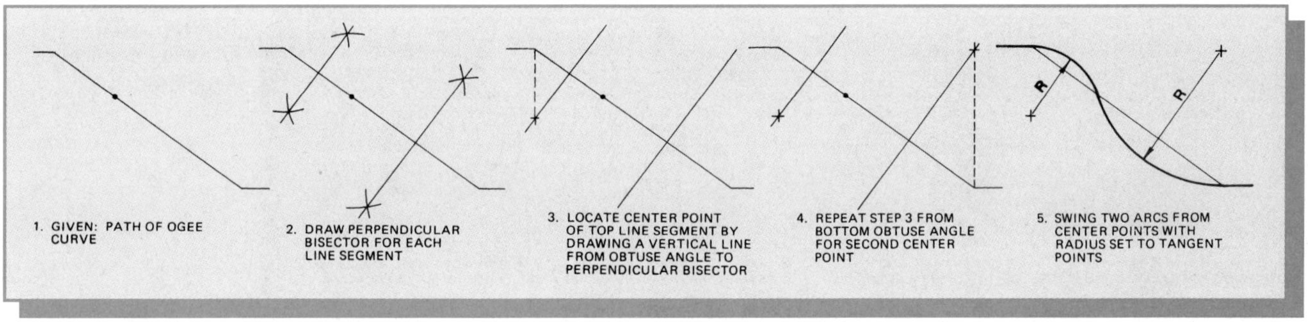

Figure 7-22
Drawing an **ogee curve** (also called a reverse curve). An ogee curve occurs in situations where a smooth contour is needed between two offset features.

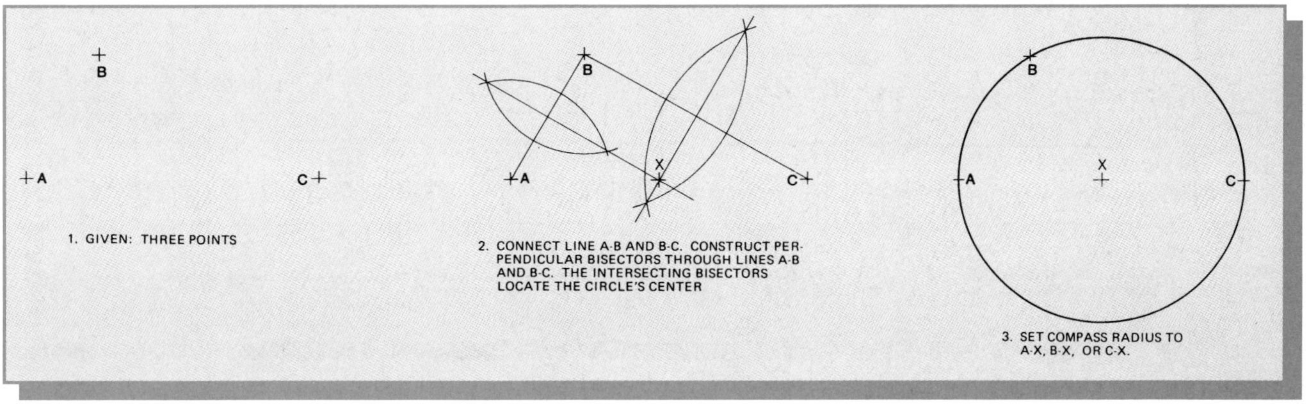

Figure 7-23
Drawing a circle through three points

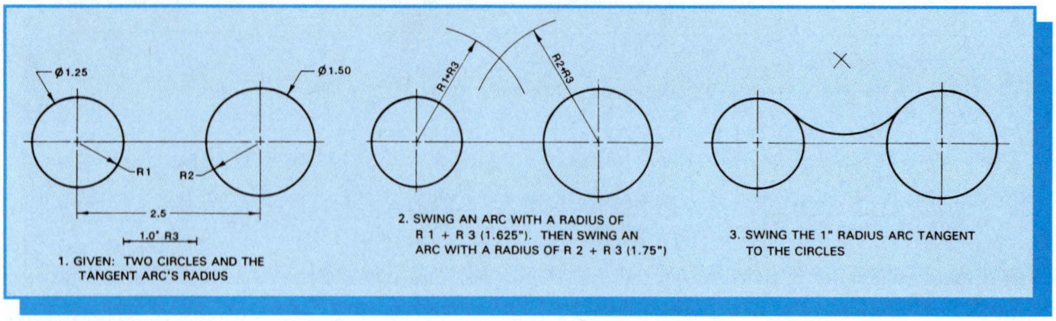

Figure 7-24
Drawing a concave arc tangent to two circles

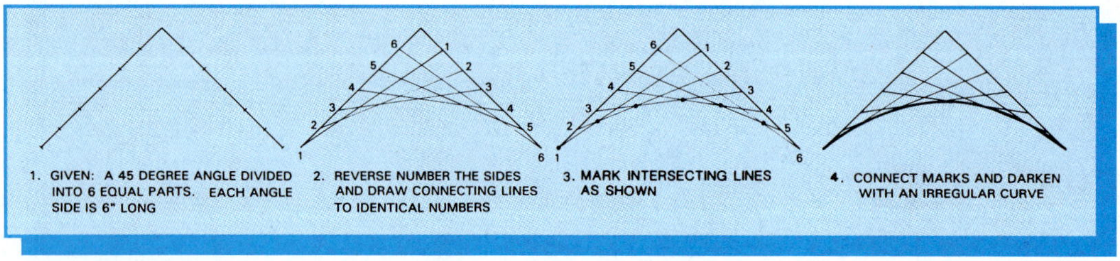

Figure 7-25
Drawing a parabolic curve within a given angle

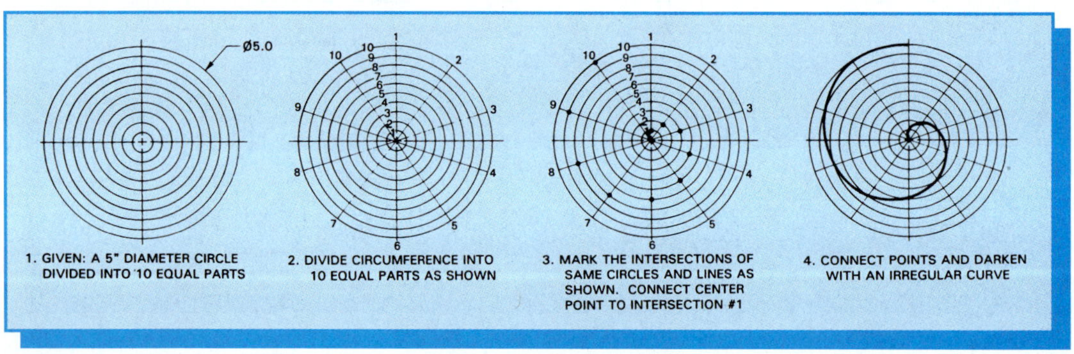

Figure 7-26
Drawing a spiral of Archimedes

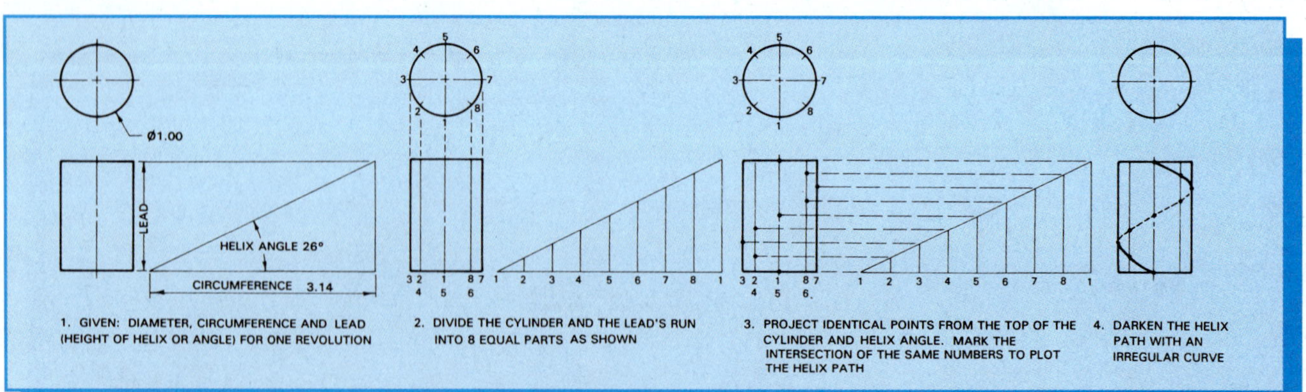

Figure 7-27
Drawing a helix

AutoCAD LT Exercises

Object Snaps

One of the most important aspects of a CAD drawing is the accurate database. A data base is the computer information that is stored in the drawing. This information can be used to do many different things. The database from an AutoCAD drawing may be used to program a computer-controlled machine tool. AutoCAD LT uses Object **Snaps** (OSnaps) to help create accurate drawings.

OSnaps allow the user to select exact locations without knowing the coordinate positions. OSnaps will pick the location according to the Osnap selected, such as Endpoint or Intersection on an existing visible object. Whenever AutoCAD LT *prompts you for a point,* an OSnap can be used.

Endpoint—snaps to the nearest end of a line or arc **(Figs. 7-28, 7-29)**. To use an OSnap, you need to be in a command that is prompting "from point" or "to point." The box that appears at the cursor when an Osnap is used needs to be placed close to the point or object you are snapping to. The center of the box does not need to be exactly on the point.

Figure 7-28
Endpoints

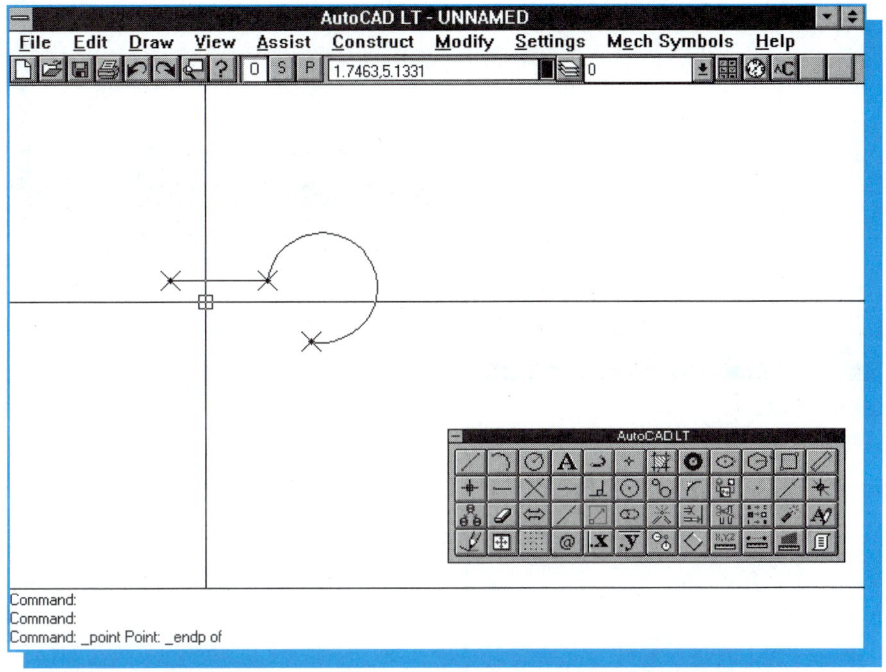

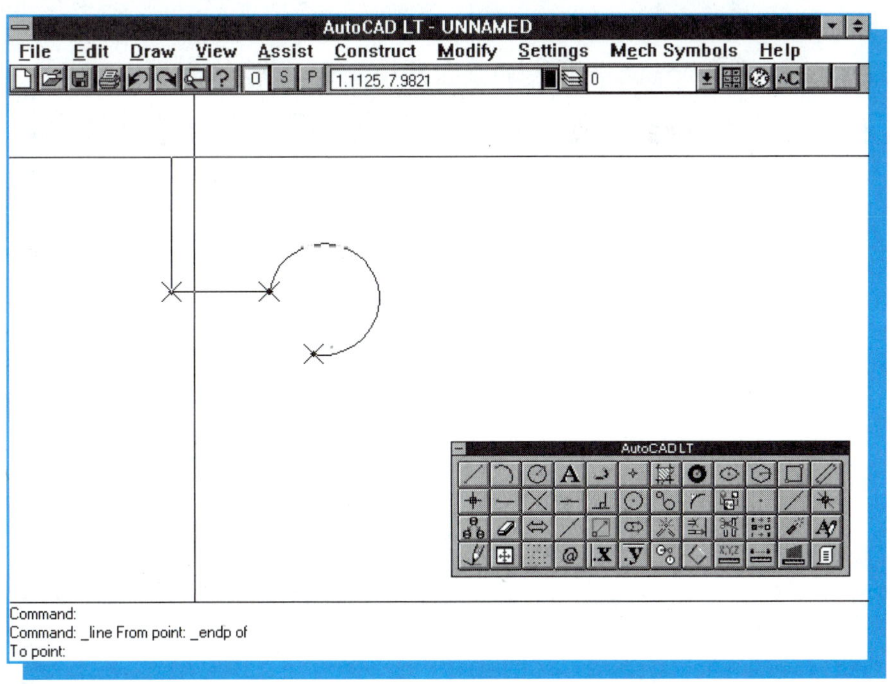

Figure 7-29
Snapping to endpoints

Intersection—snaps to the intersection of lines, circles, and arcs **(Figs. 7-30, 7-31)**.

Midpoint—snaps to the midpoint of lines and arcs **(Fig. 7-32)**.

Perpendicular—snaps a **perpendicular** from the last point entered **(Fig. 7-33)**.

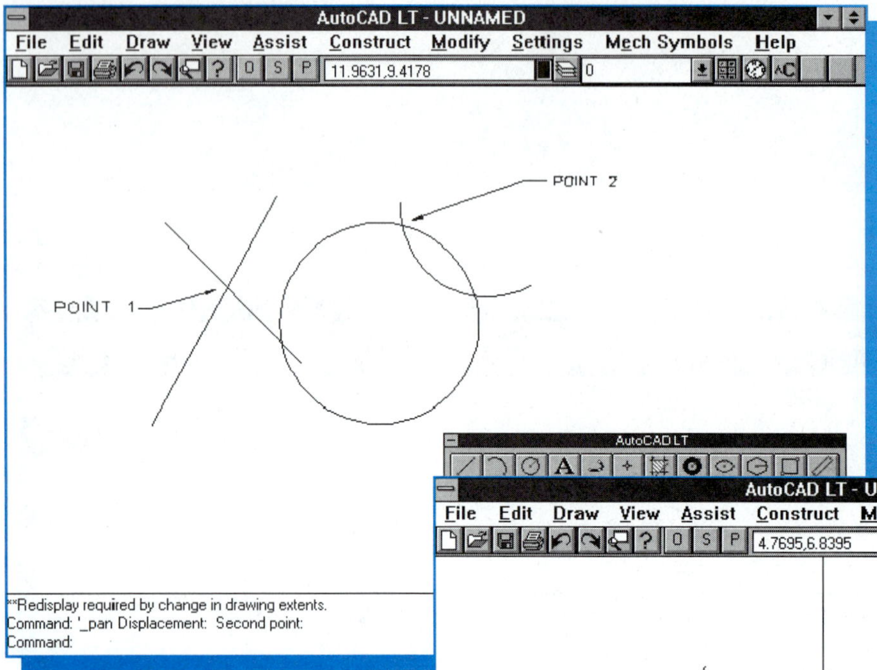

Figure 7-30
Intersections

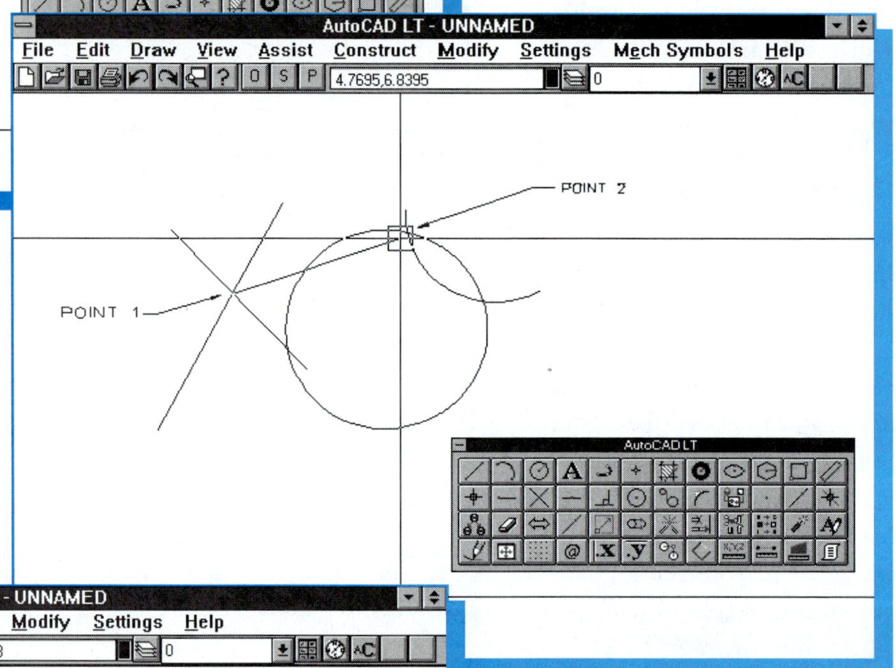

Figure 7-31
Snapping to intersections

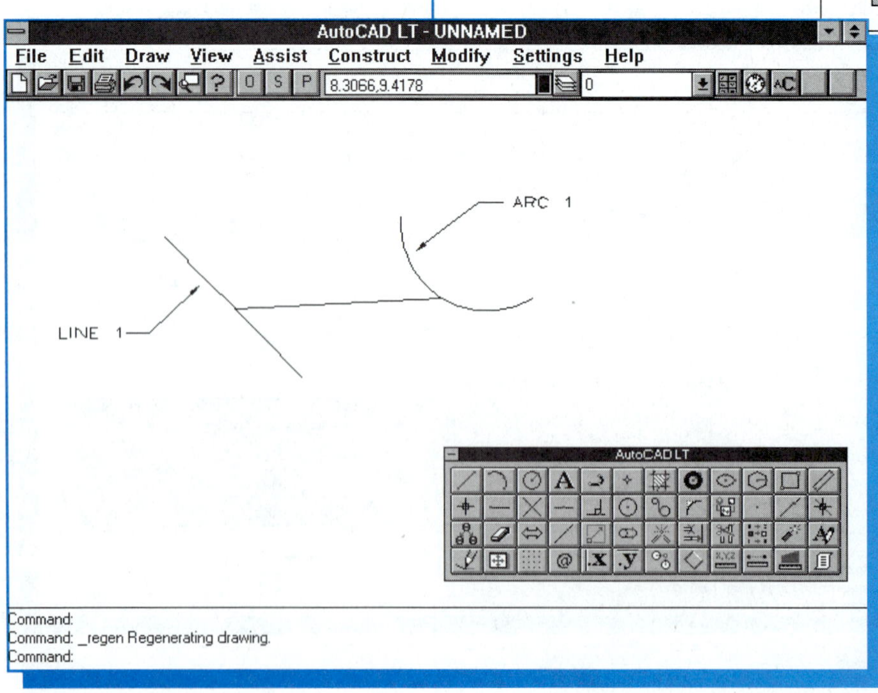

Figure 7-32
Snapping to midpoints

Center—snaps to the center of an arc or circle **(Fig. 7-34)**.

> **tip:** *When using center, select the object, not the center of the object.*

Tangent—snaps to the **tangent** point on an arc or circle from the last point **(Fig. 7-35)**.

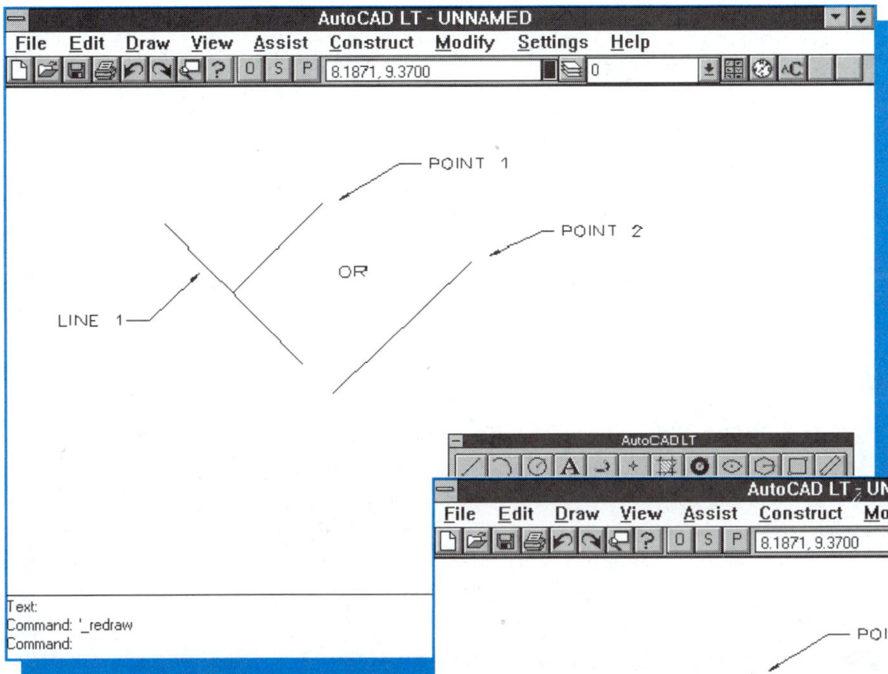

Figure 7-33
Snapping from a point perpendicular to a line

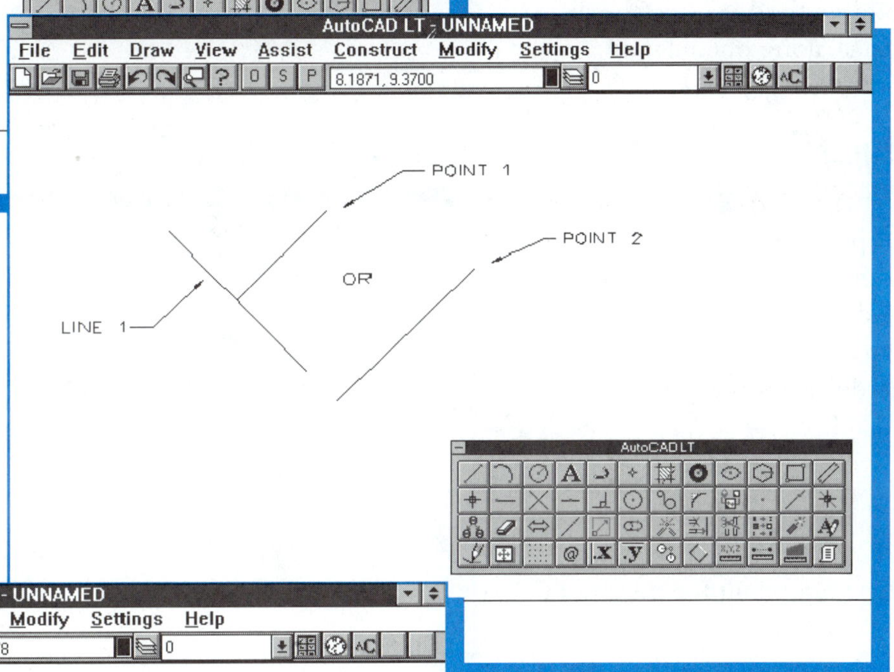

Figure 7-34
Snapping to the center of a circle or arc

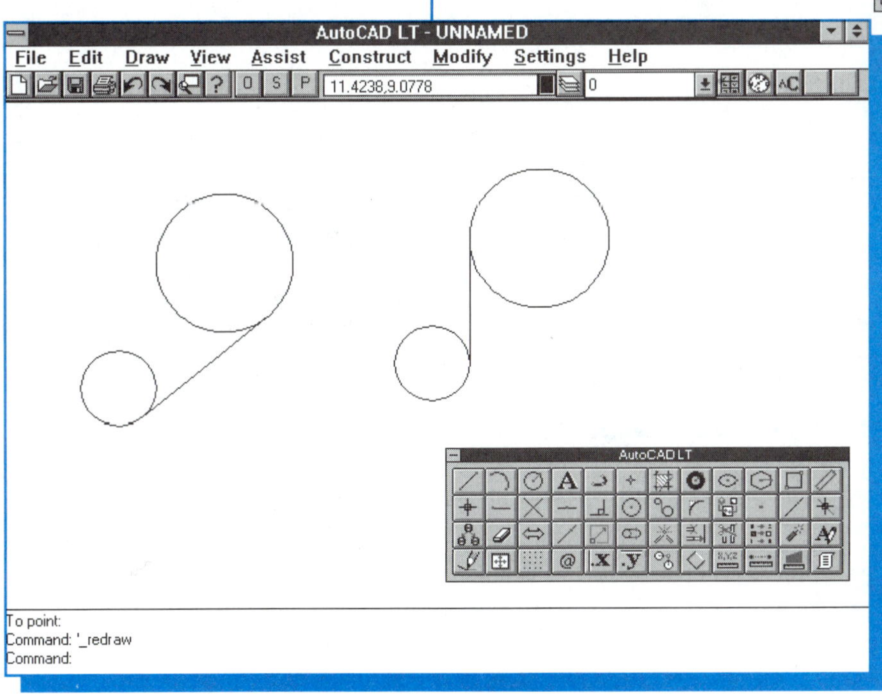

Figure 7-35
Snapping a line tangent to two circles

Polygons

Closed regular polygons with 3 to 1,024 sides can be drawn in AutoCAD LT using the polygon command or the polygon button in the toolbox. There are two methods to draw a polygon. The first method is using the radius of an inscribed circle to give the distance across the flats of the polygon **(Fig. 7-36a)**. The second method is using the distance along one edge of a polygon **(Fig. 7-36b)**.

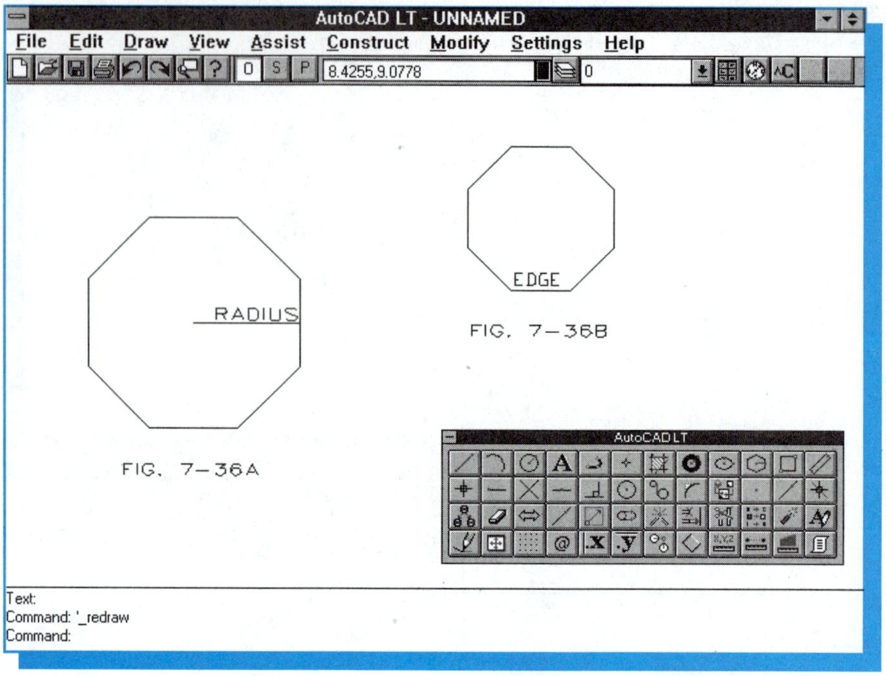

Figures 7-36a & 7-36b
Creating a polygon with a known radius or edge

Fillets

AutoCAD LT allows you to put round corners automatically between lines or lines and circles or lines and arcs. The command that allows you to do this is the **fillet** command.

1. Select the fillet button from the toolbox.
2. Type **R** for radius.
3. Type **.500** for the size of the radius.
4. Select objects that need a fillet. **Figure 7-37** shows objects before fillet. **Figure 7-38** shows objects with a R.500 fillet.

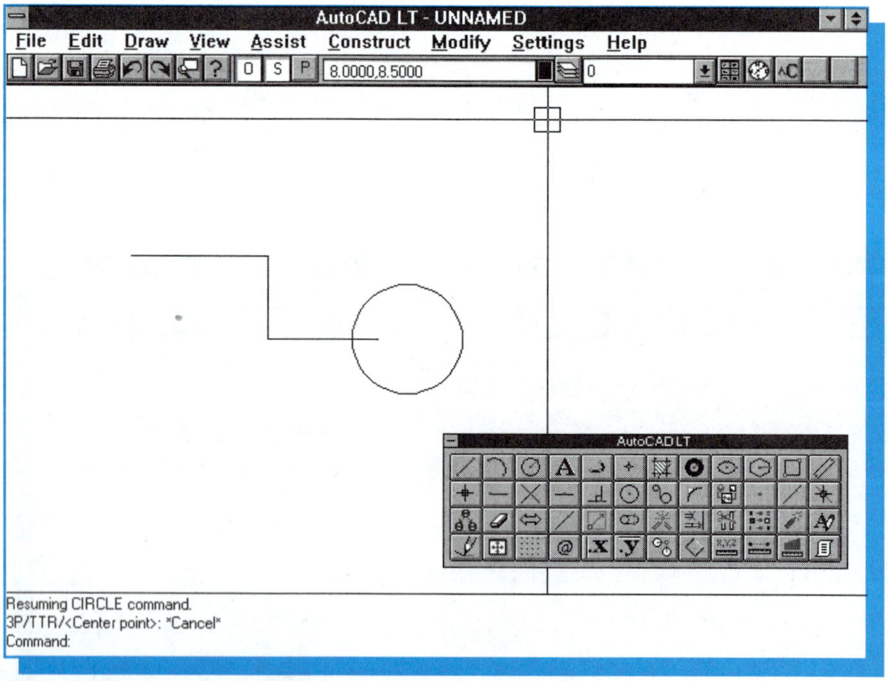

Figure 7-37
Drawing before using the fillet or chamfer command

Chamfers

The **chamfer** command (CF) puts angles between lines. The angles can be equal on each side or they may be longer one way. The first thing we need to set is the size of the chamfer.

1. Type **CF** at the command line.
2. Type **D** distance, and then **.500**, and (EB) at both prompts.
3. Select the two lines that need a chamfer **(Fig. 7-39)**.

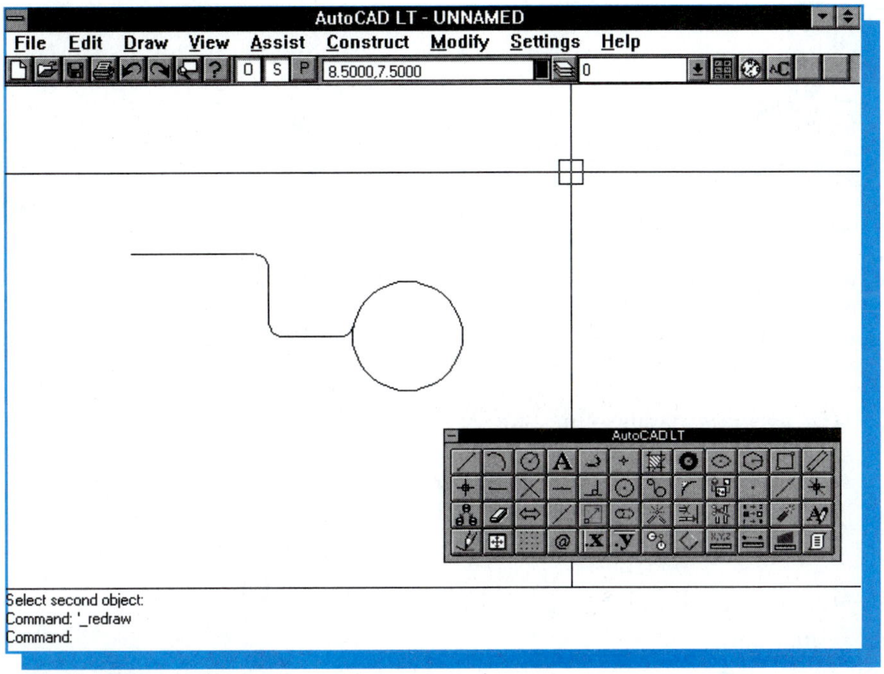

Figure 7-38
Using the fillet command

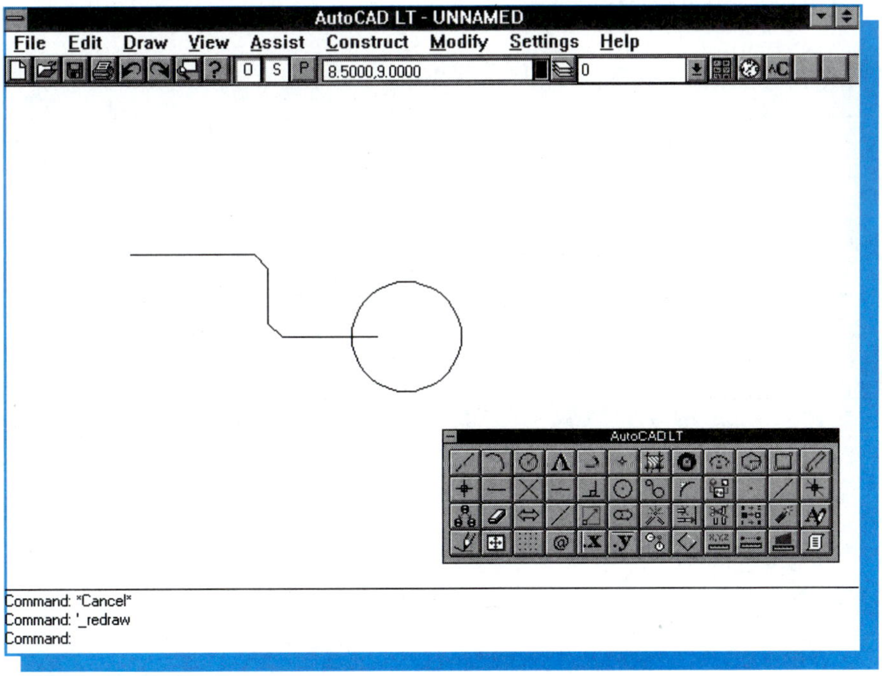

Figure 7-39
Using the chamfer command

Drafting Exercises

Do these exercises using drawing instruments or CAD.

1. Construct a perpendicular bisector for a 4.75" line.
2. Construct a parallel line to a 3.5" line set at 30°.
3. Construct a tangent line to a 1.75" diameter circle.
4. Divide a 5" line into eight equal parts.
5. Bisect a 42° angle.
6. Construct an **equilateral** triangle with 2.25" sides.
7. Construct a right triangle with the shortest side 1.5".
8. Construct a 1.375" square.
9. Construct a **pentagon** (five sides) in a circumscribed circle with a 1.5" radius.
10. Construct a **hexagon** with a 2.0" radius inscribed circle.
11. Construct an **octagon** with a 3.75" diameter inscribed circle.
12. Construct a tangent arc within a 30° angle with a .5" radius.
13. Construct a right triangle with the short side 2.0" long.
14. Draw the basic geometric constructions shown in **Figures 7-40** and **7-41**.
15. Draw the two-dimensional geometric forms shown in **Figure 7-42**.

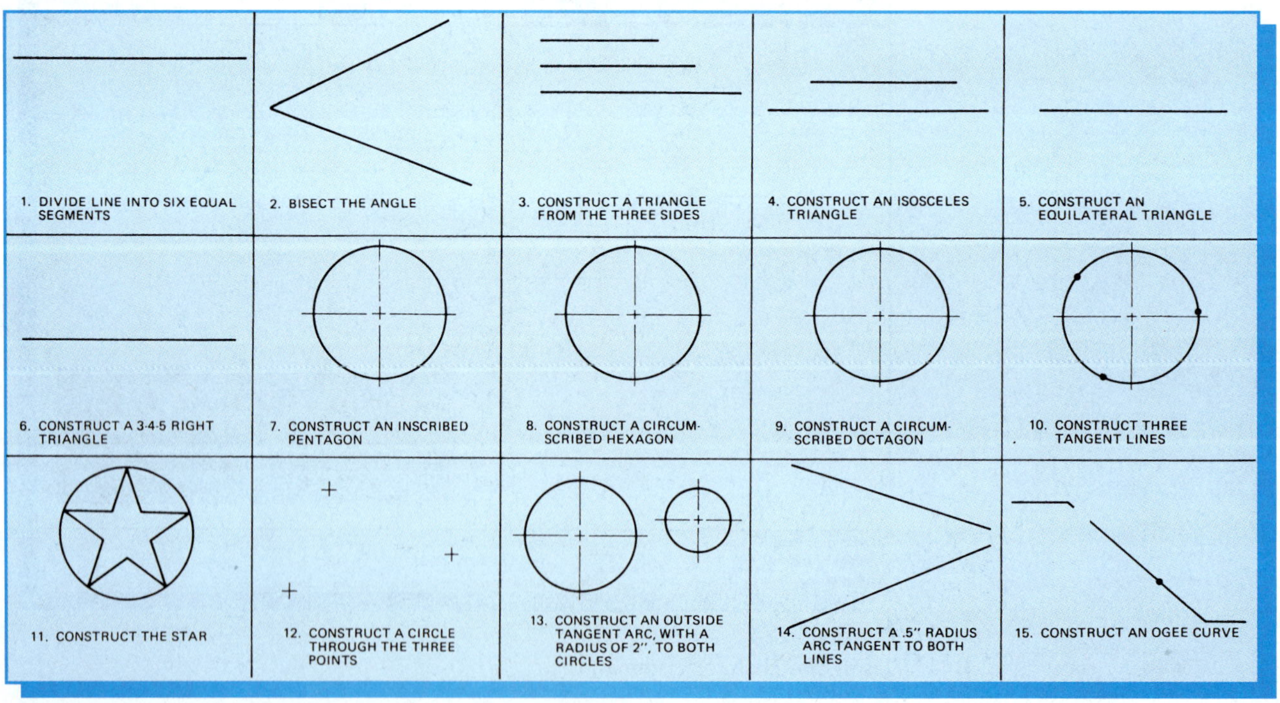

Figure 7-40
With drawing instruments or a CAD system, draw the geometric forms. Double the size of each exercise.

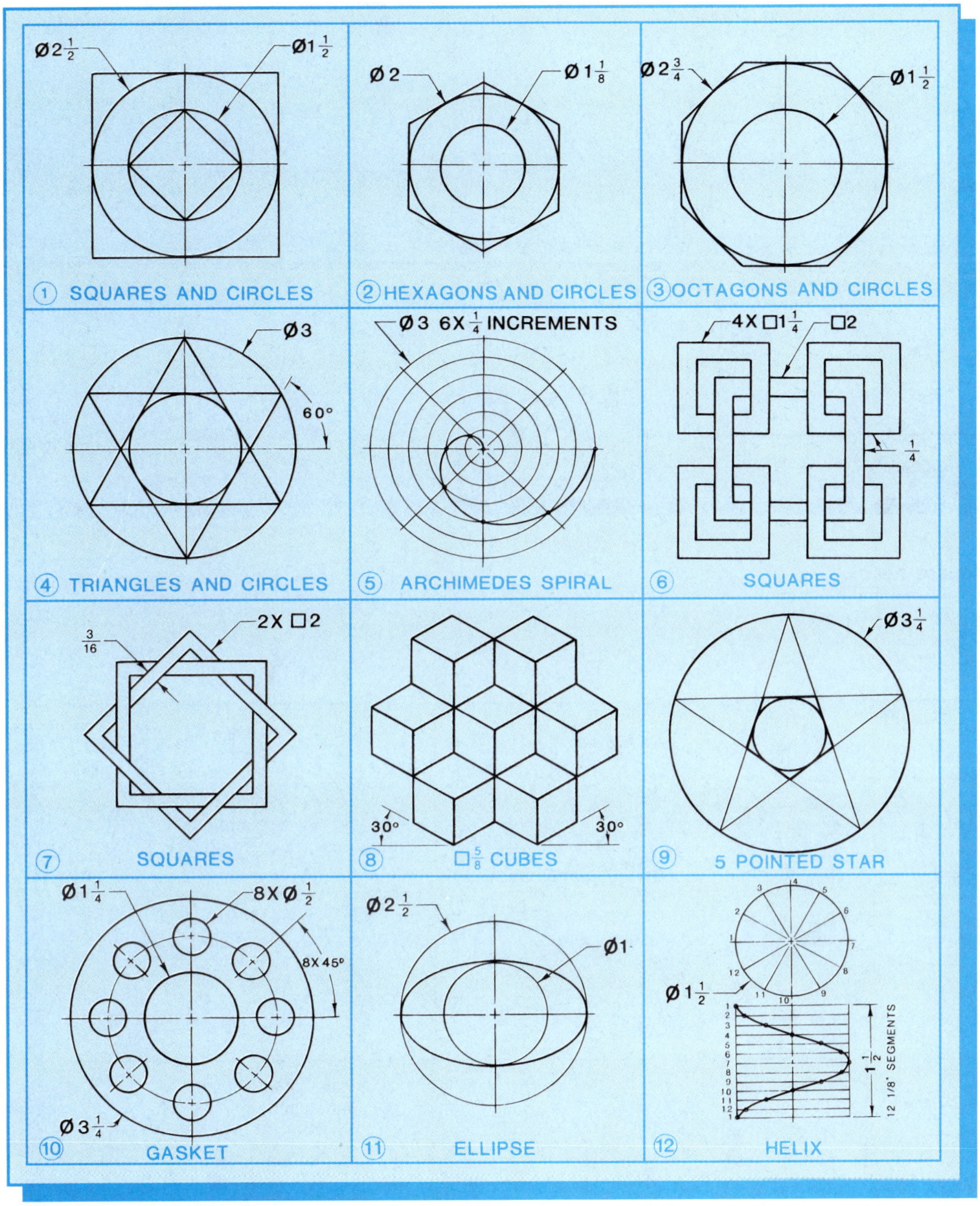

Figure 7-41
With drawing instruments or a CAD system, draw the geometric exercises.

16. Draw the geometric forms shown in **Figure 7-43** with a CAD system.

17. Draw the plates shown in **Figure 7-44**.

18. Draw the lever shown in **Figure 7-45**.

19. Draw the gasket shown in **Figure 7-46**.

20. Draw the prop shown in **Figure 7-47**.

21. Draw the geometric figures in **Figure 7-48** with metric dimensions (millimeters).

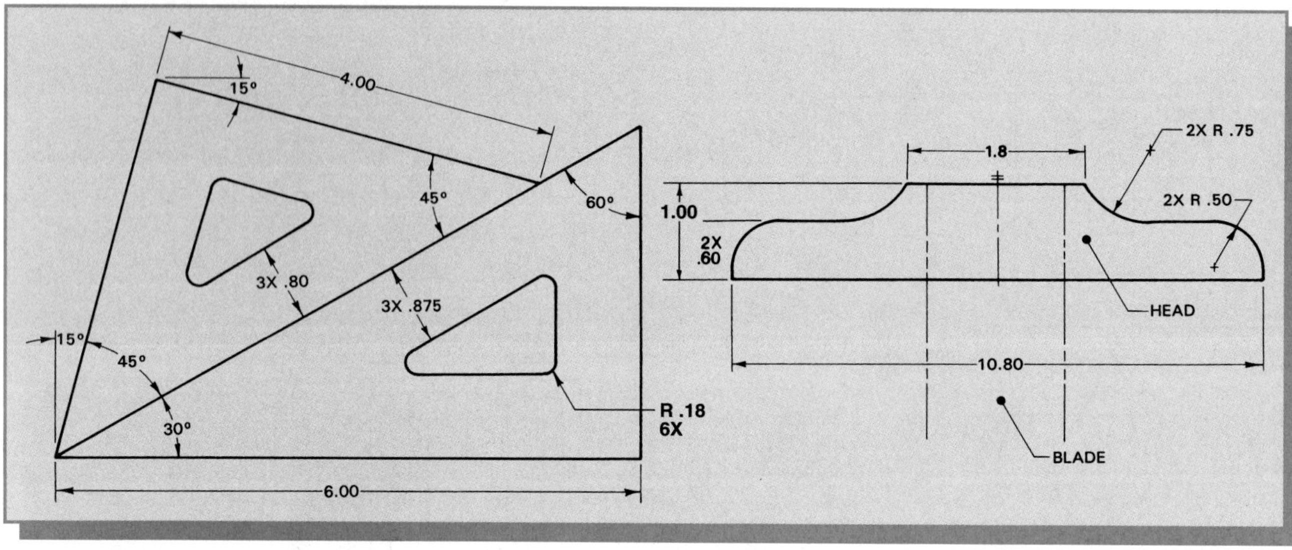

Figure 7-42
Draw the triangles and T square head

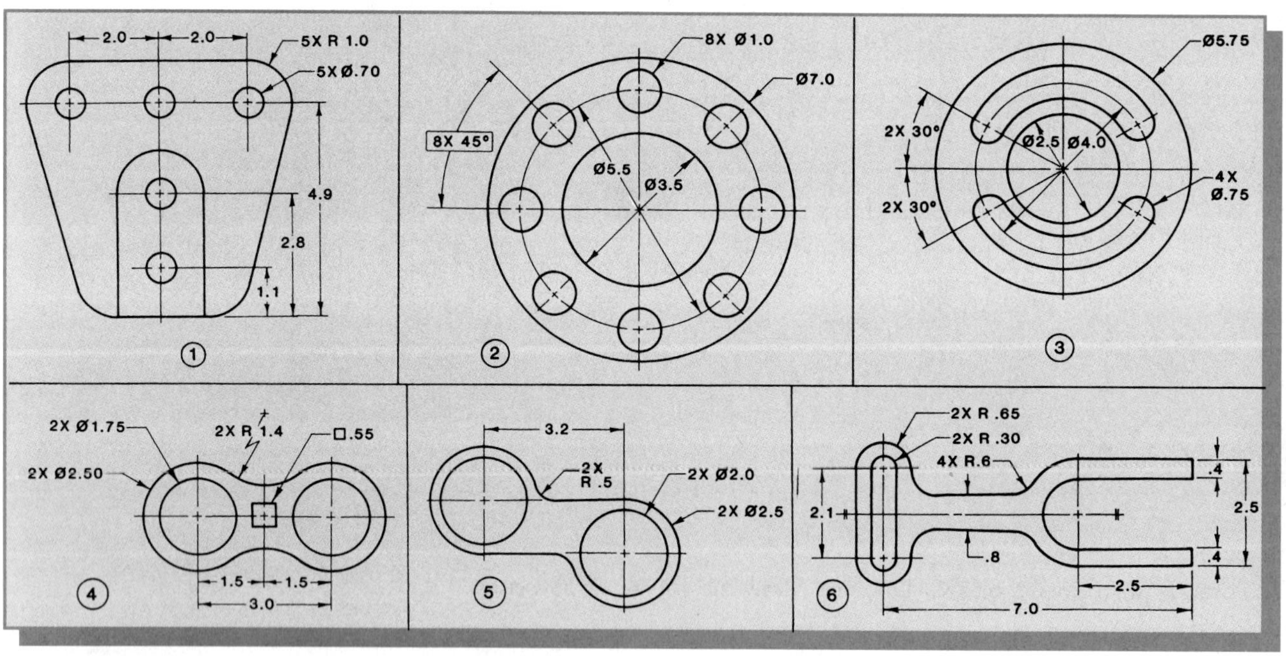

Figure 7-43
With drawing instruments or a CAD system, draw the geometric forms.

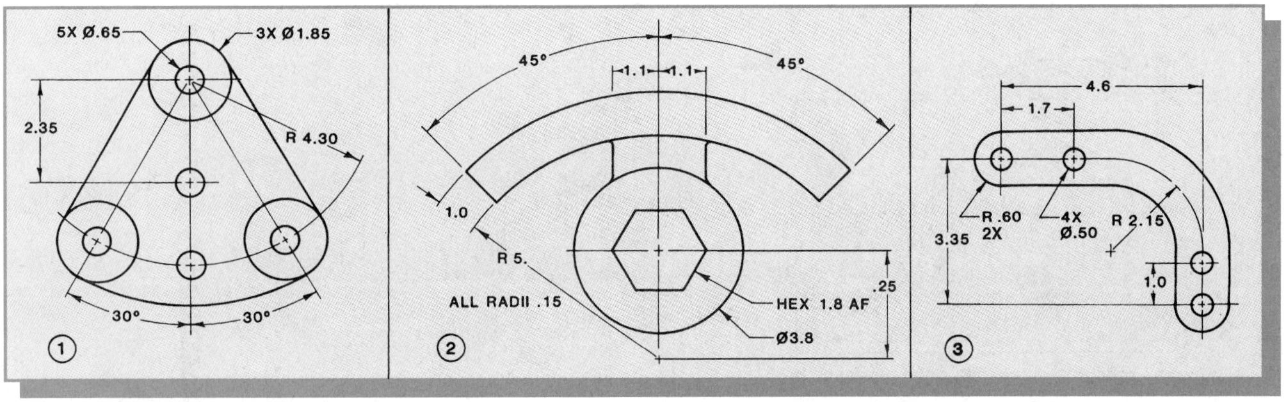

Figure 7-44
With drawing instruments or a CAD system, draw three plates.

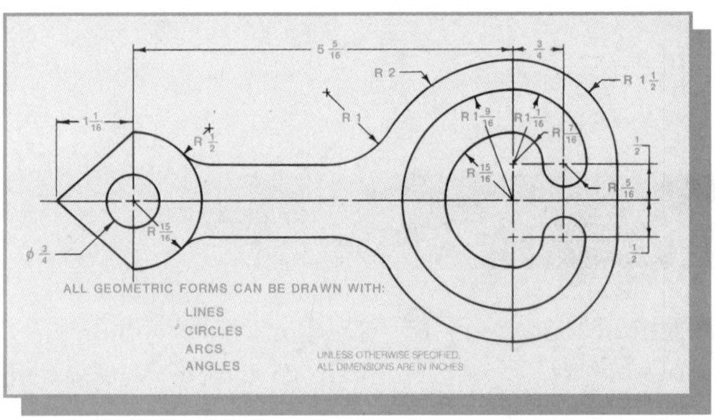

Figure 7-45
Lever

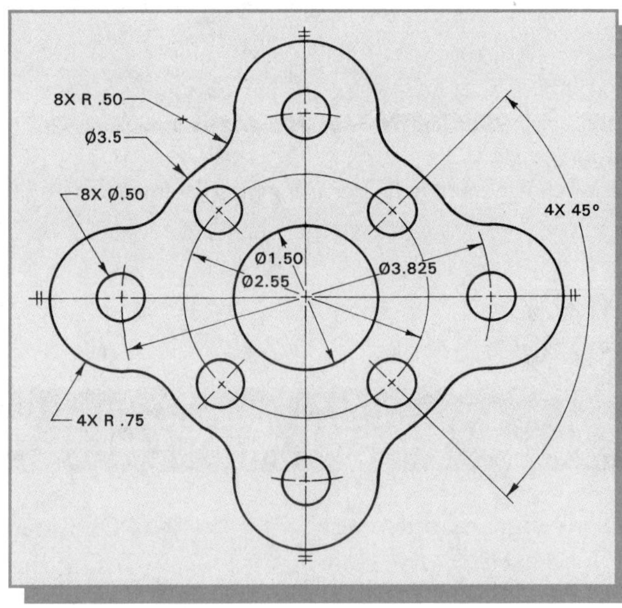

Figure 7-46
Gasket

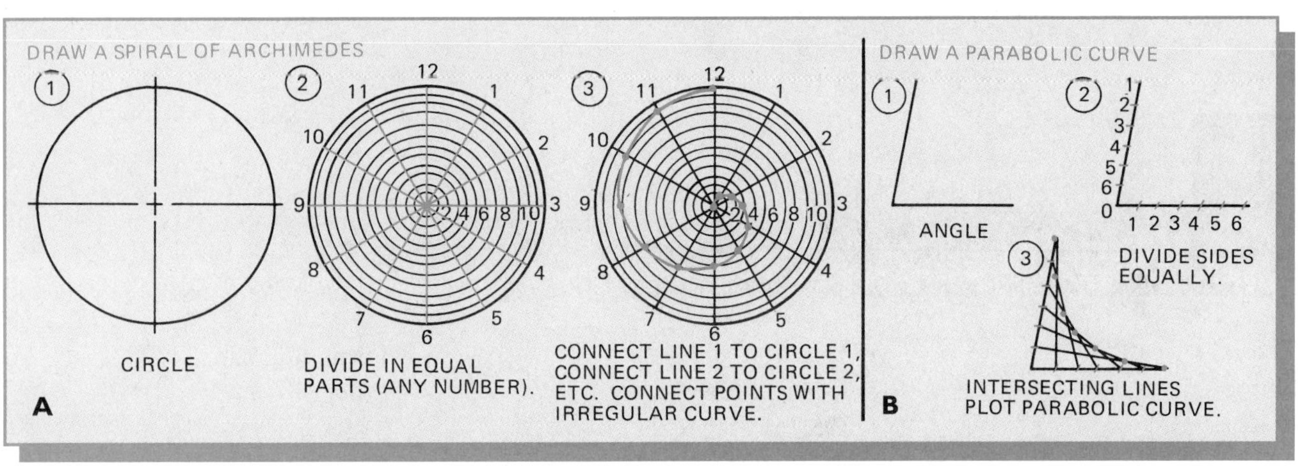

Figure 7-47
Draw Figures A and B

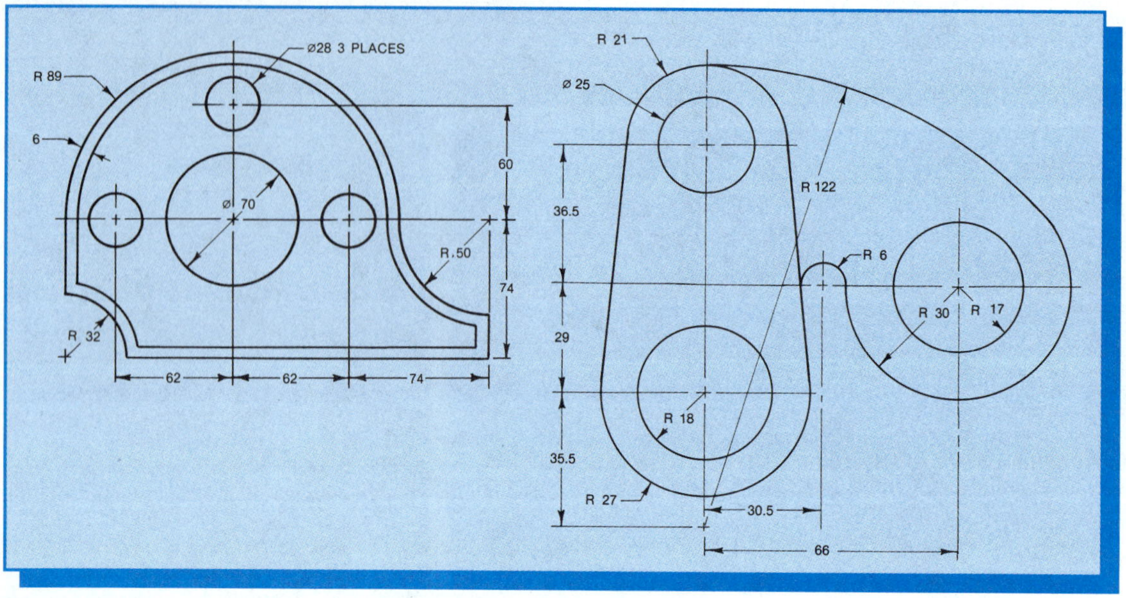

Figure 7-48
Draw the geometric figures using metric dimensions (millimeters).

Design Exercises

1. Design a family crest for yourself and/or your family.
2. Design a company logo for a business called SPACE & AERONAUTICS INC.
3. Design a new school logo using the school's colors.
4. Design a business card for your mother and/or father.
5. Design a new geometric pattern with a compass and straightedge.
6. Design and draw an object using as many basic lines and curved surfaces as you can.

Key Terms

Chamfer
Ellipse
Equilateral
Fillet
Geometric construction
Hexagon
Octagon
Ogee curve
OSnap
Parallel
Pentagon
Perpendicular
Polygon
Tangent
Three-dimensional
Two-dimensional

Chapter 8

Multiview Drawings

Objectives

The student will be able to:

- visualize the planes of projection
- select the front view for a multiview drawing
- draw third-angle projection multiview drawings
- draw first-angle projection multiview drawings
- center a multiview drawing
- draw runouts on a multiview drawing
- project an incline surface on a multiview drawing
- project an ellipse on a multiview drawing
- project an oblique surface on a multiview drawing
- generate multiview drawings with a CAD system

Multiview Orthographic Projections

Orthographic projection is a projection of a single view of an object on a drawing surface that is perpendicular to both the view and the lines of projection. Horizontal orthographic projection lines are used to align views with the same height dimensions. Vertical orthographic projection lines are used to align views with the same width dimensions **(Fig. 8-1)**. The technique of orthographic projection is used to create **multiview** drawings of a single object. The related views of the object appear as if they were all in the same plane. Multiview drawings describe the exact size and shape of an object. **Figure 8-2** shows a CAD-generated multiview drawing of a machine part.

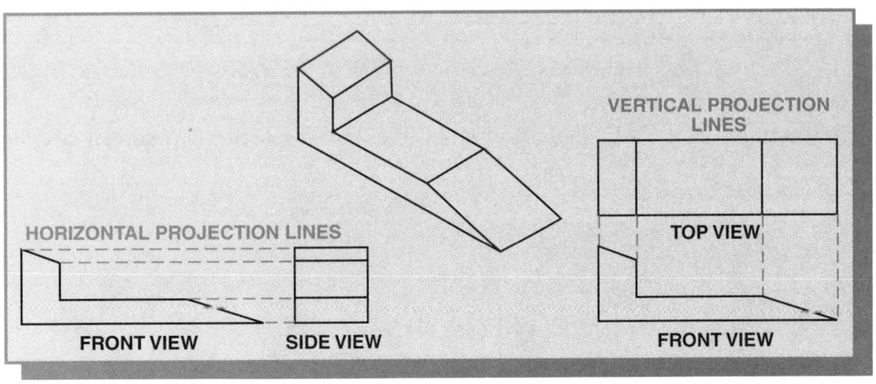

Figure 8-1
An example of the orthographic projection of the side and top views from the front view

Selection of Views

Describing the shape of an object with a multiview drawing should always be accomplished with the minimum number of views. The first view chosen is the **front view** because it best shows the form of an object. **Figure 8-3** shows the selection of the front view

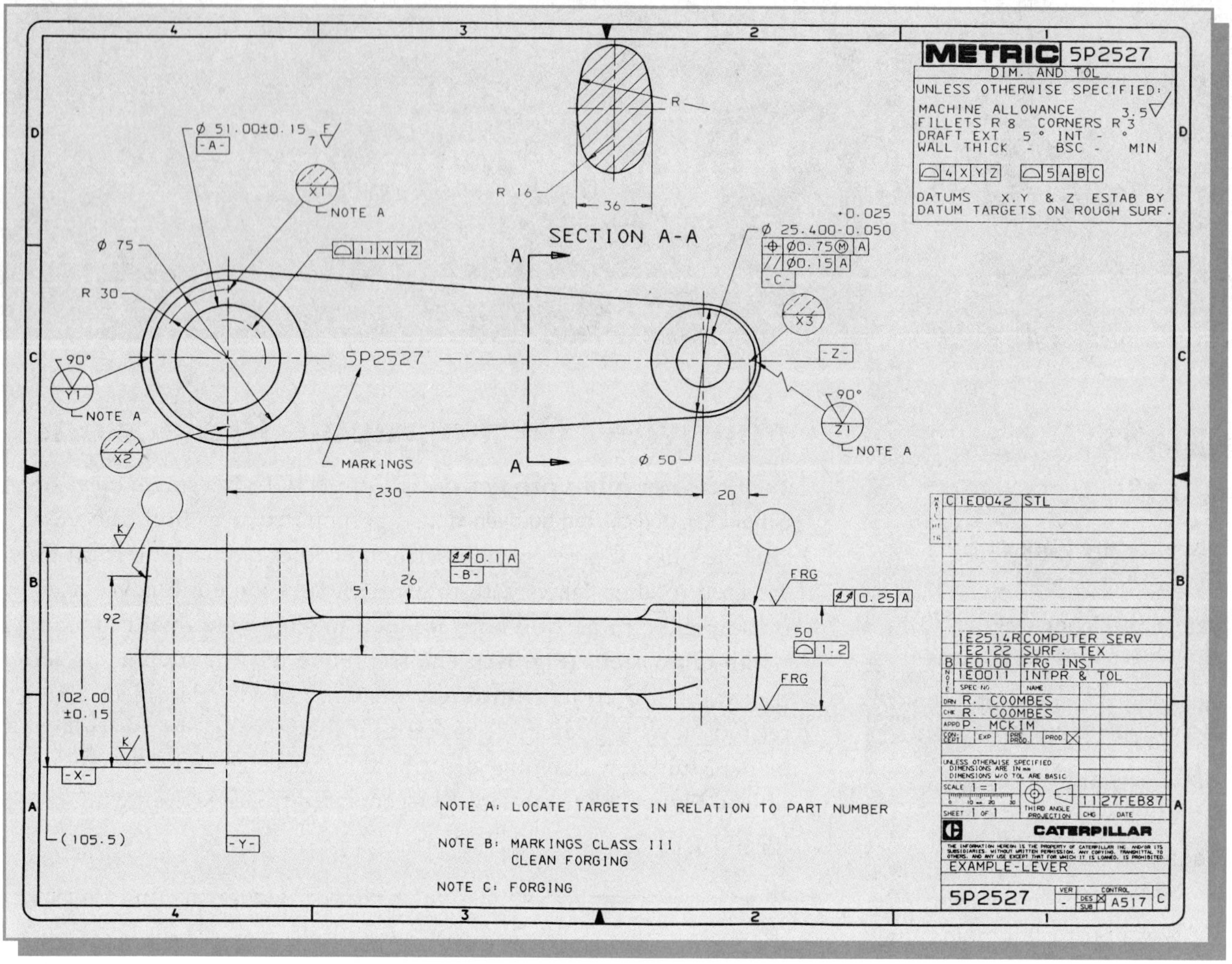

Figure 8-2
An example of a CAD-generated multiview working drawing *(Courtesy of Caterpillar, Inc.)*

Figure 8-3
The front view selection depends on which view best describes the shape of the item being drawn.

of several objects. In each case, the front view reveals the most unique and distinguishing shape feature of the object.

Simple flat objects, such as a gasket **(Fig. 8-4)**, require only one view, since the thickness is uniform and can be described in a note.

Symmetrical objects may need only one or two views. It is not necessary to draw another view if it is identical to an existing view. In a multiview drawing of a solid sphere (ball), all views are identical so only one view is needed. Only two views are required to describe the object in **Figure 8-5**, since the front and side views are the same.

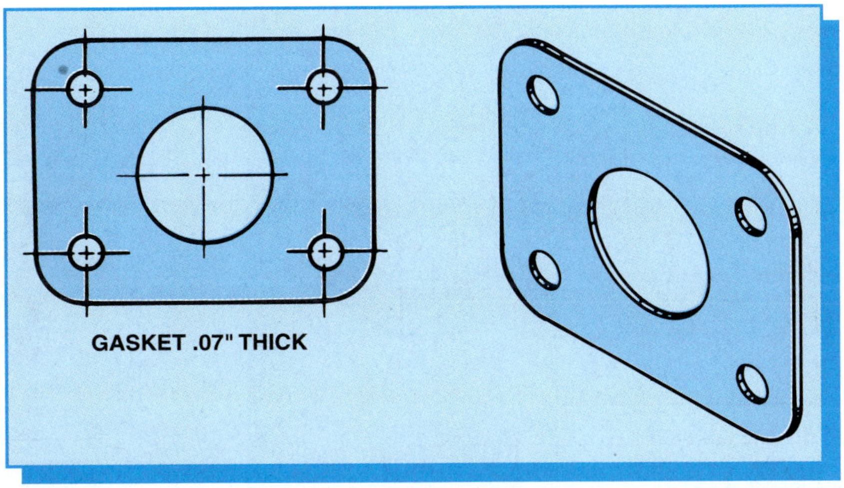

Figure 8-4
Simple flat objects can be defined with a single view.

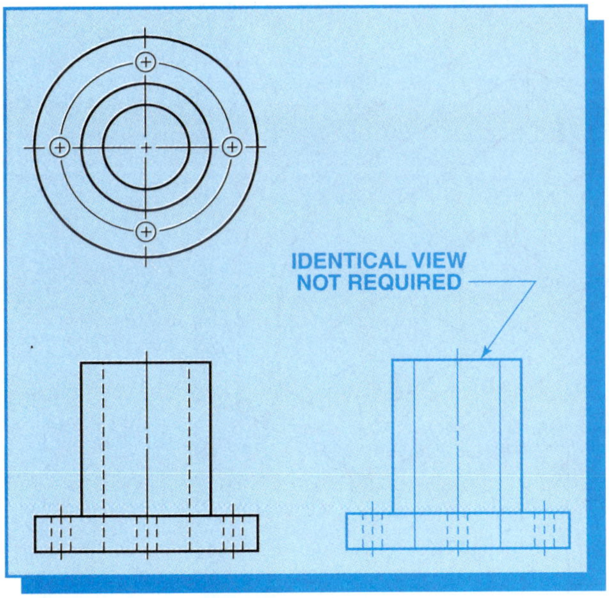

Figure 8-5
A symmetrical object requires only two views for a full description.

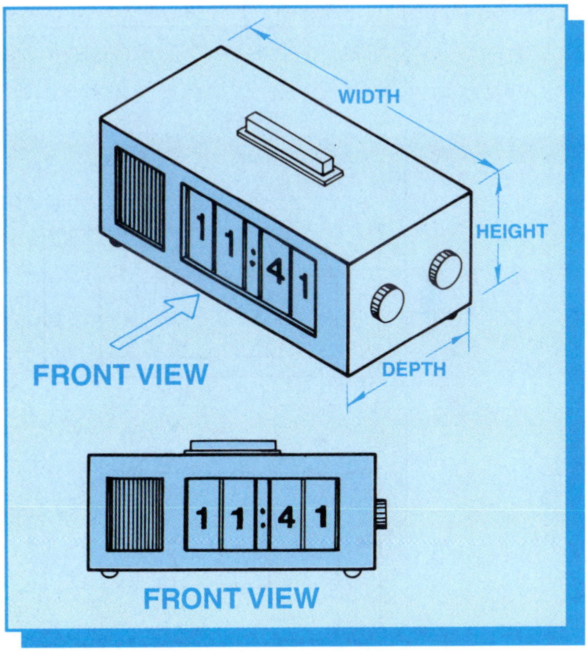

Figure 8-6
A multiview drawing's front view of a digital clock

Three views are used to describe a clock radio. The front view **(Fig. 8-6)** shows the height and width of the clock radio; the **top view (Fig. 8-7)** shows the depth and width; and the **right-side view (Fig. 8-8)** shows the height and depth. The size and shape of most objects can usually be described in three views **(Fig. 8-9)** when the bottom, left-side, and rear views would not help to further clarify any aspect of the object.

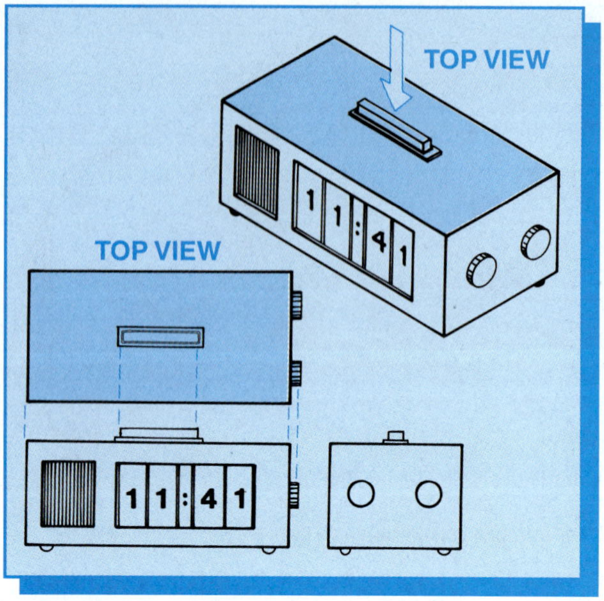

Figure 8-7
The digital clock's top view

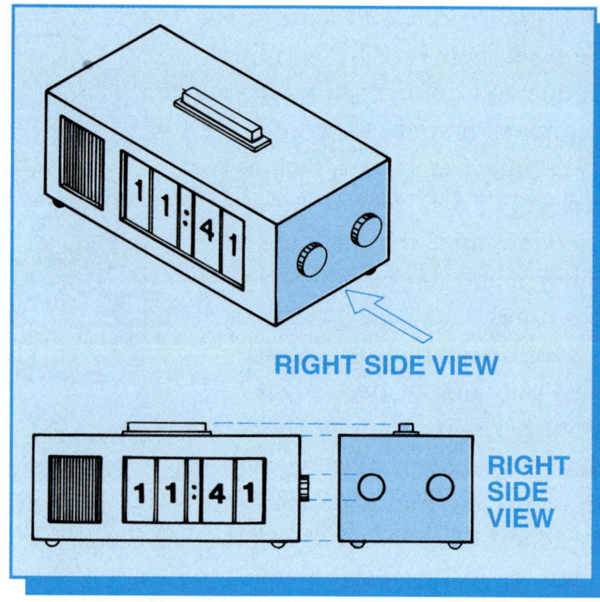

Figure 8-8
The digital clock's right-side view

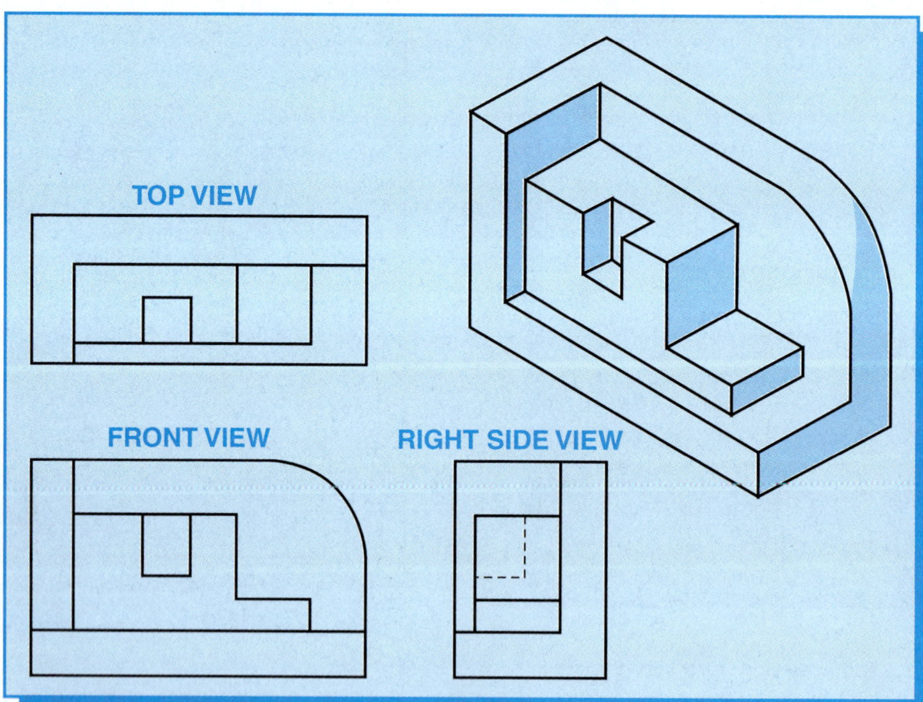

Figure 8-9
Most objects can be fully described with three views.

Other objects may require up to six views to show the exact size and shape of each side **(Fig. 8-10)**. **Figure 8-11** shows a six-view drawing of the same clock radio shown in the three views of **Figures 8-6** through **8-8**. Although the three views have identical dimensions, the details of each of the six views shown in **Figure 8-11** are different. Six views, therefore, are needed to show all the details that may be necessary for manufacturing the clock radio case. Many additional multiview working drawings are needed for the manufacture of the interior of the clock radio.

Planes of Projection

Each view in a multiview drawing is drawn as it appears on an imaginary **plane of projection**. Imagine a sheet of glass (plane of projection) placed between a viewer and an object **(Fig. 8-12)**. Now imagine the viewer's lines of sight as parallel and connected through the glass to each line and corner of the object. The shape of the object can be visualized where the lines of sight intersect the plane of projection.

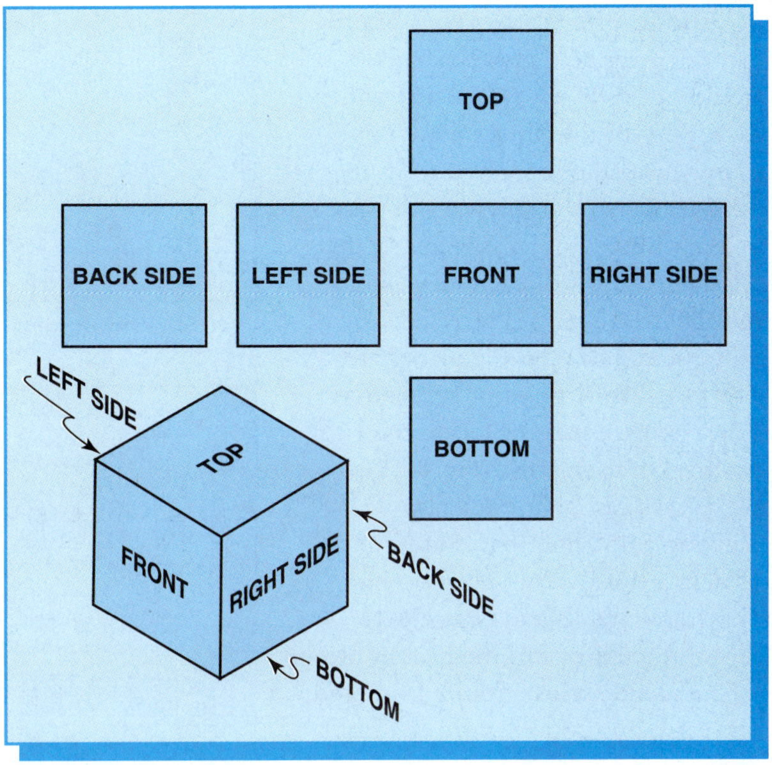

Figure 8-10
Orthographic projection of six views using third-angle projection for the placement of views

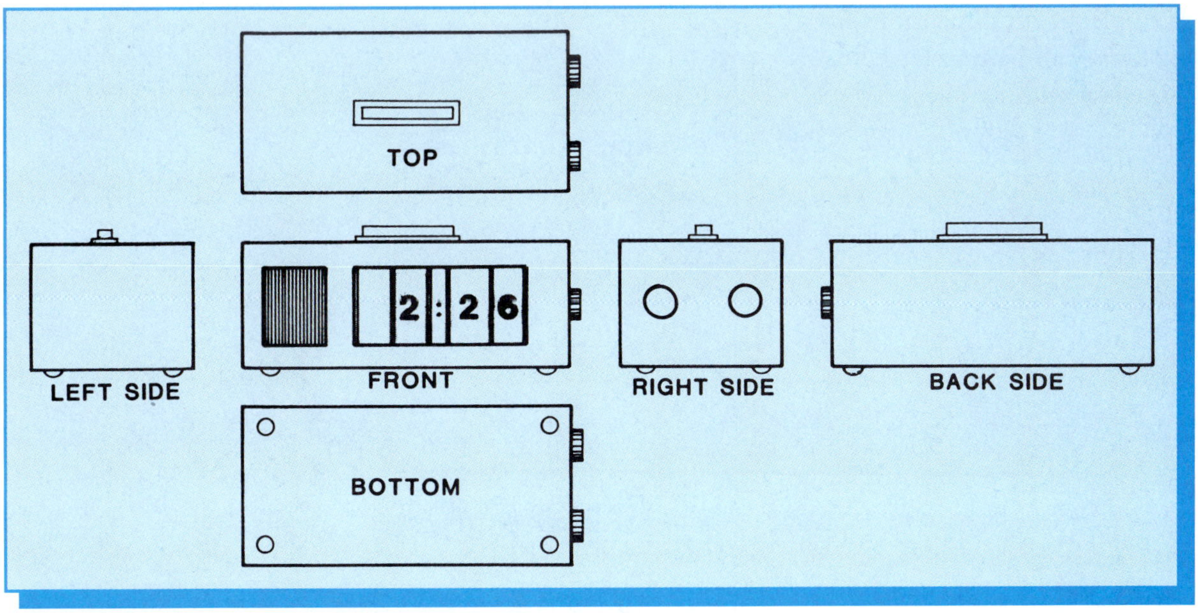

Figure 8-11
A multiview drawing may have a maximum of six orthographic views.

To visualize all six planes of projection, imagine a glass box surrounding an object **(Fig. 8-13)** with each view of the object drawn on each glass plane as viewed from the outside. When the sides of the glass box are hinged open **(Fig. 8-14)**, the relative position of each view in a multiview drawing is revealed. The hinge lines between the planes are also called **fold lines** or **reference lines**, and are used to locate the position of each view **(Fig. 8-15)**.

Views can be located also by drawing a 45° line that is used to project a third view once two base views are established. **Figure 8-16** shows the use of this method in projecting a **side view** from a front and top view, or a top view from a front and side view.

Another method of drawing a third view from two established views is the use of dividers to transfer distances from completed views. For example, depth dimensions from the top view can be transferred to the side view with dividers.

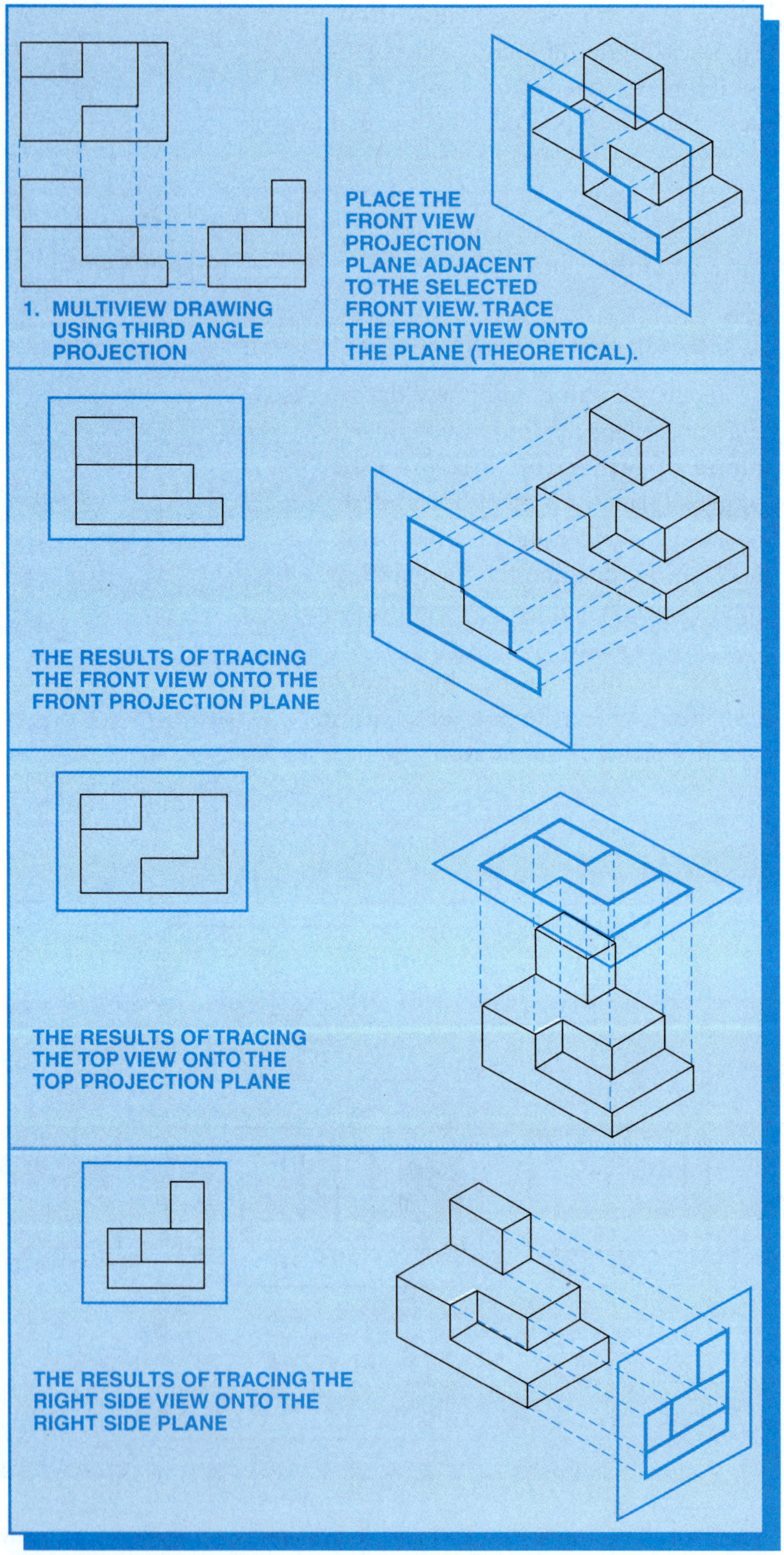

Figure 8-12
Planes of projection

Figure 8-13
Views are projected perpendicular to the six planes.

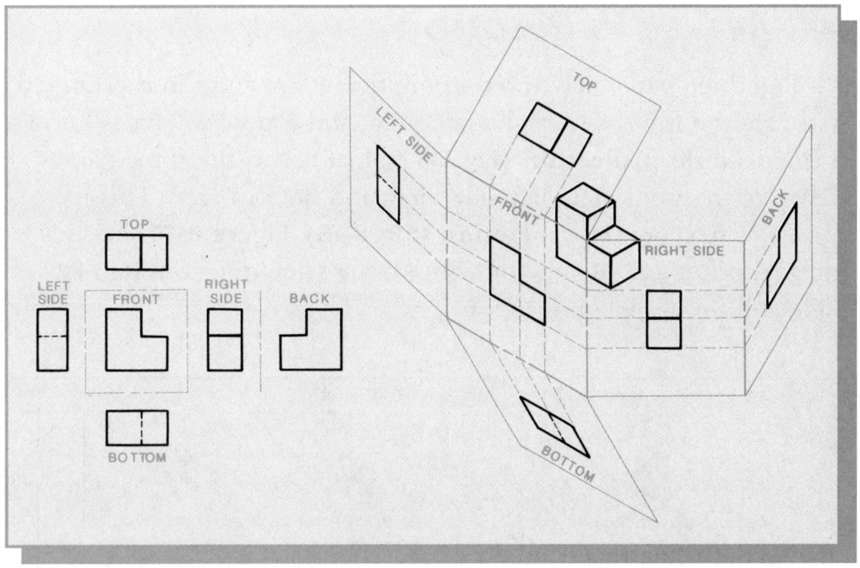

Figure 8-14
Opening the projection box on its fold lines (reference lines)

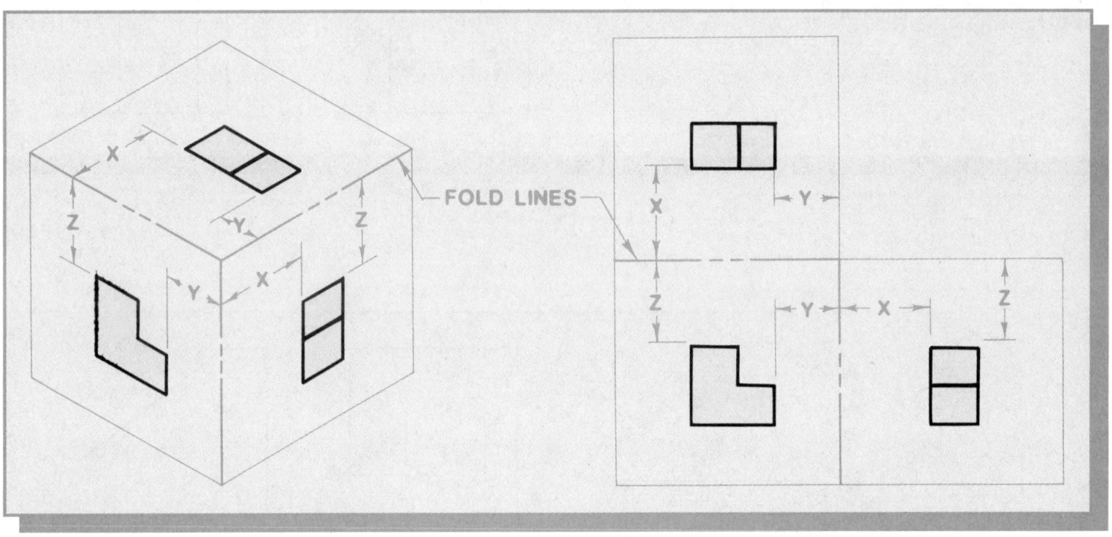

Figure 8-15
Locating views from the fold lines (reference lines)

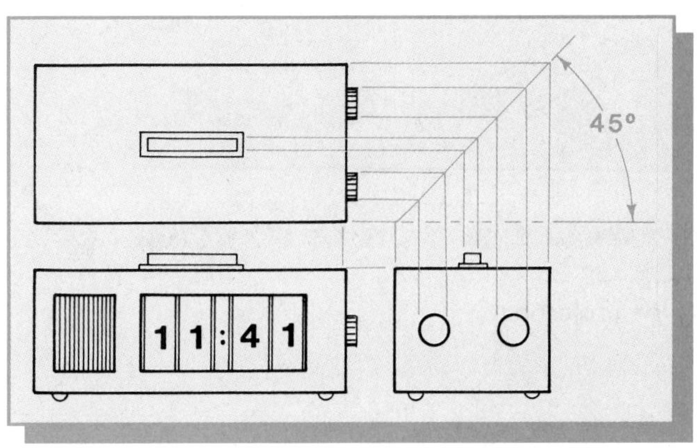

Figure 8-16
A third view can be projected from two established views.

Angles of Projection

The three basic views used in multiview drawings in the United States are the front view, right-side view, and top view. This is known as **third-angle projection (Fig. 8-17)**. In Europe, the three basic views are the front view, left-side view, and bottom view. This is known as **first-angle projection (Fig. 8-18)**. **Figure 8-19** shows a comparison of a multiview drawing of the same object using first-angle and third-angle projection.

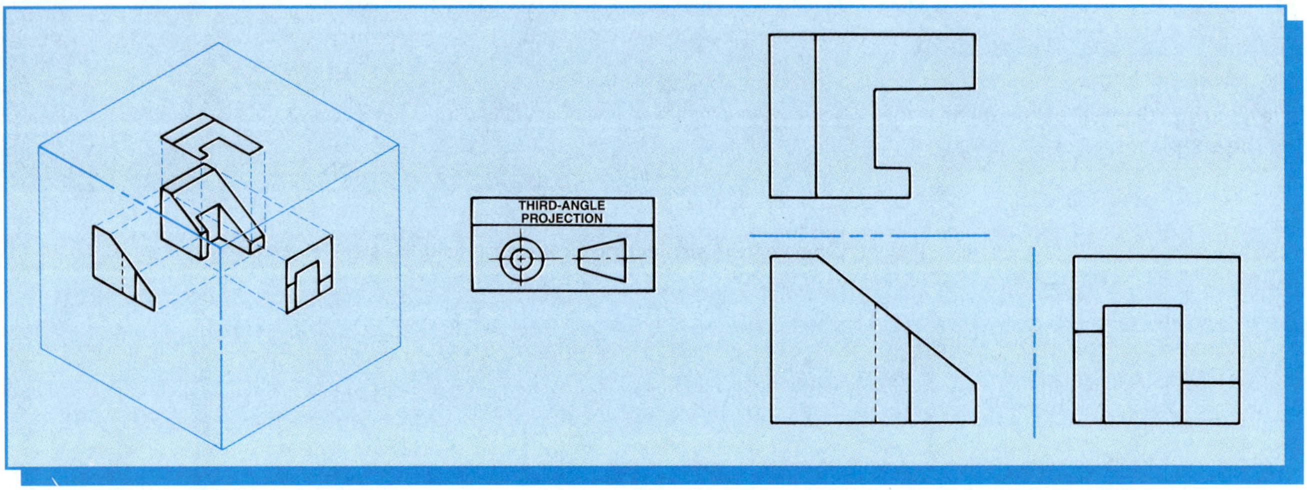

Figure 8-17
Third-angle projection

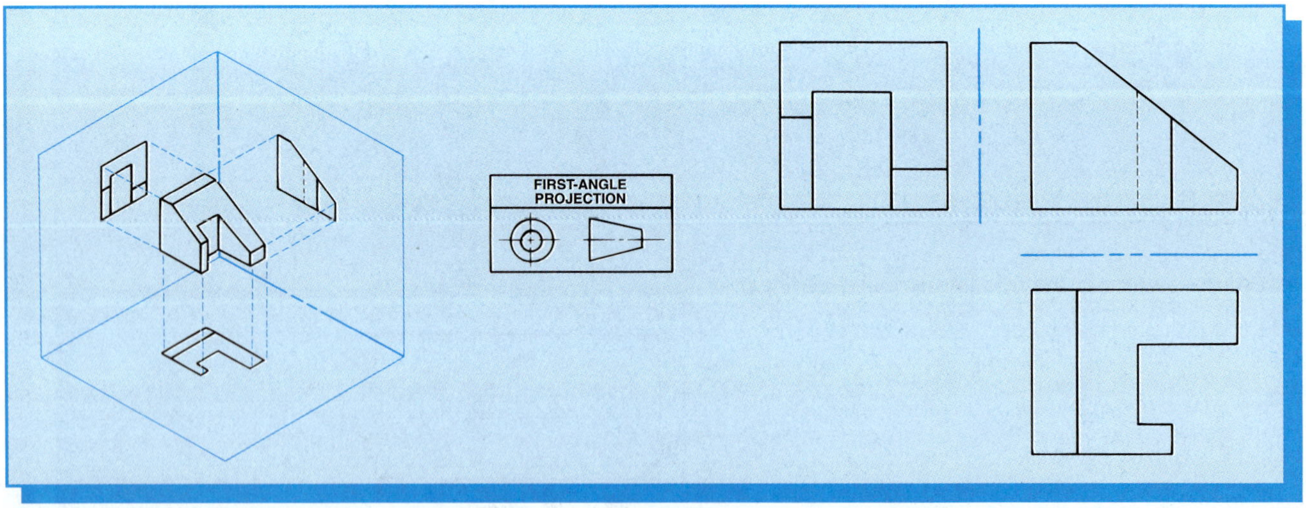

Figure 8-18
First-angle projection

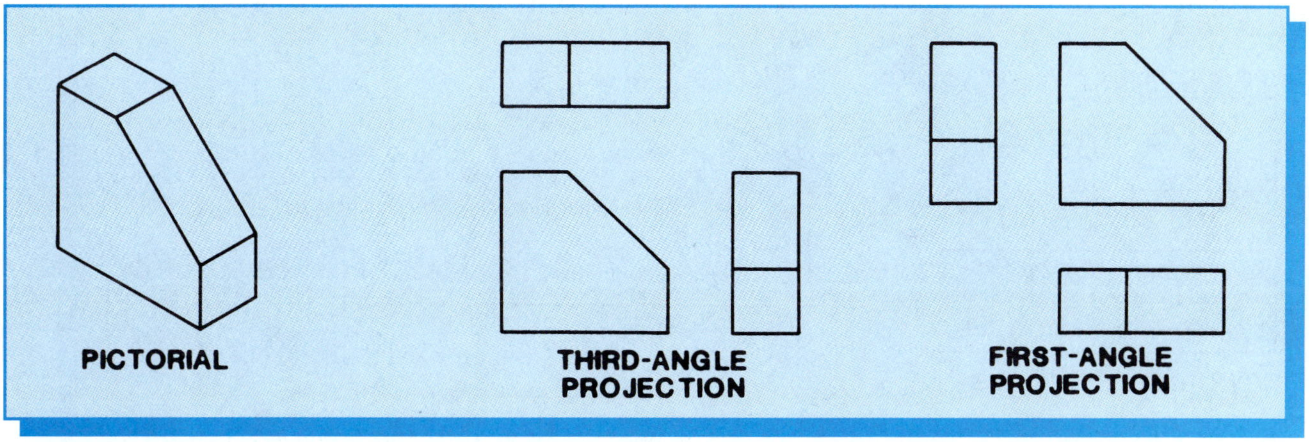

Figure 8-19
A comparison of first-angle and third-angle projection

Line Precedence

When objects are viewed through a plane of projection, some lines fall on an identical perpendicular plane, although one line may be further from the viewer than the other. When lines overlap in this manner, the darkest line takes precedence (called **line precedence**) on the drawing **(Fig. 8-20)** in the following order:

1. Cutting plane line
2. Object line
3. Hidden line
4. Centerline

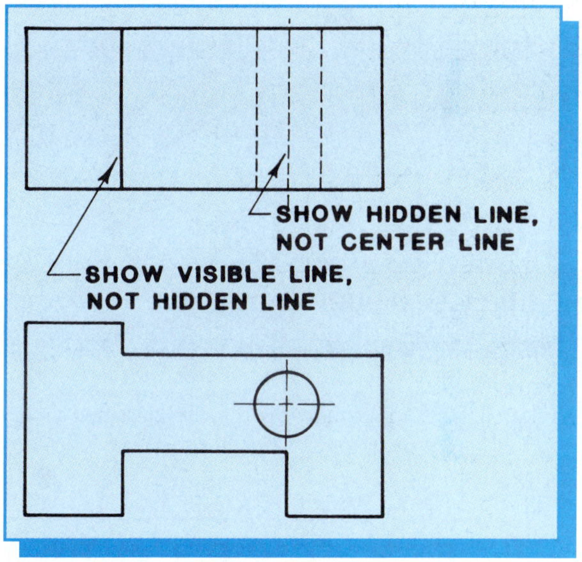

Figure 8-20
Line precedence for overlapping lines

Visualization

All flat surfaces in multiview drawings are either normal, inclined, or oblique **(Fig. 8-21)**. **Normal surfaces** are **true size surfaces**, and are parallel or per pendicular to the plane of projection. **Inclined surfaces** are angular; their planes are not parallel or perpendicular to the plane of projection. Inclined surfaces are slanted to normal surfaces. **Figure 8-22** shows the relationship between normal and inclined surfaces. **Oblique surfaces** are not parallel or perpendicular to any normal surface. **Figure 8-23** shows the relationship between normal and oblique surfaces. Numbers and letters are used on **Figures 8-22** and **8-23** to identify identical surfaces on different views.

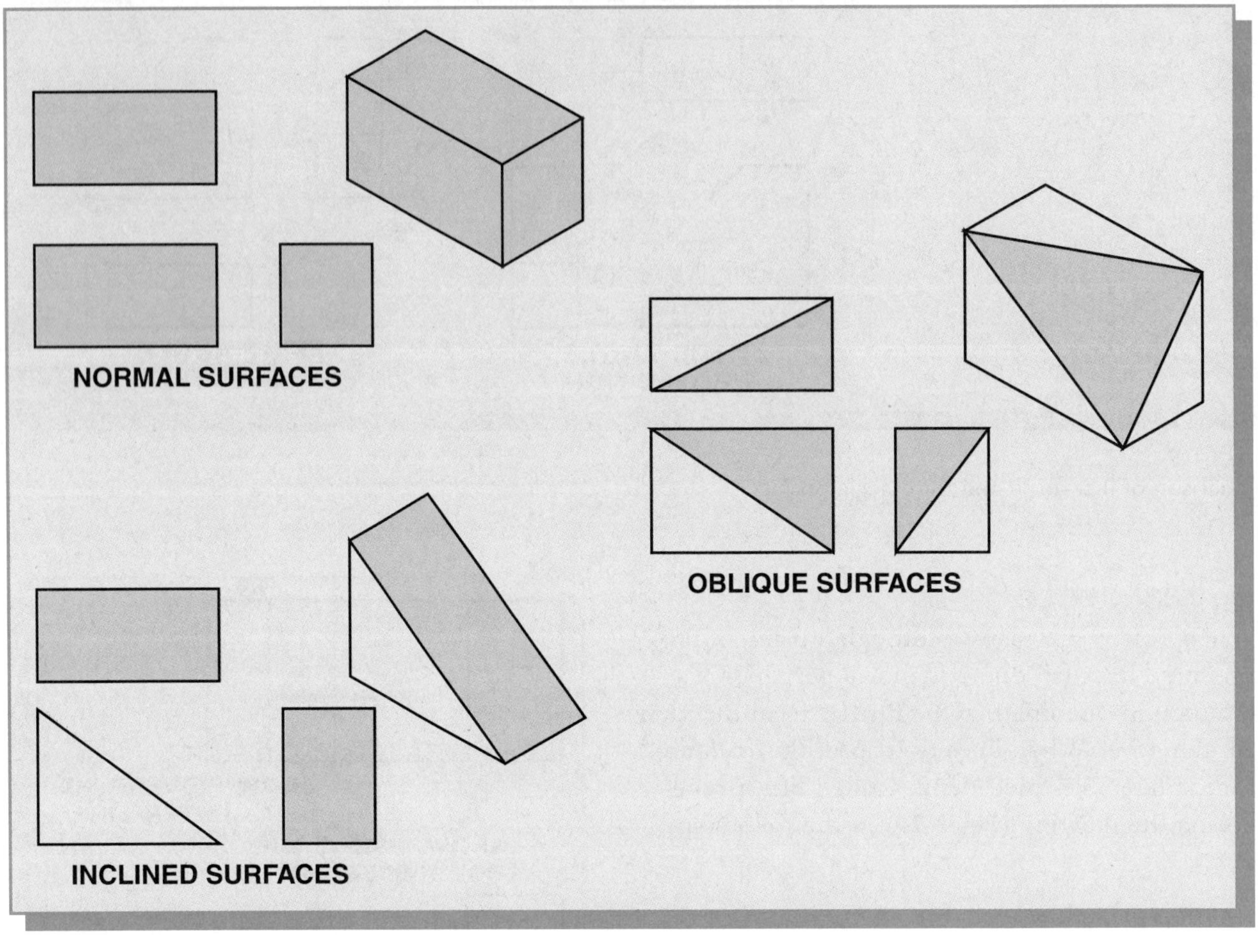

Figure 8-21
All flat surfaces in a multiview may be classified as normal, inclined, or oblique.

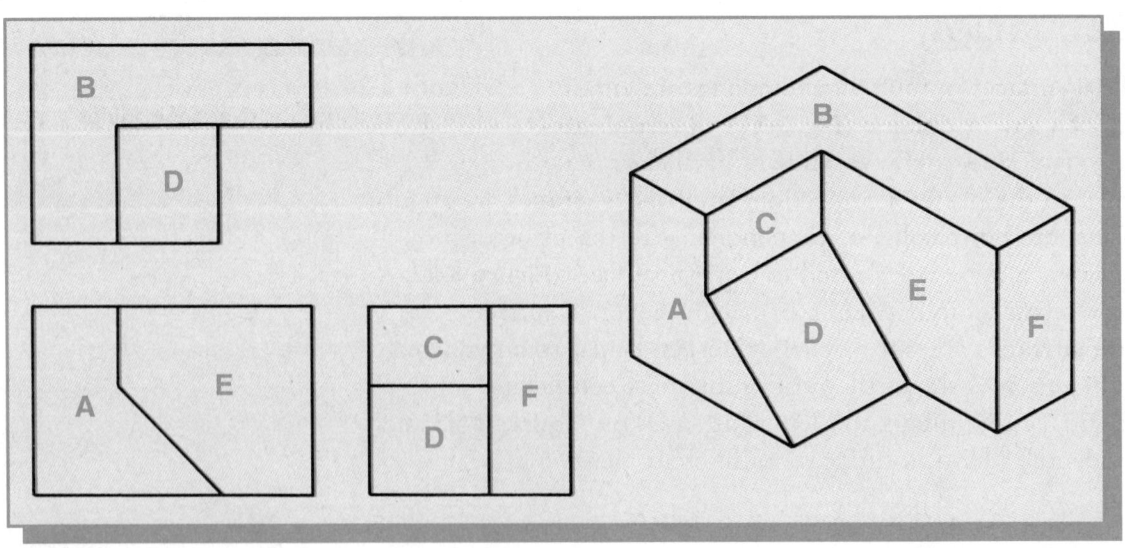

Figure 8-22
Identifying normal and inclined surfaces

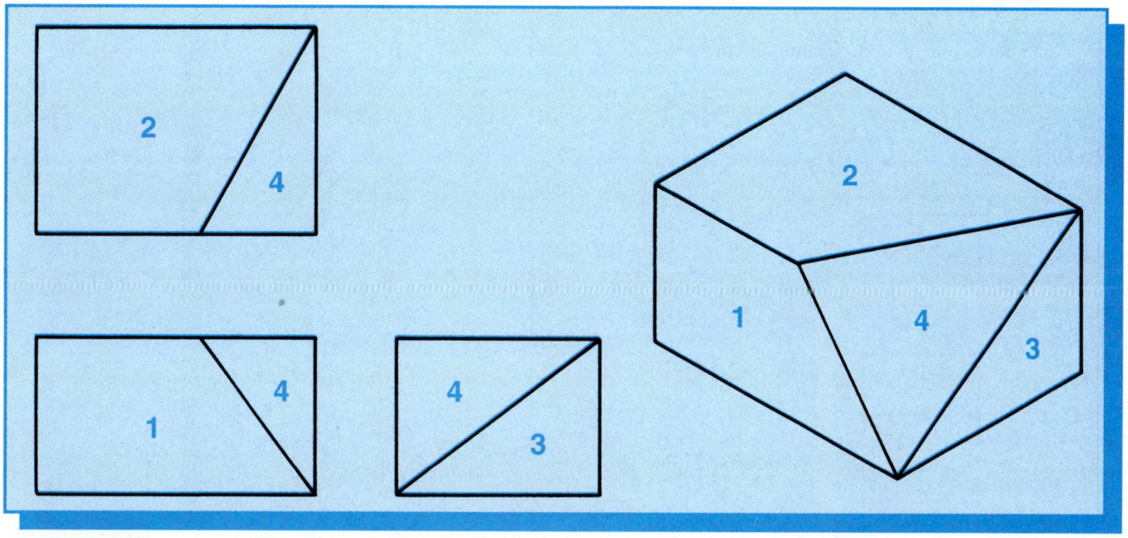

Figure 8-23
Identifying normal and oblique surfaces

Inclined and Oblique Surface Projections

Surfaces perpendicular or parallel to the plane of projection always appear as true size surface or edge on the orthographic plane. Surfaces that lie on any angle to the plane of projection appear foreshortened (not true size surfaces) on the orthographic plane. **Figure 8-24** shows a comparison of true size surfaces and foreshortened surfaces on an orthographic view.

Holes and cylinders appear as true circles on normal surfaces but appear as **ellipses** on inclined surfaces **(Fig. 8-25)**. Since inclined surfaces recede from the plane of projection at right angles, the ellipse's major diameter parallel to the plane of projection is true size. The ellipse dimension perpendicular to the plane of projection is foreshortened.

The amount of foreshortening is related to the angle of incline. Small angles produce narrow ellipses and large angles produce wide ellipses **(Fig. 8-26)**. **Figure 8-27** shows ellipse widths derived from common inclined angles.

The use of standard ellipse templates eliminates plotting ellipse layouts on inclined surfaces. The steps shown in **Figure 8-28** must be followed to insure ellipse accuracy.

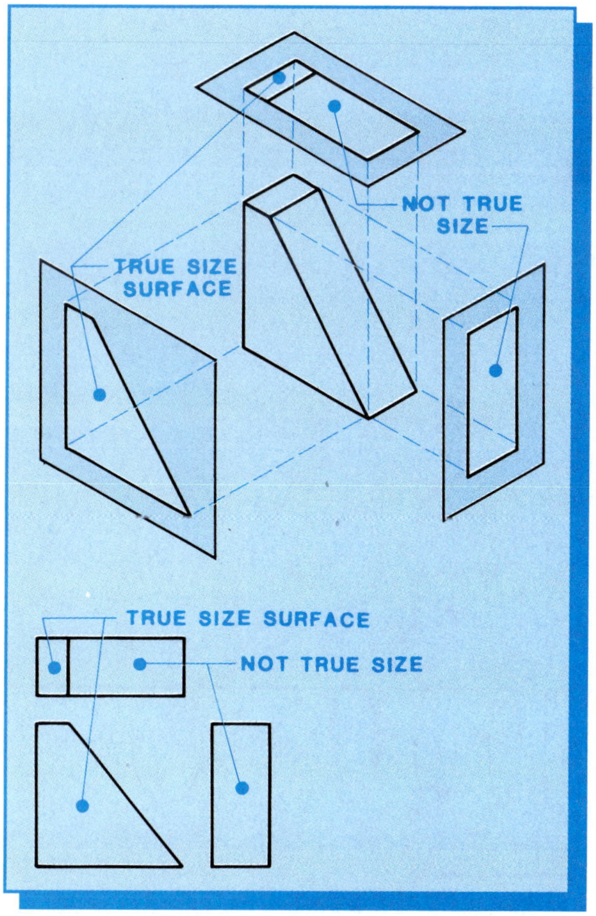

Figure 8-24
Projection of perpendicular and inclined surfaces

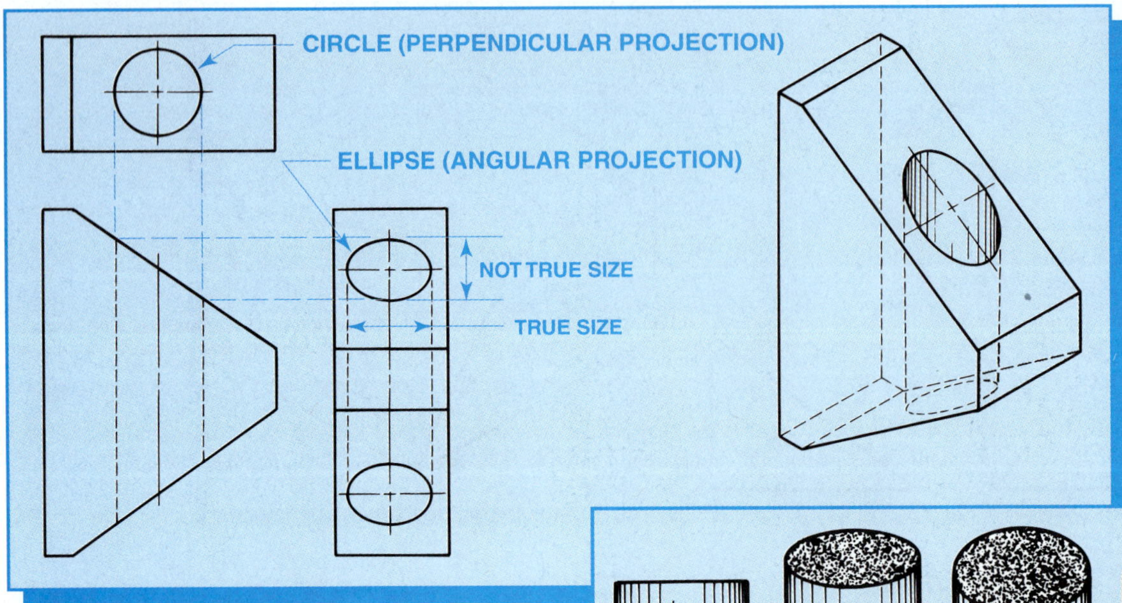

Figure 8-25
A hole will appear as an ellipse on an inclined plane.

Figure 8-26
Examples of ellipses

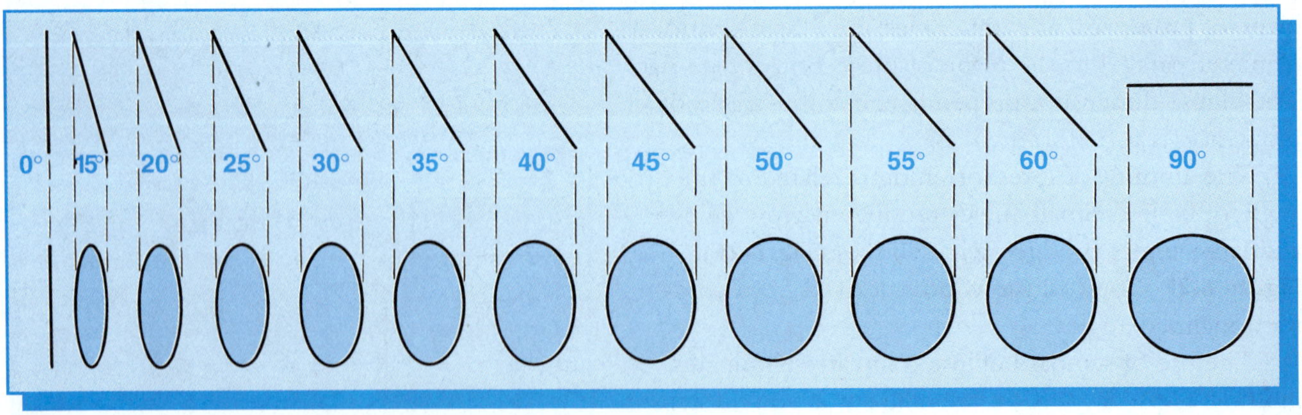

Figure 8-27
The angle of projection determines the minor diameter of the ellipse.

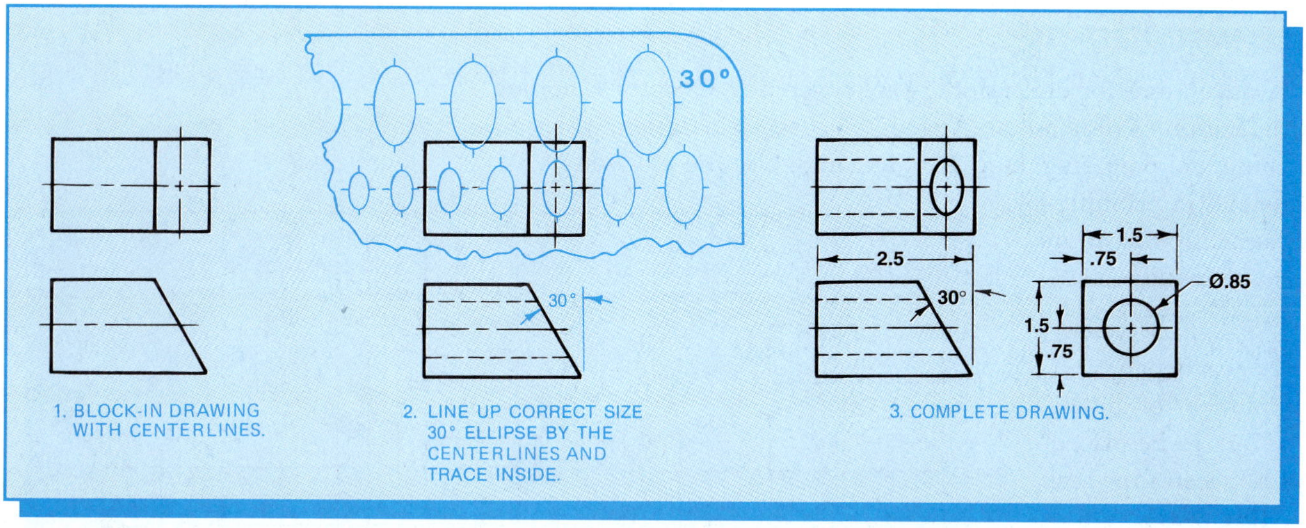

Figure 8-28
Drawing an ellipse with an ellipse template

Since oblique surfaces **(Fig. 8-29)** are not parallel to any normal plane, oblique surfaces, including circles, must be plotted from known true distances on other views.

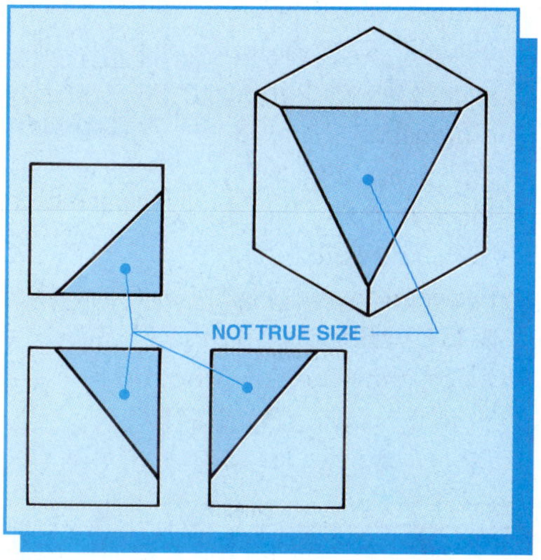

Figure 8-29
An oblique surface will not appear true size in any view.

Runouts

Surfaces with round or smooth contours do not intersect with straight lines. When surfaces blend together in this manner, the merging surface is called a **runout**. Square runouts and round runouts are shown in **Figure 8-30**.

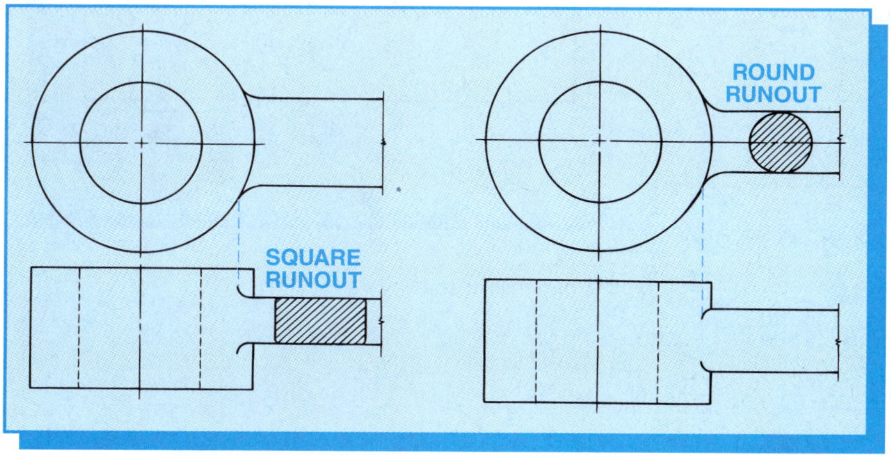

Figure 8-30
Examples of runouts on a multiview drawing

Dimensioning

Conventions for dimensioning are covered in Chapter 9. Some consideration of dimensioning needs is necessary, however, when planning and preparing multiview drawings. Use the following guidelines in planning for the dimensioning of multiview drawings:

1. Spacing between views depends on the number of dimensions to be placed between the views **(Fig. 8-31)**.
2. Place the closest dimension to the view at .375" clearance. Every additional dimension is spaced .25" apart **(Fig. 8-32)**.
3. Place as many dimensions as possible between views **(Fig. 8-33)**, rather than on the outside.
4. Place dimensions on the most descriptive view of that feature **(Fig. 8-34)**.

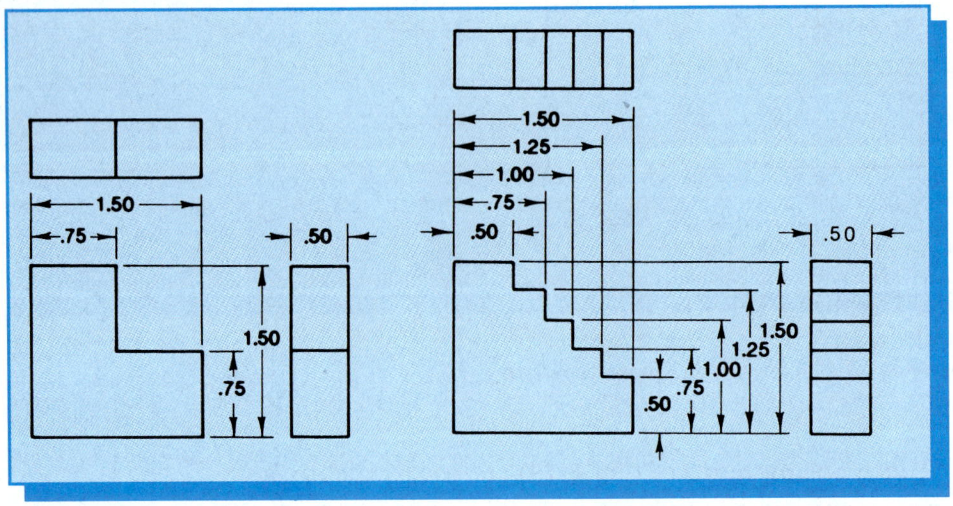

Figure 8-31
The numbers of dimensions between views will determine the space between the views.

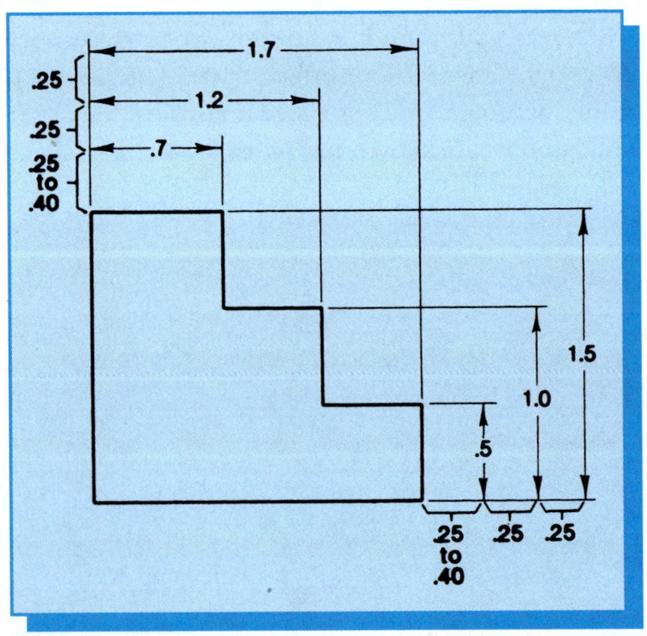

Figure 8-32
Typical dimension spacing

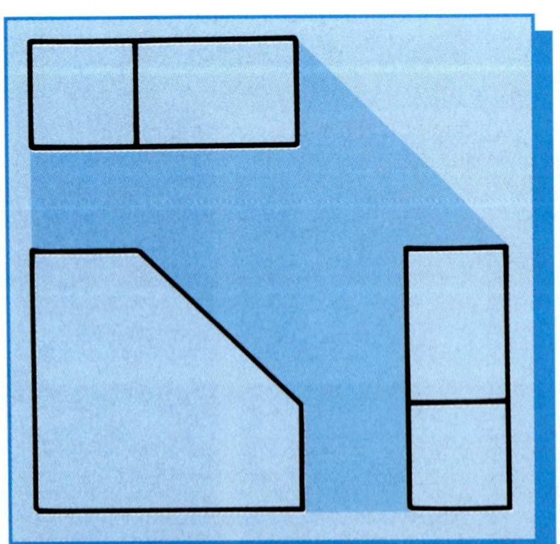

Figure 8-33
Place as many dimensions as possible between the views.

5. A working drawing must be completely dimensioned with enough information to manufacture the item without guesswork or fabrication redesign.
6. Do not duplicate the same dimensions on a different view.
7. The overall dimensions must always be given.
8. One dimension in a full chain of dimensions must be omitted.
9. Holes are always dimensioned on the circular view.
10. Shafts are dimensioned on their side view.

Most companies position the first drawing on a sheet in a corner to allow space for additional details. Nevertheless, a drafter must know how to center a drawing in a given space to be certain the finished drawing will fit into the available space. To center a drawing, follow the steps in **Figure 8-35**.

Drawing Procedures

When drawing an object, first determine the number of views needed, and then decide which surface will become the front view. Next, establish the space needed between views for dimensioning and calculate the center of the drawing. Once these preliminary decisions have been made, follow the steps in **Figure 8-36**.

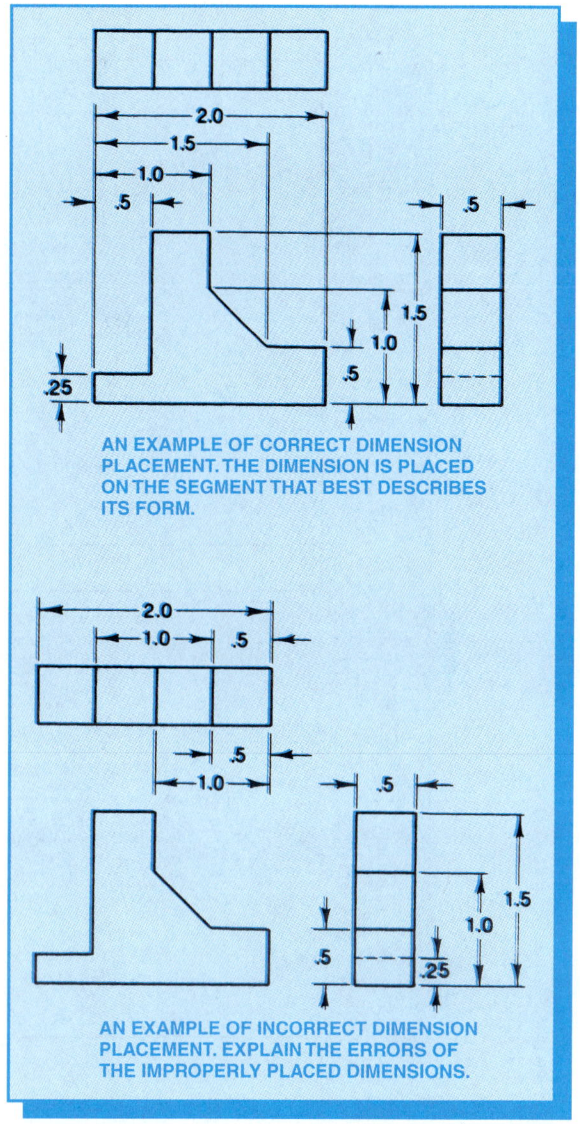

Figure 8-34
Comparison of correct and incorrect dimension placements on a multiview drawing

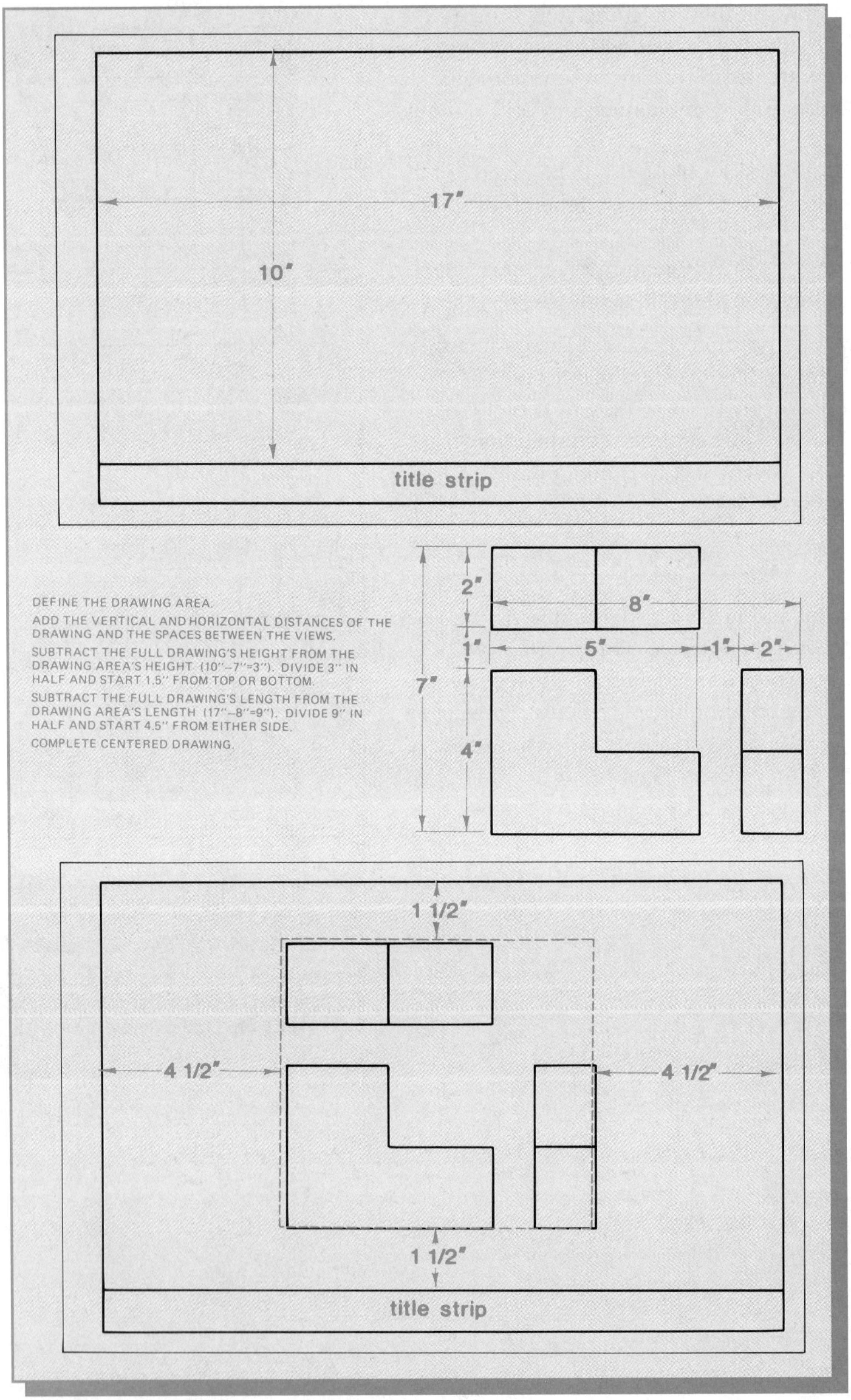

Figure 8-35
Centering a multiview drawing

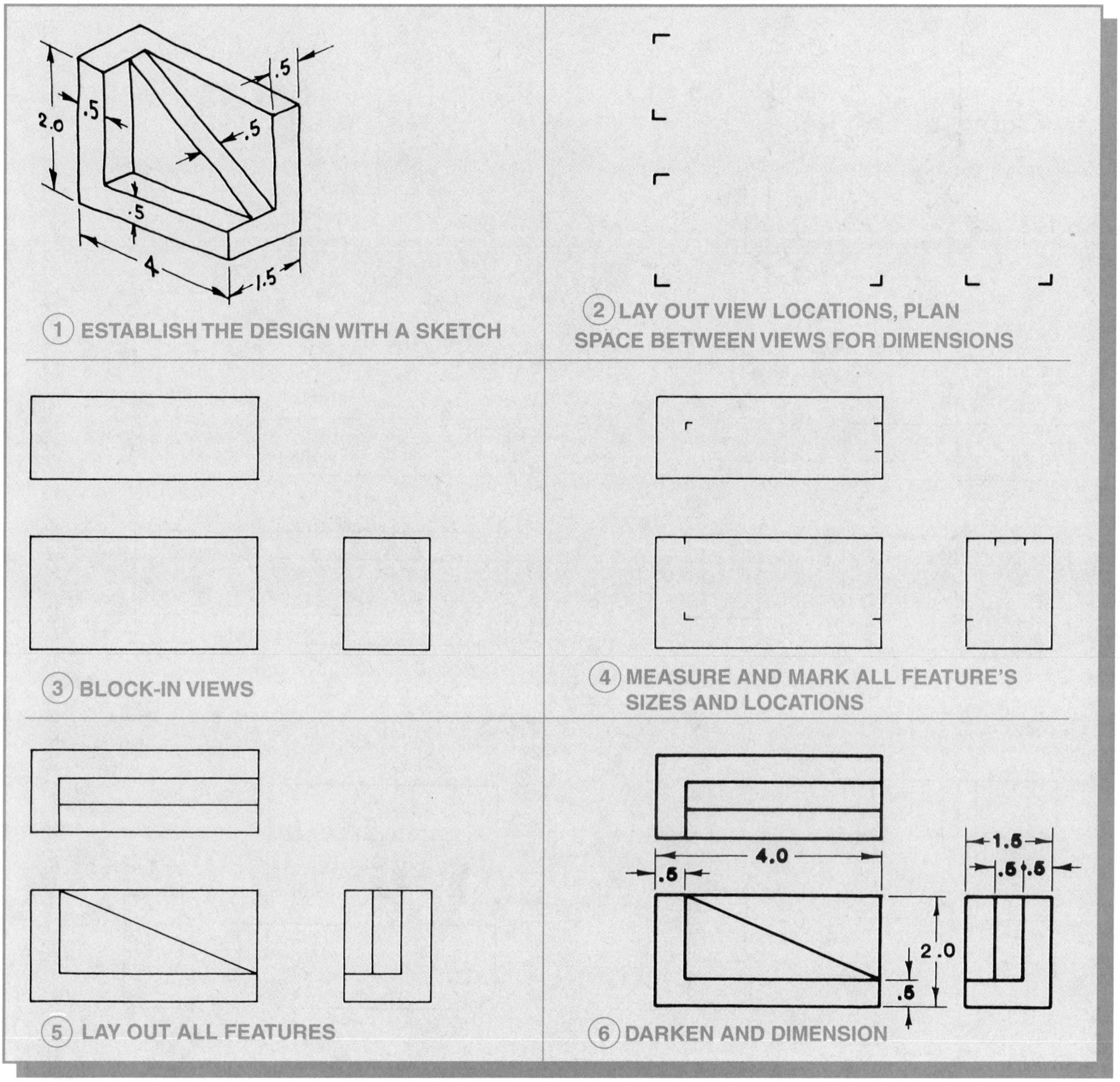

Figure 8-36
Steps to draw a multiview drawing with an inclined surface

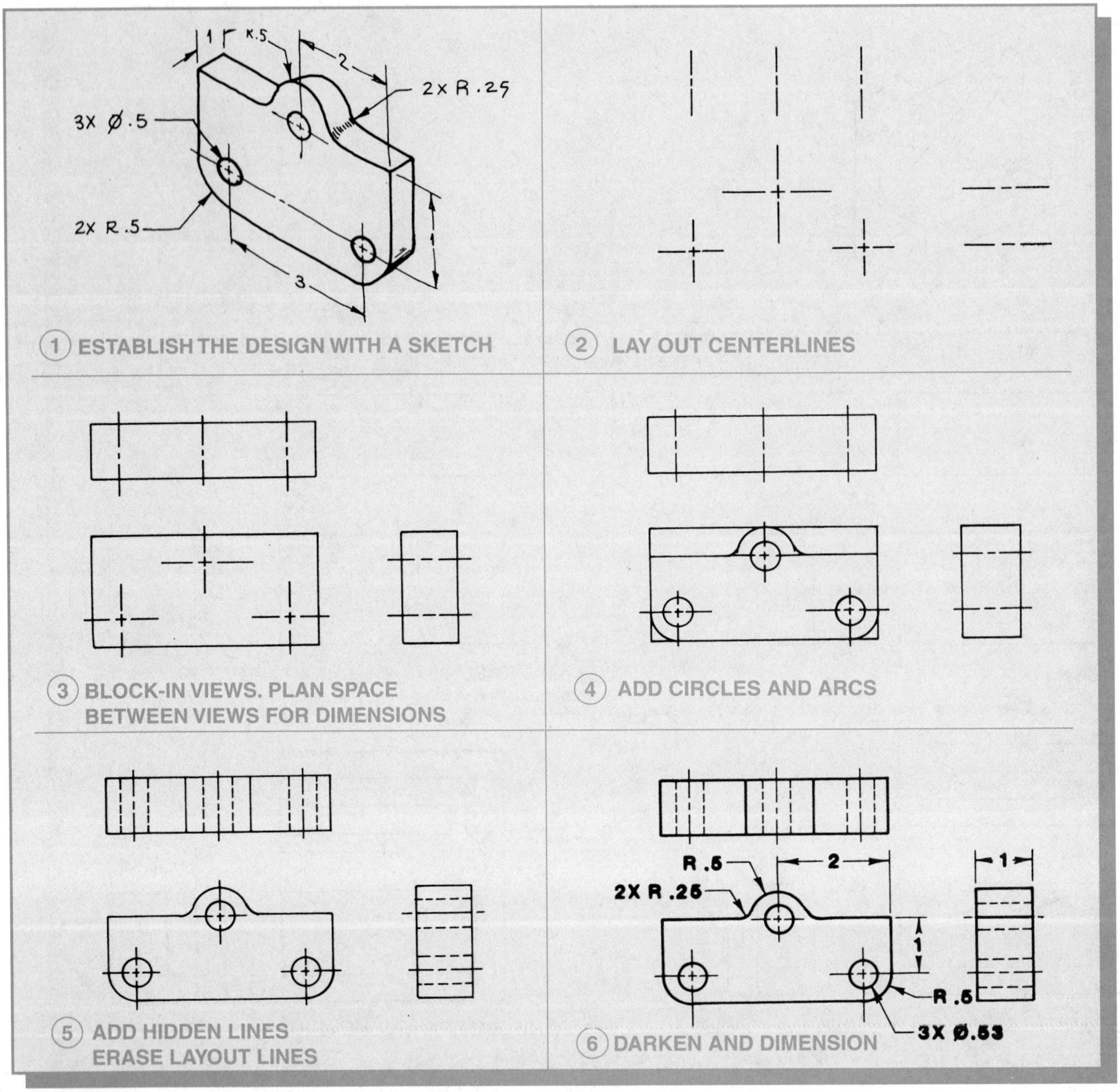

Figure 8-37
Steps to draw a multiview drawing with circles and arcs

Follow the steps in **Figure 8-37** to draw arcs and circles in a multiview drawing. Refer to **Figure 8-38** to prepare multiview drawings that contain oblique surfaces.

One of the great advantages of AutoCAD is the ability to combine drawings. In this chapter, you are going to use two drawings that were created in previous chapters. You will start the drawing with the prototype drawing created in Chapter 6 and insert the front view stated in Chapter 5.

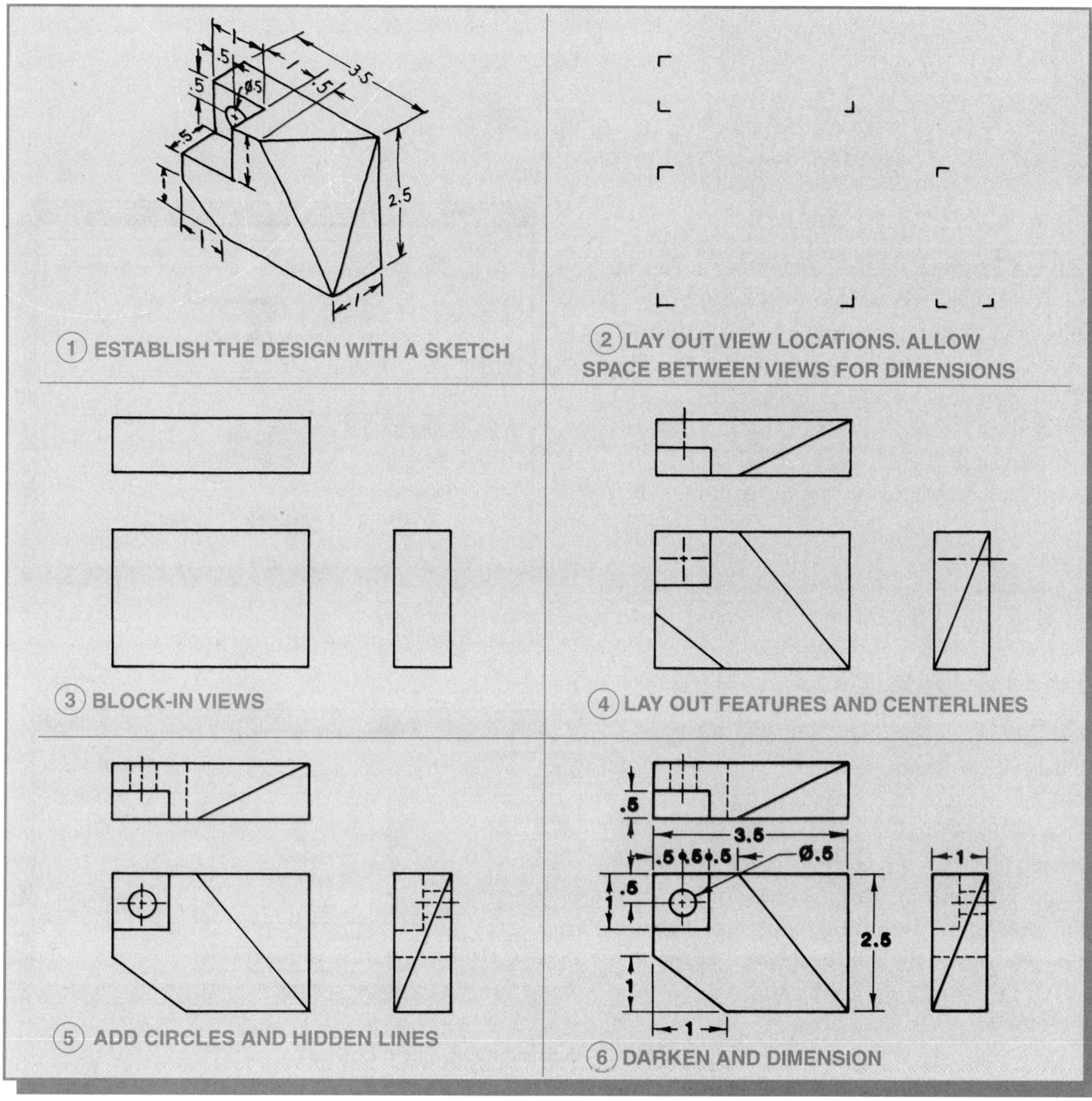

Figure 8-38
Steps to draw a multiview drawing with an oblique surface

Multiview drawings are combinations of lines, circles, and arcs to create different views of the same object. The use of grid marks, orthographic projection lines, and different layers for object, hidden, center, and construction lines simplifies and increases the speed of preparing multiview drawings.

The following is one method of preparing a multiview drawing with a CAD system.

AutoCAD LT Exercises

Using a Prototype Drawing

1. Select New from the File pull-down menu in AutoCAD Drawing Editor **(Fig. 8-39)**.
2. Select the Prototype button and use the scroll button to find the PRO6.dwg **(Fig. 8-40)**. The PRO6.dwg was created in Chapter 6.
3. Select the PRO6.dwg using the (PB) and select OK using the (PB).
4. Turn Snap and Grid on. Set Snap and Grid to .250.
5. After selecting OK, your screen should like **Figure 8-41**. Check to see if all the layers are in the drawing.

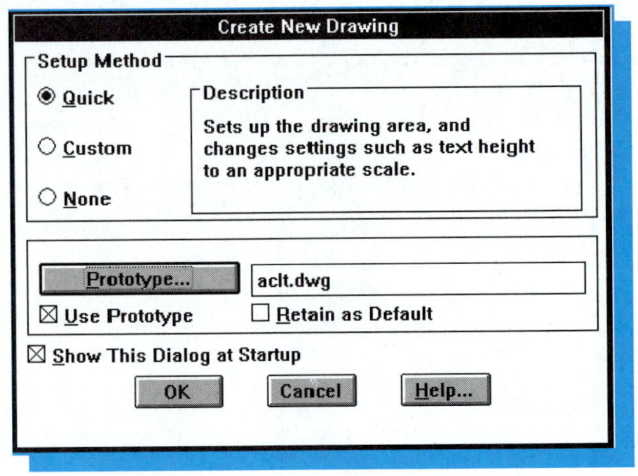

Figure 8-39
New Drawing dialog box

Inserting a Drawing

1. Select Draw and Insert Block **(Fig. 8-42)**.
2. Select File—use the scroll arrow to find CHP5-1.dwg, which was created in Chapter 5.
3. Insert at 0,0 default all the choices (default means hit the Enter key).
4. The drawing screen should look like **Figure 8-43**. Move the toolbox to the side of the screen using the Toolbox Control button in the toolbar.

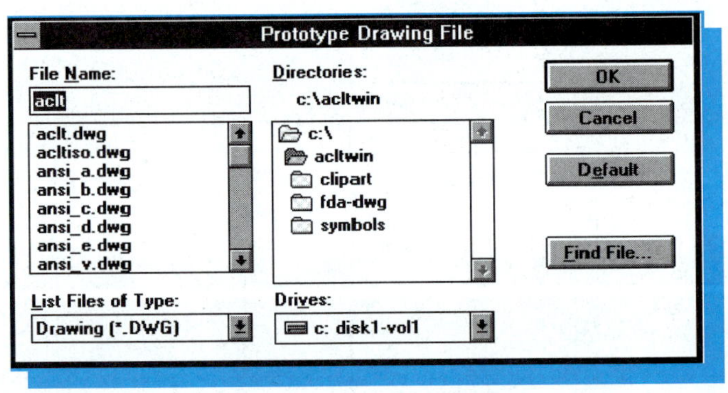

Figure 8-40
Prototype Drawing dialog box

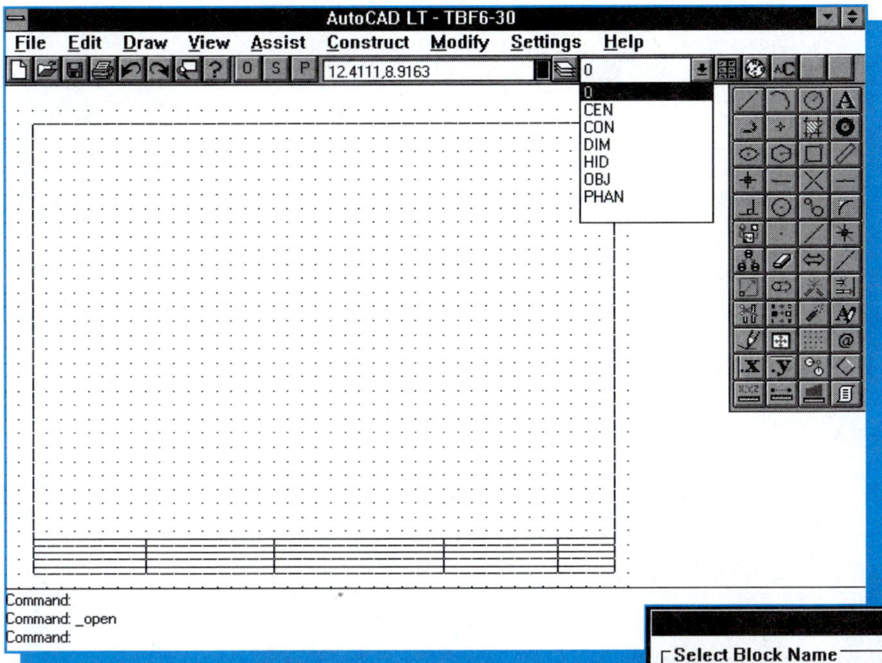

Figure 8-41
New Drawing with proto.dwg

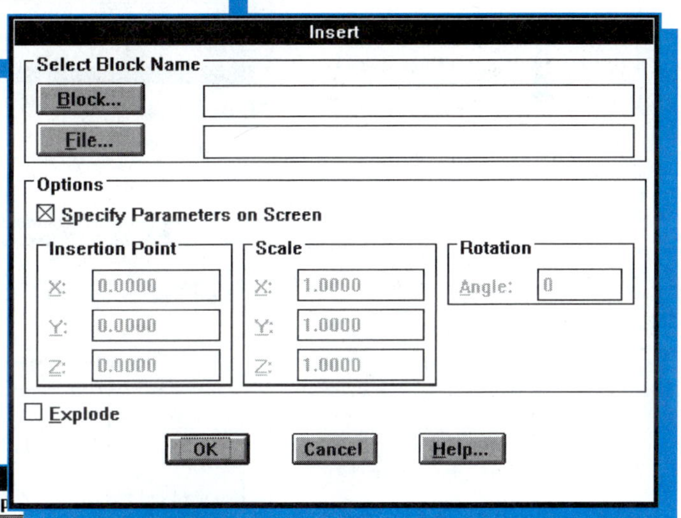

Figure 8-42
Insert dialog box

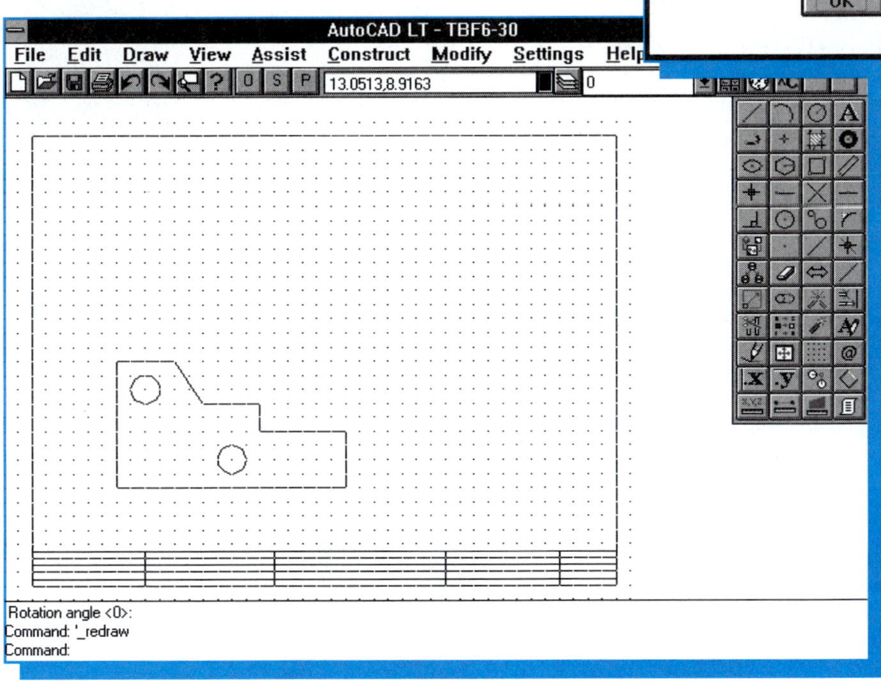

Figure 8-43
Drawing inserted into another

Drawing the Top View

1. Set the layer to OBJ layer.
2. Use Snap and the cursor to draw a horizon line 1/2" down from the top border. Make sure the line lines up with the front view **(Fig. 8-44)**.
3. Draw lines from each point on the front view where a line changes direction **(Fig. 8-45)**. Use the Perpendicular Osnap to end the line at the top line.
4. Use the Offset command under Construct to create the front line on the top view.
   a. Offset distance is 2.00" (the depth of the part).
   b. Select the back line of the top view.
   c. Pick any point on the side toward the front view **(Fig. 8-45)**.
5. Use the Trim command under Modify to get rid of the construction lines **(Fig. 8-46)**.
   a. Select the front line in the top view as the entity to trim to.
   b. Hit the Enter key. Select the construction lines between the top and front view to trim.
6. Change layer to the HID layer to get hidden lines.
7. Toggle the Snap button by selecting it with the Pick button.

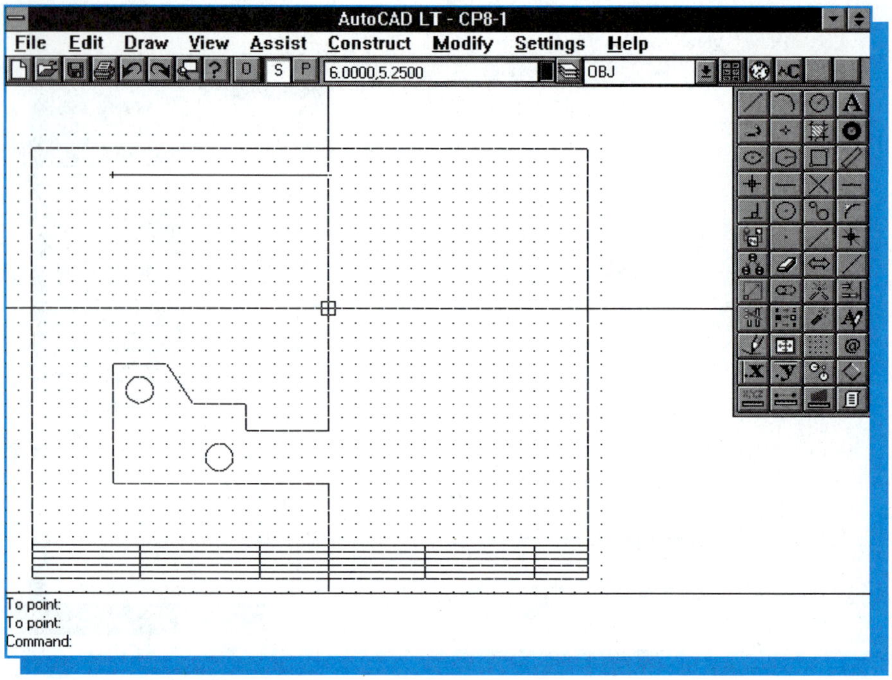

Figure 8-44
Back line top view

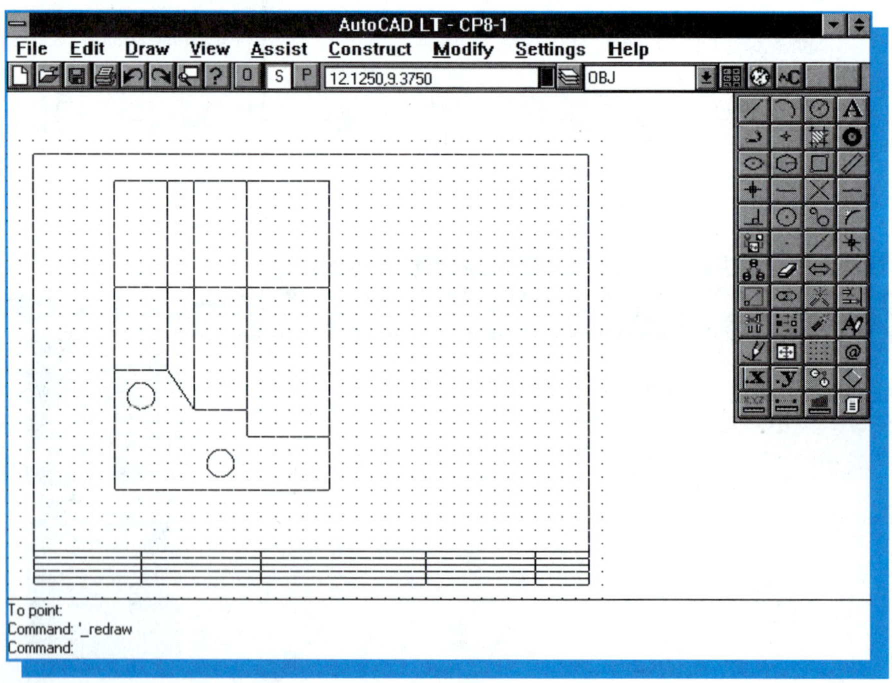

Figure 8-45
Construction lines

8. Draw lines tangent to the holes in the front view and perpendicular to the back line in the top view.
9. Use the Trim command to trim the hidden construction lines **(Fig. 8-47)**.
10. Change layer to the CEN layer to get center lines.
11. Turn the Snap back on.
12. Draw the center lines at $^1/_4$" from the edge of the part **(Fig. 8-48)**.

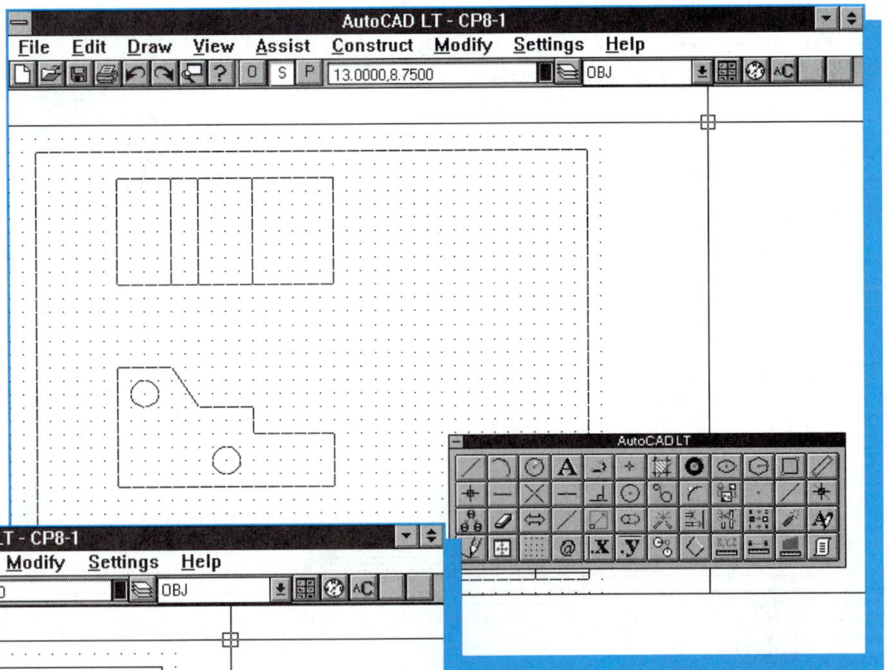

Figure 8-46
Drawing with construction lines trimmed

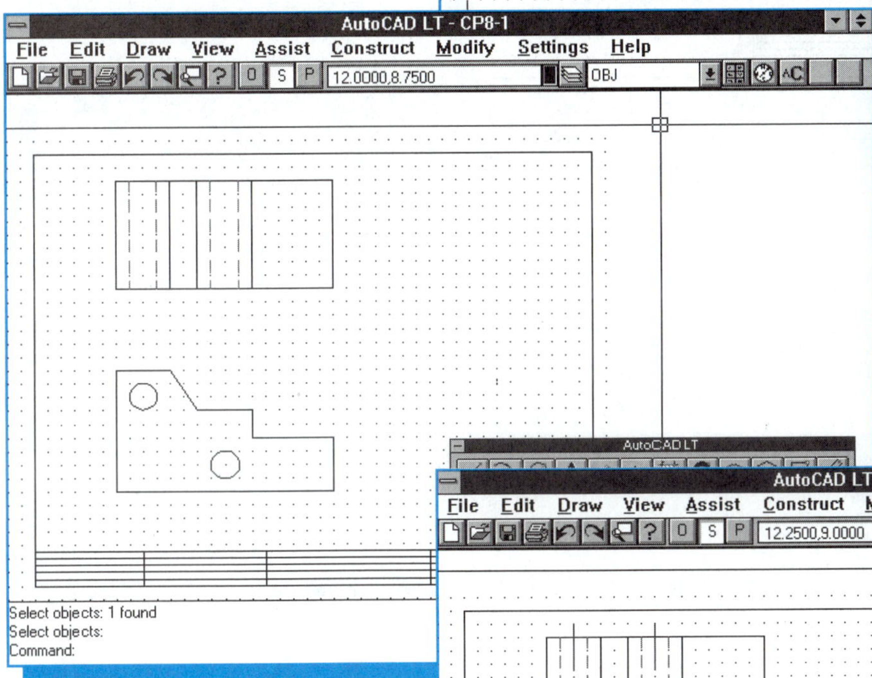

Figure 8-47
Hidden lines added and trimmed

13. Use the same techniques to draw the right-side view **(Fig. 8-49)**.

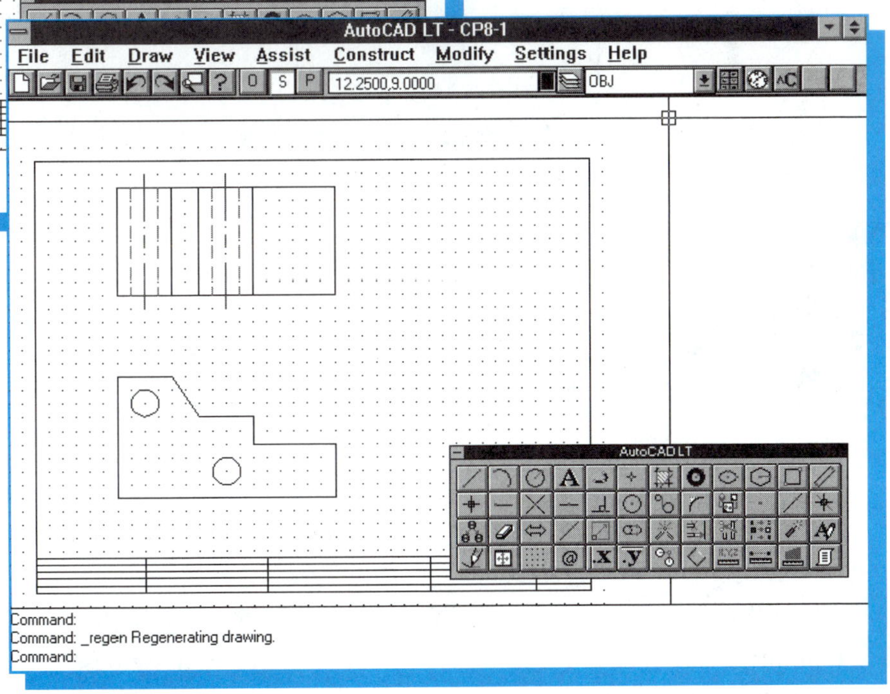

Figure 8-48
Center lines added

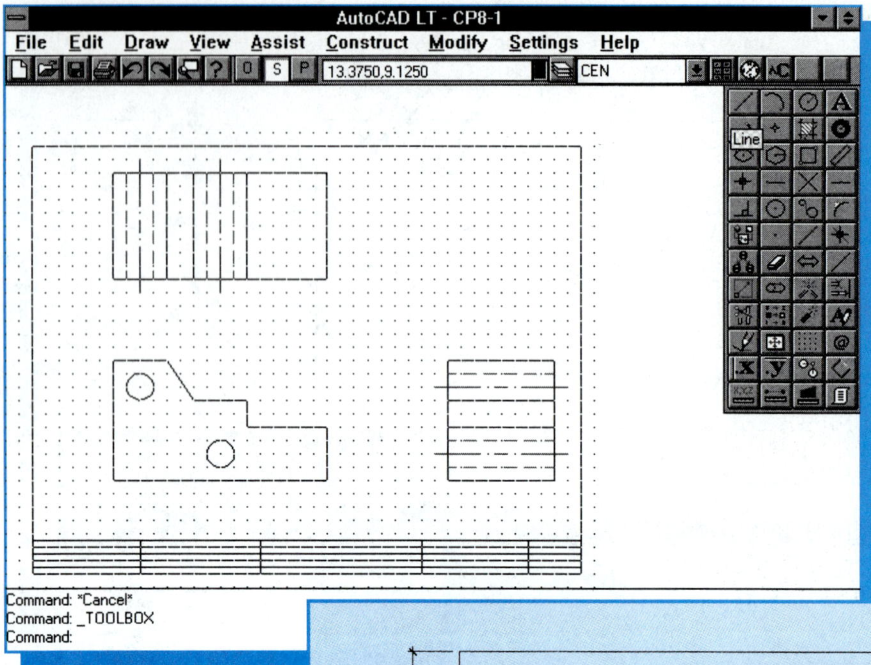

Figure 8-49
Adding the right-side view.

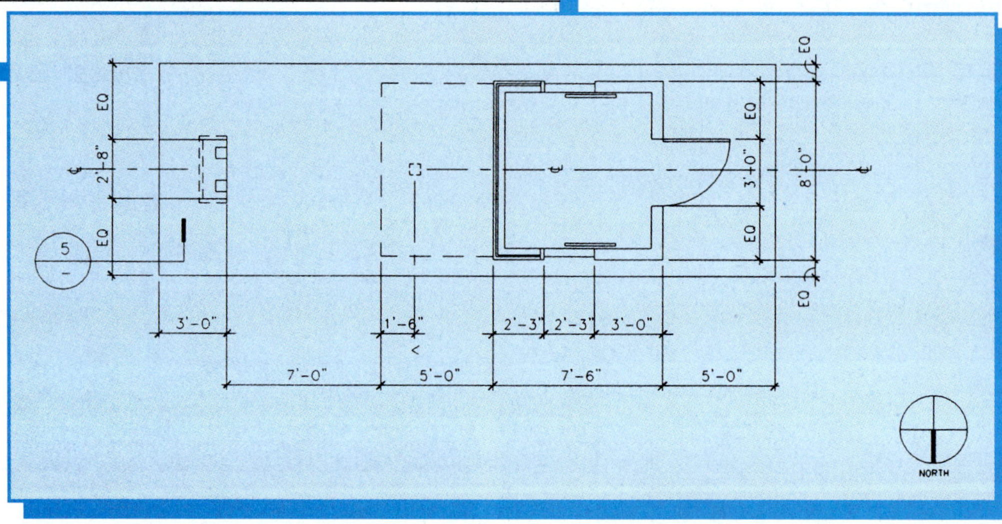

Example of a CAD generated drawing.

Exercises

1. Complete the non-drawing exercises in **Figures 8-50a** through **8-50e**. Follow the instructions for each exercise.

2. Complete the freehand sketching exercises in **Figures 8-51** through **8-54**.

All of the following in this chapter may be done by sketching, instrument drawing, and CAD.

3. Complete the two or three view multiview drawings in **Figures 8-55** through **8-64**.

4. Complete the two- or three-multiview drawings from the pictorial exercises in **Figures 8-65** through **8-94**.

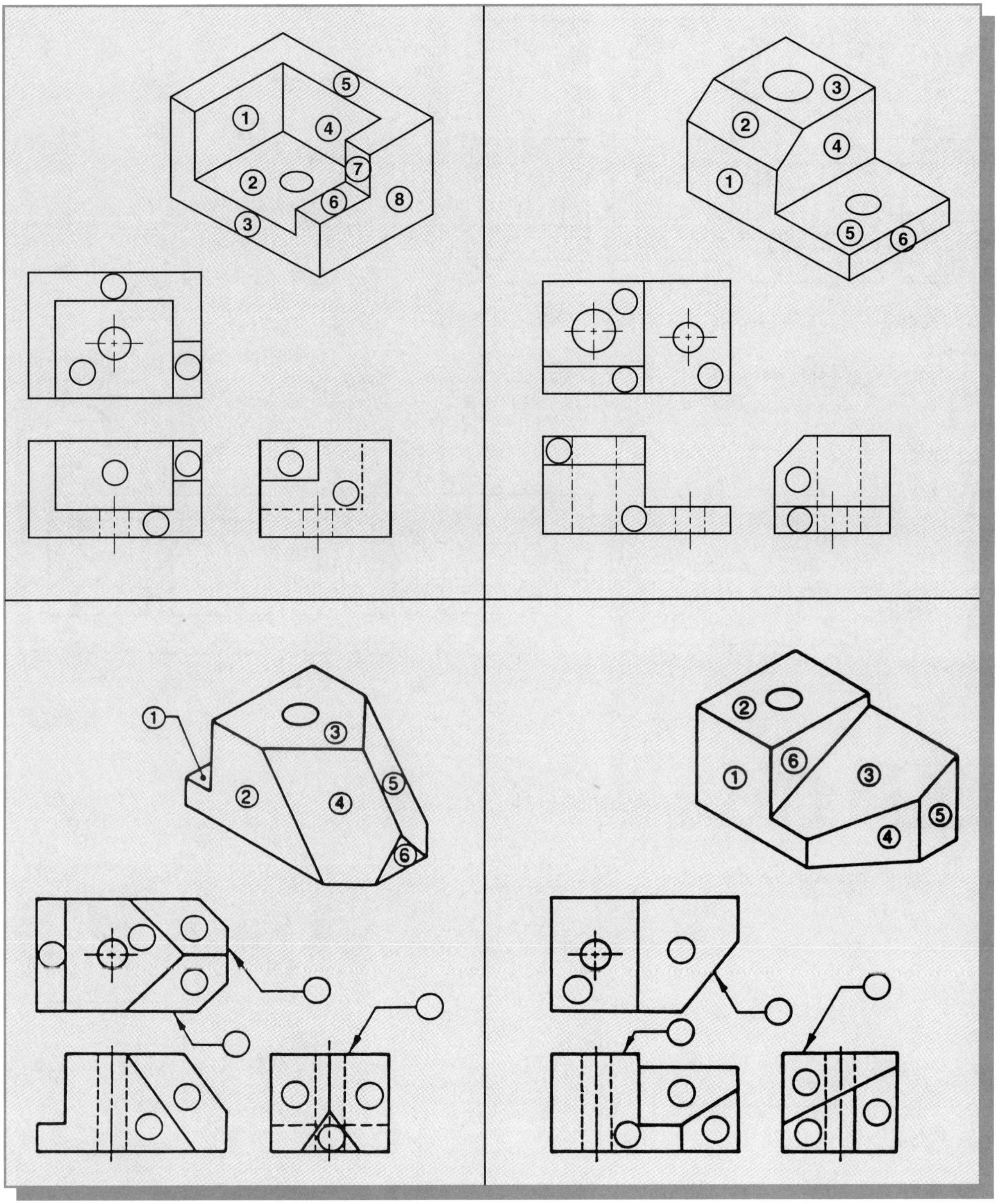

Figure 8-50a
Identify each surface with the correct number.

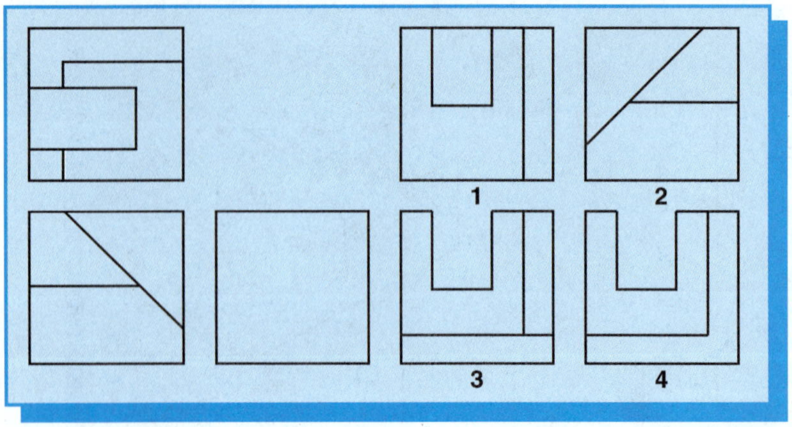

Figure 8-50b
Identify the correct right-side view(s)

Figure 8-50c
Identify the correct right-side view(s)

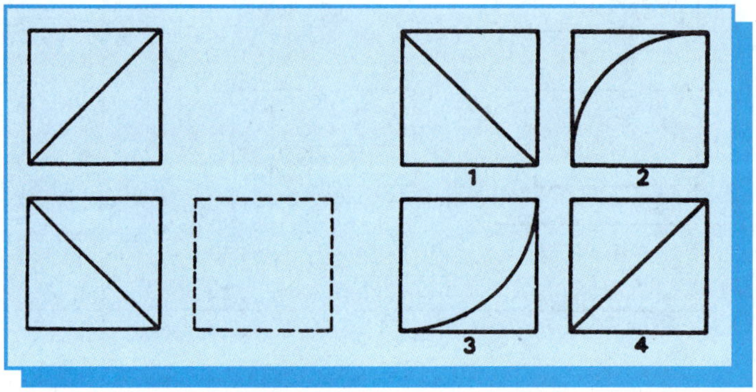

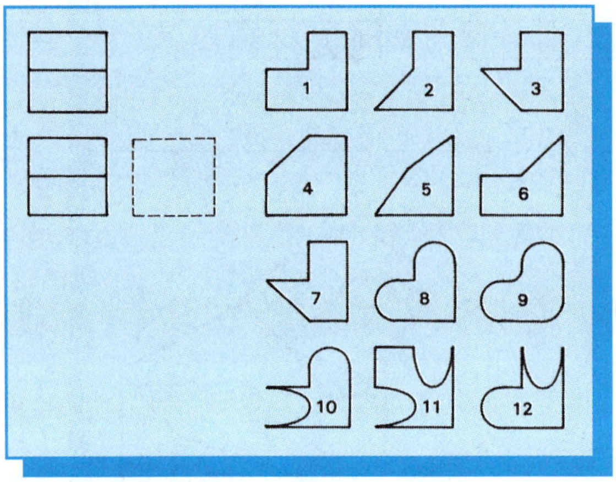

Figure 8-50d
Identify the correct right-side view(s)

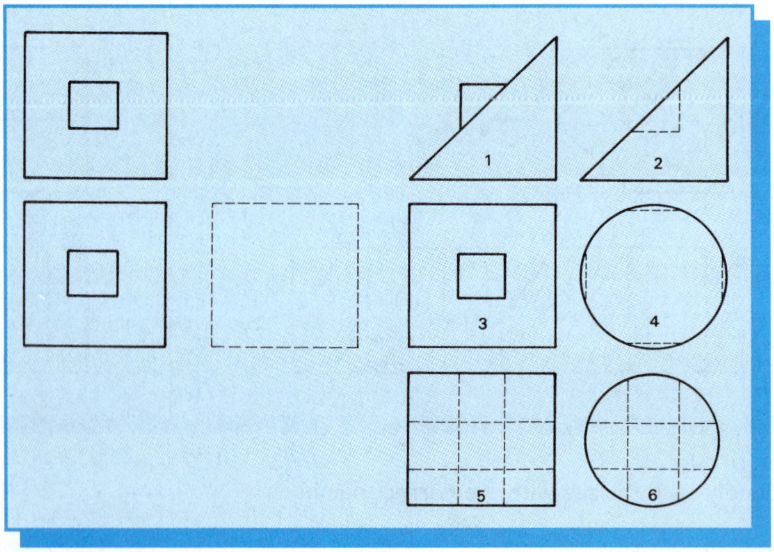

Figure 8-50e
Identify the correct right-side view(s)

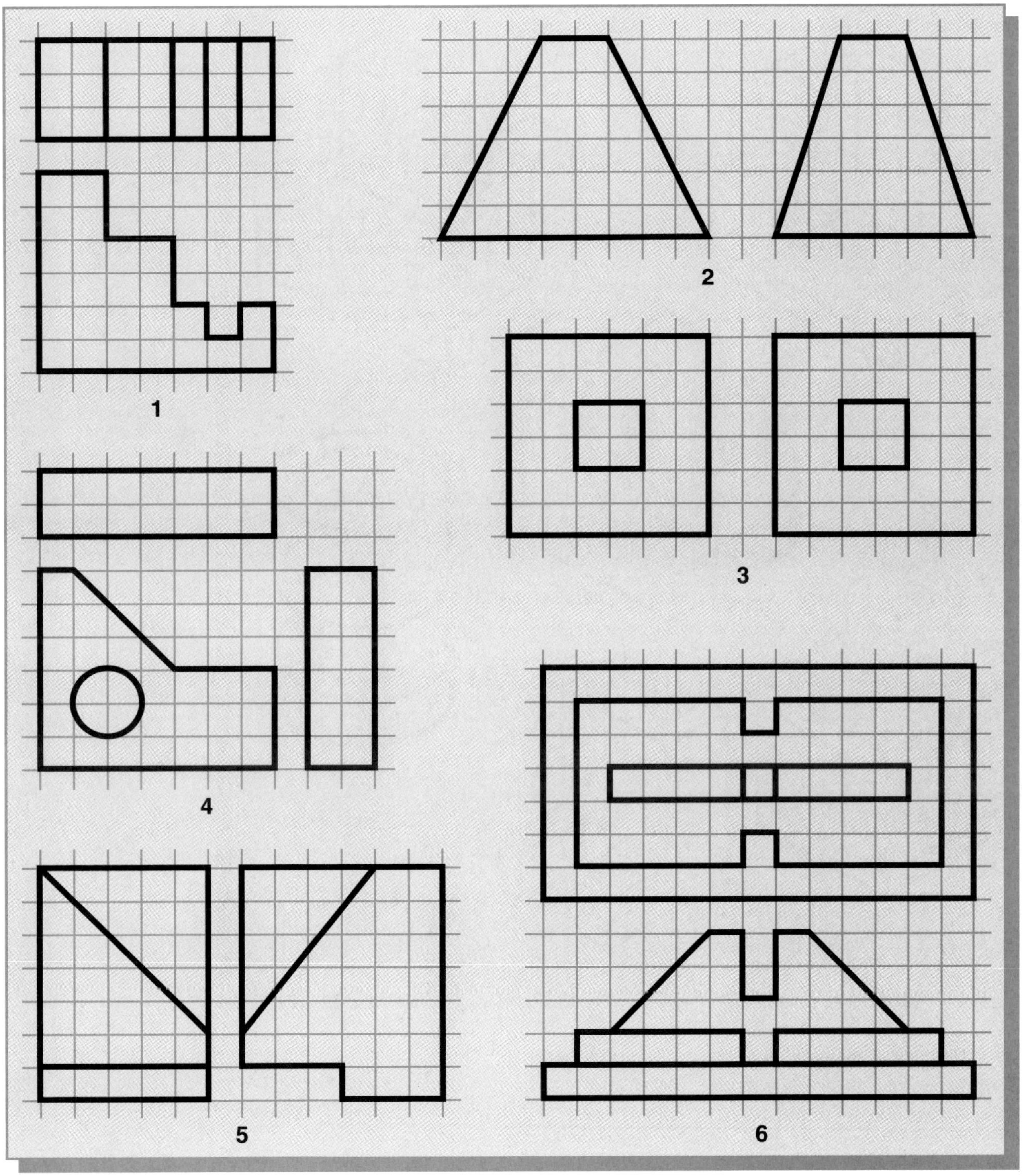

Figure 8-51
On 1/4" grid paper, sketch the multiview drawings and complete the third view. When completed, sketch an isometric view.

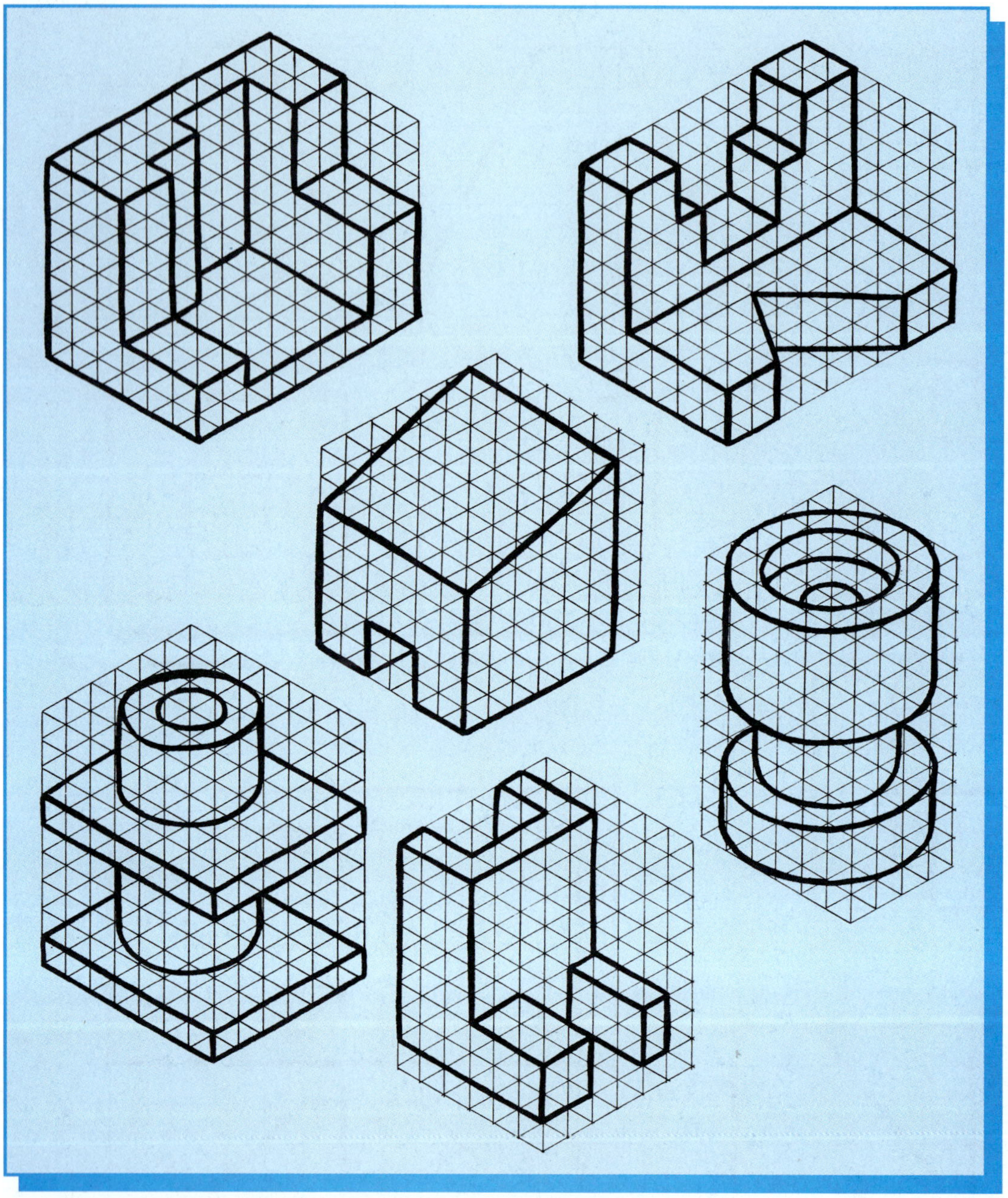

Figure 8-52
Sketch the three-view multiview drawings on 1/4" grid paper

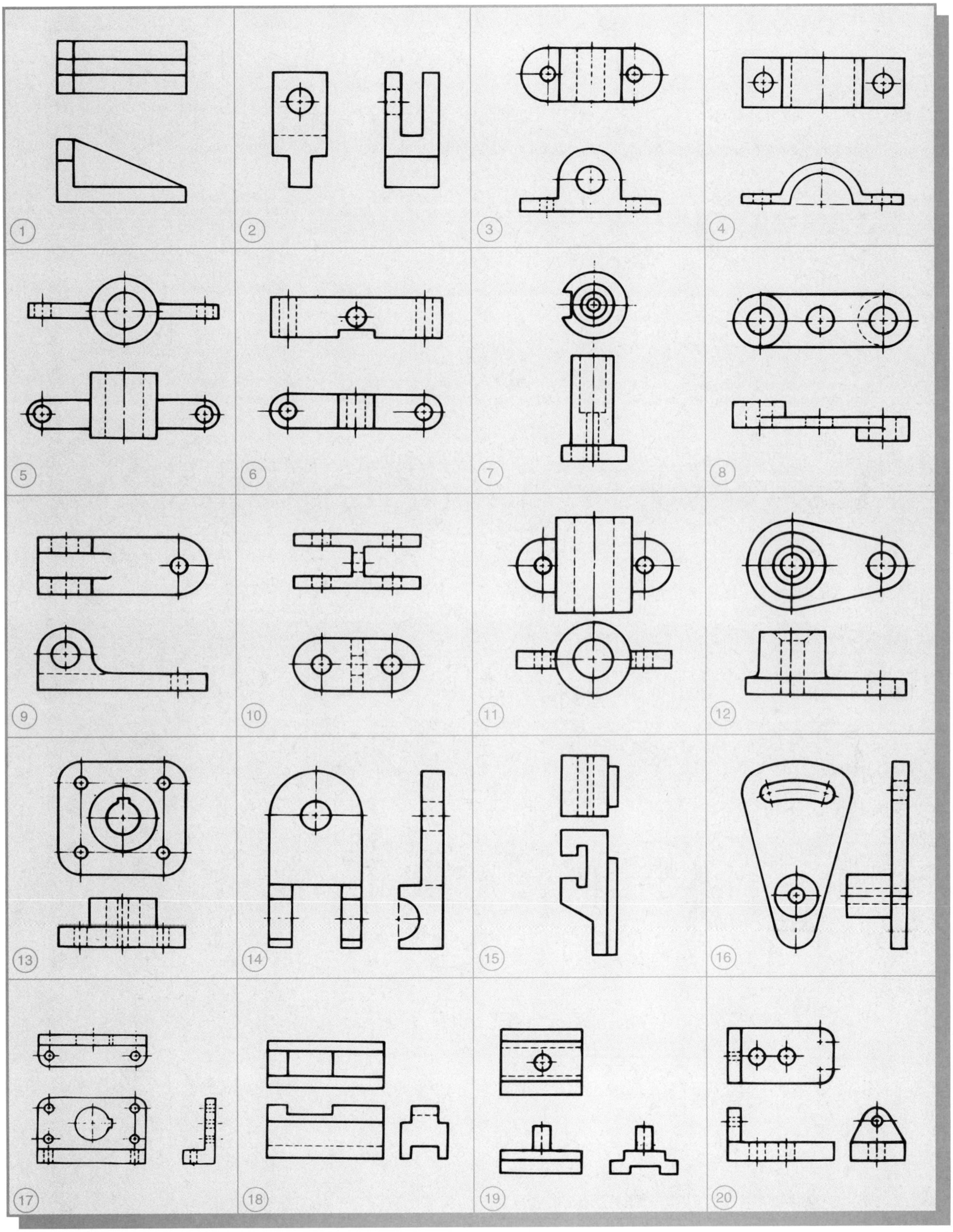

Figure 8-53
Sketch the multiview drawings.

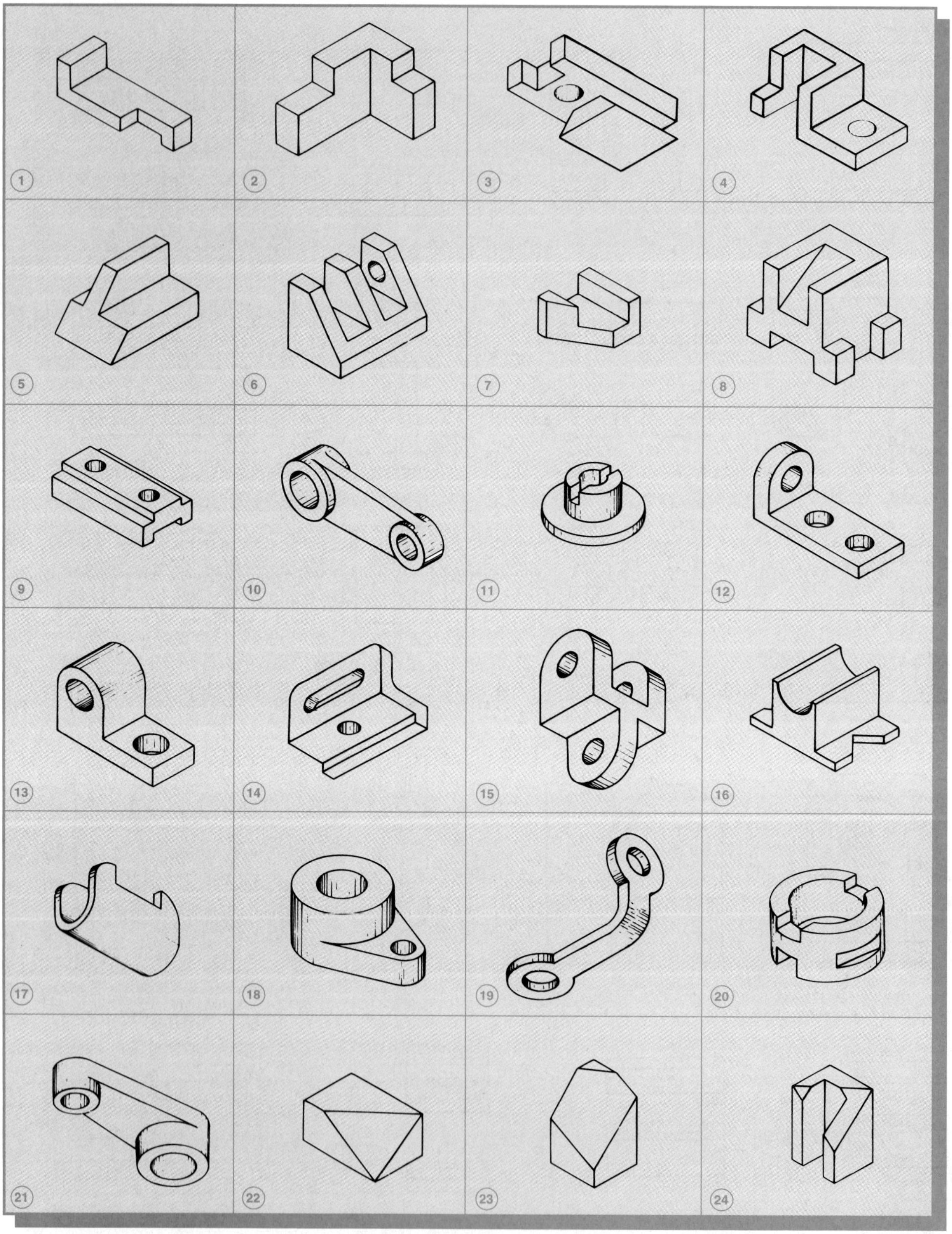

Figure 8-54
Sketch the multiview drawings for each isometric drawing.

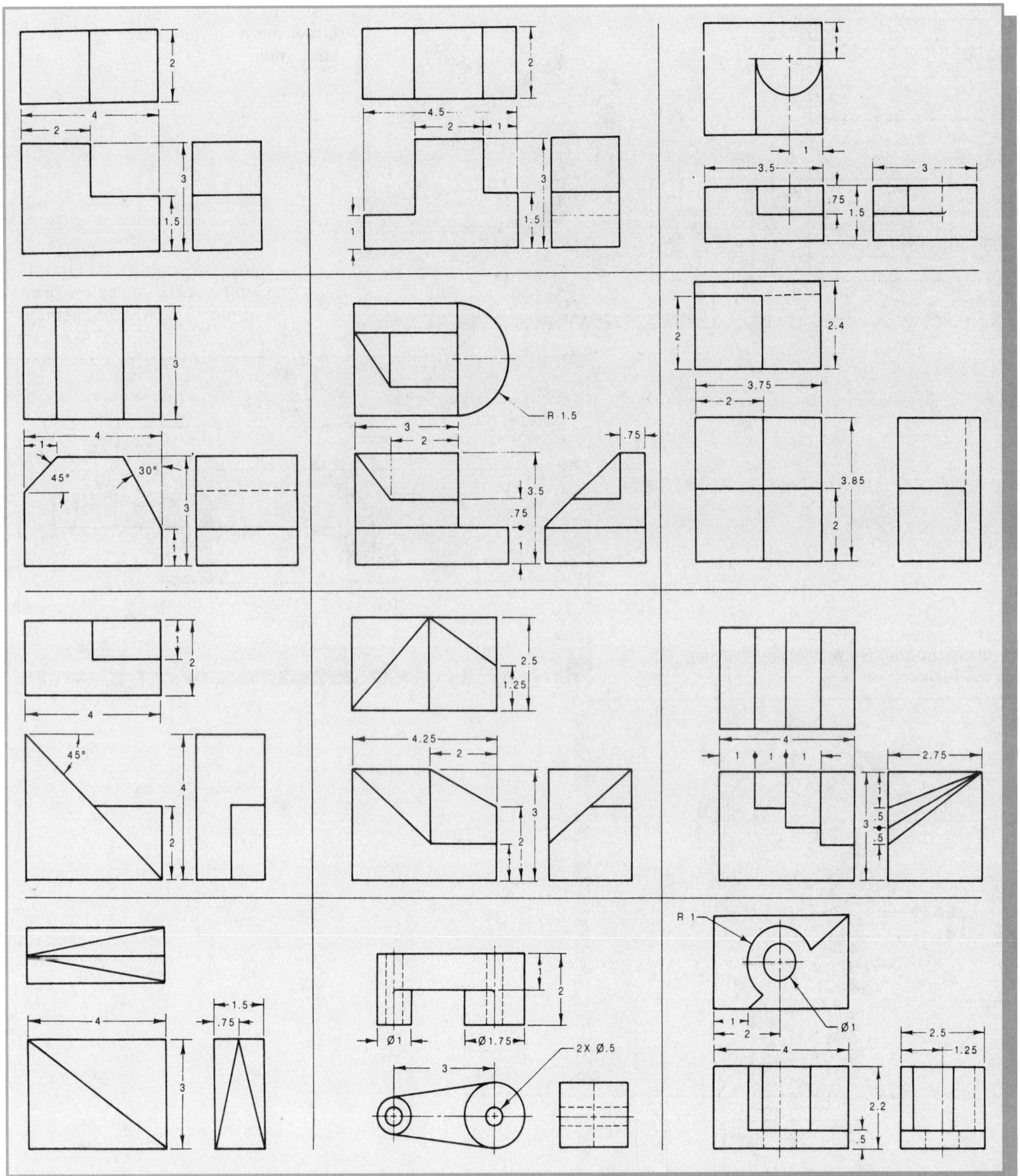

Figure 8-55
Draw the exercise with drawing instruments or a CAD system.

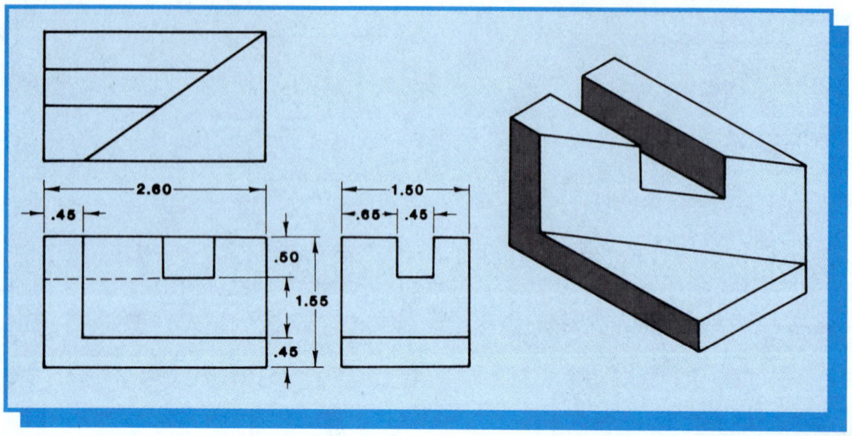

Figure 8-56
Draw the multiview drawing of the base guide.

Figure 8-57
Complete the multiview drawing (metric) with instruments or a CAD system.

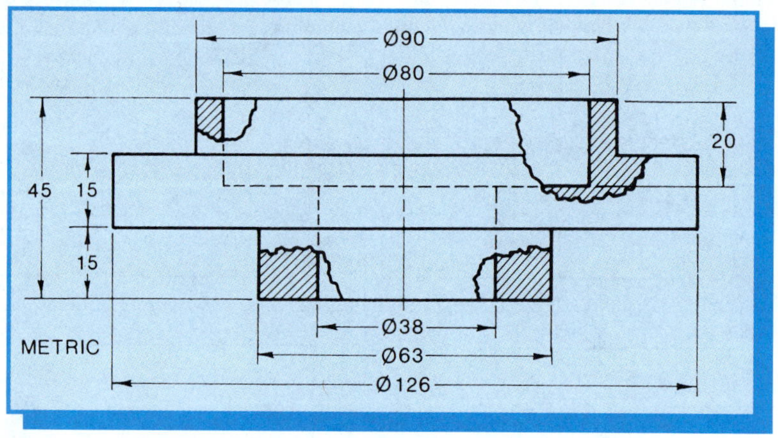

Figure 8-58
Redraw the multiview working drawing of the lock screw.

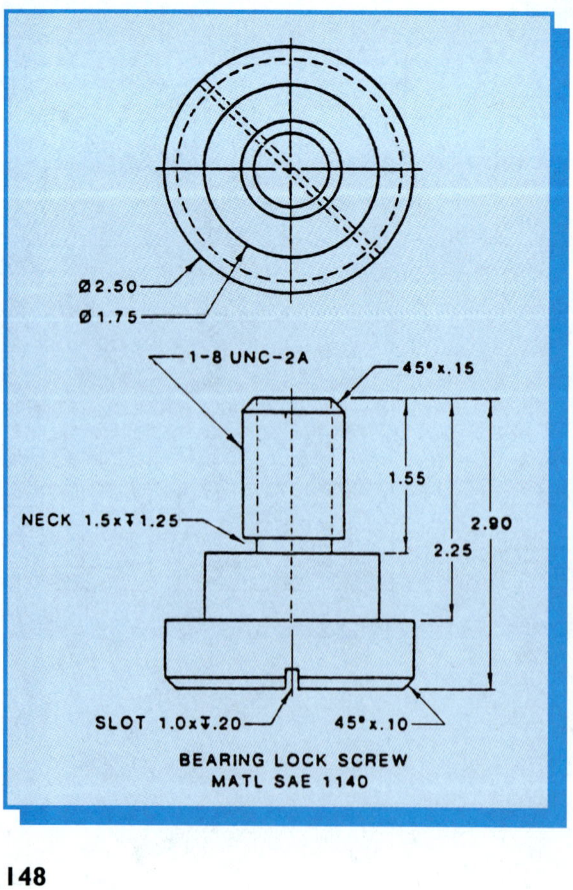

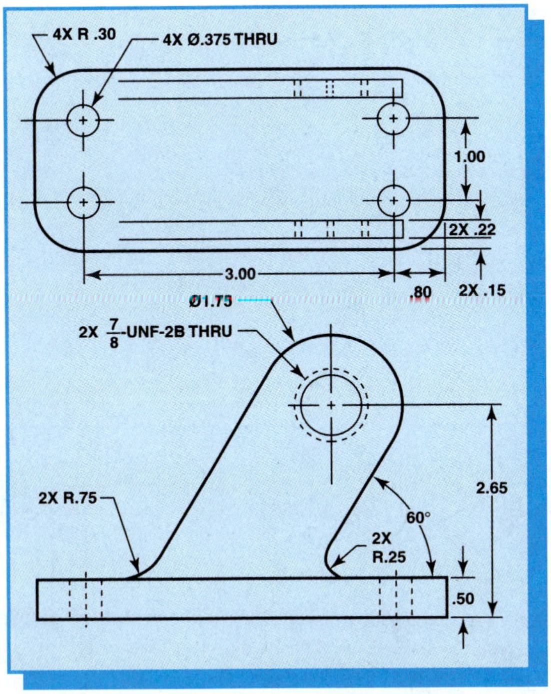

Figure 8-59
Redraw the multiview working drawing of the angle bracket. Scale: 3/4" = 1.

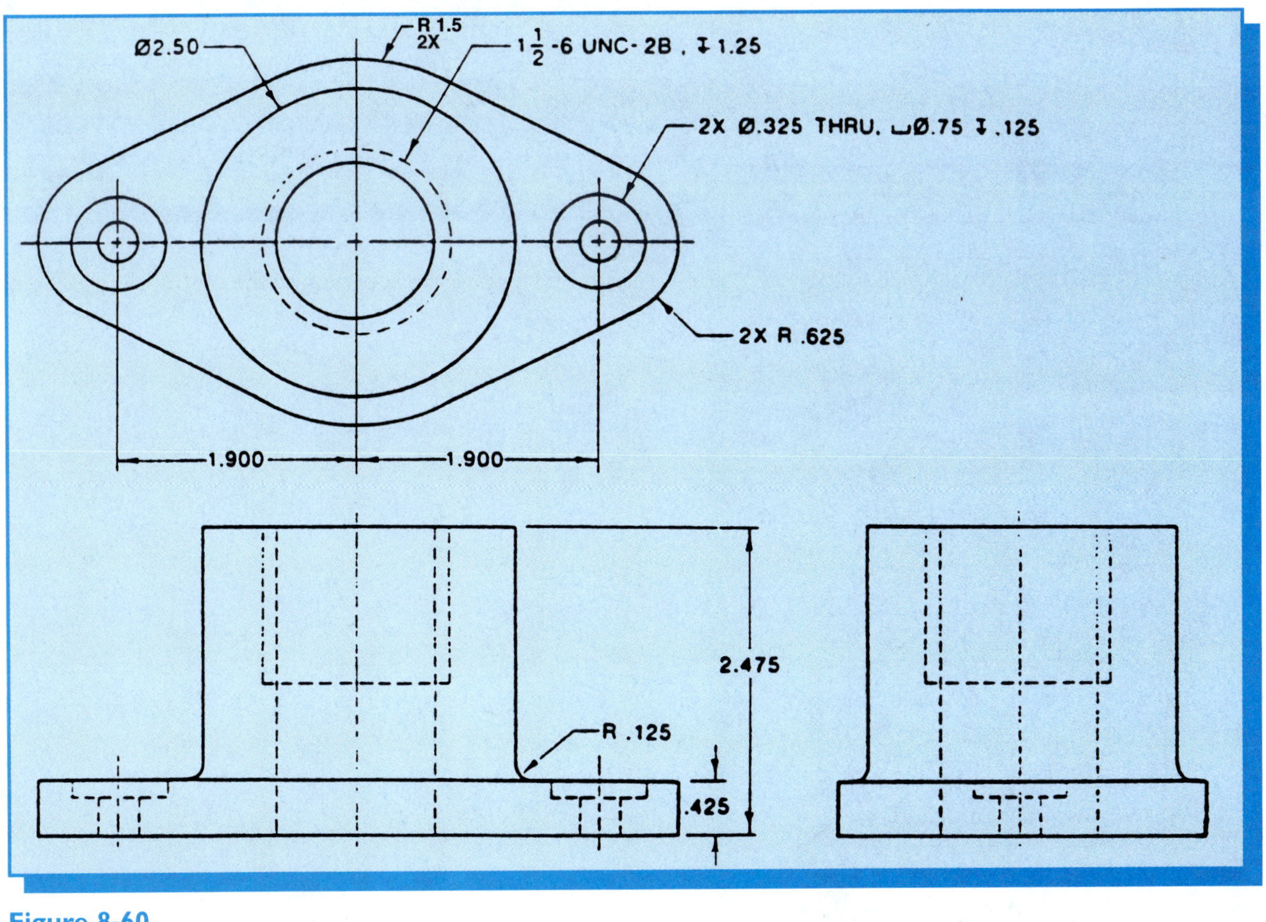

Figure 8-60
Redraw the multiview working drawing of the cap bearing. Scale: 3/4" = 1.

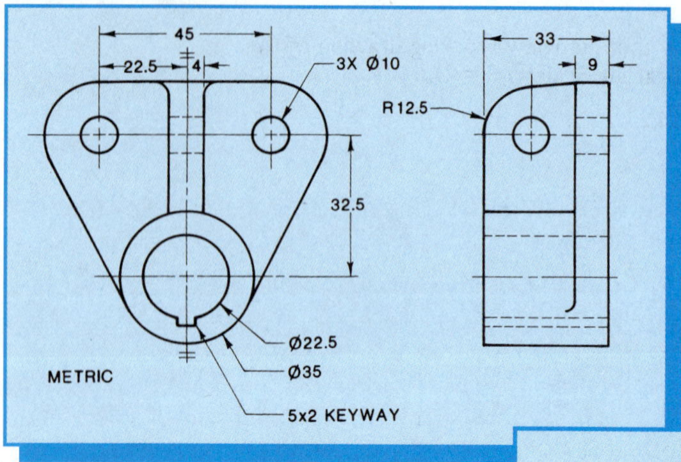

Figure 8-61
Complete the top view of the metric bearing locator fixture.

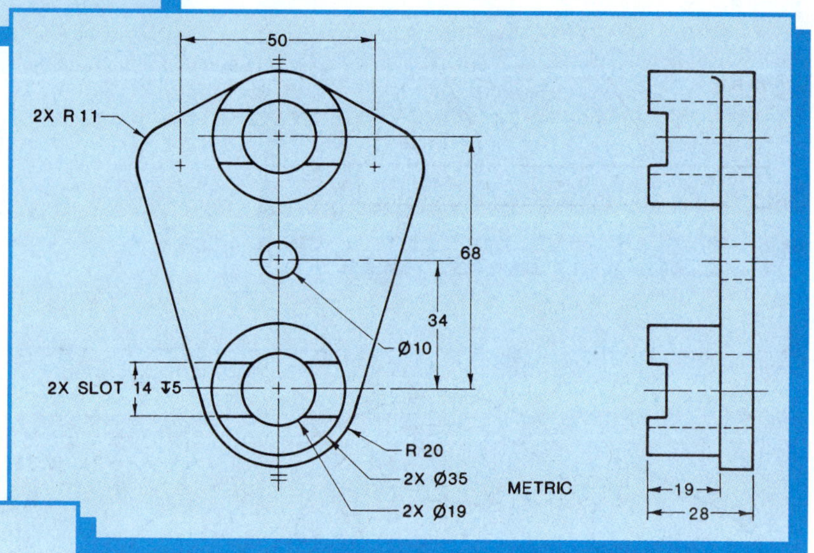

Figure 8-62
Complete the top view of the metric slide-guide jig.

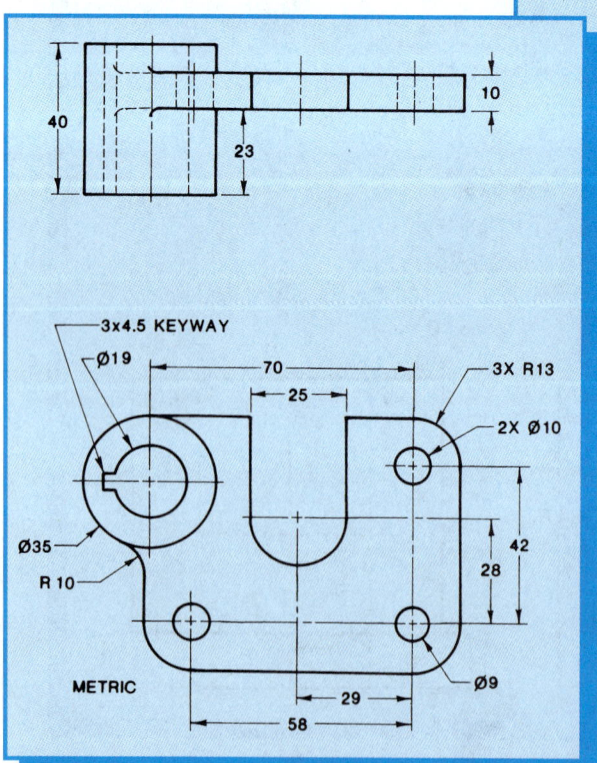

Figure 8-63
Complete the third-view of the metric bearing fixture

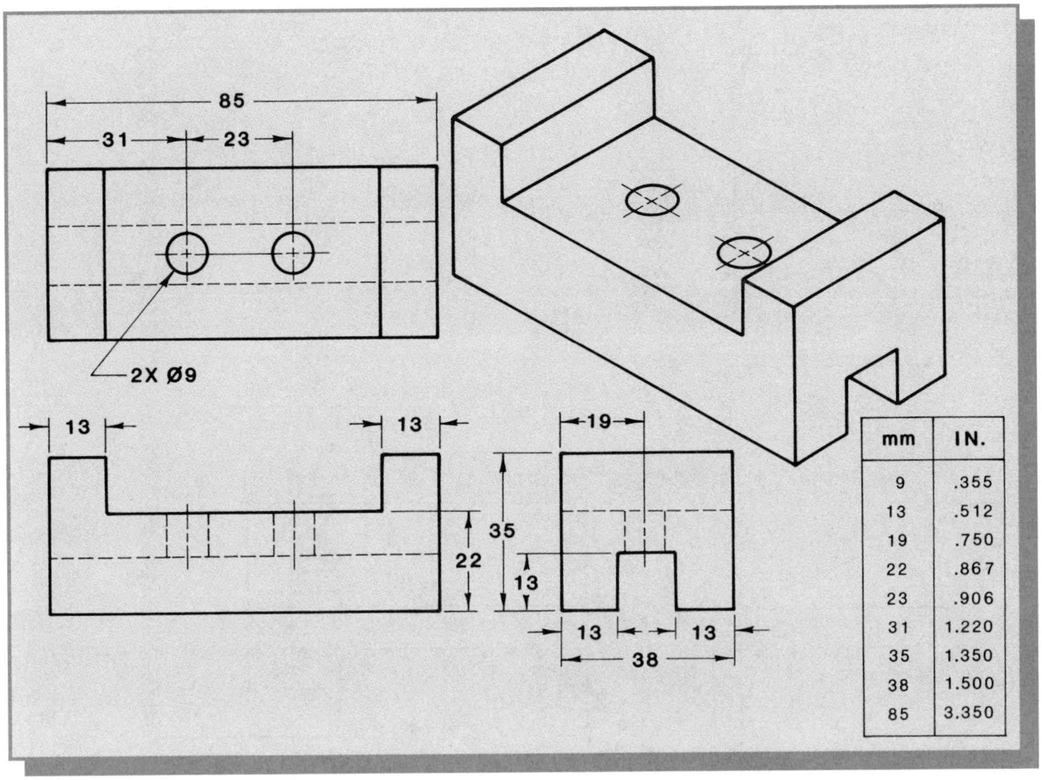

mm	IN.
9	.355
13	.512
19	.750
22	.867
23	.906
31	1.220
35	1.350
38	1.500
85	3.350

Figure 8-64a
Draw the multiview drawing.

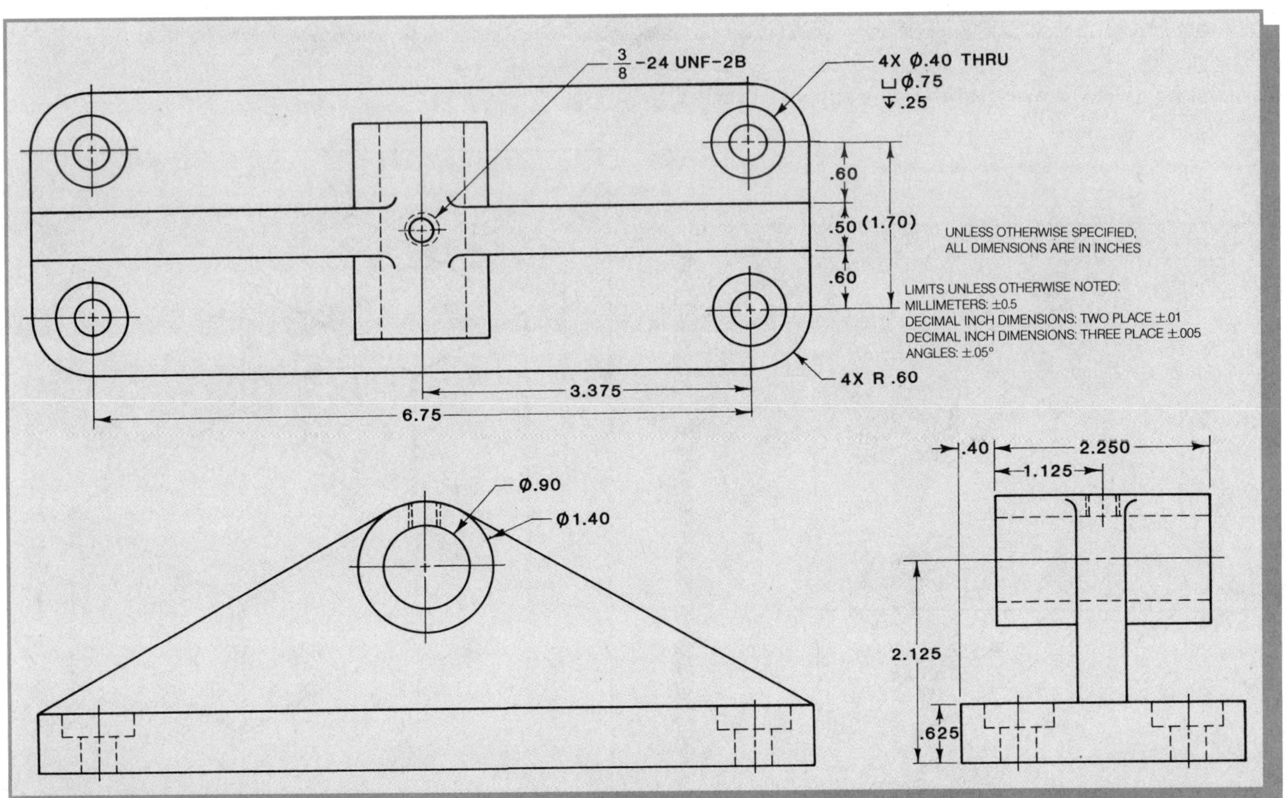

Figure 8-64b
Draw the multiview drawing.

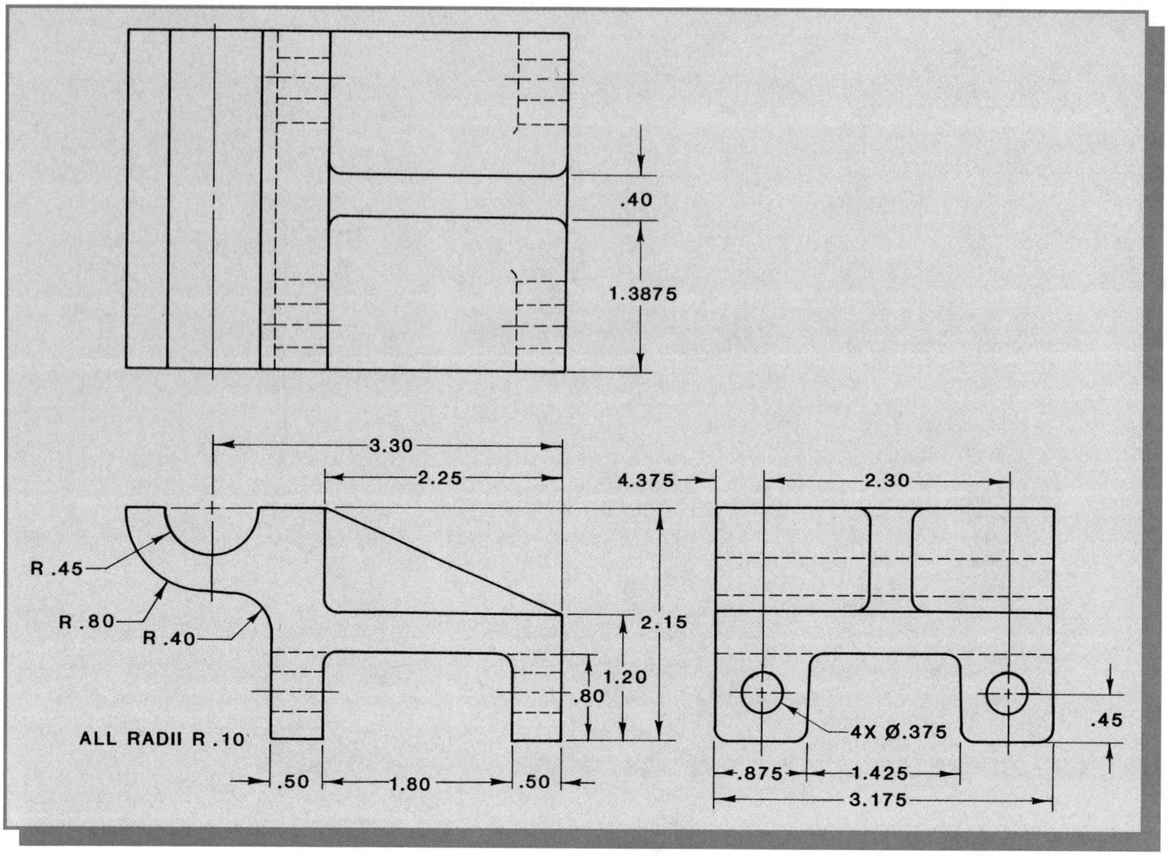

Figure 8-64c
Draw the multiview drawing of the pipe support bracket.

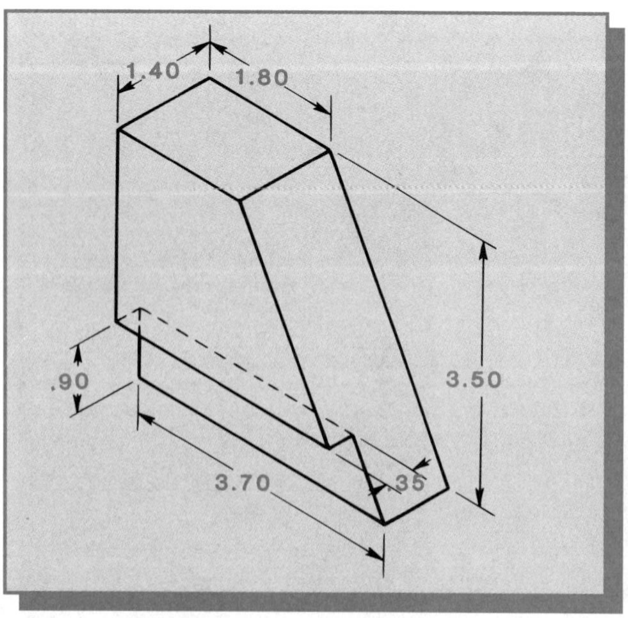

Figure 8-665a
Draw the multiview drawing.

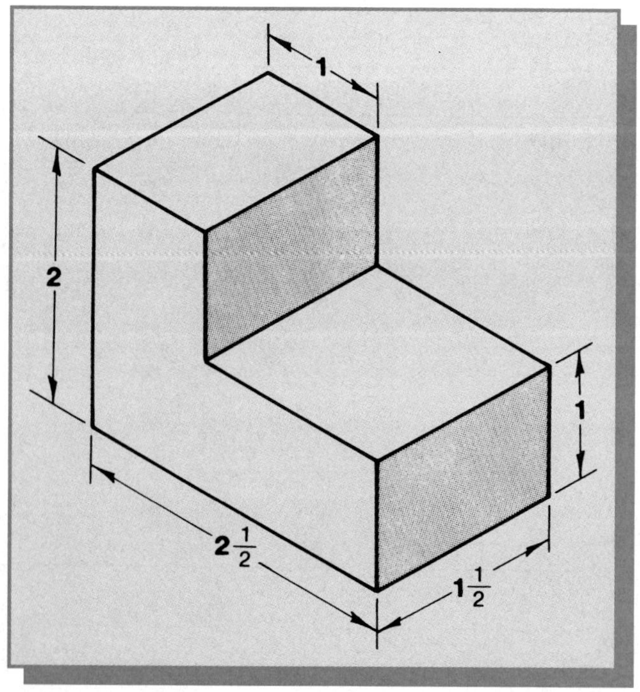

Figure 8-65b
Draw the multiview drawing of the step block.

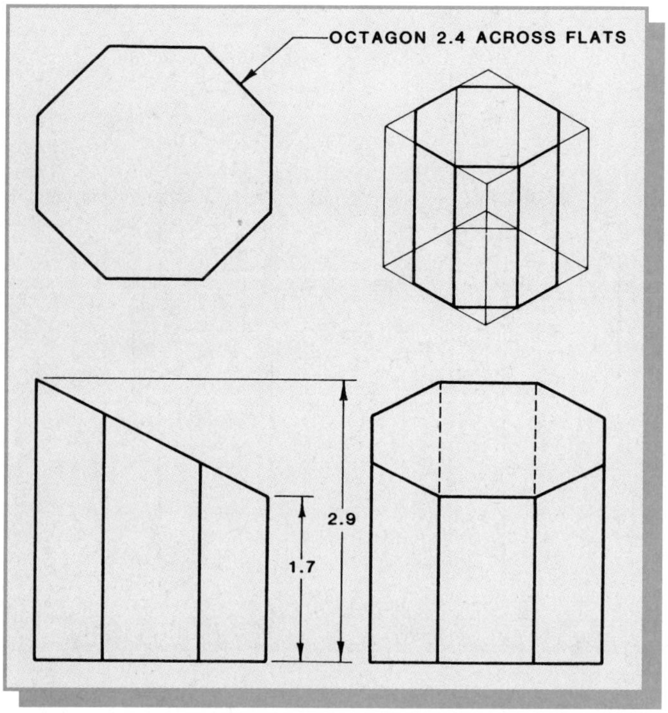

Figure 8-66
Draw the multiview drawing of the truncated octagonal prism and complete the isometric drawing.

Figure 8-67
Draw the multiview drawing (metric) with instruments or a CAD system.

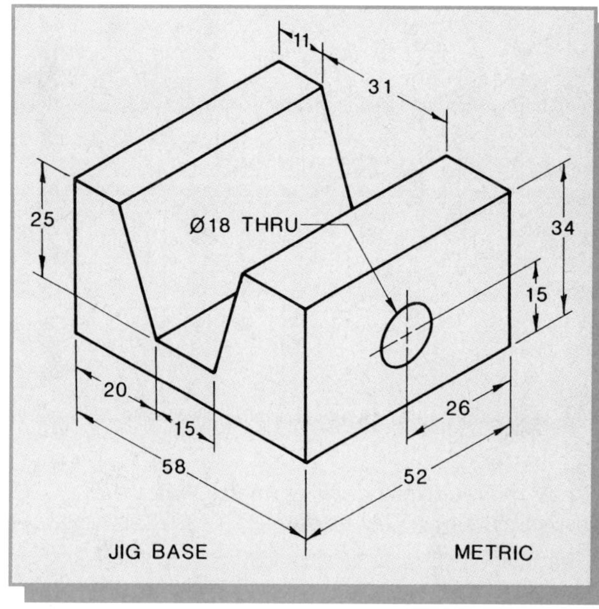

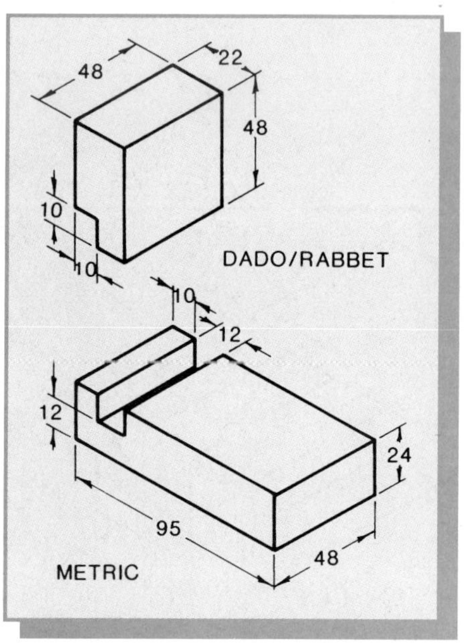

Figure 8-68
Draw the multiview drawing (metric) with instruments or a CAD system.

Figure 8-69
Draw the multiview drawing (metric) with instruments or a CAD system.

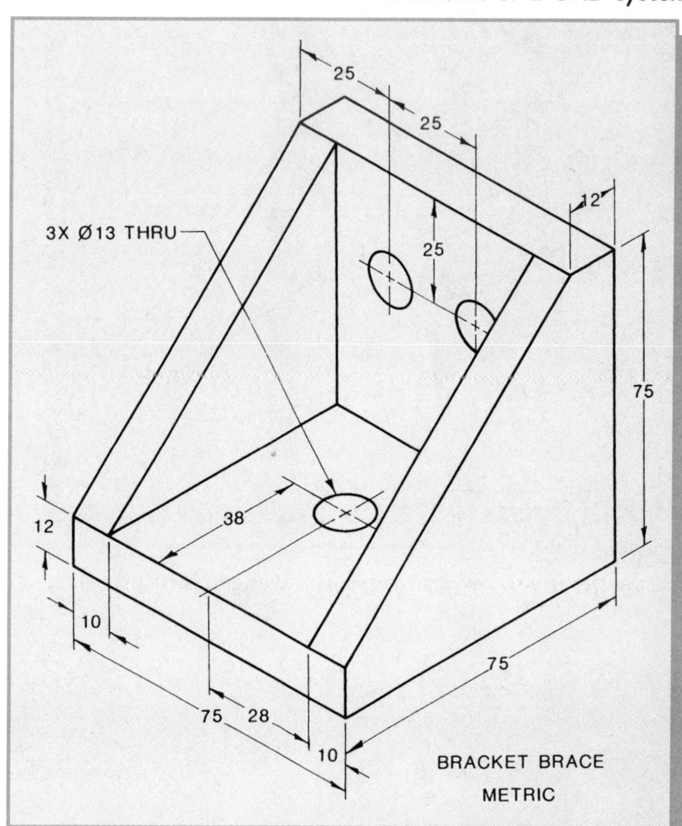

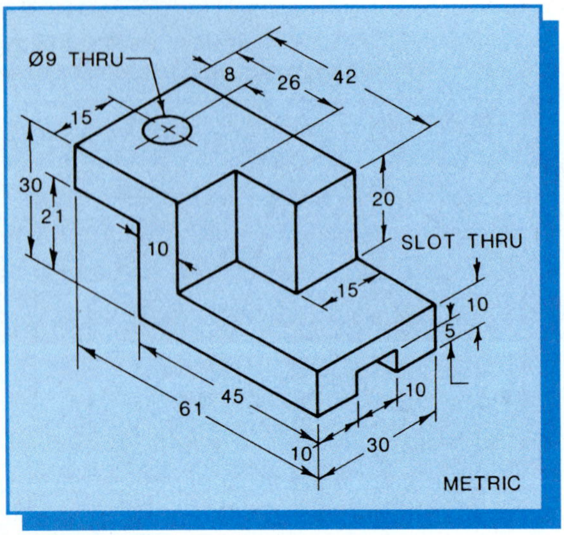

Figure 8-70
Draw the multiview drawing (metric) with instruments or a CAD system.

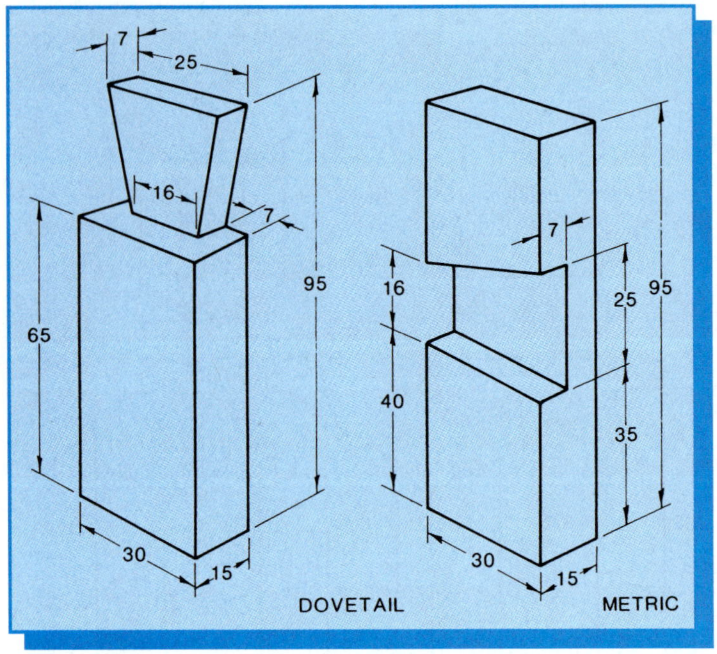

Figure 8-71
Draw the multiview drawing (metric) with instruments or a CAD system.

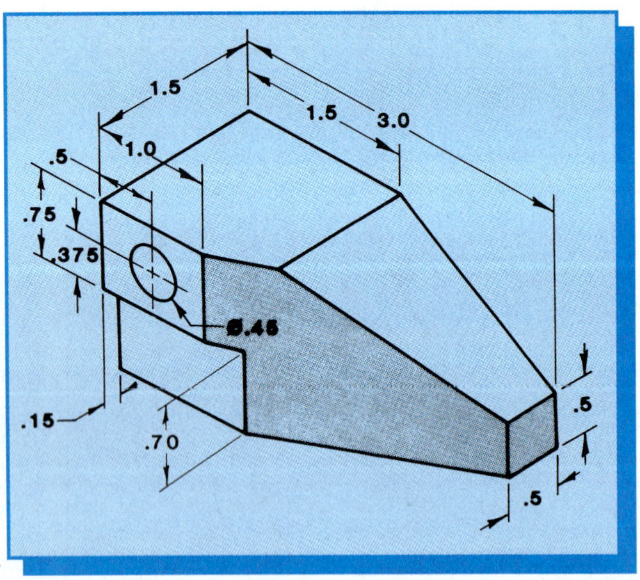

Figure 8-72
Draw the multiview drawing of the wedged bearing guide.

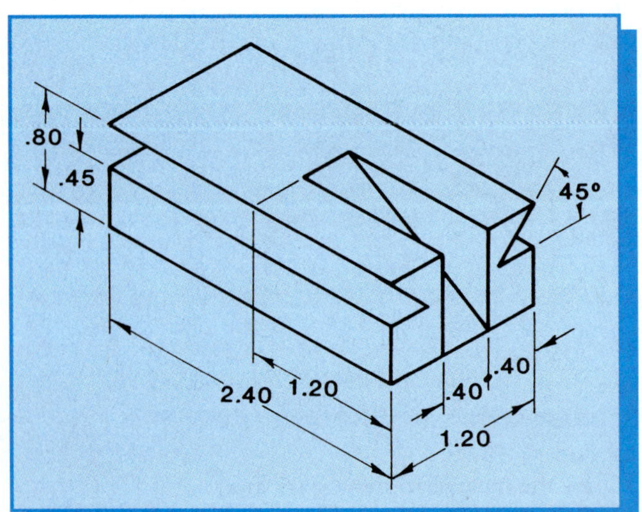

Figure 8-73
Draw the multiview drawing of the wedge support.

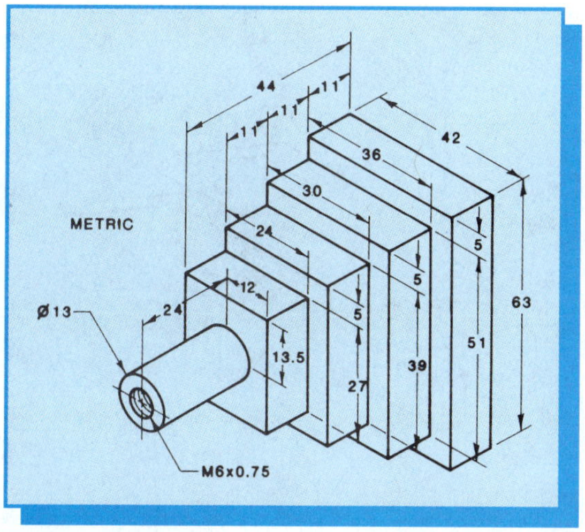

Figure 8-74
Draw a two-view multiview drawing of the centering guide.

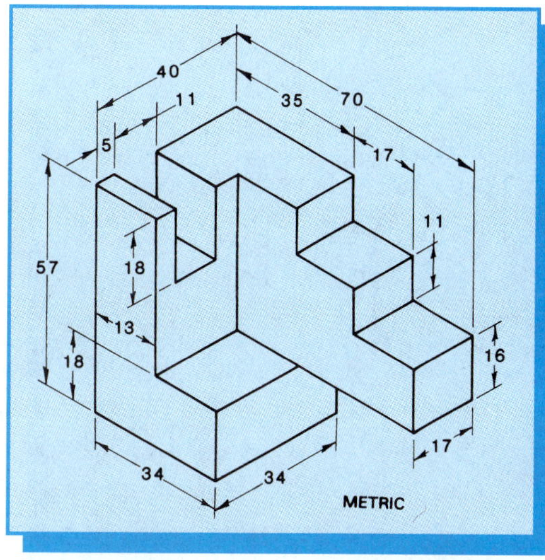

Figure 8-75
Draw a two-view multiview drawing of the metric step block.

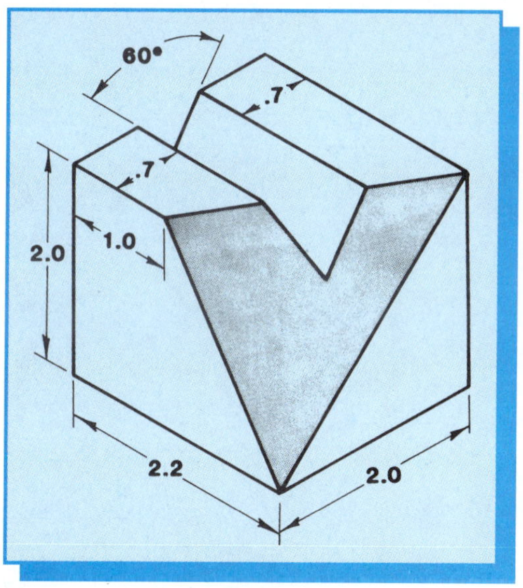

Figure 8-76
Draw a multiview drawing of the V-groove corner block.

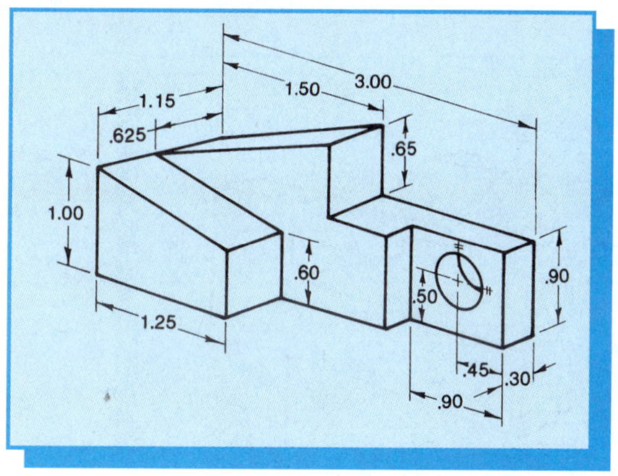

Figure 8-77
Draw a two-view multiview drawing of the double incline guide.

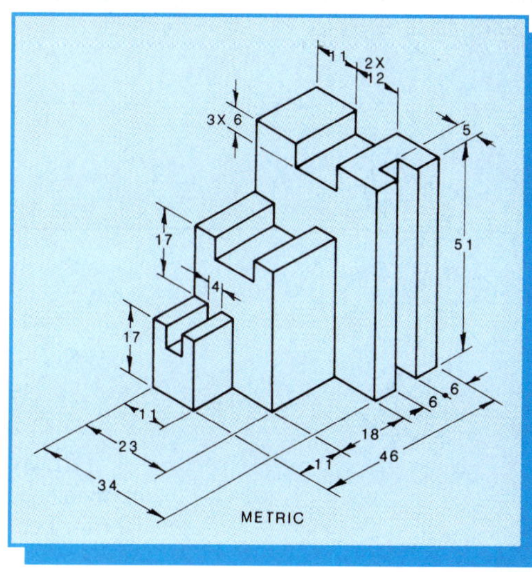

Figure 8-78
Draw a three-view multiview drawing of the slotted-guide jig.

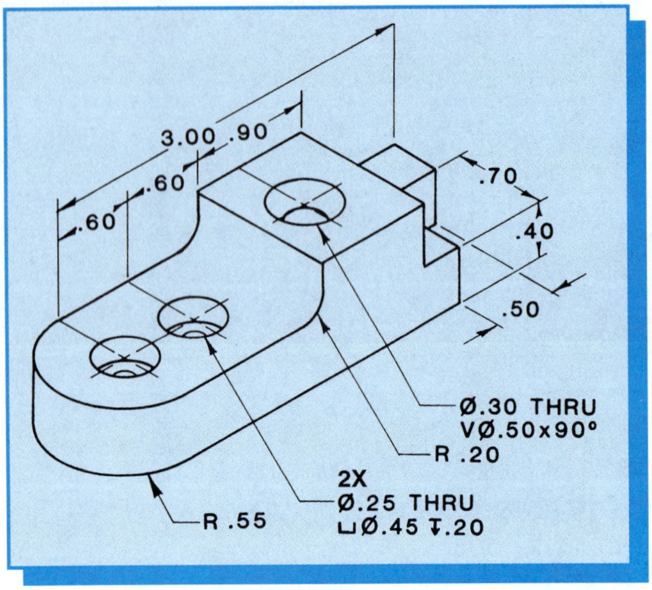

Figure 8-79
Draw the multiview drawing of the locator bracket.

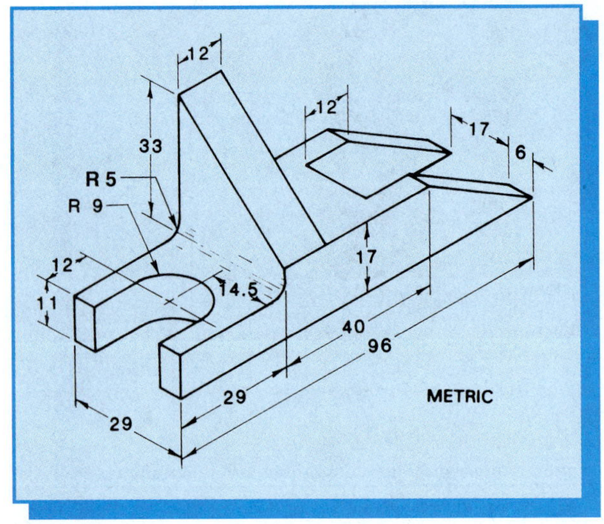

Figure 8-80
Draw a three-view multiview drawing of the wedge gauge aligner.

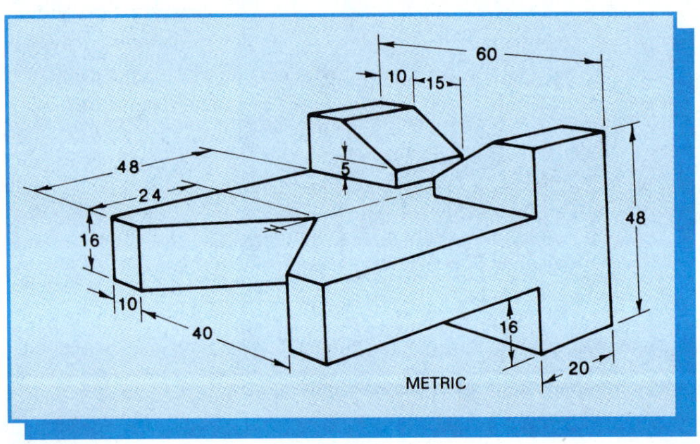

Figure 8-81
Draw a two-view multiview drawing of the metric V-guide fixture.

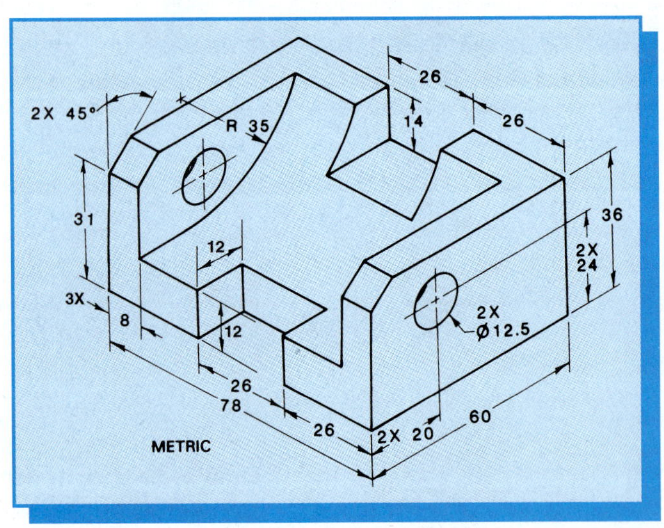

Figure 8-82
Draw a three-view multiview drawing of the 12 mm bearing support.

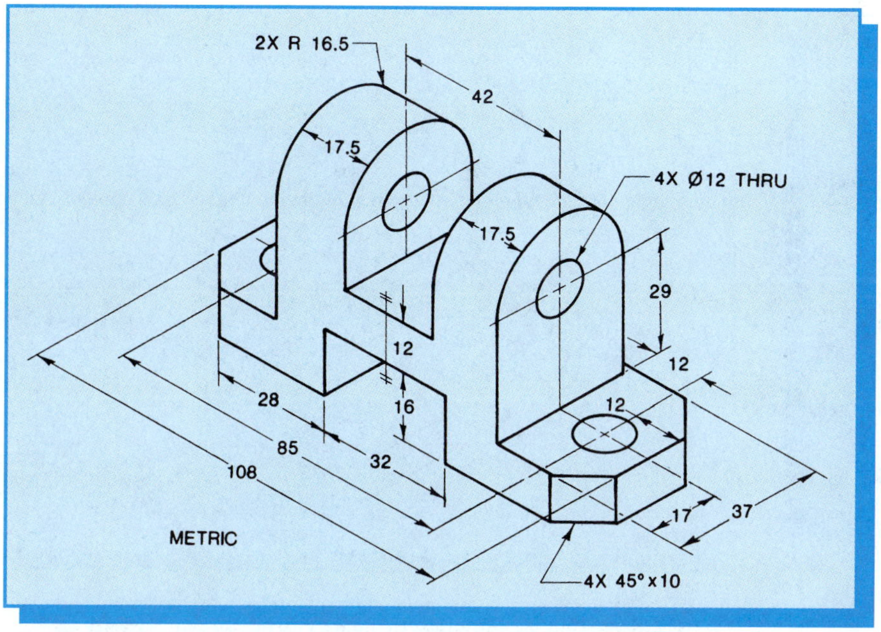

Figure 8-83
Draw a two-view multiview drawing of the metric bearing support.

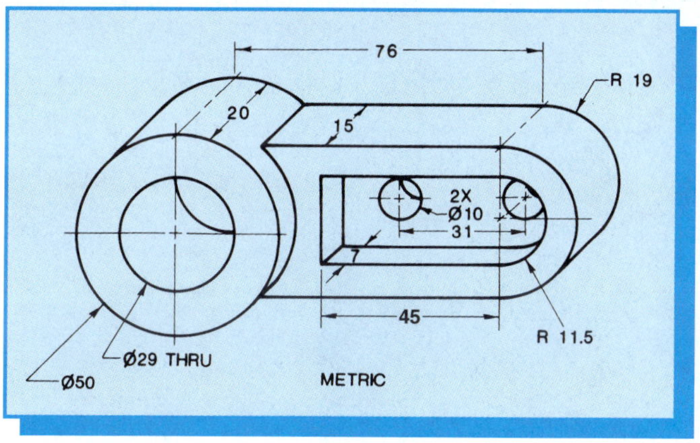

Figure 8-84
Draw a two-view multiview drawing of the metric bearing support guide.

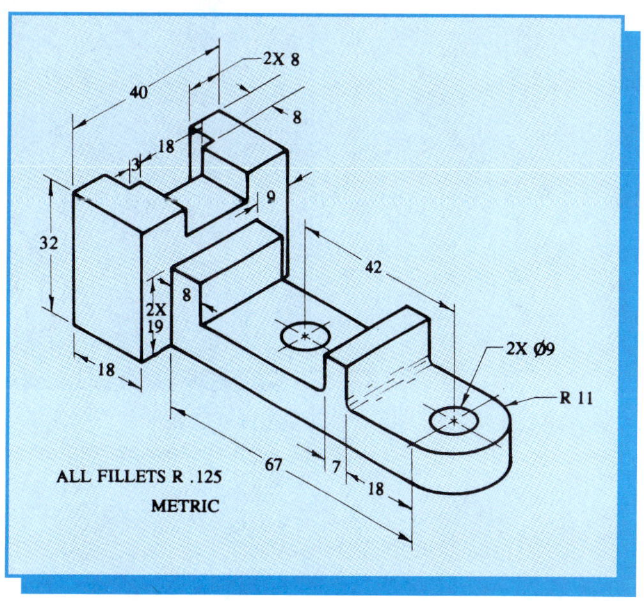

Figure 8-85
Draw a three-view multiview drawing of the location jig

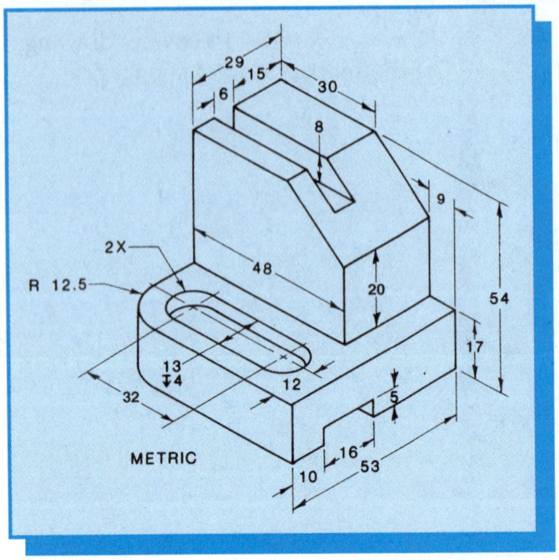

Figure 8-86
Draw a three-view multiview drawing of the metric slip guide.

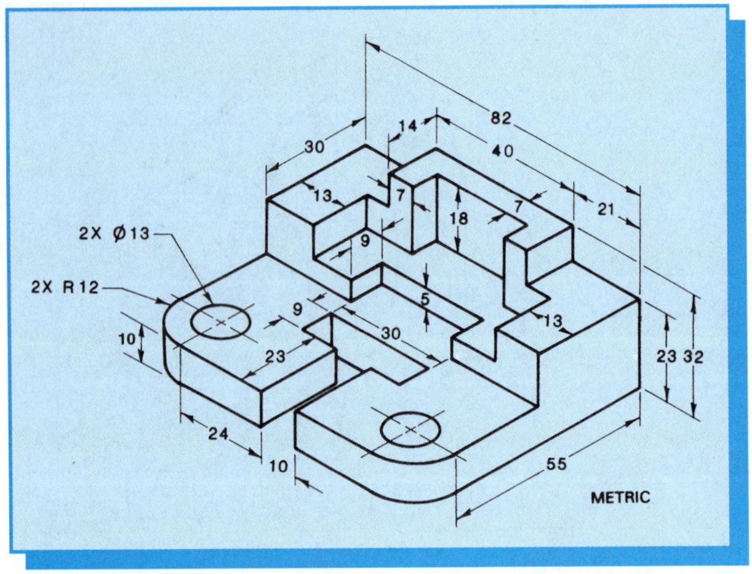

Figure 8-87
Draw a three-view multiview drawing of the metric flatbed alignment jig.

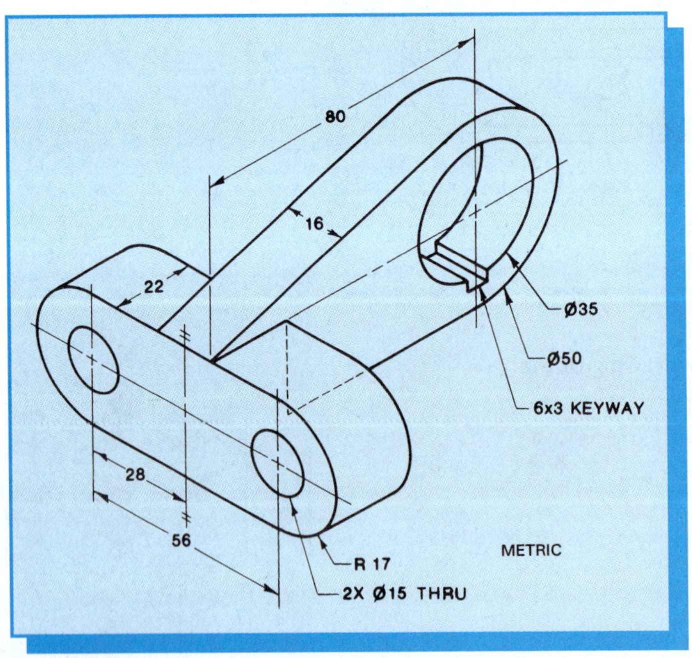

Figure 8-88
Draw a three-view multiview drawing of the metric single-bearing support.

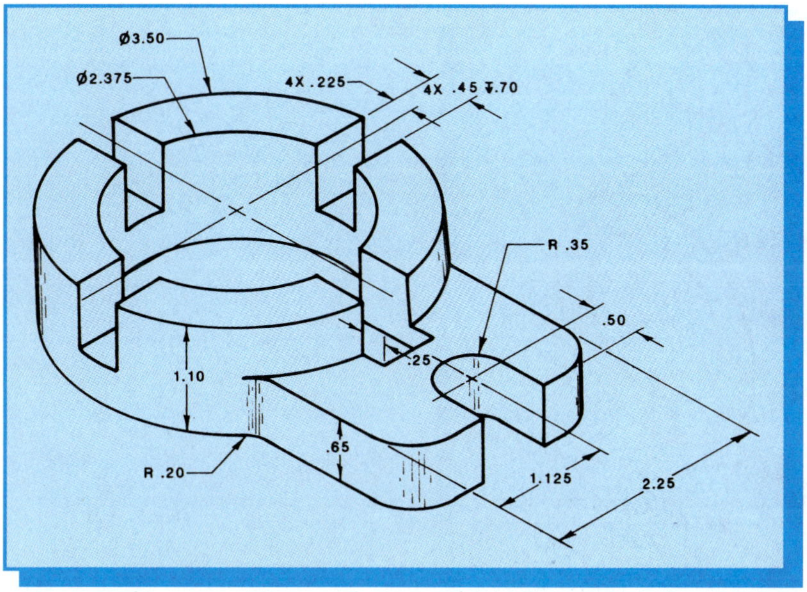

Figure 8-89
Draw a two-view multiview drawing of the ring alignment jig.

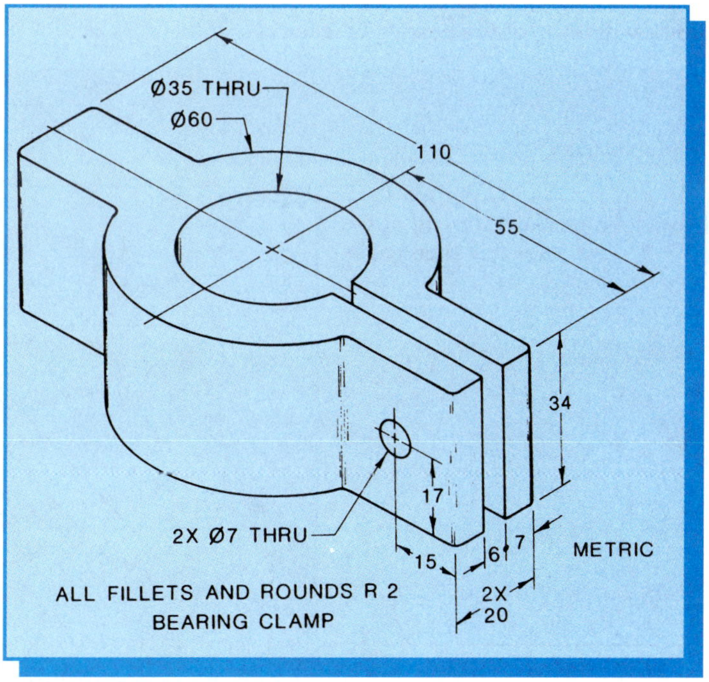

Figure 8-90
Draw the multiview drawing (metric) with instruments or a CAD system.

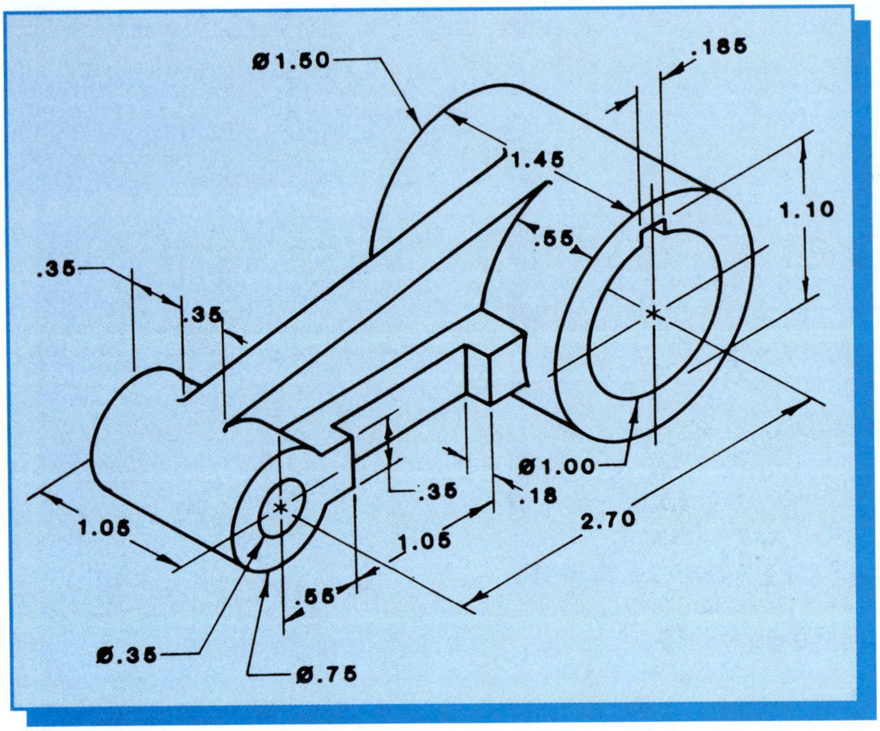

Figure 8-91
Draw the multiview drawing of the bearing bracket.

Figure 8-92
Draw the multiview drawing of the dovetail slide guide.

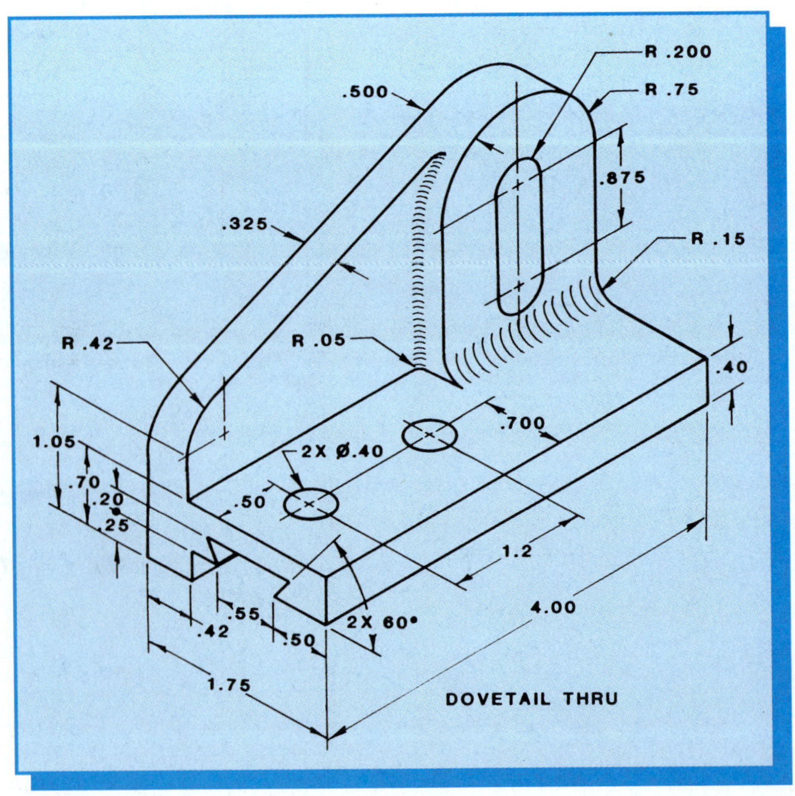

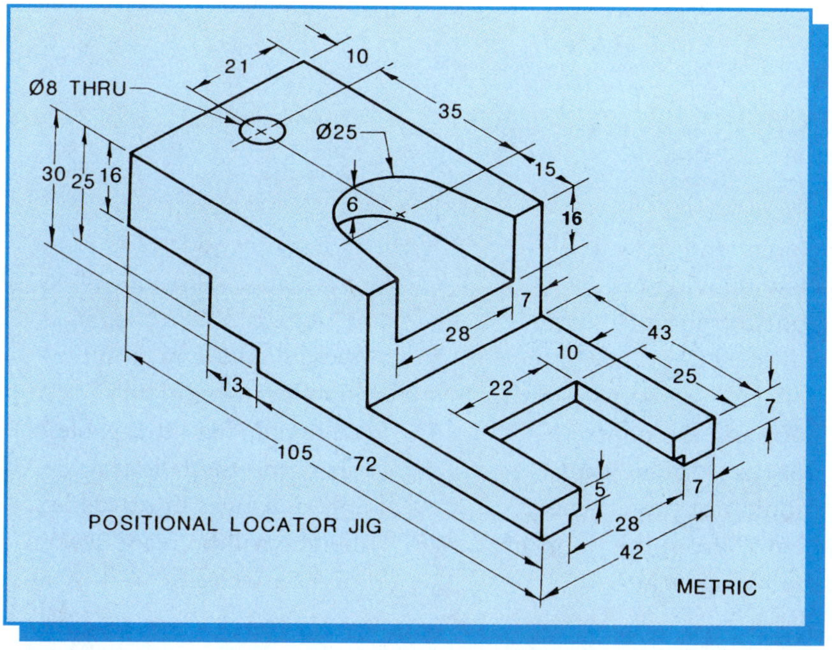

Figure 8-93
Complete the multiview drawing (metric) with instruments or a CAD system.

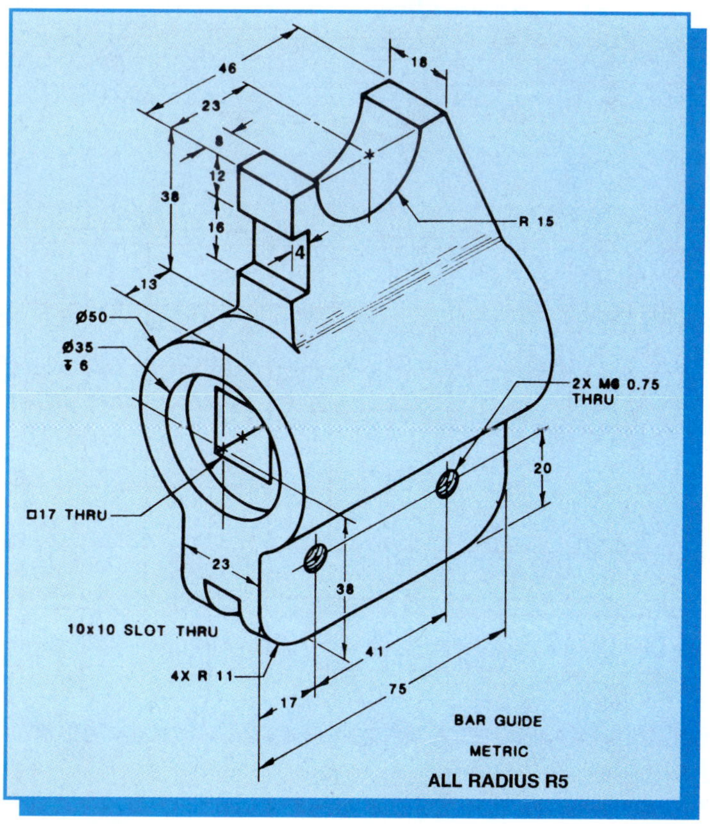

Figure 8-94
Draw a three-view multiview drawing of the bar guide.

Design Exercises

1. Draw the multiview drawings for the industrially produced vehicles in **Figures 8-95** through **8-101**. Redesign each vehicle to your own specifications.

2. Design and draw a working drawing for a small storage cabinet that will store music cassettes, VCR tapes, and CD-ROM disks.

3. Design and draw a multiview drawing for a doorstop for a factory's 10' x 8' steel door. The space between the bottom of the door and the concrete floor is .5". The stop should not project more than 2" from the door into the passageway.

4. A heavy wood 10' x 3' worktable, with four 3" x 3" legs, is 30" in height. Design and draw a multiview drawing of an adjustable leg lift capable of changing the height to 34". Changes in the table's height will be made several times a week.

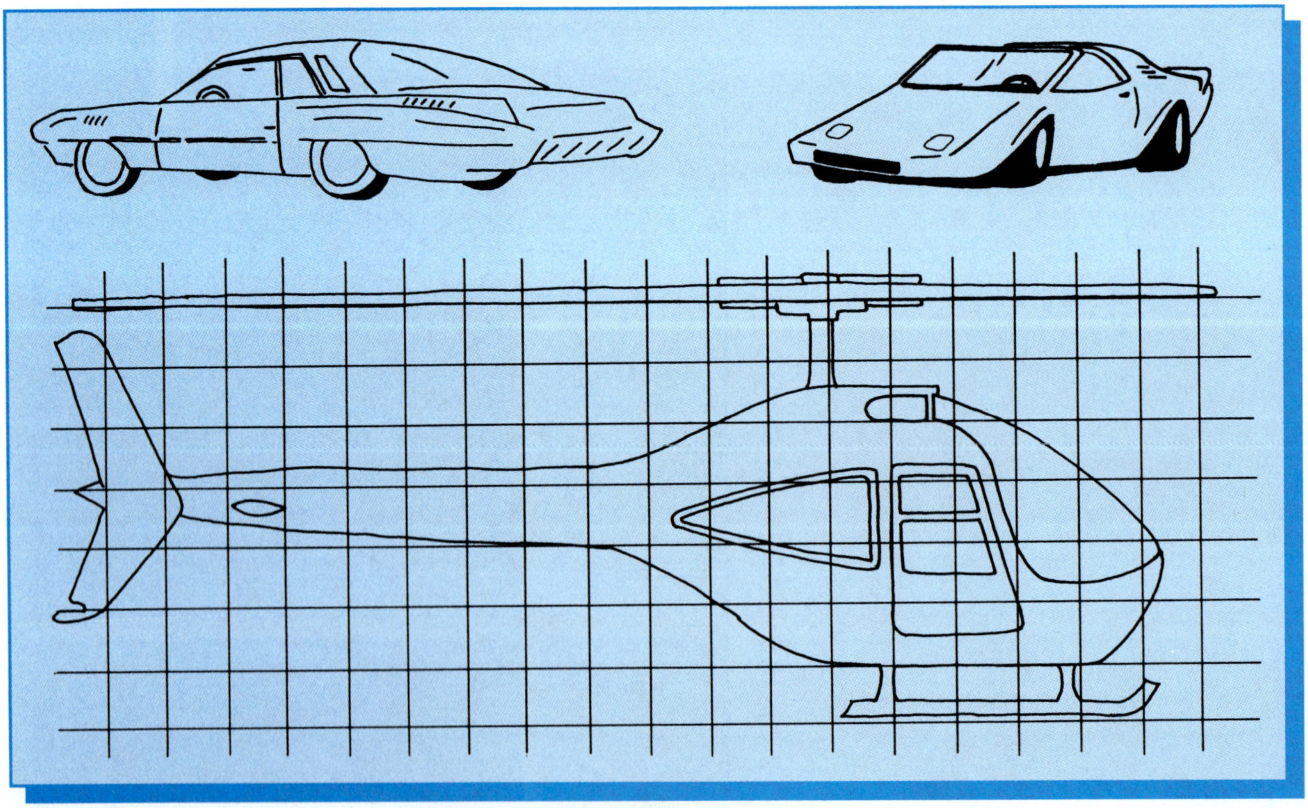

Figure 8-95
Draw the futuristic auto and complete the top, left-side, and right-side views.

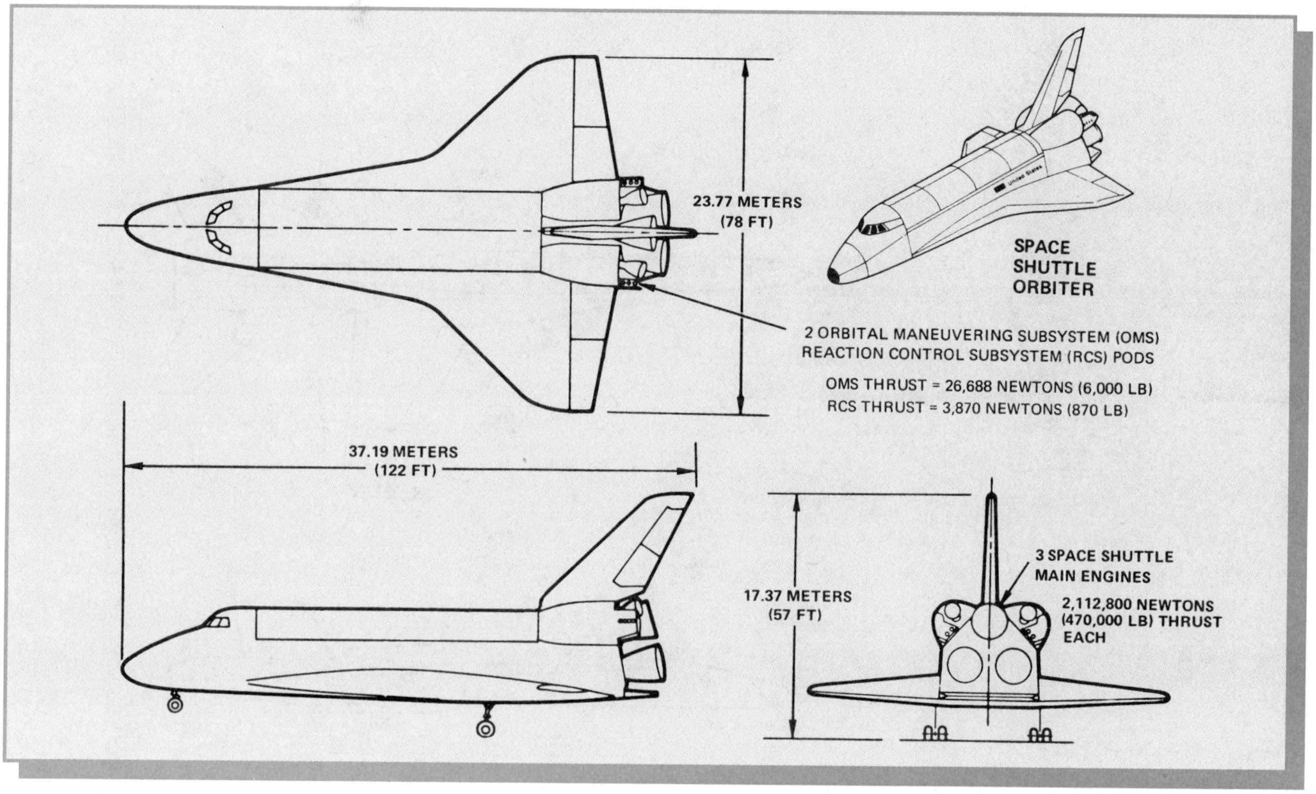

Figure 8-96a
Draw the multiview drawing of the space shuttle. *(Courtesy of NASA)*

Figure 8-96a
Draw the multiview drawing of the spacecraft.

Figure 8-97
Draw the multiview drawing of an airplane.

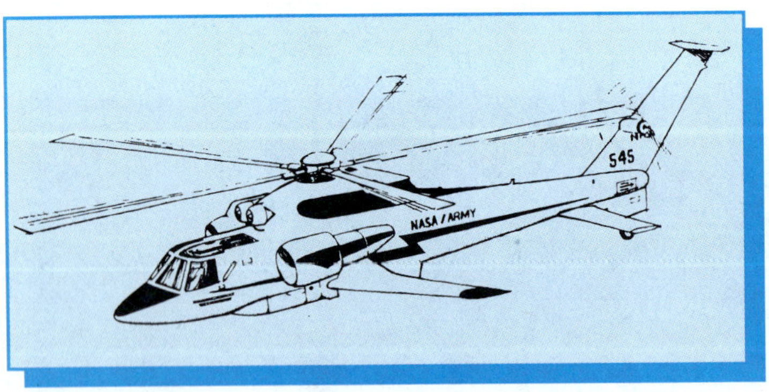

Figure 8-98a
Draw the multiview drawing of a helicopter. *(Courtesy of NASA)*

Figure 8-98b
Draw the multiview drawing of the World War II German airplane.

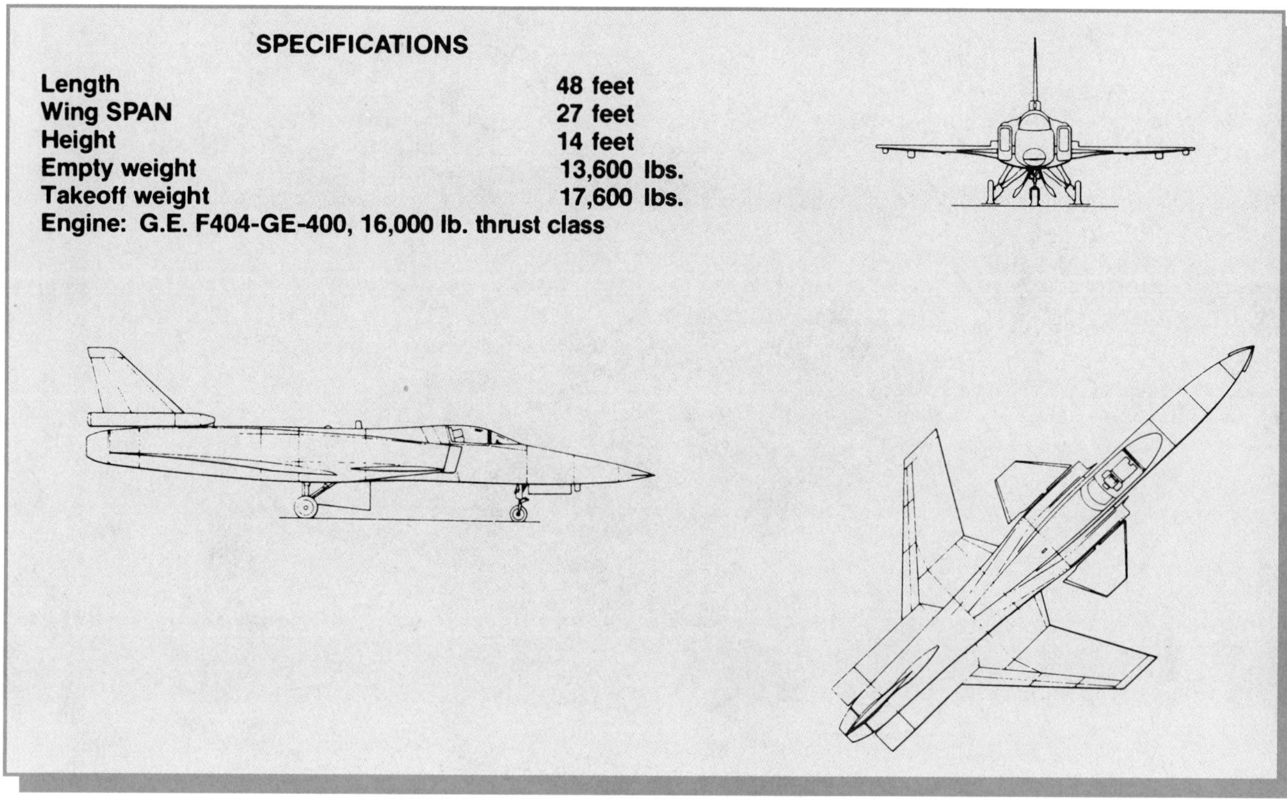

Figure 8-99
Draw the multiview drawing of the X-29 aircraft. Reposition the views so they can be drawn with orthographic projection. *(Courtesy of NASA)*

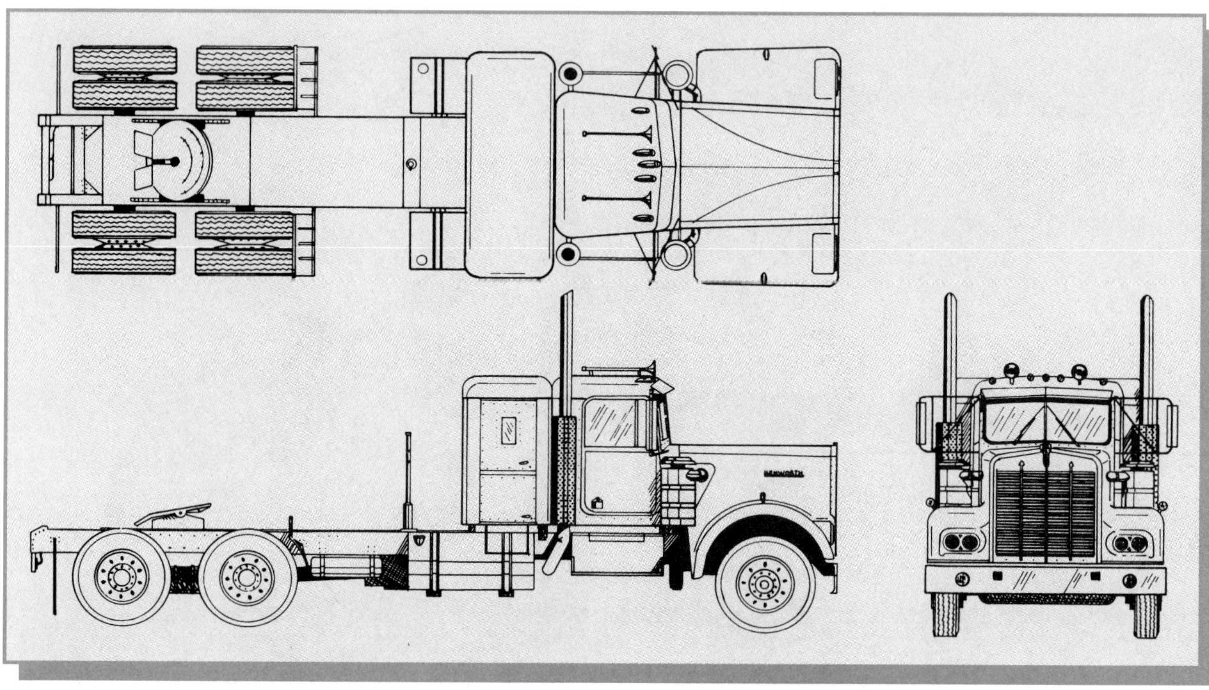

Figure 8-100
Draw the three views of the diesel truck. Use dividers to transfer dimensions.

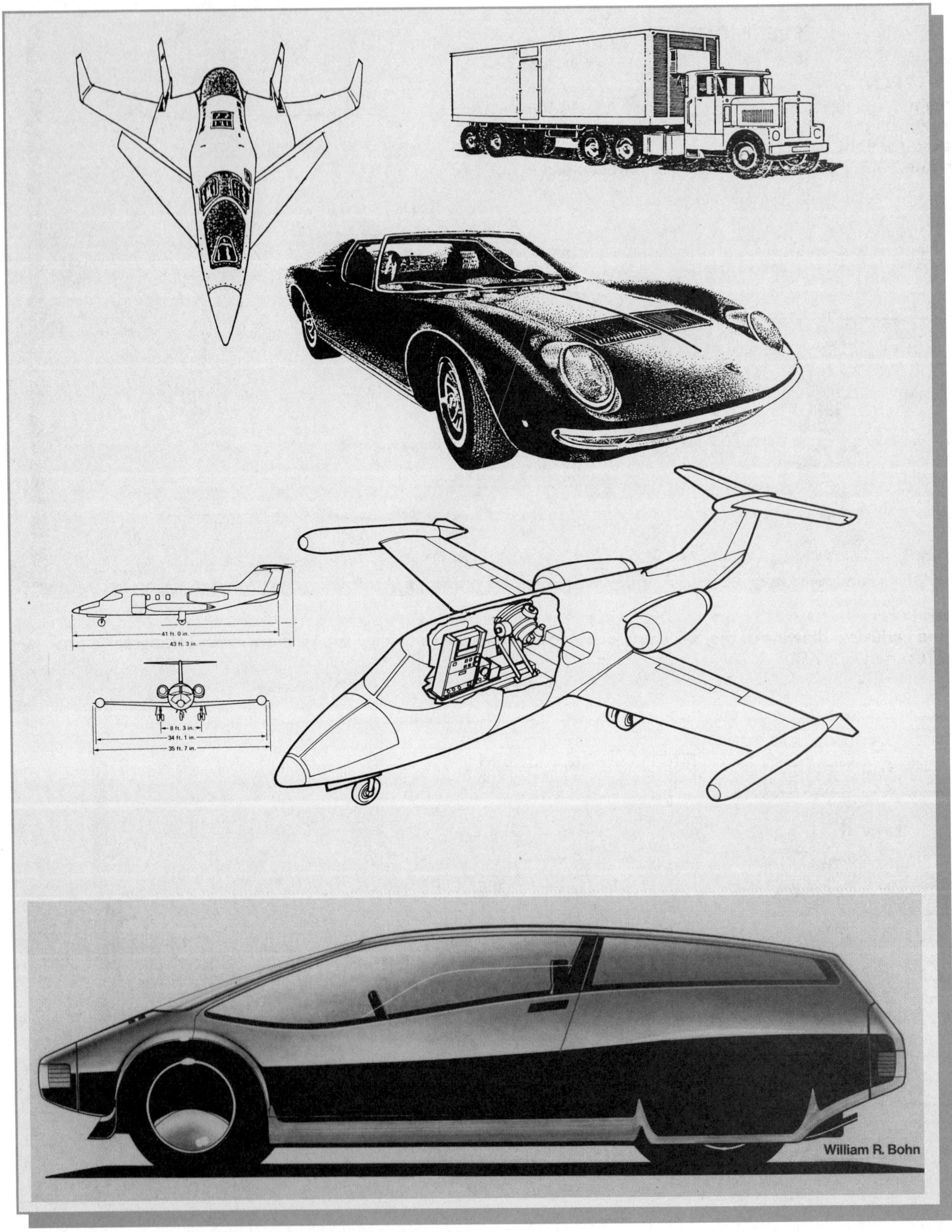

Figure 8-101
Draw the multiview drawings of each vehicle.

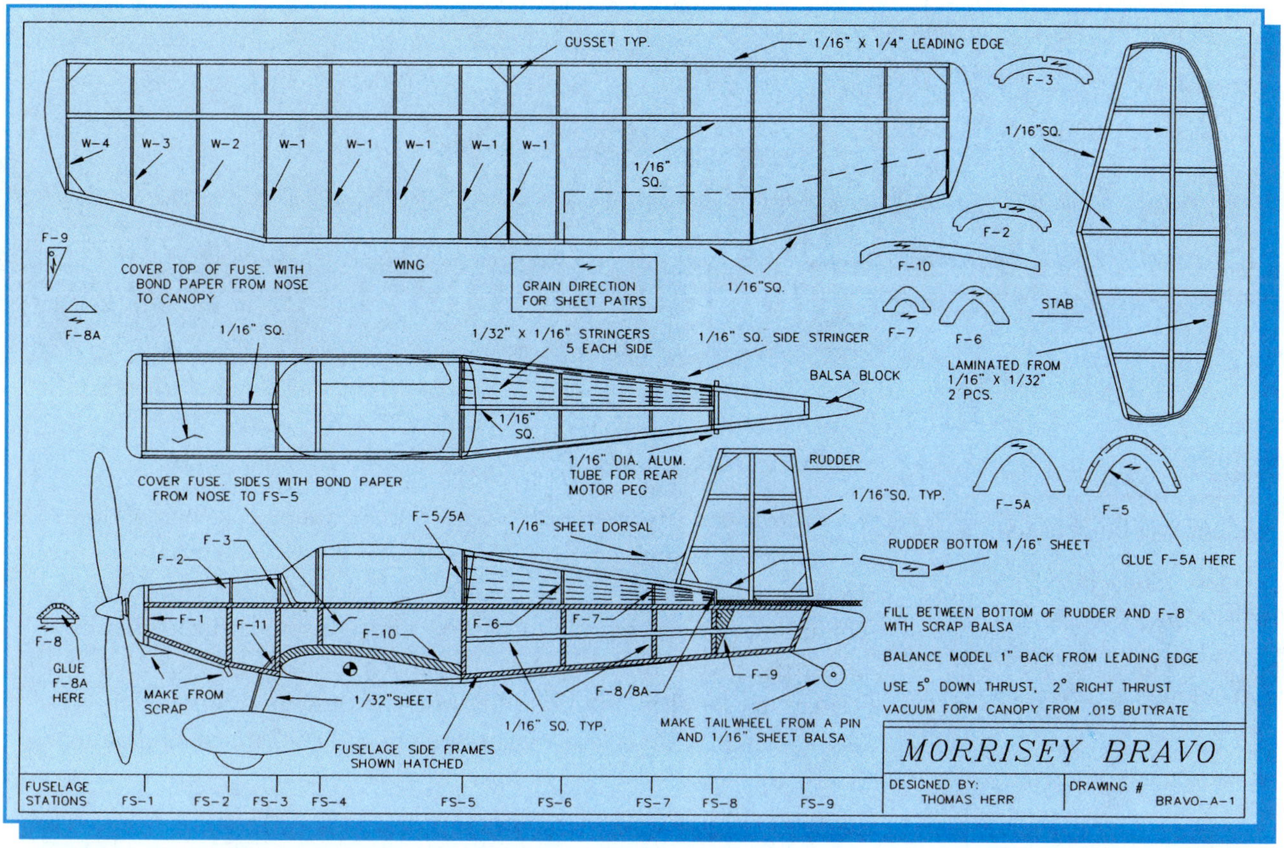

Figure 8-104
AutoCAD drawing for a model airplane

Key Terms

- Ellipses
- First-angle projection
- Fold line
- Front view
- Inclined surface
- Line precedence
- Multiview
- Normal surface
- Oblique surface
- Orthographic projection
- Plane of projection
- Reference line
- Right-side view
- Runout
- Side view
- Symmetrical
- Third-angle projection
- Top view
- True size surface

Chapter 9

Dimensioning Conventions and Surface Finishes

Objectives

The student will be able to:

- recognize standardized dimensioning symbols
- apply correct dimensioning techniques to a working drawing
- dimension all geometric forms
- dimension machine operations
- dimension an isometric drawing
- add surface control symbols to a working drawing

Introduction

Engineering working drawings must contain all information necessary for manufacturing. This includes exact descriptions of the size, shape, and material specifications of every part of a product. These descriptions are accomplished through standardized **dimensioning** conventions **(Fig. 9-1)**. The machine part in **Figure 9-1** can be manufactured in any plant because the drawing contains universally accepted ASME (American Society of Mechanical Engineers)

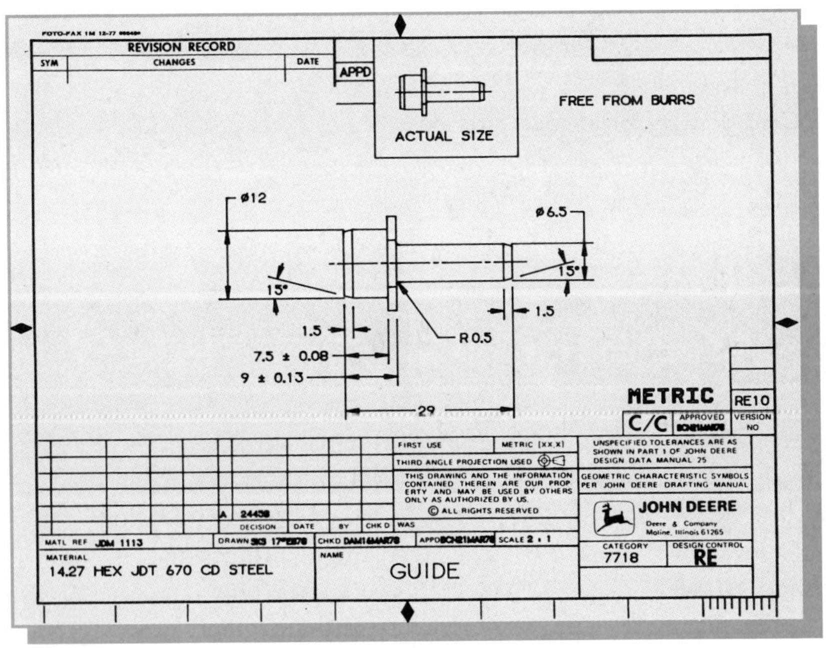

Figure 9-1
Dimensions and notations will assure accurate manufacturing. *(Courtesy of Deere & Company)*

dimensional symbols and notations. Without these symbols and notations, the drawing **(Fig. 9-2)** is meaningless and the part could not be manufactured without extensive guesswork. Product quality and accuracy would depend on the interpretation of the manufacturer.

Adhering to established standards is important in preparing working drawings manually or with a CAD system.

Working drawing dimensions are of two types; **size dimensions** and **location dimensions.** Size dimensions describe the size of each geometric form **(Fig. 9-3)**. Location dimensions provide the exact location of each geometric part of a drawing **(Fig. 9-4)**. In addition to size and location dimensions, **notations** are needed to specify materials **(Fig. 9-5)** and any other information necessary for manufacturing. If any dimension or notation is omitted, costly production errors and delays could result.

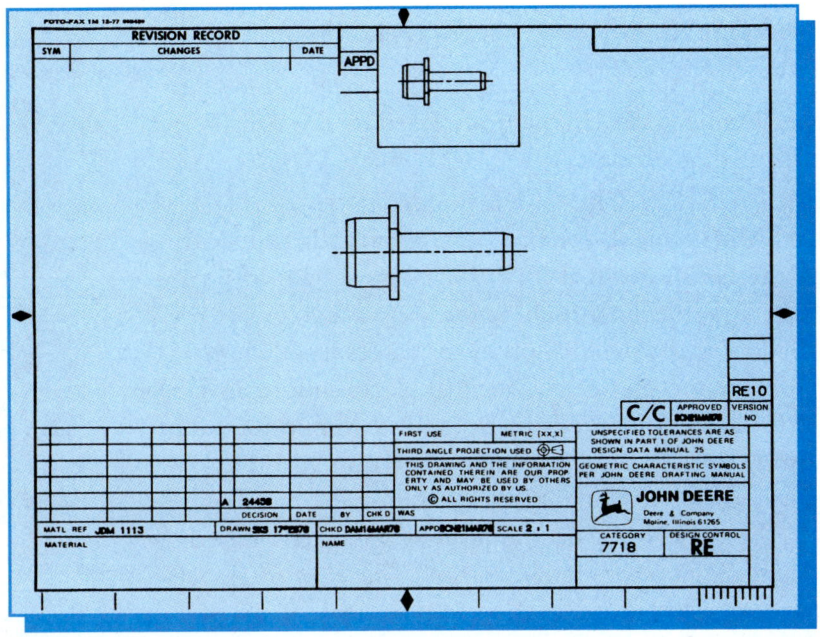

Figure 9-2
Without dimensions and notations, manufacturing of this part is impossible.

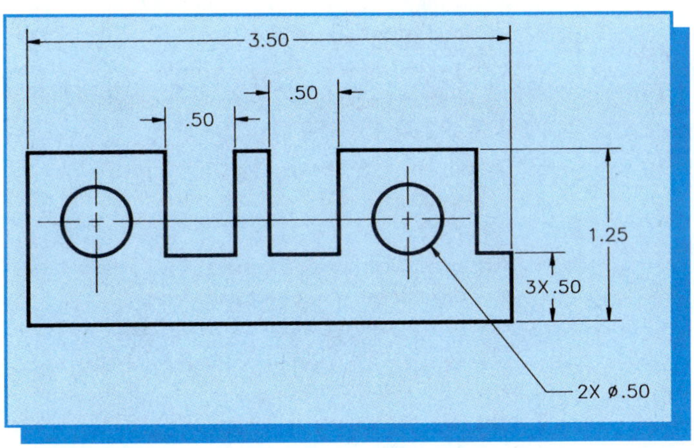

Figure 9-3
Size dimensions

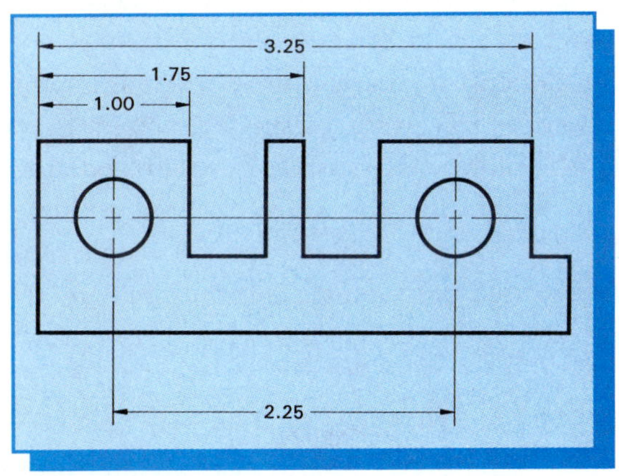

Figure 9-4
Location dimensions

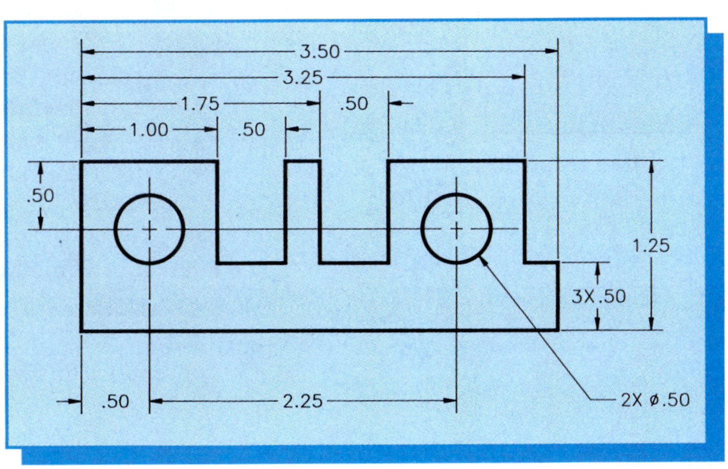

Figure 9-5
A fully dimensioned drawing

Systems of Dimensioning

The principles and rules for dimensioning, including dimensioning systems, are standardized by the ASME.

U.S. Customary Dimensioning

ASME standards for **U.S. customary dimensioning,** with the use of decimal-inch values, are shown in **Figure 9-6**. In using decimal-inch dimensions, a zero is not used to the left of the decimal point for values less than one inch. The same number of decimal places should always be used for both dimensions in a tolerance dimension range. **Tolerance** is the total acceptable variation within a dimension (see Chapter 10). Establishing the number of decimal places represents the level of tolerance desired. Three decimal places implies a higher level of tolerance than two.

1. Use the decimal inch for dimension values (5.75).
2. Do not use a zero before the decimal point for values less than one inch (.50).
3. The dimension value is expressed to the same number of decimal places as its tolerance (5.500 ± .002).
4. The number of decimal places may represent the tolerance:
   - 5.5 = large tolerance
   - 5.50 = average tolerance
   - 5.500 = small tolerance

Figure 9-6
ASME conventions for U.S. customary dimensions

Metric Dimensioning

ASME standards for **metric dimensioning (Fig. 9-7)** require all dimensions to be expressed in millimeters. Therefore, the millimeter symbol (mm) is not needed on each dimension, but it is used when a dimension is used in a notation. In this case, a single space separates the numeral and the millimeter symbol. Zeros precede the decimal point when the value is less than one millimeter; zeros are not used when the dimension is a whole number. Commas are not used between thousand units. A metric dimensional note should be displayed in or near the title block.

1. All dimensions will be in millimeters.
2. Do not use millimeter symbol mm with dimensions (25 not 25 mm).
3. When a dimension value is less than one millimeter, a zero will precede the decimal point (0.55 not .55).
4. When the dimension is a whole number, do not use a zero or decimal point (150 not 150.0).
5. When the dimension uses a decimal value, do not end it with a zero (75.5 not 75.50).
6. An exception to convention No. 5 is: if the dimension value is followed by a tolerance, the number of decimal places must be equal (75.50 ± 0.05 not 75.5 ±0.05).
7. Do not use commas or spaces (1575 not 1,575 or 1 575).
8. When specifying millimeters of the drawing, as in a document, leave a space between the number and the symbol (2585 mm not 2585mm).
9. Display the word "METRIC" with large letters near the title block.
10. Display note: "UNLESS OTHERWISE SPECIFIED, ALL DIMENSIONS ARE MILLIMETERS."
11. The abbreviation for system international (metric system) is SI.

Figure 9-7
ASME conventions for metric dimensions

Dual Dimensioning

Working drawings are normally prepared with all U.S. customary or all metric dimensions. However, when the object is to be manufactured using both metric and U.S. customary measuring systems, **dual dimensions** may be necessary. Dual dimensioning may also be necessary when converting a metric drawing to a U.S. customary drawing and vice versa. When dual dimensions are used, the metric dimensions are placed under the dimension line in parentheses. Sometimes a metric conversion chart is used **(Fig. 9-8)**.

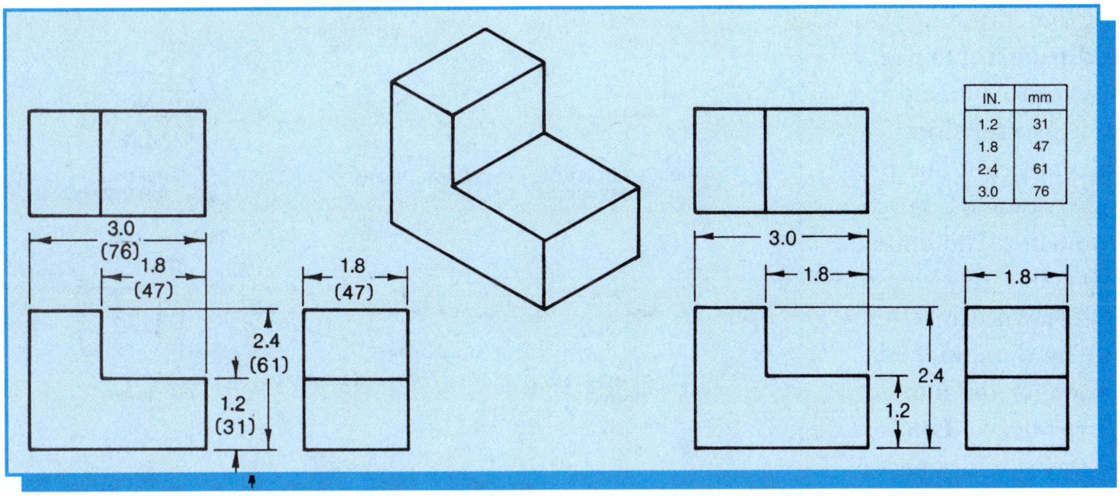

Figure 9-8
Two procedures used for dual dimensioning

Dimensioning Elements

ASME has also standardized the use of lines, numerical values, symbols, notations, and their placement on working drawings. **Figures 9-9** and **9-10** show the application of basic ASME **dimensioning symbols.** They illustrate the positioning of dimensioning lines, leaders, and notes to define the size and shape of common geometric forms and angles.

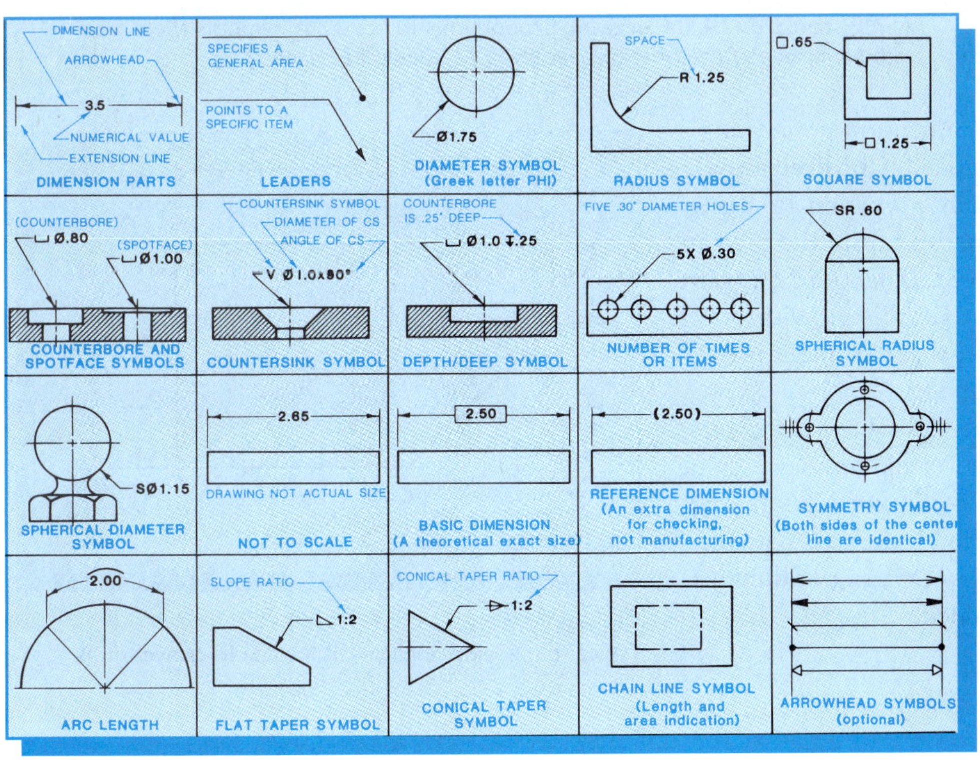

Figure 9-9
ASME dimensioning conventions

ASME dimensioning standards use **dimension lines** with arrowheads, dimension numerals, and **extension lines.** The extension line projects from the drawing but does not touch it. The dimension line touches the extension lines and is drawn parallel to the part being dimensioned. Leave a space for the numbers and add arrowheads. **Leaders** can also be used to connect a specific area of a drawing to a dimension or note. **Figure 9-11** shows the application of these standards to a typical view. **Figure 9-12** shows the correct method of drawing and spacing these lines. **Figure 9-13** shows the accepted proportion of a standard dimension arrowhead.

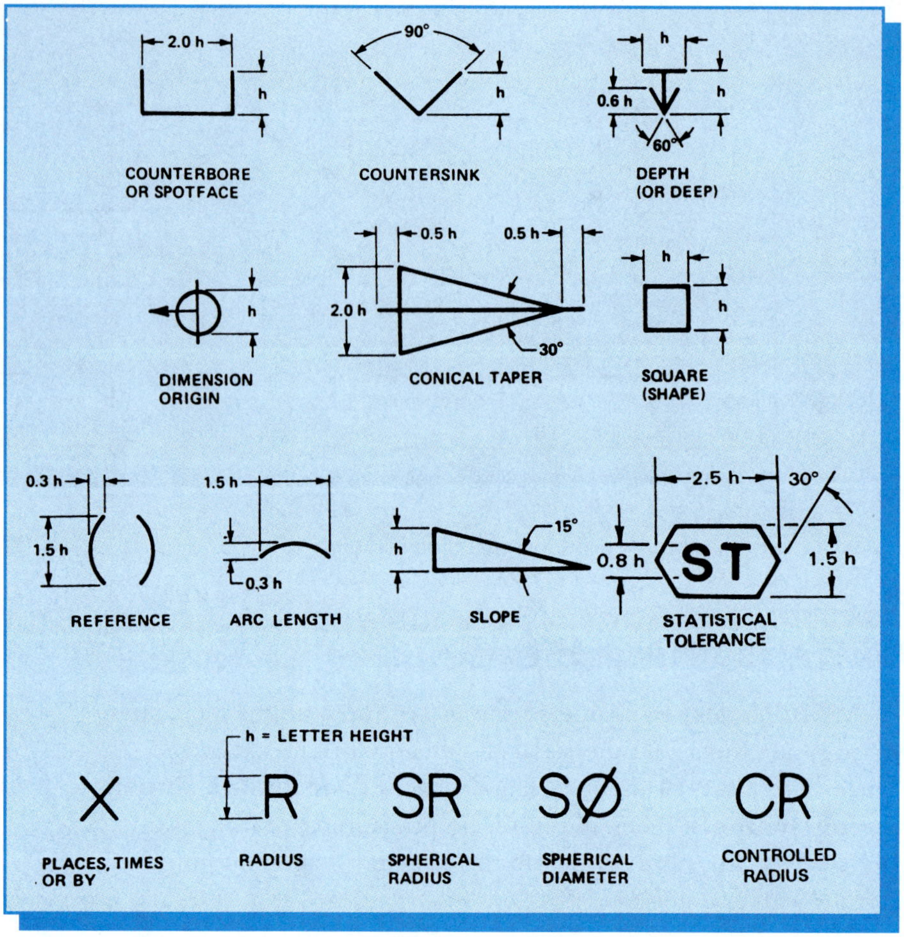

Figure 9-10
ASME standard Y14.5M, drawing proportions for drawing symbols *(Reprinted with permission of the American Society of Mechanical Engineers)*

Dimensioning Guidelines

Figure 9-14 illustrates basic dimensioning types. **Figure 9-15** shows the two standard dimension positioning systems: unidirectional and aligned. **Unidirectional dimensions** all read from the bottom and are easier to read. **Aligned dimensions** read from the bottom or right of the drawing. Aligned dimensions conserve space on a drawing since they are aligned with dimension lines. Either system is acceptable if used consistently throughout a drawing.

If dimensions are not spaced properly, the drawing surface area will be wasted or dimensions may be too close to be read without confusion. Closely spaced dimensions can also inhibit the easy reading of object lines. For these reasons, spacing of dimensions **(Fig. 9-16)** is recommended.

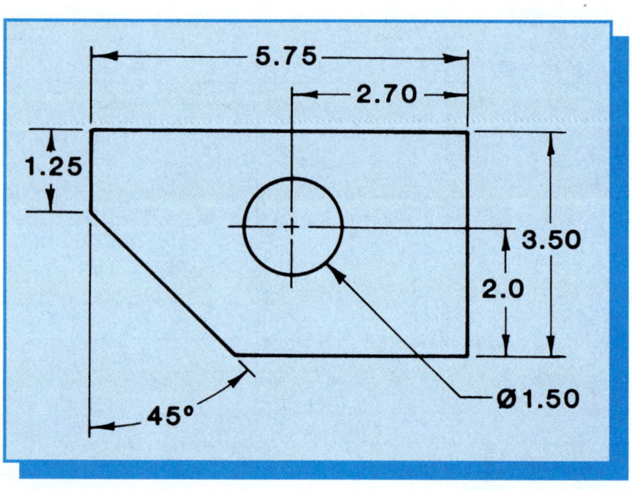

Figure 9-11
Typical dimension application for ASME conventions

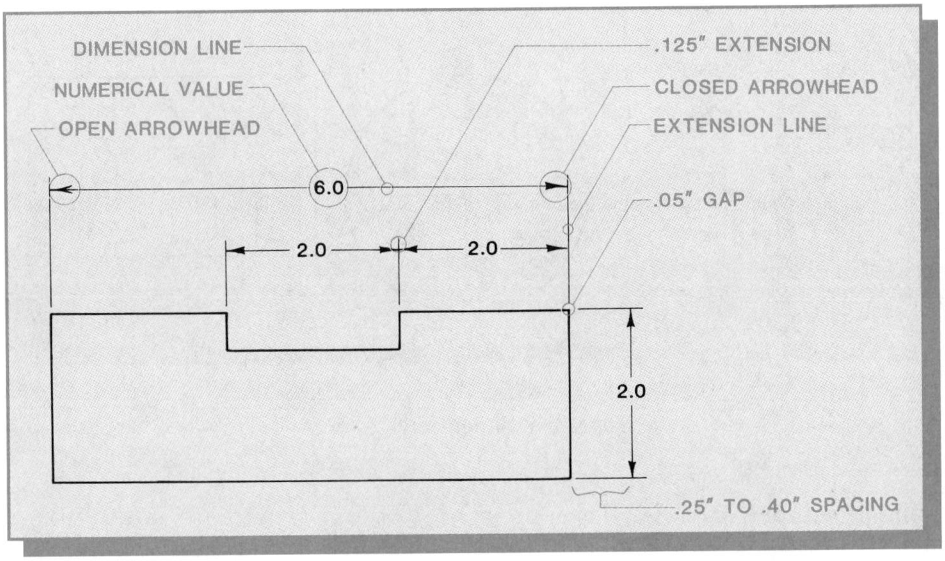

Figure 9-12
General dimensioning conventions

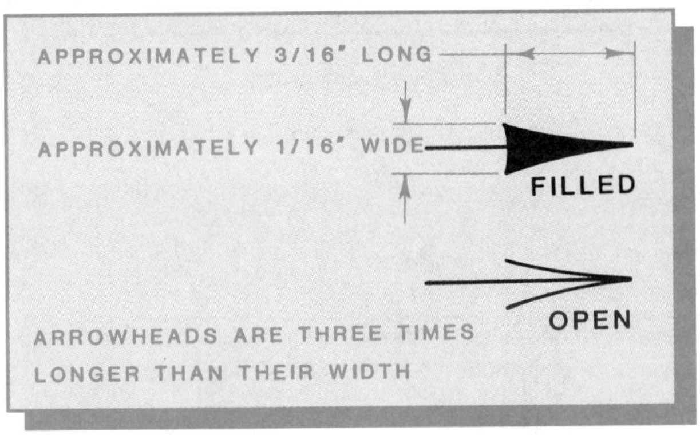

Figure 9-13
The freehand arrowhead and its proportions

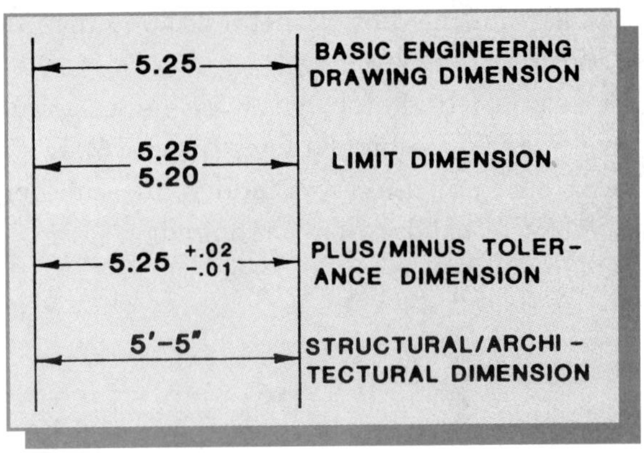

Figure 9-14
Dimension types

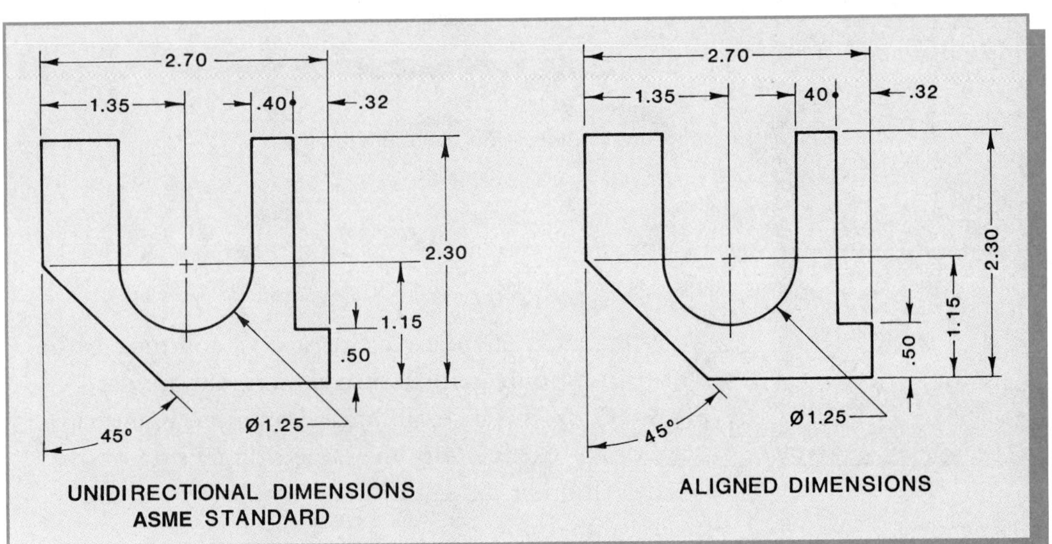

Figure 9-15
Positioning systems for reading dimensions

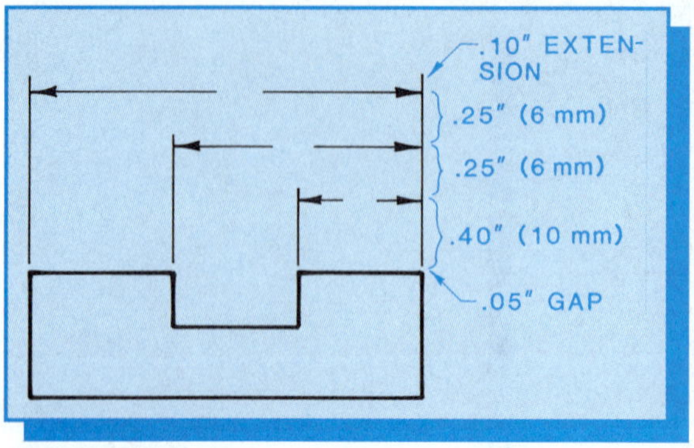

Figure 9-16
Dimension line spacing

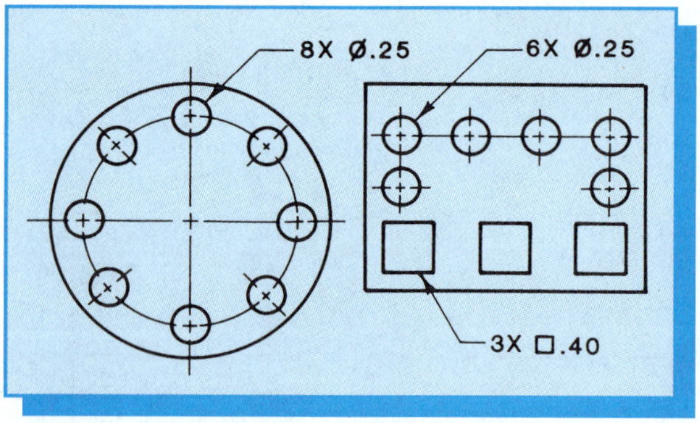

Figure 9-17
Repetitive dimensions

Many manufactured objects contain geometric forms or parts that are identical. When this occurs, only one dimension is needed **(Fig. 9-17)**. A dimension that covers many identical (repetitive) parts is always accompanied by a label indicating the number of parts of the same size.

Regardless of the object, always dimension to visible lines, not to hidden lines **(Fig. 9-18)**. If it is necessary to dimension to a hidden line, another view or sectional drawing is required.

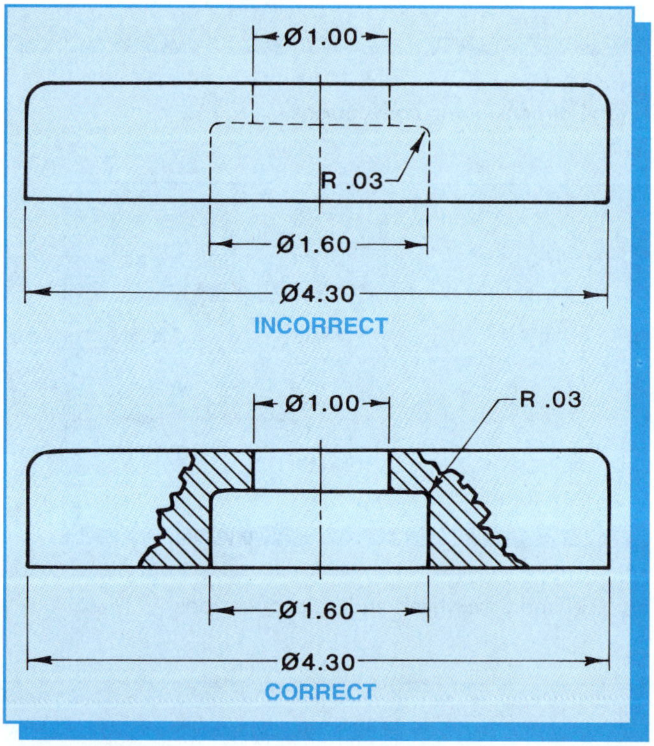

Figure 9-18
Always dimension to visible lines.

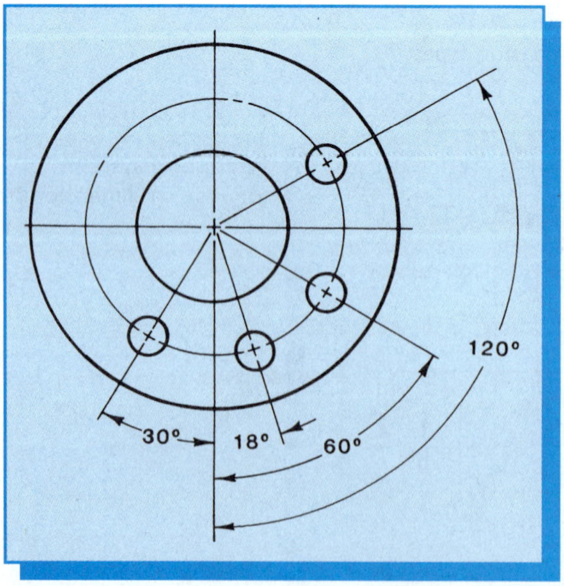

Figure 9-19
Example of polar coordinate dimensioning

Polar Coordinate Dimensioning

When parts of a drawing align on a common **radial centerline, polar coordinate** dimensions are used **(Fig. 9-19)**. In this case, all radial dimension lines originate from a baseline with the largest dimension on the outside. All other dimensions progress inward to the smallest dimension.

Rectangular Coordinate Dimensioning

When a large number of parts, projections, holes, or complex contours are contained in a small area, **rectangular coordinate** dimensions may be used in lieu of conventional dimension practices. Rectangular coordinate systems eliminate the need for dimension lines by establishing distances from base intersections, usually a corner **(Fig. 9-20)**. The baseline is assigned a value of zero. The dimension to each extension line represents the distance from zero. To eliminate size dimensions, each standard hole or projection is assigned a letter that relates to the size specified in an accompanying chart.

Rectangular coordinate grid systems are used extensively to define and locate irregular contour lines **(Fig. 9-21)**. In this application, the grid lines are spaced evenly so each square can be increased (or decreased) in size to

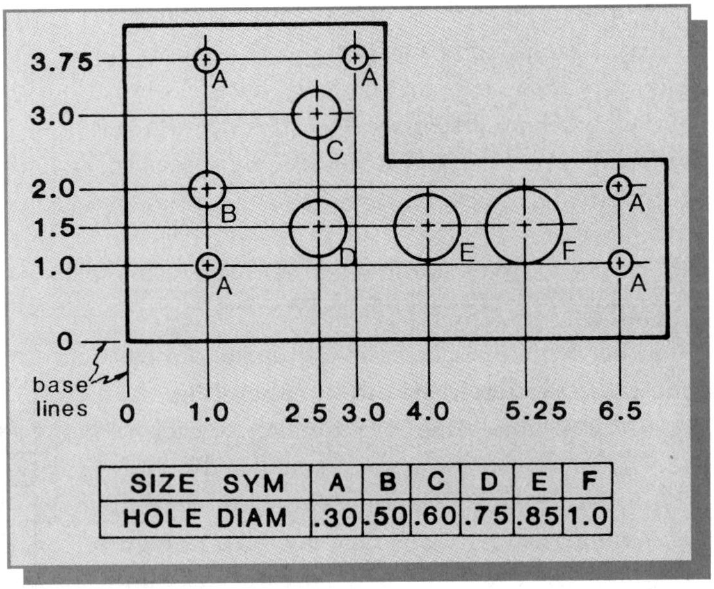

SIZE SYM	A	B	C	D	E	F
HOLE DIAM	.30	.50	.60	.75	.85	1.0

Figure 9-20
Rectangular coordinate dimensioning without dimension lines

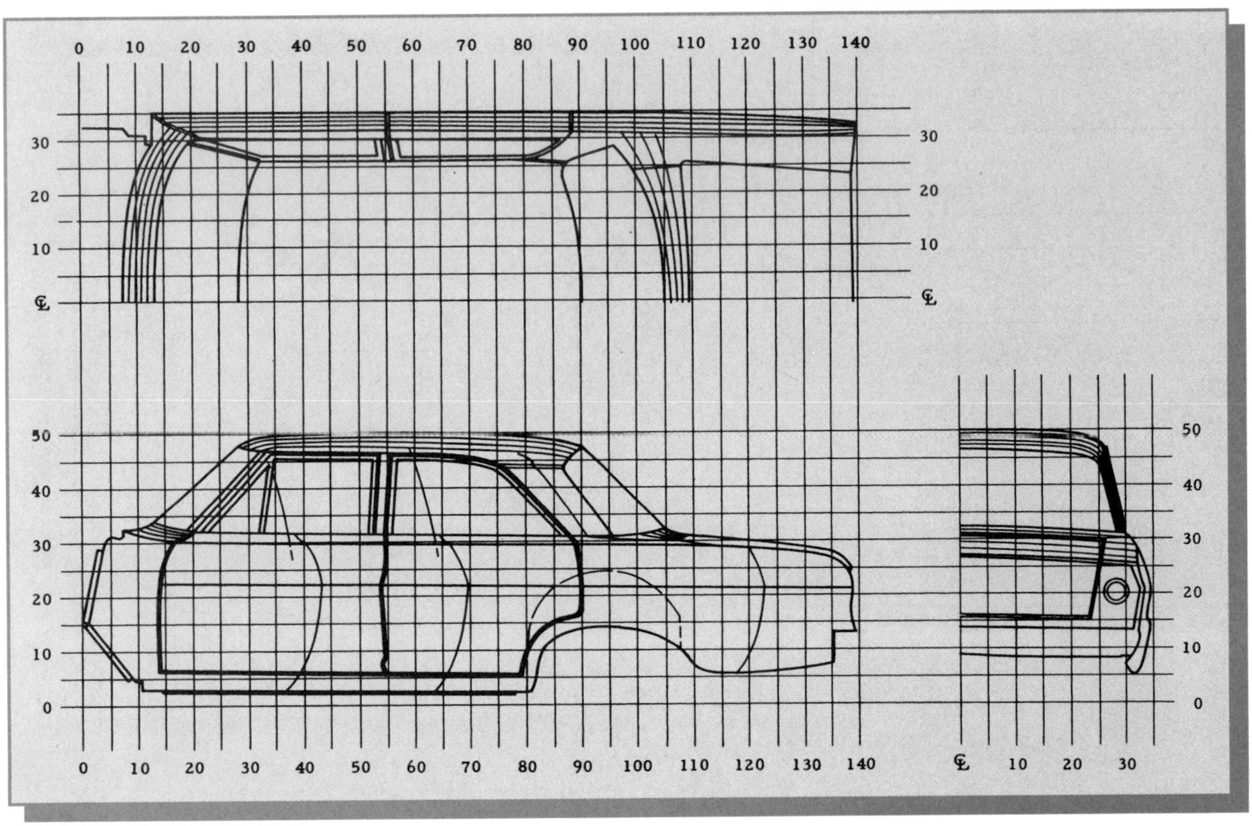

Figure 9-21
Rectangular coordinates form zones for area locations. *(Courtesy of General Motors Corporation Engineering Standards)*

duplicate the outline of the design. Some rectangular coordinate systems completely eliminate all detail dimensions with a chart that contains distances from established baselines. In **Figure 9-22**, the location and depth of holes on X, Y, and Z surfaces are shown in the top chart. The bottom chart shows the number and diameter of each hole.

Tabulated Outline Dimensioning

Another type of rectangular coordinate dimensioning is the **tabulated outline** method **(Fig. 9-23)**. In this example, the X-Y distance of each plotting point on a contour line is tabulated on an X-Y coordinate chart. The conventional method of dimensioning a curve of this type is shown in **Figure 9-24**. Notice all dimensions originate from baselines.

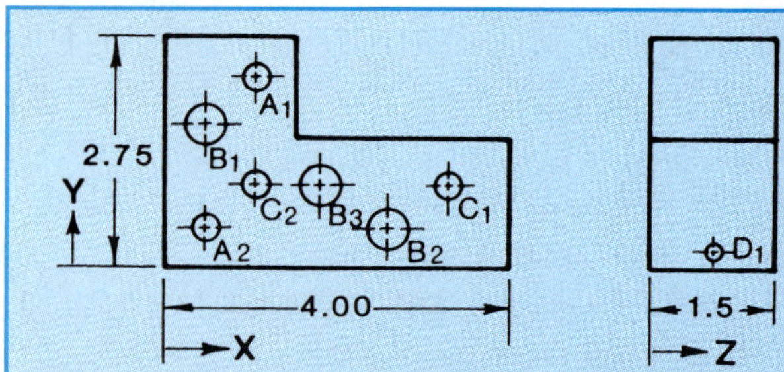

HOLE	FROM SURF	X	Y	Z
A_1	X,Y	1.0	2.25	THRU
A_2	X,Y	.5	.5	THRU
B_1	X,Y	.5	1.7	.50
B_2	X,Y	2.6	.5	.50
B_3	X,Y	1.8	1.0	.50
C_2	X,Y	1.0	1.0	THRU
C_1	X,Y	3.3	1.0	THRU
D_1	Z,Y	THRU	.1	.75

HOLE	DESCRIPTION	QUANTITY
A	Ø.30	2
B	Ø.45	3
C	Ø.28	2
D	Ø.20	1

Figure 9-22
Rectangular coordinate dimensioning in tabular form

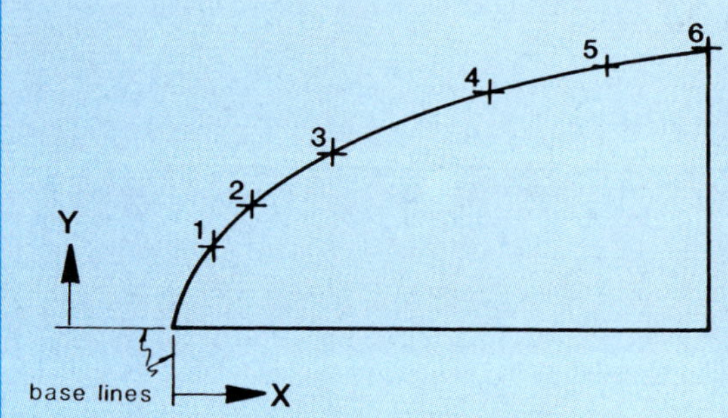

COORDINATE	1	2	3	4	5	6
X	.5	1.0	2.0	4.0	5.5	6.5
Y	1.0	1.5	2.25	3.0	3.3	3.5

Figure 9-23
Tabulated outline

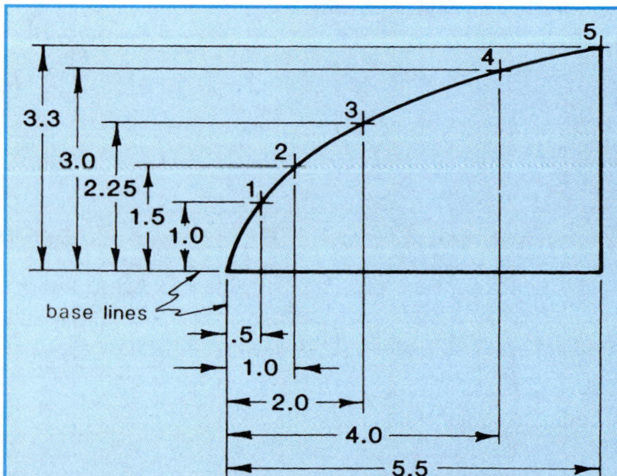

Figure 9-24
Coordinate or offset outline dimensioning

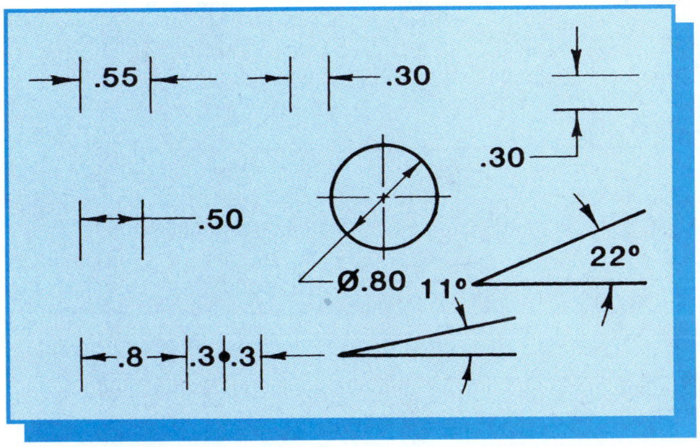

Figure 9-25
Dimensioning conventions for crowded areas

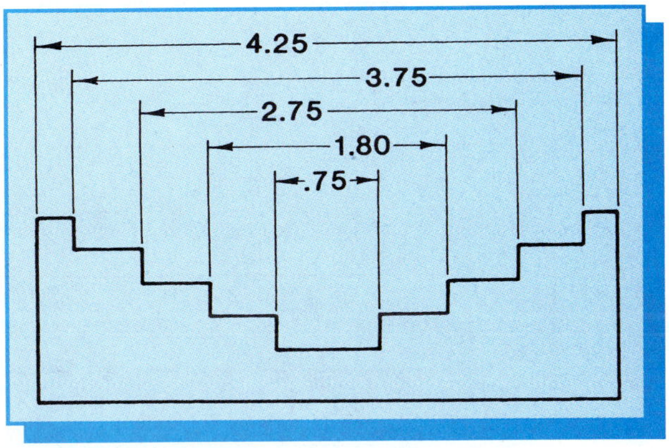

Figure 9-26
Examples of staggered dimensions

Special Dimensioning Procedures

Since some object configurations or areas are difficult to dimension, special dimensioning procedures are standardized.

Dimensioning Crowded Areas

Figure 9-25 shows examples of dimensioning practices for crowded areas. **Figure 9-26** shows how dimensions should be staggered for easier reading. In rare cases, oblique dimensioning **(Fig. 9-27)** is used to remove a dimension from a crowded area. Often, a corner or area to be dimensioned is outside the object outline or may be eliminated during manufacturing. In this case, extension lines are drawn to imaginary intersections. Extension and dimension lines are then added **(Fig. 9-28)**.

Since a wide variety of machined holes are used in manufactured products, dimensioning standards include ASME symbols **(Fig. 9-29)**. From these notation symbols, the print reader can visualize the information shown in the accompanying sectional view.

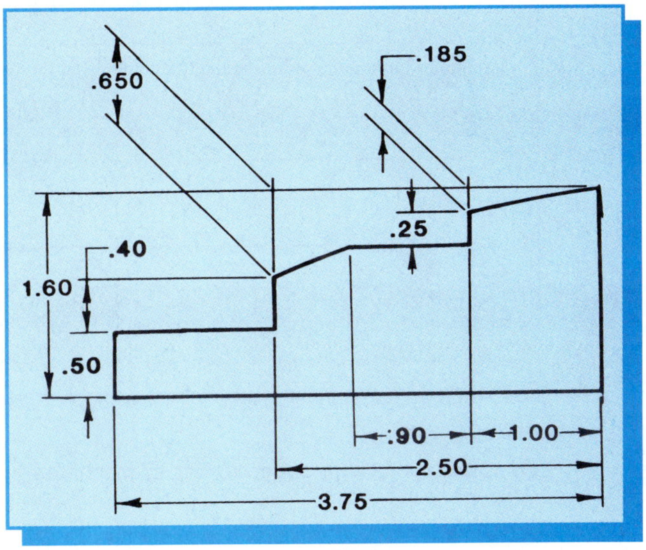

Figure 9-27
Examples of oblique dimensions

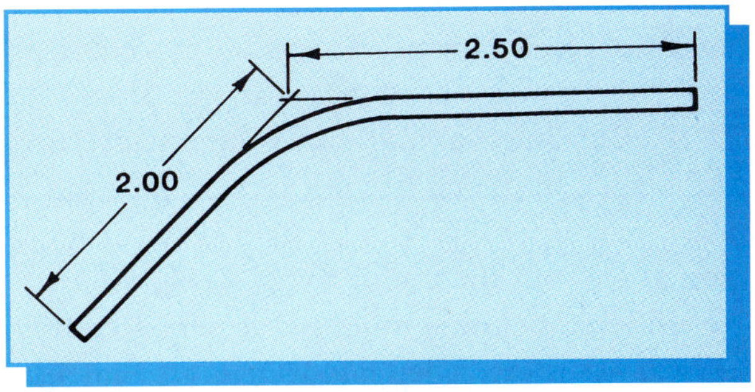

Figure 9-28
Locating a point for dimensioning

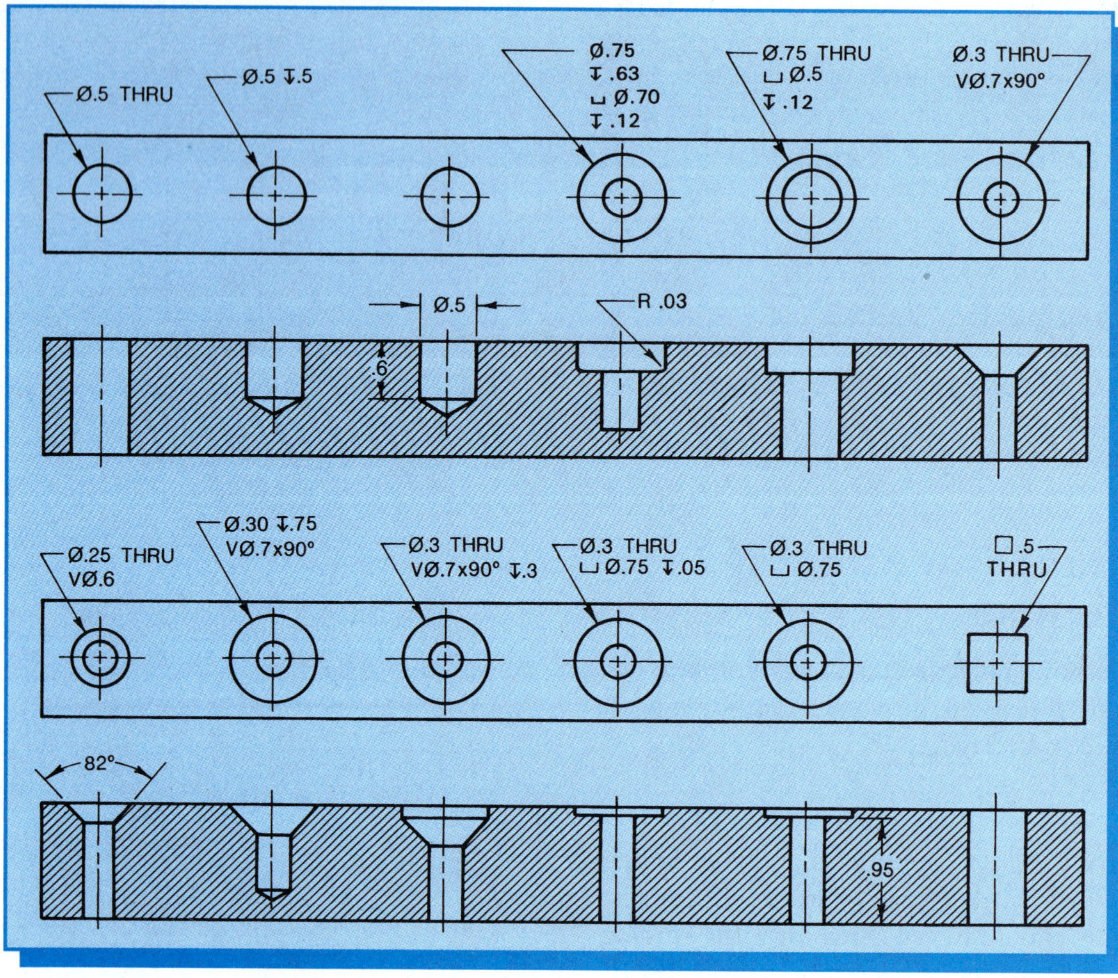

Figure 9-29
Conventions for the dimensioning of machined holes

Dimensioning Slots, Radii, Fillets, Rounds, and Chamfers

Often, the dimensioning of rounded ends **(Fig. 9-30)**, slots, radii, fillets, and chamfers offers an acceptable opportunity to shortcut normal dimension procedures. **Figure 9-31** shows three methods of dimensioning **slots.**

Radii dimensioning shows the distance from apex to arc. If the apex of the radii is not located near the drawing, an offset dimension line can be used. Very small radii are dimensioned without the use of an apex **(Fig. 9-32)**.

Fillets are inside rounded corners. **Rounds** are outside rounded corners. Fillets and rounds are dimensioned in a similar manner **(Fig. 9-33)**.

Chamfers are flat corners and are either external, internal, or oblique **(Fig. 9-34)**.

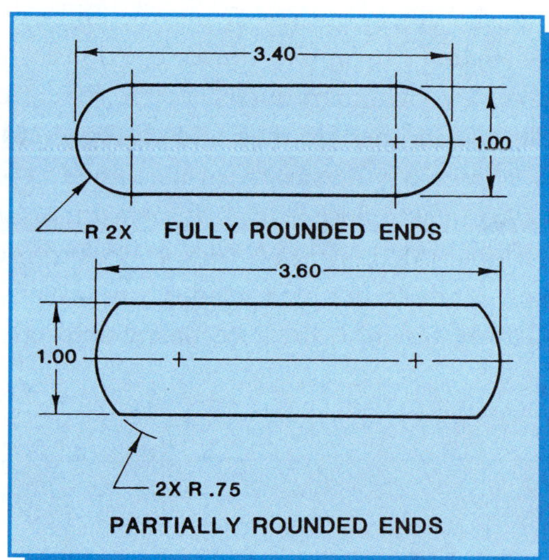

Figure 9-30
Conventions for the dimensioning of rounded ends

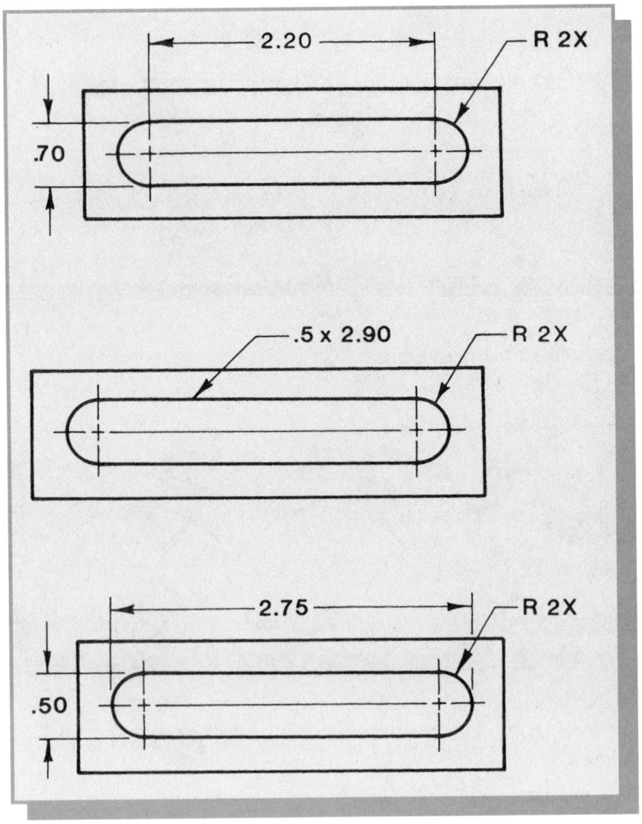

Figure 9-31
Conventions for the dimensioning of slotted holes

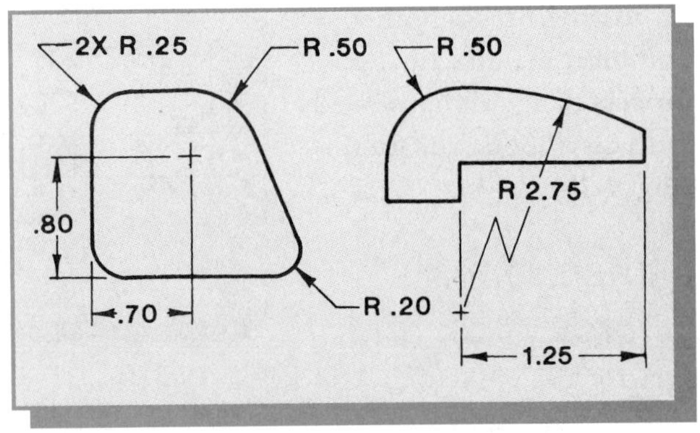

Figure 9-32
Conventions for the dimensioning of radii

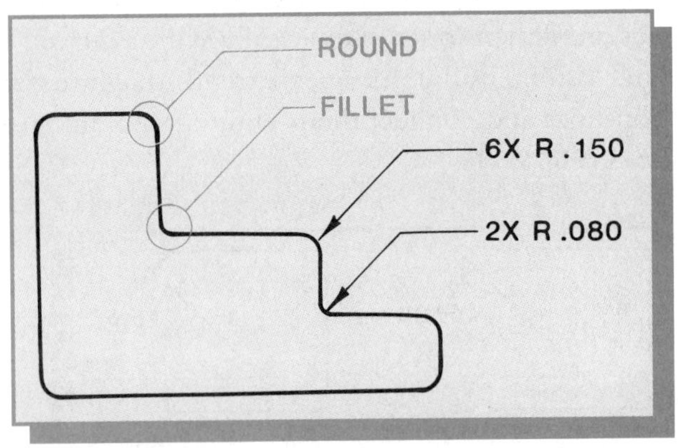

Figure 9-33
Conventions for the dimensioning of fillets and rounds

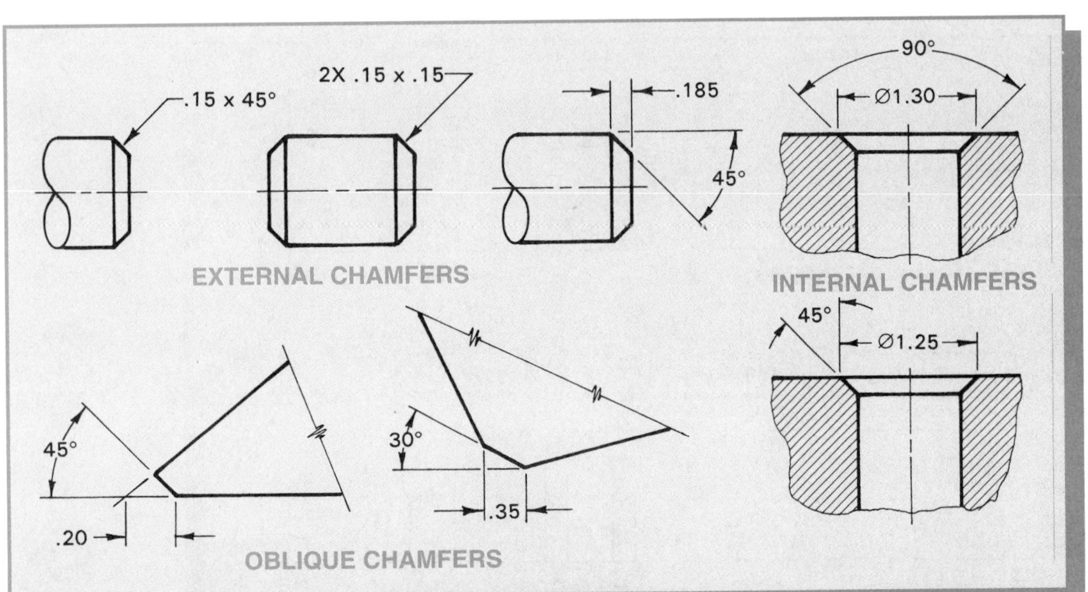

Figure 9-34
Conventions for the dimensioning of chamfers

Knurling is another common manufacturing process that is dimensioned with a symbolic notation **(Fig. 9-35)**.

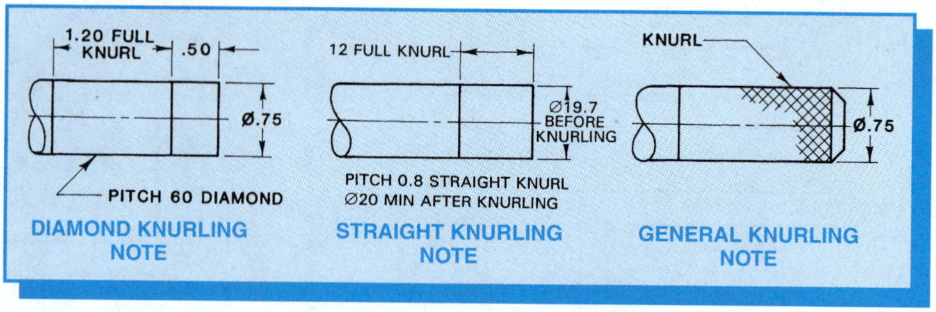

Figure 9-35
Conventions for knurling dimensions and notations

Dimensioning Chords, Arcs, and Angles

Chords, arcs, and angles may appear similar in many drawings, but should be dimensioned differently **(Fig. 9-36)**. Chords are dimensioned numerically with a straight line, and arcs are dimensioned numerically with a curved line. In dimensioning angles, extend straight extension lines and connect them with curved dimension lines **(Fig. 9-37)**.

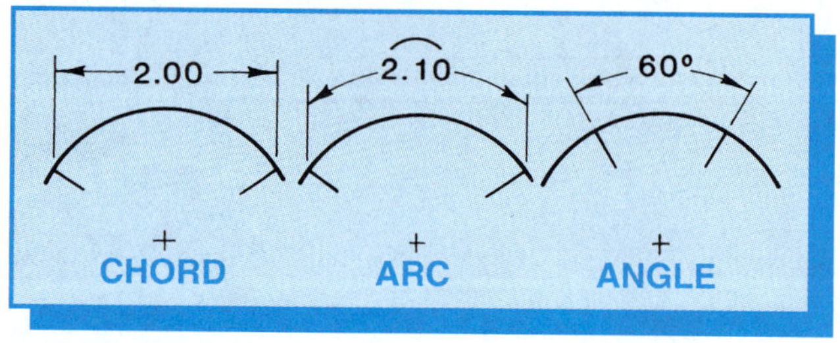

Figure 9-36
Conventions for the dimensioning of chords, arcs, and angles

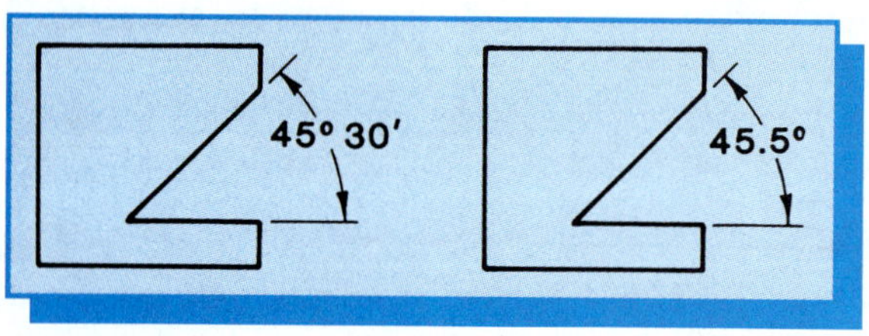

Figure 9-37
Conventions for dimensioning angles

Dimensioning Prisms, Cylinders, Cones, Pyramids, and Spheres

Other basic forms, such as prisms, cylinders, cones, pyramids, and spheres, are dimensioned by ASME standards **(Fig. 9-38)**.

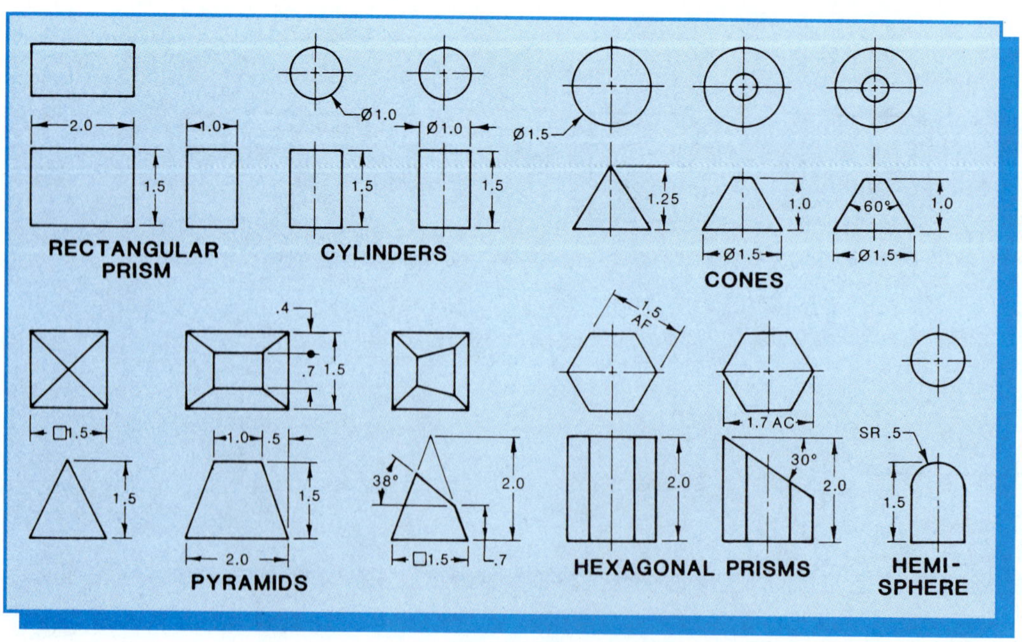

Figure 9-38
Examples of dimension placement on basic forms

Pictorial Dimensioning

Pictorial drawings are rarely used as working drawings and are usually accompanied by a fully dimensioned multiview drawing. If a pictorial drawing must be used as a working drawing, it should be dimensioned as shown on the isometric drawing in **Figure 9-39**. (See Chapter 15 for pictorial drawing details.)

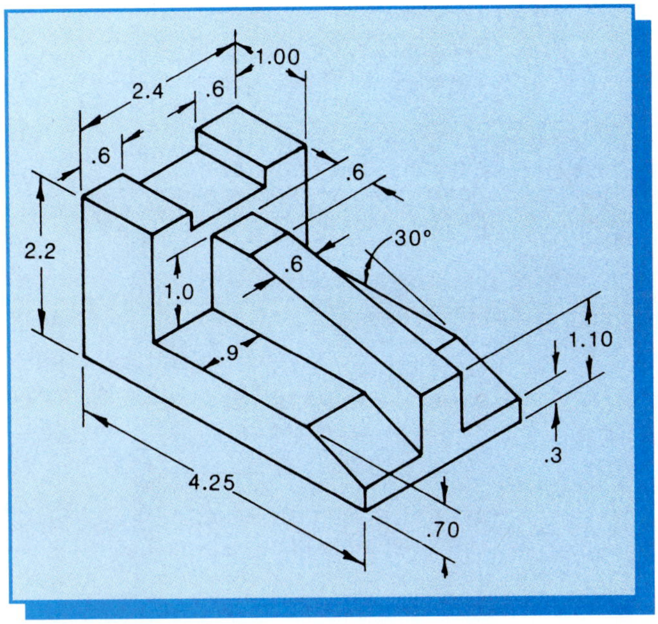

Figure 9-39
Example of a dimensioned isometric drawing

Surface Control

The **surface texture** for manufacturing processes is specified on working drawings through **finish symbols (Fig. 9-40)**. The surface texture value is controlled with a symbol or notation on the working drawing **(Fig. 9-41)**. The point of the symbol is placed on the surface for which a surface finish is specified **(Fig. 9-42)**. Different forms of the surface finish symbol are used to designate different surface finishes **(Fig. 9-43)**.

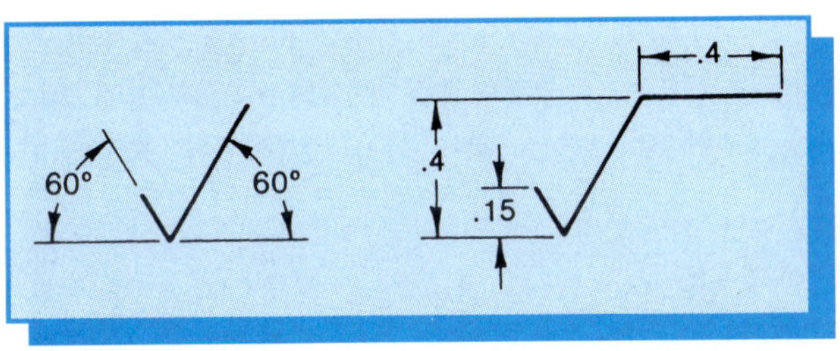

Figure 9-40
Typical size dimensions for the surface finish symbol

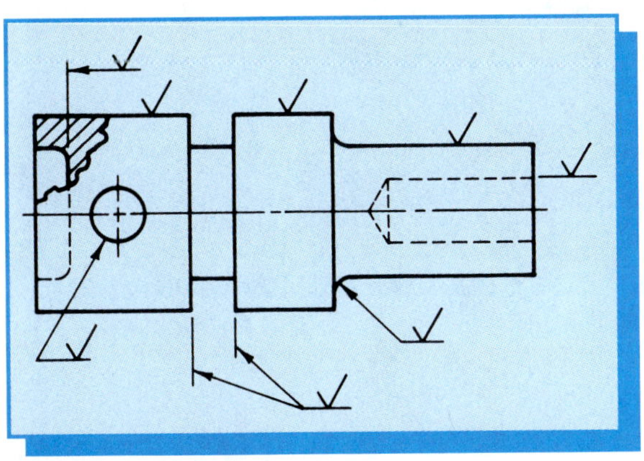

Figure 9-42
Examples for various positions of the surface finish symbols

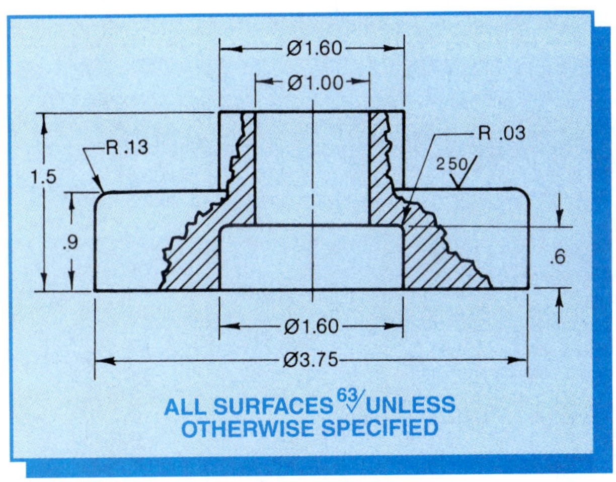

Figure 9-41
The surface texture value may be controlled with symbols or a notation placed on the drawing.

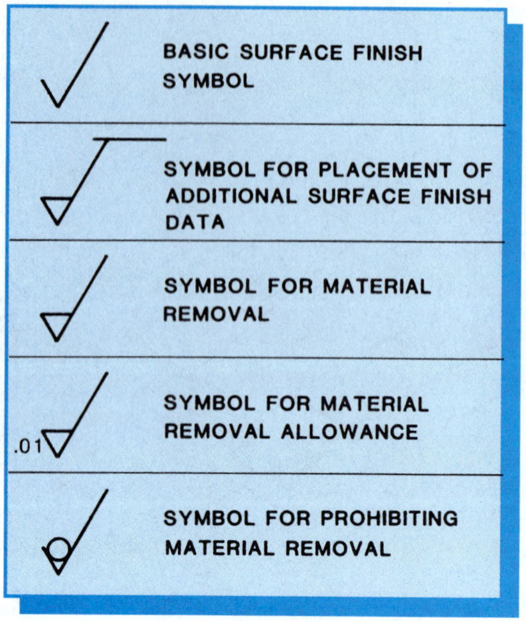

Figure 9-43
Various types of surface finish symbols

Surface **smoothness** is a relative term. No surface appears smooth under a microscope. Some surfaces require a higher degree of smoothness than others. Bearings, seals, gears, pistons, and most matching and moving parts require a high degree of smoothness. Other surfaces, such as outside walls of machine parts, can be as rough as the final castings or forgings permit. It is therefore important to correctly specify the required degree of smoothness or **roughness** on working drawings. Roughness is the opposite of smoothness. **Figure 9-44** shows the basic characteristics of surface textures that are used to define degrees of roughness.

In specifying degrees of roughness, the drafter must specify values for roughness height **(Fig. 9-45)**, roughness width **(Fig. 9-46)**, waviness height **(Fig. 9-47)**, and waviness width **(Fig. 9-48)**.

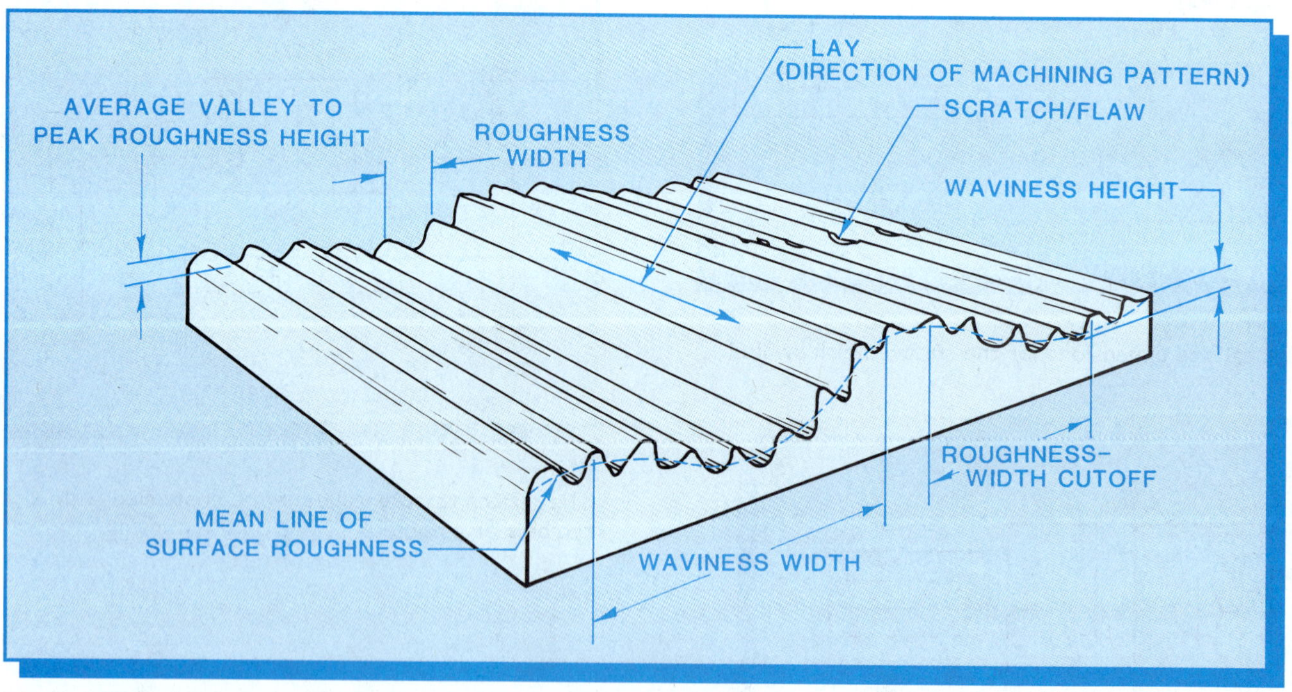

Figure 9-44
Characteristics of surface texture

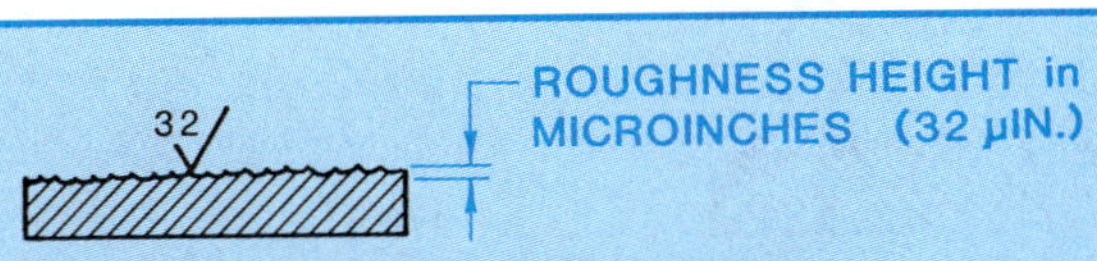

Recommended Roughness Average Rating Values Microinches, μin. [Micrometers, μm]*

μin.	μm	μin.	μm	μin.	μm	μin.	μm
1	[0.025]	**10**	[0.25]	50	[1.25]	250	[6.3]
2	**[0.050]**	13	[0.32]	**63**	**[1.6]**	320	[8.0]
3	[0.075]	**16**	**[0.40]**	80	[2.0]	400	[10.0]
4	**[0.100]**	20	[0.50]	100	[2.5]	**500**	**[12.5]**
5	[0.125]	25	[0.63]	**125**	**[3.2]**	600	[15.0]
6	[0.15]	**32**	**[0.80]**	160	[4.0]	800	[20.0]
8	**[0.20]**	40	[1.00]	200	[5.0]	1000	[25.0]

*Boldface values preferred

Figure 9-45
The roughness height values noted on the finish symbol. The prefix micro is one-millionth (.000001 or 10^{-6}).

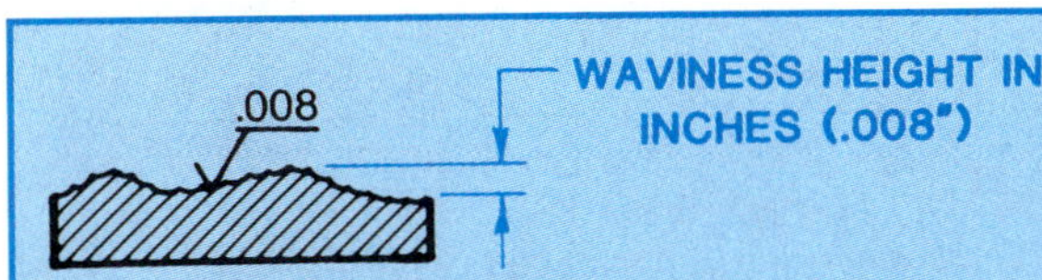

Recommended Waviness Height Values Inches, in. Millimeters, mm*

in.	mm	in.	mm	in.	mm
.00002	**[0.0005]**	.00003	[0.008]	**.005**	**[0.12]**
.00003	[0.0008]	**.00005**	**[0.021]**	.008	[0.20]
.00005	**[0.0012]**	.00008	[0.020]	**.010**	**[0.25]**
.00008	[0.0020]	**.00010**	**[0.025]**	.015	[0.38]
.00010	**[0.0025]**	**.002**	**[0.05]**	**.020**	**[0.50]**
.0002	**[0.005]**	.003	[0.08]	.030	[0.80]

*Boldface values preferred

Figure 9-47
The waviness height values noted on the finish symbol

Figure 9-46
The callout for the roughness width noted on the finish symbol

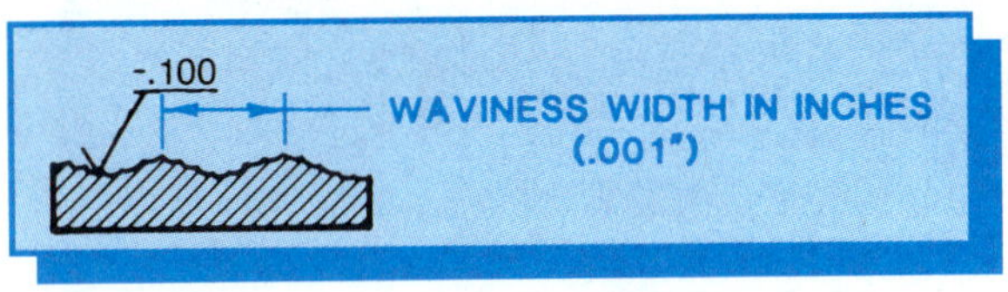

Figure 9-48
The waviness width noted on the finish symbol

Lay is the primary direction of the surface pattern made by machine tool marks. Seven symbols are given for the lay patterns produced by different manufacturing processes **(Fig. 9-49)**.

Roughness-width cutoff is the greatest spacing of repetitive surface irregularities included in the average roughness height. It is specified in inches on working drawings **(Fig. 9-50)**.

Figure 9-51 shows a surface control symbol with all its controlling features. **Figure 9-52** shows the method used for specifying minimum **contact area** for mating parts.

SYM	DESCRIPTION OF LAY	TOOL MARKS AND FINISH SYMBOLS
=	Lay parallel to the line representing the surface to which the symbol is applied.	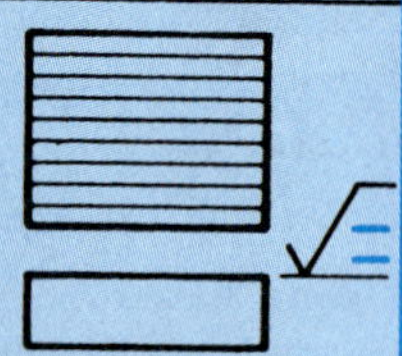
⊥	Lay perpendicular to the line representing the surface to which the symbol is applied.	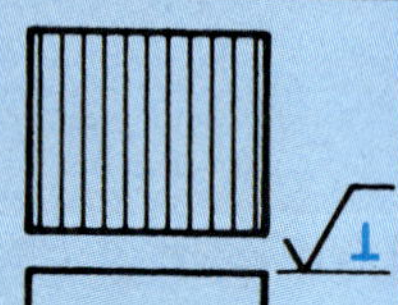
X	Lay angular in both directions to line representing the surface to which symbol is applied.	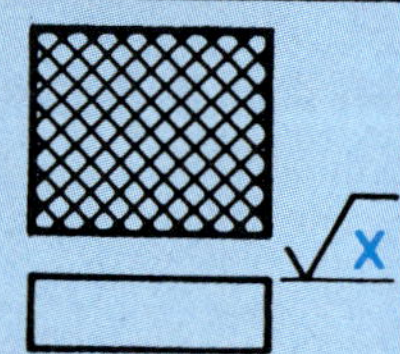
M	Lay multidirectional/ random pattern.	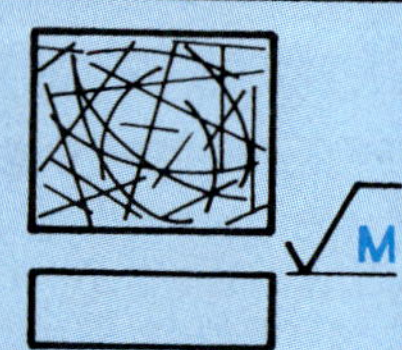
C	Lay approximately circular relative to the center of the surface to which the symbol is applied.	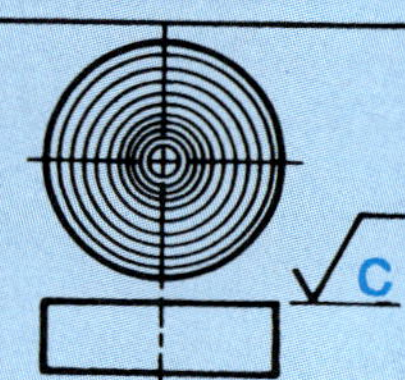
R	Lay approximately radial relative to the center of the surface to which the symbol is applied.	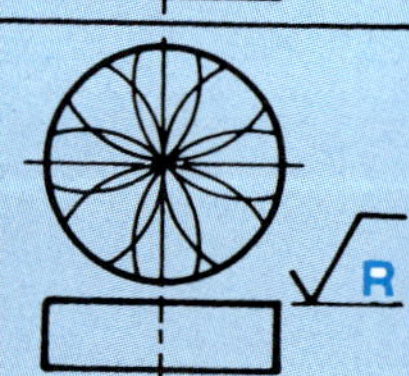
P	Lay particulate, nondirectional, or protuberant.	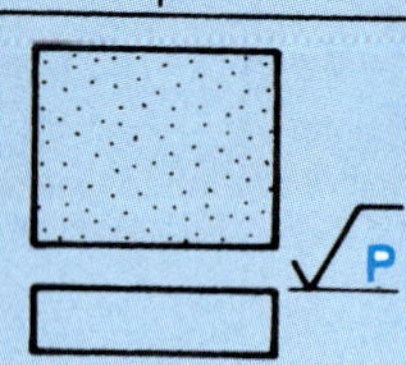

Figure 9-49
The lay symbols and their characteristics

STANDARD ROUGHNESS-WIDTH CUTOFF VALUES*						
mm	0.075	0.250	0.750	2.500	7.500	25.000
in	0.003	0.010	0.030	0.100	0.300	1.000

*When no value is specified, the value .030 in. (0.750 mm) is assumed.

Figure 9-50
The roughness-width cutoff noted on the finish symbol. (The greatest spacing of repetitive surface irregularities, due to machining, that is included in the average roughness height.)

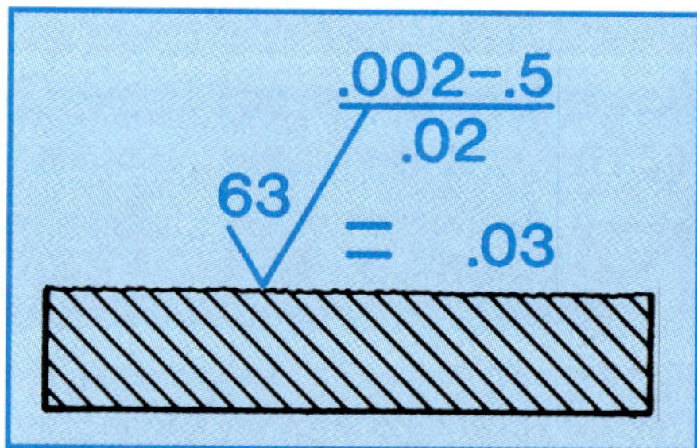

Figure 9-51
Surface control symbol in full use

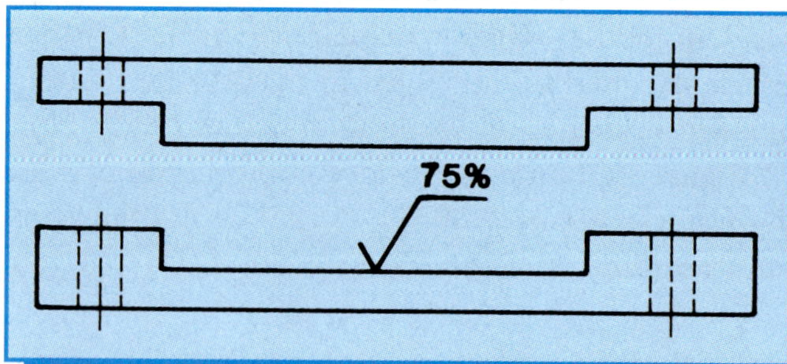

Figure 9-52
Specifying the minimum contact area

Roughness Tolerance

Different manufacturing processes, such as depth of cut, amount of feed, and machining repetition, result in varying degrees of surface smoothness. Engineers must use design processes that produce the proper degree of smoothness with the least amount of manufacturing processing. Producing surfaces that are smoother than necessary may drastically increase production time and costs. Producing surfaces that are too rough may result in an inferior-quality product. In specifying degrees of smoothness—that is, tolerance for roughness—refer to the charts shown in **Figures 9-53** and **9-54**. These charts show the degree of roughness height produced by various manufacturing and machining processes.

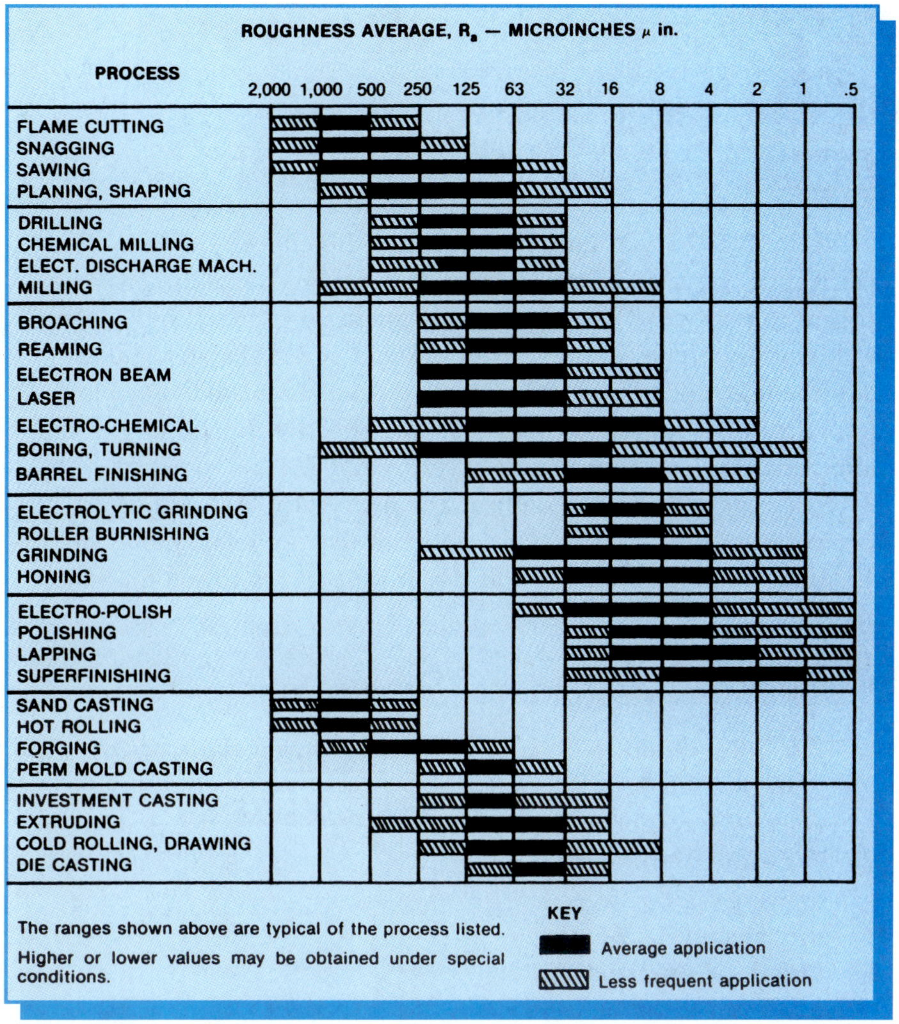

Figure 9-53
The roughness height ratings for surface finishes produced by common production procedures *(Reprinted from* Blueprint Reading For Manufacturing *by Hoffman and Wallach)*

ROUGHNESS HEIGHT MICROINCHES (μ INCH)	SURFACE DESCRIPTION	MACHINING PROCESSES
1000√	Very rough	Saw or torch cutting, forging, or sand casting
5000√	Rough machining	Coarse feeds and heavy cuts in machining
250√	Coarse	Coarse surface grind, medium feeds, and average cuts in machining
125√	Medium	Sharp tools, light cuts, fine feeds, high speeds with machining
63√	Good finish	Sharp tools, light cuts, extra fine feeds, light cuts with machining
32√	High grade finish	Very sharp tools, very fine feeds, and cuts
16√	Higher grade finish	Surface grinding, coarse honing, coarse lapping
8√	Very fine machine finish	Fine honing and fine lapping
3√	Extremely smooth machine finish	Extra fine honing and lapping

Figure 9-54
Typical roughness height values

AutoCAD LT Exercises

AutoCAD LT can automatically place dimensions on parts. Horizontal, vertical, oblique, aligned, and unidirectional dimensioning are all possible. The ability to dimension arcs, circles, and angles is also an important feature of AutoCAD. The CAD system calculates the dimensions from the geometry that was created. Drawing accurately is an absolute necessity for the effective functioning of the automatic dimensioning feature of AutoCAD LT.

Since the geometry of the part is where AutoCAD gets its data to dimension the part, it is very important that we select points on the part carefully. In order to do this, it is necessary to use the object snap commands.

Vertical and Horizontal Dimensions

1. Use **File, Open** to load drawing CP5-1.
2. Select **View, Zoom, Window** to make the drawing appear larger on the screen **(Fig. 9-55)**.
3. Use the **Construct, Fillet** to add an R.500 radius to the part.
   a. Type **R** and give radius size of .5. Then press select to bring back the fillet command.
   b. Select lines A and B.
4. Select **Settings, Dimension Styles, Text Location**; double click on the 0.180 in the Text Height box; and type in **.12. Select OK**, **OK**.
5. Select **Settings**, **Units Style**, and use the scroll arrow to set the precision to a two-place decimal. Select the 0.00 **(Fig. 9-56)** and select **OK**.

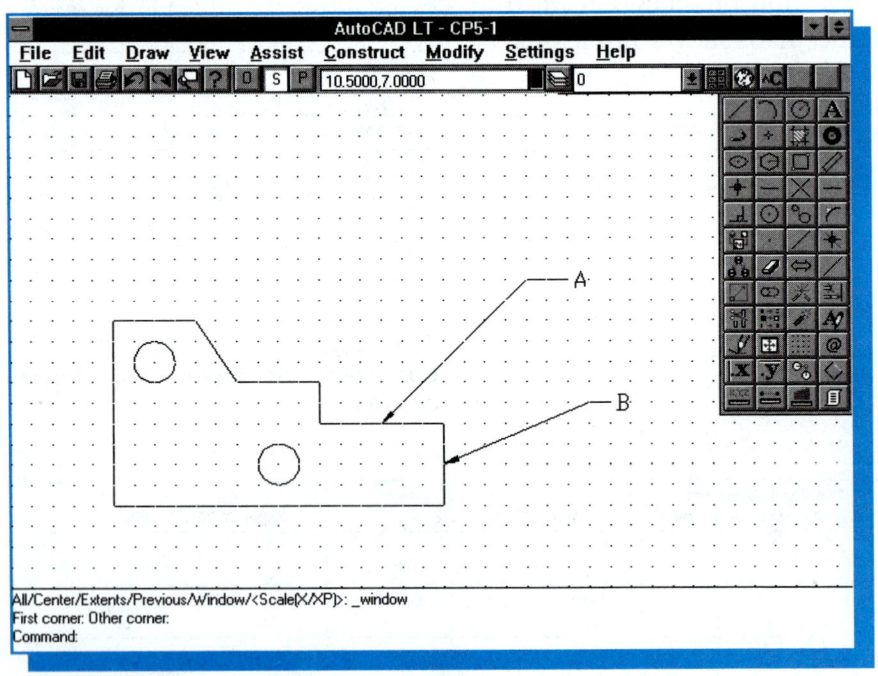

Figure 9-55
Drawing CP5-1 zoomed using Zoom—Window command

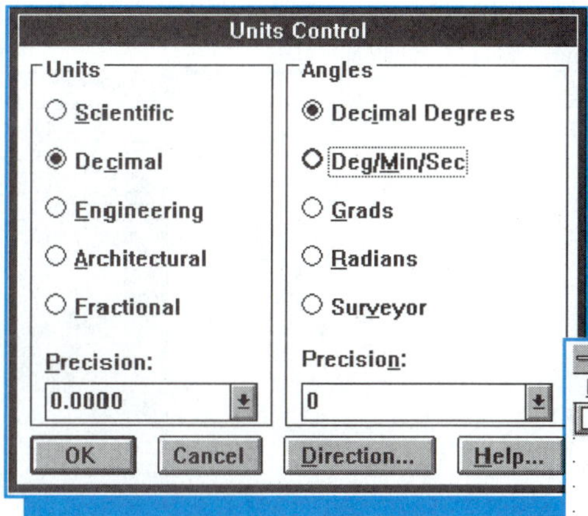

Figure 9-56
Units Control dialog box for setting the number of decimal places

Figure 9-57
Placement of the vertical dimension

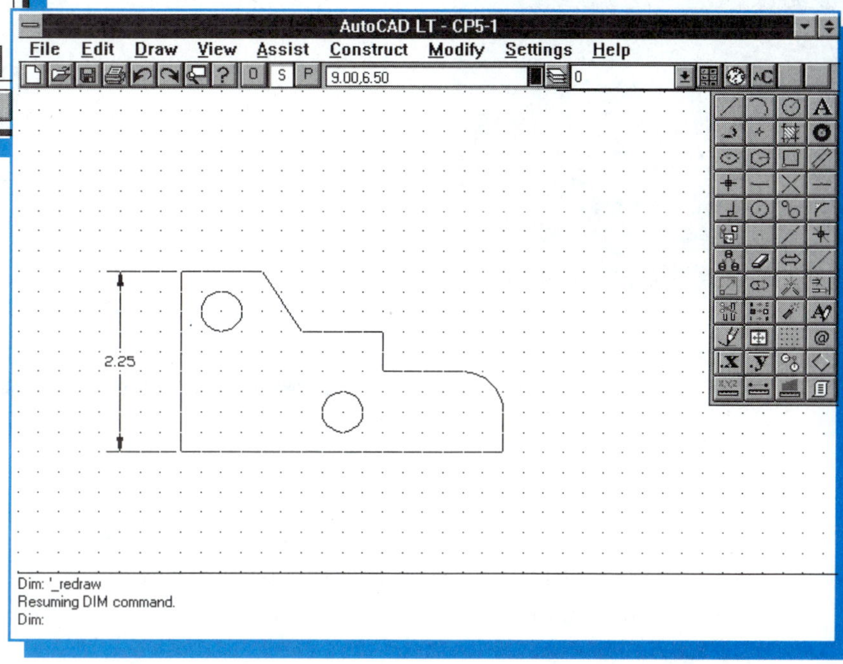

6. Select **Draw, Linear Dimensions, Vertical**, and use the intersection Osnap to select the two corners (Pts. 1 and 2) on the left side of the drawing.

   Place the dimension 1.00 inch away from the part. Use Snap and select Pt. 3. Select Enter to complete the dimension **(Fig. 9-57)**.
7. Using the same procedure, place the **horizontal** dimension using Pts. 1, 2 and 3 **(Fig. 9-58)**.

Angular Dimensions

1. Select **Draw**, **Angular Dimension.** Select lines A and B **(Fig. 9-59)**. Place arc line location as shown using Pick button. To select Pt. 1, press (E8) for the Default size. Place the angle text as shown. Pick Pt. 2

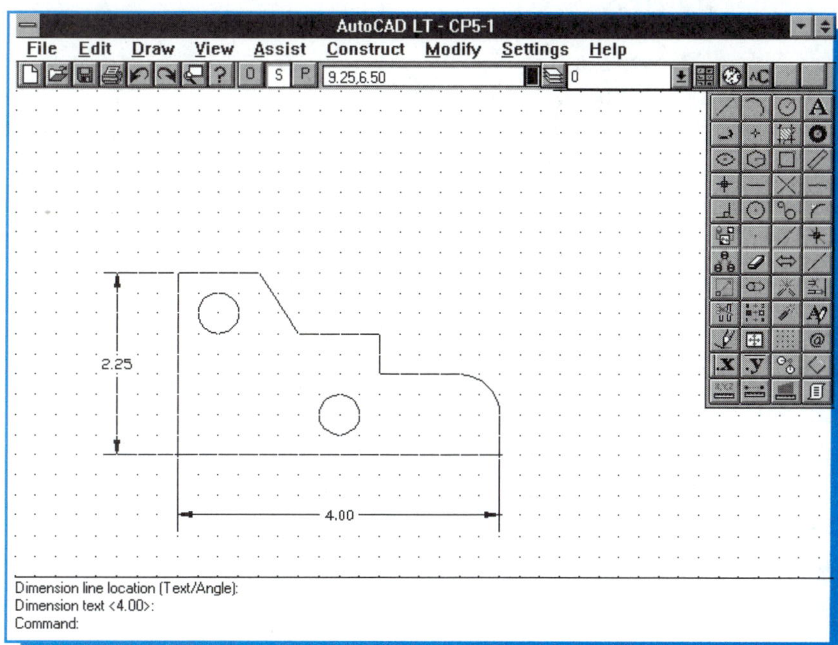

Figure 9-58
Placement of the horizontal dimension

Diameter Dimensions

1. Select **Draw**, **Radial dimension**, **Diameter**. Select the circle to dimension.

> **tip:** *The arrowhead of the leader will be placed where you select the circle (Pt. 1). Select the location of the dimension text* **(Fig. 9-60)**. *Press (E8) for the default size. Select the leader length.*

Radius Dimensions

1. Select **Draw**, **Radial dimension**, **Radius**. Select the radius to dimension. Press (E8) for the Default size. Select the location of the dimension text **(Fig. 9-60)**.

Center Mark

1. Select **D-aw**, **Radial dimension**, **Center mark.** Select the circle without a center mark.

These are the basics of dimensioning with AutoCAD LT. Experiment with dimensioning using the other dimensioning commands under DRAW. The AutoCAD LT manual that came with the software is an excellent source for more detailed dimensioning instructions.

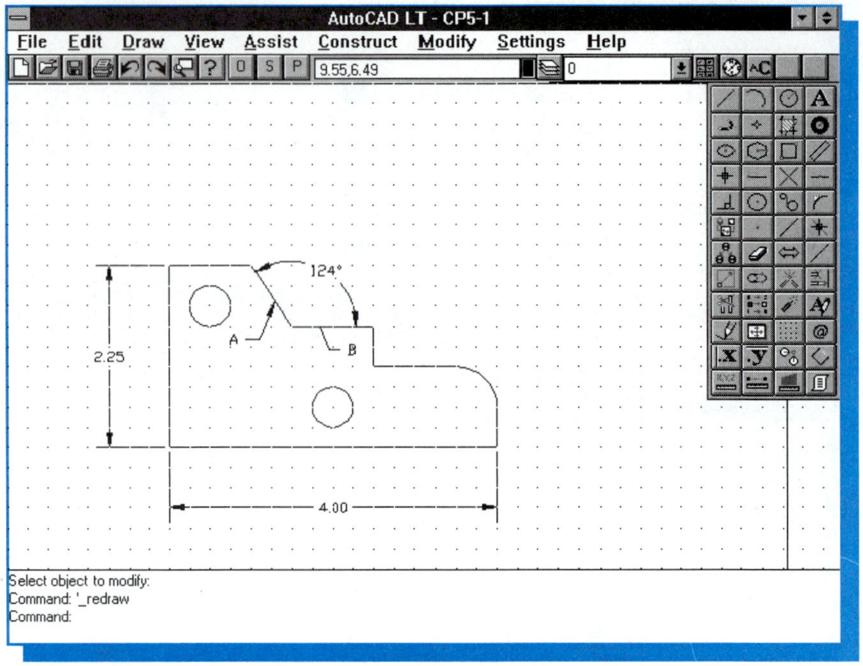

Figure 9-59
Angular dimensioning

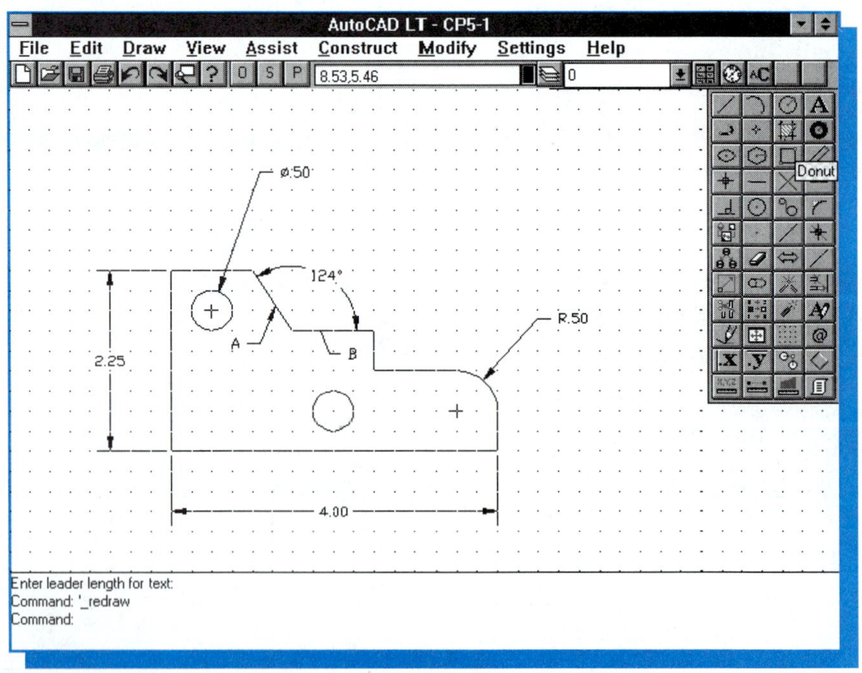

Figure 9-60
Radial dimensioning—diameter and radius

Drafting Exercises

1. Sketch the single-view drawings in **Figure 9-61** on grid paper and add dimensions.
2. With drafting instruments, or using CAD, draw and dimension the drawings in **Figures 9-62** and **9-63.**
3. With a CAD system, draw and dimension the multiview drawings in **Figure 9-64**.
4. List the dimensioning errors in **Figure 9-65**. Redraw and dimension it correctly.

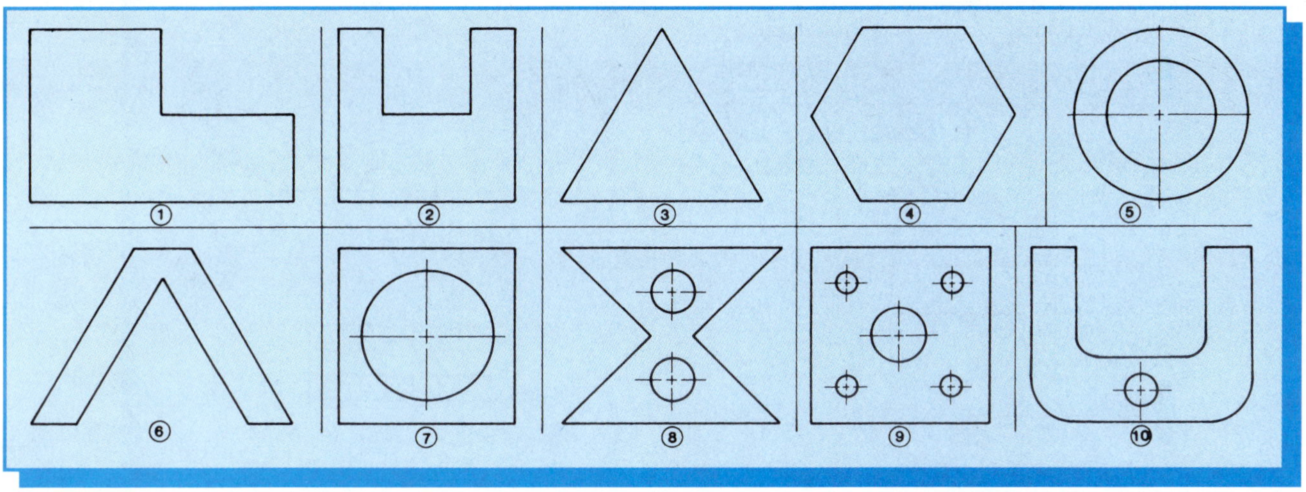

Figure 9-61a
Sketch and dimension each example. Double the size of each drawing with a scale or dividers.

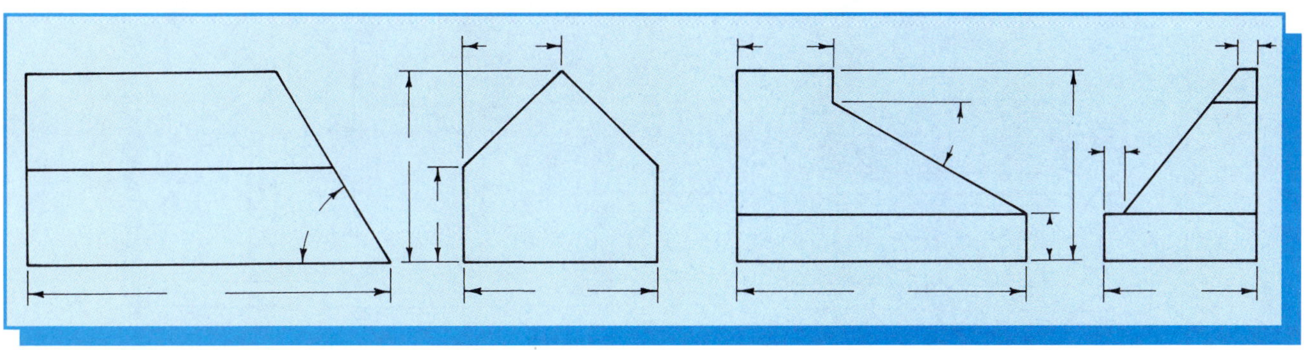

Figure 9-61b
Draw the multiview drawings 3x's larger and add dimensions.

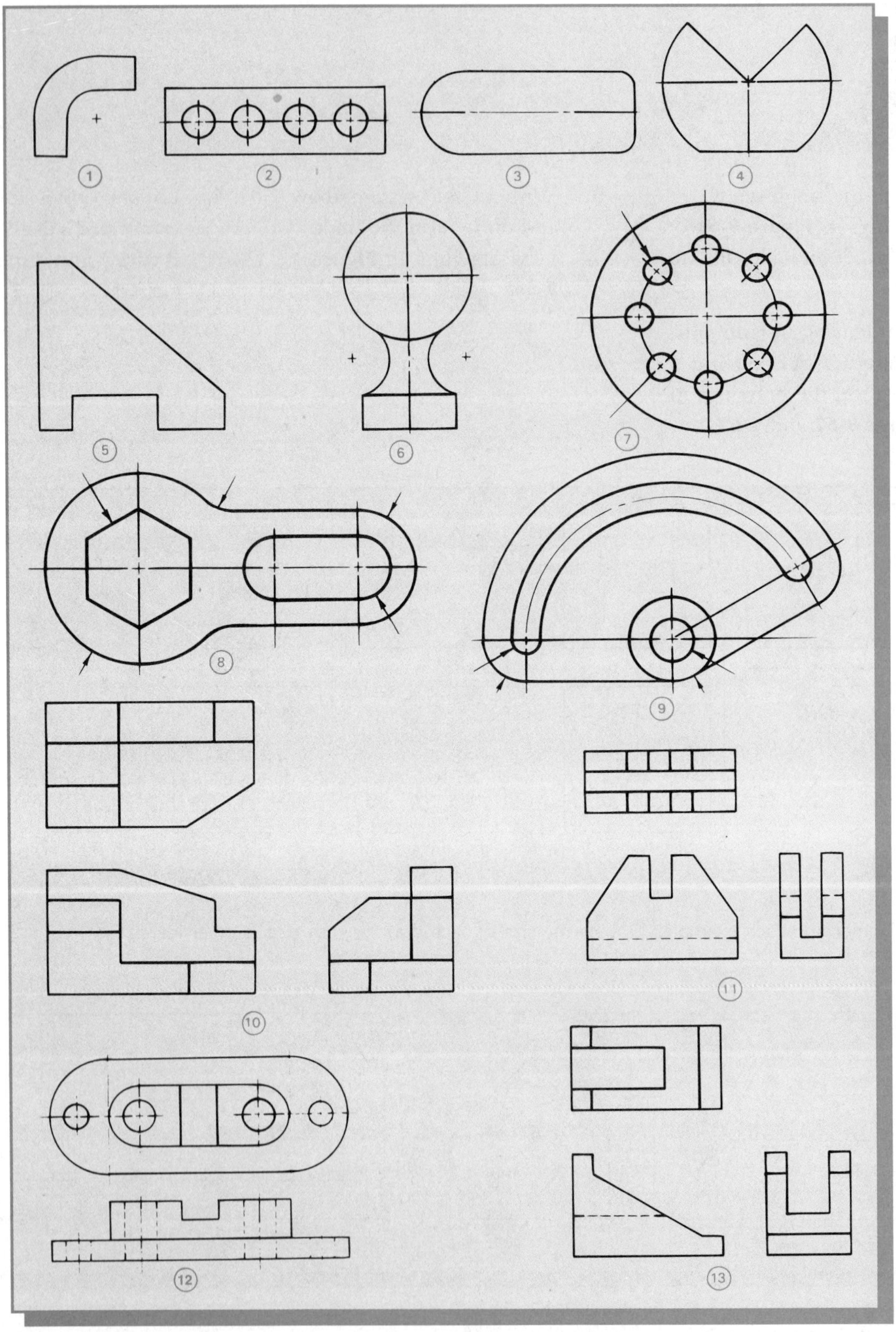

Figure 9-62
Double or triple the size of each object with a scale or divider. Use a CAD system and/or manual drafting.

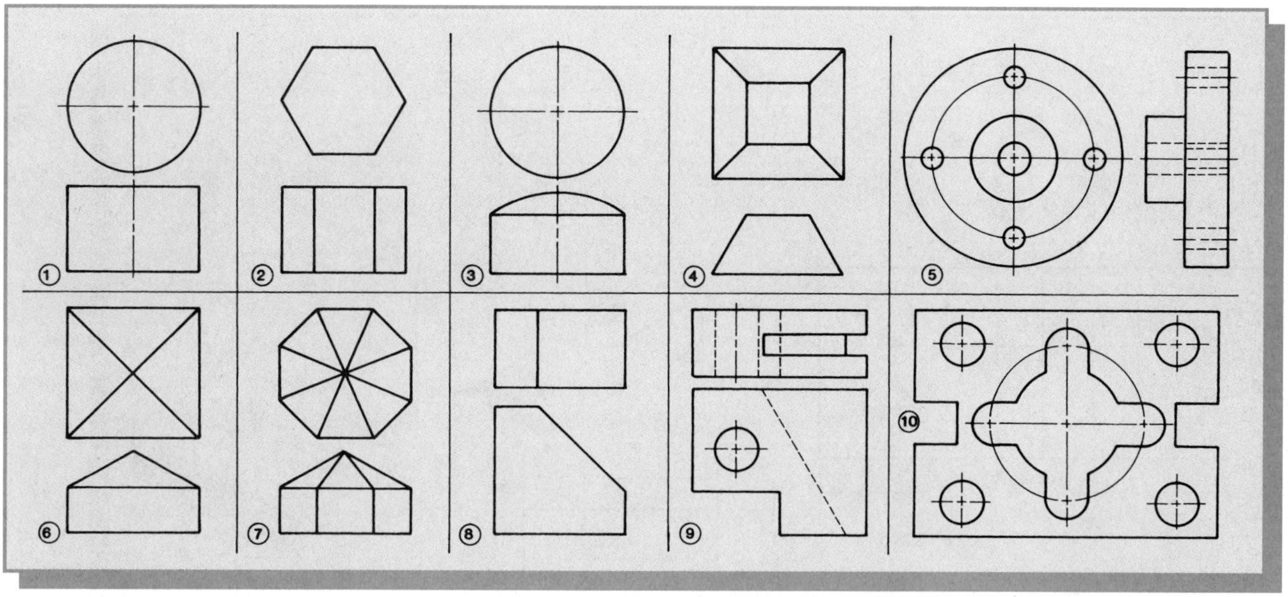

Figure 9-63
With drafting instruments, draw and dimension each example. Double the size of each multiview drawing with a scale or dividers, and dimension. Allow additional space between views as needed for dimensioning.

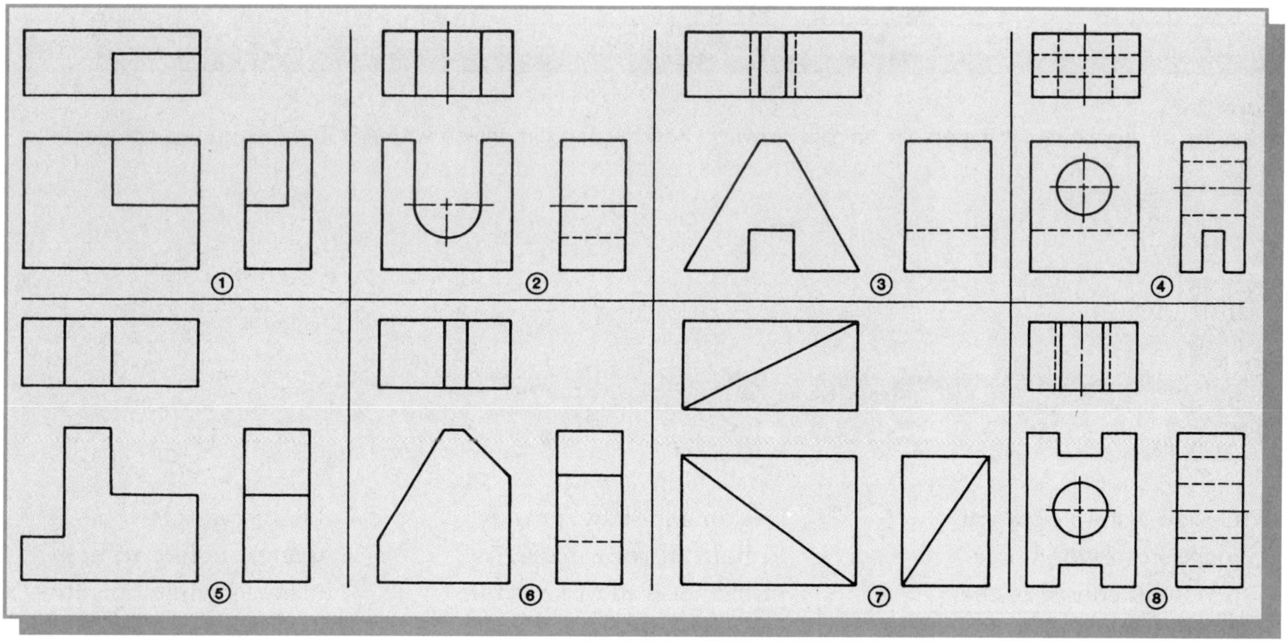

Figure 9-64
With a CAD system, draw and dimension each example. Measure dimensions directly from the page. Allow additional space between views as needed for dimensioning.

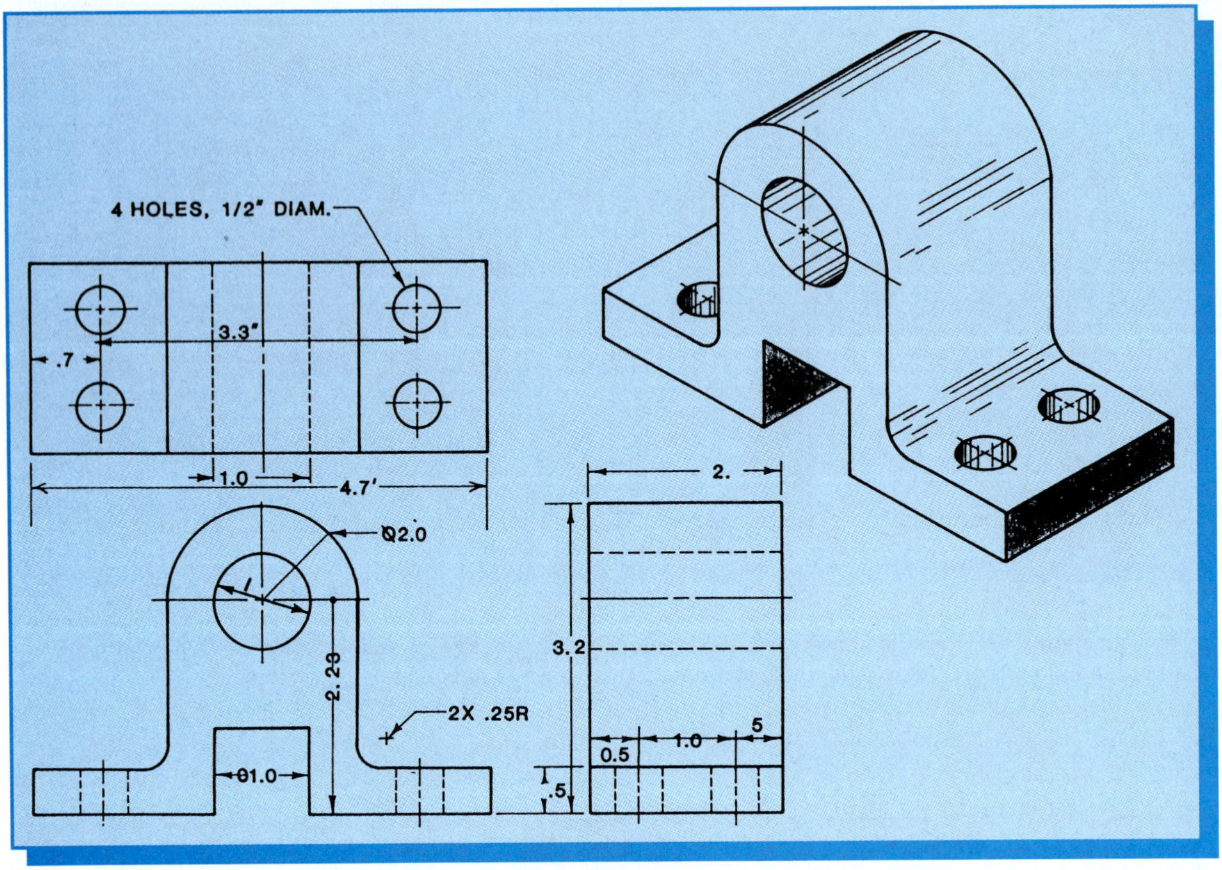

Figure 9-65
Make a list of the dimensioning errors on this drawing. Redraw and dimension with ASME dimensioning conventions.

Design Exercises

1. Design a simple toy car made from wood. The parts will consist of the body, two wood dowel axles, and four wood wheels. Draw an assembly drawing and a detailed drawing of each part fully assembled.

2. Design and draw a caddy to hold all your manual instruments in order while you are drafting.

3. Design and draw a CAD manual holder to hold manuals while operating a CAD system.

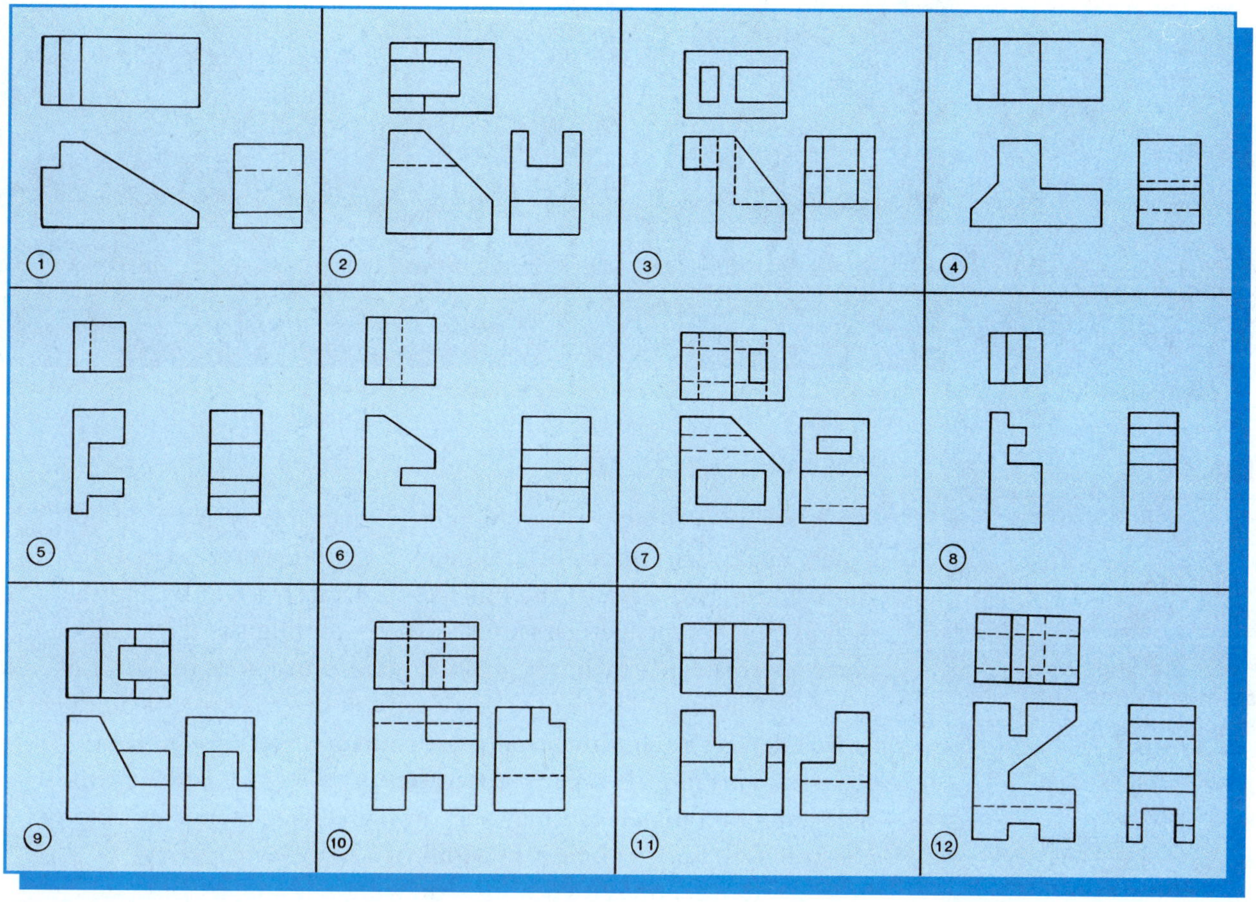

Figure 9-66
Redraw the multiview drawings 3x larger and redimension.

Key Terms

- Aligned dimensions
- Chamfer
- Contact area
- Dimensioning
- Dimension line
- Dimensioning symbols
- Dual dimensioning
- Extension line
- Fillets
- Finish symbol
- Knurling
- Lay
- Leaders
- Location dimensions
- Metric dimensioning
- Notations
- Polar coordinates
- Radial centerline
- Radii
- Rectangular coordinates
- Roughness
- Roughness-width cutoff
- Rounds
- Size dimensions
- Slots
- Smoothness
- Surface texture
- Tabulated outline dimensioning
- Tolerance
- Unidirectional dimensions
- U.S. customary dimensioning

Chapter 10

Tolerancing and Geometric Tolerancing

Objectives

The student will be able to:

- place limit dimensions on a working drawing
- place plus and minus tolerance dimensions on a working drawing
- recognize geometric tolerancing symbols
- design various fits for mating parts
- dimension a working drawing with form tolerance symbols
- understand the concepts for each form tolerancing symbol

Introduction

Absolute accuracy is not completely attainable in any field. An expert marksman cannot hit the bull's-eye dead center every time. But is hitting the edge of the bull's-eye tolerable? Or is hitting the second ring acceptable occasionally? And is hitting the third ring never acceptable? Whichever goal is tolerated becomes the limit of accuracy. There are limits of accuracy in the manufacture of products as well. In fact, each dimension must contain a working limit of accuracy. This limit is called a **tolerance.** A tolerance is the permissible amount a dimension may vary in size during manufacturing or construction. For example, the width of a highway can vary 1/4" from the dimension shown on the drawing without serious consequences. However, if a watch part varies only a thousandth of an inch, the part will not fit or function. To control these variances (tolerances), manufacturers are assigned degrees of accuracy required in each product. This is done through a tolerance range to each dimension, or through a general tolerance notation.

Tolerance Dimensions

The use of **basic (simple) dimensions** that do not include tolerance limits allows manufacturers to establish their own acceptable limits that may not produce the desired result. **Figure 10-1** shows a comparison of a drawing dimensioned with basic

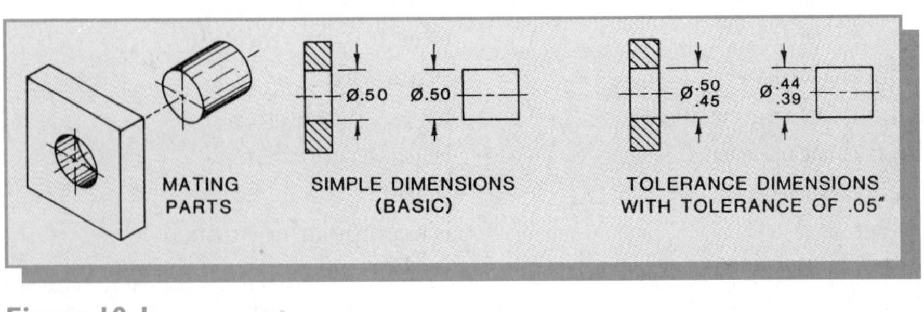

Figure 10-1
Comparison of basic dimensions and tolerance dimensions

dimensions and with tolerance dimensions of .05". The tolerance dimensions shown on the right assures the shaft will fit into the hole with the clearance desired. Basic dimensions **(Fig. 10-2)** show only the general size of a part without any tolerance limitations.

Tolerance is the difference between two dimensions. If a 190'-6" sidewalk is acceptable when constructed at 190'-5" or 190'-7", that is one inch over and one inch under the basic dimension. But the tolerance range is actually two inches (190'-7" minus 190'-5" = 2").

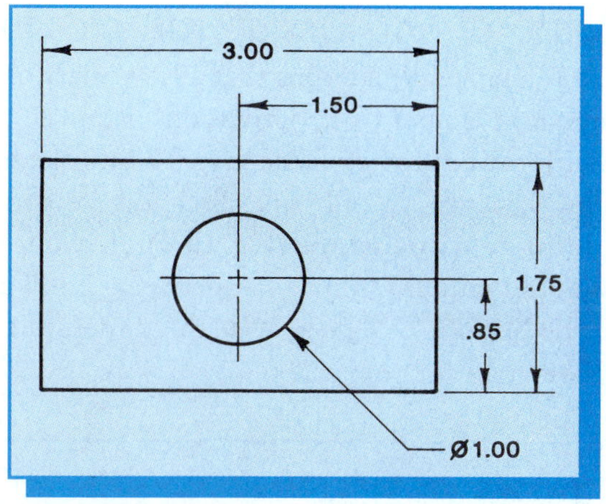

Figure 10-2
The basic dimension is the theoretical exact size.

Tolerance Practices

A **limit tolerance** is a statement of the variations that can be permitted from a given dimension. When tolerance limits are aligned vertically, the larger dimension is placed on top. When aligned horizontally, the smaller dimension is placed to the left of the larger one. The number of decimal places in the basic dimension and in the tolerancing dimension must be equal **(Fig. 10-3)**.

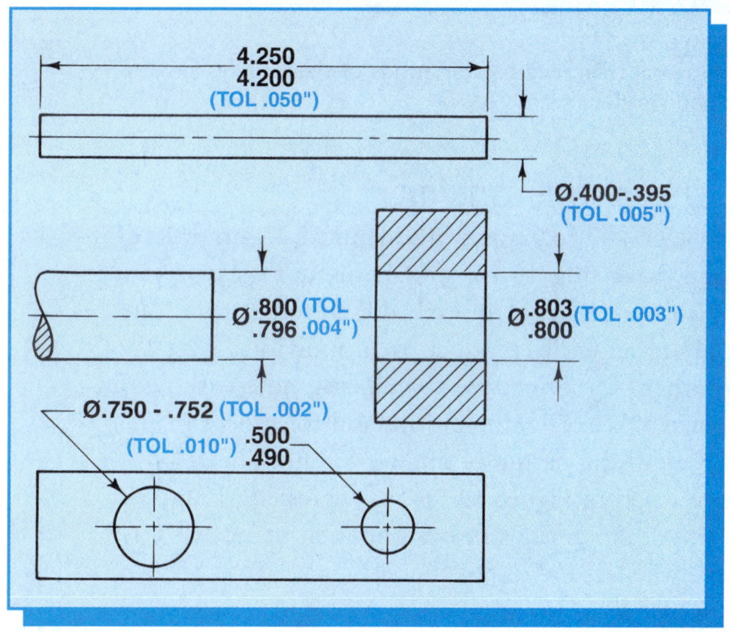

Figure 10-3
A limit tolerance is a precise statement of the variations that can be permitted within a dimension. Note the large dimension is always on the top of the vertical tolerances, and to the right of the horizontal tolerances.

Plus and minus tolerance dimensions (Fig. 10-4) are used to indicate the tolerance range above and below the basic dimension. The tolerance level higher than the basic dimension is marked with a plus sign (+). The tolerance level less than the basic dimension is marked with a minus sign (–). Plus and minus tolerance dimensions are placed to the right of the basic dimension, with the plus dimension on top of the minus dimension. If the plus tolerance dimension is the same as the minus tolerance dimension, a single number may be used preceded by a combined ± sign. The tolerance of 3.00" ± .03 **(Fig. 10-4)** indicates the actual dimension may be as large as 3.03" or as small as 2.97". This represents a tolerance range of .6".

Bilateral and Unilateral Tolerancing

Tolerance dimensions that allow variation in both directions from a basic dimension (+ and –) are **bilateral tolerance dimensions (Fig. 10-5)**. Tolerances that allow variances in only one direction are **unilateral tolerance dimensions (Fig. 10-6)**. In the top of **Figure 10-5**, the dimension may fall between 5.19" and 5.05". In the bottom of **Figure 10-6**, the dimension may fall between 1.000" and 1.003".

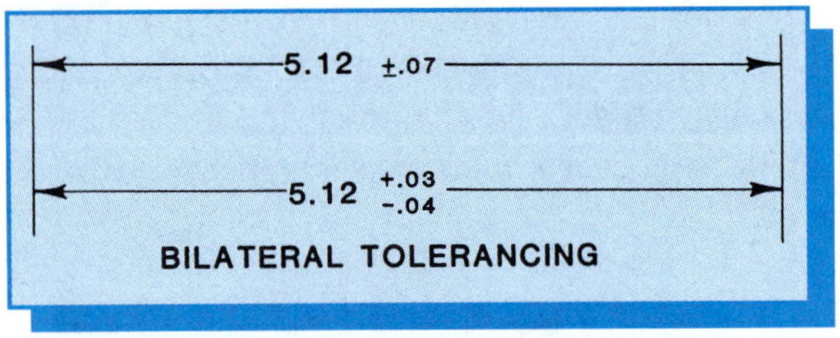

Figure 10-5
A bilateral tolerance has variations of size of larger and smaller (+ and –).

General Tolerancing

When all tolerances are identical, a **general tolerance note (Fig. 10-7a)** may be used. This saves repeating tolerance limits on each dimension. A note can apply to an entire product or an individual part. Another type of general tolerancing note is shown in **Figure 10-7b**. The dimensions and tolerances may vary with the design of different items. The tolerance levels in **Figure 10-7b** are expressed as follows:

- All dimensions below one inch may have a tolerance of ±.001".
- All dimensions between 1.01 and 6 inches may have a tolerance of .004".
- All dimensions above six inches may have a tolerance of .008".

TOLERANCES UNLESS OTHERWISE SPECIFIED:
0 TO 1.00 ARE ± .001
1.01 TO 6.00 ARE ± .004
OVER 6.00 ARE ± .008

Figure 10-7b
General tolerance note

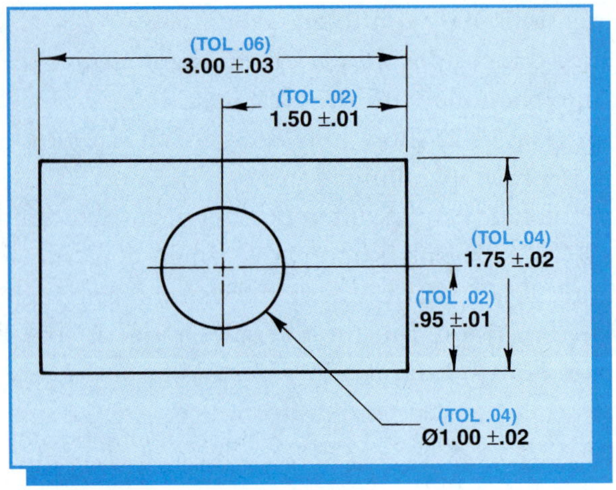

Figure 10-4
Plus and minus tolerance dimensions

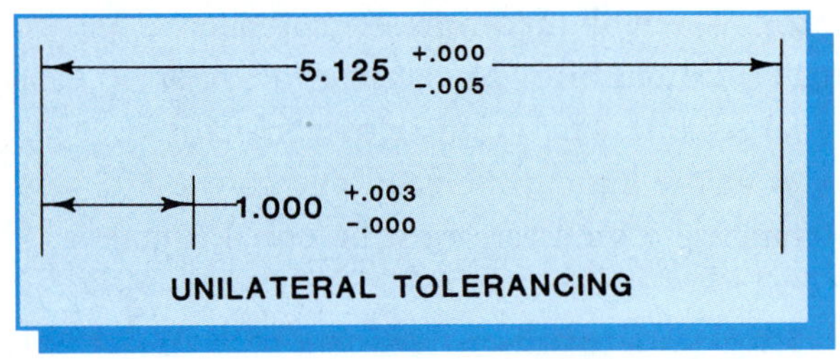

Figure 10-6
A unilateral tolerance only has a variation in one direction—larger or smaller from the basic dimension.

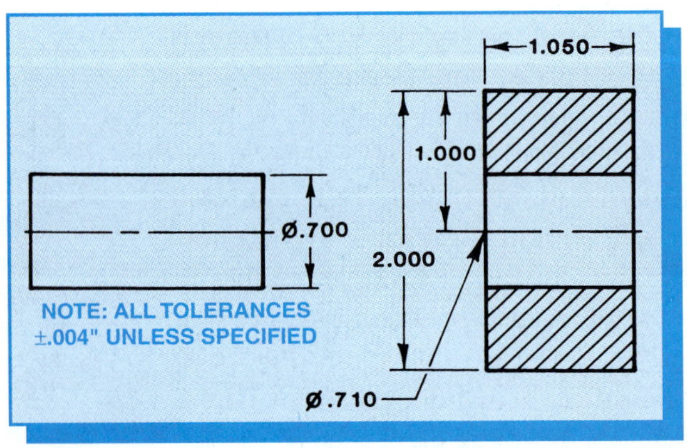

Figure 10-7a
A general tolerance notation may be placed near the drawing or in the title block.

Chain Dimensioning

Tolerance dimensions on **chain dimensions** are not recommended. Chain dimensioning accumulates tolerances. For example, the cumulative tolerance for the length of the part in **Figure 10-8** can be plus .05" (.02 + .01 + .01 + .01) or minus .05", a total tolerance range of .10", or anywhere in between.

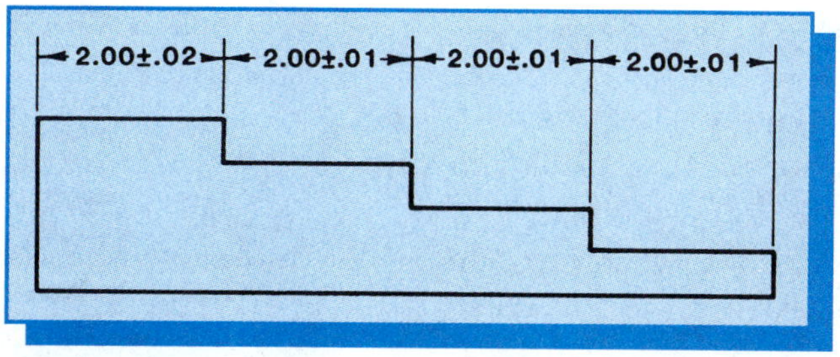

Figure 10-8
Chain dimensions accumulate tolerances. Accumulation in this drawing can be plus or minus .05". This method is not recommended.

Datum Dimensioning

In assigning tolerances to chain dimensions as in **Figure 10-8**, a dimensioned 8" length could be as short as 7.95" or as long as 8.05". For this reason **datum dimensioning** is recommended where tolerance limits are critical. In datum dimensioning, all dimensions originate from a single point and do not allow tolerances to accumulate **(Fig. 10-9)**. The total length in **Figure 10-9** cannot deviate more than +.03".

When the distance between any two points is critical, a **direct dimension** should be used to insure the least tolerance buildup. Points A and B in **Figure 10-10** should be within .04" of 6.00". Therefore, a tolerance dimension is used to connect these two points even though the remainder of the drawing is datum dimensioned.

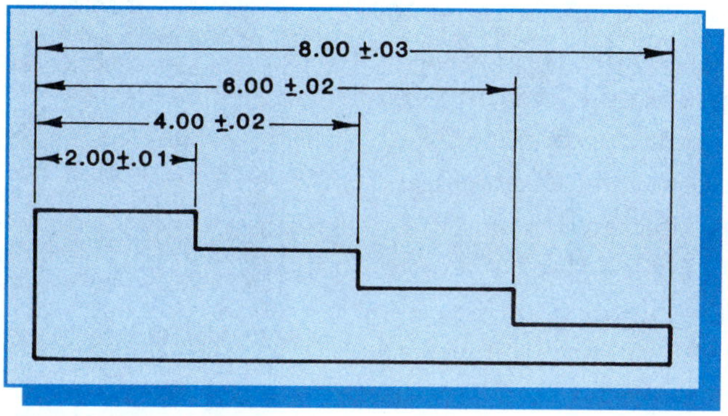

Figure 10-9
Datum dimensioning will not accumulate tolerances The maximum tolerance in this drawing is plus or minus .03".

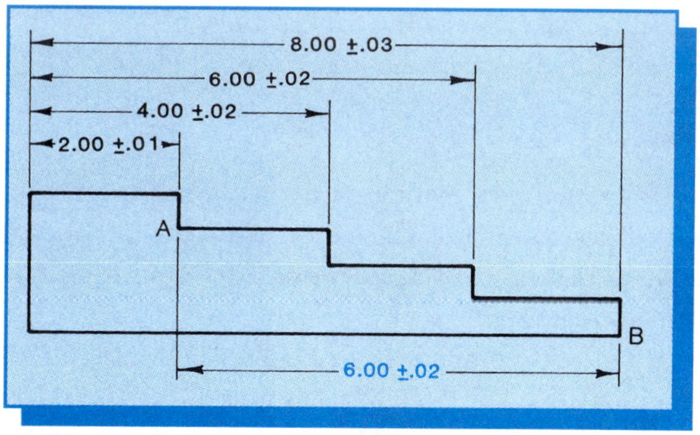

Figure 10-10
Direct dimensioning for a specific dimension will have the most accuracy.

Angle Tolerancing

While most tolerances relate to linear measurements, angle tolerances are also needed on many working drawings. **Figure 10-11** shows the correct method of specifying angle tolerances in degrees, minutes, and seconds, and with degrees and a decimal part of a degree.

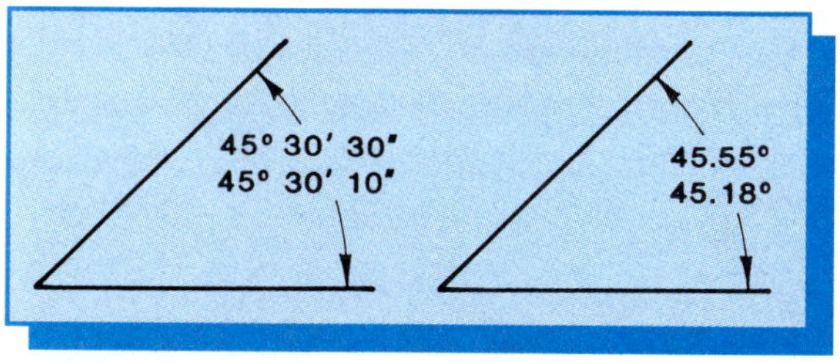

Figure 10-11
Specifying angle tolerances with degrees, minutes, and seconds, and with degrees and a decimal part of a degree.

Mating Parts

Appropriate tolerance levels for **mating parts** are very critical. If too much space is allowed between parts, the fit may wobble or fall apart. If too little space is allowed, parts may not fit or excessive friction may occur when moved. Very close tolerances also require more precise manufacturing methods and are expensive to produce. For these reasons, it is important that the appropriate fit be designed, dimensioned, and manufactured.

Figure 10-12 is a table showing the appropriate tolerances for various industrial applications at different sizes. The tolerance grade range extends from Grade 4 to Grade 13, with the lower number representing the closest tolerance limits.

STANDARD TOLERANCES										
Nominal Size Range in Inches	Slip Gauges, Production and Wear Tolerance of Gauges and Measuring Instruments		Parts Subject to Very Tight Tolerances, Precision Bearings, Precision Assemblies		Precision Engineered Designs (General)		Engineering Work in General, Giving Scope for Wider Tolerances		Rough Work, Steel Structures, Castings, Agricultural Machinery	
Over–To	Grade 4	Grade 5	Grade 6	Grade 7	Grade 8	Grade 9	Grade 10	Grade 11	Grade 12	Grade 13
0.04–0.12	.00015	.00020	.00025	.0004	.0006	.0010	.0016	.0025	.004	.006
0.12–0.24	.00015	.00020	.0003	.0005	.0007	.0012	.0018	.0030	.005	.007
0.24–0.40	.00015	.00025	.0004	.0006	.0009	.0014	.0022	.0035	.006	.009
0.40–0.71	.0002	.0003	.0004	.0007	.0010	.0016	.0028	.0040	.007	.010
0.71–1.19	.00025	.0004	.0005	.0008	.0012	.0020	.0035	.0050	.008	.012
1.19–1.97	.0003	.0004	.0006	.0010	.0016	.0025	.0040	.006	.010	.016
1.97–3.15	.0003	.0005	.0007	.0012	.0018	.0030	.0045	.007	.012	.018
3.15–4.73	.0004	.0006	.0009	.0014	.0022	.0035	.005	.009	.014	.022
4.73–7.09	.0005	.0007	.0010	.0016	.0025	.0040	.006	.010	.016	.025
7.09–9.85	.0006	.0008	.0012	.0018	.0028	.0045	.007	.012	.018	.028
9.85–12.41	.0006	.0009	.0012	.0020	.0030	.0050	.008	.012	.020	.030
12.41–15.75	.0007	.0010	.0014	.0022	.0035	.006	.009	.014	.022	.035
15.75–19.69	.0008	.0010	.0016	.0025	.004	.006	.010	.016	.025	.040
19.69–30.09	.0009	.0012	.0020	.003	.005	.008	.012	.020	.030	.050
30.09–41.49	.0010	.0016	.0025	.004	.006	.010	.016	.025	.040	.060
41.49–56.19	.0012	.0020	.003	.005	.008	.012	.020	.030	.050	.080
56.19–76.39	.0016	.0025	.004	.006	.010	.016	.025	.040	.060	.100
76.39–100.9	.0020	.003	.005	.008	.012	.020	.030	.050	.080	.125
100.9–131.9	.0025	.004	.006	.010	.016	.025	.040	.060	.100	.160
131–171.9	.003	.005	.008	.012	.020	.030	.050	.080	.125	.200
171.9–200	.004	.006	.010	.016	.025	.040	.060	.100	.160	.250

Figure 10-12
Standardized grades of tolerances

Gauge instruments .04" to .12", Grade 4, have the smallest acceptable tolerance, .00015". Castings, 171.9" to 200", Grade 13, have the largest acceptable tolerance, .250".

Clearance Fits

Tolerancing of mating parts relies on the control of clearance desired between parts. Clearance (allowance) is the **tolerance zone** between mating parts. **Clearance fits** require separate tolerancing limits for each part.

The shaft's dimensions in **Figure 10-13** are .50"/.45" with a tolerance of .05". The hole's dimensions are .53"/.51" with a tolerance of .02". When the largest shaft diameter of .50" mates with the smallest hole with a diameter of .51", the difference or allowance will be .01". When the smallest shaft (.45' diameter) mates with the largest hole (.53" diameter), the allowance is .08". With mass production, the mating of parts may have any allowance in the range of .01" to .08".

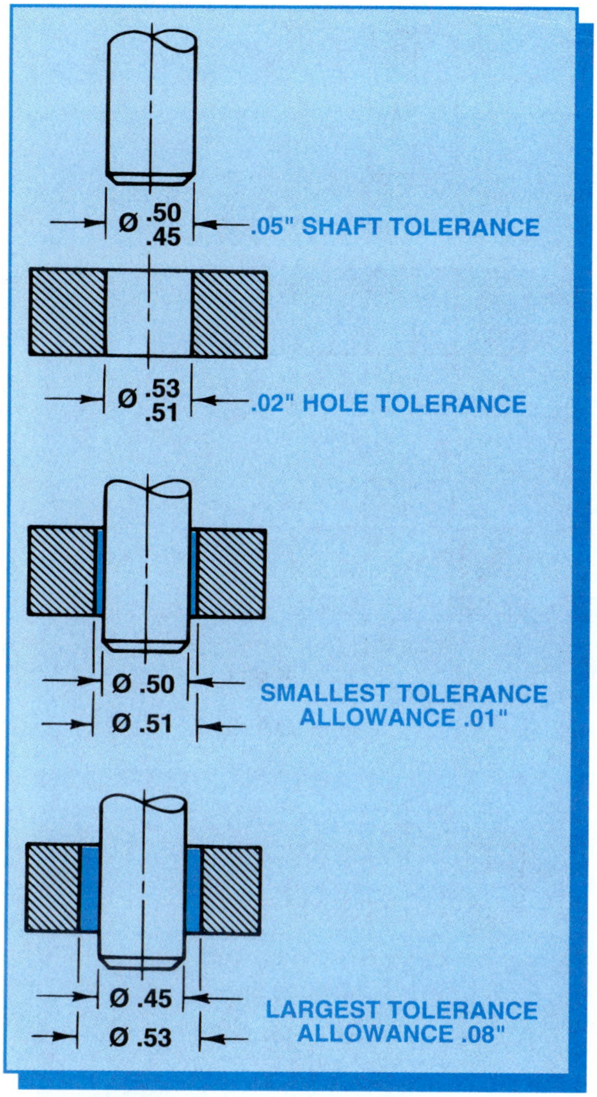

Figure 10-13
Tolerance zones for clearance fits

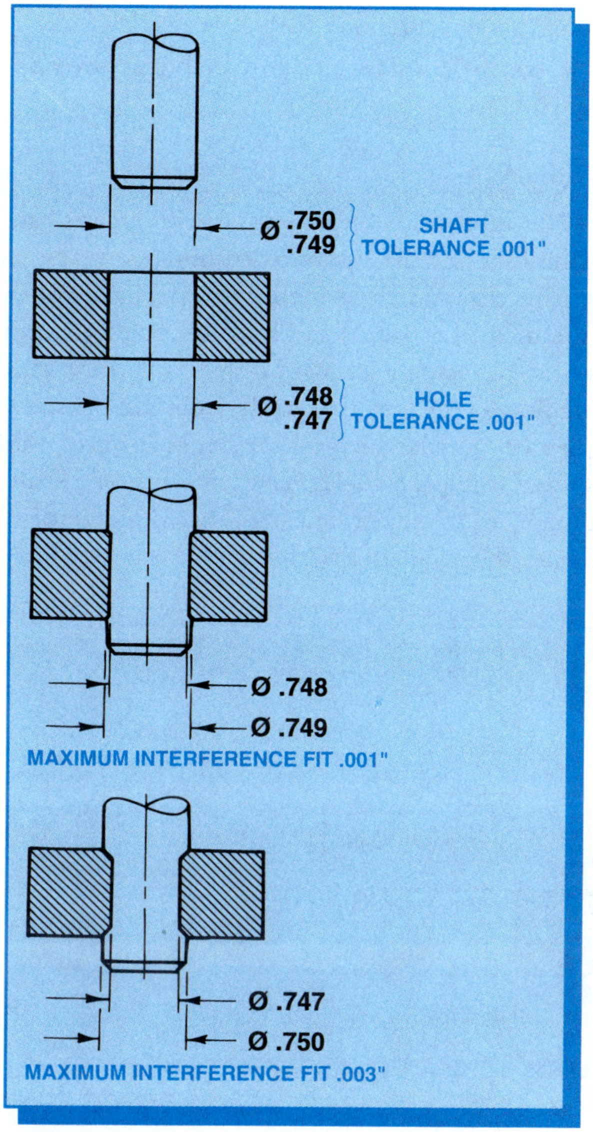

Figure 10-14
Interference fits (force fits)

Interference Fits

When mating parts are designed to be permanently attached and not movable, an **interference fit** may be specified. Interference fits (force fits) have negative clearances so that the parts must be forced together **(Fig. 10-14)**. The amount of interference tolerance depends on the flexibility and hardness of the material. Soft materials force together more easily than hard materials.

Transition Fits

A **transition fit** is one in which parts overlap in size **(Fig. 10-15)**.

Basic Sizes

The appropriate matching of mating parts can be accomplished with custom machining; however, standard-size parts should be specified when possible. For example, in the mating of a hole and shaft, the shaft size can be adjusted to match a standard hole size or vice versa. Matching a shaft size to a **basic hole size** is shown in **Figure 10-16**. Matching a hole size to a **basic shaft size** is shown in **Figure 10-17**. The basic size of a shaft refers to the standard manufactured size. The basic size of a hole refers to the standard drill bit size.

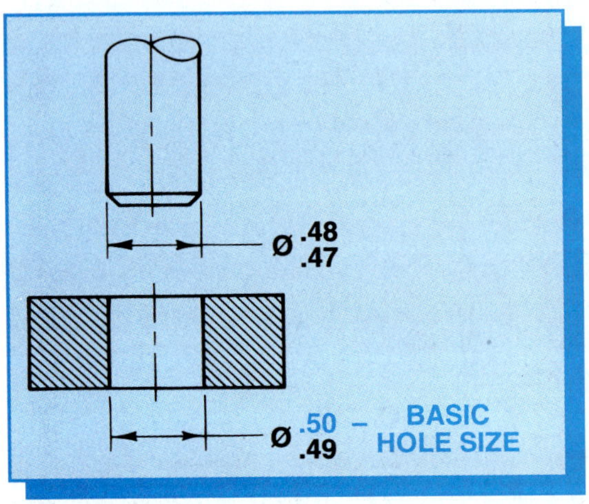

Figure 10-16
System of basic hole tolerances

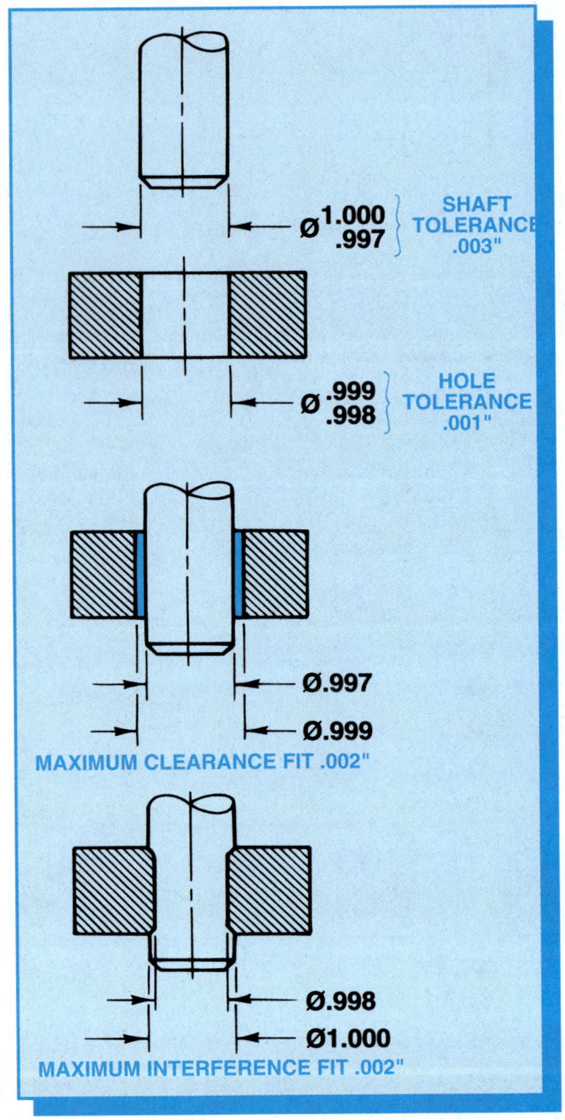

Figure 10-15
Transitional fits may be interference and/or clearance fits

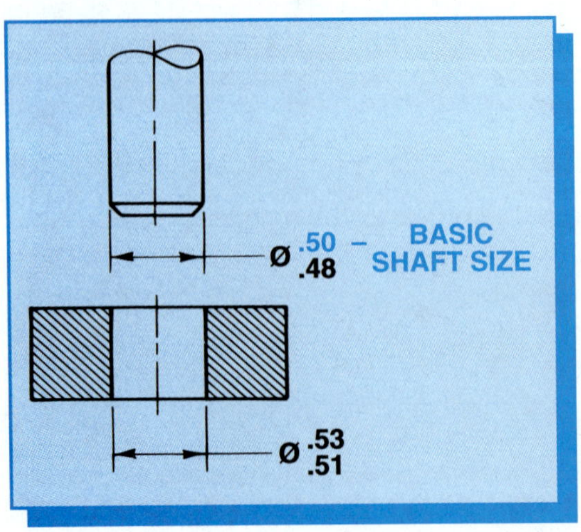

Figure 10-17
System of basic shaft tolerances

Material Condition

When establishing hole tolerance sizes of mating parts, material conditions such as **maximum material condition (MMC)** and **least material condition (LMC)** must be considered. The MMC for a hole is the smallest allowable diameter that leaves the maximum amount of material. The MMC for a shaft is the largest allowable diameter that leaves the maximum amount of material. The LMC for a shaft is the smallest allowable diameter that leaves the minimum amount of material. The MMC and LMC for a shaft are shown in **Figure 10-18a**, and the MMC and LMC for a hole are shown in **Figure 10-18b**.

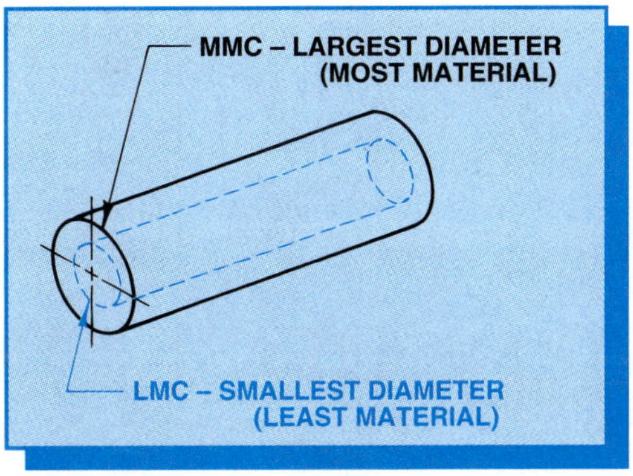

Figure 10-18a
Maximum and minimum material conditions for a shaft

RFS means **regardless of feature size.** This indicates the tolerance must be met regardless of where the feature lies within its size tolerance.

Basic dimensioning controls only the shape and size of an item on a technical drawing. **Geometric tolerancing** is an extension of the dimensioning process that describes more precisely how an item is to be manufactured. Geometric tolerancing includes form tolerancing and positional tolerancing. **Form tolerancing** controls the geometric shape of objects **(Fig. 10-19)**. **Positional tolerancing** controls the exact location of product features such as holes, depressions, and projections **(Fig. 10-20)**.

Geometric characteristic symbols are used to control the tolerance limits of form, profile, orientation, location, and runouts **(Fig. 10-21)**. These symbols are a form of dimensioning shorthand, each representing a geometric feature to be controlled. Other modifying symbols used for geometric tolerancing are summarized in **Figure 10-22**.

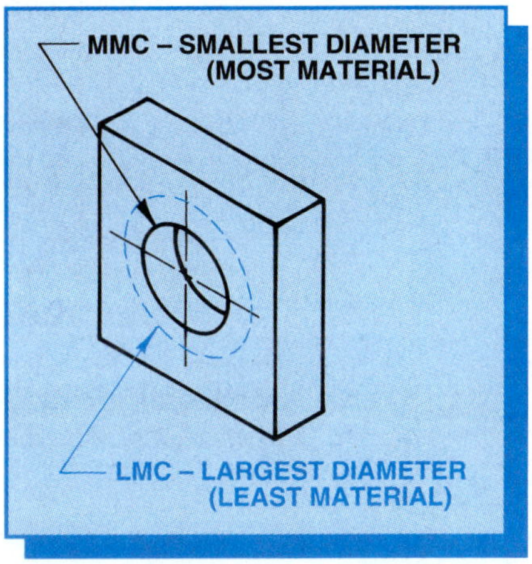

Figure 10-18b
Maximum and minimum material conditions for a hole

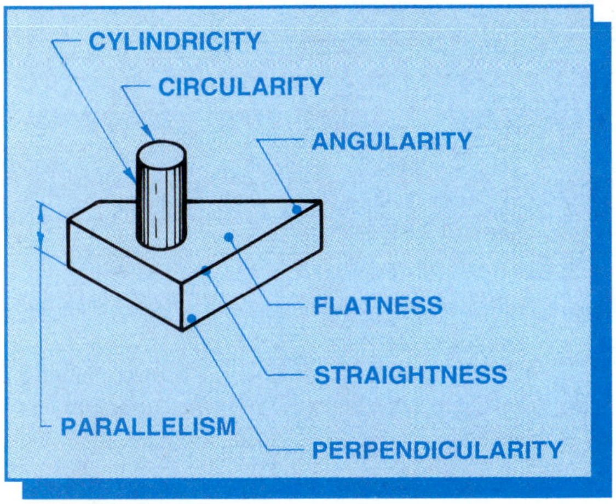

Figure 10-19
Form tolerancing controls

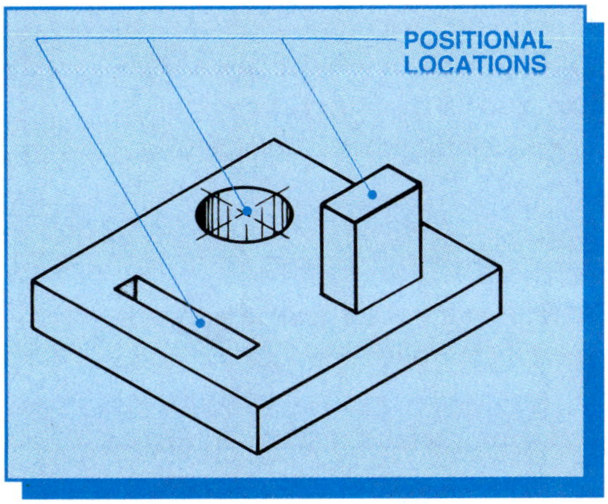

Figure 10-20
Positional tolerancing controls

	TYPE OF TOLERANCE	CHARACTERISTIC	SYMBOL
FOR INDIVIDUAL FEATURES	FORM	STRAIGHTNESS	⏤
		FLATNESS	⏥
		CIRCULARITY (ROUNDNESS)	○
		CYLINDRICITY	⌭
FOR INDIVIDUAL OR RELATED FEATURES	PROFILE	PROFILE OF A LINE	⌒
		PROFILE OF A SURFACE	⌓
FOR RELATED FEATURES	ORIENTATION	ANGULARITY	∠
		PERPENDICULARITY	⊥
		PARALLELISM	//
	LOCATION	POSITION	⌖
		CONCENTRICITY	◎
		SYMMETRY	⌯
	RUNOUT	CIRCULAR RUNOUT	↗*
		TOTAL RUNOUT	⌰*

*Arrowhead(s) may be filled in.

Figure 10-21
ASME standard Y14.5M-1994, geometric characteristic symbols. *(Reprinted with permission of the American Society of Mechanical Engineers)*

Datums

All geometric tolerancing begins with the establishment of a datum. A **datum** is a reference location easy to locate and measure. Theoretically, it is a perfect plane, point, line, or axis. The location of all geometric characteristics originate from a datum. On a technical drawing, the **datum identification box** or **datum target** is used to locate specific surfaces for reference **(Fig. 10-23)**.

Feature Control Frame

The **feature control frame** contains data for each controlled specification and includes the characteristic symbol, tolerance value, and datum reference letter. In **Figure 10-24**, the marked surface is to be parallel within .003" of datum A.

TERM	SYMBOL
AT MAXIMUM MATERIAL CONDITION	Ⓜ
AT LEAST MATERIAL CONDITION	Ⓛ
PROJECTED TOLERANCE ZONE	Ⓟ
FREE STATE	Ⓕ
TANGENT PLANE	Ⓣ
DIAMETER	Ø
SPHERICAL DIAMETER	SØ
RADIUS	R
SPHERICAL RADIUS	SR
CONTROLLED RADIUS	CR
REFERENCE	( )
ARC LENGTH	⌒
STATISTICAL TOLERANCE	⟨ST⟩
BETWEEN	↔

Figure 10-22
ASME standard Y14.5M-1994, modifying symbols used in specifying geometric tolerancing *(Reprinted with permission of the American Society of Mechanical Engineers)*

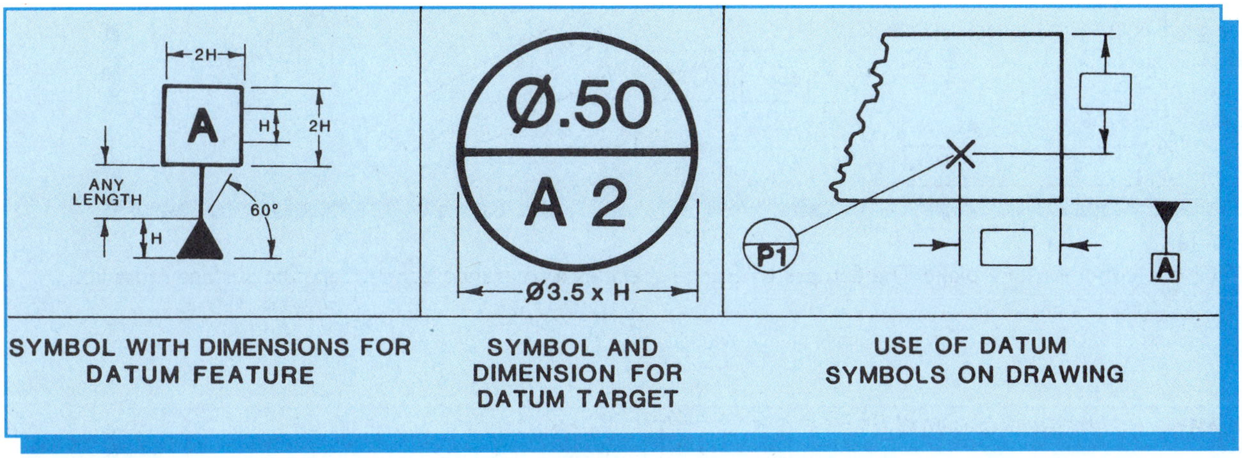

Figure 10-23
ASME Standard Y14.5M-1994 Reference datum symbols and their drawing sizes

Geometric Characteristic Symbols

As previously mentioned, each geometric characteristic symbol represents a geometric feature that must be controlled to the level specified. Most symbols are referenced to a datum; however, **straightness**, **flatness**, **circularity** (**roundness**), and **cylindricity** may stand alone. **Figures 10-25** through **10-37** show characteristic symbols accompanied with an explanation and a working drawing example.

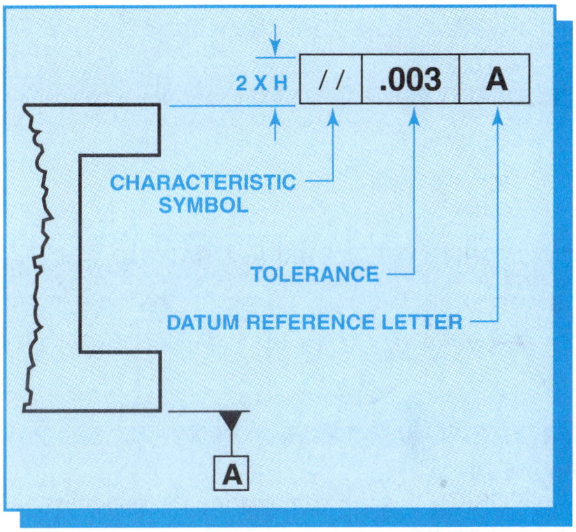

Figure 10-24
The feature control frame's drawing size. It states "The surface is to be parallel within .003" of surface -A-."

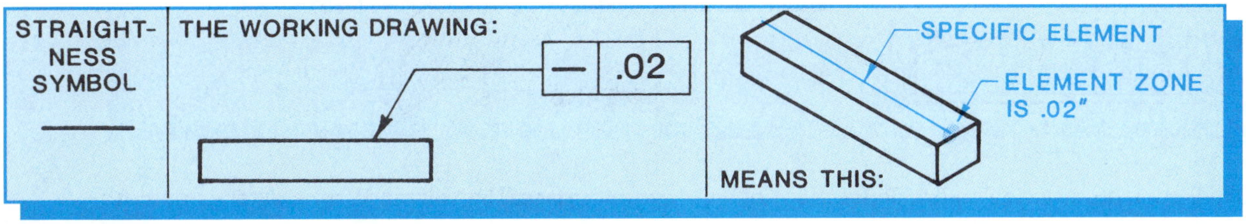

Figure 10-25
STRAIGHTNESS refers to an element of a surface or an axis that is a straight line. The straightness tolerance specifies a tolerance zone where the element or axis lies.

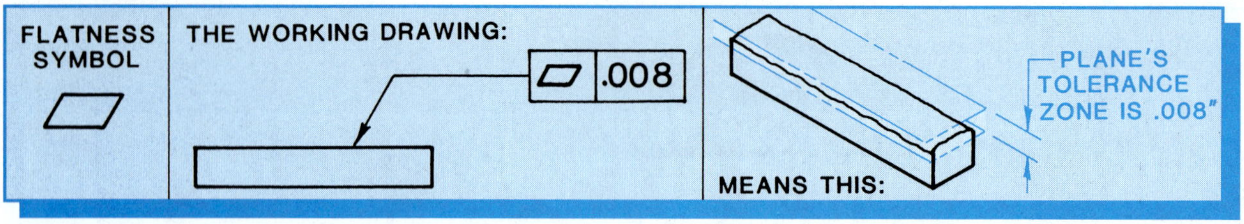

Figure 10-26
FLATNESS is the condition of a plane. The flatness tolerance specifies a tolerance zone where the surface must lie.

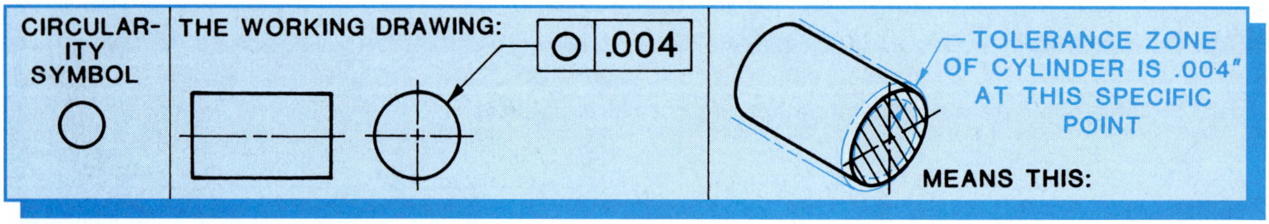

Figure 10-27
CIRCULARITY (ROUNDNESS) specifies a tolerance zone bounded by two concentric circles at a single specific point.

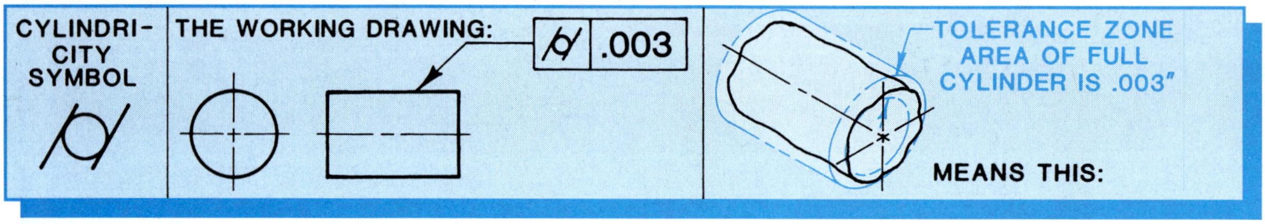

Figure 10-28
CYLINDRICITY is the combination of roundness and straightness for the whole cylinder. The tolerance zone is bounded by two concentric circles.

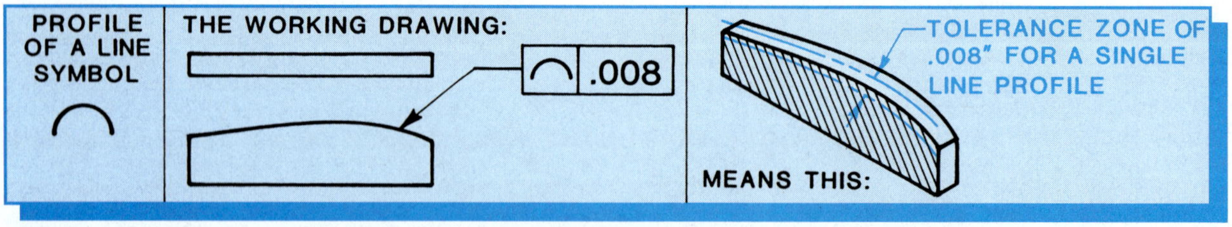

Figure 10-29
LINE PROFILE controls a single line profile (section) of an irregular surface. The tolerance zone extends on both sides of the profile along its full length.

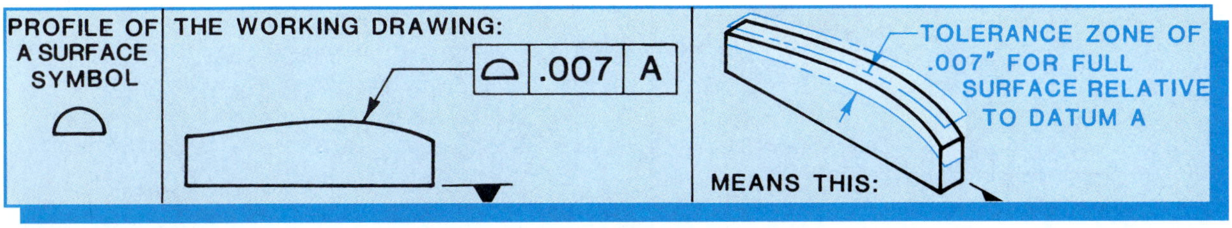

Figure 10-30
SURFACE PROFILE controls the entire surface for an irregular area. The tolerance zone extends the length and width of the surface.

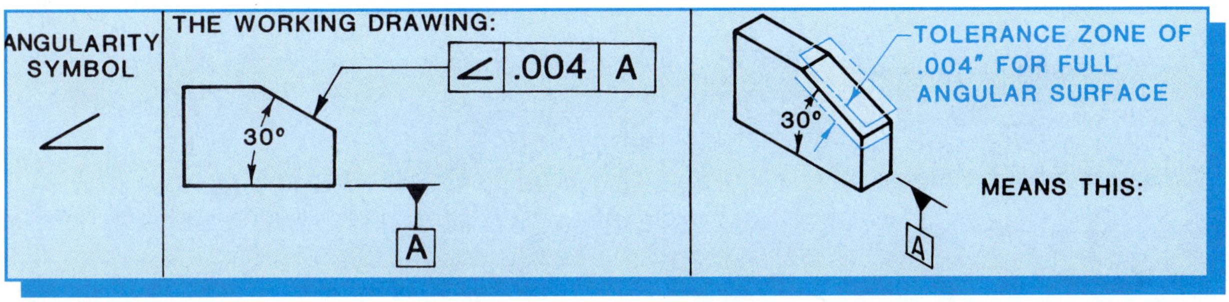

Figure 10-31
ANGULARITY is the condition of a surface or axis at an angle, other than 90°, from a datum plane or axis. The tolerance zone is defined by two parallel lines at the specific angle from the datum.

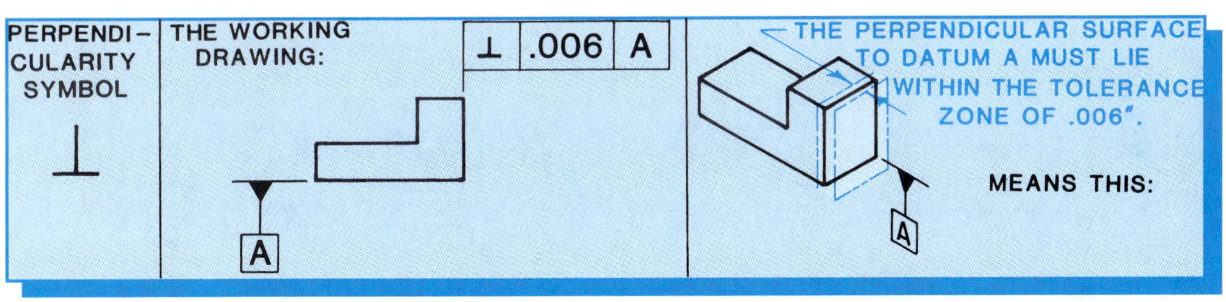

Figure 10-32
PERPENDICULARITY is the condition of a surface, plane, or axis, at a right angle to a datum plane or axis.

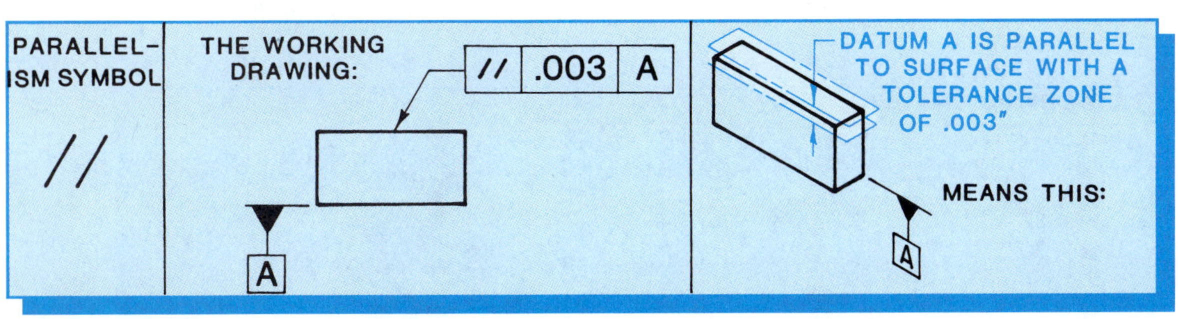

Figure 10-33
PARALLELISM is the condition of a surface equidistant at all points from a datum plane or axis. The tolerance zone is defined by two planes or lines parallel to the datum.

Figure 10-34
FEATURE POSITION is the theoretical exact location of its axis or center plane from the features from which it is dimensioned. The positional tolerance is the permissible error for the location of a feature relative to its other features.

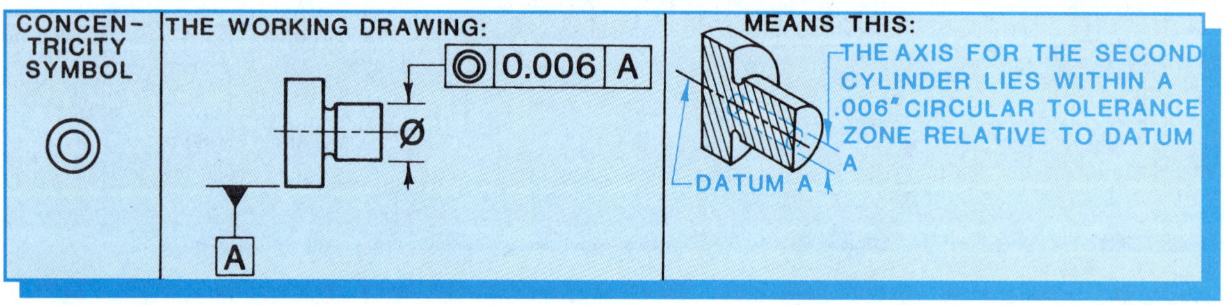

Figure 10-35
CONCENTRICITY is the condition where the axis of a cylinder aligns to a common axis datum feature. The tolerance zone forms a cylinder where the axis must lie.

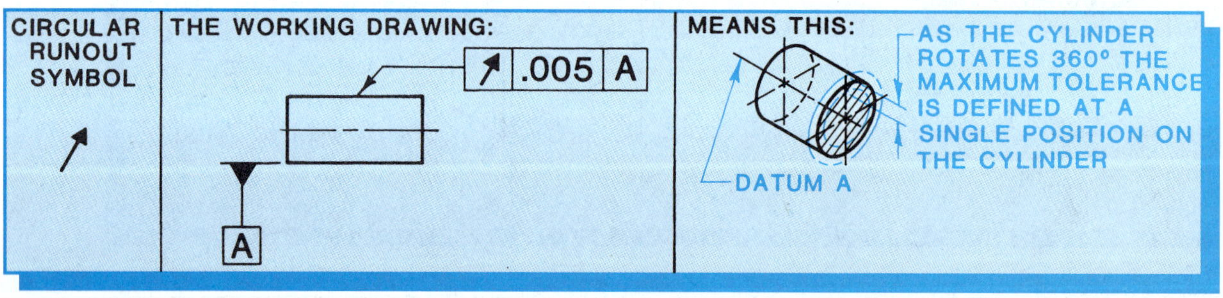

Figure 10-36
CIRCULAR RUNOUT controls a circular element at one position on a surface. The tolerance zone lies within two circles as the surface rotates one revolution.

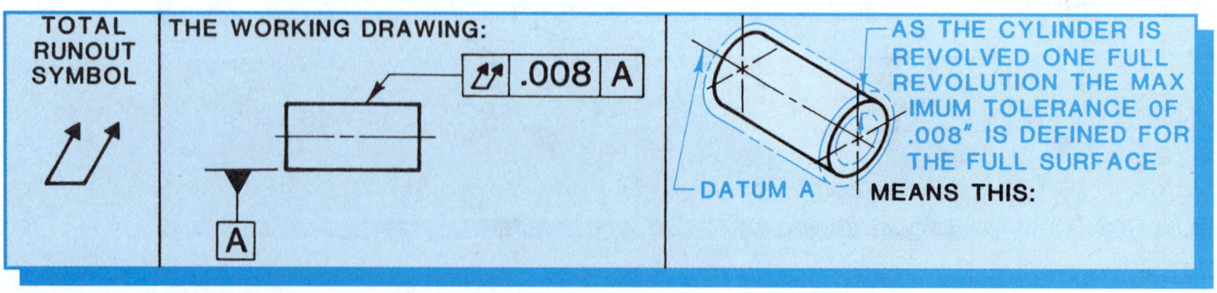

Figure 10-37
TOTAL RUNOUT controls the entire circular surface as the part is rotated 360°.

An example of a fully dimensioned metric working drawing, complete with dimensional and geometric tolerancing notations, is shown in **Figure 10-38**. Interpret each dimension and feature control frame in this illustration.

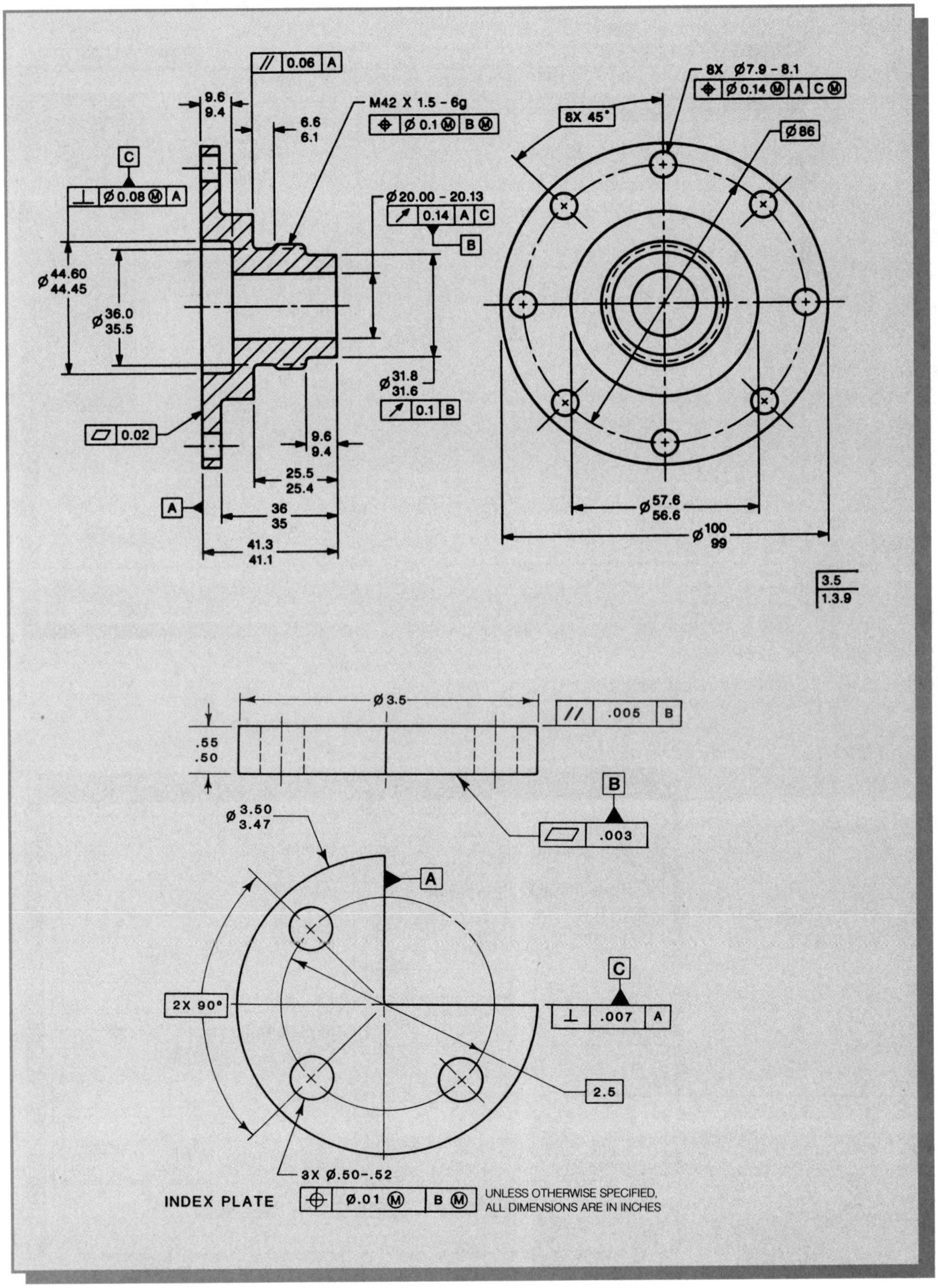

Figure 10-38
Examples of a full-dimensioned metric working drawing with tolerances and geometric tolerancing ASME Y14.5M-1994. *(Reprinted with permission of the American Society of Mechanical Engineers)*

AutoCAD LT Exercises

Tolerancing with AutoCAD LT

With AutoCAD, tolerances can be added to the dimensions automatically by changing the dimension variables. Since tolerancing is a common drafting function, changing the dimensioning variables has been made easy for the CAD operator.

1. In AutoCAD, select Settings from the menu bar.
2. Select Dimension Style **(Fig. 10-39)**.
3. Select Text Format to change the variables to get the appropriate tolerance form **(Fig. 10-40)**.

   Notice the terminology used in the tolerances box (None, Variance, Limits).

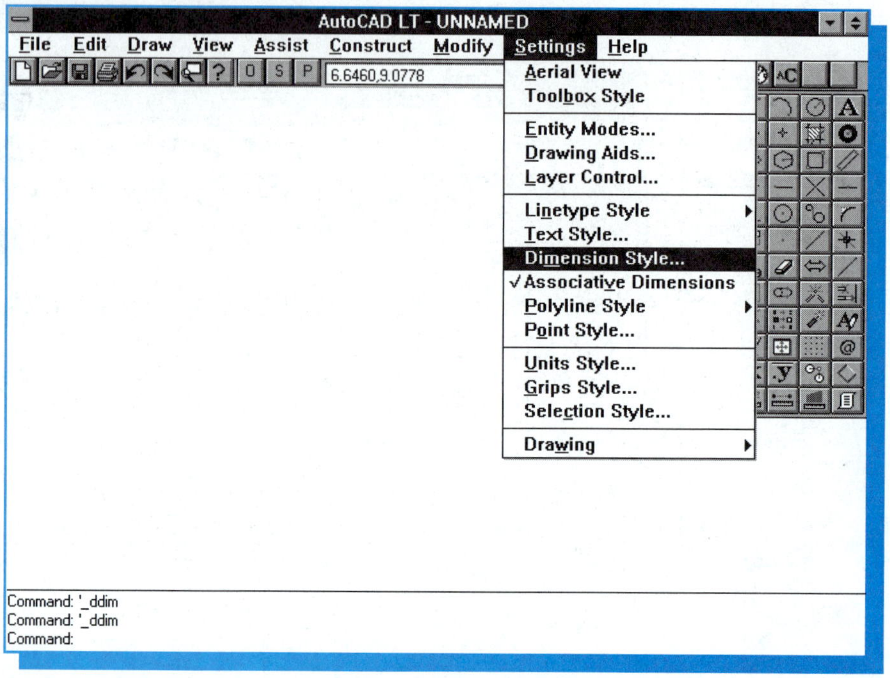

Figure 10-39
Settings dialog box: Select Dimension Style

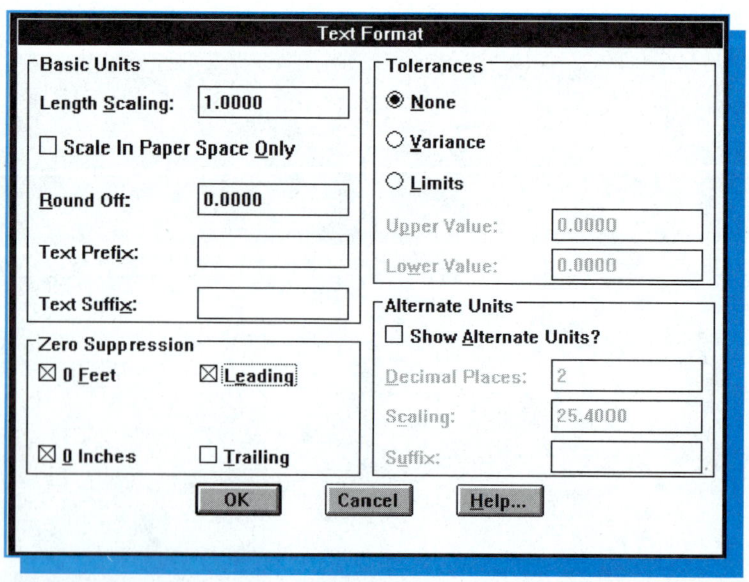

Figure 10-40
Text Format dialog box: settings for no specified tolerance

Figure 10-41 illustrates six methods that can be used to show the same 4.000" dimension on a CAD drawing using different dimensioning variables. The different methods are

A: None—no specified tolerance
B: Variance—bilateral plus and minus tolerance
C: Variance—unilateral with all the tolerance on the plus side
D: Variance—unilateral with all the tolerance on the minus side
E: Variance—bilateral with different plus and minus tolerance
F: Limit—limit tolerance

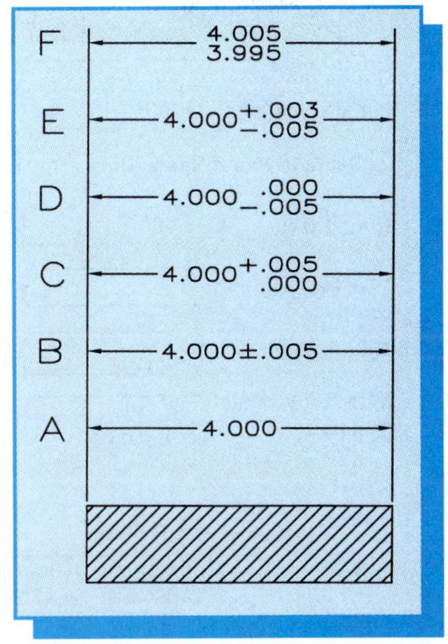

Figure 10-41
Examples of tolerances generated with AutoCAD LT.

Tolerancing

1. Draw the shape in **Figure 10-42** without dimensions. Do not worry about the tolerances. Be extremely careful to draw it accurately. Remember, AutoCAD will use the size of the shape created for the dimensions.
2. Setting a ± .005 tolerance:
   a. Select **Settings**, **Dimension Style**, **Text Format**.
   b. In the Tolerance box, select **Variance**.
      Type **.005** in the **Upper** and **Lower Value** boxes **(Fig. 10-43)**.
   c. Select **OK**.
3. Dimension the length of the part. It should look like the top dimension in **Figure 10-42**. Experiment with tolerancing by selecting limits in the tolerance box and setting the Upper and Lower Values to various decimal fractions and then dimensioning part features.
4. Finish dimensioning the part so it looks like **Figure 10-44**.

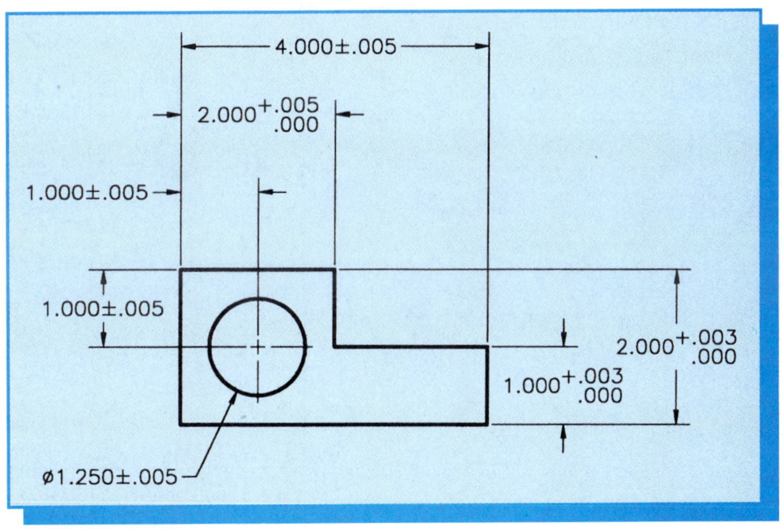

Figure 10-42
Dimension drawing

tip: *When dimensioning using tolerances, dimension all the features with similar tolerances before changing the tolerances for the next set of toleranced dimensions.*

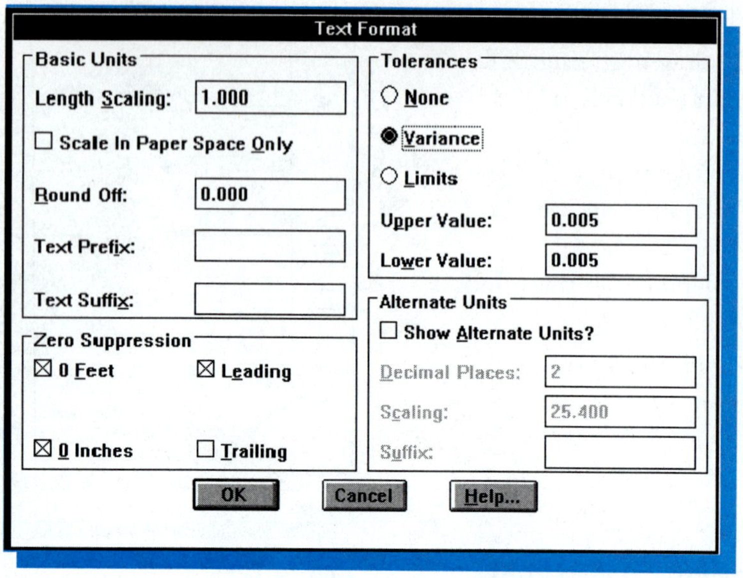

Figure 10-43
Text Format dialog box: settings for bilateral plus and minus tolerance

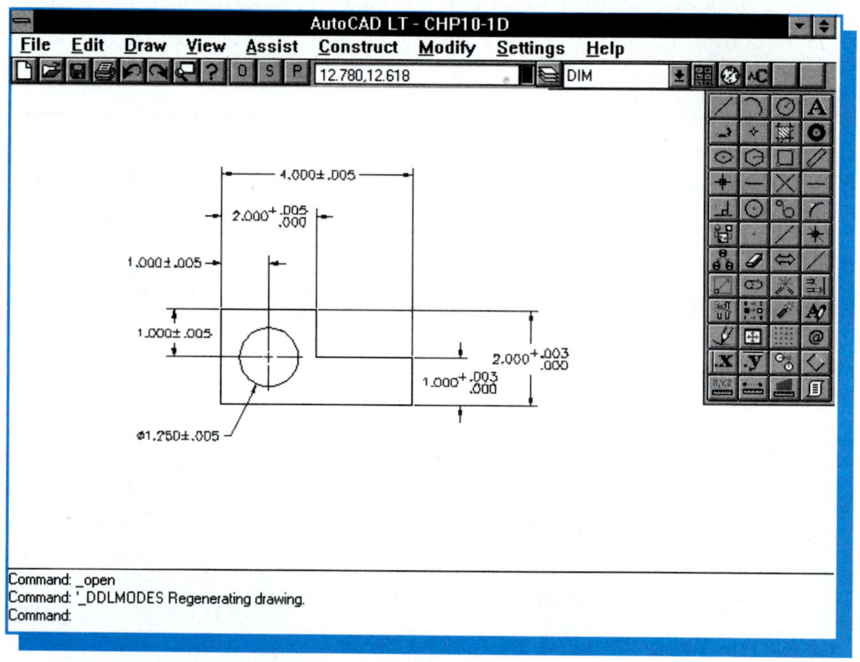

Figure 10-44
An AutoCAD-generated drawing with unilateral and bilateral tolerances

Geometric Dimensioning and Tolerancing

Applying geometric and positional tolerances on a CAD drawing is more difficult than simple tolerancing. ASME Y14.5M is the national standard for tolerancing. The symbols have been standardized and symbol libraries have been developed. However, AutoCAD LT does not come with these symbols loaded. A Geometric Dimensioning and Tolerancing symbols library may be purchased and loaded into the program. The symbol libraries include all of the standard geometric symbols, as well as the reference blocks and other dimensioning aids. To use the symbols, the CAD operator uses the insert command under DRAW. The drawings in **Figure 10-45** are

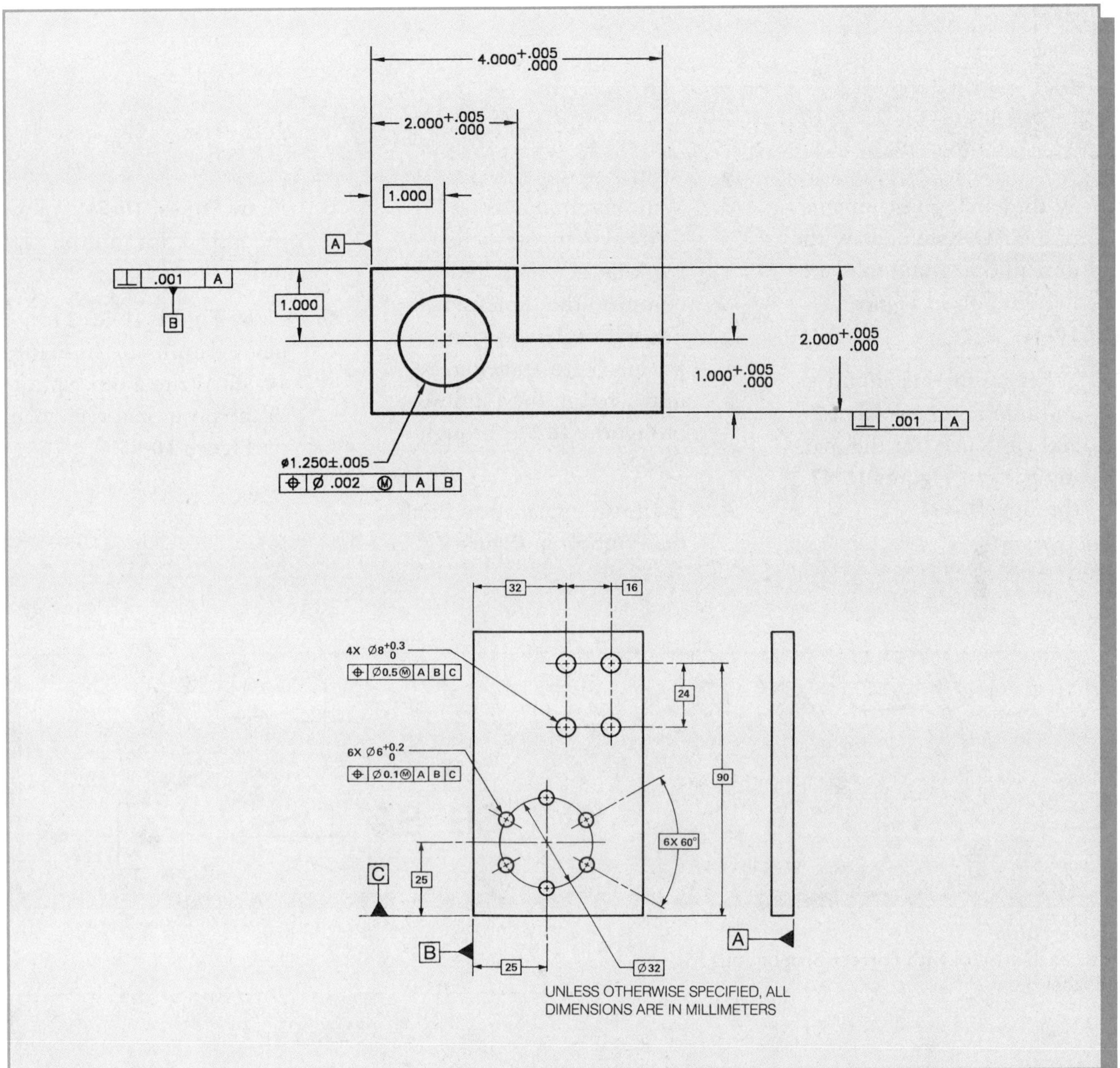

Figure 10-45
AutoCAD-generated drawings with tolerancing and geometric tolerancing symbols

examples of drawings where a symbol library was used to add geometric dimensioning and tolerancing to the drawing. If a symbol library is not available, the symbols may be created using TEXT, RECTANGLE as well as other drawing commands.

The following have been added:

1. datums A and B
2. form tolerances of perpendicularity
3. hole locational dimensions have been changed to basic sizes
4. the positional tolerance of true position

Drafting Exercises

1. With drawing instruments or a CAD system, draw the dimensioning and tolerancing symbols in **Figure 10-46**.

2. What are the maximum and minimum tolerances and clearances for the mating parts in **Figures 10-47** through **10-49**?

3. With instruments or a CAD system, draw the working drawings and complete the dimensioning, with tolerances and geometric tolerancing, as instructed in the captions of **Figures 10-50** through **10-52**.

4. Write an explanation of all the symbols in **Figure 10-53**.

5. Draw **Figure 10-54**. Follow the instructions in the caption.

6. Draw **Figure 10-55**. How many datums are in **Figure 10-55**? Write a description of the positional tolerance on **Figure 10-55**.

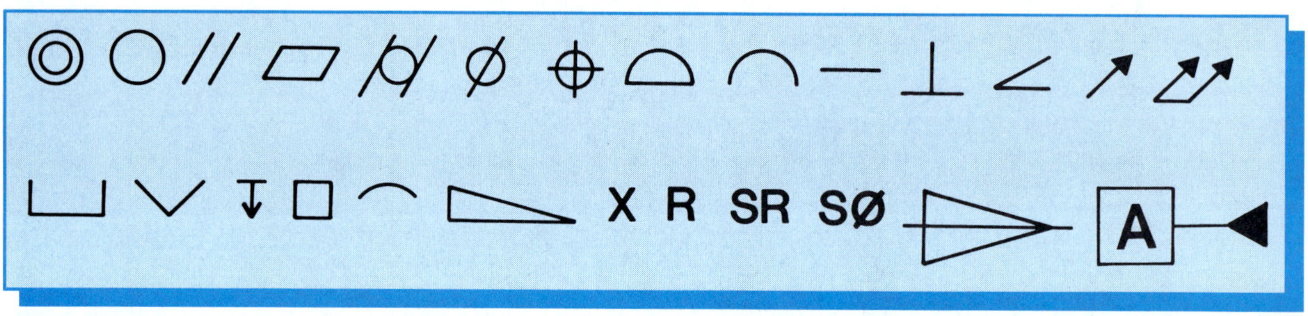

Figure 10-46
Draw each symbol with correct proportions and label them.

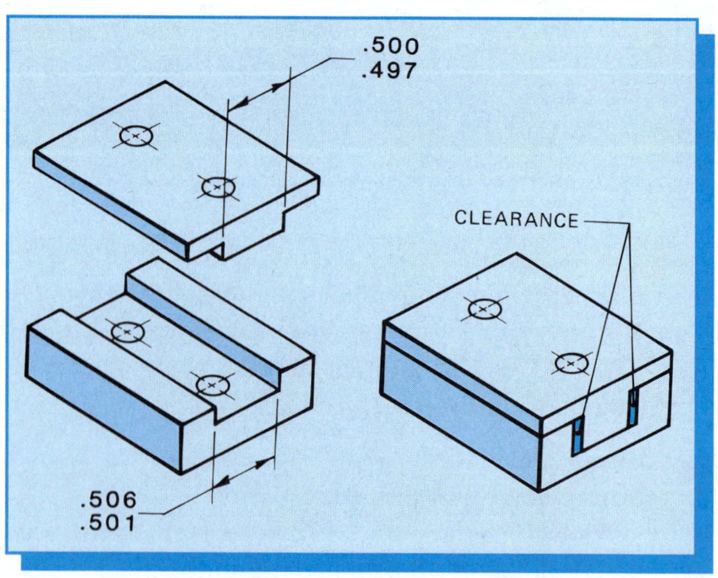

Figure 10-47
What are the maximum and minimum tolerances and clearances for the mating parts?

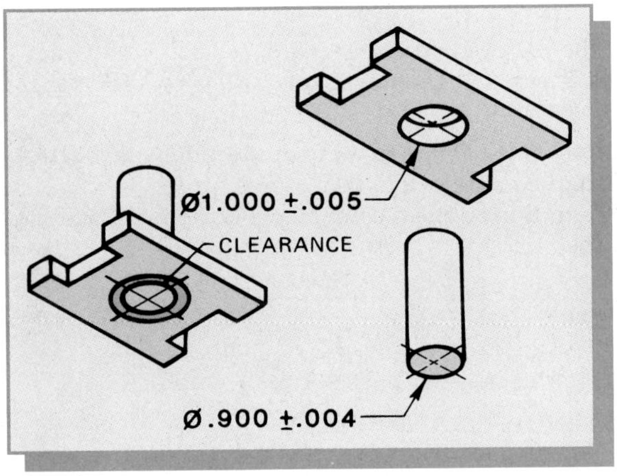

Figure 10-48
What are the maximum and minimum tolerances and clearances for the hole and shaft?

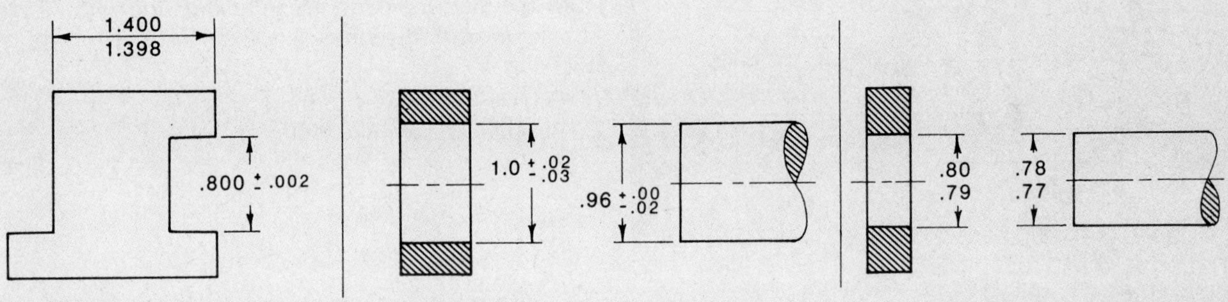

1. What is the tolerance at the top of the drawing?
2. What is the largest possible size of the slot?
3. What is the smallest possible size of the slot?
4. What is the basic size of the slot?

1. What is the tolerance of the hole?
2. What is the tolerance of the shaft?
3. What is the MMC of the hole?
4. What is the MMC of the shaft?
5. What is the LMC of the hole?
6. What is the LMC of the shaft?
7. What is the allowance of the tightest fit?
8. What is the allowance of the loosest fit?

1. What is the tolerance of the hole?
2. What is the tolerance of the shaft?
3. What is the MMC of the hole?
4. What is the MMC of the shaft?
5. What is the LMC of the hole?
6. What is the LMC of the shaft?
7. What is the allowance of the tightest fit?
8. What is the allowance of the loosest fit?

Figure 10-49
Answer the following questions on tolerancing.

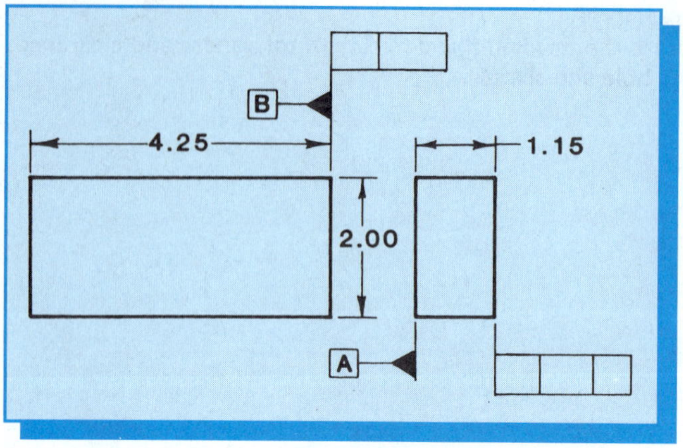

Figure 10-50
Draw the multiview drawing and dimension with Grade 12 tolerances. Add geometric tolerancing symbols for the following:

1. Datum A is parallel to its opposite side within .01".
2. Datum B is flat within 005".
3. Datum B is perpendicular to the bottom surface within .02".

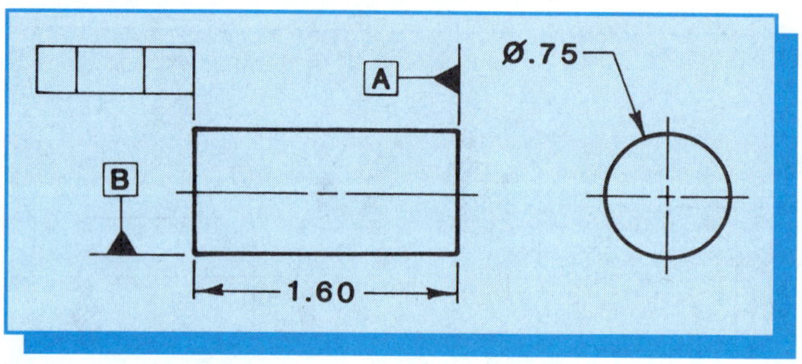

Figure 10-51
Draw the multiview drawing and dimension with Grade 11 tolerances Add geometric tolerancing symbols for the following:

1. Datum A is parallel to the opposite end within .004".
2. Roundness is within .008".

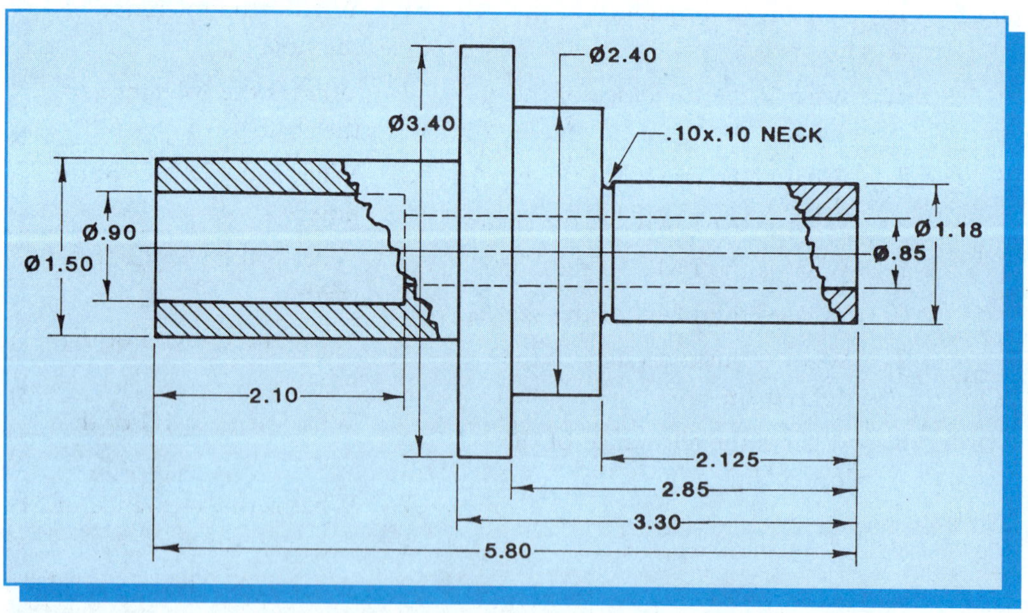

Figure 10-52
As a designer, add tolerances of 002" and 008" to the appropriate dimensions. Add geometric characteristic symbols of flatness, straightness, perpendicularity, and circularity where they apply.

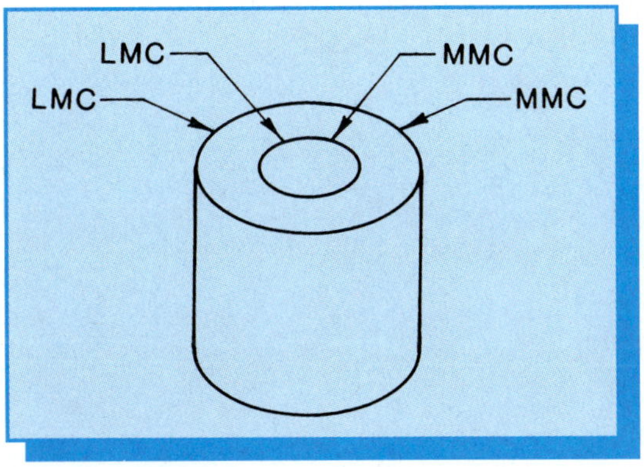

Figure 10-53
Write a short description how the geometric tolerance symbol affects the cylinder and the hole.

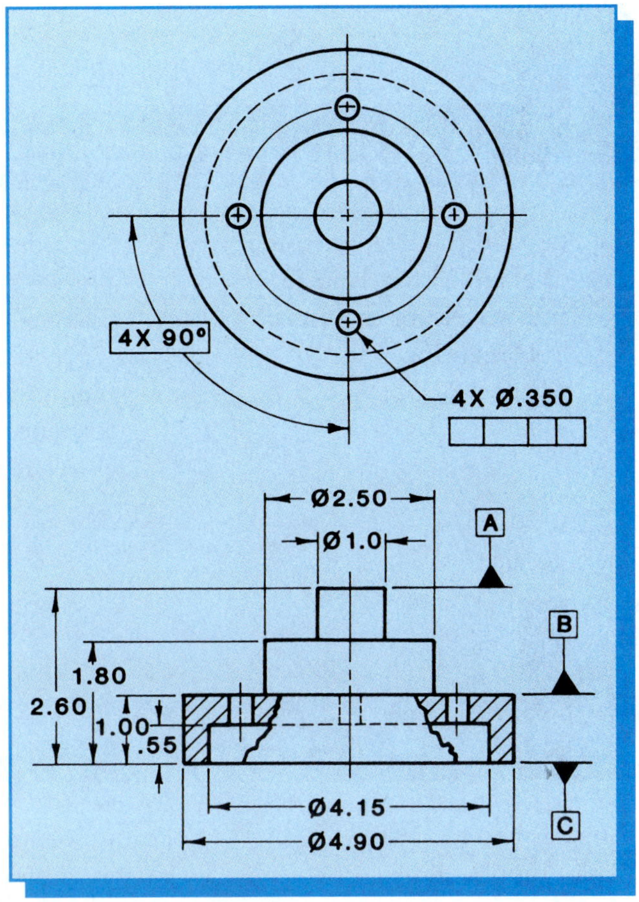

Figure 10-54
Draw the working drawing and dimension with Grade 8 tolerances. Add geometric tolerancing symbols for true position within .005" for the holes, and flatness within .003" for datums A, B, and C.

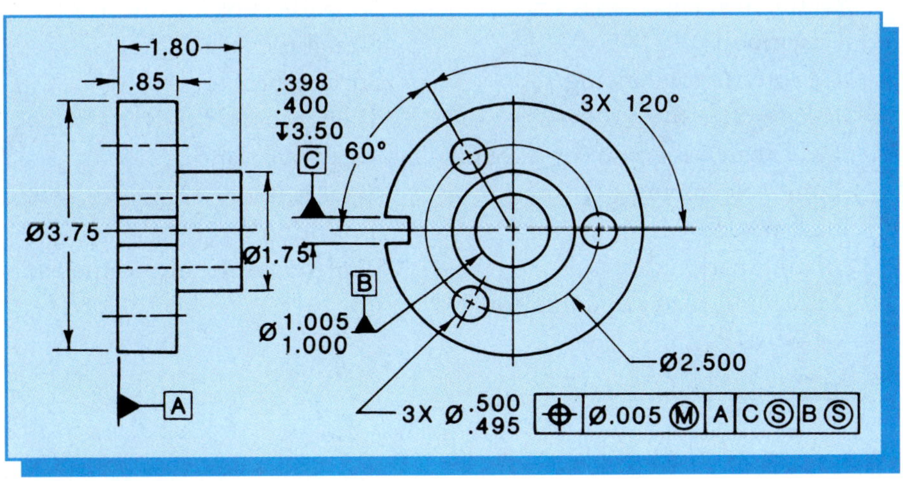

Figure 10-55
Write a description for each geometric tolerance symbol.

Design Exercises

1. Use a tolerance dimension of grade 13 for the axle and wheels for the toy car in Design Exercise 1 in Chapter 9.

2. Do a drawing of two steel pipes 6" long. The inside diameter of one pipe is .75". Design the second pipe to fit inside with a tolerance grade of 7. Add all required dimensions.

Key Terms

Angularity
Basic (simple) dimensions
Basic hole size
Basic shaft size
Bilateral tolerance dimensions
Chain dimensions
Circular runout
Circularity (roundness)
Clearance fit
Concentricity
Cylindricity
Datum
Datum dimensioning
Datum identification box
Datum target
Direct dimension
Feature control frame
Feature position
Flatness
Form tolerancing
General tolerance note
Geometric characteristic symbols
Geometric tolerancing
Interference fit
Least material condition (LMC)
Limit tolerancing
Line profile
Mating parts
Maximum material condition (MMC)
Minus tolerancing dimensions
Parallelism
Perpendicularity
Plus tolerancing dimensions
Positional tolerancing
Regardless of feature size (RFS)
Straightness
Surface profile
Tolerance
Tolerance zone
Total runout
Transition fit
Unilateral tolerance dimensions

Chapter 11

Sectional Views

Objectives

The student will be able to:

- draw a full section drawing
- draw a half section drawing
- draw a broken-out section drawing
- draw an offset section drawing
- draw an assembly section drawing
- draw a removed section drawing
- draw a revolved section drawing
- draw a thin material section drawing
- draw a pictorial section drawing
- rotate and align features for a section drawing

Introduction

Multiview drawings are adequate for describing the size and shape of an object when all important features can be seen on the normal orthographic views. However, if the internal surfaces require the extensive use of hidden lines, the drawing may be difficult to understand. When this occurs, a sectional view is usually prepared to show the size and shape of the interior more clearly **(Fig. 11-1)**. A **sectional view**, or section, is a drawing of an object as it would appear if cut in half or quartered. Sectional views do not contain

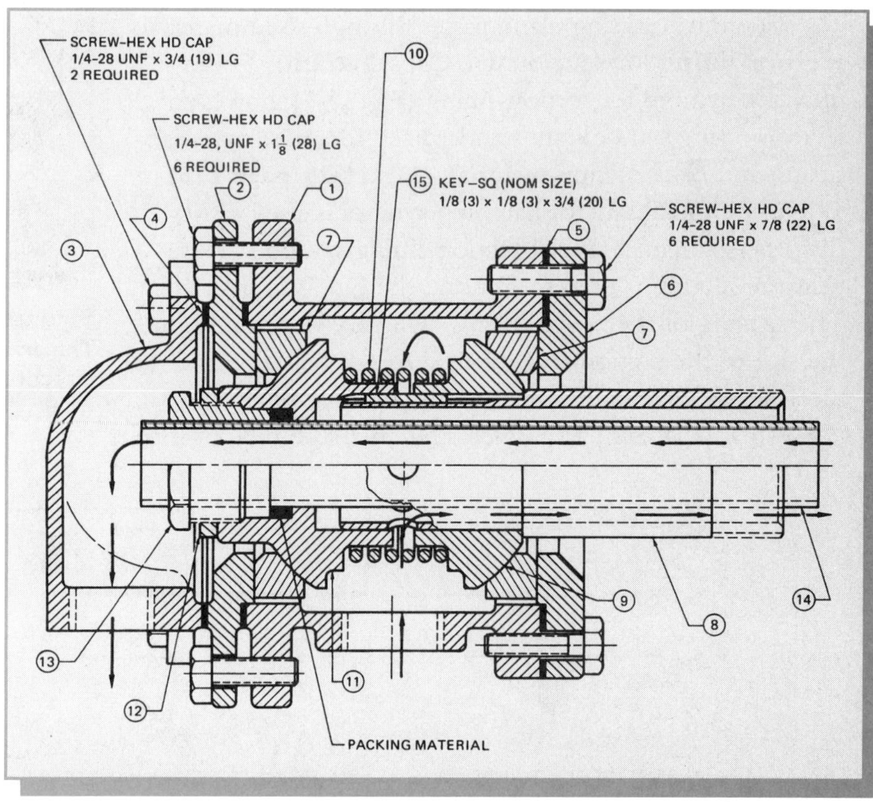

Figure 11-1
This working drawing view would be difficult to read if not drawn as a section. *(Reprinted from* Technical Drawing and Design *by Goetsch and Nelson, © 1986 Delmar Publishers Inc.)*

hidden lines because dashed lines can confuse the clarity of the sectional view **(Fig. 11-2)**.

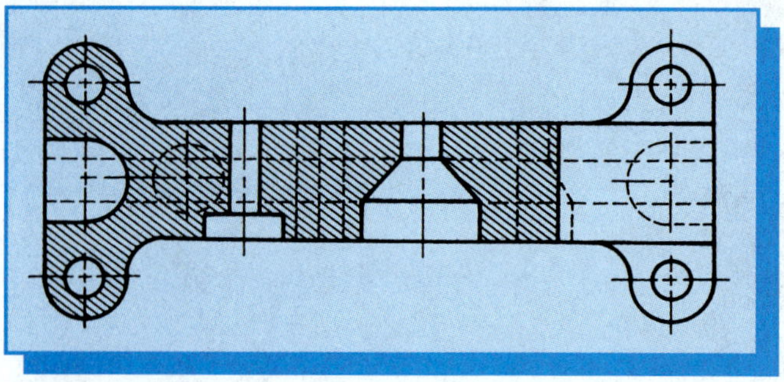

Figure 11-2
Hidden lines will confuse the clarity of a sectional view.

Cutting Plane

To understand sectional views, imagine a saw cutting an object in half. Replace the position of the saw with an imaginary plane and remove the nearest part of the object. What remains is a sectional view.

To define the location of the imaginary sectional plane, a **cutting plane line** is drawn on the view where the cut was made. Cutting plane lines are very wide and dense, with arrows placed on the ends to indicate the line of sight direction for the sectional drawing **(Fig. 11-3)**. Several types of cutting plane lines are used, depending on the type of drawing, size, and location of the section **(Fig. 11-4)**. **Figure 11-5** shows the relationship between the cutting plane, the cutting plane line, and the resulting sectional view.

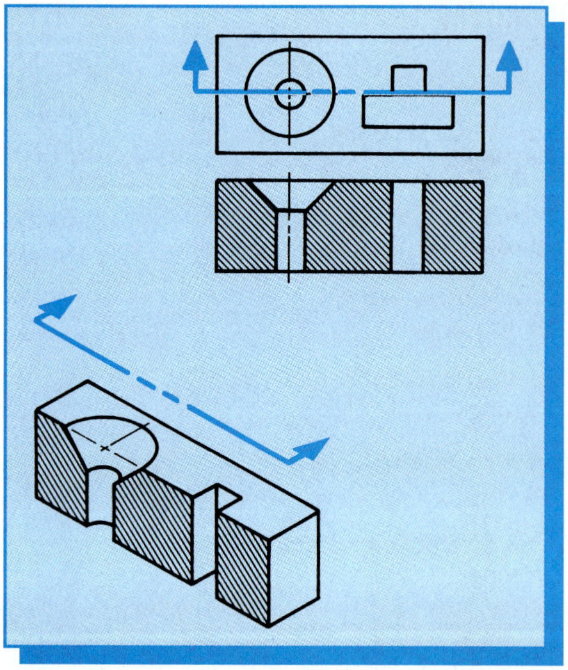

Figure 11-3
The arrows on the cutting plane indicate the direction of the line of sight for the sectional view.

Section Lining

Where the cutting plane passes through solid material, **section-lining** lines are drawn. **Crosshatching** is the universal symbol for section-lining **(Fig. 11-5)**; however, section-lining symbols are standardized for most manufacturing and construction materials **(Fig. 11-25, page 226)**. When the universal crosshatching symbol is used rather than the standard material section-lining symbol, a note near the section is used to indicate the type of material. The spacing of the crosshatching will vary depending on the size of the section **(Fig. 11-6)**. Do not take the time to measure the spacing. Estimate with your eye when drawing. When adjacent parts are crosshatched, the direction of the

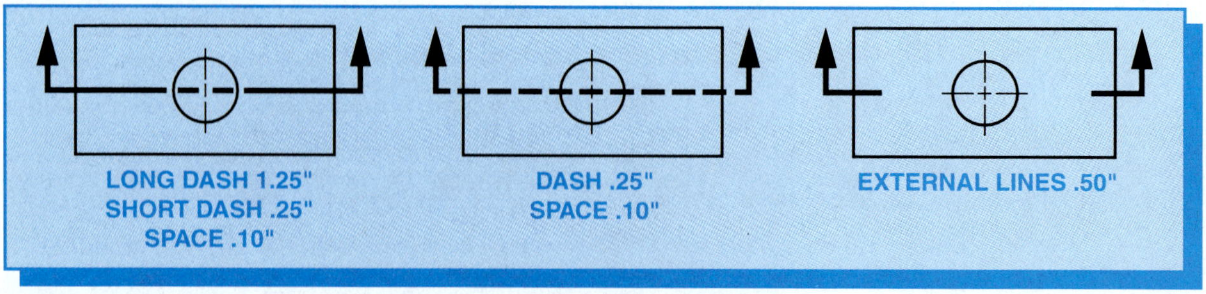

Figure 11-4
Average size and spacing recommended for the cutting plane lines

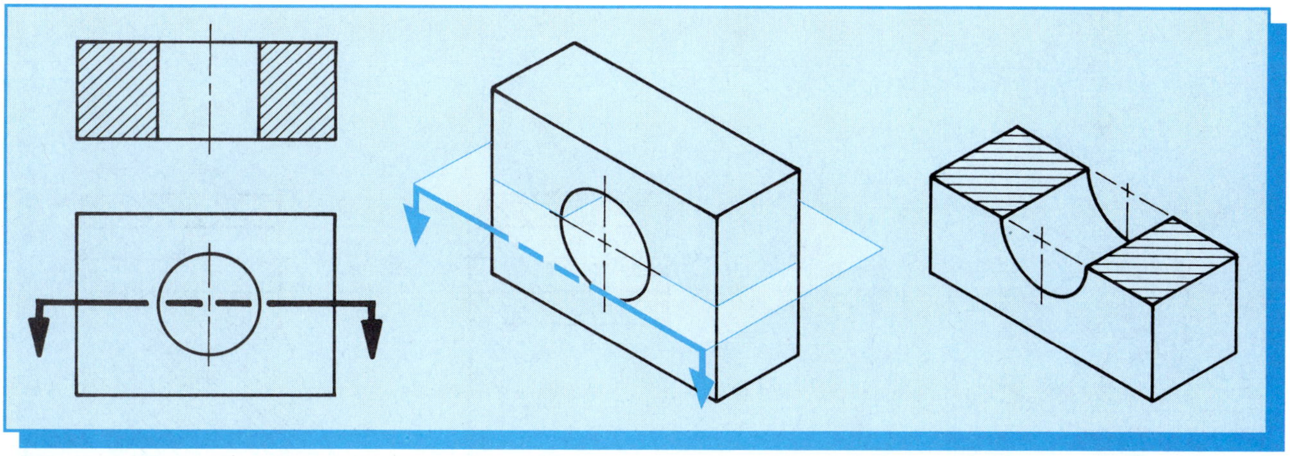

Figure 11-5
The cutting plane defines the surface to be sectioned.

lines is reversed or the angle of the crosshatching is changed to differentiate the separate pieces **(Figures 11-7** and **11-1)**.

Figure 11-8 shows the method of interrupting the section lining to place dimensions and notes. This should only be done in large sectioned areas. It is preferable to place notes and dimensions on the outside of the view.

Section-Lining Exceptions

Although section lining is added to areas intersected by the cutting plane, there are some exceptions to this principle.

Ribs, webs, spokes, gear teeth, lugs, and ball bearings are not section-lined, because they are either comparatively thinner than the remainder of the object or they appear intermittently **(Fig. 11-9)**.

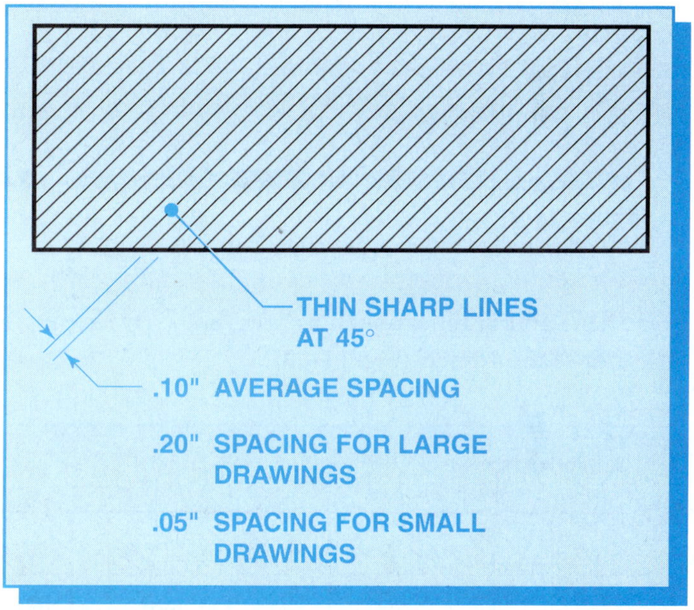

Figure 11-6
Crosshatch spacing will vary according to the drawing size. Space the crosshatching by "eye" when drawing.

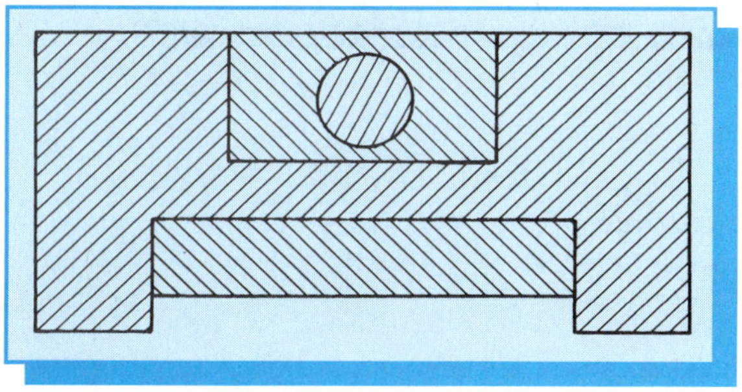

Figure 11-7
Reverse the direction of the crosshatching for adjacent parts.

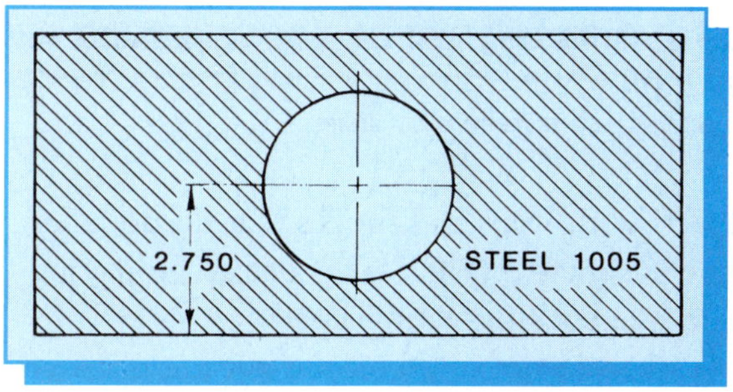

Figure 11-8
If necessary, notes and dimensions may be placed on the sectional view as shown.

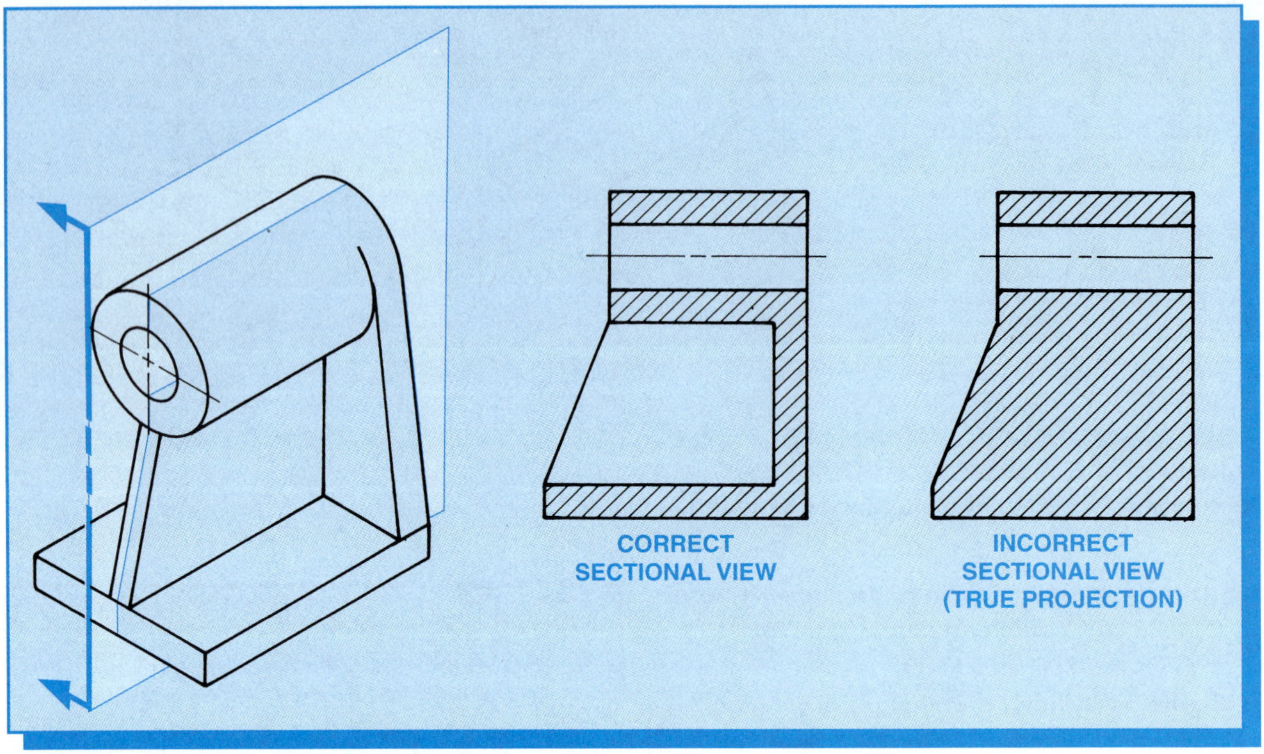

Figure 11-9
Do not crosshatch thin supporting materials.

Shafts and fasteners, such as nuts, bolts, rivets, pins, and screws, are also not section lined **(Fig. 11-10)**.

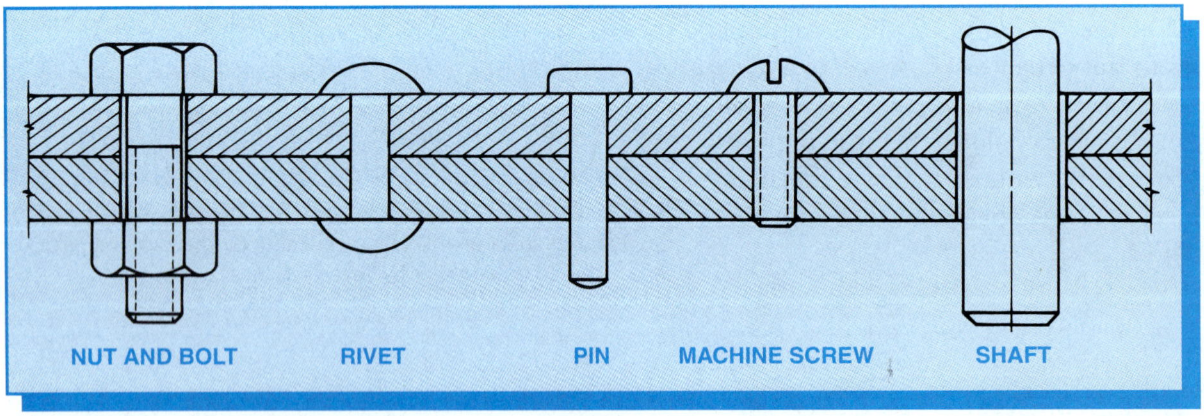

Figure 11-10
Do not crosshatch fasteners or shafts.

Objects with webs and spokes that do not align with the cutting plane line are rotated **(Fig. 11-11)** before projecting the sectional view. The spoke is not section lined.

Holes that do not intersect the cutting plane are also rotated to reveal a hole through the section view **(Fig. 11-12)**. Objects with protrusions are rotated and sectioned in the same manner **(Fig. 11-13)**.

When objects containing curved surfaces are sectioned, the true projection may reveal curved lines outlining the section lined areas. When this occurs, the section is drawn using straight line separations **(Fig. 11-14)**.

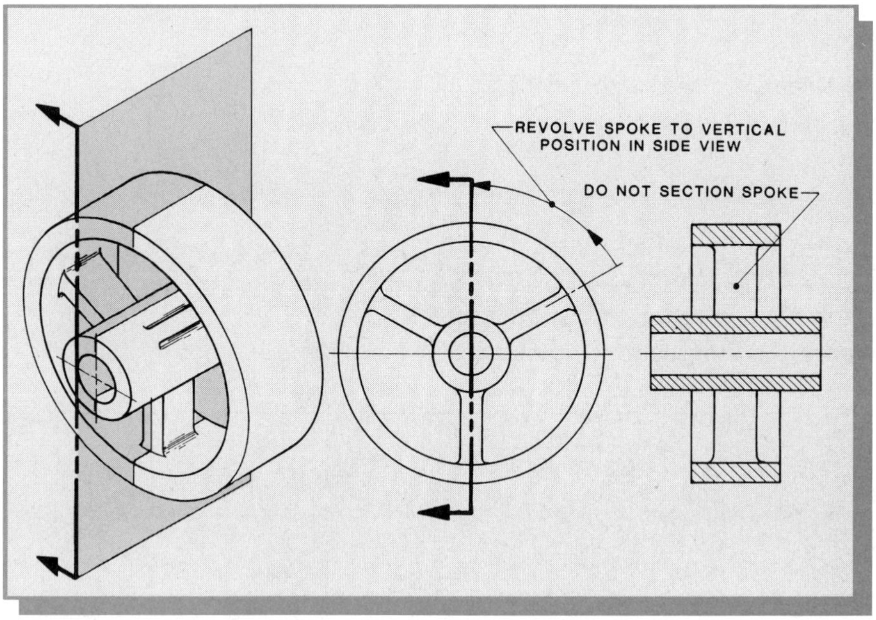

Figure 11-11
Revolve a spoke to a vertical position in a sectional view.

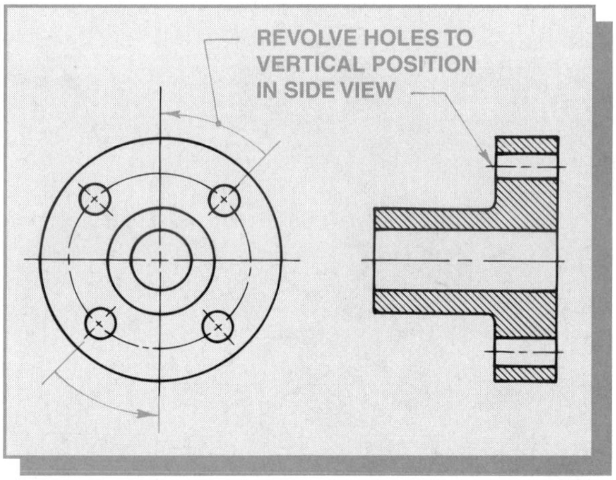

Figure 11-12
Revolve holes to a vertical position in a sectional view.

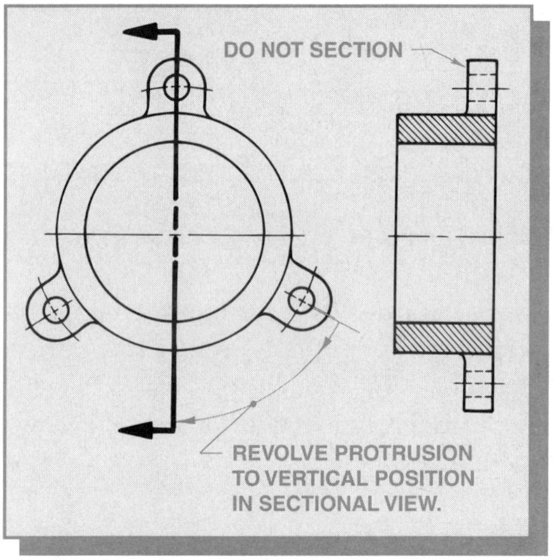

Figure 11-13
Revolve protrusions to a vertical position in a sectional view

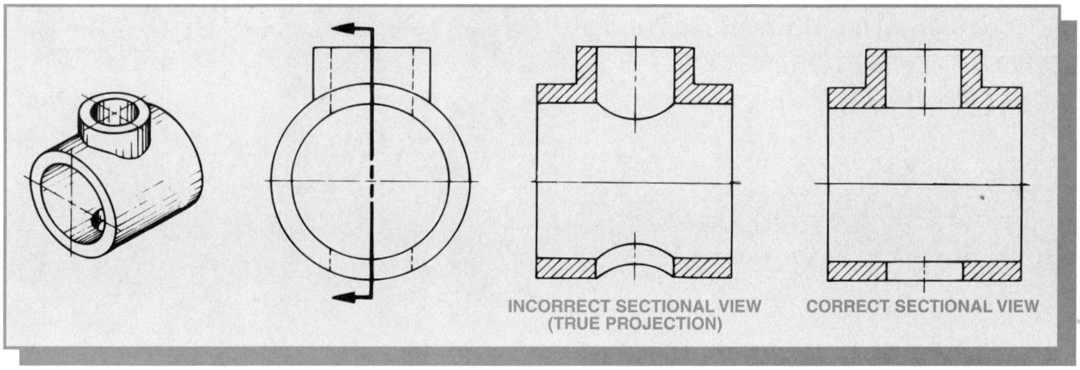

Figure 11-14
Drafting conventions simplify intersections for curved surfaces.

Sectional View Types

The type of sectional view selected depends on the size, scale, and complexity of the area to be sectioned.

Full Sections

When the cutting plane extends entirely through an object, the resulting section is a **full section. Figure 11-15** shows a multiview drawing with a cutting plane on the top view, and the corresponding full sectional view in the front-view position. Full sections of a front view will place the cutting plane on either horizontal (top) views or vertical (profile) views. In architectural drawings, a full section top view is called a plan view or floor plan. A full section for a profile view is an elevation section.

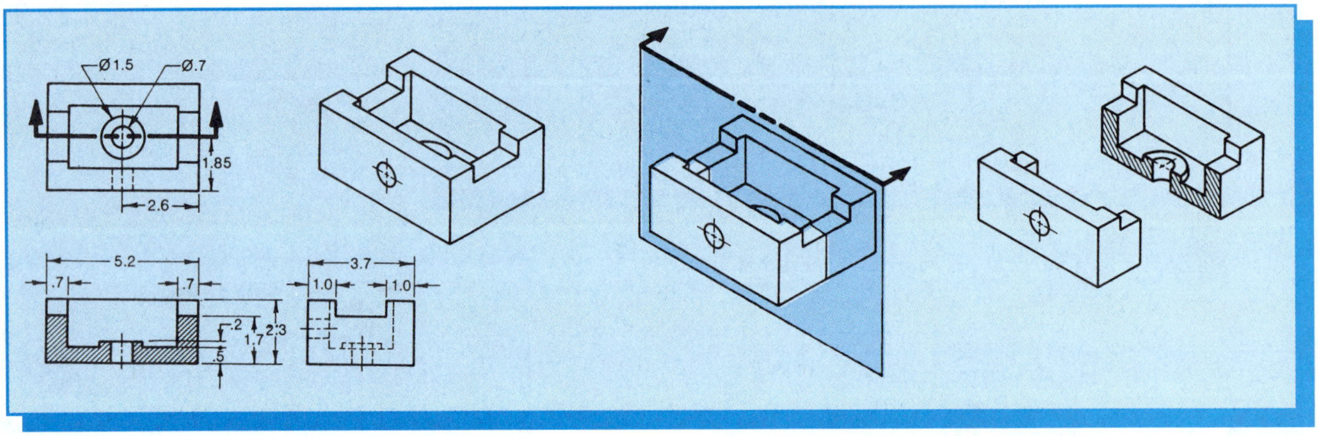

Figure 11-15
Full-section working drawings for a jig base

Half Sections

A **half section** is one-half of a full section. Half sections are prepared for symmetrical objects where the details on both sides of a center line are identical. Two perpendicular cutting plane lines, one on the vertical plane and one on the horizontal plane, define the section area. These align with center lines and remove an imaginary quarter of the object **(Fig. 11-16)**.

Broken-Out Sections

When only a small part of an object needs to be sectioned, a **broken-out section** of that part is prepared.

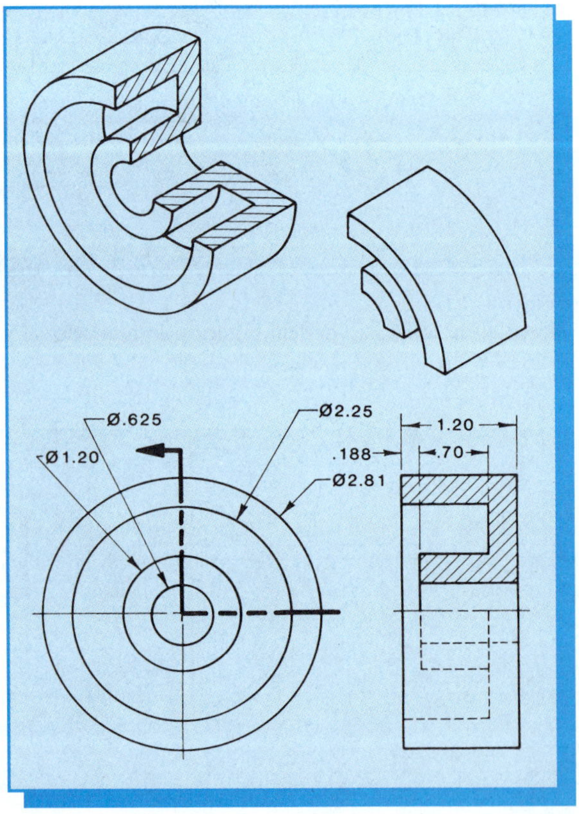

Figure 11-16
Half-section working drawing of a belt pulley

An irregular short break line separates the sectioned area from the remainder of the drawing. **Figure 11-17a** shows a broken-out section in both pictorial and orthographic views. Note how a broken-out section facilitates the dimensions to visible lines in **Figure 11-17b**.

Offset Sections

A cutting plane normally extends through an object in a straight line. However, a straight cutting plane may miss key features that should be shown for maximum clarity. When this occurs, the cutting plane is offset to align with special features. The cutting plane line in **Figure 11-18a** is **offset** to align with a countersunk hole, a counterbored hole, and a slot, none of which fall on a straight line. When offsetting cutting plane lines, always bend the line at right angles to the cutting plane and perpendicular to the plane of projection. Note that the bends in the cutting plane are never shown in the sectioned view **(Fig. 11-18b)**.

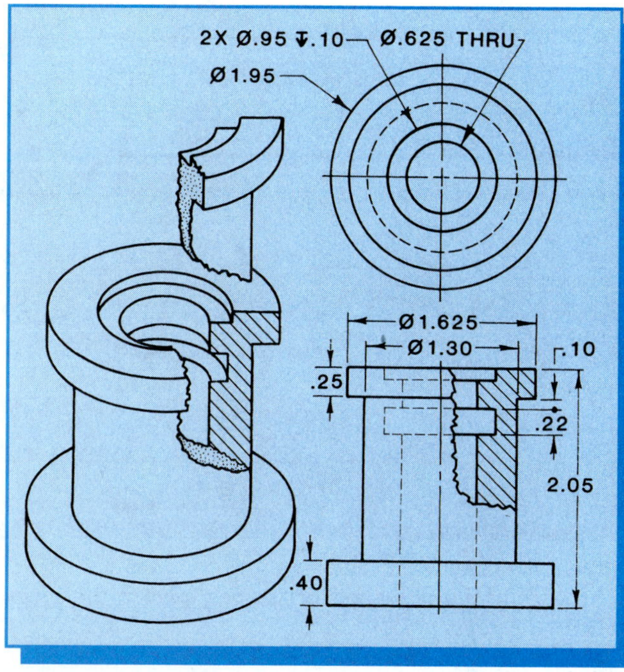

Figure 11-17a
Broken-out section working drawing of a sleeve bearing

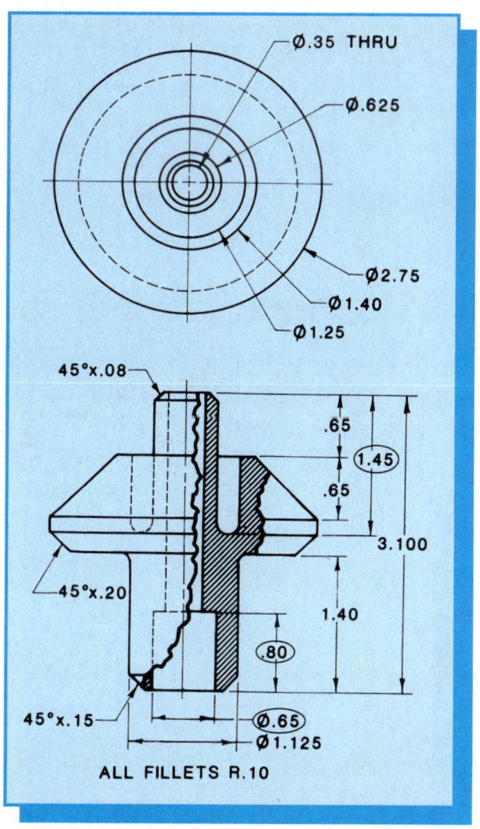

Figure 11-17b
A broken-out section will permit dimensioning to object lines instead of hidden lines.

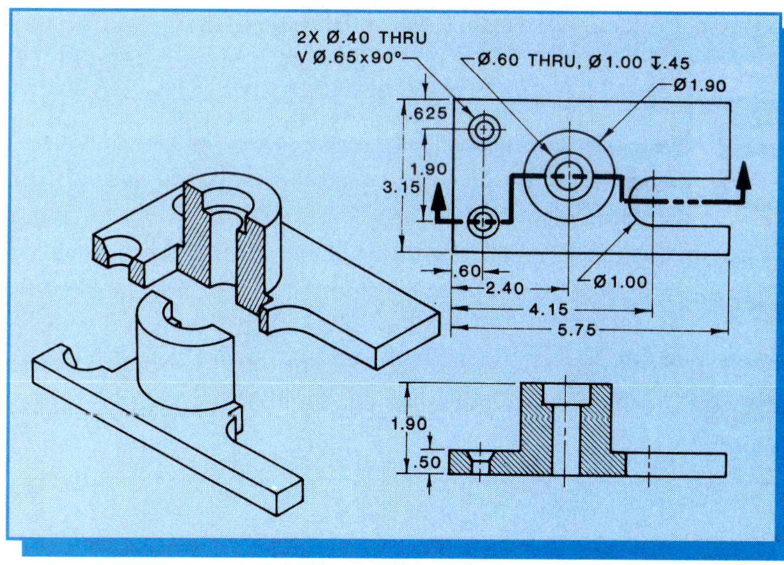

Figure 11-18a
Offset-section working drawing of a jig bearing locator

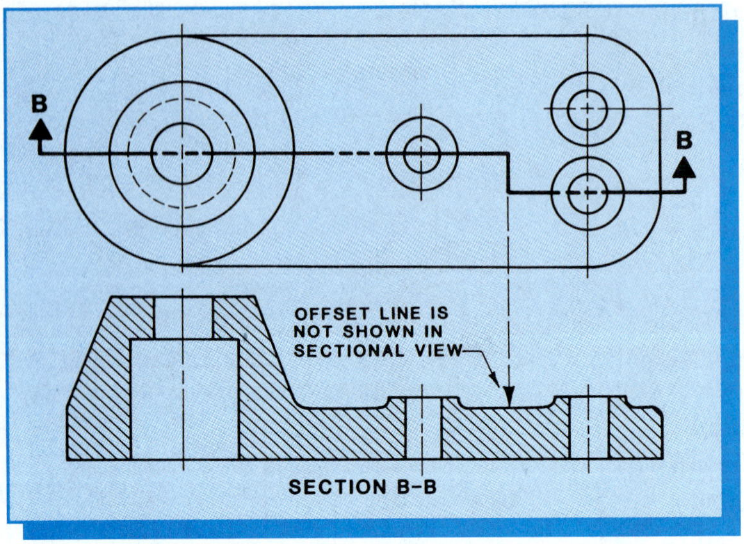

Figure 11-18b
The edge of the offset section is not shown in the sectional view.

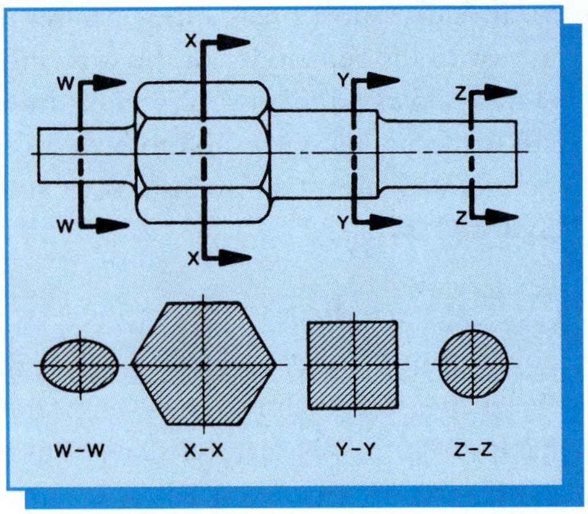

Figure 11-19
Removed sections in a drawing of a spindle arbor

Removed Sections

When sections are drawn in a location other than in one of the six multiview positions, they are **removed sections.** Sections are often removed to be drawn at a different scale or to dimension some portion in detail. When sections are removed, they must be indexed to a cutting plane line containing identifying letters **(Fig. 11-19)**.

Revolved Sections

Sectional views are often revolved 90° to reveal a section of an object perpendicular to the plane of projection. In visualizing a revolved section, imagine a cutting plane passing through an object creating a slice of the object. Now imagine rotating the slice 90°. The rotated slice is a **revolved section.** Some revolved sections break the object to show the section, and others show the section directly on the view **(Figures 11-20a** and **b)**.

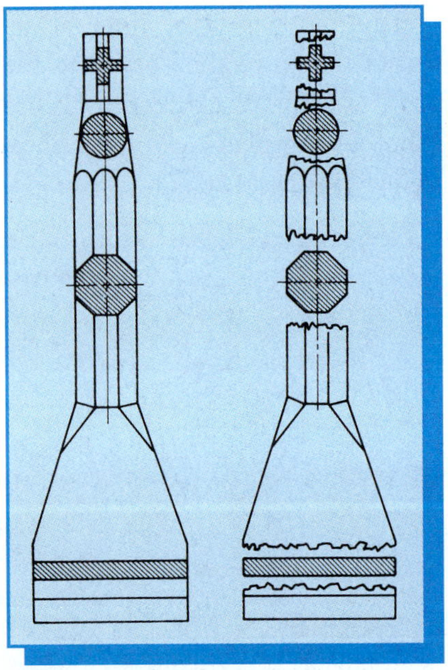

Figure 11-20b
Two methods for drawing revolved sectional views for a jackhammer chisel

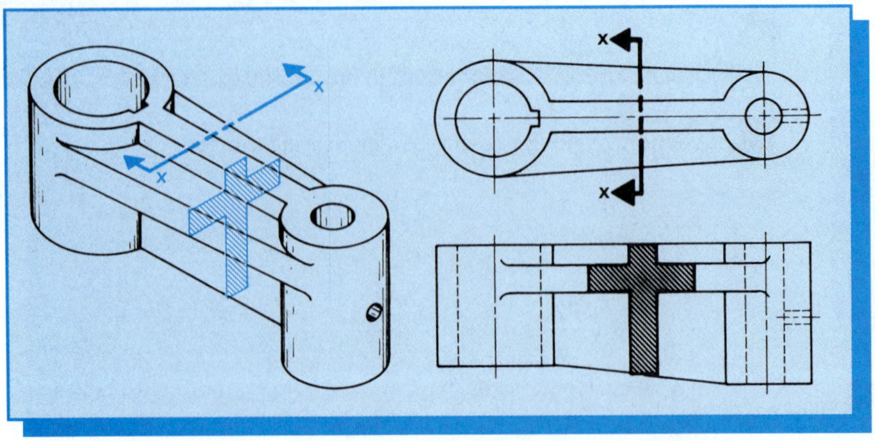

Figure 11-20a
Example of a revolved section in a working drawing and an isometric drawing

Assembly Sections

One common use for sectional drawings is to show the assembly of parts. When multiple-part assembly drawings are sectioned, the adjoining parts are section-lined at different angles. **Figure 11-21** shows an example of an **assembly section** with multiple parts sectioned.

Thin Wall Sections

The outlines of ribs, spokes, and other thin walls are drawn on sectional views. However, section lining is not added to these parts. **Figure 11-22** shows an example of a typical **thin wall section** for a corner bracket.

Materials too thin to section with cross-hatching, such as paper and sheet metal, are sectioned solid.

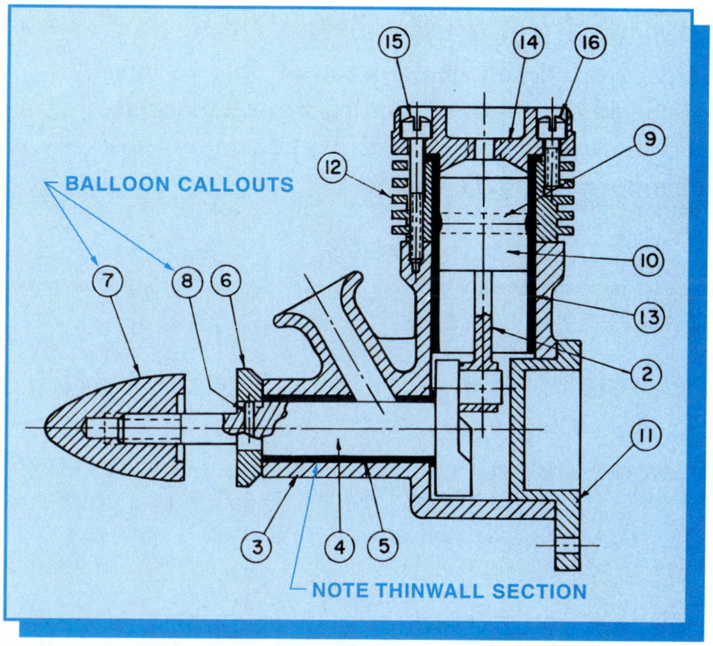

Figure 11-21
An assembly section drawing depicting multiple sectioned parts of a model airplane motor *(Reprinted from* Technical Drawing and Design *by Goetsch and Nelson, © 1986 Delmar Publishers Inc.)*

Auxiliary Sections

Most sectional views are aligned with the normal planes of projection. When a section must be drawn of an auxiliary view, follow the guidelines covered in Chapter 12. **Auxiliary sections** should be aligned with the projection lines of the auxiliary views **(Fig. 11-23)**.

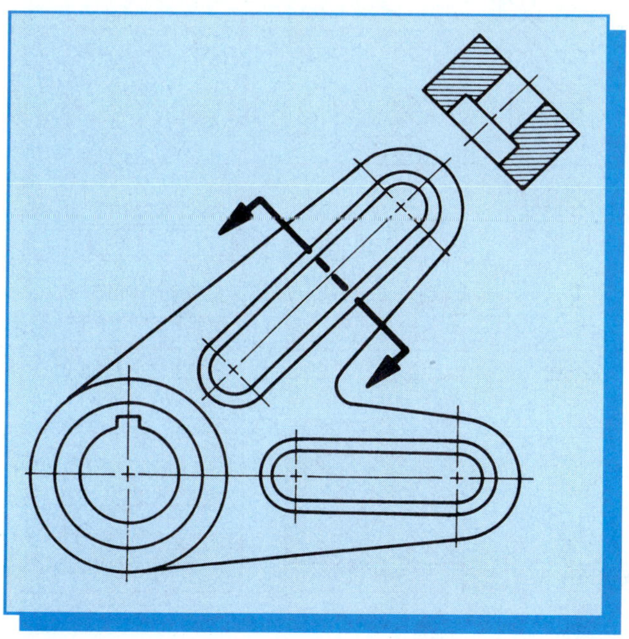

Figure 11-23
Auxiliary section for a locking lever arm

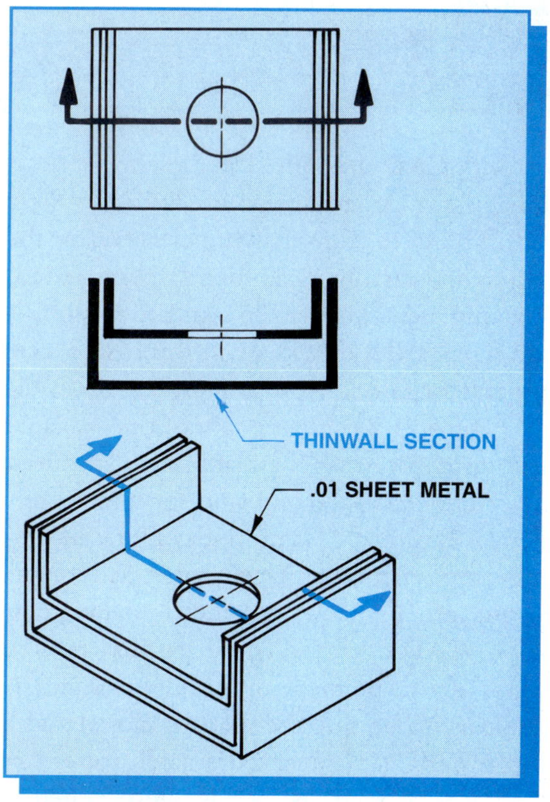

Figure 11-22
Thin wall section for a corner bracket

Enlarged Sections

Sectional details are often too small to be interpreted and dimensioned. In such a case, **enlarged sections** are prepared to clarify details and for ease of dimensioning **(Fig. 11-24)**.

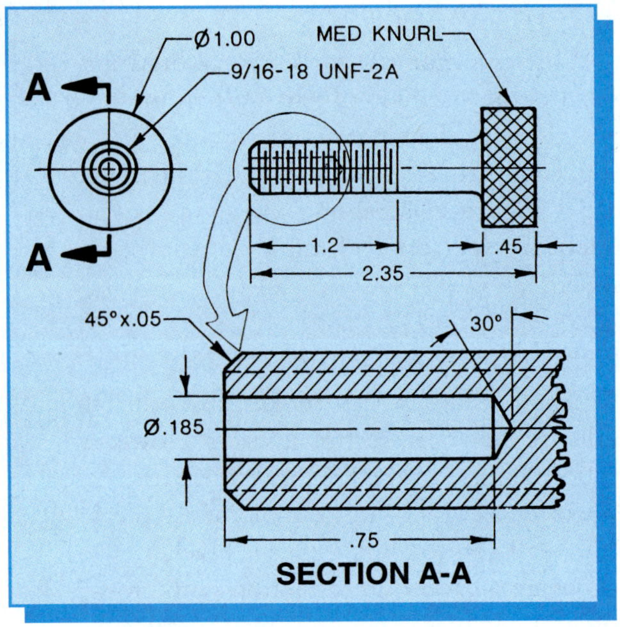

Figure 11-24
Enlarged section of a screw adjuster

AutoCAD LT Exercises

AutoCAD simplifies the drawing of sectional views. An advantage of all CAD systems is the ability to store drawing elements in the system and rapidly recall them when needed. Drawing elements may be standard symbols, line types, styles of text, or in this case, **sectioning patterns**.

Figure 11-25 shows a few of the available sectioning patterns. These are called **hatches.** The AutoCAD LT command for inserting them is Boundary Hatch under Draw in the pull-down menu bar. These hatch patterns may be put into any closed geometric shape drawn by the CAD operator. The shape may be a circle, or any shape made up of lines and arcs. It is necessary to be sure the shape is closed and has no gaps. If a gap appears, the hatch pattern will not insert correctly. A layer is often created in the drawing for the construction of the hatch patterns.

Figure 11-25
Examples of CAD-generated sectioning (hatch) patterns

Sectioning Steps

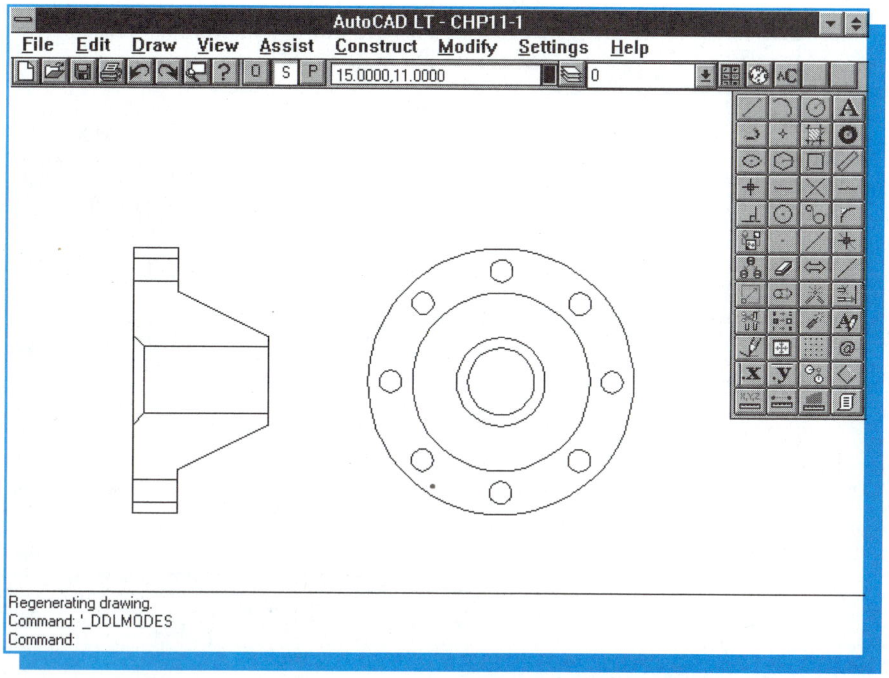

Figure 11-26
Draw these views; use continuous lines and circles (no hidden lines).

1. Draw the desired views in **Figure 11-26**. In the view to be sectioned, use solid lines in the place of hidden lines. The views have been created with line and circle commands. *For the purpose of this sectioning exercise, estimate the sizes for the views.*
2. Select **Draw** from the pull-down menu.
3. Select **Boundary Hatch**.
4. From the Boundary Hatch dialog box **(Fig. 11-27)**, select **Pick Points**.
5. Pick the four areas shown in **Figure 11-28** and when finished picking areas, press EB.
6. Select **Preview Hatch** from the Boundary hatch dialog box to see if you selected the correct areas **(Fig. 11-29)**.
7. Select Continue.
8. Select **Apply** to finish hatching.

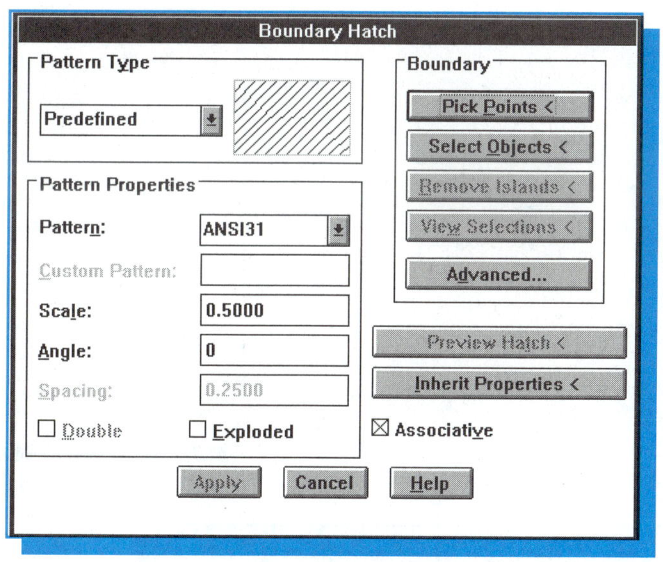

Figure 11-27
Boundary Hatch dialog box

Polylines

Polyline is AutoCAD's way to draw lines that have a given width. To do this:

1. Select **Draw**, and then **Polyline.**
2. Pick a point to start a line.
3. Type **W** (for width of the polyline).
4. Type **.01.**
5. Press EB to accept the ending width of .01 and continue to draw lines; draw several.
6. Press EB to end the command.

   Erase the polyline (note that it is one continuous line). The beginning and the end of polylines do not have to be the same width. Figure 11-30 shows some of the different things that can be done with the polyline command. Experiment with it. One of the uses of polyline is to create the extra wide lines for the cutting plane line in sectioning drawings.

tip: *Use a width of zero for the beginning or ending of a polyline to create an arrowhead.*

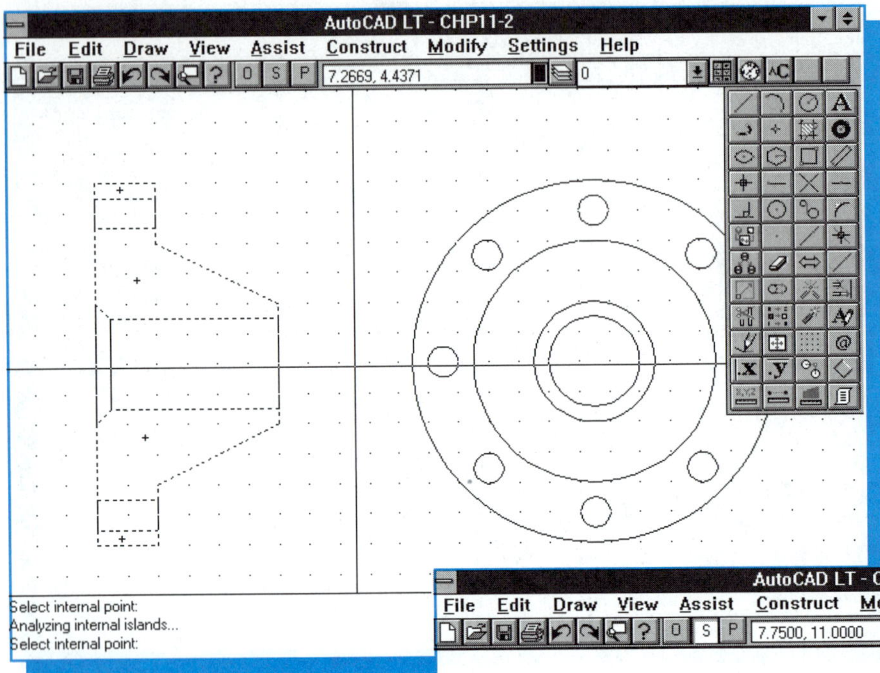

Figure 11-28
Use the Pick Points to select the areas to be hatched.

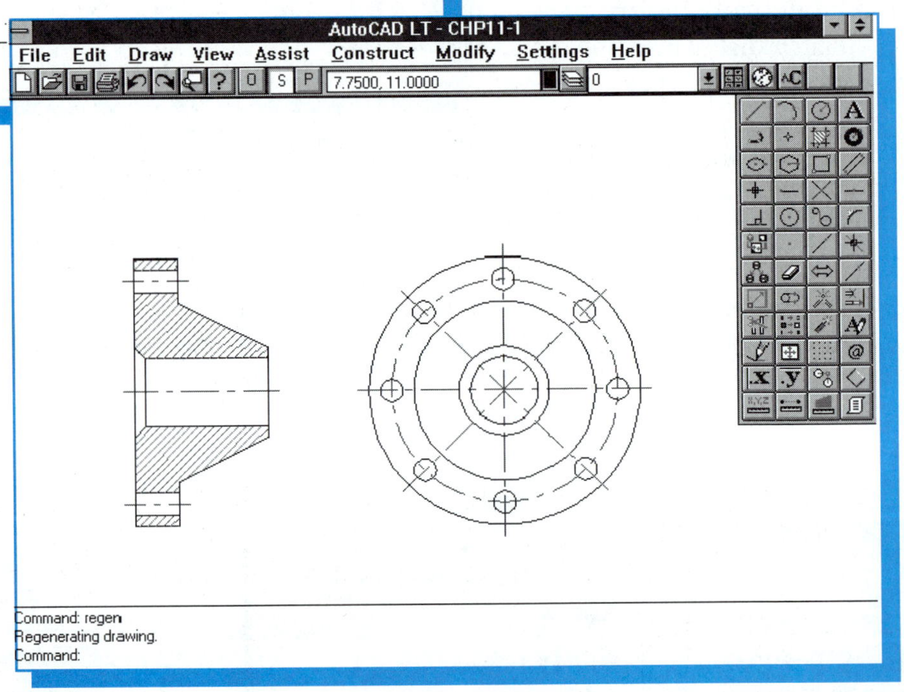

Figure 11-29
Layout of a new drawing or call-up of an existing drawing from the AutoCAD data bank to be sectioned

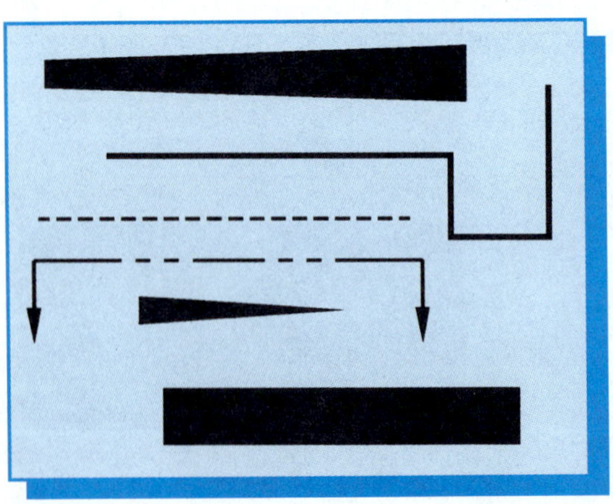

Figure 11-30
Examples of polylines

Drafting Exercises

1. Sketch the sections in **Figure 11-31**.
2. With instruments, draw the sections noted in **Figures 11-32a** and **11-32b**. Have your instructor add new cutting planes for additional sections to be drawn in the working drawings.
3. Draw the working drawings and the isometric sections for **Figures 11-33** and **11-34**.
4. Make the following ECOs (refer to "Engineering Change Orders," in Chapter 17) to **Figure 11-35**:
   - **a.** Change the base dimension 1.60" to 1.50".
   - **b.** Change the base dimension 2.10" to 2.00".
   - **c.** Change the counterbore depth to .200".
   - **d.** Change the vertical section to a horizontal section through the bearing.
5. With a CAD system, draw the sections in **Figure 11-35**.
6. Draw the twin-engine Cessna airplane and the removed sections in **Figure 11-36a**.
7. Complete the multiview sectional drawings.
8. Draw a front and a side view of the Rod Hanger in **Figure 11-37**. Make the side view a full section.
9. Draw the views of the Bearing Hanger in **Figure 11-38**. Change the right side view to a full section.

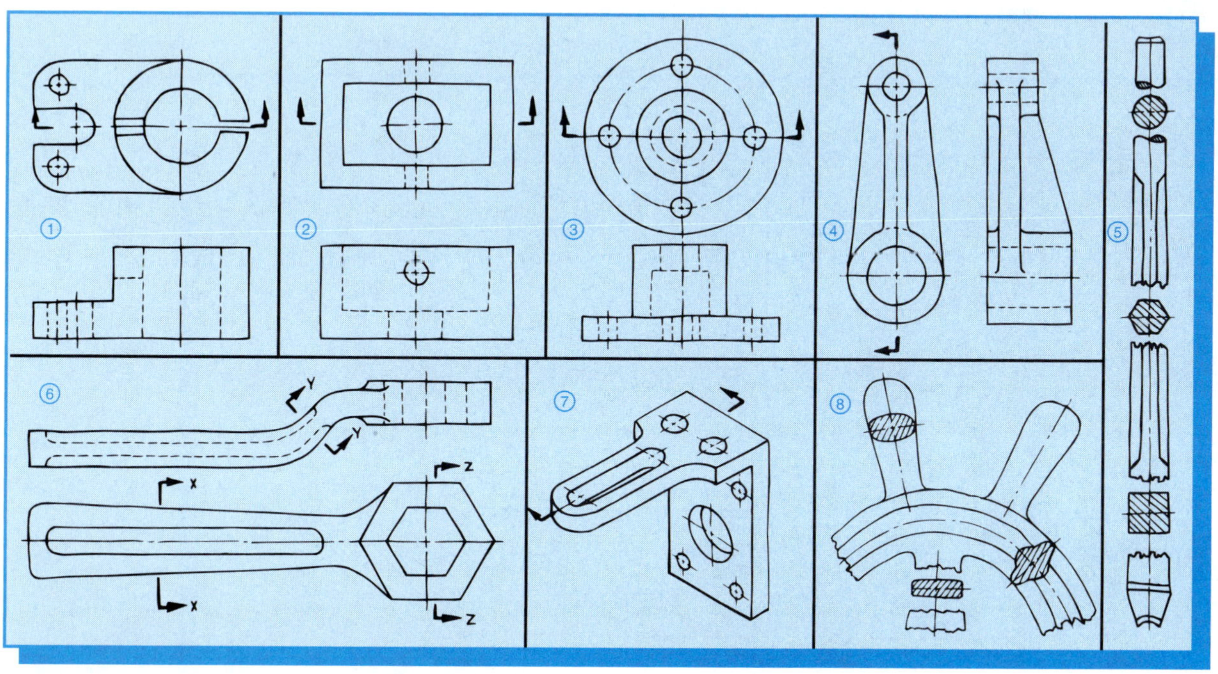

Figure 11-31
Sketch the sectional views for each exercise.

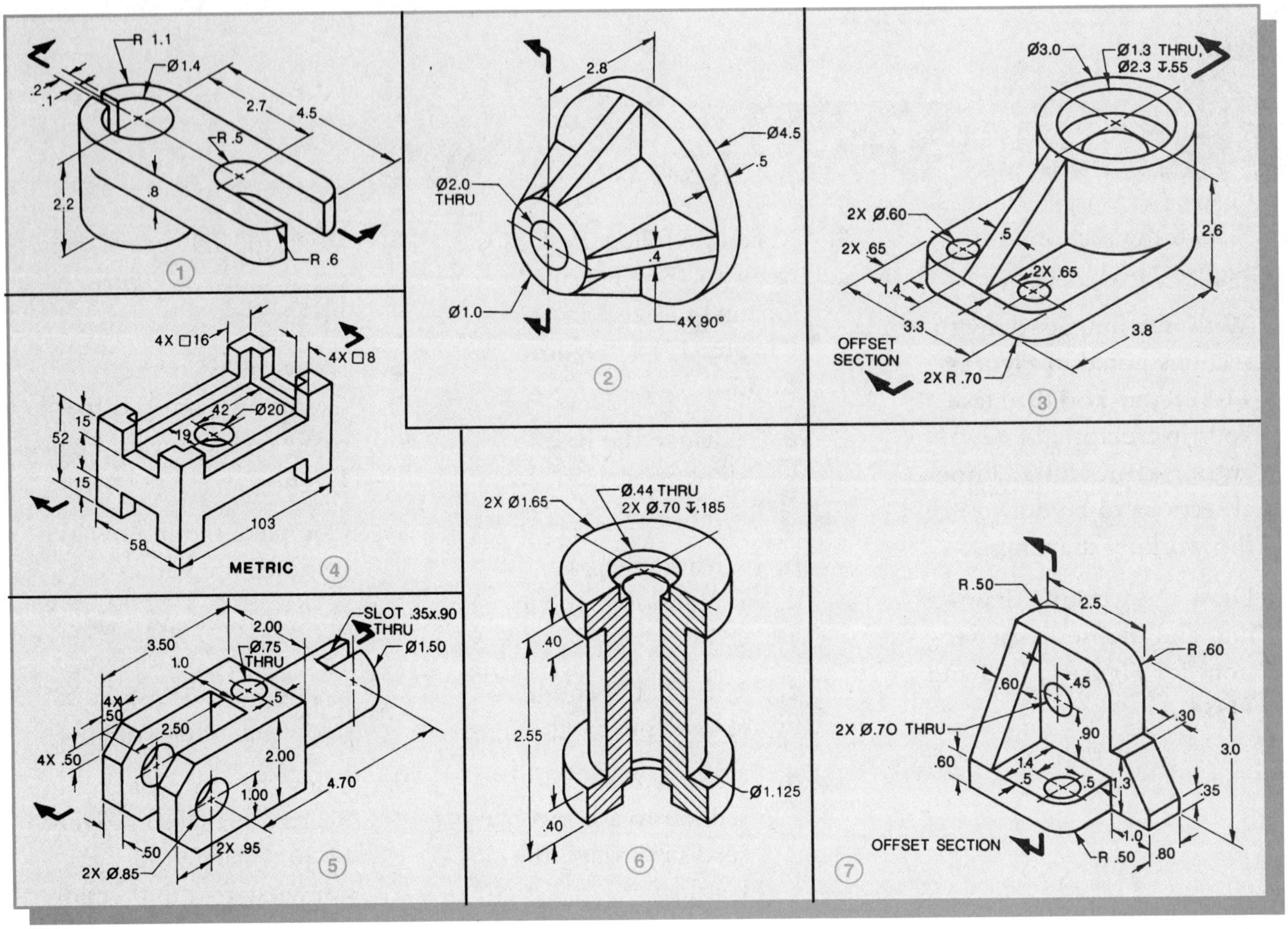

Figure 11-32a
With drawing instruments, draw the sectional views as required.

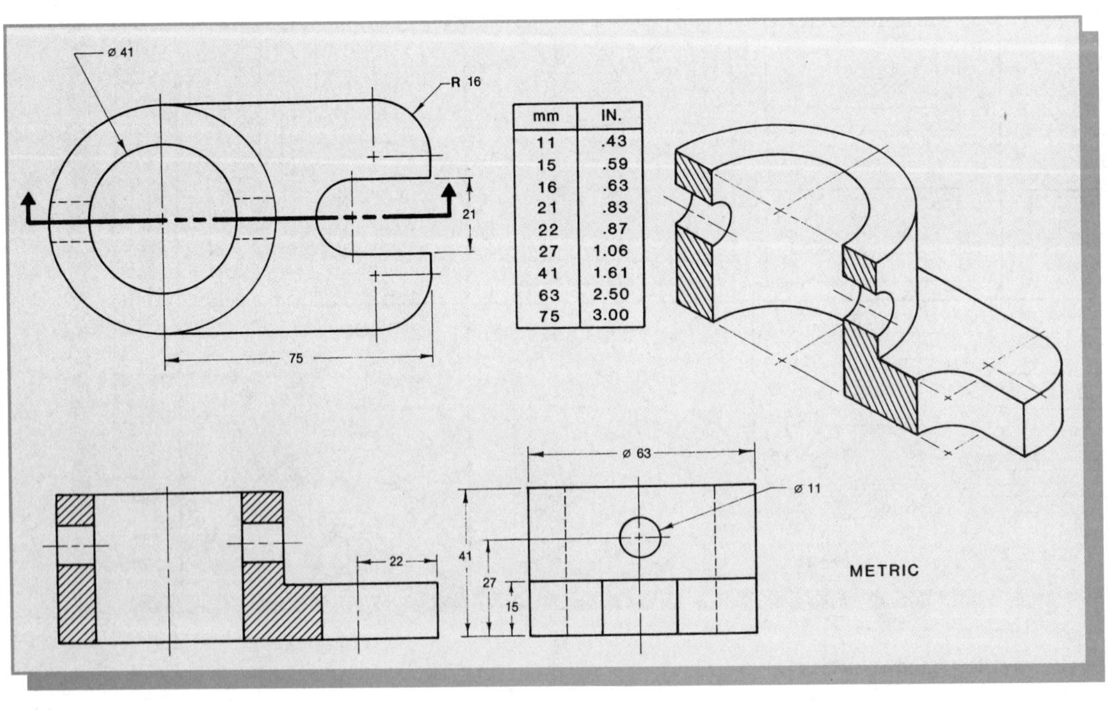

mm	IN.
11	.43
15	.59
16	.63
21	.83
22	.87
27	1.06
41	1.61
63	2.50
75	3.00

Figure 11-32b
Draw the sectional drawings for the metric bearing guide.

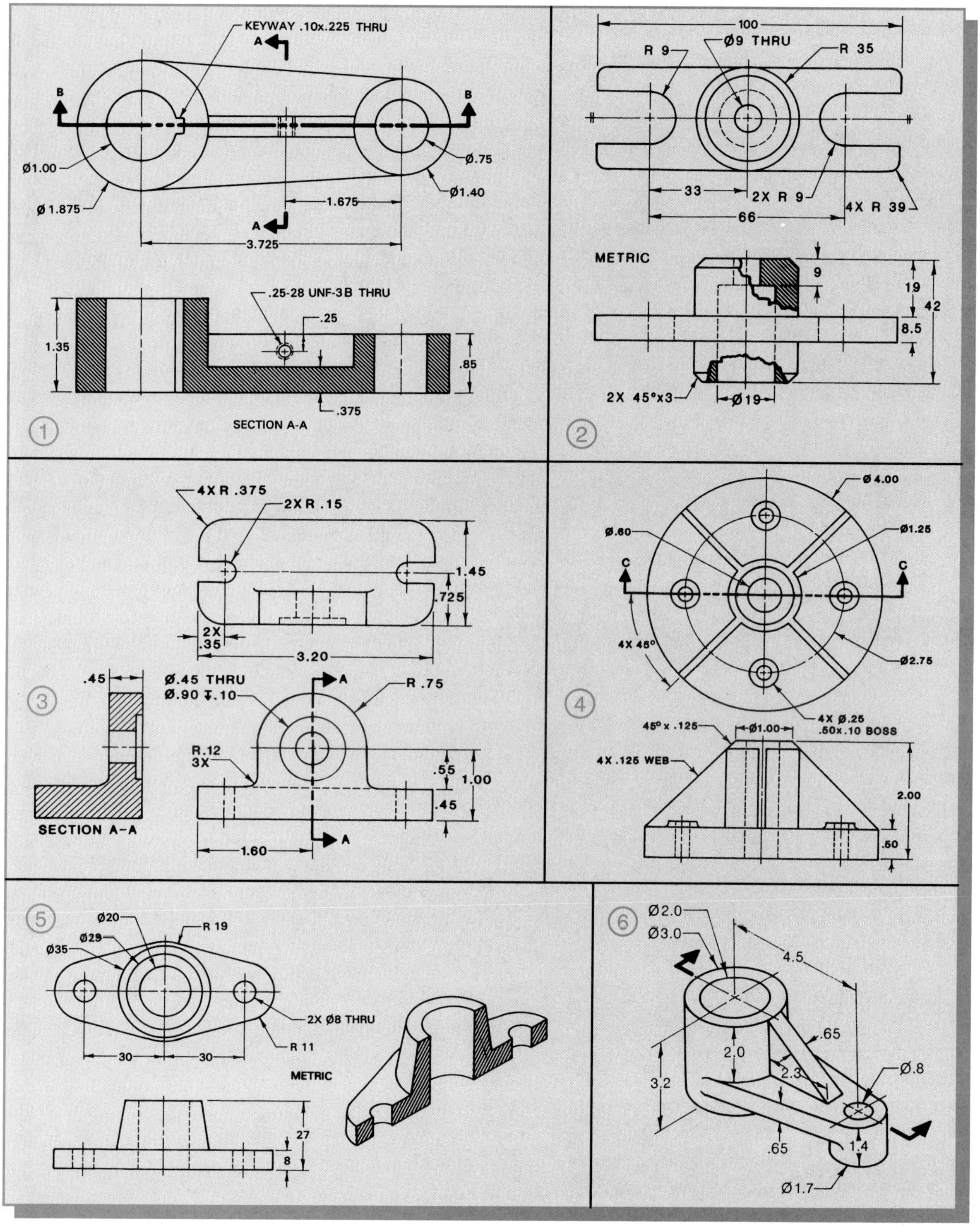

Figure 11-33
Complete the sectional drawings.

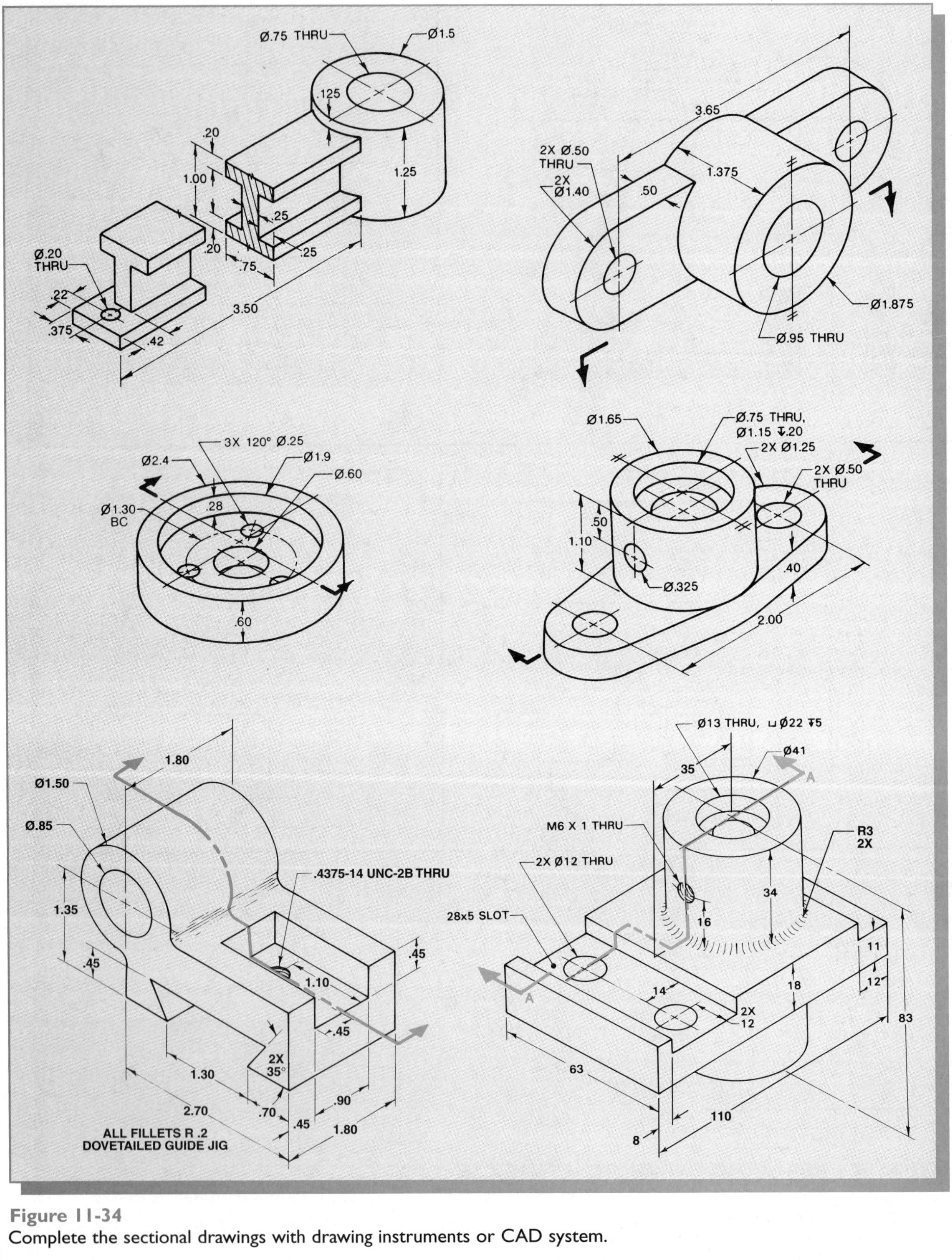

Figure 11-34
Complete the sectional drawings with drawing instruments or CAD system.

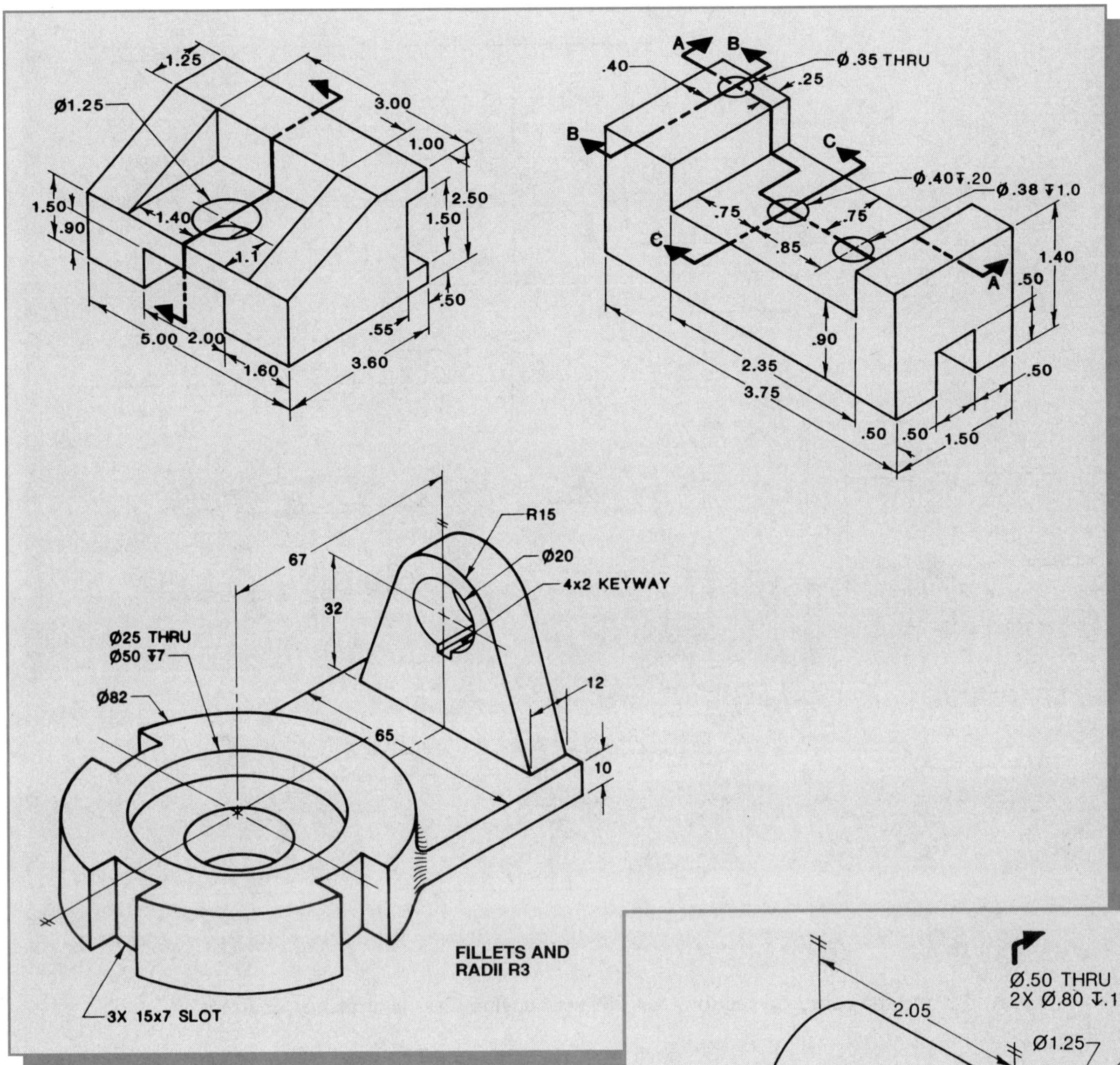

Figure 11-34
Complete the sectional drawings with drawing instruments or CAD system.

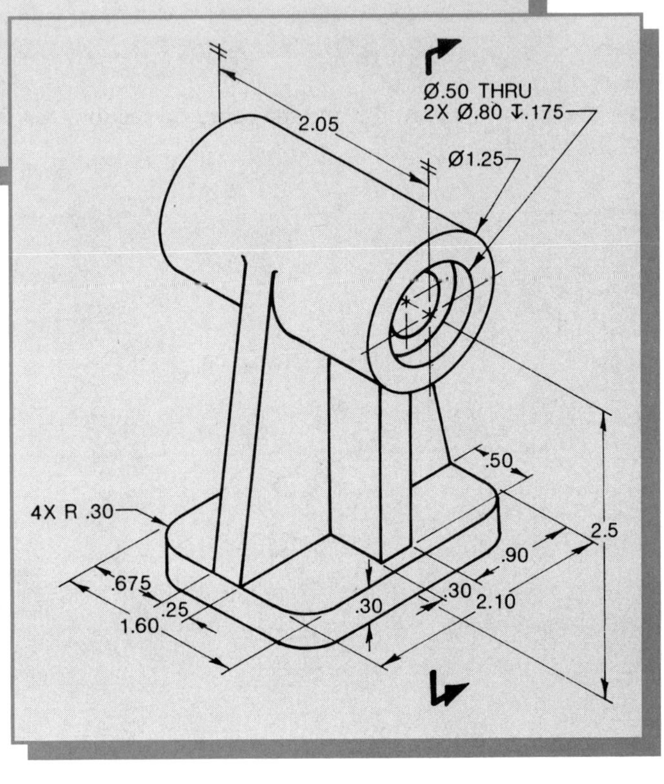

Figure 11-35
Make the ECOs (refer to "Engineering Change Orders" in Chapter 17) listed in Exercise 4.

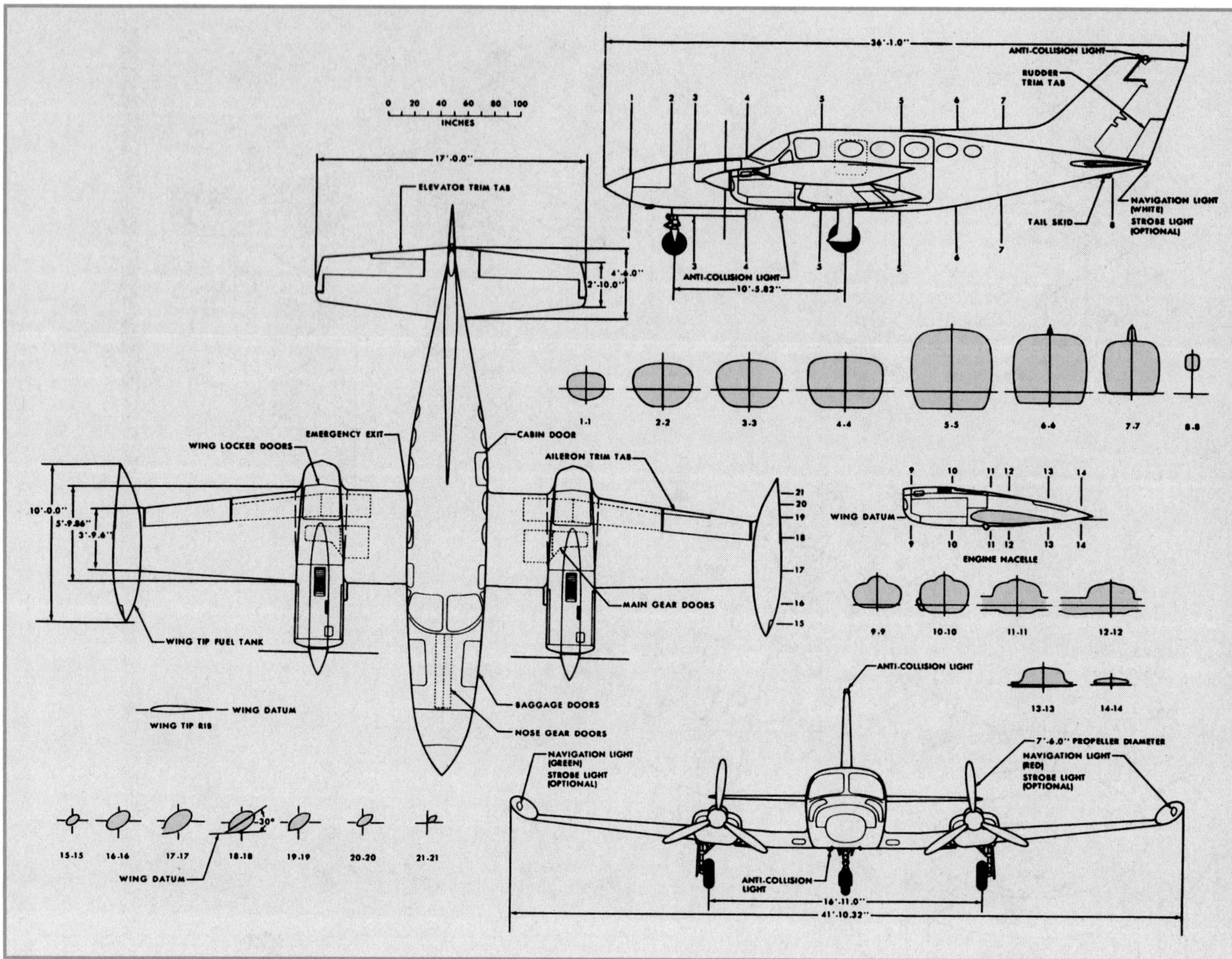

Figure 11-36a
Draw the multiview drawings and the removed sections for the twin engine Cessna airplane. *(Courtesy of Cessna Aircraft Company)*

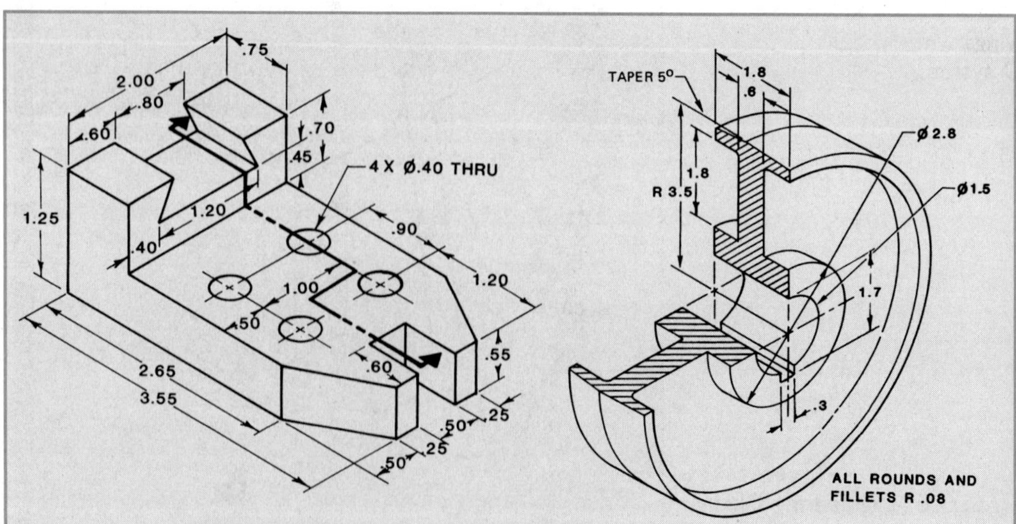

Figure 11-36b
Complete the multiview sectional drawings.

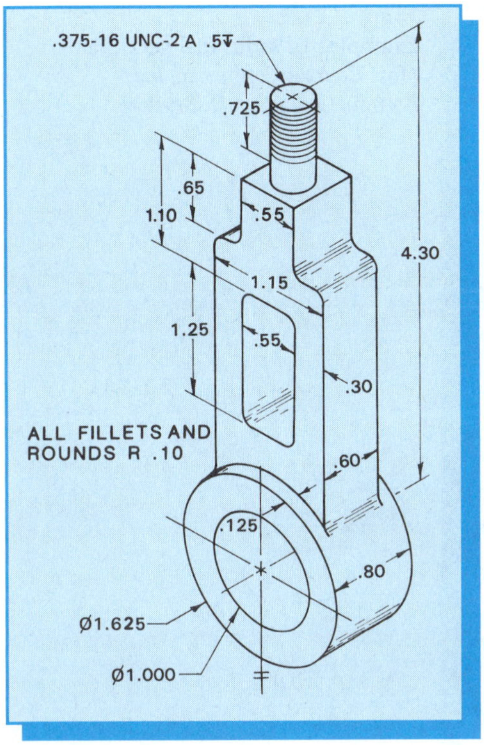

Figure 11-37
Rod Hanger

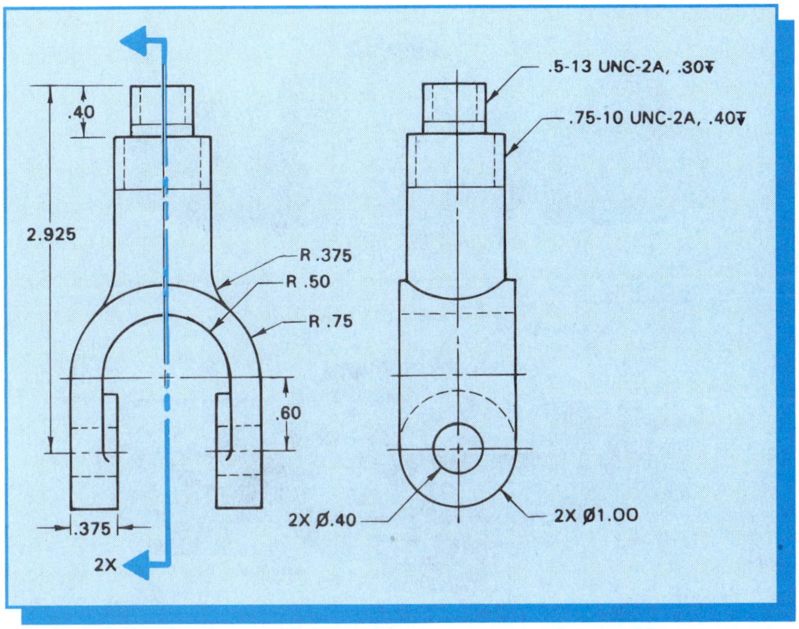

Figure 11-38
Bearing Hanger

Design Exercises

1. Design a tape dispenser for a standard-size roll of Scotch tape. Draw a set of working drawings, including a full sectional drawing.
2. Design a coffee cup. Draw the working drawings, including a sectional drawing.
3. Design a stand to support a video monitor over a microcomputer. Research the dimensions for any computer and monitor. Draw a set of working drawings, including sectional drawings.
4. Design a wheel cover (hubcap) for a sports car. Include a section drawing.
5. Design a steering wheel for a sports car.
6. Redesign the bracket bearing holder in **Figure 11-37** so it will support a 1.5" pipe. Add strength with the design to the screw support. Draw an isometric section of your new design.
7. Redesign the bearing support in **Figure 11-38** so it will support a .5" steel rod. Add a new bracket support for the top. Draw an isometric section of your new design.

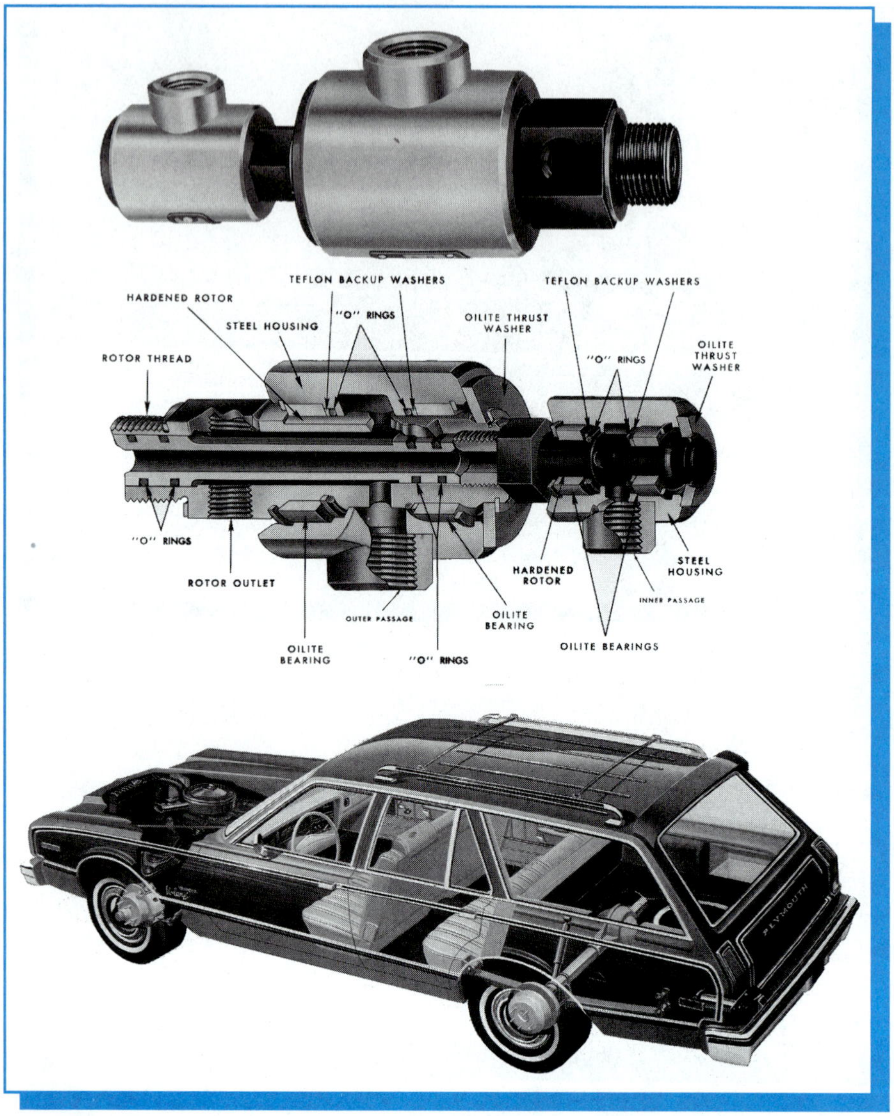

Figure 11-39
Examples of industrial sections
(Top: Courtesy of Neoco, Inc. Bottom: Courtesy of Chrysler)

Key Terms

- Assembly section
- Auxiliary section
- Broken-out section
- Crosshatching
- Cutting plane line
- Enlarged section
- Full section
- Half section
- Offset section
- Removed section
- Revolved section
- Section lining
- Sectional view
- Sectioning patterns
- Thin wall section

Chapter 12

Auxiliary Views & Revolutions

Objectives

The student will be able to:

- project foreshortened lines
- project an auxiliary view
- project an auxiliary view from the front view
- project an auxiliary view from the top view
- project an auxiliary view from the side view
- draw primary and secondary auxiliary views
- draw a simple auxiliary view on a CAD system

Introduction

In Chapter 8, "Multiview Drawings", you learned that when each surface of an object is either parallel or perpendicular (90°) to other surfaces, one or more of the six normal orthographic views (front, top, right side, left side, back, or bottom) can be used to accurately describe the object. **Figure 12-1** shows an object of this type in which all angles are perpendicular. The three views are drawn as viewed through the normal orthographic planes of projection. Each projection plane is parallel to the corresponding face of the object. In this drawing, all lines and surfaces are shown as true size.

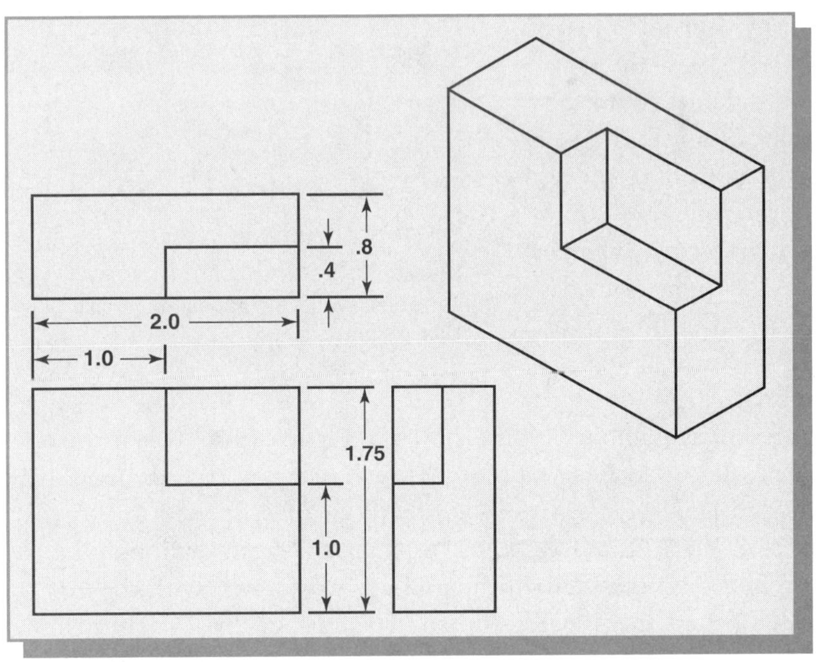

Figure 12-1
All surfaces and edges (lines) that are perpendicular and parallel to the planes of projection will appear true size.

When an inclined surface is viewed from a point perpendicular to the surface plane, the true shape of the inclined surface will appear **(Fig. 12-2)**. These views are known as **auxiliary views**.

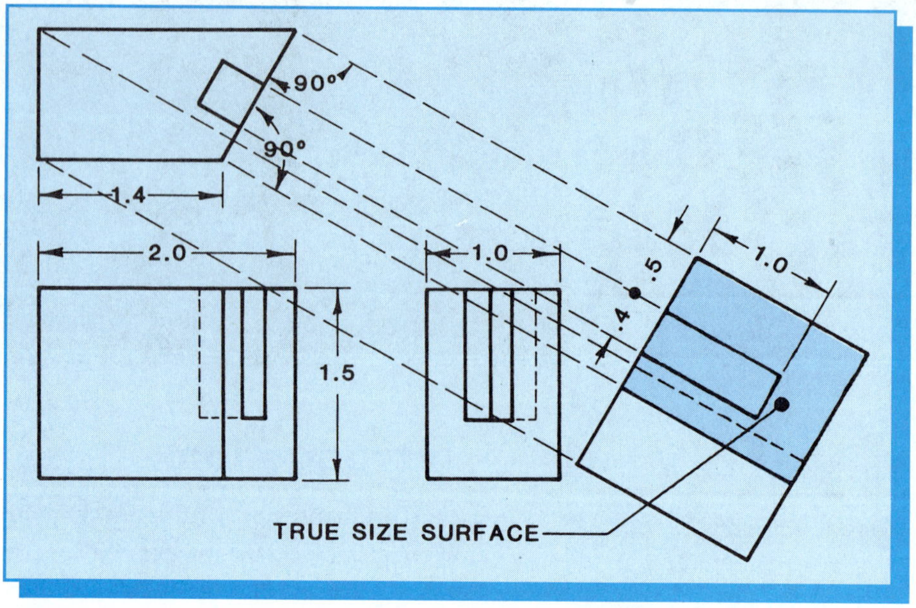

Figure 12-2
An auxiliary view is projected perpendicular to its inclined surface.

Foreshortening

Many objects contain surfaces that are not parallel or perpendicular to other surfaces. When objects of this type are drawn using the normal planes of projection, any inclined or sloped surface will be **foreshortened.** A foreshortened line or surface is one that appears shorter than actual size, because the viewing angle is not perpendicular to the line of sight. For example, hold a 12" ruler perpendicular to your line of sight **(Fig. 12-3)**. Note the true vertical distance covered by the ruler. Now, tilt the ruler from the bottom to move the top of the ruler away from you. The ruler no longer appears 12" long, nor does it cover the same amount of vertical distance. It appears foreshortened.

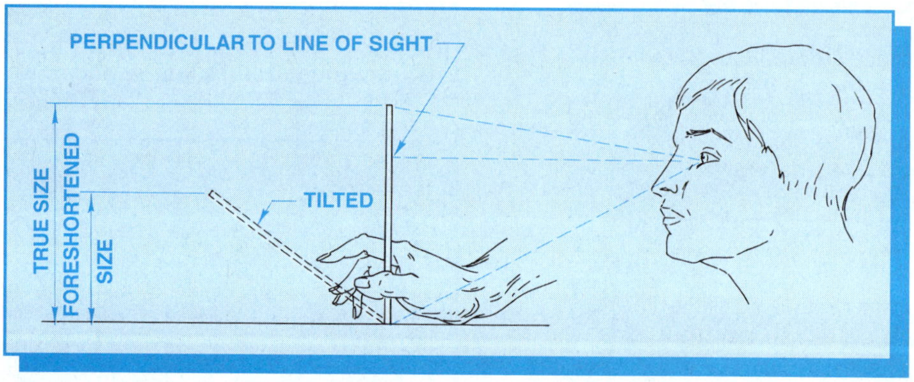

Figure 12-3
An object will only appear true size when the line of sight is perpendicular to that object.

Foreshortened lines are also shown in **Figure 12-4**. When the music box lid is closed (A), the front view shows all true dimensions (width and height). Likewise, when the lid is fully opened to 90° (B), the 2" x 4" lid measures exactly 2" x 4" on the front orthographic view. When the front view of the music box is drawn with the lid partially opened, the 4" width doesn't change, but the 2" depth is foreshortened (C). You can also partially open your book cover and observe the same effects.

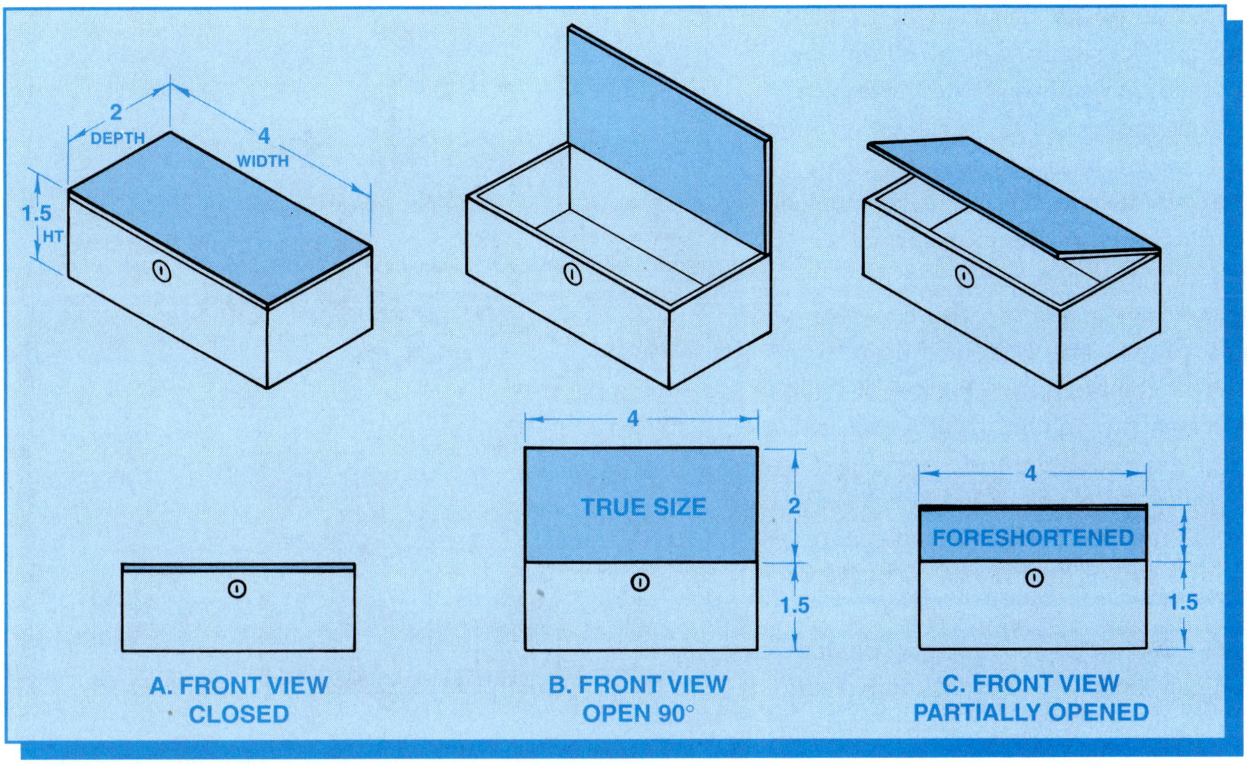

Figure 12-4
The apparent size of a surface will change when viewed at different angles.

To understand this concept better, look at **Figure 12-5**. Surface D is foreshortened in both the top and right-side views because this surface is not perpendicular to either the top or right-side planes of projection. Surfaces A, B, and C are shown in true size because these surfaces are perpendicular to the normal orthographic planes of projection. Only the true depth of surface D is shown on the top and left-side views. Only the true length of surface D is shown, as an inclined line representing the edge, on the front view. Even when all six orthographic planes are used **(Fig. 12-6)**, the true shape of surface D in **Figure 12-5** will not appear. This surface appears only, in foreshortened form, on the top and right-side views. It appears hidden, in foreshortened form, in the bottom and left-side views.

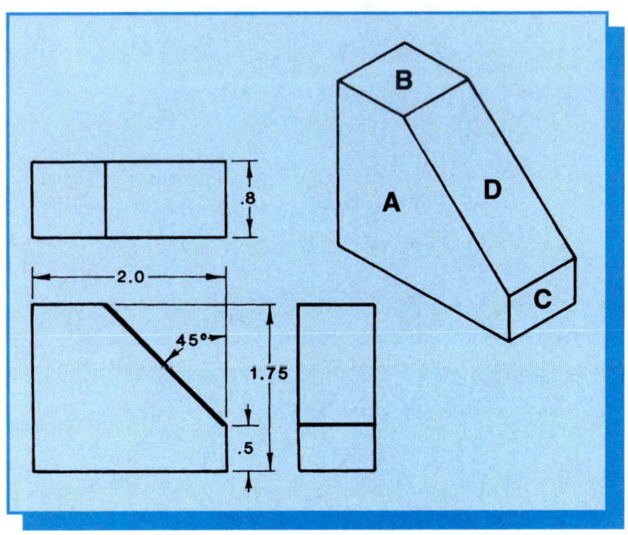

Figure 12-5
Inclined surfaces will appear foreshortened (not true size).

Auxiliary Planes

The plane through which the inclined surface is viewed is the **auxiliary plane.** The auxiliary plane must always be parallel to the inclined surface to be drawn. **Figure 12-7** shows the use of an auxiliary plane compared to the normal orthographic planes of projection. You can see that viewing a surface through a parallel auxiliary plane eliminates foreshortened lines and reveals the true geometric shape and size of the surface. Viewing an inclined surface through an auxiliary plane is similar to viewing the sides of a right-angled object through one of the six planes of an orthographic projection box.

The auxiliary planes can also be envisioned as **hinged planes** that are folded from the normal orthographic planes. **Figure 12-7** shows the auxiliary plane on a projection box, and an auxiliary plane folded from the normal orthographic planes. Both methods reveal the true shape of the inclined surface which cannot be seen through any of the normal orthographic planes.

Figure 12-8 shows an orthographic drawing containing an auxiliary view of a manufactured jig guide. In this drawing, the auxiliary views

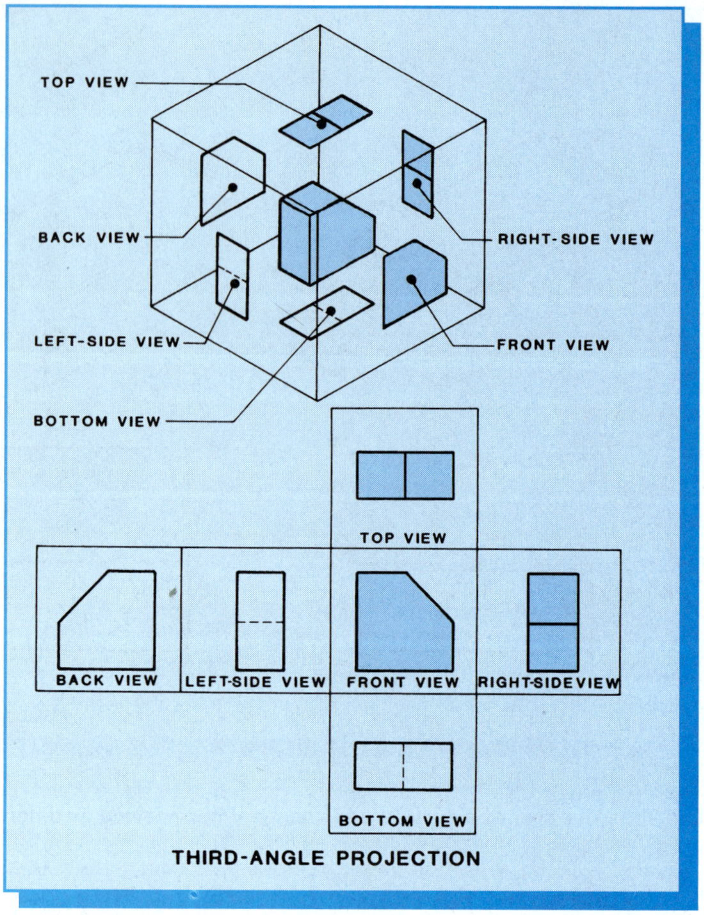

Figure 12-6
The true size of the inclined surface is not shown in any of the six orthographic projections.

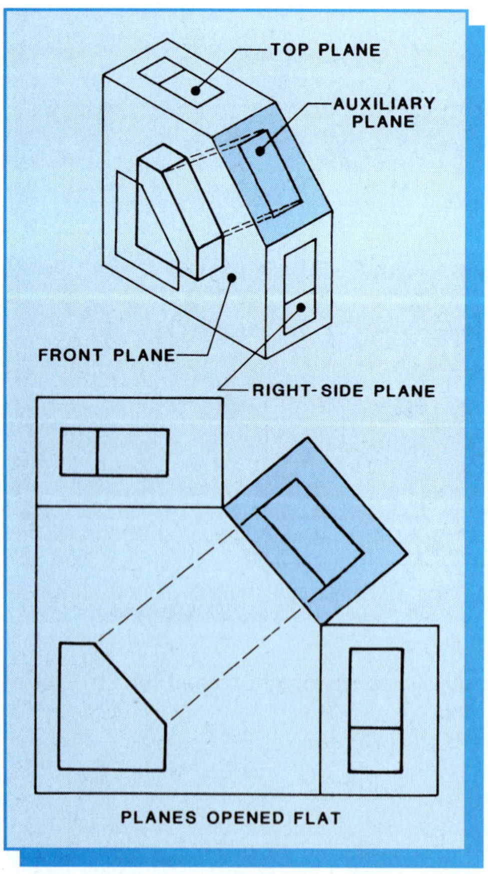

Figure 12-7
Auxiliary projection plane

are needed because the true shape of the angled surfaces does not show in any of the orthographic views. The front view does show the exact shape and size of this part. The top view shows round holes as ellipses rather than true circles. It also shows foreshortened overall lengths. An auxiliary view can show the true shape of this part with the proper relationship between holes and radii.

Projection

When drawing auxiliary views, always project perpendicularly from a **true size edge** on a normal view. In **Figure 12-9**, the true length of the inclined surface is shown as an inclined line representing the surface edge on the front view. The length of this line is true (not foreshortened); therefore, the auxiliary view length can be projected from this line. The true depth of the inclined surface is shown on the top and side views. This depth can be transferred at 90° angles from the

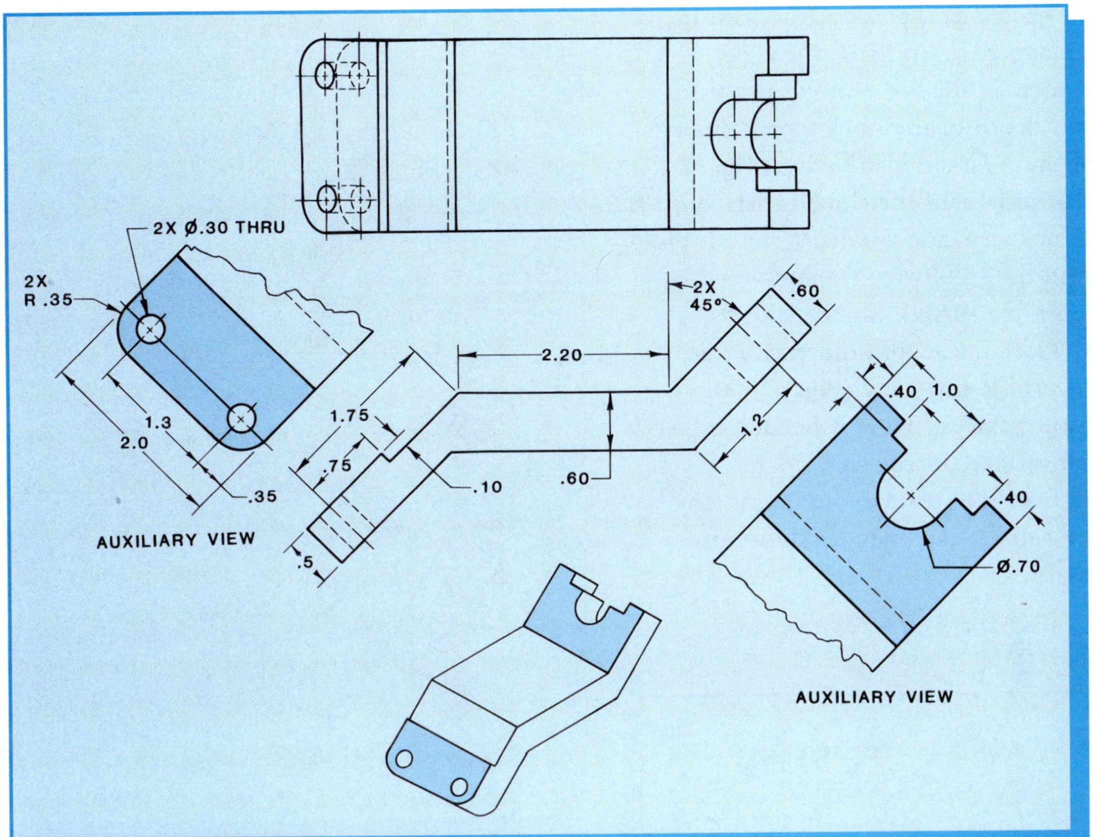

Figure 12-8
An auxiliary view shows the offset surface in true size and form.

length projection line to form the outline of the auxiliary view. Two procedures for transferring width dimensions for an auxiliary view are shown in **Figure 12-10**.

Primary Auxiliary Views

Primary auxiliary views are drawings projected directly from any of the six normal orthographic views. The three most common auxiliary views are projected from the front, side, or top normal views. **Primary auxiliary views** are used to define either depth, height, or width details not found in true form in one of the principal views. The true height and width of the object, plus the true length of the inclined plane edge, are shown on the orthographic front view in **Figure 12-11**. The true size surface of the sloped surface is not shown; therefore, a primary front auxiliary view is projected from the front view to show the sloped surface in true size. This

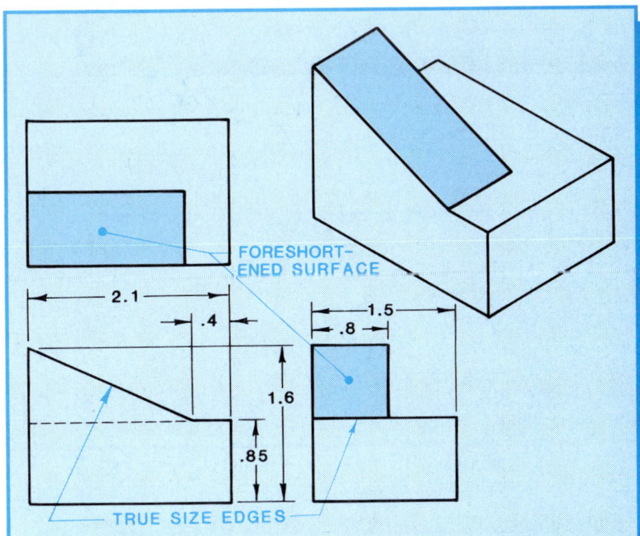

Figure 12-9
Only the edge view of an inclined surface will appear true length.

view is projected perpendicularly from the inclined plane edge on the front view.

The normal side view shows the true height and depth of the object, plus the true length of the inclined plane edge **(Fig. 12-12)**. A true size surface of the sloped surface is not shown. A primary-side auxiliary view is projected from the side orthographic view to show the sloped surface in true size.

True width and depth dimensions are on the orthographic top view **(Fig. 12-13)**. A primary top auxiliary view is needed to show the sloped surface in true size.

More than one primary auxiliary may be drawn off a multiview view as shown in **Figure 12-14**.

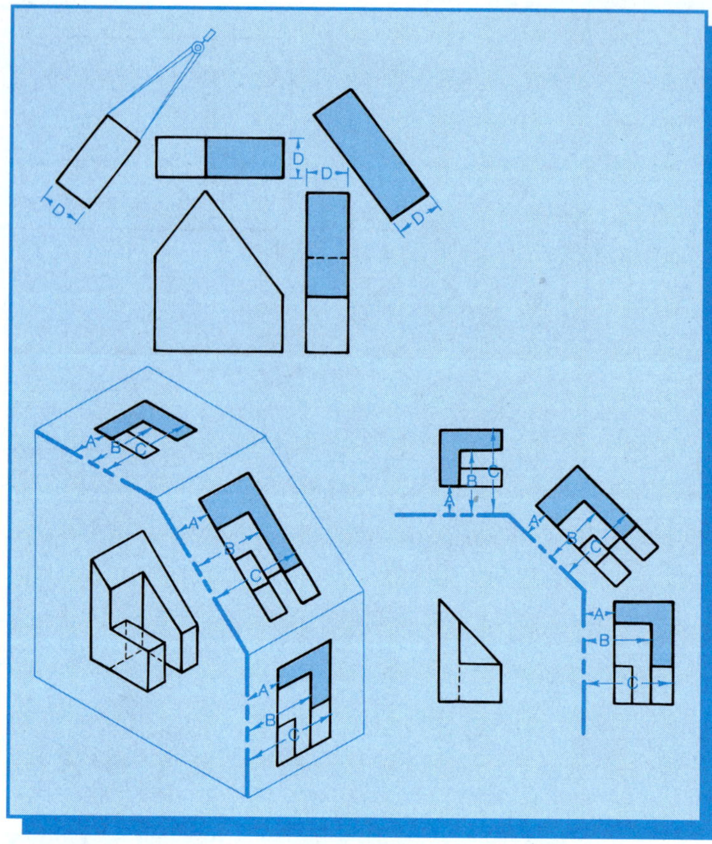

Figure 12-10
Procedure for depth measurements of an auxiliary view. A divider or scale may also be used to transfer measurements.

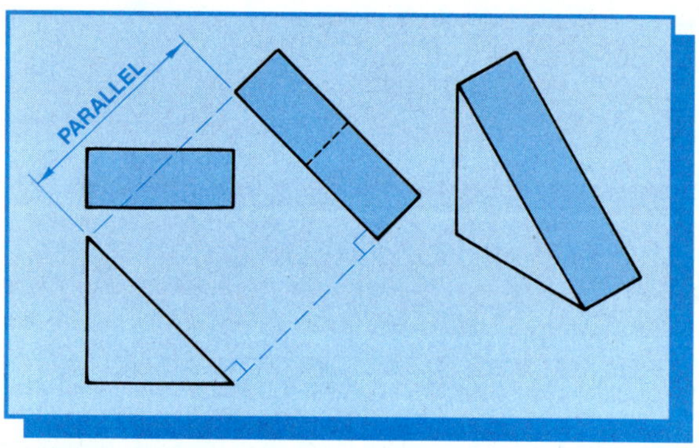

Figure 12-11
Example of a front-view auxiliary

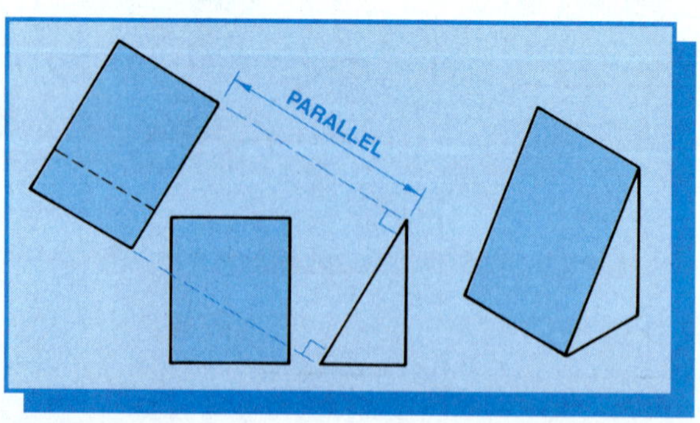

Figure 12-12
An example of a side-view auxiliary

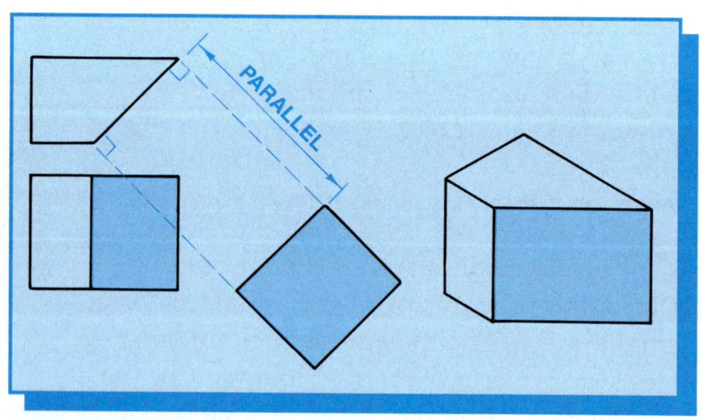

Figure 12-13
An example of a top-view auxiliary

Secondary Auxiliary Views

When surfaces are sloped in several primary auxiliary views, a secondary auxiliary view may be necessary to show the true shape of the inclined surfaces. A

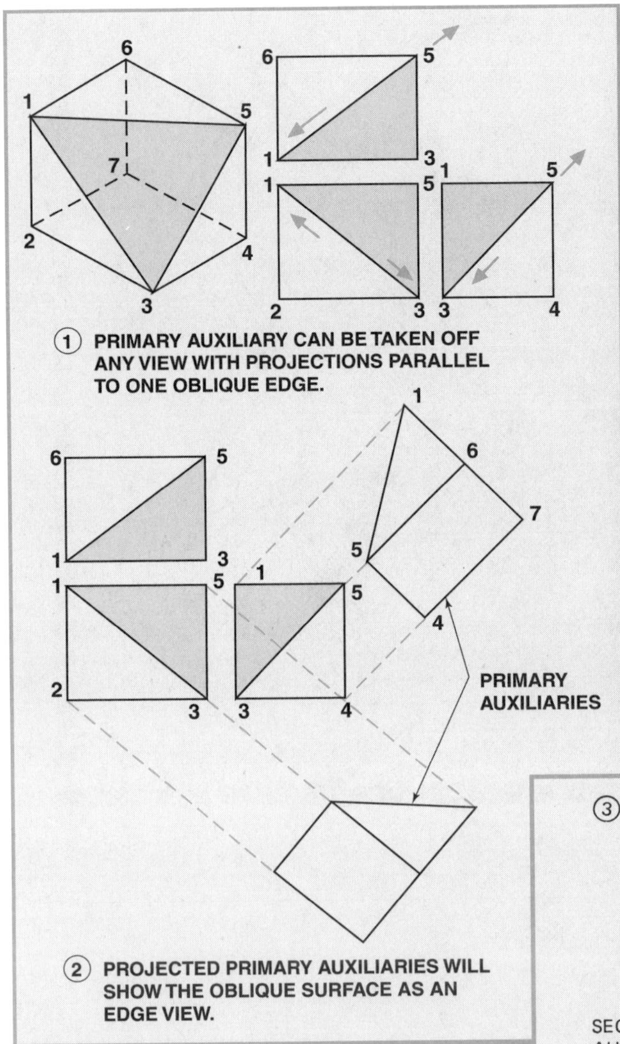

Figure 12-14
The first projected auxilary is called a primary auxilary view.

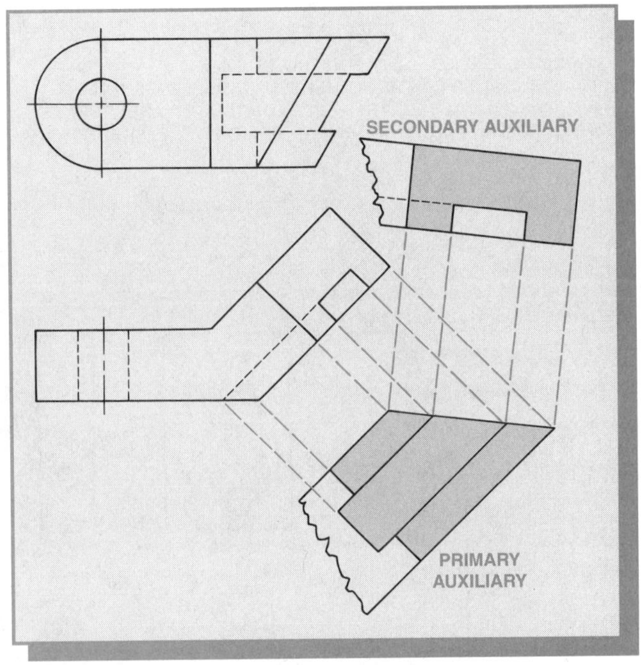

Figure 12-15
Multiple auxiliary views may be drawn.

secondary auxiliary view is a view projected from a primary auxiliary view. To draw a secondary auxiliary view, a primary auxiliary view must first be drawn. The secondary auxiliary view is then projected perpendicularly from a selected true size edge on the primary auxiliary view **(Figs. 12-15** and **12-16)**.

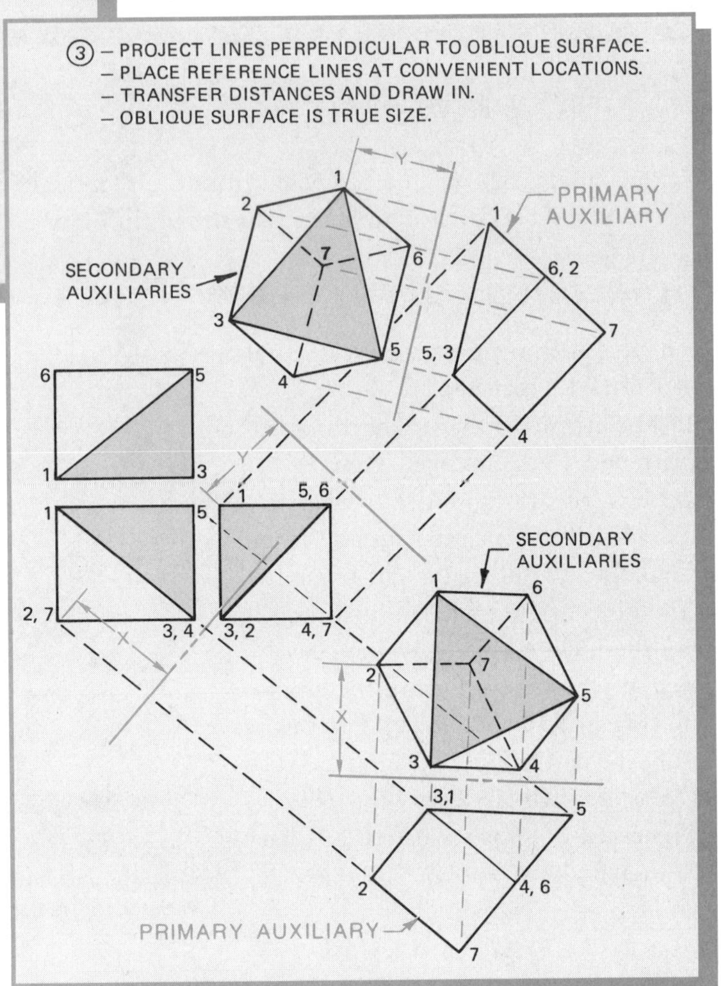

Figure 12-16
The auxiliary view projected from a primary auxiliary is called a secondary auxiliary.

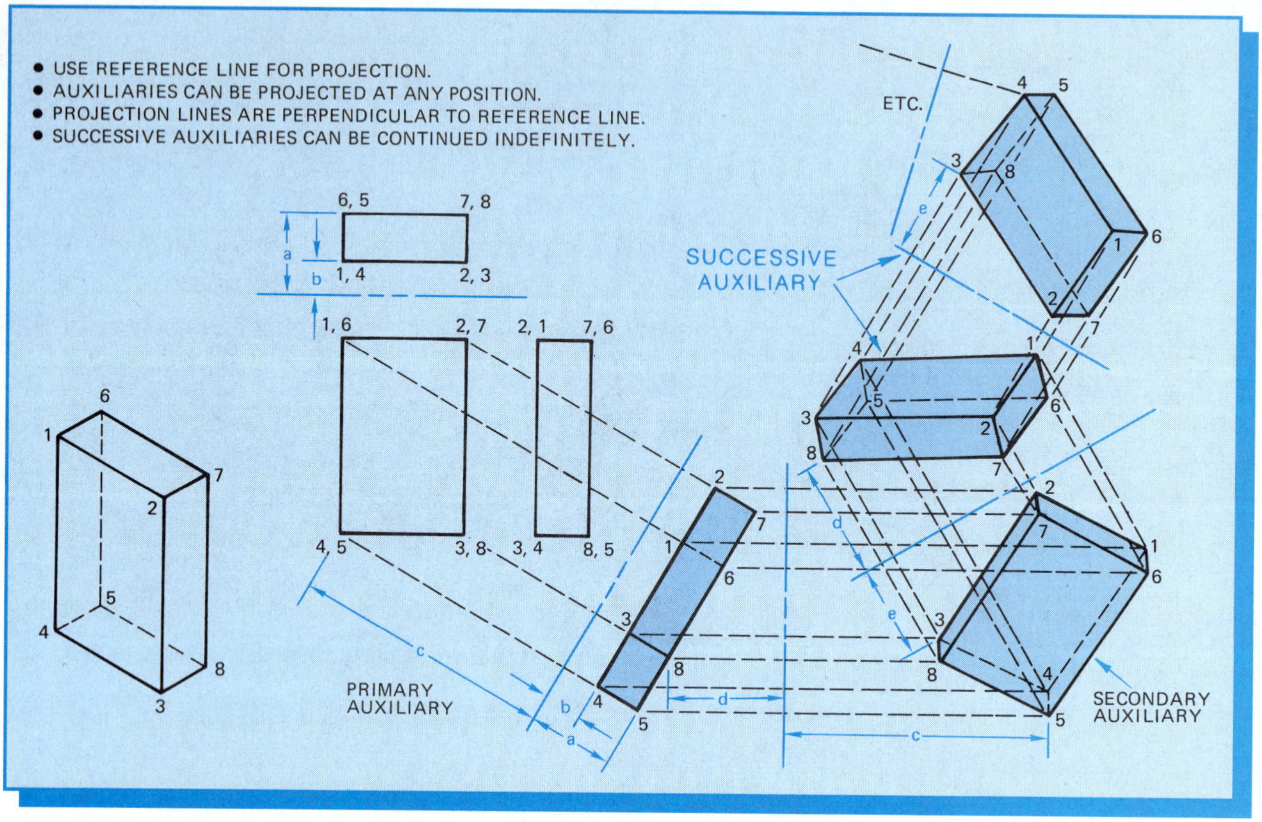

Figure 12-17
Examples of successive auxiliary drawings

Additional auxiliaries may be projected from the secondary auxiliary. They are called successive auxiliaries, as shown in **Figure 12-17**.

Completeness of Auxiliary Views

In principal orthographic views, all parallel or perpendicular surfaces, are shown in true scale, but inclined surfaces are distorted or foreshortened. In an auxiliary view, just the opposite is true. Because the auxiliary plane is parallel to the inclined surface, it will not be parallel to the normal right-angle surfaces of the object. Therefore, all surfaces, other than the inclined plane, will appear foreshortened when viewed from the auxiliary plane **(Figs. 12-18** and **12-19)**. For this reason, only the actual inclined surface is usually shown as a partial auxiliary view **(Fig. 12-20)**. **Figure 12-21** shows a partial auxiliary view of a digital clock.

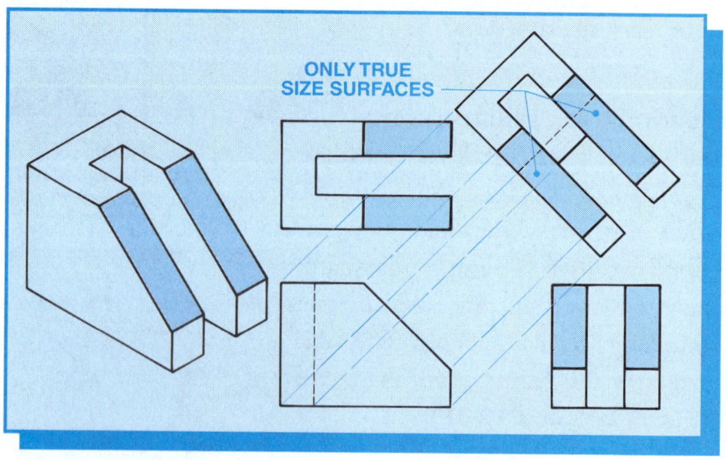

Figure 12-18
Only the inclined surface will appear true size in the auxiliary view, because the auxiliary is viewed perpendicular to its inclined surface.

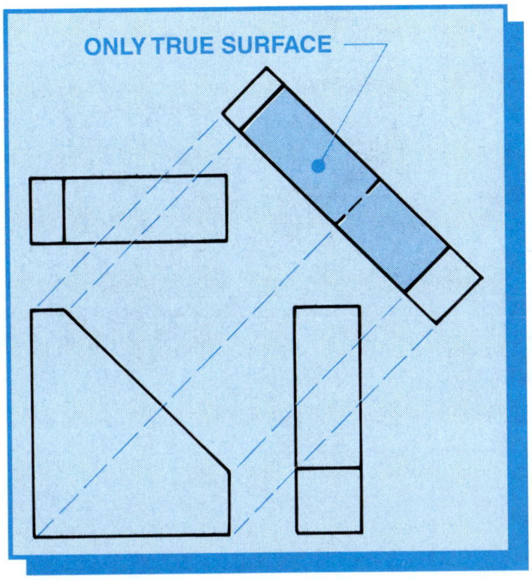

Figure 12-19
A full auxiliary drawing

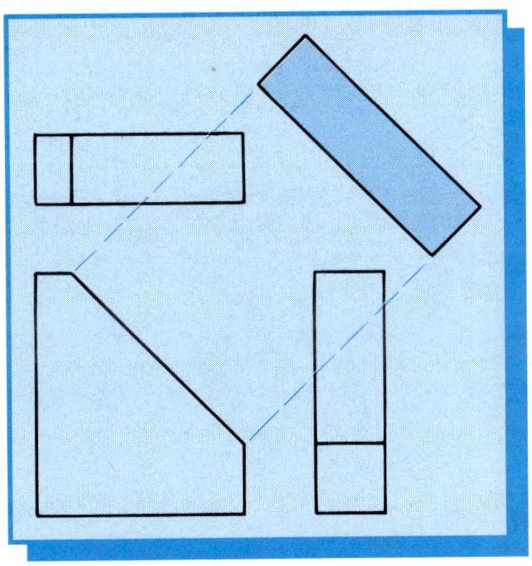

Figure 12-20
A partial auxiliary drawing

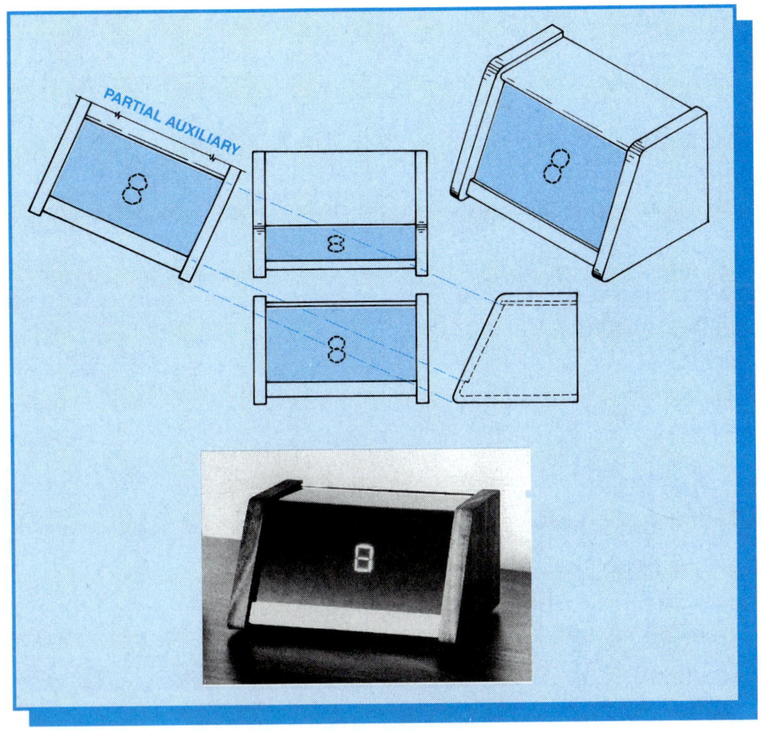

Figure 12-21
An auxiliary drawing of a single-number digital clock *(Courtesy of Michael Robbins)*

Other Angles

In addition to being projected from normal orthographic views or other auxiliary views, auxiliary views can be projected from any surface. In all cases, the view must be projected perpendicular to the surface to be shown to insure a true shape description. **Figure 12-22** shows a nut with auxiliary views projected from surfaces that are not normal orthographic views.

Notice that the auxiliary planes of projection are all parallel to the surfaces to be drawn. The auxiliary projections are perpendicular to the inclined surface.

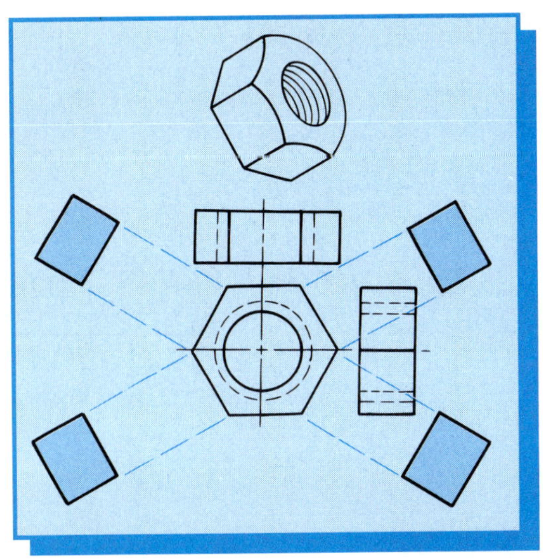

Figure 12-22
The number of auxiliary projections is unlimited.

Hidden Lines

Normally, hidden (dashed) lines are omitted from auxiliary views as their use may clutter the drawing and make interpretation difficult. However, if needed for clarity, hidden lines can be added to auxiliary views.

Holes and Ellipses

When a hole is projected from an inclined plane on an orthographic drawing, it will appear as an ellipse **(Fig. 12-23)**. To show the hole in its true form, an auxiliary view must be drawn.

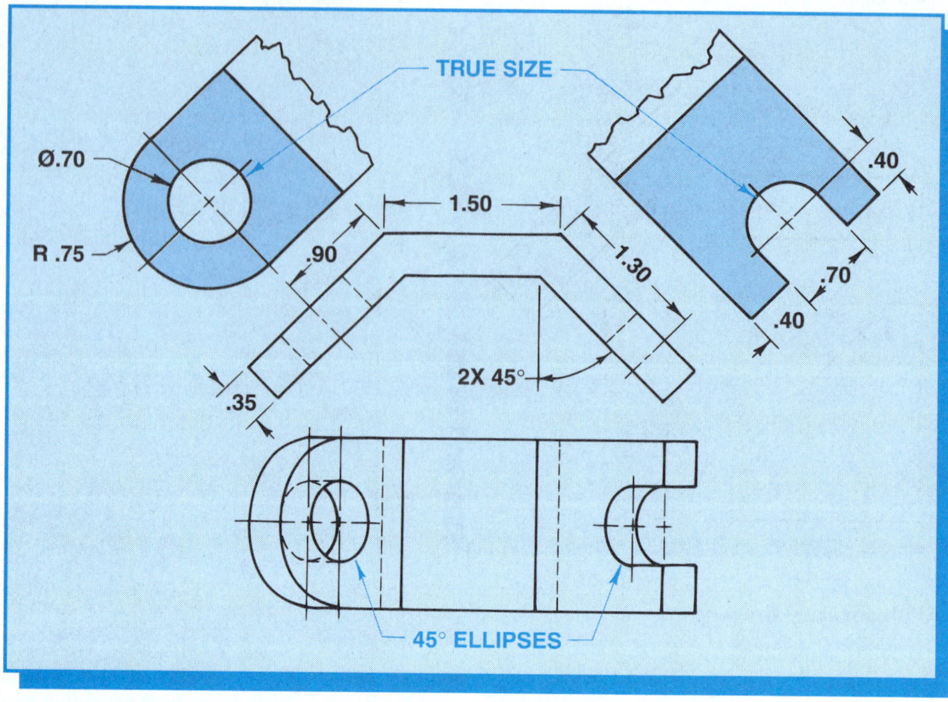

Figure 12-23
Projecting circles and ellipses

Dimensioning

A basic rule for dimensioning is to dimension only true sizes. Never dimension a foreshortened line or surface. Draw the auxiliary view, and place the required dimensions on the true size auxiliary view **(Fig. 12-24)**.

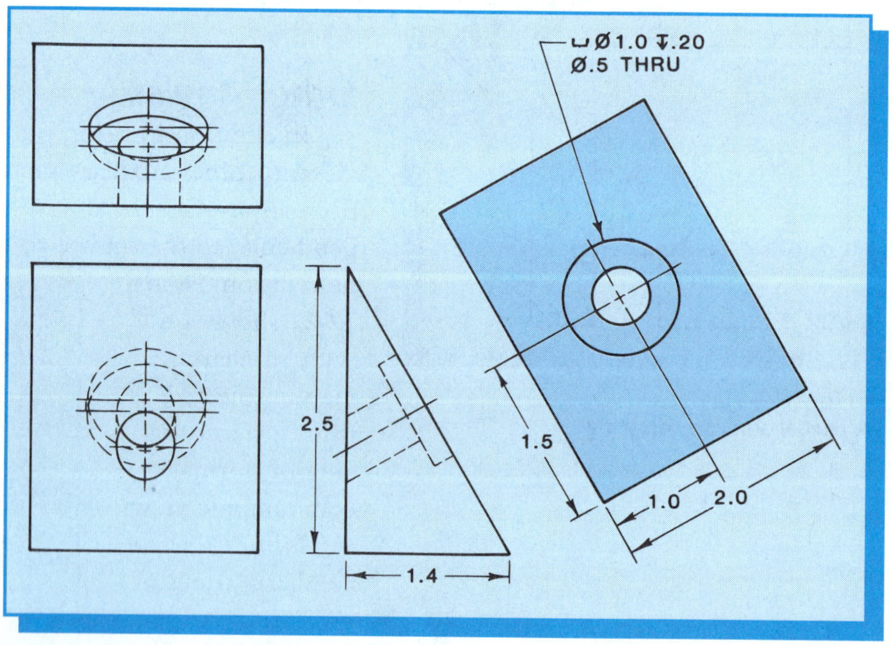

Figure 12-24
Dimension only true sizes.

Procedures

The following illustrations will show the procedure to develop several types of auxiliary views. **Figure 12-25** shows the steps for an auxiliary view of a wedge stop.

Figure 12-26 develops the auxiliary view of one surface of a rectangular pyramid.

Figure 12-27 develops the auxiliary view of a truncated hexagonal prism.

Figure 12-28, the last example, shows the steps to draw the auxiliary view of a truncated cylinder.

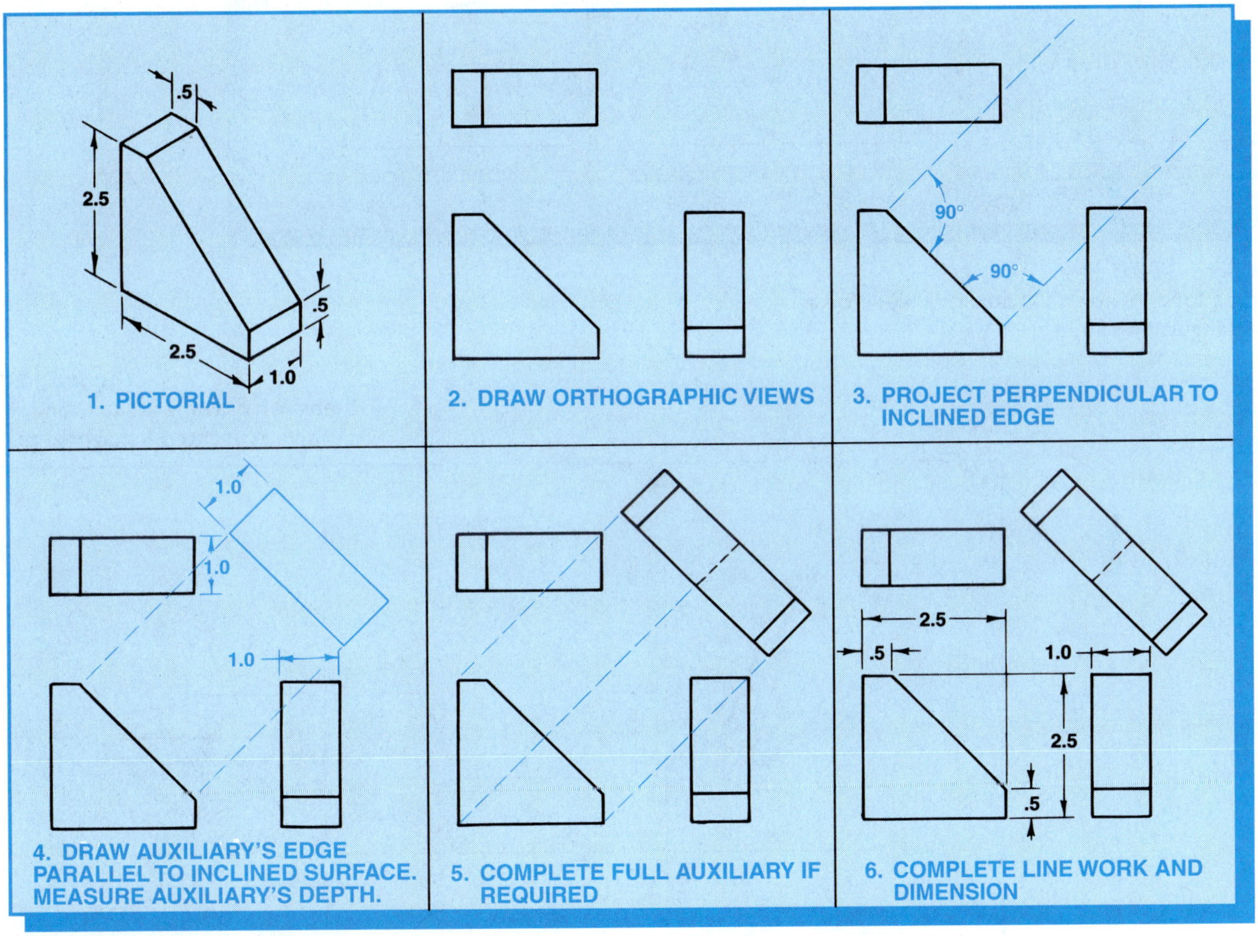

Figure 12-25
Steps for drawing an auxiliary view for a wedge stop

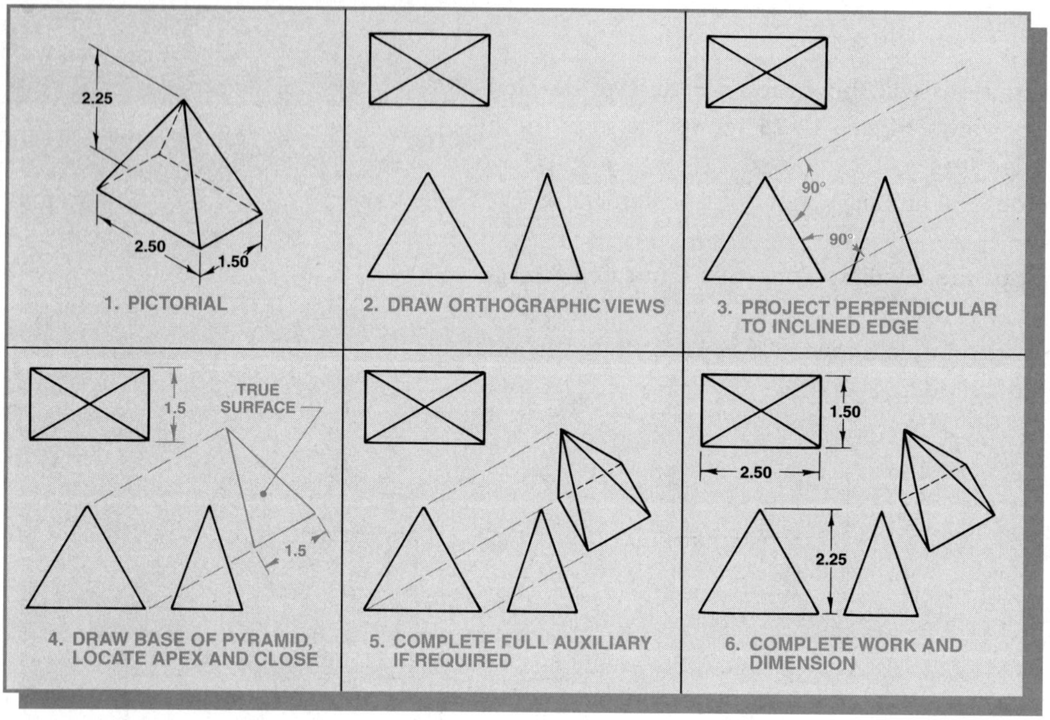

Figure 12-26
Steps for drawing a full auxiliary view for a rectangular pyramid

Figure 12-27
Steps for drawing an auxiliary view for a truncated hexagonal prism

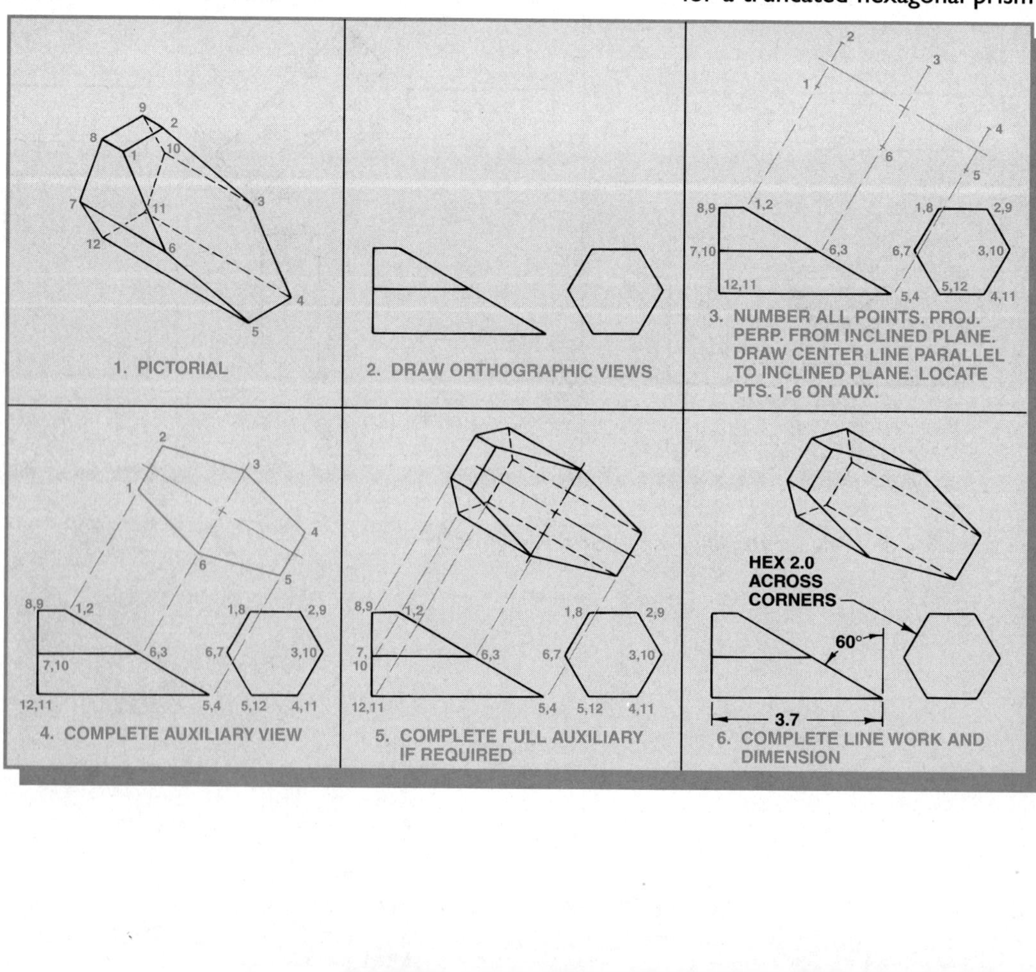

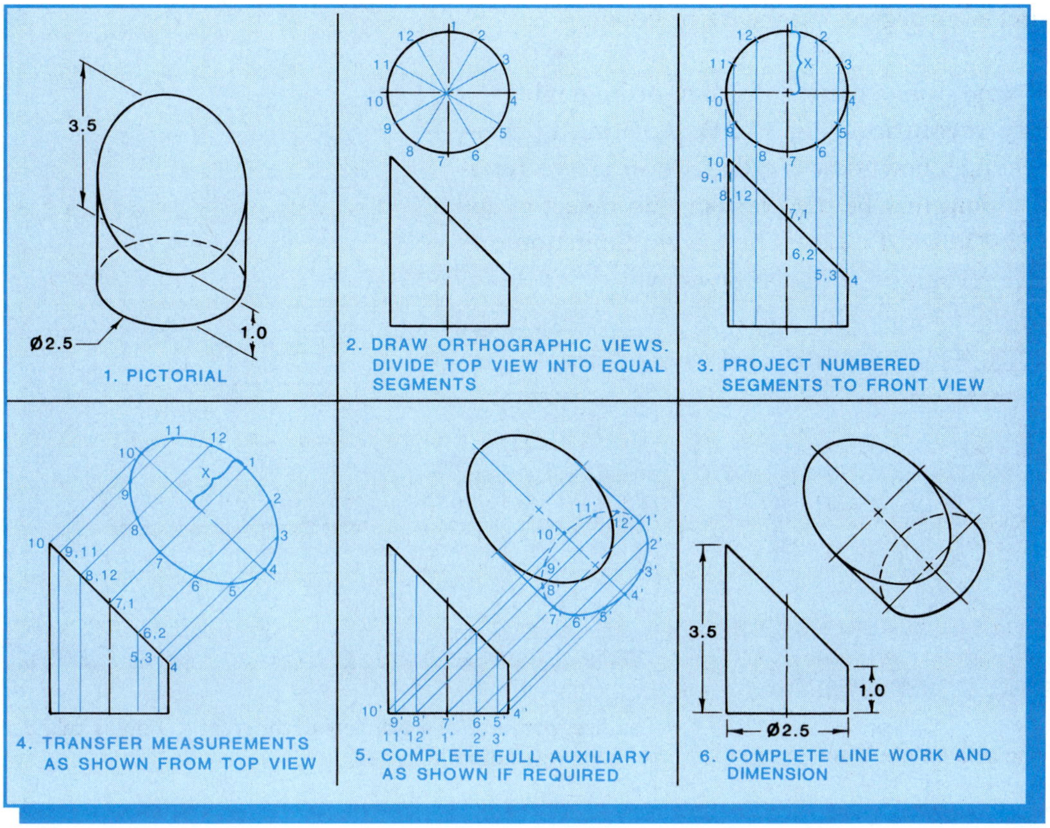

Figure 12-28
Steps for drawing an auxiliary view for a truncated cylinder

Revolutions

The positioning of an object's multiview drawing for clarity may be done with three procedures. Orthographic projection is the basis for all multiview drawings (Chapter 8). All projections are perpendicular between adjacent views. Descriptive geometry is the theory of doing graphical solutions of points, lines, and planes in space using the principles of orthographic projection (Chapter 13). Auxiliary projections can be drawn at any angle off a primary orthographic view or auxiliary view, as explained earlier in this chapter.

A **revolution drawing** is a fourth procedure used to reposition a view from the primary orthographic drawing, and project its repositioned view **(Fig. 12-29)**. The drafter must decide the most advantageous viewing positions.

Any view can be rotated about a theoretical axis placed on the view **(Fig. 12-30)** or on a corner of the view **(Fig. 12-29)**. The views

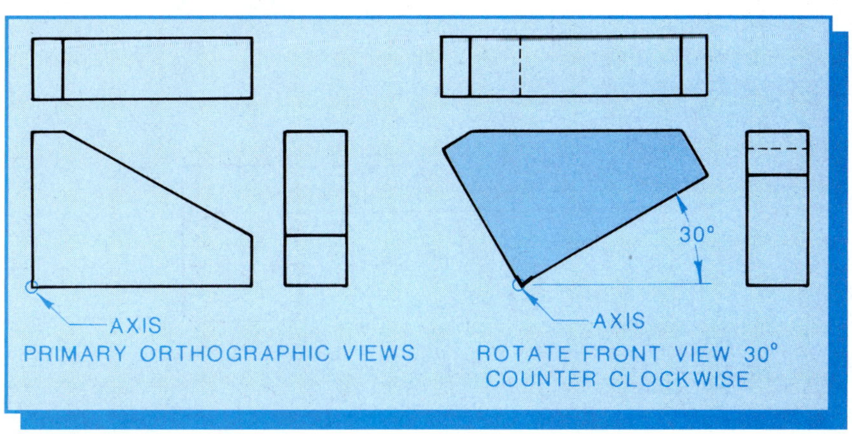

Figure 12-29
Rotation of a view using a corner as an axis

may be rotated any number of degrees clockwise or counter-clockwise.

The first view projected from a rotated primary orthographic view is called a **primary revolution (Fig. 12-31)**. Rotating a primary revolution and projecting a new view is called a **successive revolution.** Successive revolutions may be used to rotate an object to any desired position **(Figs. 12-32a** and **12-32b)**. There is no limitation on the number of successive revolutions that can be drawn.

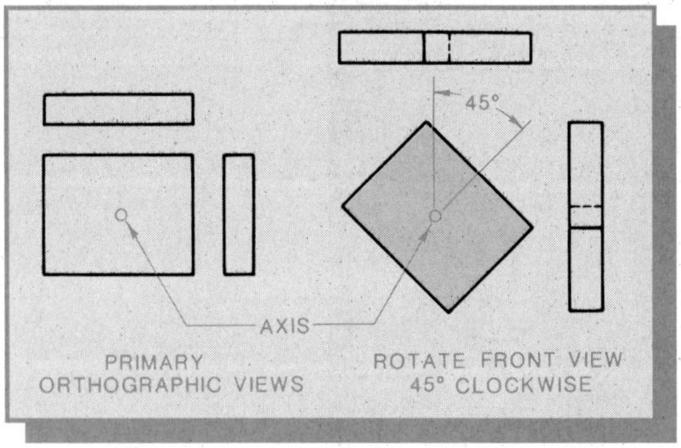

Figure 12-30
Rotation of a view with the axis on the view

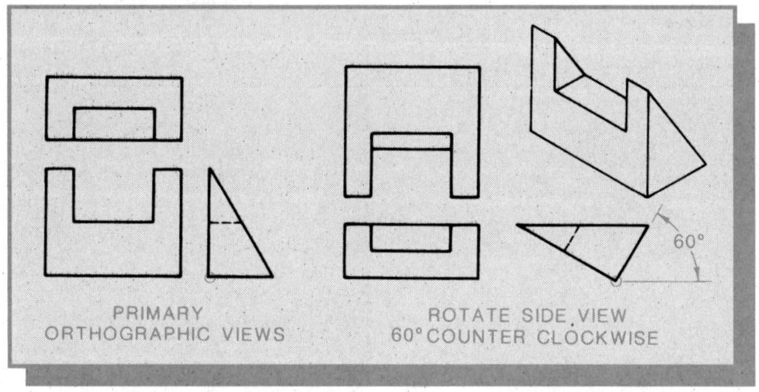

Figure 12-31
The first rotation of a primary orthographic view is called a primary revolution.

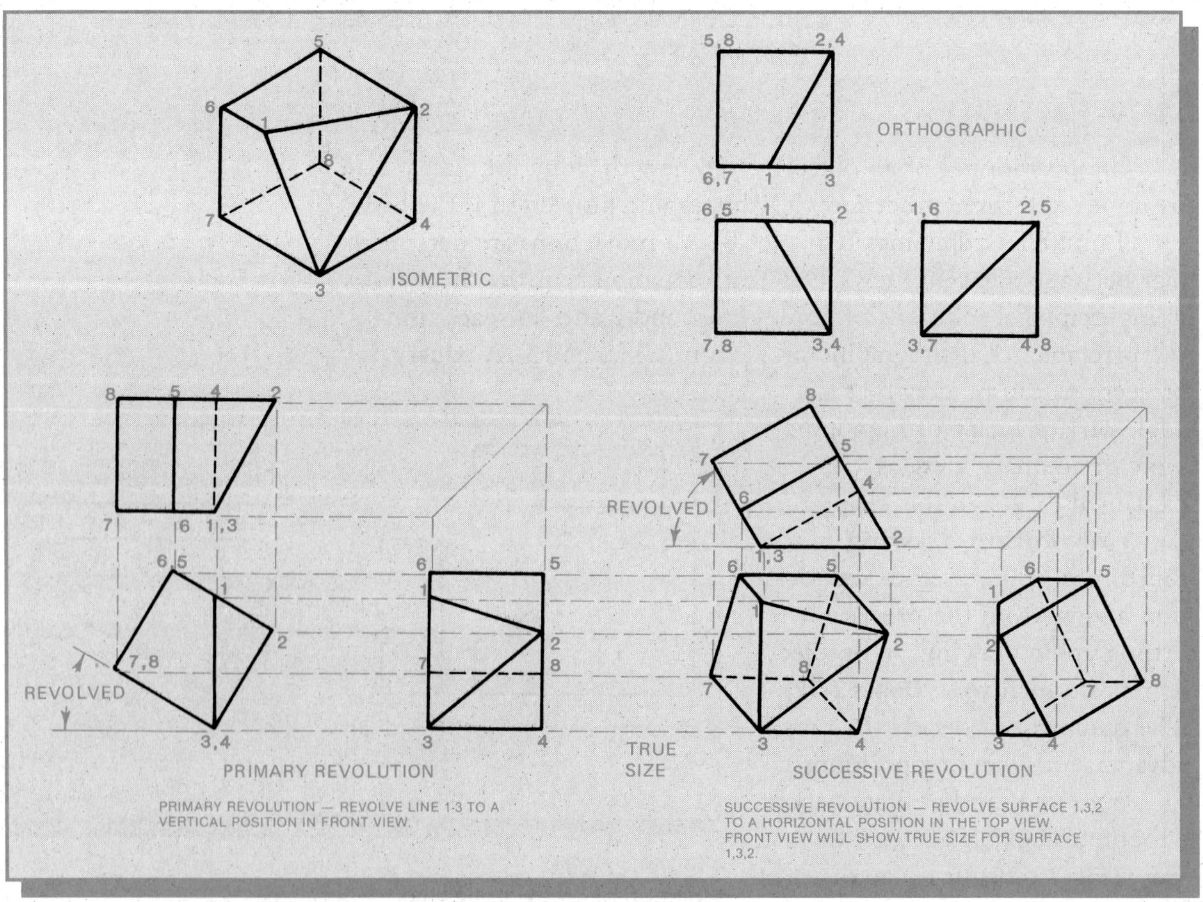

Figure 12-32a
Examples of successive revolutions

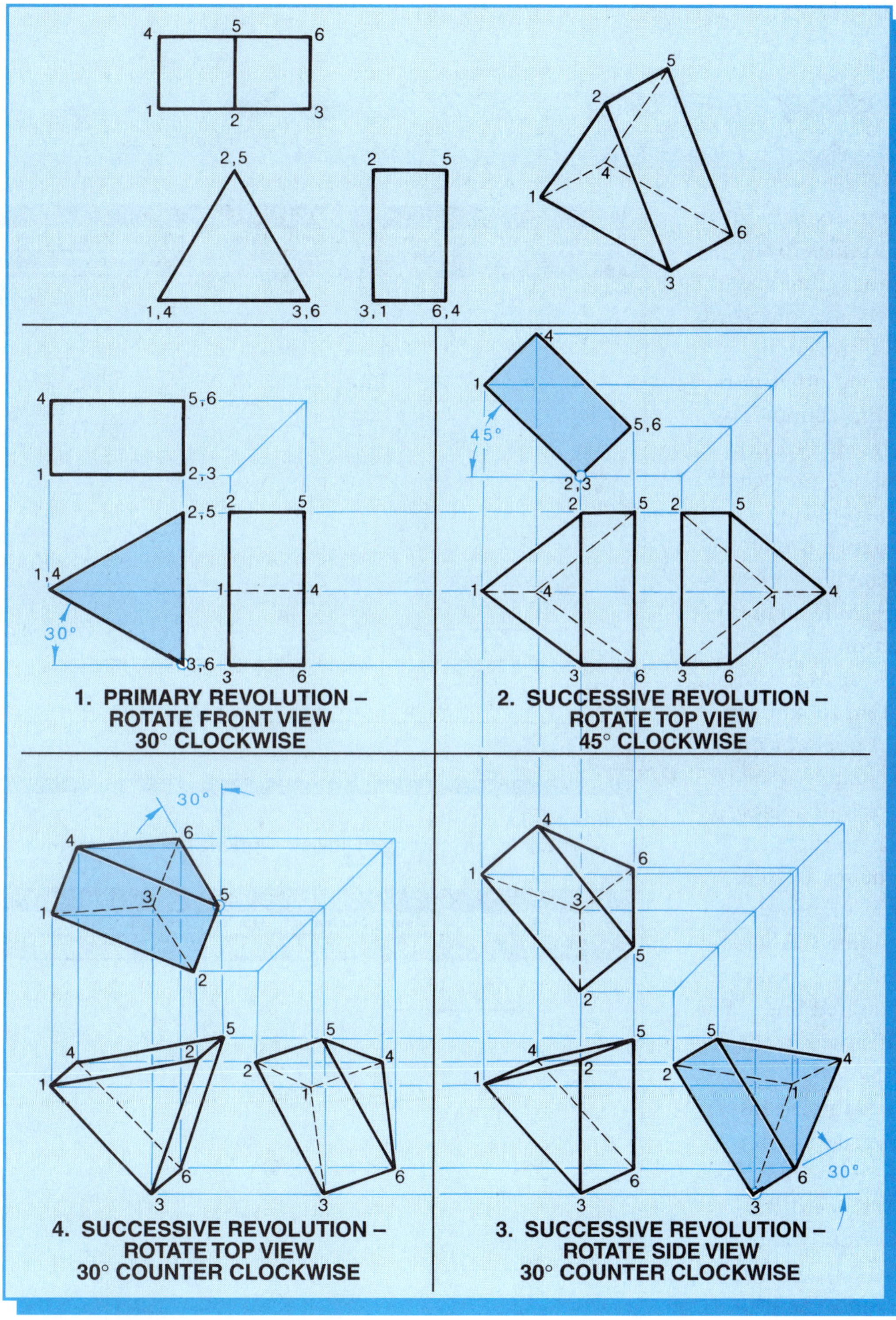

Figure 12-32b
When drawing successive revolutions, any view may be rotated any number of degrees to achieve specific placement of revolved views.

AutoCAD LT Exercises

Developing auxiliary views with CAD is similar to the layout of orthographic views. Auxiliary views are composed of lines, circles, arcs, and dimensions, as are other normal views of projection. The major difference is that the auxiliary views are projected from an inclined plane. Another difference is that it is common for auxiliary view drawings to include ellipses. The construction of ellipses on CAD systems varies greatly from one system to another.

The CAD operator must be able to accurately draw parallel lines at various angles (perpendicular to the edge view of the inclined plane). One way to accurately draw inclined lines on a CAD system is to establish a grid at the angle of the desired lines. This grid method is used in the following examples.

The first example illustrates the construction of a simple auxiliary drawing. The starting point is the orthographic projection of the object **(Fig. 12-33)**. Note that in this illustration, the grid pattern is horizontal and vertical. The next step is to establish the inclined grid to match the inclined plane of the object **(Fig. 12-34)**.

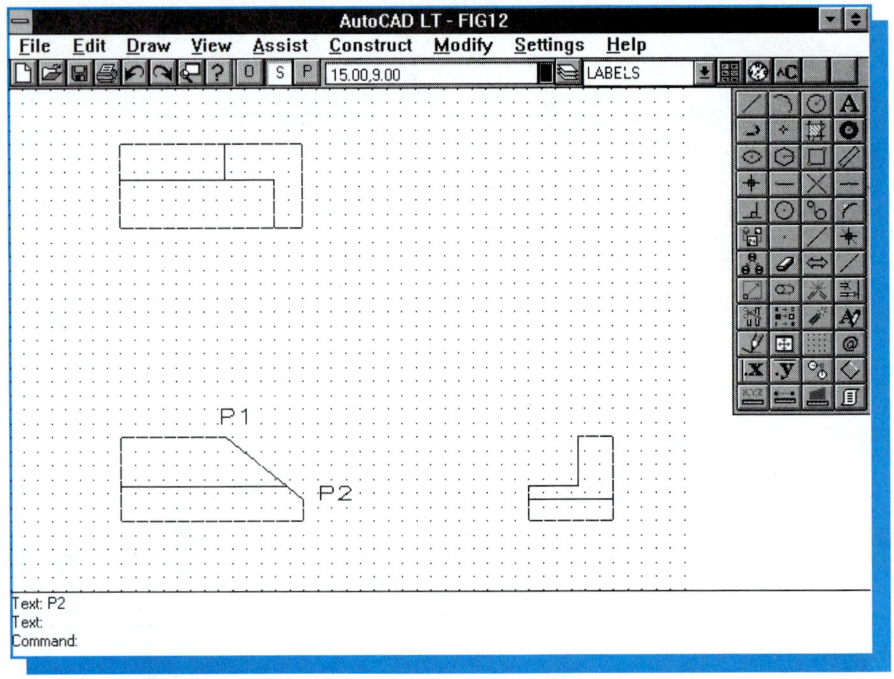

Figure 12-33
Draw or call up an existing multiview working drawing stored in the database.

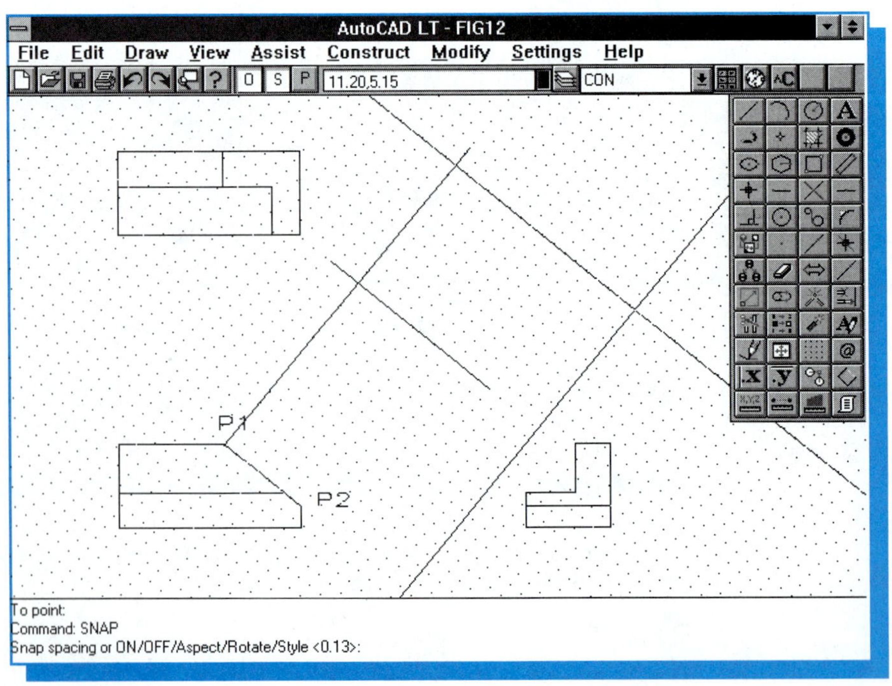

Figure 12-34
Set the grids to be parallel to the inclined surface by using the Snap, Rotate commands. Note that the cursor lines will match the grids.

Rotate the Grid

tip: *Use Osnap for accuracy when selecting the points.*

1. Type **Snap** at the command line
2. Type **R** for Rotate.
3. In response to the prompt "Base Point", select one end of the edge view of the inclined plane (P1)
4. Select the other end of the line (P2).

The result, rotated grid aligned with the inclined plane, should look like **Figure 12-34**.

The auxiliary view is completed by projecting from the inclined plane using the new grid **(Fig. 12-35)**. This makes the drawing of the auxiliary view as easy as a normal view. To finish the drawing replace the inclined grid with a normal grid by using the drawing aids under **Settings (Fig. 12-36)**.

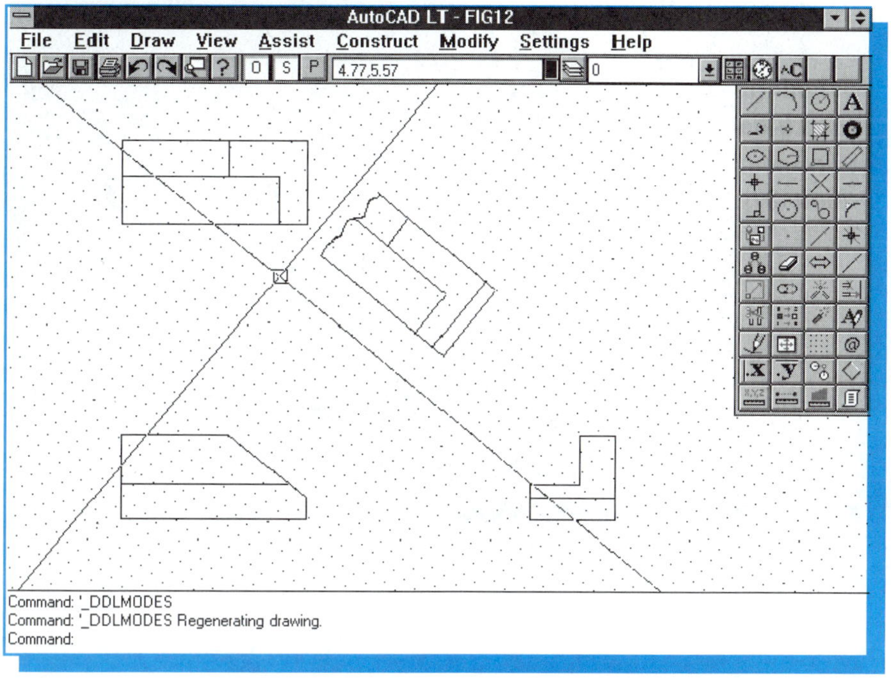

Figure 12-35
Transfer the depth dimensions, and with the line command, complete the auxiliary view.

A more complex auxiliary drawing is illustrated in **Figure 12-42**. The auxiliary view is projected from the top view, and the object contains a hole. The hole will project as ellipses on two of the views. The procedure is as follows:

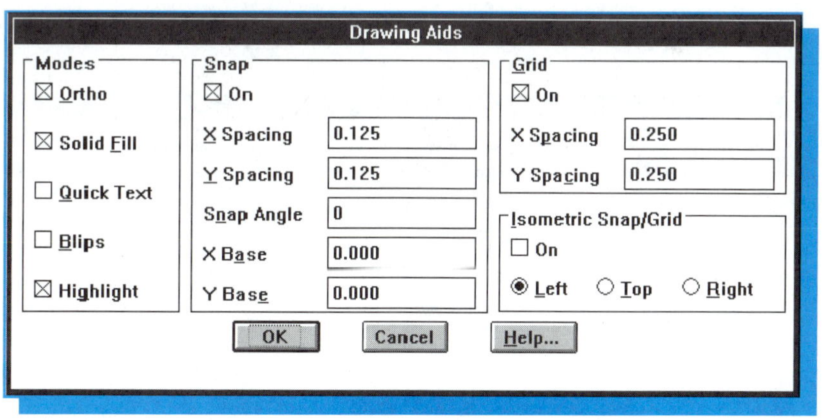

Figure 12-36
Drawing Aids dialog box. Set snap and grid.

1. Construct the normal orthographic views. Use all the available CAD aids. These include Grid, Snap, and Ortho, as well as the basic entities **(Fig. 12-38)**.
2. Change the grid to match the angle of the inclined plane. Note that the screen cursor also changed and matches the new grid angles **(Fig. 12-39)**.

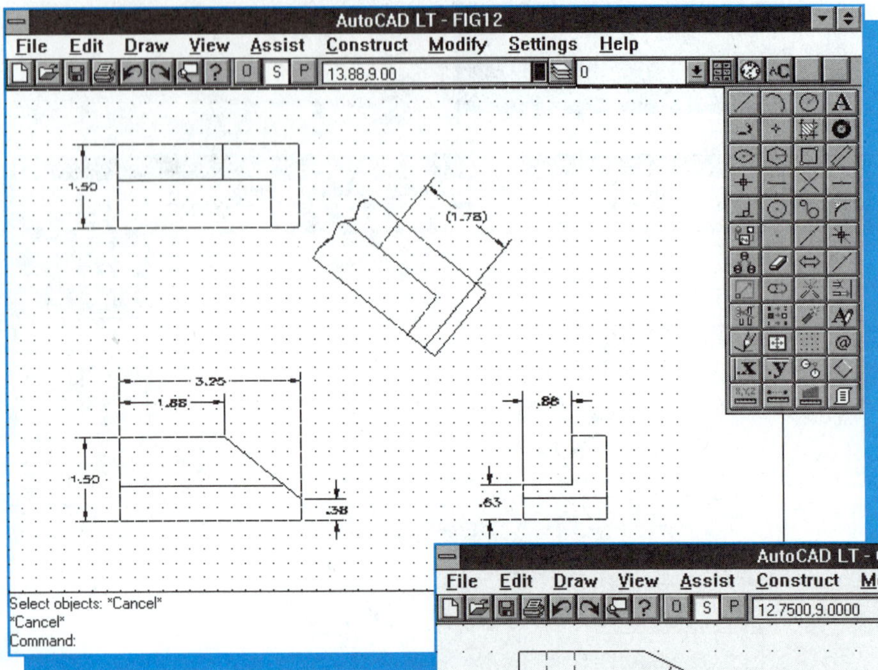

Figure 12-37
Reset grid to normal and add dimensions with the dimension commands.

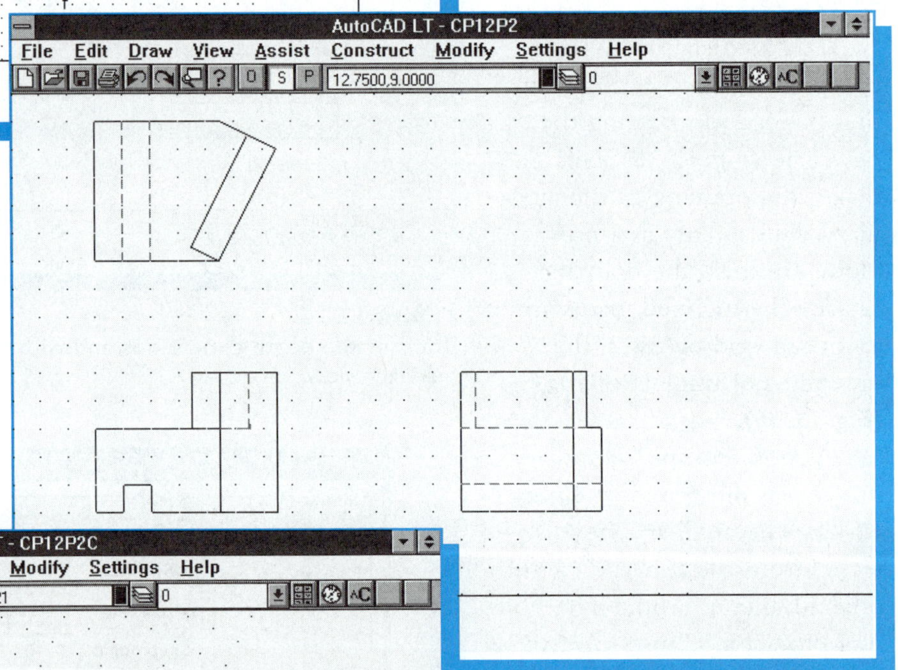

Figure 12-38
Draw or call up the stored multiview drawing.

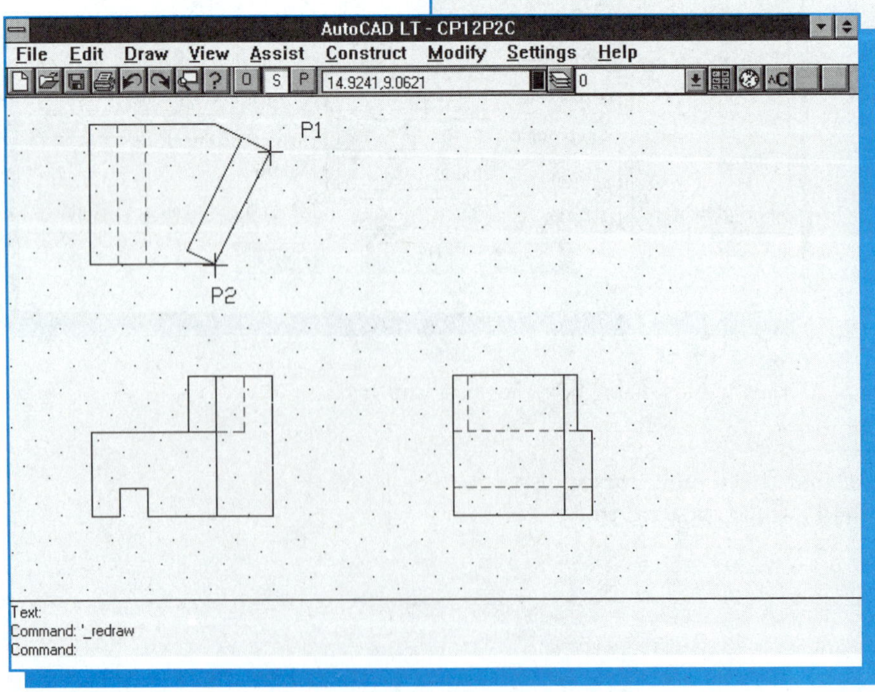

Figure 12-39
Rotate the grid to be parallel to the inclined surface.

3. Construct the auxiliary view and add details to it, including hidden lines and the hole **(Fig. 12-40)**.
4. Project the hole (as ellipses) back to the normal views **(Fig. 12-41)**. Use the **ellipse** command under **Draw** to construct the ellipses from the construction lines.
5. Turn off the construction lines used to place the ellipses. Return to the normal grid and add the necessary dimensions to the drawing **(Fig. 12-41)**. In **Figure 12-41**, the grid has also been turned off to make the drawing easier to read. The final drawing will appear as in **Figure 12-42** when plotted. Turn off the grids. Check the drawing and make a hard copy.

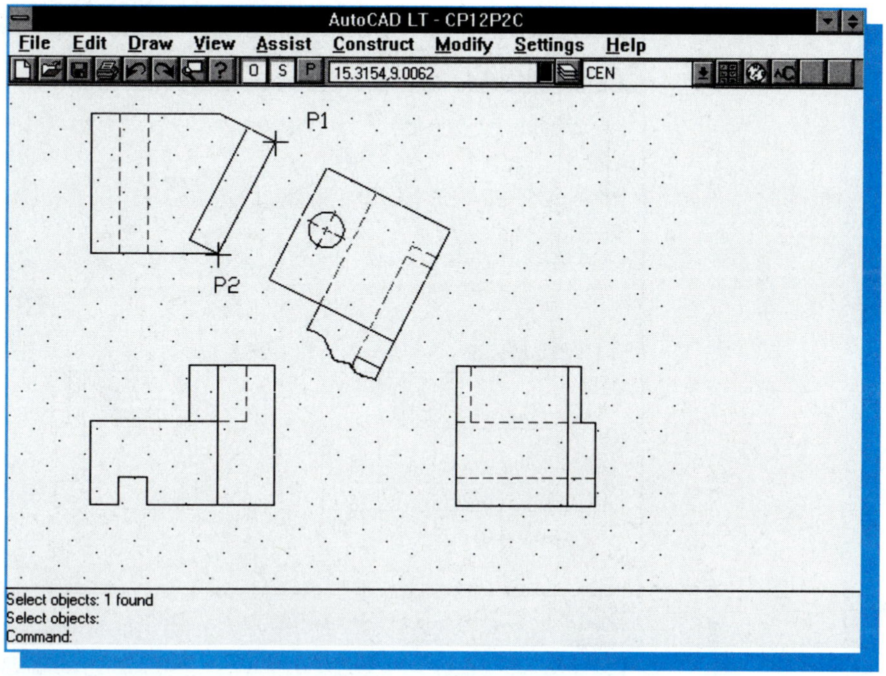

Figure 12-40
Construct the auxiliary view with the Circle and Line commands.

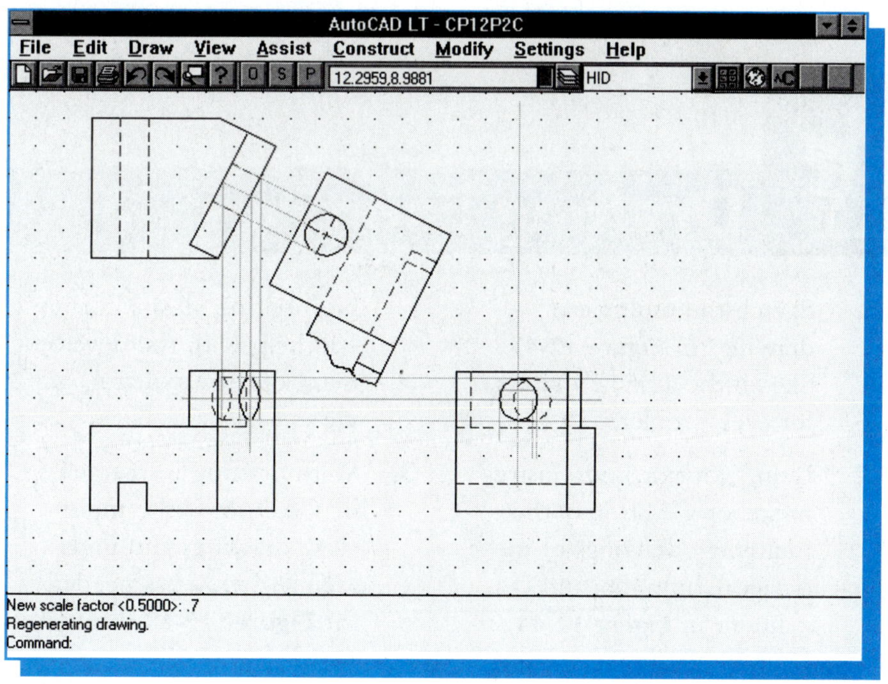

Figure 12-41
Project the hole back to multiviews. Use the Ellipse command to place the hole on the top and side views.

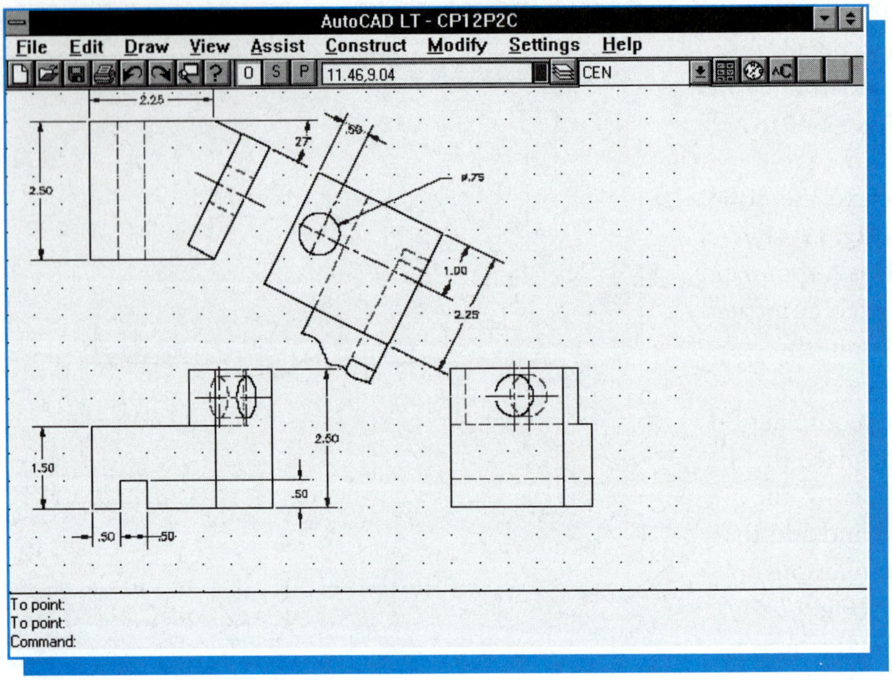

Figure 12-42
Remove the projection lines. Dimension the part. Turn off the grid. Check the drawing and make a hard copy.

Drafting Exercises

1. Sketch the multiview drawings in **Figure 12-43**. Sketch the auxiliary views for each problem. Numbering all the corners will help with the development of the auxiliary views.

2. Using your drawing instruments or CAD, draw the multiview drawings of the octagon, hexagon, and cylinder in **Figure 12-44**. Draw their auxiliary views.

3. With drafting instruments or CAD, draw the multiview drawings and their auxiliary views as needed for **Figures 12-43** through **12-62**.

4. With a CAD system, duplicate as closely as possible the procedures in **Figures 12-63** through **12-66**.

5. Follow the instructions for the revolution exercises in **Figures 12-67** and **12-68**.

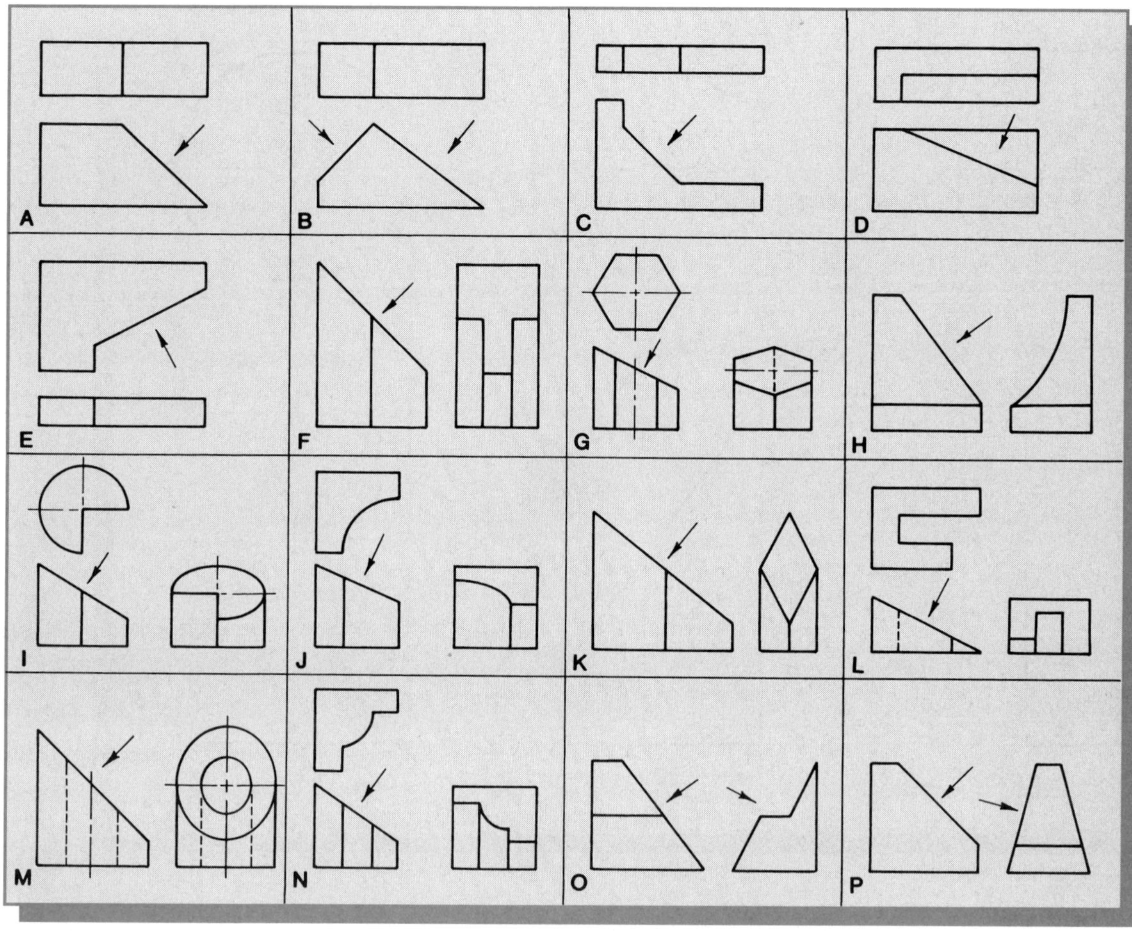

Figure 12-43a
Sketch the multiview drawings and the auxiliary solution for each exercise.

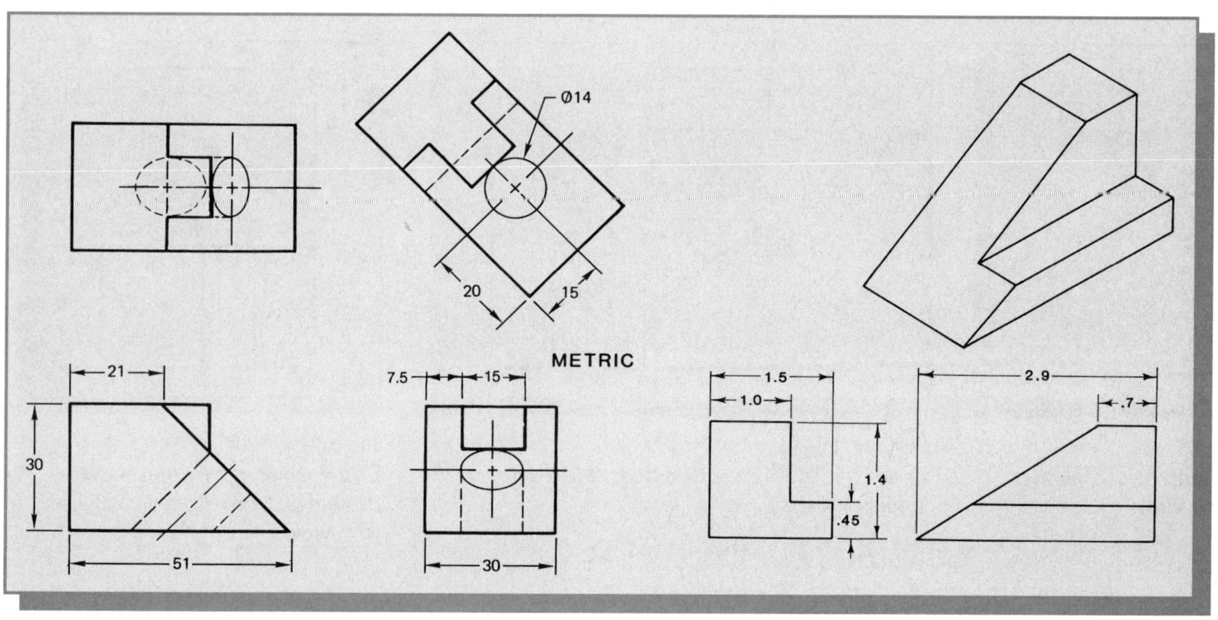

Figure 12-43b
Complete the multiview drawing and the auxiliary view for each problem.

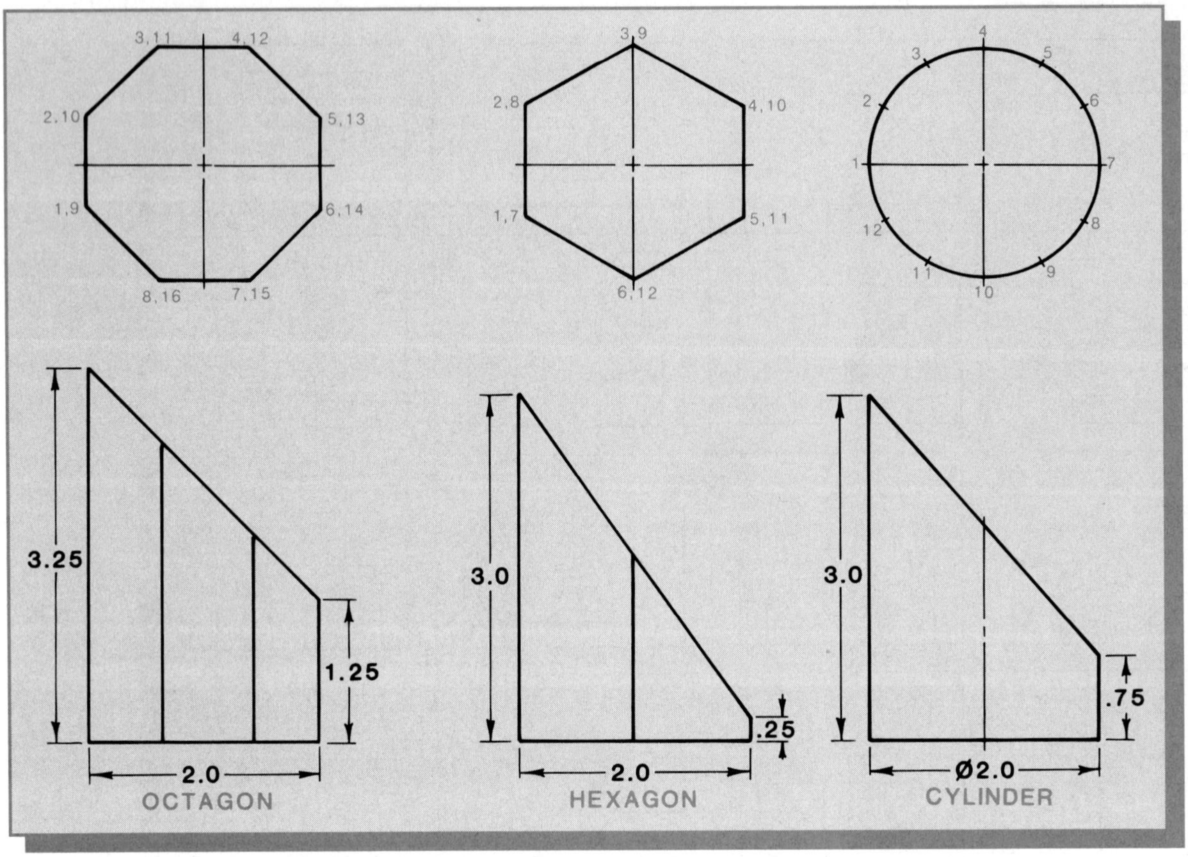

Figure 12-44
Draw each truncated prism and its auxiliary. Number all corners (points). Note that the first number is the visible corner or point.

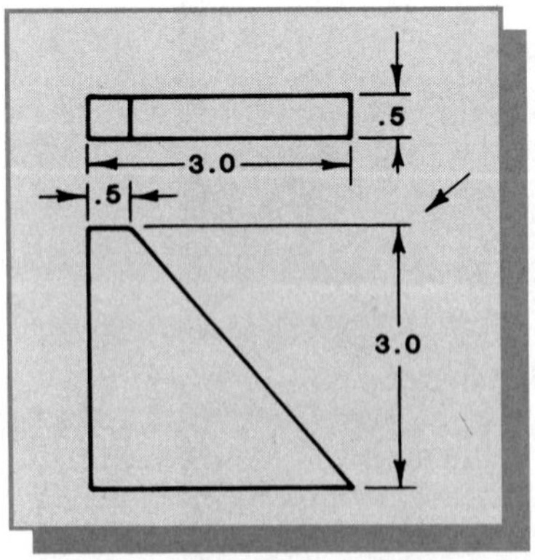

Figure 12-45
Draw the multiview drawings and auxiliary view.

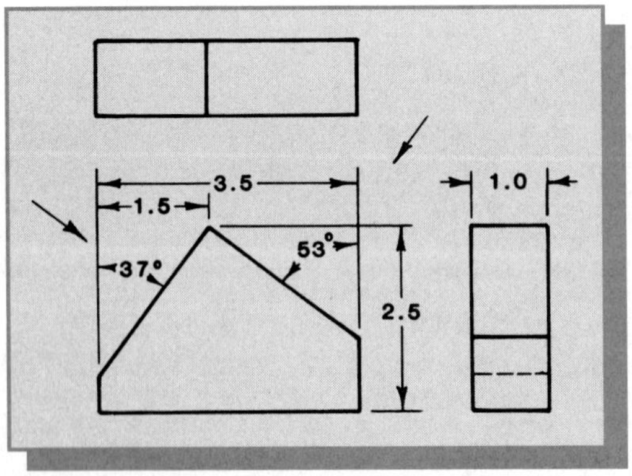

Figure 12-46
Draw the multiview drawings and two auxiliary views.

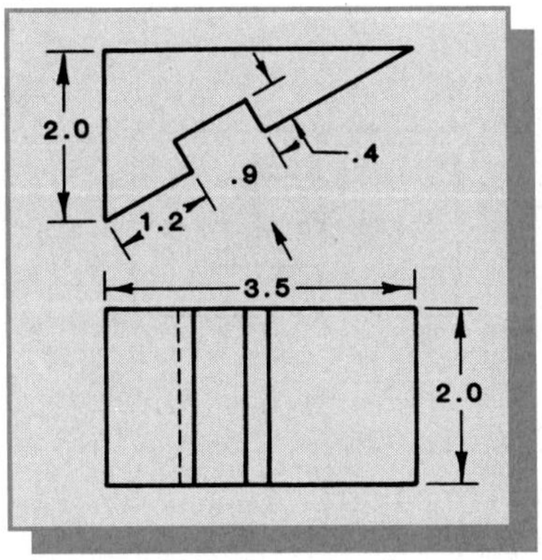

Figure 12-47
Draw three-view multiview drawings, isometric, and auxiliary views.

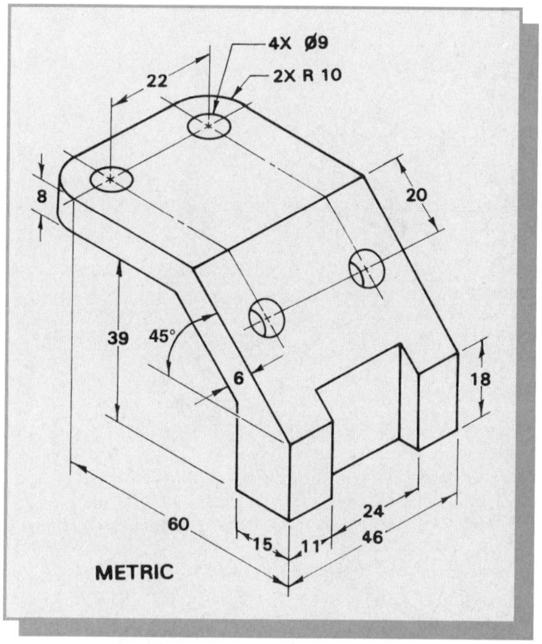

Figure 12-48
Draw the multiview and auxiliary drawings.

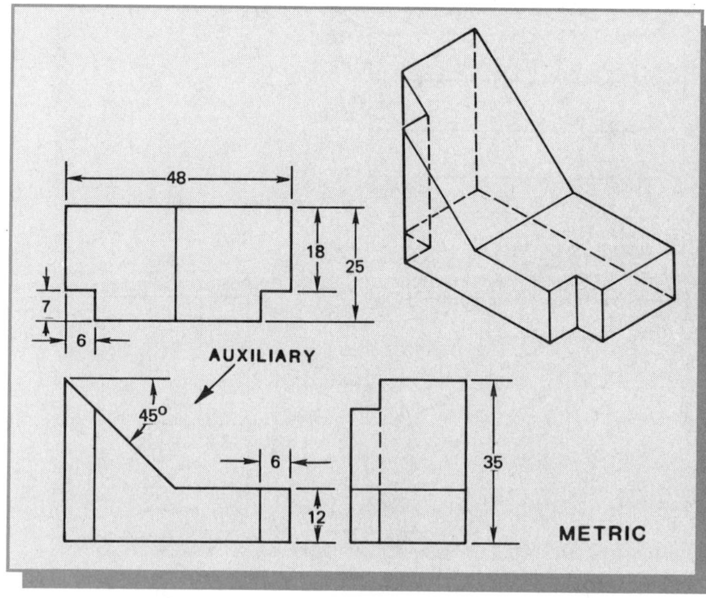

Figure 12-49
Draw the multiview, isometric, and auxiliary drawings.

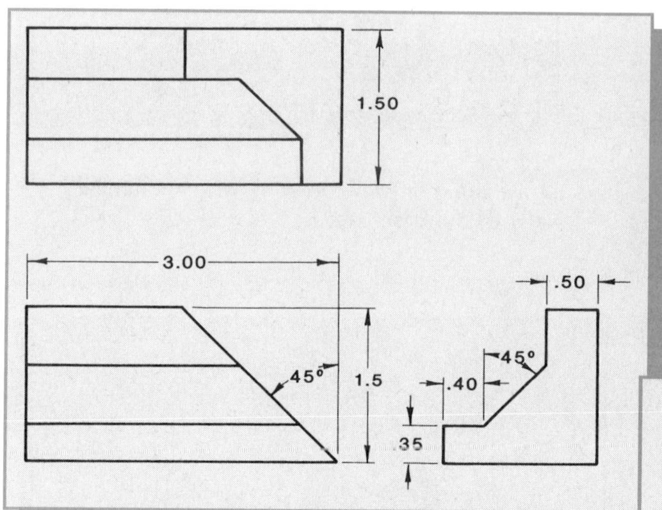

Figure 12-50
Draw the multiview drawing and auxiliary views of the front and side views.

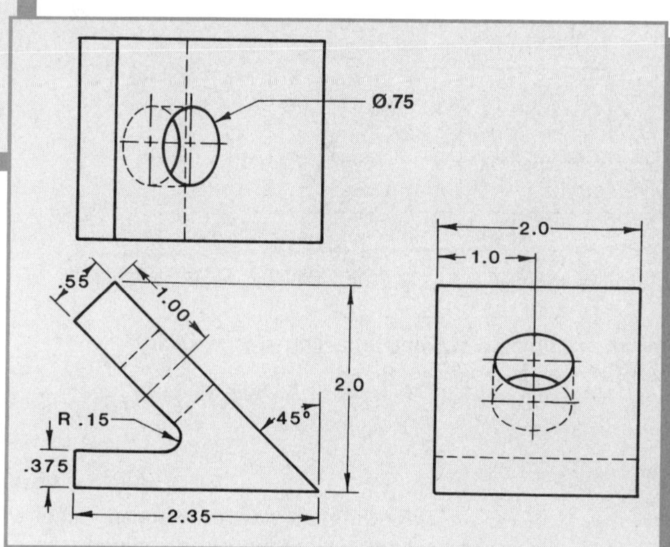

Figure 12-51
Draw the multiview drawing and an auxiliary view of the inclined surface.

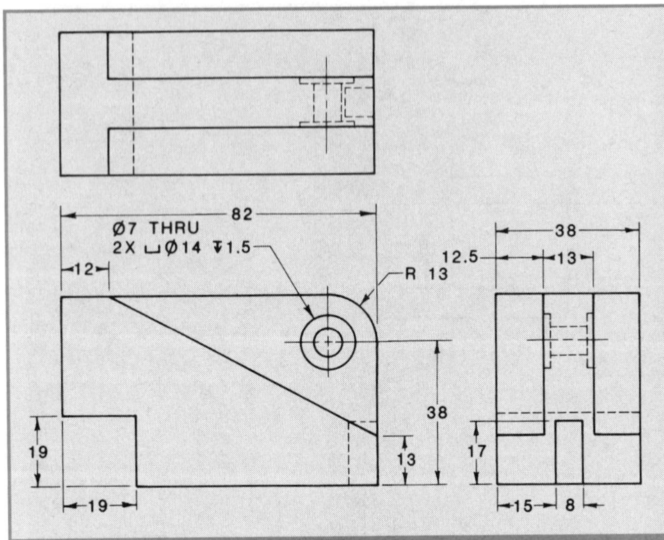

Figure 12-52
Draw the multiview drawing and an auxiliary view to show the true size of the inclined surfaces.

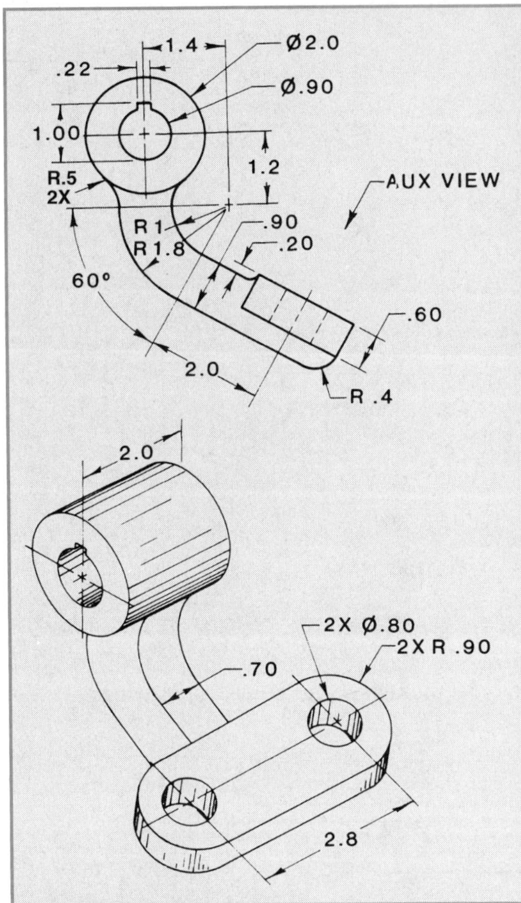

Figure 12-53
Draw the multiview drawing with an auxiliary view of the base pad.

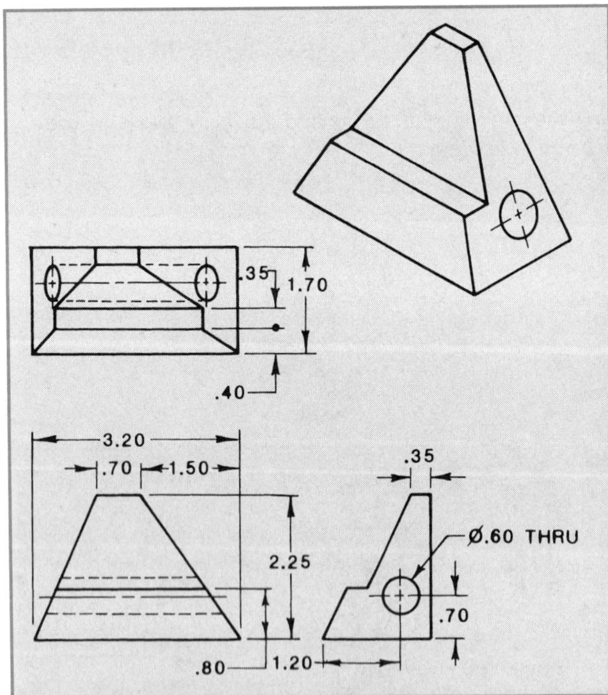

Figure 12-54
Draw the multiview drawing and right-side auxiliary view of the angle insert guide.

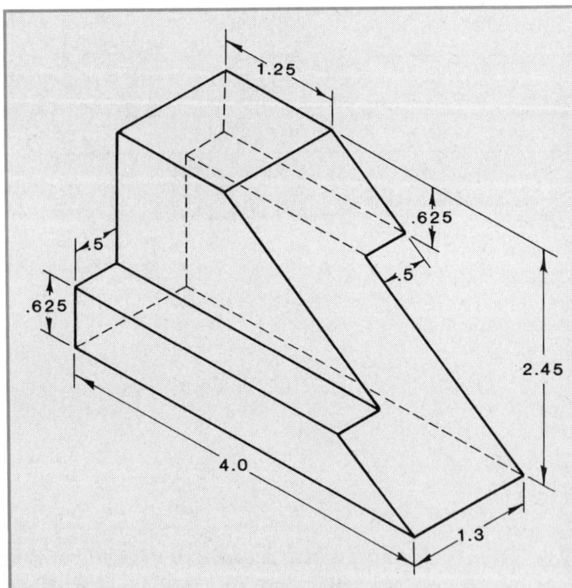

Figure 12-55
Draw the multiview drawing and auxiliary view of the inclined surface.

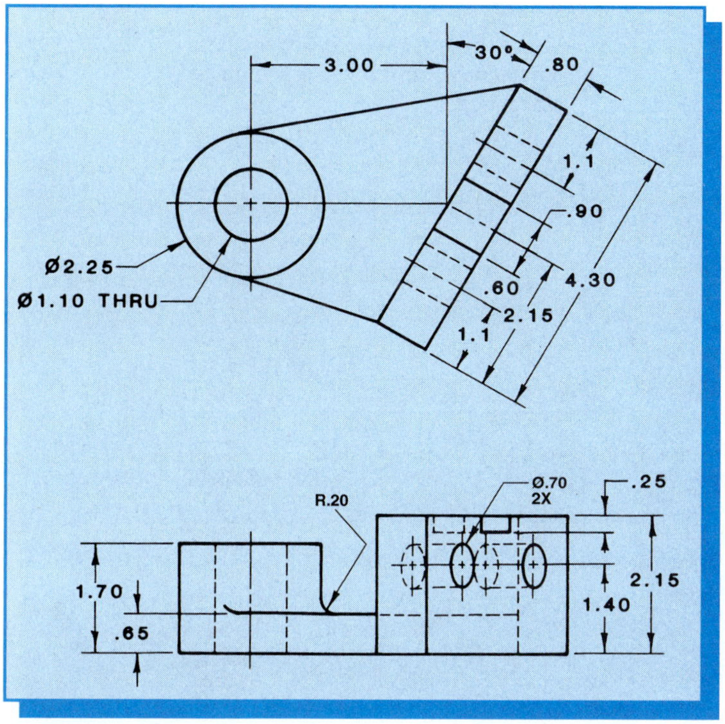

Figure 12-56
Draw the multiview drawing with an auxiliary view of the angled base. Select the best direction for auxiliary view.

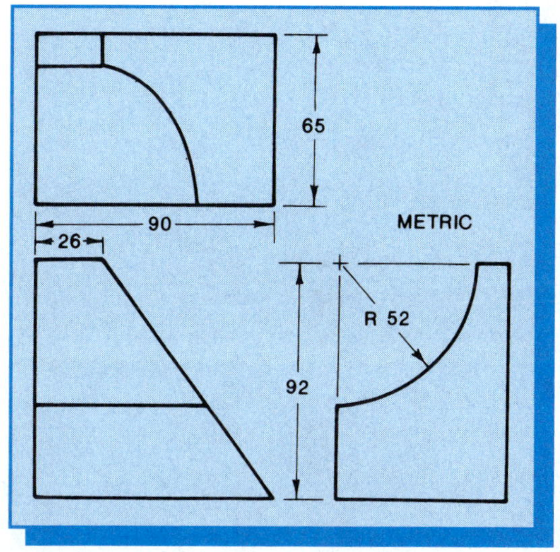

Figure 12-58
Draw the multiview, isometric, and auxiliary view.

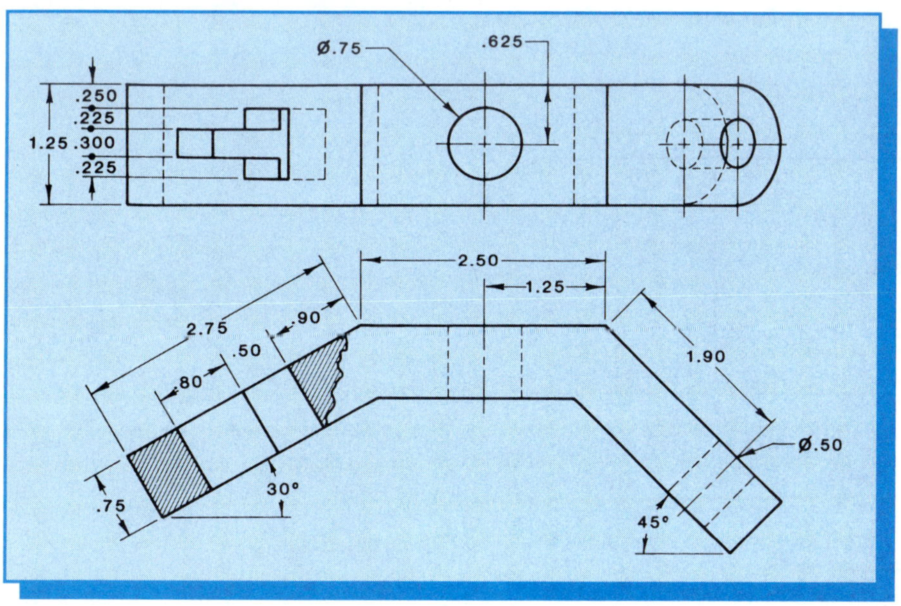

Figure 12-57
Draw the multiview drawing and two partial auxiliary views for the angled clamping jig.

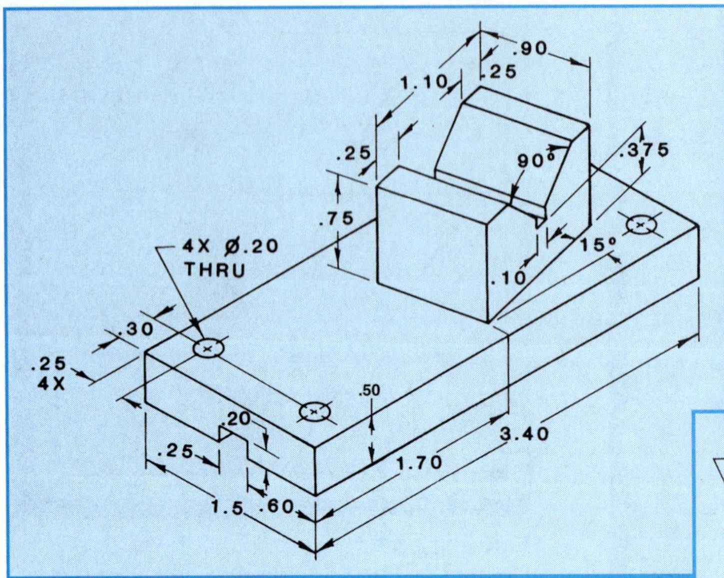

Figure 12-59
Draw the multiview drawing and a partial auxiliary view for the angle slot's face.

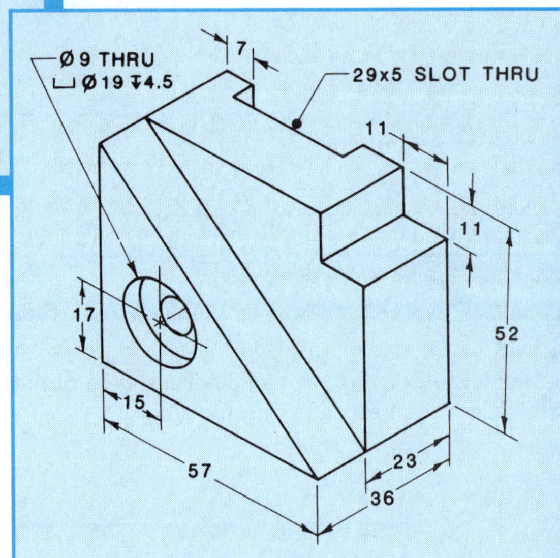

Figure 12-61
Draw the multiview, isometric, and auxiliary view.

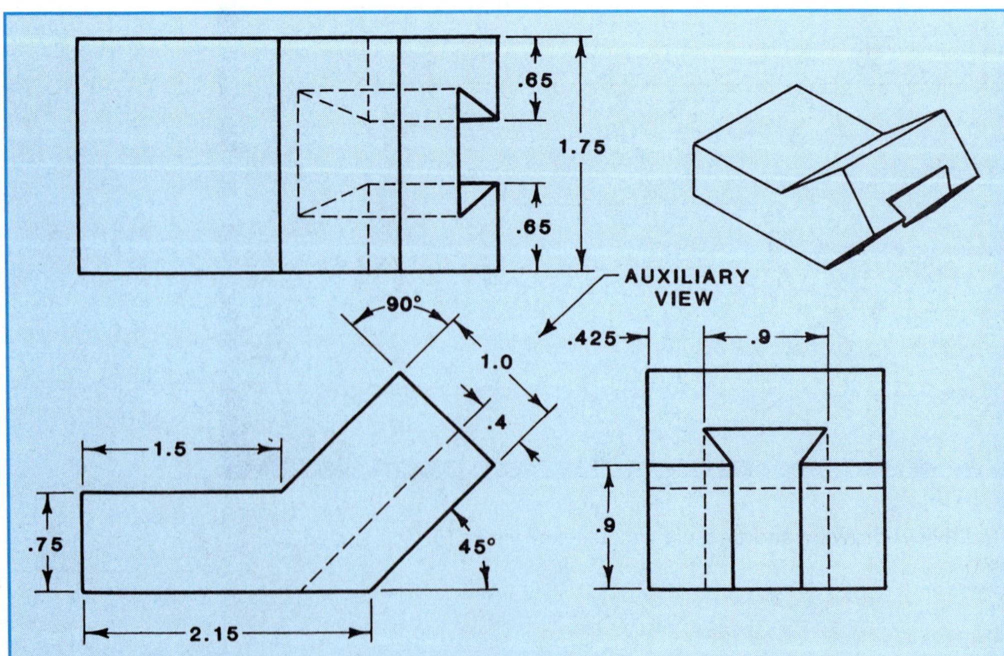

Figure 12-60
Draw the multiview, isometric, and auxiliary view.

Figure 12-62
Draw the multiview drawing with an auxiliary view of the inclined surface.

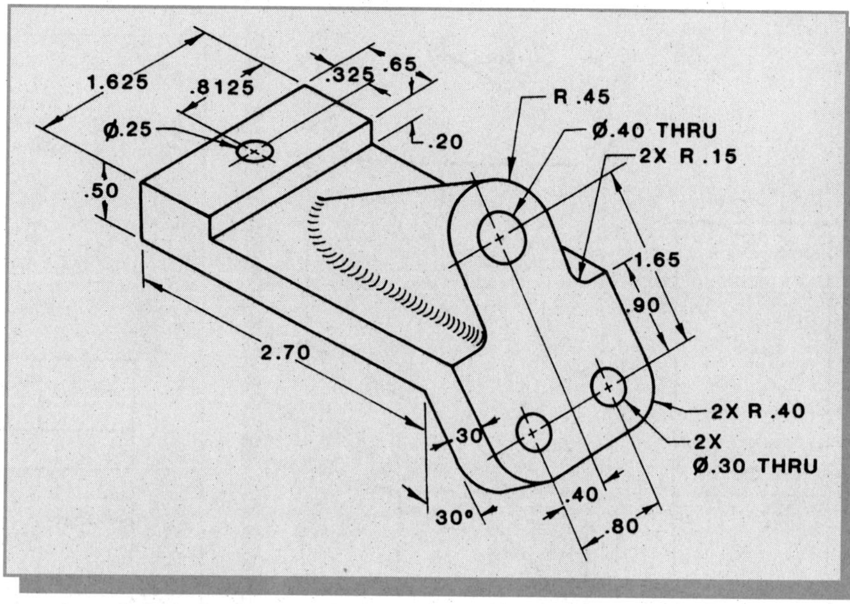

Figure 12-63
Draw the orthographic, isometric, and front-view revolution as shown. Revolve the side view 15° counter-clockwise and complete the front and top views. (See Chapter 15 for isometric procedure.)

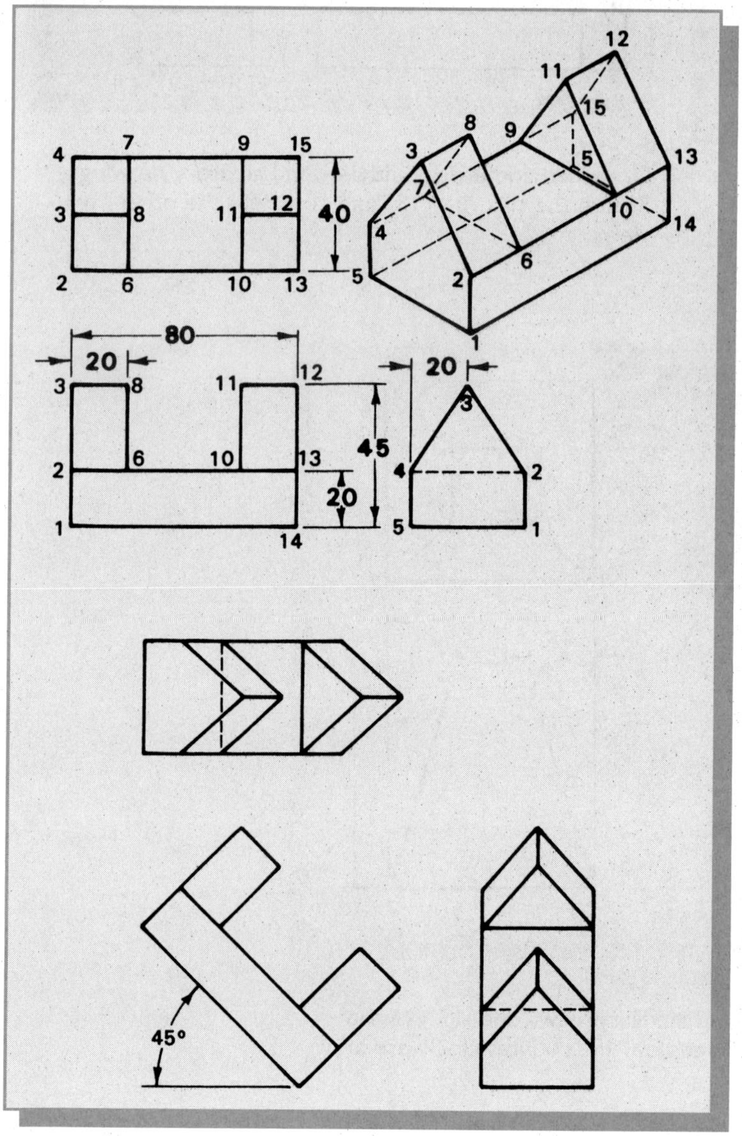

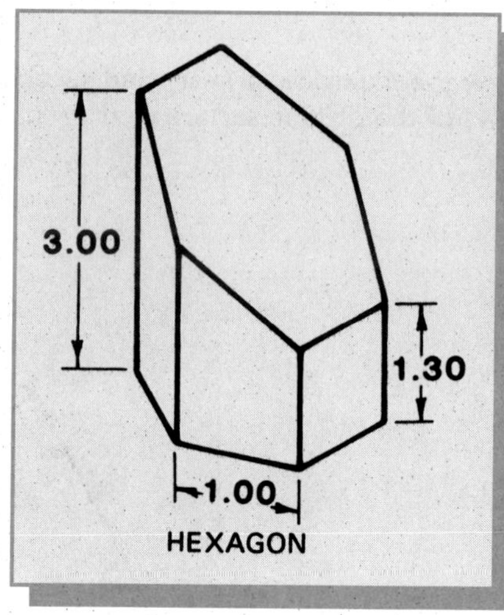

Figure 12-64
Draw the isometric, multiview, and auxiliary view drawings. Rotate the front view 45° clockwise and complete the other view(s).

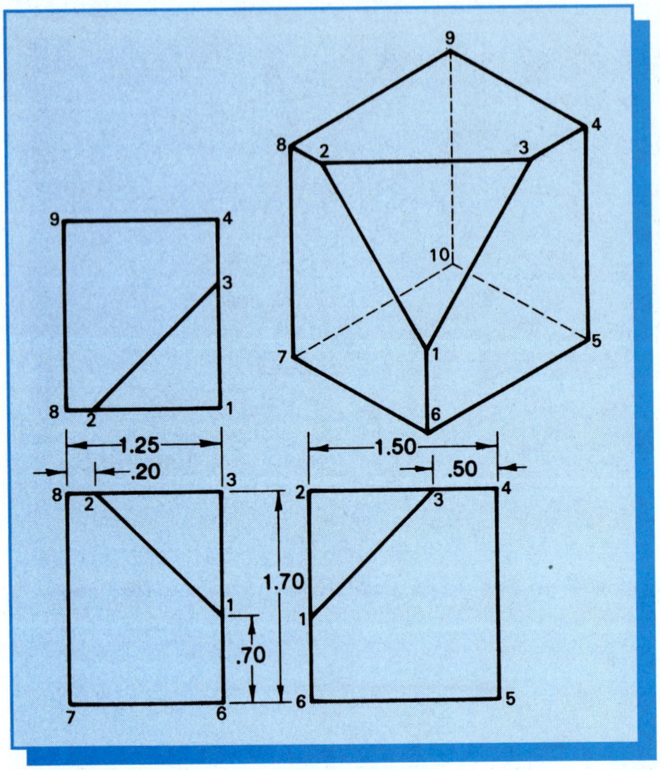

Figure 12-65
Draw the multiview, primary, and secondary auxiliary views of the oblique surface to draw its true size.

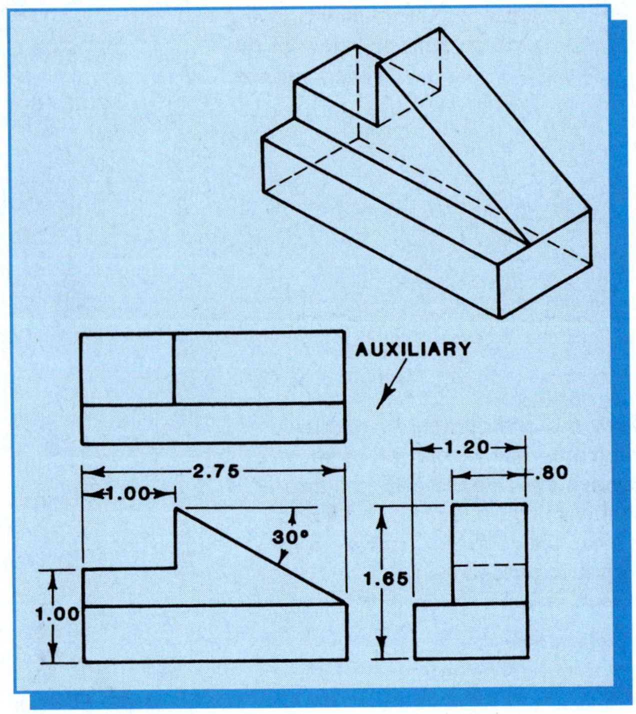

Figure 12-66
Draw the isometric, multiview, and auxiliary drawings. Rotate the top view 15° and complete the other two views.

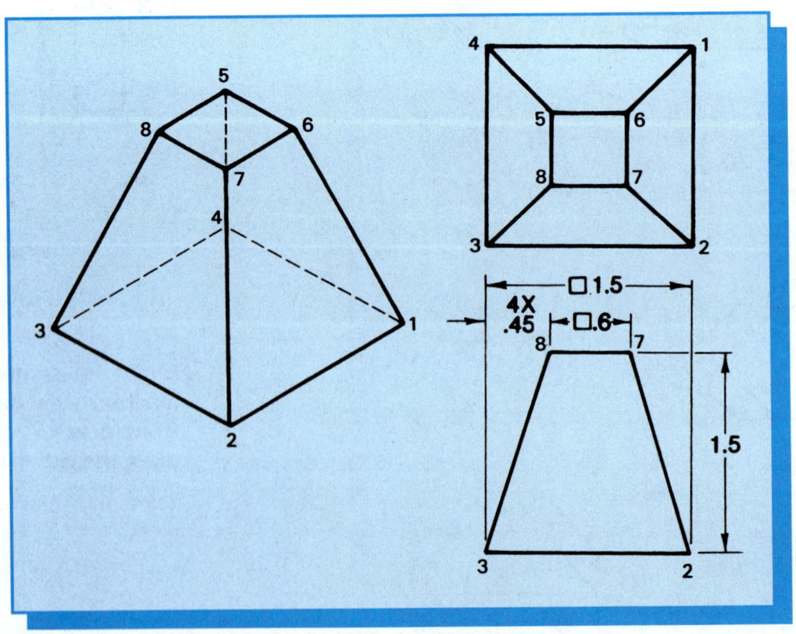

Figure 12-67
Draw the isometric, multiview, and auxiliary view (one side) of the frustum pyramid. Revolve the front view 45° counter-clockwise and complete the other two views.

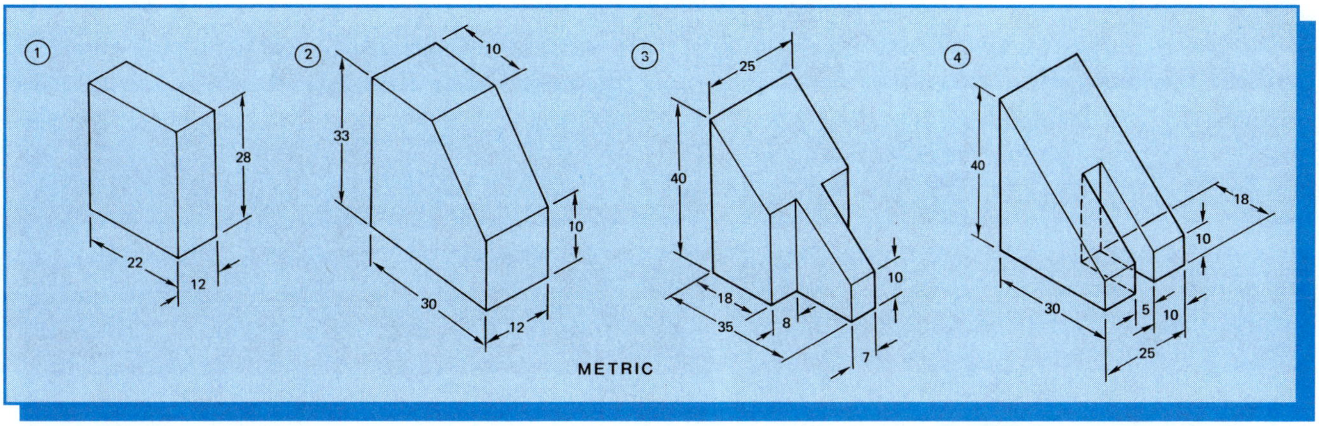

Figure 12-68

1. Draw isometric and number all corners.
2. Draw orthographic and number all corners.
3. Select and draw a primary auxiliary (use reference plane).
4. Select and draw a secondary zuxiliary (use reference plane).

Design Exercises

1. Design a recipe book holder for use when cooking in the kitchen. Draw a multiview working drawing and an auxiliary view of the slanted position of the slanted support for the cookbook.
2. Design a wall-mounted mirror and frame holder that will reflect a back view of your head from the mirror in your bathroom medicine cabinet.

Key Terms

Auxiliary plane
Auxiliary views
Foreshortened
Hinged planes
Primary auxiliary views
Primary revolution
Revolution drawing
Secondary auxiliary view
Successive revolutions
True size edge

Chapter 13

Descriptive Geometry

Objectives

The student will be able to:

- draw a point in space onto various planes of projection
- draw a line in space onto various planes of projection
- draw a point view of a line
- draw the true length of a line
- draw an edge view of a surface
- draw a true size surface
- lay out and solve a descriptive geometry exercise on a CAD system

Introduction

Descriptive geometry is a study of the three-dimensional relationship of points, lines, angles, and surfaces as drawn on two-dimensional surfaces. The principles of descriptive geometry are used to graphically solve problems and gain information about geometric forms and their position in space. Since descriptive geometry is directly related to orthographic and auxiliary projection, a basic understanding of these concepts is a vital prerequisite to understanding the material presented in this chapter. **Figure 13-1** graphically defines the geometric terms commonly used in the development of graphic solutions using descriptive geometry. Most graphic problems can be solved by finding the true length and point projection of lines, and the true size and edge view of planes.

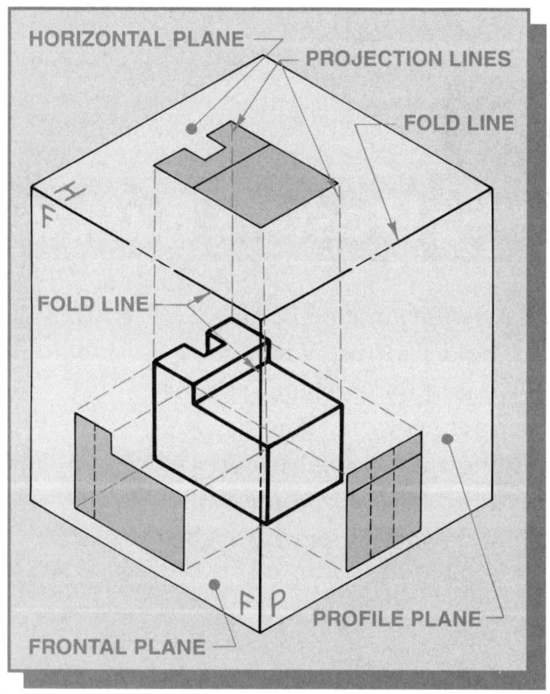

Figure 13-1
Graphic terms for descriptive geometry

Points in Space

Geometric forms are all constructed from a series of points. Understanding the precise location of a single point is the first step in understanding the principles of descriptive geometry.

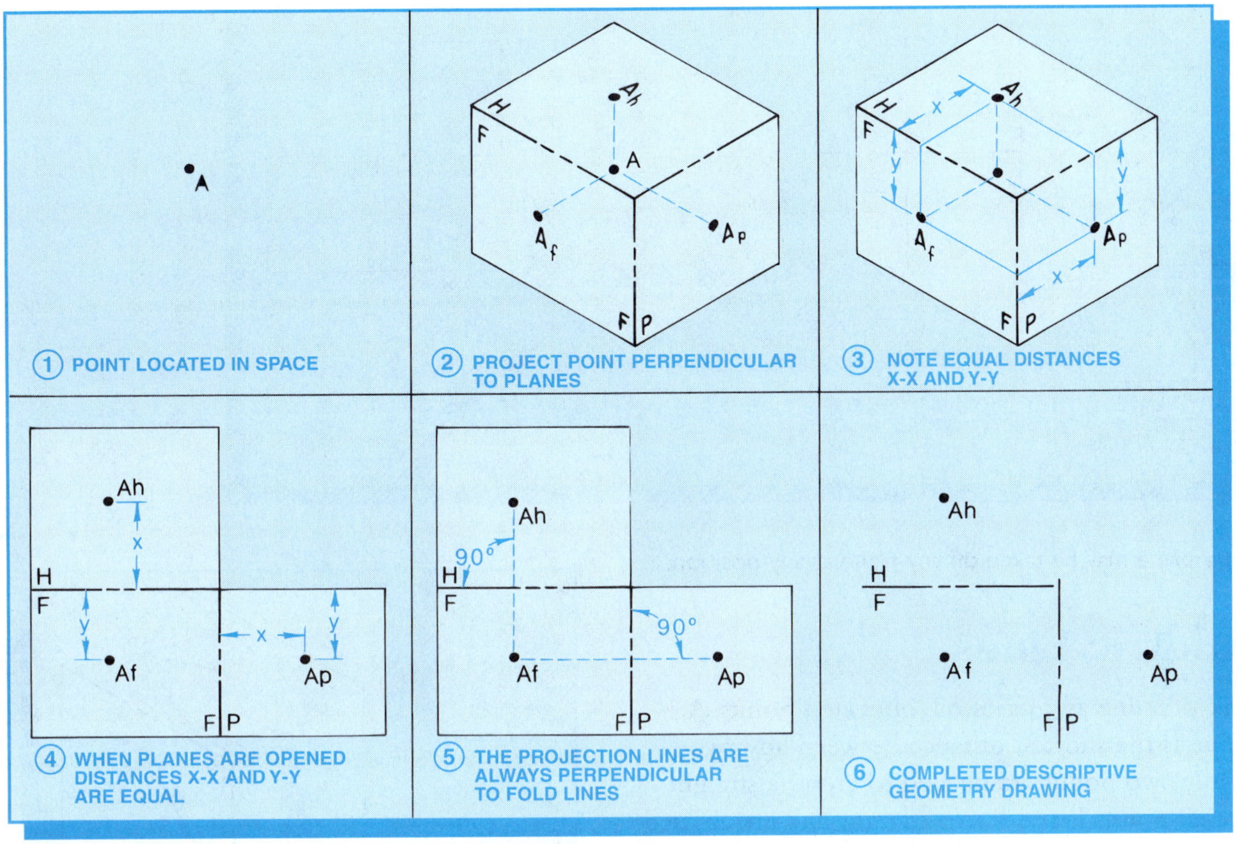

Figure 13-2
Descriptive geometry graphics layout for a single point in space

When a **point in space** is theoretically surrounded by the **frontal plane**, **horizontal plane**, and **profile plane**, it may be projected perpendicularly to the face of the planes **(Fig. 13-2)**. It is extremely important to understand why X = X and Y = Y. Study **Figure 13-2** until it is clear. When the planes are opened flat onto a two-dimensional surface, the distances of X = X and Y = Y remain the same. It is possible to solve the location of a point to a third plane when the point's location is known on two planes. This principle in **Figure 13-2** is the basis of descriptive geometry. When you can solve for one point, you can solve for many points, and points make up drawings.

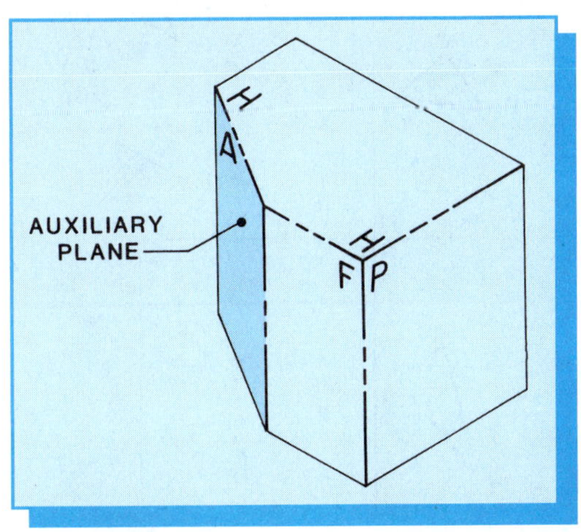

Figure 13-3
An auxiliary plane

Geometric Planes

Planes used in descriptive geometry include the normal orthographic planes, plus auxiliary planes. There are only six normal orthographic planes; however, the number of auxiliary planes is unlimited. An **auxiliary plane (Fig. 13-3)** is a plane that lies on any angle other than the normal orthographic planes. **Figure 13-4** shows several examples of auxiliary planes projected from orthographic planes. Any angle may be used with a **fold line** of an auxiliary plane.

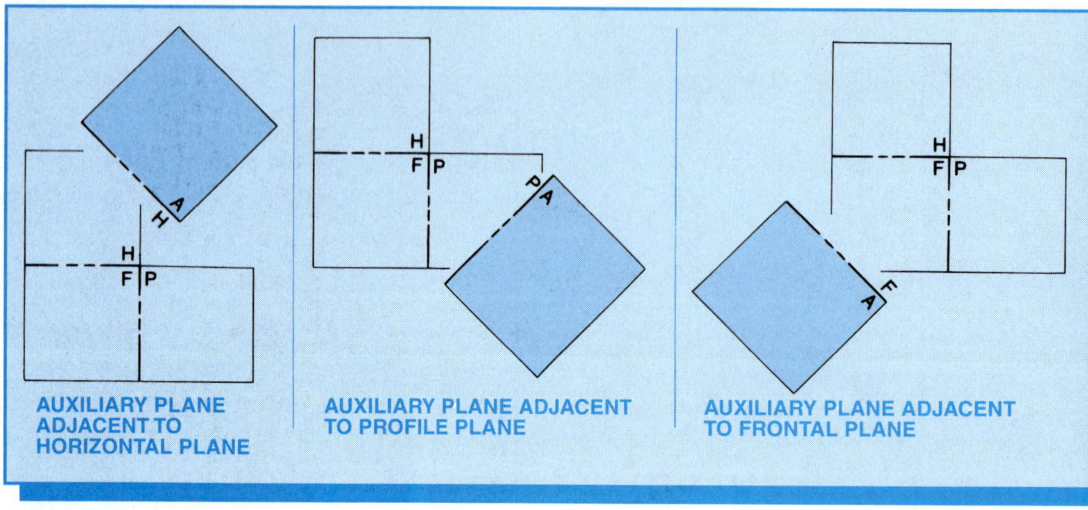

Figure 13-4
An auxiliary plane may be taken off any plane at any position.

Geometric Lines

Think of a line as a series of connected points. A straight line is the shortest distance between any two points. Only two points are needed to create a straight line, since all points located on a straight line are aligned.

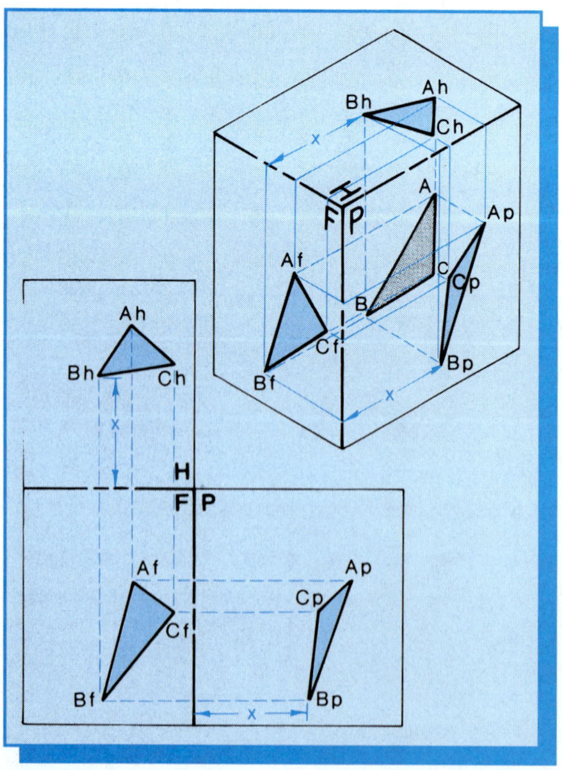

Figure 13-6
Descriptive geometry layout for a triangular surface in space (three points)

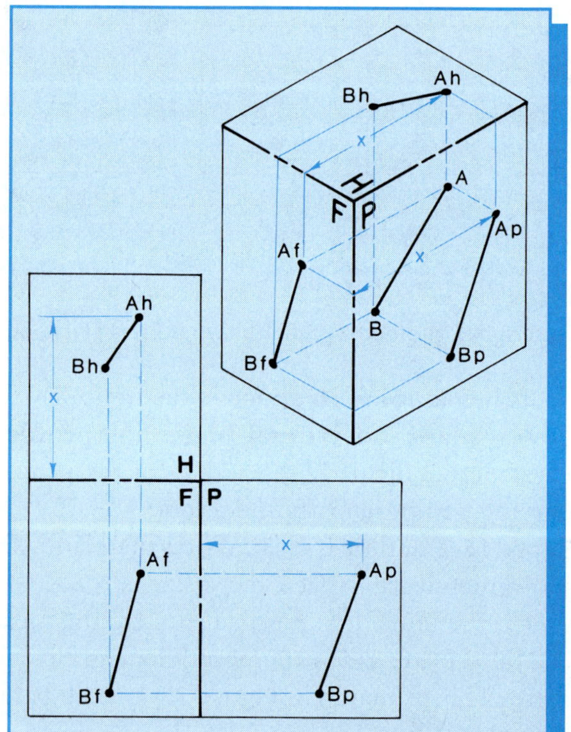

Figure 13-5
Descriptive geometry layout for a line in space (two points)

Figure 13-5 shows the projection of two points in space to create a straight line.

The connection of three points in space creates a **triangle (Fig. 13-6)**. It logically follows that any two-dimensional plane surface or three-dimensional solid object can be created by the projection and connection of multiple points in space **(Fig. 13-7)**.

Curved lines are comprised of a series of unaligned points. A curved line is actually a series of very short straight lines connecting closely spaced points. Closely spaced points create smooth curves, and widely spaced points create ragged curves.

True Line Length

Lines perpendicular and parallel to normal orthographic planes are all **true length**. Lines on auxiliary planes that are oblique or inclined are not true length when viewed from a normal plane of projection. However, lines on auxiliary planes are true length when viewed perpendicular to the adjacent plane.

When a fold line is parallel to a line in space, the line on that plane will be true size. Therefore, to find the true length of a line on any plane, locate the fold line parallel to the line. Project the line to the adjacent plane where the lines become true length **(Fig. 13-8)**.

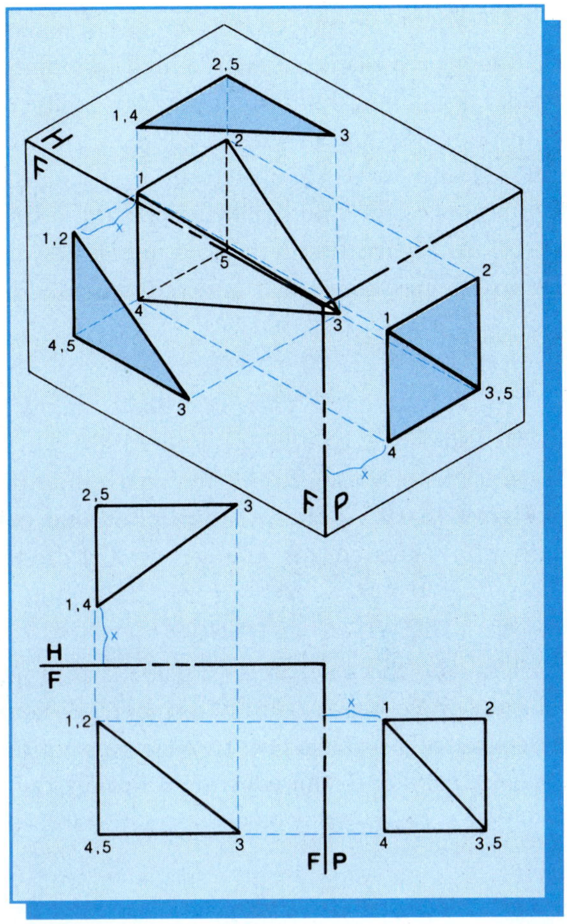

Figure 13-7
Descriptive geometry layout for a five-point solid in space planes.

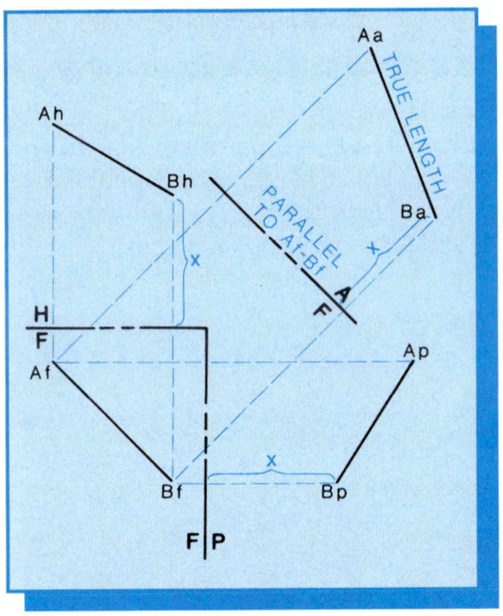

Figure 13-8
When the fold line is parallel to a line in space, the line on that projected plane will be true size.

A second procedure that may be used to find the true length of a line is to revolve the line on the horizontal plane until it is parallel to the reference line. Then project the end of the line to the frontal plane as shown in **Figure 13-9**.

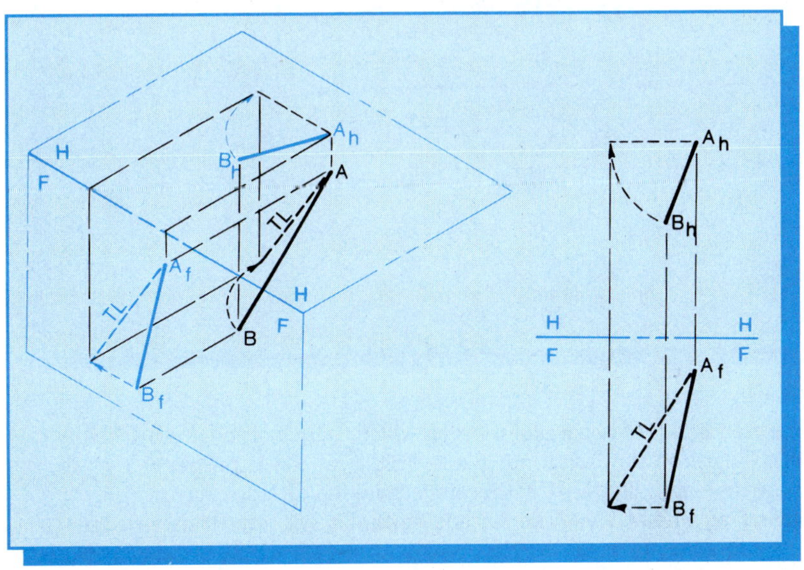

Figure 13-9
Finding the true length of a line by revolving it on the horizontal plane and projecting it to the frontal plane

Point View of a Line

Lines on a plane are represented by points on all other perpendicular planes. Think of a point on a line as a cross section of the line. The **point view** of a line can be drawn by placing the fold line perpendicular to the true length line. Project the two points of the line that will appear as a single point **(Fig. 13-10)**. In **Figure 13-10**, the fold line A/B is perpendicular to the true length line. All X and Y distances are equal.

Edge View of a Plane

To find the **edge view** of a plane, follow the steps in **Figure 13-11**. Establish a true length line within the plane's surface. Locate a point view of the true length line using the procedures in **Figure 13-10**. Locate the third point and connect the points with a line. This creates an edge view of the triangular plane.

True Size of a Surface

Drawing the true size of a plane is an important graphic solution procedure used in descriptive geometry. To draw the **true size** of a surface, find an edge view of the surface. Draw a fold line parallel to the edge view. Project the surface to the new plane and it will measure true size and shape **(Fig. 13-12)**.

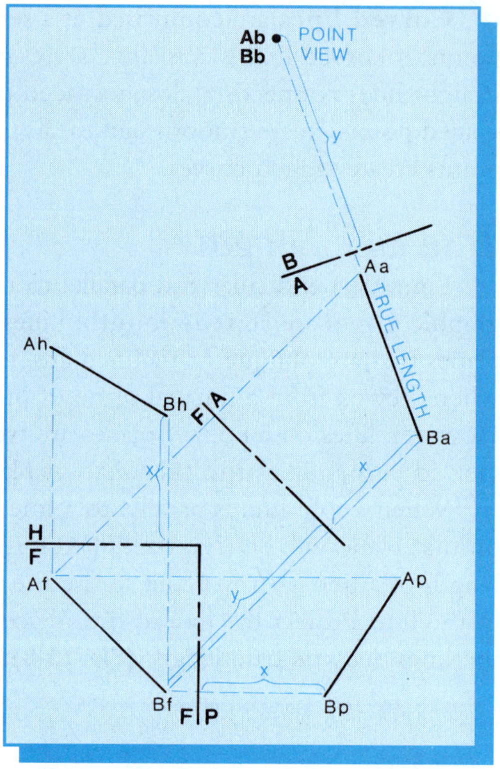

Figure 13-10
When the fold line is perpendicular to the true length line, the projection will be a point view. Note that all X distances are equal and all Y distances are equal.

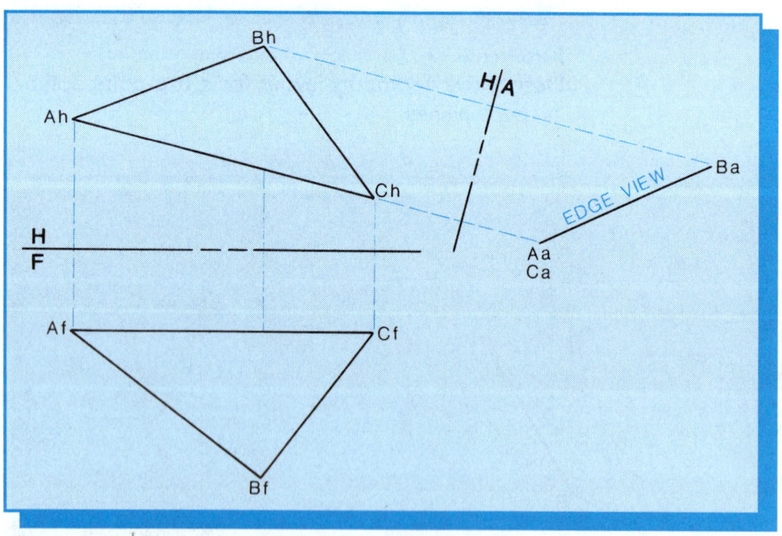

Figure 13-11
The fold line F-H is parallel to edge Af-Cf; therefore, edge Ah-Ch is true length on the horizontal plane. Fold line H-A is perpendicular to the true length line Ah-Ch; therefore, true length edge Aa-Ca will appear as a point view. The surface Aa-Ba-Ca will appear as an edge view on auxiliary plane A.

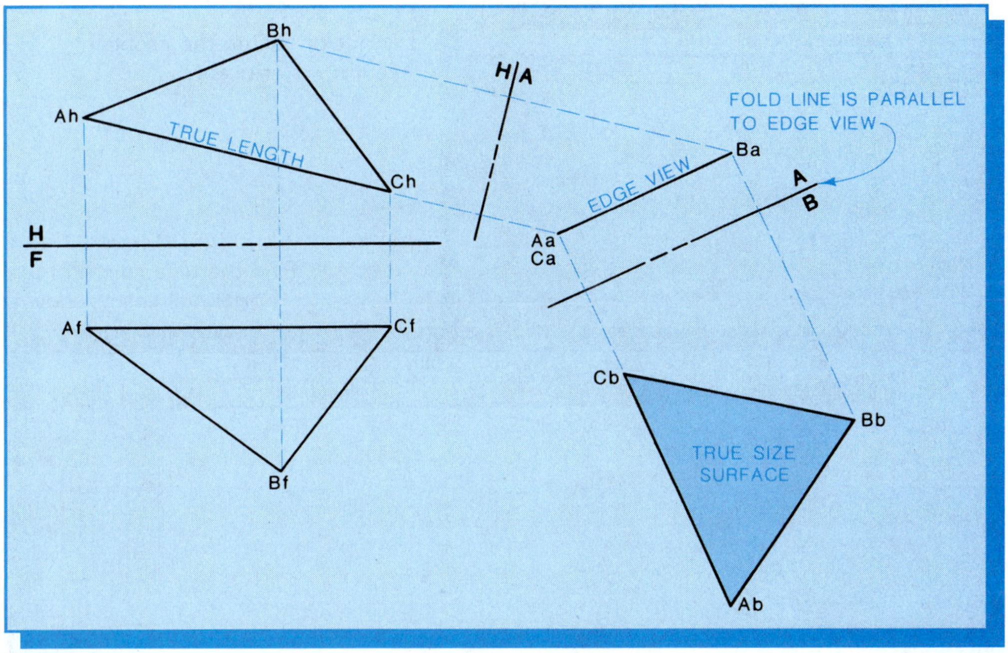

Figure 13-12
Once the edge view is drawn. the placement of a fold line parallel to it will show the true size of the surface on auxiliary plane B.

AutoCAD LT Exercises

Since solving descriptive geometry problems consists primarily of drawing a series of auxiliary views, CAD procedures are similar to those used in manually drawing auxiliary views. CAD is an effective tool for solving descriptive geometry problems because of the accuracy that can be achieved. CAD operators must be extremely careful in drawing projections to insure that all geometric data is accurate. If the CAD operator's accuracy is correct, the results will be accurate. To aid in the construction of the various geometric projections, the grid should be rotated to align with the reference (fold) line **(Figs. 13-15** and **13-16)**. Note that the grid is aligned with the reference line (fold line).

To find the true size of a plane on a CAD system, follow these steps:

Label your drawing using the Text command.

1. Draw the problem. Extreme care must be taken to insure the problem is set up accurately. Label the views **(Fig. 13-13)**.
2. Draw a horizontal line in the frontal view, and find its resulting true length line in the horizontal view **(Fig. 13-14)**.

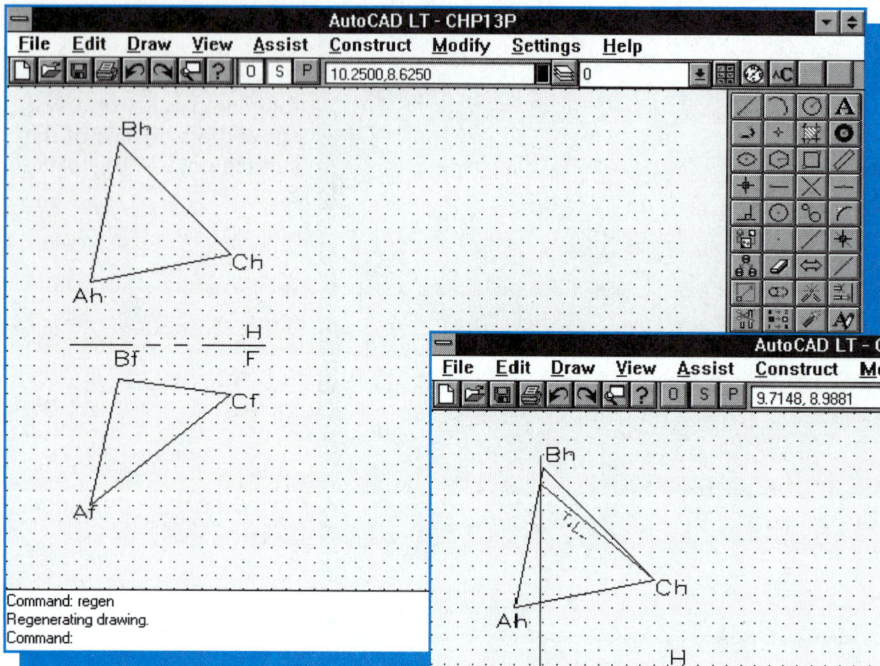

Figure 13-13
Lay out or call up the problem. Accuracy is critical.

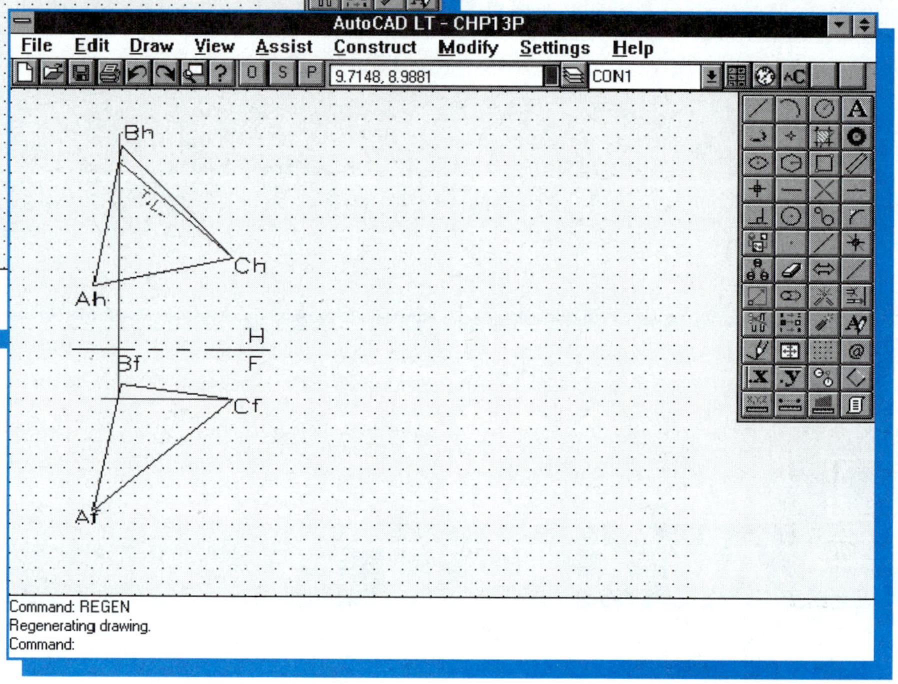

Figure 13-14
Find the true length of the horizontal line as shown.

3. Draw a reference line perpendicular to the true length line. Rotate the grid so it is aligned with the new reference line. Using projection lines perpendicular to the reference line and transferring points to the new view, construct and label the edge view **(Fig. 13-15)**.

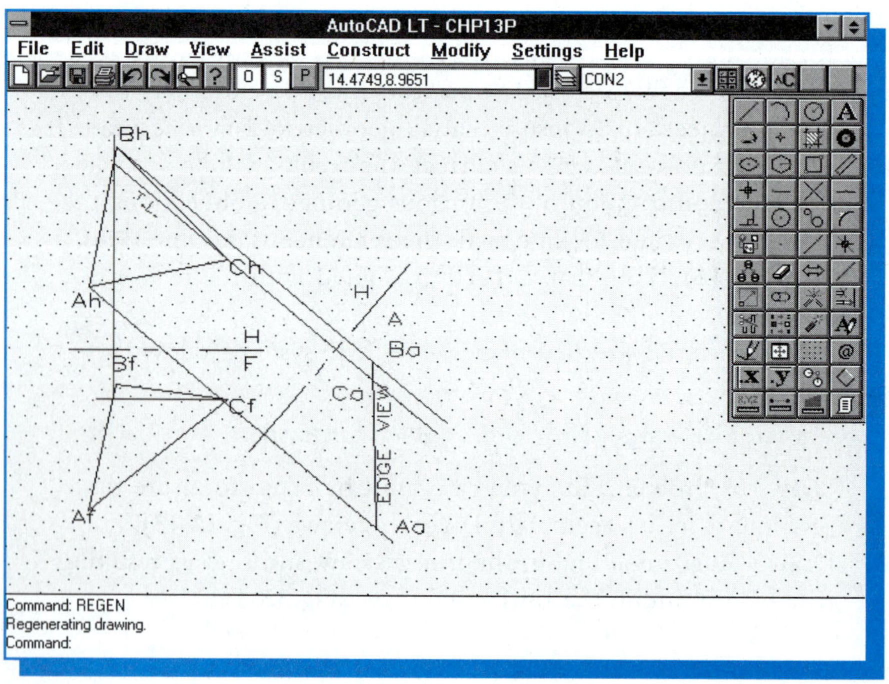

Figure 13-15
The edge view is drawn by placing the fold line perpendicular to the true length line and projecting the plane.

4. Draw a reference line parallel to the edge view. Rotate the grid so it is aligned with the new reference line. Draw projection lines and transfer points to construct the true size of the plane. Label the true size view **(Fig. 13-16)**.

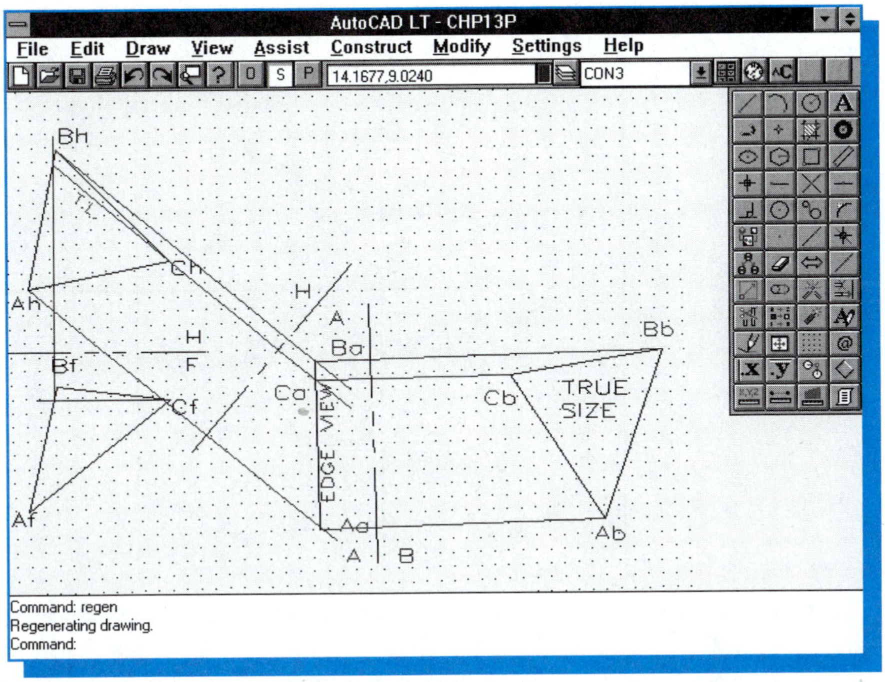

Figure 13-16
The true size of the surface is drawn by placing the fold line parallel to the edge view.

Drafting Exercises

1. Sketch the graphic solutions for the exercises shown in **Figure 13-17**. Sketch the exercises two or three times larger than shown.

2. With instruments, find the graphic solutions for the problems in **Figure 13-18**. Triple the size of the exercises with a scale or dividers.

3. Solve the problems in **Figure 13-18** using a CAD system.

4. Find the point in space on the profile plane if the point on the frontal plane is in the exact center of the plane, and the point on the horizontal plane is .75" above the fold line. Draw all planes 2" square.

5. Complete the horizontal plane for **Figure 13-19**. Find the edge view of the plane and then the true size.

6. Complete the profile plane for **Figure 13-20**. Find the edge view of the plane and then the true size.

7. Note in **Figure 13-21** on the frontal plane the surface A, B, C, and point P. Find the shortest distance from surface A, B, C, to point P.

8. Complete the auxiliary planes A, B, C, and E for the rectangular prism in **Figure 13-22**.

9. Complete the true shape drawing for **Figure 13-23**.

10. Follow the instructions on page 277 for **Figure 13-24**.

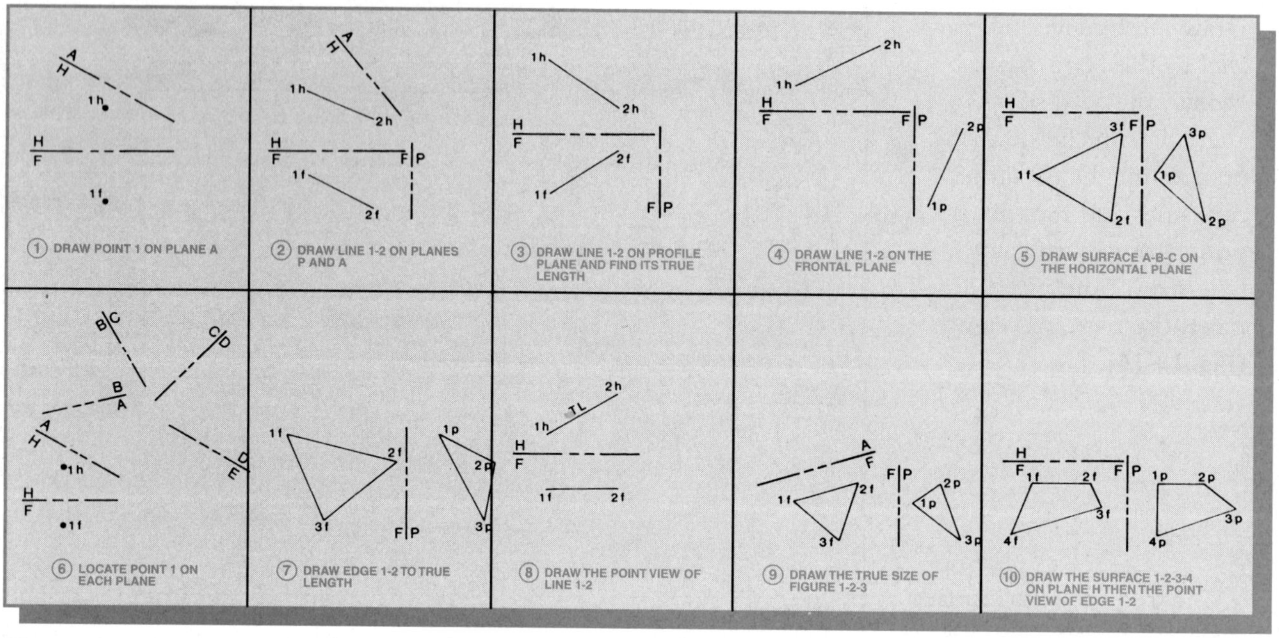

Figure 13-17
Triple each exercise with a scale or dividers on a separate drawing format. Sketch or draw the solutions with instruments.

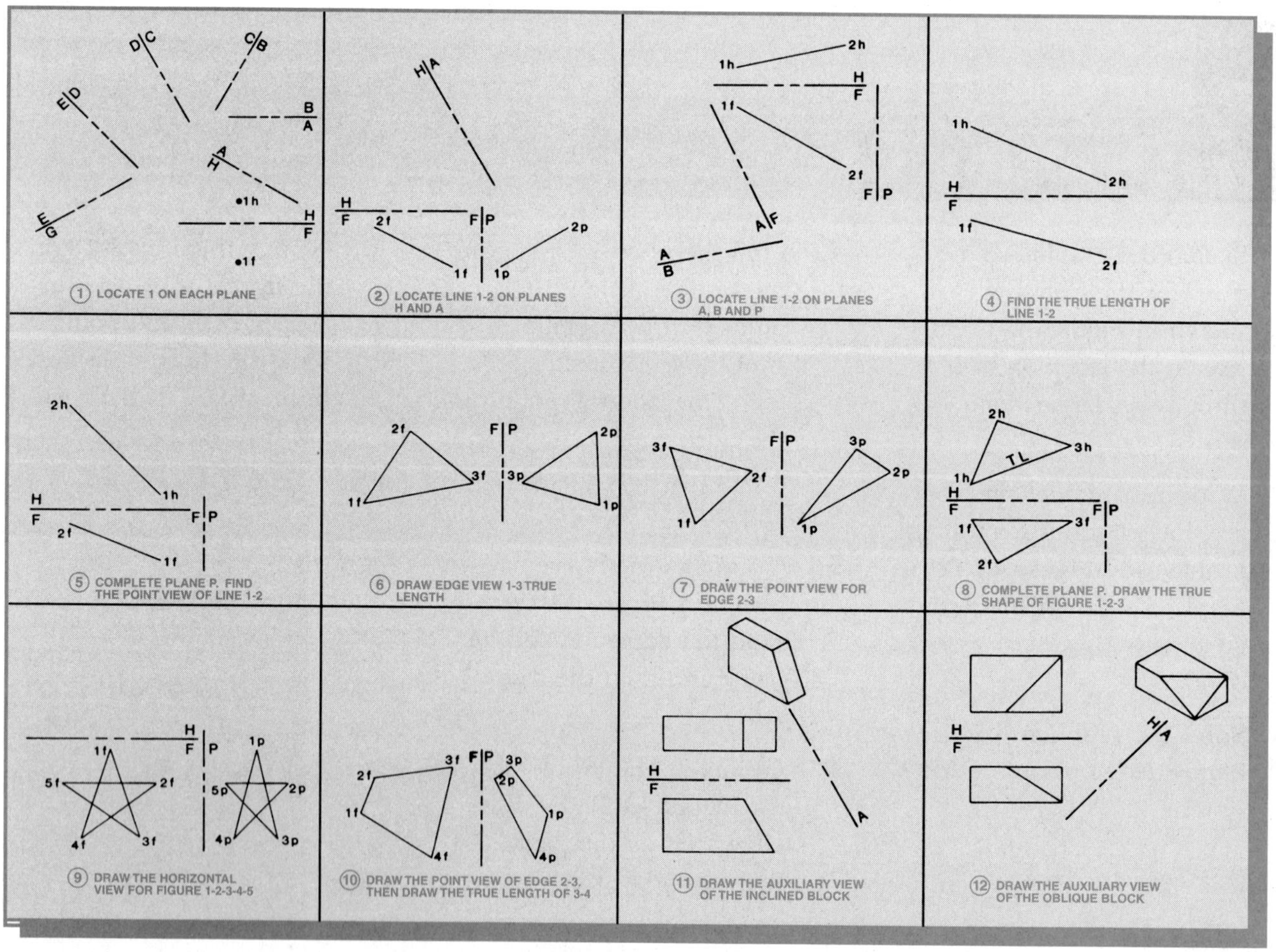

Figure 13-18
With drawing instruments, triple the size of each exercise and solve.

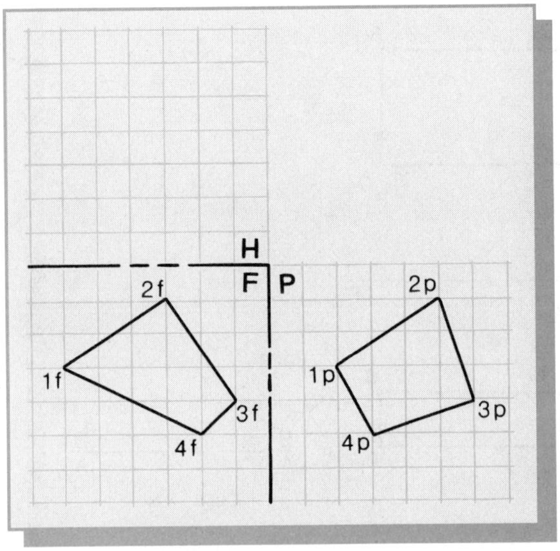

Figure 13-19
Complete the drawing on the horizontal plane, and then find the edge view and true size of the plane.

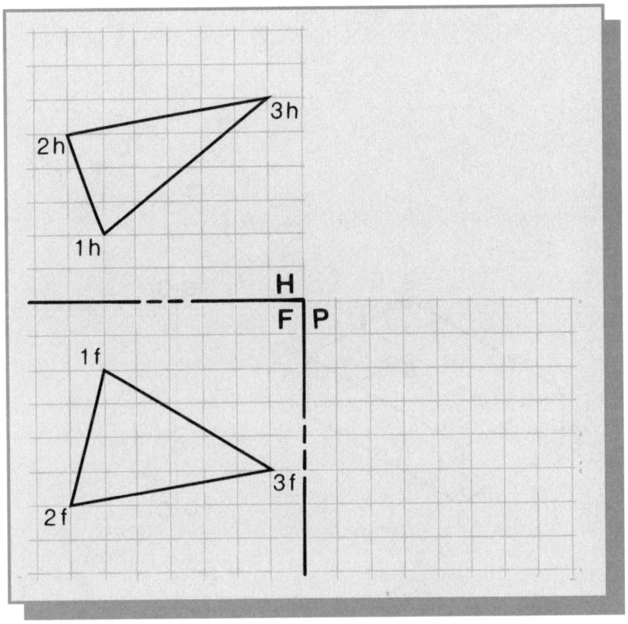

Figure 13-20
Complete the drawing on the profile plane. Find the edge view of the plane and then the true size.

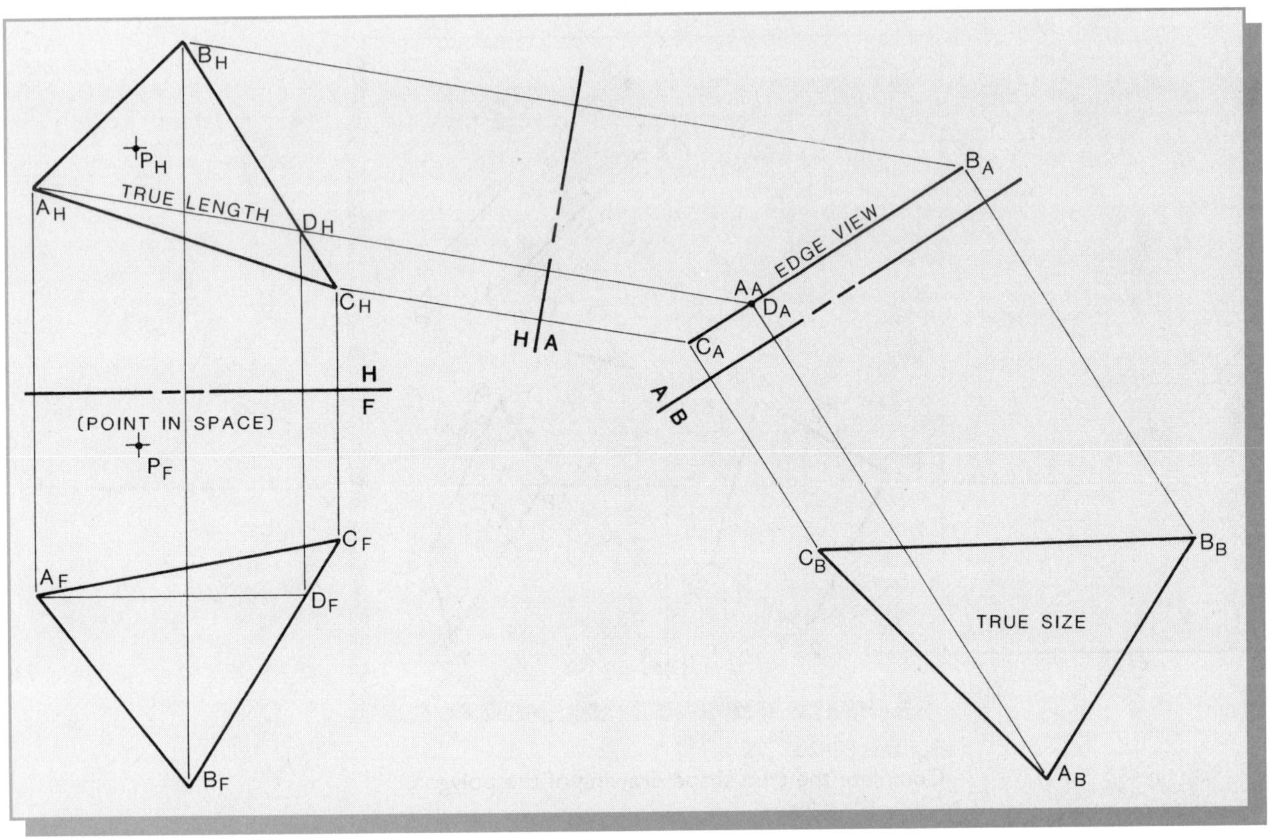

Figure 13-21
1. Locate point P in views A and B.
2. Find the shortest distance of point P from the surface.

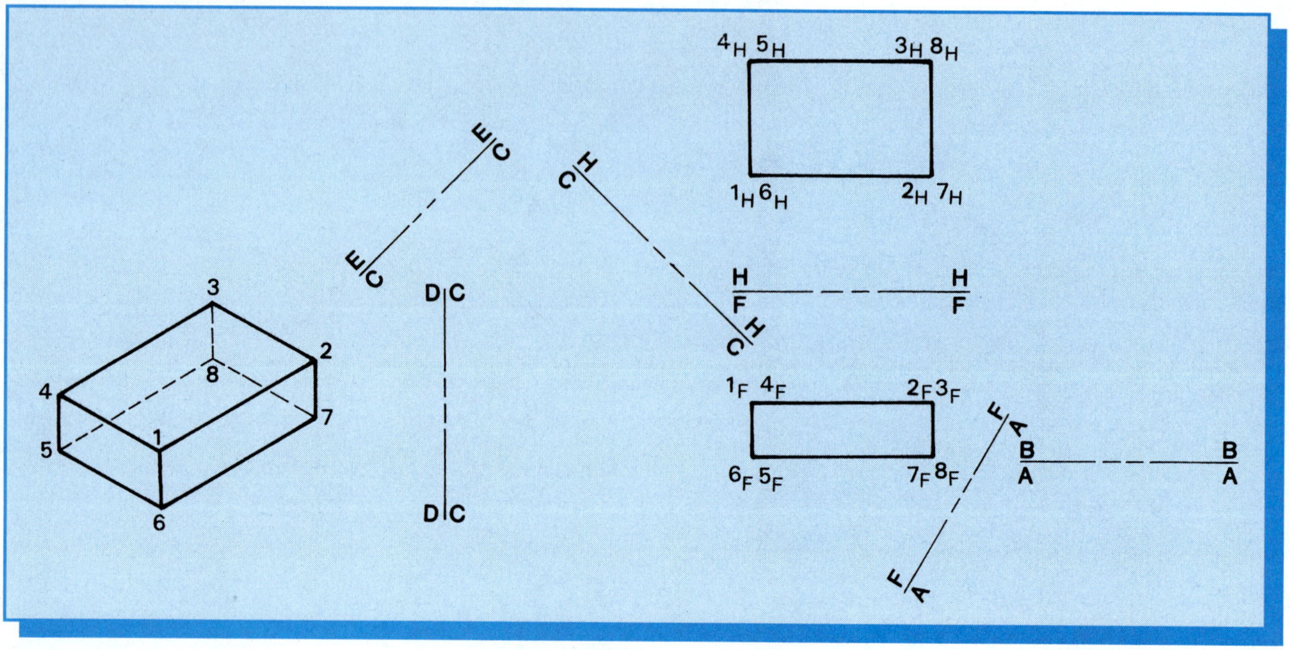

Figure 13-22
Double the size of this exercise (dividers or photocopying) and complete all the auxiliary planes A, B, C, and E.

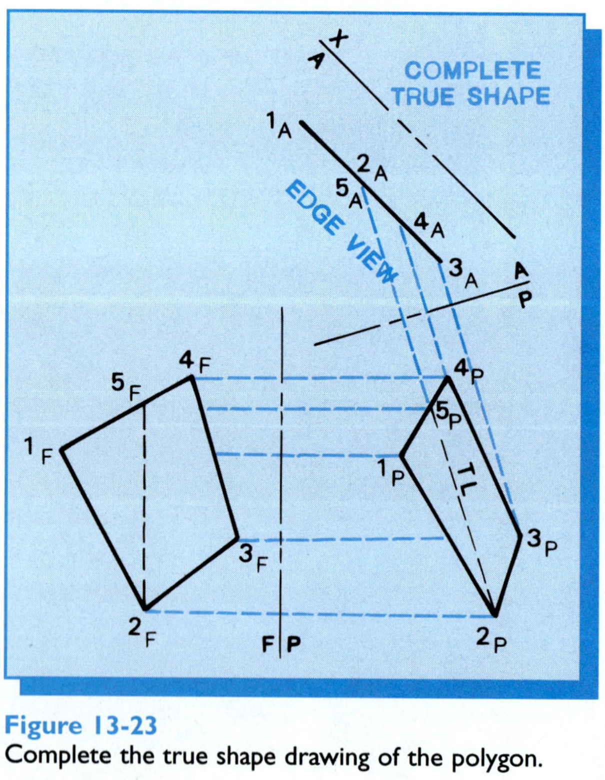

Figure 13-23
Complete the true shape drawing of the polygon.

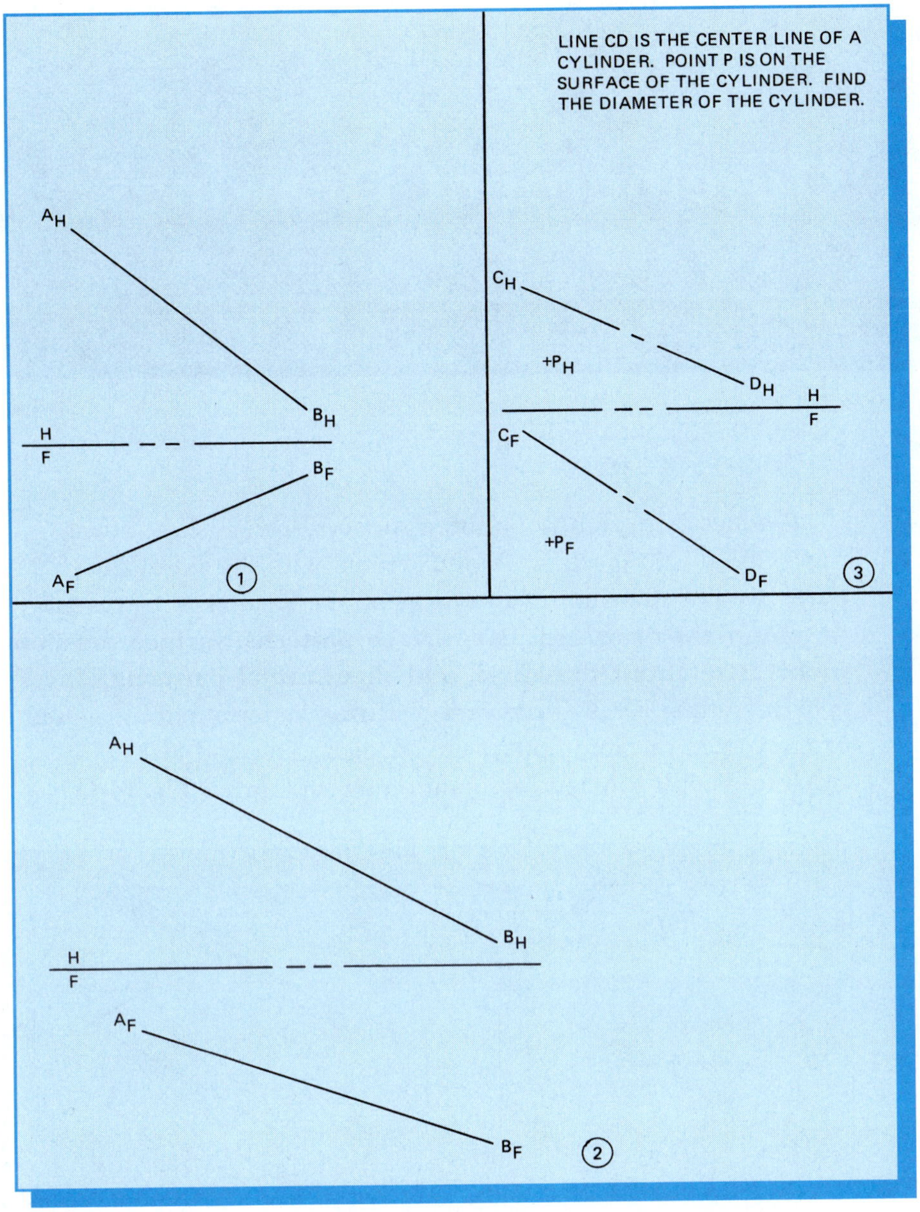

Figure 13-24
Find the point views for exercises 1 and 2. Find the diameter of the cylinder in exercise 3. (Redraw or photocopy this page.)

Key Terms

- Auxiliary plane
- Curved line
- Descriptive geometry
- Edge view
- Fold line
- Frontal plane
- Horizontal plane
- Point in space
- Point view
- Profile plane
- Triangle
- True length
- True size

Chapter 14

Development Drawings

Objectives

The student will be able to:

- recognize the basic geometric shapes
- find true length lines
- draw patterns using the parallel line method
- draw patterns using the radial line method
- draw patterns using the triangulation method
- draw patterns using a CAD system

Introduction

Products with thin, continuous surfaces, such as sheet metal, paper, and plastic sheets, are manufactured with the aid of flat **patterns**. The drawings used to develop these patterns are called **development drawings, flat surface patterns, surface developments, stretchout drawings, and sheet metal patterns.** This field of drafting is called **pattern drafting**. Pattern drafting is used extensively in the automotive, aircraft, packaging, and HVAC (Heating, Ventilation and Airconditioning) industries **(Fig. 14-1)**.

Figure 14-1
Many products are manufactured from pattern developments.

Although finished products are three-dimensional, development drawings (patterns) are drawn on a two-dimensional surface. All dimensions on a pattern drawing must be true size, since the pattern is used as a template to cut material to an exact size and shape. The material, once cut in the shape of the pattern, is formed into a three-dimensional shape and fastened into its final form. **Figure 14-2** shows a container box pattern outlined as a flat surface. **Figure 14-3** shows the container material cut to the outline of the pattern. **Figure 14-4** shows how the material is folded into a three-dimensional form and fastened into position **(Fig. 14-5)**.

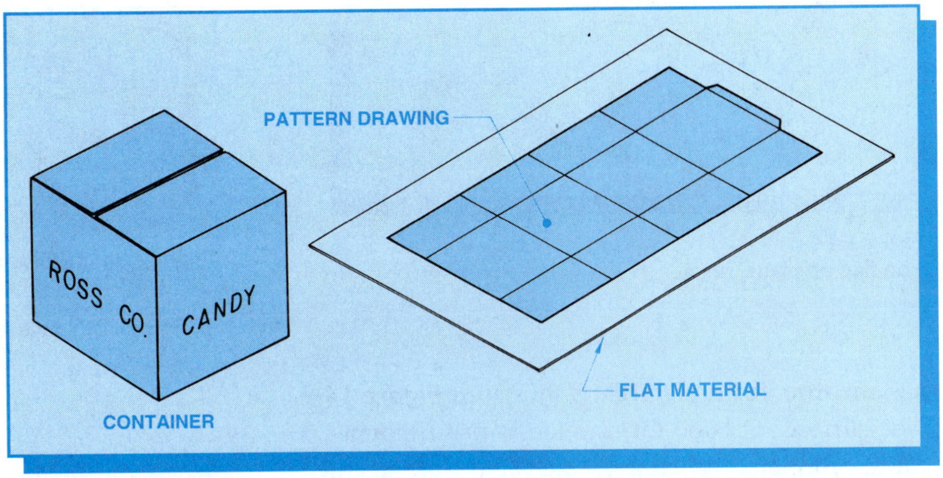

Figure 14-2
A container box and its pattern on a flat material. All dimensions are true size on a pattern drawing.

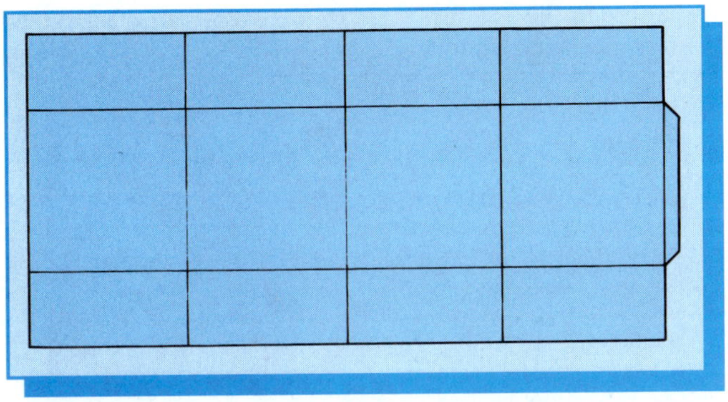

Figure 14-3
Cut-out pattern

Surface Forms

Surface planes, for which development drawings are prepared, include flat, curved, warped, regular solid, and basic geometric forms. Plane (flat) surfaces are simple to lay out because there are no inclined or oblique surfaces to project **(Fig. 14-6)**.

Surfaces with a single curve include cylinders and cones **(Fig. 14-7)**. The development of pattern drawings for **single-curve surfaces** is also simple, since all lines are either parallel or radiate from a common apex. **Warped surfaces** are curved in two directions, which makes pattern layout more complex. Warped surfaces include spheres, paraboloid, and hyperboloid forms **(Fig. 14-8)**. The surfaces of regular solids, such as the tetrahedron, hexahedron, octahedron, dodecahedron, and

Figure 14-4
Pattern is formed by folding.

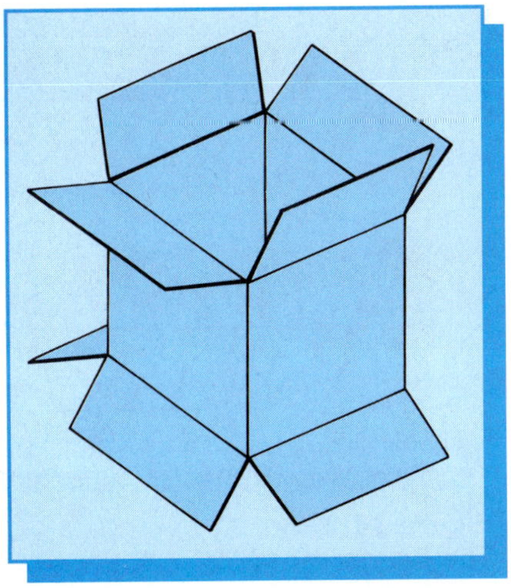

Figure 14-5
Fasten together by taping or gluing.

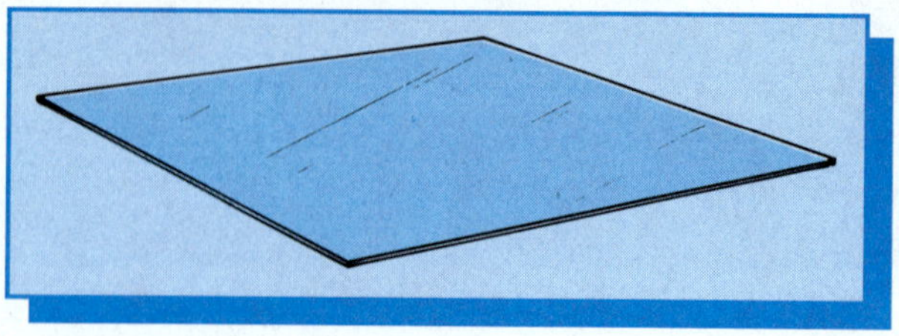

Figure 14-6
The flat or plane surface

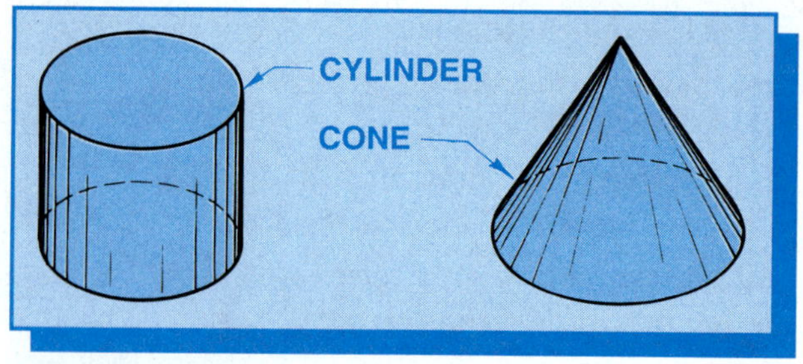

Figure 14-7
Examples of single-curve surfaces

icosahedron, form patterns as shown in **Figure 14-9**. The surfaces of basic forms, such as prisms, pyramids, cylinders, and cones, are created through the development of pattern shapes as shown in **Figure 14-10**.

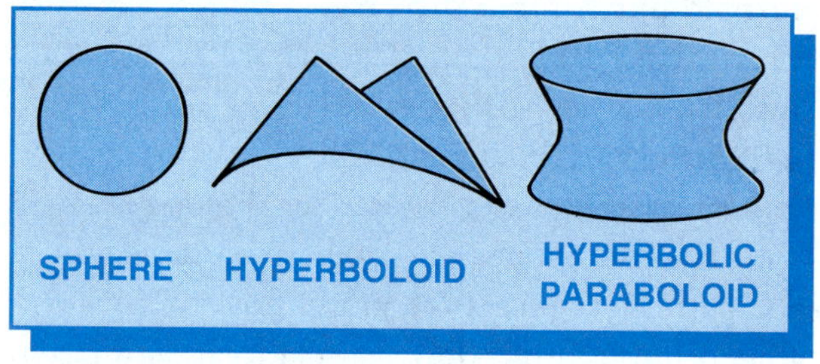

Figure 14-8
Examples of warped (double-curve) surfaces

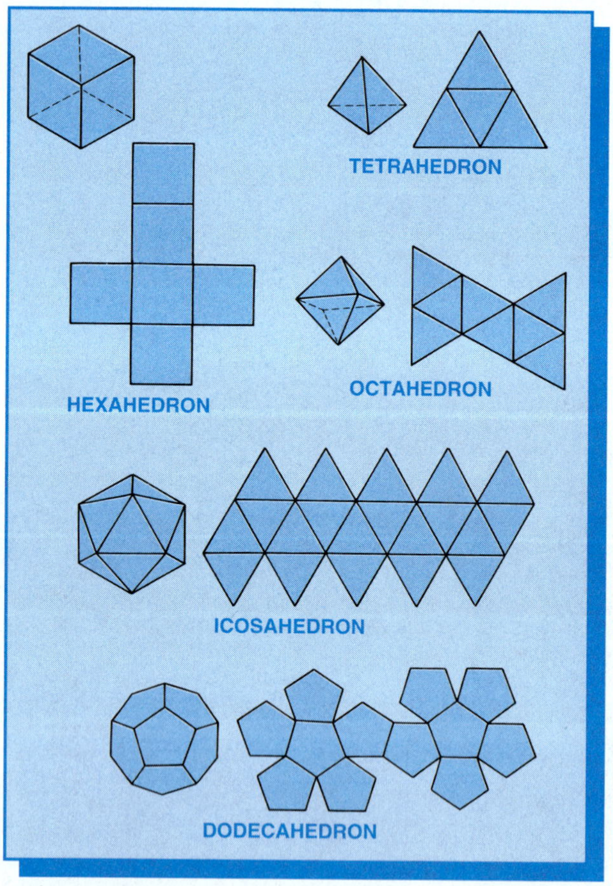

Figure 14-9
The five regular solids

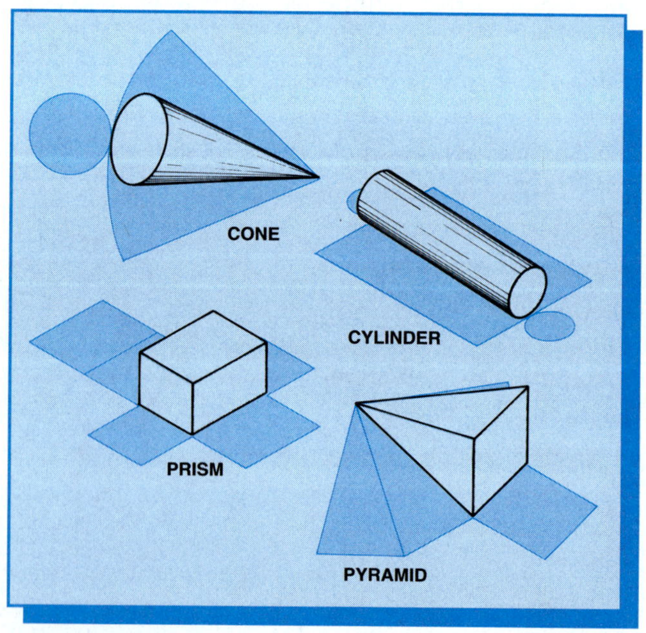

Figure 14-10
Basic forms and their patterns

Pattern Drawing Terminology

Pattern drawings are used as the guide for cutting the surface outline of an object from a sheet. Pattern drawings are also used to show the location of lines for bending, folding, or rolling flat forms into three-dimensional shapes. In addition, they are used for outlining the position of seams and hems for fastening. This is all accomplished through **stretchout lines**, **fold (bend) lines**, and **tab lines (Fig. 14-11)**. Tabs are used to form hems and seams. **Hems** are formed along the edges of sheet metal pieces to seal in the sharp edges. **Seams** are used to connect two or more pieces of material together **(Fig. 14-12)**.

In addition to the basic outline of the pattern, further material must be added to provide a base for permanent connections. The type of connection depends on the size and type of material, plus the fastening method. Common methods for fastening flat material include soldering, welding, riveting, gluing, stapling, and sewing.

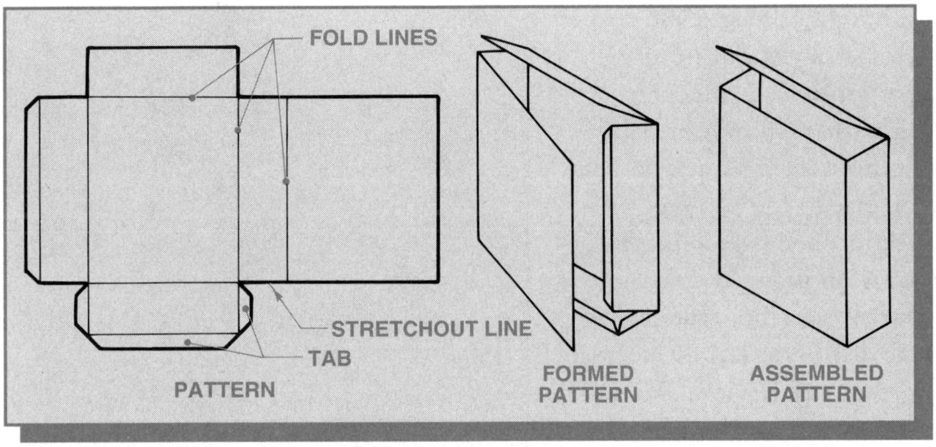

Figure 14-11
Terminology for a pattern drawing

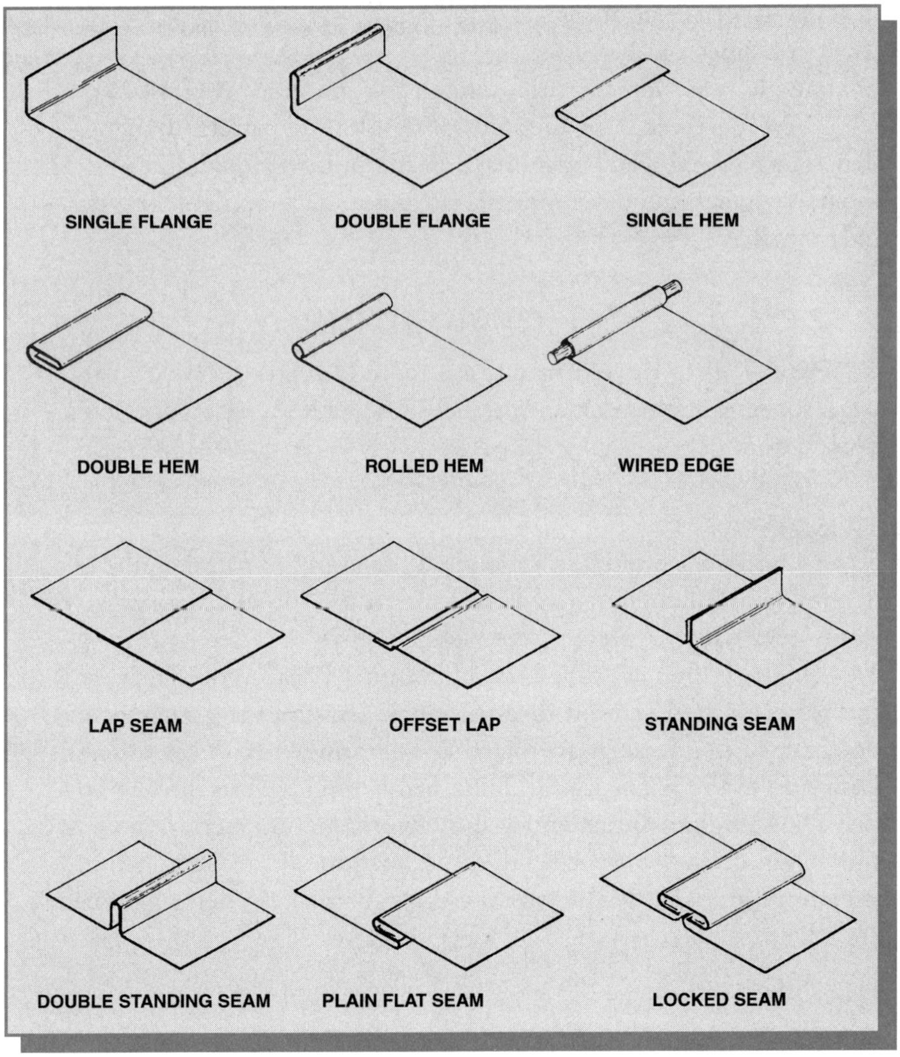

Figure 14-12
Various types of seams and hems *(Reprinted from* Technical Drawing and Design *by Goetsch and Nelson, © Delmar Publishers Inc.)*

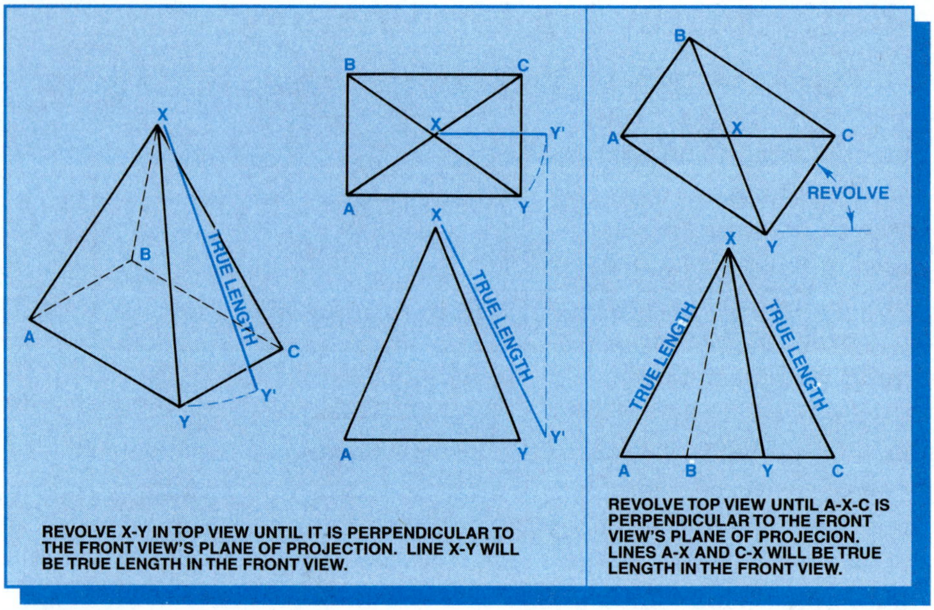

Figure 14-13
Two procedures used to find the true length of a line

All lines on a pattern drawing must be true size; otherwise the size and shape of the object cut from the pattern will be incorrect. All lines that fall on normal orthographic planes, perpendicular to the plane of projection, are **true length**. All lines on inclined or oblique surfaces are not true length (**foreshortened**) on normal orthographic views. The true length of these foreshortened lines must be determined before they can be used in a pattern drawing. The true length of foreshortened lines can be obtained by revolvement, using one of two methods **(Fig. 14-13)**.

There are three basic methods of developing pattern drawings depending on the geometric shape of the finished form. These include parallel line development, radial line development, and triangulation.

Parallel Line Development

Parallel line development is a method used to develop patterns for objects that contain parallel lines, such as prisms and cylinders.

Prisms

All lines in a multiview drawing of a prism are true length. Therefore, parallel lines are extended from the top and bottom of the front view, which represents the true height of the pattern **(Fig. 14-14)**. These lines are called stretchout lines because they represent the pattern's stretched out form. The true length of the pattern is determined by measuring, projecting, or transferring the width of each side to the pattern. A fold line marks the location of each corner. The fold lines are extended upward using construction lines, and the height of each corner is projected from the front view to the stretchout lines. Lastly, the tabs necessary for making hems and seams are added.

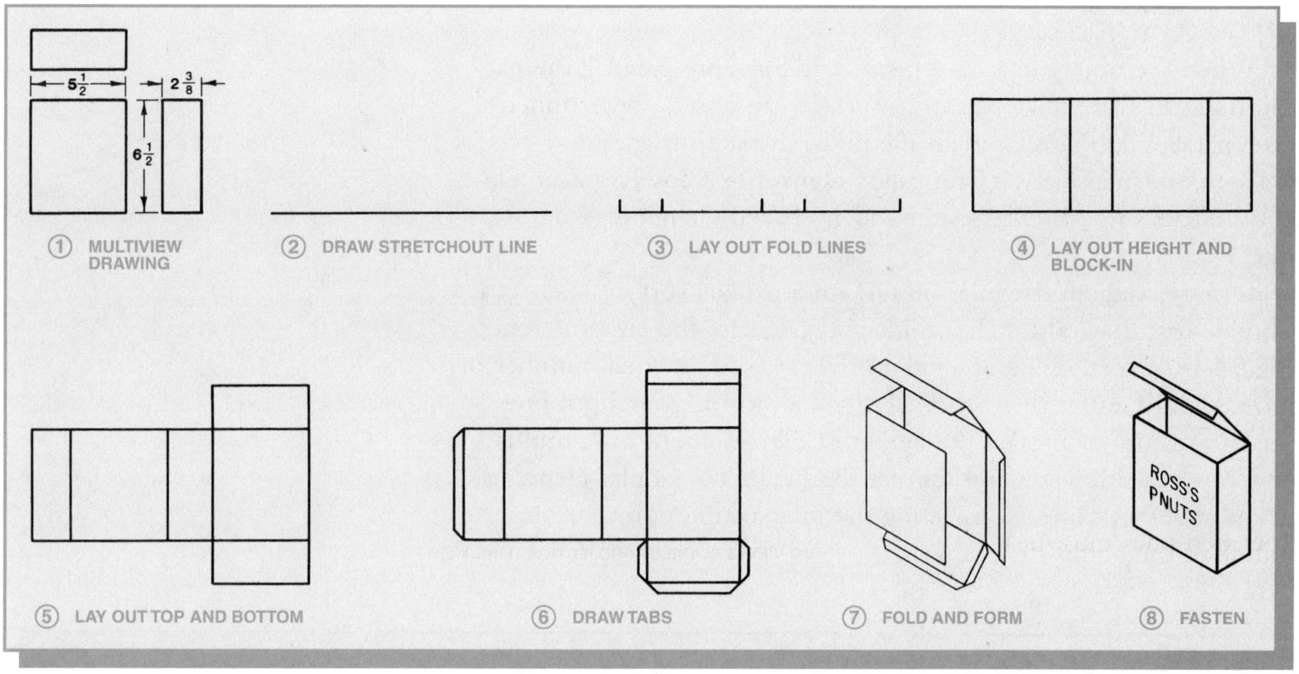

Figure 14-14
Steps for parallel line development of a rectangular prism

Cylinders

In developing a pattern for a cylinder, parallel stretchout lines are projected from the top and bottom of the front view **(Fig. 14-15)**. The length of the stretchout line is equal to the circumference of the cylinder. The circumference is determined by multiplying the diameter by 3.1416 (π), or by stepping off the **chordal distances** around the circumference of the top view.

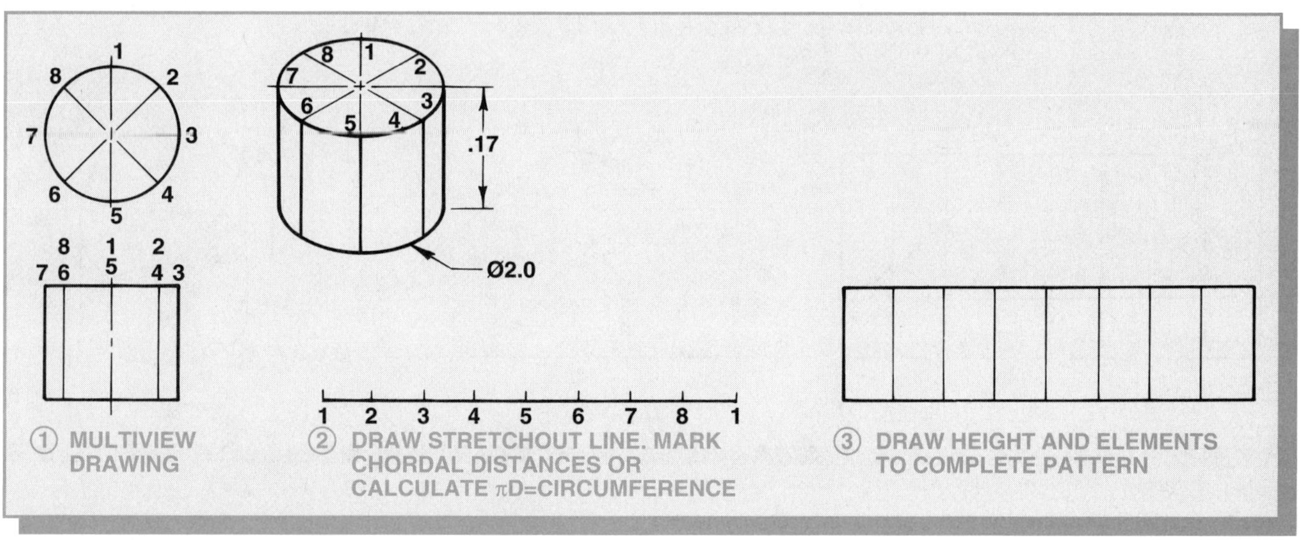

Figure 14-15
Steps for parallel line development of a cylinder

Curved Surfaces

When the object lines of a geometric form are parallel, the pattern outline of irregular cuts in its surface can also be determined using parallel line development methods. Curved surfaces may consist of a series of small flat surfaces called **elements**. Closely spaced elements create smooth curves, and widely spaced elements create flat-sided curves.

In projecting an irregular outline on a parallel surface, such as a cylinder, first draw a stretchout line and calculate the circumference **(Fig. 14-16)**. Next, divide the circumference into an equal number of **chords**, and transfer these chordal distances to the stretchout line. Finally, measure or project the height of each element on a multiview front or side view, and transfer the heights to similar elements on the stretchout line. Connecting the intersecting points creates the outline of the completed pattern.

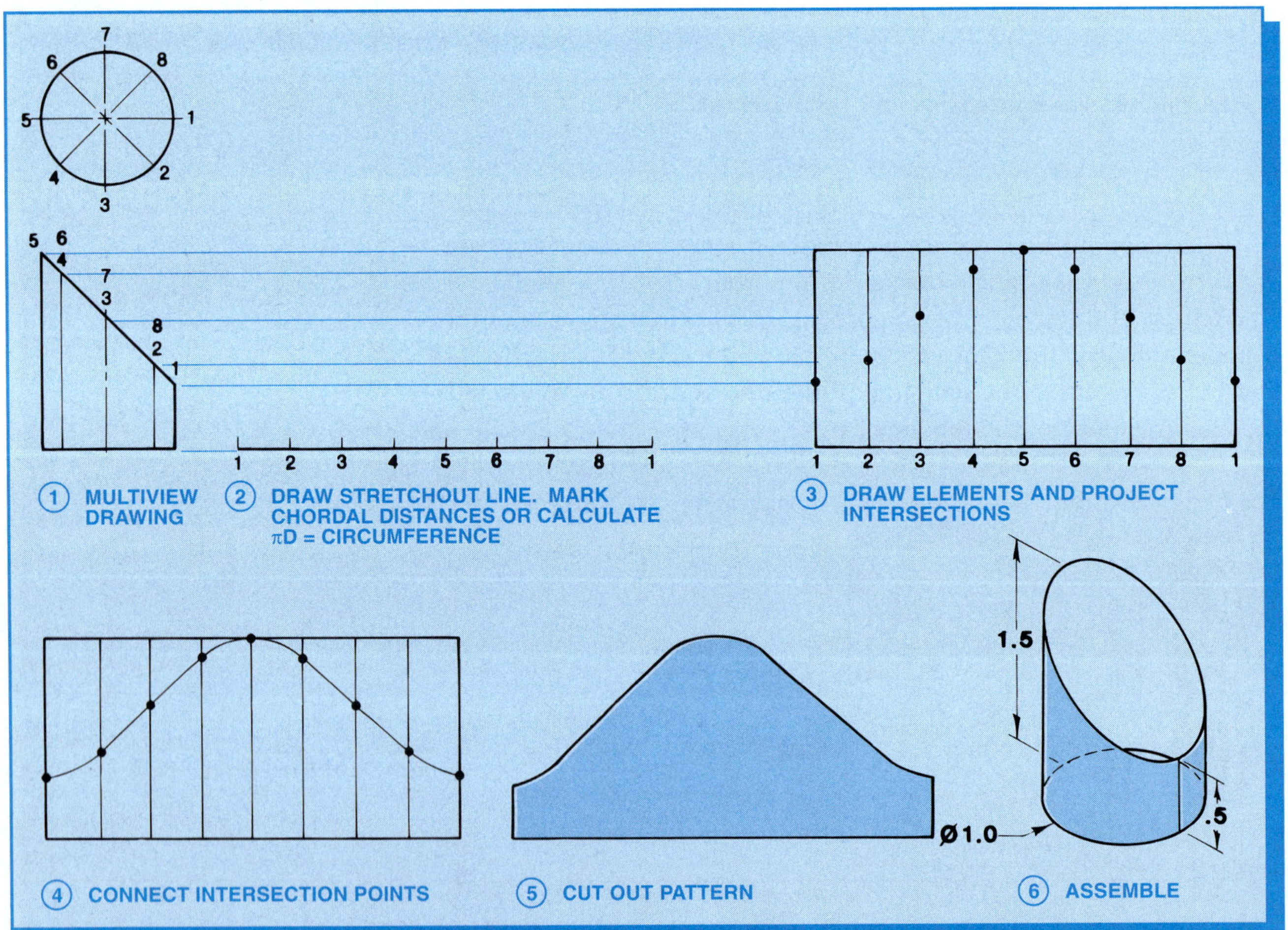

Figure 14-16
Steps for parallel line development of a truncated cylinder

Radial Line Development

Geometric forms, such as pyramids and cones, do not contain parallel lines. These forms contain lines that radiate to a common apex and are developed through the use of **radial lines**. Since the inclined vertical edges of cones and pyramids are not parallel, stretchout lines are not parallel. In radial line developments, stretchout lines are arcs, and element lines become triangles that radiate from an apex to the **stretchout arc (Fig. 14-17)**.

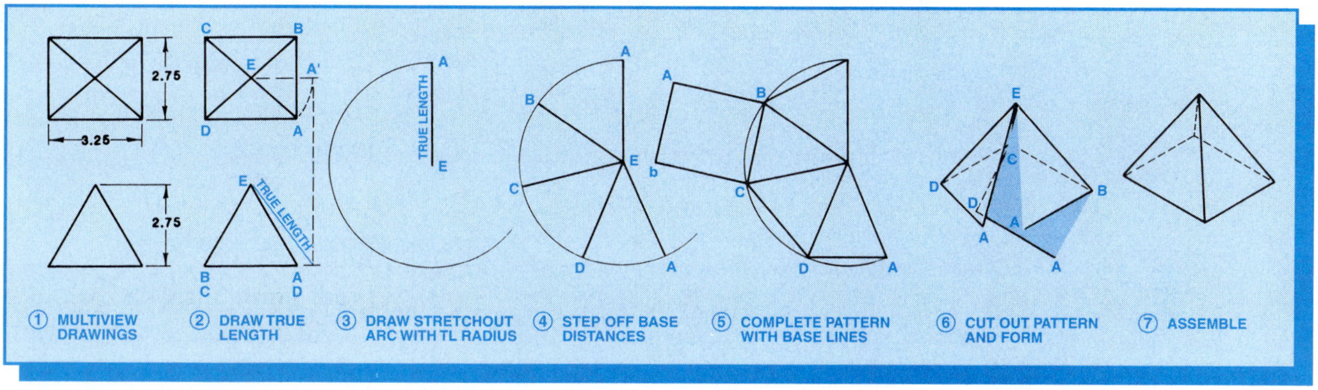

Figure 14-17
Steps for radial line development of a rectangular pyramid

Pyramids

The orthographic top view of a pyramid reveals true width and length measurements. The orthographic front or side view contains receding object lines that are not true length, depending on the orientation of the view. To find the true length of a pyramid's side, the pyramid's corners must be revolved to convert inclined lines into normal orthographic true height lines. This is shown in **Figure 14-17** by rotating corner A to position A', aligning corner A-E on a normal orthographic plane. This true length line (A'-E) is the true length of the radius between the apex of the stretchout line. Once the stretchout arc is drawn, the base distances (which are true length in the top view), are located on the arc and connected to the apex. The pattern is then completed with the addition of the base.

Truncated and Triangular Pyramids

To develop a pattern for a truncated (cut) pyramid, follow the first five steps demonstrated in **Figure 14-17**. Then determine the true length of the lines connecting the base corners with the corners of the cutoff by revolving a corner **(Fig. 14-18)**. This involves finding the true length of lines H-C, G-D, F-E, and I-B. These lengths are projected to the true length line A-H', then transferred to finish the outline of the truncated portion of the pyramid.

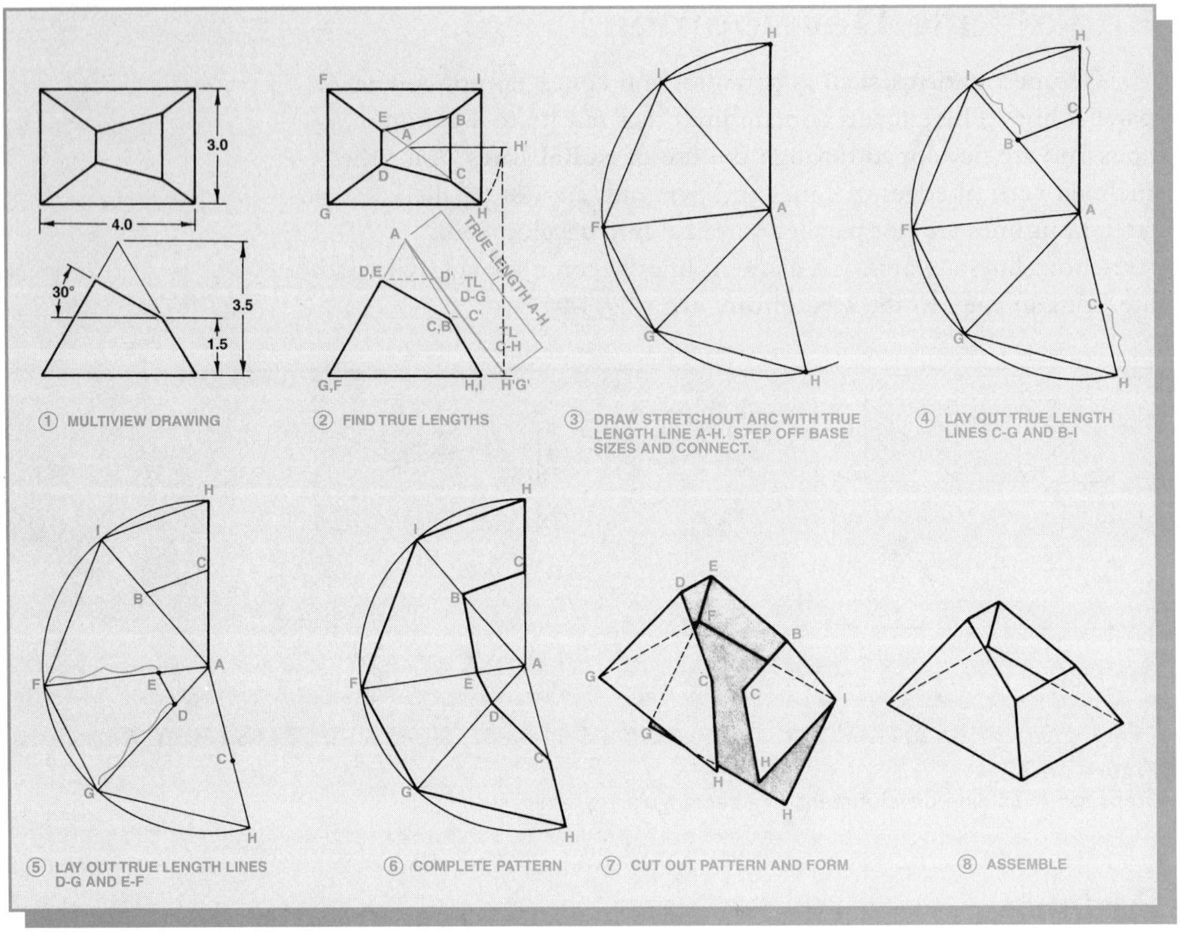

Figure 14-18
Steps for radial line development of a truncated pyramid

The development of patterns for triangular pyramids is similar to square pyramids. The exception is that only three base sides are projected on the stretchout line **(Fig. 14-19)** after the true length for the stretchout arc is found.

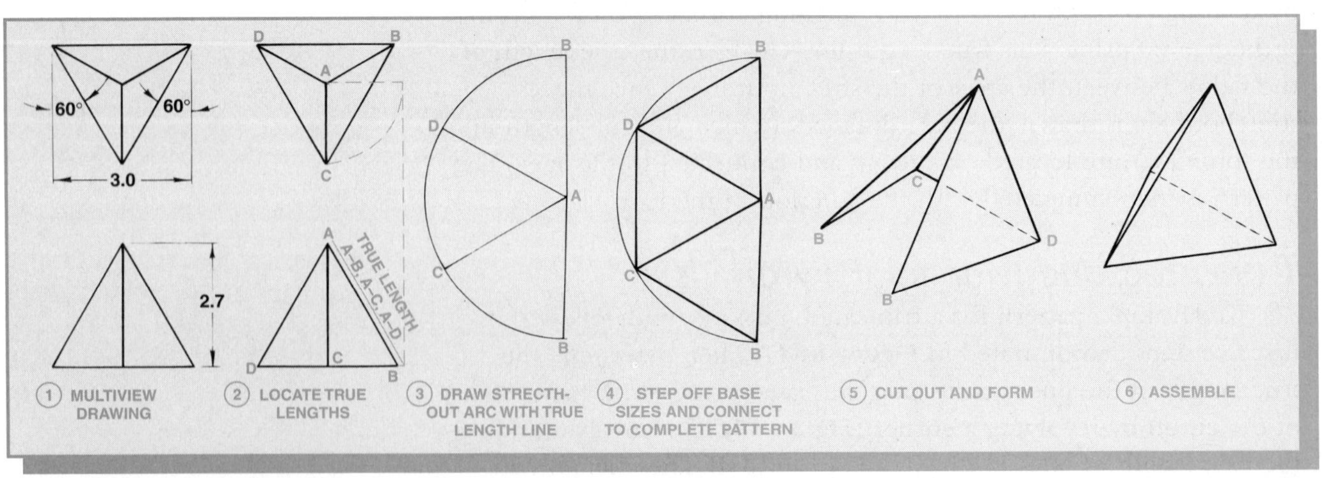

Figure 14-19
Steps for radial line development of a triangular pyramid

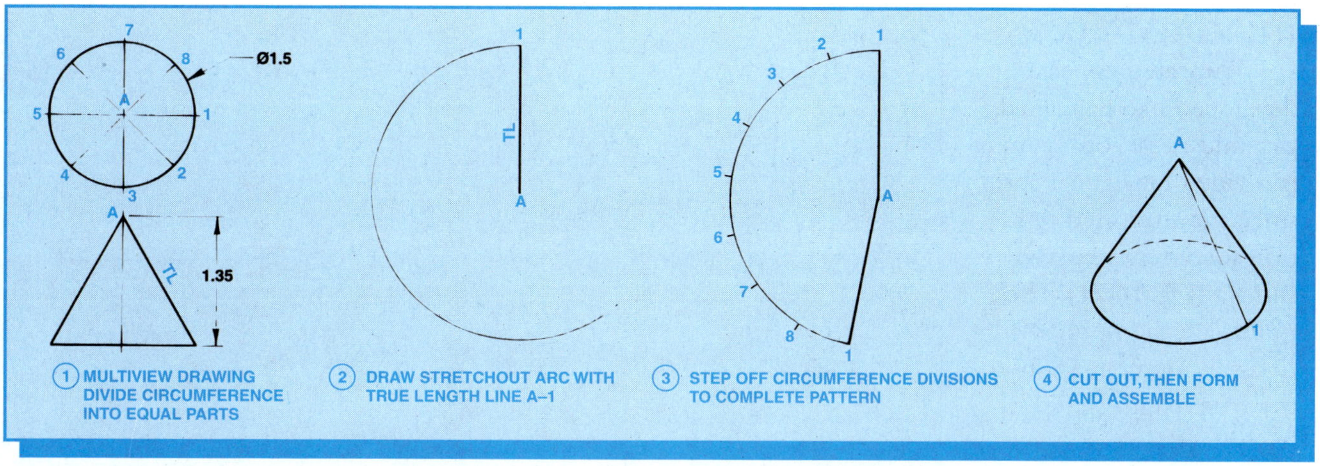

Figure 14-20a
Steps for radial line development of a cone

Cones

Unlike pyramids, the elements of a cone are all equal in length from base to apex. The side and front view of a cone are identical, and contain true length object lines **(Fig. 14-20a)**. The radial stretchout arc for the development of a cone pattern is equal to the distance from the base to the apex, as found on the side view. The top view of a cone reveals the true circumference, but the element lines are foreshortened.

To complete a pattern drawing of a cone, draw an arc with a radius equal to the distance from the base to the true length apex **(Fig. 14-20a)**. Divide the circumference of the base into evenly spaced distances, and transfer these distances to the radial stretchout line. These points are then connected to the apex, forming triangular elements. There are eight elements in **Figure 14-20a**. More elements will create a smoother and more accurate cone.

A simple, symmetrically shaped cone, without cuts or intersections, can also be plotted mathematically by determining the angle of the stretchout and the circumference of the base. The angle of the stretchout is determined by multiplying the base diameter by 180, divided by the true length of the side. The length of the stretchout line is determined by multiplying the base diameter by 3.1416 (π) **(Fig. 14-20b)**.

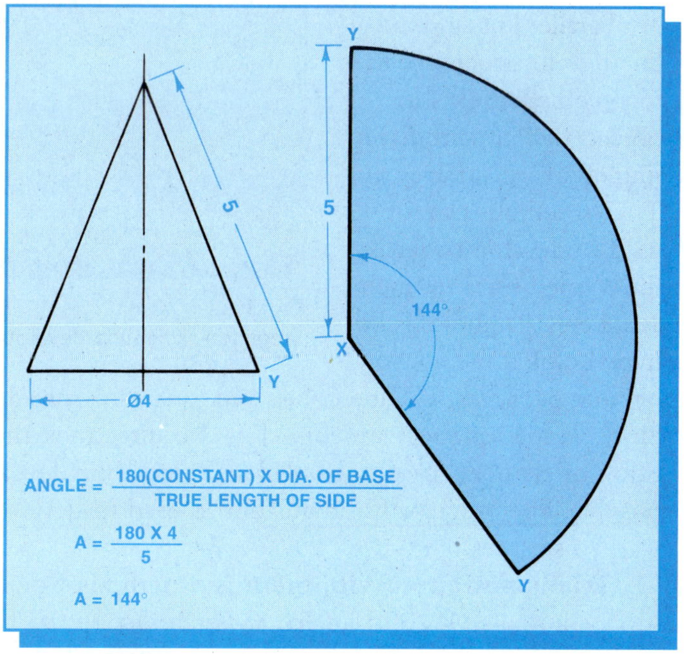

Figure 14-20b
Computing the stretchout angle for the development of a cone's pattern

Truncated Cones

Truncated cones are developed like truncated pyramids. The top opening is divided into equal parts, since the truncated portion intersects a curved surface **(Fig. 14-21)**. These points are then projected to a side element to show the true lengths. After a true-length radial stretchout is drawn, each true distance from base to the intersection point is transferred to equivalent points on the stretchout arc. Connecting these points completes the pattern.

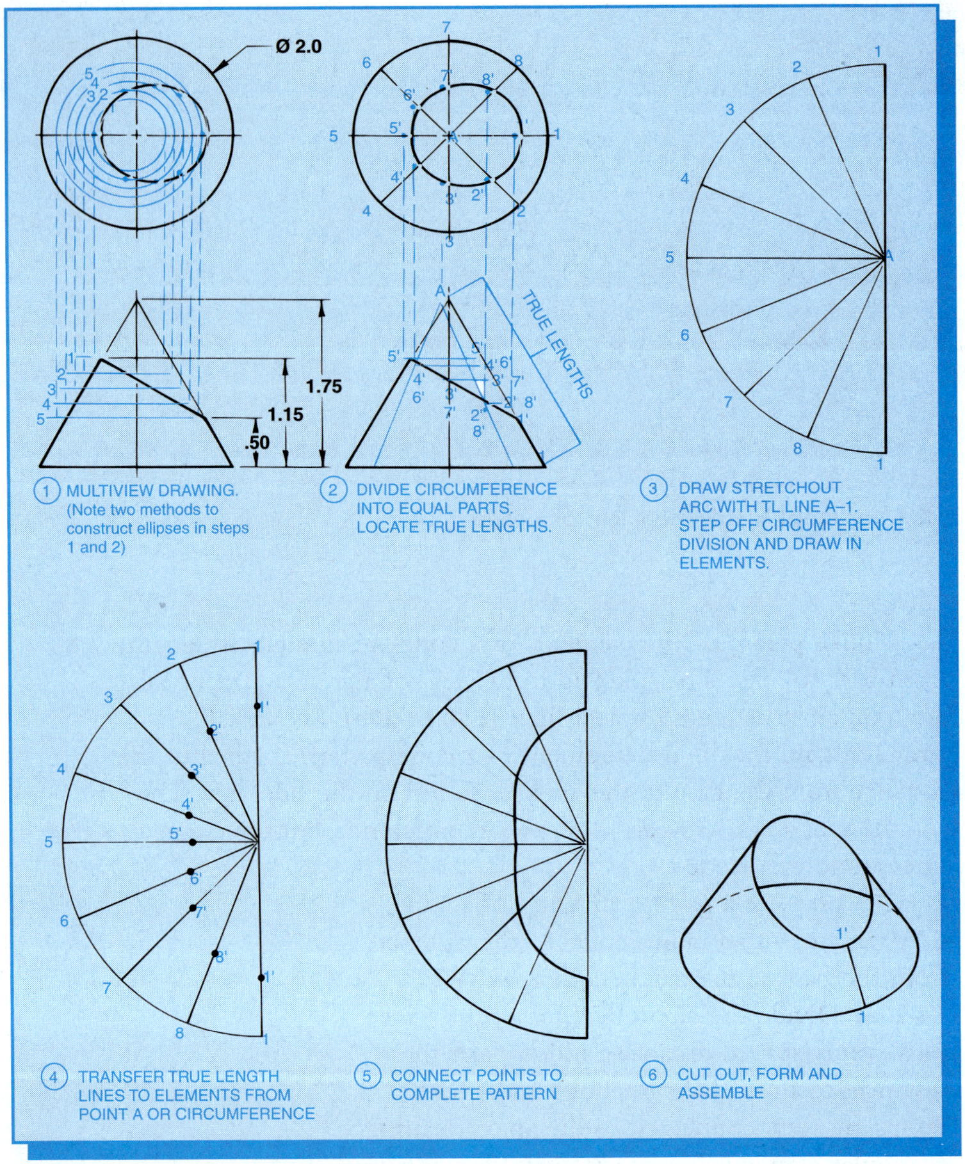

Figure 14-21
Steps for radial line development of a truncated cone

Triangulation Development

Parallel line and radial line developments are used to develop patterns for objects that contain flat or single-curved surfaces. These methods cannot be used to develop patterns for warped surfaces, such as spheres, paraboloids, or hyperboloids, including oblique pyramids, oblique cones, and square-to-round intersections. Since warped surfaces are curved in two directions, the even distribution of elements used in parallel and radial line development is not possible. The process of triangulation is used to develop patterns for warped surfaces.

Triangulation development is a method of dividing a warped surface into a series of triangles, and transferring the true size of each triangle to a flat pattern. When flat triangulated patterns are bent or folded into a three-dimensional form, the desired warped form is created. Triangulation is an approximate method of pattern development. The use of more and smaller triangles creates a closer approximation of the desired form than a few large triangles.

The process of triangulation involves dividing the top and front views of an object into a series of triangles. This is done by dividing the top and bottom outlines into a series of parts and connecting them with element lines on both views. The intersections of element lines are evenly spaced on the top view. They converge at common points on the bottom view, creating a succession of triangles **(Fig. 14-22)**. Next, true lengths are determined for each element line. These distances are transferred progressively to the stretchout. In the development of triangulated patterns, one triangle at a time is added to the stretchout. Each element line is connected with the chordal divisions laid out on the circumference of the cone's base in the top view.

Figure 14-22
Steps for triangulation development of an oblique cone

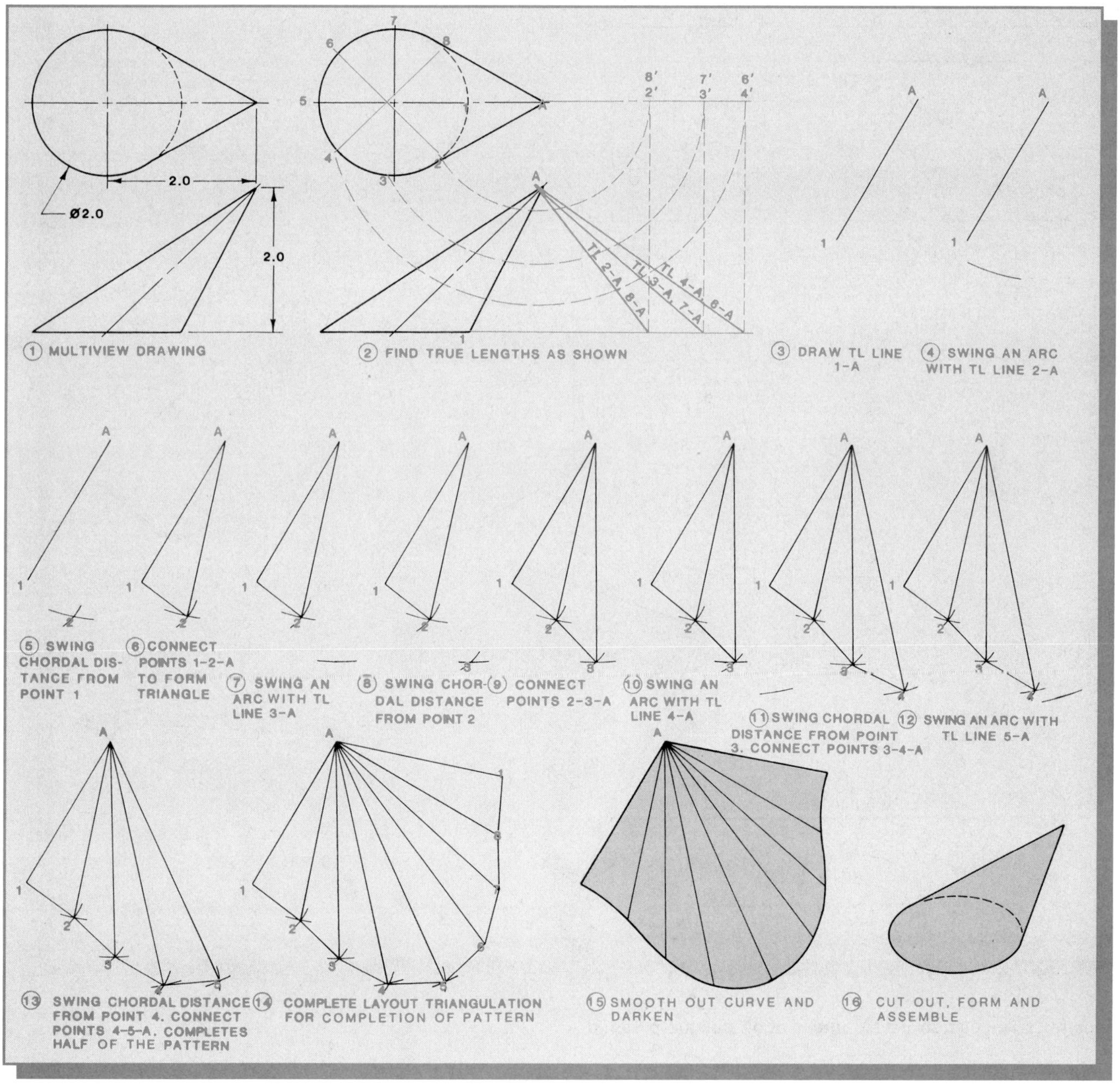

Triangulation can also be used to develop flat or single-curved surfaces. Since the process is very time consuming, triangulation is normally used only where necessary. **Figure 14-22** shows the development of an oblique cone pattern through triangulation methods. **Figure 14-23** shows the triangulation of an oblique pyramid pattern.

When flat surfaces meet and the intersection lines are straight, the related pattern can be developed using parallel line development.

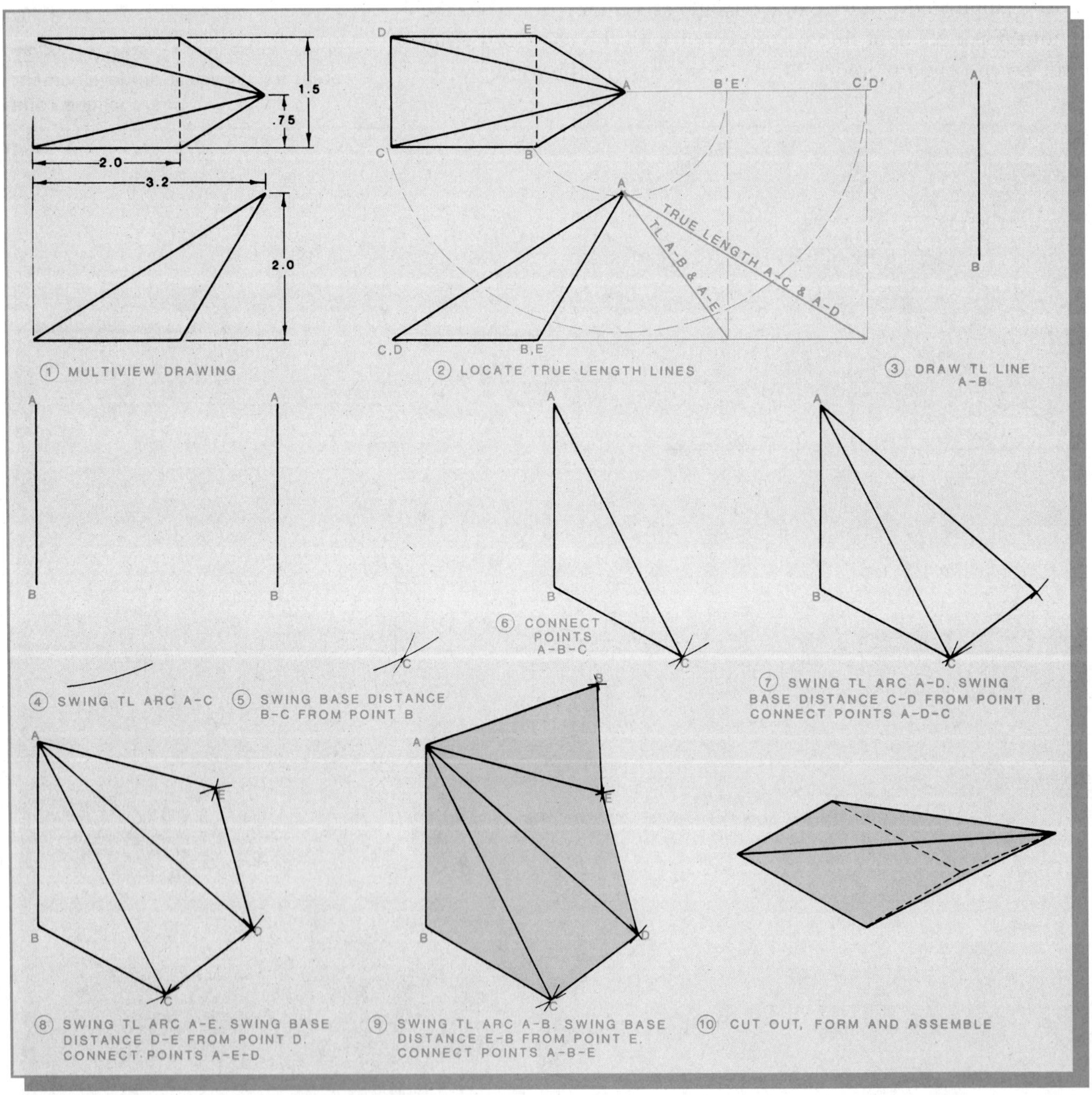

Figure 14-23
Steps for triangulation development of an oblique pyramid

When single-curved surfaces meet, the intersection lines can be developed using radial line development methods. However, when warped surfaces meet, or when different geometric forms intersect at varying angles, triangulation must be used to develop the pattern for both parts. **Figure 14-24** shows the development of a pattern drawing for a square form intersecting a round form. This same procedure can be used to develop patterns for any intersecting geometric forms.

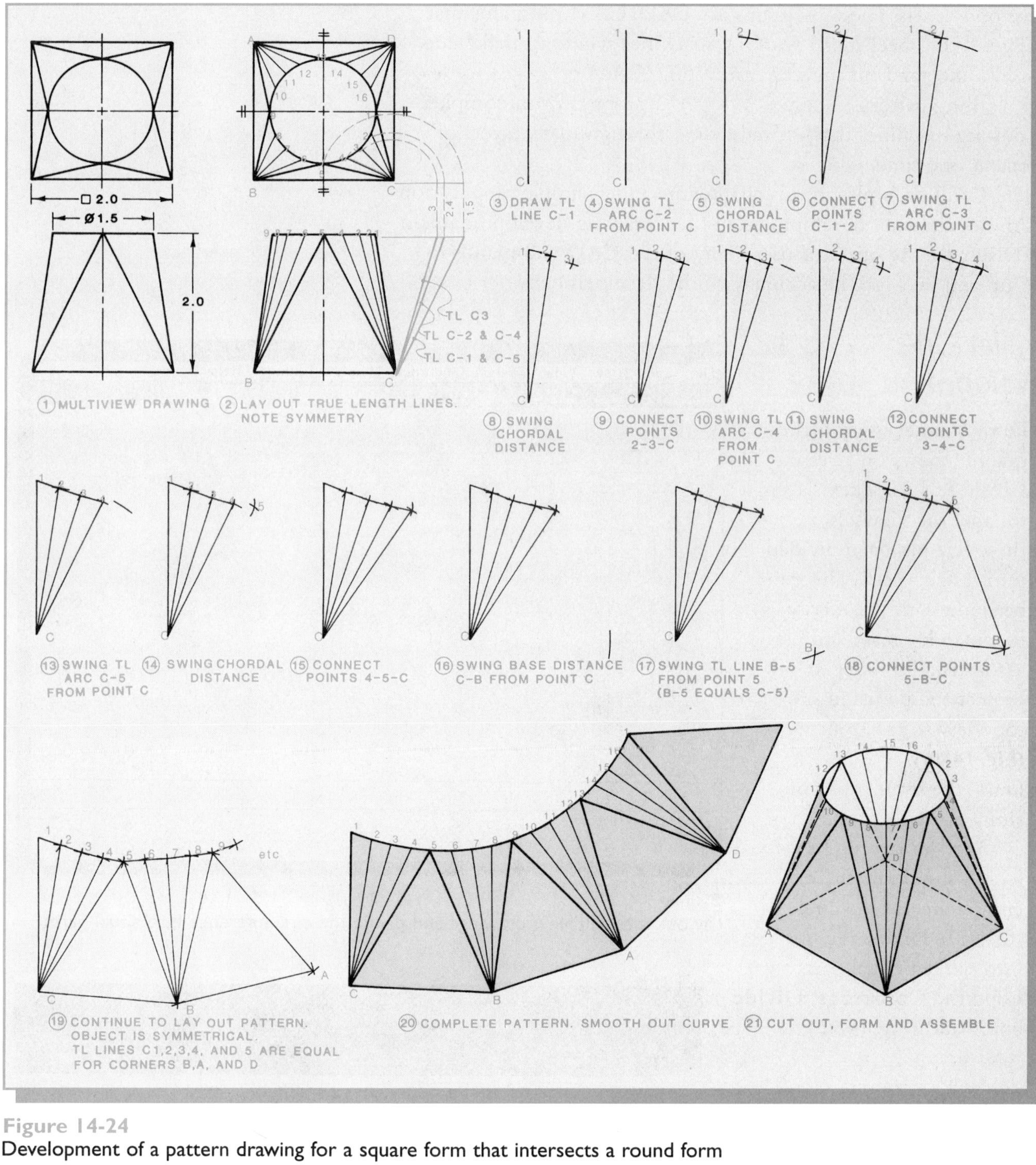

Figure 14-24
Development of a pattern drawing for a square form that intersects a round form

AutoCAD LT Exercises

The application of CAD to draw developments is widespread. The automotive industry uses special CAD software to develop automotive body parts. These programs are CAD/CAM programs that actually tell the sheet metal working machines where to make cuts and folds. The garment industry uses CAD/CAM programs to develop clothing patterns. These CAD/CAM programs run complex laser cutting machines that accurately cut through many layers of material at one time.

The CAD examples in this chapter are much simpler applications of CAD for drawing developments. When drawing developments on a CAD system, the operator uses many of the CAD commands to speed up and increase the accuracy of the developments.

Parallel Line Development

1. Draw the necessary orthographic views on layer 0. Create and use a text layer to label the views. Note the circle has been divided into sixteen parts. This can be done with the **Array** command. Create and use a construction layer to project the points in the top view to the front view **(Fig. 14-25)**.
2. Draw the stretchout line along the base of the cylinder on layer 0. Its length should be the circumference of the circle (C = D). Divide the line into sixteen equal parts using the **Construct, Divide** command **(Fig. 14-26)**.

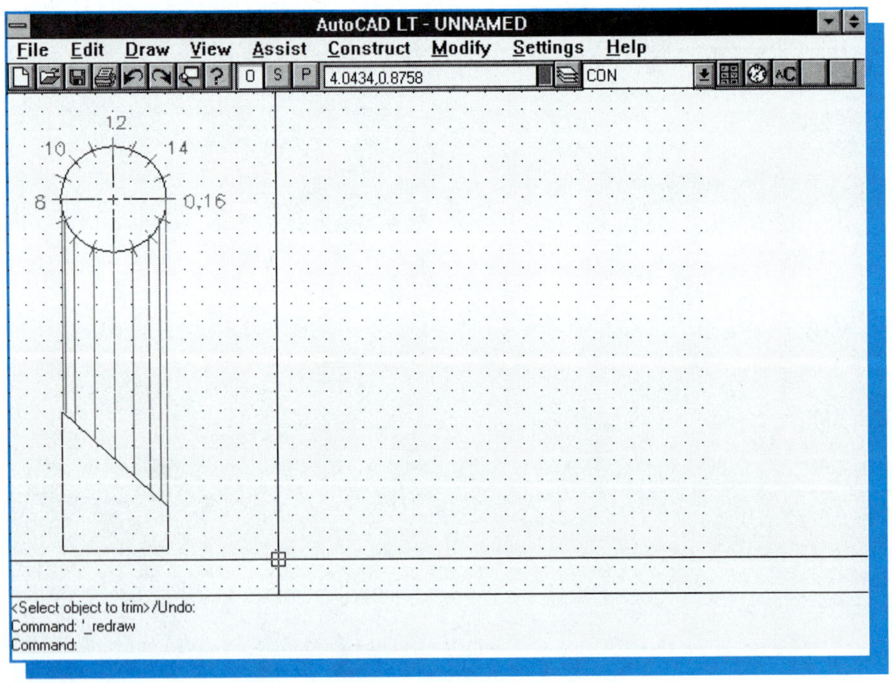

Figure 14-25
Lay out the multiview drawings and divide the circumference into equal parts.

tip: *Use the Point Style under Settings to pick a point style that can be seen.*

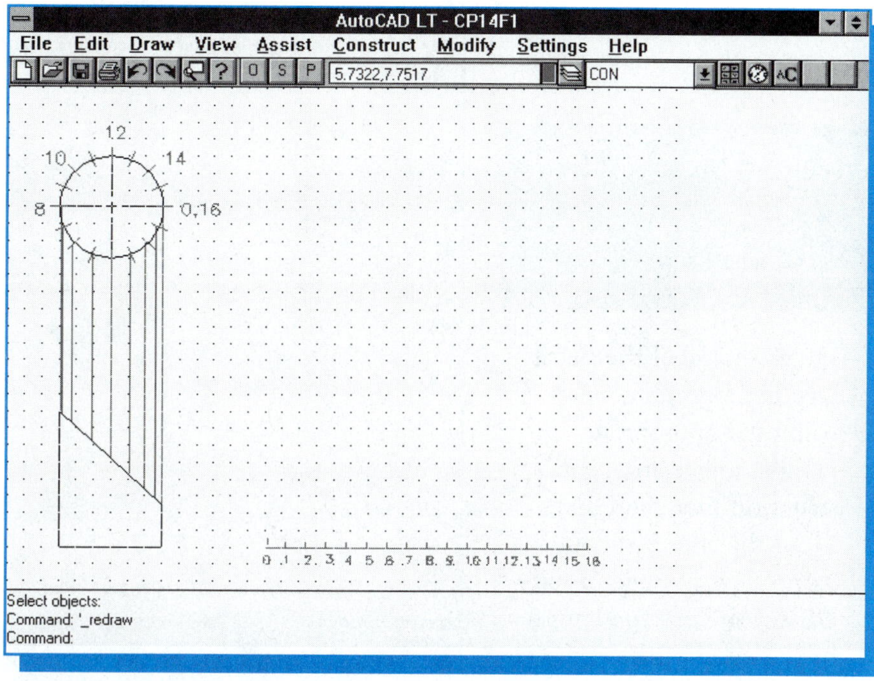

Figure 14-26
Lay out the stretchout line for the circumference's length and divide into sixteen equal parts.

3. Project each segment of the development view using the construction layer.
4. Switch to layer 0 and use the **Draw, Polyline** command to connect the proper intersections.
5. Use **Modify, polyline edit, fit curve** to have the polyline automatically smooth out **(Fig. 14-27)**.
6. Under layers, freeze, the construction layer and the development is clearly shown **(Fig. 14-28)**.

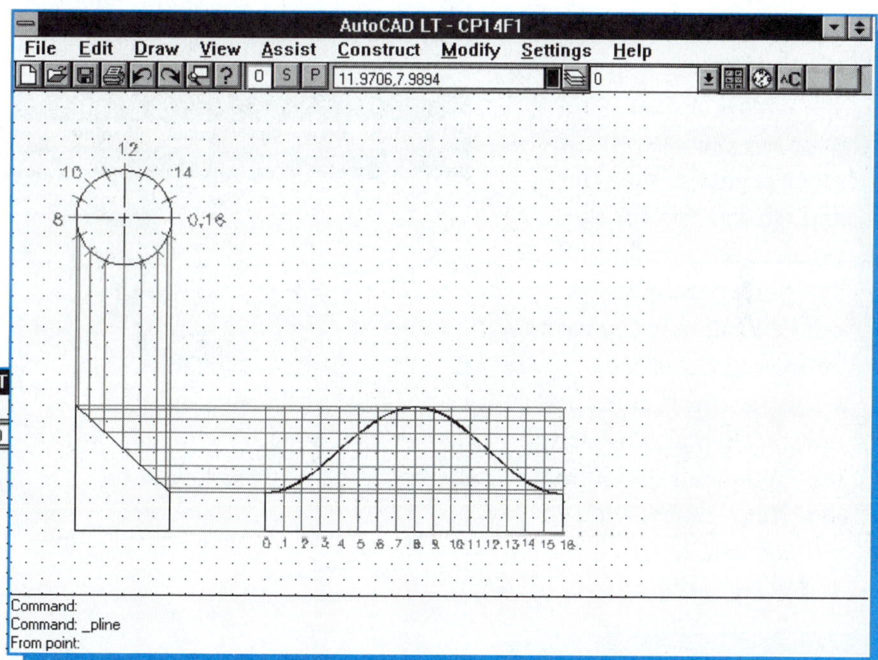

Figure 14-27
Use polyline to connect the intersections of the construction lines.

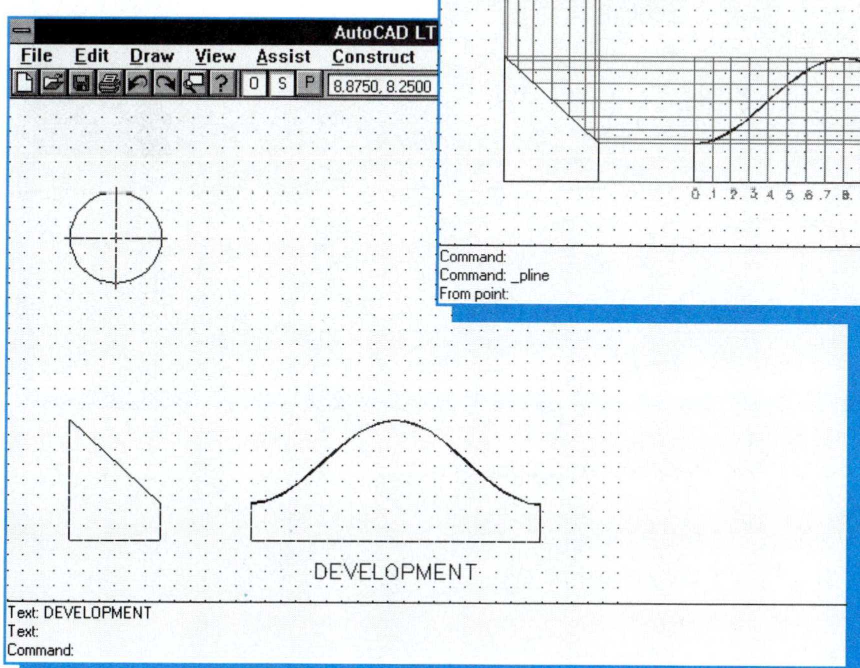

Figure 14-28
Completed pattern—freeze the construction line layers.

Radial Line Development

tip: *It is important that all construction is accurate, so be sure to use object snaps when selecting points.*

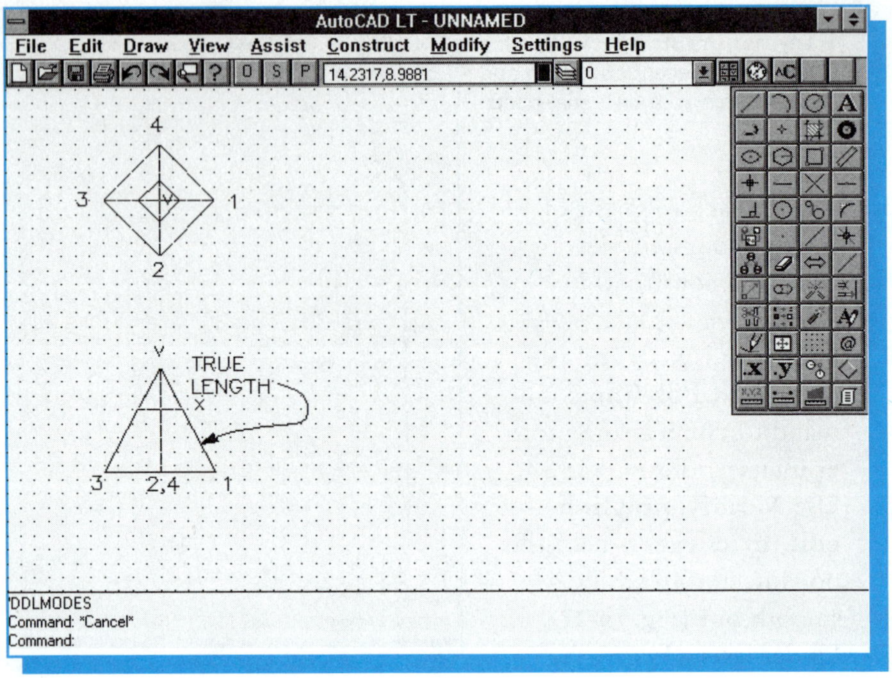

Figure 14-29
Draw and label multiview drawings.

1. Draw and label the necessary orthographic views, using the appropriate layers for object lines, construction lines, and text. Note that the horizontal (top) view has been rotated so that the true length of the edge of the pyramid is shown in the frontal view as true size **(Fig. 14-29)**.
2. Use **Draw**, **Point** command to pick any convenient point on the screen, and label it "V" for the vertex of the pyramid. The location of the point is not critical, as it can easily be moved if needed.
3. Using a construction layer, draw stretch arcs by drawing circles using "V" on the front view as the center and true length V-X and V-1 as radii.

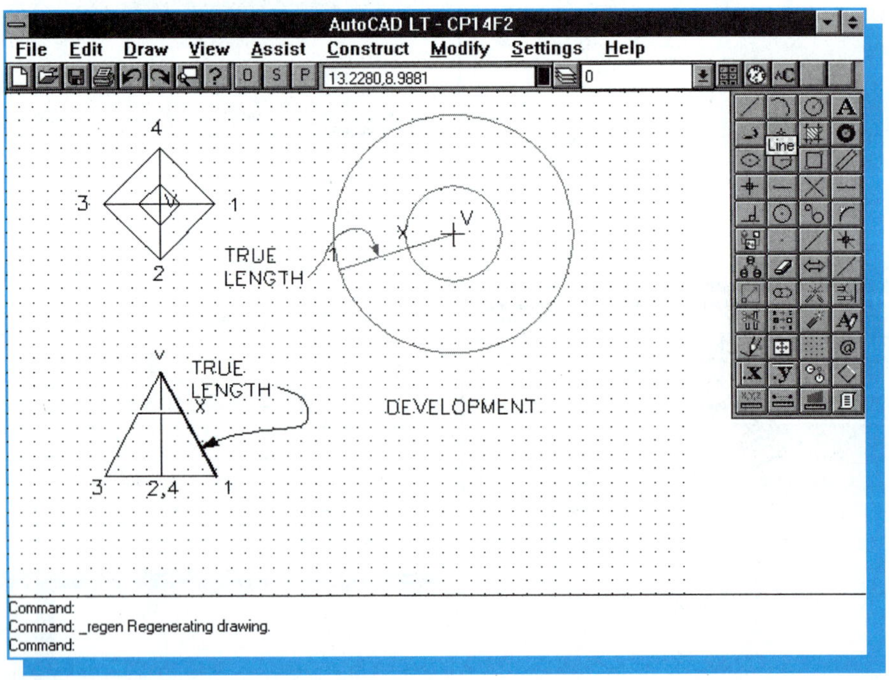

Figure 14-30
Lay out the two stretchout arcs using circles.

tip: *As you draw a development, it may be easier to have several construction layers. This will allow you to freeze the ones that get in the way.*

4. Use the **Modify**, **Move** command to relocate the center of the circles to point "V" **(Fig. 14-30)**.

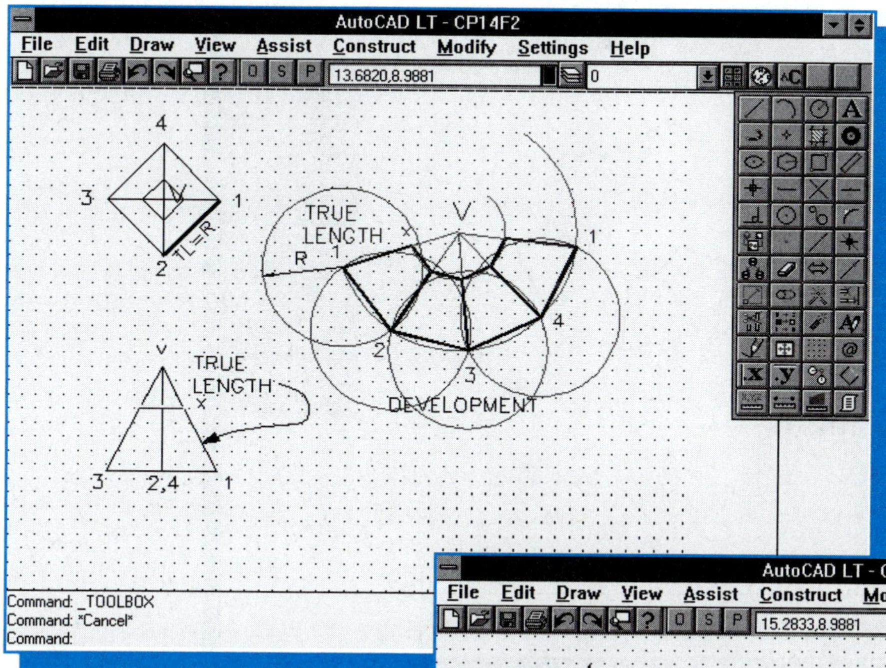

Figure 14-31
Lay out base segments on stretchout arcs and connect to apex V.

5. Draw and label a line from point "V" to the larger circle **(Fig. 14-30)**.
6. Lay out the segments 1-2, 2-3, 3-4, and 4-1 along the larger arc. Draw a circle using 1-2 in the top view as its radius. Use the **Copy** command to copy this circle to find the points of intersection for the development line segments **(Fig. 14-31)**.
7. Using Osnap (intersection), draw straight lines connecting 1-2, 2-3, 3-4, and 4-1. Draw all other needed lines for the development **(Fig. 14-32)**.
8. Freeze the construction line layers, and the development is complete **(Fig. 14-33)**.

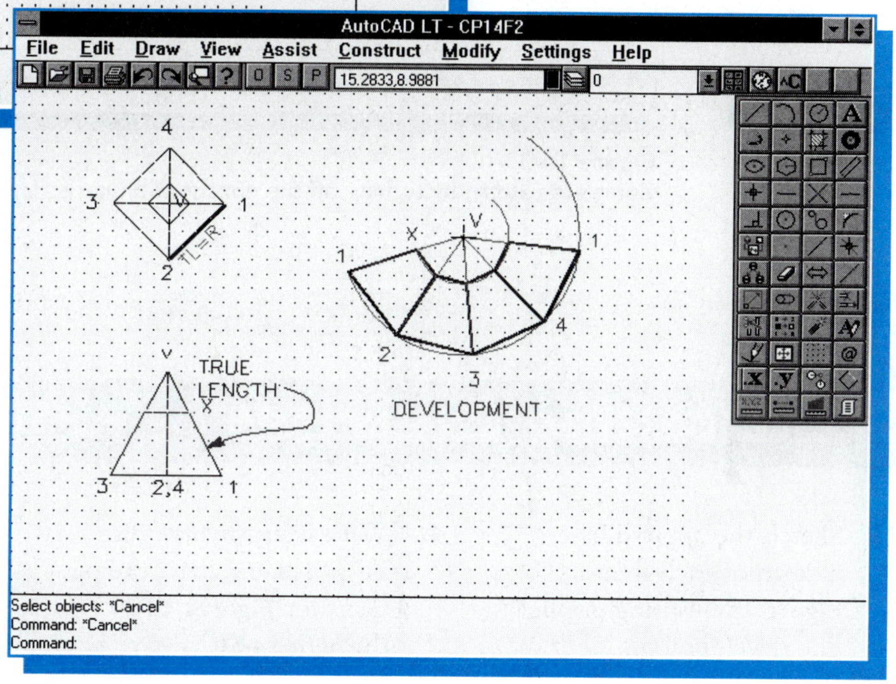

Figure 14-32
Connect the base points.

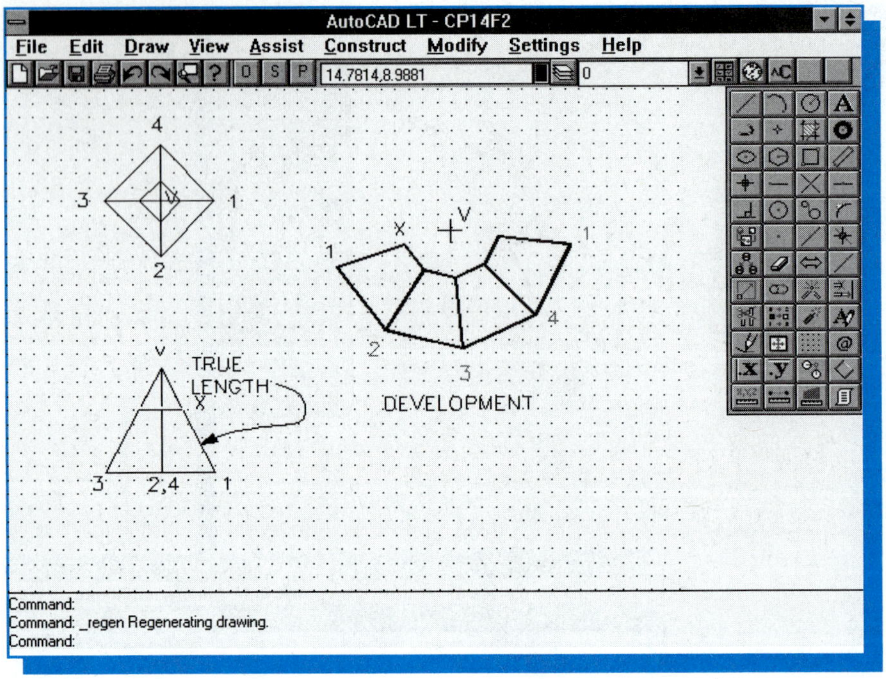

Figure 14-33
Complete pattern by turning off the construction layout layers.

Drafting Exercises

1. Sketch the approximate pattern of each item in **Figure 14-34** using parallel line development.

2. Sketch the approximate pattern of each item in **Figure 14-35** using radial line development.

3. Sketch the approximate pattern of each item in **Figure 14-36** using triangulation development.

4. With instruments or a CAD system, draw the patterns for **Figures 14-37** through **14-50**.

5. Make the following ECO (refer to "Engineering Change Order" in Chapter 17) changes for **Figure 14-51**.

- **a.** Change the vertex offset from 1.1" to 1.25".
- **b.** Increase the base diameter to 4.00".
- **c.** Change the overall height to 2.5".

6. Design a two-piece sheet metal pattern for a transitional fit in **Figure 14-52**.

7. Develop the patterns 1-6 by parallel line development..

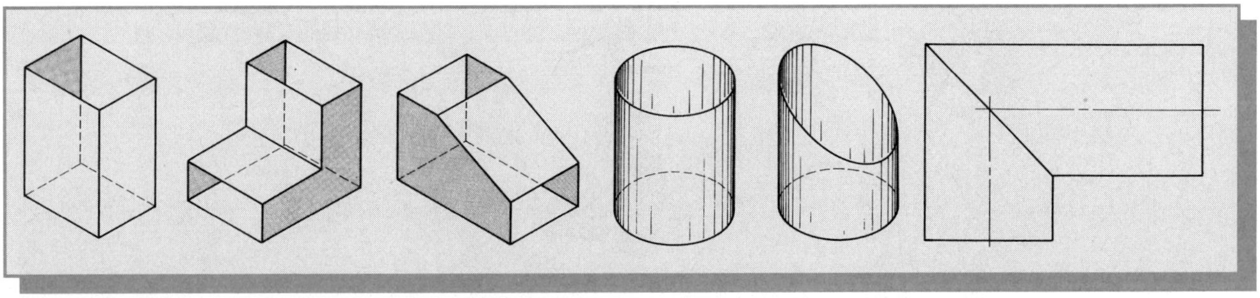

Figure 14-34
Sketch an approximate parallel line pattern for each item.

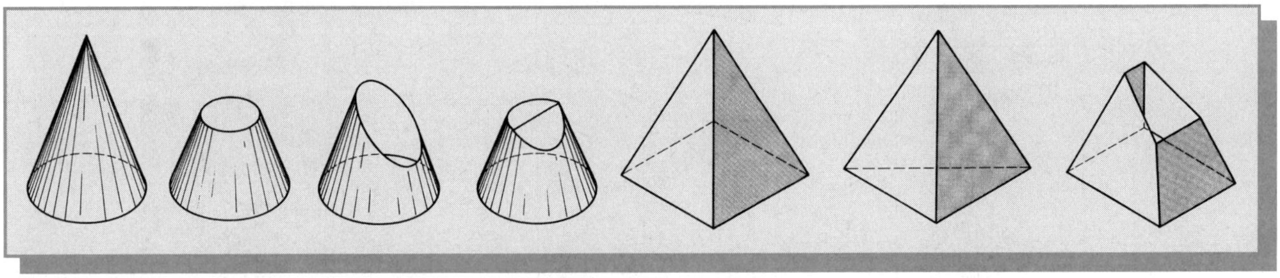

Figure 14-35
Sketch an approximate radial line pattern for each item.

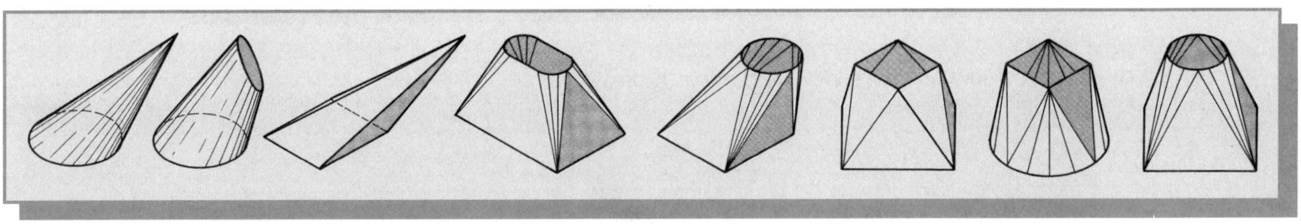

Figure 14-36
Sketch an approximate triangulation pattern for each item.

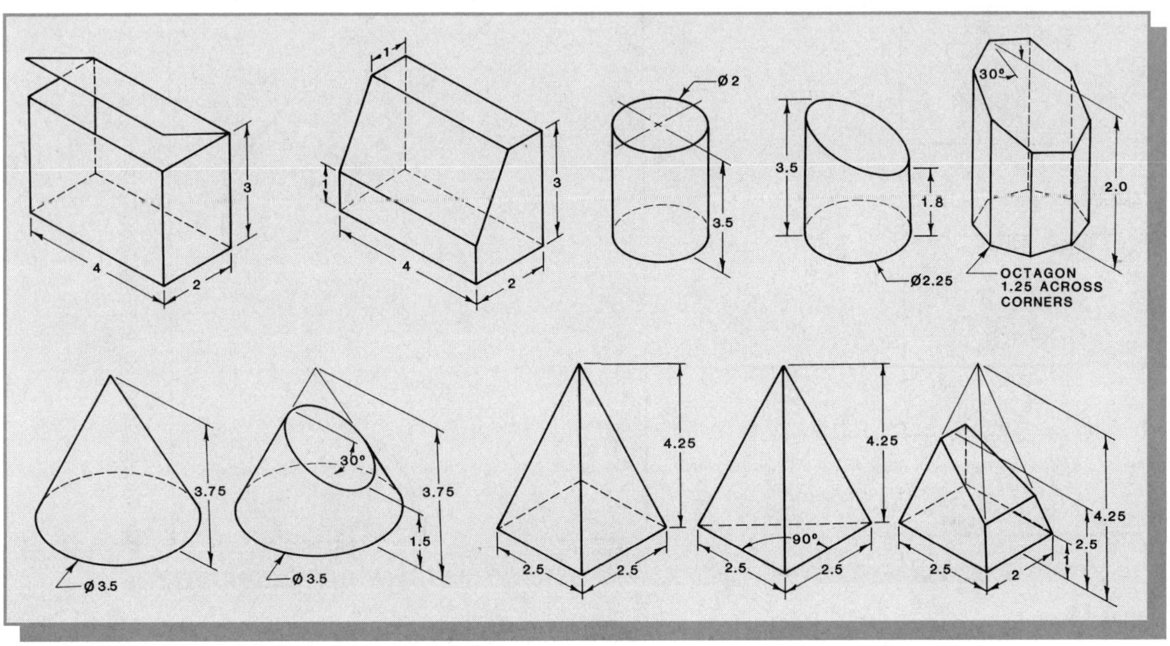

Figure 14-37
Draw the pattern for each item.

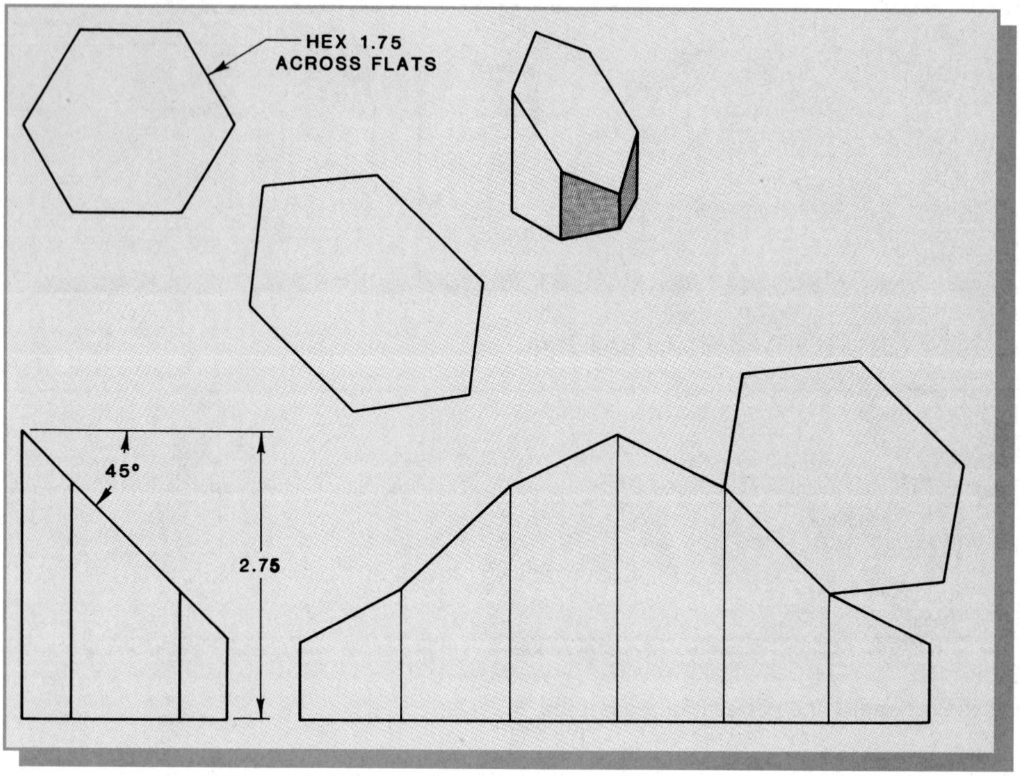

Figure 14-38
Draw the multiview, auxiliary view, and pattern drawings.

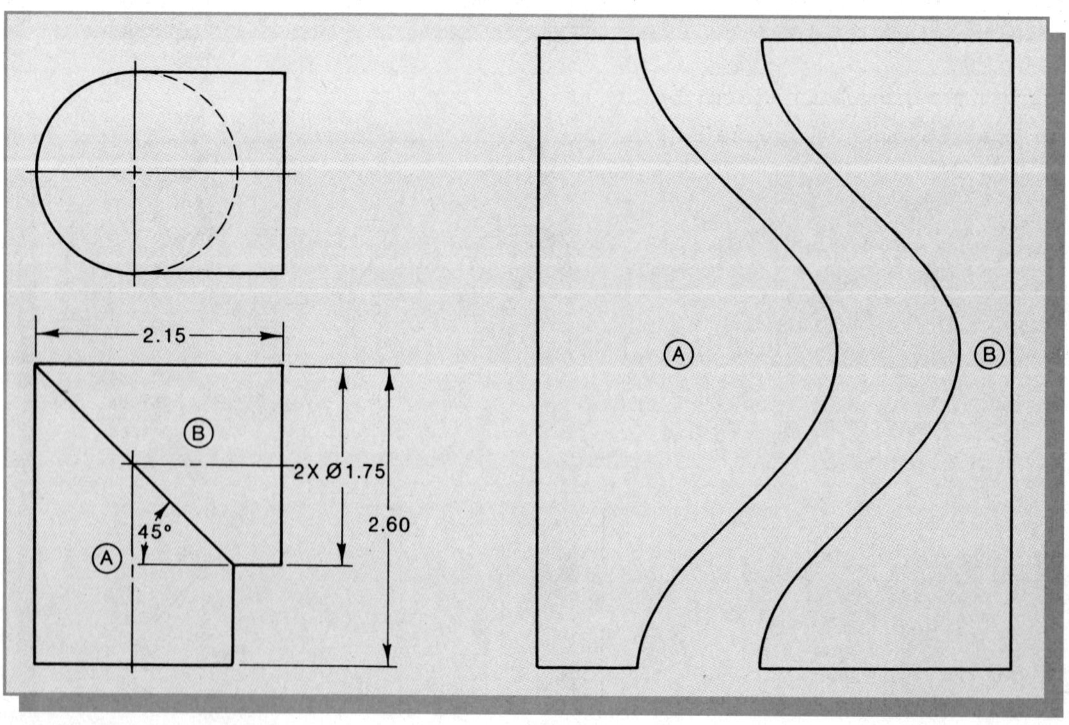

Figure 14-39
Redraw the multiview drawing and patterns for the 90° sheet metal elbow.

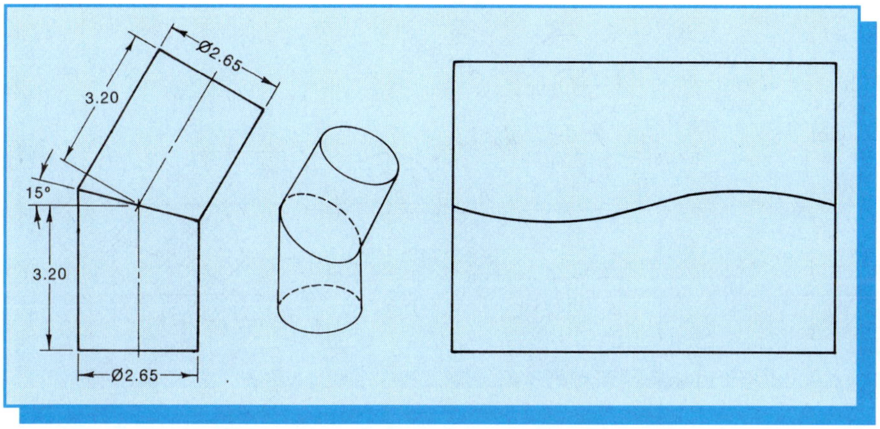

Figure 14-40
Draw the multiview drawing and patterns for the elbow.

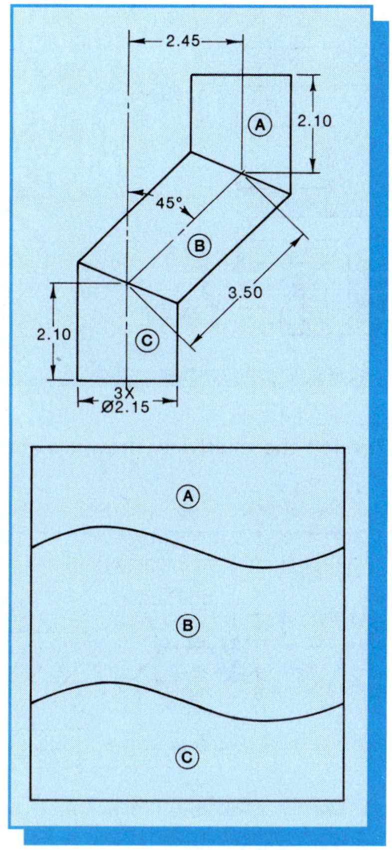

Figure 14-41
Draw the working drawing and patterns for the three-piece elbow.

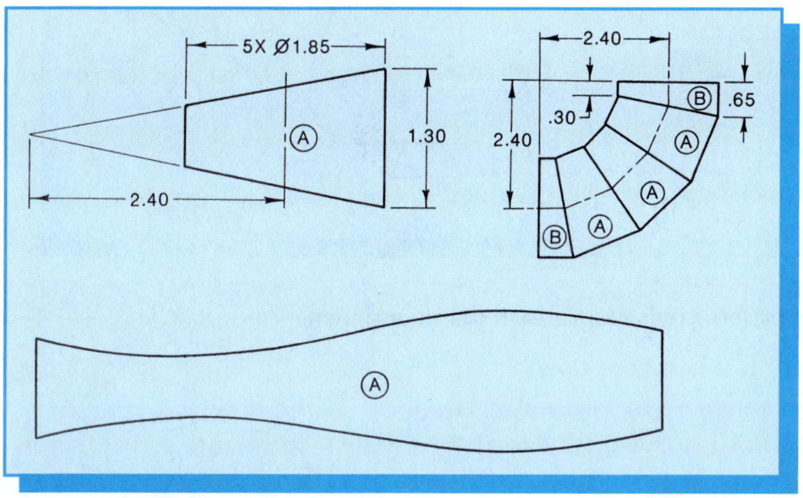

Figure 14-42
Draw the working drawing and patterns A and B to make the five-piece elbow. Pattern A is already drawn.

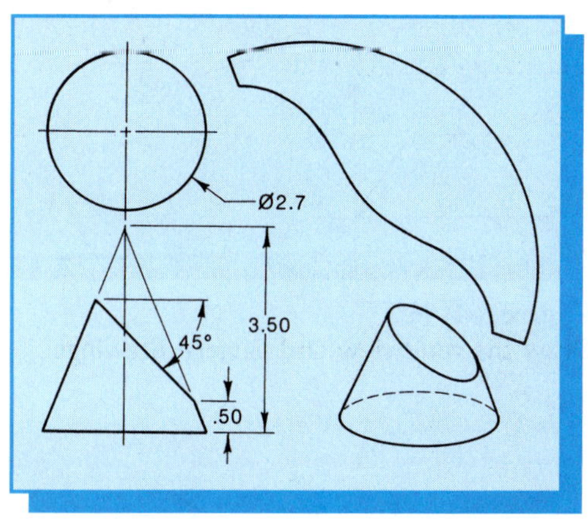

Figure 14-43
Draw the multiview and pattern drawings.

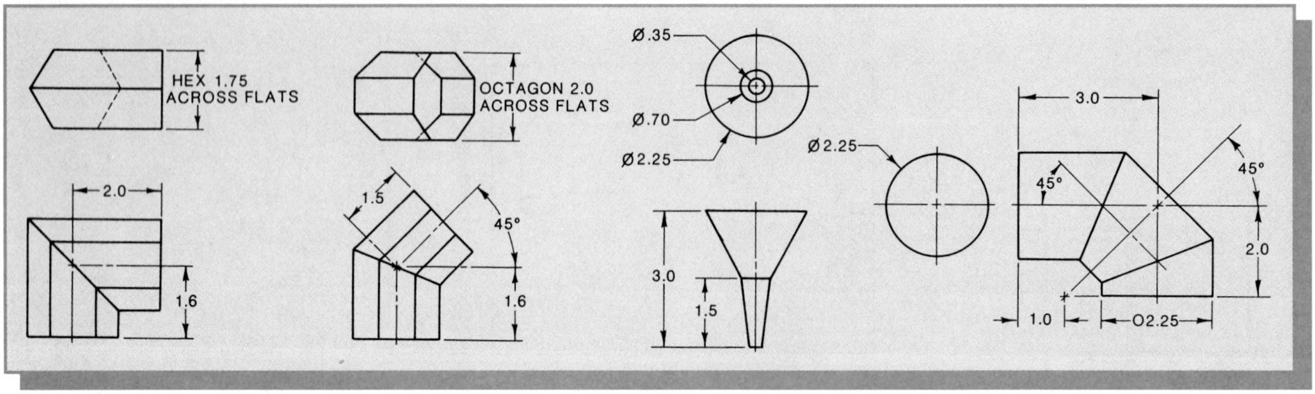

Figure 14-44
Develop the multiview drawings and the multiple patterns required for each item.

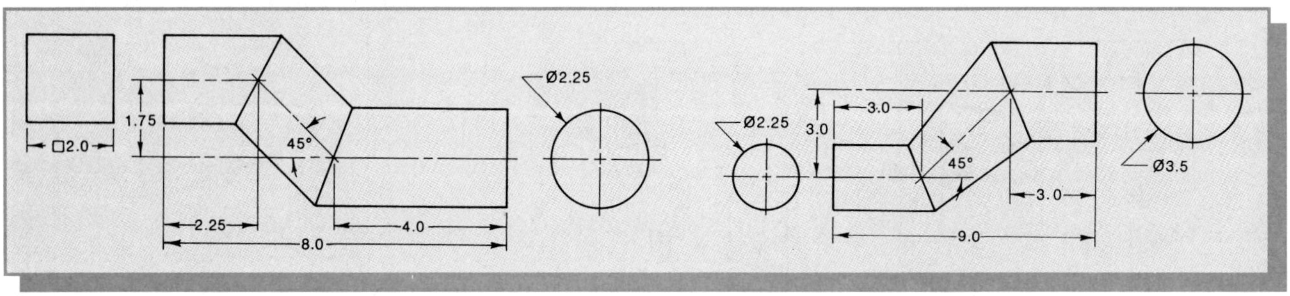

Figure 14-45
Develop the multiview drawings and the multiple patterns required for each transitional item.

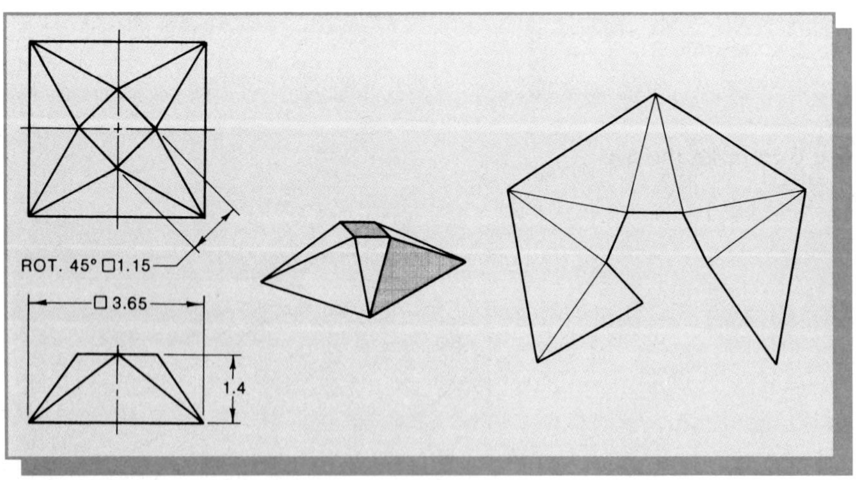

Figure 14-46
Draw the multiview and pattern drawings.

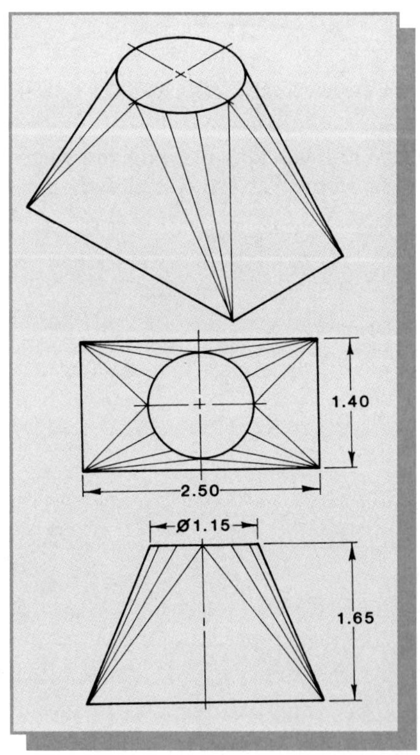

Figure 14-47
Redraw the multiview and pattern drawings.

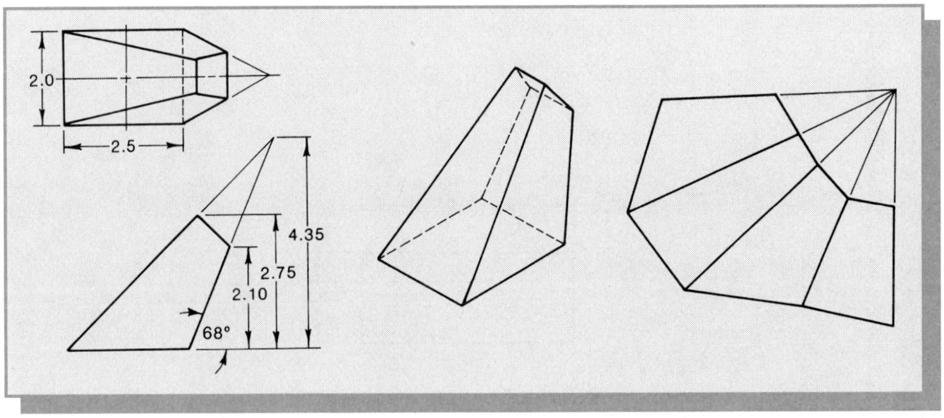

Figure 14-48
Redraw the multiview and pattern drawings.

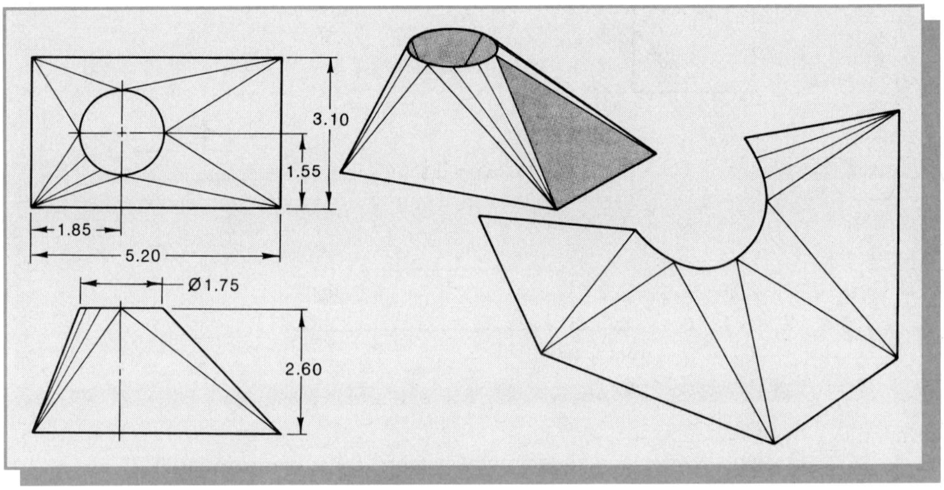

Figure 14-49
Redraw the multiview and pattern drawings.

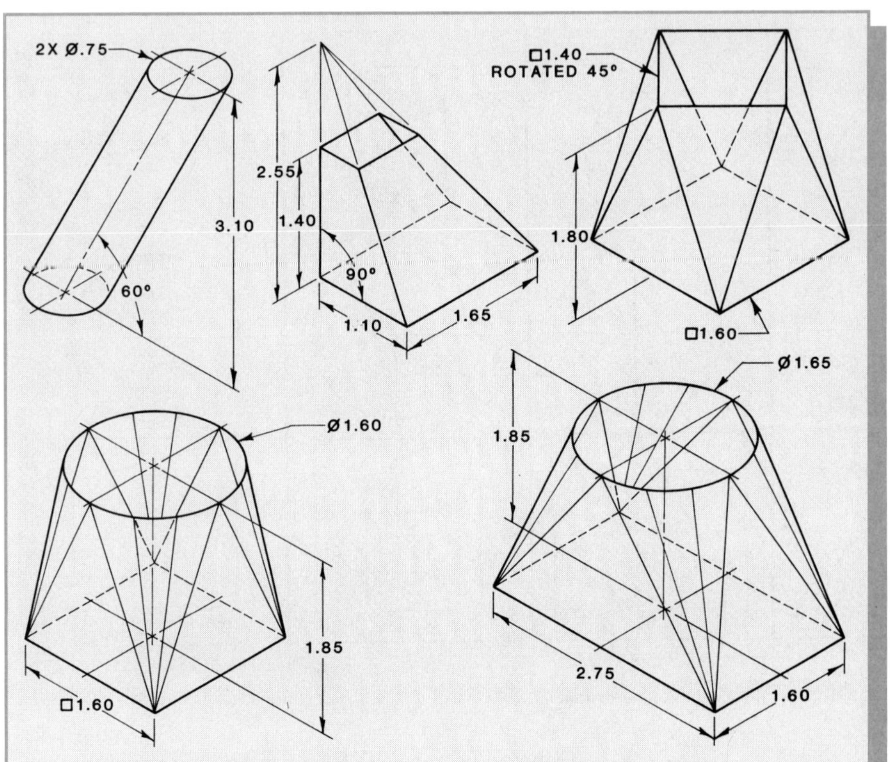

Figure 14-50
Develop the multiview and pattern for each item.

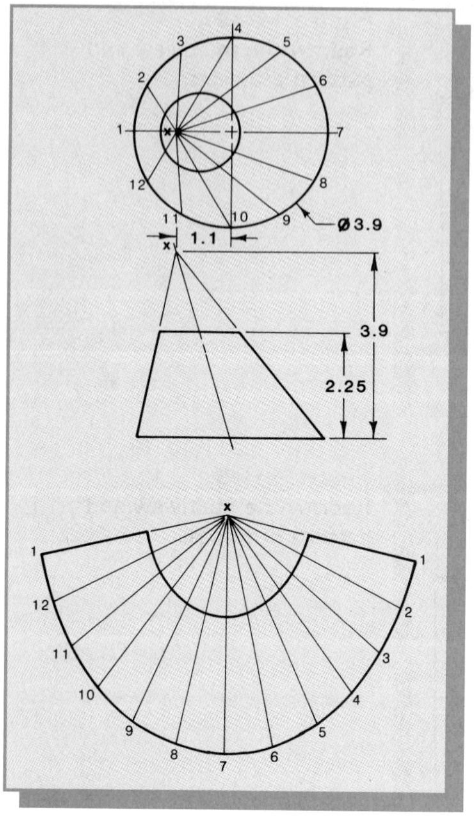

Figure 14-51
Make the ECOs (refer to "Engineering Change Orders") as specified in Drafting Exercise 5, page 296, and complete the multiview and pattern.

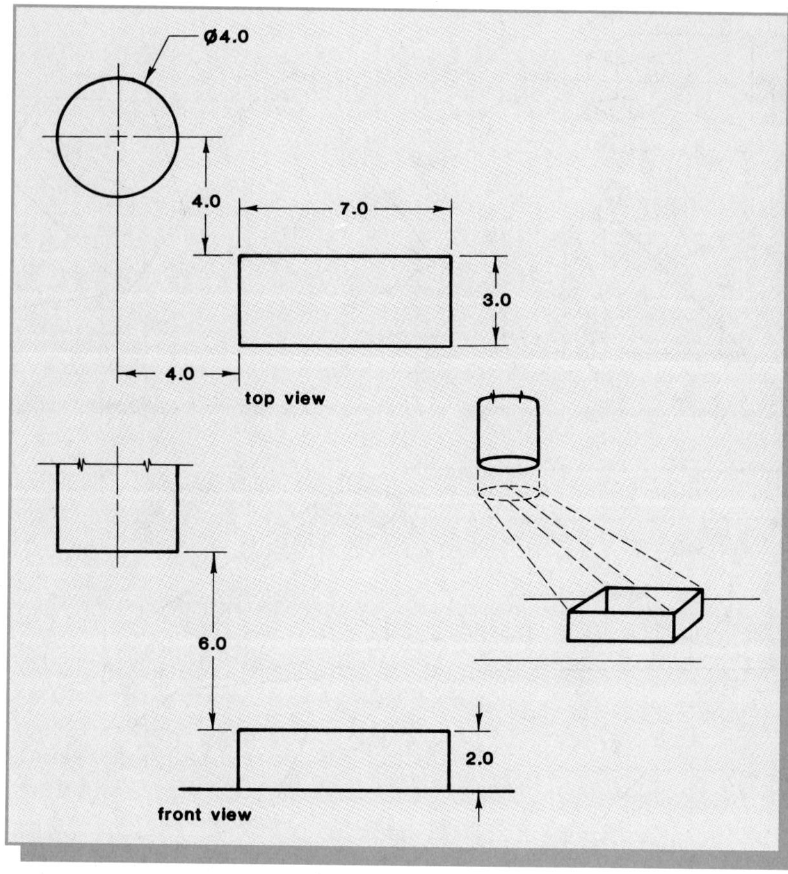

Figure 14-52
Design a two-piece sheet metal pattern for a transitional fit.

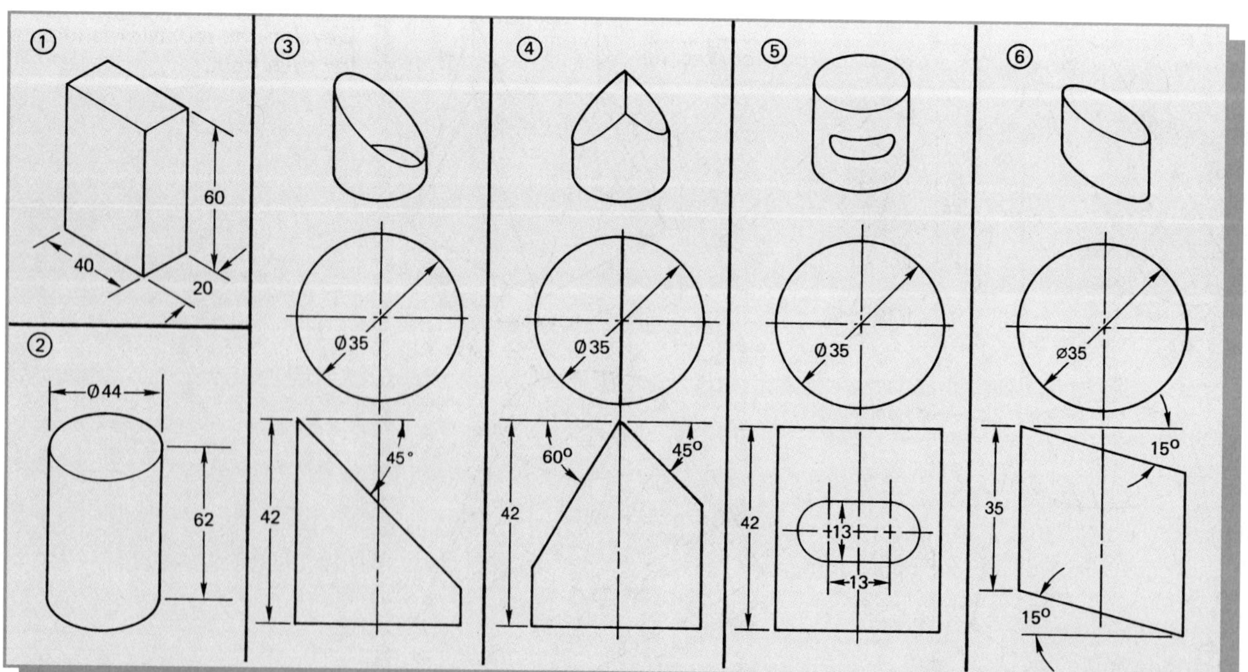

Figure 14-53
Develop patterns 1–6 by parallel line development.

Design Exercises

1. Design a one-piece cardboard candy box with the dimensions of 4" x 2" x 6". Add the following design information to the outside surfaces:
   Company name
   Logo design
2. Design a heavy cardboard storage case to hold 100 5.25" square floppy disks with a thickness of .075". Draw and cut out a paper test pattern. The disks should be loosely packed.
3. Design a cardboard carrying case for 8 standard-size cans of soda.
4. Design a cardboard carrying case or envelope to carry an 18" x 24" drafting board, a 24" T square, and assorted drafting supplies, such as pencils, erasers, triangles, irregular curves, and compasses.
5. Design a sheet metal mailbox for the front curb. Draw a multiview working drawing and the required pattern drawings.
6. Design an automatic signaling device for a mailbox that will alert the residents that the mail has been delivered.

Key Terms

Chordal distances
Chords
Development drawings
Elements
Fold (bend) lines
Foreshortened
Hems
Parallel line development
Patterns
Pattern drafting
Radial line development
Single-curve surfaces
Seams
Stretchout arc
Stretchout lines
Tab lines
Triangulation development
True length
Warped surfaces

Chapter 15

Pictorial Drawings

Objectives

The student will be able to:

- block in an isometric square
- draw nonisometric lines
- construct isometric circles
- draw isometric circles with a template
- draw isometric drawings with axes in various positions
- center an isometric drawing
- plot an isometric curve
- draw isometric radii and rounds
- draw various types of isometric sections
- draw an isometric assembly drawing
- draw an isometric exploded-view drawing
- dimension an isometric drawing with the aligned system
- dimension an isometric drawing with the unidirectional system
- sketch an isometric drawing on isometric grid paper
- draw a dimetric drawing
- draw a trimetric drawing
- draw an oblique cabinet drawing
- draw an oblique cavalier drawing
- draw circles and curves on an oblique drawing
- draw simple pictorial drawings on a CAD system
- draw a one-, two-, and three-point perspective drawing

Introduction

Pictorial drawings have been used throughout history to communicate ideas that cannot be described easily in words. Early attempts to show depth and detail were crude and often distorted the appearance of objects. **Figure 15-1** shows primitive pictorial drawings that lack correct proportion and realism. Methods and techniques used to prepare realistic pictorial drawings have been continually refined and improved through the years. **Figure 15-2** shows a contemporary pictorial drawing that is difficult to distinguish from a photograph.

Figure 15-1
Primitive drawings usually appeared two-dimensional and distorted.

In addition to their function as an art form, **pictorial drawings** are used today by engineers, architects, and all types of designers and drafters to illustrate the size and shape of an object. Pictorial drawings are multidimensional; that is, they show more than one side of an object **(Fig. 15-3)**. Pictorial drawings can show width, height, and length on one drawing. For this reason, they are easier to understand and interpret than orthographic multiview drawings, which show only one face of an object on each view. Since pictorial drawings are

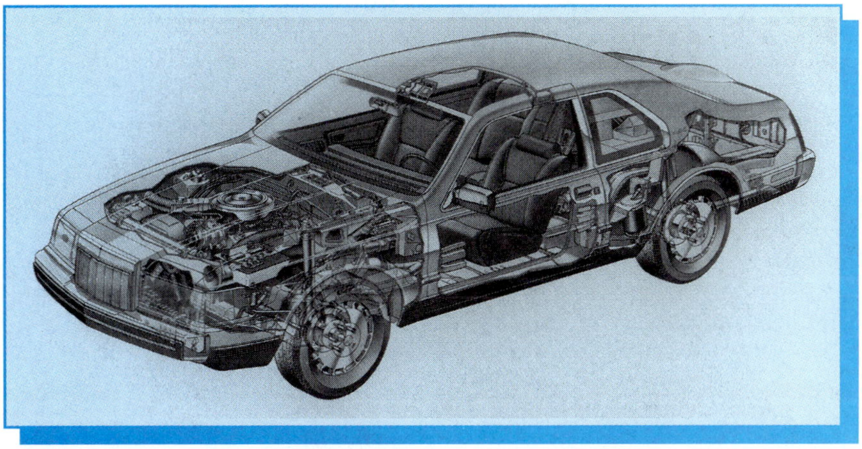

Figure 15-2
An artistic version gives a three-dimensional effect and realistic appearance. *(Courtesy of Ford Motor Company)*

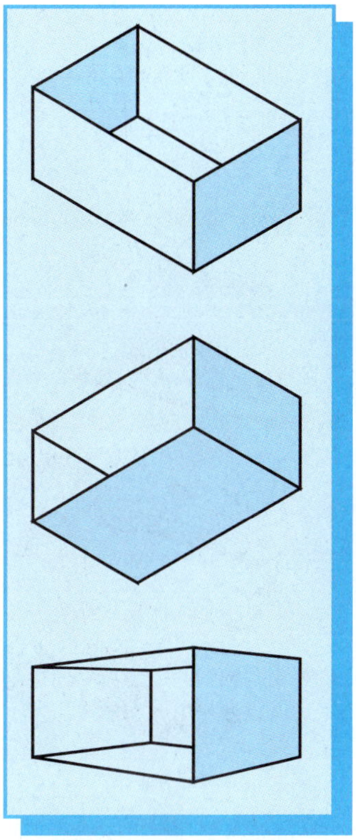

Figure 15-3
Pictorial drawings will show two or more sides on each drawing.

not always dimensionally accurate, they are rarely used as detailed working drawings for manufacturing or construction purposes.

In engineering and architecture, pictorial drawings are used to improve the visualization skills of nontechnical personnel who may have difficulty reading orthographic drawings. Pictorial drawings are used extensively:

- as idea recorders in the design process
- to clarify assembly and repair instructions
- for new product sales training
- for sales promotion

Once a product is manufactured, a photograph may be used for these purposes. Only pictorial drawings or models can be used to show the complete appearance of a product before manufacture. Unlike photographs, pictorial drawings can be used to create sectional or **exploded views** which show details or the interior of an object. **Figure 15-4a** shows a photograph of a technical pen set. **Figure 15-4b** shows a drawing of an inking tip with sections removed to heighten clarity.

Figure 15-4c shows a photograph of a technical pen and an exploded-view line drawing of the same pen.

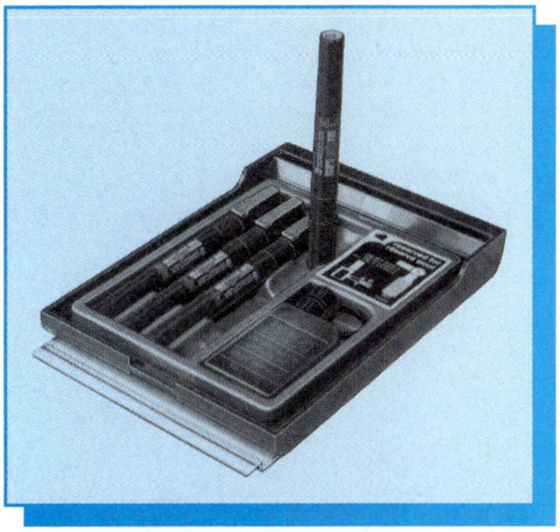

Figure 15-4a
Photograph of a technical pen set
(Courtesy of J. S. Staedtler, Inc.)

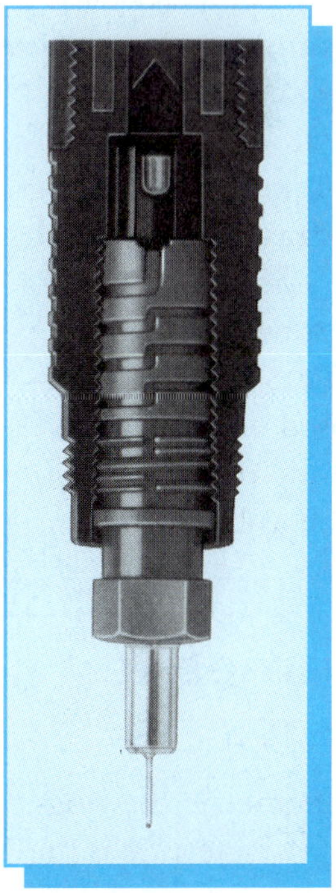

Figure 15-4b
Sectional drawing of a technical pen
(Courtesy of J. S. Staedtler, Inc.)

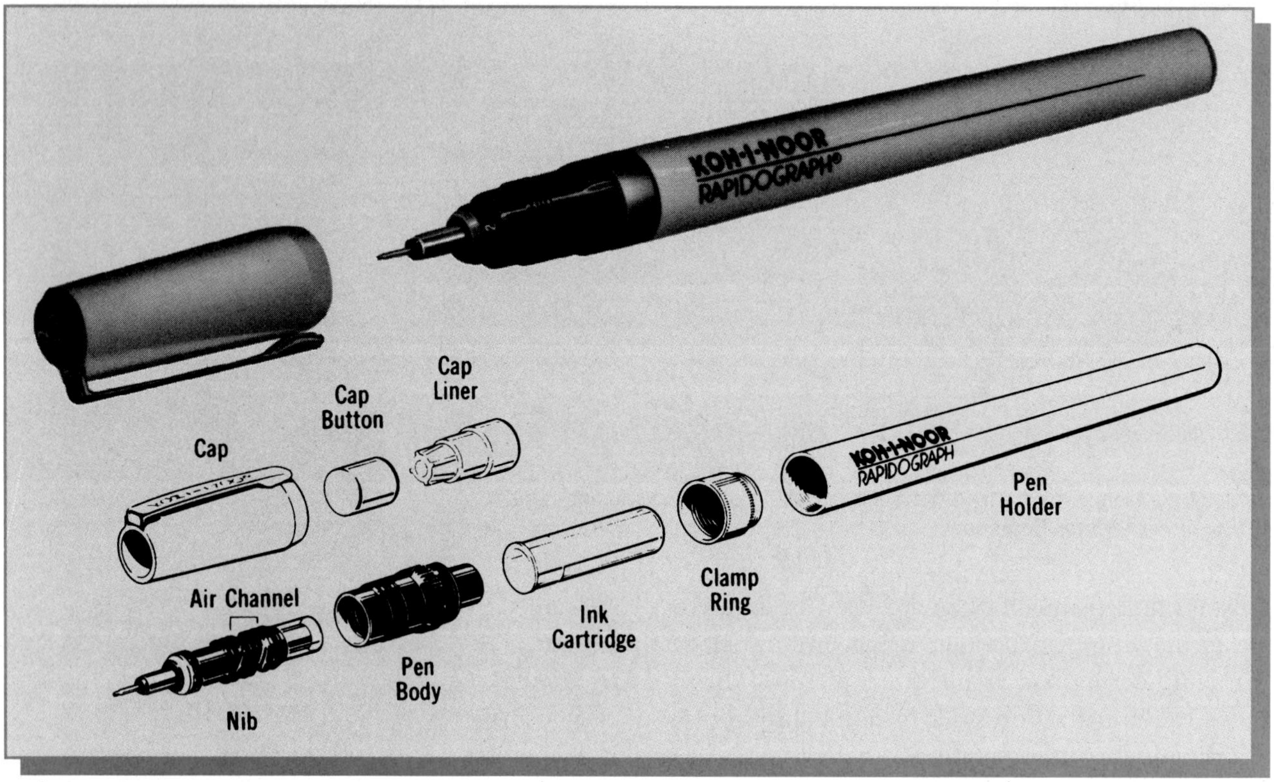

Figure 15-4c
A photograph and an exploded pictorial line drawing
(Courtesy of Koh-I-Noor)

In the fields of engineering and architecture, there are three basic types of pictorial drawings—axonometric, oblique, and perspective. **Figure 15-5a** compares these three types applied to the same object. **Figure 15-5b** shows four pictorial drawings for a single-number digital clock. **Axonometric drawings** are most frequently used for engineering drawings. **Oblique drawings** are used mostly for simple thin objects or for progressive design sketches. **Perspective drawings** are used extensively to create realistic renderings of objects such as buildings, cars, and boats.

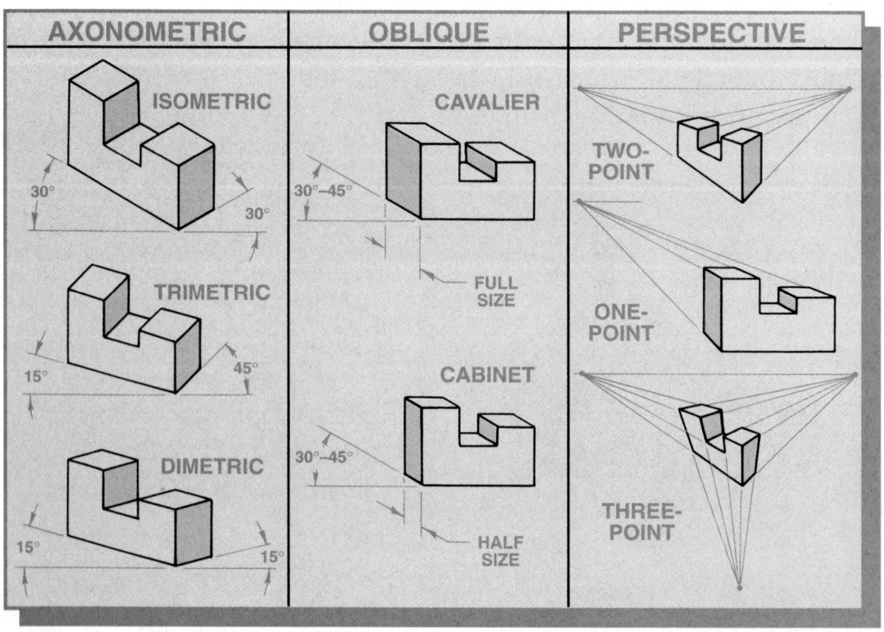

Figure 15-5a
Various types of pictorial drawings

Axonometric Drawings

In axonometric drawings, the major characteristic is the receding lines that are always parallel. Axonometric drawings

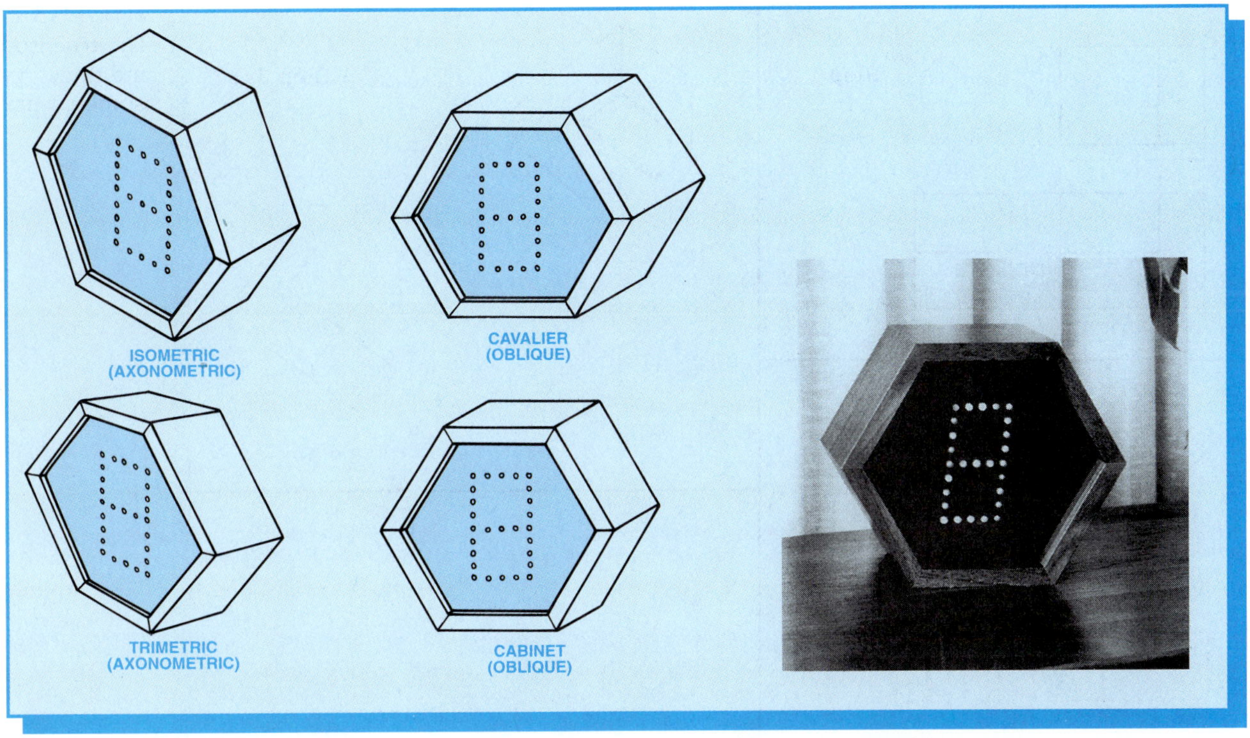

Figure 15-5b
Various types of pictorial representation of a digital clock *(Courtesy of Michael Robbins)*

show all three faces (front, side, and top) of an object in one view. There are three types of axonometric drawings—**isometric**, **dimetric**, and **trimetric (Figs. 15-5a** and **15-5b)**.

Isometric Drawings

The most popular form of axonometric drawing used in engineering is the isometric. Isometric drawings are easier to draw, detail, and dimension than any other type of pictorial drawing. This is because the three planes are shown true size **(Fig. 15-6)**. To visualize an object in true isometric form, the following steps are necessary **(Fig. 15-7)**.

Step 1. Draw the three orthographic views.

Step 2. Revolve the top view 45° and draw the remaining two views.

Step 3. Revolve the side view in Step Two 35°. Project and draw the remaining two views.

The front view will be a true isometric projection. Note that the vertical lines are not the true length. The dimensions will appear foreshortened because of the 35° revolution. Regular isometric drawings will use true size dimensions for speed and convenience.

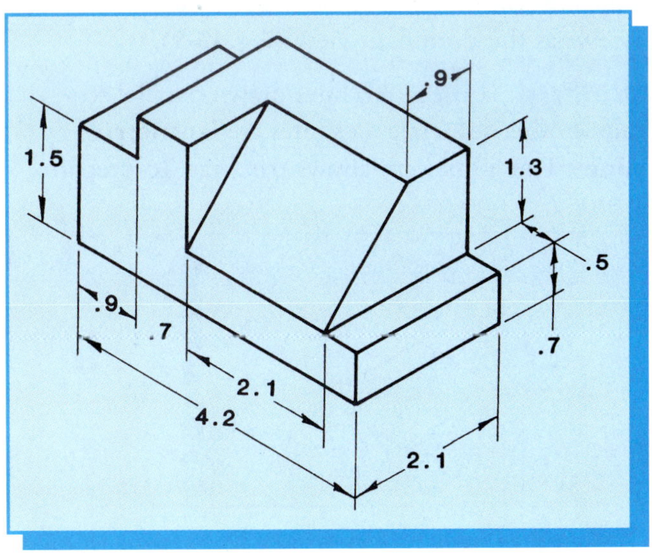

Figure 15-6
This isometric drawing shows the front, top, and right sides.

Isometric Axis. The **isometric axis** is the basis for all isometric drawings. This axis consists of three lines, 120° apart. These three

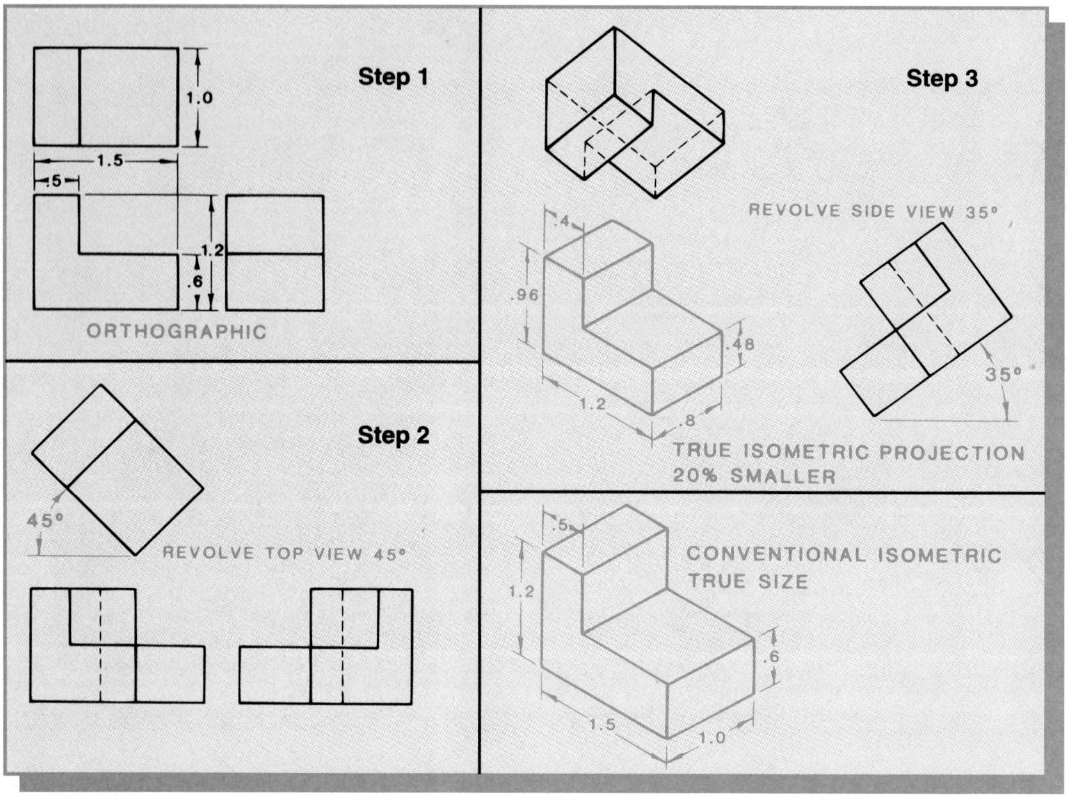

Figure 15-7
The true isometric and the conventional isometric drawings.

lines represent the front corner and the two receding edges of an object **(Fig. 15-8)**. Although the three-axis lines are always 120° apart, the axis can be positioned to show either the top, bottom, or side views as the dominant view **(Fig. 15-9)**.

Isometric Lines. All lines drawn parallel to any of the isometric axis lines are **isometric lines**. These lines are always true size. To prepare

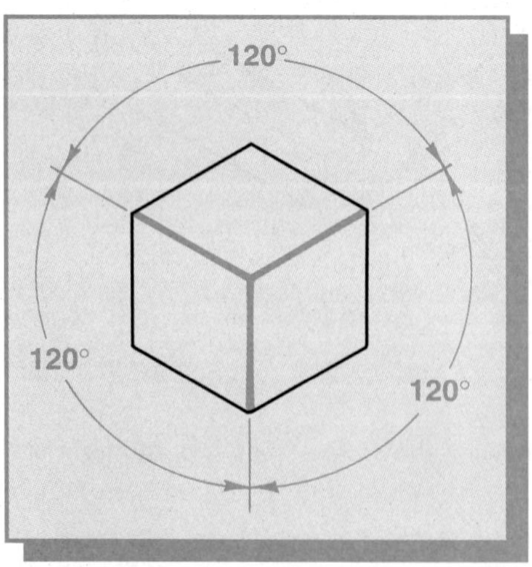

Figure 15-8
The major axes for an isometric drawing are spaced at 120°.

Figure 15-9
Alternate positions for isometric axes

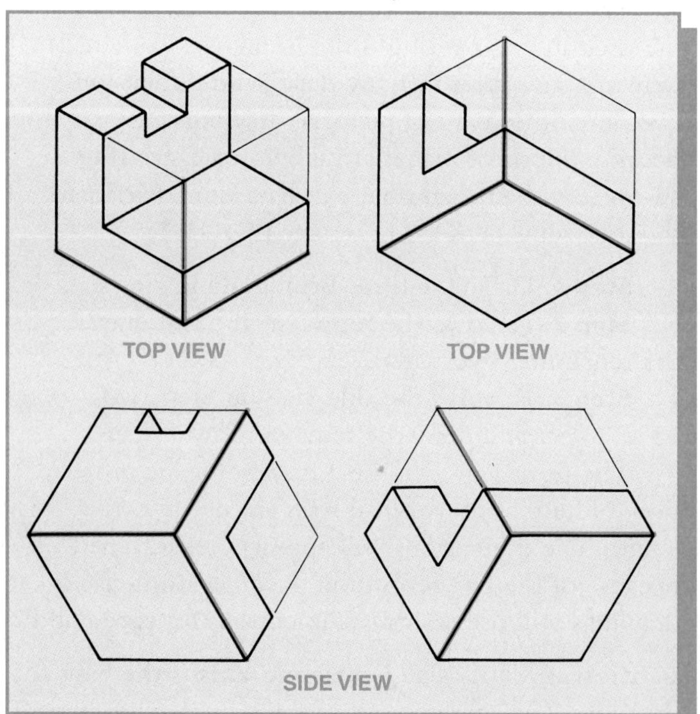

an isometric drawing, all isometric lines are drawn 30° or vertical from the horizontal **(Fig. 15-10)**.

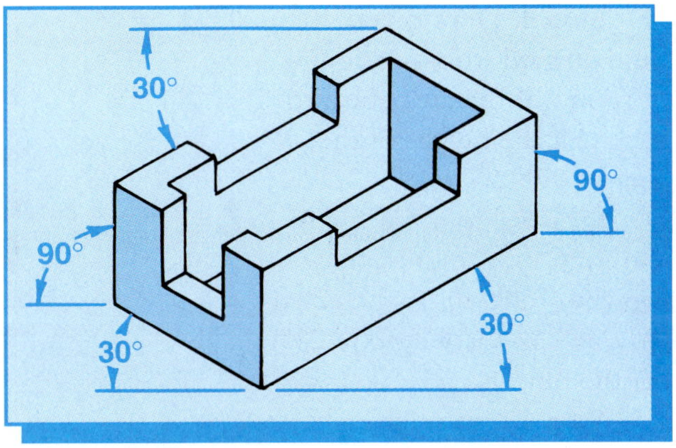

Figure 15-10
Isometric lines are 30° from the horizontal.

Nonisometric Lines. As you learned in developing auxiliary views, objects may contain lines that are not parallel to the major axes of the object. These lines, representing sloping surfaces, are called **nonisometric lines**. Nonisometric lines are lines that are not parallel to isometric lines. The angle and length of nonisometric lines can only be established by locating points of intersection with isometric lines. This is because actual measurements can only be made on true size isometric lines.

Before drawing nonisometric lines, first lightly draw all isometric lines, completing the outline of the object as it would appear if no surfaces were slanted **(Fig. 15-11)**. Then transfer the distances of each corner from an orthographic view to one of the faces of the isometric drawing. Connect these points on the isometric lines to form the nonisometric lines. Project 30° isometric lines from these points until the lines intersect the lines representing the back surface of the object. Erase the isometric construction lines and complete the isometric drawing as shown.

Centering Isometric Drawings. To center an isometric drawing, follow the steps in **Figure 15-12a**:

Step 1. Check the overall dimensions of the object to be drawn.
Step 2. Outline the area needed for the drawing.
Step 3. Locate the center of the area.

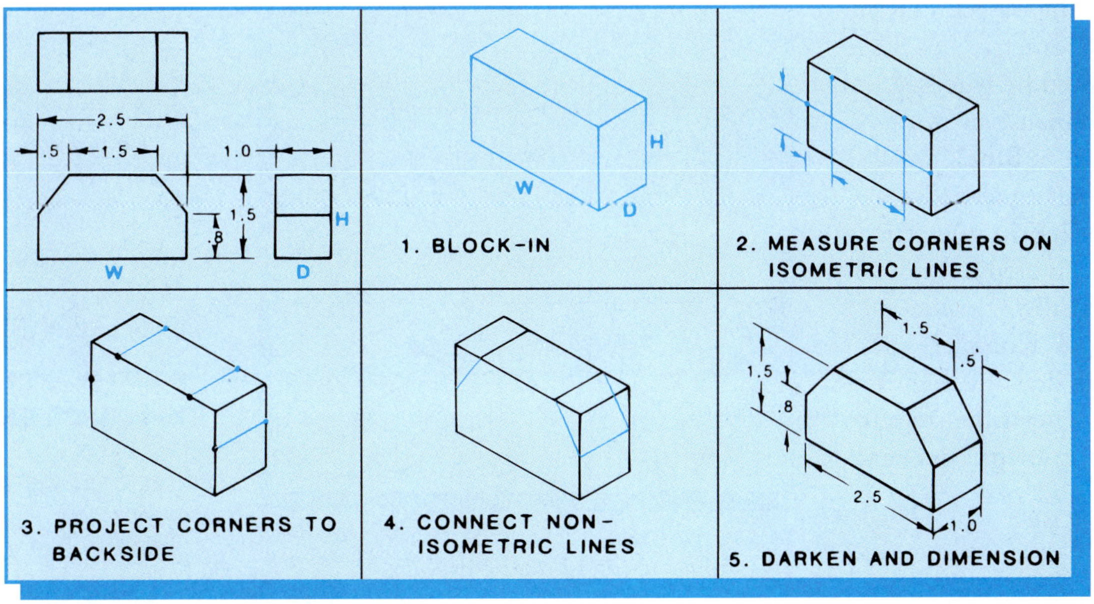

Figure 15-11
Steps in drawing an isometric drawing

Step 4. Draw a vertical line upward from the counterpoint of the area equal to one half the height of the object (.75").

Step 5. From the top of this line, draw a line 30° downward to the right, equal to one half the length (1.25") of the object.

Step 6. From the right end of this line, draw a line 30° downward to the left, equal to one half the thickness of the object (.5"). The left end of this line represents the front, top corner (X) of the object.

Step 7. Extend a light vertical line down from corner X to represent the front edge of the object. The length of this line is the height of the object at its highest point (1.5"). Next, extend 30° lines upward to the left and right to represent the top right and top left edges of the object. This creates the isometric axis. These lines should be true size overall dimensions of the object. Now, **block in** all remaining isometric lines. The blocked-in isometric will be centered horizontally and vertically.

Step 8. Complete the isometric drawing. **Figure 15-12b** reviews the steps to draw an isometric drawing.

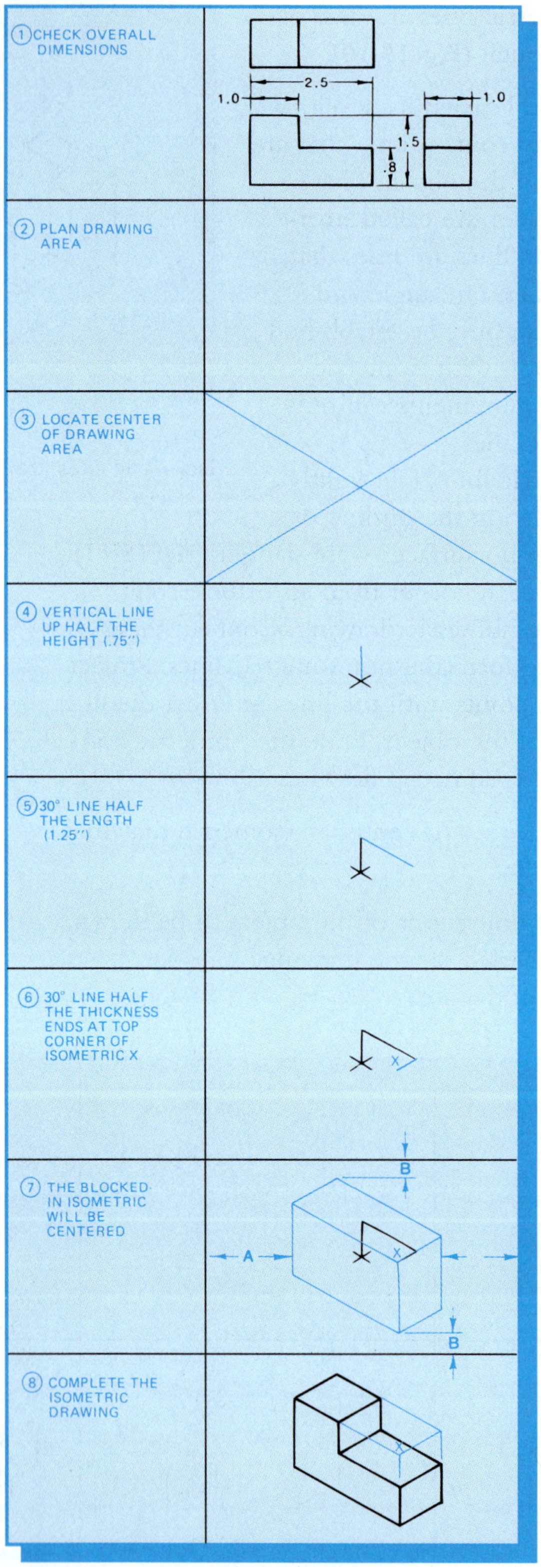

Figure 15-12a
Steps to center an isometric drawing

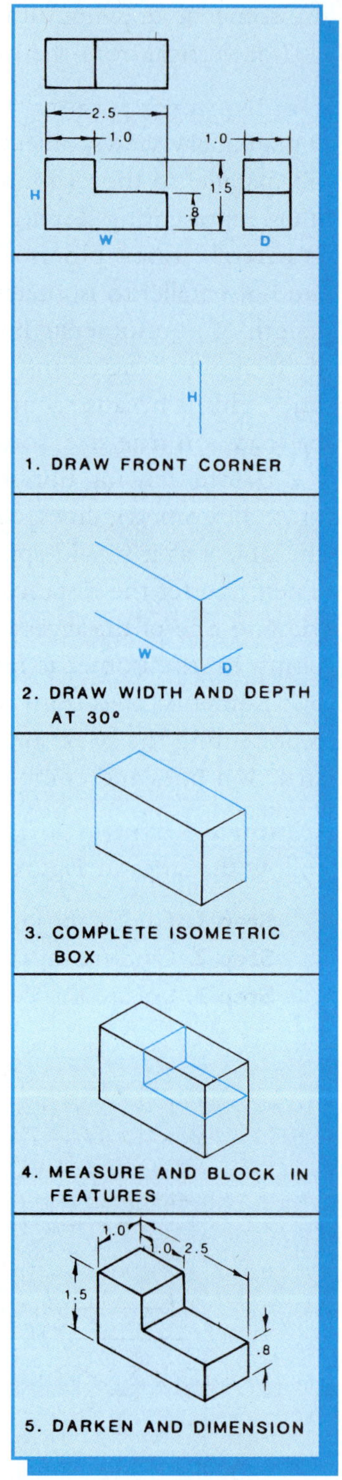

Figure 15-12b
Steps in drawing an isometric drawing

Isometric Circles. Circles will appear as **ellipses** on isometric drawings because they are located on **receding planes** that are not perpendicular to the line of sight. Ellipses are oriented differently on each of the three isometric planes **(Fig. 15-13).** To draw **isometric circles** (which are actually 35° ellipses), follow the steps outlined in **Figure 15-13**:

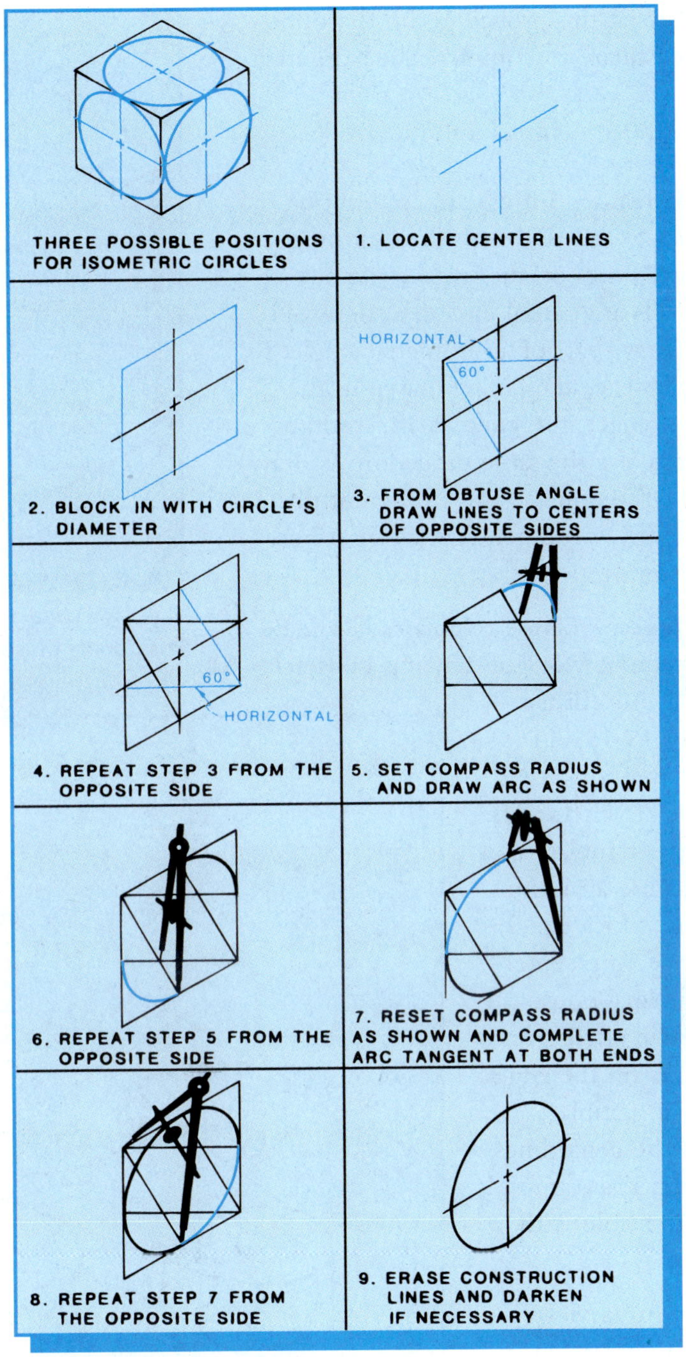

Figure 15-13
Steps in constructing an isometric circle

Step 1. Locate the center of the circle. Then draw centerlines that are parallel to the isometric lines of the object.

Step 2. Establish the size of the circle by drawing an isometric square with sides equal to the diameter of the circle.

Step 3. Draw two lines from the vertex of the obtuse angle of the square to the point where the center lines intersect the opposite edges of the square. One line will be horizontal and the other will be 60°.

Step 4. Repeat the above procedure in reverse from the obtuse angle on the opposite side of the square.

Step 5. Place the point of a compass on the intersection of these two lines. Set the compass opening to align with the intersection of the centerline and the outer edge of the square. Draw an arc from this point to the corresponding point on the adjacent side of the square.

Step 6. Repeat Step 5 on the opposite end of the square.

Step 7. Place the compass point on the obtuse angle corner of the square. Draw a large arc on the opposite side to connect with the edges of the small axis drawn in Steps 5 and 6 above. Be certain to draw smooth tangent points.

Step 8. Repeat this procedure on the opposite side of the square to complete the isometric circle.

Step 9. Erase all construction lines.

Fillets and Rounds. **Fillets** are continuous internal (concave) arcs. **Rounds** are continuous external (convex) arcs on the surface of an object. Since arcs are partial circles, to draw an isometric arc, draw part of an isometric circle. To visualize this, observe the circles drawn in the corners of the object in **Figure 15-14**. To draw **isometric arcs**, use the same procedure as drawing isometric circles on the three different planes. Draw only the part of the circle that describes the fillets or rounds.

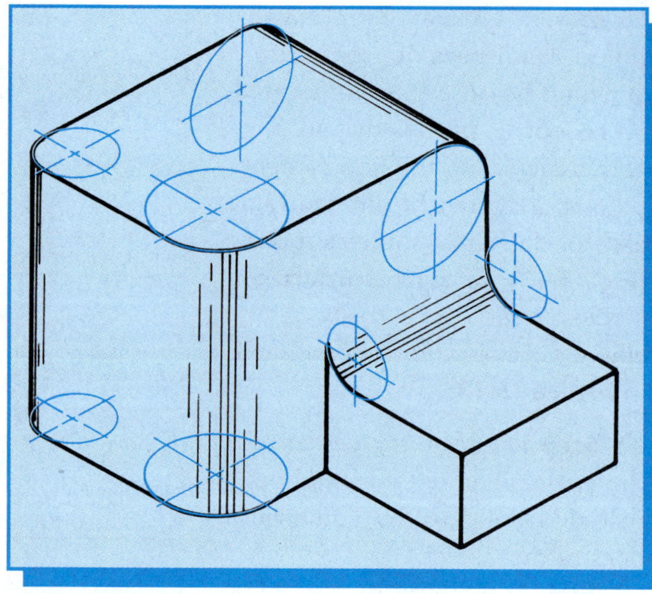

Figure 15-14
Positioning of isometric circles

Ellipse Templates. Ellipse templates eliminate much of the tedious work of drawing isometric circles and arcs. An **ellipse template (Fig. 15-15)** is the most efficient method of drawing isometric circles. The isometric ellipse template uses 35° ellipses. To use an isometric template, first locate and draw the intersecting centerlines of the circle. Choose the correct isometric circle size. Align the marks on the edge of the isometric template opening with the centerlines on the drawing. Trace around the inside of the hole with a pencil.

Ellipse on Nonisometric Planes. The template and construction method of drawing ellipses can only be used on isometric planes that are at 30°. These methods cannot be used on nonisometric planes that do not align with the isometric axis.

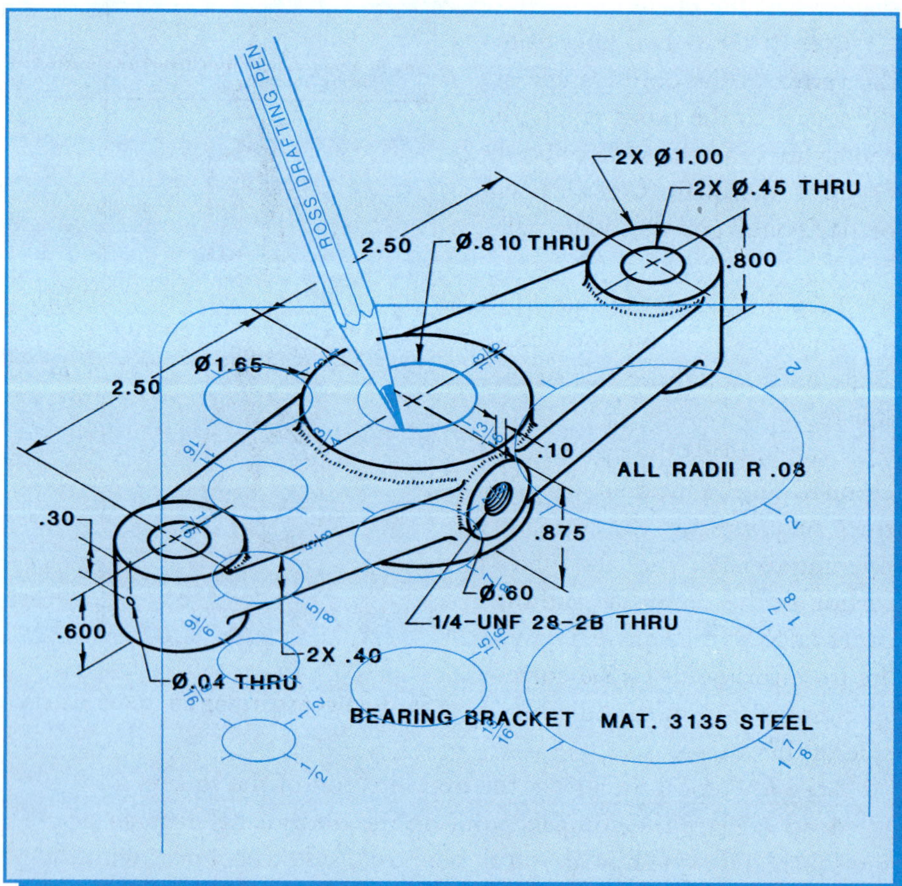

Figure 15-15
Line up the diagonals on the isometric template's circle on the drawing's isometric center lines, and trace around to complete the isometric circle.

Ellipses, or any projection or depression on a nonisometric plane, may be plotted directly from a normal orthographic view.

To plot ellipses on nonisometric surfaces, first prepare an orthographic two-view drawing. Show the sloped surface profiled on one view, and the true shape of the circle on the other. Follow the steps outlined in **Figure 15-16**:

Step 1. Divide the circle into eight or more equal parts as shown on the front view. Number each equally spaced mark on the circumference of the true circle. The accuracy and smoothness of the final ellipse will increase as the number of circle divisions is increased. Project these points horizontally to intersect the line representing the edge of the inclined plane on the front view. Label the points of intersection on the profile plane. Draw the isometric form. Draw a centerline representing the vertical center of the ellipse on the inclined surface.

Step 2. Measure the X and Y distances to find the position of the horizontal centerline on the inclined surface. Label the intersection of the vertical and horizontal centerline with a 0.

Step 3. Measure the distance from 0 to 3, and from 0 to 7 on the front view. Transfer these distances to the horizontal centerline on the inclined surface as shown.

Step 4. Determine the distance from 0 to 1, and from 0 to 5 on the side view sloping edge. Transfer or project these distances to the vertical centerline on the inclined surface as shown. At this point, much time could be saved if you had an ellipse that would fit points 1-5 to 3-7 on the isometric drawing.

Step 5. Measure the horizontal and vertical X-Y distances from the inclined surface for points 8 and 2, and then 6 and 4. If more points are used, lines representing all of the points would be drawn at this time. Draw a dot where these lines intersect with corresponding numbered lines measured from the

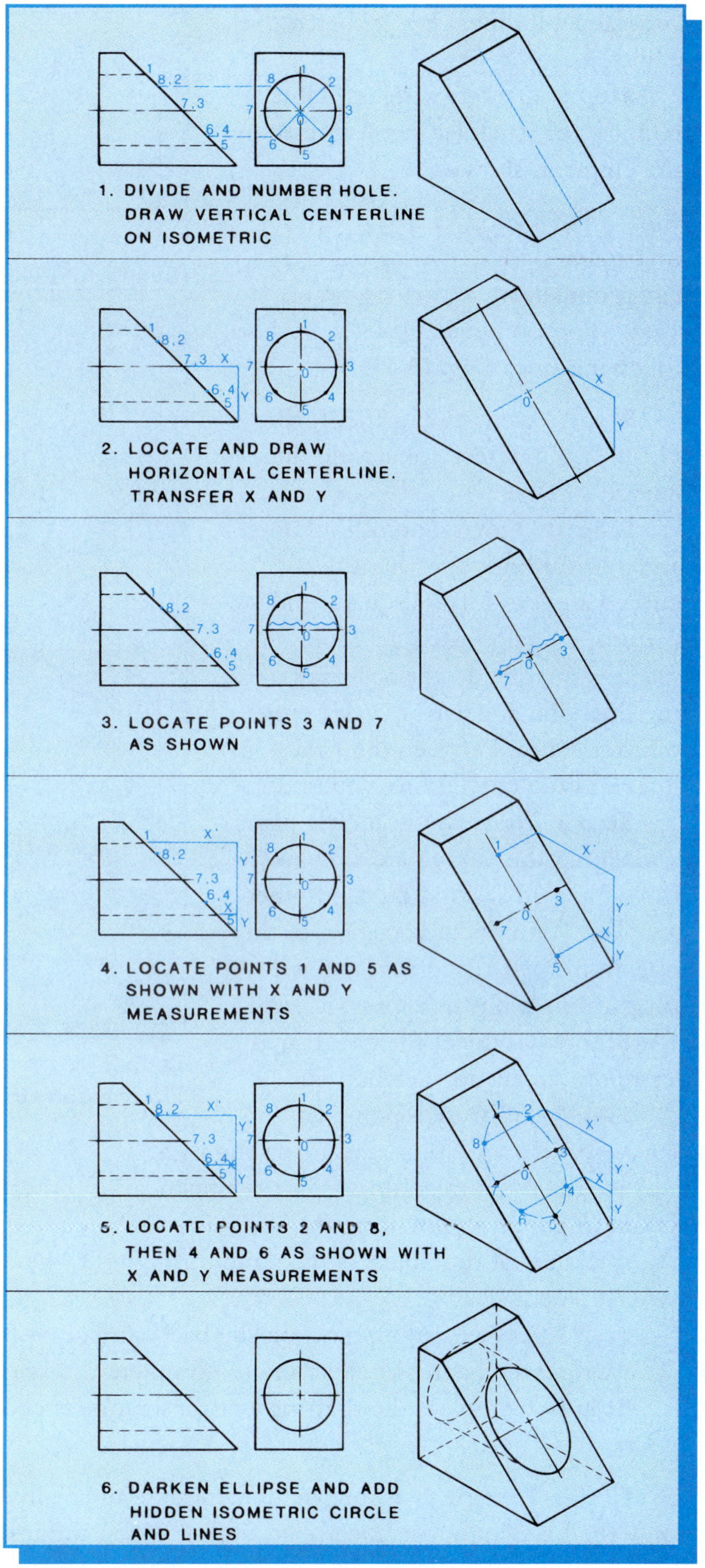

Figure 15-16
Plotting a nonisometric circle

vertical centerline. Lightly sketch the ellipse.

Step 6. Darken with an ellipse template or irregular curve to form the ellipse as shown.

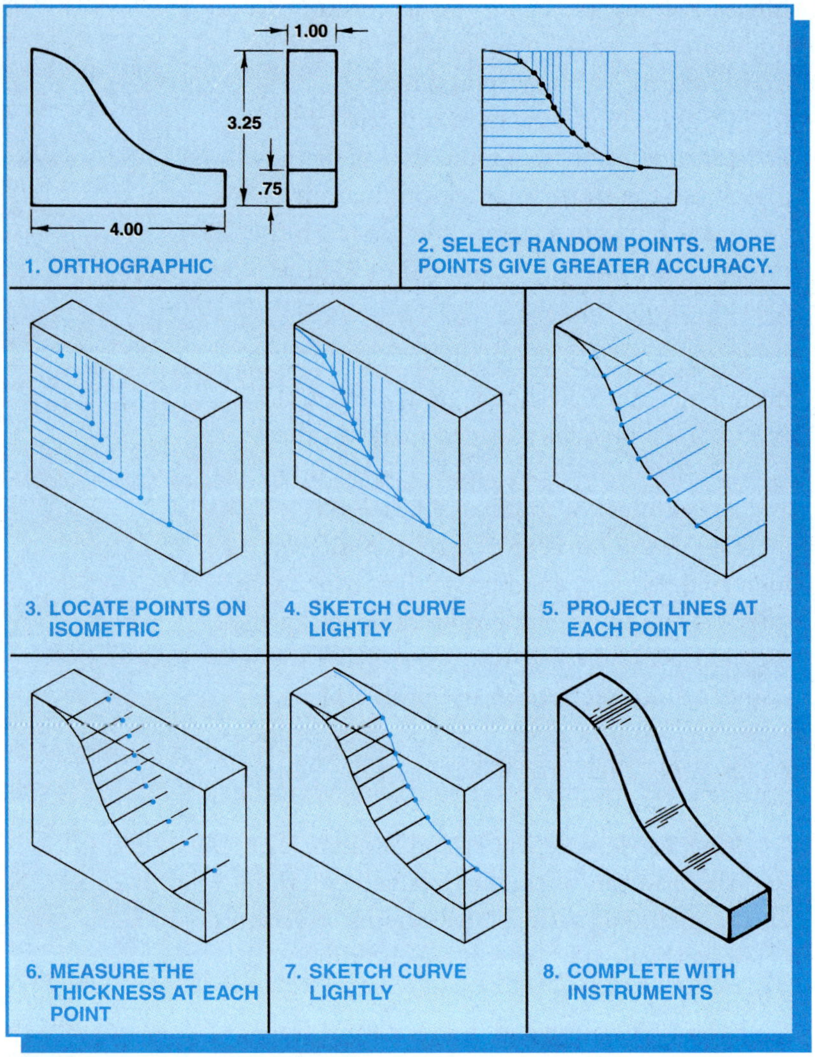

Figure 15-17
Plotting isometric irregular curves

Plotting Curves. Curved nonisometric lines are drawn by transferring points from an orthographic view to an isometric plane, as in the following steps **(Fig. 15-17)**:

Step 1. Draw two orthographic views of the object containing the curved surface.

Step 2. Draw random horizontal and vertical lines covering the curved areas to form a square grid pattern. The closer the grid line spacing, the greater the accuracy of the final drawing. Mark each point of intersection between the grid lines and the curved line with a dot.

Step 3. Draw the isometric plane containing the curved areas. Draw the same grid spacing on the isometric plane. Transfer the position of the dots from the orthographic view to similar points on the isometric grid.

Step 4. Connect the points with a smooth, freehand-sketched line.

Step 5. From each point on the curve, project a 30° line upward to the right.

Step 6. Measure and transfer the thickness of the object at each point, to points representing the back side of the curved surfaces. If the thickness of the object is the same throughout, these distances will all be equal.

Step 7. Connect the points on the back surface of the curved areas with a freehand-sketched line to complete the layout.

Step 8. Complete the drawing with instruments and an irregular curve.

Isometric Sections. Isometric drawings are normally used to show only the exterior appearance of an object. When the inside of an object must be shown in pictorial form, an isometric section is also prepared. The two basic types of isometric sections are the full isometric section and the half isometric section.

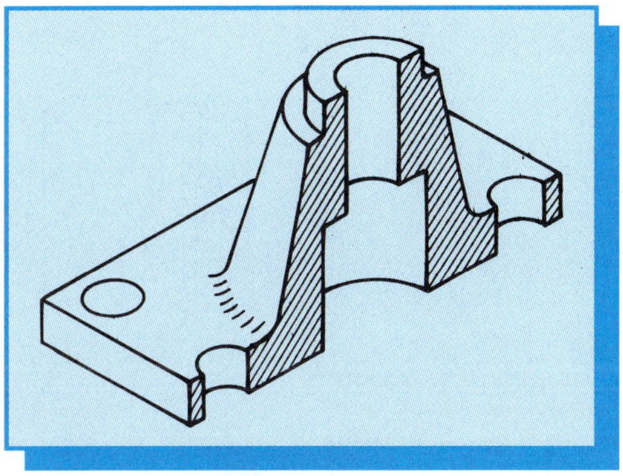

Figure 15-18
Full isometric section

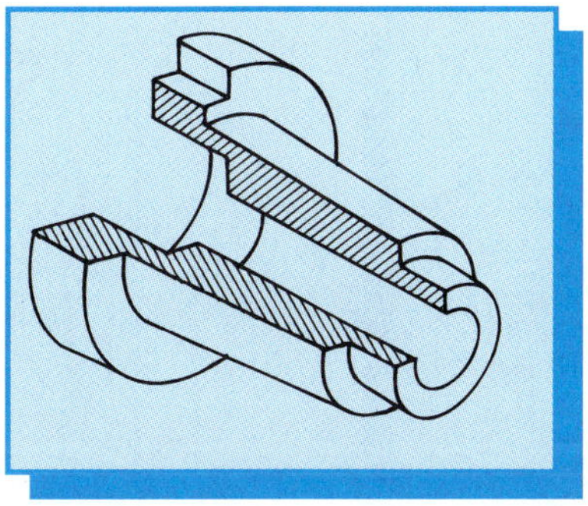

Figure 15-19
Half isometric section

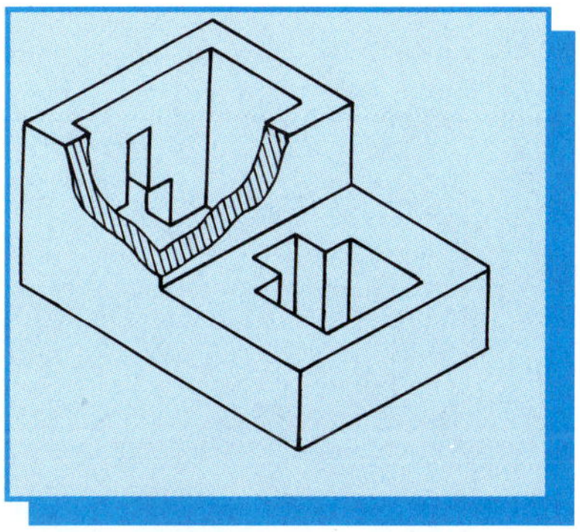

Figure 15-20
Broken-out isometric section

Isometric sections, like orthographic sections, follow a cutting plane line that defines the edge of the area to be sectioned. In a **full isometric section**, the cutting plane extends across the entire width or height of an object **(Fig. 15-18)**. A full isometric section cuts the object in half to show its inside half. In a **half isometric section**, the cutting plane extends only halfway through the object. Half sections show only a one-fourth cutaway view of the object **(Fig. 15-19)**.

Symmetrical objects are usually half-sectioned to their centerline. When objects are not symmetrical, or do not contain a centerline, a separate broken-out isometric section may be prepared. In **Figure 15-20**, part of the wall is removed to reveal details on the back surface. **Broken-out sections** are also used to show the shape of an object that may be difficult to interpret from the isometric drawing. The drawing of the nuclear reactor vessel includes a broken-out section to expose the shape of the interior of the vessel **(Fig. 15-21)**.

Sometimes a sectional view is removed to show the shape of an object without changing the form of the isometric drawing **(Fig. 15-22)**. There are two methods of drawing isometric sections. The first involves drawing the complete object lightly **(Fig. 15-23)**. The cutting plane line or outline of the cutaway area is drawn. Lines representing the cutaway are erased to show the section. In the second method, only the sectioned area part is drawn **(Fig. 15-24)**.

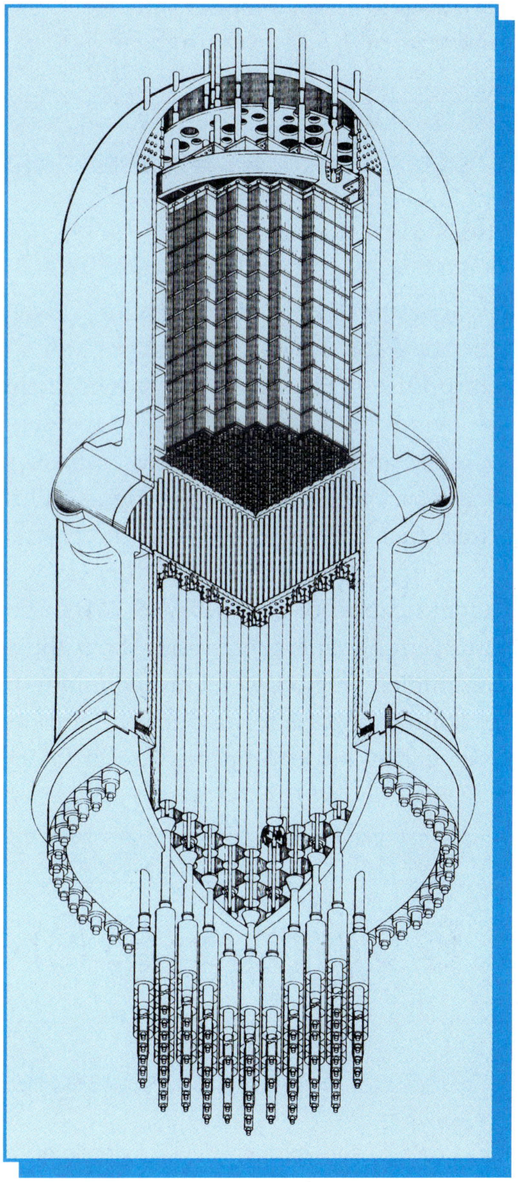

Figure 15-21
Isometric cutaway sectional drawing of a nuclear reactor vessel *(Courtesy of Combustion Engineering, Inc.)*

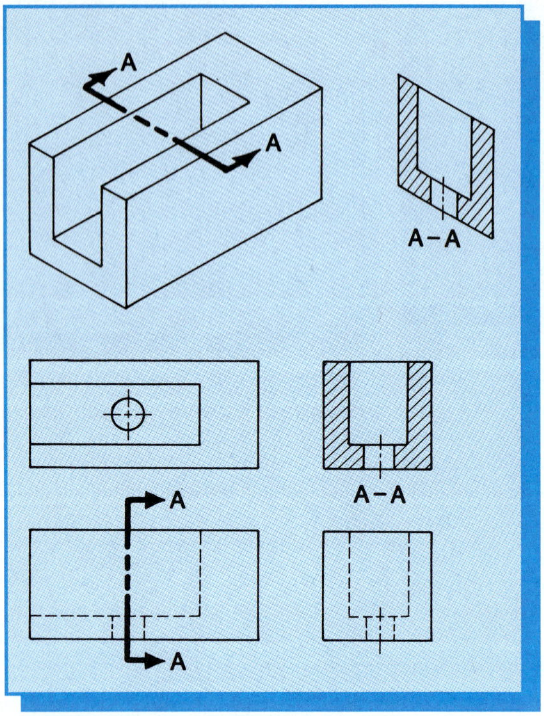

Figure 15-22
Isometric removed section

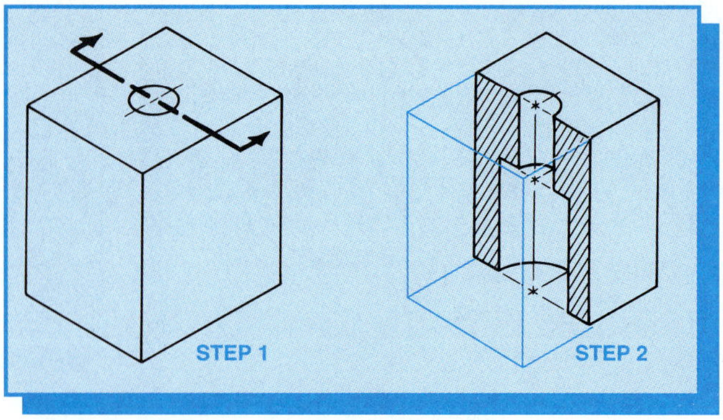

Figure 15-23
Isometric section

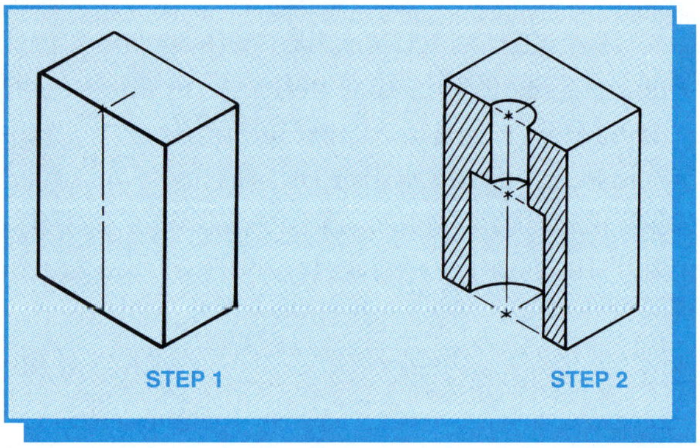

Figure 15-24
Lay out only the isometric section.

Isometric Exploded Drawings. Isometric drawings are ideal for showing various parts of a multipart component in related position **(Fig. 15-25)**. Drawings of this type are exploded-view drawings because the parts are separated (disassembled). The parts are aligned and related to their assembled position. Exploded views allow each part to be drawn showing its full shape. This eliminates hidden areas that do not show when the parts are assembled.

Exploded-view assembly drawings also allow for the labeling and identification of parts **(Fig. 15-26)**. In very complex assembly drawings, connecting center lines show the relationship of parts when assembled **(Fig. 15-27)**. This technique provides a guide for the assembly of parts to produce the finished component. Exploded views may be accompanied by an assembly drawing of the

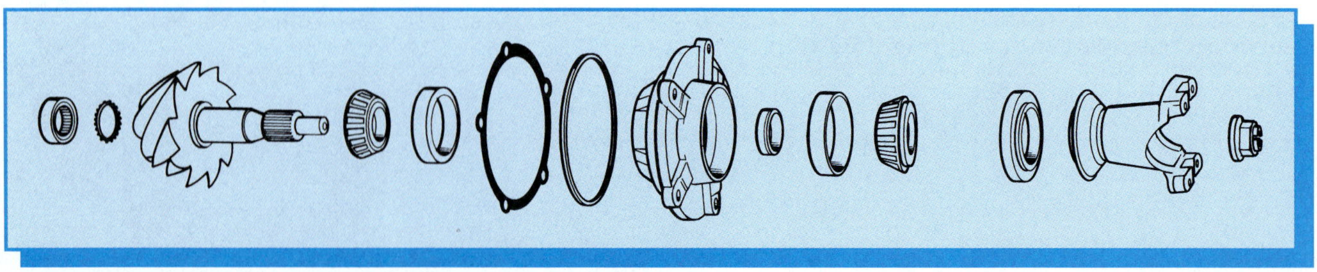

Figure 15-25
Example of an exploded-view isometric drawing used for assembly instruction

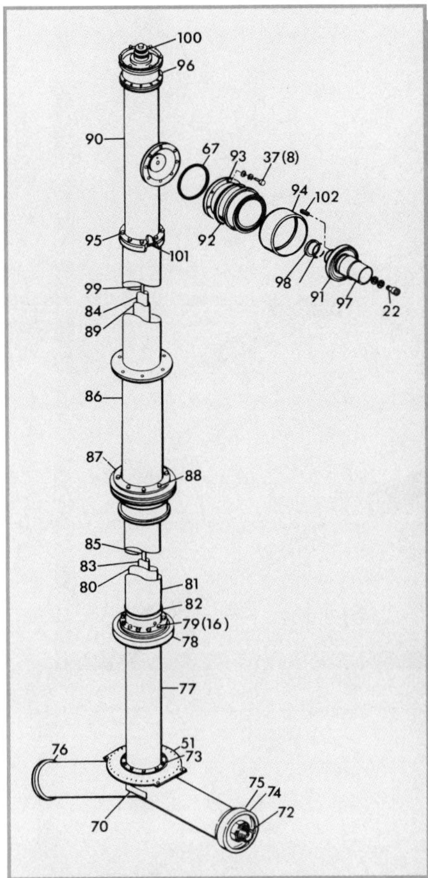

Figure 15-26
Exploded-view drawings are used to label parts for identification.

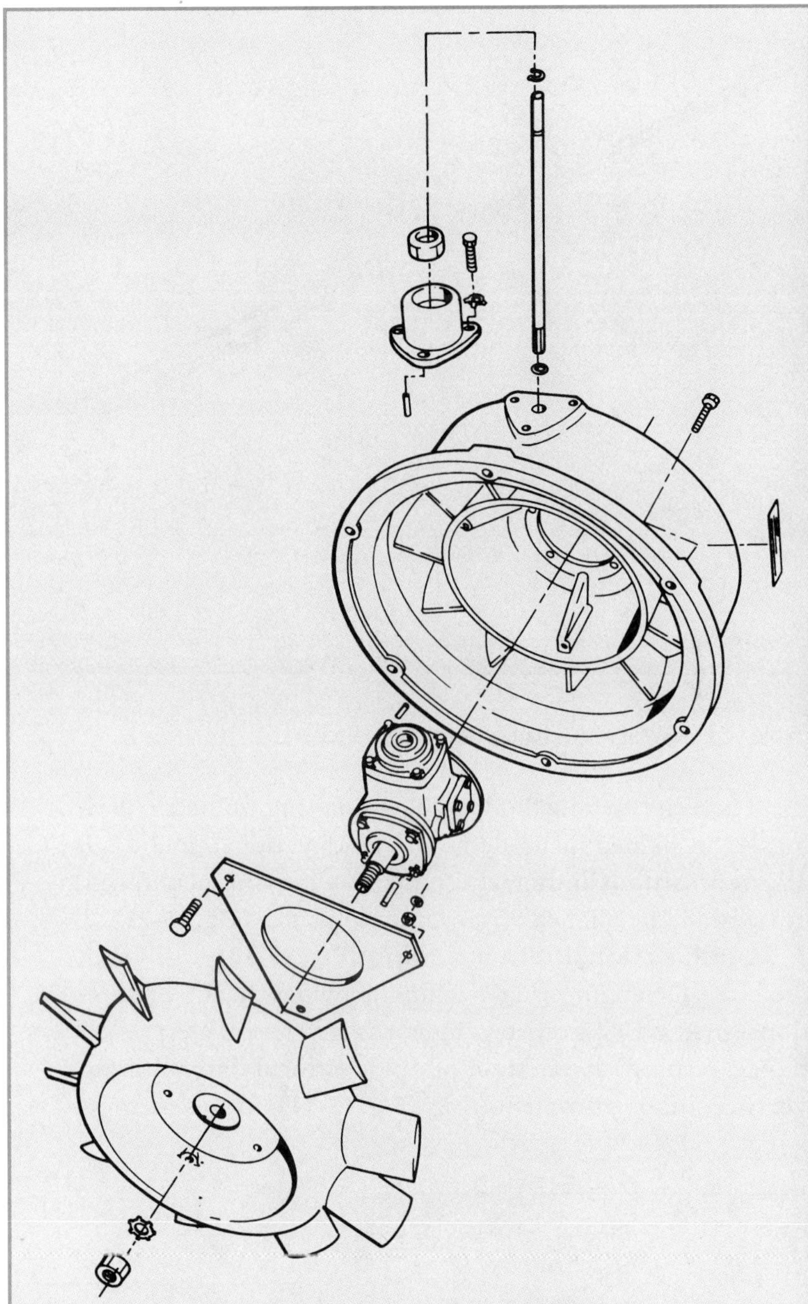

Figure 15-27
Centerlines are used to align mating parts of an assembly.

component with the parts all assembled **(Fig. 15-28)**.

Isometric Dimensioning. Since isometric drawings are not normally used as working drawings, detailed dimensions are not found on most of them. However, when isometric drawings of very basic objects are used for manufacturing or construction purposes, dimensions, notes, and specifications are included.

Dimension lines, extension lines, arrowheads, letters and numerals on isometric drawings are similar to those on orthographic drawings. Isometric dimensions and extension lines all follow the isometric axis. Arrowheads are also drawn in the same plane as dimension and extension lines **(Fig. 15-29)**.

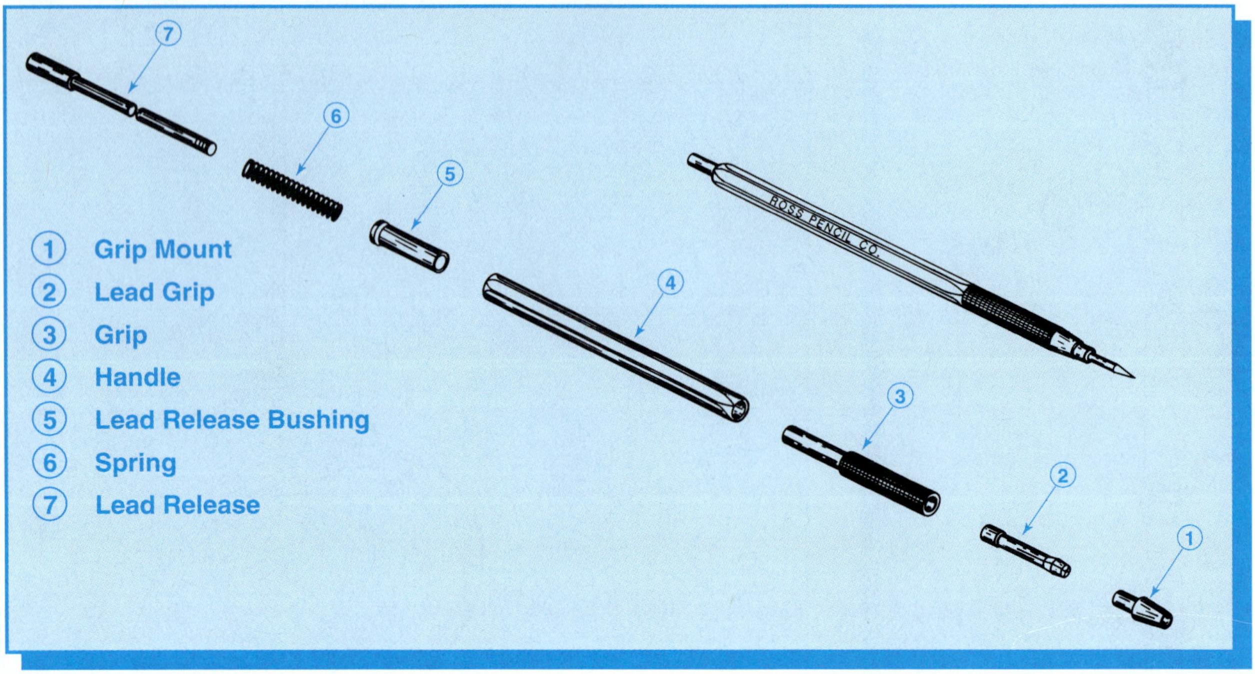

Figure 15-28
Exploded view of a mechanical drafting pencil

There are two methods of dimensioning isometric drawings—the aligned method and the unidirectional method. In **aligned dimensioning**, all isometric letters, numerals, and arrowheads are aligned with isometric planes **(Fig. 15-29)**.

Unidirectional dimensioning (Fig. 15-30) is generally preferred for engineering working drawings. In the unidirectional method, all isometric numerals and letters are positioned vertically. A variation of unidirectional dimensioning is **vertical plane dimensioning (Fig. 15-31)**. In this method,

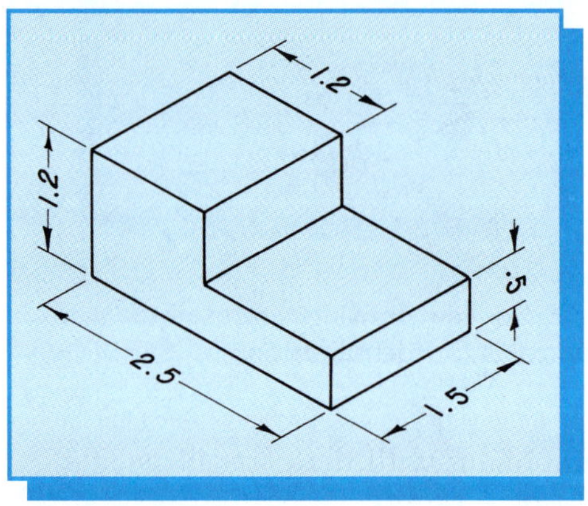

Figure 15-29
Aligned isometric dimensions

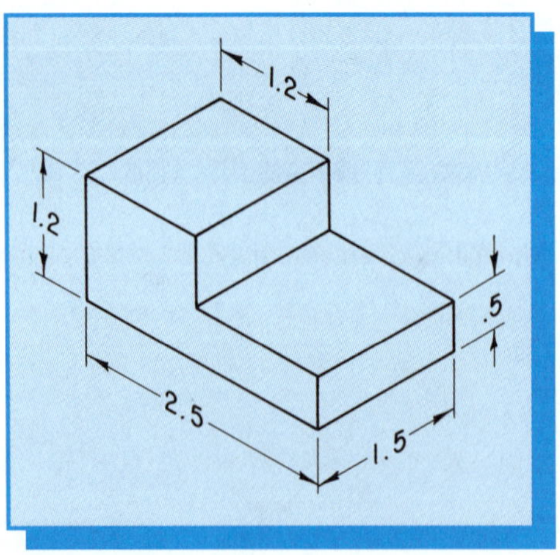

Figure 15-30
Unidirectional isometric dimensions

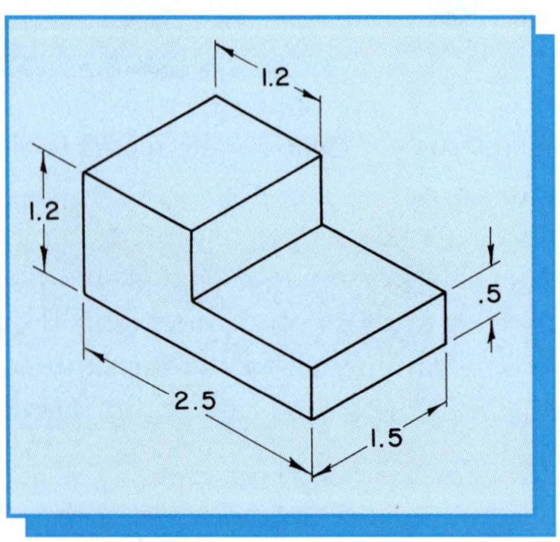

Figure 15-31
Vertical plane dimensioning on an isometric drawing

all dimension and extension lines are drawn in the isometric plane, with the numerals placed horizontally. Straight horizontal lettering is simpler and faster to use on isometric drawings.

Isometric Grids. To save time in preparing isometric drawings, isometric grid paper is often used **(Fig. 15-32)**. The grid lines are printed on 30° angles representing the isometric axis. Grid lines are often printed in nonreproducible blue. This enables an isometric drawing to be prepared directly on the grid paper without the grid lines appearing on the final print. The grid paper may be used as an underlay guide in preparing isometric drawings. The use of grids is a great time-saver for sketching and isometric layout work.

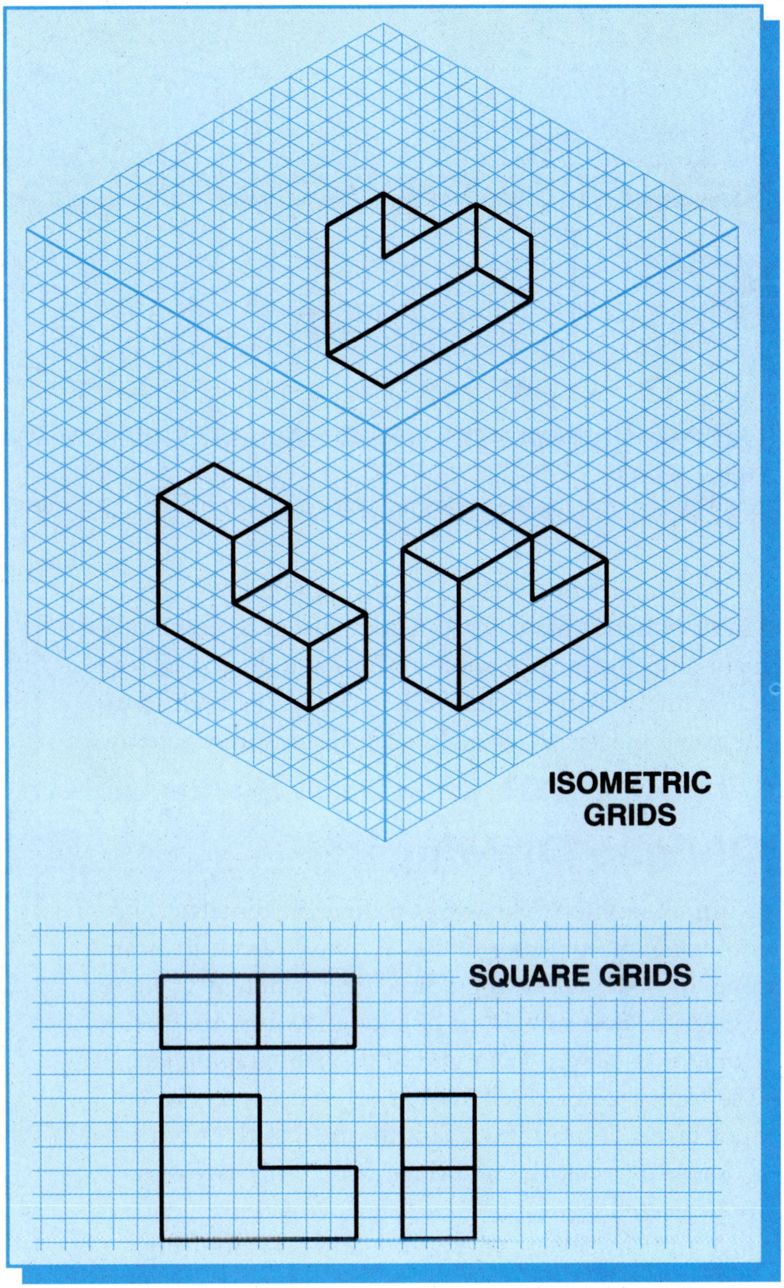

Figure 15-32
Grid papers

Dimetric Drawings

In dimetric drawings, the receding angles may vary, but two of the three faces and axes are always equal **(Fig. 15-33)**. The method of drawing dimetric pictorials is similar to preparing isometric drawings. Once the axis angles are established, all lines extending in the same direction must be kept parallel.

Trimetric Drawings

In isometric drawings, all three angles around the isometric axis are the same. In dimetric drawings, two of the angles are the same. In trimetric drawings, all three of the angles around the axis are different **(Fig. 15-34)**. Trimetric drawings are prepared like isometric and dimetric drawings, except that each pictorial plane is projected on a different angle. All lines in each of the three planes must be parallel. This three-angle difference makes the preparation of trimetric drawings very time-consuming. Trimetric drawings are also difficult to dimension since each dimension and extension line must be drawn parallel to each of the three different planes of projection.

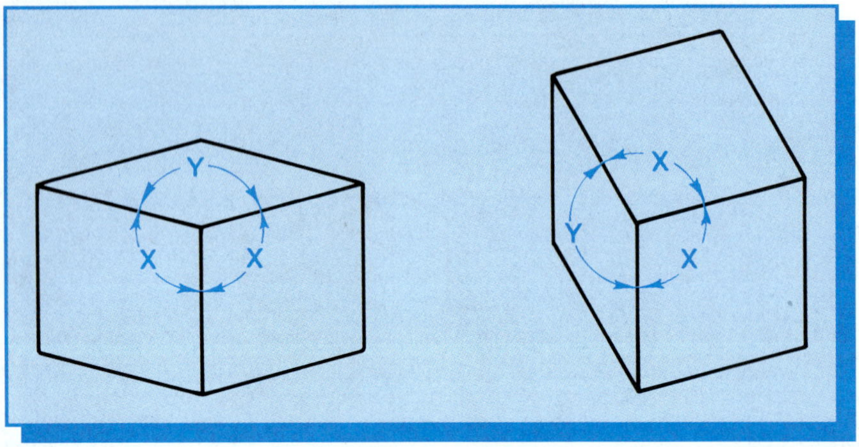

Figure 15-33
In a dimetric drawing the angles may vary, but two of the angles must be equal.

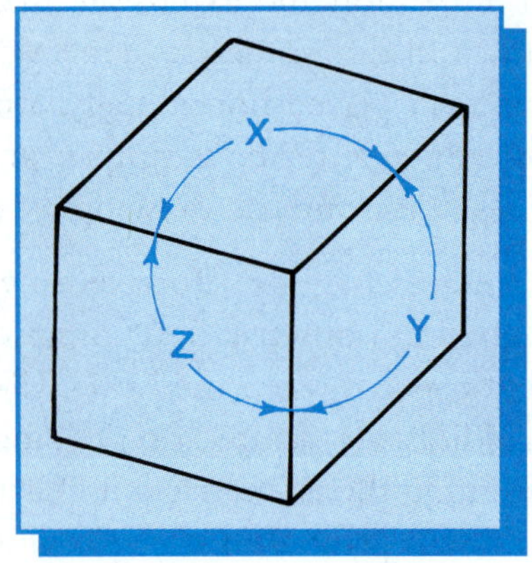

Figure 15-34
In a trimetric drawing angles may vary, but no angles will be equal.

The angles of trimetric drawings, and to a lesser extent dimetric drawings, can be adjusted to any convenient angle to provide more realism and emphasize one side while making another side recede. This is closer to the way objects are normally viewed in isometric drawings.

Oblique Drawings

In axonometric drawings (isometric, dimetric, and trimetric), the object is drawn as viewed from its nearest edge. In oblique drawings, the object is drawn as viewed from its nearest surface. In oblique drawings, the object is drawn as viewed from one of its orthographic faces. The front face of an oblique drawing is parallel to the front plane of projection. This means that the front face is drawn to exact size and shape. The other two surfaces are projected from this front face at any convenient angle (usually 30° or 45°) and in any direction **(Fig. 15-35)**. The angle and direction depend on the nature of the object and the planes to be emphasized. There are two types of oblique drawings—cavalier and cabinet.

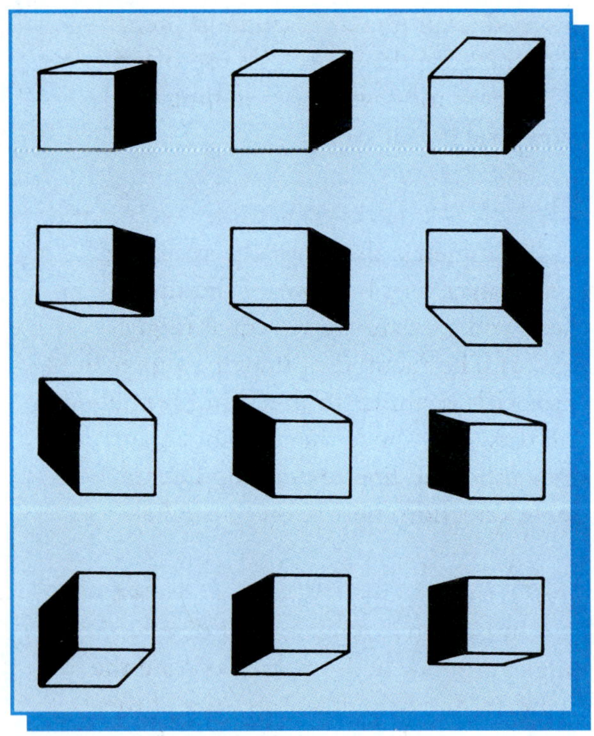

Figure 15-35
Alternate positions for oblique drawings. Receding lines are usually at 15°, 30°, or 45°.

Cavalier Drawings

When the lengths of the receding angles of an oblique drawing are drawn to full scale, the drawing is a **oblique cavalier drawing (Fig. 15-36)**. If the length of the receding lines is excessive, a cavalier drawing will appear distorted.

Because the front face is both geometrically and dimensionally accurate, the front face can be dimensioned the same as in an orthographic drawing. The receding planes can be dimensioned as in isometric drawings with the dimension and extension lines drawn parallel to the angle of the plane.

Cavalier drawings are especially appropriate for objects with a shallow depth dimension, since depth distortion is minimized. The largest area is usually selected as the front plane to avoid as much distortion as possible on the receding planes. Since the front plane of a cavalier drawing is geometrically accurate, the front face should contain the greatest amount of detail or irregularity.

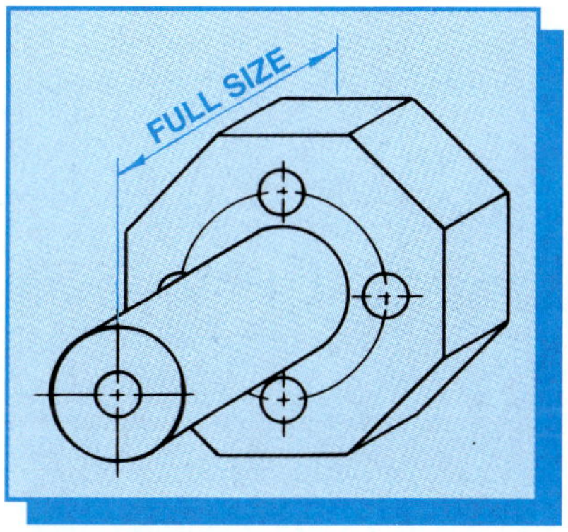

Figure 15-36
Example of an oblique cavalier drawing. All dimensions are full size.

Cabinet Drawings

To reduce the amount of visual distortion, long receding lines are sometimes reduced by one half, but dimensioned true size. When this is done, the drawing is known as a **oblique cabinet drawing (Fig. 15-37)**. This creates a more realistic drawing, but eliminates the use of true scale dimensions on the receding planes.

Figure 15-38 shows a comparison of the types of pictorial drawings. The drafter must be familiar with all pictorial drawing methods in order to select the best possible type for a set of working drawings.

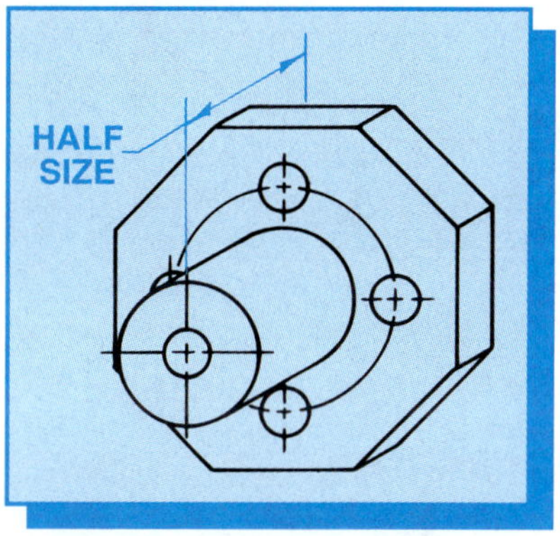

Figure 15-37
Example of an oblique cabinet drawing. Only receding lines are drawn at half size.

Oblique Circles

Because the front face of an oblique drawing is not distorted, angles, circles, and arcs drawn on this surface are true and accurate. For this reason, the front face is chosen to show the circular end view of a cylinder hole or other curved surface **(Fig. 15-39a)**. Circles and arcs shown on the receding planes of an oblique drawing will appear as ellipses. This is similar to the way they appear on axonometric receding planes **(Fig. 15-39b)**.

Both isometric circles and oblique circles on receding planes are drawn as ellipses. When plotting circles on oblique receding planes, follow the steps in **Figure 15-40**.

The accuracy of the ellipse on an oblique drawing is not critical. A faster method of drawing an ellipse by estimation is shown in **Figure 15-41**.

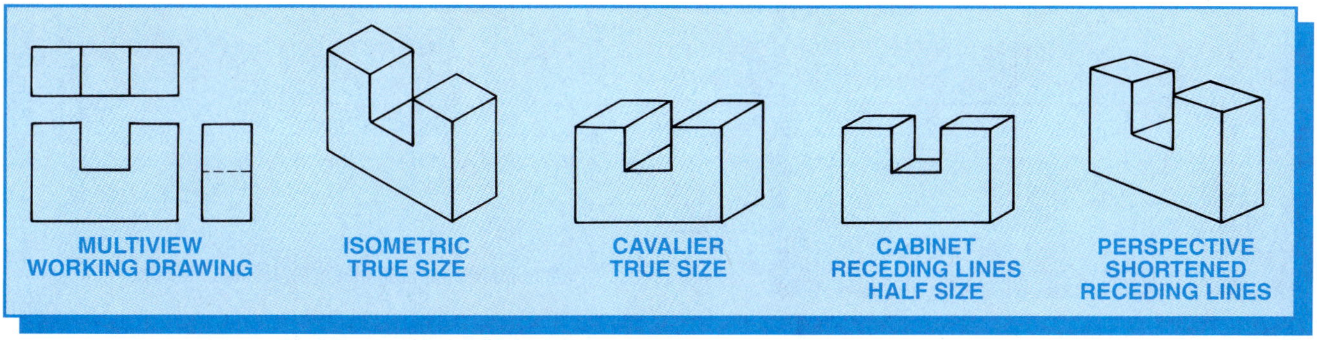

Figure 15-38
Comparison of pictorial drawings

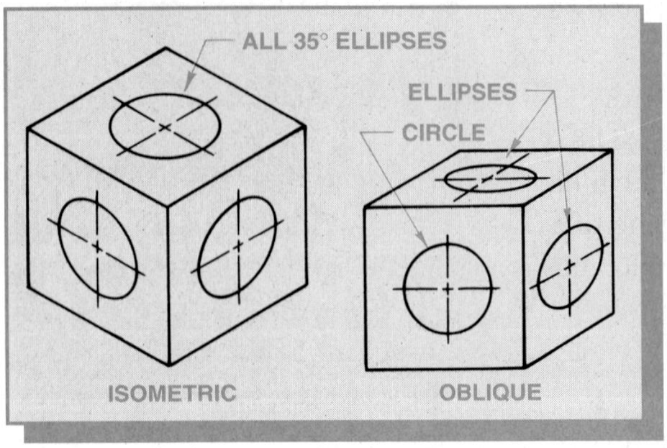

Figure 15-39a
Isometric and oblique circles

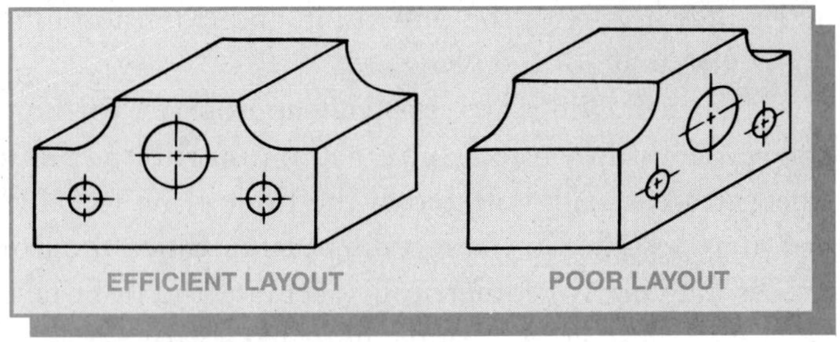

Figure 15-39b
Layout for isometric and oblique circles

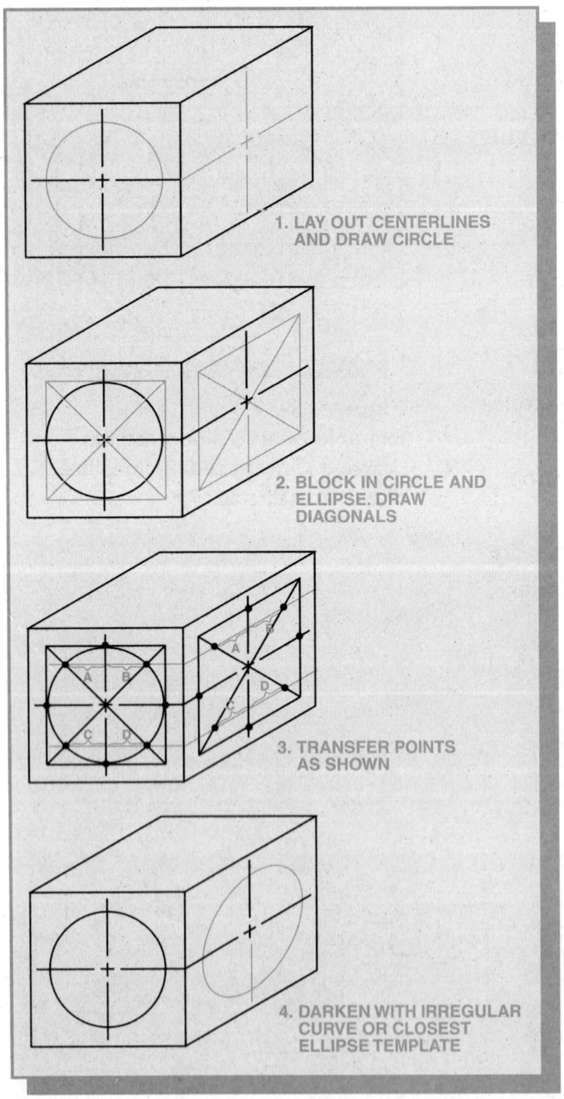

Figure 15-40
Constructing oblique circles

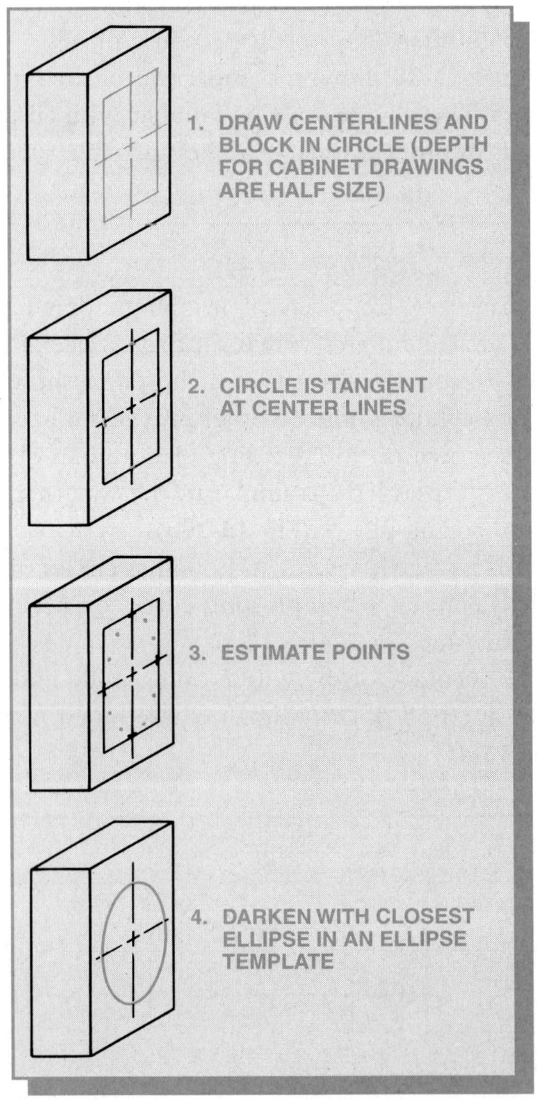

Figure 15-41
Estimation procedure for drawing an ellipse

Oblique Sections

Oblique full and half sections are drawn in a similar fashion as isometric sections. When oblique sections are drawn, they are usually cut through the receding plane **(Fig. 15-42)**. To draw an oblique section, draw the front orthographic view. Project the receding lines, beginning at the points of intersection between the cutting plane line and front plane lines. Extend these lines on the established angle to the depth of each part of the object.

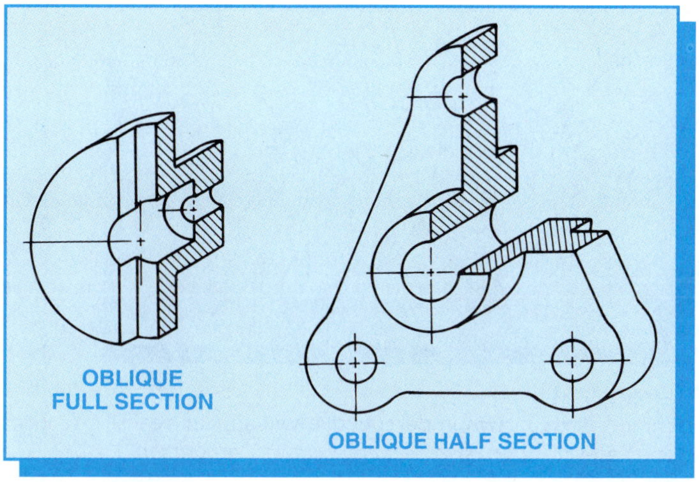

Figure 15-42
Oblique sections

Perspective Drawings

The receding sides of axonometric and oblique drawings are parallel. This creates visual distortion because the human eye does not perceive objects in this manner. It sees close-up objects in true size, detail, and color. Distant objects are seen as smaller, less detailed, and paler in color. The combination of these factors gives the eye the ability to perceive the dimension of depth. Look at the perspective drawing in **Figure 15-43**. Notice how the lines of the road are not parallel in the picture. Yet, in reality the two sides of the road are actually parallel. The camera, like the eye, sees objects in perspective. That is, parts of the object further from the eye appear to get progressively smaller as the viewing distance is increased. The reason objects appear to grow smaller and closer is the curvature of the earth. Therefore, on a perspective drawing the receding lines will converge to a vanishing point on the horizon. Perspective drawings, unlike axonometric drawings, are prepared with the more distant areas proportionally smaller to create the illusion of depth **(Fig. 15-44)**.

Figure 15-43
A perspective view gives the illusion of depth.

Perspective Uses

As previously mentioned, axonometric and oblique drawings of large objects appear greatly distorted because the large receding sides are parallel. For this reason, perspective drawings are often prepared for large objects, such as buildings, cars, boats, and machinery. **Figure 15-45** is a perspective drawing of a building housing a steam generator. Notice the increased realism found in the perspective drawing.

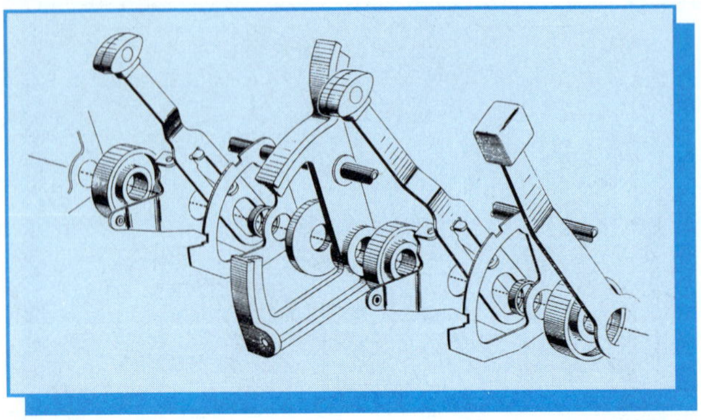

Figure 15-44
Machined parts drawn in perspective will appear realistic to the eye. *(Courtesy of General Motors Corporation. Engineering Standards)*

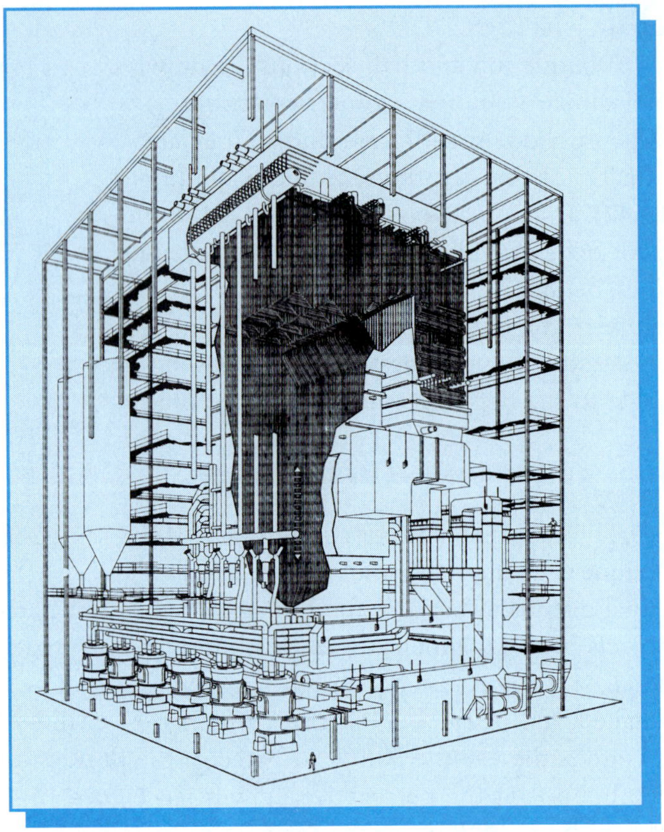

Figure 15-45
Perspective drawings show depth with the diminishing lines *(Courtesy of Combustion Engineering, Inc.)*

Perspective Types

There are three basic types of perspective drawings: one-, two-, and three-point perspective **(Fig. 15-46)**. The type of drawing used depends on the size and shape of the object and the preferred viewing angle.

One-Point Perspective

A **one-point perspective** (parallel) drawing is similar to an oblique drawing, except the receding lines are not parallel. The receding lines of a one-point perspective drawing meet at a distant location called a **vanishing point (Fig. 15-47)**. Like oblique drawings, the front face is drawn at true scale. The receding planes can be projected in any direction from this true front plane.

The vanishing point can be located any distance from the horizon. When the distance from the horizon to the vanishing point is short, the angles of the receding sides will be more acute **(Fig. 15-48)**.

Before preparing a one-point perspective drawing, the concepts of station point, picture plane, and lines of vision must be understood **(Fig. 15-49)**. The **station point** represents the location of the observer. The final drawing will therefore appear as viewed from the station point. The **picture plane** represents the vertical plane upon which the object is viewed. The distance from the station point to the picture plane determines the size of the drawing. An object

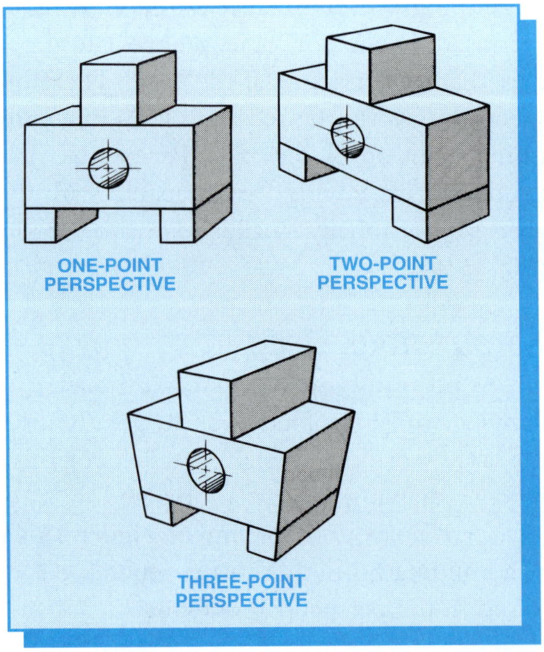

Figure 15-46
Examples of one-, two-, and three-point perspective drawings

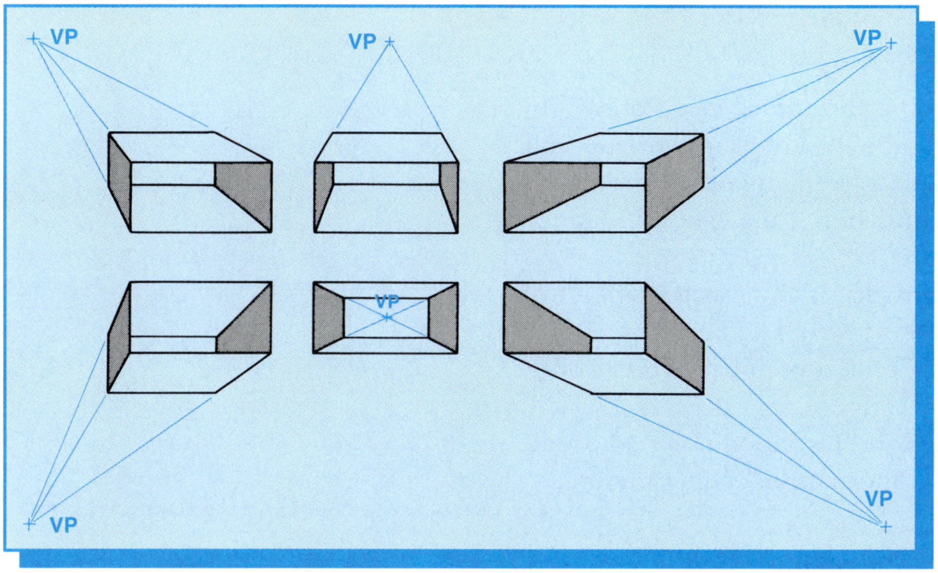

Figure 15-47
Receding lines can be projected to vanishing points at any position.

Figure 15-48
The distance or position that the vanishing point is placed from the drawing will affect its final form.

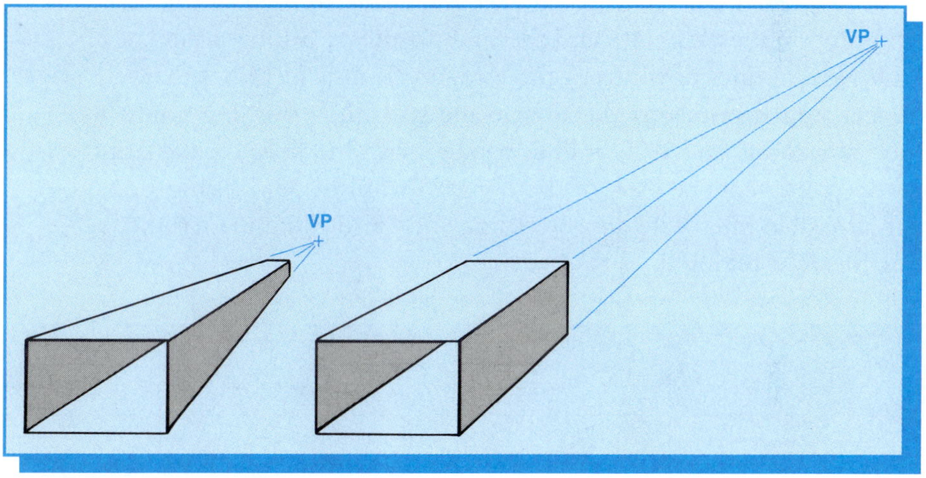

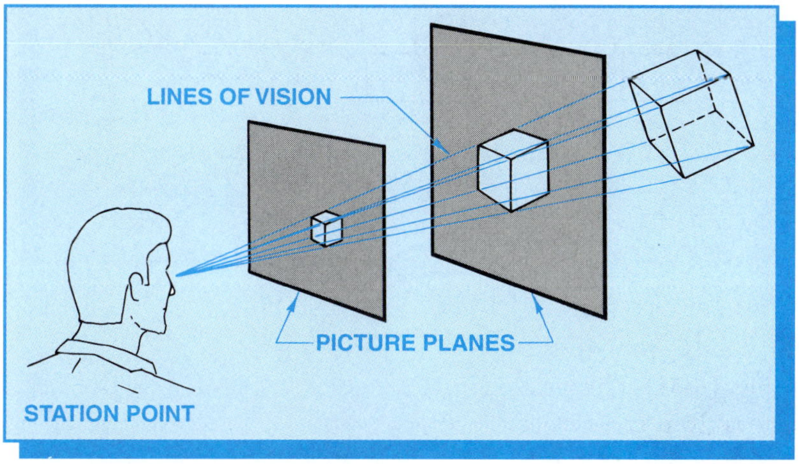

Figure 15-49
The location of the picture plane controls the size of the perspective drawing.

on a picture plane closer to the station point will appear smaller than the same object on a picture plane further from the station point. **Lines of vision** are imaginary lines connecting the station point with the picture plane and object.

To draw a one-point perspective of the exterior of an object, follow the steps outlined in **Figure 15-50**:

Step 1. Draw the front and top orthographic views of the object. Draw horizontal lines representing the picture plane, horizon, and ground line. The **horizon** is a horizontal line upon which the vanishing points are placed. The **ground line** is the base line upon which the front plane is viewed.

Step 2. Locate the station point; that is, the exact position from which the object (top view) will be viewed. The final drawing will appear as viewed from this point. Connect each back corner of the top view with the station point.

Step 3. Project a vertical line from the station point up to intersect the horizon line. This point of intersection is the vanishing point.

Step 4. Connect each corner of the front view with lines to the vanishing point.

Step 5. Establish points of intersection between each of the station point connecting lines and the lines to the vanishing point. Draw vertical lines down from these points to intersect lines that connect similar points on the front plane with the vanishing point. Draw horizontal and vertical lines parallel with the lines on the front plane to represent the back of the object. To finish the drawing, connect all visible lines with an object line and erase all construction lines. Shade if required.

Figure 15-50
Steps for drawing with one-point perspective

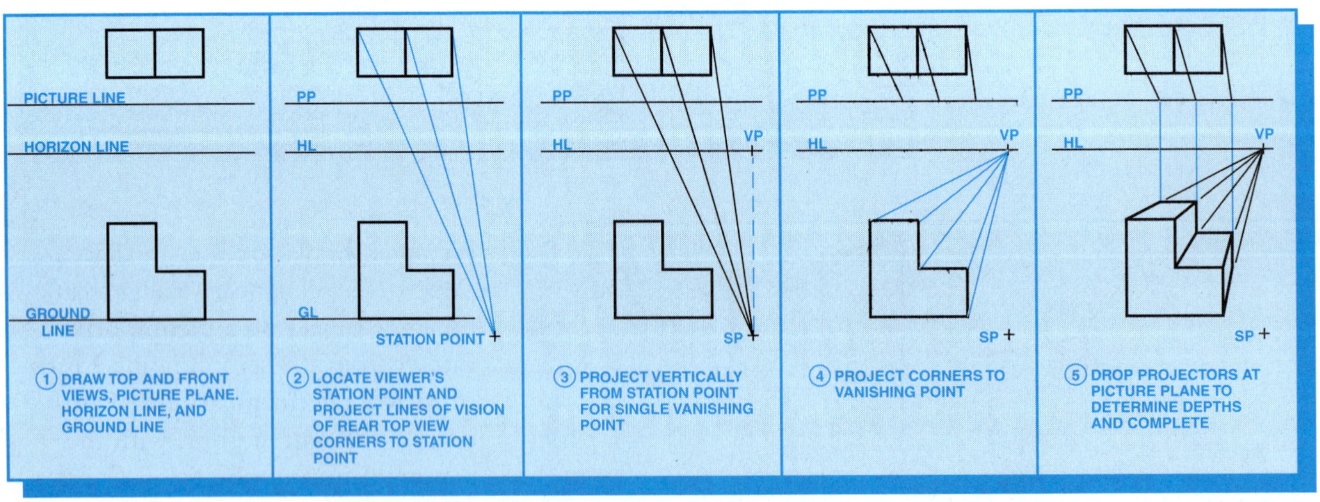

If part of the front plane's detail is missing, lightly block in the entire front plane, including the empty areas **(Fig. 15-51)**. This provides a base area from which to project light lines from the corners of the front face to the vanishing point. Draw the outline of the object on the receding sides using heavy lines. Erase the construction lines to complete the drawing.

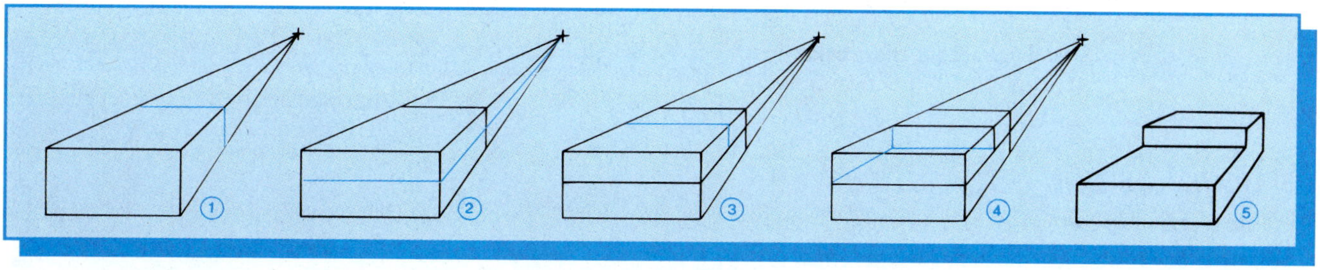

Figure 15-51
Laying out and blocking in a one-point perspective

One of the most common uses for a one-point perspective drawing is to show the interior view of a building **(Fig. 15-52)**.

Two-Point Perspective

Two-point perspective (angular) drawings **(Fig. 15-53)** are most popular because they provide the greatest amount of realism with a minimum amount of complex projection. Two-point perspective drawings are prepared when viewed from a corner, not from a front orthographic face. Each of the two sides are projected to separate vanishing points on the horizon. **Figure 15-54** shows how the two vanishing points relate to the object, picture plane, horizon, and station point.

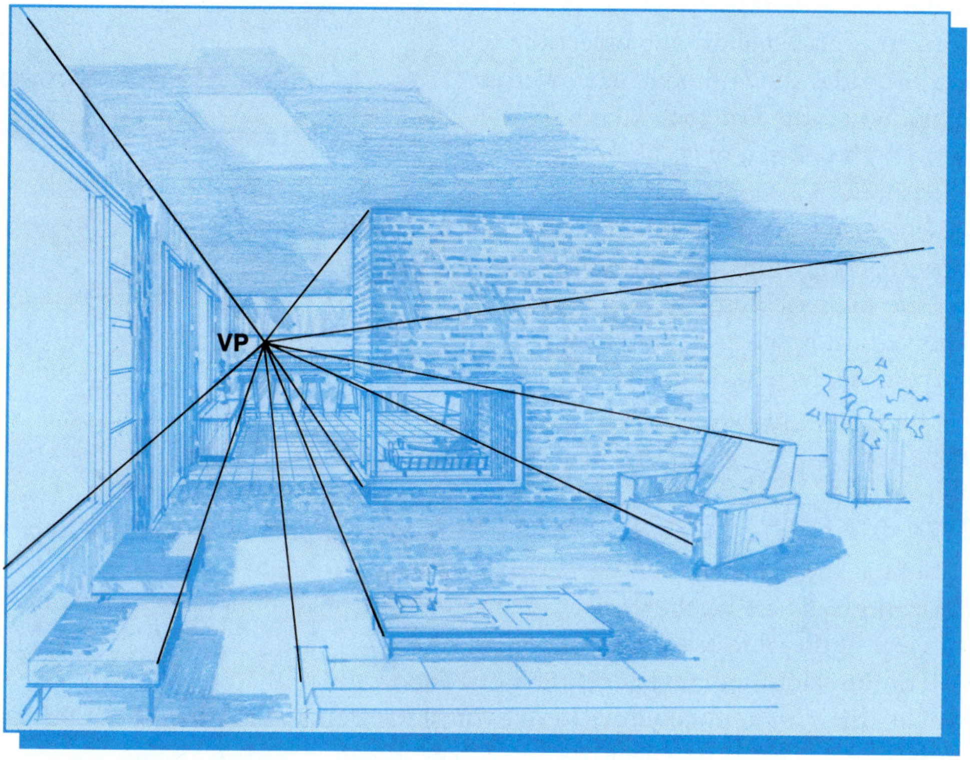

Figure 15-52
Example of a one-point interior perspective *(Courtesy of Home Planners, Inc.)*

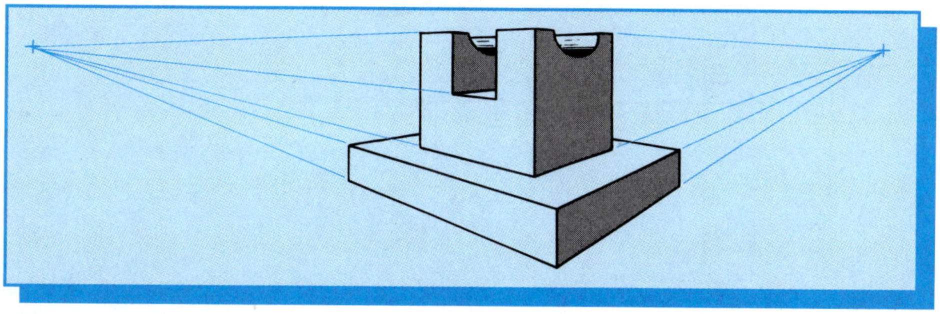

Figure 15-53
Example of a two-point perspective

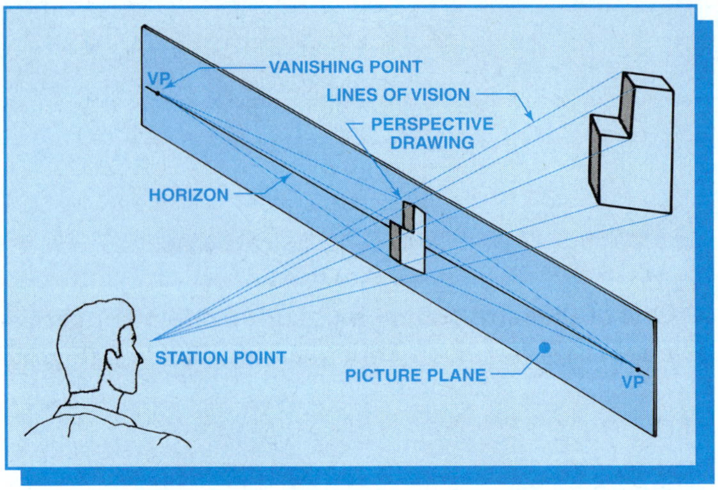

Figure 15-54
Deciding factors that make up a two-point perspective drawing

Vertical Position Alternatives

The vertical position of the horizon line in relation to the object affects the amount of each surface on the final drawing **(Fig. 15-55)**. When the horizon is placed below the object, the bottom and two sides show (producing a worm's-eye view). When the horizon is placed above the object the top and two sides show (producing a bird's-eye view). When the horizon extends through the object, only the two sides show. This view more closely relates to an eye-level view, especially for large objects that are higher than eye level. The vertical position of the horizon is selected to provide the most appropriate viewing angle.

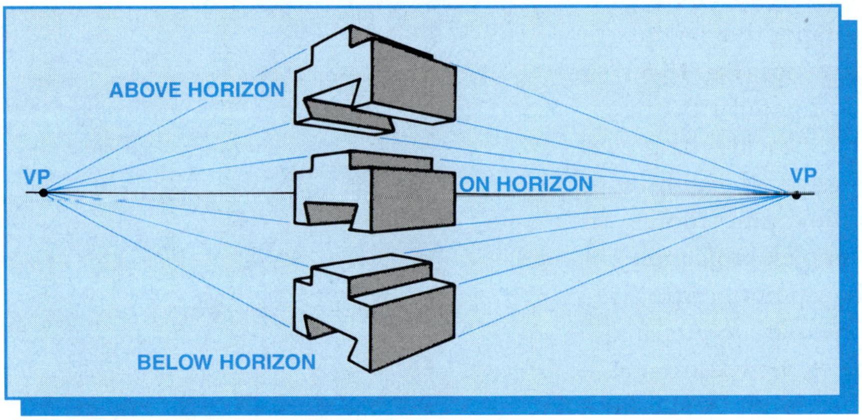

Figure 15-55
The vertical position of the viewer affects the perspective drawing.

Station Point Position

Just as the placement of the horizon greatly affects the vertical view of the object, the position of the station point changes the horizontal view of the object. Moving the station point closer to one vanishing point and farther away from the other results in the differences shown in **Figure 15-56**.

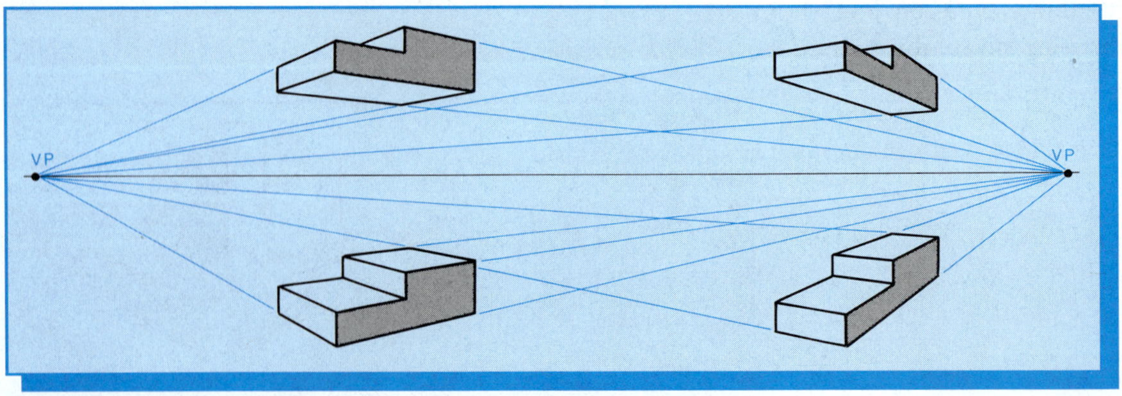

Figure 15-56
As the viewer's position changes, the perspective drawing will change.

Vanishing Point Location

In addition to the vertical location of the horizon and the horizontal location of the station point, the final appearance of the drawing is greatly affected by the placement of the vanishing points on the horizon. **Figure 15-57** shows a perspective drawing with widely spaced vanishing points compared with a drawing with closely spaced vanishing points. To insure the base angle is more than 90°, spread the vanishing points apart on the horizon line and/or draw the perspective layout closer to the horizon line **(Fig. 15-57)**. The station point should be located below the picture plane at a distance of approximately two times the width of the object **(Fig. 15-58)**.

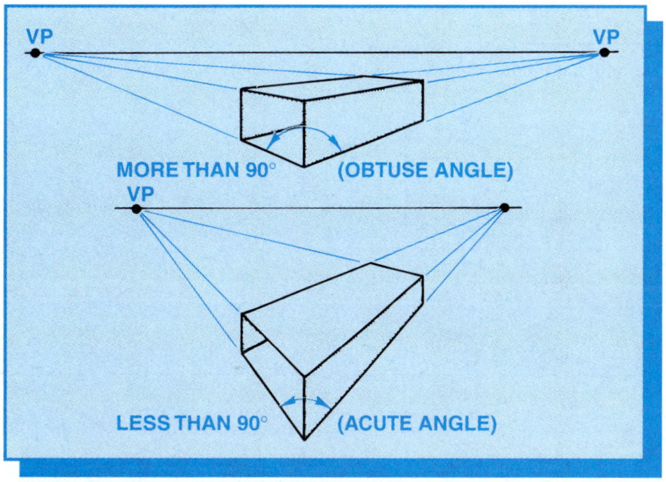

Figure 15-57
Placing the vanishing points too close together will cause the drawing to appear distorted.

Estimating Procedure

Since perspective drawings are not used as working drawings, drawing to precise dimensional limits is usually not required. When only rough proportions are sufficient, a two-point perspective drawing can be prepared by using the following estimating procedures **(Fig. 15-59)**:

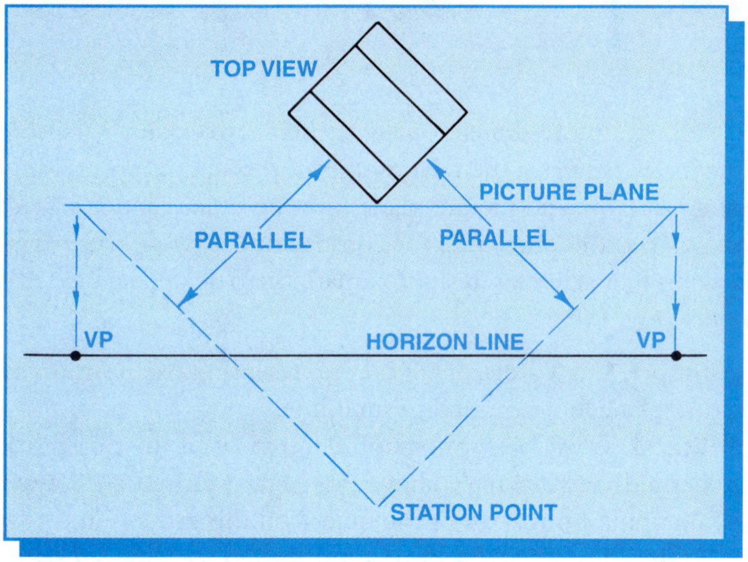

Figure 15-58
Locating the vanishing point's position on the horizon

Step 1. Draw a horizontal line and locate the position of two vanishing points. Draw a vertical line representing the leading edge of the object above or below the horizon, depending on the view required.

Step 2. Draw lines connecting the front corner ends with each vanishing point.

Step 3. Estimate the length of the object compared to the height of the front corner line. Mark the estimated length and depth with vertical lines drawn between the receding lines.

Step 4. Draw lines connecting the back receding corners to the opposite vanishing point.

Step 5. Draw details by estimating the position of features on the receding planes.

Step 6. Connect intersecting points with lines extending to the opposite vanishing point.

Step 7. Complete by erasing the construction lines, darkening the object lines, and shading.

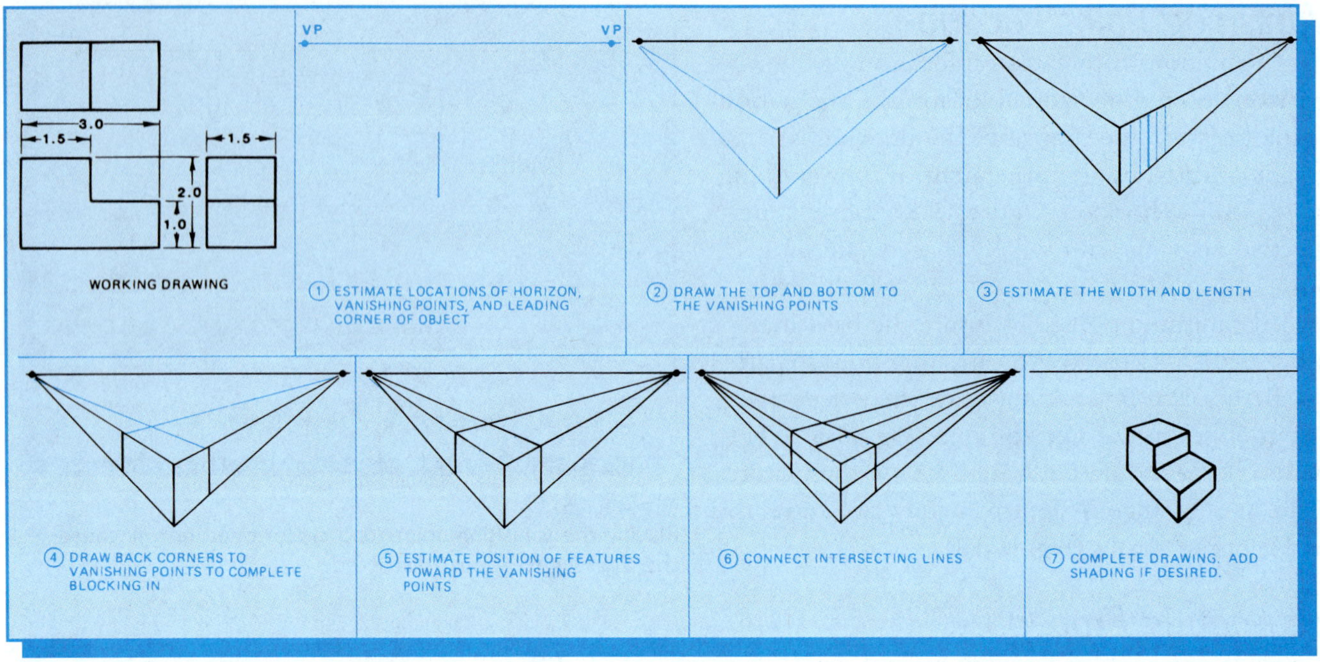

Figure 15-59
Drawing a two-point perspective drawing using an estimating procedure

Construction Procedure

When a more dimensionally accurate two-point perspective drawing is required, the relationship of the station point, vanishing point, horizon, and picture plane must be precisely controlled. The following is the procedure for constructing a two-point perspective drawing that will closely approximate the true appearance of the object **(Fig. 15-60)**:

Step 1. Draw horizontal lines representing the position of the picture plane, horizon, and ground line.

Step 2. Draw the orthographic top view of the object, rotated to the desired viewing angle above the picture plane line. Draw the orthographic front view of the object on the ground line.

Step 3. Locate the station point, and project lines of sight to the picture plane parallel to the sides of the top view. Draw vertical lines downward from the picture plane to intersect the horizon to locate the vanishing points.

Step 4. Draw lines connecting all corners of the object with the station point.

Step 5. From the points of intersection between the projected lines and the picture plane line, project vertical lines downward to intersect the ground line. Mark the line representing the front corner line.

Step 6. From the side view, project the object's height with horizontal lines to intersect the front corner line.

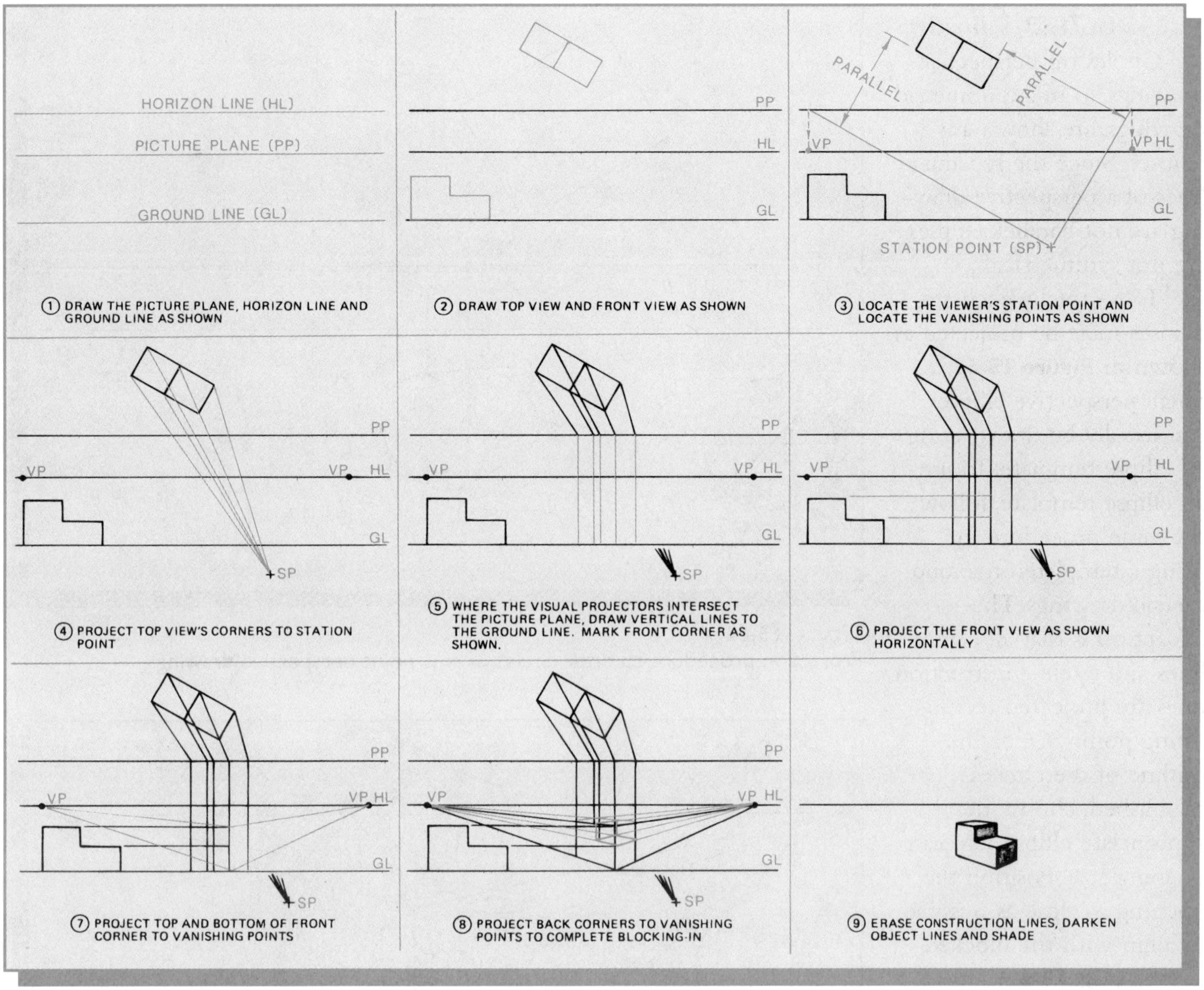

Figure 15-60
Projection procedure for a two-point perspective drawing

Step 7. Establish the height of the front corner line by marking the intersection between the vertical and horizontal projection lines. Project the lines from the bottom and top of the front corner line to each vanishing point. Draw vertical lines representing the width and depth of the object between the receding lines.

Step 8. Project lines from the back corners to the opposite vanishing points. Locate the position of depressions, cutaway areas, extensions, and holes on the planes. Project lines to the vanishing points to establish the final shape. Adjust the edges for any fillets and rounds.

Step 9. Darken the major object lines and erase all construction lines. Shade and render if desired.

Perspective Ellipses

Circles on perspective drawings, as in axonometric drawings, are shown as ellipses. Since the receding lines of a perspective drawing are not parallel, ellipses are not symmetrical.

To be totally accurate, ellipses must be projected as shown in **Figure 15-61**. Small perspective ellipses can usually be drawn using an ellipse template. To use an ellipse template, follow the same procedure for using a template on axonometric drawings. The exception is that all centerlines and circle construction lines are projected to vanishing points. Once the outline of the circle is established, choose the appropriate ellipse template opening size. Position the opening as close as possible to align with the blocked-in area **(Fig. 15-62)**.

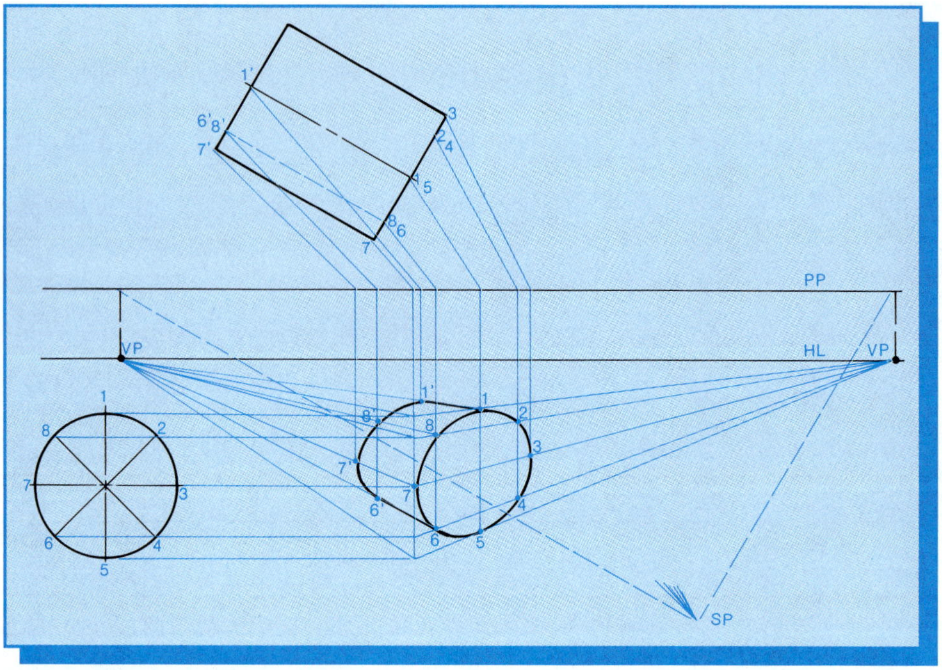

Figure 15-61
Projection procedure to draw circles in two-point perspective drawing

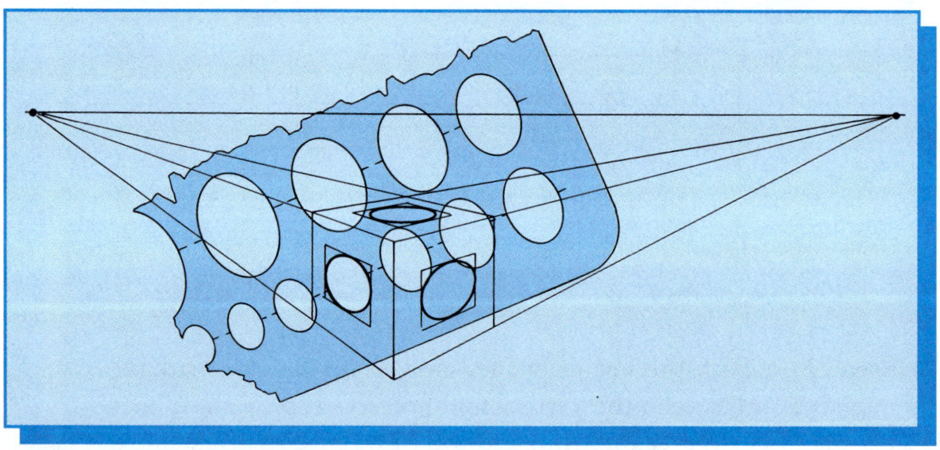

Figure 15-62
To draw an estimated perspective circle, block in a perspective square and find the closest ellipse on an ellipse template.

Perspective Grids

Perspective grid paper, like isometric grid sheets, are often used as a guide in preparing perspective drawings. Grid sheets are available for one-, two-, and three-point perspective drawings, although two-point grids are the most popular. Grids are also available for either interior or exterior perspective drawings. Grids **(Fig. 15-63)** are plotted on specific preset positions of the vanishing point, station point, and picture plane. This limits the options in selecting the exact vertical and horizontal position of the view and angles of the receding sides. Although excellent time-savers, the theory of perspective projection must be thoroughly understood to effectively use perspective grids.

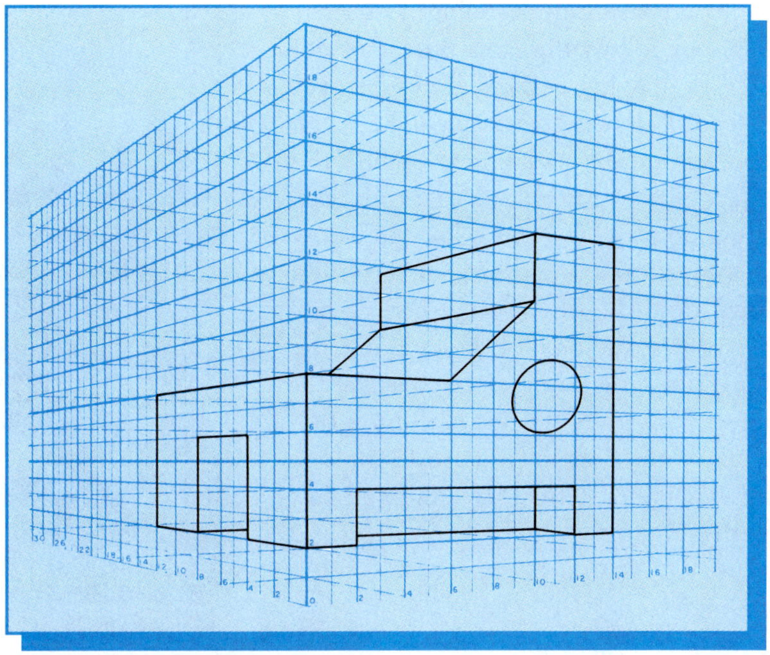

Figure 15-63
Drawing on a perspective grid is fast and will help produce well-proportioned perspective drawings.

Three-Point Perspective

When a two-point perspective drawing contains long, narrow, vertical planes that extend far from the horizon, the object will appear distorted. This optical illusion is created because the vertical lines are parallel. This is not consistent with the normal way the eye sees the object. **Figure 15-64** shows a distorted two-point perspective drawing of this type. **Three-point perspective** drawings are used to correct this illusion by providing a third vanishing point **(Fig. 15-65)**. There are no parallel lines on a three-point perspective drawing since all lines connect to one of the three vanishing points. Lines representing horizontal surfaces are connected to vanishing points one and two, just as in two-point perspective drawings. All lines representing vertical planes are connected to vanishing point three.

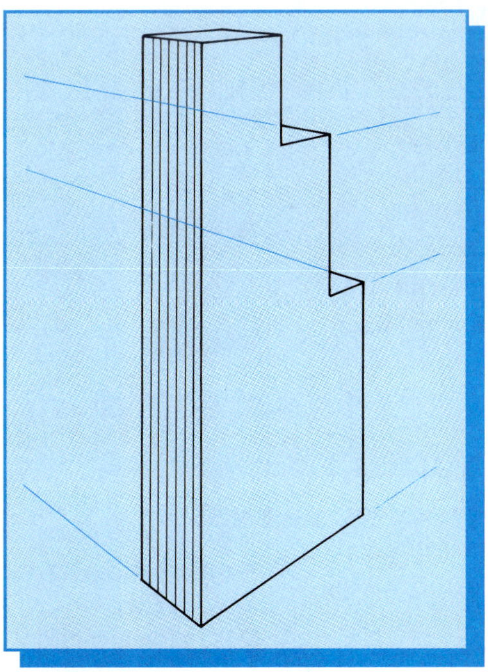

Figure 15-64
Tall structures will appear distorted with a two-point perspective drawing.

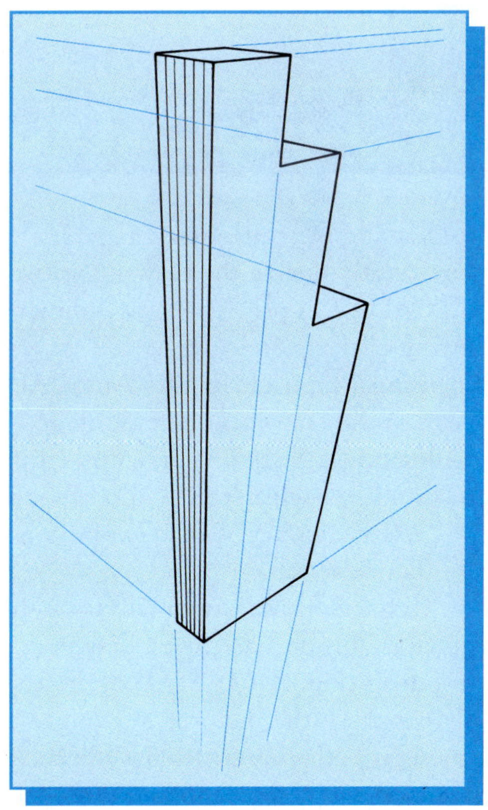

Figure 15-65
Three-point perspective drawings for tall structures add realism.

In preparing three-point perspective drawings, follow the same procedures outlined in drawing two-point perspectives. Instead of drawing vertical parallel lines representing the width and depth of the object, draw vertical lines to a third vanishing point. This point is located below the object on a line projected vertically down from the front corner line. If the viewer is at the base of the building, the third vanishing point is placed above the object. This gives the added illusion of depth to the drawing. **Figure 15-66** is an industrial perspective drawing of an automobile engine.

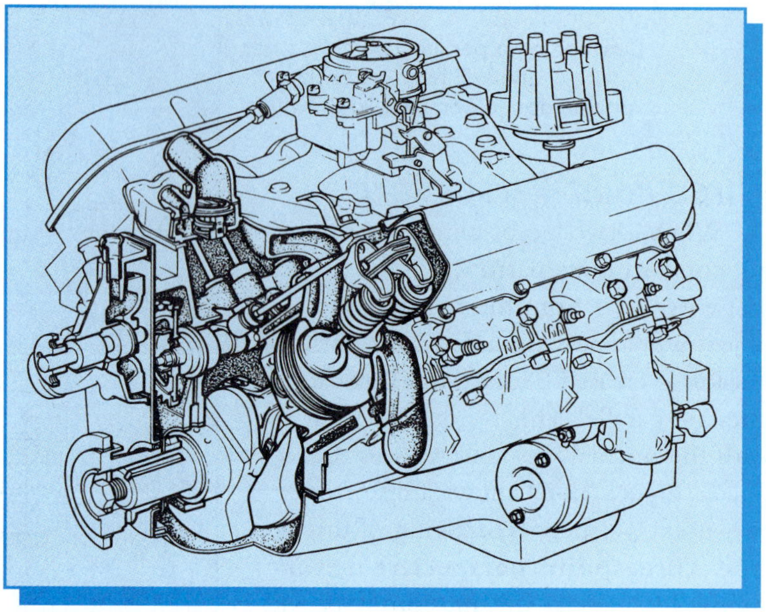

Figure 15-66
A perspective drawing of a V-8 automobile engine *(Courtesy of General Motors Corporation, Engineering Standards)*

AutoCAD LT Exercises

Drawing pictorial drawings with a CAD system varies greatly due to the drawing software available. AutoCAD LT is primarily a two-dimensional (2D) drawing package; however, it does have some three-dimensional (3D) capabilities. AutoCAD LT can draw in 3D using the concepts of elevation and depth, and 3D coordinates (X,Y,Z). The drawings created in this manner are considered **wire frame** 3D drawings. **Figure 15-67** is an example of a twelve-button puck drawn as a wire frame model.

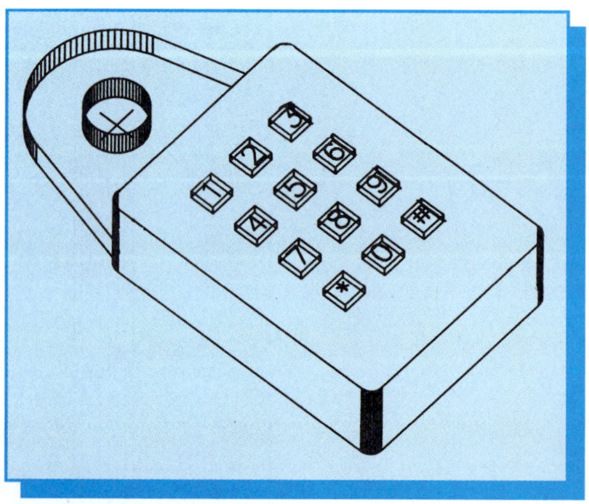

Figure 15-67
3D wire frame model

More expensive systems are truly three-dimensional. These systems do wire frame drawings as well as **solid modeling (Fig. 15-68)**. Some of these systems automatically construct the pictorial drawing as the orthographic drawing is being constructed. Others create the orthographic views from the solid 3D model.

One way to simulate 3D drawings is to draw isometric views. Isometric drawing sometimes may be the quickest way to show a 3D representation. The following shows how to draw a simple isometric drawing.

Note: An isometric drawing may look 3D, but it is only 2D. Therefore none of AutoCAD's 3D capabilities such as viewpoints, hide, and shade, apply to isometric drawings.

Figure 15-68
3D solid model

Isometric Object with Circles

1. From the pull-down menus select **Settings, Drawing Aids**.
   a. Pick the **On** box under the **Isometric snap/grid area**.
   b. Set the **Grid** to **On** and **spacing** to **.25**.
   c. Set the **Snap** to **On** and spacing to **.25**.
   Note: When the Isometric snap/grid is on, only the Y spacing can be set, which sets both the X and the Y spacing on the screen.
   d. The settings should look like **Figure 15-69**. Select **OK**. The screen should look like **Figure 15-70**.
2. It is necessary to change the cursor to the different isometric planes to draw an isometric figure. Two convenient ways to do this are by using the **F5** func-

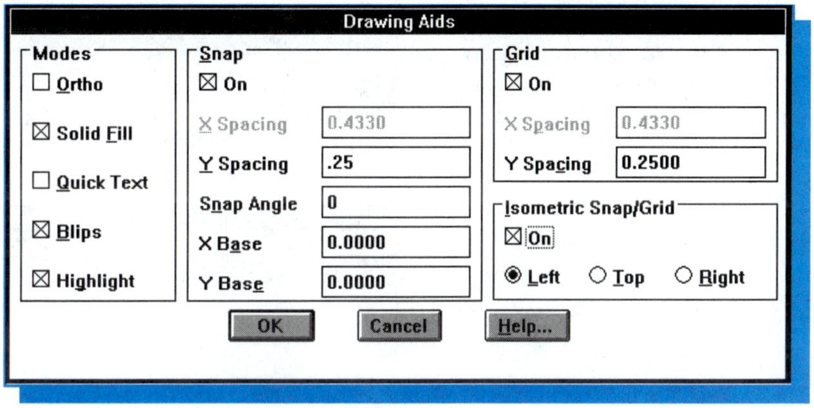

Figure 15-69
Dialog box to set the Isometric Snap/Grid

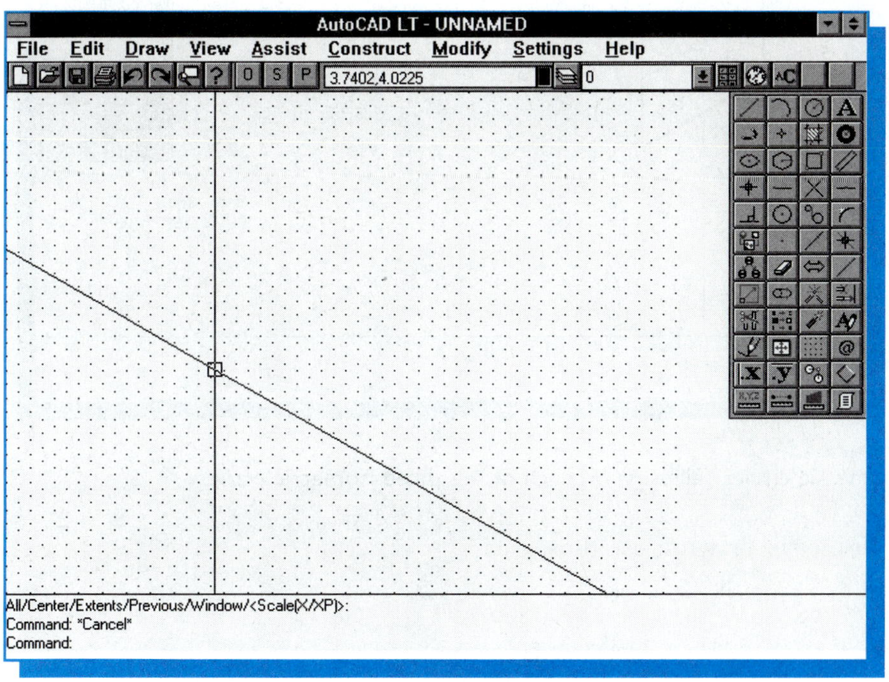

Figure 15-70
Isometric set-up. Note that the cursor will match the isometric grids.

tion key or by pressing **CTRL + E.** Try either of these methods and watch the cursor on the screen.

3. Using **Draw, Line**, draw the Isometric figure in **Figure 15-71**. (Your object will not have the dimensions).
   a. Start at any convenient point.
   b. Make sure that your grid and snap are set to **.250**, and that the **Snap** and **Ortho** commands are on.
   c. As you draw, watch the coordinate display at the top of the screen to get the proper size. To see the length of the line along the isometric axes use the F6 function key to view polar coordinates. Example: 2.000<30.

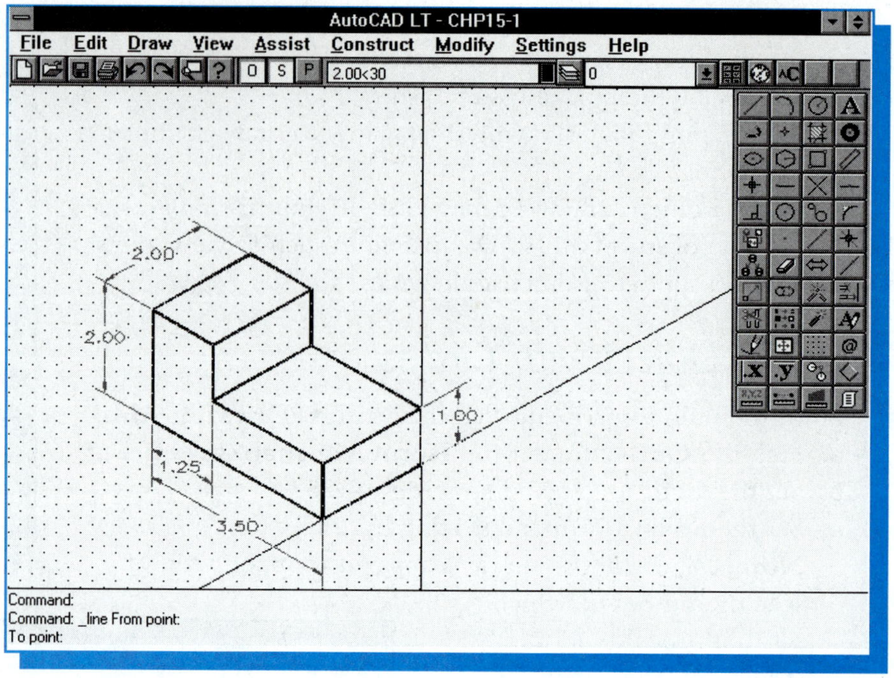

Figure 15-71
Draw the Isometric drawing using the grid Snap. (Each line end Is on a grid point.) Do not dimension.

Note: Polar coordinates will only work when in the Line command.

Isometric circles (ellipses) are unique, and AutoCAD has an isometric circle command to draw them. **Figure 15-72** shows a box containing an isometric ellipse in each of the three isometric planes. When drawing the isometric circles, it is necessary to be in the proper isometric plane—remember that F5 changes the planes.

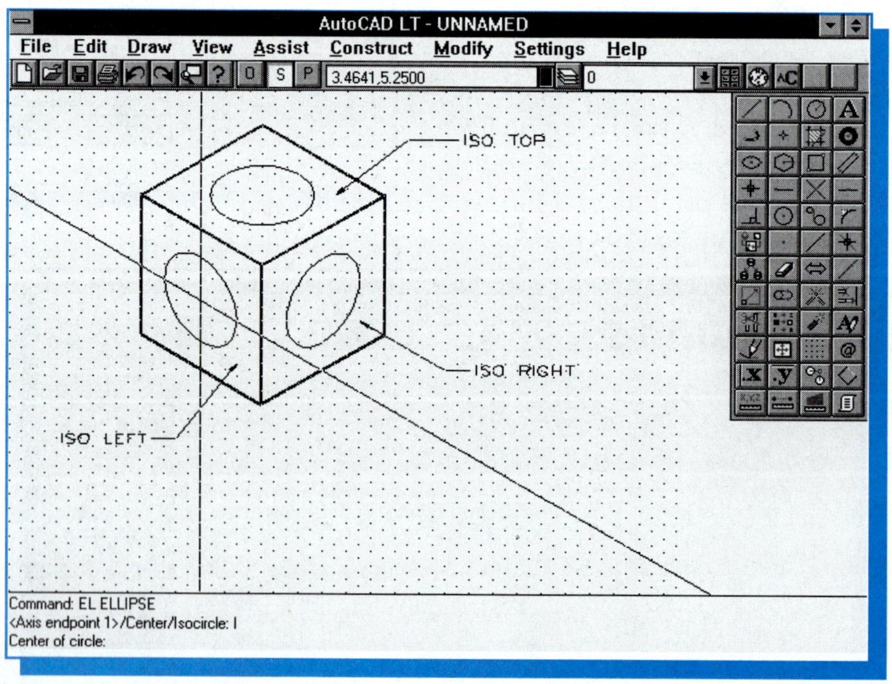

Figure 15-72
Isometric circles (ellipses) on each of the three isometric planes

To add circles and arches to our Isometric drawing use the **Draw, Ellipse, Isocircle** commands.

Note: The Fillet command *does not work* correctly with the isometric drawings.

1. Select **Draw, Ellipse, Isocircle** commands.
   a. Using the F5 function key, change the cursor until it appears to be on the surface where an isocircle is to be placed.
   b. Pick the desired center location of the ellipse.
   c. Type in or snap to the desired radius.
   d. Make your drawing look like **Figure 15-73**.

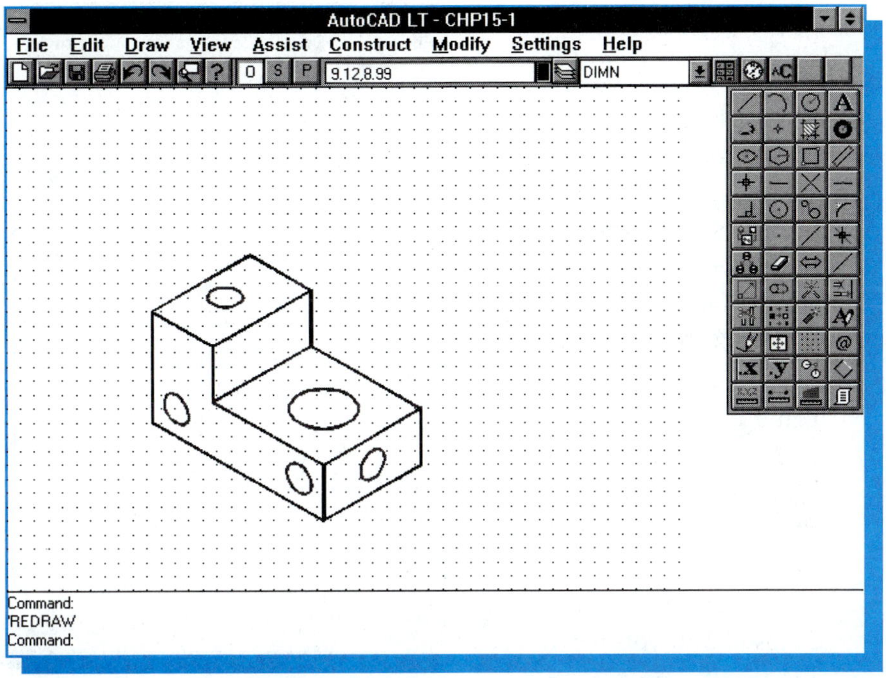

Figure 15-73
Circles (ellipses) placed on different isometric planes, using the **Ellipse, Isocircle** command

Figure 15-74
Placing isocircles to construct fillets

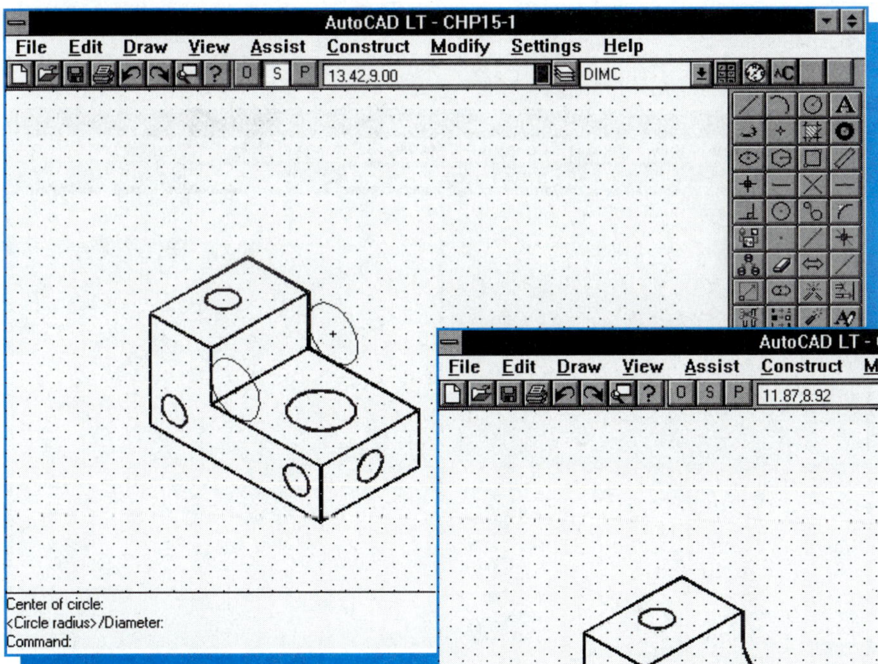

2. To draw fillets, place isocircles tangent to the lines where the fillets are to appear, as shown in **Figure 15-74**.
3. Use the **Trim and Erase** commands to complete your isometric drawing **(Fig. 15-75)**.

Figure 15-75
Complete isometric drawing.

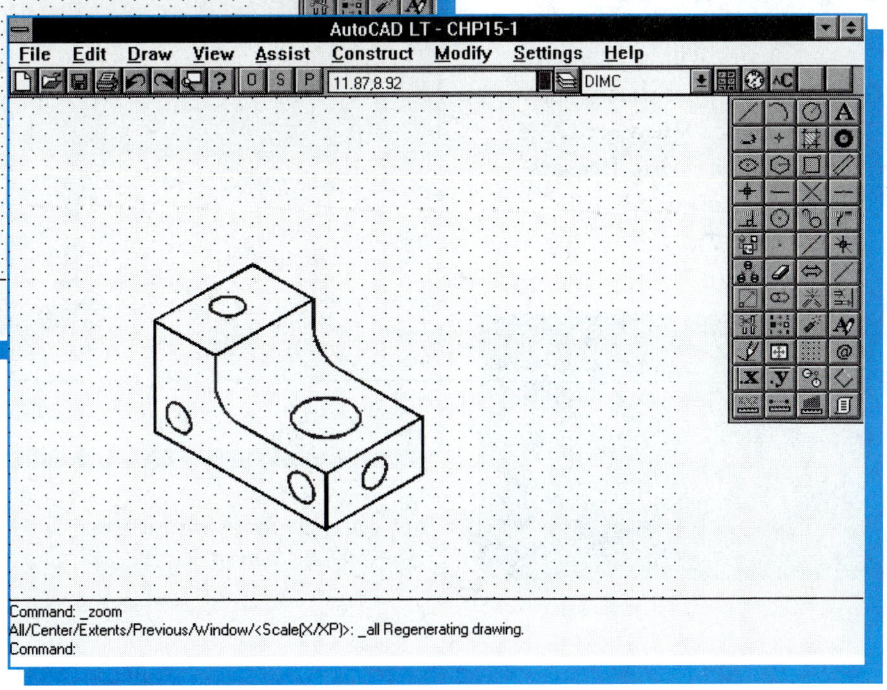

3D Wire Frame

1. Start a **New** drawing.
2. Set the **Snap** and **Grid** to **.250** and turn them both on.
3. Set UCS icon. This shows which viewing plane you are in. Select **Assist, UCSICON**, and type **On**. Select **Enter, Enter** (to bring back the last command). Type **A** and then type **Or**. This will locate the UCS icon in the later views.
4. Create two view screens. (This allows you to view the part from different angles.) Select **View, Viewports, 2 viewports, vertical (hit Enter for default) (Fig. 15-76)**.
5. Set the view in the right viewport.
   a. Move cursor to the right viewport and pick. **Note:** The cursor crosshairs will appear in the viewport that is active.
   b. Select **View—3D Viewpoint Presets—Iso view SE (Fig. 15-77)**.

tip: *Use Fillet command to draw radius and Zoom to locate the part on screen. As you draw the profile in the left view you will see it being created in the right view.*

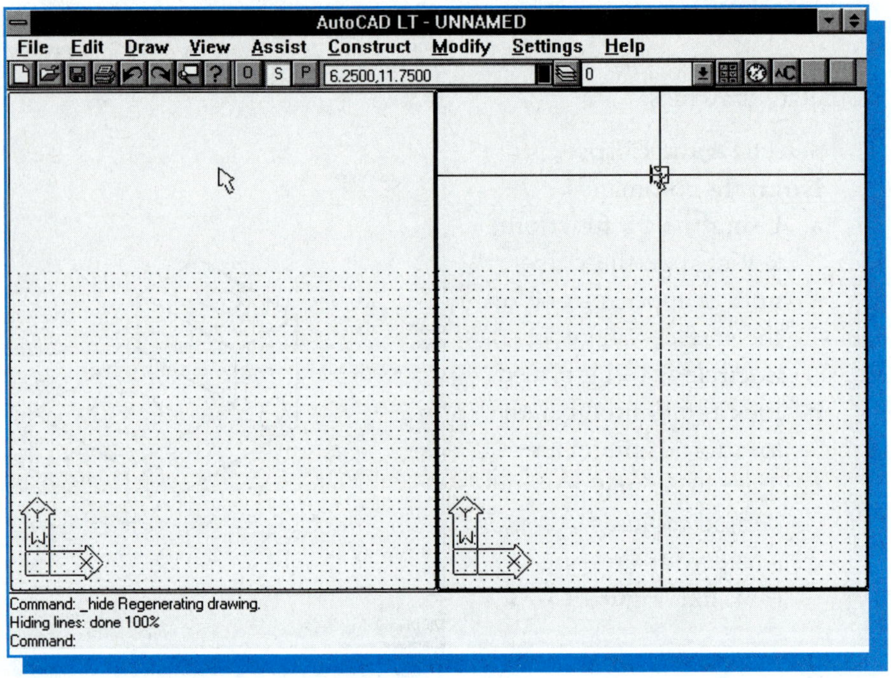

Figure 15-76
Two vertical viewports with UCS icons on. The active viewport is the one with crosshairs.

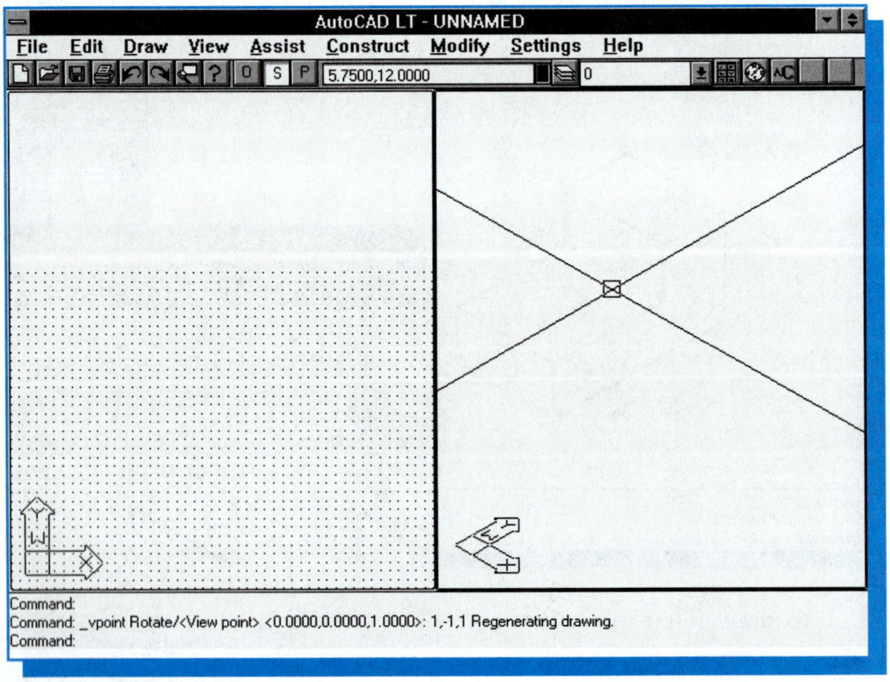

Figure 15-77
The right viewport set to iso view SE.

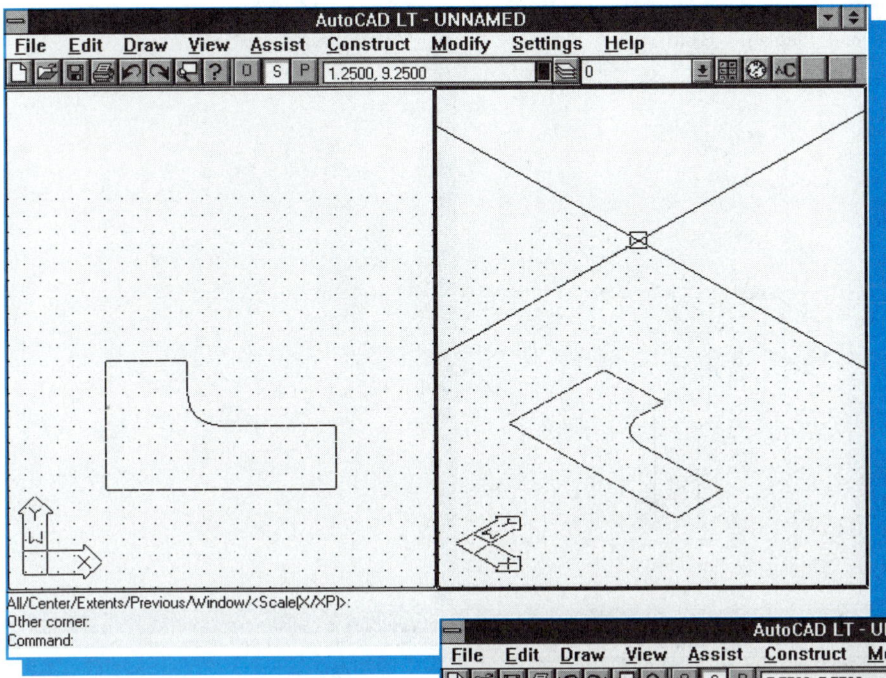

Figure 15-78
Profile of part using the Zoom command to show larger on screen.

6. Select the left viewport and draw the profile on the part to look like **Figure 15-78**.
7. Copy the profile in the Z direction.
   a. Make the left Viewport active.
   b. Select the **Construct, Copy** command.
   c. Use Window to select all the lines or type **all** and press **EB**.
   d. Select the upper-left corner of the part (be sure to use Snap or Osnap).
   e. At the (Second point of displacement) prompt select **Assist, xyz Filters, xy**.
   f. Select the same point
   g. At the prompt (need Z), type **2**.
   The screen should look like the **Figure 15-79**.
8. Draw lines connecting the two sides.
   a. Select **Assist, Object snap** and check the intersection box **OK**.
   b. Move the cursor to the viewport.
   c. Select Line and draw a line between each corner on the two surfaces **(Fig. 15-80)**.

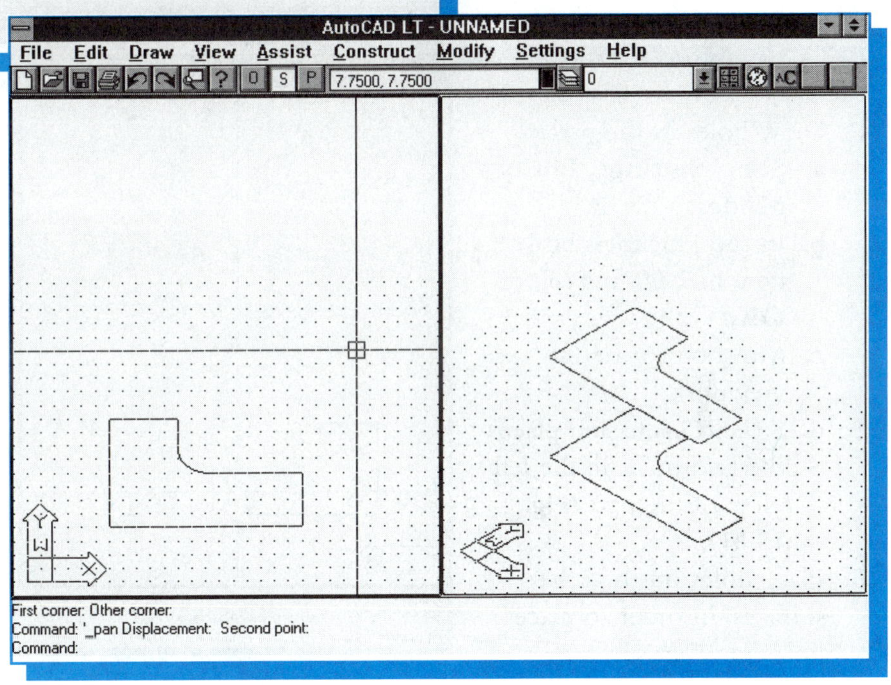

Figure 15-79
Copy the profile of part in the Z direction.

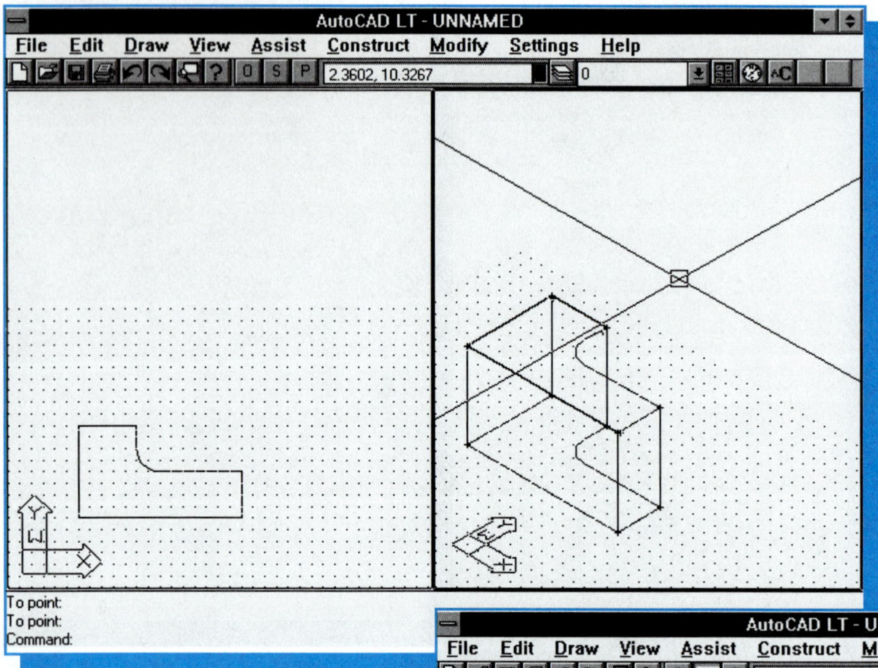

Figure 15-80
Wire frame model

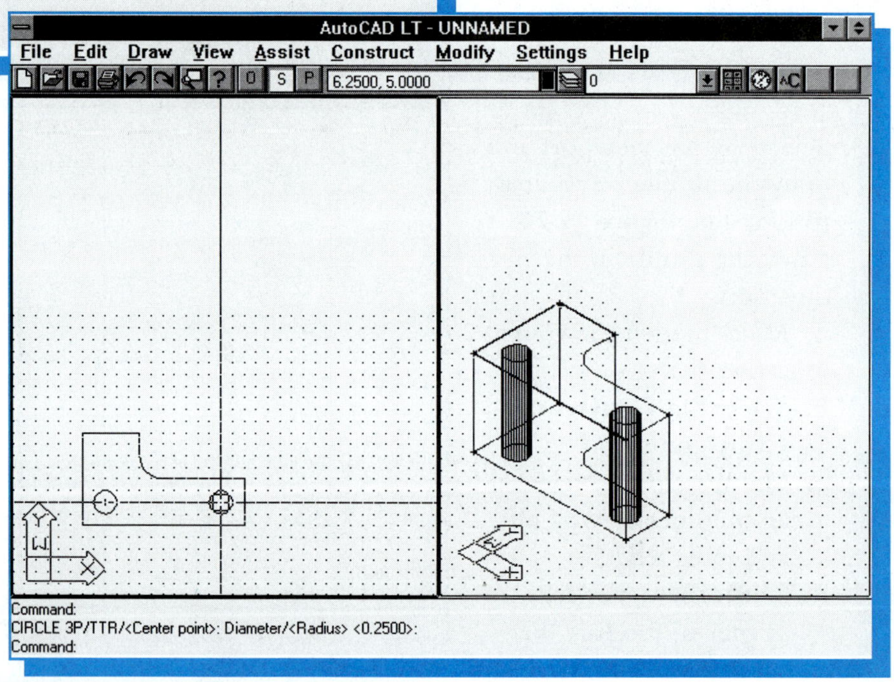

Figure 15-81
Wire frame with holes

9. Draw holes in the part.
   - a. Select **Settings, Entity Modes**.
   - b. In the Thickness box type in **2.00** and select **OK.**
   - c. Move cursor to left viewport.
   - d. Select **Circle** and place the two holes in the left view of the part **(Fig. 15-81)**.
10. Place holes on the other surfaces. In order to place holes in other surfaces, it is necessary to be on that surface. AutoCAD uses the user coordinate system (UCS) to move to different surfaces.
    - a. Select **Assist, Set UCS, origin**.
    - b. Pick the upper-left corner in the 3D view **(Fig. 15-82)**.
    - c. Select **Assist, Set UCS, X axis rotate** type **-90**. Press **(EB)**. The UCS is now on the top surface.
    - d. Activate the left view by clicking in that viewport.
    - e. Select **View, 3D Plan View, Current UCS.**

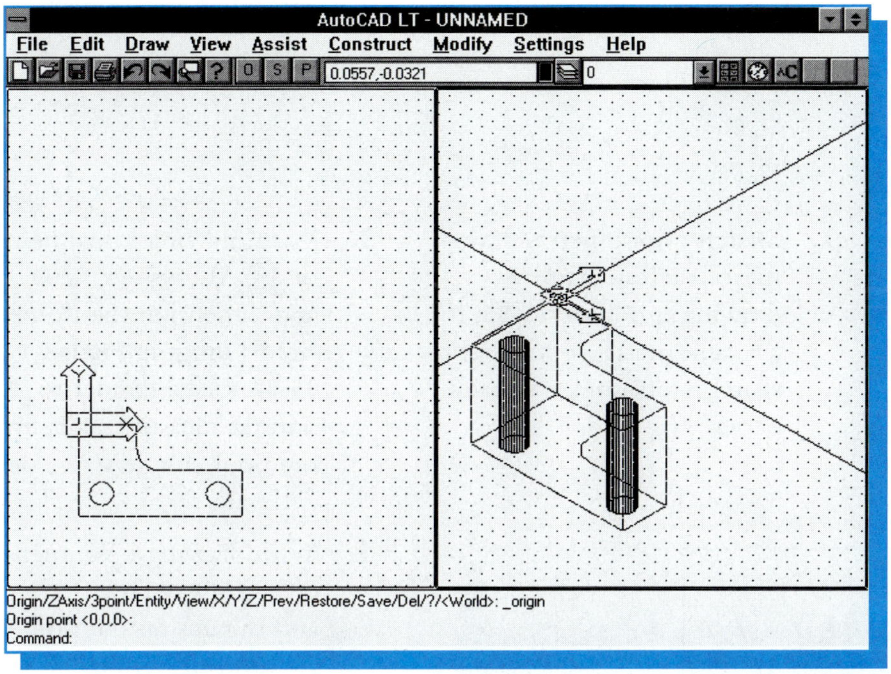

Figure 15-82
Moving the UCS

f. Select **View, Zoom, Center**. Pick the center of that part. Type **8**.

g. Change the Thickness of the entity to -2 (see above).

h. Draw a circle on this surface **(Fig. 15-83)**.

All of AutoCAD's 3D depends on the ability to create and view user coordinate systems. It would be a good idea for the student to create and view different UCS planes as demonstrated in the previous example.

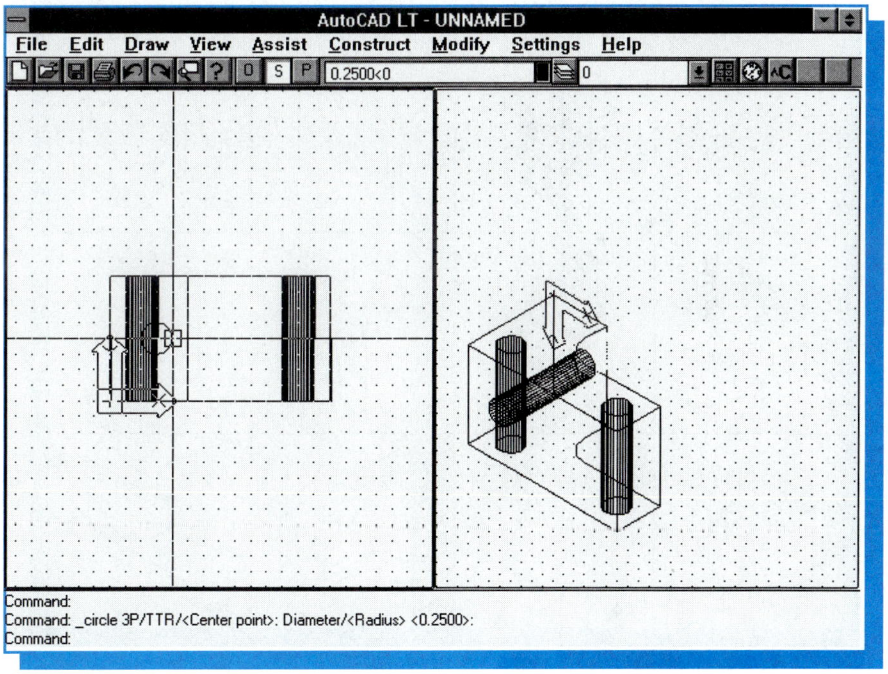

Figure 15-83
Placing holes on other surfaces by locating a new UCS

Drafting Exercises

1. Sketch the isometric drawings in **Figure 15-84**. Make your sketches approximately two times larger. If available, make your sketches on isometric grid paper.

2. Sketch the isometric drawing of each multiview drawing in **Figure 15-85**. If available, make your sketches on isometric grid paper.

3. With drawing instruments or a CAD system, draw and dimension the isometric drawings in **Figure 15-86**. Use a divider to transfer measurements from the book. Draw at 1:1 or 2:1 scale.

4. With drawing instruments or a CAD system, draw and dimension the isometric drawings in **Figure 15-87**.

5. With drawing instruments or a CAD system, draw and dimension the multiview and isometric drawings in **Figure 15-88**. Use a divider to transfer dimensions from the book.

6. With drawing instruments or a CAD system, draw objects in **Figure 15-89** as cabinet drawings.

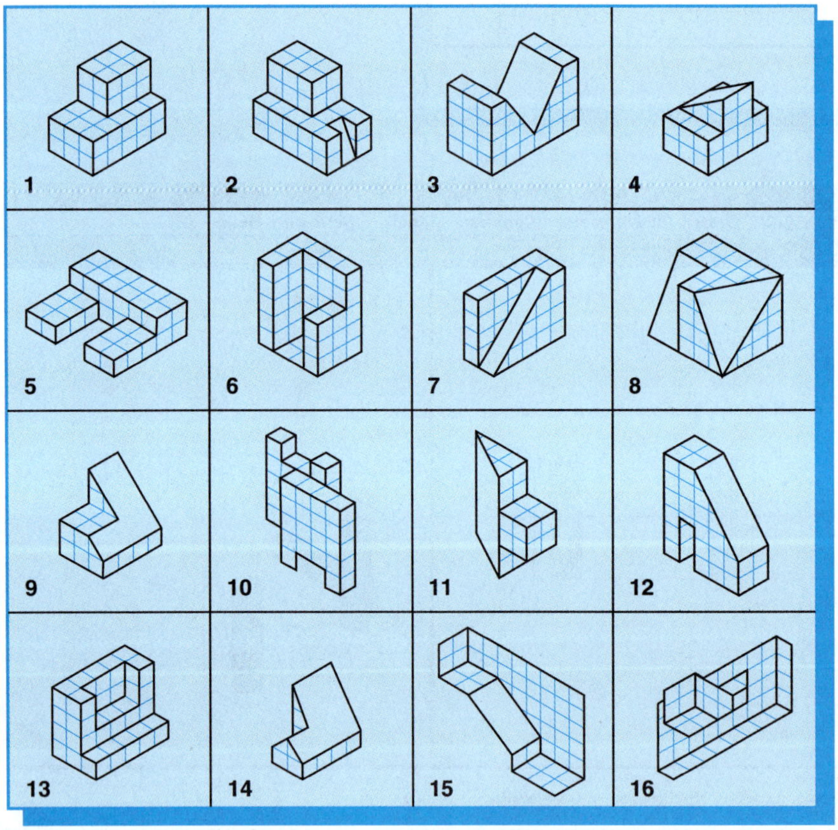

Figure 15-84
Sketch each isometric exercise. Each square is 1/2"

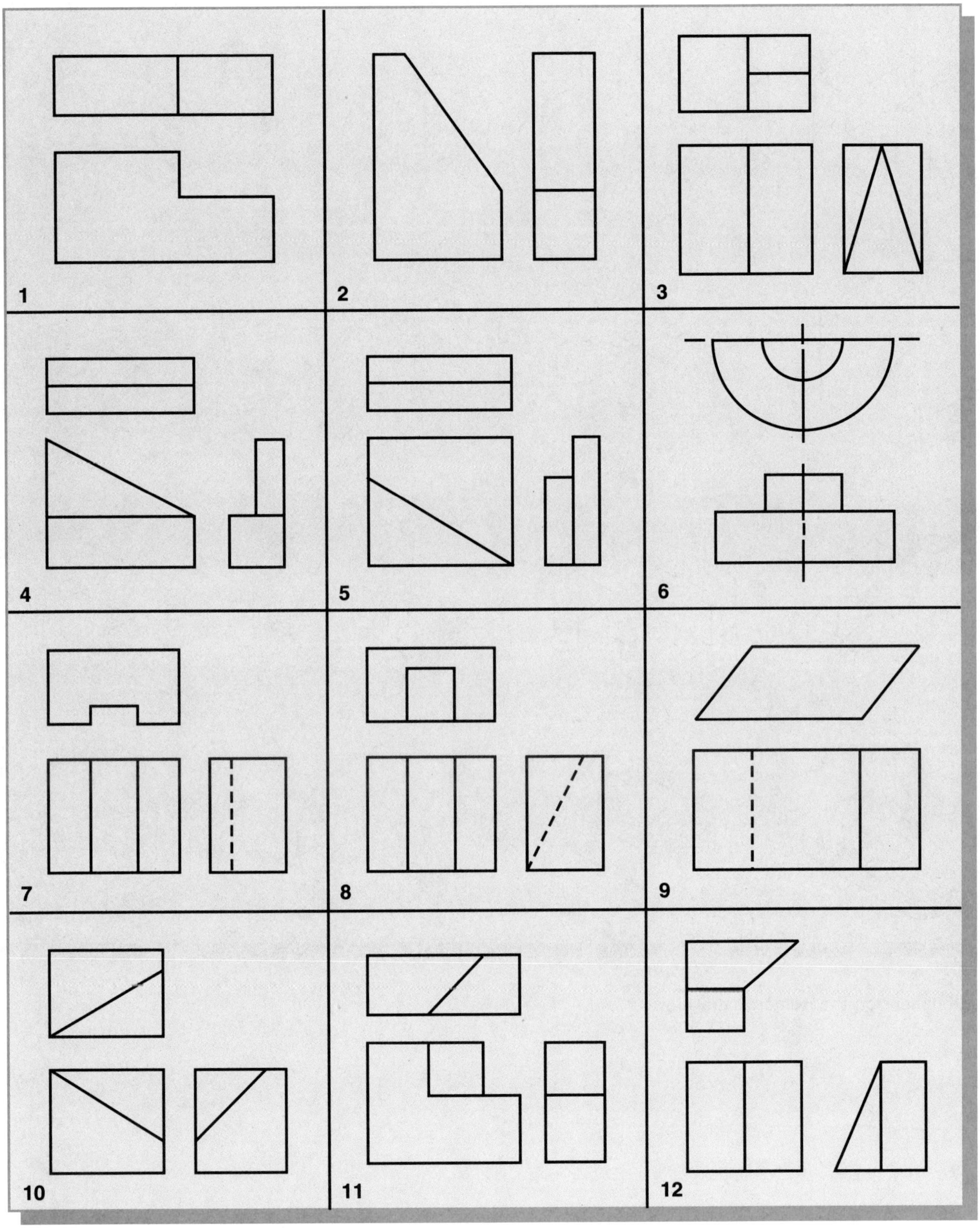

Figure 15-85
Sketch the isometric for each multiview drawing.

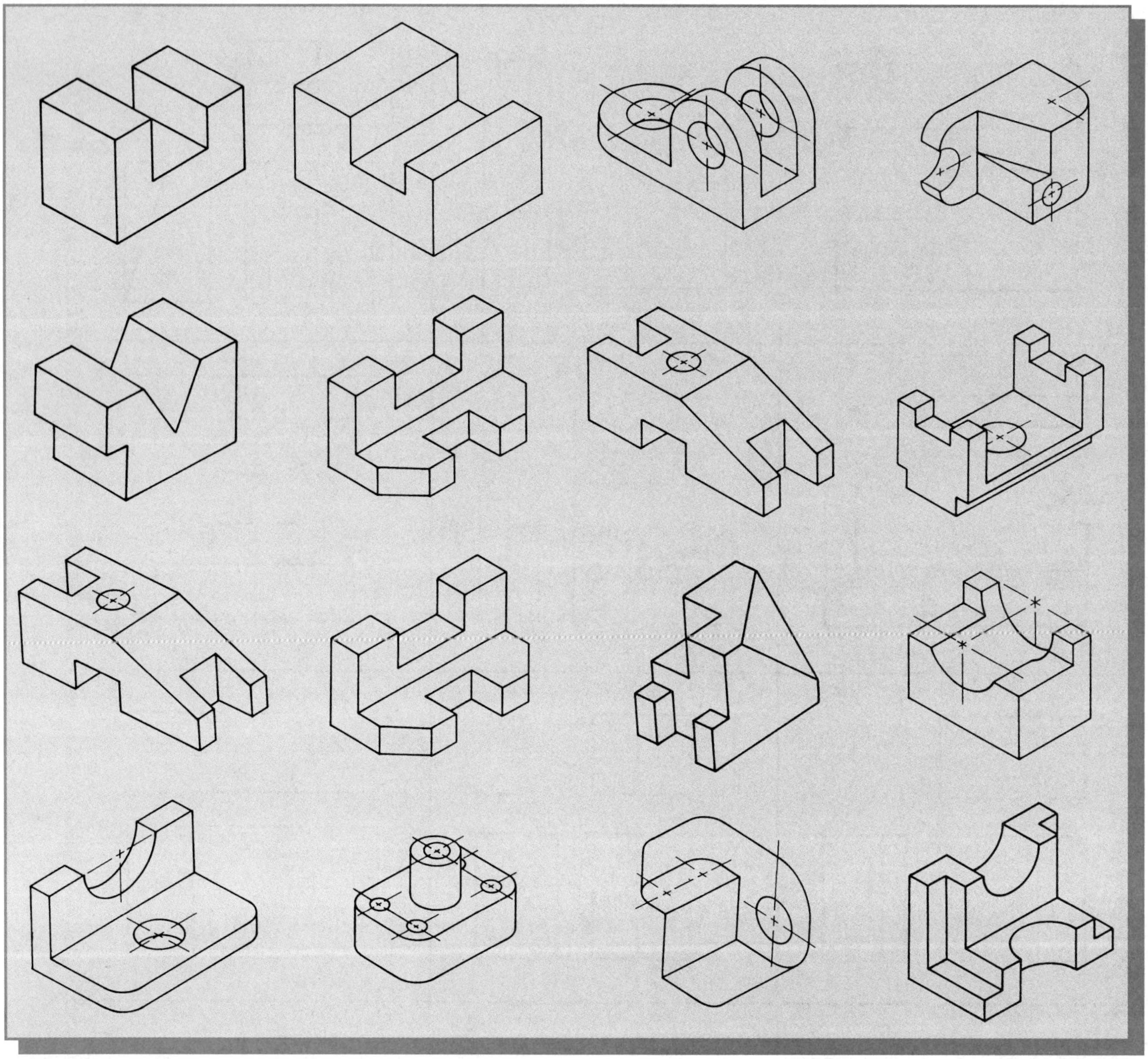

Figure 15-86
Draw and dimension the isometric drawings

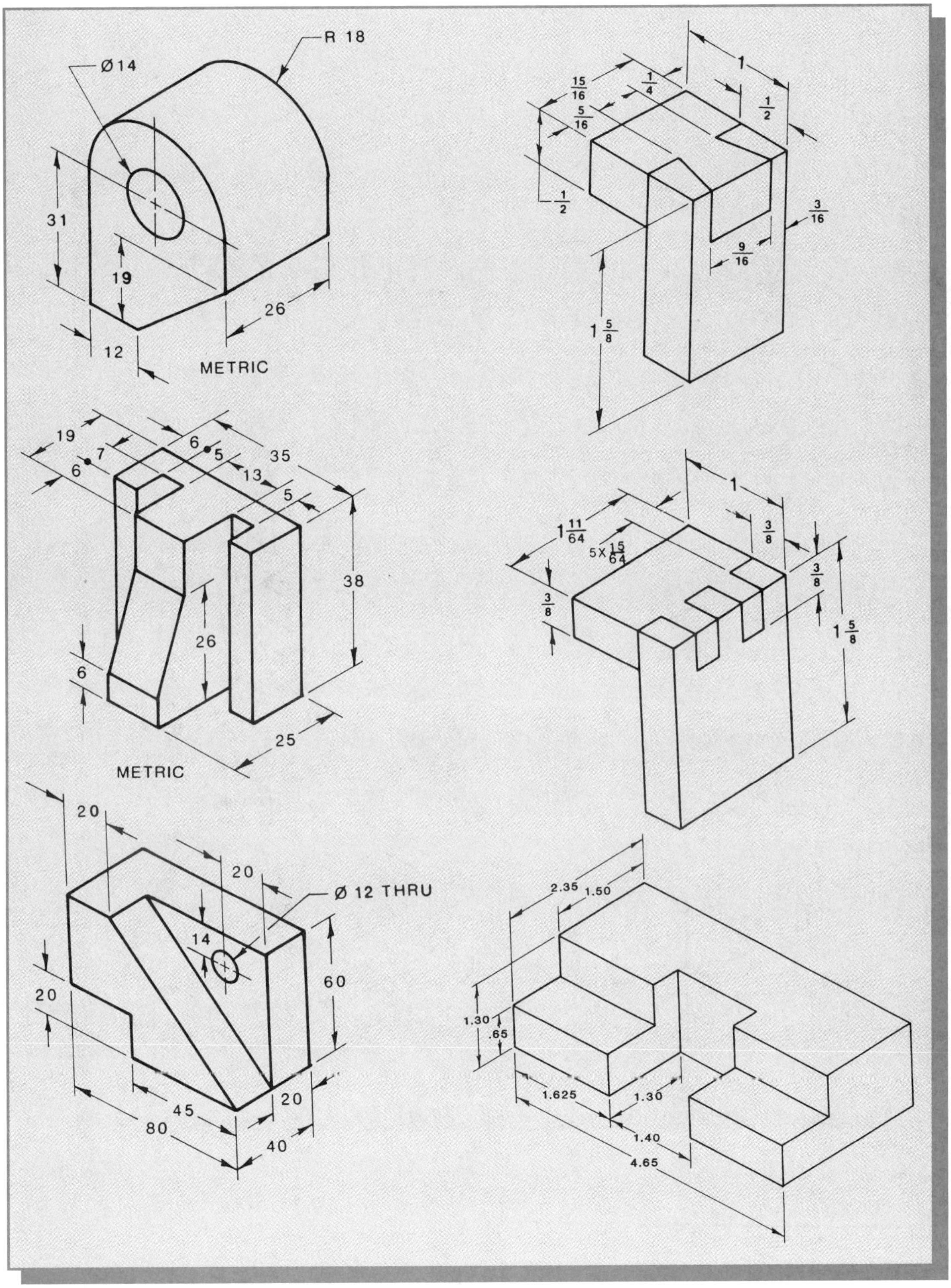

Figure 15-87
Draw the isometric drawing of the jig guide support. Make the following ECOs (refer to "Engineering Change Orders" in Chapter 17):

1. Reduce .20" web to .15".
2. Reduce 6 holes to .25".
3. Reduce four-corner radii to .25".
4. Increase height from 1.15" to 1.25".
5. Increase base thickness to .25".

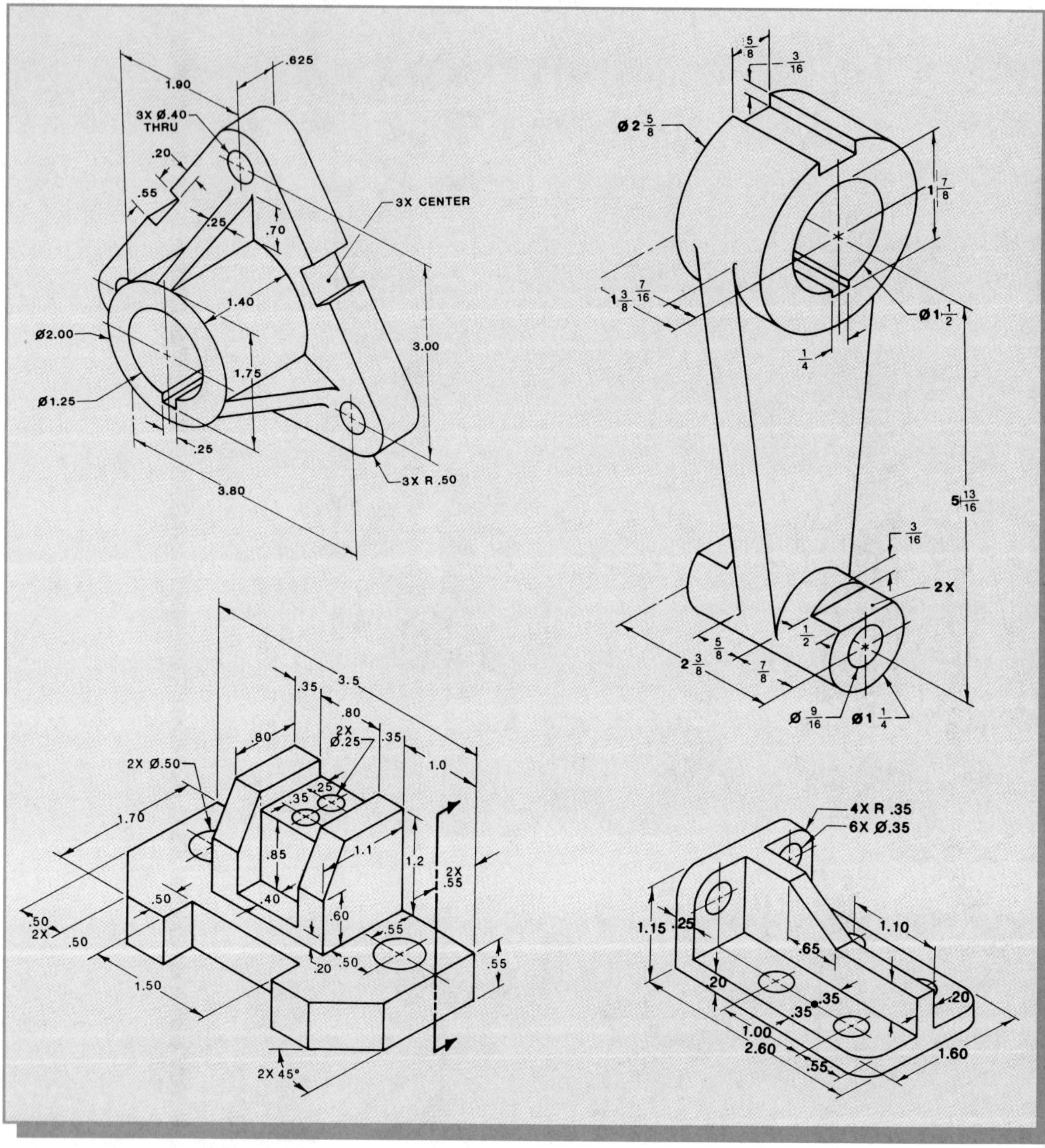

Figure 15-87
continued

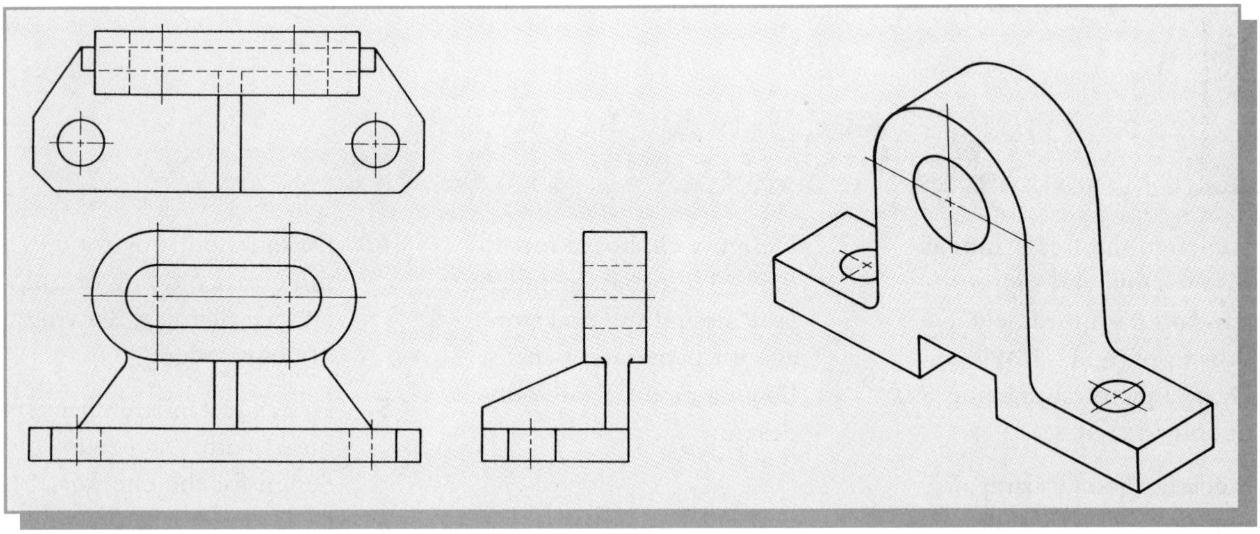

Figure 15-88
Draw and dimension the multiview and isometric drawings.

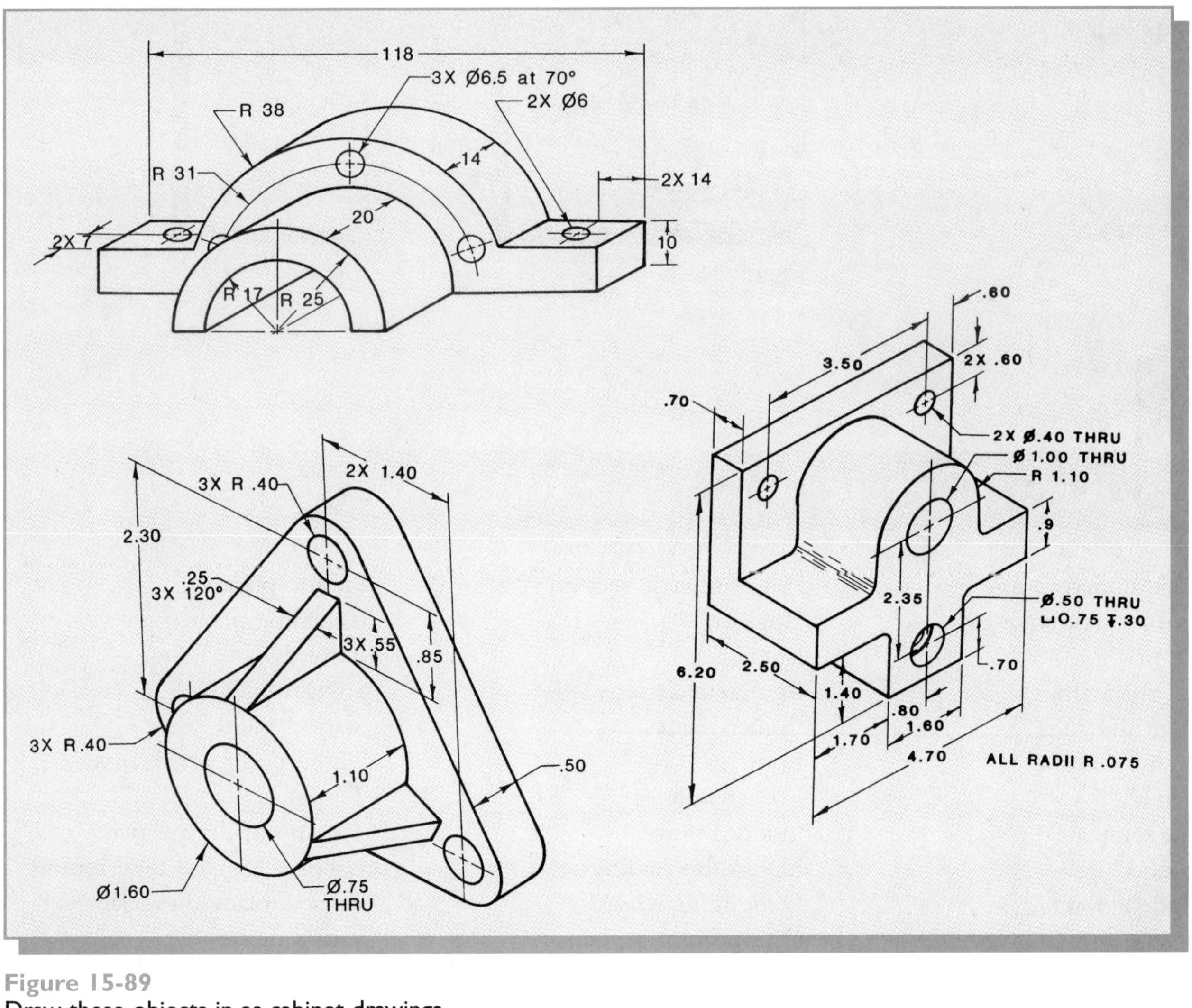

Figure 15-89
Draw these objects in as cabinet drawings.

Design Exercises

1. Redesign the free-standing base to hold and support a standard 35mm camera as shown in **Figure 15-90**. Draw a pictorial drawing of your design.
2. Redesign the C-clamp, in **Figure 15-91** so it will attach to a drafting table and support a tray for your drafting instruments and supplies.
3. Design a clipboard for 8.5"x 11" paper. Include in your design internal storage for paper and pencils. Draw a pictorial of your design.
4. Design a pull-toy for a three year old boy or girl. Draw a pictorial drawing of your design.
5. Design a standard checkerboard with a stylized design for the checkers. Draw a pictorial drawing of your design.

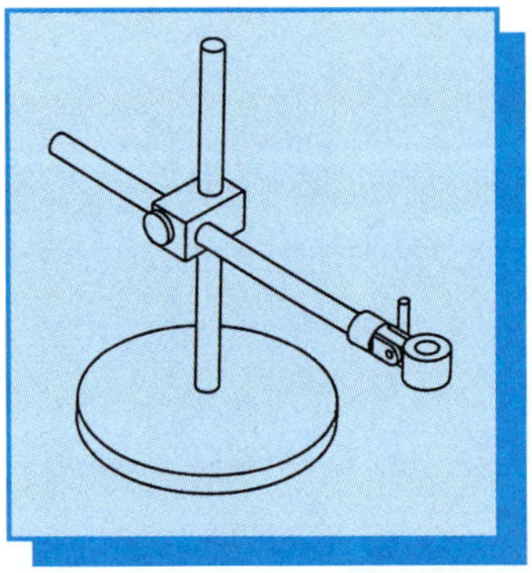

Figure 15-90

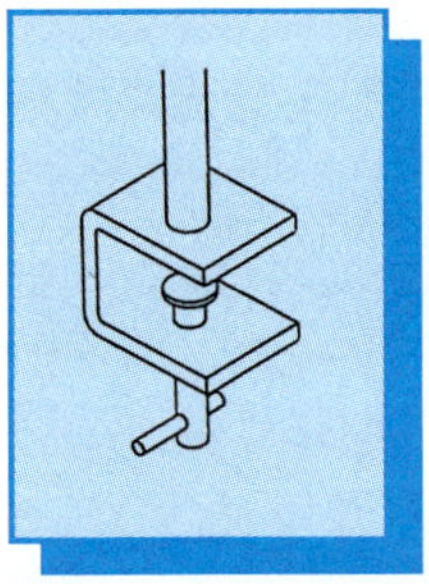

Figure 15-91

Key Terms

Aligned dimensioning
Axonometric drawings
Block in
Cabinet drawing
Cavalier drawing
Cutaway views
Dimetric
Ellipse template
Ellipses
Exploded views
Fillets
Full isometric section
Ground line
Half isometric section
Horizon
Isometric
Isometric circles
Isometric arcs
Isometric axis
Isometric lines
Lines of vision
Nonisometric lines
Oblique drawings
One-point perspective
Perspective drawings
Pictorial drawings
Picture plane
Receding planes
Rounds
Solid modeling
Station point
Three-point perspective
Trimetric
Two-point perspective
Unidirectional dimensioning
Vertical plane dimensioning
Vanishing point
Wire frame drawing

Chapter 16

Fasteners

Objectives

The student will be able to:

- draw American standard internal and external thread symbols (simplified, schematic, and detailed)
- draw ISO metric internal and external thread symbols (simplified, schematic, and detailed)
- read and understand fastener tables
- recognize the various types of fasteners used in industry

Introduction

A fastener is any device used to join separate parts or materials. There are two general types of fasteners—removable and permanent. The use of removable fasteners, such as screws, pins, keys, nuts, and bolts, allows attached parts or materials to be separated and reattached. Permanent fasteners such as rivets, nails, and staples, and fastening methods, such as soldering, welding, and gluing, affix parts or materials that never need separating. The **American Society of Mechanical Engineers (ASME)**, the **International Standards Organization (ISO)**, and the **Industrial Fasteners Institute (IFI)** establish the standards for fastening methods and devices.

Threaded Fasteners

The functioning of all threaded fasteners is based on the principle of the inclined plane. An **inclined plane** is a simple machine that accomplishes work by reducing force, by increasing distance. A screw thread is a spiral inclined plane. An evenly spaced spiral curve around a cylinder is called a **helix. Figure 16-1** shows how an inclined plane is applied to a cylinder to create a helix. The **lead** is the lateral distance a screw moves

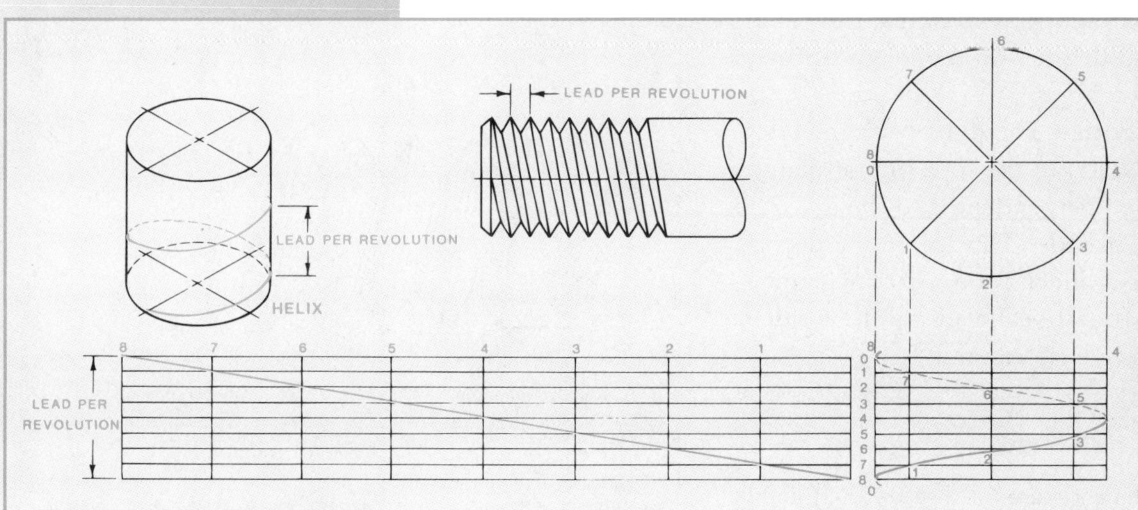

Figure 16-1 The thread's form is a helix. One revolution equals the lead.

along the axis with one 360° revolution. The **lead per revolution** of the inclined plane is equal to the circumference of the cylinder.

Threaded Fastener Nomenclature

To work effectively with threaded fasteners using conventional drafting instruments or a CAD system, you must understand the following standard nomenclature **(Figs. 16-2** and **16-3)**:

- **Angle of thread**—The included angle formed between the major diameter and the depth of a thread.
- **Axis of Fastener**—The cylindrical center of a fastener shaft.
- **Screw Thread Series**—Screws with similar combinations of pitches and diameters are divided into several related series. These are specified in the American National Standard Unified screw threads for screws, bolts, nuts, and other fasteners in ASME B1.1 (UN) and in ISO for metric threads. The following is the symbol designation for thread series:

  Coarse thread series—UNC, NC
  Fine thread series—UNF, NF
  Extra-fine thread series—UNEF, NEF
- **Thread Fit Classes**—Threads are divided into the following three tolerance fit classes, 1A, 2A, and 3A (for external threads) and 1B, 2B, and 3B (for external threads):

  Class 1A and 1B—loose fits
  Class 2A and 2B—general purpose fits
  Class 3A and 3B—tight fits
- **Crest**—The point of widest diameter of a screw thread that lies on the major diameter plane.
- **Thread Depth**—The distance from crest (major diameter) to the root (minor diameter) measured perpendicular to the axis.
- **External Thread**—Threads on the outside surface of a cylinder as on a bolt or screw.
- **Internal Thread**—Threads on the inside surface of a hole as in a nut.

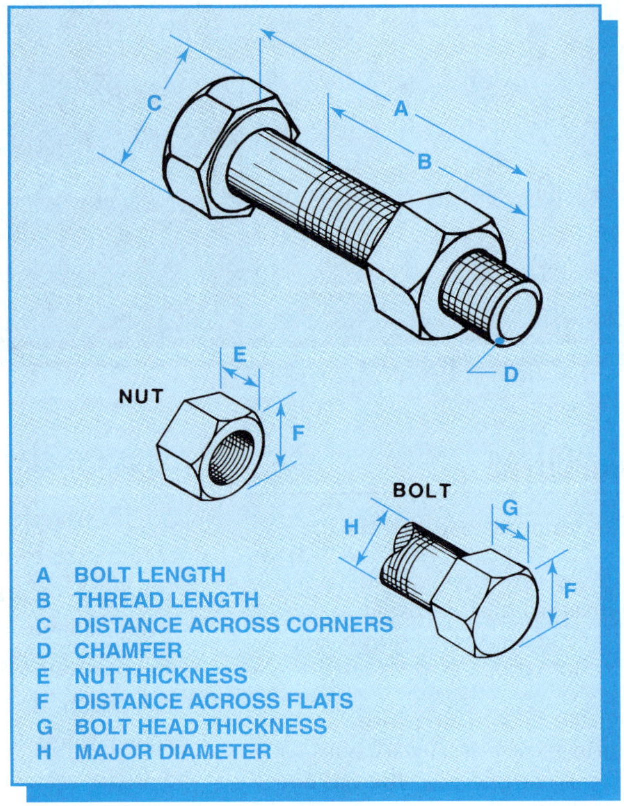

Figure 16-2
Nomenclature for nuts and bolts

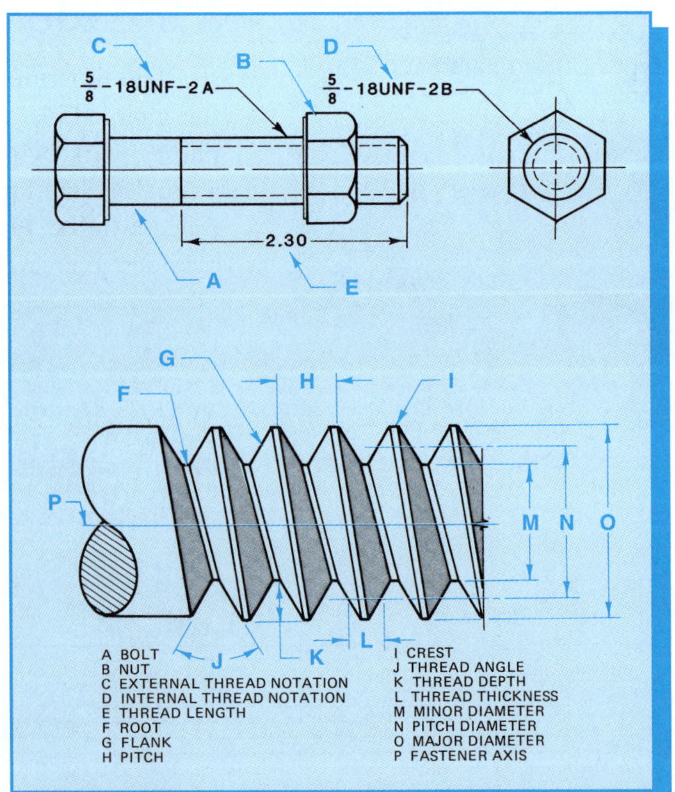

Figure 16-3
Nomenclature for threaded nuts and bolts

- **Lead—**The lateral distance the axis of a fastener moves during one 360° revolution of the fastener.
- **Major Diameter**—The outside diameter of the threaded portion of a fastener measured from crest to crest.
- **Minor Diameter**—The diameter of the thread measured from root to root perpendicular to the axis.
- **Root—**The deepest point or valley of a thread.
- **Thread Forms**—The shape of the thread in profile section **(Fig. 16-4)**.
- **Right-Hand Thread**—As viewed from the head, a right-hand thread turns clockwise into an object or threaded opening.

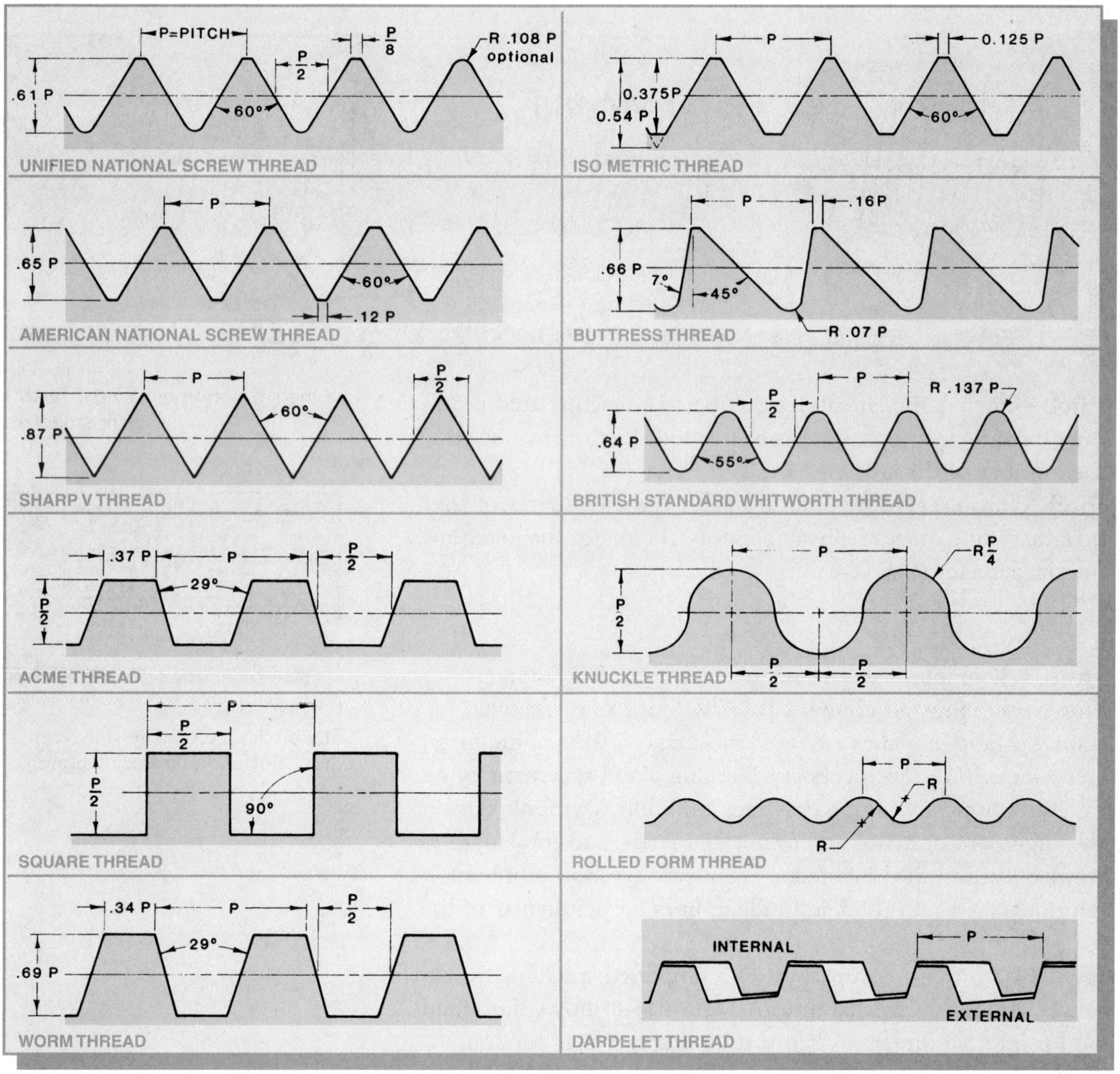

Figure 16-4
Thread forms

- **Left-Hand Thread**—As viewed from the head, a left-hand thread turns counterclockwise into an object or threaded opening.
- **Single Thread**—Threads with one continuous helical pattern in which the lead (one full turn) is equal to the single-thread pitch **(Fig. 16-5)**.
- **Multiple Threads**—Threads that contain more than a single helical pattern side-by-side. For example, a double thread **(Fig. 16-5)** includes two side-by-side helical threads; the lead is two times the pitch. In a triple-threaded fastener, the lead is three times the pitch. Multiple-threaded fasteners are used where few turns are desired to move the axis a maximum distance.

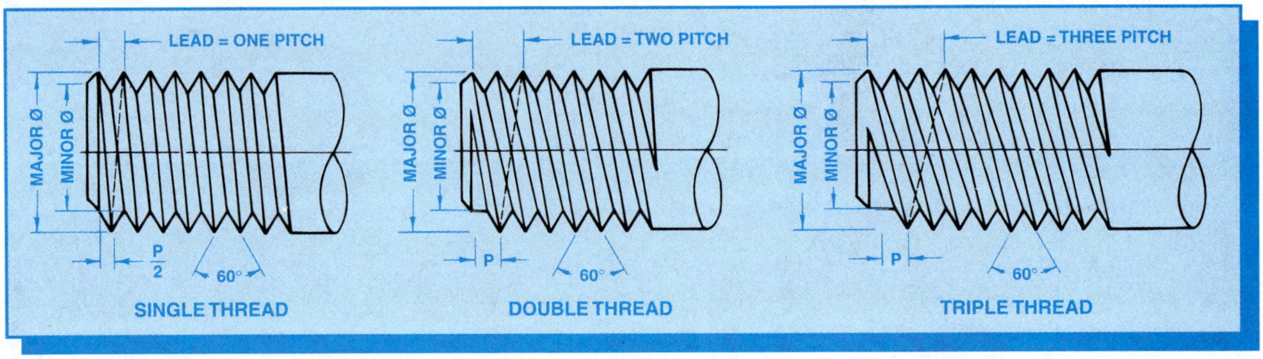

Figure 16-5
One revolution equals the pitch of the threaded rod.

- **Pitch**—Pitch is the lateral distance between two related points on a thread (crest to crest or root to root). Pitch is expressed in the number of **threads per inch (TPI).**
- **Pitch Diameter**—The pitch diameter is the diameter of an imaginary line drawn halfway between the major diameter and minor diameter **(Fig. 16-6)**.

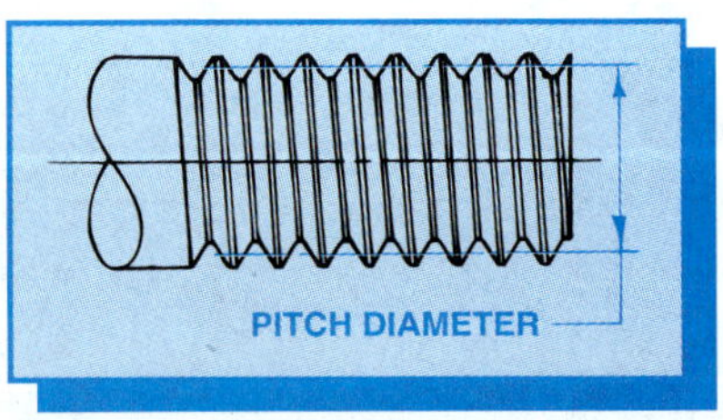

Figure 16-6
The pitch diameter lies between the major and minor diameters.

Thread Symbolism

A true orthographic view or section of a threaded fastener is very time consuming and complex to draw. Since most fasteners shown on working drawings are standard, a true orthographic or pictorial representation is not necessary. Simplified and schematic symbols are used on most working drawings. Simplified symbols show only the alignment of major and minor diameters, and the length of the threaded portion of the fastener. Schematic symbols represent crest and root lines with solid and hidden lines perpendicular to the axis.

Figure 16-7 shows a comparison of a simplified, a schematic, and a pictorial symbol for external threads. **Figure 16-8** shows simplified symbols for internal threads. When threads are shown in section, the three types of symbols are drawn **(Fig. 16-9)**; however, the simplified symbol is most widely used and recommended.

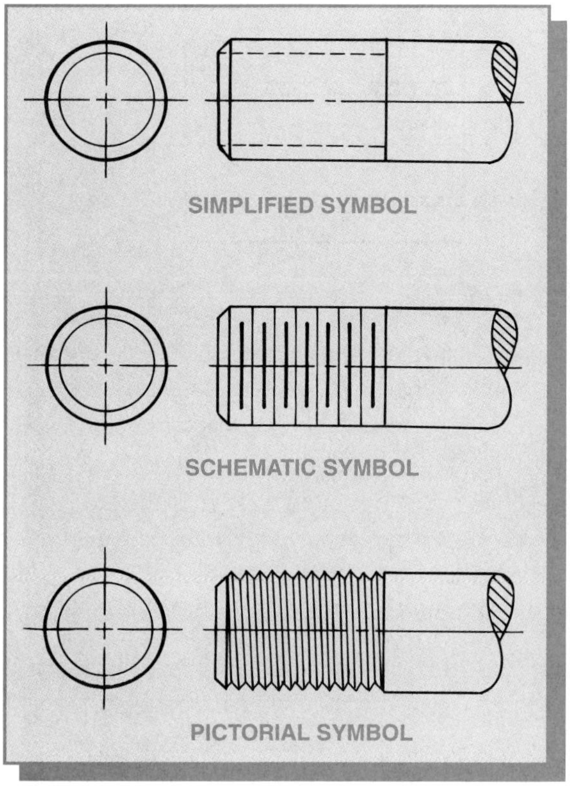

Figure 16-7
Symbols for drawing external threads. The simplified symbol is recommended for working drawings.

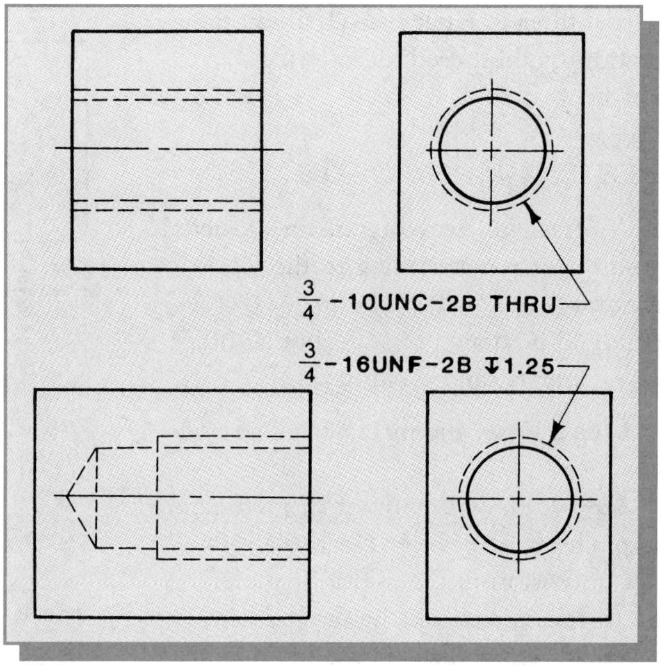

Figure 16-8
Simplified symbols for drawing internal threads

Figure 16-9
Drawing symbols for sectioned internal threads

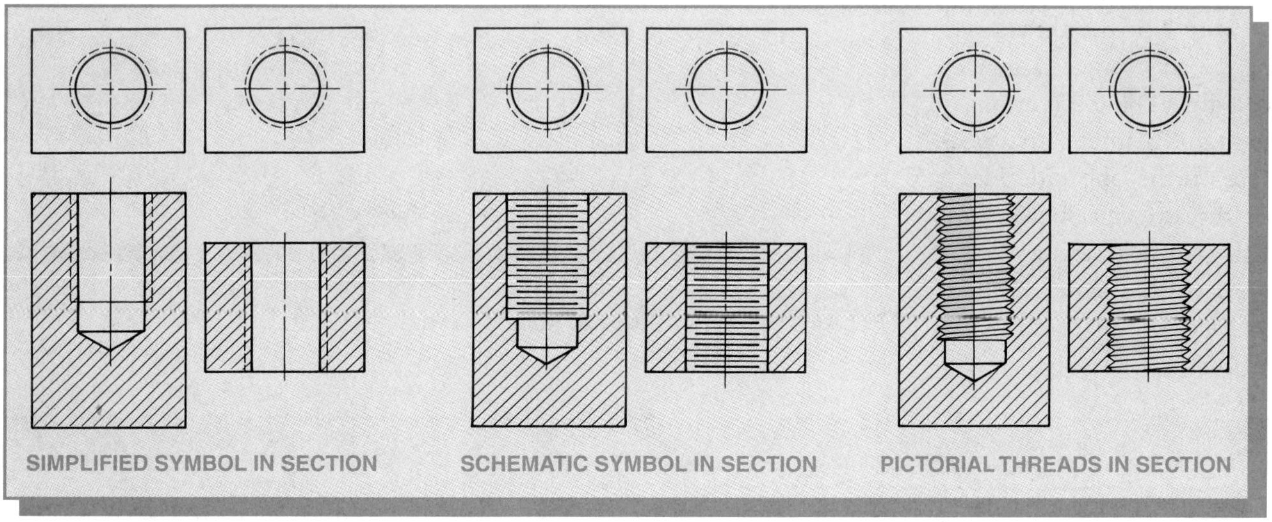

Thread Notations

Regardless of the thread symbol used, all essential information cannot be shown graphically or with normal dimensioning methods effectively. Notations are used to show the type, class, series, form, TPI, and major diameter of fasteners on working drawings. **Figure 16-10** shows the notation method used to specify these features for internal threads. **Figure 16-11** shows the notation method used for external threads.

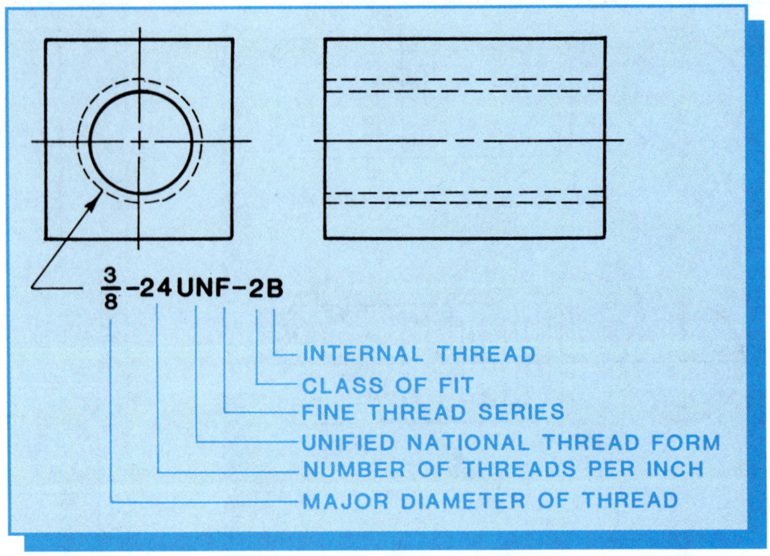

Figure 16-10
Placement of internal thread notation's data

Drafting Procedures

A **pictorial drawing of an external thread** is drawn according to the following steps **(Fig. 16-12)**. (The major diameter should be drawn to scale, but all other measurements can be estimated.)

Step 1. Draw the major diameter and axis.

Step 2. Draw the minor diameter and the pitch diameter lines. Draw two 60° lines upward from the minor diameter line (root). As a check, the distance between the 60° lines on the pitch circle line should be about one-half the space of the 60° lines at the major diameter (crest).

Step 3. Continue to draw the 60° V threads for the full length of the threaded rod at the top and bottom.

Step 4. If the thread is a single lead, connect a line from a peak angle (vertex) on the major diameter to a peak angle on the opposite side, one thread over. Repeat for all the major (crest) and minor (root) diameters to complete the pictorial thread layout.

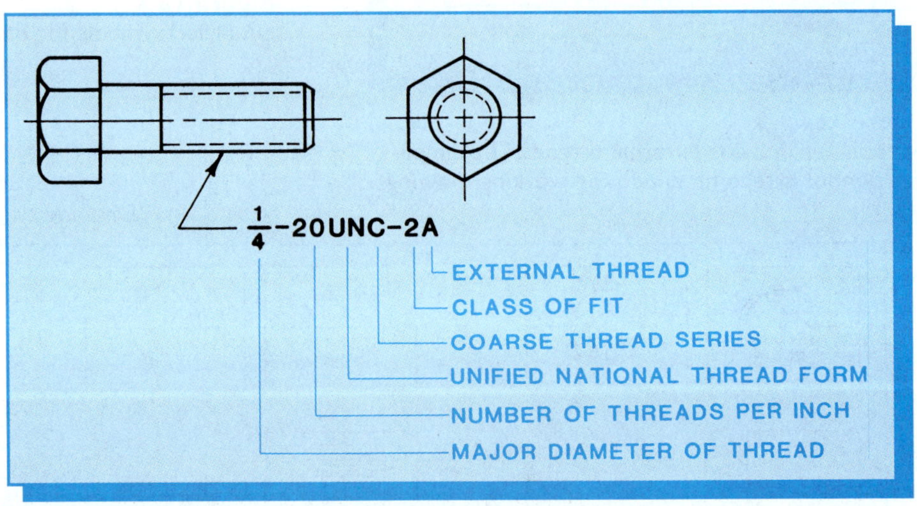

Figure 16-11
Placement of external thread notation's data

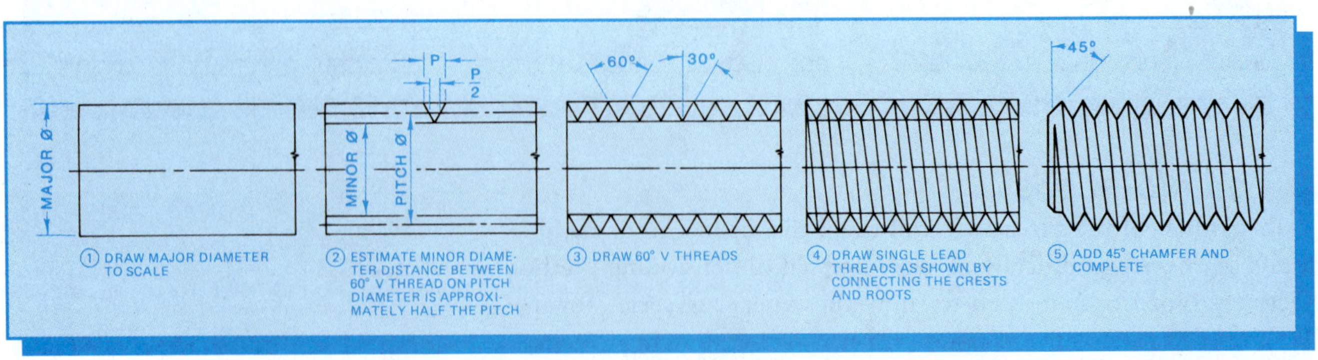

Figure 16-12
Steps to draw external pictorial (detailed) V thread symbol

Step 5. Add a 45° chamfer and erase all construction lines. Darken to complete.

Simplified internal thread symbols are drawn according to the following steps **(Fig. 16-13)**:

Step 1. Draw the drill depth.
Step 2. Draw the major diameter and depth of the thread.

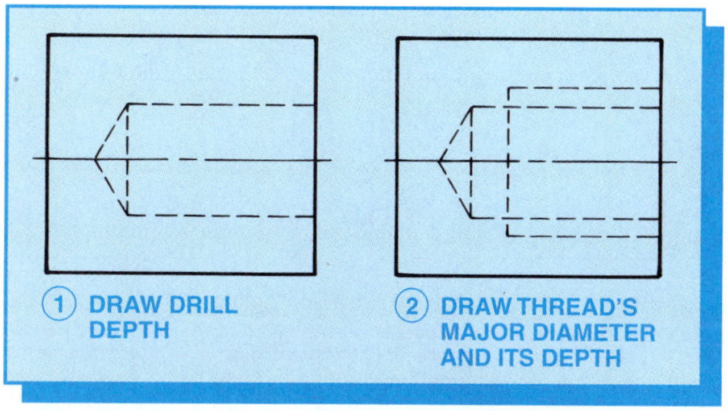

Figure 16-13
Steps to draw simplified internal thread symbols

Simplified external thread symbols are drawn according to the following steps **(Fig. 16-14)**:

Step 1. Draw the outline and centerline of the fastener. The object lines, which parallel the axis centerline, represent the major diameter and are drawn to scale. Draw a line representing the threaded length.
Step 2. Draw a 45° chamfer on the opposite end. With construction lines, project a 60° included angle from the major diameter (crest) in order to determine the thread depth (root). At this point (thread depth), draw dashed lines parallel to the axis to represent the minor diameter.

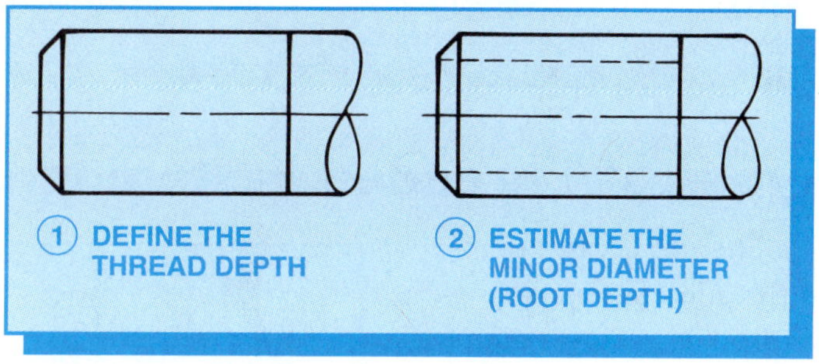

Figure 16-14
Steps to draw simplified external thread symbols

Schematic internal thread symbols are drawn according to the following steps **(Fig. 16-15)**:

Step 1. Draw the outline of the tap drill hole's diameter to the required depth.
Step 2. Draw the outline of the thread's major diameter to the required depth.
Step 3. Estimate the pitch spacing and draw lines to the major diamter (crest to crest).
Step 4. Draw the lines for the minor diameter lines from root to root (use the tap drill line).
Step 5. Darken and add cross hatching as shown in **Figure 16-15**.

External schematic thread symbols are drawn following the five steps in **Figure 16-16**.

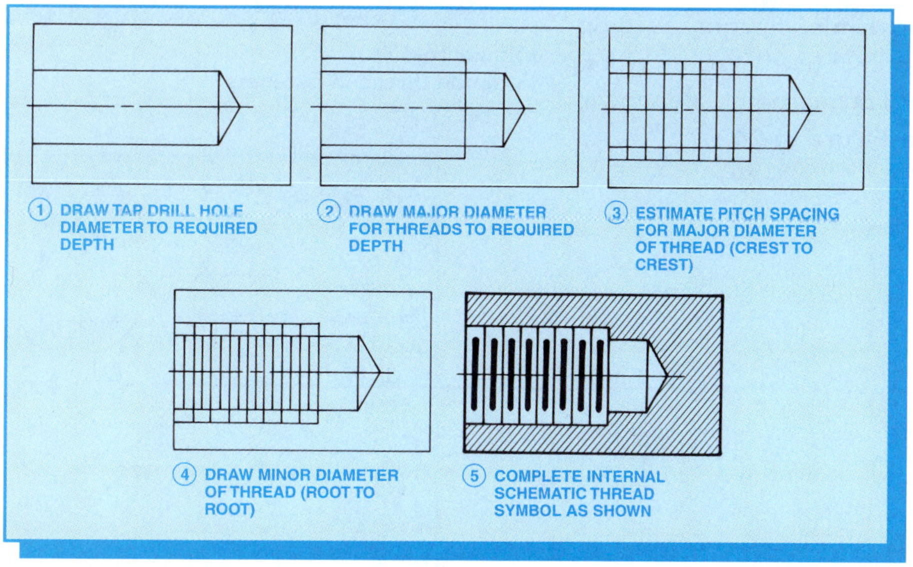

Figure 16-15
Steps to draw schematic internal thread symbols

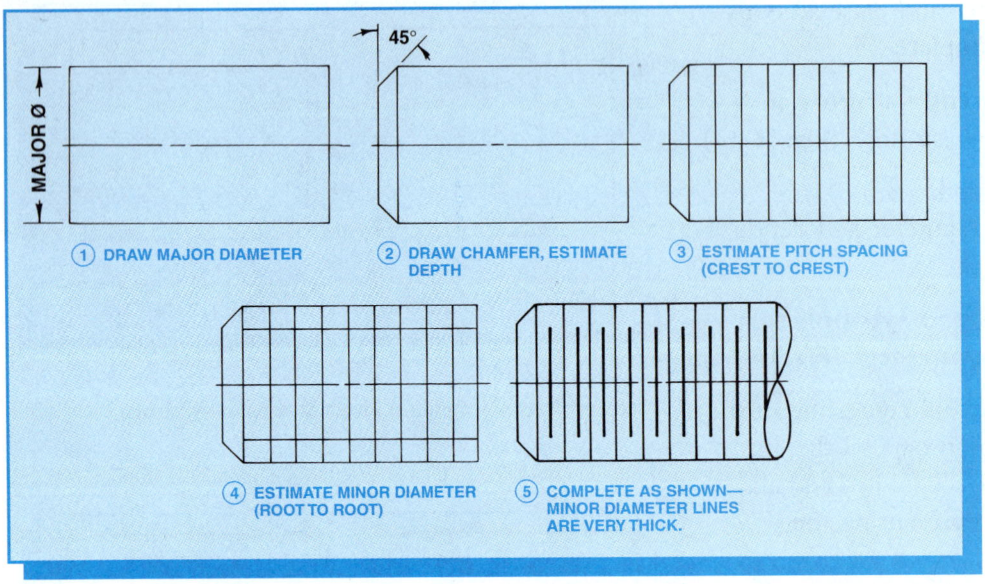

Figure 16-16
Steps to draw schematic external thread symbols

Types of Threaded Fastening Devices

There is a vast number of threaded fasteners ranging from the most common **(Fig. 16-17)** to very specialized. Threaded fasteners consist of a body, head, and point. **Figure 16-18** shows a variety of fastener heads, and **Figure 16-19** shows drive configurations used on fastener heads. Various types of fastener points are shown in **Figure 16-20**.

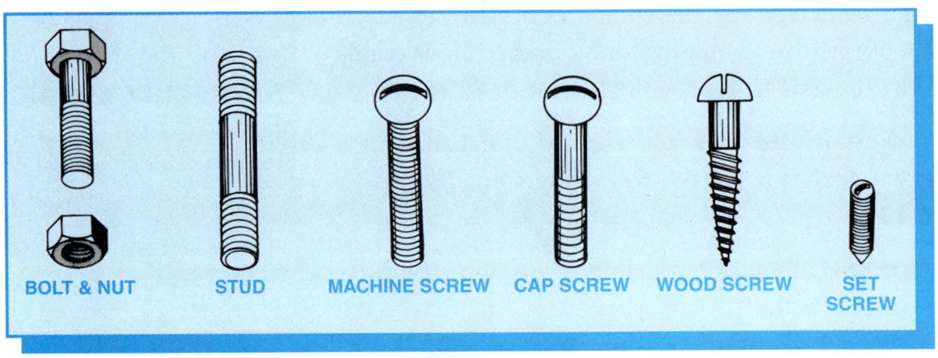

Figure 16-17
Common threaded fasteners

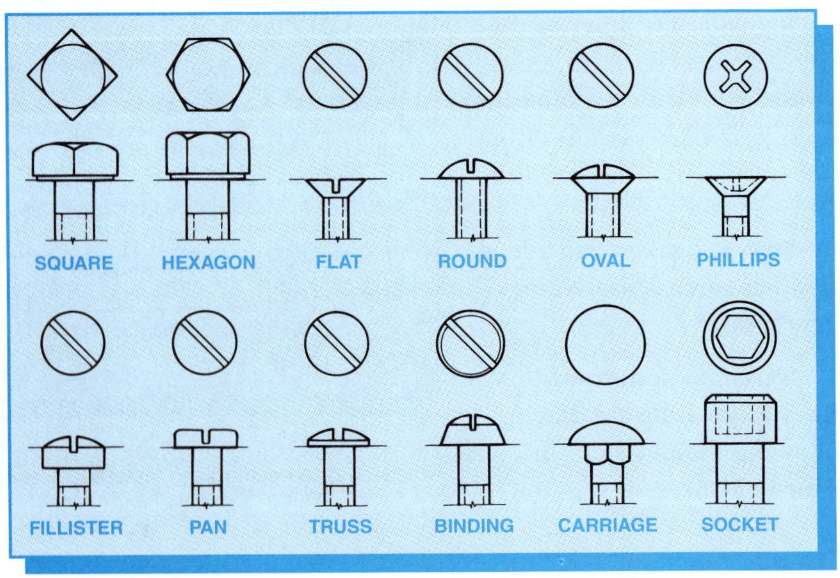

Figure 16-18
Fastener heads

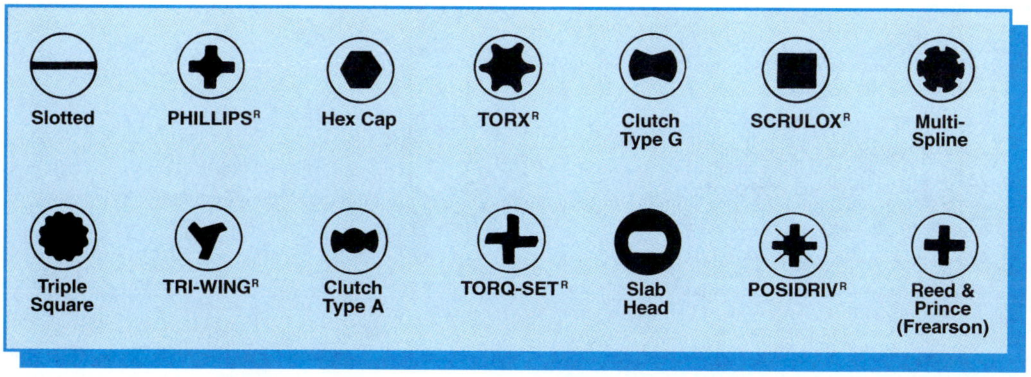

Figure 16-19
Various types of drive configurations for threaded fasteners *(Reprinted from* Technical Drawing andDesign *by Goetsch and Nelson, © 1986 Delmar Publishers, Inc.)*

In designing the type and size of fasteners for a project, information concerning the class and dimensions is necessary to insure the fastener will fit and perform the desired function. To find this detailed data, ASME, ISO, IFI, and manufacturer's charts are used. **Tables 16-1, 16-2,** and **16-3** show samples of this type of data for threaded fasteners.

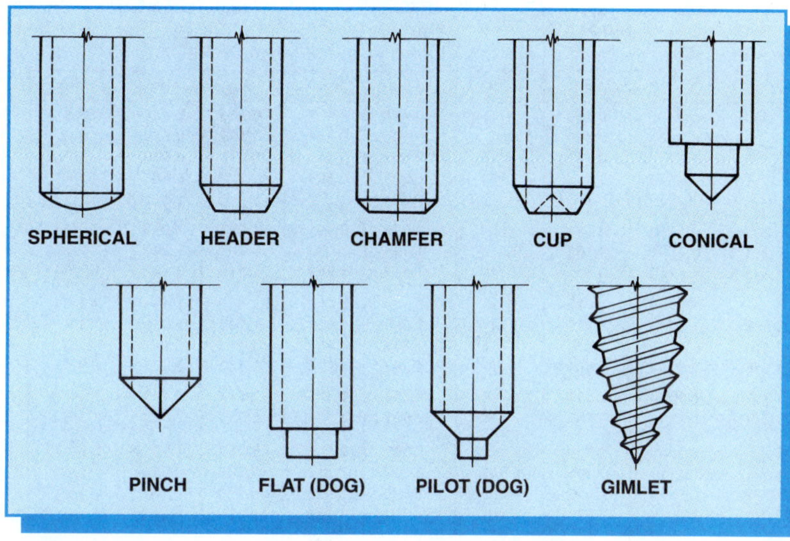

Figure 16-20
Fastener points

Major Diameter	Number of Threads per Inch	
	Course UNC	Fine UNF
0 (.060)		80
1 (.073)	64	72
2 (.086)	56	64
3 (.099)	48	56
4 (.112)	40	48
5 (.125)	40	44
6 (.138)	32	40
8 (.164)	32	36
10 (.190)	24	32
12 (.216)	24	28
1/4	20	28
5/16	18	24
3/8	16	24
7/16	14	20
1/2	13	20
9/16	12	18
5/8	11	18
3/4	10	16
7/8	9	14
1	8	12
1 1/8	7	12
1 1/4	7	12
1 3/8	6	12
1 1/2	6	12
1 3/4	5	—
2	4 1/2	—
2 1/4	4 1/2	—
2 1/2	4	—
2 3/4	4	—
3	4	—
3 1/4	4	—
3 1/2	4	—

Table 16-1
Thread table for commonly used ASME fasteners

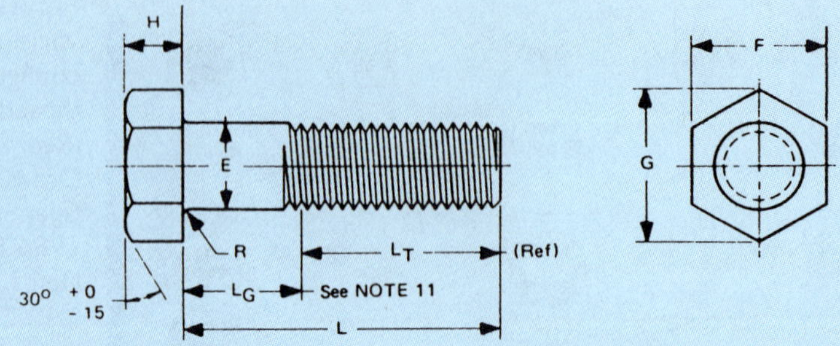

Nominal Size or Basic Product Dia (17)		E	F			G		H			R		L_T	
		Body Dia (7)	Width Across Flats (4)			Width Across Corners		Height			Radius of Fillet		Thread Length For Bolt Lengths (11) 6 in. and shorter	over 6 in.
		Max	Basic	Max	Min	Max	Min	Basic	Max	Min	Max	Min	Basic	Basic
1/4	0.2500	0.260	7/16	0.438	0.425	0.505	0.484	11/64	0.188	0.150	0.03	0.01	0.750	1.000
5/16	0.3125	0.324	1/2	0.500	0.484	0.577	0.552	7/32	0.235	0.195	0.03	0.01	0.875	1.125
3/8	0.3750	0.388	9/16	0.562	0.544	0.650	0.620	1/4	0.268	0.226	0.03	0.01	1.000	1.250
7/16	0.4375	0.452	5/8	0.625	0.603	0.722	0.687	19/64	0.316	0.272	0.03	0.01	1.125	1.375
1/2	0.5000	0.515	3/4	0.750	0.725	0.866	0.826	11/32	0.364	0.302	0.03	0.01	1.250	1.500
5/8	0.6250	0.642	15/16	0.928	0.906	1.083	1.033	27/64	0.444	0.378	0.06	0.02	1.500	1.750
3/4	0.7500	0.768	1 1/8	1.125	1.088	1.299	1.240	1/2	0.524	0.455	0.06	0.02	1.750	2.000
7/8	0.8750	0.895	1 5/16	1.312	1.269	1.516	1.447	37/64	0.604	0.531	0.06	0.02	2.000	2.250
1	1.0000	1.022	1 1/2	1.500	1.450	1.732	1.653	43/64	0.700	0.591	0.09	0.03	2.250	2.500
1 1/8	1.1250	1.149	1 11/16	1.688	1.631	1.949	1.859	3/4	0.780	0.658	0.09	0.03	2.500	2.750
1 1/4	1.2500	1.277	1 7/8	1.875	1.812	2.165	2.066	27/32	0.876	0.749	0.09	0.03	2.750	3.000
1 3/8	1.3750	1.404	2 1/16	2.062	1.994	2.382	2.273	29/32	0.940	0.810	0.09	0.03	3.000	3.250
1 1/2	1.5000	1.531	2 1/4	2.250	2.175	2.598	2.480	1	1.036	0.902	0.09	0.03	3.250	3.500
1 3/4	1.7500	1.785	2 5/8	2.625	2.538	3.031	2.893	1 5/32	1.196	1.054	0.12	0.04	3.750	4.000
2	2.0000	2.039	3	3.000	2.900	3.464	3.306	1 11/32	1.388	1.175	0.12	0.04	4.250	4.500
2 1/4	2.2500	2.305	3 3/8	3.375	3.262	3.897	3.719	1 1/2	1.548	1.327	0.19	0.06	4.750	5.000
2 1/2	2.5000	2.559	3 3/4	3.750	3.625	4.330	4.133	1 21/32	1.708	1.479	0.19	0.06	5.250	5.500
2 3/4	2.7500	2.827	4 1/8	4.125	3.988	4.763	4.546	1 13/16	1.869	1.632	0.19	0.06	5.750	6.000
3	3.0000	3.081	4 1/2	4.500	4.350	5.196	4.959	2	2.060	1.815	0.19	0.06	6.250	6.500
3 1/4	3.2500	3.335	4 7/8	4.875	4.712	5.629	5.372	2 3/16	2.251	1.936	0.19	0.06	6.750	7.000
3 1/2	3.5000	3.589	5 1/4	5.250	5.075	6.062	5.786	2 5/16	2.380	2.057	0.19	0.06	7.250	7.500
3 3/4	3.7500	3.858	5 5/8	5.625	5.437	6.495	6.198	2 1/2	2.572	2.241	0.19	0.06	7.750	8.000
4	4.0000	4.111	6	6.000	5.800	6.928	6.612	2 11/16	2.764	2.424	0.19	0.06	8.250	8.500

Table 16-2
Hex bolt table for commonly used ASME standard hex bolts *(Reprinted from The American Society of Mechanical Engineers, ASME B18.2.1-1981)*

Basic Major DIA & Pitch	INTERNAL THREADS Tap Drill DIA	Minor DIA MAX	Minor DIA MIN	EXTERNAL THREADS Major DIA MAX	Major DIA MIN	Clearance Hole
M1.6 × 0.35	1.25	1.321	1.221	1.576	1.491	1.9
M2 × 0.4	1.60	1.679	1.567	1.976	1.881	2.4
M2.5 × 0.45	2.05	2.138	2.013	2.476	2.013	2.9
M3 × 0.5	2.50	2.599	2.459	2.976	2.870	3.4
M3.5 × 0.6	2.90	3.010	2.850	3.476	3.351	4.0
M4 × 0.7	3.30	3.422	3.242	3.976	3.836	4.5
M5 × 0.8	4.20	4.334	4.134	4.976	4.826	5.5
M6 × 1	5.00	5.153	4.917	5.974	5.794	6.6
M8 × 1.25	6.80	6.912	6.647	7.972	7.760	9.0
M10 × 1.5	8.50	8.676	8.376	9.968	9.732	11.0
M12 × 1.75	10.20	10.441	10.106	11.966	11.701	13.5
M14 × 2	12.00	12.210	11.835	13.962	13.682	15.5
M16 × 2	14.00	14.210	13.835	15.962	15.682	17.5
M20 × 2.5	17.50	17.744	17.294	19.958	19.623	22.0
M24 × 3	21.00	21.252	20.752	23.952	23.577	26.0
M30 × 3.5	26.50	26.771	26.211	29.947	29.522	33.0
M36 × 4	32.00	32.270	31.670	35.940	35.465	39.0
M42 × 4.5	37.50	37.799	37.129	41.937	41.437	45.0
M48 × 5	43.00	43.297	42.587	47.929	47.399	52.0
M56 × 5.5	50.50	50.796	50.046	55.925	55.365	62.0
M64 × 6	58.00	58.305	57.505	63.920	63.320	70.0
M72 × 6	66.00	66.305	65.505	71.920	71.320	78.0
M80 × 6	74.00	74.305	73.505	79.920	79.320	86.0
M90 × 6	84.00	84.305	83.505	89.920	89.320	96.0
M100 × 6	94.00	94.305	93.505	99.920	99.320	107.0

Table 16-3
Thread table for commonly used ISO metric fasteners

Metric Thread Series

ASME conventions cover inch standards. They are not compatible with the ISO metric series. Today the United States is the only country that uses an inch-based thread system, although the IFI has established U.S. metric standards for some fasteners. A comparison of U.S. standard (UN) threads and metric threads is shown in **Figure 16-21**. ISO metric notations differ from ASME notations **(Fig. 16-22)**. Metric notations include pitch in millimeters, major diameter in millimeters, and the ISO thread symbol.

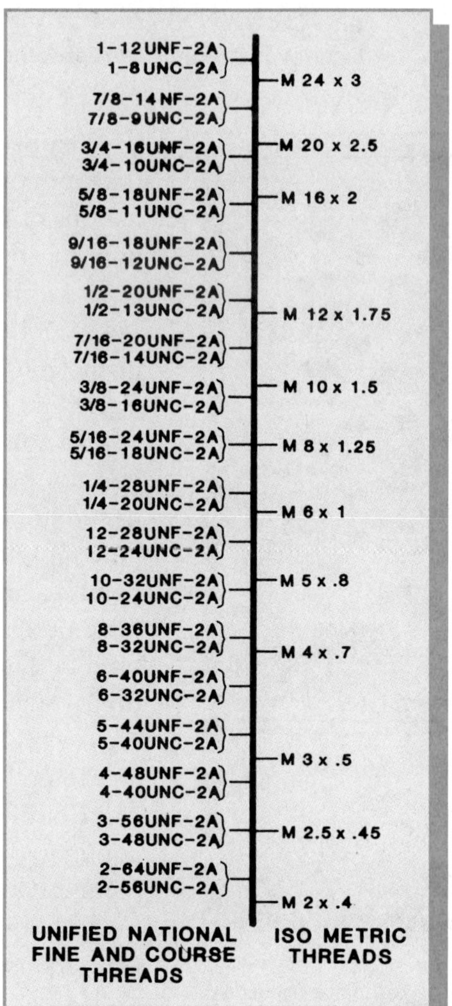

Figure 16-21
Comparison of typical ASME's UN thread series to ISO metric threads

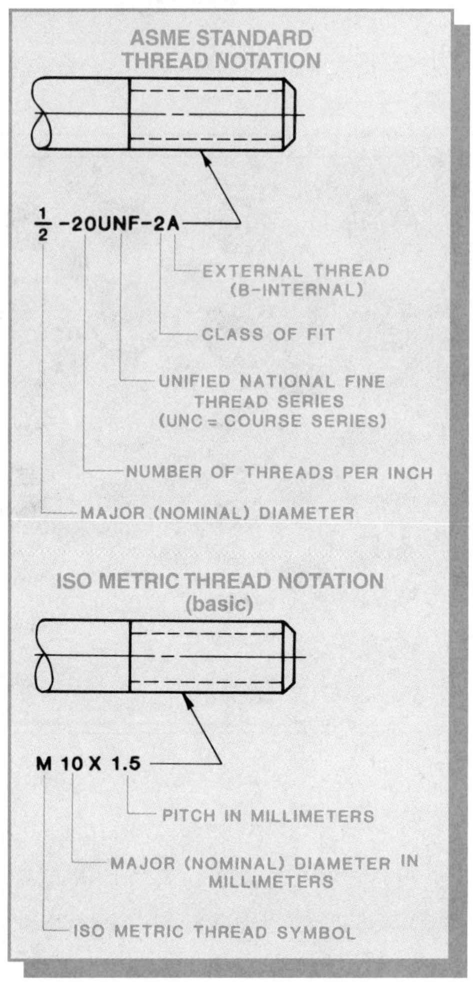

Figure 16-22
Comparison of ASME's standard inch thread note to ISO metric thread note

Pipe Threads

Pipes have either straight or tapered threads **(Fig. 16-23)**. They are drawn with either detailed or simplified symbols. In either case, the pipe thread notation includes the nominal diameter of the pipe in inches, threads per inch, standard used, pipe designation, and an entry indicating whether the pipe is straight or tapered.

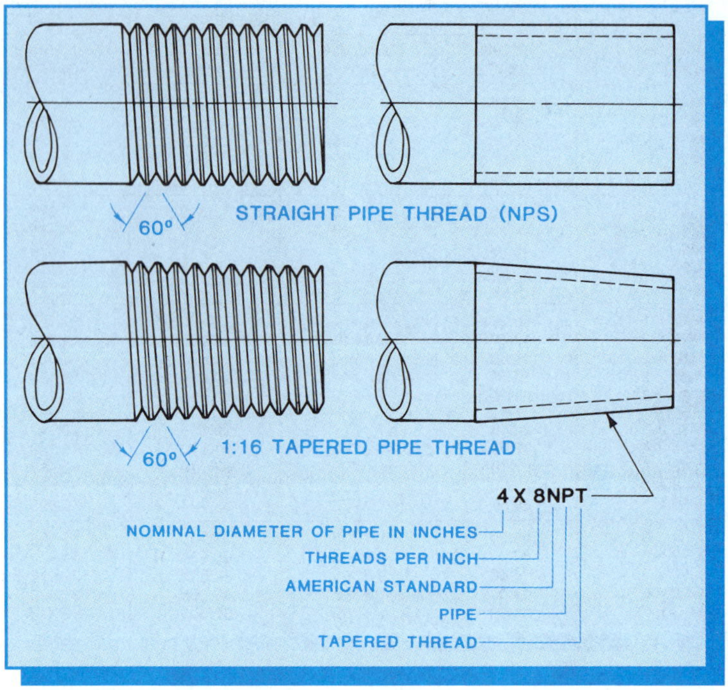

Figure 16-23
Pictorial and simplified representation of pipe threads and notations

Fastener Templates

The normal laying out and drawing of fasteners with instruments is a long and costly process. Consequently, most fasteners are drawn with templates using simplified fastener symbols. Templates provide outlines for two- and three-dimensional fastener drawings **(Figs. 16-24** and **16-25)**.

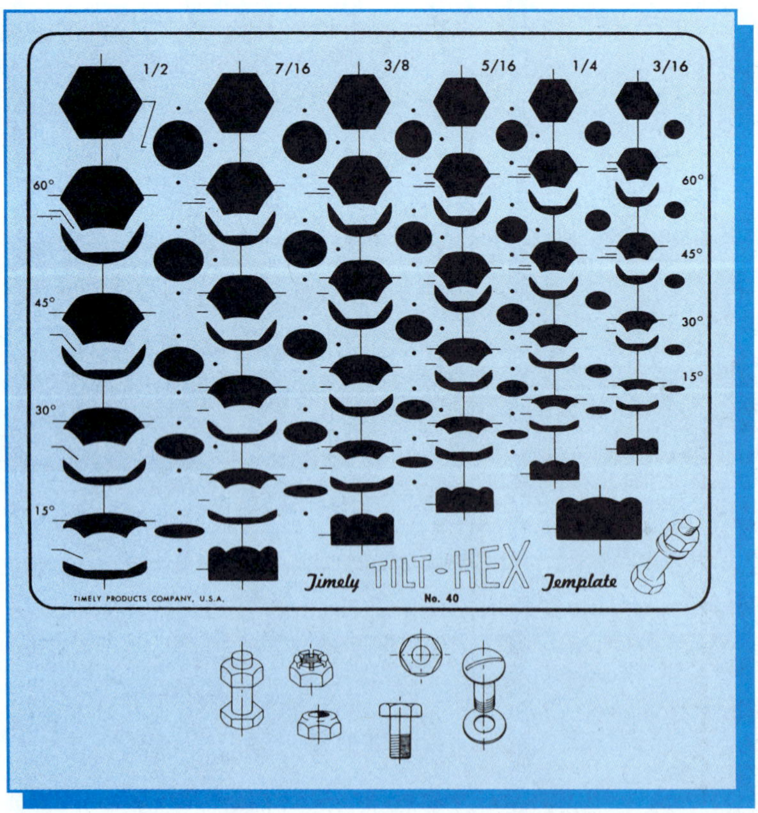

Figure 16-24
Two- and three-dimensional bolt and nut drawing template *(Courtesy of Timely Products)*

Templates provide the outline of fastener components; however, combinations of template openings must be used to draw all the features of a fastener symbol. **Figure 16-26** shows the steps used to draw a bolt using a template. To use a template, first establish centerline **(Fig. 16-27a)**. In **Figure 16-27a** a centerline positions the hexagon nut opening, providing a guide for the addition of concentric circles with a hole template. The use of some template openings may require reversing a template outline to draw identical sides of a symbol. This procedure is illustrated in the layout of the jamb nut shown in **Figure 16-27b**.

Templates do not always provide the exact outline of every fastener size and configuration. Some contain the most common sizes with the most common heads, bodies, and points. Most templates contain separate heads that can be

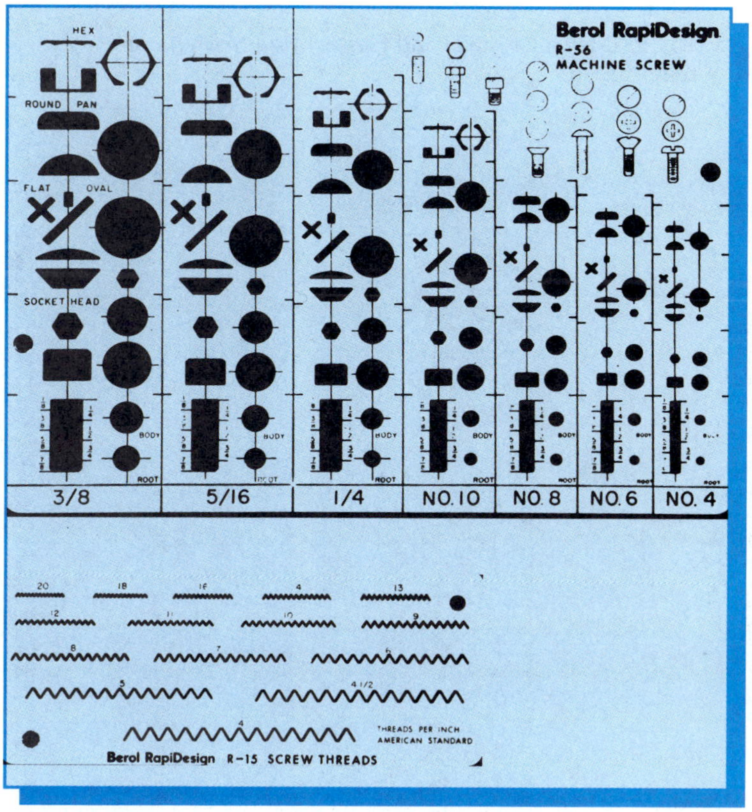

Figure 16-25
Various types of working drawing templates for screw threads and machine screws *(Courtesy of Berol USA/RapiDesign)*

combined with different lengths and points to produce the symbol needed. A machine screw template is used to draw the flat head and fillister machine screws in **Figure 16-28**.

Pictorial representations of fasteners are also accomplished with the use of templates **(Fig. 16-29)**. In using pictorial templates, the drafter must make sure that the template angle of projection is the same as the angle in the drawing. Most technical drawing pictorial templates include isometric angles, although ellipse templates cover a wide range of angles. Isometric fastener templates contain symbol outlines designed for the receding 30° planes and the top plane. These templates provide isometric views of unique features that, used with centerlines and isometric circles (35°), produce the final drawing.

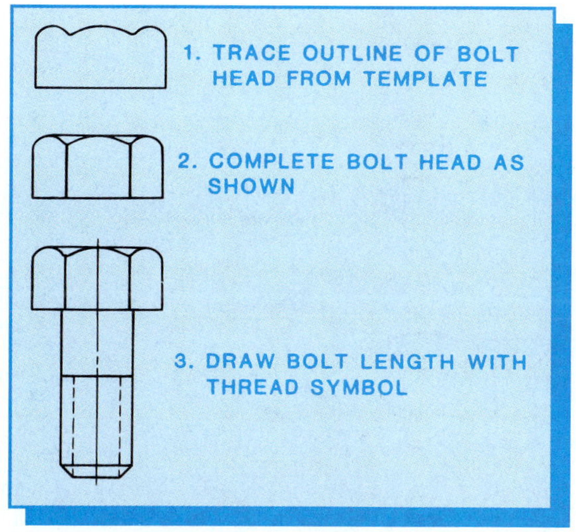

Figure 16-26
Steps to draw a bolt with a template

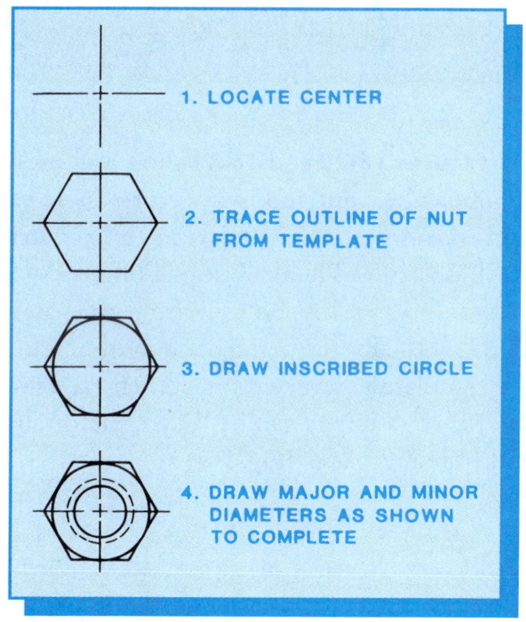

Figure 16-27a
Steps to draw the face view of a nut with a template

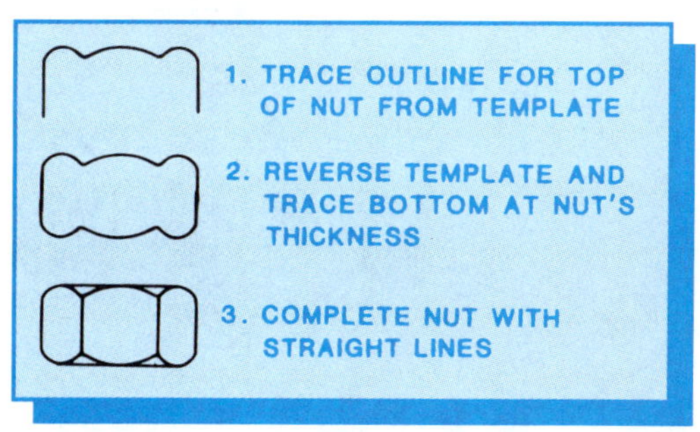

Figure 16-27b
Steps to draw a side view of a jamb nut with a template

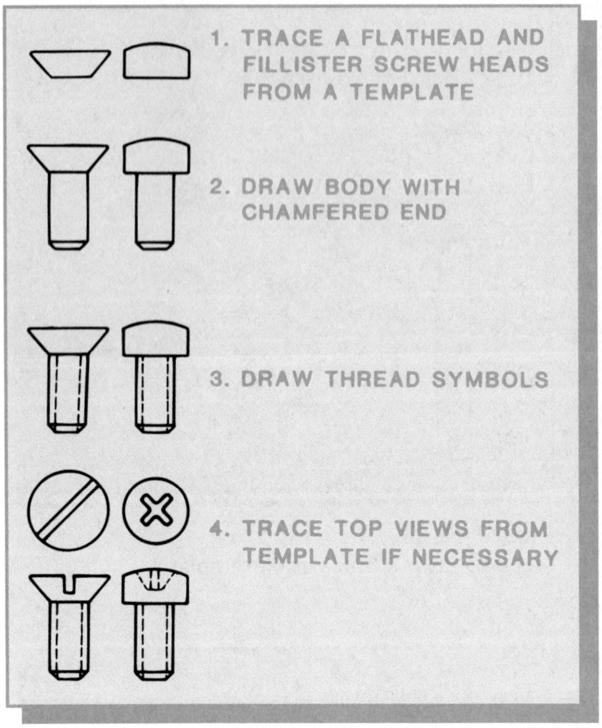

Figure 16-28
Steps to draw a flat head and fillister head machine screw with a template

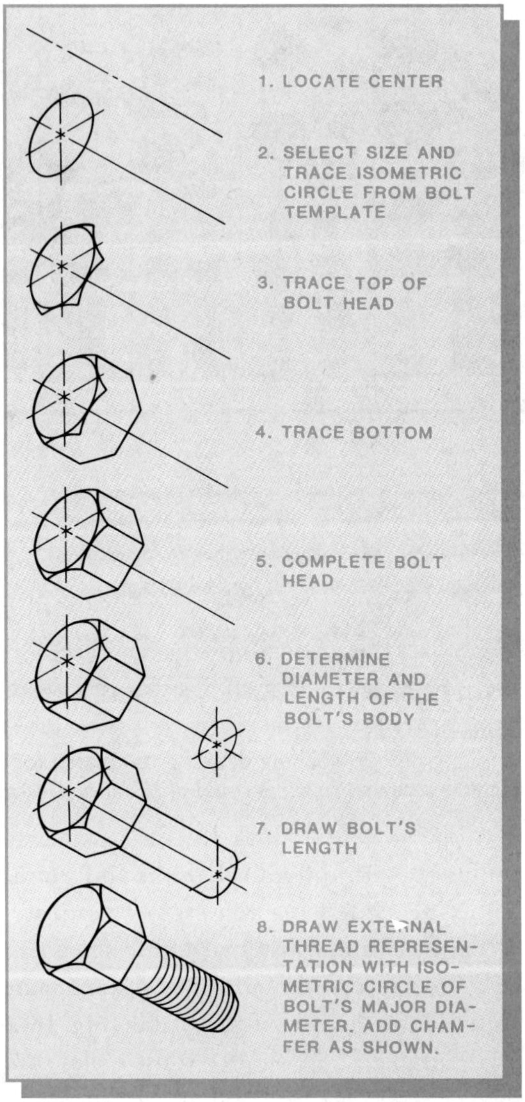

Figure 16-29
Steps to draw an isometric hex bolt with a template

Figures 16-29 and **16-30** show the procedure to draw an isometric hex nut and bolt. **Figure 16-31** shows the application of this method to drawing an isometric flat head machine screw. There are four isometric planes available on most templates: top, bottom, right side, and left side **(Fig. 16-32)**. Usually, the right-plane outlines are provided, and the template is reversed for the left receding plane. A wide variety of fasteners can be drawn pictorially with fastener and hole templates **(Fig. 16-33)**.

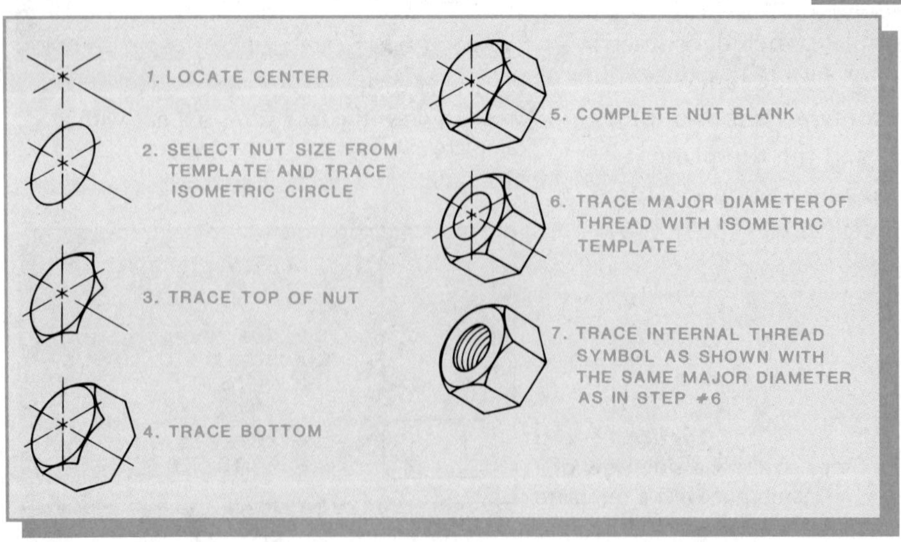

Figure 16-30
Steps to draw an isometric hex nut with a template

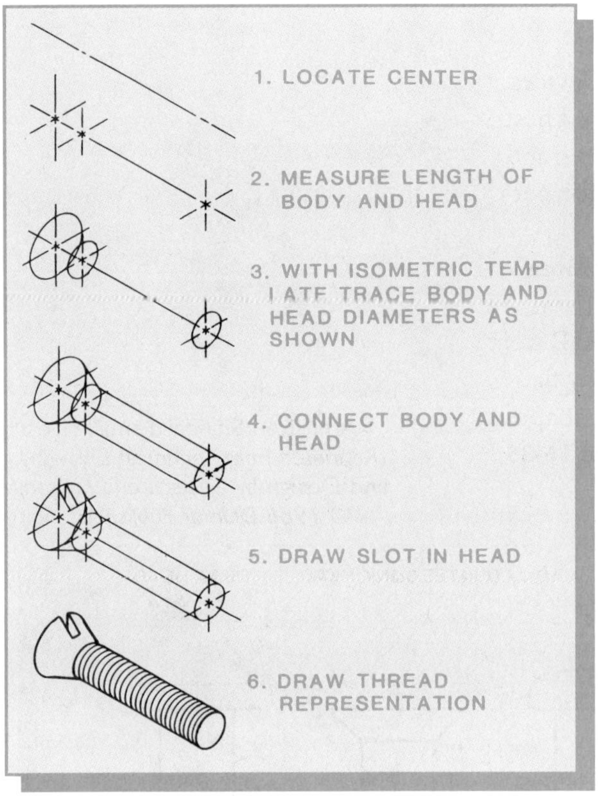

Figure 16-31
Steps to draw an isometric flat head machine screw with a template

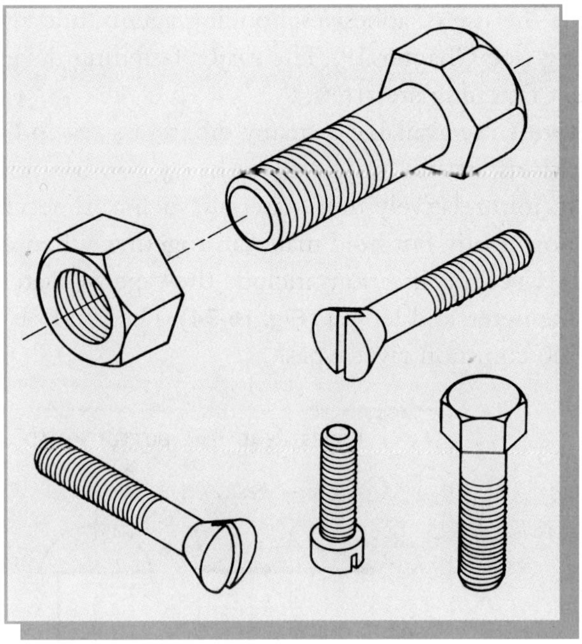

Figure 16-32
Various positions of isometric planes

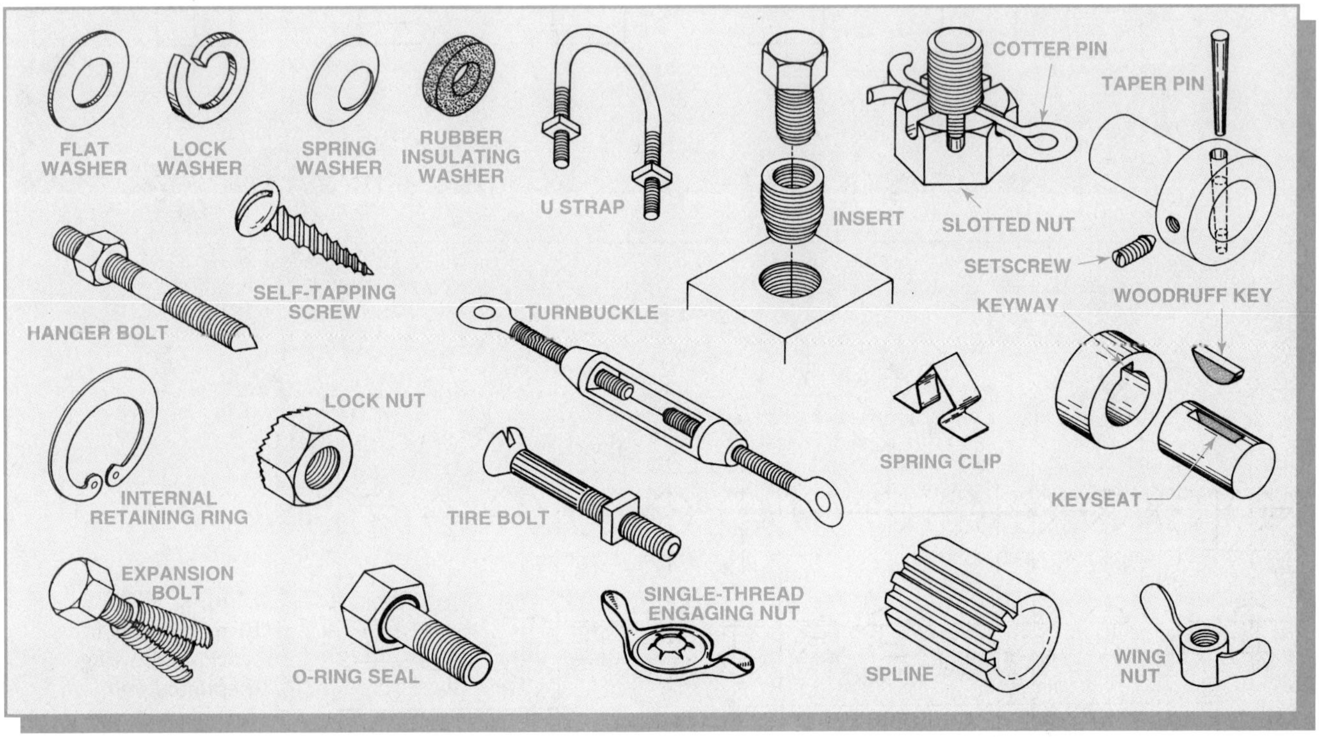

Figure 16-33
Miscellaneous types of fasteners

Threaded Permanent Fasteners

Most permanent fastening systems do not use fastening devices, but bond materials through heat, pressure, or chemical action. These include the use of adhesives, bonding agents, brazing, soldering, and welding (see Chapter 19). The major fastening devices used for permanent fastening are rivets.

Rivets are available in many different sizes and configurations. Large rivets are used primarily in heavy construction. Small rivets are used to join relatively thin materials, such as sheets and plates. Rivets have no threads, but hold materials together when one head is flattened. Rivet specifications include the type of material, head shape, shaft diameter, and length **(Fig. 16-34)**. The symbols in **Figure 16-35** describe common rivet types.

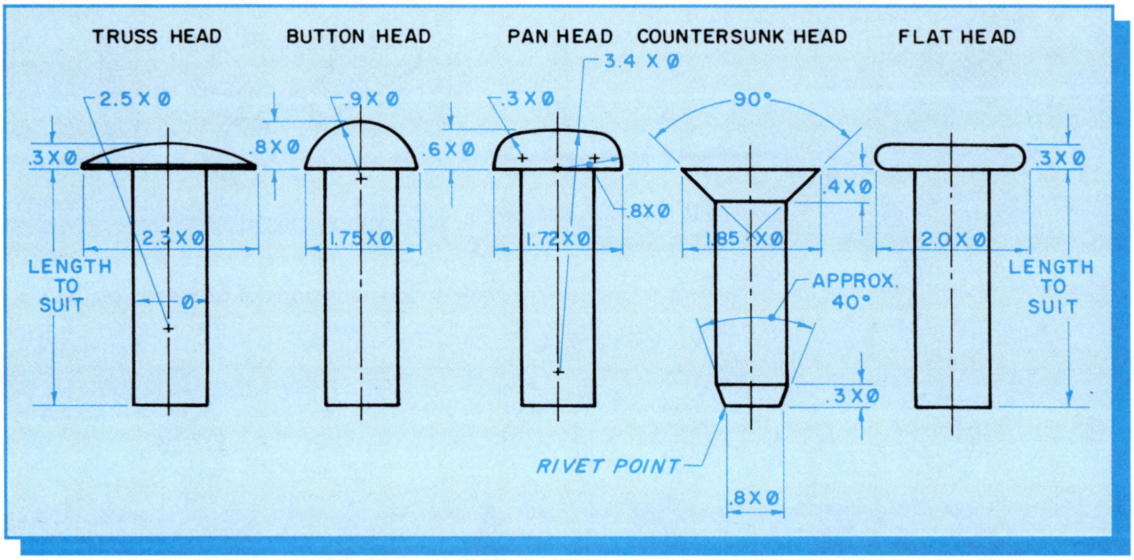

Figure 16-34 American Standard small rivets *(Reprinted from* Technical Drawing and Design *by Goetsch and Nelson, © 1986 Delmar Publishers, Inc.)*

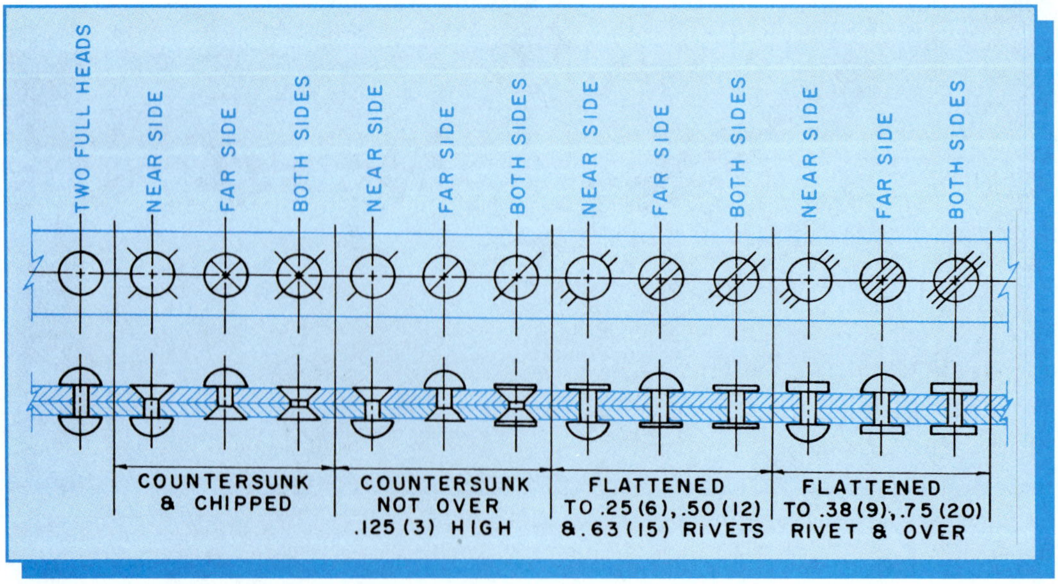

Figure 16-35 Rivet symbols for working drawings *(Reprinted from* Technical Drawing and Design *by Goetsch and Nelson, © 1986 Delmar Publishers, Inc.)*

Springs

A **spring** is a mechanical device, often in the form of a helical coil, that yields by expasmeon or contraction due to pressure, force, or stress applied. Springs are made to return to their normal form when the force or stress is removed. Springs are designed to store energy for the purpose of pushing or pulling machine parts by reflex action into certain desired positions. Improved spring technology provides springs with the ability to function a long time under high stresses. The effective use of springs in machine design depends upon five basic criteria, including material, application, functional stresses, use, and tolerances.

Continued research and development of spring materials have helped improve spring technology. The spring materials most commonly used include high-carbon spring steels, alloy spring steels, stainless spring steels, music wire, oil-tempered steel, copper-based alloys, and nickel-based alloys. Spring materials, depending upon use, may have to withstand high operating temperatures, and high stresses under repeated loading.

Spring design criteria are generally based on material gauge, kind of material, spring index, direction of the helix, type of ends, and function. Spring-wire gauges are available from several different sources ranging in diameter from number 7/0 (.490 in.) to number 80 (.013 in.). The most commonly used spring gauges range from 4/0 to 40. There are a variety of spring materials available in round or square stock for use depending on spring function and design stresses. The spring index is a ratio of the average coil diameter to the wire diameter. The index is a factor in determining spring stress, deflection, and the evaluation of the number of coils needed and the spring diameter. Recommended index ratios range between 7 and 9, although other ratios commonly used a range from 4 to 16. The direction of the helix is a design factor when springs must operate in conjunction with threads or with one spring inside of another. In such situations, the helix of one feature should be in the opposite direction of the helix for the other feature. Compression springs are available with ground or unground ends. Unground, or rough, ends are less expensive than ground ends. If the spring is required to rest flat on its end, then ground ends should be used. Spring function depends upon two basic factors, compression or extension. Compression springs release their energy and return to the normal form when compressed. Extension springs release their energy and return to the normal form when extended **(Fig. 16-36)**.

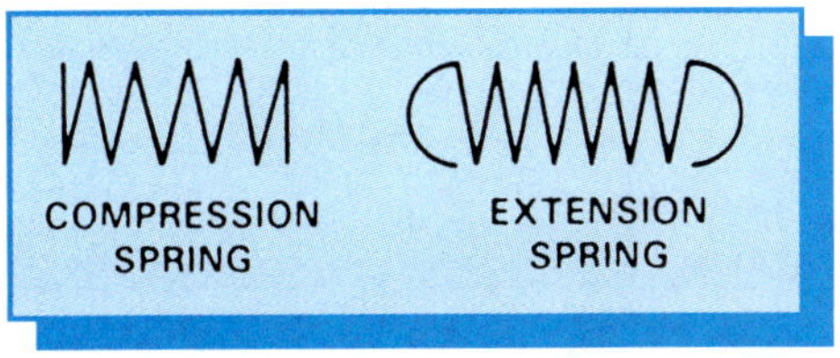

Figure 16-36
Compression and extension spring

Spring Terminology

The springs shown in **Figure 16-37** show some common characteristics, as follows.

Ends. Compression springs have four general types of ends: open or closed ground ends and open or closed unground ends, shown in **Figure 16-38**. Extension springs have a large variety of optional ends, a few of which are shown in **Figure 16-39**.

Helix Direction. The helix direction may be specified as the right hand or the left hand. **(Fig. 16-38)**.

Free Length. The length of the spring when there is no pressure or stress to affect compression or extension is known as free length.

Compression Length. The compression length is the maximum recommended design length for the spring when compressed.

Solid Height. The solid height is the maximum compression possible. The design function of the spring should not allow the spring to reach solid height when in operation unless this factor is a function of the machinery.

Loading Extension. The extended distance to which an extension spring is designed to operate is the loading extension length.

Pitch. The pitch is one complete helical revolution, or the distance from a point on one coil to the same corresponding point on the next coil.

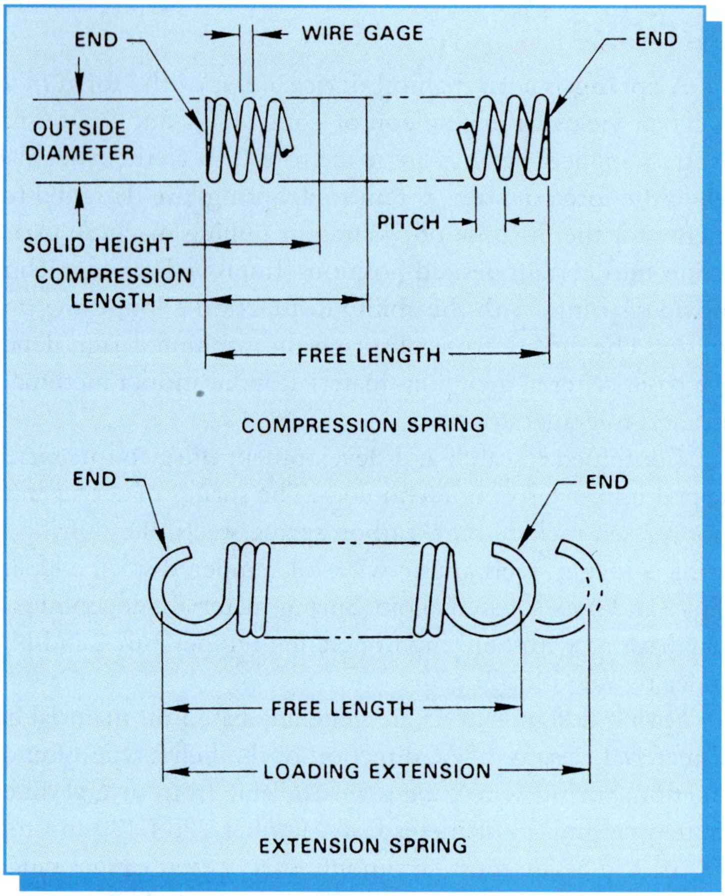

Figure 16-37
Spring characteristics

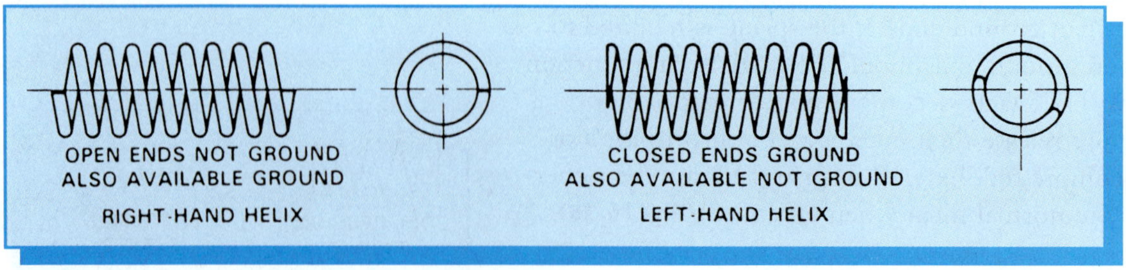

Figure 16-38
Helix direction and compression spring end types

Torsion Springs

Torsion springs are designed to transmit energy by a turning or twisting action. Torsion is defined as a twisting action that tends to turn one part or end around a longitudinal axis while the other part or end remains fixed. Torsion springs are often designed as antibacklash devices or as self-closing or self-reversing units **(Fig. 16-40)**.

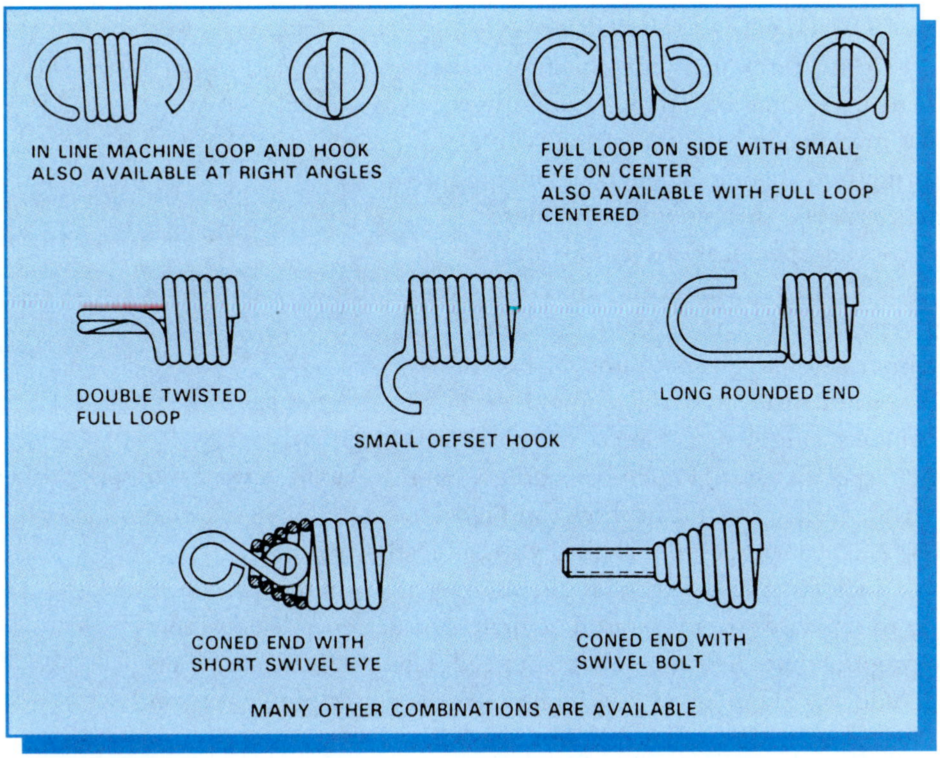

Figure 16-39
Extension spring end types

Flat Springs

Flat Springs are arched or bent flat-metal shapes designed so that when placed in machinery, they cause tension on adjacent parts. The tension may be used to level parts, provide a cushion, or position the relative movement of one part to another. One of the most common examples of a flat spring are leaf springs on an automobile.

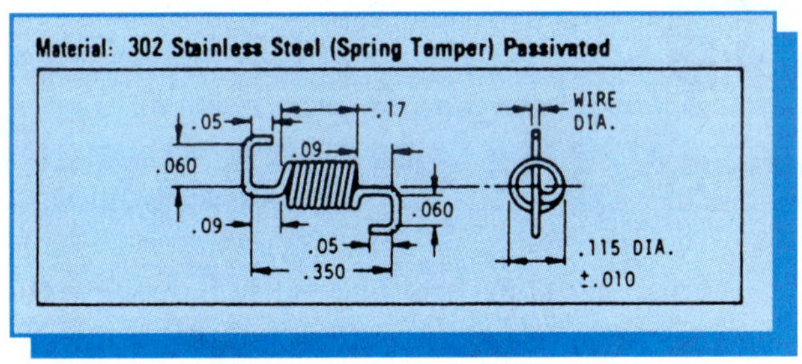

Figure 16-40
Torsion spring. *(Courtesy NORDEX Inc.)*

Spring Representations

There are three types of spring representations, detailed, schematic, and simplified as seen in **Figure 16-41**. Detailed spring drawings are used in situations that require a realistic representation, such as vendors' catalogs, assembly instructions, or detailed assemblies. Schematic spring representations are commonly used on drawings. The single-line schematic symbols are easy to draw and clearly represent springs without taking the additional time required to draw a detailed spring. The use of simplified spring drawings is limited to situations where the clear resemblance of a spring is not necessary. While very easy to draw, the simplified spring symbol must be accompanied by clearly written spring specifications. The simplified spring representation is not very useful in assembly drawings or other situations that require a visual comparison of features.

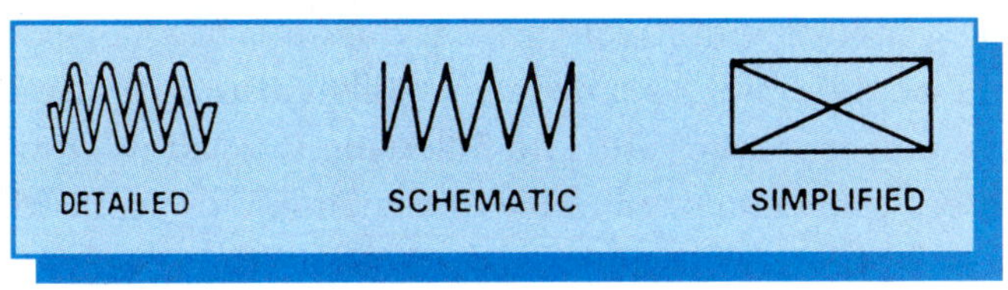

Figure 16-41
Spring representations

Spring Specifications

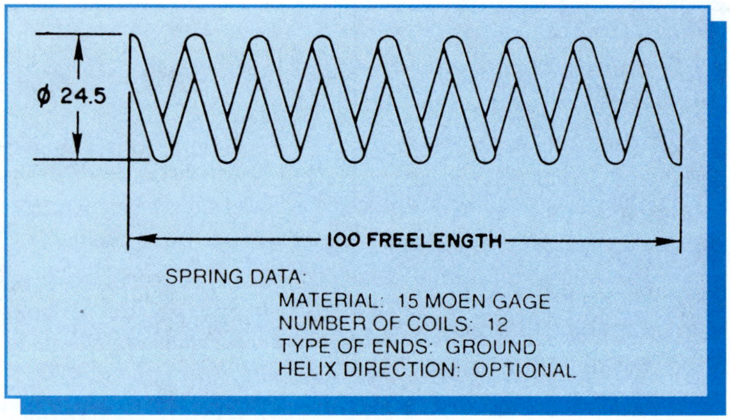

Figure 16-42
Spring drawing with spring data chart

No matter which representation is used, there are several important specifications that must accompany the spring symbol. Spring information is generally lettered in the form of a specific or general note.

Spring specifications include outside or inside diameter, wire gauge, kind of material, type of ends, surface finish, free and compressed length, and number of coils. Other information, when required, may include spring design criteria and heat treatment specifications. The information is often provided on a drawing as shown in **Figure 16-42**. The material note is usually found in the title block.

Since fasteners are standard items, they are manufactured according to standard specifications. The drafter or designer selects the appropriate fasteners for the design need. These standard fasteners include the full range of nuts, bolts, washers, cotter pins, rivets, and other fastening devices.

AutoCAD LT Exercises

Because fasteners are standardized, drawing them on CAD is easy. Each type of fastener is drawn only once and saved in the library as a block for insertion or coping as needed. Standard mechanical symbols libraries may be purchased and added to AutoCAD LT. They make drawing detailed fasteners as easy as picking a menu item. If your system does not have a custom symbols library, the CAD operator may easily create and save the standard symbols for future use. The following steps show how to create and insert a standard symbol into a drawing. This standard fastener is a 1/4-20 hex bolt, 1.25 long.

Creating a Symbol

1. Obtain accurate information about the fastener. Three sources may be used: a text, such as this one; the *Machinery's Handbook*; or a vendor's catalog. Once the fastener is found, locate the size for a 1" hex bolt.

2. Draw the 1" hex bolt **(Fig. 16-43)**.
   a. Select **Draw**, **Polygon**, and type **6** for the number of sides. Pick a point. Radius of Circle: **.5**
   b. Rotate the hexagon 90°.
      Select **Modify**, **Rotate**. Select the hexagon and press Enter. Base Point: pick the center of the hexagon. Rotation: **90**. Press **EB**.
   c. Draw a 1" circle inside the polygon.
   d. Create a hidden line layer and construction layers as needed.
   e. Add hidden line circles to the polygon to form the top view.
   f. Draw the side view **(Fig. 16-43)**. Use construction lines as a drawing aid.
   g. Copy the top view to the right side of the drawing.
   h. Change the hidden layer to layer 0. Select **Modify**, **Change properties**, and select the hidden lines in the bottom view.
      Select **Layer, Layer 0,** and press **OK** twice. The drawing should look like **Figure 16-44** without the text.

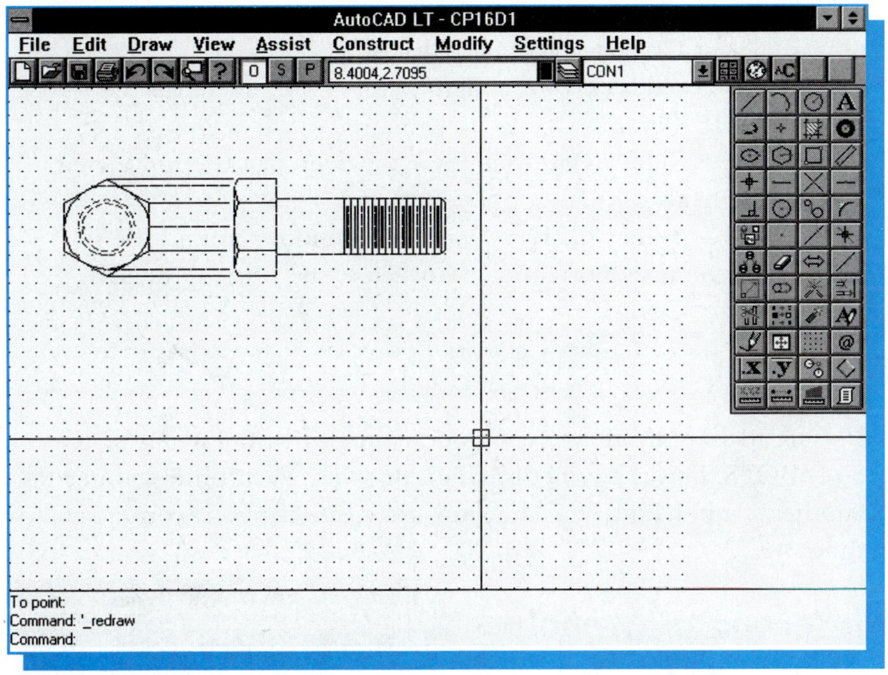

Figure 16-43
Drawing the bolt to specifications

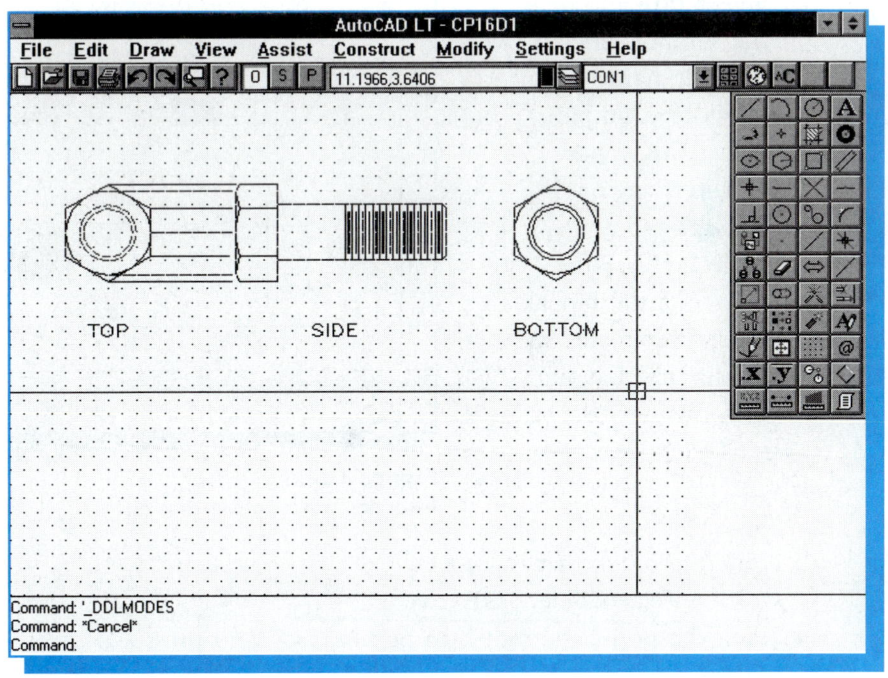

Figure 16-44
Parts ready to be made into blocks

3. Create blocks (symbols) of the drawings.
   a. Select **Construct**, **Make Block.**
   b. Give the block a meaningful, easily remembered name (such as HEXTOP).
   c. Select **Select Point** and pick the center of the top view of the hexagon.
   d. Select **Select Objects** and put a window around the top view of the hexagon. Do not select the name. Press **EB**.
   e. Select **OK.**
   f. Do the same process for the other two views. Name them HEXSIDE and HEXBOT.

Be sure to save this drawing. A good name for this drawing is SYMBOLS. This is a symbols library drawing. Each time you need another symbol, go to this drawing and create it just as you did these three.

Inserting a Symbol

1. Start a new drawing.
2. Select **Draw**, **Insert Block**.
   a. Select **File** and pick the symbols library drawing name (from previous example). Select **OK** twice.
   b. At the insertion prompt, type **CTRL+C.** This inserts all the blocks that were created in the library drawing into this new drawing.
3. Select **Draw**, **Insert block**.
   a. Select **Block**.
   b. Pick the block that is to be inserted. Select **OK** twice.
   c. Pick the point where it is to be inserted , and press **EB** three times.
4. Insert the other symbols.

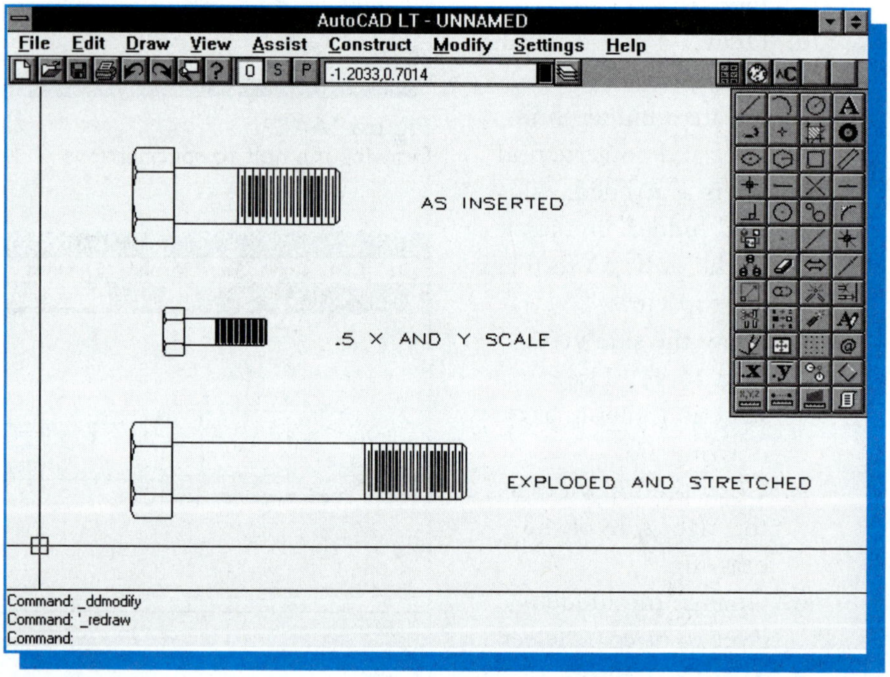

Figure 16-45
Inserting parts into a drawing

NOTE: When inserting a block, watch the command line. The program will prompt you to change the scale or rotate the block as it is inserted. Experiment with this. If no change in the block is necessary, hit the Enter key to default the prompts. **Figure 16-45** is an example of modified blocks.

If it is necessary to change a block after it is inserted, it will be necessary to explode the block.

1. Select **Modify**, **explode** and select the block to explode. Press **EB**.
2. Select **Modify**, **stretch**—to make the part longer after you have exploded it.
   a. Select area of the part to stretch with the crossing window.
   b. With Ortho on, pick a point on the area to be stretched and move it until the required length is obtained.

This process of drawing commonly used parts or symbols and making blocks of them is very useful and will be used in some of the following chapters. Standard symbols may be purchased and added to AutoCAD LT. **Figure 16-46** is an example of some standard symbol libraries available.

Figure 16-46
AutoCAD LT Mechanical Symbols

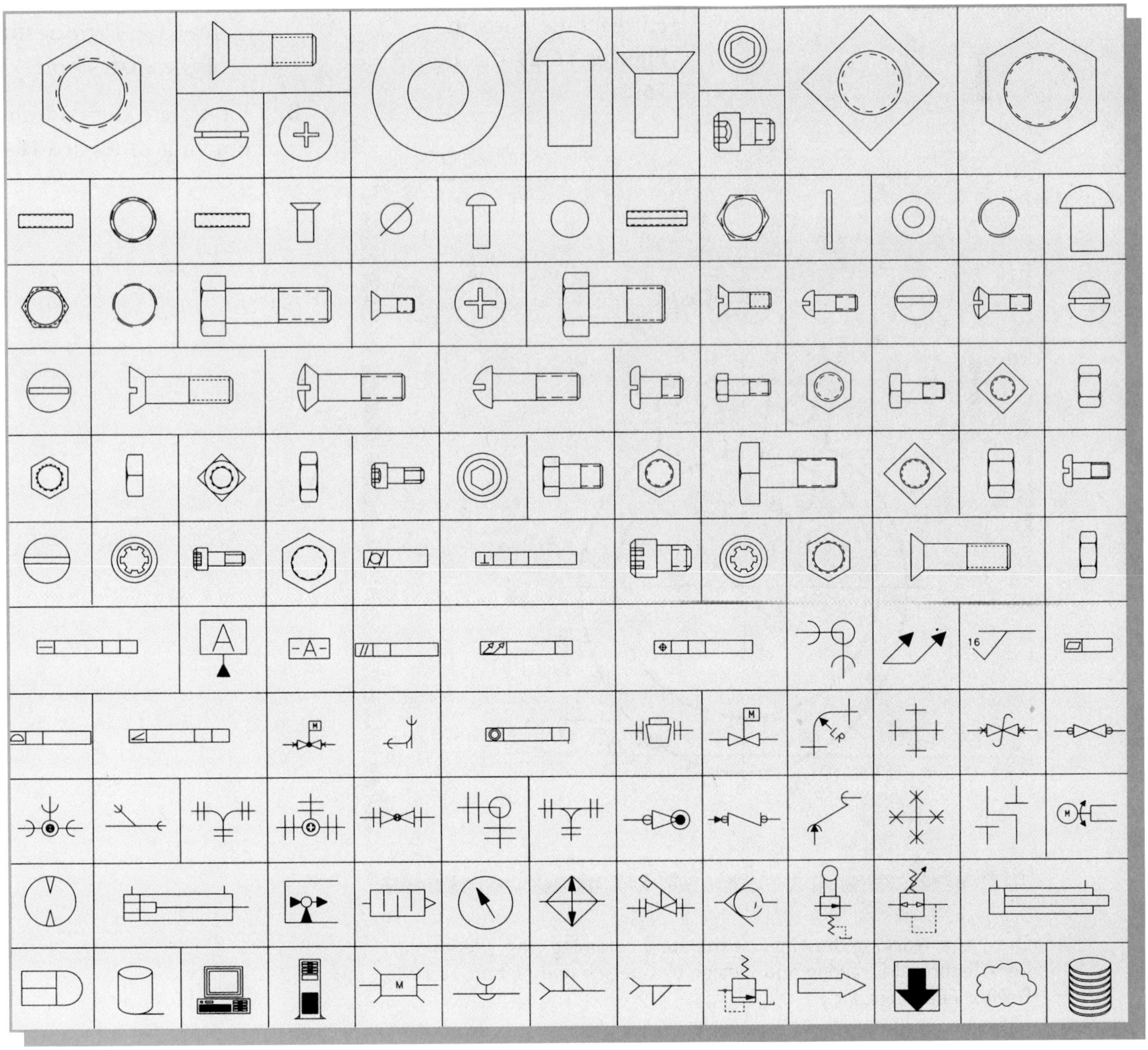

Drafting Exercises

1. Sketch **Figures 16-47** through **16-49**.
2. With an isometric fastener template, draw the various nuts in **Figure 16-50**.
3. With drawing instruments or a CAD system, draw the fasteners in **Figures 16-51** through **16-55**.
4. With drawing instruments or a CAD system, draw the metric fasteners in **Figures 16-56** and **16-57**.
5. Draw and label the twenty-three fastener symbols in **Figure 16-58**.
6. Draw with instruments or a CAD system the threaded machine parts in **Figures 16-59** through **16-61**.
7. Perform the following ECOs for **Figure 16-59**:
   - Change all UNF threads to UNC.
   - Add a fifth threaded fastener to the flange.
   - Add an external UNC thread 1.5" from the top of the shaft.
   - Add an internal UNF thread 2.0" from the top of the shaft's hole.
8. Complete the coordinate drawing in **Figure 16-61**.

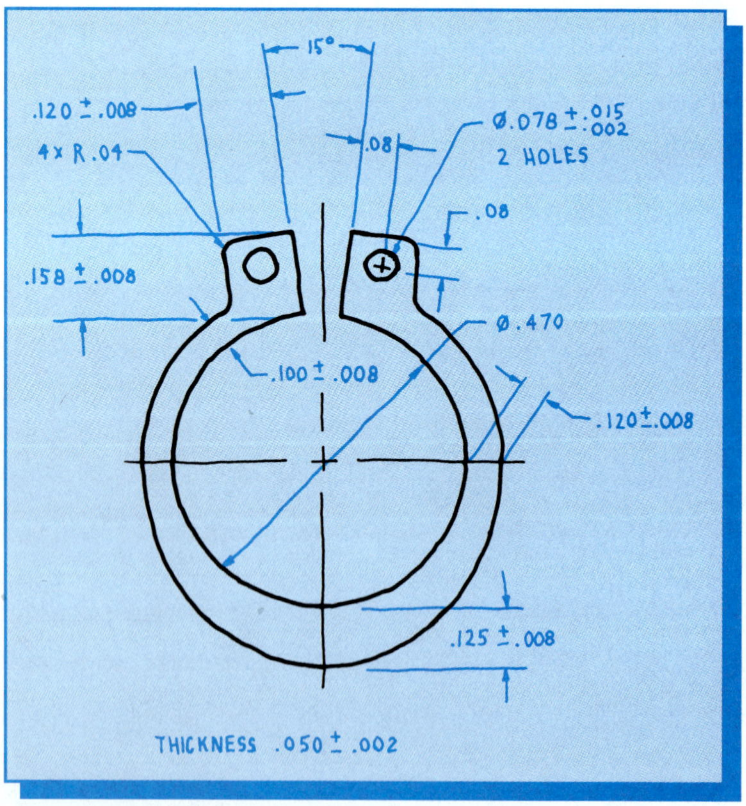

Figure 16-47
Draw the working drawings of the steel retaining ring. *(Reprinted from* Technical Drawing and Design *by Goetsch and Nelson, © 1986 Delmar Publishers, Inc.)*

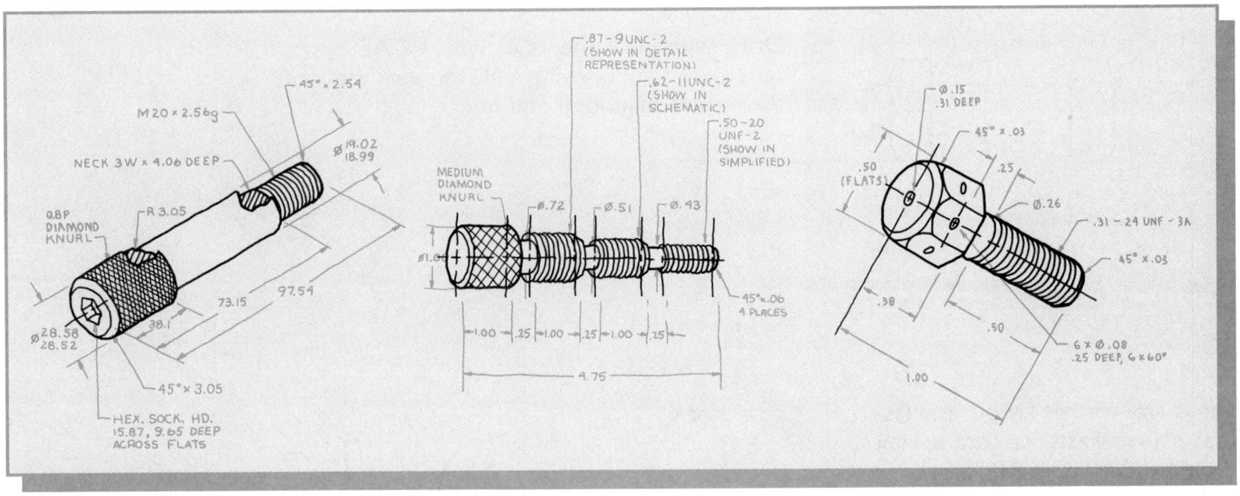

Figure 16-48
Sketch a pictorial of each item freehand or on a CAD system. Draw a multiview working drawing with metric or inch/decimal dimensions. *(Reprinted from* Mechanical Drafting *by Madsen, Shumaker, and Stewart, © 1986 Delmar Publishers, Inc.)*

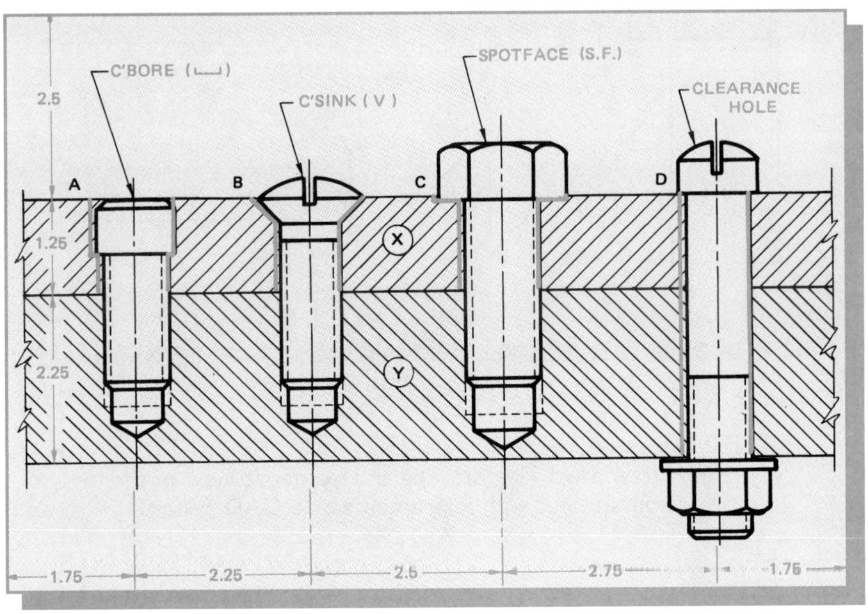

Figure 16-49
Draw or sketch each fastener and add a thread notation. *(Reprinted from* Technical Drawing and Design *by Goetsch and Nelson, © 1986 Delmar Publishers, Inc.)*

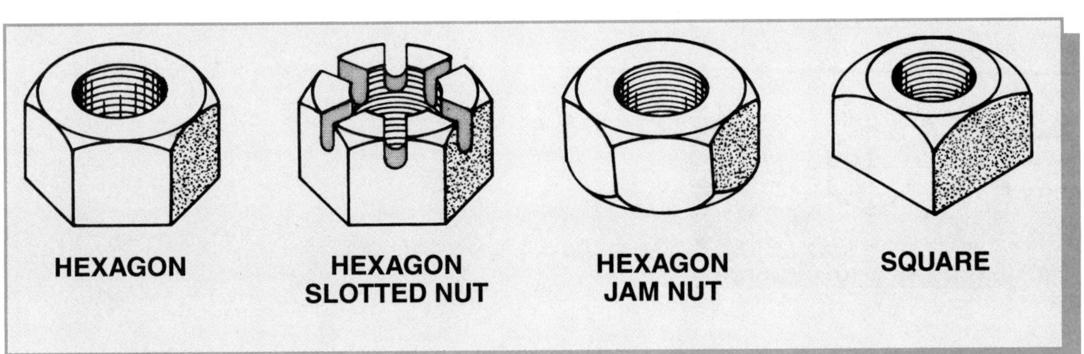

Figure 16-50
Draw an example of each nut as a detail drawing and a pictorial drawing.

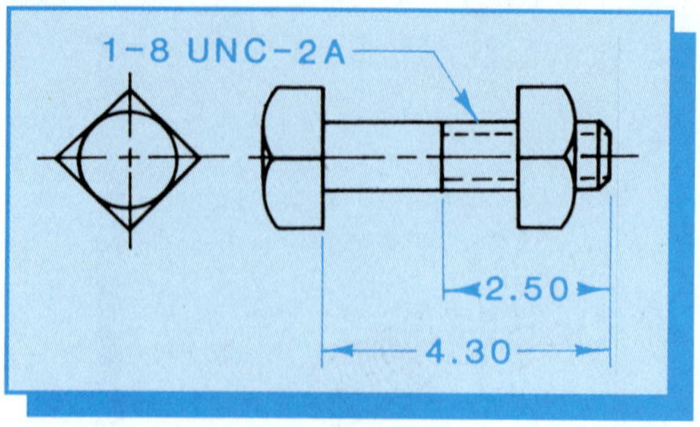

Figure 16-51
Draw with instruments or with a CAD system a working drawing and isometric view of the square head bolt and nut.

Figure 16-52
On an A-size format lay out the fasteners as shown. Fastener data is: Hex bolt 1-3/8-UNC x 3.0 long; Lock washer fits a 7/8" diameter cap screw Square nut 1-12 UNF 2B; Square head cap screw 3/4-10-UNC x 3.0 long. *(Reprinted from* Technical Drawing and Design *by Goetsch and Nelson, © 1986 Delmar Publishers, Inc.)*

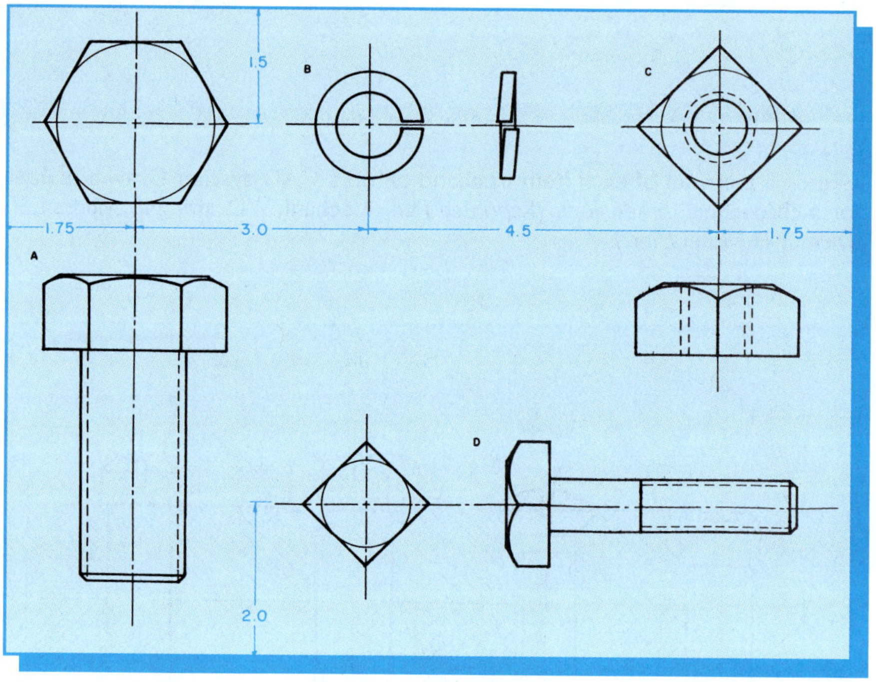

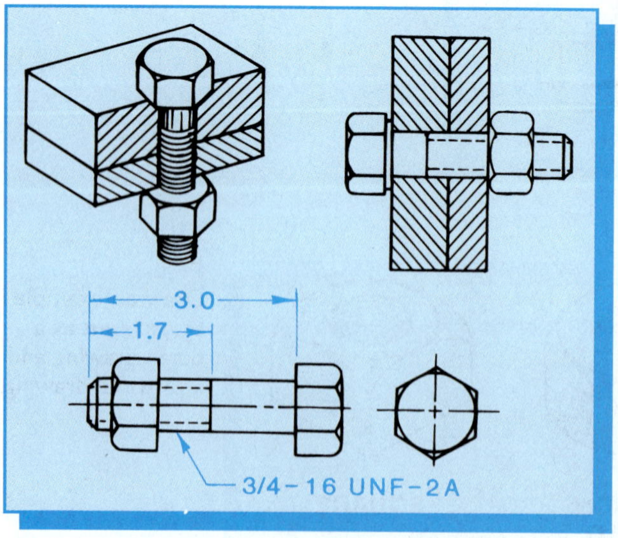

Figure 16-53
Draw a working drawing and isometric view of the hex bolt and nut with instruments or a CAD system.

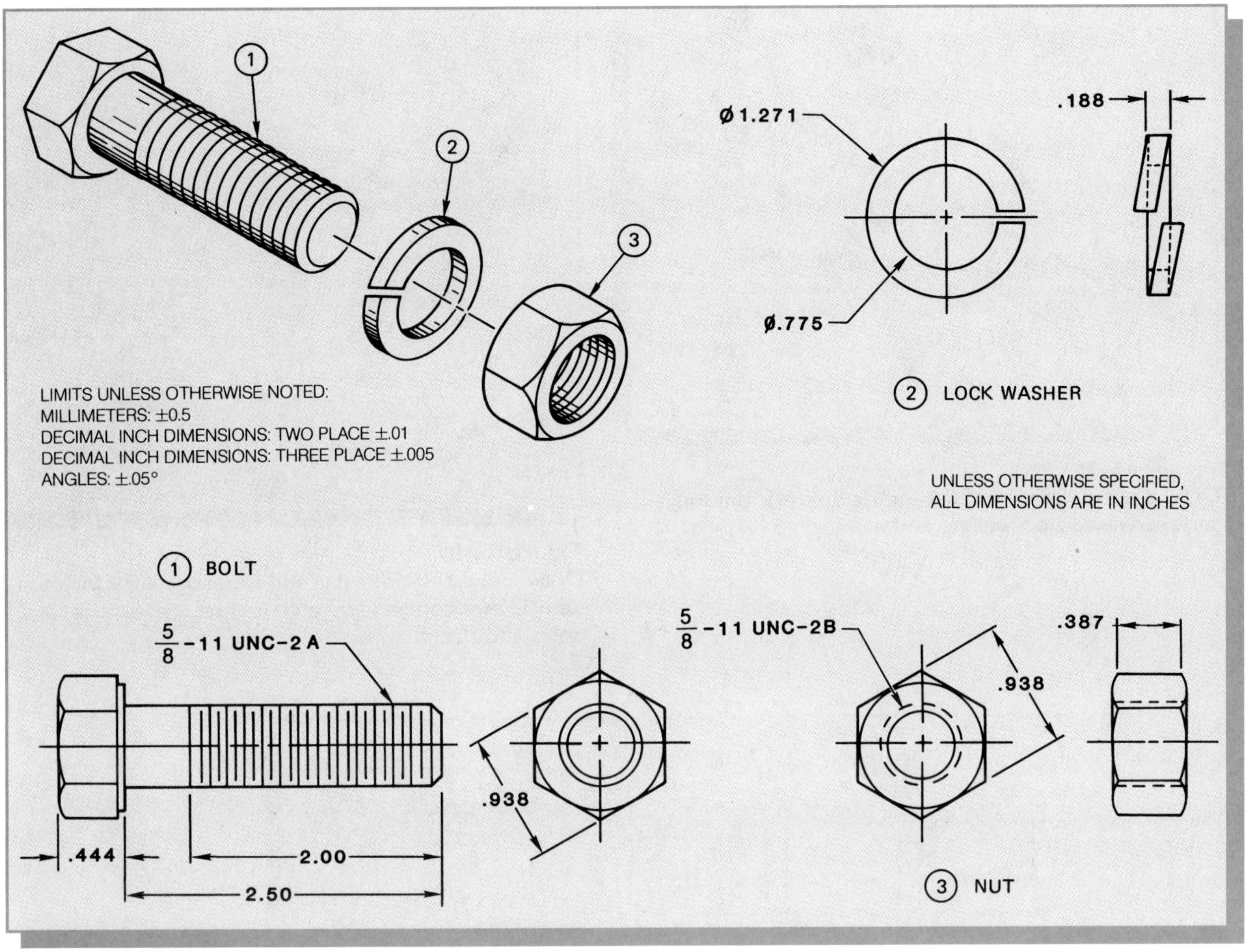

Figure 16-54
Redraw the fasteners.

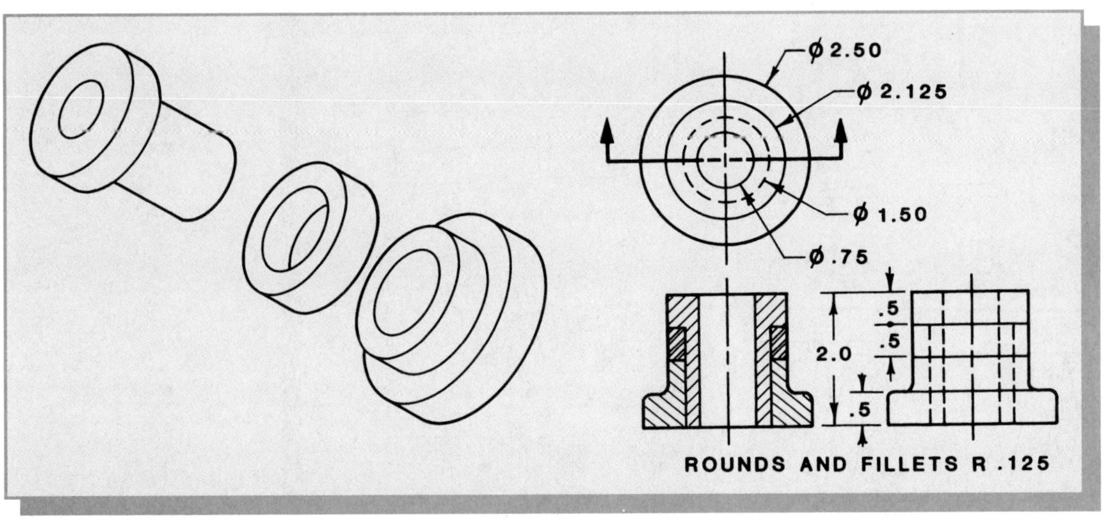

Figure 16-55
Draw the working drawing of the bearing assembly. Plan a thread system and add all thread notes and thread symbols.

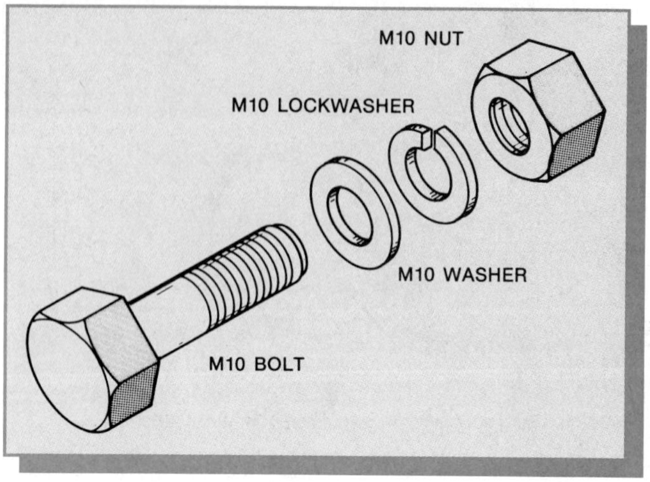

Figure 16-56
Draw a detail drawing of the metric bolt (45 mm long) and the washer, lockwasher, and nut.

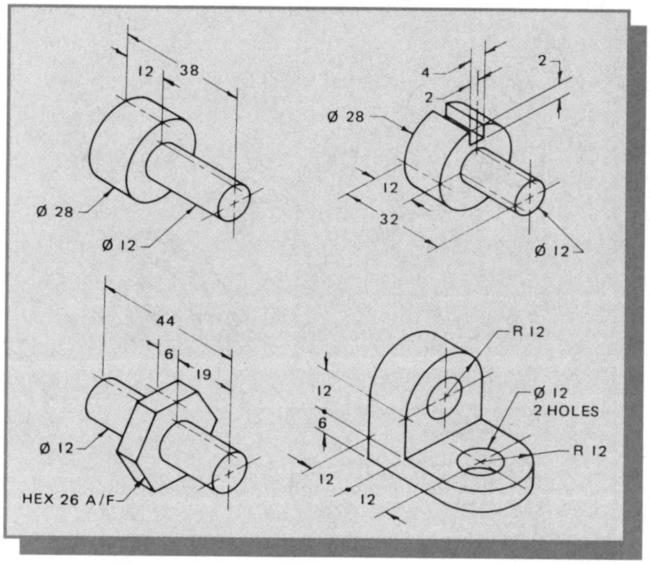

Figure 16-57
Draw a multiview drawing and isometric view for each item. Dimension and add metric thread notations for all holes and shafts.

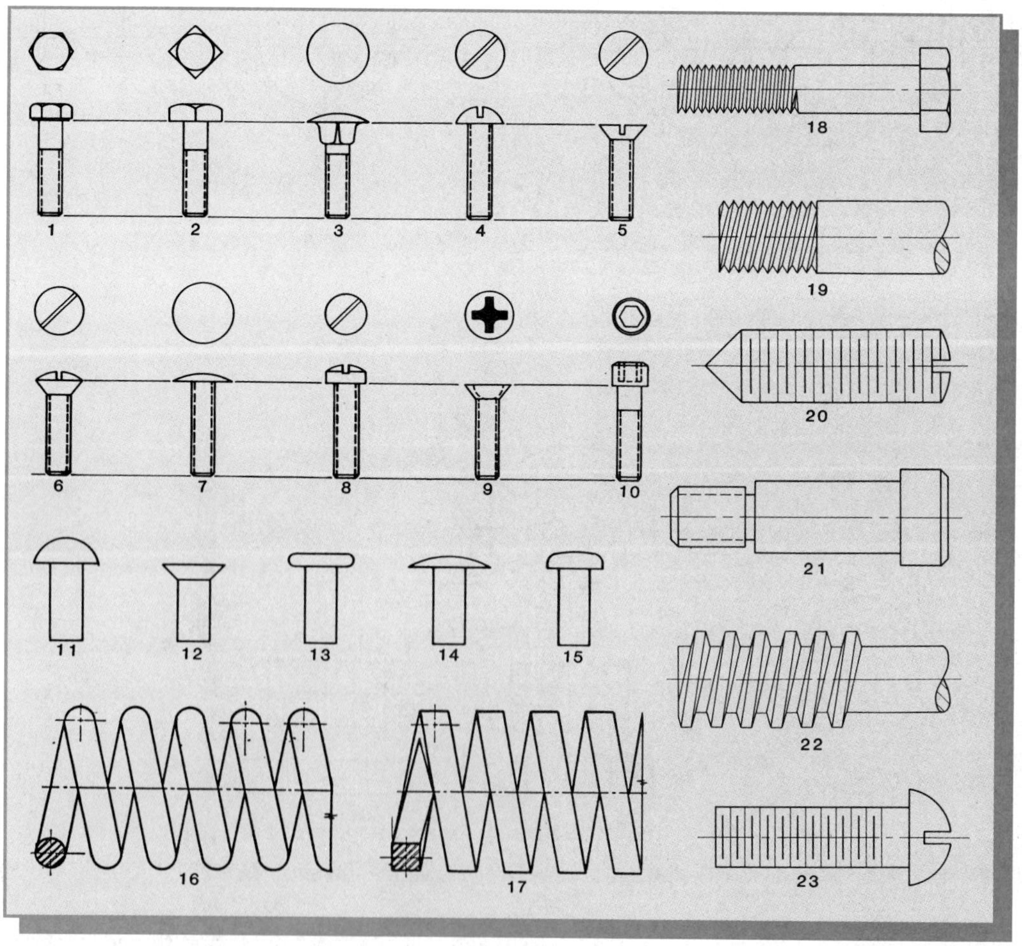

Figure 16-58
Draw each fastener symbol to a scale of 2:1 and name each symbol.

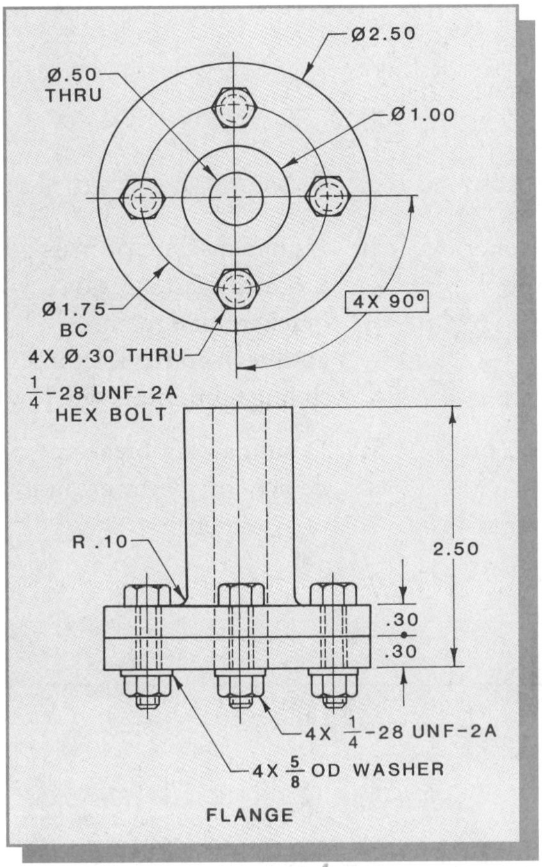

Figure 16-59
Draw accurately with instruments or CAD.

Figure 16-61
Draw a multiview drawing for the metric hexagon cap

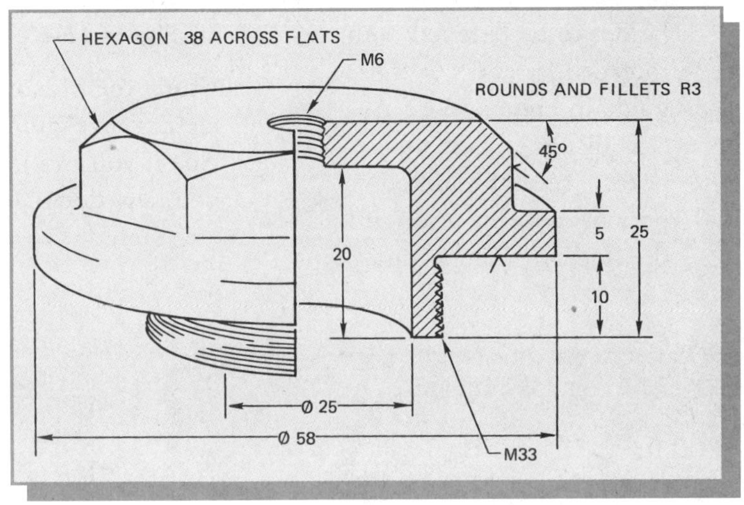

Figure 16-60
Draw the detail drawings of the jeweler's clamp. Then complete the assembly drawing, noting all metric threads.

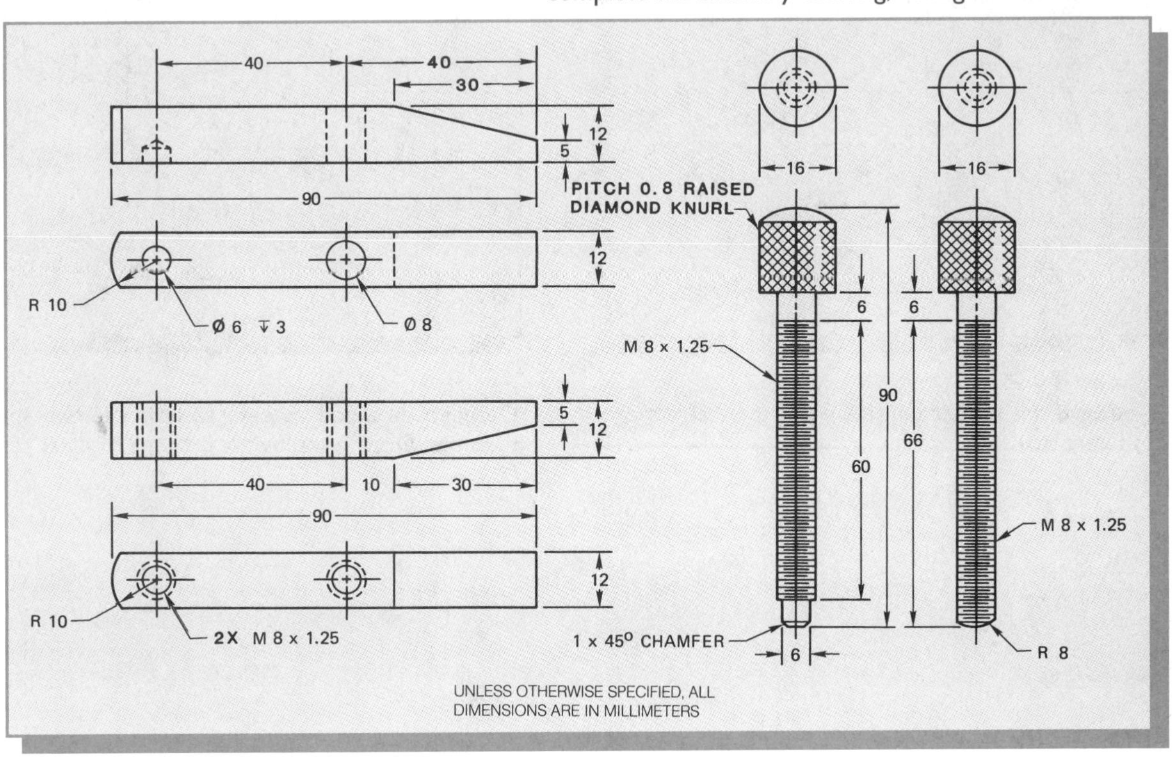

Design Exercises

1. Redesign and complete the plate with a hook or holder to be fastened with threaded fasteners to the guide in **Figure 16-62**. A .75" diameter rope supporting several hundred pounds will pass through the hook or holder unit.
2. Design a threaded fastener to connect the two spools in **Figure 16-63**.
3. Study the old and new designs in **Figure 16-64**. See if you can improve on the new design.
4. Using any type of mass-produced fasteners or brackets, design an antibreak-in device for a sliding aluminum window.
5. Design an antibreak-in device for the front door of a residence.

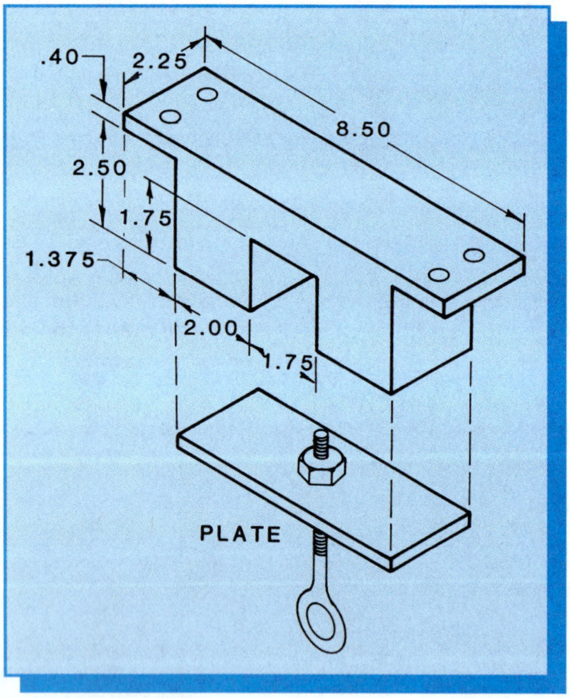

Figure 16-62
Redesign this holder unit following the instructions in Exercise 1.

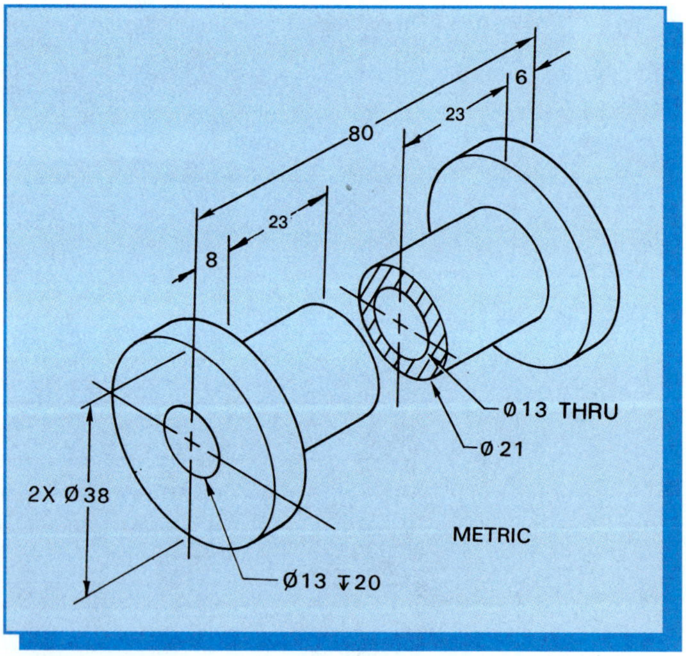

Figure 16-63
Design a threaded fastener to hold the two spool forms together. Draw a multiview drawing of each part.

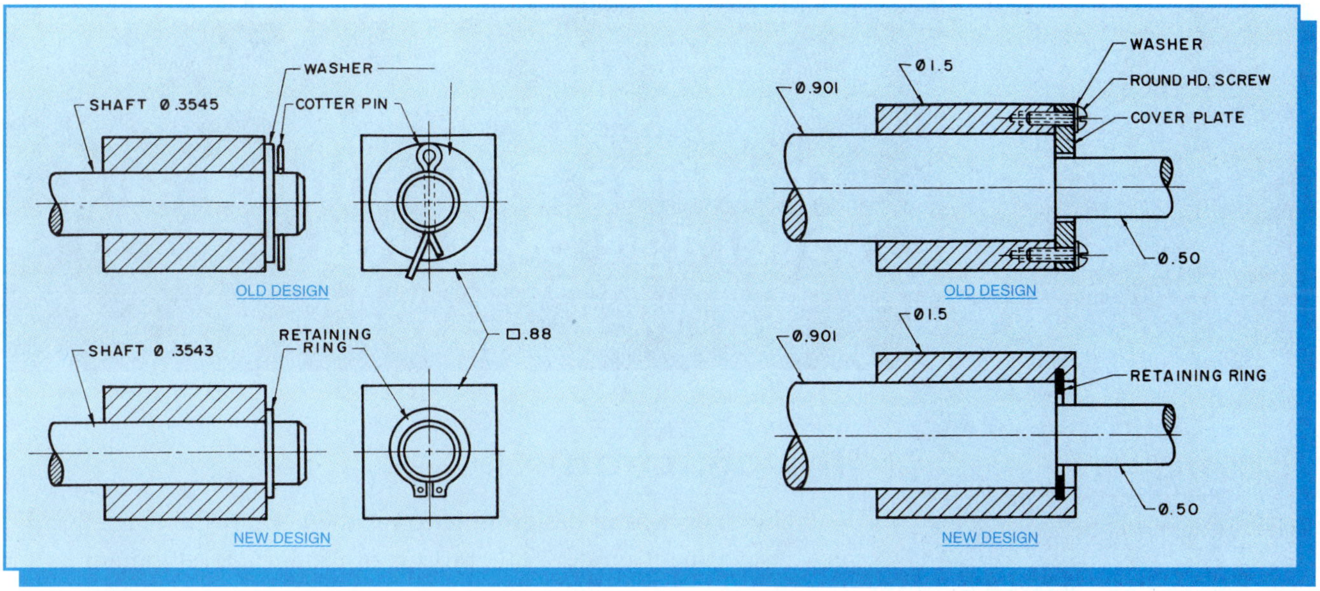

Figure 16-64
Sketch the multiview drawings and isometric views of the fastening systems. *(Reprinted from* Technical Drawing and Design *by Goetsch and Nelson, © 1986 Delmar Publishers Inc.)*

Key Terms

American Society of Mechanical Engineers (ASME)
Angle of thread
Axis of fastener
Compression length
Crest
Ends
External thread
Flat springs
Free length
Helix
Helix direction
Inclined plane
Industrial Fasteners Institute (IFI)
Internal thread
International Standards Organization (ISO)
Lead
Lead per revolution
Left-hand thread
Loading extension
Major diameter
Minor diameter
Multiple threads
Pitch
Pitch diameter
Right-hand thread
Rivets
Root
Screw thread series
Single thread
Solid height
Spring
Thread depth
Thread forms
Thread fit classes
Threads per inch (TPI)
Torsion springs

Chapter 17

Drafting Procedures and Storage Systems

Objectives

The student will be able to:

- make revisions on existing working drawings
- use a microfilm reader
- check other students' working drawings
- prepare a working drawing using several overlays
- write a progress report covering the preparation of working drawings on a specific item
- lay out a drawing format with the following information: borders, title block, parts list, revision list, drawing identification number, parts identification numbers, and drawing zones
- use a photograph as the basis for a working drawing

Introduction

The drafting and design function is the core of the entire product development process. The process of product development includes the following phases:

1. Ideas and concept generation
2. Preliminary design
3. Design revisions and refinements
4. Working drawing development
5. Checking, revision, and modifications of drawings
6. Preparation of related documents, contractor's specifications, and estimates
7. Reproduction of drawings and documents
8. Production—manufacturing or construction
9. Product marketing and distribution
10. Servicing or repair
11. Redesign

To complete the drafting functions relating to these phases, preliminary design drawings and later working drawings must be drawn, recorded, revised, checked, reproduced, stored, and retrieved through a variety of systems and procedures.

Functional Drafting

The preparation of the most accurate and complete set of technical working drawings in the least amount of time is a primary goal in the development of every product. To accomplish this, traditional instrument and board systems or CAD systems may be used, depending on the availability of equipment and the complexity of the product. Regardless of the standard system, many procedures are used to accelerate the production of drawings by organizing data and eliminating repetitive tasks. In manual drafting, these include the use of pin graphics, scissor drafting, reformatting, photodrafting, and standardized formats.

Pin Graphics

Pin graphics is a system of binding transparent drawing film or paper together with a **pin bar** so the various layers of the drawings will **register** (align) **(Fig. 17-1)**. This function is known as **layering** when performed on a CAD system.

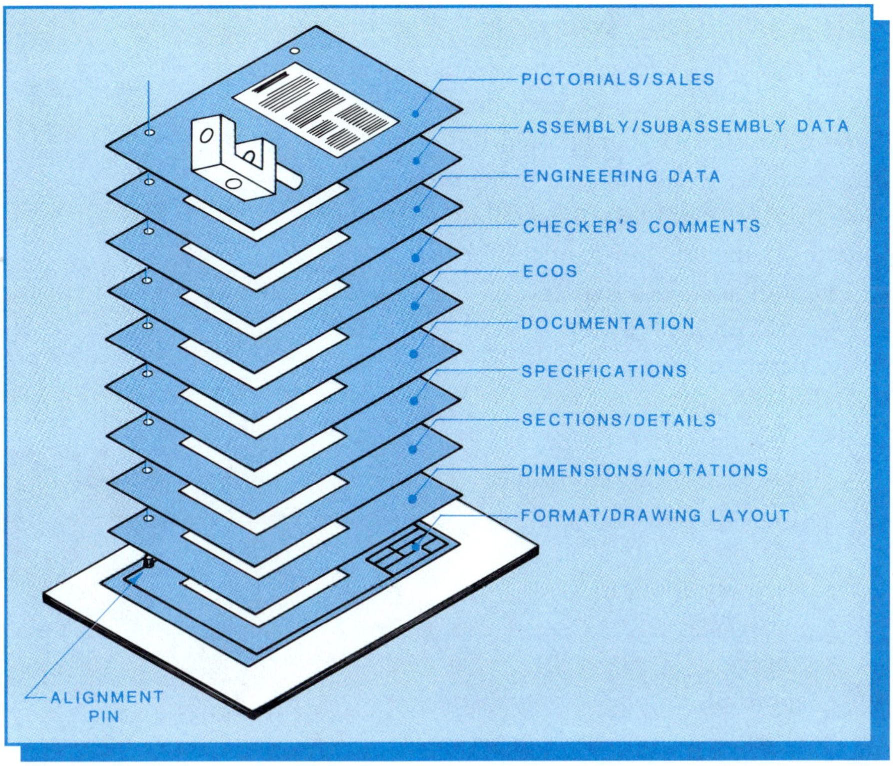

Figure 17-1
This is an example of how pin graphics (layering) may split up a working drawing into its separate elements of engineering, design, drawing, and documentation.

Scissor Drafting

Scissor drafting is a system of cutting out unwanted areas of a drawing, making a copy, and then redrawing or pasting the new drawing segments in the blanked-out area **(Fig. 17-2)**. The revised copy is then used as the master copy.

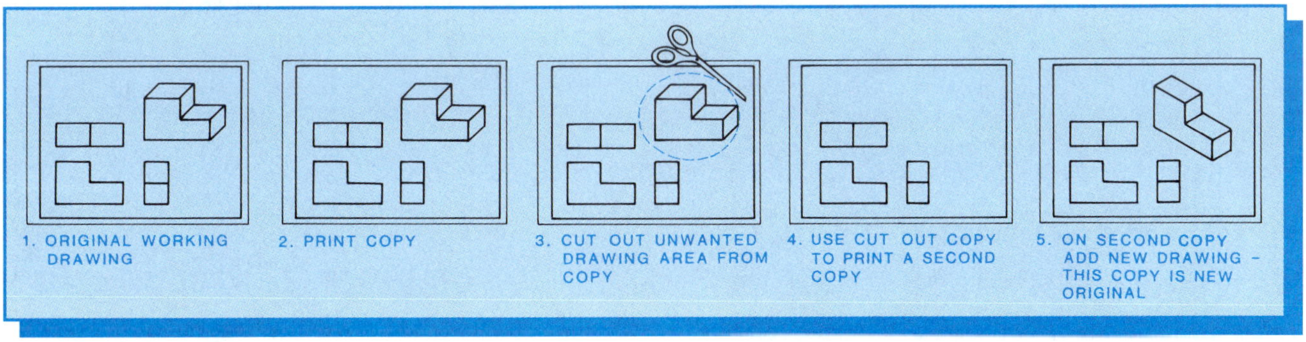

Figure 17-2
Procedures for scissor drafting

Reformatting

Reformatting is an intermediate copy made where changes may be dry erased, wet erased, or cut out for revision. The revised copy is then used as the master copy.

Photodrafting

Photodrafting is a system of copying a photograph with a high-resolution copy machine, and adding dimensions, linework, and notes for use as a working drawing **(Fig. 17-3)**.

Standardized Formats

Preprinted formats eliminate the repetition of lines and headings on each drawing **(Fig. 17-4)**. This format can be preprinted on drawing surfaces or applied as a preprinted transparent overlay **(Fig. 17-5)**. The following items are normally included in standard formats:

- title block and borders
- revision list
- parts list
- company's name
- drawing identification number
- part identification number
- drawing zones
- tolerances
- assembly application
- material
- finish
- approvals
- scale
- microfilm targets
- format size

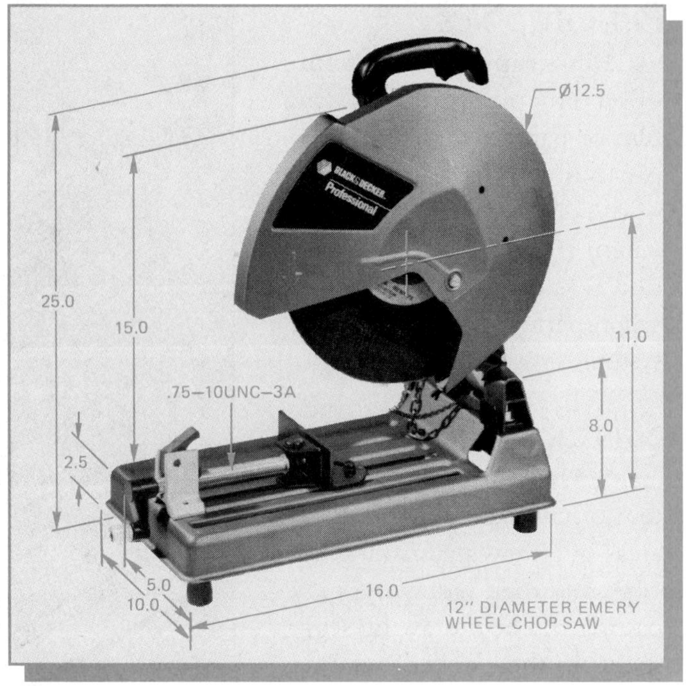

Figure 17-3
Photodrafting uses a photograph as the base for a working drawing. *(Courtesy of Black and Decker U.S. Power Tools Group)*

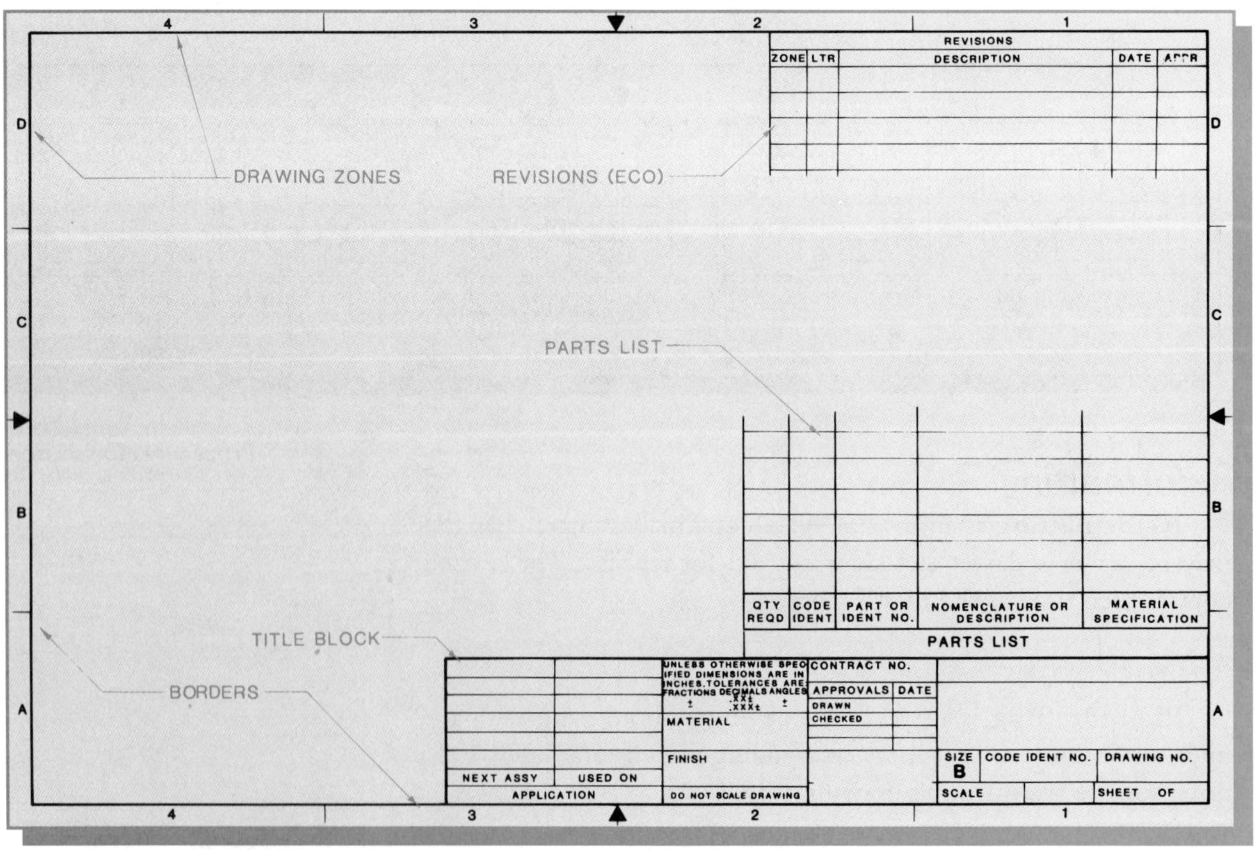

Figure 17-4
Example of a standardized preprinted format of a sheet layout

Revisions

Revisions to working drawings are made for a variety of reasons, including design changes **(modifications)** and the correction of errors. Revisions may involve the change of a simple line, dimension, or note, or it may require a redraw of an entire part or component. In making changes and corrections, only those areas of a drawing that are affected are revised.

Each revision must be recorded in the title block. **Figure 17-6** shows the entry of five changes, A through E. Changes begin at the bottom of the **revision list,** and successive changes are added on top of the previous change. Changes are also accompanied with a document requesting and authorizing each change. These are called **Engineering Change Orders (ECOs), Engineering Change Requests (ECRs),** or **Engineering Change Notices (ECNs).**

Figure 17-5
Transparent film stick-ons are available in many formats, or may be designed to fit the drafter's needs.

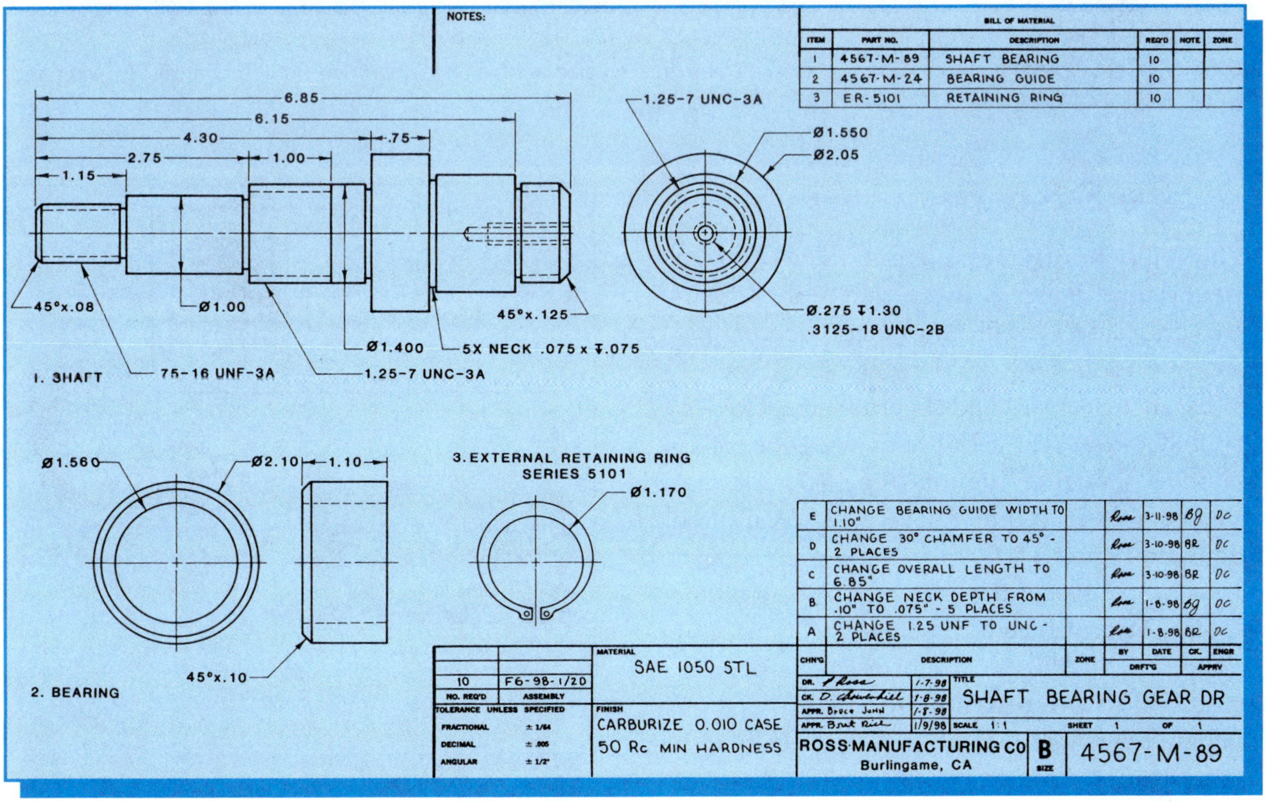

Figure 17-6
Engineering change orders (ECOs) listed on a working drawing

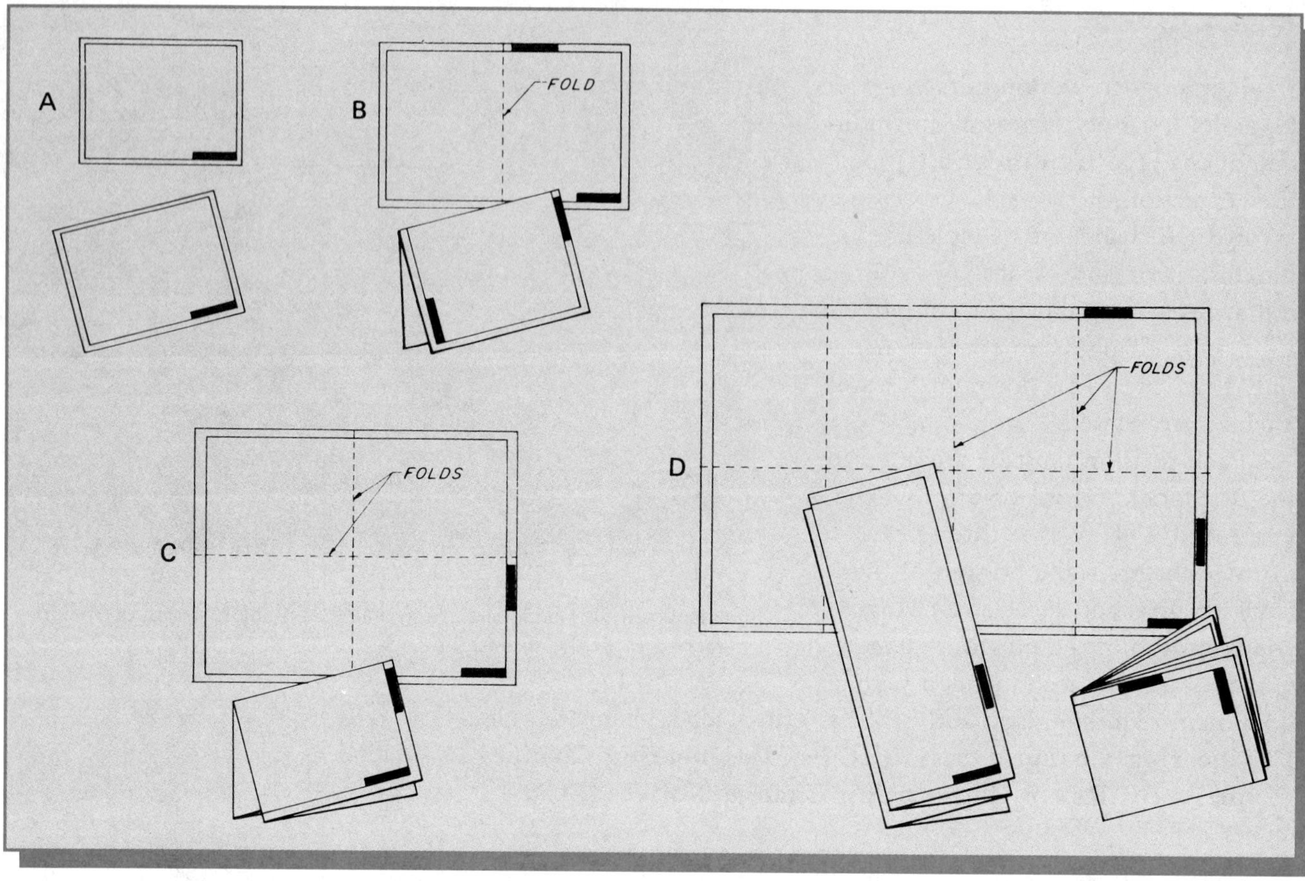

Figure 17-7
Procedure for folding standardized drawing formats of prints to size A for filing *(Reprinted from* Technical Drawing and Design *by Goetsch and Nelson, © 1986 Delmar Publishers, Inc.)*

Storage Systems

The ability to organize, store, and quickly retrieve original drawings, prints, and related documents is critical to the efficient functioning of any drafting operation. Original drawings should be stored flat in a cool, dry, and fireproof area. Copies can be folded to A size with the drawing identification number shown on the top **(Fig. 17-7)**. All standard drawing sheets can be folded to size A using the procedure in **Figure 17-7**.

To conserve space, drawings may be microfilmed for storage. **Microfilmed** drawings can be placed in individual aperture filing cards or rolls, or placed in flat sheets, called **microfiche (Fig. 17-8)**. Microfilmed drawings may be stored manually in file cabinets or entered into a computerized automatic storage and retrieval system.

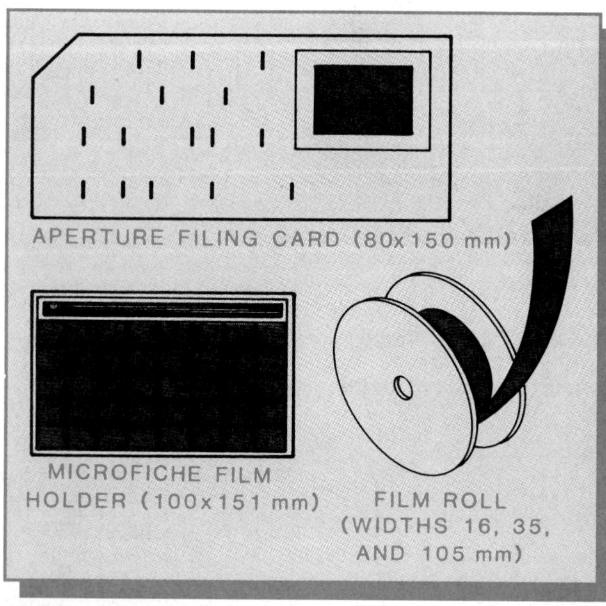

Figure 17-8
Microfilmed drawings and documents may be stored in various ways.

AutoCAD LT Exercises

CAD Storage Systems

Drawings completed on a CAD system may be stored on a floppy disk, hard disk, drum, or magnetic tape. For hard-copy drawings and computer-stored CAD drawings, a backup copy should be stored in a separate location to insure against loss. The CAD operator should save work every fifteen minutes to a backup disk in case of a power loss or surge that might erase all the work done. This way, only part of the work would need to be repeated. Because of the CAD software's special capabilities for block storage, many types of working drawings and pictorial drawings can be saved for instantaneous retrieval for reference or utilization.

CAD Databases

In addition to performing CAD functions, the same computer is capable of doing word processing, organizing and manipulating data into **spreadsheets**, calculating, accounting and management, engineering calculations, as well as a host of other tasks that help with the documentation of engineering drawings. Many of those functions require the existence of a **database**.

As a CAD drawing is being generated, the computer is developing a database. A database is a body of information that is gathered and sorted to be used for future tasks. In drafting, the database created by a CAD drawing may be used to accurately identify and describe segments of a drawing; calculate areas, volumes, angles, and clearances; create parts lists; calculate product cost; and write programs for automated machining. The following material covers some of the data accumulated as a CAD operator produces a drawing and how this data may be used. As technology advances, more uses will be discovered for CAD and its database.

Recording Elements

Whenever a point, line, circle, arc, dimension, or any other element is drawn on a CAD drawing, the element is accurately recorded in a database for that drawing **(Fig. 17-9)**. Note that **Figure 17-9** contains three elements: two circles and a line. The circles were drawn using the CIRCLE command and specifying their centers and radii. The line was drawn tangent to the circles, and its length and angle are not known. By using the LIST command, selecting the three elements (the small box on the drawing is the selector box), and pressing the Enter key, the properties of each of the elements are displayed on the monitor **(Fig. 17-10)**. This is the data for these three elements. The data for the circles include the location of the center point, radius, circumferences, and areas. The line data includes location of the start and end points of the line, its length, angle, and the change (**delta**) in the X and Y directions. This is more information than the CAD operator knew when drawing the elements. It may be valuable design data needed in the future.

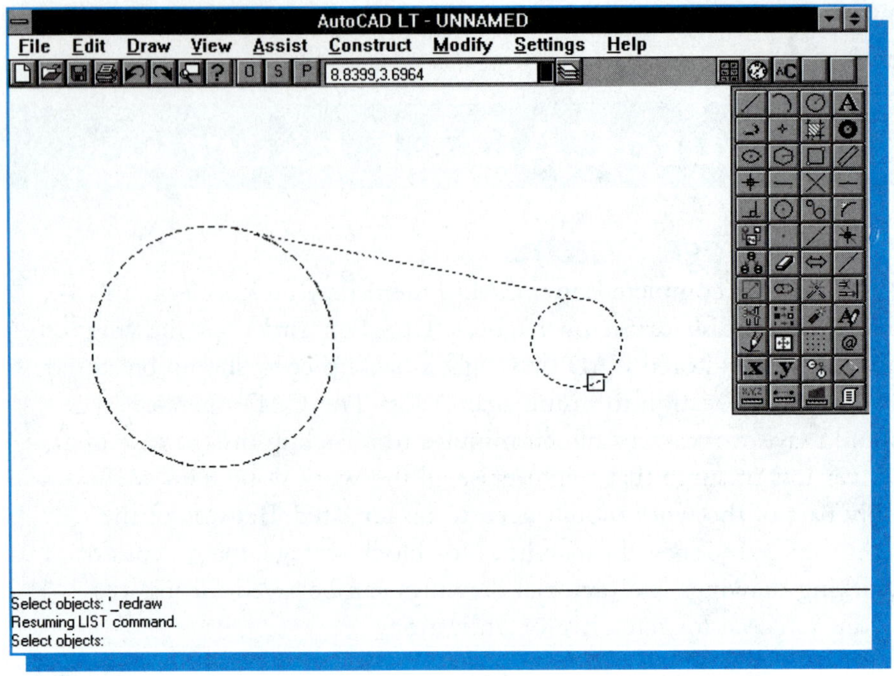

Figure 17-9
Using the LIST command to display the database of each drawing entity

Calculating

It was noted above that a computer may be used for calculating. The fact is that it is a very powerful calculator. Built into most CAD software are formulas for doing standard calculations. The CAD operator gives the computer the information by accurately generating the geometry **(Fig. 17-11)**. Using this information and the built-in formulas, the computer can very rapidly perform valuable calculations. An example of this is the AREA command **(Fig. 17-11 through 17-13)**. The drawing includes three shapes—a rectangle and a triangle, plus a more complex polygon. Dimensions were put on the drawings so the reader of this text may manually do the calculations to verify the data. The computer does not use the dimensions when calculating. It uses the database for each element of geometry. In **Figure 17-11**, the rectangle has been selected (note the square selector boxes). The area and perimeter of the feature are instantly displayed in the prompt area below the drawing. The same

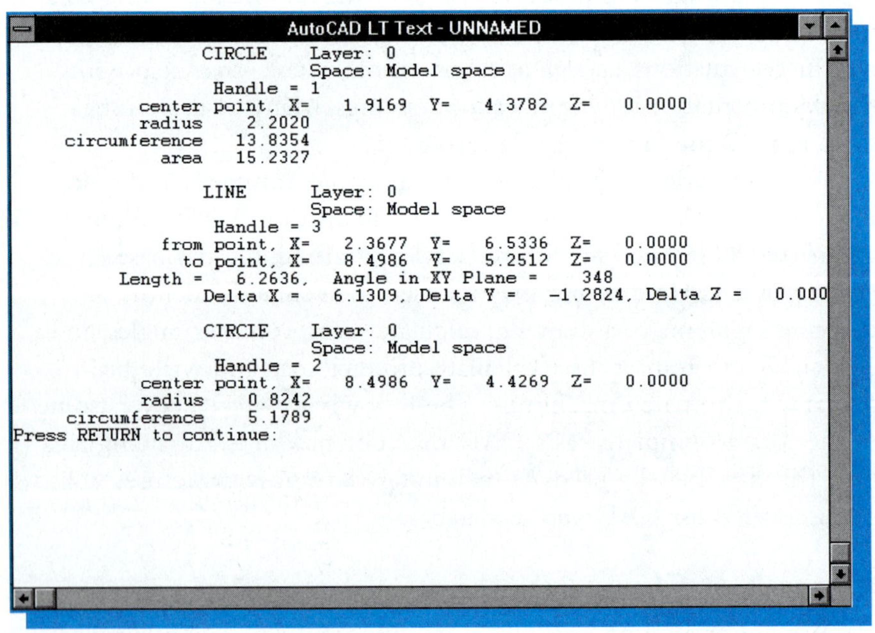
```
AutoCAD LT Text - UNNAMED
          CIRCLE    Layer: 0
                    Space: Model space
           Handle = 1
   center point, X=   1.9169  Y=   4.3782  Z=   0.0000
         radius     2.2020
  circumference    13.8354
           area    15.2327

          LINE      Layer: 0
                    Space: Model space
           Handle = 3
     from point, X=   2.3677  Y=   6.5336  Z=   0.0000
       to point, X=   8.4986  Y=   5.2512  Z=   0.0000
     Length =    6.2636,  Angle in XY Plane =     348
          Delta X =   6.1309, Delta Y =    -1.2824, Delta Z =    0.000

          CIRCLE    Layer: 0
                    Space: Model space
           Handle = 2
   center point, X=   8.4986  Y=   4.4269  Z=   0.0000
         radius     0.8242
  circumference     5.1789
Press RETURN to continue:
```

Figure 17-10
The database for each entity in Figure 17-9

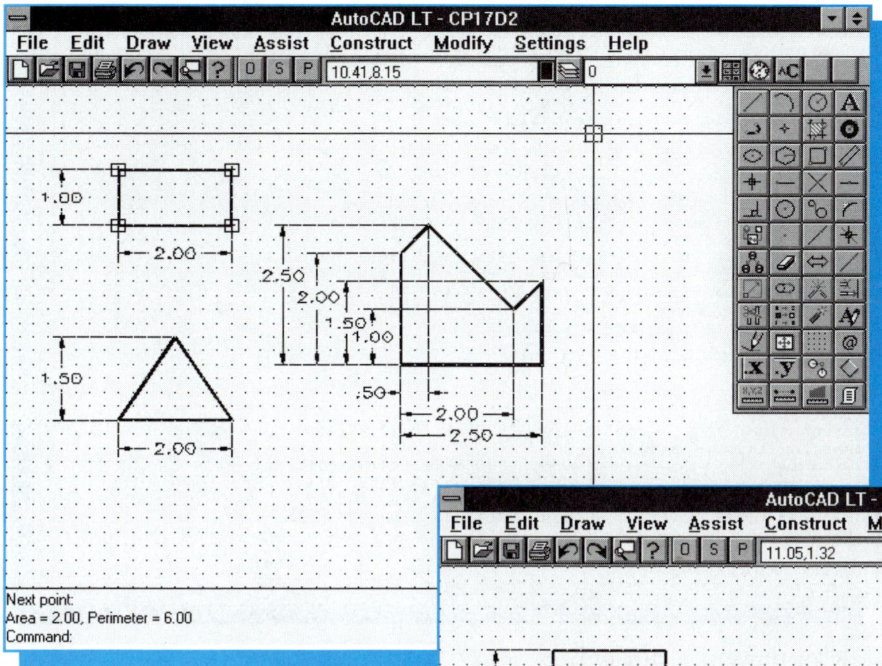

Figure 17-11
The AREA command calculates the area and perimeter for the rectangle.

Figure 17-12
The AREA command calculates the area and perimeter for the triangle.

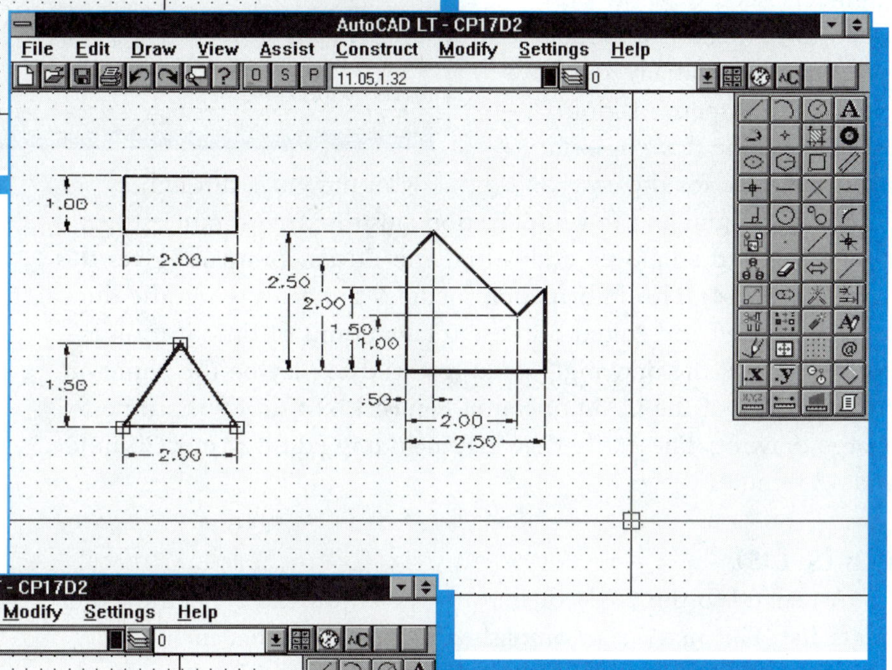

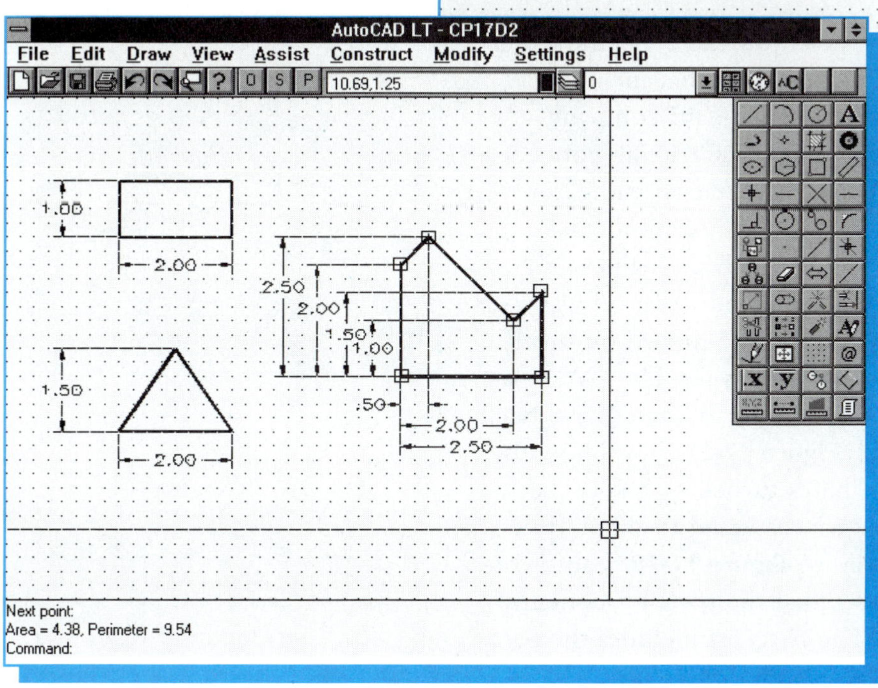

Figure 17-13
The AREA command calculates the area and perimeter for the irregular polygon.

procedure was used to get the area and perimeter of the triangle and polygon in **Figures 17-12** and **17-13**. The calculation for the area and perimeter for **Figure 17-13** involves geometry and trigonometry, and the computer does the calculations instantly.

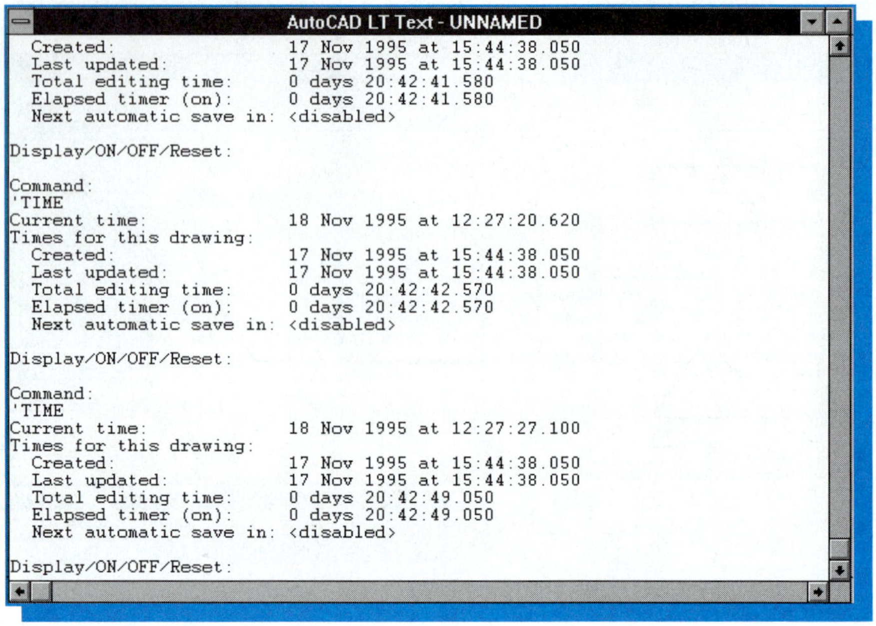

Figure 17-14
The TIME command keeps records on the drawings in progress.

Time Records

Managers, designers, and engineers need to know how much time it takes to produce a CAD drawing initially and how much time is spent making revisions to the drawings. They are responsible for the costs and the flow of drawings through a company's system and this information aids them with cost estimating, billing, and employee evaluation. The TIME command gives this data **(Fig. 17-14)**. This information includes the dates when the drawing was created and edited, and the total amount of time it took to create and edit the drawing. Added to this data may be the name or code number of the CAD operator. **Figure 17-14** shows the time for a large drawing. The entire drawing, including editing, took 20 hours and 42 minutes.

Parts List

A real aid to the CAD operator is the automatic generation of a **parts list**. This may be accomplished for drawings that include several of the same items, such as fasteners or standard company parts. The example given here is a floor plan of a CAD lab containing seven CAD stations **(Fig. 17-15)**. Each CAD station is the same. By the use of attributes, data may be added to the components. **Figure 17-16** shows one of the CAD stations with the names and part numbers superimposed over the parts. This information was placed on the drawing in an attribute that may be turned on or off. It will be used as a database for generating a parts list. The CAD station has been inserted into **Figure 17-15** and the attributes have been turned off (it was inserted seven times). **Figure 17-17** shows how the drawing would look with the data attributes turned on. Note that the bookcase and the cabinet also have data assigned to them. By using the data in the blocks, the parts list in **Figure 17-18** is automatically generated. It is important to notice that the parts list contains more information than the attributes. The parts list includes the part

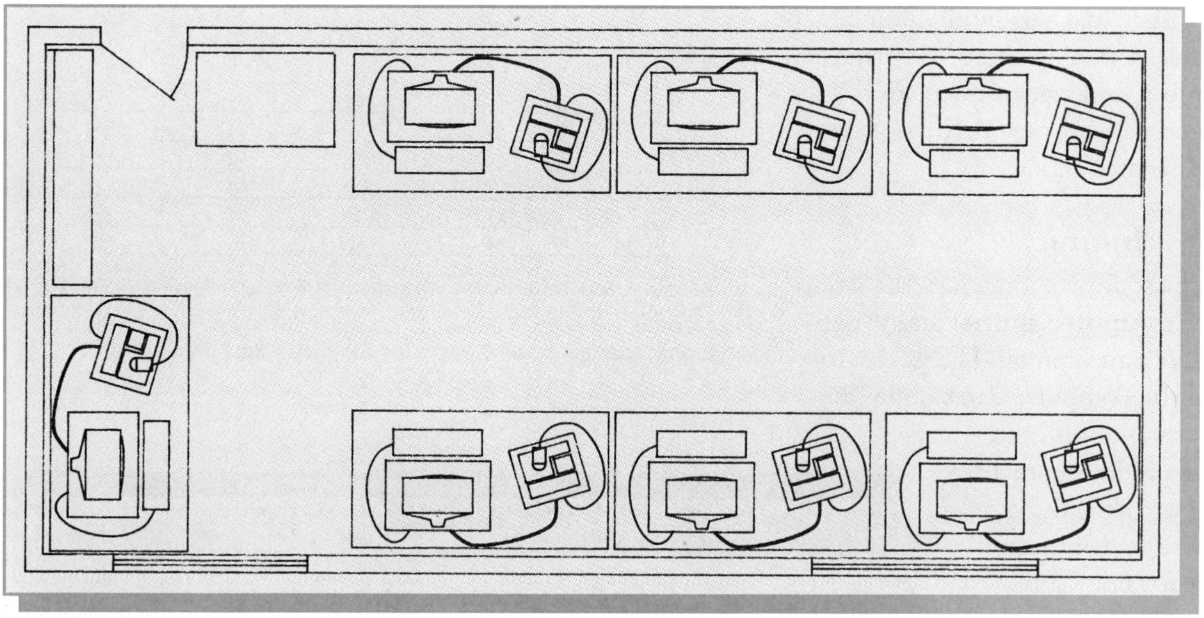

Figure 17-15
Seven-station, computer-aided drafting lab. Note that one lab was drawn and duplicated, revolved, and placed into the other six positions.

Figure 17-16
Single CAD workstation with its parts data displayed

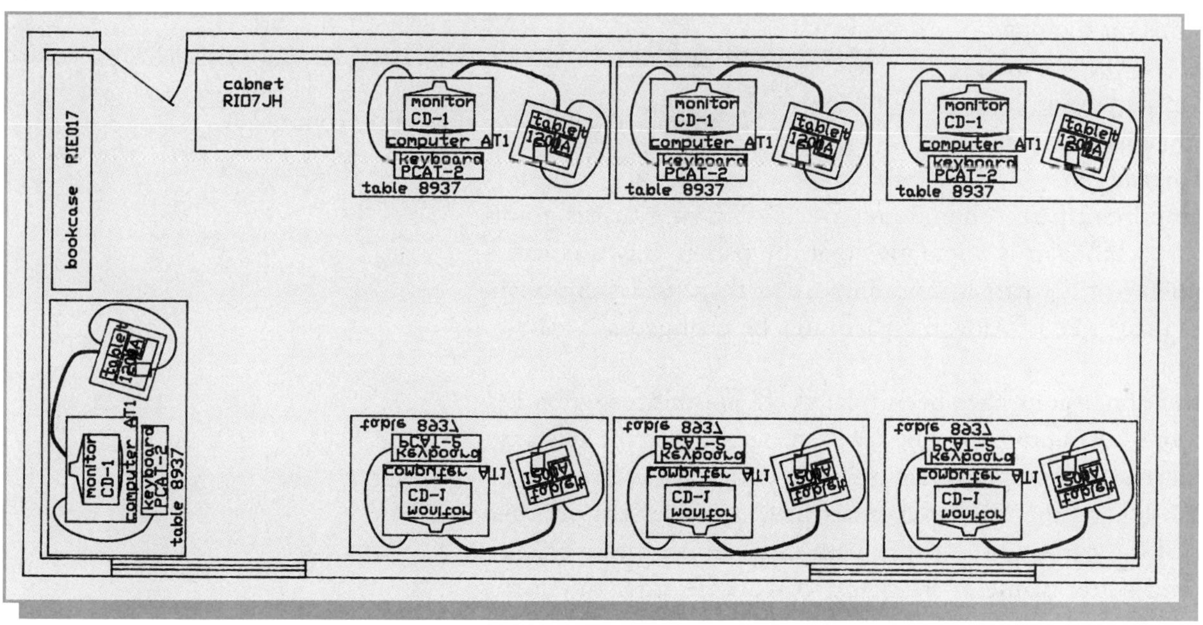

Figure 17-17
Full CAD lab with its parts data displayed

number, quantity (derived by the number of times the blocks were inserted), description, vendor number, and stock information. This computerized part list may be part of the company's inventory data.

PART NO.	QTY	DESCRIPTION	VENDOR NO.	IN STOCK
RIE017	1	BOOKCASE 5'X8'	107	YES
R107JH	1	CABNET,LOCKING	107	YES
8937	7	TABLE 3'X5'	107	5
AT1	7	COMPUTER, 386	84	YES
CD-1	7	MONITOR, COLOR	84	YES
PCAT-2	7	KEYBOARD, W/FKEYS	84	YES
1200A	7	GRAPHIC TAB W/PUCK	85	NO

Figure 17-18
Computer generated parts list for CAD lab

CNC Machining

Another use for the drawing database is to program **computer numerically controlled (CNC)** machining. The geometry recorded in the computer database by the drawing, as shown with the LIST command in **Figure 17-10**, may be sent directly to a CNC center for manufacturing of the part. The CNC center uses this CAD-generated geometry to select the material from inventory, and to cut, mill, or turn the part as required. **Figure 17-19** is an example of a part to be CNC machined.

Before sending the data to a CNC center, the CAD operator may use a CNC program to check the machining process **(Fig. 17-20)**. The CNC program is not going to use the dimensions on the drawing, but will use the geometry; therefore, the dimension layer is turned off **(Fig. 17-21)**.

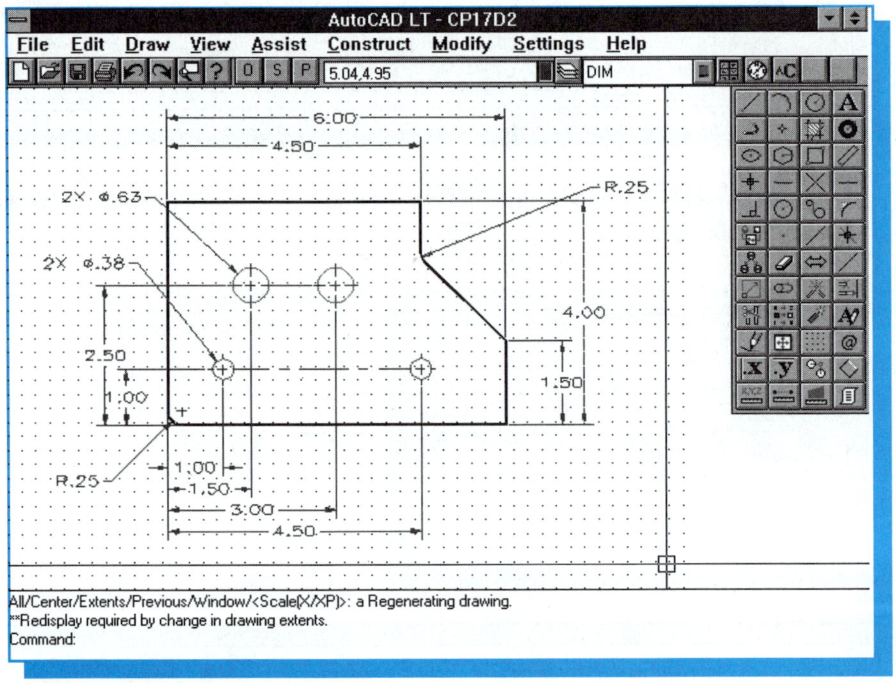

Figure 17-19
Part to be CNC machined (computer numerically controlled)

This brings up an extremely important part. When doing a CAD drawing that is going to be CNC machined, it is imperative that the part is drawn correctly. The geometry of the part is more important than its dimensions. If a change to a part size is made, the part must be changed as well as the dimension.

Once the dimensions have been turned off, another program is loaded into the CAD program. The CNC program uses the database and creates a machine program that guides a cutter around the part. **Figure 17-20** displays the simulated cutter path on the monitor. This visual view of the part and the cutter path indicates to the operator if the path and cutter diameter are going to work in order to cut this part.

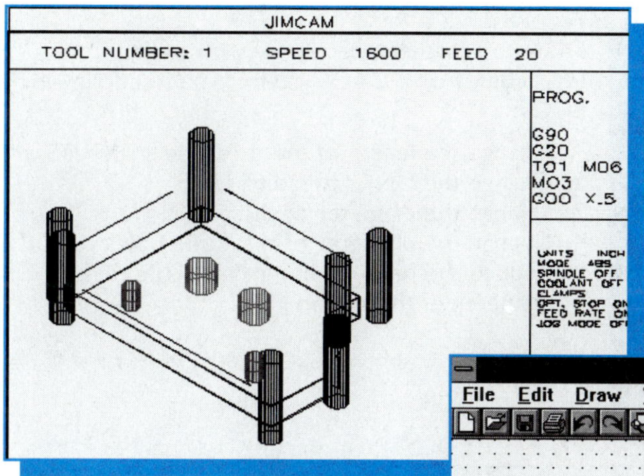

Figure 17-20
Third-party CNC simulation of part being cut.

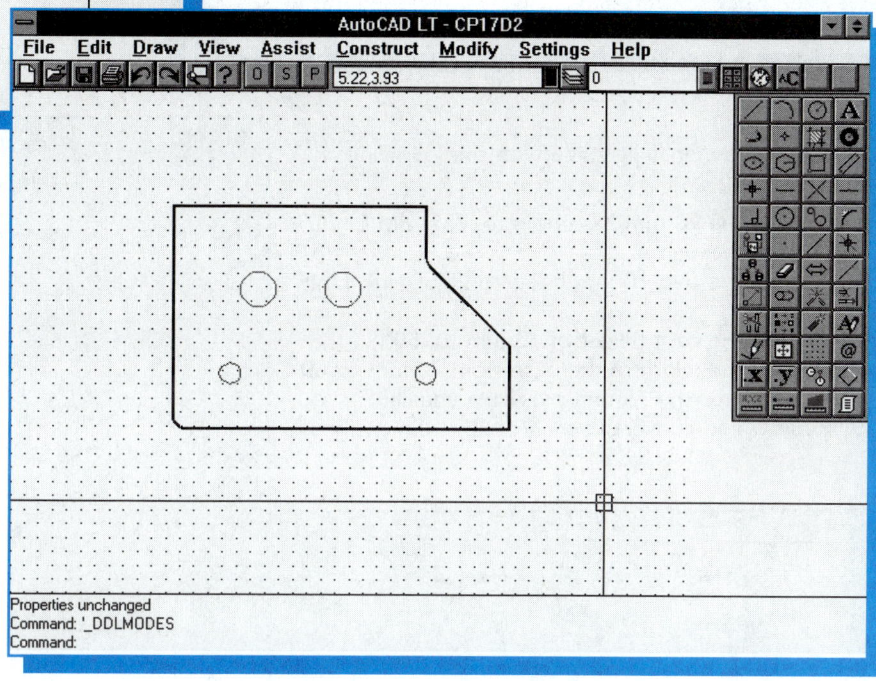

Figure 17-21
CNC drawing part with the dimension layer turned off.

Drafting Exercises

1. Draw the working drawings in **Figures 17-22** through **17-24** with drafting instruments or with a CAD system. Make the revisions as noted in each caption.
2. Design a slip-proof handle for the threaded rod (note ECOs) in **Figure 17-25**.
3. Draw **Figure 17-26**, making the ECO changes. Write a progress report covering the drafting time and any specific design recommendations.
4. Write the ECOs for the U-bolt in **Figure 17-27** to improve its design functions. Redraw the drawing, listing the ECOs in a revision list.
5. Take black-and-white photographs with your camera of a few simple machine parts (use a very light background). Dimension the prints with technical ink pens on a vellum overlay.

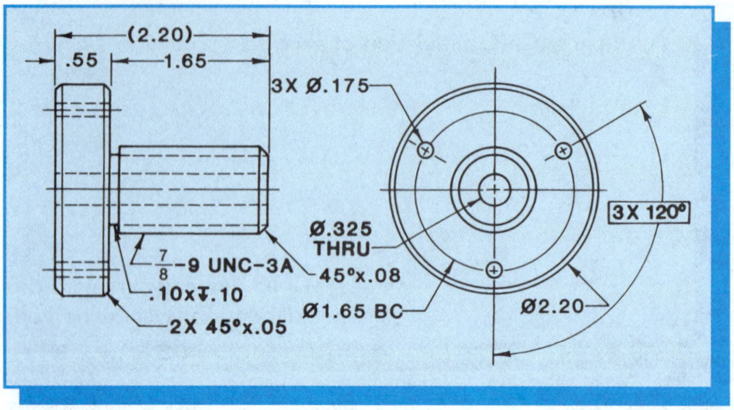

Figure 17-22
Draw the threaded set screw with drafting instruments or a CAD system. Make the following ECOs:

1. Change the length of the threaded shaft 1.75".
2. Change the UNC thread to UNF.
3. Change the diameter of the bolt circle to 1.50".
4. Change the diameter of the three holes to .20".
5. Reduce the neck diameter below the minor diameter of the thread (.15).

Figure 17-23
Redesign the metric jig guide with the following ECOs:

1. Change the 20 mm base hole to .75" with UNC threads through.
2. Change the two 10 mm holes to .375" with UNF threads.
3. Change the base height of 10 mm to .50".
4. Change the width of the base from 70 mm to 60 mm. Reduce the 15 mm overhang equally.
5. Redraw and convert to inch/decimal dimensions.

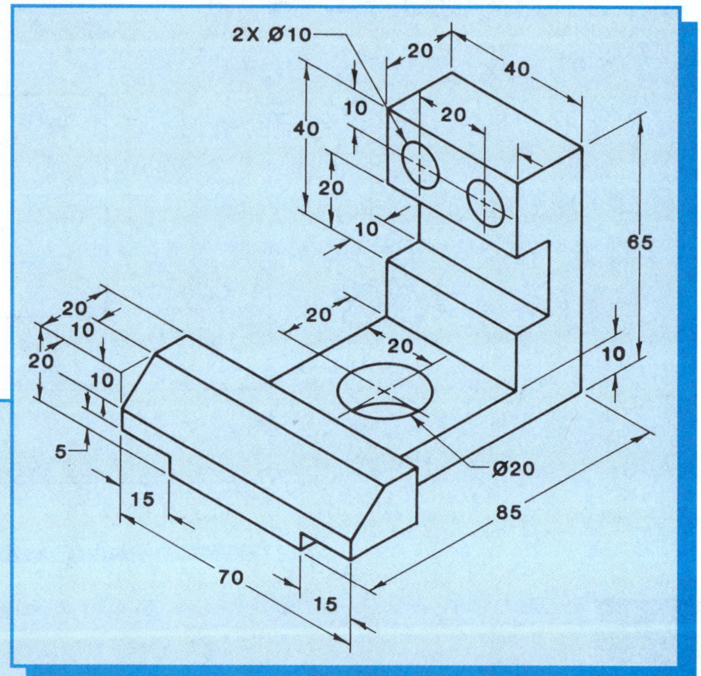

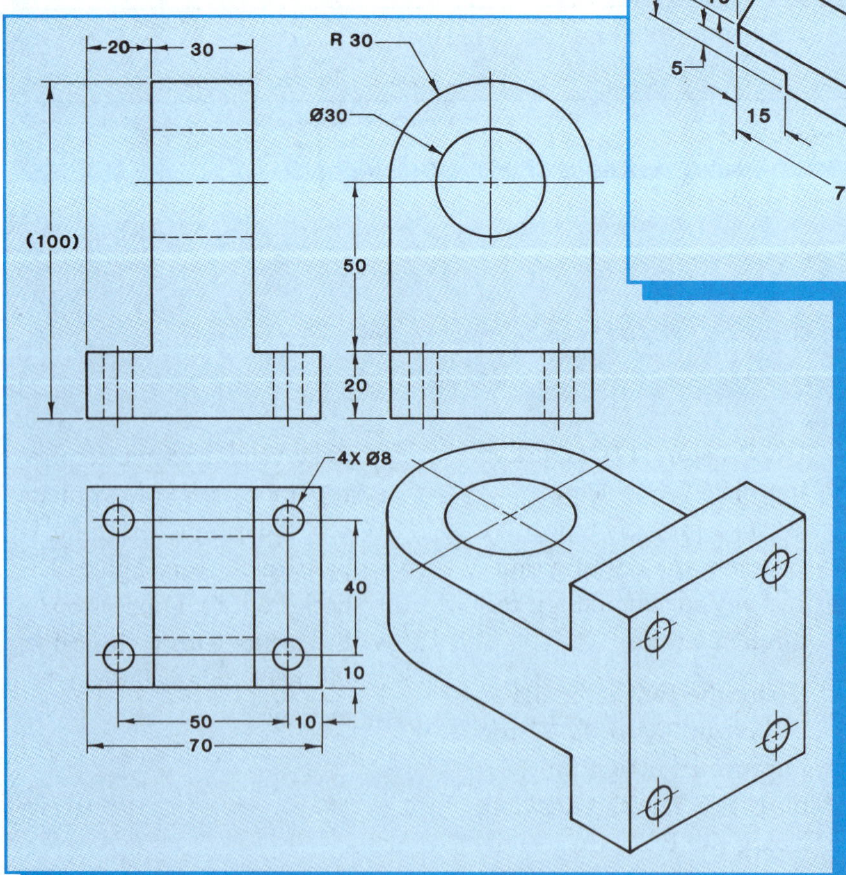

Figure 17-24
Redraw the (first-angle projection) T support working drawing and make the following ECOs:

1. With the scissor drafting technique, change to third-angle projection.
2. Convert metric dimensions to inch/decimal.
3. Change 30 mm to 25 mm. Add UNC threads through.
4. Add UNF threads through the four 8 mm holes.
5. Change stem of T support to 25 mm.

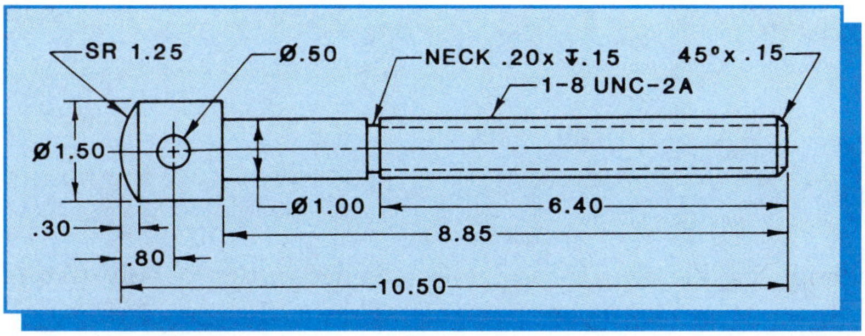

Figure 17-25
Redraw the threaded rod, make the following ECOs, and record them on the format's revision list:

1. Change the spherical radius to 1.00".
2. Change the thread length to 6.50".
3. Change the neck depth to .175".
4. Change the thread fit to class 3.
5. Change the .50" hole in the handle to .30" diameter through.
6. Convert dimensions to metric.

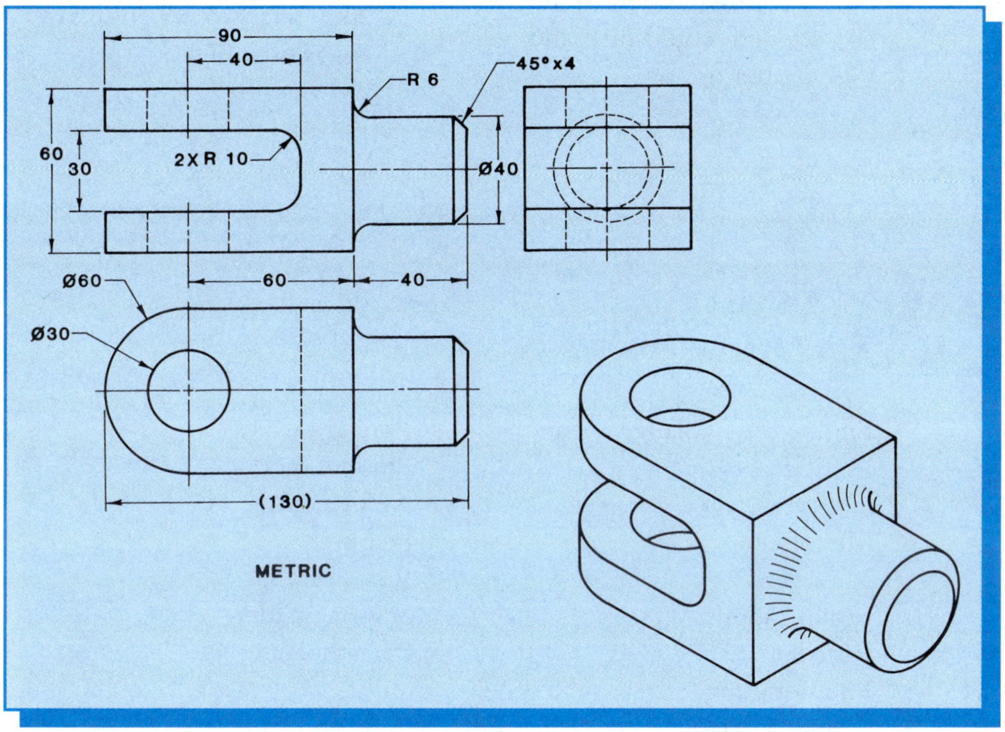

Figure 17-26
Redraw the (first-angle projection) bearing bracket, the isometric drawing, and make the following ECOs:

1. With the scissor drafting technique, change to a third-angle projection.
2. Convert the metric dimensions to inch/decimal.
3. Add 1" depth of UNC threads to round extrusion.
4. Add UNC threads through to all internal holes.
5. Reduce all fillets by 50 percent.
6. Reduce 45° chamfer by 50 percent.

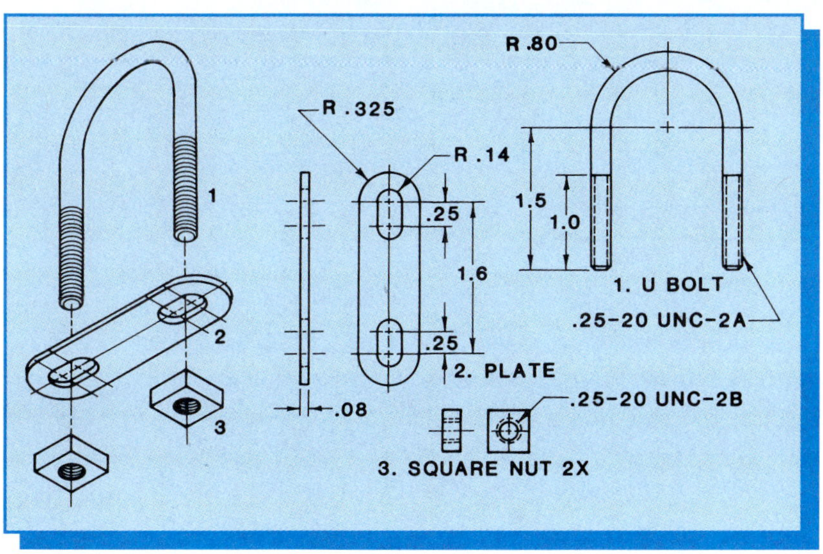

Figure 17-27
Redesign the U-bolt so it can strap a 1.5" diameter pipe. Convert it to metric dimensions.

Design Exercises

1. Redesign the set screw in **Figure 17-22** so it may be tightened with a $^3/_8$" hex allen wrench.
2. Redesign the T support with a dovetail fit for the base.
3. Redesign the handle for the threaded rod in **Figure 17-25** so that a person impaired with very arthritic hands and with little strength could turn and tighten the rod.
4. Redesign the U-bolt strap without the threaded fasteners.
5. Redesign the handle for the emerywheel chop saw in **Figure 17-3**.

Key Terms

Computer Numerically-Controlled (CNC)
Database
Delta
Engineering Change Orders (ECOs)
Engineering Change Notices (ECNs)
Engineering Change Requests (ECRs)
Layering
Microfiche
Microfilm
Modifications
Parts list
Photodrafting
Pin bar
Pin graphics
Reformatting
Register
Revision list
Scissor drafting
Spreadsheets

Chapter 18

Working Drawings

Objectives

The student will be able to:

- prepare detailed drawings manually
- prepare assembly drawings manually
- prepare detailed drawings on a CAD system
- prepare assembly drawings on a CAD system
- select the proper assembly drawings for a working drawing set

Introduction

Working drawings provide specific information and instructions needed to manufacture or construct products and structures. This includes detailed graphics of precise shapes, and dimensions of exact sizes. They also contain other notations for manufacturing, such as material descriptions, finishing methods, and tolerances. Working drawings are used in all phases of manufacturing, including fabrication and assembly. All information needed to manufacture a product must be included in a working drawing. The manufacturer must not be forced to guess about any detail to produce a product as designed.

Working Drawings

There are two general categories of working drawings—detail drawings and assembly drawings. **Working drawings** usually include fully dimensioned multiview drawings of the assembled product, plus detail drawings for each part and component **(Fig. 18-1)**. **Detail drawings** contain information relating to the manufacture of parts. **Assembly drawings** describe the relationship of parts and the product assembly.

Pictorial drawings may be used as working drawings if the exact shape and necessary dimensions can be shown with clarity **(Fig. 18-2)**. Pictorial drawings are usually used to illustrate product assembly, while multiview drawings are used to convey manufacturing details.

Detail Drawings

A product may consist of a single part or contain hundreds or thousands of parts **(Figs. 18-3a** and **18-3b)**. The manufacture of each part requires a working drawing containing specifications for its manufacture. These include dimensions, tolerances, scale, material, machining and casting details, fastening methods, finishing instructions, identification numbers, parts and standard purchase lists, and

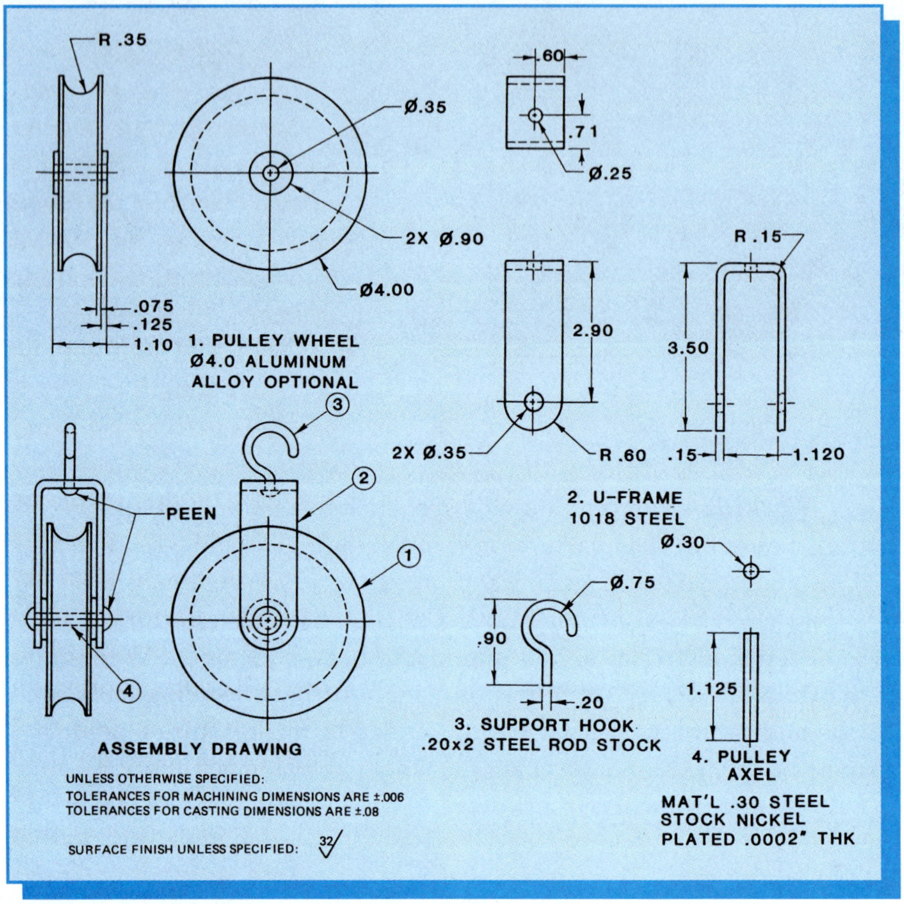

Figure 18-1
Detail drawings

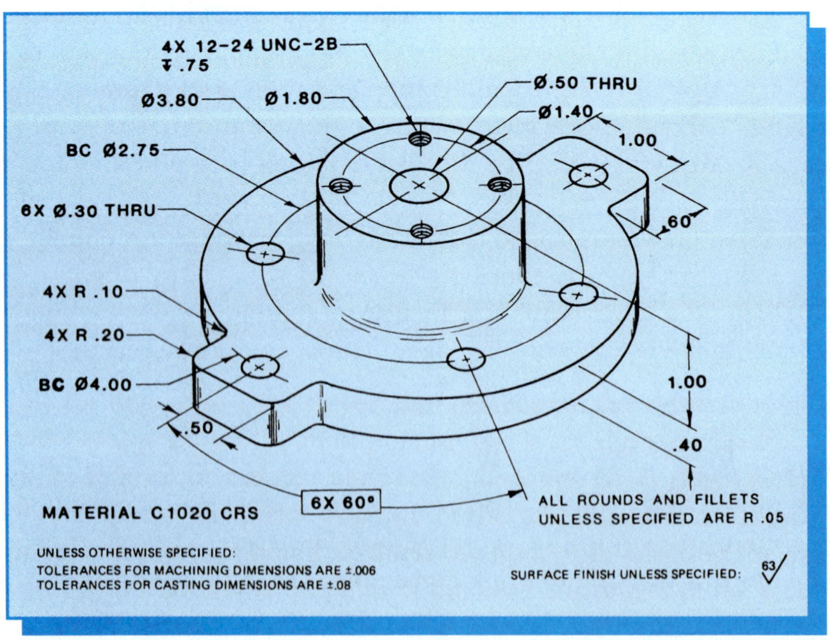

Figure 18-2
Pictorial working drawing of a plate cover

Figure 18-3a
Thousands of working drawings are required to manufacture this airplane. *(Courtesy of Cessna Aircraft Company)*

Figure 18-3b
Hundreds of drawings were required to manufacture this valve.

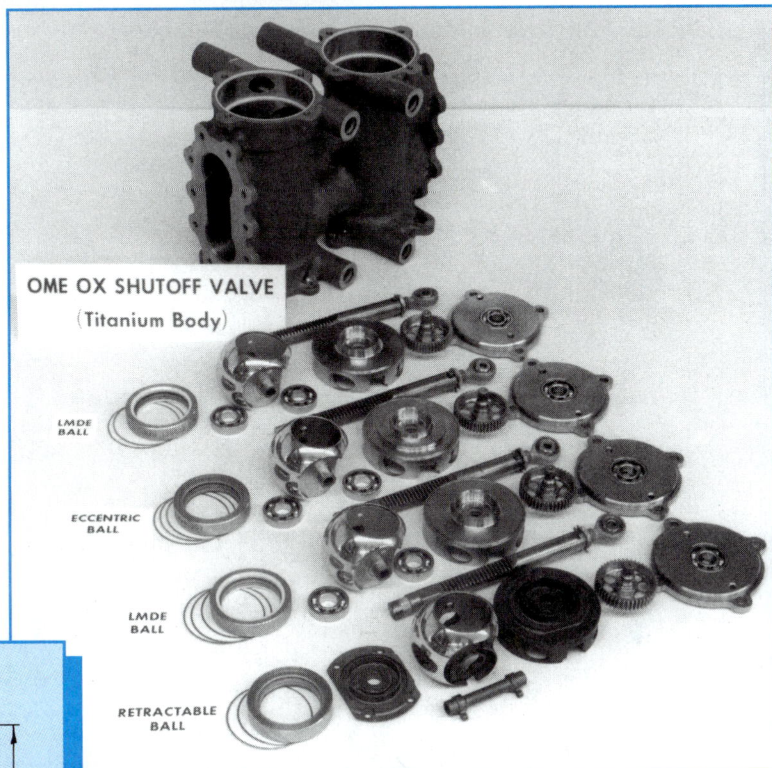

indexes to related part drawings. A typical working drawing of a single part is shown in **Figure 18-4**.

Since different parts may be manufactured and/or assembled in different locations, it is imperative that uniform standards, such as ASME, be used in the preparation of all detail drawings. Unless the entire manufacturing operation is contained in one location, detail drawings for one part should not be placed on the same sheet as drawings for another part. Even then, careful indexing and cross-indexing are necessary to avoid confusion. Drawing sheet layout standards, which outline indexing methods, are covered in Chapter 6.

Specialized detail drawings are often prepared in separate sets for different manufacturing operations. Specialized sets may be prepared for foundry use, rough machining, finish machining, and surface finishing. The detail drawings show only the information needed for the particular operation. When this is done, great care must be taken to insure the individual sets are totally compatible, especially in the assignment of tolerances.

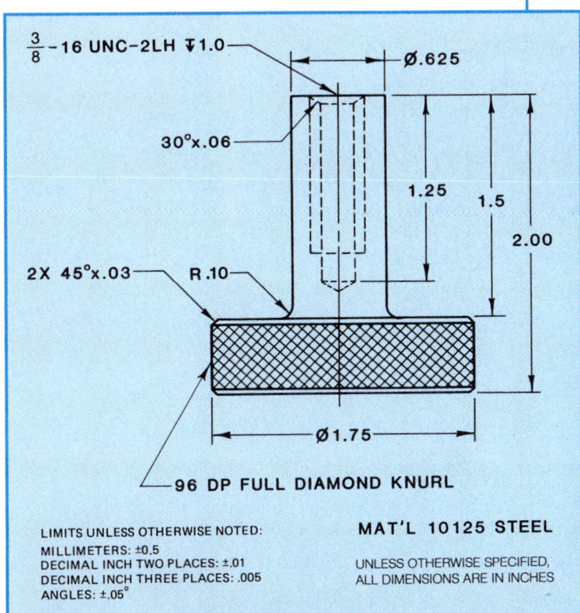

Figure 18-4
Detail working drawing for an adjustment knob from a lathe assembly

Assembly Drawings

A drawing that shows the relationship of two or more parts of a product is an assembly drawing **(Fig. 18-5).** There are many different types of assembly drawings. The amount of detail and drawing type depend on its use during manufacturing. Assembly drawings do not contain as much detail as in detail drawings. Most are prepared without hidden lines to promote clarity. The following are the most common types of assembly drawings.

- **General assembly drawings** include orthographic drawings of assembled products, parts lists, and related data needed in the manufacturing process **(Fig. 18-6).**

Figure 18-6
General assembly drawing and parts list for a gear puller *(Reprinted from* Technical Drawing and Design *by Goetsch and Nelson, © 1986 Delmar Publishers, Inc.)*

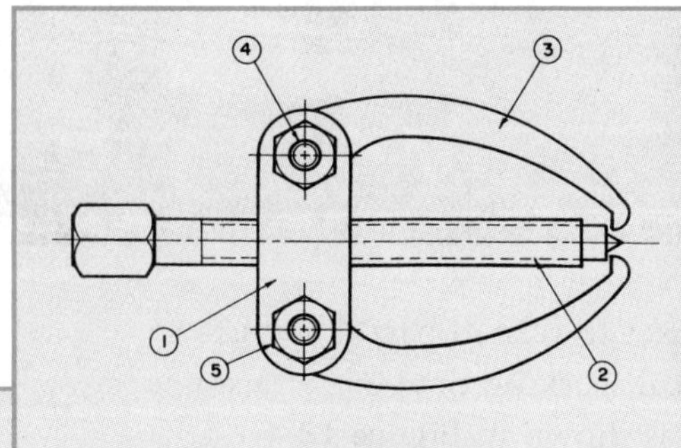

NO.	DRAWING NO.	ITEM	REQ'D.
-	-	1/4-20 UNC	-
5	PURCH	LOCK NUT-HEX	2
-	-	1/4-20 UNC X 1 1/2 LG.	-
4	PURCH	SCREW-CAP HEX HD	2
3	A661982	HOOK	2
2	A661981	SCREW-CENTER	1
1	A661983	YOKE	1

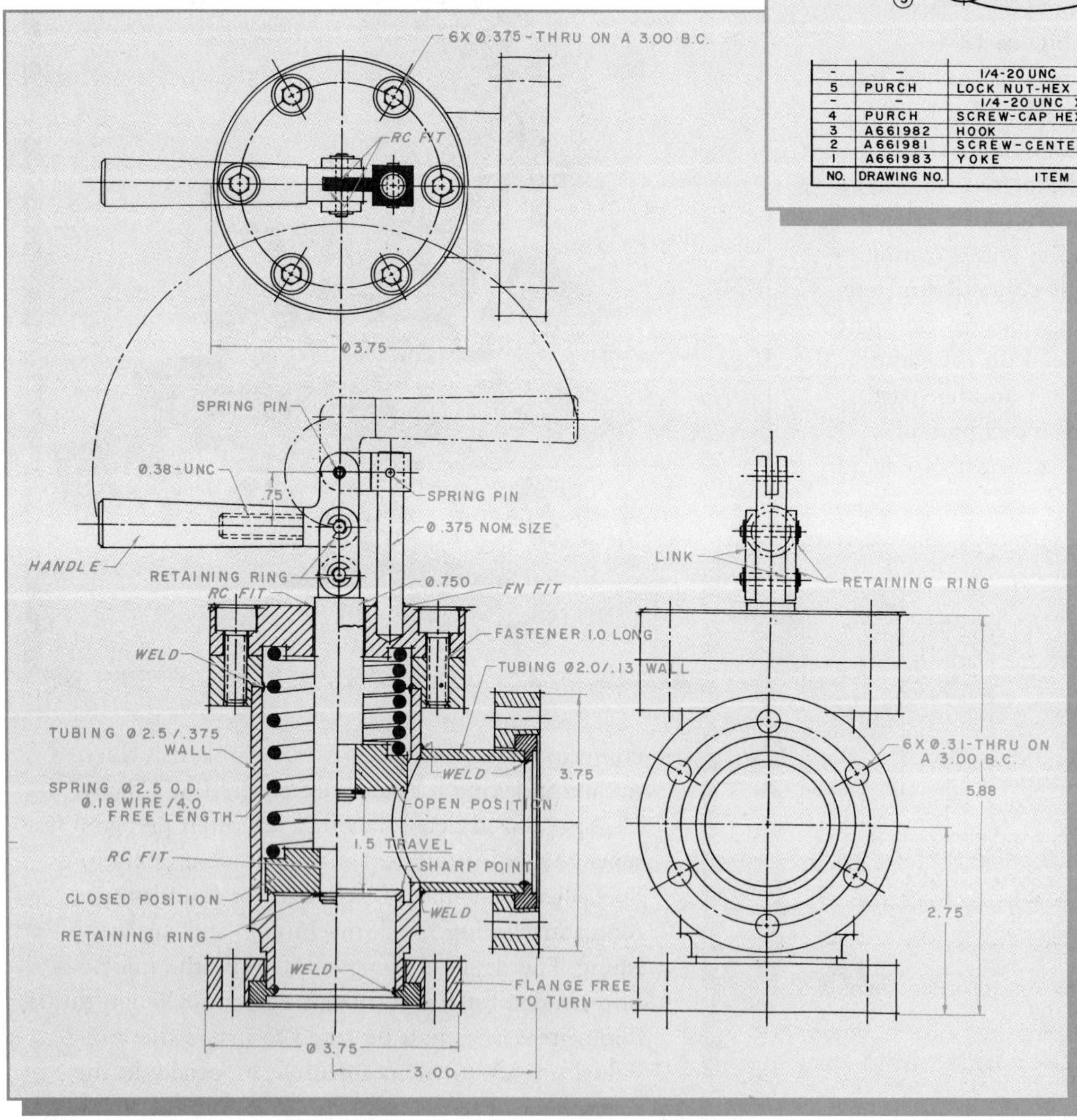

Figure 18-5
Assembly drawing for a pipe valve *(Reprinted from* Technical Drawing and Design *by Goetsch and Nelson, ©1986 Delmar Publishers, Inc.)*

- **Layout assembly drawings** show rough preliminary design ideas, sometimes in sketch form. They show overall size, form, location, and relationship of only major components **(Fig. 18-7)**.

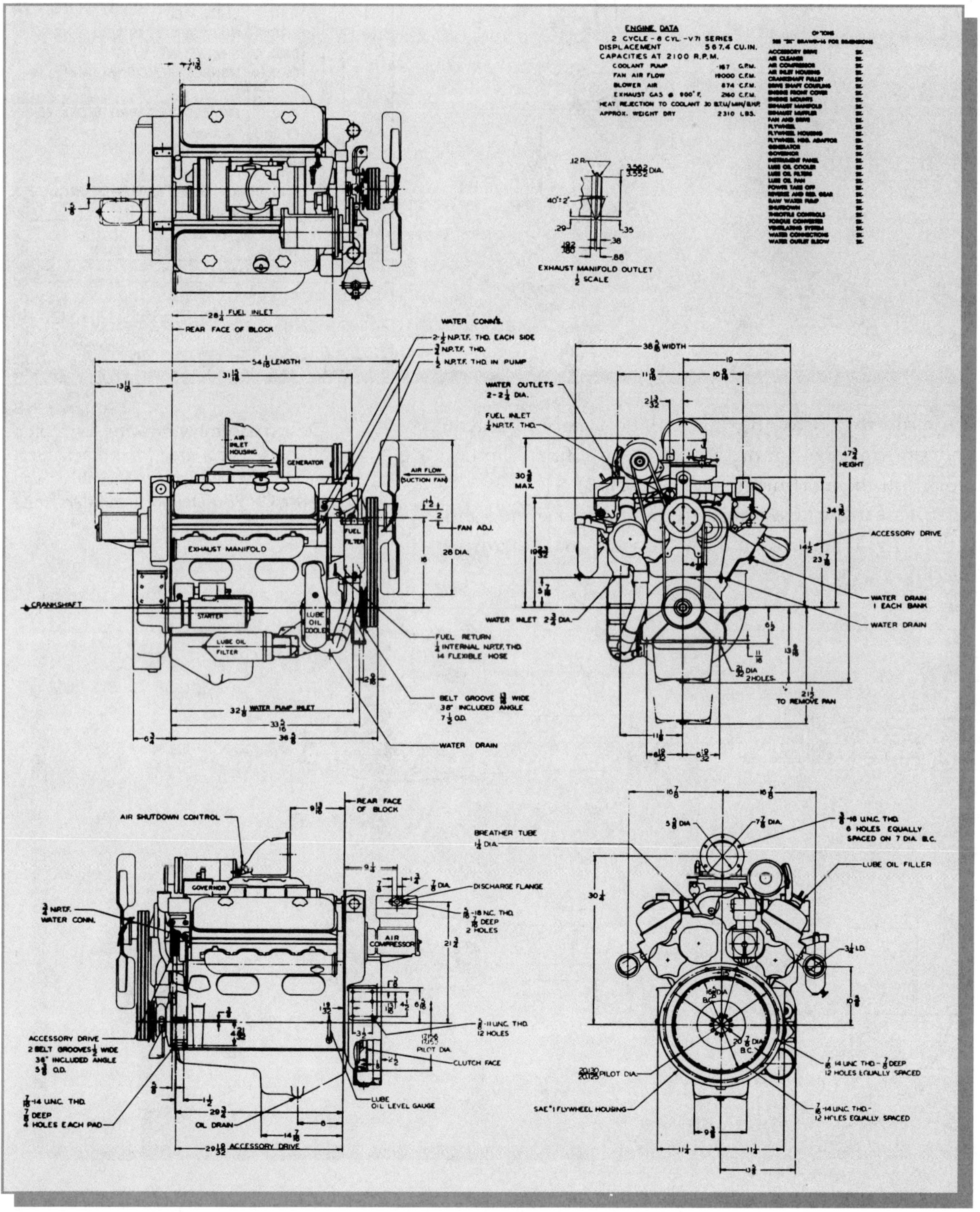

Figure 18-7

Layout assembly drawing for an automobile engine *(Courtesy of General Motors Corporation, Engineering Standards)*

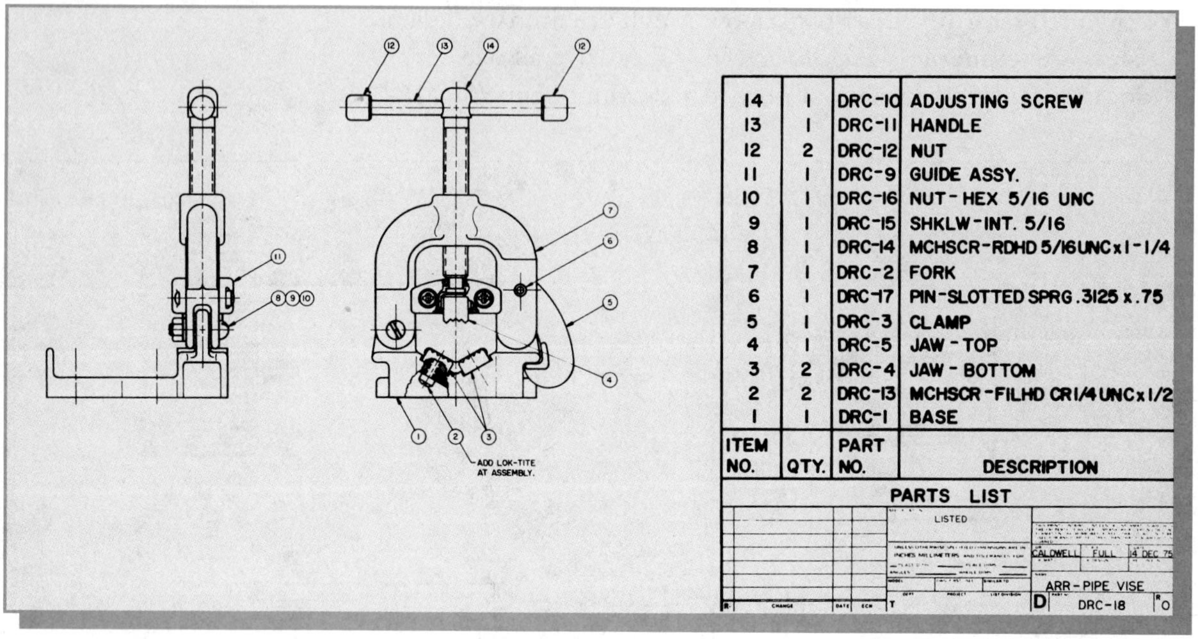

Figure 18-8
Design assembly drawing for a pipe vise *(Reprinted from* Mechanical Drafting *by Madsen, Shumaker, and Stewart, ©1986 Delmar Publishers, Inc.)*

- **Design assembly drawings** are refined versions of layout assembly drawings with more accurate and refined dimensions, line work and manufacturing data **(Fig. 18-8)**.
- **Working assembly drawings** are fully dimensioned and noted drawings that are used directly in the manufacturing process **(Fig. 18-9)**.

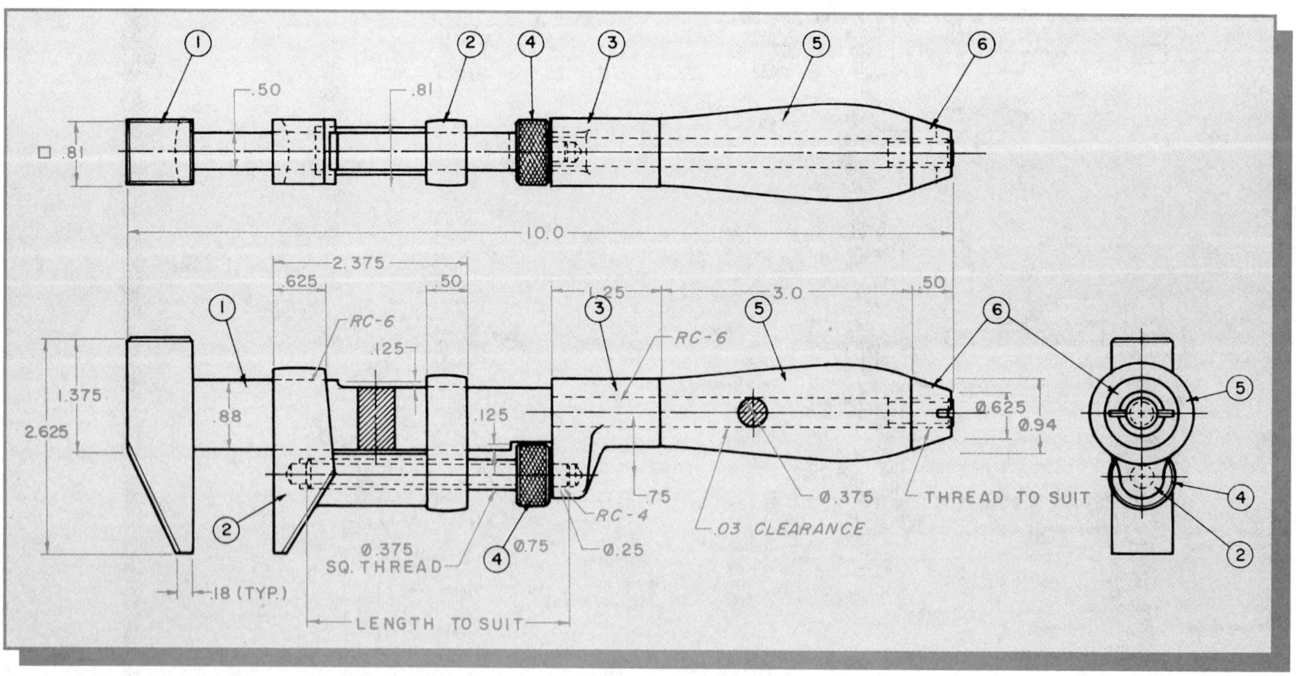

Figure 18-9
Working assembly drawing for a wrench *(Reprinted from* Technical Drawing and Design *by Goetsch and Nelson, ©1986 Delmar Publishers, Inc.)*

- **Erection assembly drawings** are used in the construction industry and include all information necessary for the erection of structural framework. Detailed information relating to structural members is omitted **(Fig. 18-10)**.

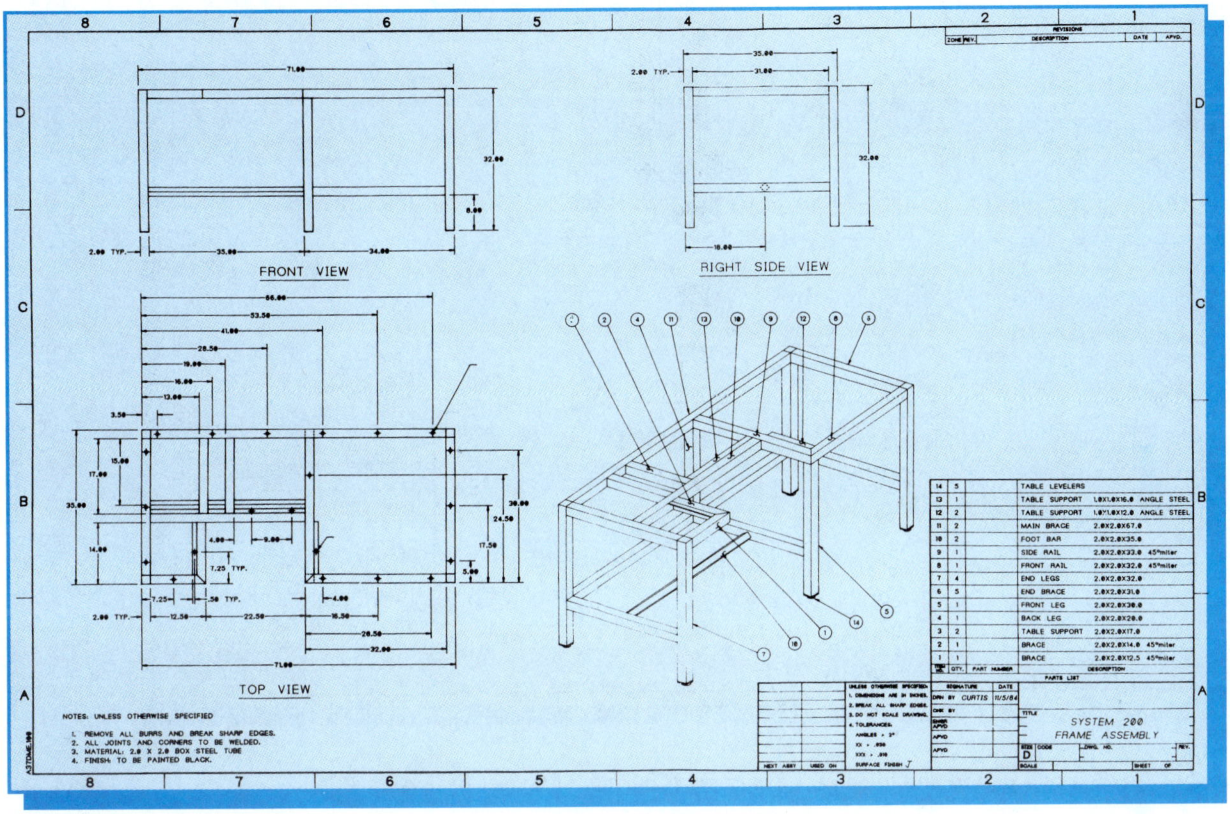

Figure 18-10
Erection assembly drawing of a bench frame *(Courtesy of EFT Systems)*

- **Subassembly drawings** show the smallest assembled components that are part of a larger component. Working drawings may include many layers of subassembly drawings, depending on the size and complexity of the finished product **(Fig. 18-11)**.
- **Pictorial assembly drawings** are substitutes for multiview assembly drawings if the product is not complex or if it is intended for nontechnical interpretation **(Fig. 18-12)**. Many pictorial assembly drawings are sectional views designed to show the interior structure and relationship of static and moving parts, and components **(Fig. 18-13)**. Pictorial assembly drawings are often drawn in exploded-view form (see Chapter 15) to show the relationship of parts before assembly.

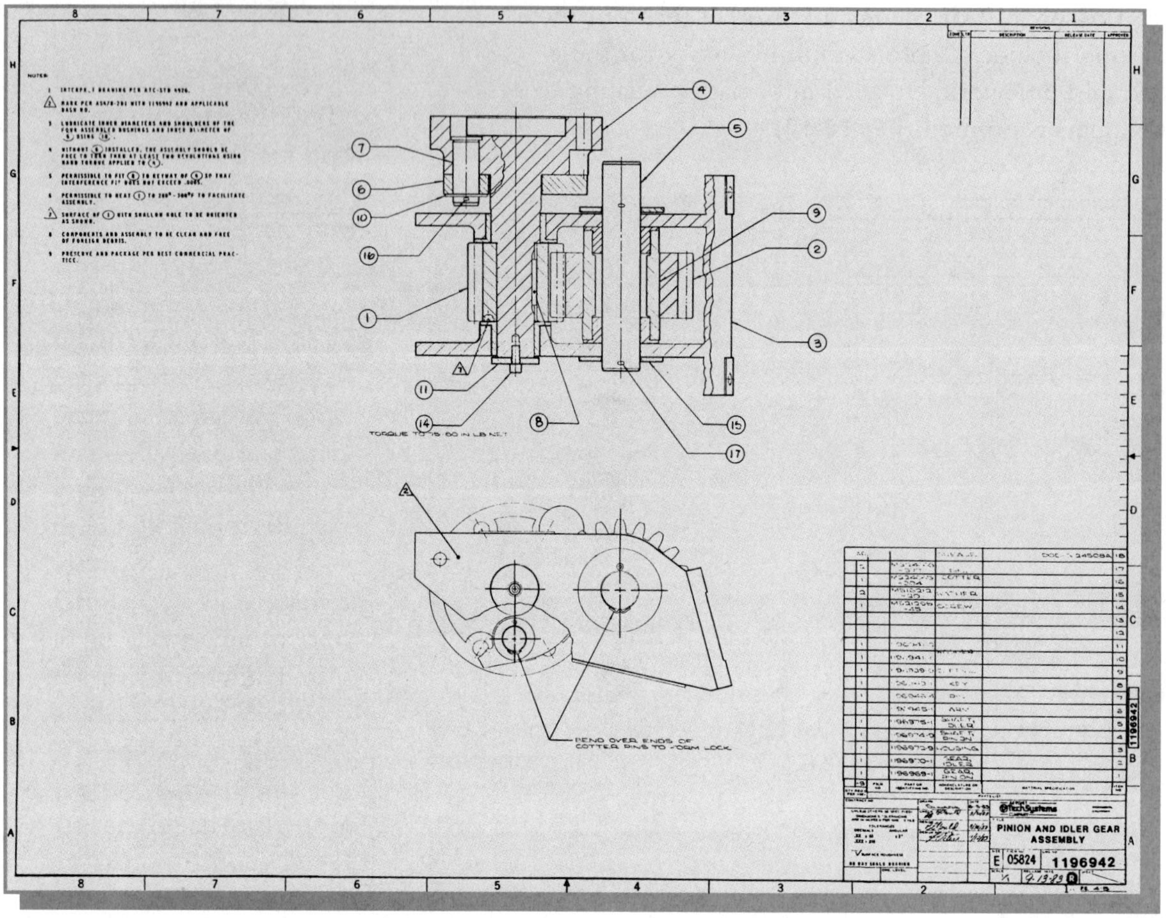

Figure 18-11
Subassembly drawing of a pinion and idler gear assembly *(Courtesy of Aerojet Techsystems Co.)*

Figure 18-12
Pictorial assembly drawing *(Courtesy of General Motors Corporation, Engineering Standards)*

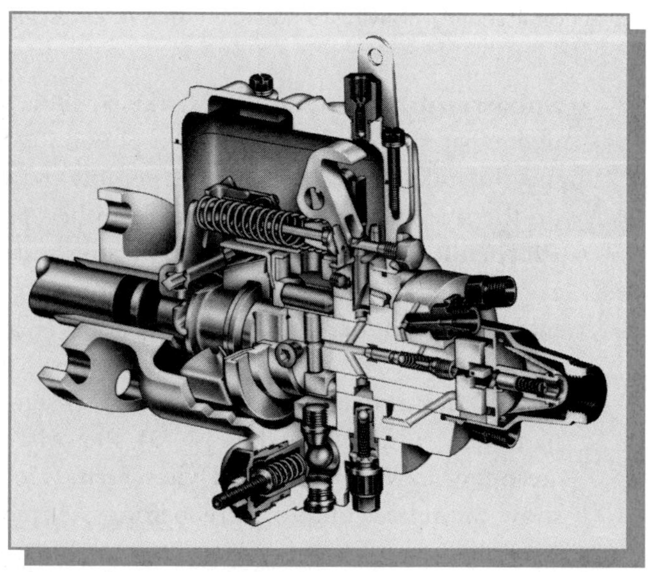

Figure 18-13
Cutaway pictorial assembly drawing *(Courtesy of Stanadyne Diesel Systems)*

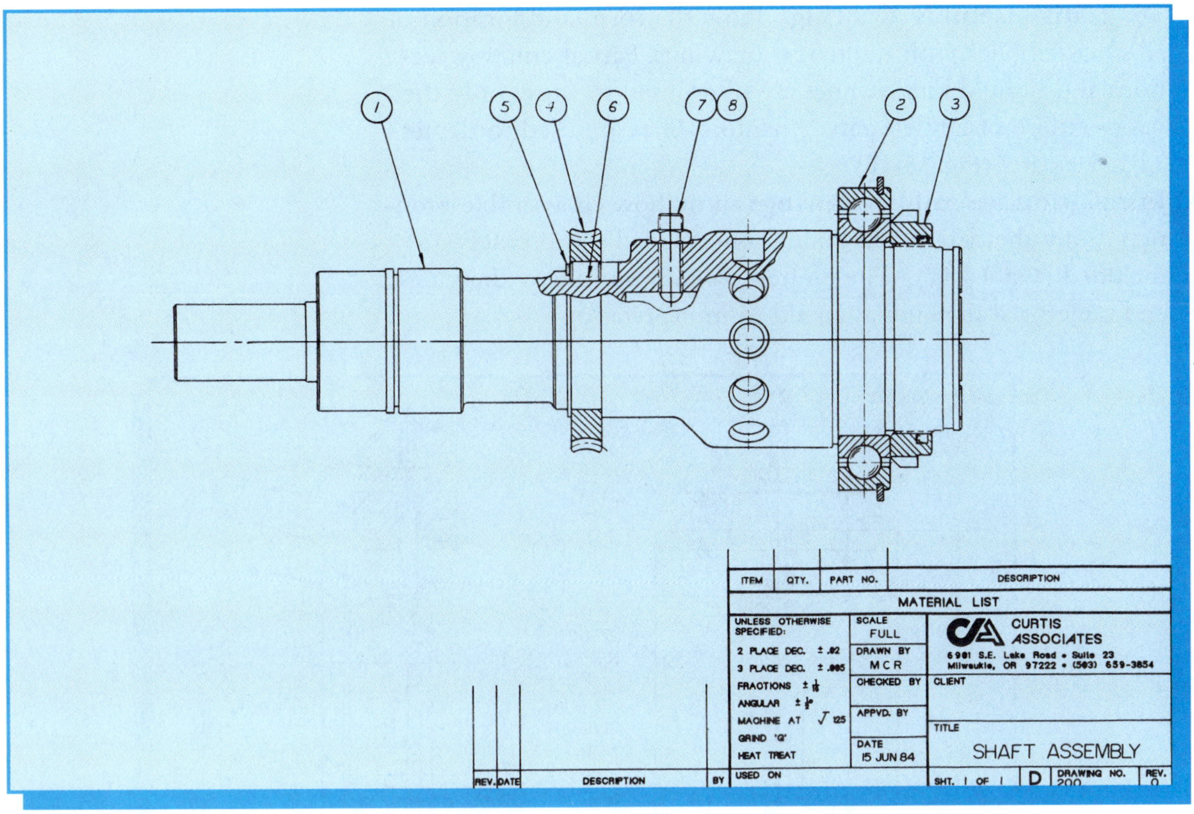

Figure 18-14
Outline assembly drawing with a broken-out section of a shaft *(Courtesy of Curtis Associates)*

- **Outline assembly drawings** show only the major outline of an assembly or subassembly and include only a few basic notations. Symmetrical objects are often sparsely detailed on one half with only major outlines shown on the other half **(Fig. 18-14)**. They are also used to show only the basic positioning of subassemblies. Catalog publications often use outline assembly drawings.
- **Operation assembly drawings** illustrate the mechanical operation. Multiple positions of moving parts are often shown with phantom lines, sections, and enlarged details **(Fig. 18-15)**.

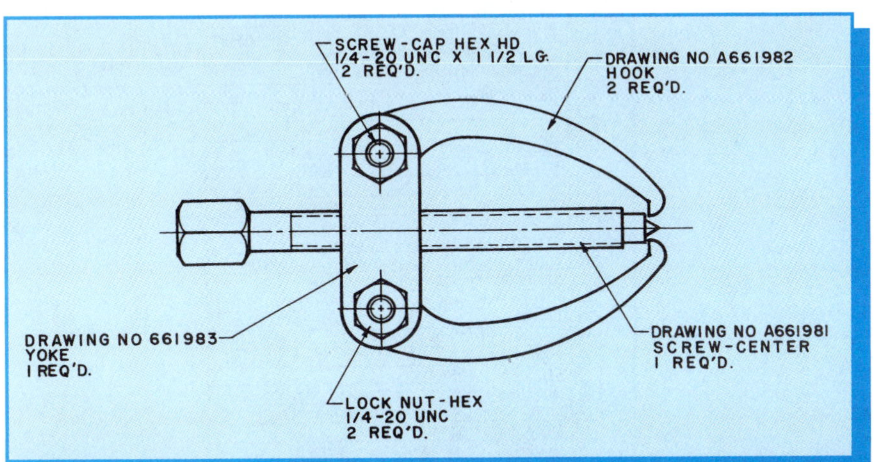

Figure 18-15
Operation assembly drawing for a gear puller *(Reprinted from* Technical Drawing and Design *by Goetsch and Nelson, Delmar Publishers, Inc.)*

- **Diagram assembly drawings** show the form and location of all subassemblies with multiview drawings. Partial cutaway sections and limited hidden lines are often used to show only the basic outline of hidden parts. Phantom lines are used to depict adjacent parts **(Fig. 18-16)**.
- **Installation assembly drawings** show how to assemble products. Only the details and dimensions needed for assembly are included **(Fig. 18-17)**. These often contain multiview drawings and a pictorial drawing as an aid to interpretation.

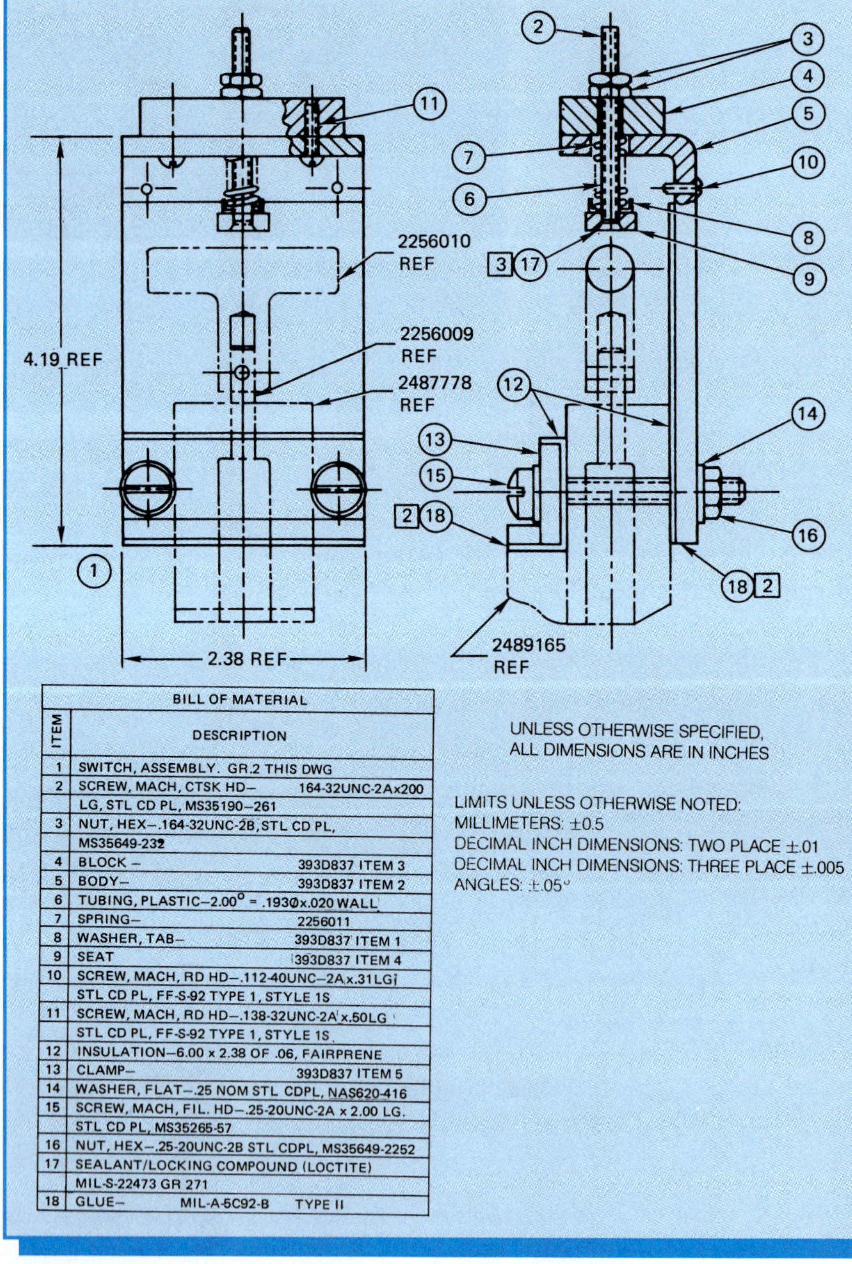

BILL OF MATERIAL	
ITEM	DESCRIPTION
1	SWITCH, ASSEMBLY. GR.2 THIS DWG
2	SCREW, MACH, CTSK HD– 164-32UNC-2Ax200
	LG, STL CD PL, MS35190–261
3	NUT, HEX–.164-32UNC-2B, STL CD PL,
	MS35649-232
4	BLOCK – 393D837 ITEM 3
5	BODY– 393D837 ITEM 2
6	TUBING, PLASTIC–2.00° = .193Øx.020 WALL
7	SPRING– 2256011
8	WASHER, TAB– 393D837 ITEM 1
9	SEAT 393D837 ITEM 4
10	SCREW, MACH, RD HD–.112-40UNC–2Ax.31LG
	STL CD PL, FF-S-92 TYPE 1, STYLE 1S
11	SCREW, MACH, RD HD–.138-32UNC-2Ax.50LG
	STL CD PL, FF-S-92 TYPE 1, STYLE 1S
12	INSULATION–6.00 x 2.38 OF .06, FAIRPRENE
13	CLAMP– 393D837 ITEM 5
14	WASHER, FLAT–.25 NOM STL CDPL, NAS620-416
15	SCREW, MACH, FIL. HD–.25-20UNC-2A x 2.00 LG.
	STL CD PL, MS35265-57
16	NUT, HEX–.25-20UNC-2B STL CDPL, MS35649-2252
17	SEALANT/LOCKING COMPOUND (LOCTITE)
	MIL-S-22473 GR 271
18	GLUE– MIL-A-5C92-B TYPE II

UNLESS OTHERWISE SPECIFIED,
ALL DIMENSIONS ARE IN INCHES

LIMITS UNLESS OTHERWISE NOTED:
MILLIMETERS: ±0.5
DECIMAL INCH DIMENSIONS: TWO PLACE ±.01
DECIMAL INCH DIMENSIONS: THREE PLACE ±.005
ANGLES: ±.05°

Figure 18-16
Diagram assembly drawing of a valve adjuster

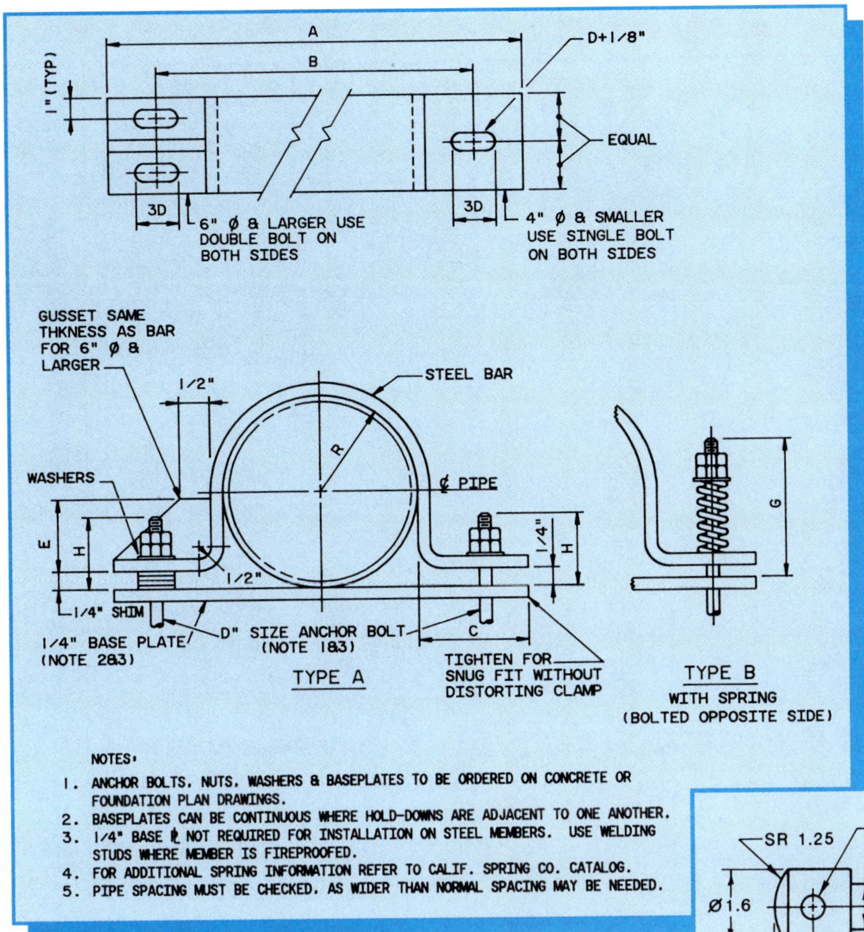

Figure 18-17
Installation assembly drawing of a base plate for pipes

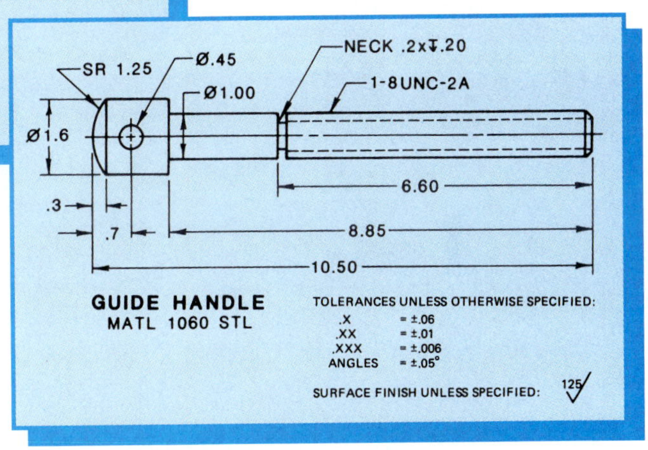

Figure 18-18
Working drawing of a guide handle with U.S. customary dimensions (ASME)

Working Drawing Dimensions

Dimensions on working drawings are used when manufacturing products to an exact specified size. Dimensions must be selected and placed to insure that no further calculations are required during any stage of the manufacturing process. All related drawings must use the same standards for tolerancing to insure proper fitting in the final assembly. **Figure 18-18** shows a summary of the use of the U.S. customary system of measure. **Figures 18-19** through **18-21** show the metric system of measure.

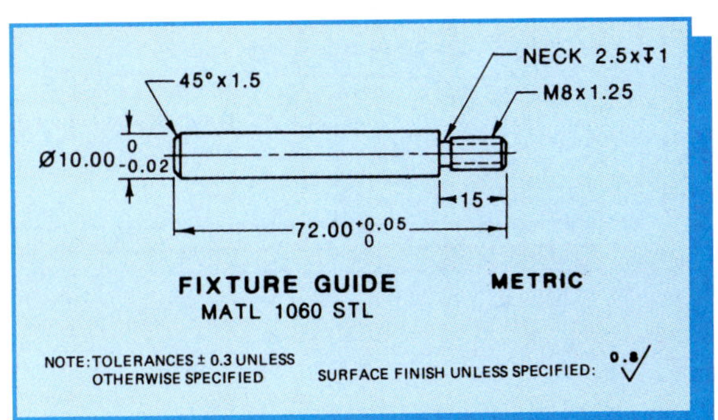

Figure 18-19
Working drawing of a fixture guide with metric dimensions (ASME)

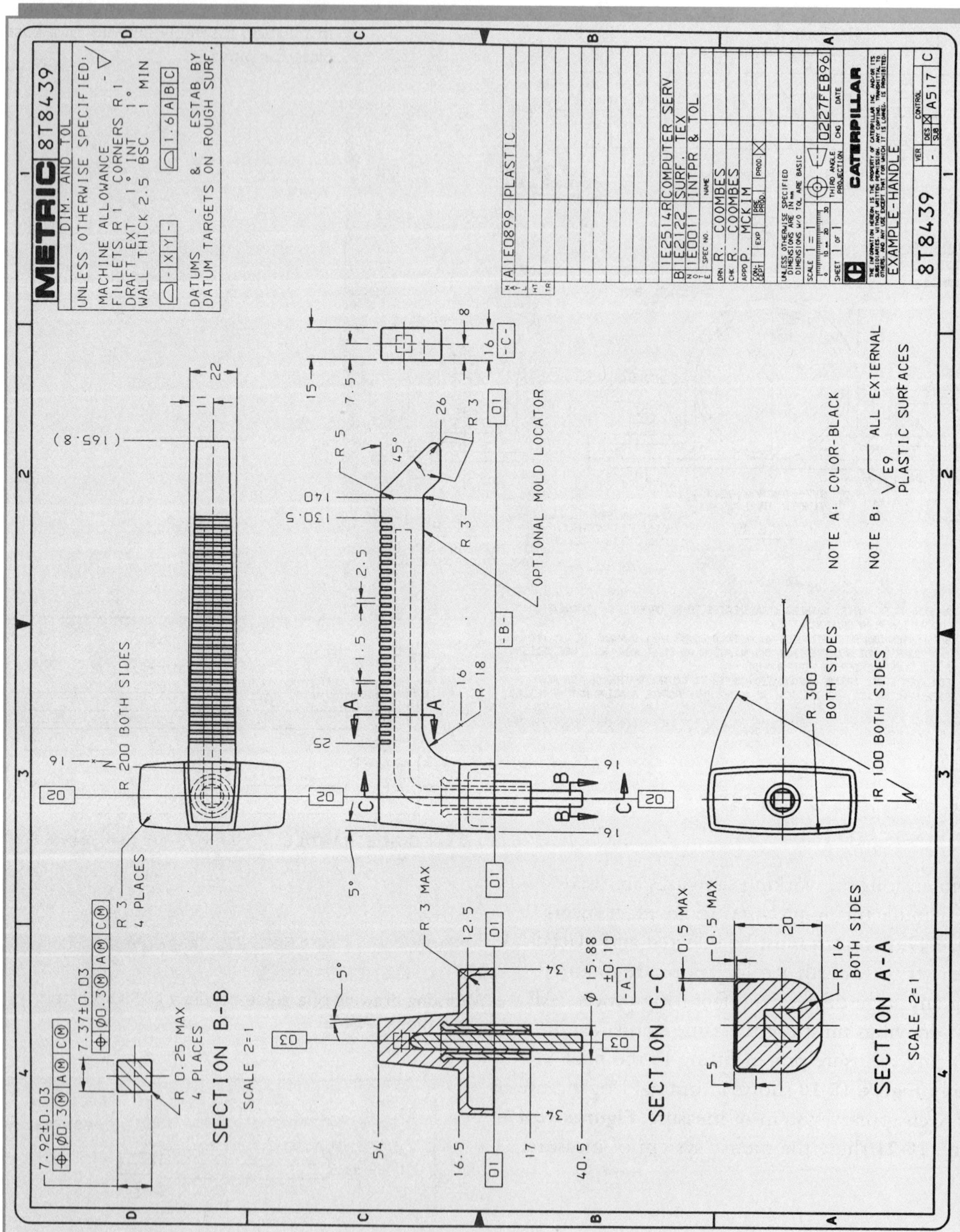

Figure 18-20
Working drawing with metric dimensions *(Courtesy of Caterpillar Inc.)*

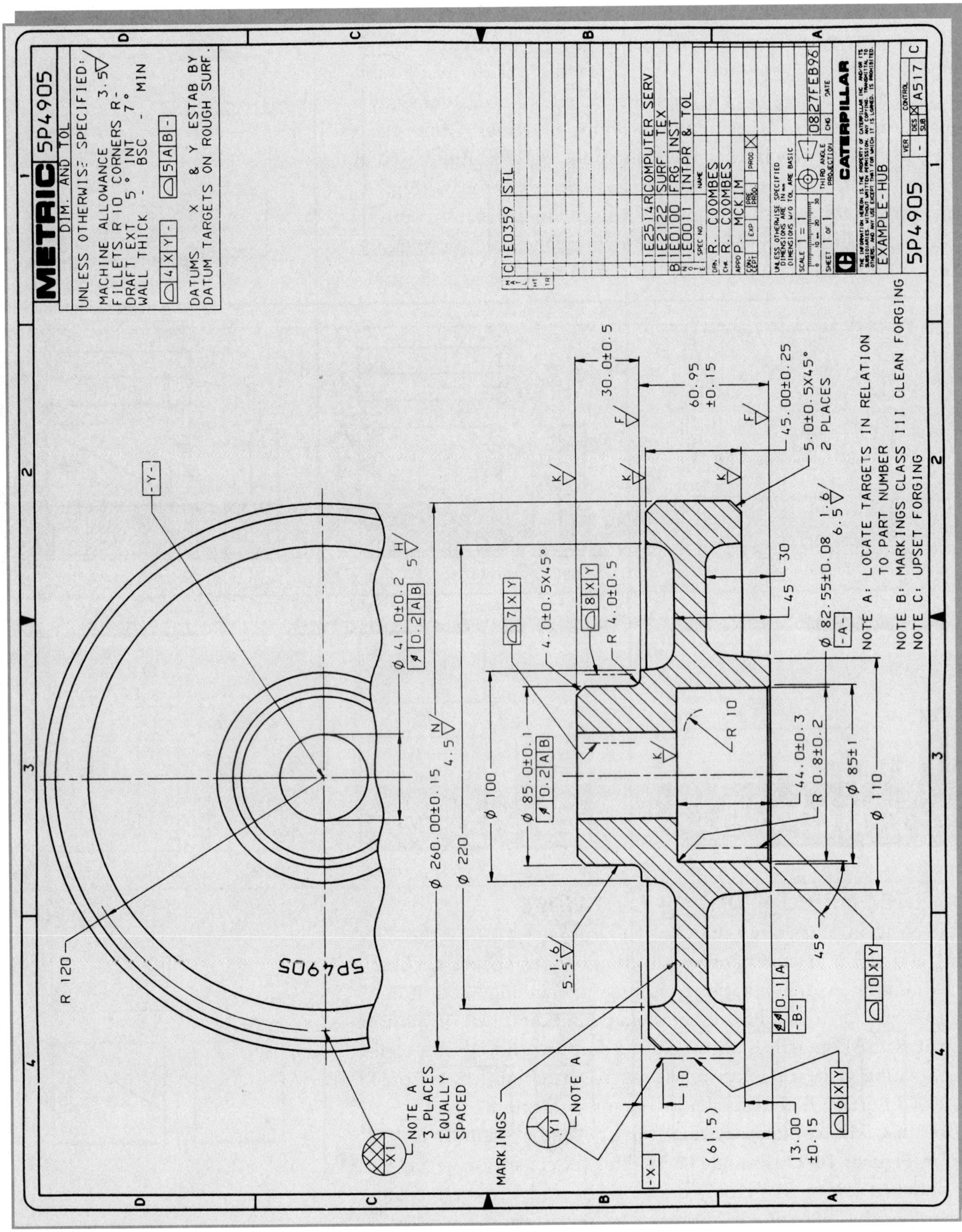

Figure 18-21
Working drawing with metric dimensions *(Courtesy of Caterpillar Inc.)*

Although working drawing dimensions must be extremely accurate and consistent, specific manufacturing processes should not be specified. Exactly how the product is manufactured is the responsibility of the manufacturer, not the engineer, designer, or drafter. Only the final desired result is specified on working drawings. Outcomes may be specified by machine process notations on working drawings. In these cases, the end result is also the goal, not the process **(Fig. 18-22)**. Since there are other manufacturing methods that can produce the same result, only the size and form of the final product are interpreted from the drawing—not the specific process.

Figure 18-22
Typical types of machining operations depicted on working drawings

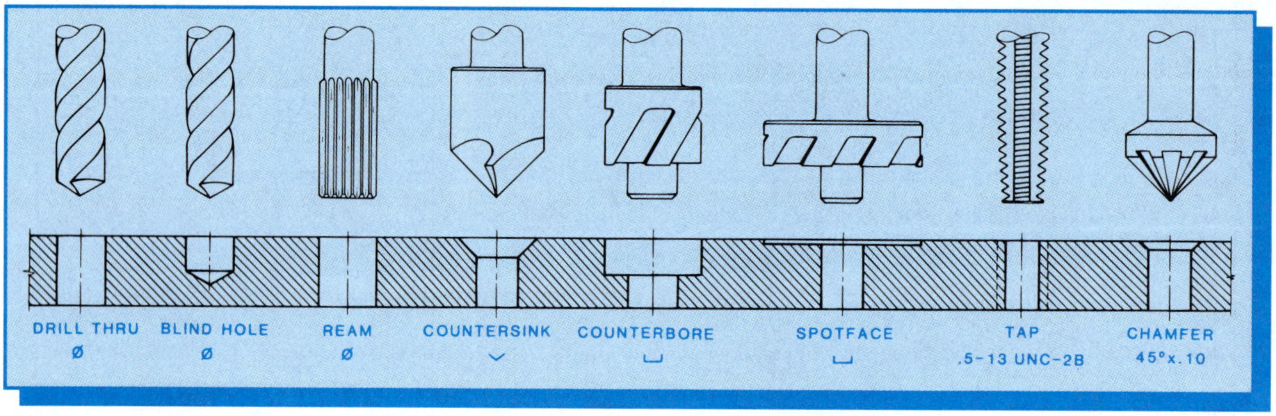

AutoCAD LT Exercises

Computer-Aided Working Drawings

The creation of working drawings incorporates most of the drafting and CAD practices covered in the previous chapters. These included multiview drawing, dimensioning, tolerancing, sectioning, auxiliary views, pictorial drawings, fasteners, and drafting conventions and formats. CAD practices also included drawing and editing techniques, rotating grids for auxiliary views, crosshatching for sectional views, and inserting fasteners, standard items, and title blocks.

A set of CAD working drawings for a collector assembly is shown in **Figures 18-23** through **18-31**. This set consists of seven detail drawings and two assembly drawings. In studying these examples, many of the following CAD practices become apparent:

1. Multiview drawing in all of the details and one of the assembly drawings **(Figs. 18-23** through **18-30)**.
2. Extensive dimensioning in **Figures 18-23** and **18-24**.
3. Tolerancing used in **Figure 18-23**. Note that both bilateral and limit tolerancing have been used.
4. Sectioning, using the HATCH command as described in Chapter 11, has been used in **Figures 18-23** and **18-24**.
5. An auxiliary view **(Fig. 18-24)** was done by rotating the grid, as described in Chapter 12.
6. Text has been used extensively to insert notes in **Figures 18-23** and **18-24**, and in the parts lists assembly drawing. Note that in **Figure 18-23**, two styles of text have been used. The standard text style has been used for dimensioning and notes, and a larger style for the SECTION A-A callout.
7. A CAD shortcut was used in creating the subassembly drawing in **Figure 18-26**. Note that in Part A, the shoe was detailed in **Figure 18-25**. The CAD operator copied it to **Figure 18-26**, mirrored it, and added Part B. This eliminated redrawing the part and saved drafting time, which is one of the advantages of CAD.
8. The spring in **Figure 18-27** is a standard fastener used by a company. The CAD operator inserted the standard spring onto the company's standard drawing format and added the necessary dimensions. The drawing was completed in minutes.
9. The parts lists are identical on the two assembly drawings. The CAD operator created it once and copied it onto the drawings. This is also true for drawing borders and title blocks for each of the drawings. These were done in the same manner as described in Chapter 6.
10. An isometric drawing, described in Chapter 15, was used to create **Figure 18-31**.

The advantages of CAD become more apparent by further studying these drawings. In addition to the timesaving advantage of preparing the working drawings initially on a CAD, there is a big timesavings when making revisions. It is much easier to change CAD-generated drawings than to change those that have been manually drawn. Many drawing changes may simply involve correcting a dimension or a tolerance, or moving the location of a note. A change may necessitate the removal or addition of a view. Simple editing commands, such as ERASE, MOVE, and COPY, make changes a snap.

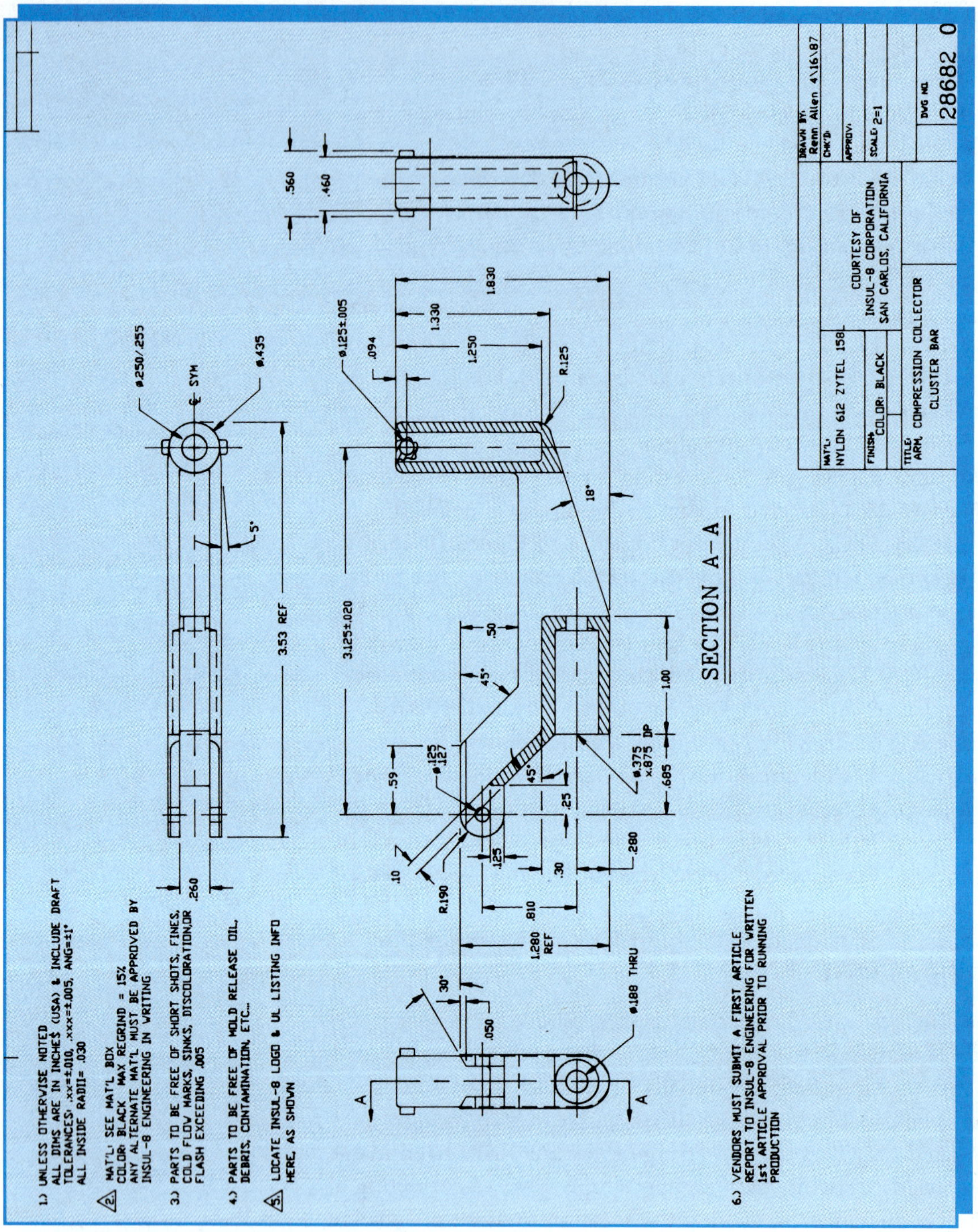

Figure 18-23
Arm-compression collector cluster bar (Sheet 1 of 9) *(Courtesy of Ms. Renn Allen, Draftsperson, Insul-8 Corporation)*

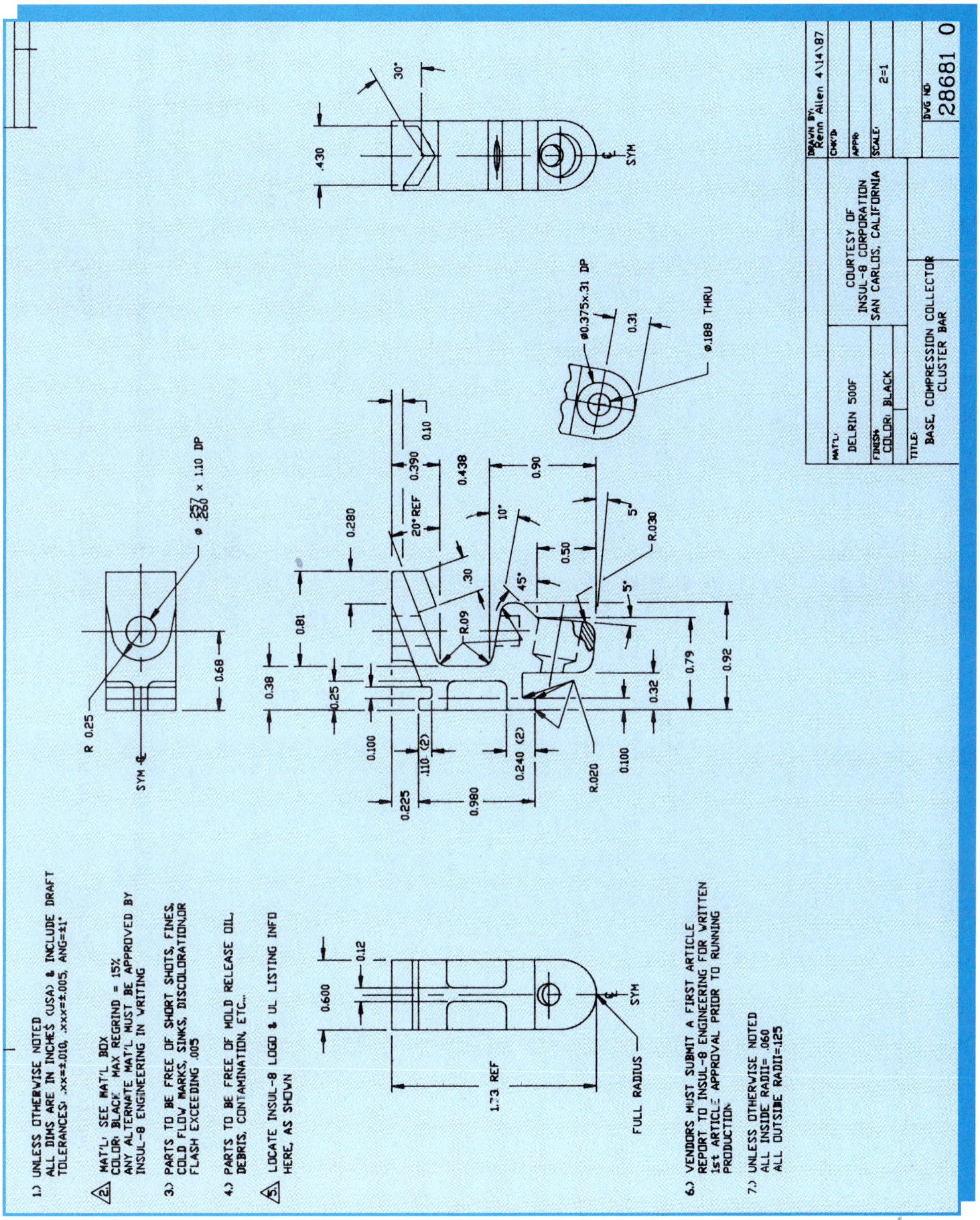

Figure 18-24 Base-compression collector cluster bar (Sheet 2 of 9) *(Courtesy of Ms. Renn Allen, Draftsperson, Insul-8 Corporation)*

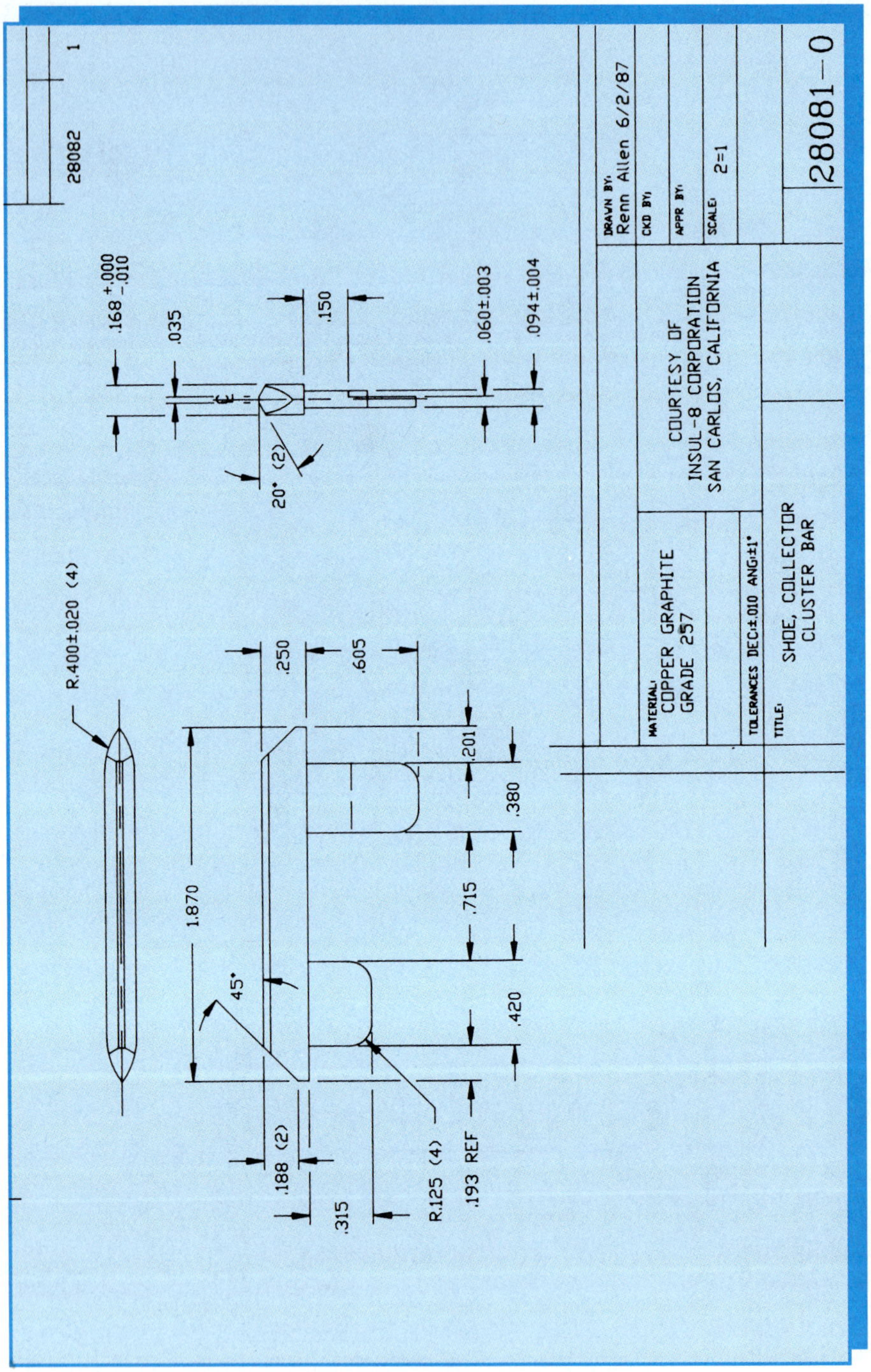

Figure 18-25
Shoe-collector cluster bar (Sheet 3 of 9) *(Courtesy of Ms. Renn Allen, Draftsperson, Insul-8 Corporation)*

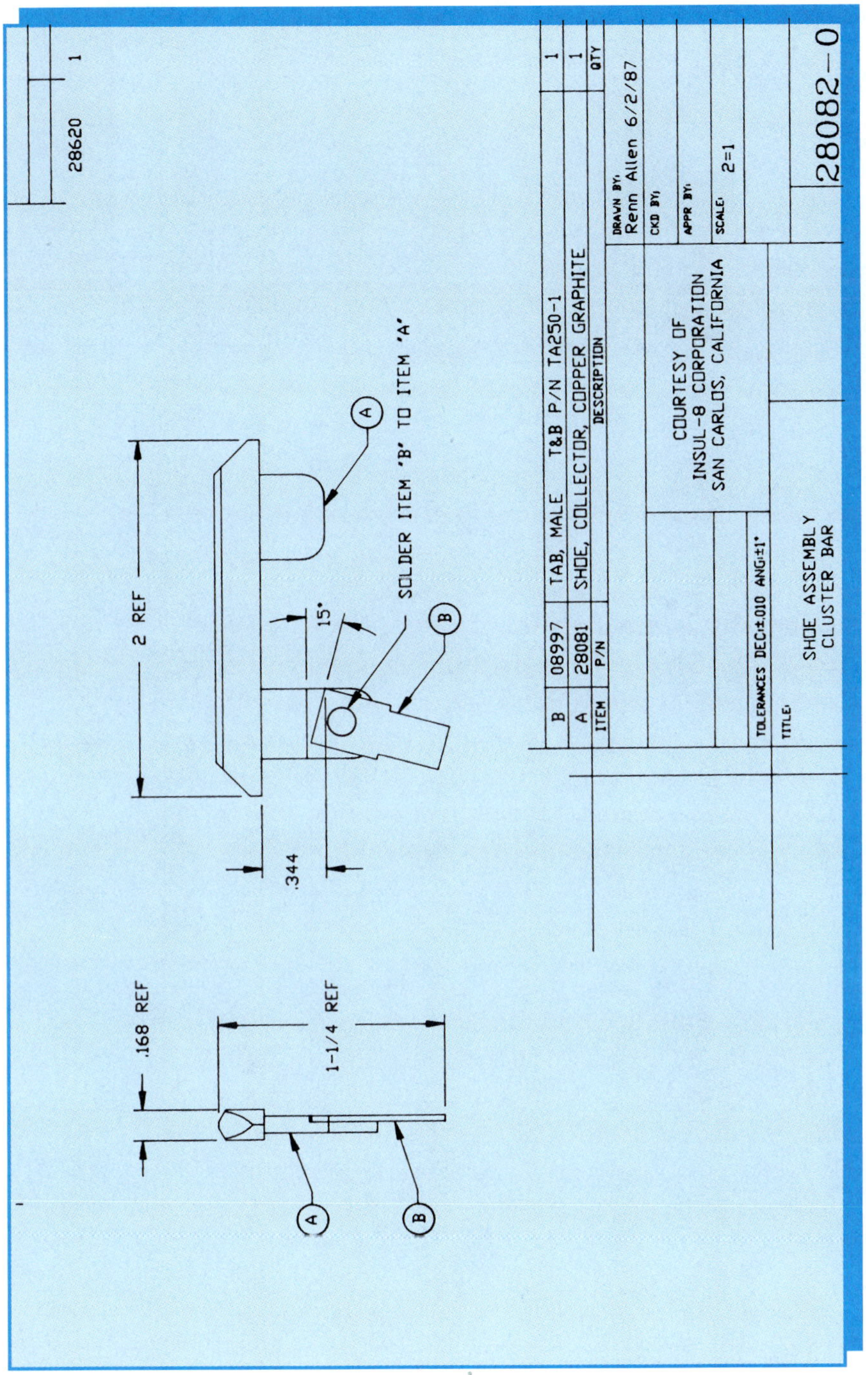

Figure 18-26
Shoe assembly-cluster bar (Sheet 4 of 9) *(Courtesy of Ms. Renn Allen, Draftsperson, Insul-8 Corporation)*

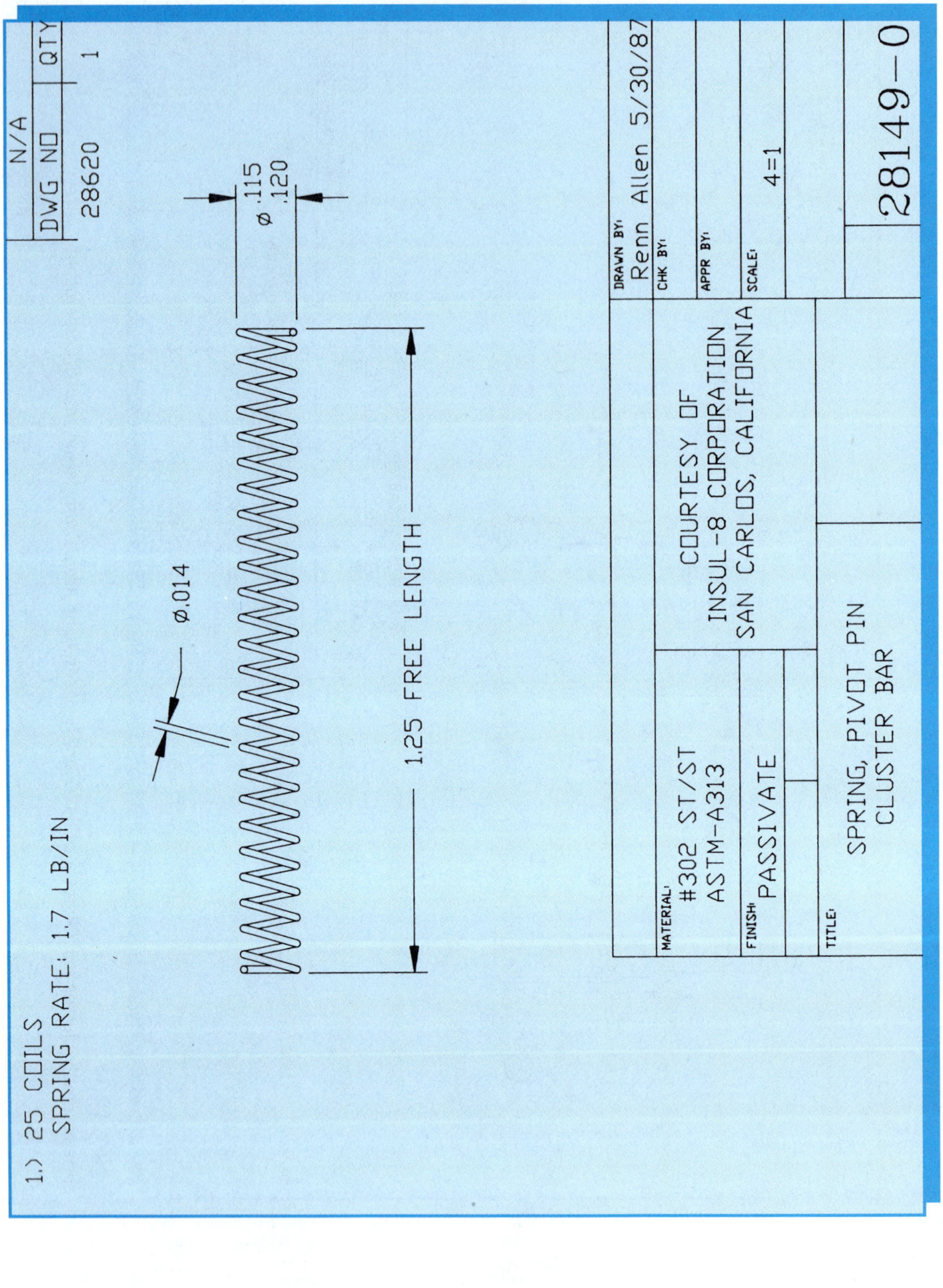

Figure 18-27
Spring-pivot cluster bar (Sheet 5 of 9) *(Courtesy of Ms. Renn Allen, Draftsperson, Insul-8 Corporation)*

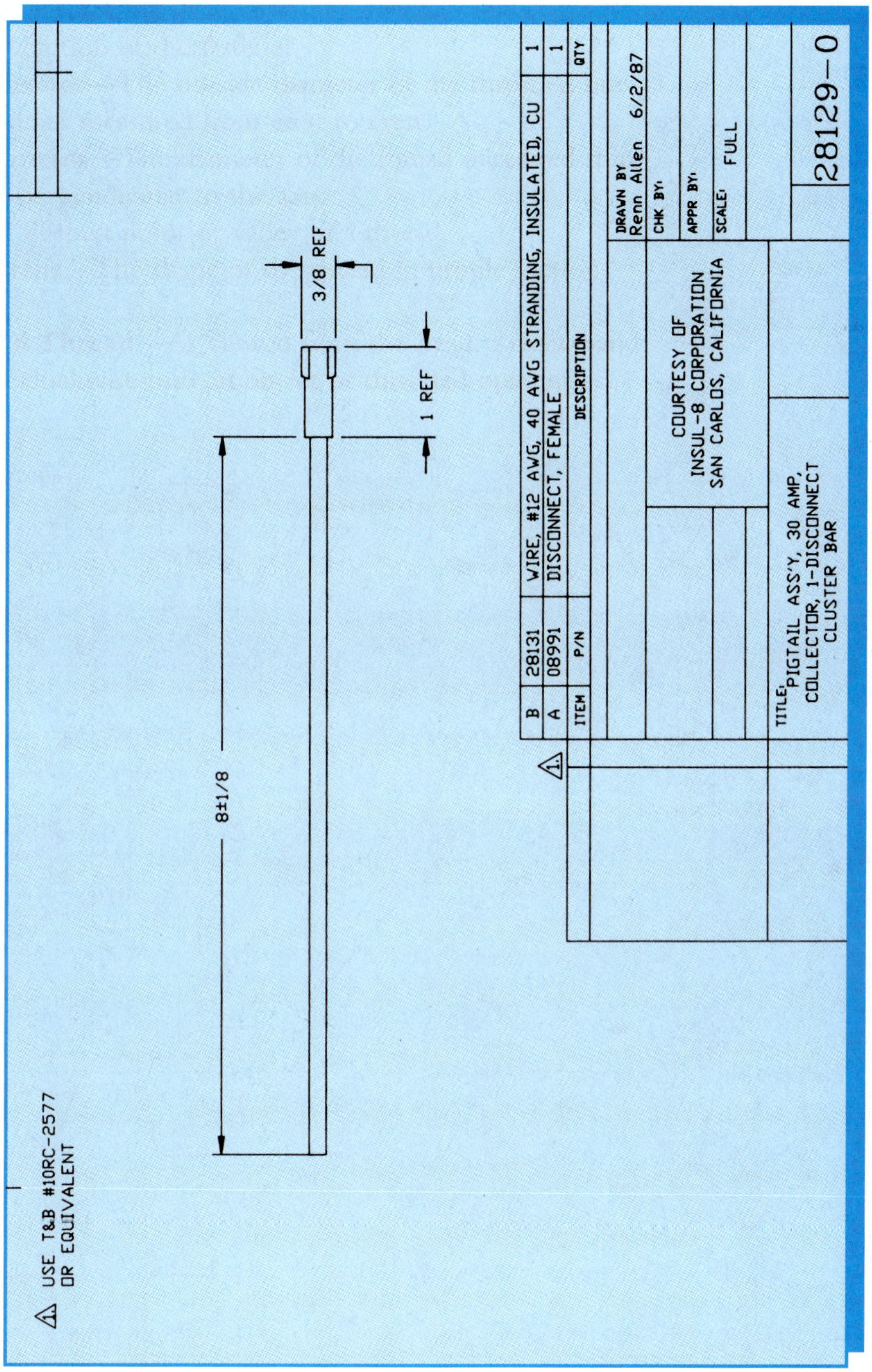

Figure 18-28
Pigtail assembly-30 amp, collector, cluster bar (Sheet 6 of 9) *(Courtesy of Ms. Renn Allen, Draftsperson, Insul-8 Corporation)*

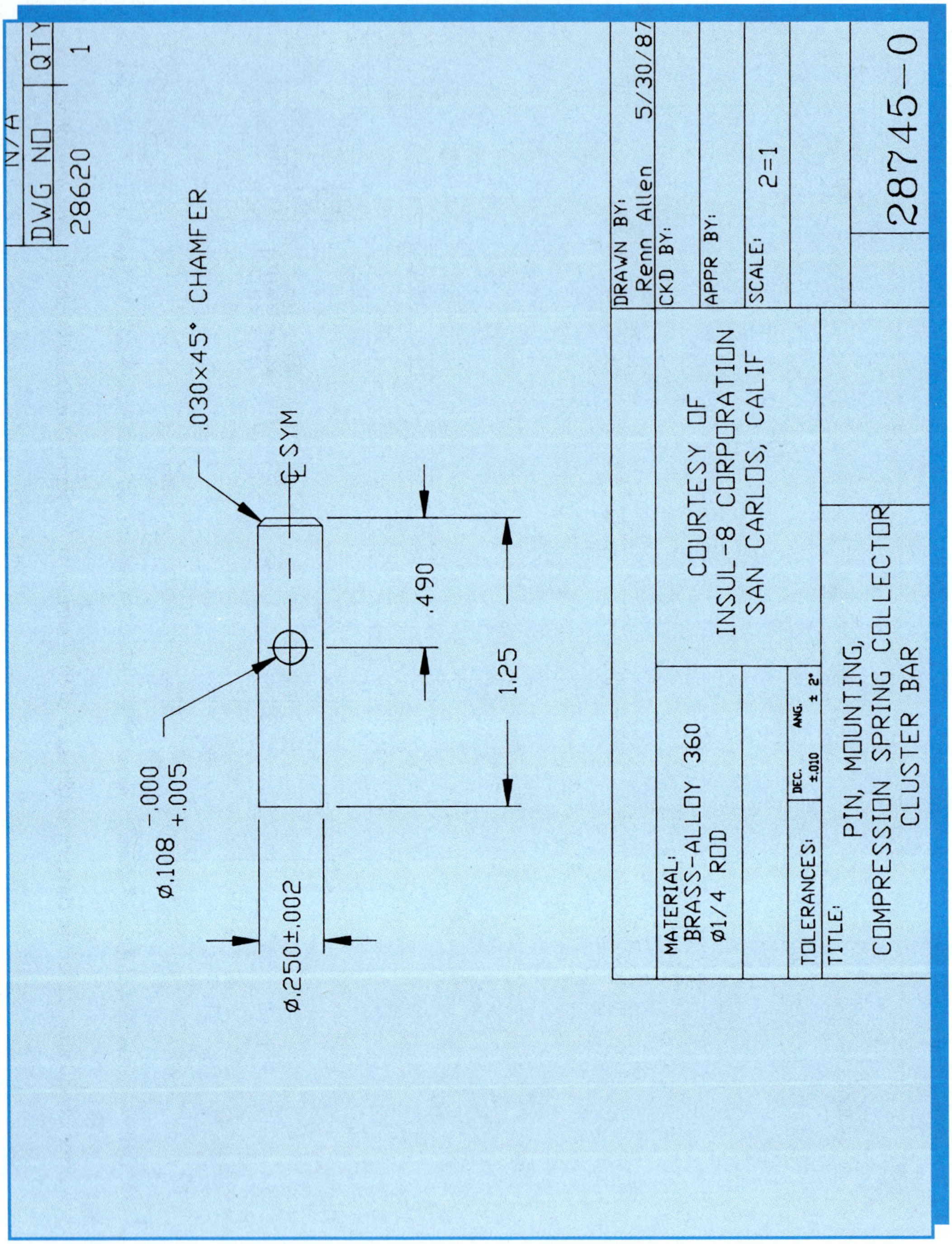

Figure 18-29
Mounting pin-compression collector cluster bar (Sheet 7 of 9) *(Courtesy of Ms. Renn Allen, Draftsperson, Insul-8 Corporation)*

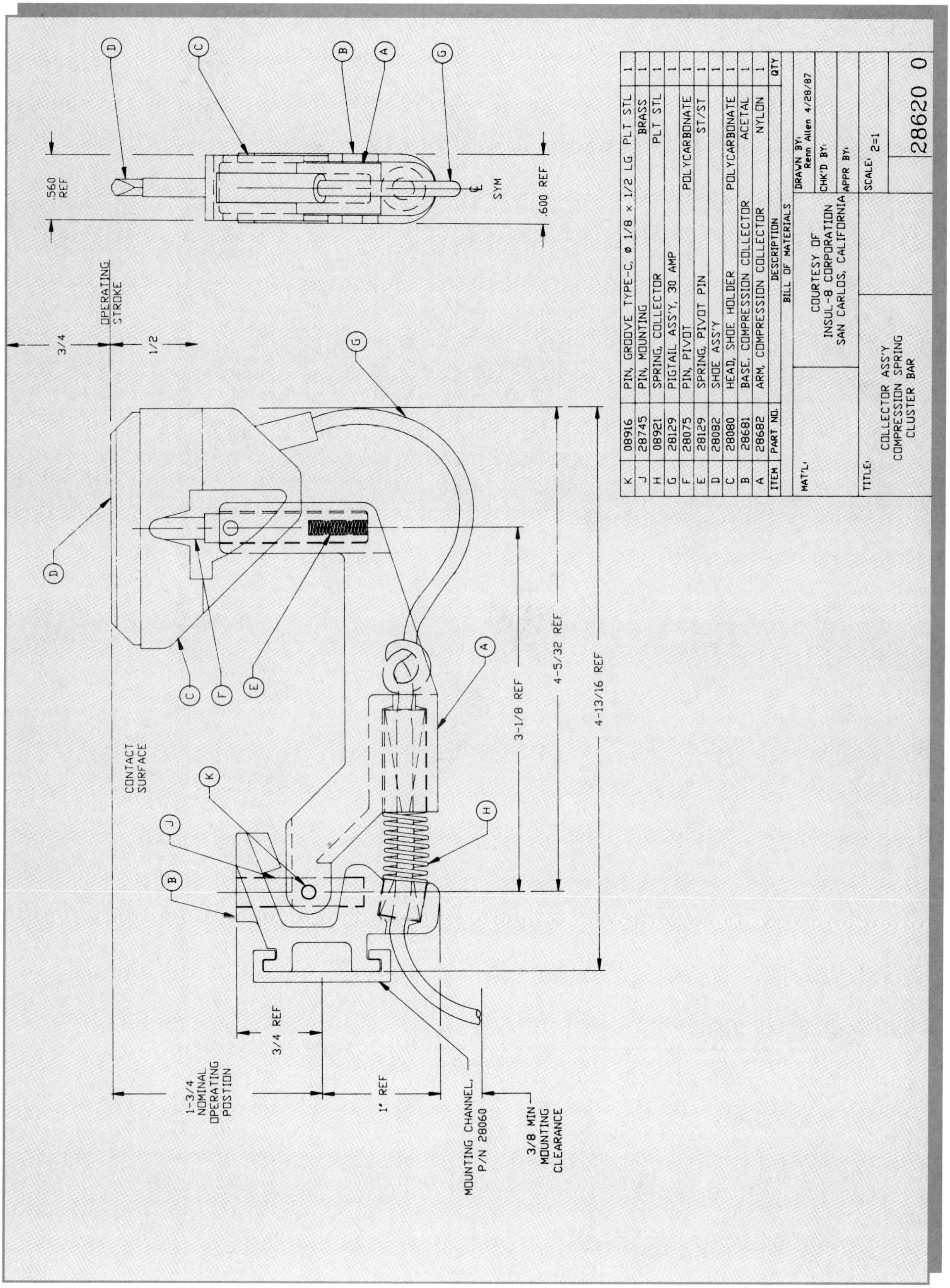

Figure 18-30 Collector assembly-compression spring cluster bar (Sheet 8 of 9) *(Courtesy of Ms. Renn Allen, Draftsperson, Insul-8 Corporation)*

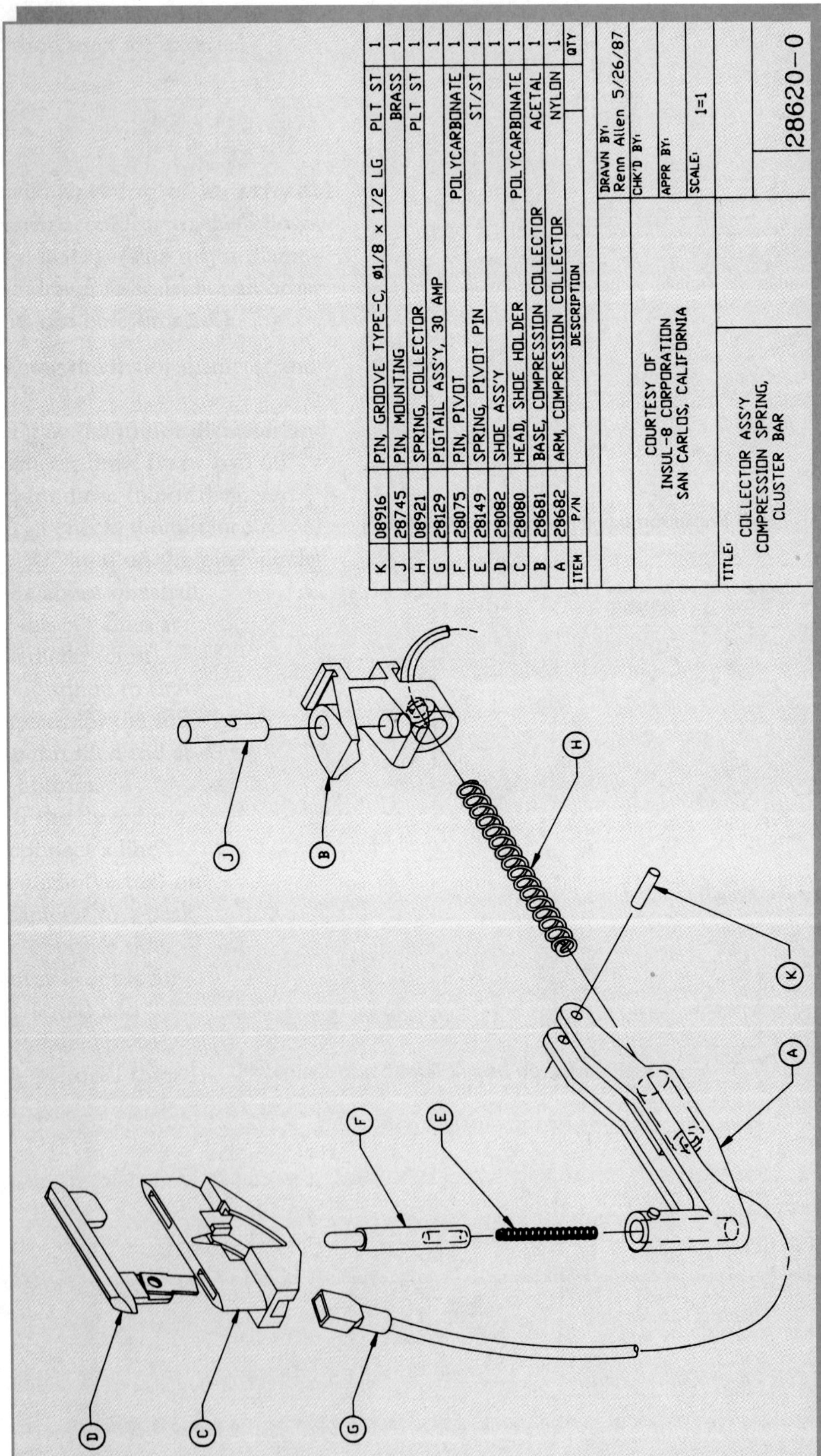

Figure 18-31
Pictorial collector assembly-compression spring cluster bar (Sheet 9 of 9) *(Courtesy of Ms. Renn Allen, Draftsperson, Insul-8 Corporation)*

Drafting Exercises

1. Sketch the drawings for **Figures 18-32** and **18-33**.

2. With drawing instruments or a CAD system, draw the multiview and isometric drawing for each of the single-item machined parts in **Figures 18-34** through **18-53**.

3. With drawing instruments or a CAD system, draw the detail and assembly drawings for items in **Figures 18-54** through **18-65**.

4. With drawing instruments or a CAD system, draw the working drawings in **Figures 18-66** through **18-73** and make all the required ECOs as noted. Record all ECOs onto the proper listing on your drawing format.

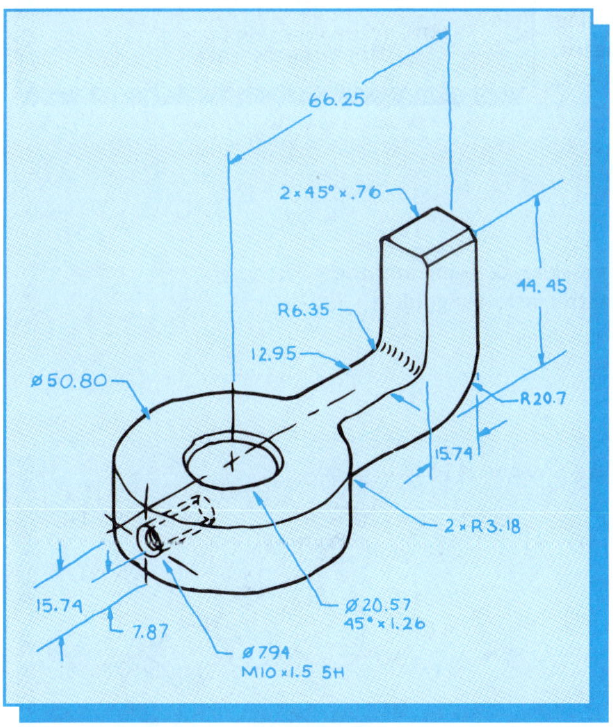

Figure 18-32
Sketch a multiview working drawing. With instruments, complete the scaled drawing for the lathe dog. *(Reprinted from* Mechanical Drafting *by Madsen, Shumaker, and Stewart, ©1986 Delmar Publishers, Inc.)*

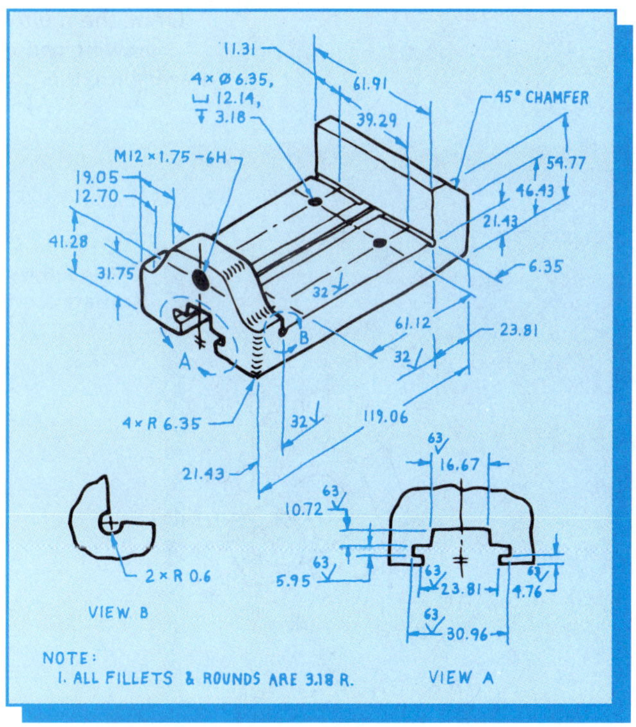

Figure 18-33
Draw the multiview working drawing and the isometric drawing for the dovetail guide.

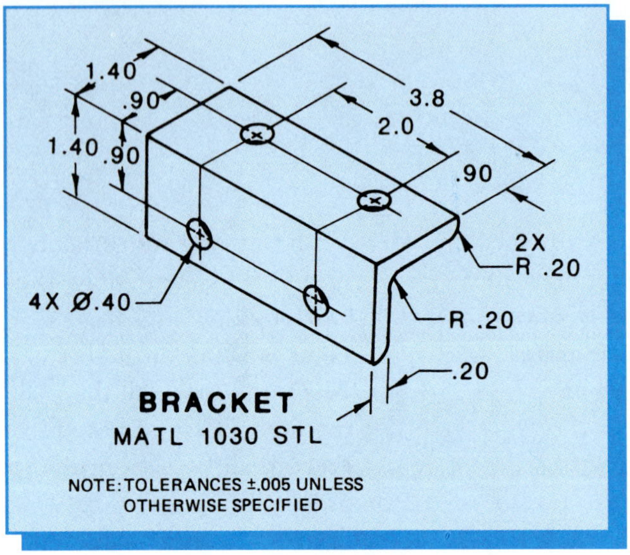

Figure 18-34
Sketch the multiview working drawing. With instruments or CAD, complete the scaled drawing for the vise base. *(Reprinted from* Mechanical Drafting *by Madsen, Shumaker, and Stewart, ©1986 Delmar Publishers, Inc.)*

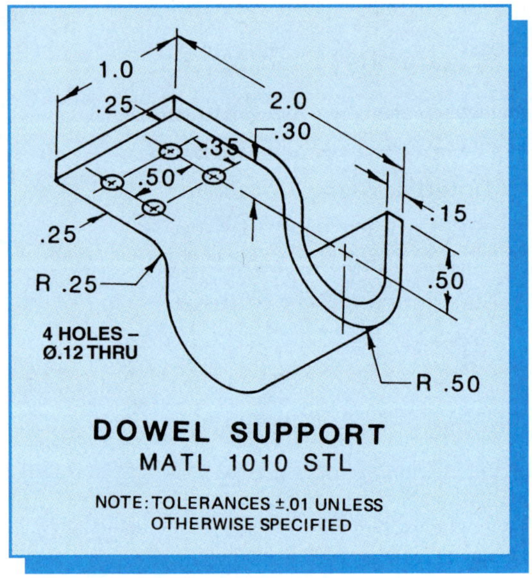

Figure 18-35
Draw the multiview working drawing and the isometric drawing for the dowel support.

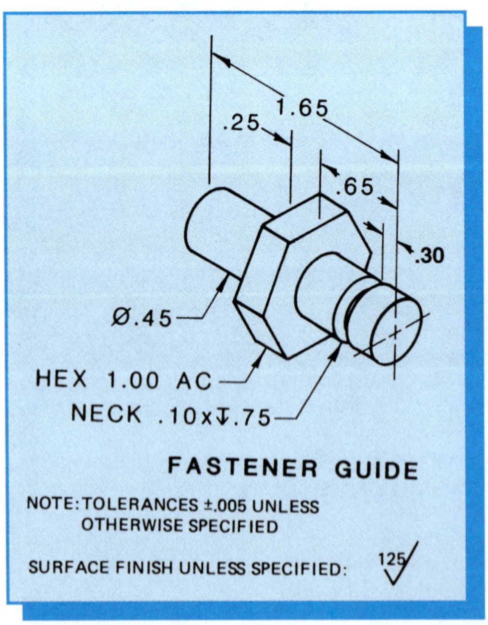

Figure 18-36
Draw the multiview working drawing and the isometric drawing for the fastener guide.

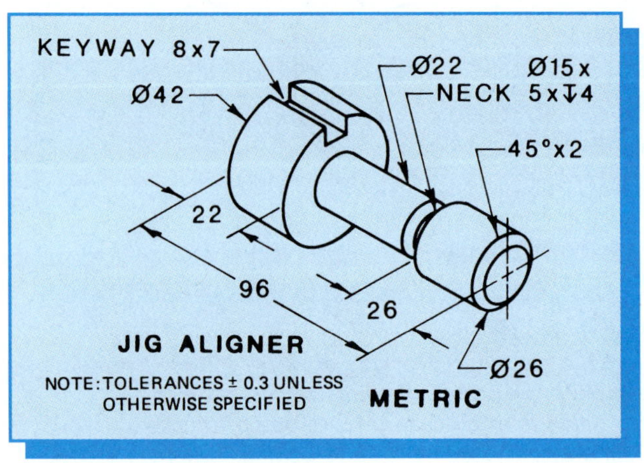

Figure 18-37
Draw the multiview working drawing and the isometric drawing for the jig aligner.

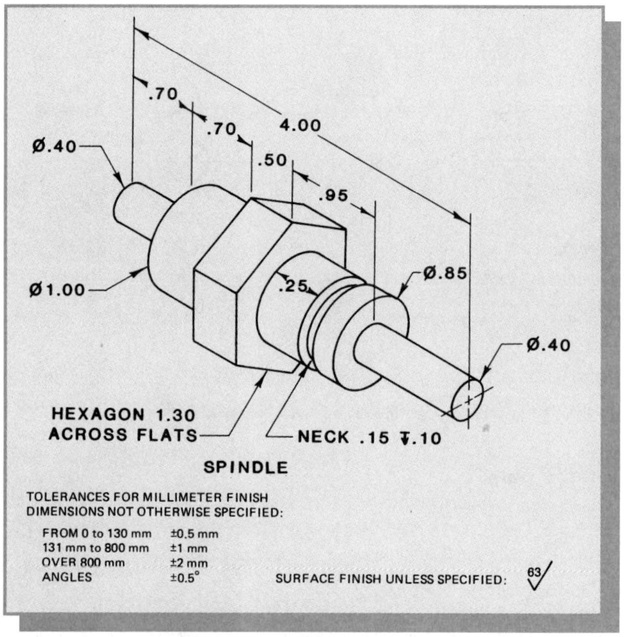

Figure 18-38
Draw the multiview working drawing of the spindle with a CAD system.

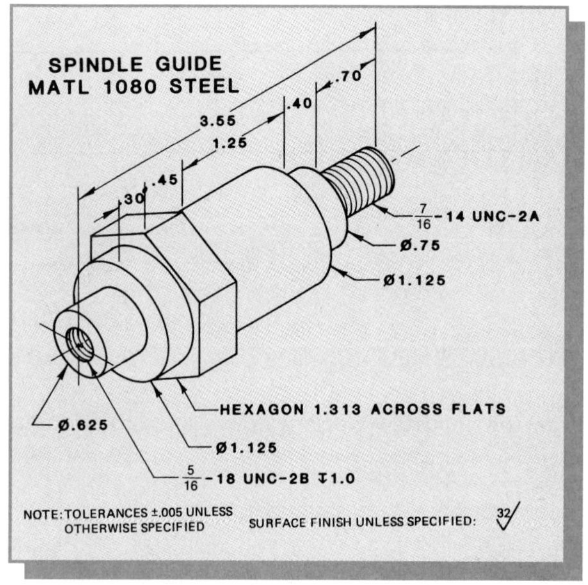

Figure 18-39
Draw the multiview working drawing of the spindle guide with a CAD system.

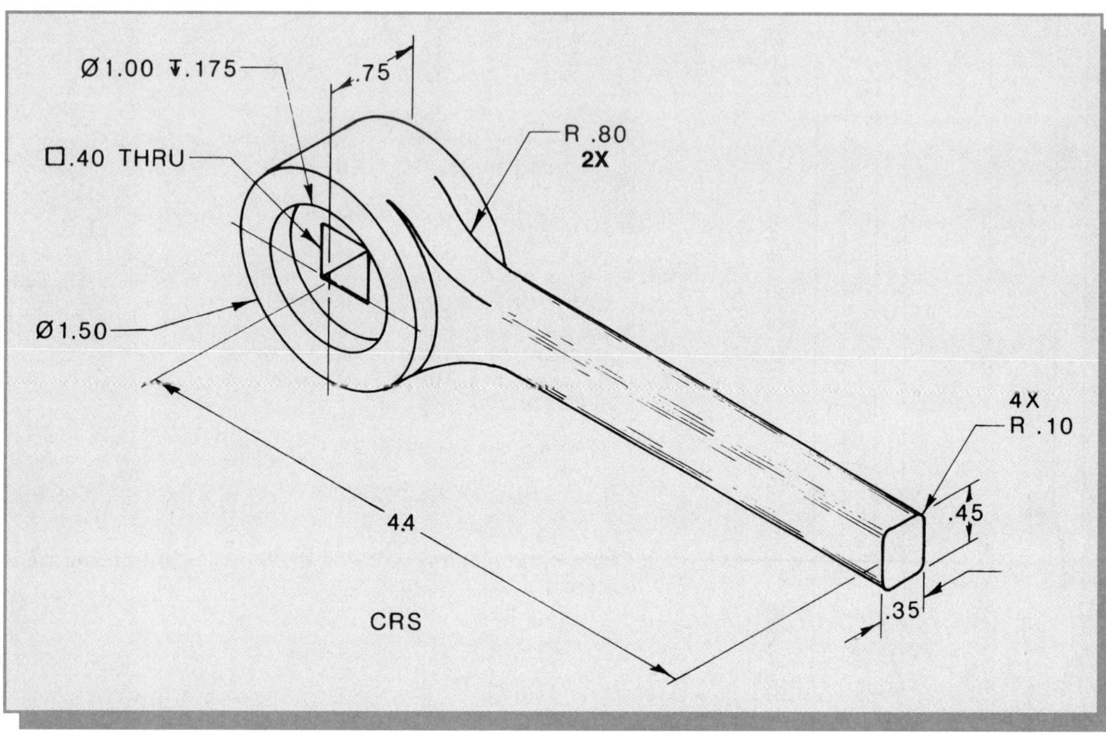

Figure 18-40
Draw the working drawings for the locking handle.

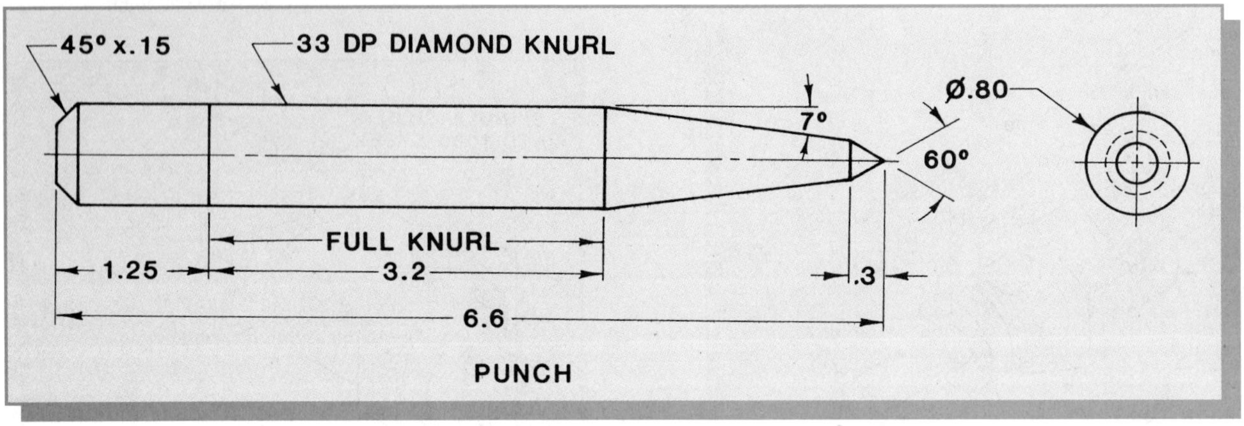

Figure 18-41
Draw the multiview working drawing and the isometric drawing for the punch.

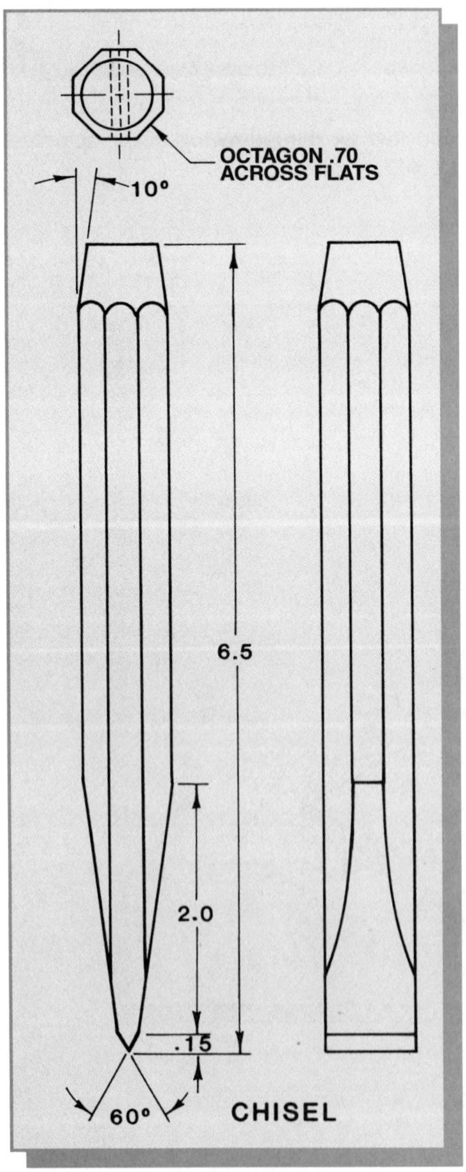

Figure 18-42
Draw the multiview working drawing and the isometric drawing for the chisel.

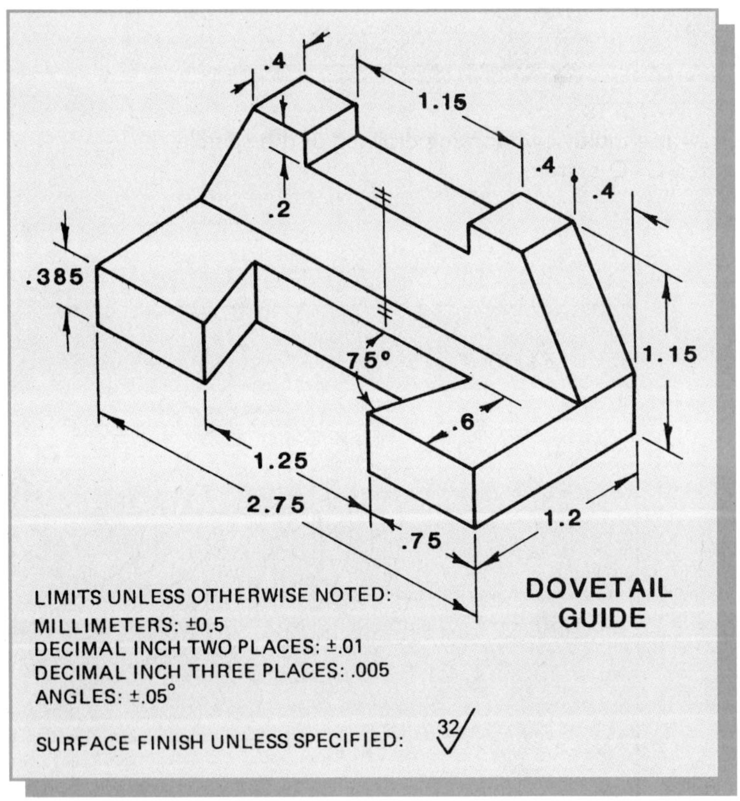

Figure 18-43
Draw the multiview working drawing and the isometric drawing for the dovetail guide.

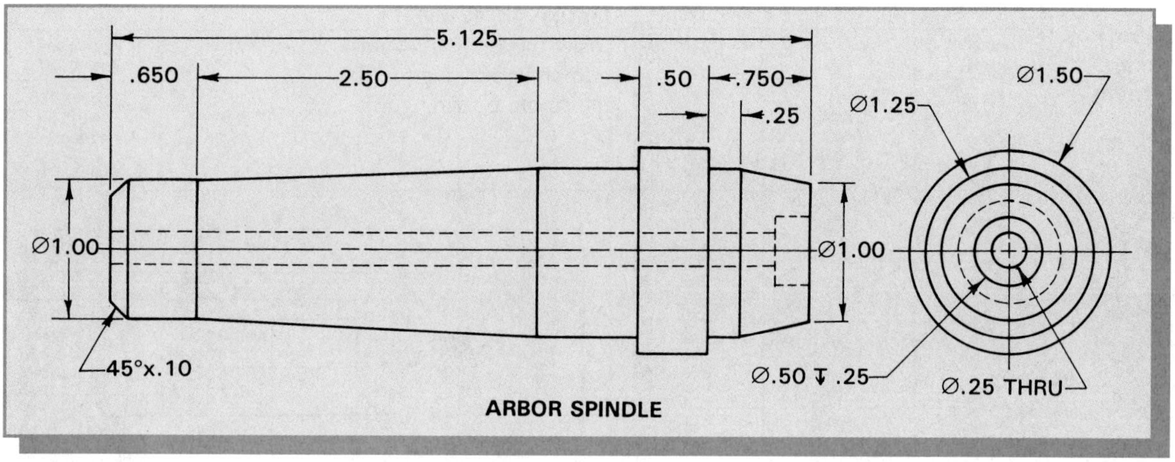

Figure 18-44
Arbor spindle

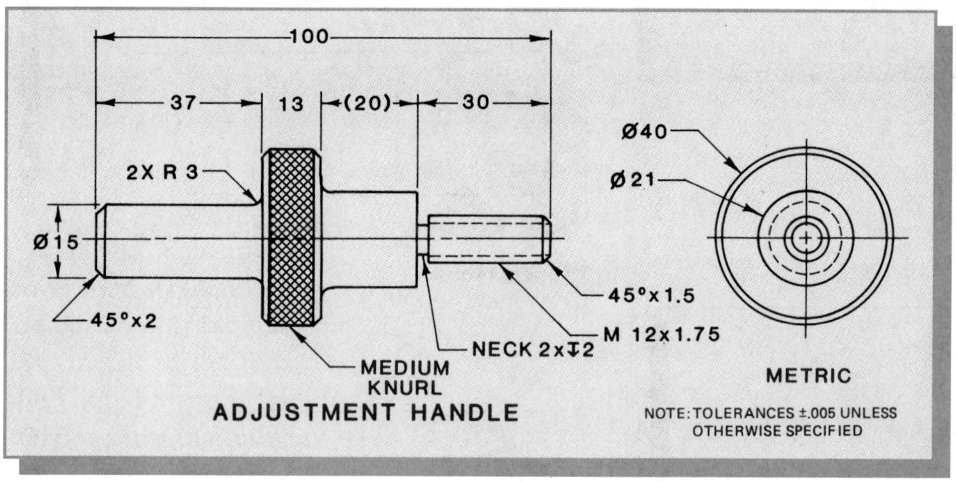

Figure 18-45
Draw the multiview working drawing and the isometric drawing for the adjustment handle.

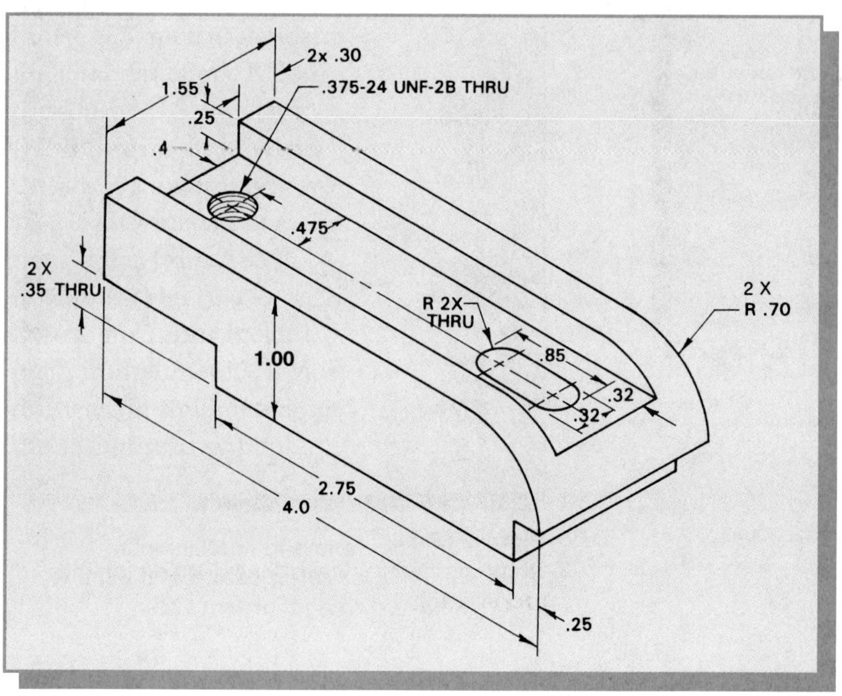

Figure 18-46
Complete the working drawing for the slot guide.

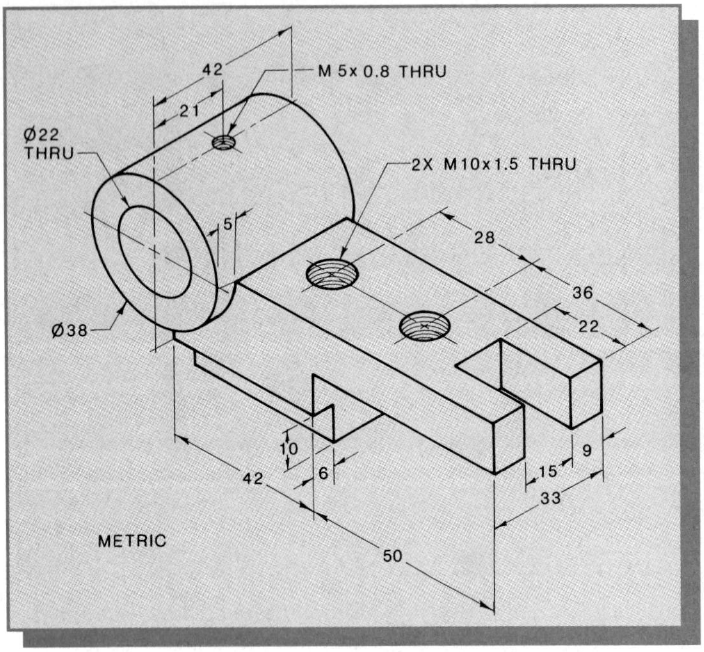

Figure 18-47
Draw the multiview and isometric drawing for the adjustable bearing guide

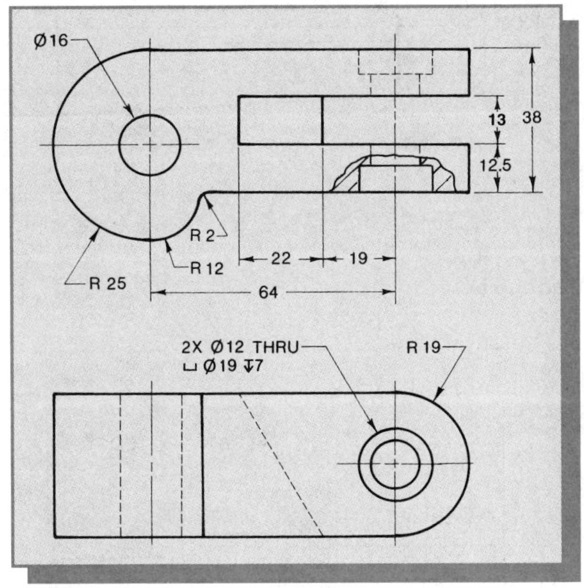

Figure 18-48
Draw multiview with CAD or instruments.

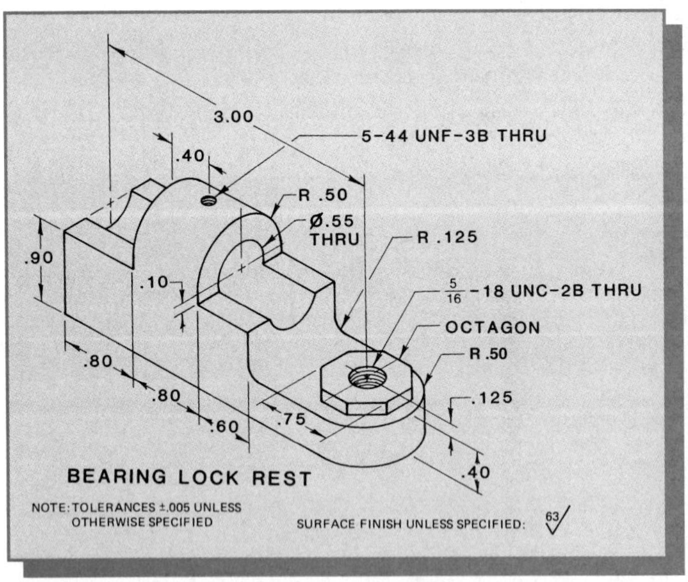

Figure 18-49
Draw the multiview working drawing for the bearing lock rest.

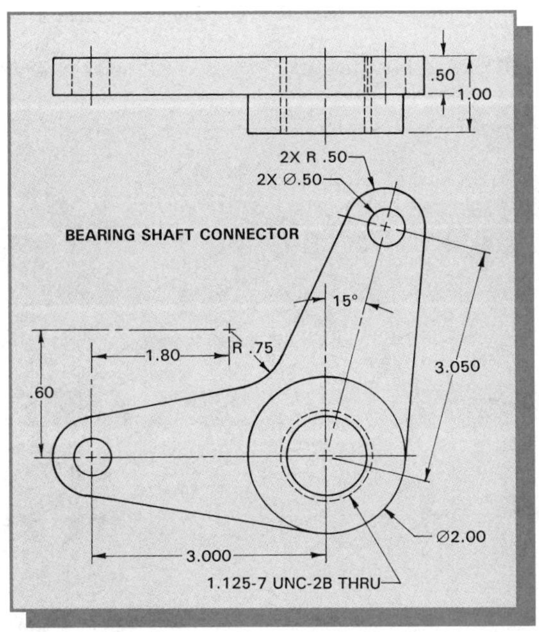

Figure 18-50
Draw the multiview working drawing for the rocker arm.

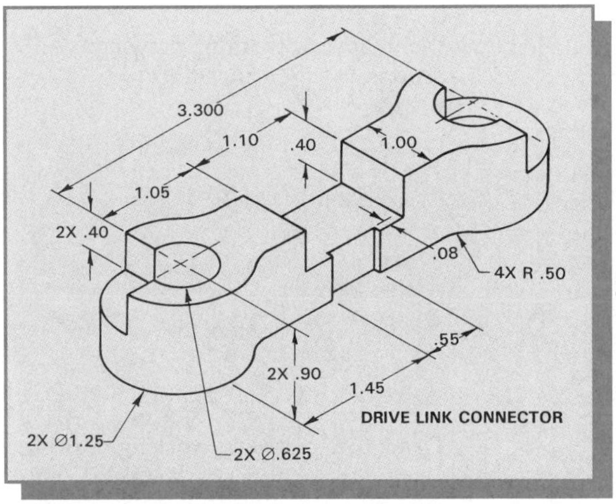

Figure 18-51
Draw the multiview working drawing for the bearing linkage.

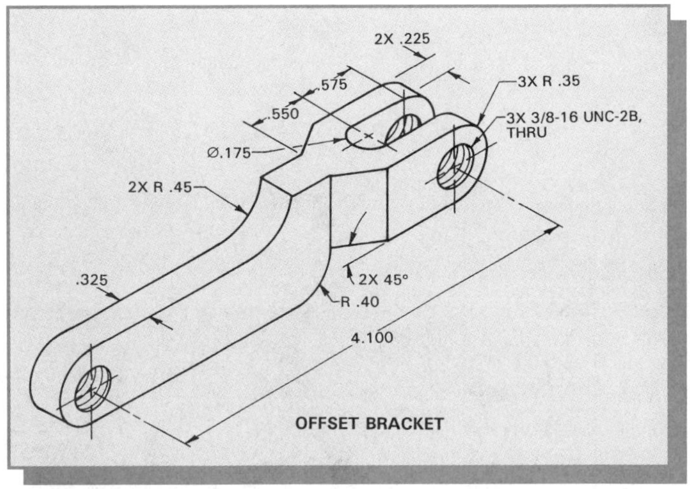

Figure 18-52
Draw the multiview working drawing for the jig alignment arm.

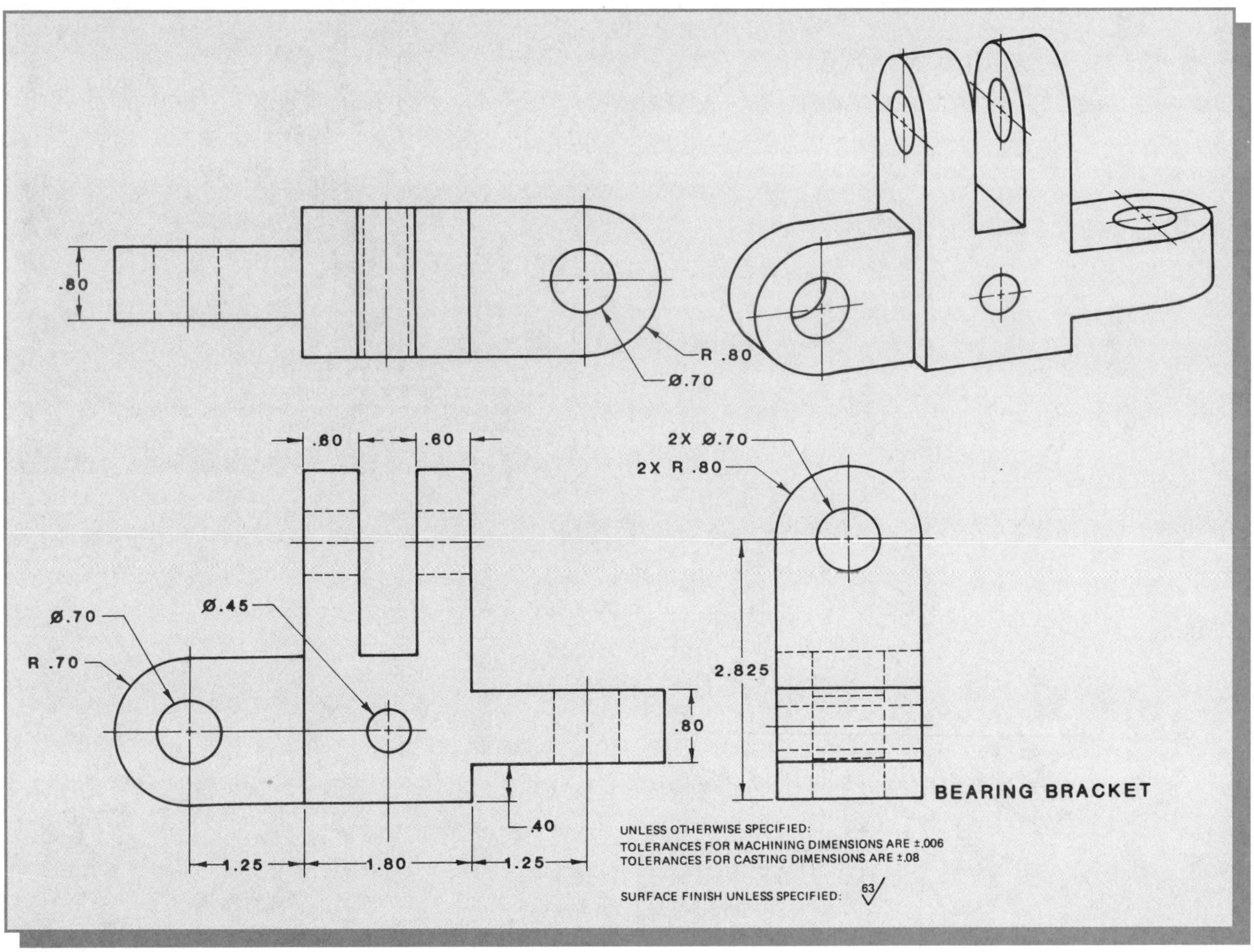

Figure 18-53
Redraw the multiview and isometric working drawings of the bearing bracket with a CAD system.

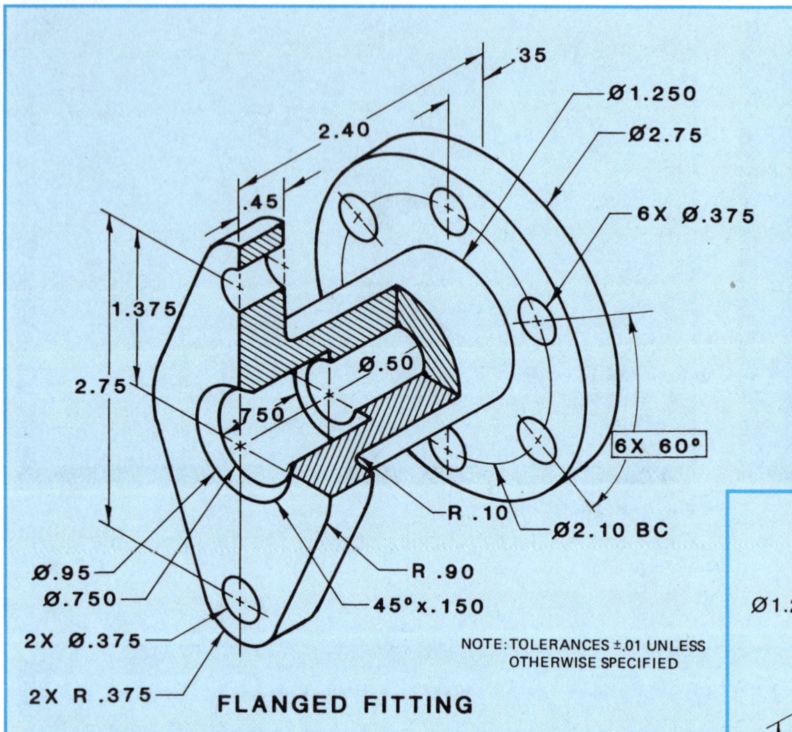

Figure 18-54
Draw the multiview working drawing, with sections, for the flanged fitting.

Figure 18-55
Draw the multiview working drawing, with sections, for the valve shell.

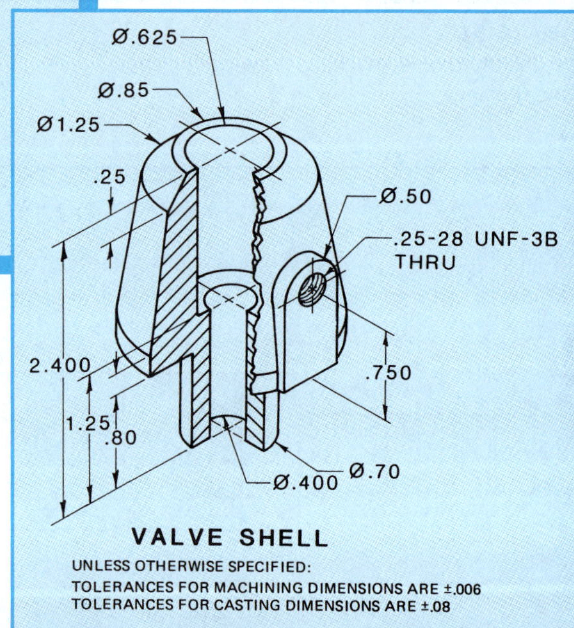

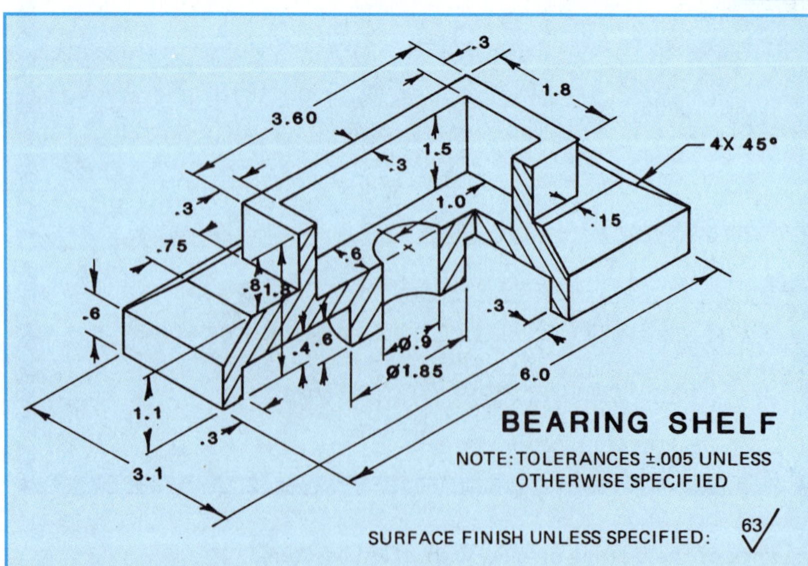

Figure 18-56
Draw the multiview working drawing, with sections, for the bearing shelf.

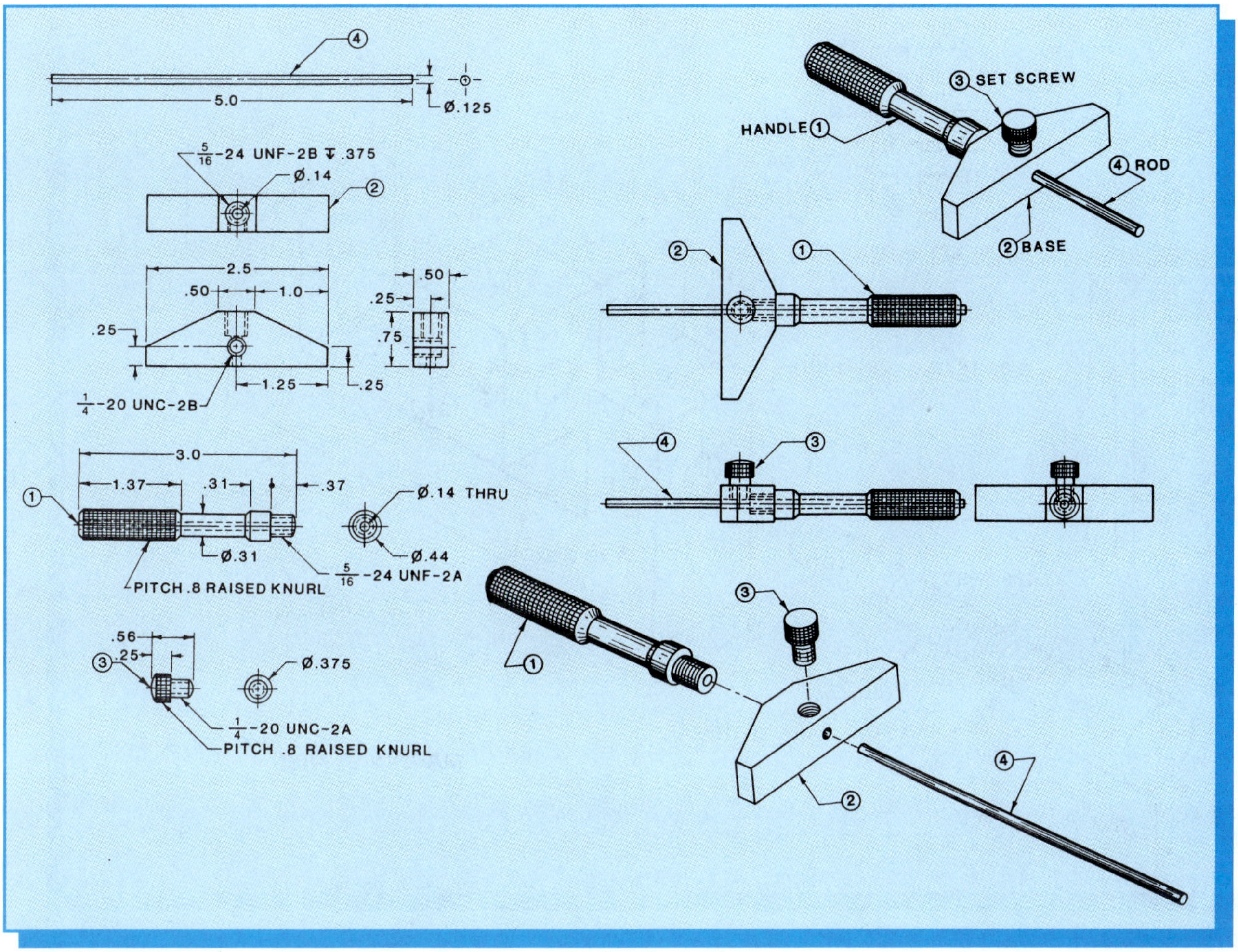

Figure 18-57
For the depth gage draw the multiview, detail, assembly and pictorial drawings.

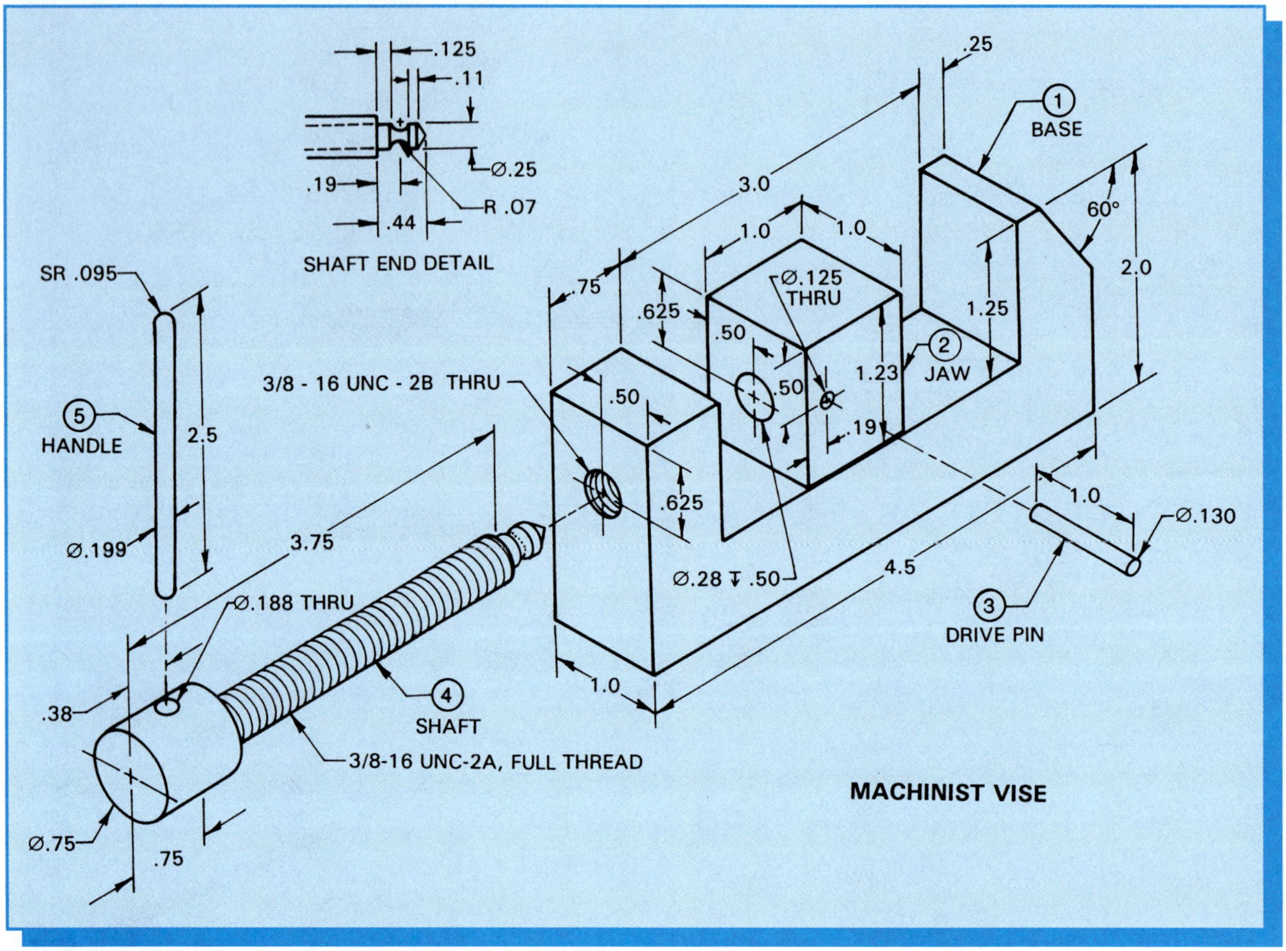

Figure 18-58
Draw a multiview drawing for the clamp vise.

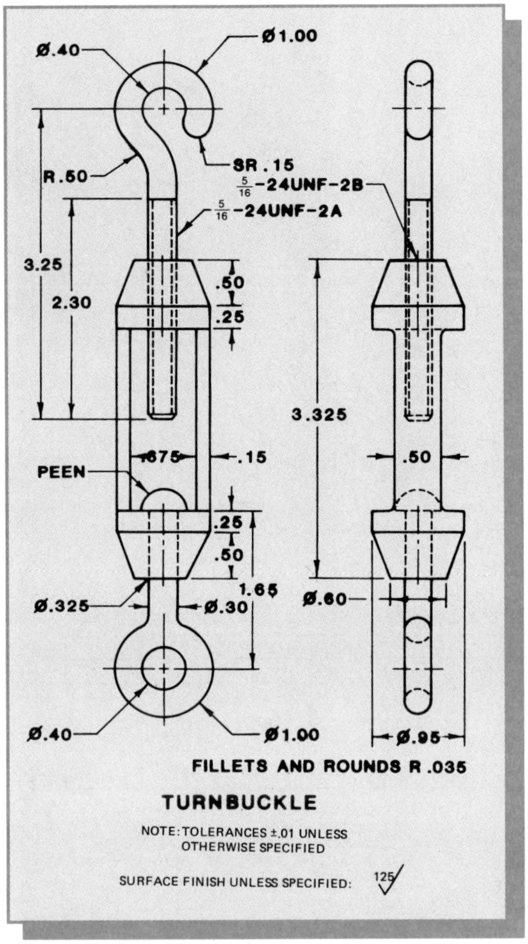

Figure 18-59
Draw the assembly and the detailed working drawings for the turnbuckle.

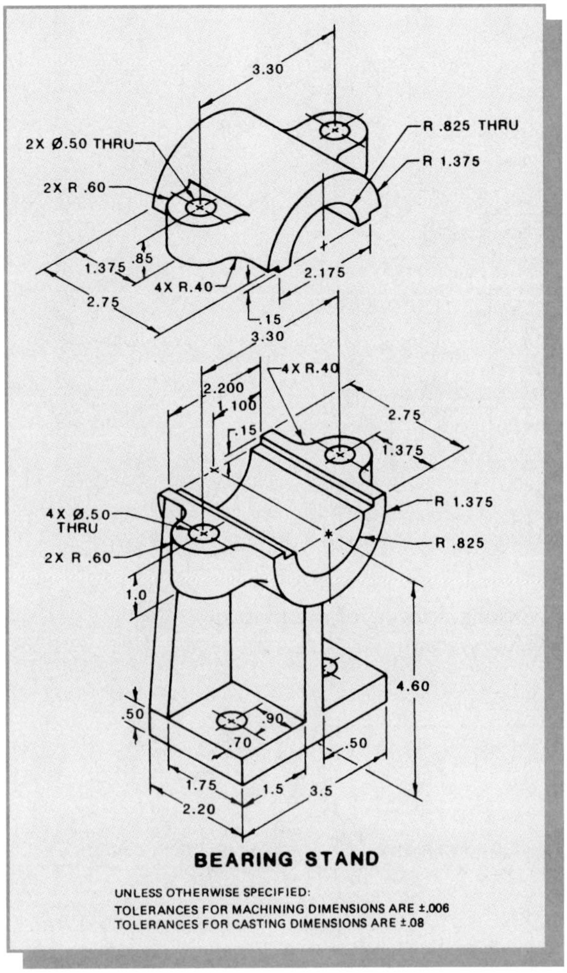

Figure 18-60
Draw the assembly working drawing for the bearing stand. Add bolts and nuts as needed.

Figure 18-61
Draw the assembly and the detailed working drawing for the clamp.

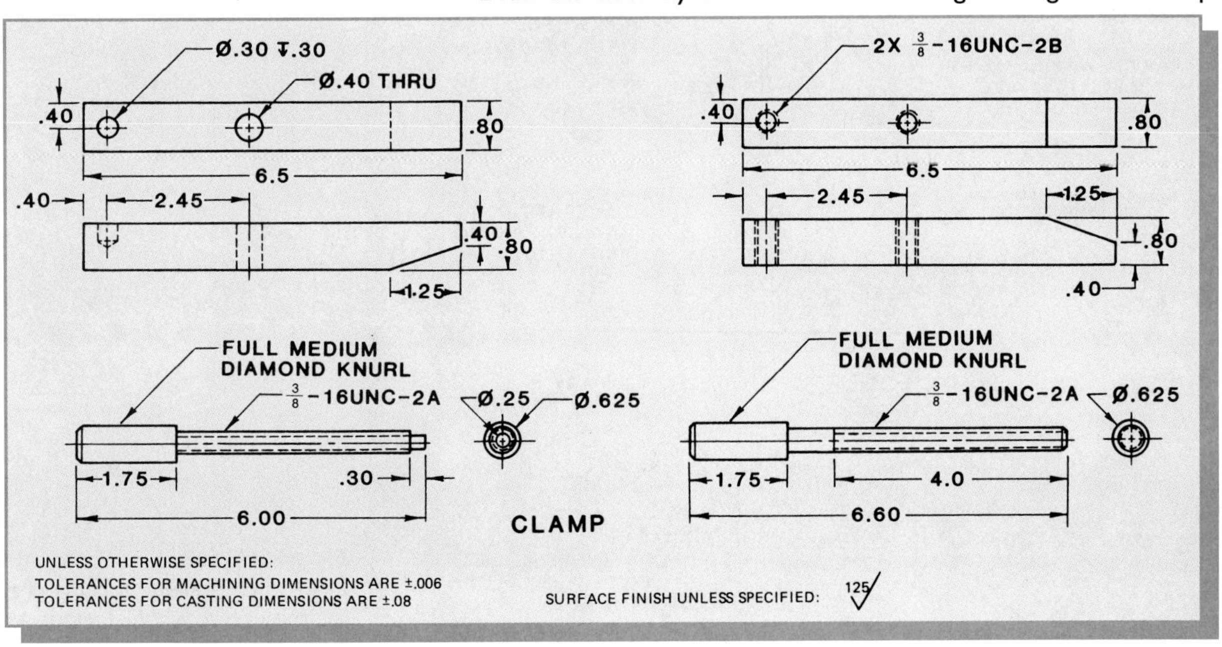

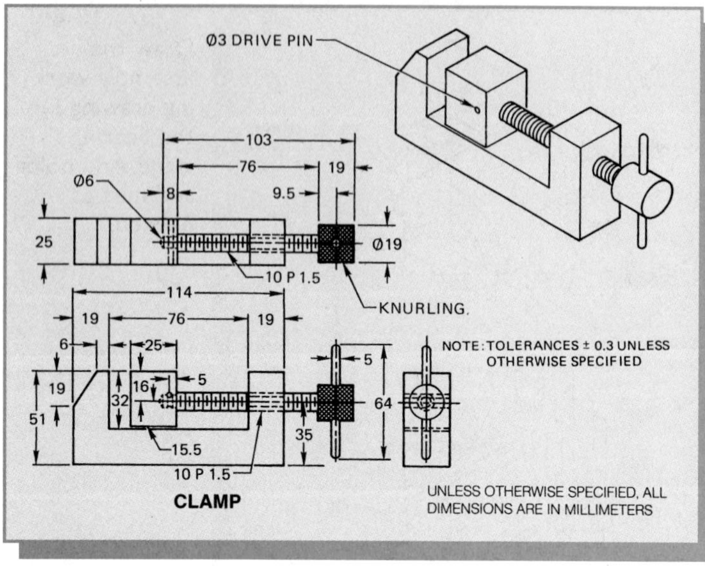

Figure 18-62
Draw the multiview working drawing of the metric clamping vise with a CAD system.

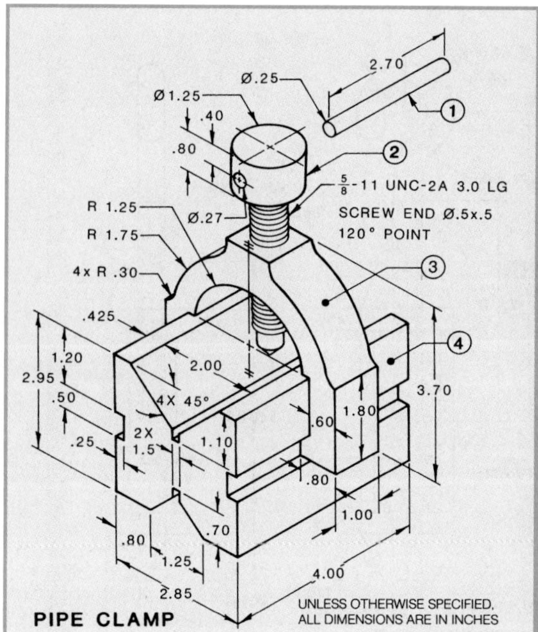

MATERIAL LIST			
PART NO.	PART	MATERIAL	AMT
1	HANDLE	1025 STEEL	1
2	ADJ SCREW	1125 STEEL	1
3	CLAMP	CAST STEEL	1
4	BASE	1025 STEEL	1

LIMITS UNLESS OTHERWISE NOTED:
MILLIMETERS: ±0.5
DECIMAL INCH DIMENSIONS: TWO PLACE ±.01
DECIMAL INCH DIMENSIONS: THREE PLACE ±.005
ANGLES: ±.05°

Figure 18-63
Draw the detailed working drawings for the pipe clamp. Include a materials list.

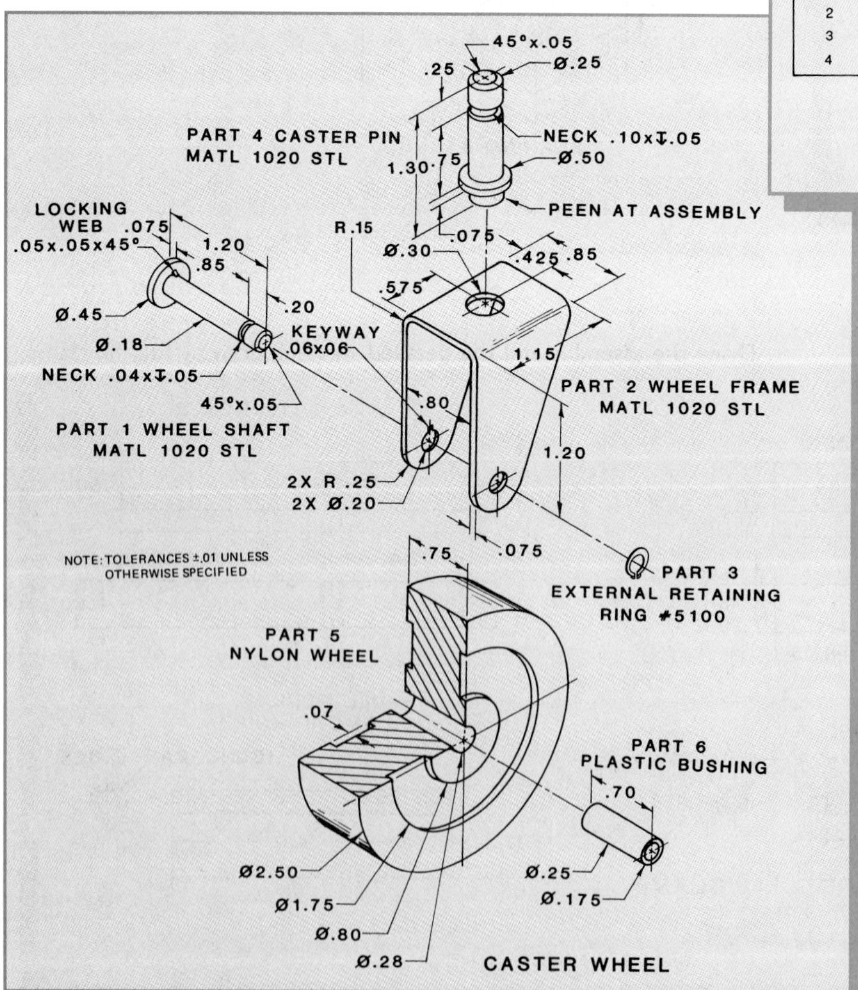

Figure 18-64
Draw the detailed working drawings for the caster wheel. Include a materials list.

Assembly Name: Machine Vice
PARTS LIST:

ITEM	QTY	NAME	DESCRIPTION	MATERIAL
1	2	HANDLE CAP		MS
2	1	HANDLE		MS
3	1	BODY		SAE 4320
4	1	SCREW		SAE 4320
5	1	MOVABLE JAW		SAE 1020
6	1	MOVABLE JAW PLATE		SAE 4320
7	1	FIXED JAW PLATE		SAE 4320
8	1	GUIDE		SAE 1020
9	2	MACHINE SCREW	.25-20UNC-2 × .500 SLOT FIL HD	STL
10	2	MACHINE SCREW	.190-32UNF-2 × .875 SLOT FIL HD	STL
11	1	SET SCREW	.25-20UNC-2 × .250 FULL DOG POINT	STL
12	2	MACHINE SCREW	.190-32UNF-2 × 6 SLOT FIL HD	STL

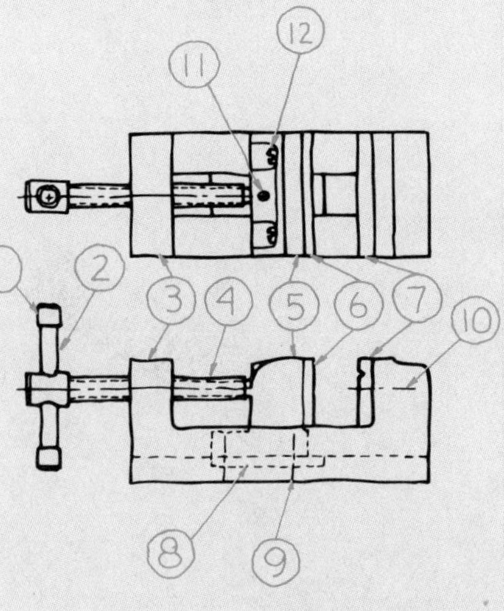

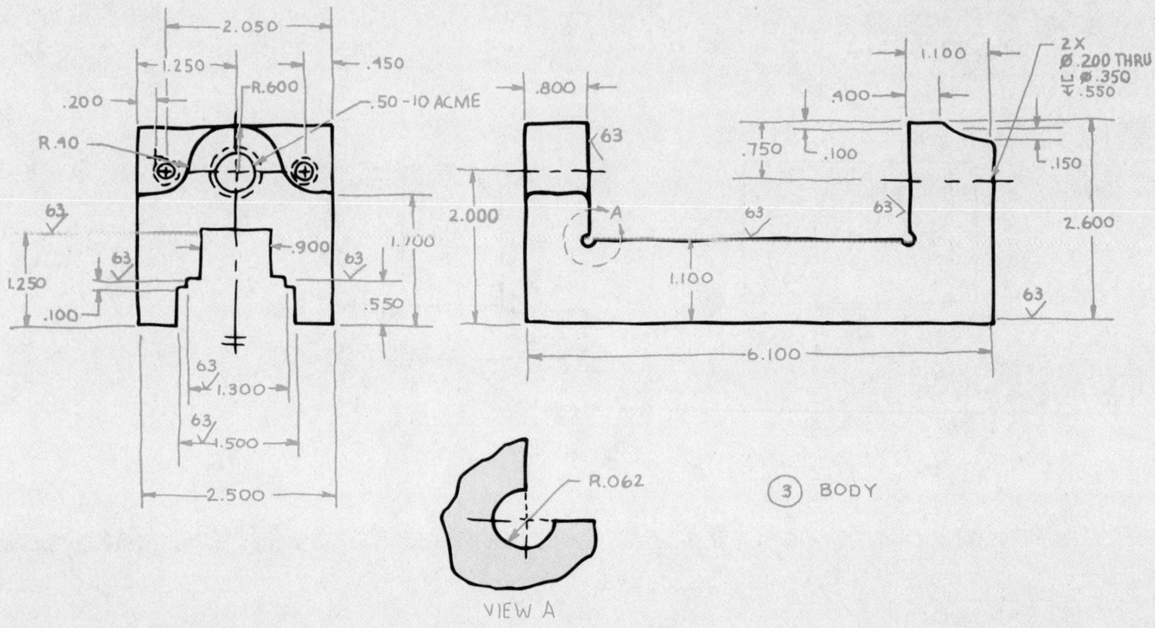

Figure 18-65
Draw the full set of working drawings for the machine vise (2-part drawing). *(Reprinted from* Mechanical Drafting *by Madsen, Shumaker, and Stewart, © 1986 Delmar Publishers, Inc.)*

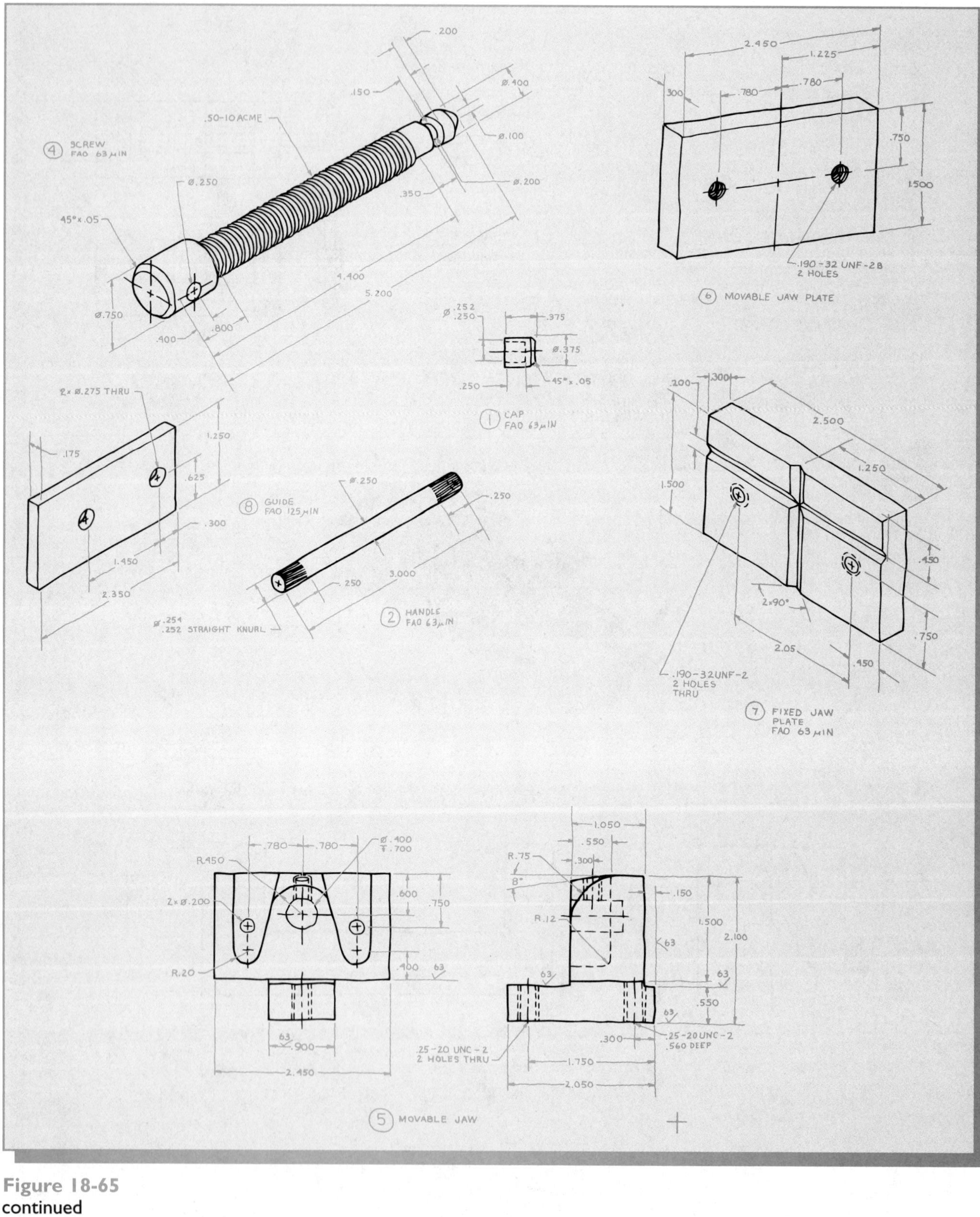

Figure 18-65 continued

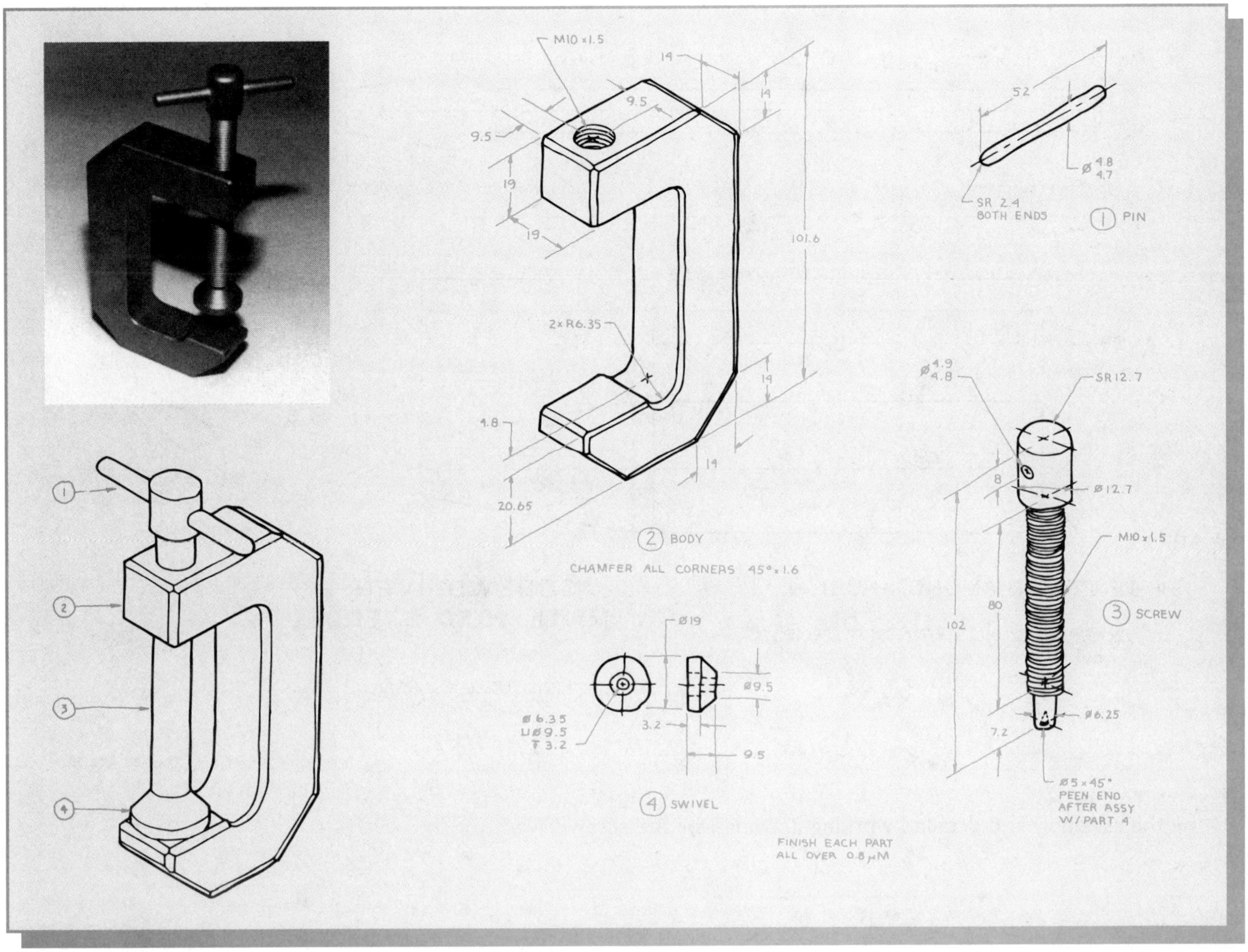

Figure 18-66
Sketch the detailed working drawings for the C-clamp. With instruments, complete the scaled drawings. *(Reprinted from* Mechanical Drafting *by Madsen, Shumaker, and Stewart, ©1986 Delmar Publishers, Inc.)*

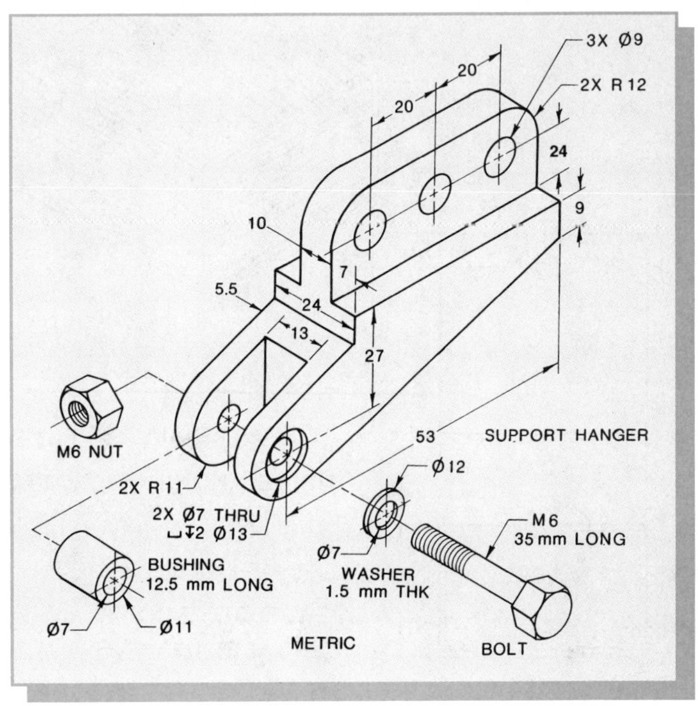

Figure 18-67
Draw the working drawings (assembly and detail) for the metric support hanger.

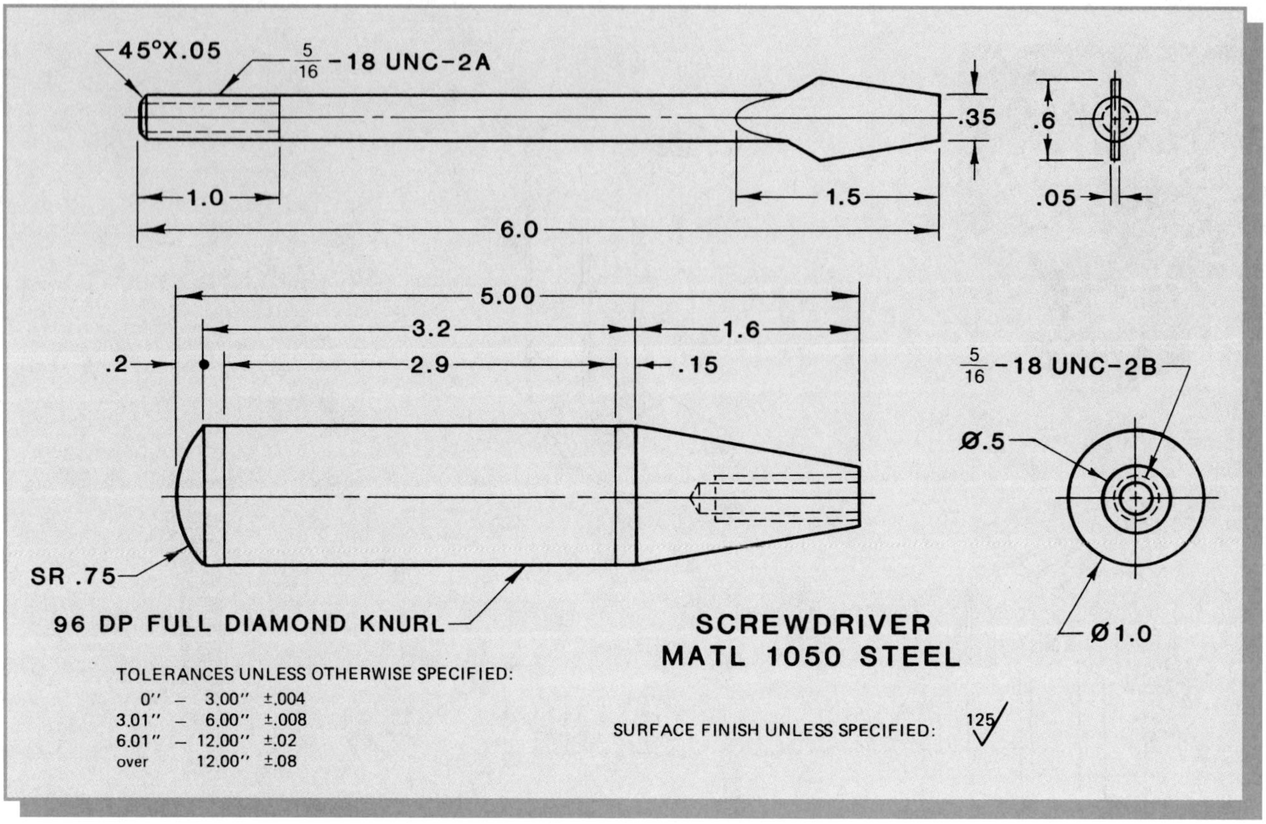

Figure 18-68
Draw the assembly and detailed working drawings for the screwdriver.

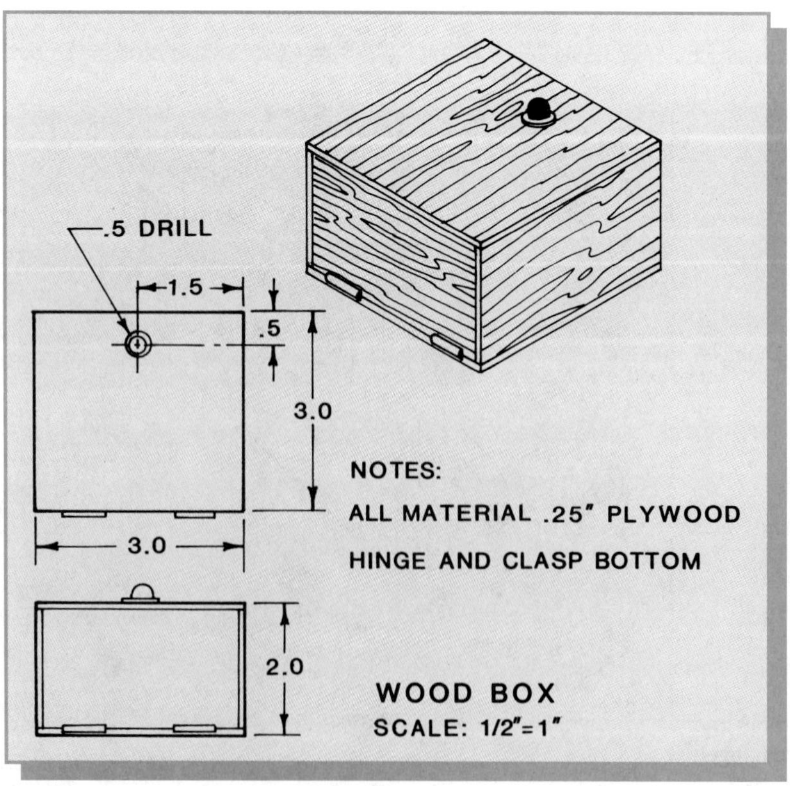

Figure 18-69
Draw the working drawings for the wood box container for a red flasher light, and make the following ECOs:

1. Change the material to 3/8" plywood.
2. Change the height to 1.75".
3. Move the hole center for light from .5" to .65".
4. Specify the hinges and clasp.
5. Specify the assembly procedures.

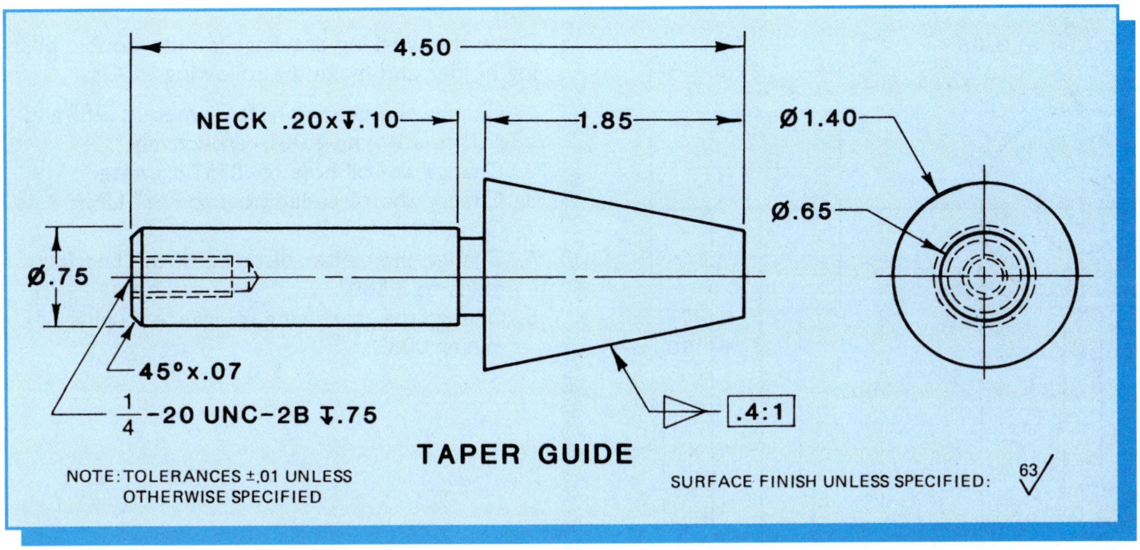

Figure 18-70

Draw the working drawings for the taper guide, and make the following ECOs:

1. Change the taper length to 2.00".
2. Change the large diameter of taper to 1.50".
3. Change the small diameter of taper to .50".
4. Calculate a new taper: TAPER = $\frac{\text{LARGER DIAM.} - \text{SMALLER DIAM.}}{\text{LENGTH}}$

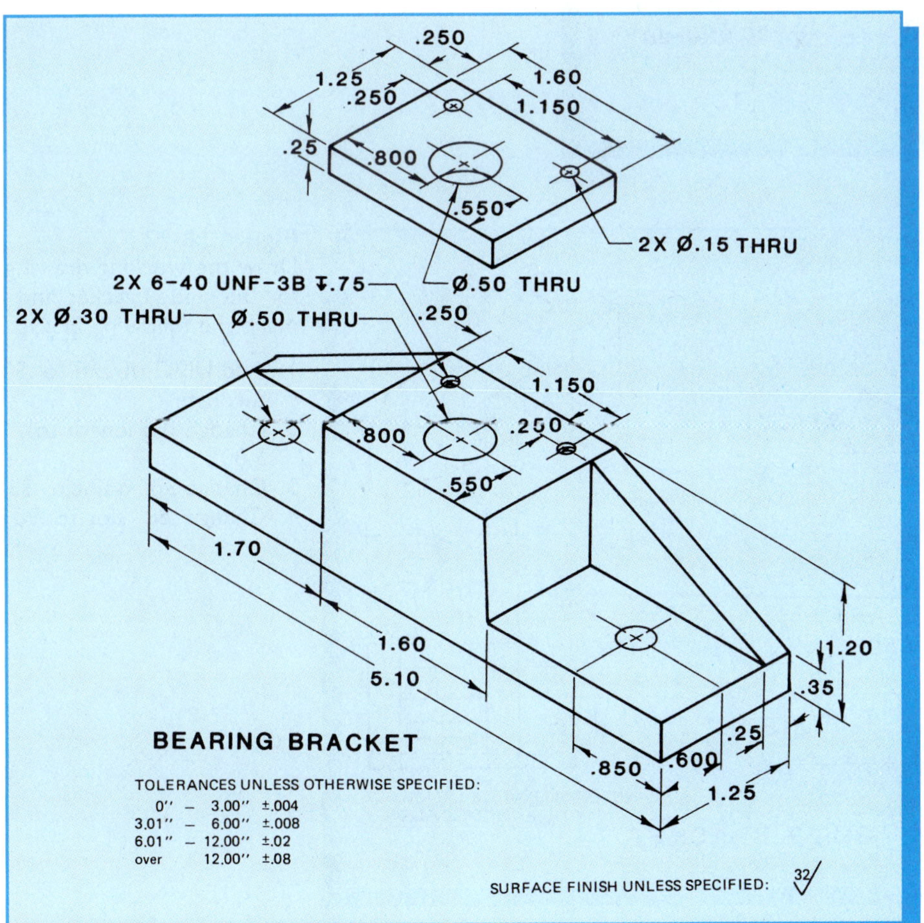

Figure 18-71

Draw the working drawings for the bearing bracket, and make the following ECOs:

1. Add .05" radius fillets and rounds to the support webs.
2. Add a boss to both .30" diameter holes I/M .05" high, .50" diameter.
3. Change cap's thickness to .50".
4. Add .05" cork gasket seal.

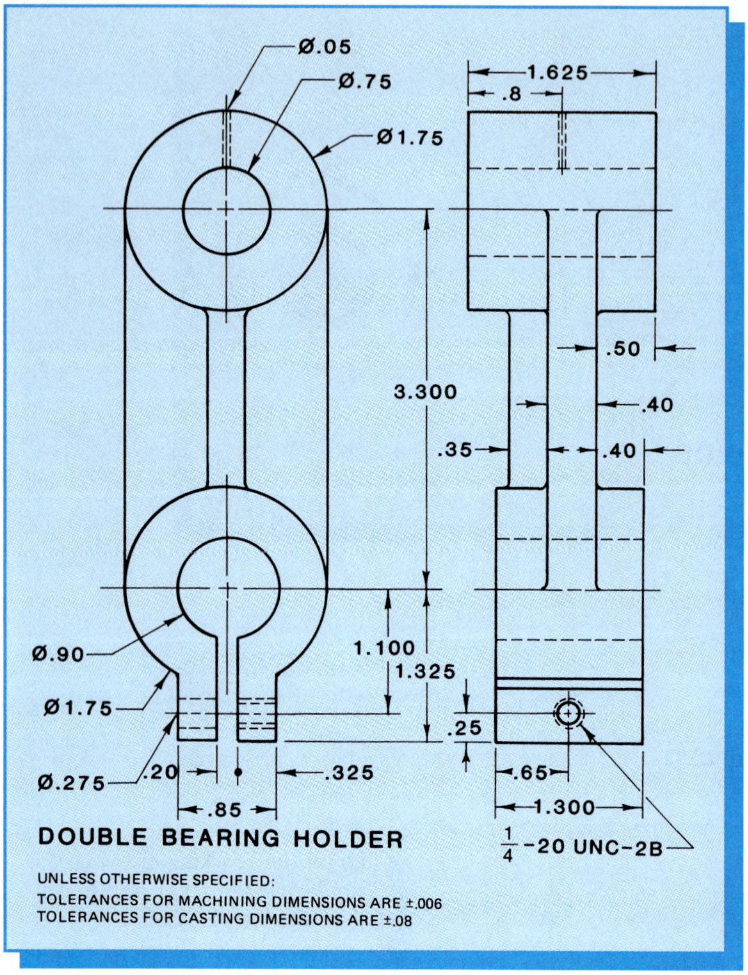

Figure 18-72
Draw the working drawings for the double bearing holder, and make the following ECOs:

1. Change the bearing hole diameters .75" and .90" to .80" and 1.00", respectively.
2. Change the oil hole to .075" diameter.
3. Change the threaded hole to 3/16" UNF, Class 3 fit.
4. Change the center distance of the bearing holes to 3.500".
5. Change the machining tolerance to plus or minus .008".

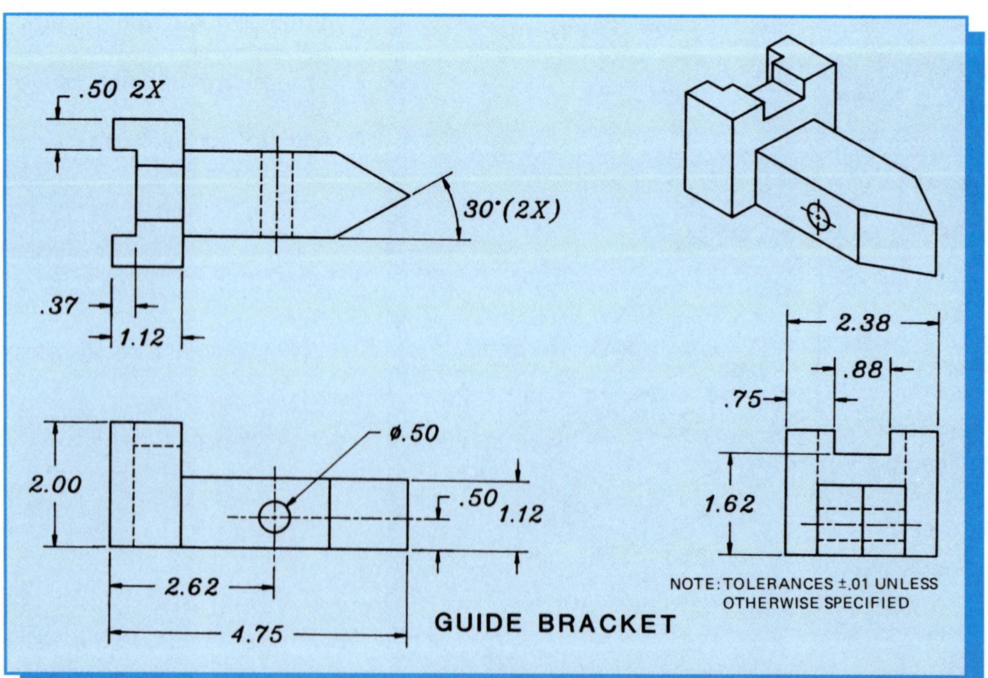

Figure 18-73
Draw the working drawings for the guide bracket, and make the following ECOs:

1. Add UNC thread to .50" through.
2. Change the length to 5.00".
3. Change 30° point to 35°.
4. Change .88" slot to .90".

Design Exercises

1. Redesign the items in **Figures 18-74** through **18-77** so they will perform their tasks more efficiently.
2. Design a more efficient type of kitchen cabinet door clasp.
3. Design an esthetic drawer pull for kitchen cabinet drawers.
4. Design a childproof latch for bathroom medicine cabinets.
5. Design an esthetic hub cap for an automobile. Use standard dimensions.
6. Design a furniture module to hold a CAD system's hardware.
7. Design a removable clasp that may be used to lift heavy 2' x 2' x 3' boxes of books.
8. Redesign the caster wheel in **Figure 18-64**.

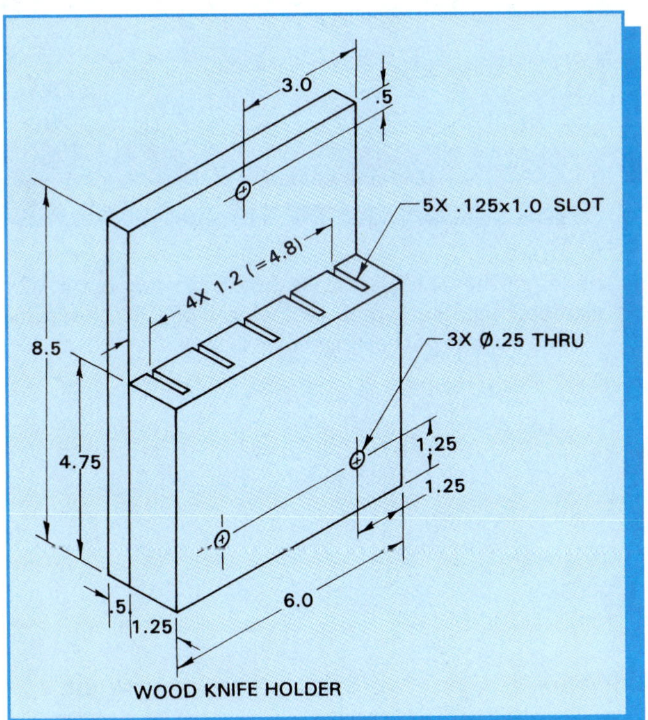

Figure 18-74
Redesign the knife holder to hold three more large knives (maximum blade width of 1.75" and a sharpening steel of .5" diameter).

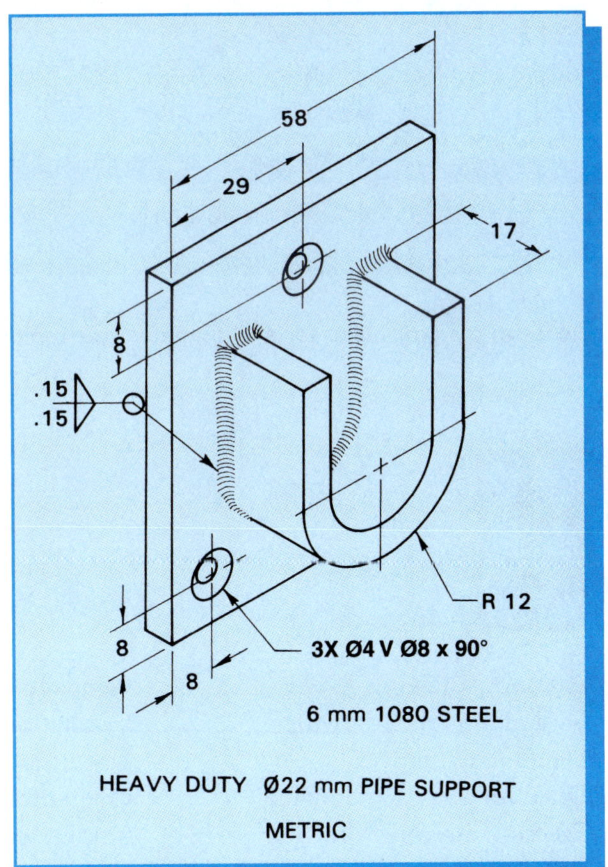

Figure 18-75
Design a locking mechanism for the steel pipe's installation.

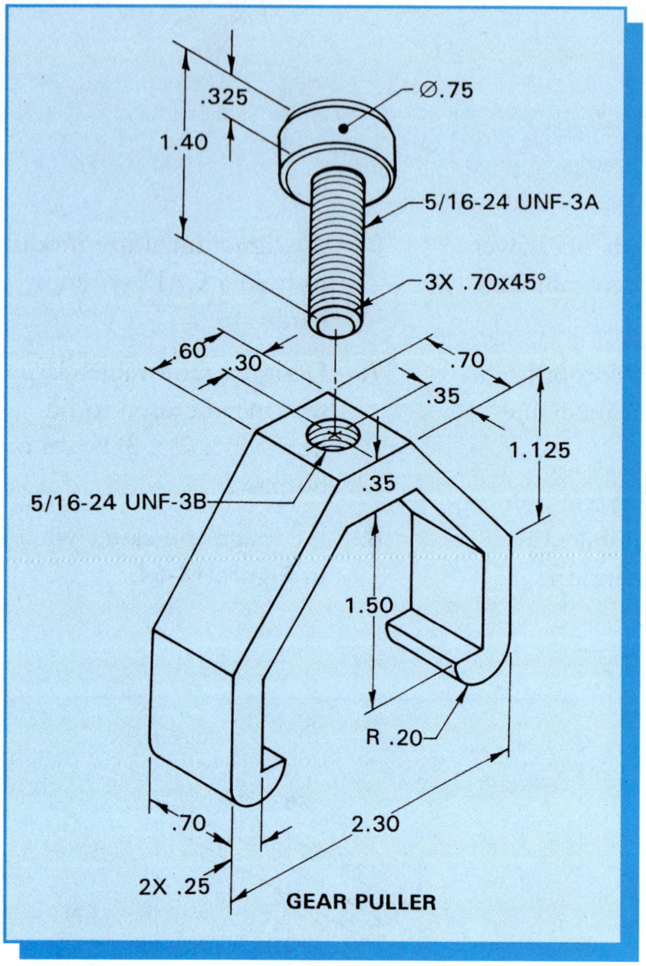

Figure 18-76
Redesign the gear puller for a 4" diameter gear removal.

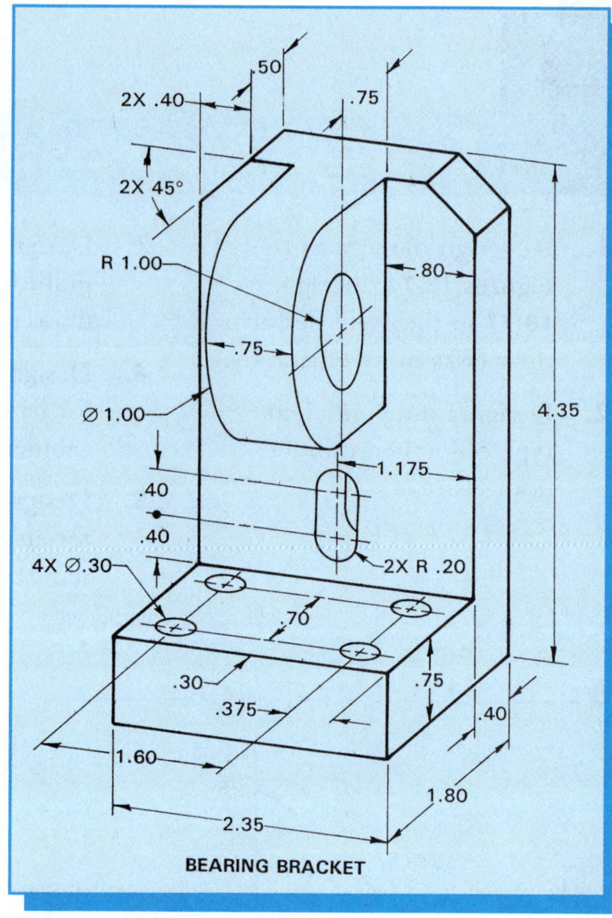

Figure 18-77
Redesign the bearing bracket holder to support a 1.25" diameter bearing with a total weight of 1,500 lbs. (double the weight of the original design).

Key Terms

Assembly drawing
Design assembly drawing
Detail drawing
Diagram assembly drawing
Erection assembly drawing
General assembly drawing
Installation assembly drawing
Layout assembly drawing
Operation assembly drawing
Outline assembly drawing
Pictorial assembly drawing
Subassembly drawing
Working assembly drawing
Working drawing

Chapter 19

Welding Drawings

Objectives

The student will be able to:

- read welding abbreviations
- select the proper type of weld
- read welding symbols
- draw a weld reference symbol
- select the proper weld finish
- specify contour symbols
- specify the size and strength of a weld
- specify the length of a weld
- specify the pitch of a weld
- draw a weld symbol with a CAD system

Introduction

When designing products with multiple parts, designers have several options for joining parts together, including mechanical devices, adhesive bonding, and welding. The joining method depends on the material to be joined, the strength of the joint required, and the need for disassembly. Mechanical devices include nails, screws, bolts, rivets, keys, and splines. Adhesive bonding methods range from cement and epoxy glues to plastic bonding solutions. Welding is a method of permanently joining metals and plastics through the application of heat. Welding is usually chosen when great strength, rigidity, permanence, or liquid-proofing is required.

Welding began over 4,000 years ago after the discovery of iron, but the science and technology of welding are relatively new. Early welders simply heated two pieces of iron in a forge, dipped them into a flux, and then pounded the pieces together with a hammer. Later, oxygen and acetylene gas were added to increase the amount of heat produced. This enabled harder metals to be welded. Nevertheless, welding had been restricted to iron and iron alloys until the twentieth century. Today, hundreds of metallurgical processes are used to weld thousands of different metals and alloys. Vehicles, aircraft, ships, and machinery contain thousands of welds of different types and categories. These modern welds are usually stronger, lighter, more durable, and more economical than other methods of assembly **(Fig. 19-1)**.

Welding Processes

The specific metallurgic combination of heat, energy, and materials used to create a weld between two metals is called a **welding process.** There are hundreds of processes and combinations of processes now available. **Figure 19-2** shows some of the most commonly used processes organized under the major categories of welding. The abbreviations shown next to each process are used to specify the welding process on working drawings. All welding processes join metals together using either solid-state welding, fusion welding, brazing, or soldering.

Figure 19-1
Welding is a major procedure used to bond metals and plastics. *(Courtesy of Arvin Industries Inc.)*

Figure 19-2
Categories of welding processes *(Courtesy of The American Welding and Manufacturing Company)*

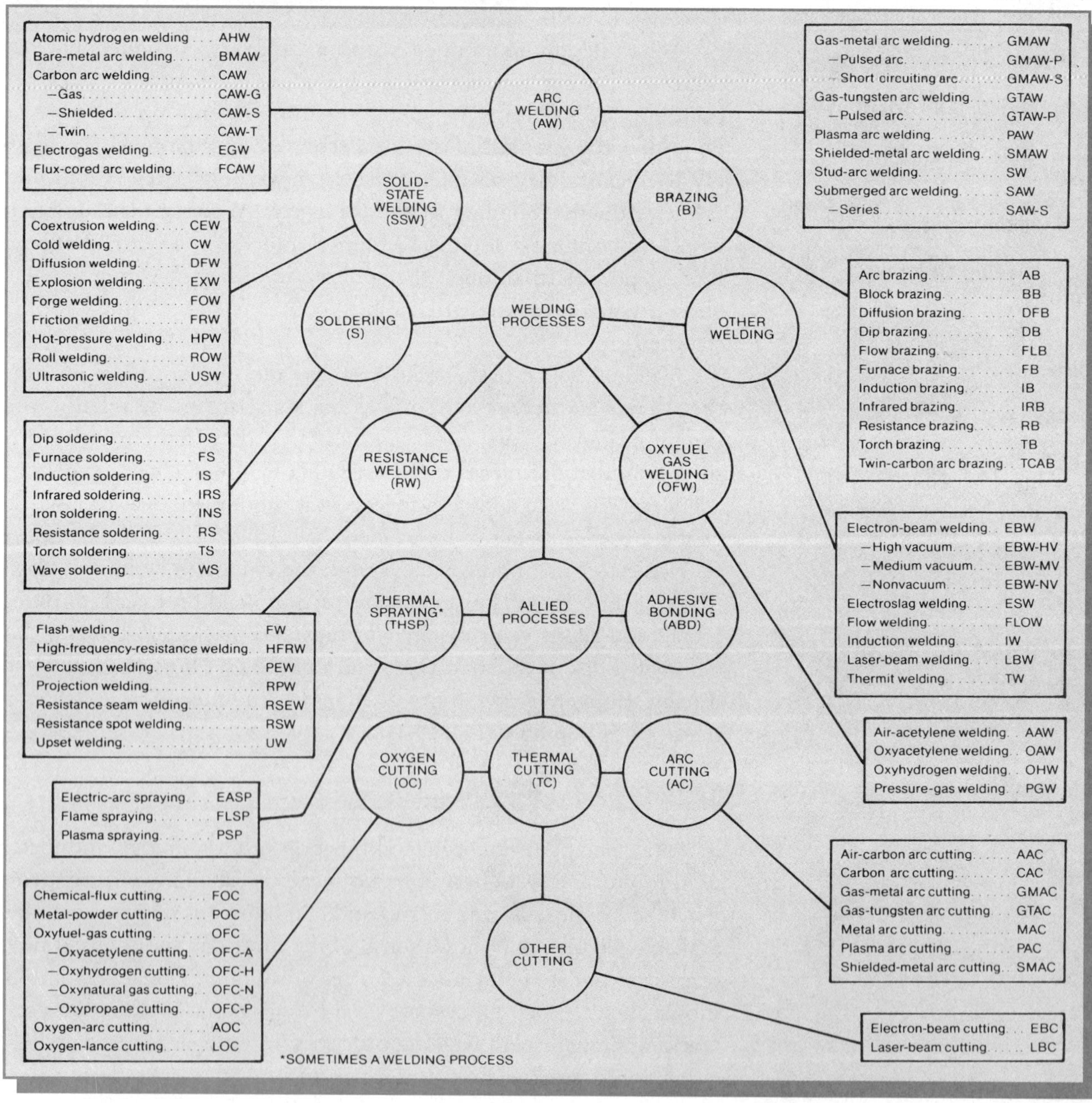

Solid-State Welding

Solid-state welding involves heating two workpieces and impacting them together until the hot material on the adjoining surfaces becomes mixed. Solid-state welding processes include forge welding, friction welding, explosive welding, and roll welding. Thin, soft, malleable metals such as wrought iron, copper, and brass are well suited to solid-state welding processes.

Fusion Welding

Fusion welding involves the common edge of two metal parts being melted to a molten state. The molten parts, combined with a filler material from a welding rod, form a molten puddle. When the puddle cools and solidifies, the separate pieces are permanently fused as one. The most common fusion welding processes include **arc welding, gas welding**, and **resistance welding.**

Arc Welding. This process uses an electric arc struck between a workpiece and an electrode. The extreme heat from the arc fuses the metals together with material from the consumable electrode. Metals well suited to arc welding include copper, brass, soft iron, aluminum, low-carbon steel, cast steel, stainless steel, and nickel alloys.

Gas Welding. The combination of gas (usually acetylene or hydrogen) with air (or oxygen) is known as **oxyacetylene welding.** Oxyacetylene welding can produce temperatures that will weld most metals, except very high carbon steel and chromium.

Resistance Welding. Both heat and pressure are used in resistance welding. A concentrated electrical current is passed through adjoining metals. The resistance to the electrical charge melts the metal while pressure is applied to the areas to be joined. When the metal cools and the pressure is removed, the separate pieces are permanently joined together. When the electrical charge is concentrated on a very small area, the type of welding is **spot welding.** When the charge is moved along the edge of a joint, the type of welding is **seam welding.** Resistance welding is widely used to spot-or seam-weld sheet metal parts together.

Brazing and Soldering

These processes use a filler rod that melts at a lower temperature, leaving the workpiece metals solid. If the filler rod melts at less than 840°F (450°C), the process is called **soldering.** If the filler rod melts at a temperature higher than 840°F, but lower than the melting temperature of the base metal, the process is called **brazing.** When the molten filler rod cools and hardens, it adheres to both parts, forming a metallurgical bond. Since brazing or soldering temperatures are not hot enough to melt and mix the two workpiece metals, the joints are not as strong as those created by fusion or solid-state welding processes. Therefore, brazed or soldered joints are not used in areas of severe impact, high heat, shear, torsion, or tension stress.

Welding Symbols

When welders read working drawings, they need to know which welding process should be used. They also need to know the size and type of each weld, the location, the type of joint, and the finish to be used. It is impossible or impractical to draw, dimension, or label all of this information on each weld location on complex working drawings. **Figure 19-3** shows a drawing of a weld joint fully dimensioned at **A**. A weld symbol containing the same information is shown at **B**. Notice how much less space the symbol information requires.

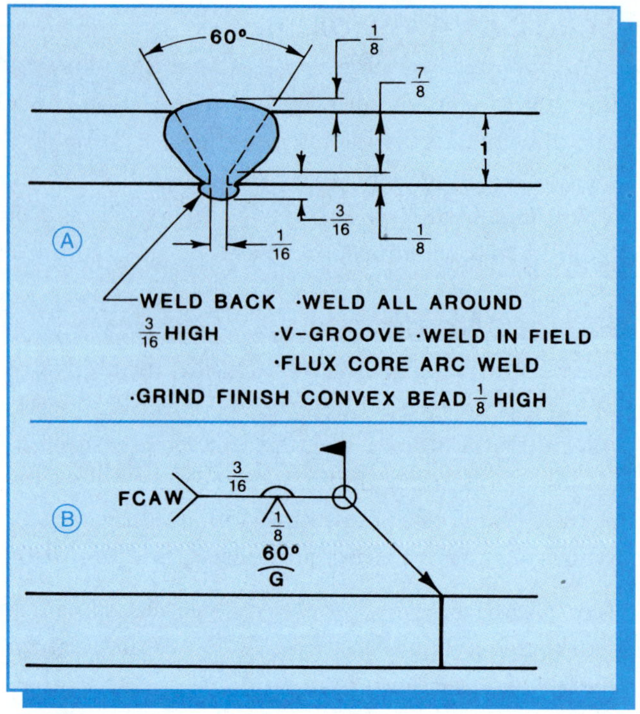

Figure 19-3
All the weld data in drawing A is included in the weld symbol in drawing B.

Anatomy of a Weld Symbol

Because of limited dimensioning space on most drawings, the actual appearance or dimensions of a weld are not drawn. The parts are drawn as they appear after welding without the weld bead or removed metal shown. For example, the material to be welded is shown at A in **Figure 19-4** as being butted together. Examples at B and C are incorrect. A weld symbol **(Fig. 19-5)** substitutes for the drawing of the actual joint or weld. A weld symbol is a combination of graphics, text, and numerical data. The following is a list of the common weld symbols.

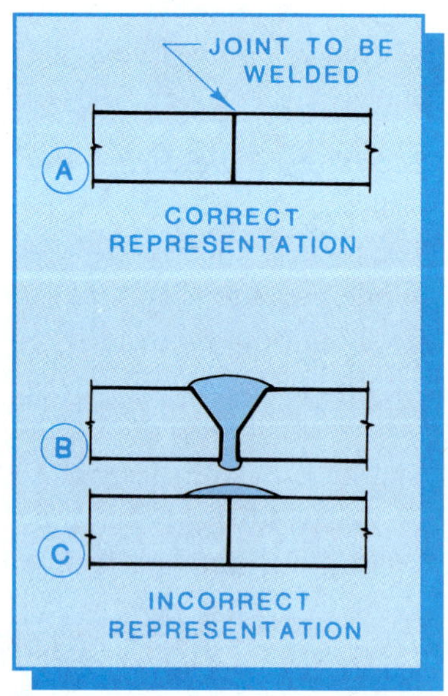

Figure 19-4
The actual welded area is not shown on the working drawing.

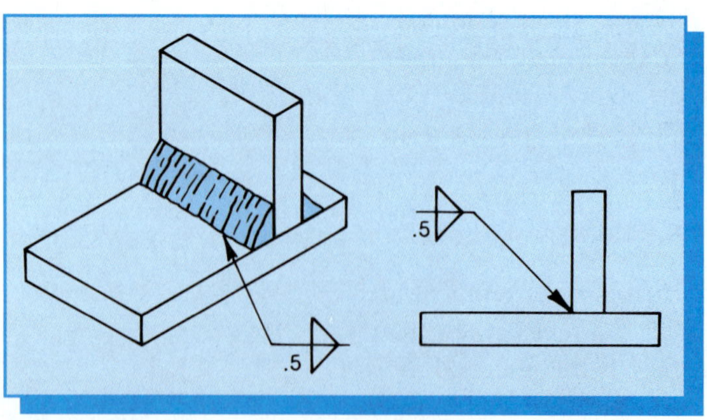

Figure 19-5
The configuration of the weld is not shown on the working drawing.

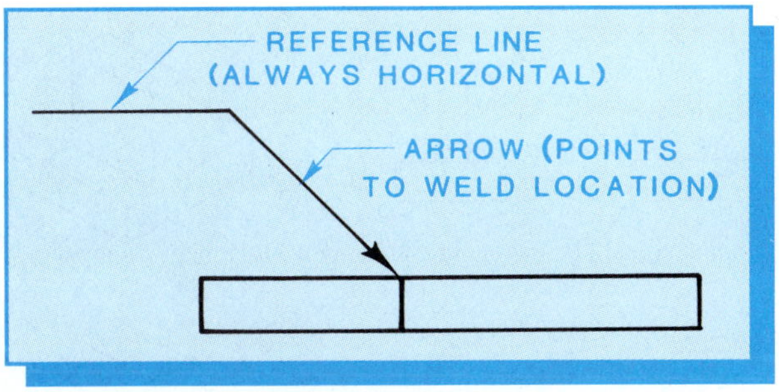

Figure 19-6
Welding reference line and arrow

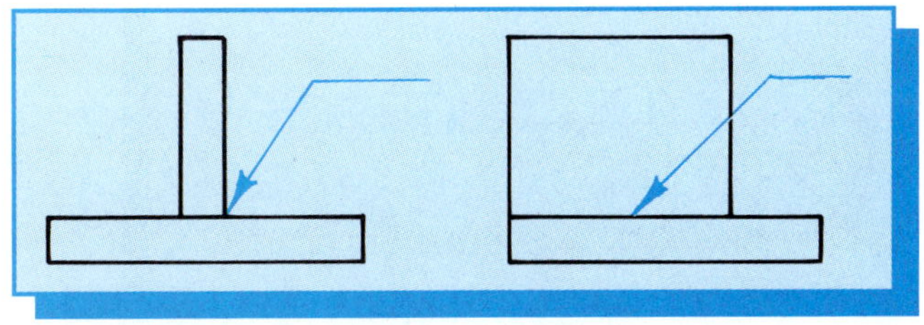

Figure 19-7
The arrow of the weld reference symbol touches the weld joint.

Reference Line and Arrow. A welding reference line is a horizontal line upon which specifications for each weld are placed. An **arrow line** always extends at an angle from the reference line to the workpiece **(Fig. 19-6)**. The arrowhead at the end of the reference line touches the location of the weld **(Fig. 19-7)**. Variations of how a weld reference symbol may be used are shown in **Figure 19-8**.

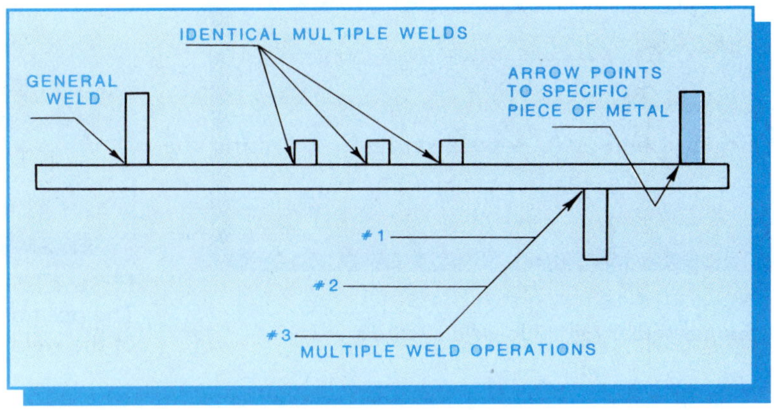

Figure 19-8
Variations of the weld reference symbol

Weld Type Graphic Symbols. Weld type graphic symbols refer to the shape and configuration of the weld. A graphic symbol resembling the shape of the weld is used on, above, or under the reference line **(Fig. 19-9)**. **Figure 19-10** shows supplementary graphic symbols that are placed on, above, or below the reference line to convey additional information about the weld. These symbols will be covered in detail later in this chapter.

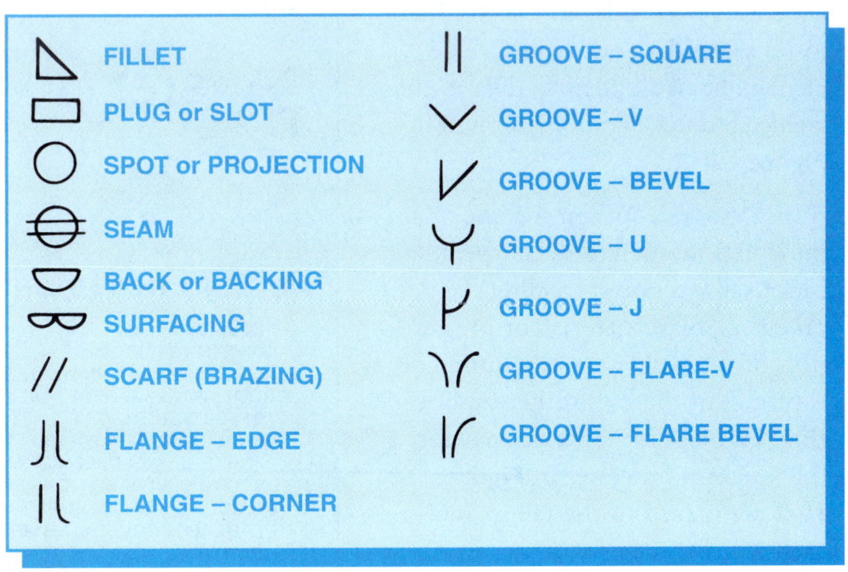

Figure 19-9
Graphic symbols for drafting

The position of the graphic symbol specifies the location of the weld **(Fig. 19-11)**. When the symbol is placed on the bottom side of the reference line, the joint is to be welded exactly where the arrowhead is placed. When the symbol is placed on the top side of the reference line, the weld is to be made on the opposite (other) side of the joint **(Fig. 19-11)** as

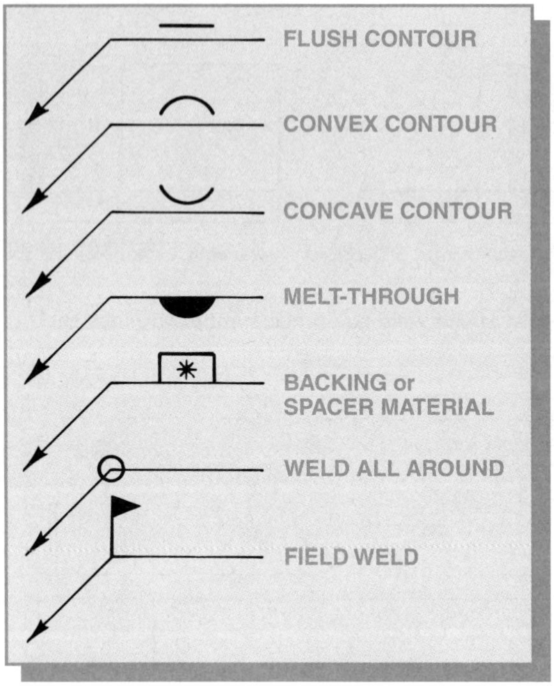

Figure 19-10
Supplementary graphic weld symbols

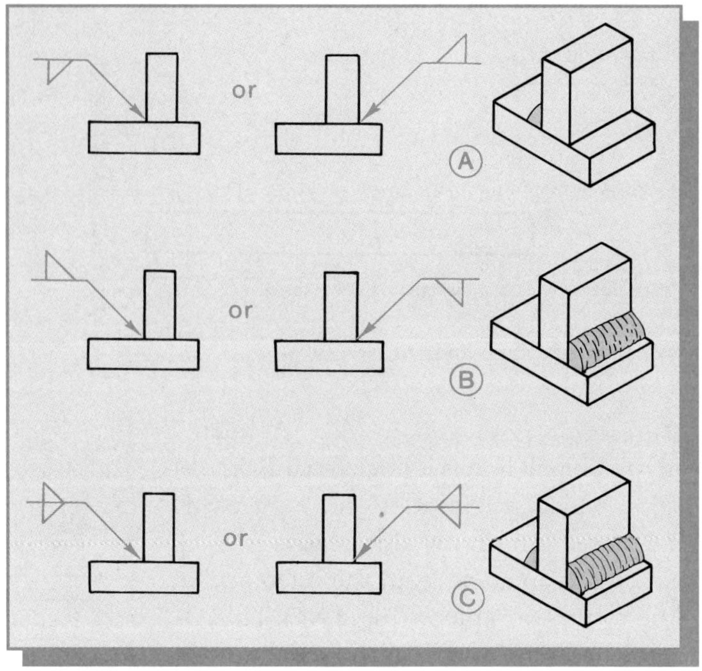

Figure 19-11
The position of the weld symbol (fillet) specifies the location of the weld.

noted in **A** and **B**. When the symbol is placed on both the arrow side and the other side of the reference line, both sides of the joint are to be welded, as noted in **C**. The placement of all possible data in the weld reference symbol is shown in **Figure 19-12**.

Tail Symbol. When specific information, such as detail specifications, procedures, welding processes, filler material, or preparation of the weld area, is required, it is placed in the tail of the symbol **(Fig. 19-13)**. The abbreviations for the welding processes in **Figure 19-2** are placed in the tail to indicate which process is to be used. If all welds use the same process,

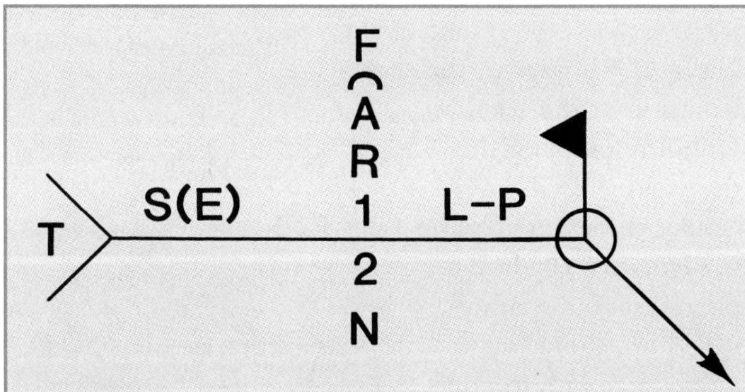

WELDING NOTATIONS:

F – FINISH SYMBOL
[contour symbol] – CONTOUR SYMBOL
A – GROOVE ANGLE: INCLUDED ANGLE OF COUNTERSINK FOR PLUG WELDS
R – ROOT OPENING: DEPTH OF FILLING FOR PLUG AND SLOT WELDS
(E) – EFFECTIVE THROAT
T – SPECIFIC PROCESS OR REFERENCE
L – LENGTH OF WELD, LENGTH OF OVERLAP (BRAZED JOINTS)
S – DEPTH OF PREPARATION, SIZE OR STRENGTH FOR CERTAIN WELDS, HEIGHT OF WELD REINFORCEMENT, RADII OF FLARE-BEVEL GROOVES, RADII OF FLARE-V GROOVES, ANGLE OF JOINT (BRAZED WELDS)
P – PITCH OF WELDS (CENTER-TO-CENTER SPACING
1 – WELD SYMBOL ON THIS SIDE MEANS THE WELD MUST BE ON THE OPPOSITE SIDE OF THE MATERIAL THE ARROW IS TOUCHING
2 – WELD SYMBOL ON THIS SIDE MEANS THE WELD MUST BE ON THE SAME SIDE OF THE MATERIAL THE ARROW IS TOUCHING
(N) – THE NUMBER OF SPOT OR PROJECTION WELDS
[field weld symbol] – WELD DONE IN THE FIELD
[weld all around symbol] – WELD ALL AROUND

Figure 19-12
Weld reference symbol with its data symbol information

this can be covered in a general note, such as "all welds to be FCAW." When this is done, the tail information may be omitted, unless there are other specific instructions.

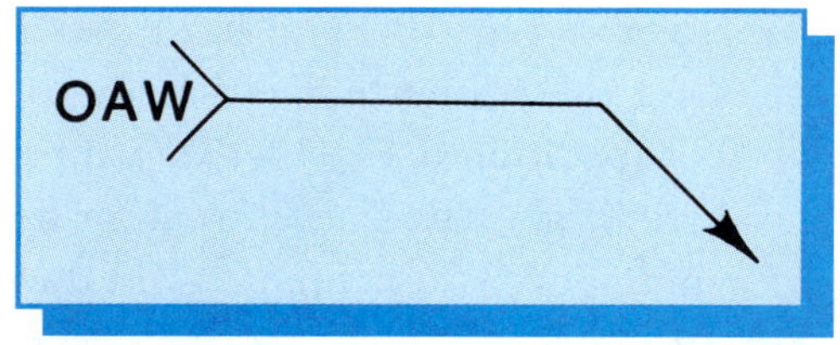

Figure 19-13
When a specific welding notation is required, a tail is added (oxyacetylene welding) as shown.

Finished Method Symbol. The method of finish for each weld is shown by a letter placed at the F position **(Fig. 19-12)**. The following letters indicate the method of finish:

C – Chipping
G – Grinding
M – Machining
R – Rolling
H – Hammering

The exact degree of finish is not shown on a weld symbol since this is part of the finishing limits established for the entire surface on the working drawing. Surface finishes are explained in Chapter 9.

Contour Symbol. The shape of the finished surface of a weld is shown directly under the finishing method symbol **(Fig. 19-12)**. A flat surface is indicated by a straight line. A rounded (concave) surface is shown by a curved line, like a smile; a depressed (convex) surface is shown by a curved line, like a frown **(Fig. 19-14)**.

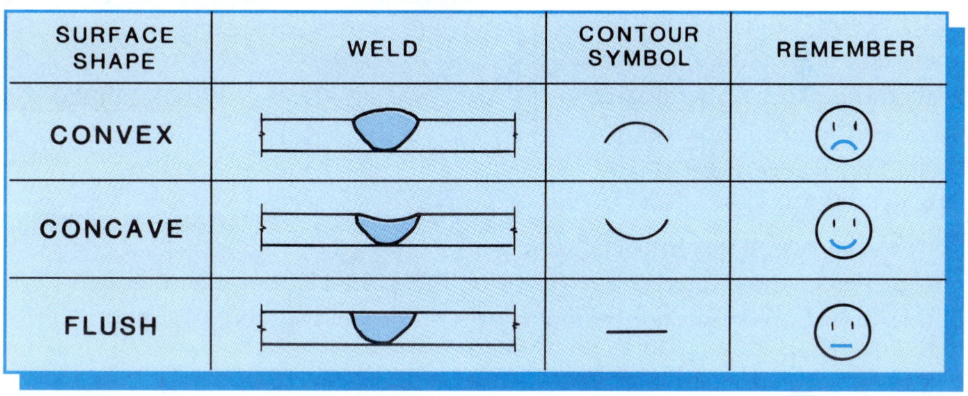

Figure 19-14
Contour symbols

Groove Angles. The letter directly below the contour symbol **(position A, Fig. 19-12)** shows the groove angle if a groove joint is specified. Groove angles are expressed as included angles **(Fig. 19-15)**. This location on the reference weld symbol is also used to show the angle of any countersinks on plug welds.

Root Opening. The gap between two workpieces before welding is known as the **root opening** or **root gap (Fig. 19-15)**. The width of the root opening is located directly under the groove angle on the welding symbol **(position R, Fig. 19-12)**. This location may also be used to show the filling depth for plug or slot welds.

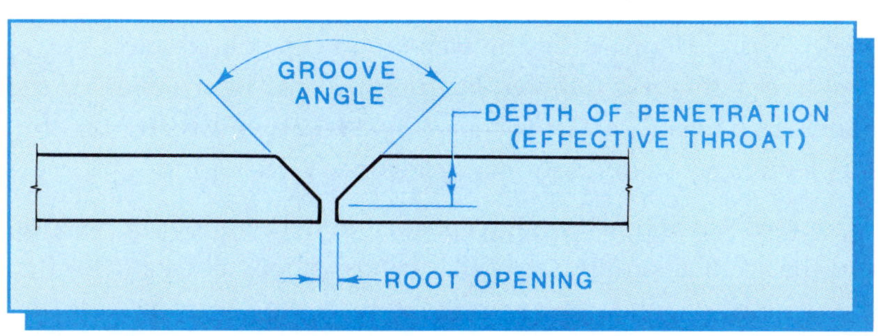

Figure 19-15
Groove angles

Size and Strength. The size or strength of the weld is shown on or under the reference line directly to the left of the graphic weld symbol **(position S, Fig. 19-12)**. Weld size can mean the height of the bead, radii of a flared bevel joint, or depth of penetration. The depth of penetration is sometimes called the **effective throat (position E, Fig. 19-12)**. For some welds, the minimum allowable strength of the joint in pounds is substituted for weld size in position S.

Length of Weld. The length of the weld is located on or below the reference line directly to the right of the weld type graphic symbol **(position L, Fig. 19-12)**. If no length is indicated on the symbol, it is assumed that the weld extends the full length of the part. This location is also used to show the length of overlap on brazed joints.

Weld Pitch. When welds are intermittent over the length of a joint, the spacing of each weld and the length of unwelded spaces between welds must be shown. The center-to-center distance between intermittent welds is called the **pitch.** The distance between intermittent welds is called the **increment (Fig. 19-16)**. The length of the pitch and increments is placed on or below the reference line **(position P, Fig. 19-12)**. For example, a dimension of 6-10 on the symbol means 6" of weld spaced 10" apart from center to center.

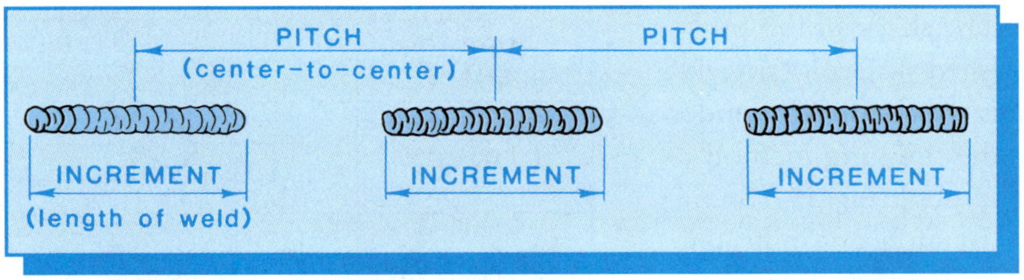

Figure 19-16
Intermittent welds (increments)

Number of Spot or Projection Welds. When spot or projection welds are used, the spacing between welds is shown below the reference line **(position N, Fig. 19-12)**.

Weld All Around. When a joint is to be welded around all of its edges, the weld all around symbol is used. This symbol is a circle drawn at the intersection of the reference line and the arrow line **(Fig. 19-12)**. This is one of the supplementary symbols shown in **Figure 19-10**.

Field Weld. If a part is to be welded on the construction site, a vertical line with a flag is drawn beginning at the intersection of the reference line and the arrow line **(Fig. 19-12)**. This is also one of the supplementary symbols shown in **Figure 19-10**.

Order of Information. Except for the tail symbol information, and the weld all around and field weld symbols, all data is located in the same horizontal order on the reference line, regardless of the position of the arrow. For example, **Figure 19-17a** includes a left arrow line symbol and a right arrow line symbol containing the same information, read in the same sequence.

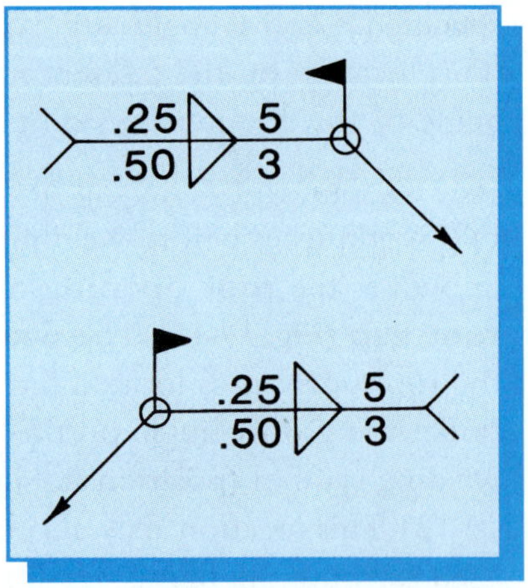

Figure 19-17a
The order of the weld data remains in the same position regardless of the arrow's direction.

Figure 19-17b
Placement of weld data depicts the location side of the weld.

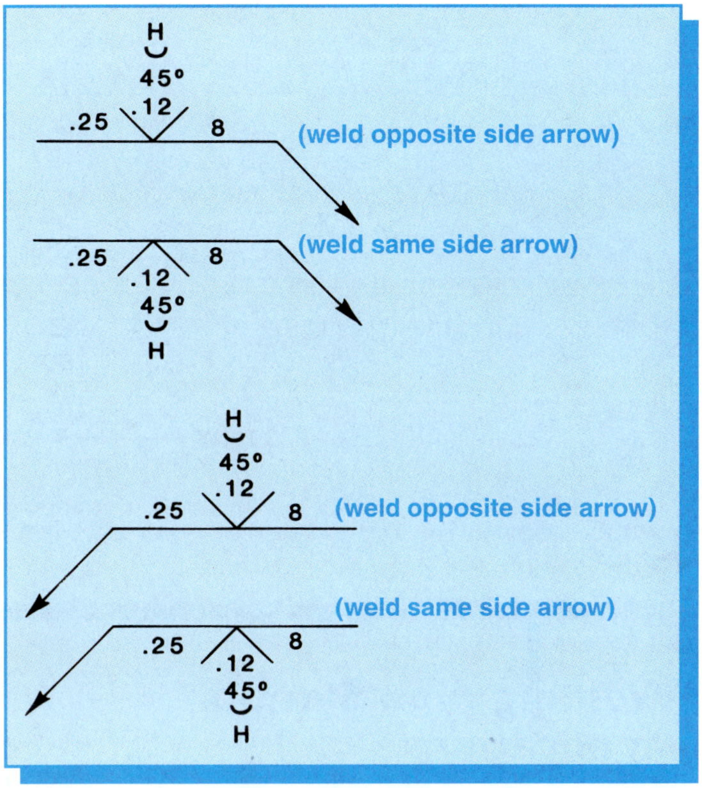

Vertically aligned data, such as the root opening, groove angle, contour symbol, and finish method symbol, are always in the same order from the reference line. Therefore, the order is reversed when these data are below the line. **Figure 19-17b** shows the order of the information applied to an arrow side symbol and another side symbol.

Data Selection

All data shown in **Figure 19-12** is not always used for each weld. Some information simply does not apply to some welds. For example, the groove angle degree does not apply to fillet welds, and the pitch distance does not apply to continuous welds.

When the data does apply to the weld, all information must be located on the symbol. If data is missing, the welder must determine the specifications of the weld without full knowledge of the design requirement. From the information supplied in **Figure 19-18**, the welder knows only the location and size of a V-groove. Not known is the welding process, finish method, surface shape, groove angle, root opening, or length of weld. In comparison, **Figure 19-19** shows a welding symbol containing complete data. From the information on this symbol, the welder knows exactly where a 1/4", 60°, V-groove weld with a 1/8" root opening is to be made. The welder also knows that the gas-tungsten arc process is to be used in the field, and that the weld is to be made all around and ground to a convex shape. The welder understands that this weld is intermittent, with 6" long welds spaced 10" apart center to center.

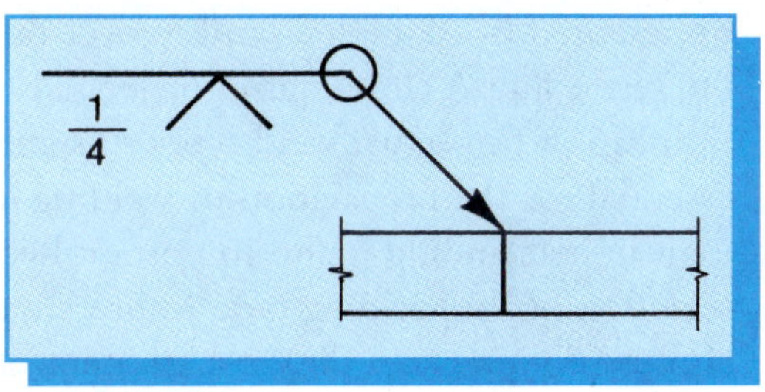

Figure 19-18
Incomplete weld information

Figure 19-20 shows another example of a complete symbol. It specifies an 8" long, 3/16" gas-metal arc fillet weld to be made all around the joint and machined to a concave surface. Although this weld symbol contains only seven items compared to eleven in **Figure 19-19**, it is complete, since the other items are not required for this type of weld.

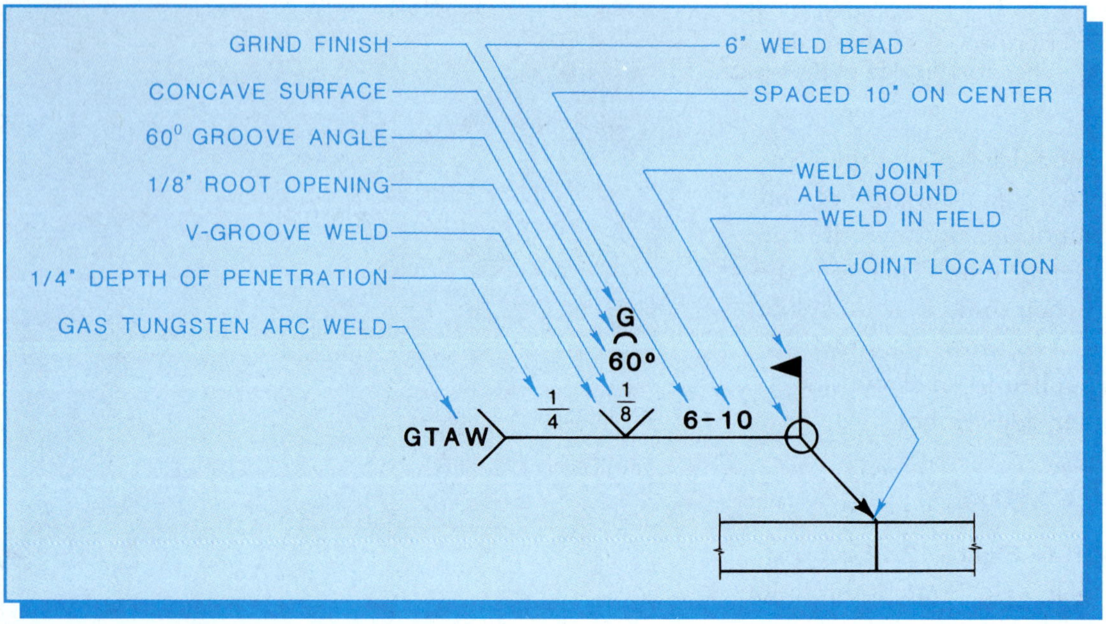

Figure 19-19
Fully noted weld instructions for a V-groove weld

Welding Type Symbol Applications

As shown in **Figure 19-9**, the type of weld is represented by a graphic symbol over or under the reference line. A thorough knowledge of the relationship of the actual weld types to symbols is essential for the preparation of welding drawings. This understanding is also important because the selection of the welding type is directly related to and usually precedes the establishment of other weld design specifications. **Figures 19-21** through **19-42** show pictorial interpretations of the welding type symbols used for each major type of weld.

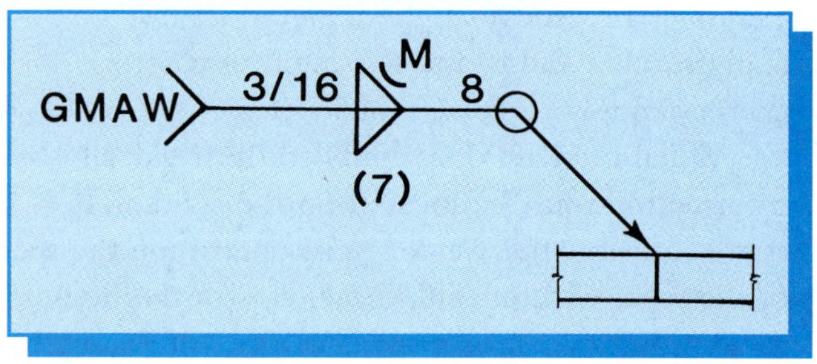

Figure 19-20
Fully noted weld instructions for a fillet weld

Fillet Welds

Fillet welds are used to weld two objects that join at right angles **(Fig. 19-21)**. Notice how the fillet weld symbol closely resembles the actual appearance of a fillet. The vertical line of the fillet weld symbol is always on the left and the slanted line on the right, regardless of the position of the weld. The size of a fillet weld is normally dimensioned as the height of the bead. If the height and width of the bead are different, both dimensions are given in parentheses, with the height entered first **(Fig. 19-22)**. Fillet weld contours are shown next to the slanted part of the fillet symbol **(Fig. 19-23)**.

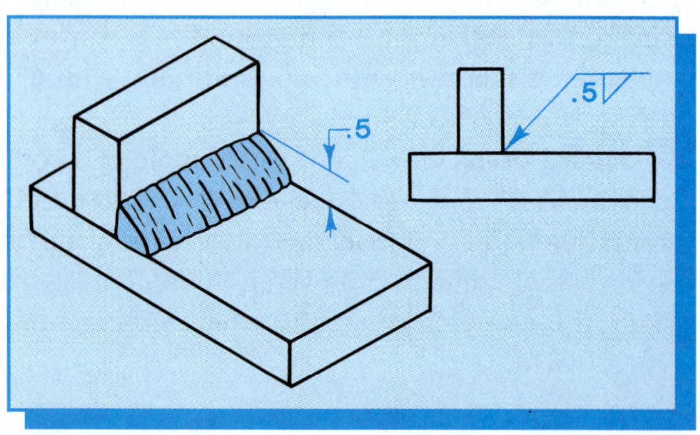

Figure 19-21
Fillet weld

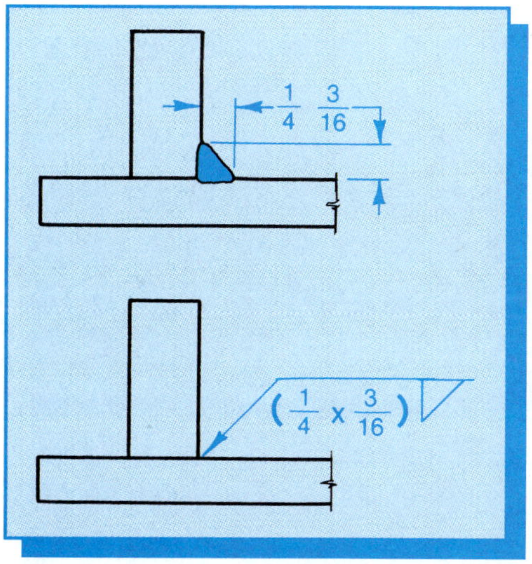

Figure 19-22
Fillet weld

Figure 19-23
Contours for a fillet weld

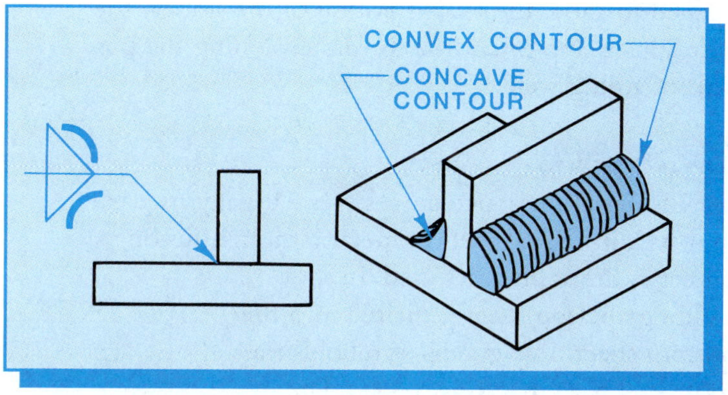

Plug Welds

Plug welds substitute for rivets. A plug weld is a hole filled with a weld filler material. Five items are necessary on plug weld symbols **(Fig. 19-24)**. The graphic symbol for a plug weld is a rectangular box roughly resembling the cross section of a plug weld. The diameter of the hole to be filled is to the left of the rectangle, and the depth of the area to be filled is inside the symbol box. If there is more than one hole, the center-to-center spacing (pitch) is to the right of the box symbol. The symbol in **Figure 19-24** specifies 1/2" holes with 60° countersinks to be spaced 2" apart (center to center) and filled to a depth of 1/4".

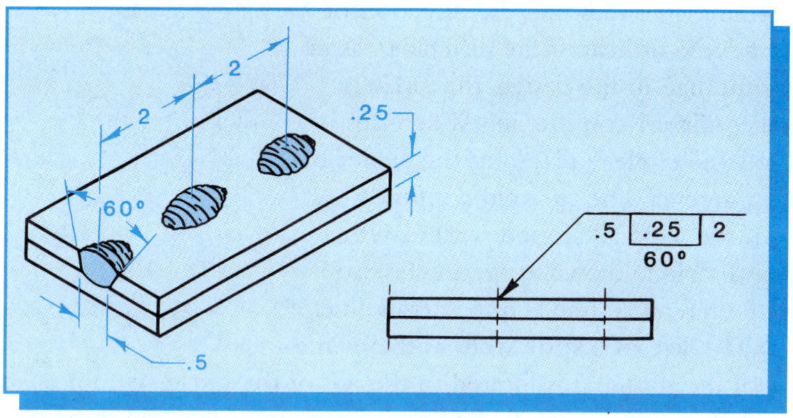

Figure 19-24
Plug welds

Slot Welds

Since **slot welds** are basically elongated plug welds, the same rectangular box symbol is used for slot welds as for plug welds. The depth of slot, depth of fill, angle of countersink, size of hole, and length of slot must be specified. Because of the amount of data required, some or all of the dimensions for a slot weld are often shown on a detail drawing **(Fig. 19-25)**. The weld reference symbol for the slot weld can also be shown as in **Figure 19-26**. In this case, a separate detail is

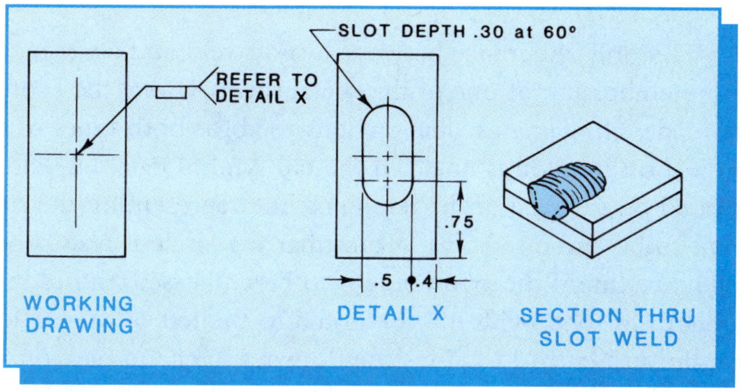

Figure 19-25
Detail drawing of slot weld

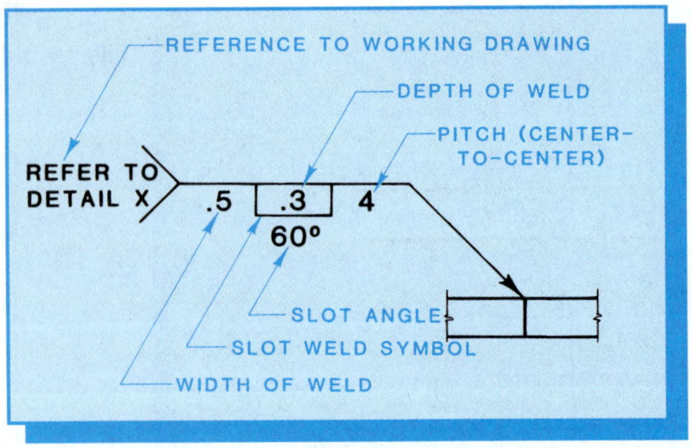

Figure 19-26
Slot weld reference symbol

needed to show the exact position of the slot on the part. This is also often true for dimensioning the position of a plug weld.

Spot Welds

Electrical resistance processes are usually used for spot welding thin sheets of metal, although fusion processes are sometimes used. In spot welding, the top sheet is melted into the bottom sheet. The graphic symbol for a spot weld is a circle **(Fig. 19-27)**. The position of the circle above or below the reference line indicates the direction of the source of heat. If the heat side has no significance in the design, the circle is placed directly on the line. When this is done, the circle is placed in the center of the reference line to avoid confusion with the weld all around symbol, which is also a circle drawn at the intersection of the reference line and the arrow line.

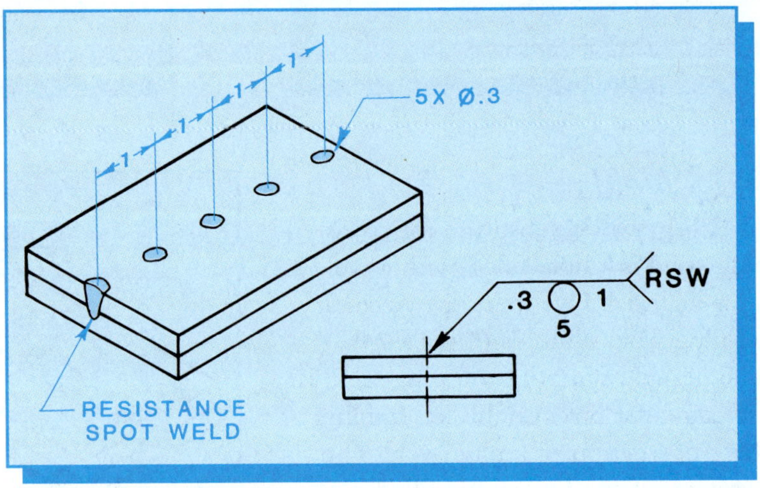

Figure 19-27
Spot welds

The size of a **spot weld** is the diameter of the spot and is located to the left of the circle symbol. A spot weld may also be classified by the minimum acceptable shear strength per spot. If this method is preferred, the strength in pounds per spot is located to the left of the circle. The pitch, located to the right of the circle, is the center-to-center spacing between each spot. **Figure 19-27** specifies five .3" diameter resistance spot welds spaced 1" apart and located on the arrow side of the reference line.

Projection Welds

The only difference between a spot weld and a **projection weld** is the embossing of one of the metal sheets where the spot weld is to be made. The same graphic circle is used for both types of welds. The projection method is noted in the tail symbol data. The circle is located on the side of the reference line representing the side of the joint to be embossed. This means that the circle is placed below the reference line if the arrow side is to be embossed. Either the size or strength of spot welds may be noted to the left of the circle. The symbol in **Figure 19-28** specifies four resistance projection welds with a joint strength of 200 lbs., spaced .8" apart on the arrow side of the joint.

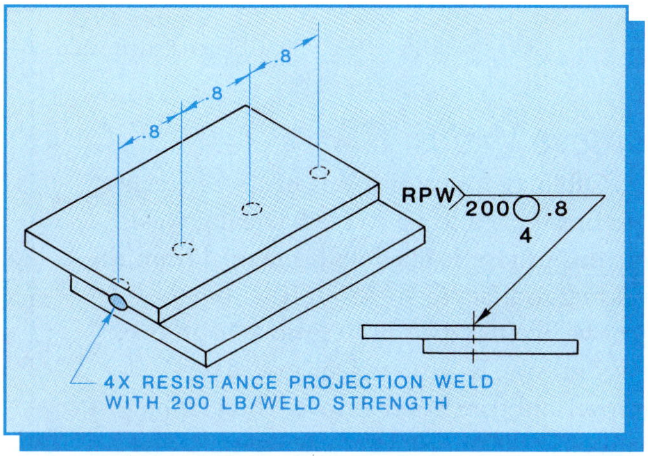

Figure 19-28
Projection welds

Seam Welds

A continuous spot weld is called a **seam weld.** Seam welds are used instead of spot welds when a tight (storage of liquids or pressurized gas) or slightly stronger than usual joint is required. The graphic symbol for a seam weld is a circle with two horizontal lines **(Fig. 19-29).** This symbol can be placed above, below, or on the reference line to show the location of the heat source, as in spot weld symbols. All other spot weld dimensioning rules apply to seam welds, except the strength is expressed in pounds per linear inch. **Figure 19-29** specifies a continuous resistance seam weld on the same side of the arrow, with a strength of 100 lbs. per linear inch.

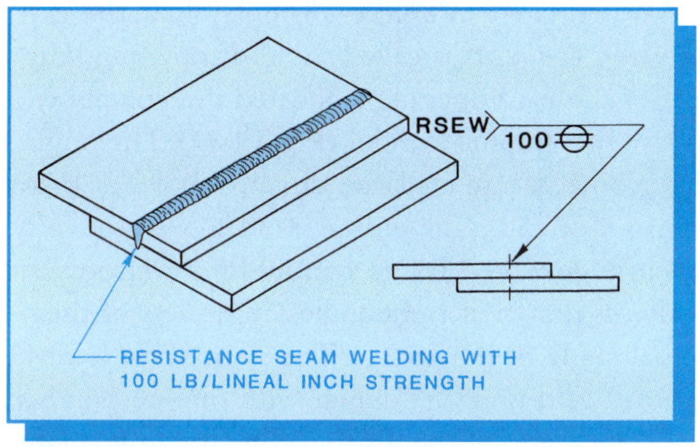

Figure 19-29
Seam weld

Flange Welds

Welding flat surfaces of two sheets of metal bent to provide a surface for welding is called **flange welding.** There are two types of flange welds—edge flange welds and corner flange welds. Edge flange welds join sheets of metal that lie in the same plane **(Fig. 19-30)**. Corner flange welds join sheets of metal that intersect at right angles **(Fig. 19-31).** The graphic symbol for each is different, but the rules for dimensioning are the same. Dimensions are shown on the reference line by using the bend radius, the distance to the point of tangency, and the height of the bead. The bend radius is located to the left of the graphic symbol, followed by a plus (+) sign and the distance from the edge of the metal to the point of tangency of the arc. The height of the bead is shown below this dual dimension. The symbol in **Figure 19-30** specifies a .10" edge flange weld with a .20" radius, and the point of tangency of the arc located .15" from the edge of the metal. **Figure 19-31** shows a $^{3}/_{16}$" corner flange weld with a $^{1}/_{8}$" radius, and a point of tangency located $^{1}/_{8}$" from the edge of the metal.

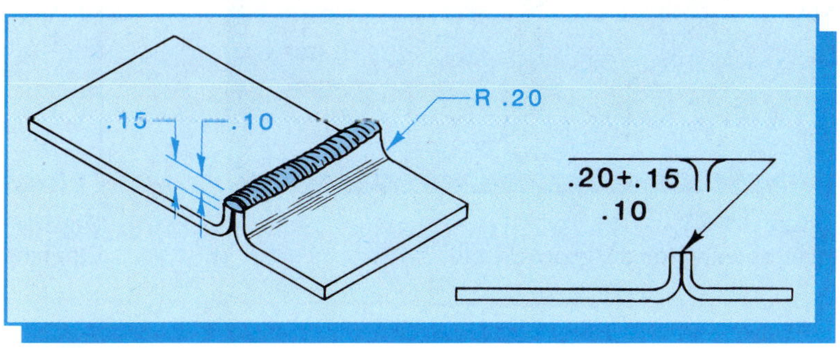

Figure 19-30
Edge flange weld

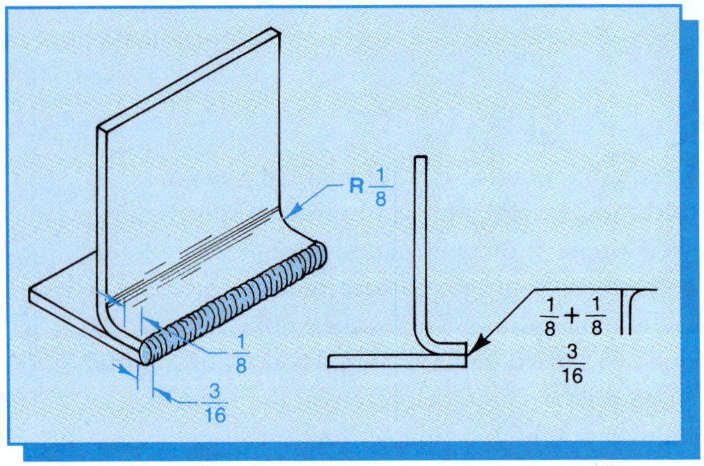

Figure 19-31
Corner flange weld

Groove Welds

Often a groove or gap is made between two metal parts to be welded. During welding, this groove is filled with material from the welding rod. There are seven basic types of **groove welds:** the square groove, V-groove, bevel groove, U-groove, J-groove, flare V-groove, and flare bevel groove. The graphic symbol for each groove weld resembles the shape of the groove.

When two workpieces do not touch, the gap between the parts is called the root opening. The size of the root opening is located directly above or below the graphic symbol **(Fig. 19-32)**. The angle of the groove, if any, is placed directly above or below the root opening dimension, as in the V-groove symbol in **Figure 19-33**. The symbol for a V-groove weld is always drawn at right angles, regardless of the actual angle of the groove. The symbol for a bevel groove weld is always drawn with one perpendicular line. The angle of the 45° bevel is shown in **Figure 19-34**.

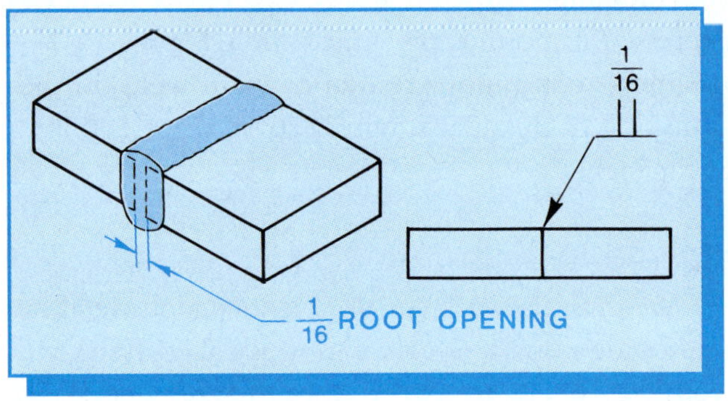

Figure 19-32
Root opening for a square groove weld

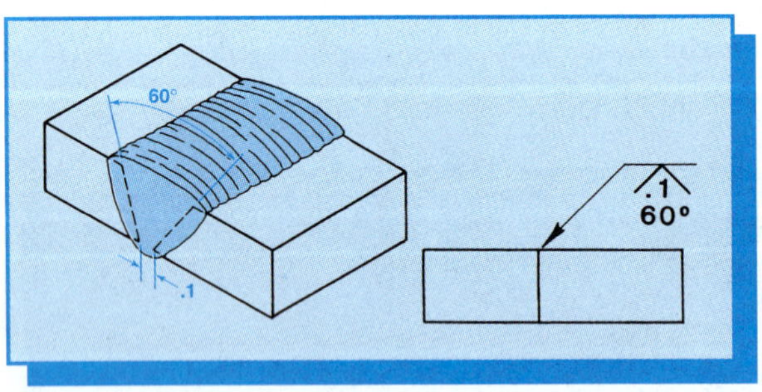

Figure 19-33
Angle opening for a V-groove weld

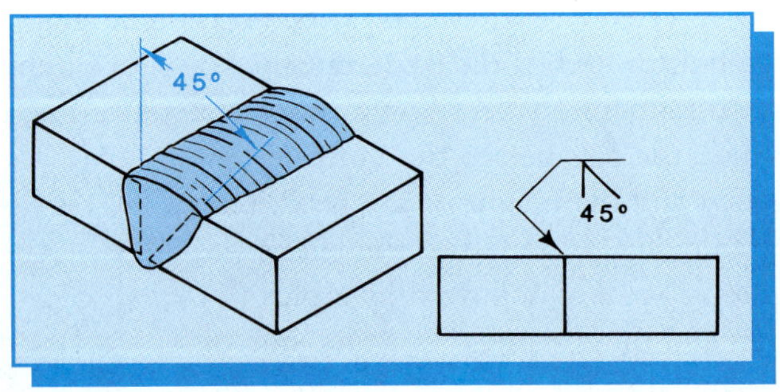

Figure 19-34
Opening for a bevel groove weld

For all groove welds, the depth of penetration is always located to the left of the graphic symbol. The size of the U-groove weld. **(Fig. 19-35)** is .5" with a .1" root opening. The size of groove welds is the depth of penetration (effective throat) of the filler material. The effective throat of the J-groove weld. **(Fig. 19-36)** is .8" and the root opening is .2".

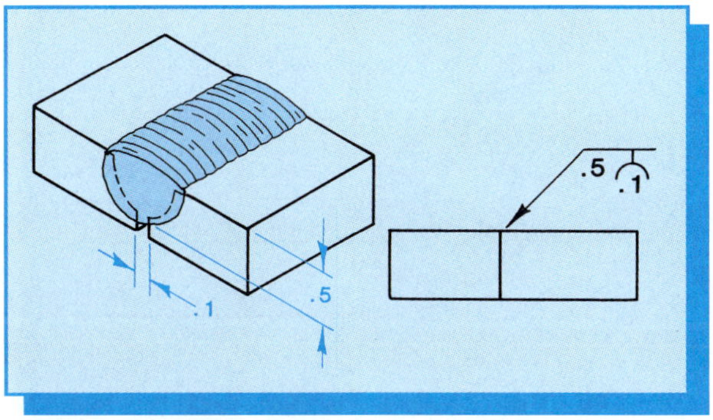

Figure 19-35
Depth of penetration for a U-groove weld

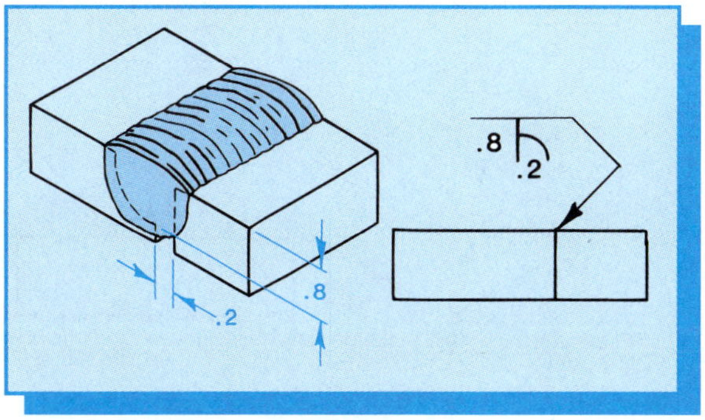

Figure 19-36
Effective throat for a J-groove weld

When two metal plates are bent to create flat adjacent surfaces for welding, the type of weld is called a flare groove weld. There are two types of flare groove welds—the flare V-groove and the flare bevel groove weld. The flare V-groove weld joins metal plates that lie in the same plane **(Fig. 19-37)**. The flare bevel groove weld joins two plates that intersect at an angle **(Fig. 19-38)**. The depth of penetration is the radius of the bevel for both types.

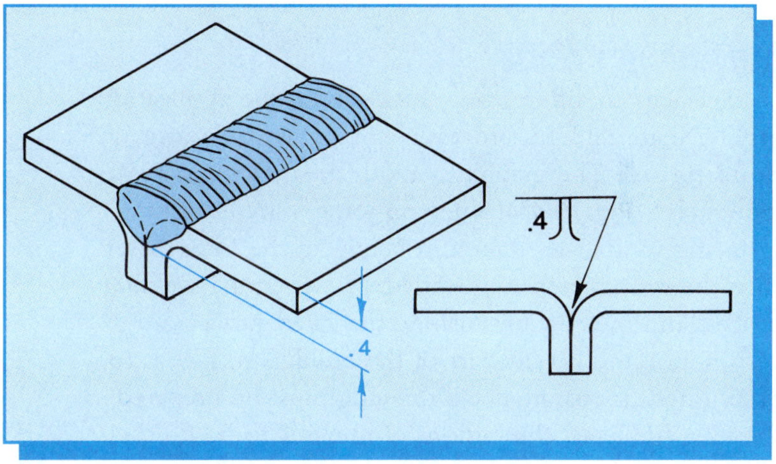

Figure 19-37
Flare V-groove weld

Backing and Melt-Through Welds

For maximum support, a weld bead is sometimes required on the back side of a joint in addition to the front. This is done in two separate operations. To specify a **backing weld,** a semicircle is drawn on the reference line directly opposite the graphic symbol. **Figure 19-39** shows this symbol applied to a V-groove weld. The depth of the backup bead is shown to the left of this semicircle.

Melt-through welds are made completely through the joint from one side with a backup bead. The semi-circle symbol is made solid **(Fig. 19-40)**. It is similar to the backing weld, but is made with one welding operation.

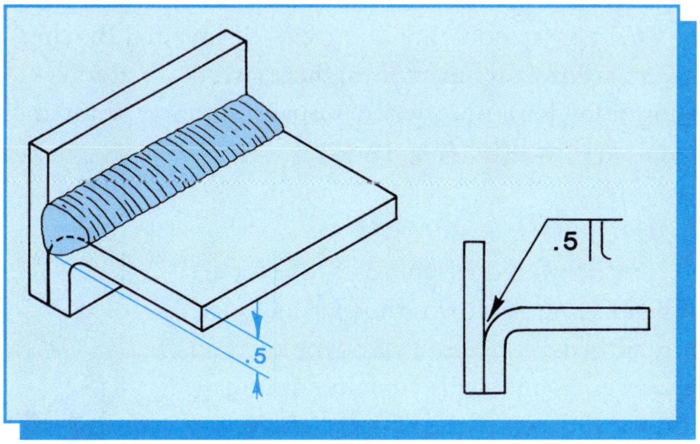

Figure 19-38
Flare bevel groove weld

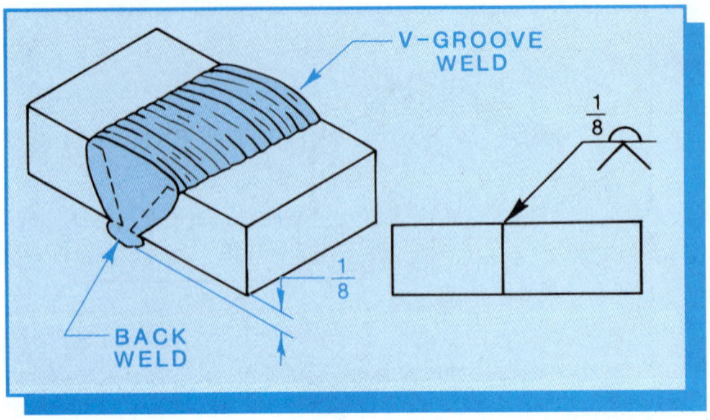

Figure 19-39
Back weld used with a V-groove weld

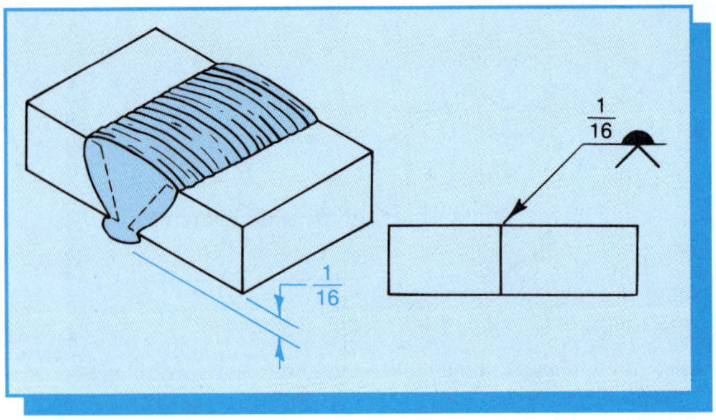

Figure 19-40
Melt-through weld

Surfacing Welds

Surfaces are often made thicker with the application of filler material. This process is known as **surfacing** or **building up.** The graphic symbol for surfacing is two semicircles **(Fig. 19-41)**. Since no joint is involved in surfacing welds, the minimum height of the filler material is the only dimension needed. However, the welding process and method of finishing the filled surface are very important. If only part of the buildup surface is to be finished, a separate detail drawing must be prepared showing the exact dimensions of the area to be finished. This information must be indexed in the welding tail symbol.

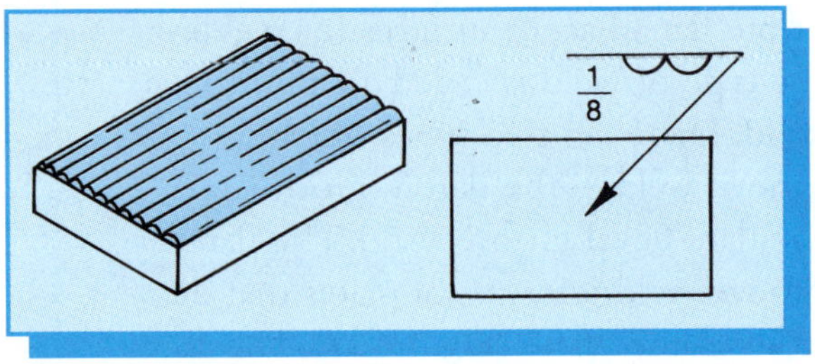

Figure 19-41
Surfacing weld

Brazing Symbols

Welding symbols are also used for brazing. Brazing occurs at lower temperatures; therefore, the materials being joined are not melted. Usually, softer or thinner metals are involved **(Fig. 19-42)**.

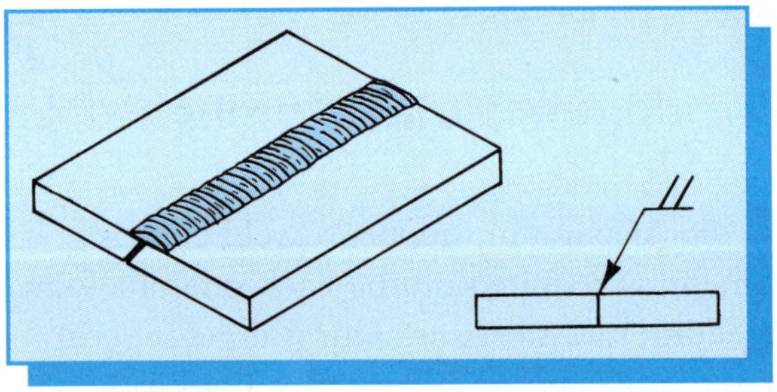

Figure 19-42
Scarf braze

Combination Welds

For instructional purposes, the types of weld symbols shown thus far included only information about one type of weld. However, many welds contain combinations of different types of welds. **Figure 19-43** shows how a combination of different welding types and dimensions are incorporated into one symbol.

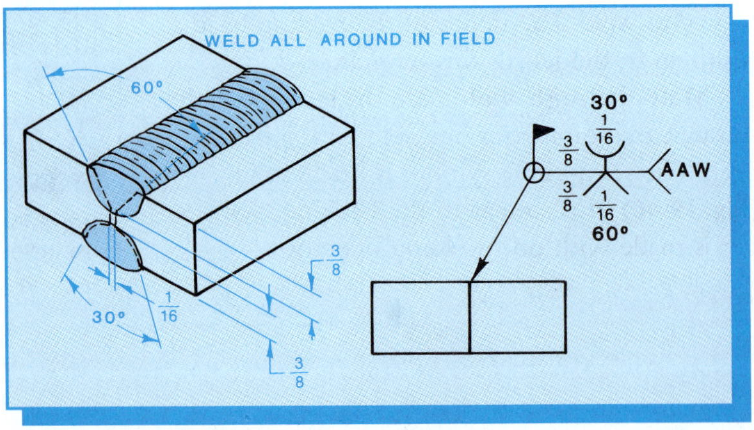

Figure 19-43
Example of combination welds

Welded Joints

In addition to knowing which type of weld is needed, the designer must know which joint is appropriate to join metal parts by welding. The common joints used in welding are shown in **Figure 19-44**. These include the butt, corner, lap, T-, and edge joints. The type of weld that can be used for each joint is also shown in **Figure 19-44**. The final selection of the joint and weld type depends on the metal specified, and the structural and aesthetic requirements of the design.

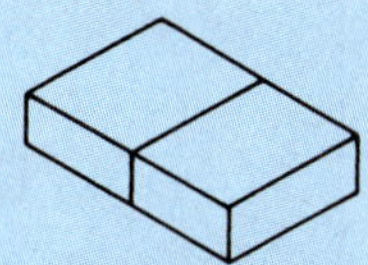

BUTT JOINT
SQUARE GROOVE
V-GROOVE
BEVEL GROOVE
U-GROOVE
J-GROOVE
FLARE V-GROOVE
FLARE BEVEL GROOVE
EDGE FLANGE
BRAZE

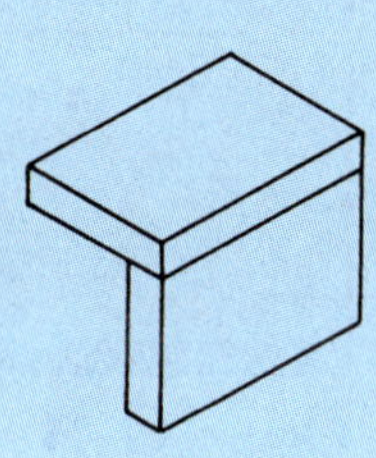

CORNER JOINT
FILLET
SQUARE GROOVE
V-GROOVE
BEVEL GROOVE
U-GROOVE
J-GROOVE
FLARE V-GROOVE
FLARE BEVEL GROOVE
EDGE FLANGE
CORNER FLANGE
SPOT
PROJECTION
SEAM
BRAZE

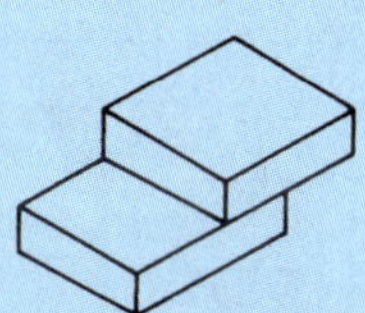

LAP JOINT
FILLET
PLUG
SLOT
BEVEL GROOVE
J-GROOVE
FLARE BEVEL GROOVE
SPOT
PROJECTION
SEAM
BRAZE

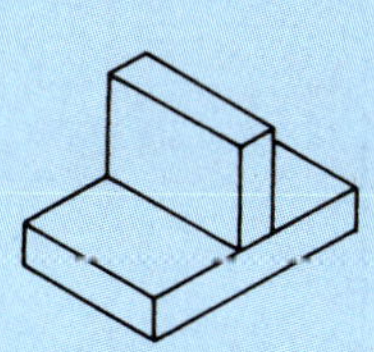

T-JOINT
FILLET
PLUG
SLOT
SQUARE GROOVE
BEVEL GROOVE
J-GROOVE
FLARE BEVEL GROOVE
SPOT
PROJECTION
SEAM
BRAZE

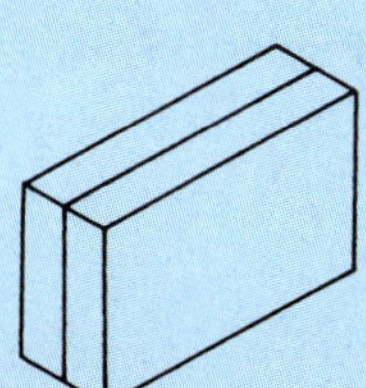

EDGE JOINT
SQUARE GROOVE
BEVEL GROOVE
V-GROOVE
U-GROOVE
J-GROOVE
EDGE FLANGE
CORNER FLANGE
SEAM

Figure 19-44
Weld joints and their recommended welds

AutoCAD LT Exercises

Weld Symbols with CAD

The application of weld symbols on working drawings is an excellent application for a CAD system because it involves drawing repetitive details. The weld symbols can be drawn once, made into blocks, and inserted without redrawing each symbol. **Figure 19-45** shows some standard weld symbols and their dimensions, and **Figure 19-46** shows the names of weld symbols to be made into blocks. To create these blocks, refer to Chapter 16, creating a symbol. Blocks can be inserted easily and rapidly on the drawing as many times as necessary. **Figures 19-47** through **19-49** show the following steps in laying out a typical working drawing with CAD welding symbols.

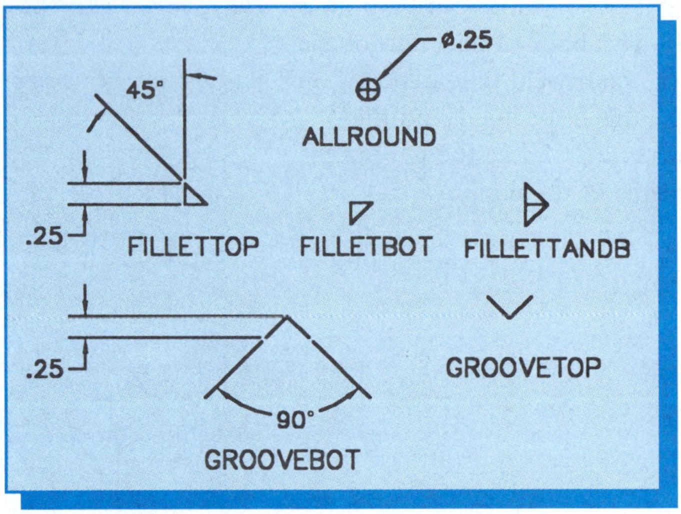

Figure 19-45
Standard weld symbols with dimensions

1. Set up the drawing parameters to include Ortho, Snap, and Grid.
2. Draw the weld symbols that you are going to use full size.
3. Make blocks of these drawings and give them appropriate names. See Chapter 16 for procedures.
4. Using the necessary entities, draw the required views of the weldment **(Fig. 19-47)**.
5. Select DRAW, LEADER to add the leaders for the weld symbols **(Fig. 19-48)**.
6. Select DRAW, INSERT to add the weld symbols. In this example, the symbols that have been inserted are GROOVE, FILLET, and ALLROUND **(Fig. 19-49)**.
7. To complete the drawing, add dimensions, a parts list, and a title block.

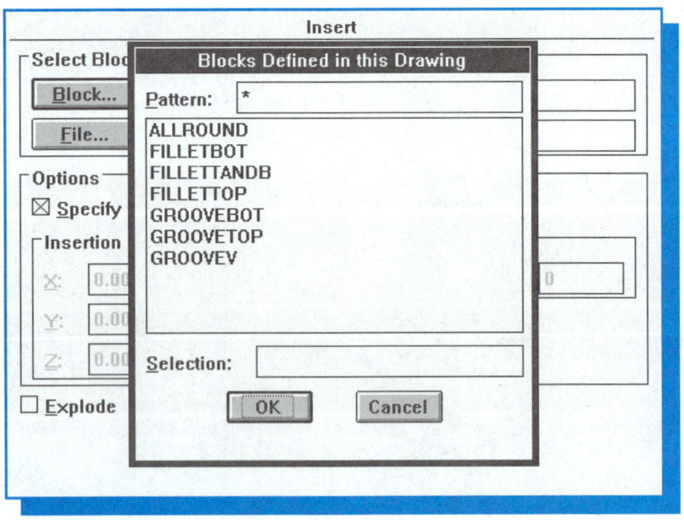

Figure 19-46
Weld symbols created using the BLOCK command

Standard symbols may be inserted on a CAD drawing in various ways depending on the system used. Another method is to use a graphics tablet with custom-made standard symbols **(Fig. 19-50)**. To use the graphics tablet, the CAD operator selects the desired symbol, moves it into place, and inserts it with a puck or mouse.

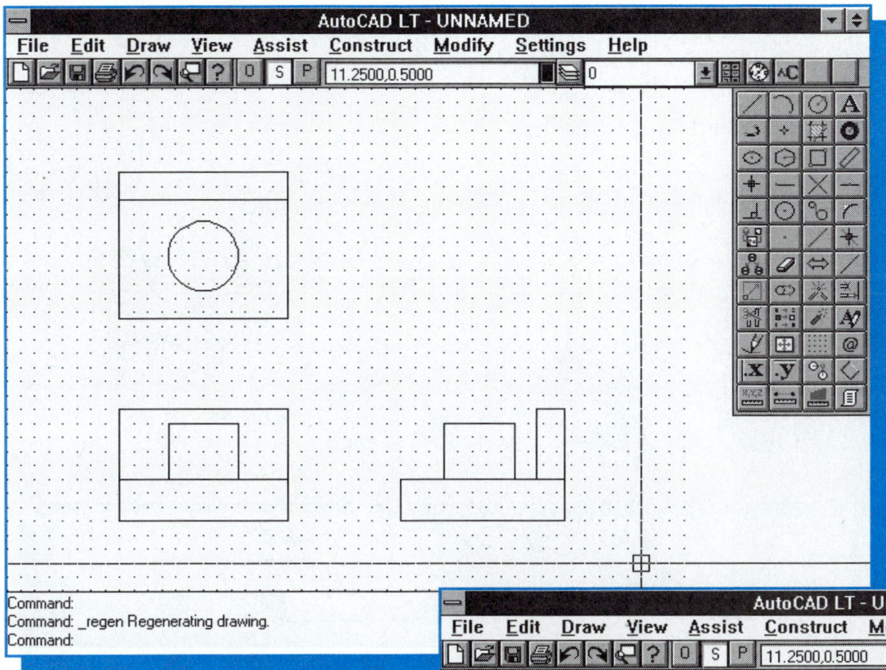

Figure 19-47
Draw or call up the weldment drawing from the CAD's storage system.

Figure 19-48
Call up the LEADER command under dimensioning and add the weld symbol leaders.

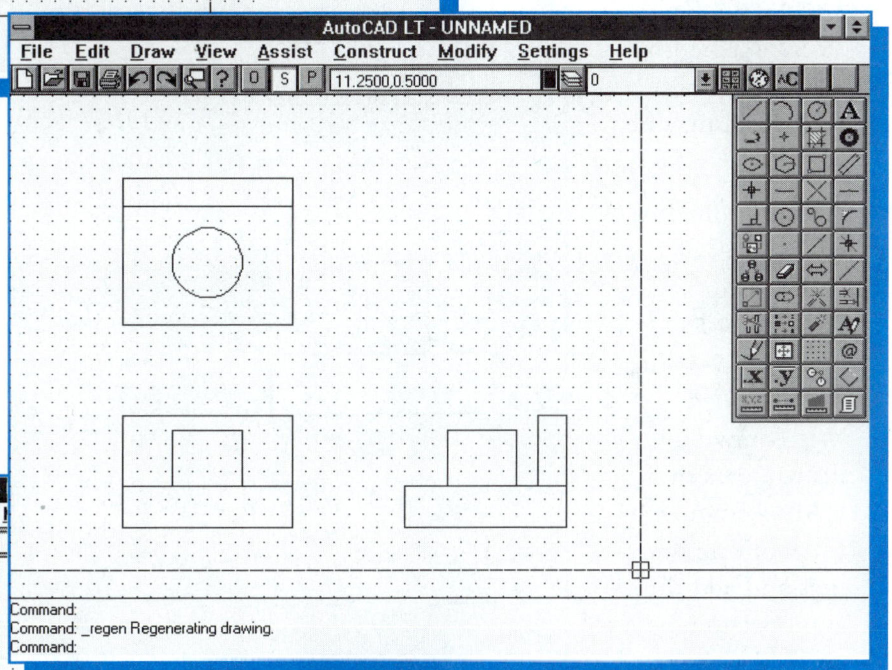

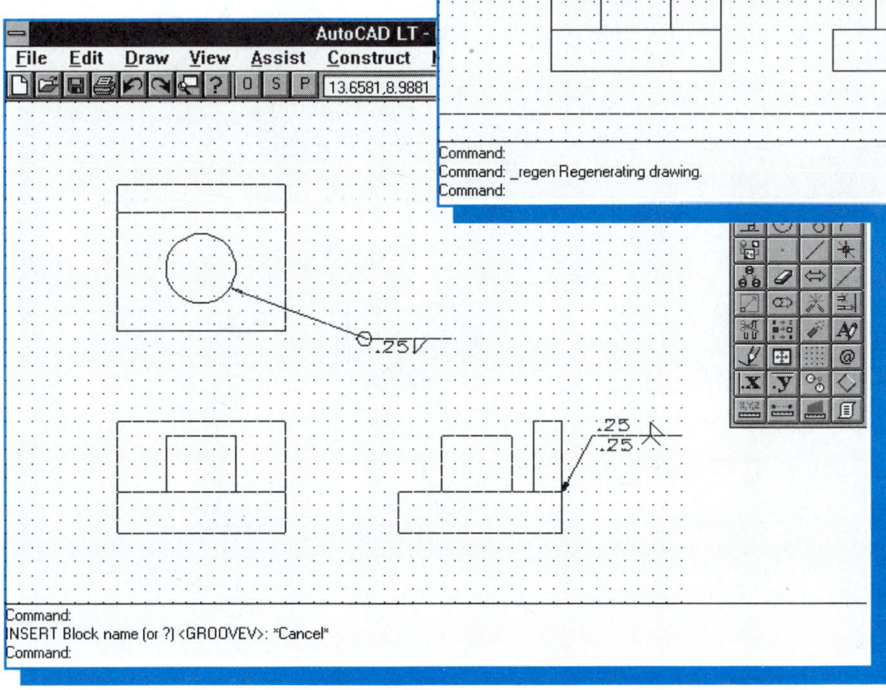

Figure 19-49
Place the required weld symbols on the drawing. Go to the DTEXT command and add the weld sizes.

Using the graphics tablet is similar to using a screen menu. The advantage is that many symbols and drafting functions may be placed on one graphics tablet. With use, the CAD operator will gain a working knowledge of their locations, thereby greatly speeding up the CAD process.

Drafting Exercises

1. Name each type of weld in **Figures 19-50, 19-51,** and **19-52**. Name one example of how each weld may be used.
2. All the individual steel pieces of materials in **Figures 19-53** through **19-66** are to be welded. With drawing instruments or a CAD system, draw the working drawings and add the appropriate type of weld symbols and notations needed.

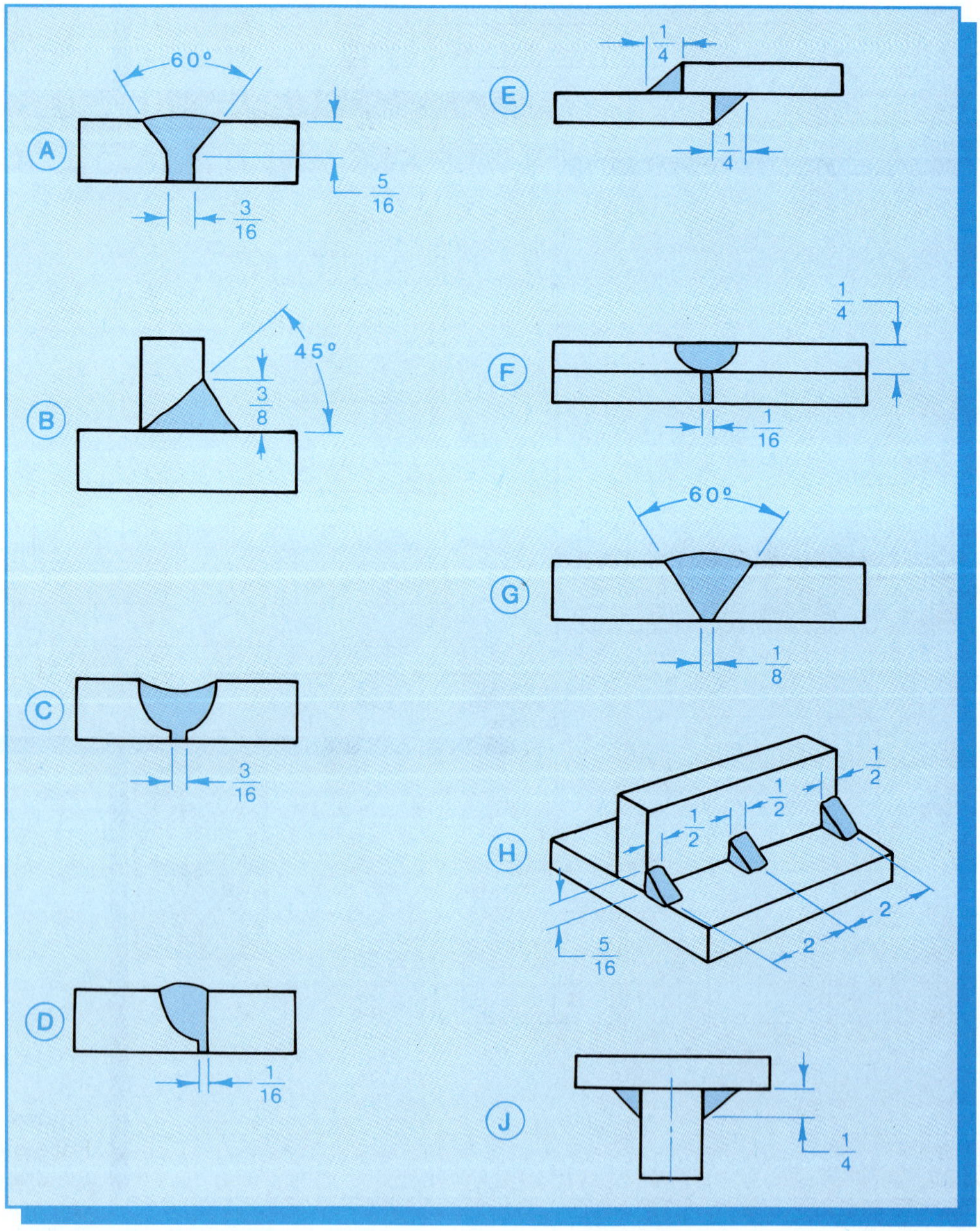

Figure 19-50
Name and sketch each weld symbol.

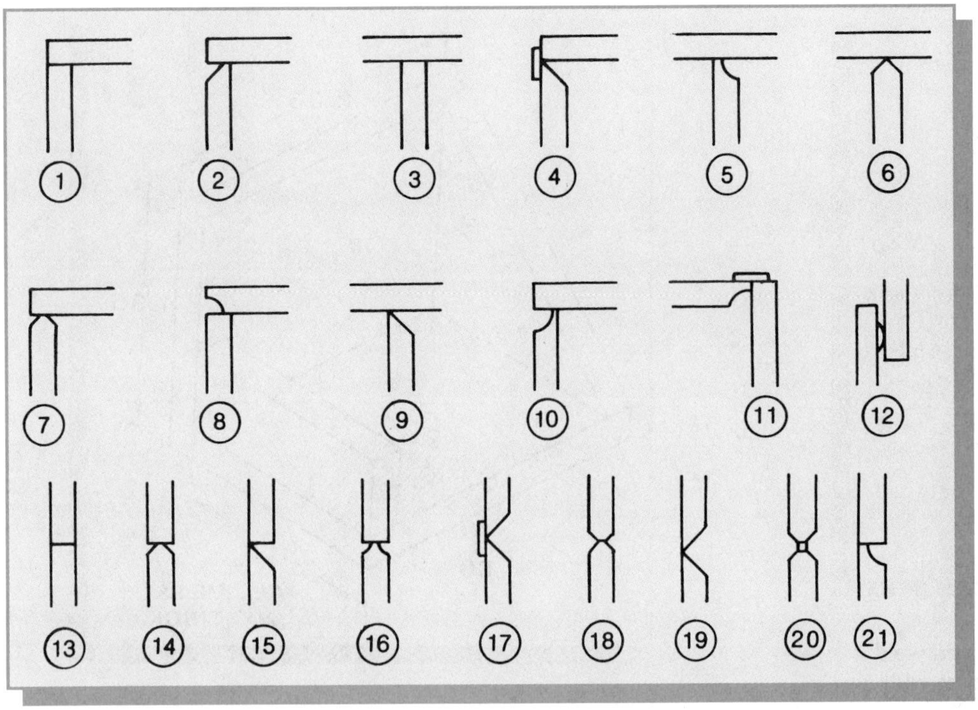

Figure 19-51
Sketch each example to be welded. Add the proper weld reference symbol and data required with another sketch as it will appear on a working drawing.

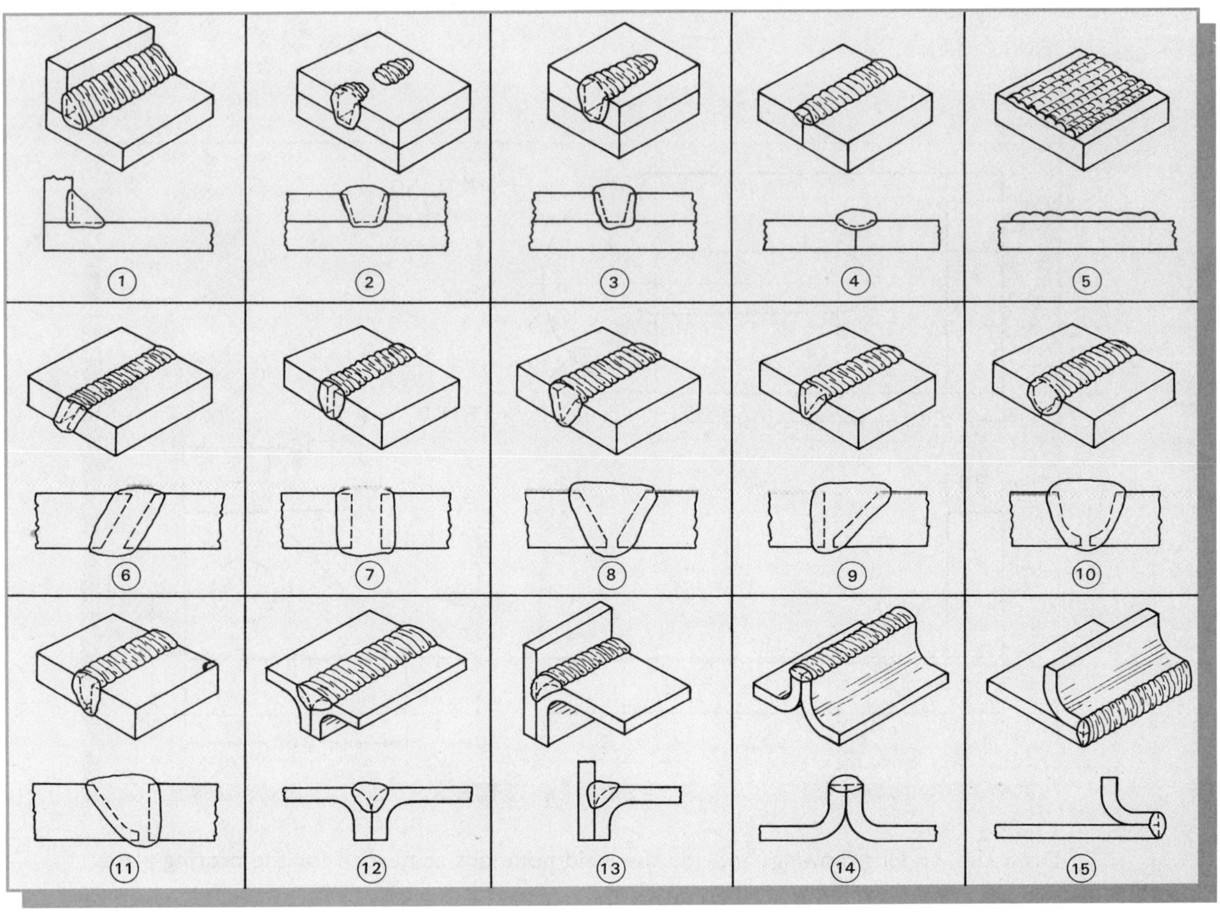

Figure 19-52
Name each type of weld.

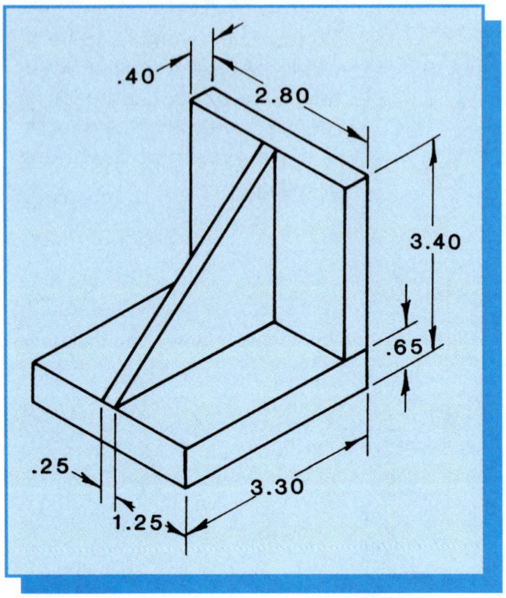

Figure 19-53
Draw the working drawings and add weld notations as needed for the bracket with a CAD system.

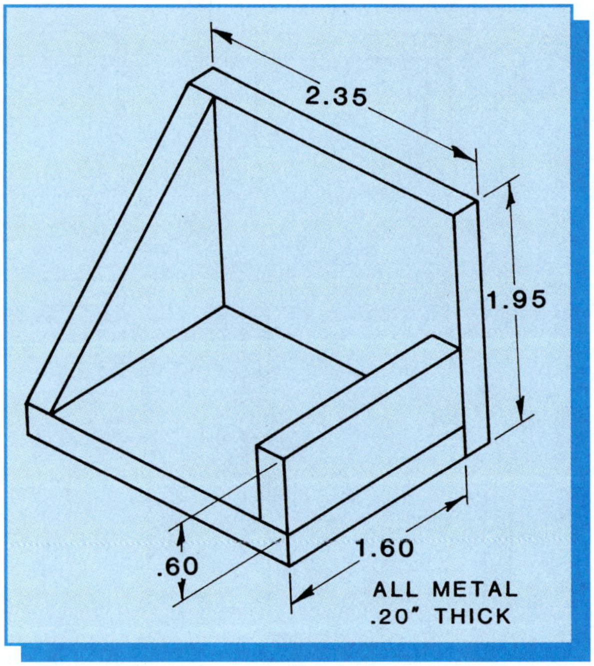

Figure 19-54
Draw the working drawings and add fillet welds with seam welds as needed.

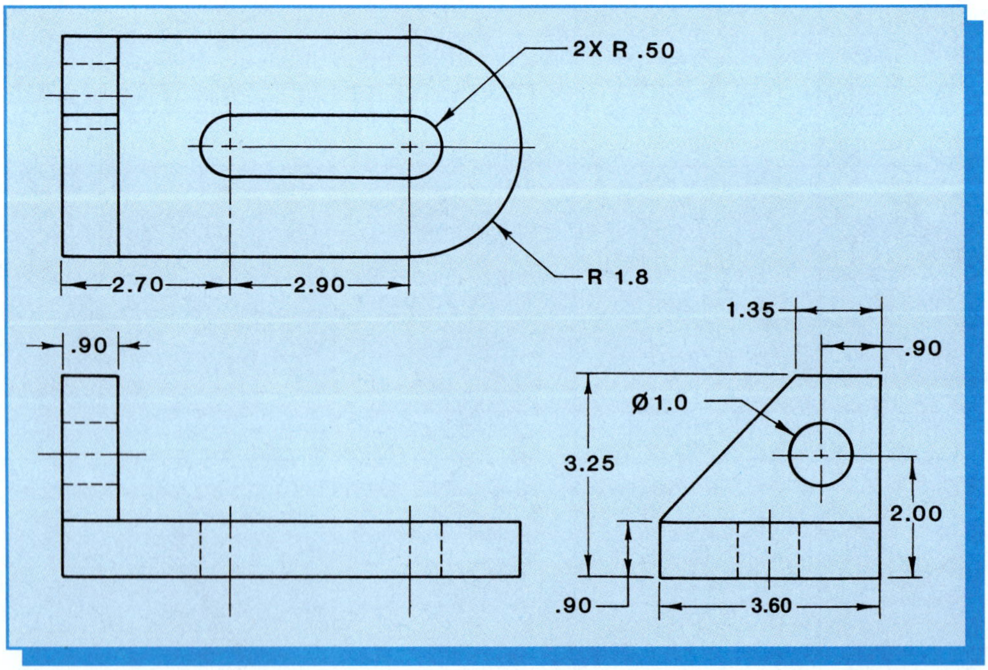

Figure 19-55
Draw the working drawings and add the weld notations as needed for the bearing plate.

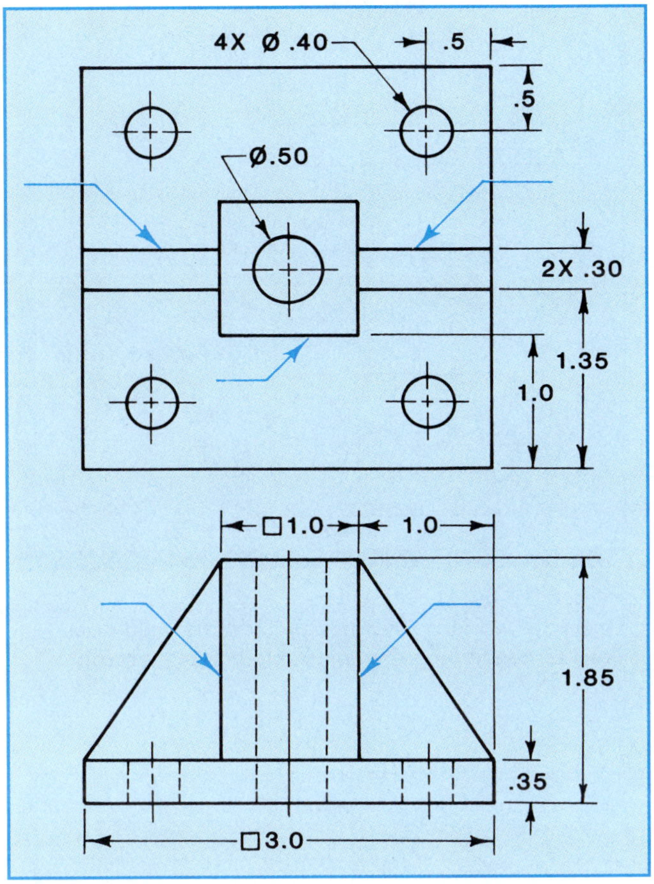

Figure 19-56
Draw the multiview drawings. Locate and indicate the proper weld notations.

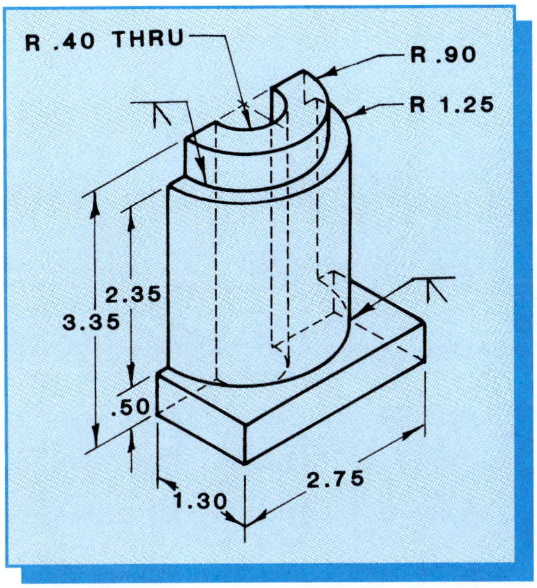

Figure 19-58
Draw working drawings of the semi-column support, and complete the weld notations with a CAD system.

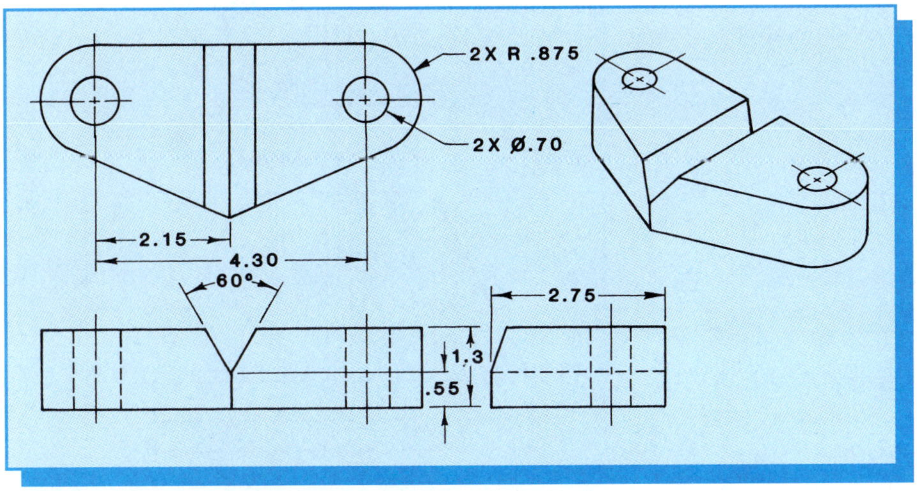

Figure 19-57
Draw the working drawings and add the weld notations as needed for the V-guide.

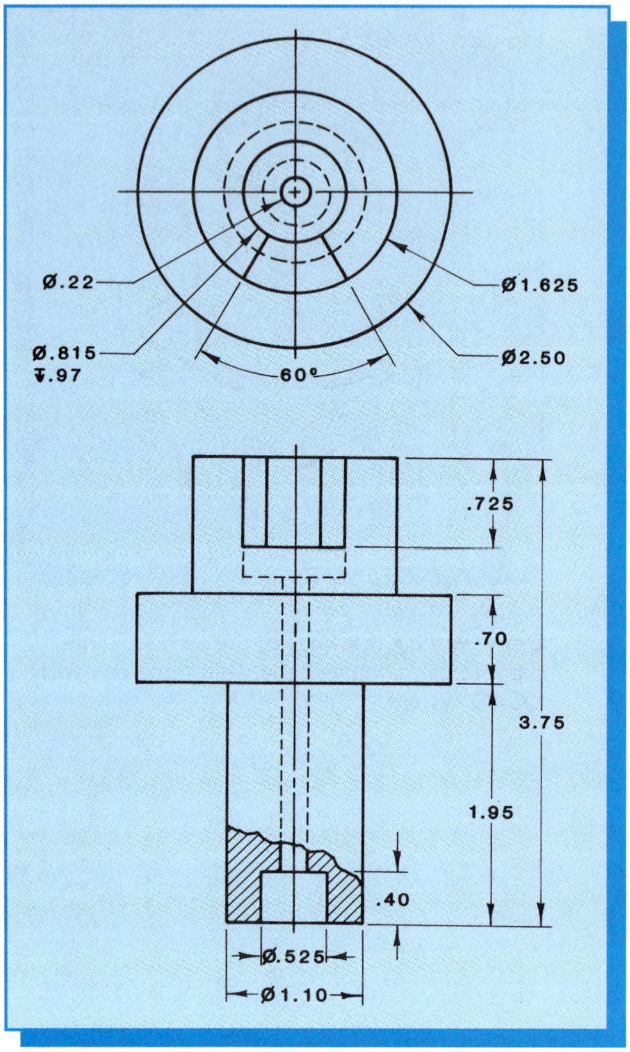

Figure 19-59
Draw the working drawings and add the weld notations as needed for the hub guide.

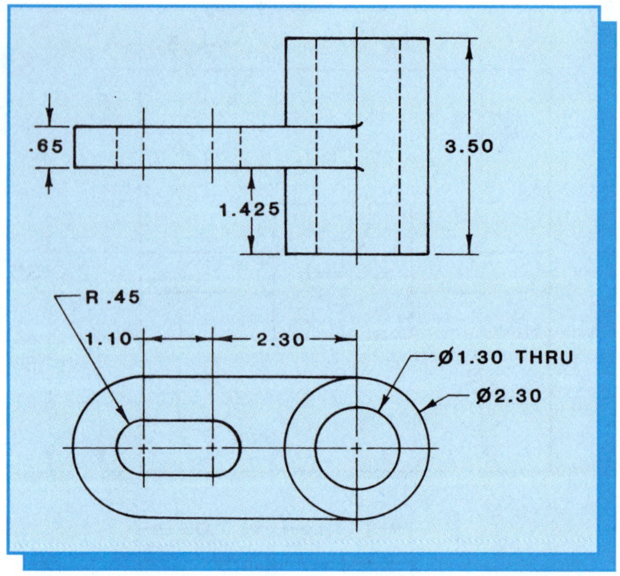

Figure 19-60
Draw the working drawings and add the weld notations as needed for the adjustable bearing locator.

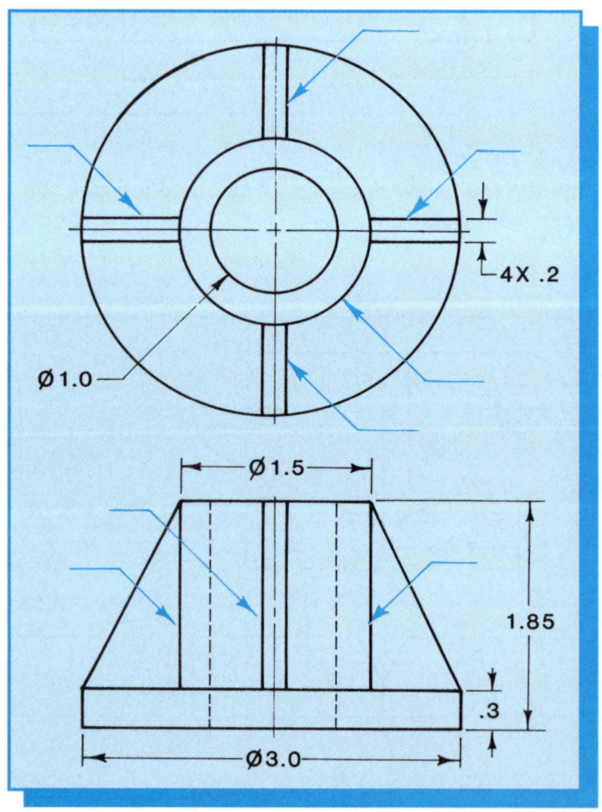

Figure 19-61
Draw the multiview drawings. Locate and indicate the proper weld notations.

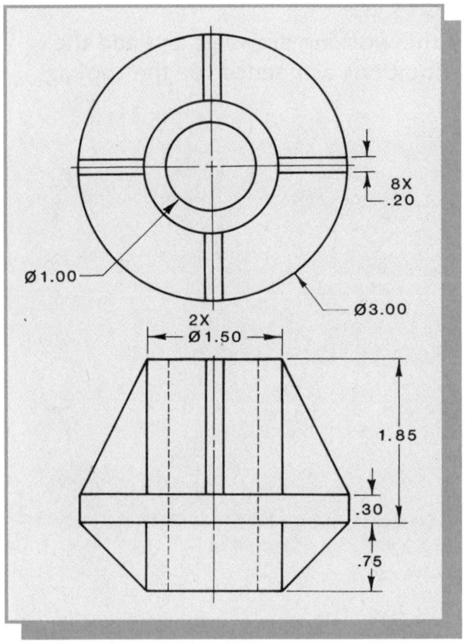

Figure 19-62
Draw the working drawings and add weld notations as needed for the cylinder and four webs with a CAD system.

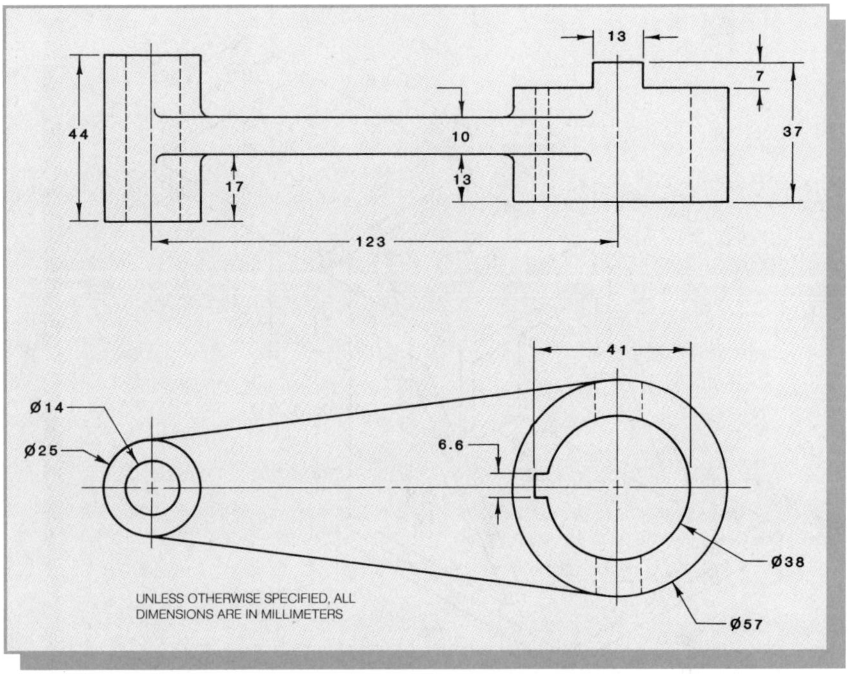

Figure 19-63
Draw the working drawings and add the weld notations as needed for the web connector.

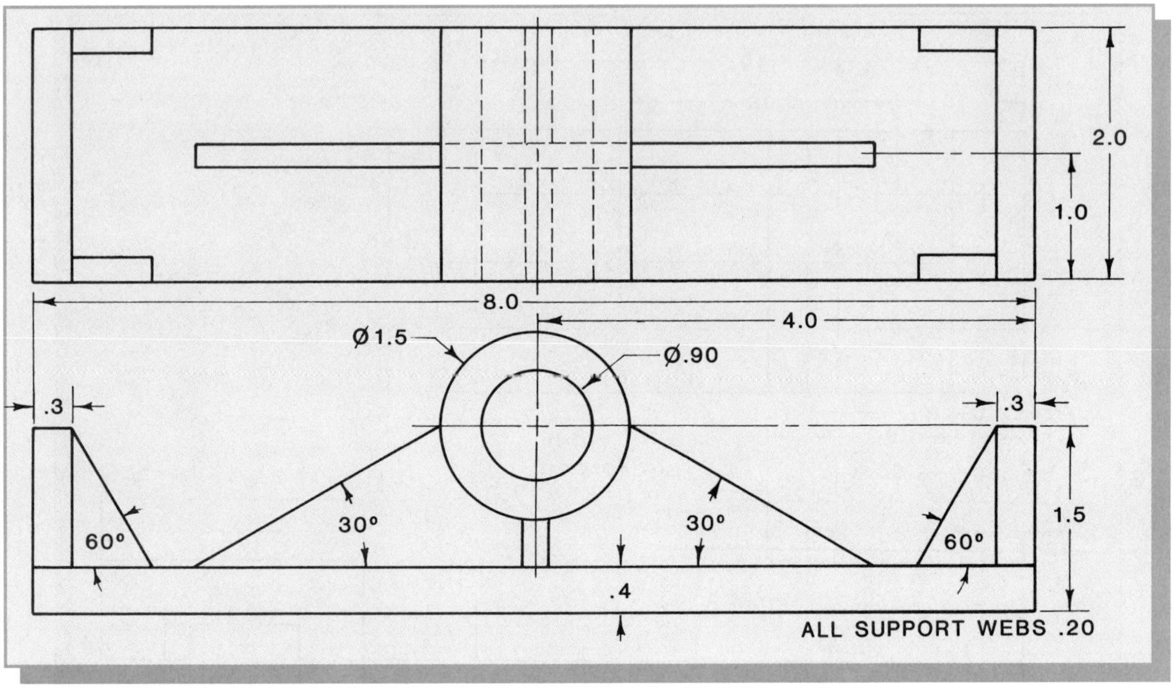

Figure 19-64
Draw the multiview drawings. Locate and indicate the proper weld notations.

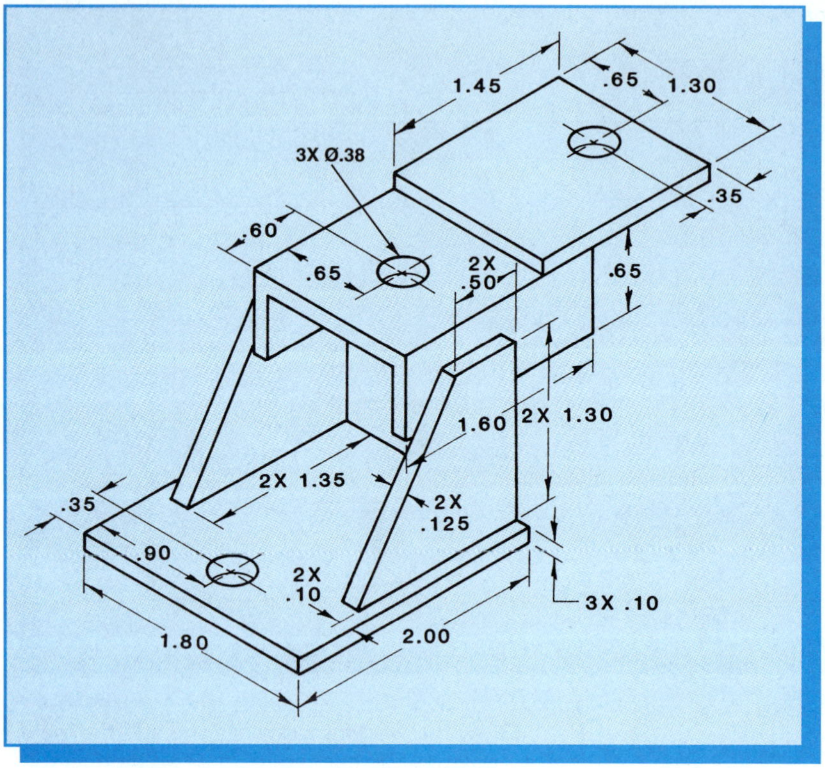

Figure 19-65
Draw the working drawings and add the weld notations as needed for the tool jig stand.

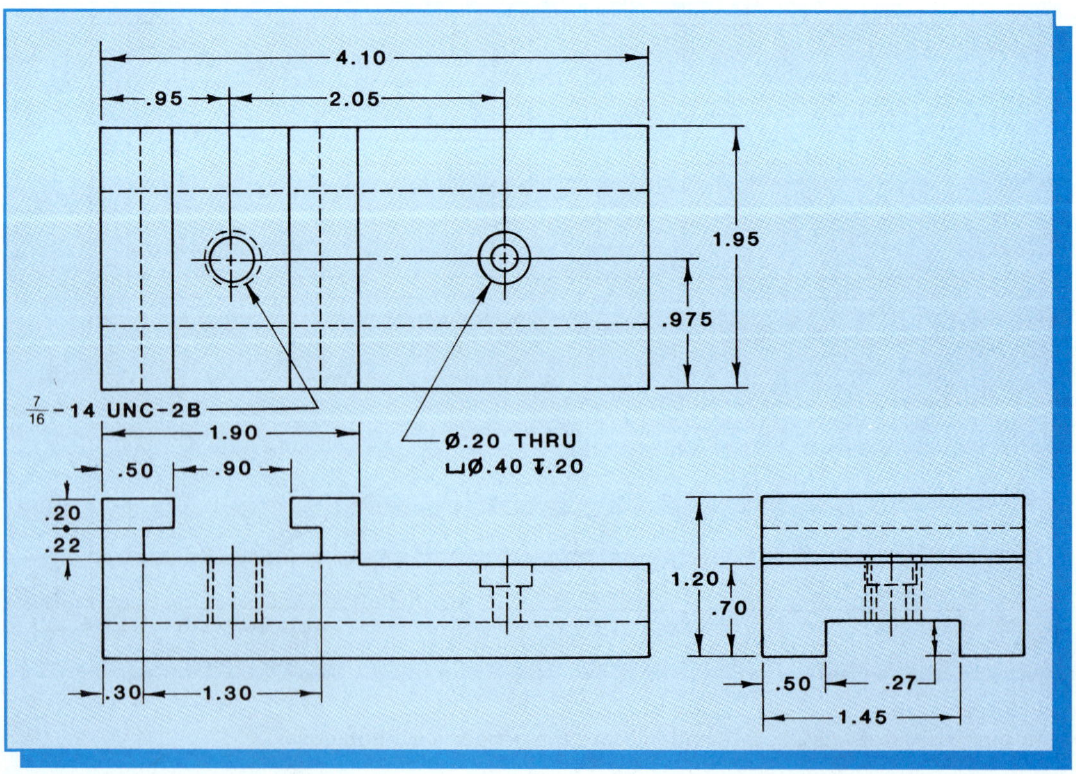

Figure 19-66
Draw the working drawings and add the weld notations as needed for the slotted fixture guide.

Design Exercises

1. Design a metal container that will hold five compressed gas tanks. Each tank has an outside diameter of 12" and a height of 36". Use weldments for the assembly. Research the safety features required for the storage of compressed gas tanks.

2. Design a "hook set-up" for a lifting crane that must pick up a 4' diameter spherical glass ball that weighs 500 pounds. Draw a working drawing showing all the components of the "hook" and attachments. Use weldments whenever possible for the assembly.

Key Terms

Arc welding
Arrow line
Backing weld
Brazing
Effective throat
Fillet welds
Flange welds
Fusion welding
Gas welding
Groove welds
Increment
Melt-through welds
Oxyacetylene welding
Pitch
Plug welds
Projection welds
Resistance welding
Root opening (root gap)
Seam welding
Seam welds
Slot welds
Soldering
Solid-state welding
Spot welding
Spot welds
Surfacing (building up)
Welding process
Weld symbols

Chapter 20

Gears and Cams

Objectives

The student will be able to:

- draw a working drawing of a spur gear
- draw the teeth of a gear with a gear template
- draw a working drawing of a rack and pinion gear
- draw a working drawing of a bevel gear
- draw a working drawing of a worm gear
- draw a plate cam from displacement diagram data
- draw a drum cam
- lay out a displacement diagram from data in a plate cam profile drawing
- lay out a displacement diagram from data in a drum cam drawing
- recognize the various types of gears and cams
- prepare a gear drawing using a CAD system
- complete a cam drawing using a CAD system

Introduction

Gears, cams, and drives are mechanical devices used to control power. These devices can transmit motion from one device to another, or can change the speed and direction of motion. To design and draw gears, cams, and drives as integral parts of machine assemblies, a basic understanding of their function is vital.

Gears and Drives

Cylinders or cones used to transmit power and motion, and change the speed and direction of motion, are known as **gears.** Gears may be smooth (friction gears) or toothed. There are many types of gears designed to perform specific transmission, directional, and speed control functions. Gears are used to transmit rotary or reciprocating (back and forth) motion between machine parts.

Gears are divided into two general classifications based on the relative position of the gear shafts. **Parallel shaft gears** are connected with spur gears, helical gears, or herringbone gears. **Perpendicular shaft gears** are connected with bevel or miter gears. If intersecting gear shafts are neither parallel nor perpendicular, connections are made with crossed helical gears or worm gears, called **angular gears.**

When one gear is rotated and brought into contact (meshed) with another gear, motion is transmitted. The second gear turns in the opposite direction of the first. If two gears are equal in circumference, both will rotate one complete revolution in opposite directions. If the second gear is twice the circumference of the first, it will rotate only one-half revolution, while the first gear rotates one full revolution. This results in the second gear rotating at one-half the speed of the first gear. The difference between the rotation of the two gears is known as the **gear ratio.** When gears are the same size and rotate at the same speed, the gear ratio is 1:1. When one gear rotates at twice the speed as the other, the ratio is 2:1. Three times rotation equals 3:1, and so forth.

The most common type of gears and drives are described below, and are illustrated in **Figures 20-1** through **20-9**:

- **Spur gears** are one of the most common types of gears in which the teeth on the circumference are all parallel to the axis and perpendicular to the sides of the wheel **(Fig. 20-1)**. The can opener has spur gears.
- **Rack and pinion gears** transmit reciprocating motion through a flat-toothed surface (rack) intersecting with a small cylindrical gear (pinion) **(Fig. 20-2)**. The pinion is the smaller gear in any gear system. Cresent wrenches use the rack and pinion gears.
- **Bevel gears** transmit power between shafts that intersect at an angle **(Fig. 20-3)**. If the intersection between shafts is a right angle, they are often called **miter gears (Fig. 20-4)**.
- **Worm gears** connect shafts that are perpendicular with shaft centerlines that do not intersect. The small gear is the worm, and it drives the larger worm gear. The worm is shaped like a screw. The worm gear is round like a wheel. The worm includes at least one thread around the pitch surface. Worm gear teeth are cut on an angle to mesh with worm teeth **(Fig. 20-5)**.
- **Helical gears** operate on parallel shafts with gear teeth cut in a twisted, oblique angle to the shaft **(Fig. 20-6)**.

Figure 20-1
Spur gears *(Courtesy of Boston Gear Division, Imo Corporation, Inc.)*

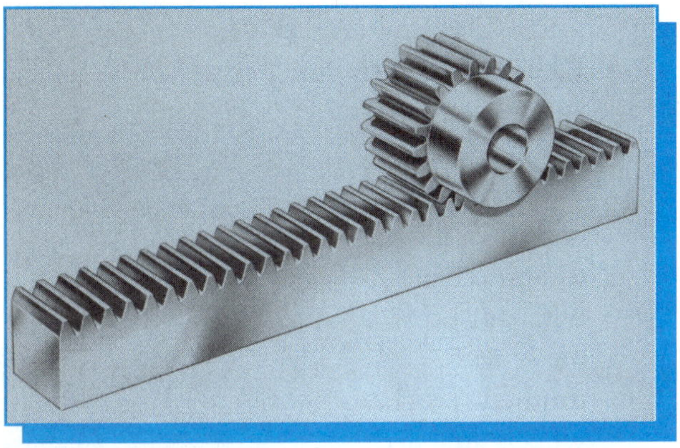

Figure 20-2
Rack and pinion gears *(Courtesy of Boston Gear Division, Imo Corporation, Inc.)*

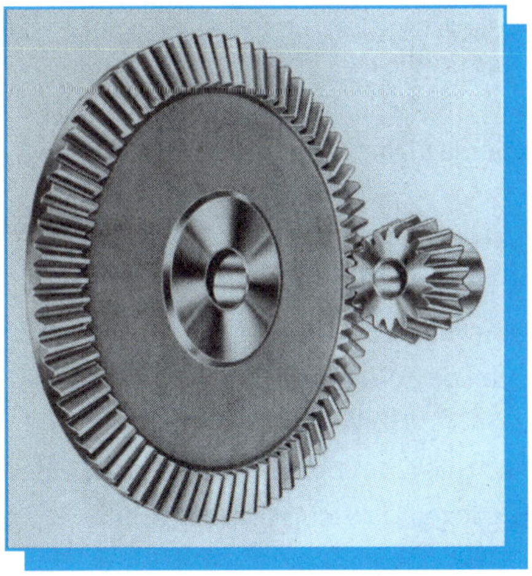

Figure 20-3
Bevel gears *(Courtesy of Boston Gear Division, Imo Corporation, Inc.)*

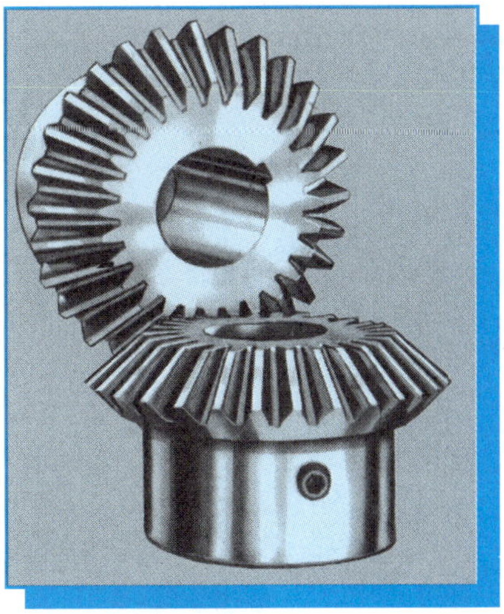

Figure 20-4
Miter gears *(Courtesy of Boston Gear Division, Imo Corporation, Inc.)*

Figure 20-5
Worm and worm gear *(Courtesy of Boston Gear Division, Imo Corporation, Inc.)*

- **Herringbone gears** contain two sets of helical gear teeth placed side by side on an opposite slant on the same gear wheel **(Fig. 20-7)**.
- **Spiral bevel gears** contain oblique, curved teeth that provide more contact between the intersecting teeth **(Fig. 20-8)**.
- **Ring gears** are used primarily in automobile differentials A propeller shaft pinion transmits power to a large ring gear that has teeth cut on the inside surface of the ring **(Fig. 20-9)**.
- **Chain and sprocket drives** are flexible power transmission devices. They consist of a continuous chain with links that mesh with toothed sprocket wheels attached to a drive shaft **(Fig. 20-10)**. Chain drives are used where ease of assembly and disassembly, and elasticity are required, or where shaft-to-shaft distances are extremely long as on a bicycle.
- **Belt drives** transmit power between shafts with a belt wrapped around the outside circumference of a pulley mounted on each shaft. Belts that connect pulleys are either flat **(Fig. 20-11)**, grooved with longitudinally

Figure 20-6
Helical gears *(Reprinted from* Mechanical Drafting *by Madsen, Shumaker, and Stewart, Delmar Publishers Inc.)*

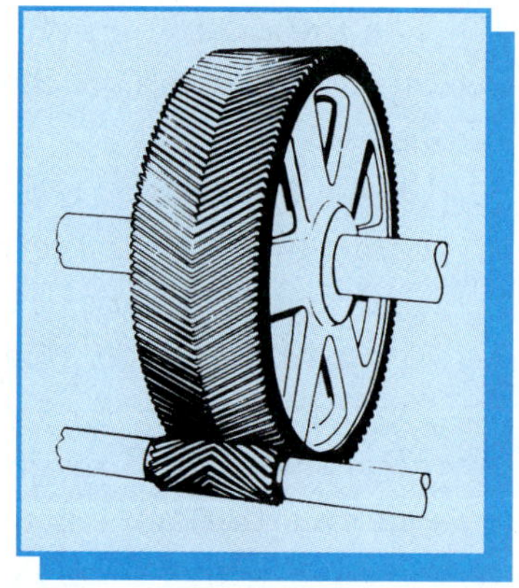

Figure 20-7
Herringbone gears *(Reprinted from* Mechanical Drafting *by Madsen, Shumaker, and Stewart, Delmar Publishers Inc.)*

Figure 20-8
Spiral bevel gears *(Courtesy of Boston Gear Division, Imo Corporation, Inc.)*

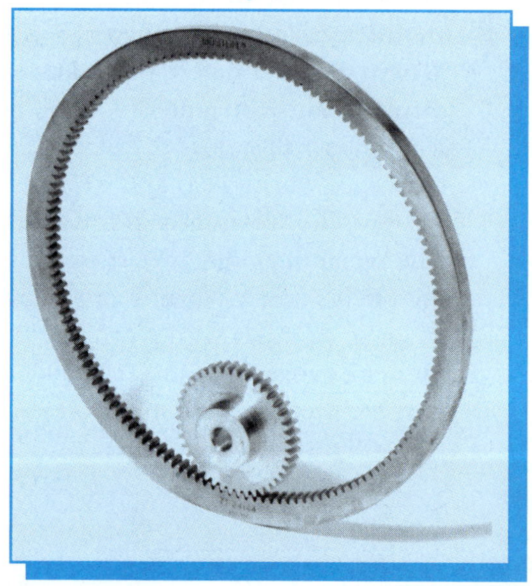

Figure 20-9
Ring gears *(Courtesy of Boston Gear Division, Imo Corporation, Inc.)*

Figure 20-10
Chain and sprocket drive *(Courtesy of Boston Gear Division, Imo Corporation. Inc.)*

ribbed undersides, or v-belts. To connect perpendicular shafts, belts can be twisted by one-quarter. Belts can also be crossed (twisted by one-half) between pulleys to reverse the direction of the second pulley.

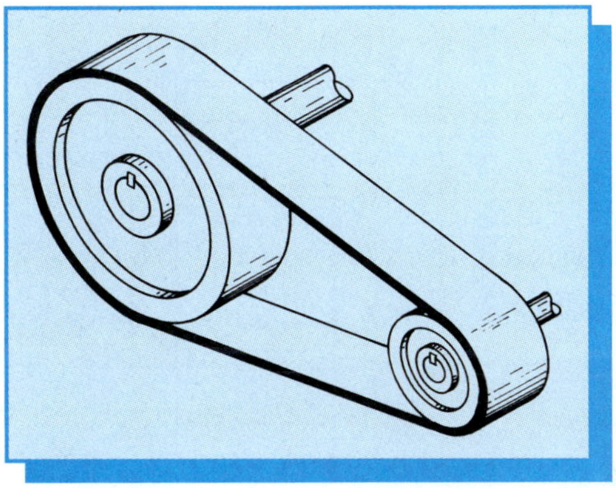

Figure 20-11
Belt drive

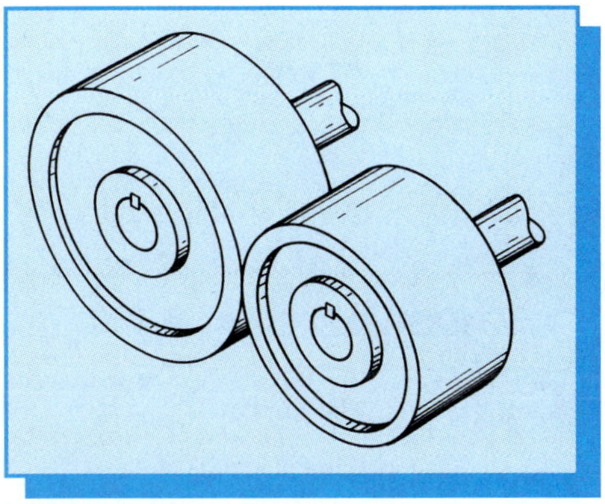

Figure 20-12
Friction drive

- **Friction wheels** are used to transmit power between shafts without using gears, chains, or belts. Friction wheels rely on heavy pressure to maintain sufficient contact to transfer power **(Fig. 20-12)**. Friction wheels engage easily but will slip when a predetermined amount of resistance is applied to the drive wheel.
- **Flanged sleeve gears** are used when an internal and powerful direct drive is required **(Fig. 20-13)**.

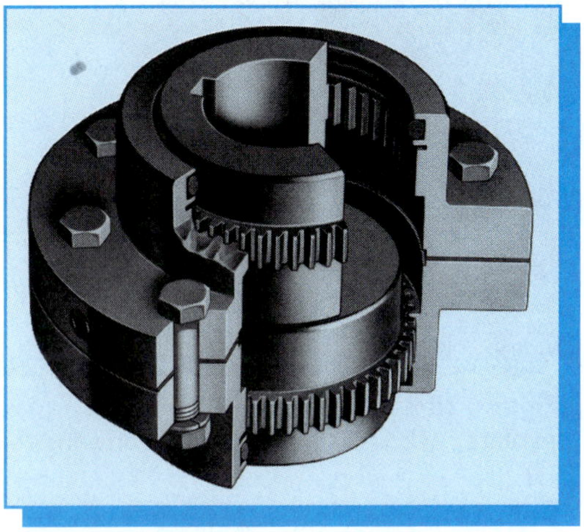

Figure 20-13
Flanged Sleeve Gears *(Courtesy of Falk Corp.)*

Gear Teeth

Two basic curves are used to define the shape of gear teeth—involute and cycloidal. An **involute curve (Fig. 20-14)** is produced by connecting the end points of a flexible and taut thread as it is unwound from a cylinder. A **cycloidal curve (Fig. 20-15)** is formed by connecting the position of a point at evenly spaced intervals on the circumference of a circle as that circle is rolled on a flat surface.

The actual shape (profile) of gear teeth is rarely used on working drawings. Simplified symbols are used to show the outline of gear teeth; however, complete cutting data notations are always necessary **(Fig. 20-16)**. When gear teeth must be shown on detail drawings, gear tooth templates of different types and sizes are used **(Figs. 20-17** and **20-18)**.

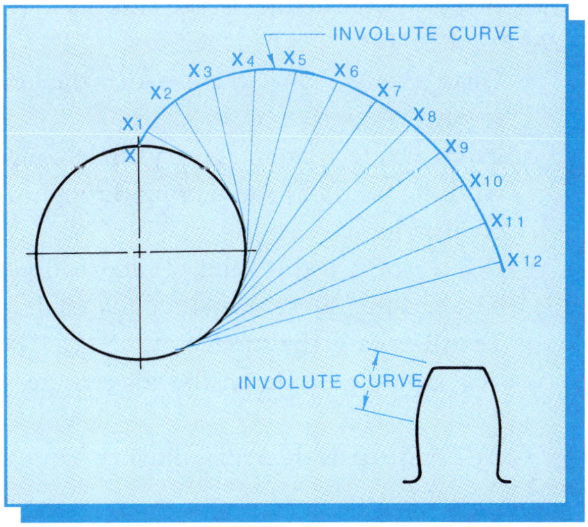

Figure 20-14
An involute curve is formed by a taut string as it unwinds from a cylinder.

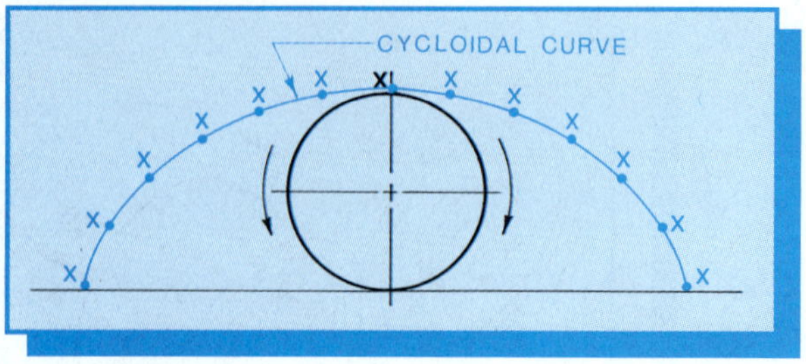

Figure 20-15
Plotting a cycloidal curve

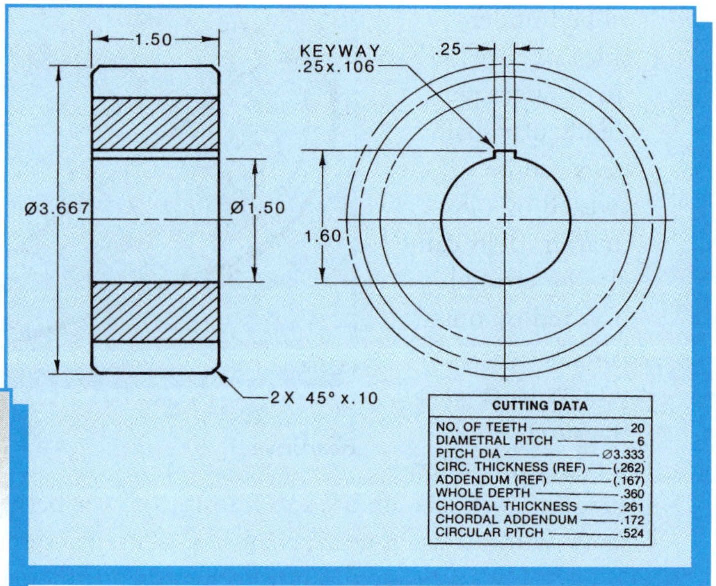

Figure 20-16
A working drawing for a gear seldom shows the actual outline of each gear tooth.

Figure 20-17
Involute rack and spur gear teeth drawing template

DIAMETRAL PITCH	TOOTH SIZE	DIAMETRAL PITCH	TOOTH SIZE
4		8	
5		10	
6		12	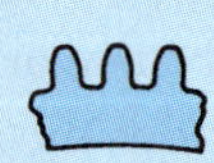

Figure 20-18
The diametral pitch determines the number and size of gear teeth on a gear. *(Reprinted from* Mechanical Drafting *by Madsen, Shumaker, and Stewart, Delmar Publishers Inc.)*

Gear Nomenclature

Before preparing working drawings involving gears, it is important to understand the following parts of a gear that define its size and shape **(Figs. 20-19** and **20-20)**:

- **Gear axis** is the central axis that the gear revolves around.
- **Pitch circle** is an imaginary circle concentric with the gear axis and passing through the pitch points.
- **Pitch point** is the point of intersection between the tooth face and pitch circle.
- **Tooth face** is the surface of the gear tooth that applies pressure on the adjacent gear's tooth.
- **Addendum** is the radial distance from the top of the tooth to the pitch circle.
- **Addendum circle (outside diameter)** is a circle connecting the outer edge of the gear teeth.

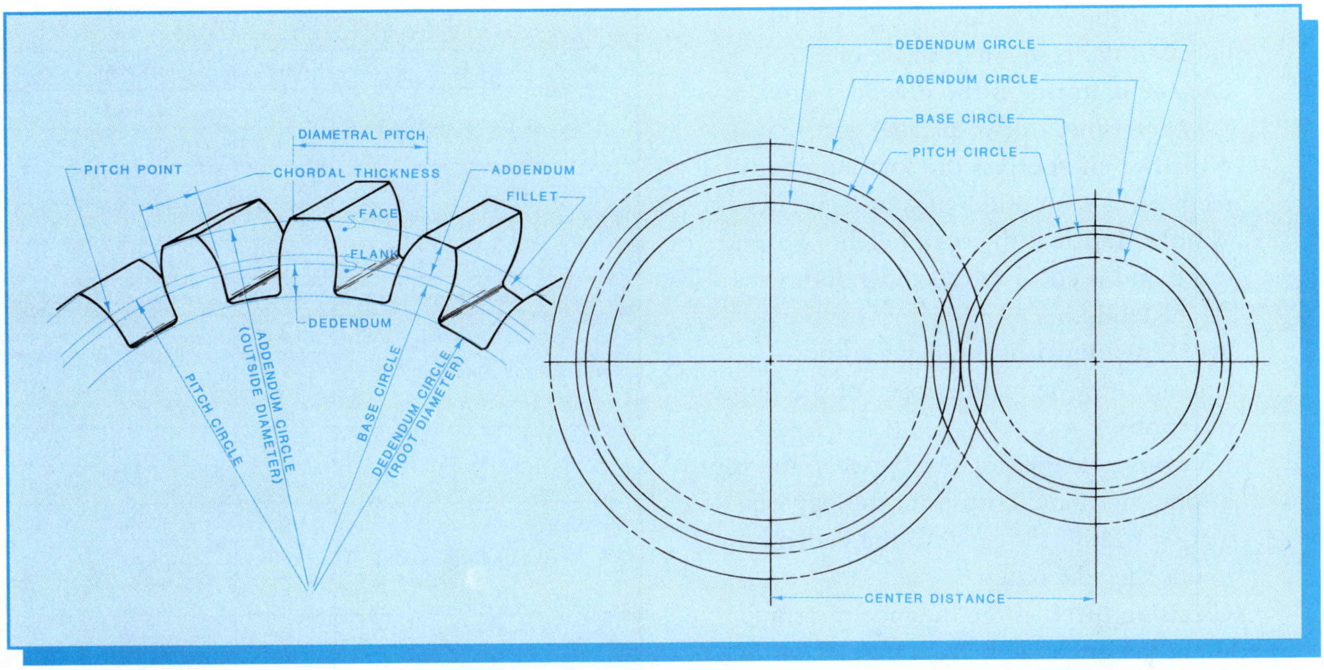

Figure 20-19
Gear terminology

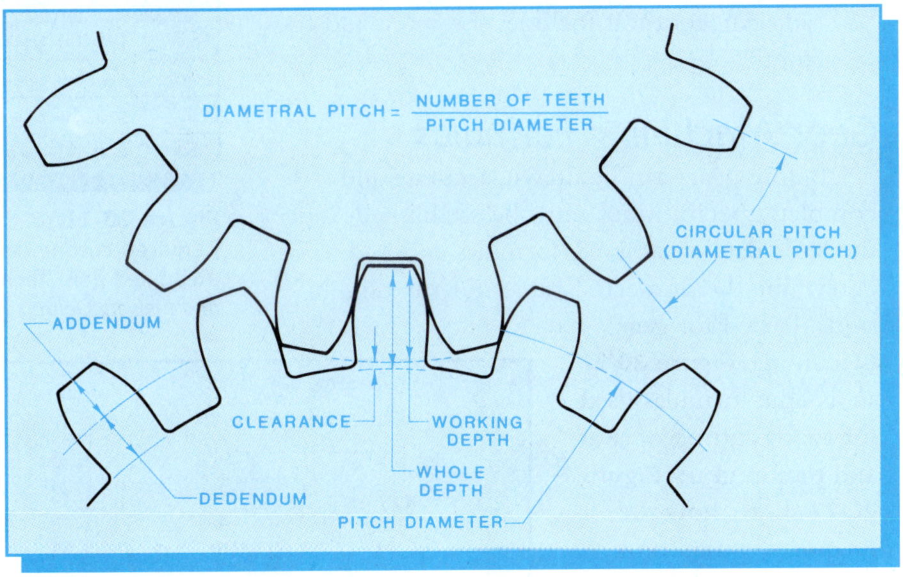

Figure 20-20
Gear terminology

- **Chordal thickness** is the thickness of a tooth measured on the pitch circle.
- **Pitch diameter** is the diameter of the pitch circle.
- **Diametral pitch** is the ratio equal to the number of teeth on the gear per inch of pitch diameter.
- **Dedendum** is the radial distance from the pitch circle to the root diameter.
- **Dedendum circle** is the diameter of the pitch diameter, minus the dedendum.
- **Root diameter** is the same as the dedendum circle. It is the smallest diameter for the base of a gear tooth.
- **Base circle** is the circle from which the involute tooth profile begins. When drawing gear teeth with a template, it is not necessary to lay out and draw the involute curve.
- **Fillet** is the concave intersection between the root and the side of a gear tooth.

- **Flank** is the lower side of a gear tooth between the base circle and root.
- **Center distance** is the distance between the centerlines of two parallel gear shafts.
- **Number of teeth** is the total number of teeth on the circumference of a gear.
- **Whole depth** is the total height of a gear tooth and is equal to the addendum, plus the dedendum.
- **Working depth** is the depth a tooth extends into the mating space of the opposite tooth.
- **Clearance** is the distance between the top of one tooth and bottom of the opposite mating tooth. It is equal to the dedendum minus the addendum.
- **Circular pitch** is the distance measured along the pitch circle from a point on one tooth to the corresponding point on the adjacent tooth. It includes the space and the tooth.

Gear Machining Formulas

Before gears can be drawn, accurate and complete specifications must be established. Once the basic data is gathered, formulas are used to determine the numerical values for the dimensions needed for gear machining. **Figure 20-21** shows the formulas used for tooth cutting of spur and pinion gears. **Figure 20-22** shows how the final data is used on a spur gear detail drawing.

Cutting data and formulas for rack and pinion gears is basically the same as for spur gears. Since the rack is a straight surface, all circular dimensions become linear, with the addendum, dedendum, whole depth, and tooth thickness identical to the mating pinion **(Fig. 20-23)**. **Figure 20-24** shows a working drawing of a rack accompanied with

REQUIRED CUTTING DATA			
ITEM	TO FIND:	HAVING	FORMULA
1	Number of teeth (N)	D & P	D x P
		DO & P	(DO x P) - 2
2	Diametral pitch (P)	p	$\frac{3.1416}{p}$
		D & N	$\frac{N}{D}$
		DO & N	$\frac{N+2}{DO}$
3	Pressure angle (Ø)	–	20° STANDARD 14°-30' OLD STANDARD
4	Pitch diameter (D)	N & P	$\frac{N}{P}$
		N & DO	$\frac{N \times DO}{N+2}$
		DO & P	$DO - \frac{2}{P}$
5	Whole depth (ht)	a & b	a + b
		P	$\frac{2.157}{P}$
6	Outside diameter (DO)	N & P	$\frac{N+2}{P}$
		D & P	$D + \frac{2}{P}$
		D & N	$\frac{(N+2) \times D}{N}$
7	Addendum (a)	P	$\frac{1}{P}$
8	Working depth (hk)	a	2 x a
		P	$\frac{2}{P}$
9	Circular thickness (t)	P	$\frac{3.1416}{2 \times P}$
10	Chordal thickness	N, D & a	$\sin\left(\frac{90°}{N}\right) \times D$
11	Chordal addendum	N, D & a	$\left[1 - \cos\left(\frac{90°}{N}\right)\right] \times \frac{D}{2} + a$
12	Dedendum (b)	P	$\frac{1.157}{P}$

Figure 20-21
Required cutting data and formulas for spur gears *(Reprinted from* Technical Drawing and Design *by Goetsch and Nelson, Delmar Publishers Inc.)*

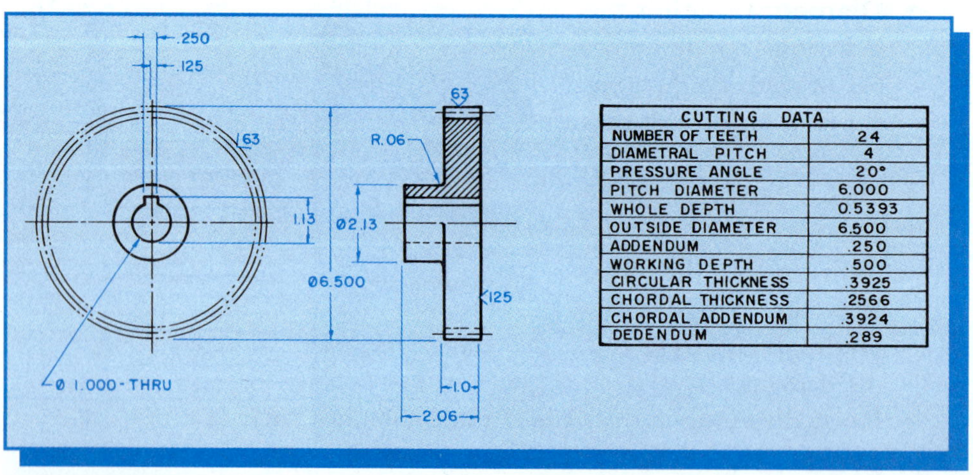

Figure 20-22
Example of a spur gear detail working drawing *(Reprinted from* Technical Drawing and Design *by Goetsch and Nelson, Delmar Publishers Inc.)*

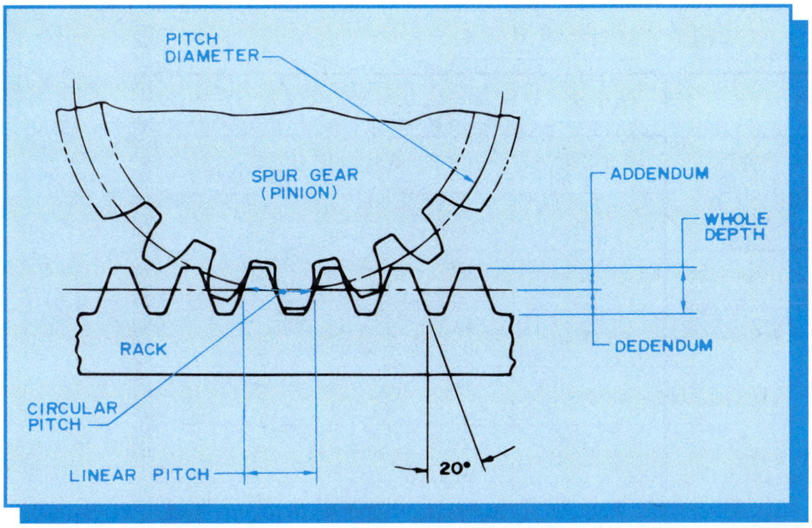

Figure 20-23
Example of rack gear terminology *(Reprinted from* Mechanical Drafting *by Madsen, Shumaker, and Stewart, Delmar Publishers Inc.)*

the data notations needed for machining.

Information for cutting bevel gear teeth is based on the same involute curve as spur gear teeth; however, bevel gear teeth are tapered in a cone shape. Bevel gear teeth are designed in pairs and are not interchangeable. Dimensions are determined by the same formulas used for spur gears, with some modifications. The pitch diameter of bevel gears is the diameter of the base of the cone. The circular pitch is measured on this base circle **(Figs. 20-25** and **20-26)**. Working drawings of bevel gears include only the outline of the gear blank. This is normally drawn as a single sectional view **(Fig. 20-27)** with all related gear tooth data provided in tabular notations.

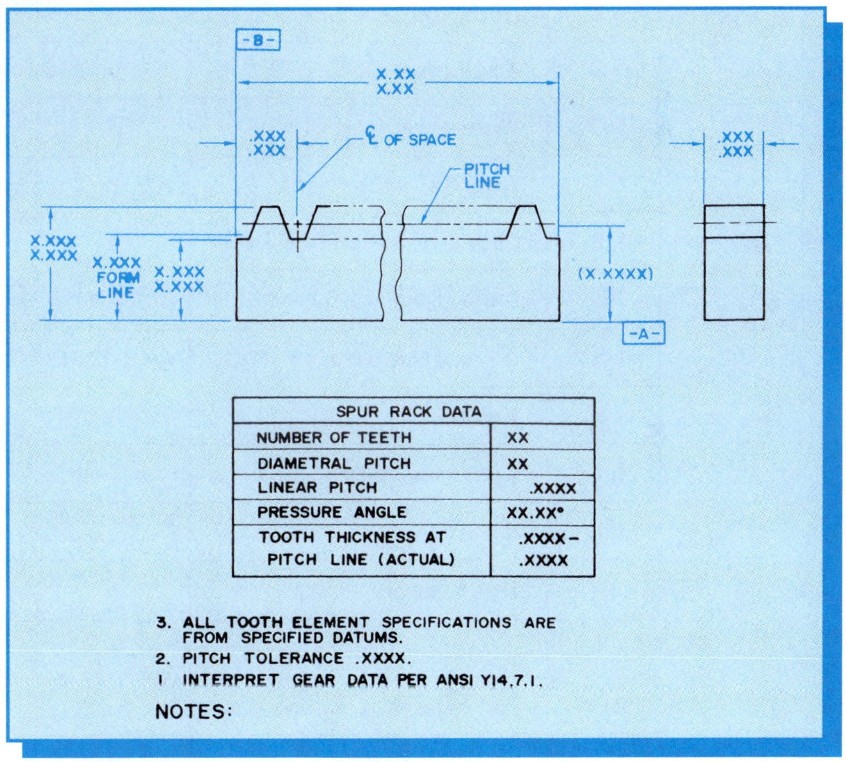

Figure 20-24
Partial example of a rack gear working drawing *(Reprinted from* Mechanical Drafting *by Madsen, Shumaker, and Stewart, Delmar Publishers Inc.)*

Worm gears consist of a worm (screw form) and a worm gear (wheel form). The worm is basically a screw with single or multiple threads. The teeth on the worm gear are curved to conform to the shape of the worm teeth. Large reductions and high mechanical advantages result from the high ratios possible between worm and worm gear (up to 50:1). One revolution of the worm may result in a lead of only one tooth width on the worm gear. The lead of the worm is the distance the worm gear will travel

REQUIRED CUTTING DATA				
ITEM	TO FIND:	HAVING	FORMULA	
			SPUR	PINION
1	Number of teeth (N)	–	AS REQ'D.	
2	Diametral pitch (P)	p	$\frac{3.1416}{p}$	
3	Pressure angle (Ø)	–	20° STANDARD 14°-30' OLD STANDARD	
4	Cone distance (A)	D & d	$\sin d \overline{\big)\frac{D}{2}}$	
5	Pitch distance (D)	p	$\frac{N}{p}$	
6	Circular thickness (t)	p	$\frac{1.5708}{p}$	
7	Pitch angle (d)	N & d (of pinion)	90°-d (pinion)	$\tan d \frac{N \text{ pinion}}{N \text{ gear}}$
8	Root angle (γR)	d & δ	d - δ	
9	Addendum (a)	p	$\frac{1}{p}$	
10	Whole depth (ht)	p	$\frac{2.188}{p} + .002$	
11	Chordal thickness (C)	D & d	$\frac{1}{2}\left(\frac{D}{\cos d}\right)$	$1-\cos\left(\frac{\frac{90°}{N}}{\cos d}\right)+a$
12	Chordal addendum (aC)	d	$\sin\left(\frac{\frac{90°}{N}}{\cos d}\right)$	
13	Dedendum (bC)	P	$\frac{2.188}{P}$ - a(pinion)	$\frac{2.188}{P}$ - a (gear)
14	Outside diameter (DO)	D, a & d	D + (2 x a) x cos. d	
15	Face	A	1/3 A (max.)	
16	Circular pitch (p)	p & N	$\frac{3.1416 \times p}{N}$	
17	Ratio	N gear & N pinion	$\frac{N \text{ gear}}{N \text{ pinion}}$	
18	Back angle (γO)	–	SAME AS PITCH ANGLE	
19	Angle of shafts	–	90°	
20	Part number of mating gear	–	AS REQ'D.	
21	Dedendum angle δ	A & b	$\frac{b}{A}$ = tan δ	

Figure 20-25
Required cutting data and formulas for bevel gears *(Reprinted from* Technical Drawing and Design *by Goetsch and Nelson, Delmar Publishers Inc.)*

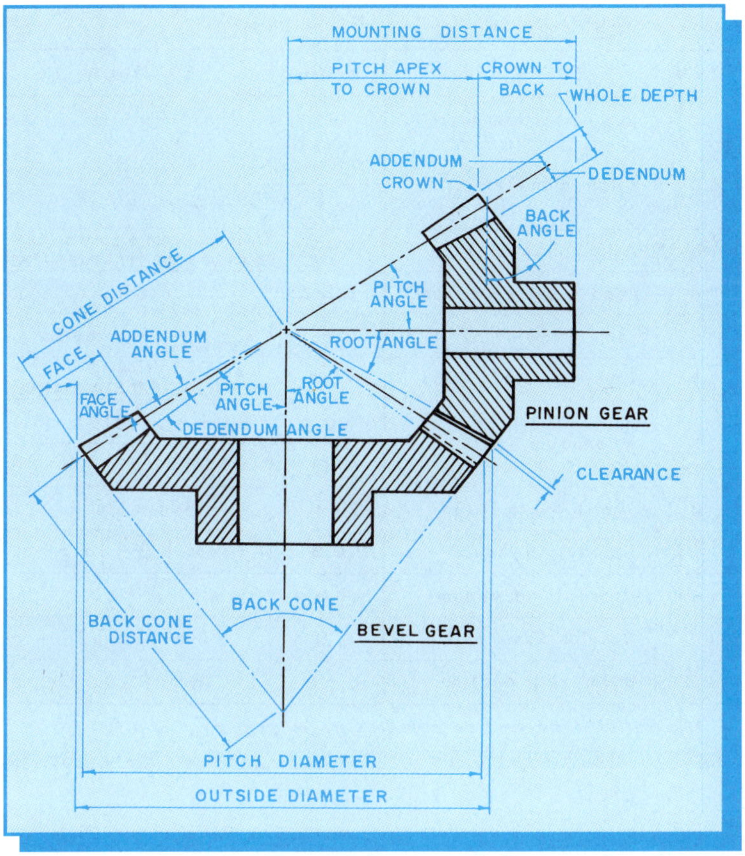

Figure 20-26
Example of bevel gear terminology *(Reprinted from* Technical Drawing and Design *by Goetsch and Nelson, Delmar Publishers Inc.)*

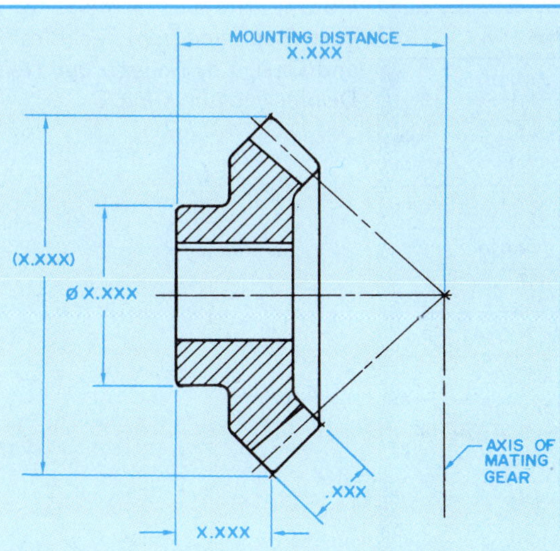

GEAR TOOTH DATA	
NUMBER OF TEETH	XX
DIAMETRAL PITCH	XX
PRESSURE ANGLE	XX.XX°
CONE DISTANCE	X.XXX
PITCH DIAMETER	X.XXX
CIRCULAR THICKNESS (REF)	.XXXX
PITCH ANGLE	XX.XX°
ROOT ANGLE	XX.XX°
ADDENDUM	.XXX
WHOLE DEPTH	.XXX
CHORDAL ADDENDUM	.XXX
CHORDAL THICKNESS	.XXX

1. INTERPRET TOOTH DATA PER ANSI Y14.7.1

NOTES:

Figure 20-27
Example of a bevel gear working drawing *(Reprinted from* Mechanical Drafting *by Madsen, Shumaker, and Stewart, Delmar Publishers Inc.)*

in one revolution of the worm. Normally, only a one-view drawing of a worm gear assembly is necessary when cutting data is supplied. **Figure 20-28** shows cutting data for a worm. **Figure 20-29** shows cutting data for a worm gear. When a second view is required, the relationship of gear teeth is shown with radial centerlines on the side view, and interlocking profiles sectioned on the front view **(Fig. 20-30)**.

REQUIRED CUTTING DATA

ITEM	TO FIND	HAVING:	FORMULA
1	Number of teeth (N)	P	$\frac{3.1416}{P}$
2	Pitch diameter (D)	Pa	(2.4 x Pa)+ 1.1
		DO & a	DO - (2 x a)
3	Axial pitch (Pa)	–	Distance from a point on one tooth to same point on next tooth
4	Lead (L) Right or Left	p & N	p x N
5	Lead angle (La)	L & D	$\frac{L}{3.1416 \times D}$ = tan La
6	Pressure angle (Ø)	–	20° STANDARD 14°-30' OLD STANDARD
7	Addendum (a)	p	p x .3183
		P	$\frac{1}{P}$
8	Whole depth (ht)	Pa	.686 x Pa
9	Chordal thickness	N,D & a	$\left[1 - \cos\left(\frac{90°}{N}\right)\right] \times \frac{D}{2} + a$
10	Chordal addendum	N,D & a	$\sin\left(\frac{90°}{N}\right) \times D$
11	Outside diameter (DO)	D & a	D + (2 x a)
12	Worm gear part no.	–	AS REQ'D.

Axial pitch (Pa) must be same as worm gear circular pitch (p)

Figure 20-28
Required cutting data and formulas for a worm *(Reprinted from* Technical Drawing and Design *by Goetsch and Nelson, Delmar Publishers Inc.)*

REQUIRED CUTTING DATA

ITEM	TO FIND:	HAVING	FORMULA
1	Number of teeth (N)	–	AS REQ'D.
2	Pitch diameter (D)	N & p	$\frac{N \times p}{3.1416}$
3	Addendum (a)	p	p x .3181
		P	$\frac{1}{P}$
4	Whole depth (ht)	p	p x .6866
		P	$\frac{2.157}{P}$
5	Lead (L) Right-Left	p & N	p x N
6	Worm part no.	–	AS REQ'D.
7	Pressure angle Ø	–	20° STANDARD 14°-30' OLD STANDARD
8	Outside diameter (DO)	Dt & Pa	Dt + .4775 x Pa
9	*Circular pitch (p)	P	$\frac{3.1416}{P}$
		L & N	$\frac{L}{N}$
10	Diametral pitch (P)	p	$\frac{3.1416}{p}$
11	Throat diameter (Dt)	D & Pa	D + .636 x Pa
12	Ratio of worm/worm gear	N worm & N worm gear	$\frac{N \text{ worm gear}}{N \text{ gear}}$
13	Center to center distance between worm & worm gear.	D worm & D worm gear	$\frac{D \text{ worm} + D \text{ worm gear}}{2}$

Circular pitch (p) must be same as worm Axial pitch (Pa)

Figure 20-29
Cutting data and formulas for a worm gear *(Reprinted from* Technical Drawing and Design *by Goetsch and Nelson, Delmar Publishers Inc.)*

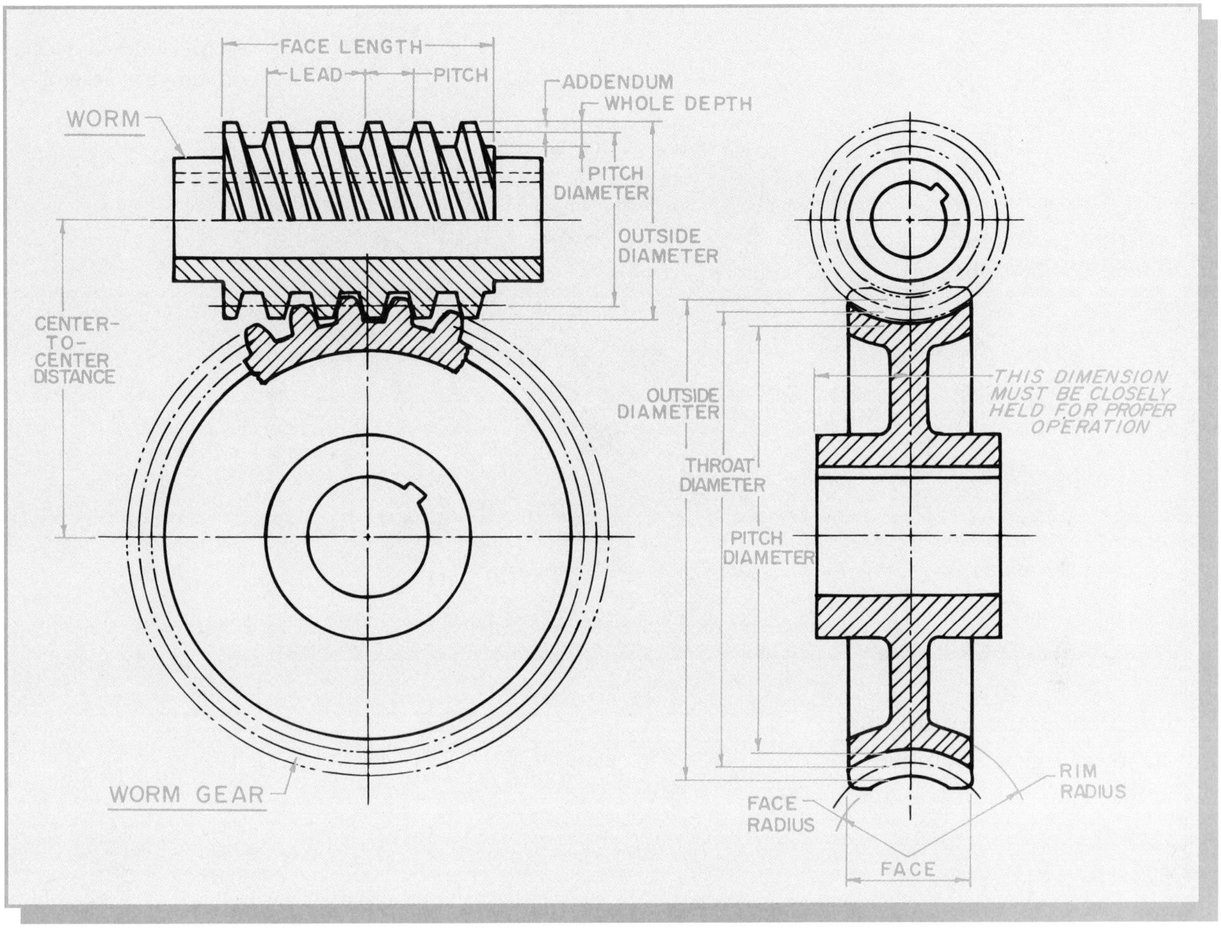

Figure 20-30
Example of worm and worm gear terminology
(Reprinted from Technical Drawing and Design *by Goetsch and Nelson, Delmar Publishers Inc.)*

Pictorial Representation

When multiple gears are connected with shafts that lie on many different planes, the positioning and relationship of gears is often difficult to visualize. This is especially true when the position of gears change as they do in an automobile transmission. Pictorial drawings (usually isometric) are used in those cases as an aid to interpretation **(Fig. 20-31)**. Pictorial drawings, unless prepared to a very large scale, do not include teeth details, since these drawings are not used for manufacturing.

Figure 20-32 is a pictorial assembly drawing designed to show the relationship of gears and other mechanical components. **Figure 20-33** is a pictorial cutaway view of a gear assembly revealing the internal

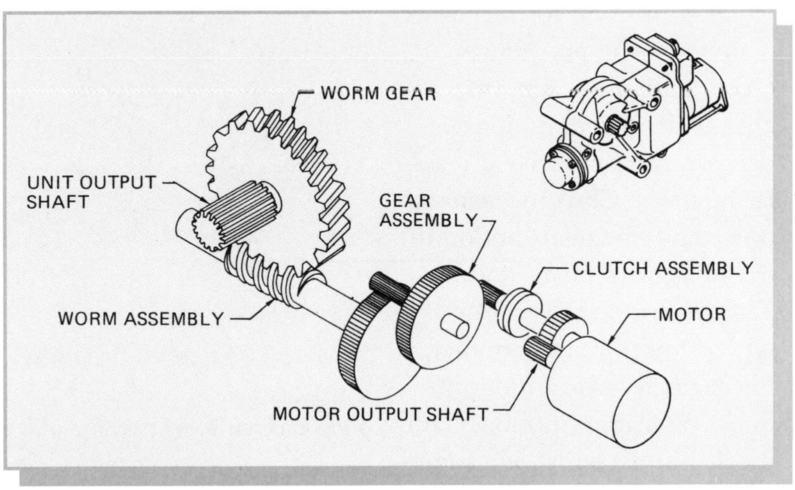

Figure 20-31
Pictorial drawing of a motor-driven gear assembly

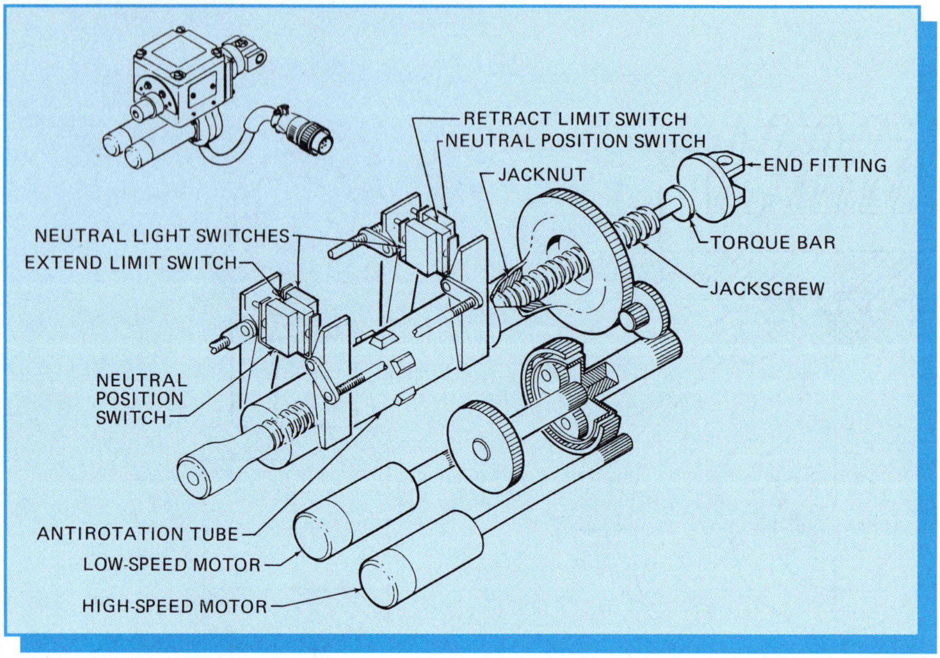

Figure 20-32
Pictorial drawing of a motor-driven gear assembly

configuration and relationship of gears and gear assemblies.

Cam Mechanisms

It is often necessary to convert the rotary motion of a motor into linear motion for many mechanical operations. It is also often necessary to convert this motion back again to rotary motion for successive manufacturing procedures. Gears and screws can only change or maintain motions that are evenly spaced at regular intervals. **Cam** mechanisms can convert regular motion to irregular or unusual motion. They also can convert one type of motion into another, for example, rotary motion to linear oscillating motion. Cam mechanisms consist of a cam, follower, and camshaft.

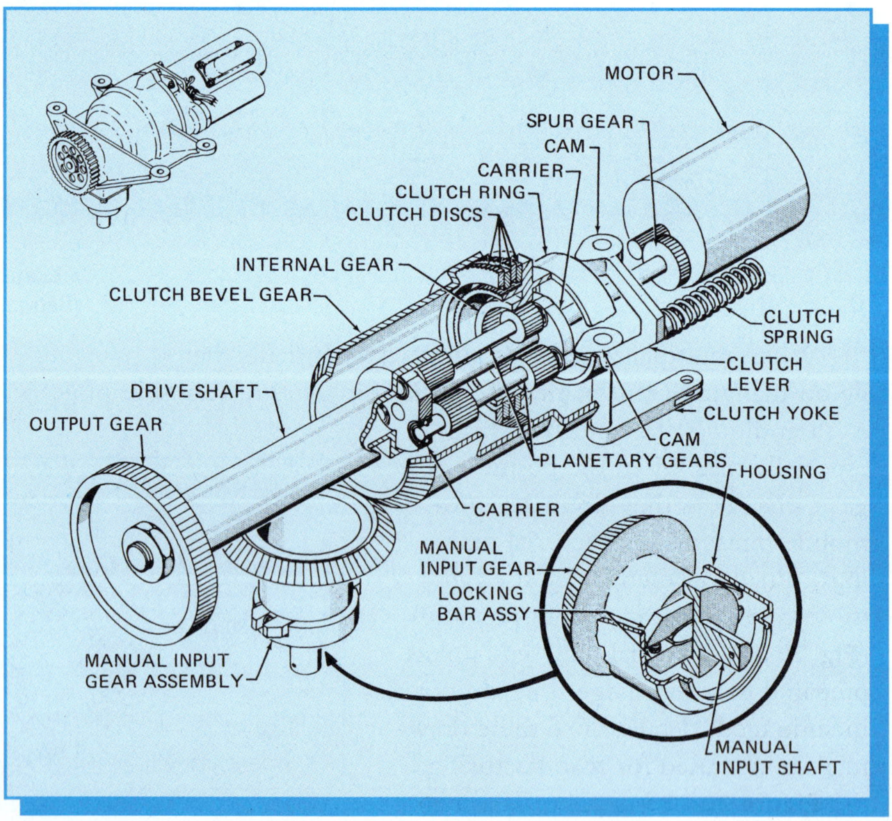

Figure 20-33
Pictorial cutaway drawing of a motor-driven gear assembly

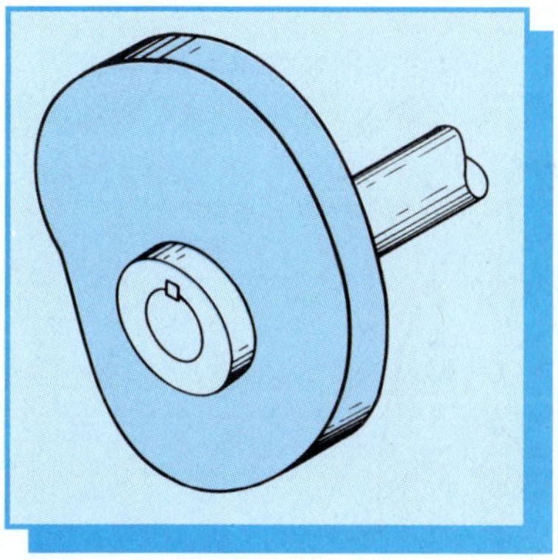

Figure 20-34
Plate cam

Figure 20-35
Drum cam

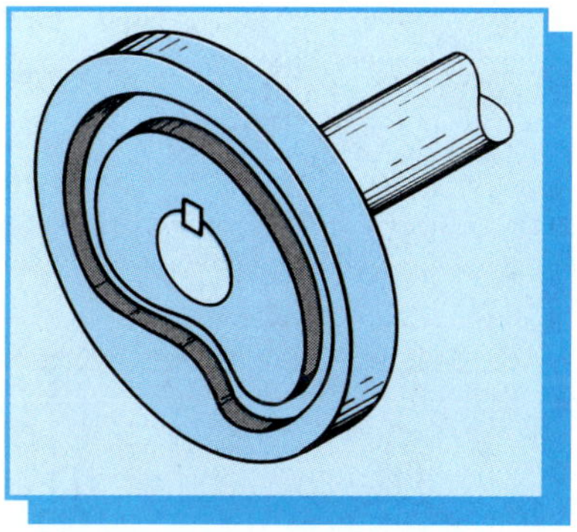

Figure 20-36
Face cam

Cams are divided into three types—**plate cams (Fig. 20-34)**, **drum cams (Fig. 20-35)**, and **face cams (Fig. 20-36)**. Cam shapes are determined by the desired motion of a follower. A plate cam follower moves vertically as the cam rotates, converting rotary motion into vertical linear (displacement) reciprocating movement **(Fig. 20-37)**. Drum cams convert circular motion to linear horizontal reciprocating movement by guiding the follower through a groove **(Fig. 20-38)**.

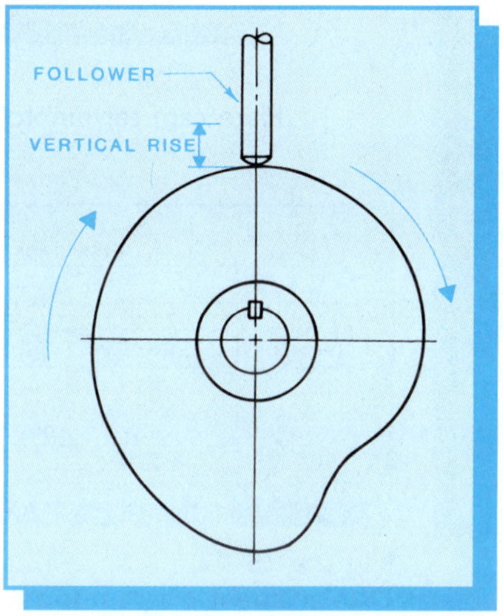

Figure 20-37
The circular motion of the plate cam is converted to a vertical linear movement by the follower.

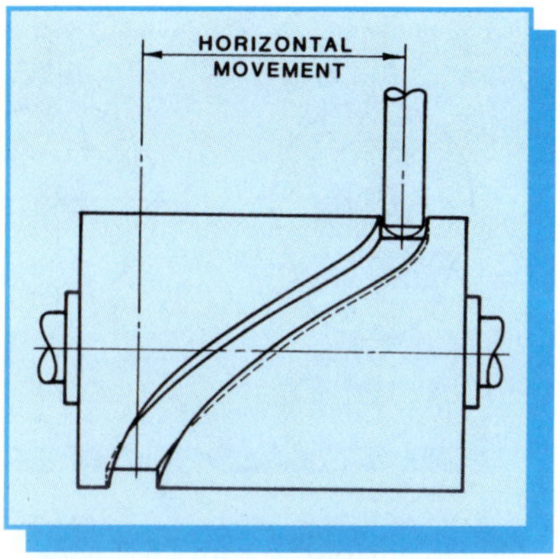

Figure 20-38
The circular motion of the drum cam is converted to a horizontal linear movement by the follower.

Cam Followers

There are four types of cam followers—**roller, pointed, round,** and **flat faced (Fig. 20-39)**. Roller followers are used the most since they operate better at higher speeds with less friction and wear. The exact shape of a cam determines the exact timing and motion limits of the follower during one cam revolution. Think of a cam as a simple inclined plane with uneven surface contours. Although **Figure 20-40** shows the follower position at only four points, it moves smoothly from one position to another. **Figure 20-41** summarizes the nomenclature used in the design and drawing of a plate cam.

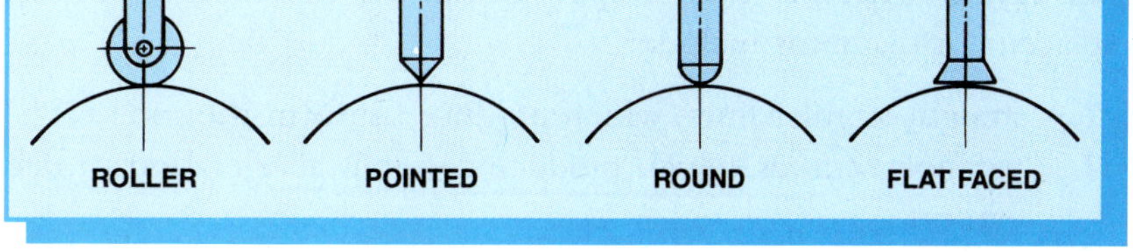

Figure 20-39
Basic types of cam followers

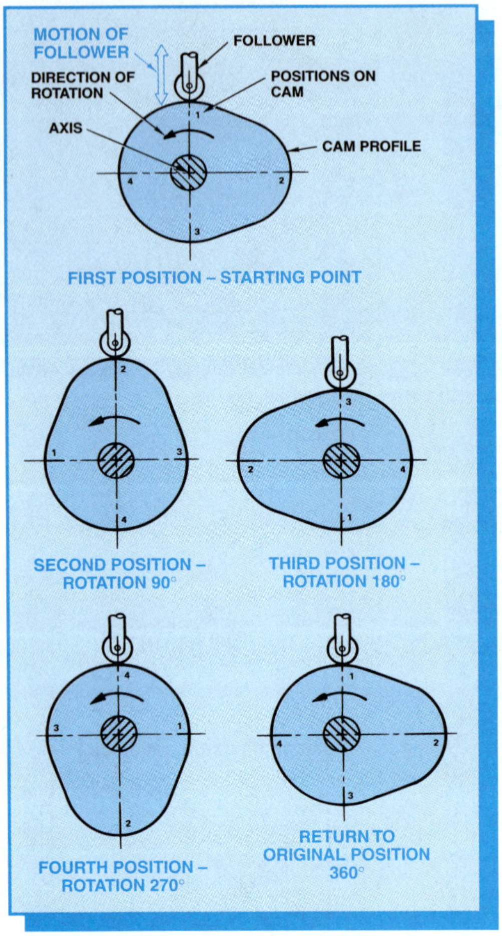

Figure 20-40
Movement of the cam's follower

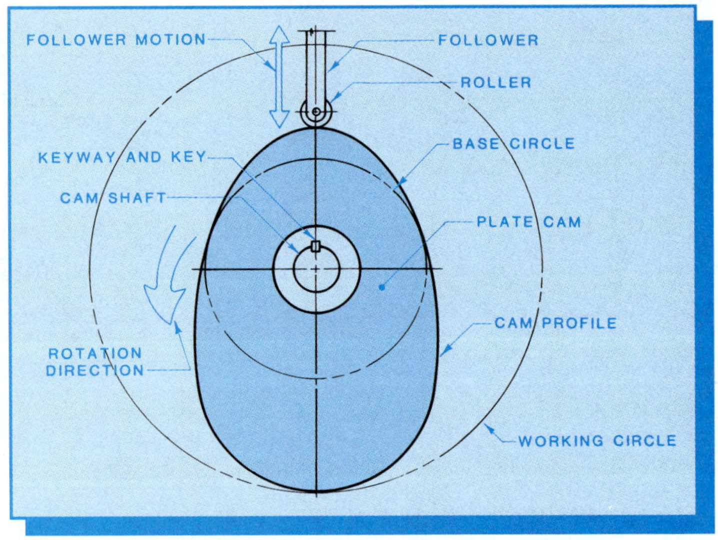

Figure 20-41
Plate cam terminology

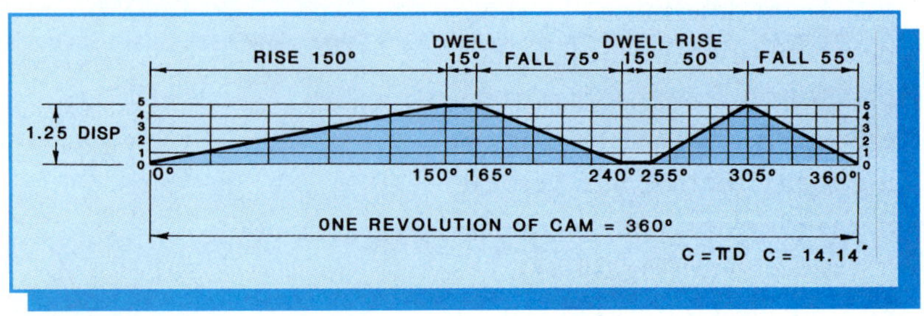

Figure 20-42
Displacement diagram for a cam

Displacement Diagram

The first step in designing and drawing a cam shape is the development of a **displacement diagram.** A displacement diagram is a curve representing the desired position (displacement) of the follower at successive intervals during one revolution of the **camshaft (Fig. 20-42)**. **Geometric line shapes** used in the development of displacement diagrams include:

1. straight angular lines, which produce uniform motion
2. parabolic curves, which produce constant acceleration or deceleration
3. harmonic curves, which produce a harmonic motion
4. straight horizontal lines, which hold the follower at a fixed **(dwell** or **rest)** position

All nonhorizontal lines result in an upward movement (**rise**) of the follower or in a downward movement (**fall**) of the follower. The displacement diagram in **Figure 20-42** illustrates two uniform rise and fall curves with two dwell (rest) periods. In **Figure 20-42**, the horizontal axis represents 360° or one revolution of the cam. The vertical grid represents the maximum distances the follower will move during one revolution of the cam. The profile line identifies the exact position of the follower required at specific degrees of rotation.

Displacement diagrams are flat, while cams are basically round. The flat profile intersections or the displacement diagram must be converted to radial distances to create a cam shape **(Fig. 20-43)**. In **Figure 20-43**, all horizontal distances are converted to degrees on concentric circles. These are spaced evenly to represent the movement range of the follower. Intersecting points are then plotted to coincide with the intersections on the displacement diagram. Connecting these points creates the outline shape of the cam that will produce the follower action plotted in the displacement diagram.

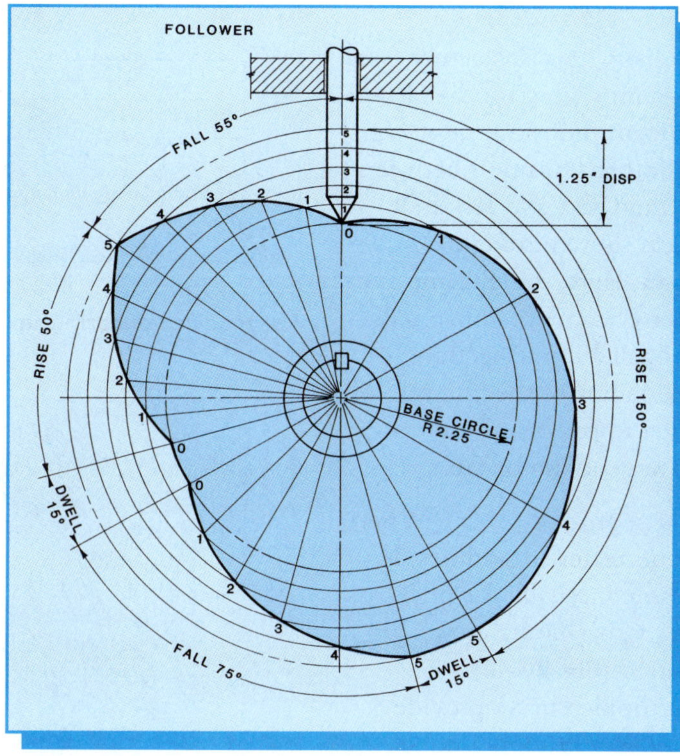

Figure 20-43
Layout of a cam drawing made from the displacement diagram in Figure 20-42

Cam Motions

Care must be taken to provide the correct cam shape, since that controls the movement of the follower. The follower must move smoothly, without jerks, at the maximum planned speed of rotation. The following displacement curves are designed to produce specific results.

Uniform Motion. A displacement diagram and resulting cam shape designed to produce a uniform motion at a constant speed are shown in **Figure 20-44**. However, the changes in direction from this curve are abrupt, and cause a jerky motion in the follower.

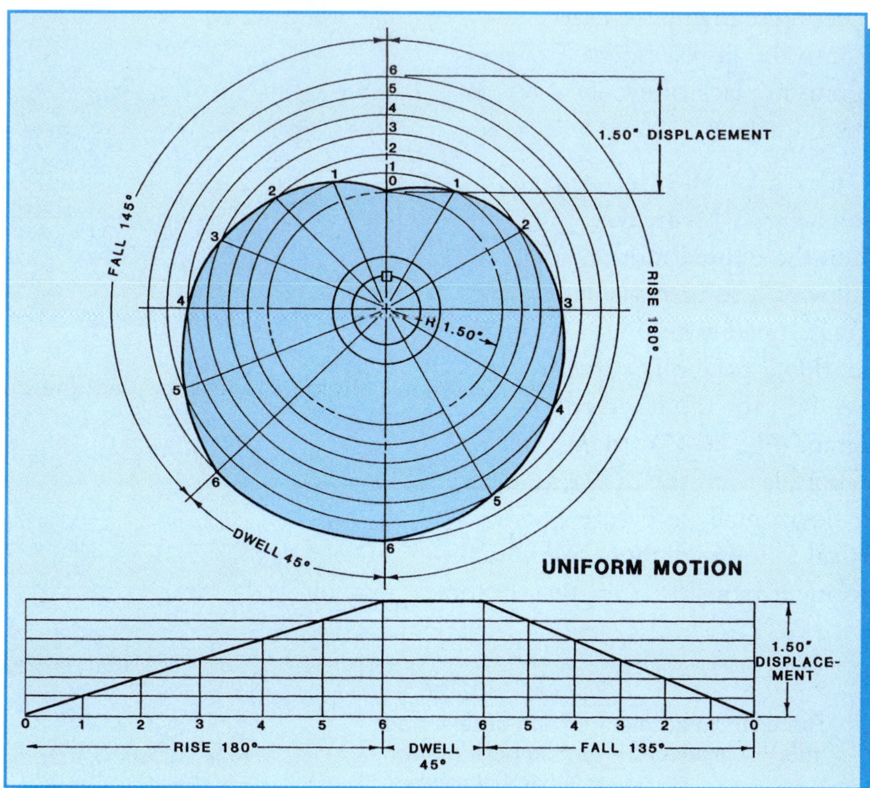

Figure 20-44
The uniform cam motion gives a jerk to the follower at the start and end of each phase.

Modified Uniform Motion. A displacement diagram with a modified uniform motion produces a smoother transition between rise and rest, and rest and fall, is shown in **Figure 20-45**. Here, straight line intersections are converted to small arc intersections. The radius of the arcs is equal to one-third the displacement rise or fall.

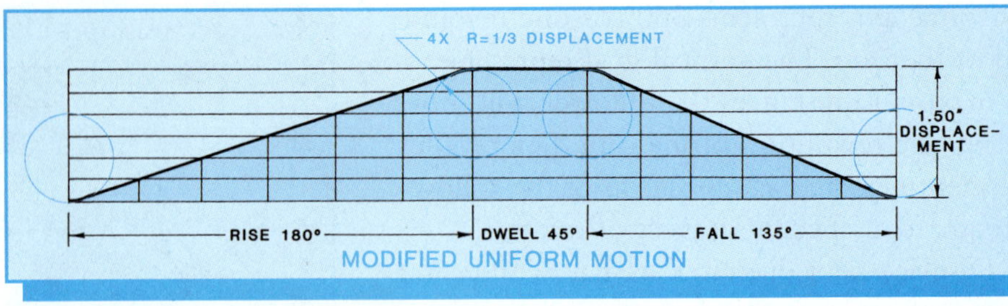

Figure 20-45
The modified uniform cam motion will smooth out the follower's jerky motion.

Harmonic Motion. When even greater smoothness is required, usually for high speeds, harmonic curves are plotted **(Fig. 20-46)**. Harmonic curves provide a smooth transition among rise, fall, and dwell. They also allow smoother start and stop movements of the follower. A harmonic curve is plotted by projecting segments of a circle onto the displacement diagram in place of evenly spaced distances.

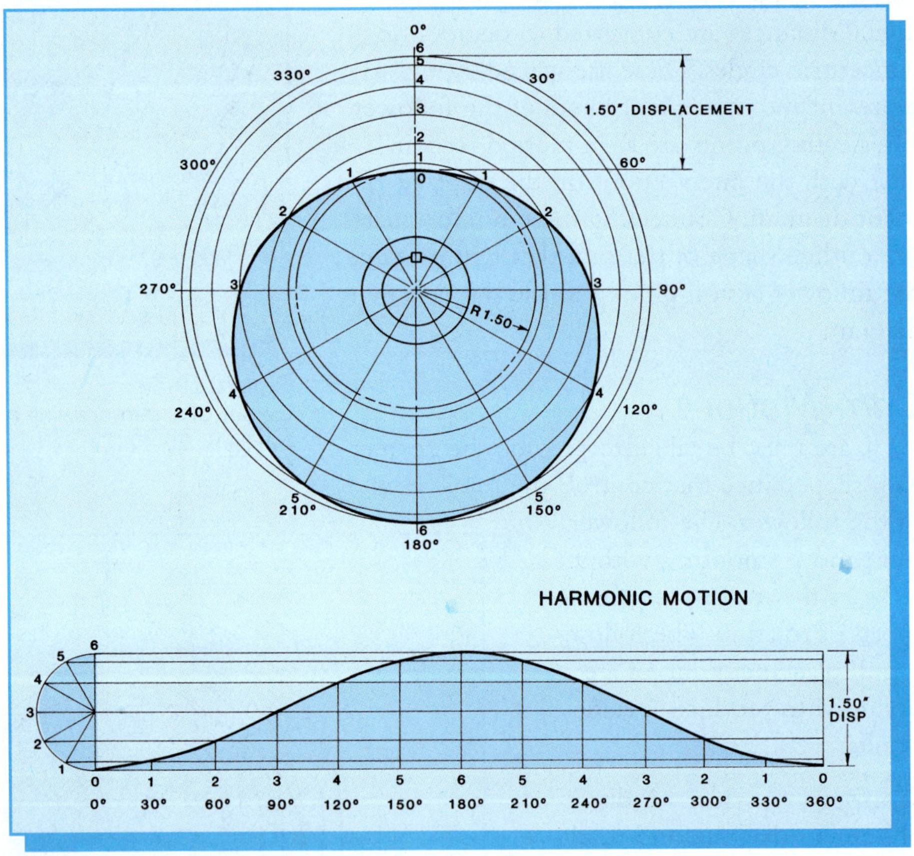

Figure 20-46
Harmonic motion provides the smoothest motion.

Uniform Accelerated and Decelerated Motion. When the desired motion of a follower is to increase or decrease steadily and smoothly, a parabolic curve is plotted on the displacement diagram **(Fig. 20-47)**. In plotting a parabola, lay out the 360° line or time line horizontally as before. Divide the vertical follower distance line into displacement ratios of 1:3:5. Project these

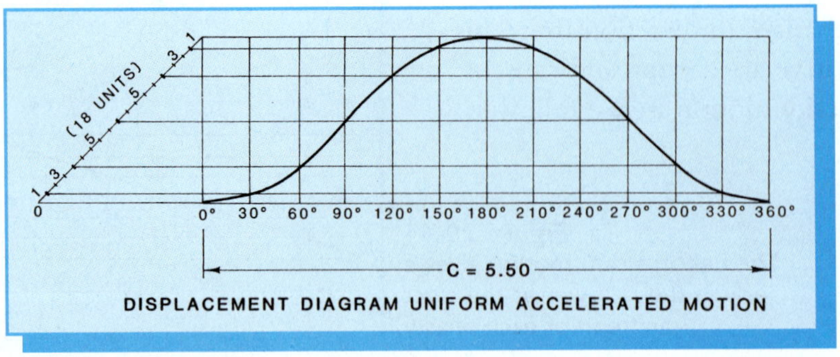

Figure 20-47
The uniform accelerated and decelerated motion produces a very smooth motion with a parabolic curve.

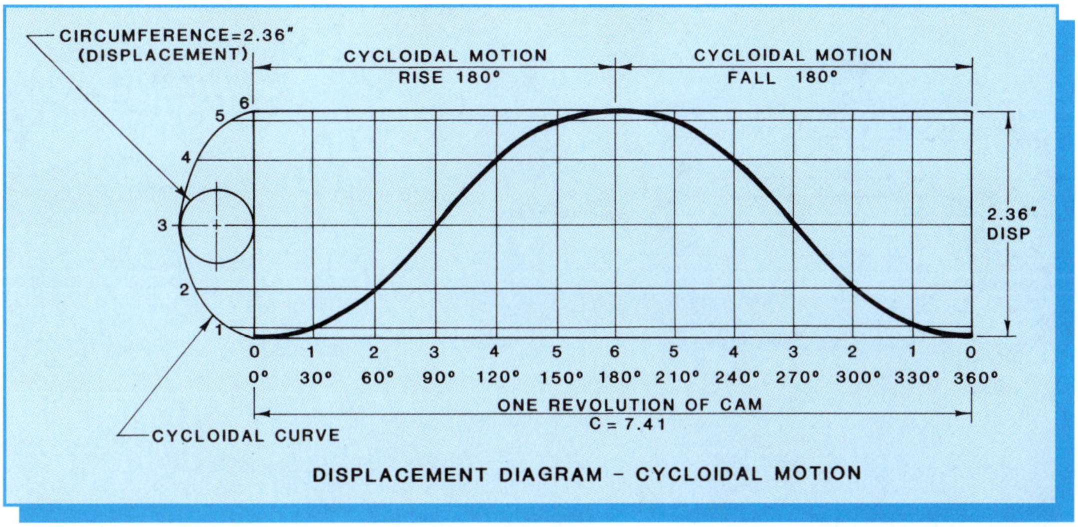

Figure 20-48
Cycloidal motion is the smoothest of the cam's motions.

lines to intersect degree lines, and connect the intersections to produce a parabola.

Cycloidal Motion. A displacement diagram designed to produce cycloidal motion in a cam mechanism is shown in **Figure 20-48**. Cycloidal curves produce extremely smooth curves by eliminating the dwell contour. Cycloidal curves are used in cams that operate at optimum efficiency at high speeds.

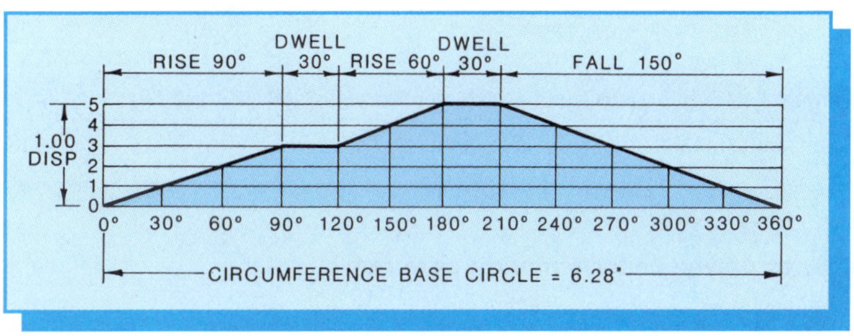

Figure 20-49
Displacement diagram designed with a uniform motion

Design and Drawing Procedures

Figures 20-49 through **20-52b** illustrate the specific steps used to design, draw, and dimension cams in detail.

A uniform motion displacement diagram **(Fig. 20-49)** is used to complete the cam profile shown in **Figure 20-50**.

The procedures used to develop and draw a uniform motion plate cam are shown in **Figure 20-51**. This involves laying out the base circle, followed by the segmenting of the circumference.

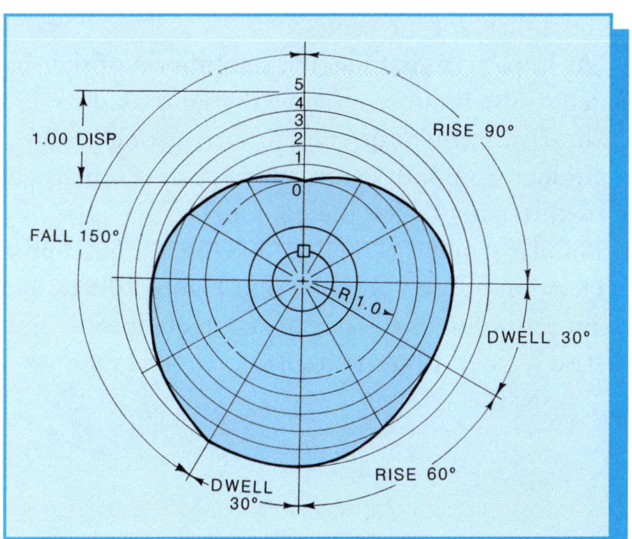

Figure 20-50
Uniform motion plate cam drawn from the displacement diagram in Figure 20-49

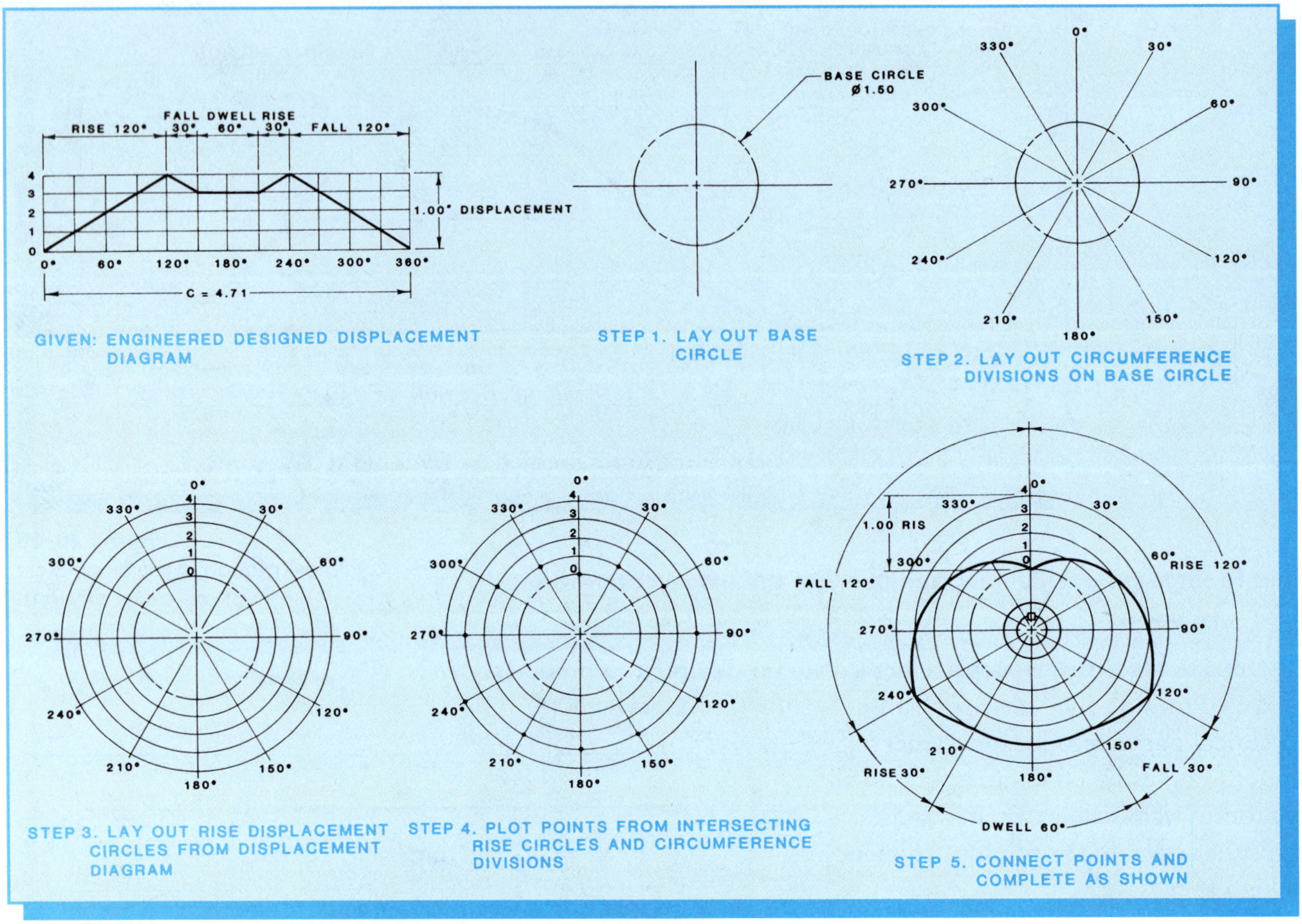

Figure 20-51
Steps to draw a uniform motion plate cam

The rise displacement circles are then added to represent the range of follower motion. Points are plotted to coincide with intersections on the displacement diagram. These points are connected to produce the final shape of the cam.

As in pattern drafting, the smoothness of the shape is directly related to the number of circle divisions used to develop the profile on the displacement diagram. If intersecting points are too far apart to produce a smooth curve, an irregular (French) curve template may be used to smooth out the contour.

Similar sequences used to develop a cam to produce a constant acceleration motion are shown in **Figure 20-52a.** The only difference in this procedure is the use of an angular profile line to produce the fall-rise segments on the displacement diagram.

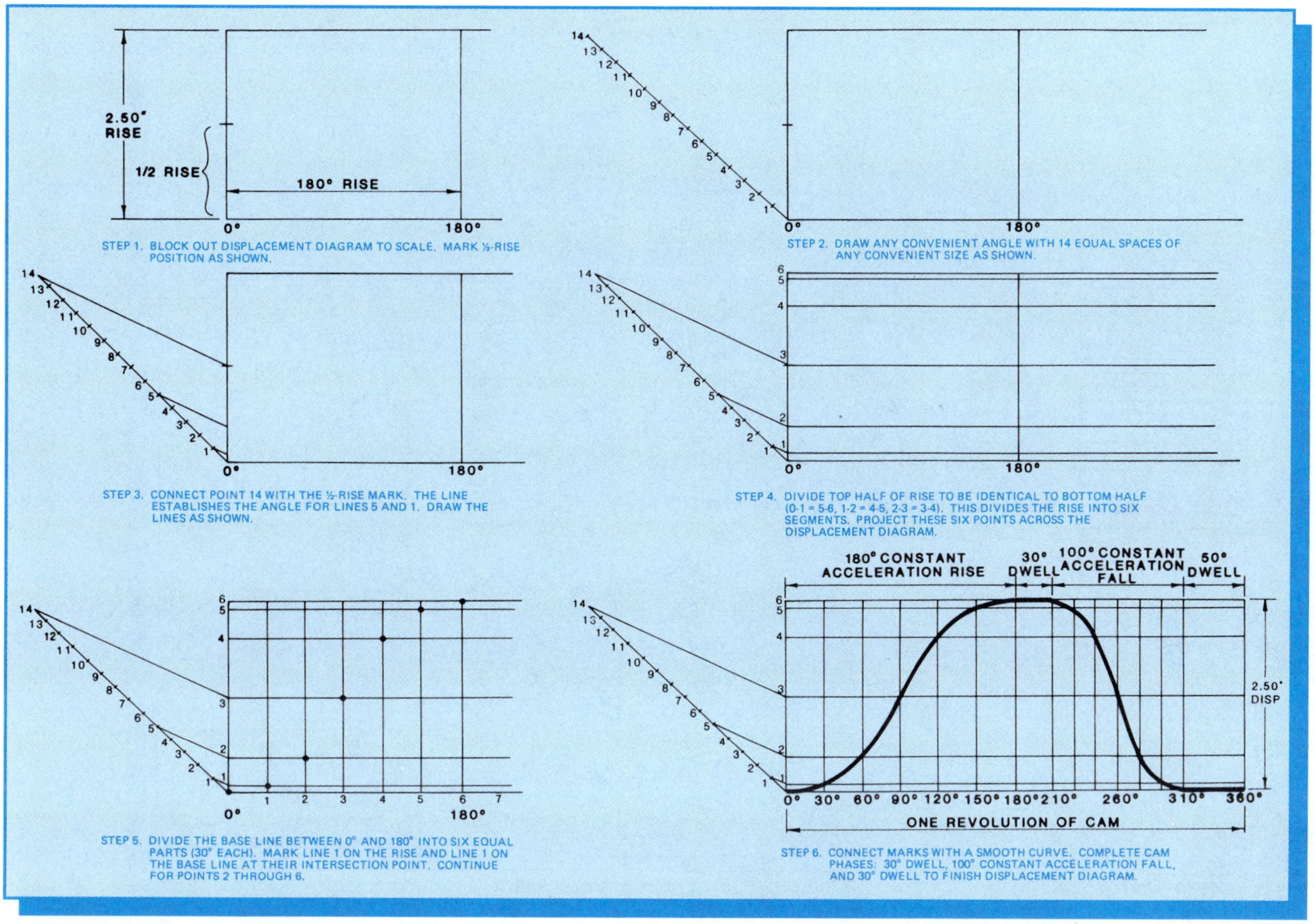

Figure 20-52a
Steps to draw a constant acceleration cam with a rise of 180°, dwell 30°, fall 100°, and dwell of 50°

Since cam shapes are irregular, grid layouts may be used to establish the exact shape of the circumference. The shape of the cam also may be defined by locating the position of the follower at various degree increments during a 360° rotation. Again, more degree divisions produce a smoother contour on the cam circumference. All other sizes are dimensioned conventionally **(Fig. 20-52b)**. **Figure 20-52b** also shows geometric tolerancing limits, which are extremely critical in cam machining.

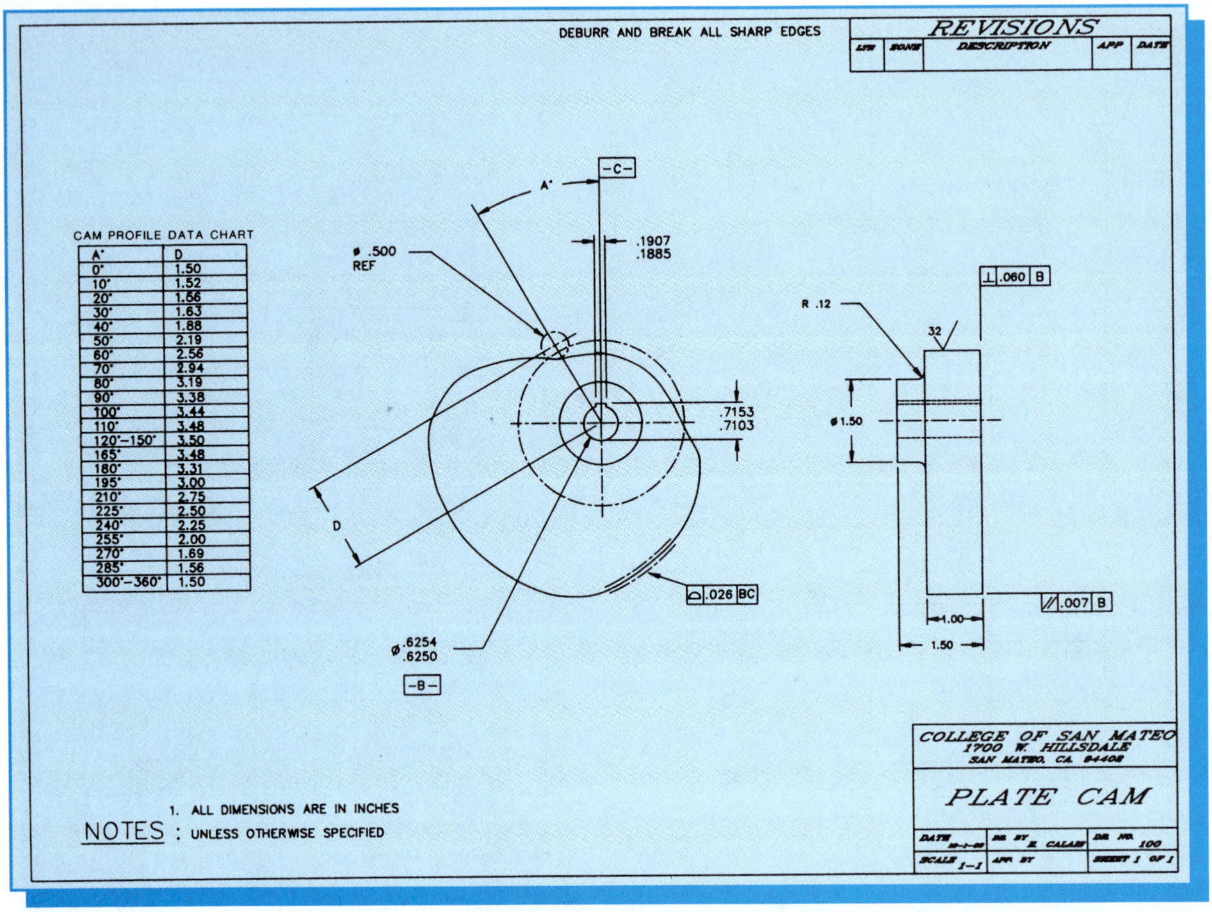

Figure 20-52b
Example of CAD-generated drawing of a plate cam layout drawn by intersecting points of the rise (D) and the angular (A) locations *(Courtesy of the College of San Mateo)*

AutoCAD LT Exercises

Gears can be accurately and rapidly drawn using a CAD system. The following information does not relate to specialized software for gear design and drawing, but to the normal AutoCAD LT software used throughout this text. The CAD command that is of great help in drawing gears is ARRAY. The Array command repeats a shape (such as a gear tooth) in a specified pattern.

The following sequence, used for drawing an approximate representation of a spur gear, utilizes the Array command to produce the individual gear teeth around the gear. The gear has a pitch diameter of 4.80" and 36 teeth.

Note: The gear calculations must be made prior to doing the drawing. These may be done with the aid of the computer, a hand calculator, or by student computations.

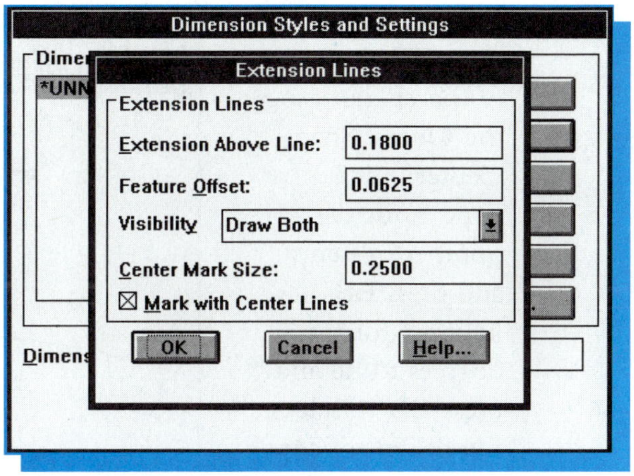

Figure 20-53
Activating and setting the size of center lines

1. The three main circles—Pitch diameter (Ø4.800), outside diameter (Ø5.067), and root diameter (Ø4.492) are drawn.
   a. Use the formulas for spur gears to verify the sizes of the circles.
   b. Use the **DRAW**, **CIRCLE**, **Center, Diameter** to get the correct size circles.
   c. Use the **SETTINGS**, **Dimension Style**, **Extension Lines** to set and activate center lines. Set the center mark size to **.250** and pick the Mark with Center Lines box, as shown in **Figure 20-53**.
   d. Select **Draw, Radial Dimensions, Center Mark.** Pick the outside diameter.

   The screen should now look like **Figure 20-54**.

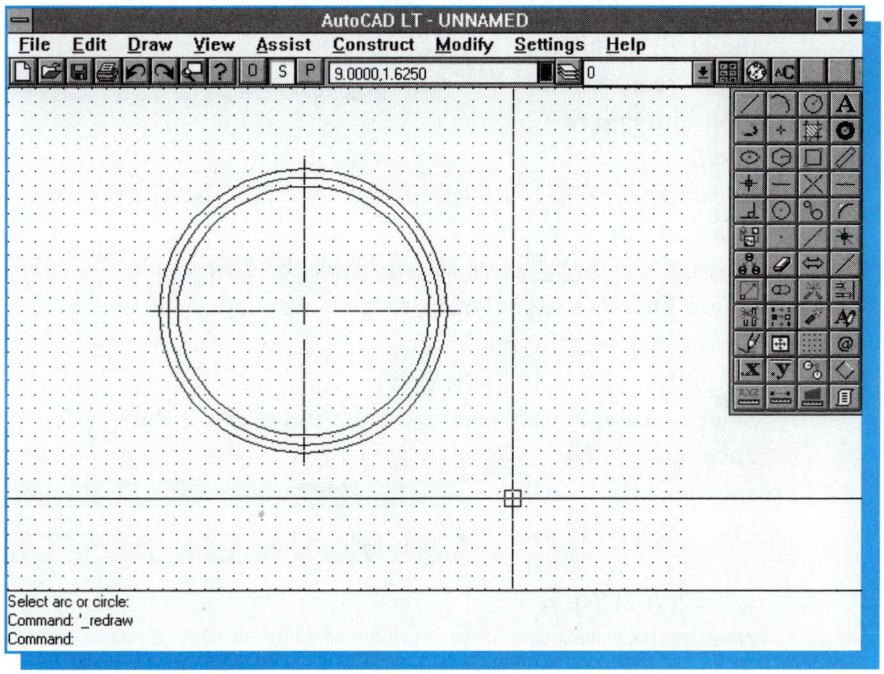

Figure 20-54
Lay out the outside diameter, pitch diameter, and root diameter after doing the gear calculations.

2. The next step is to draw the gear tooth. To do this, first use the gear tooth calculations to draw half the gear tooth and then mirror the half tooth to get a full tooth. To complete the gear drawing use the **ARRAY** command.

   The **VIEW, ZOOM, Window** command is used to see the top center of the gear clearly. By Zooming in, the operator can do fine detailed work and draw one of the gear teeth. One-half of a gear tooth is drawn using the results of the gear tooth calculations.

   a. To start the gear tooth construction, use the Array command to create radial lines.

      Select **Construct, Array,** pick the <u>Center Line</u>, and press **EB**.

Type **P** (polar) and press **EB**. Pick any of the circles using the Object Snap Center.
Type **3** for the number of items and press **EB**.
Type **-5** for the degrees to fill and press **EB** to select the default of copy items as rotated.
Repeat process for lines on other side except use **5** for the angle to fill.
The screen should look like **Figure 20-55**.

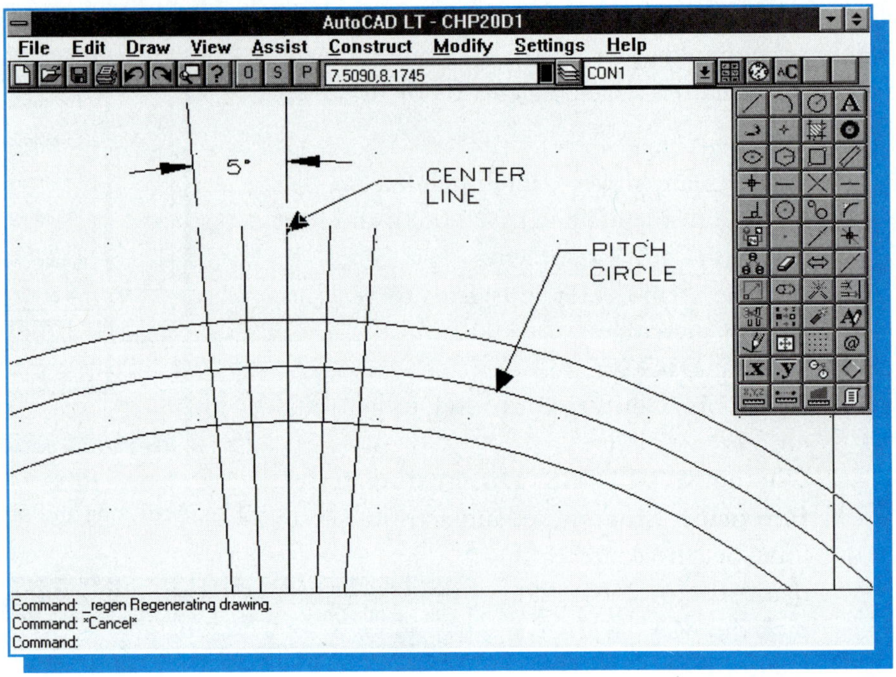

Figure 20-55
Arraying center line to get radial lines for gear tooth construction

b. Create layers CON1, CON2, and GEAR. Give the layers different colors.
Change to CON1 layer to construct the base circle.
Select **Draw**, **Line**. **Osnap** to the intersection of the center line and the pitch circle.
To point, type **@1<180b (EB)**.
Select **Draw, Line**. From the same point using Polar coordinates draw another line 1 inch line at 20 degrees.
To point, type **@1<200 (EB)** **(Fig. 20-56)**. This is the pressure angle line.
Draw a line from the center of the pitch circle, perpendicular to the pressure angle line.
Select **Draw, Line**.
Use Osnap Center to pick the pitch circle.

Figure 20-56
Constructing pressure angle

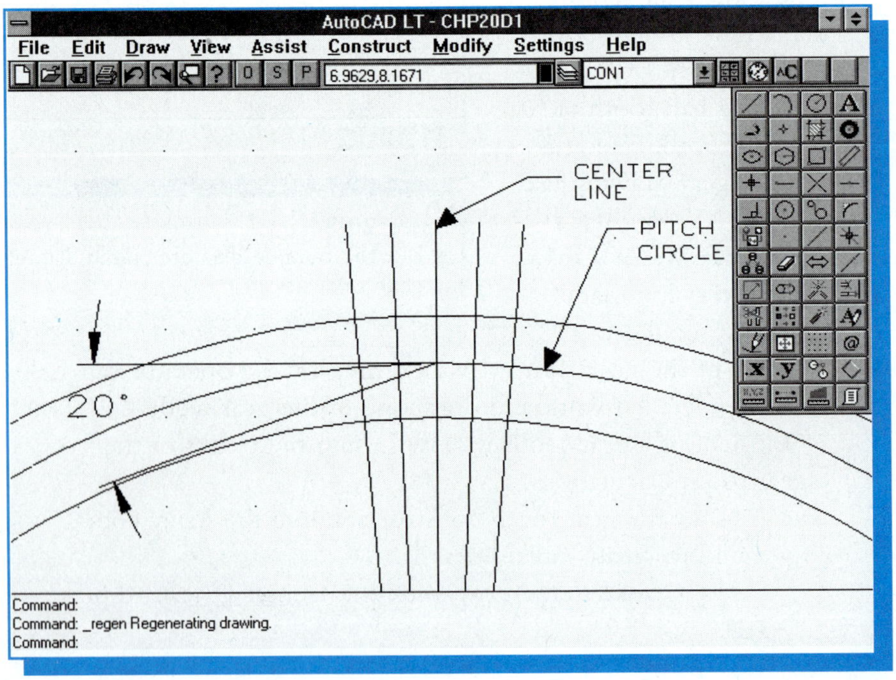

Use Osnap perpendicular to pick the pressure angle line. Using this line as the radius for the Base circle, select **Draw, Circle, Radius**.
Use the Osnap center to select the pitch circle.
Use the Osnap end to select end point of the line drawn perpendicular to the pressure angle. This is the Base circle **(Fig. 20-57)**.

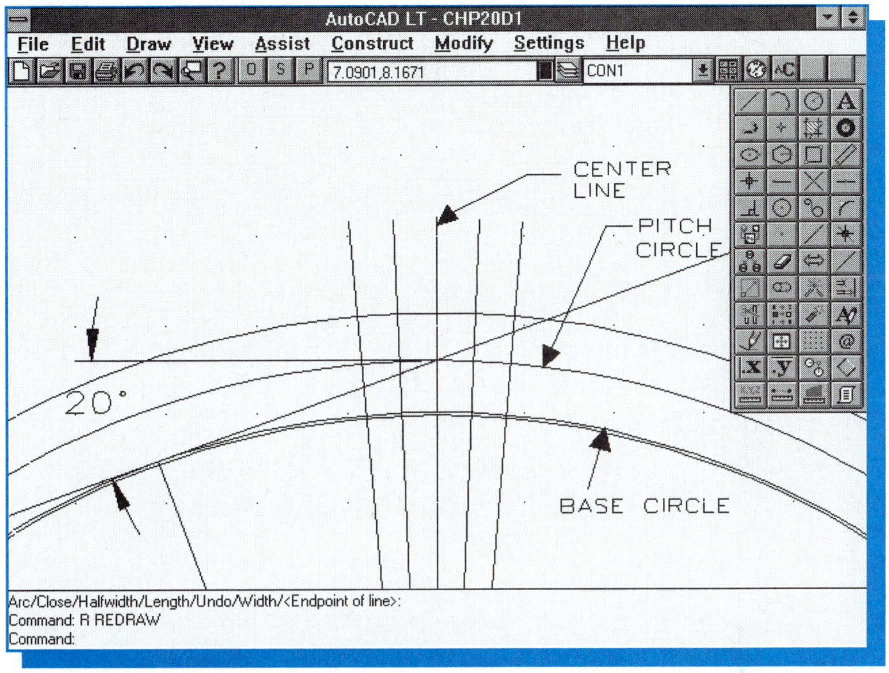

Figure 20-57
Construction of the Base Circle

c. **Zoom in** to construct half the gear tooth **(Fig. 20-58)**.
The approximate radius of a gear tooth profile equals .125 times the pitch diameter. In this example it is .125 x 4.800 = .600. To get the correct location for the radius, it will be necessary to construct a circle from the intersection of the pitch diameter and the radial line that is 2.5 degrees from the center line. Now draw the same radius circle from where the construction circle 1 crosses the base circle. This forms an approximate representation of the radius on the tooth of the gear.
Select **Draw, Circle, Radius**.
Pick the intersection marked **1** in **Figure 20-58** as the center.
Type **.600** as the radius [.125 x the pitch diameter (4.8) = gear tooth radius (.600)]. Press **(EB)**.
This is construction circle 1 as shown in **Figure 20-58**.

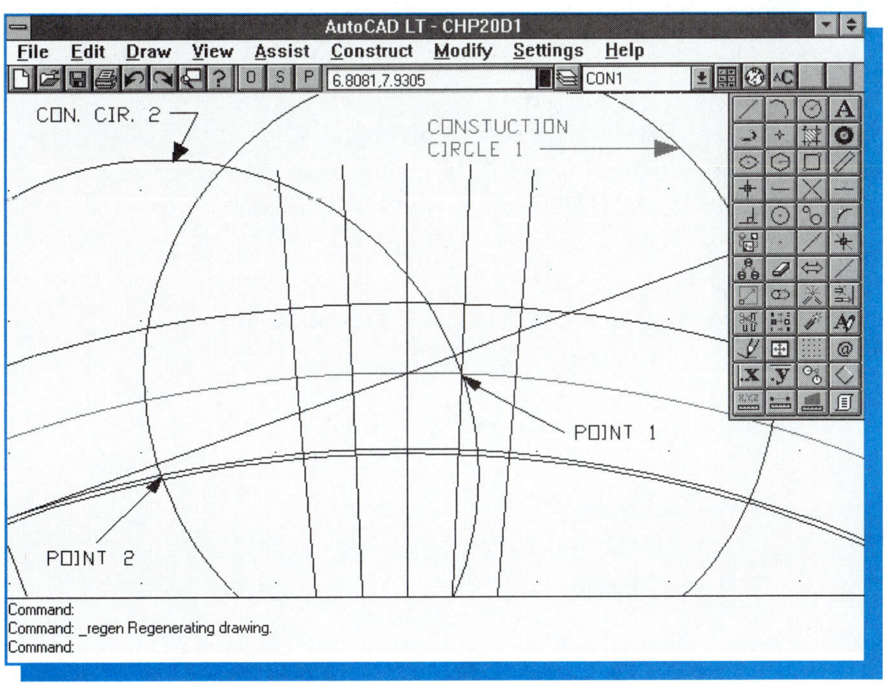

Figure 20-58
Constructing half the gear tooth shape

This circle was constructed to find the center point for the circle that forms the radius for the gear tooth. (construction circle 2).

Select **Draw, Circle, Radius**.

Pick the intersection of the construction circle 1 and the base circle, marked 2, as the center.

Type **.600** as the radius. Press **(EB)**.

This is construction circle 2 as shown in **Figure 20-58**.

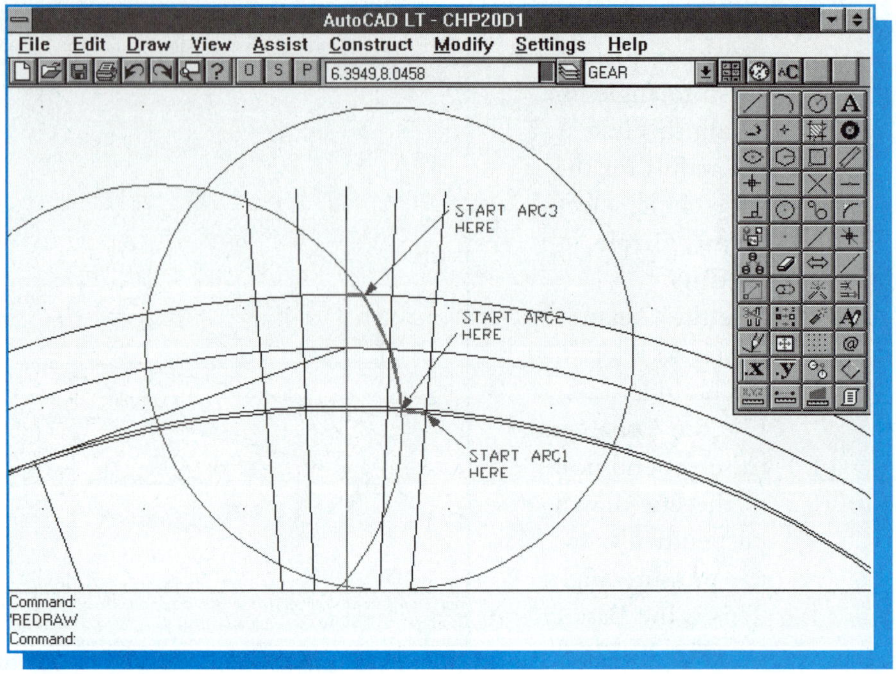

Figure 20-59
Using arc commands, draw half the tooth.

With the construction lines as guides, change layers to Gear and use the Arc command to construct half the gear tooth. This will include three arcs **(Fig. 20-59)**.

Select **Draw, Arc, Start, Center, End**. Construct half the gear tooth by drawing the three arcs.

Use Osnaps to get the intersections and centers of the radii.

Start where shown in **Figure 20-59**.

Now mirror the half tooth to get a full tooth.

Use **Construct, Mirror**. Pick only the entities of the half tooth.

Press **(EB)** when finished picking.

With Ortho on, select the endpoint of the center line.

Select any point below the endpoint of the center line.

Press (**EB**) to select the default of Not Deleting old objects.

The drawing should look like **Figure 20-60**.

Select **Construct**, **Array**. Pick all the entities of the gear tooth. Press (**EB**). when you have selected all the entities.

Select **P** (polar). Press (**EB**).

Use object snap **Center** to select the center of the pitch circle as the center point of Array.

Number of Items to array: **36**. Press (**EB**) (number of teeth in the gear).

Fill 360°. Press (**EB**).

tip: *It may be necessary to turn off some of the construction layers to select only the entities of the gear tooth.*

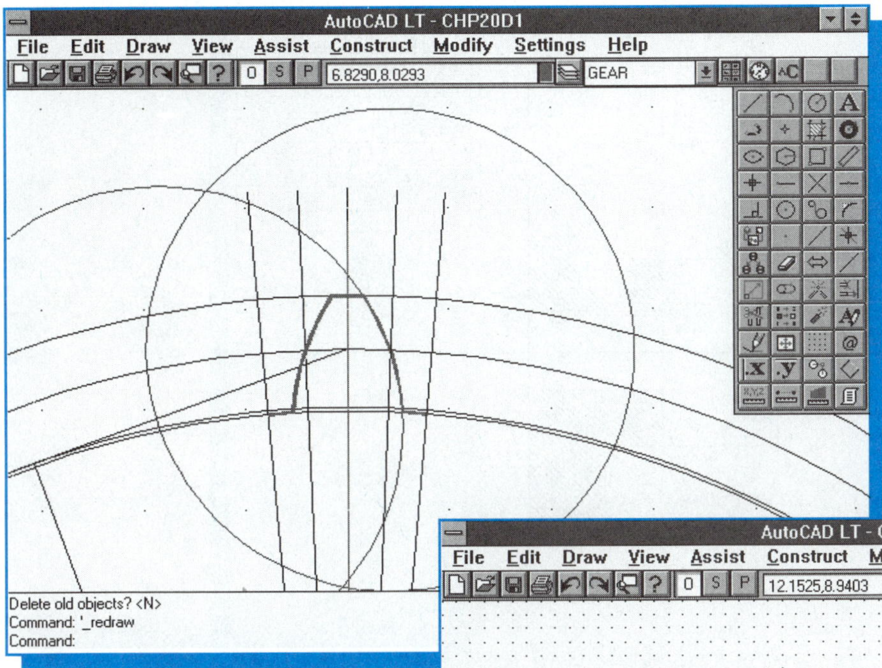

Figure 20-60
Mirror half the tooth to form the whole tooth.

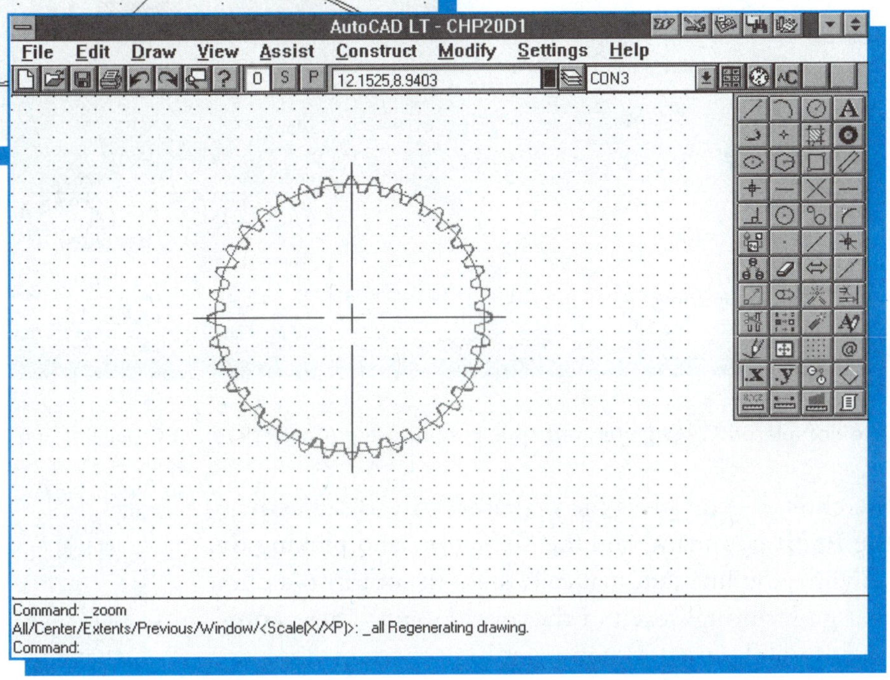

Figure 20-61
Use the Erase command to erase construction lines, or if done on a separate layer, freeze the construction layers.

Rotate items: **<y>**.
Press (**EB**).
The gear profile is completed by erasing construction lines or freezing the construction layers so your drawing will look like **Figure 20-61**.

3. To complete the spur gear drawing, add the hub, center hole, side view, dimensions, and gear data **(Fig. 20-62)**.

Although there are many types of cams, the following CAD procedures apply to a simple plate cam. These procedures can be applied easily to the design and drawing of other cam types on a CAD system. Smooth curved lines are a common feature of all cams.

When drawing cams manually, the drafter must use a French curve or flexible curve to smooth out lines. CAD's counterpart to the French curve is a command called **Pline** which stands for "polyline." The Pline command has many features, one of which is to draw a smooth curve through a series of points. This is done in three steps **(Fig. 21-63)**. First, the points are drawn. Second, the points are connected using a polyline (**Pline** command). The third step is to use a command called **Pedit**, which stands for "polyline edit." One of

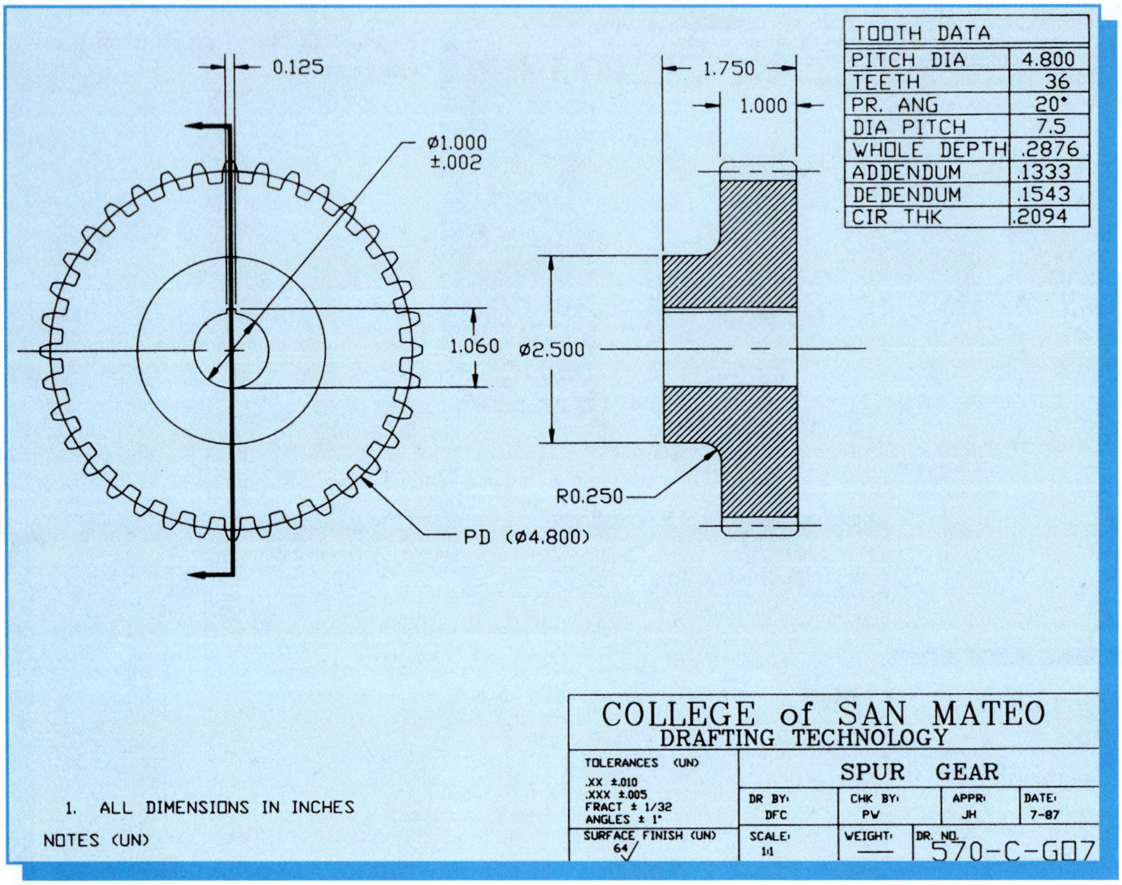

Figure 20-62
The completed CAD-generated spur gear drawing *(Courtesy of the College of San Mateo)*

the choices Pedit gives the operator is **fit**. By selecting the **Pedit** command and the **fit** option, and picking the polyline, the line automatically smooths out into a curve that goes through each of the given points. These commands will be used to construct both the cam displacement diagram and the cam.

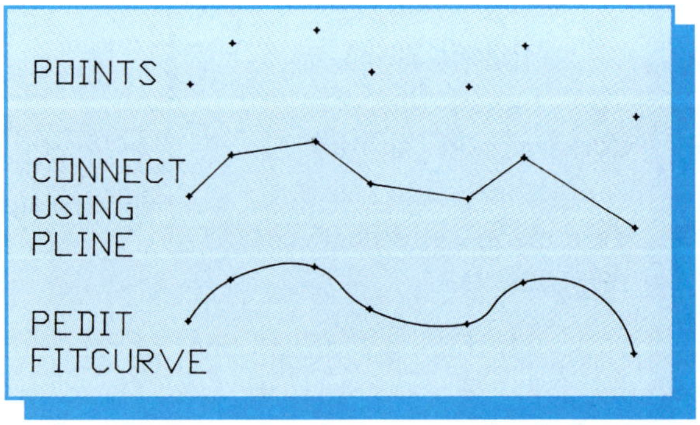

Figure 20-63
Use the Pline and Pedit commands for irregular curves.

Figure 20-64 shows a CAD-generated displacement diagram. The curve was drawn using the Pline command. This displacement diagram will be used to draw the profile of a plate cam **(Figs. 20-65** through **20-67)**.

Use CAD to draw a plate cam using the steps to draw a uniform motion plate cam shown previously in this chapter.

1. Start by accurately drawing the displacement drawing.
   Instructions for drawing a displacement diagram are shown in **Figure 20-52a**. Use polyline to make a smooth curve through the points. It should look similar to **Figure 20-64**.
2. To draw the plate cam:
   Draw the base circle and the label the angles.
   Indicate the counter-clockwise rotation of the cam.

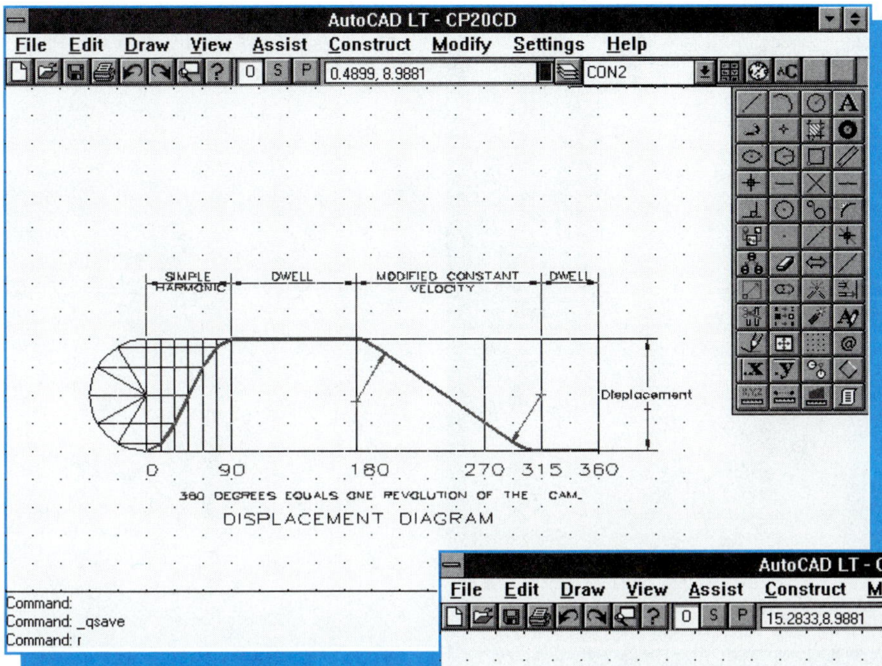

Figure 20-64
CAD-generated cam displacement diagram

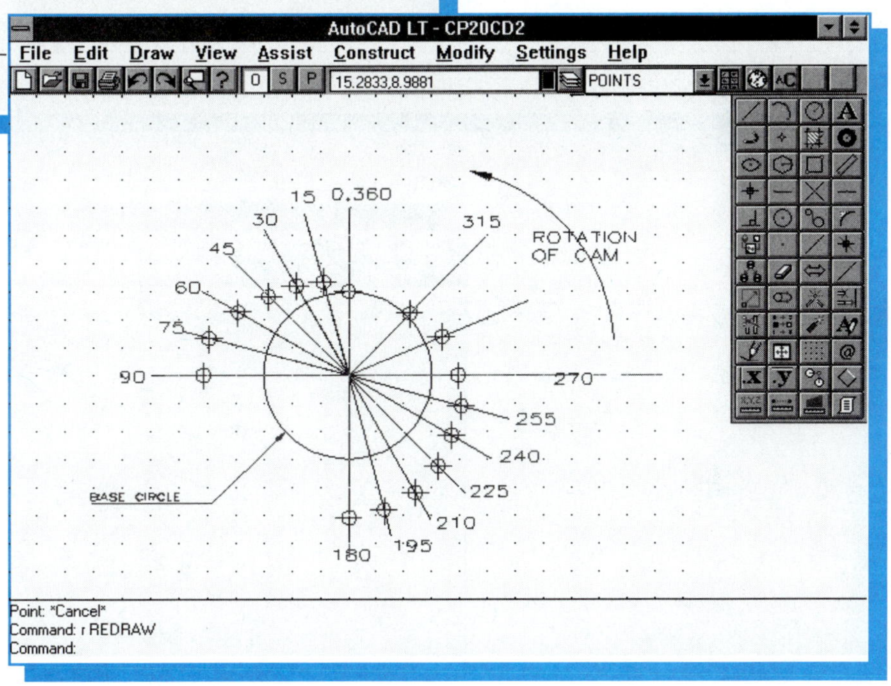

Figure 20-65
Plate cam construction layout

Draw the critical radial lines representing the fall and rise of the cam.

Note: In this example, lines were not drawn between 90° and 180°. That's because this cam dwells during this period. The placement of the radial lines were acquired from the displacement diagram.

Follow the procedures shown previously in this chapter to find the points on the radial lines that create the surface of the cam. The layout for the plate cam should look like **Figure 20-65**.

3. Create and change to a layer called CAM.

Using **Draw**, **Polyline** connect the points.

Using the **Modify**, **Polyline Edit**, **Fit** smooth out the line. This provides for a smooth, accurate cam profile **(Fig. 20-66)**.

The construction lines and points are erased or their layer frozen, depending on how they were drawn. The hub and profile view are drawn **(Fig. 20-67)**.

Dimensions, a cam profile data chart, notes and a title block have been added to complete the drawing **(Fig. 20-68)**.

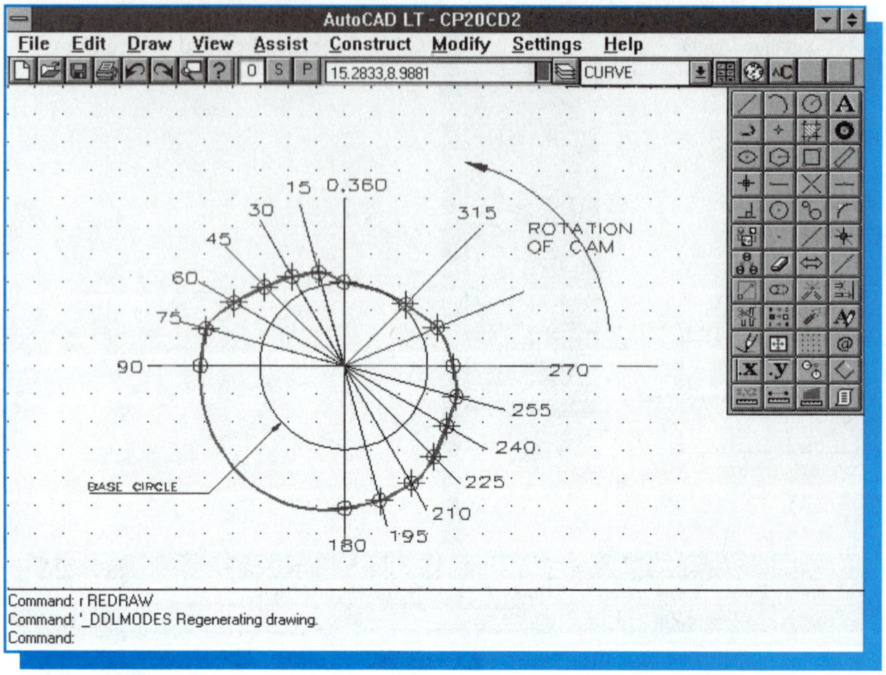

Figure 20-66
Plate cam outline drawn with Pline and Pedit commands

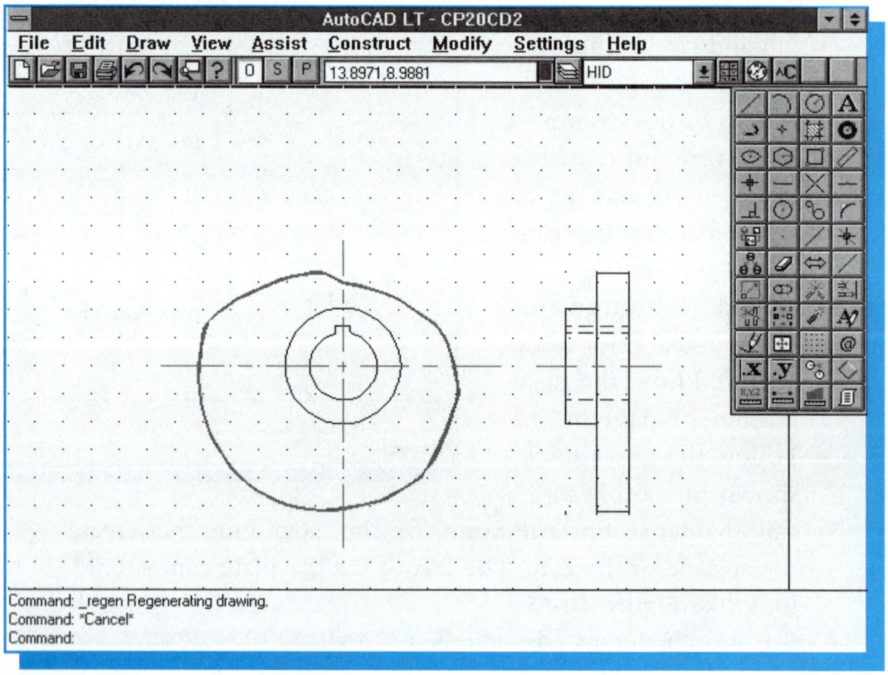

Figure 20-67
Remove the construction lines and draw the profile view.

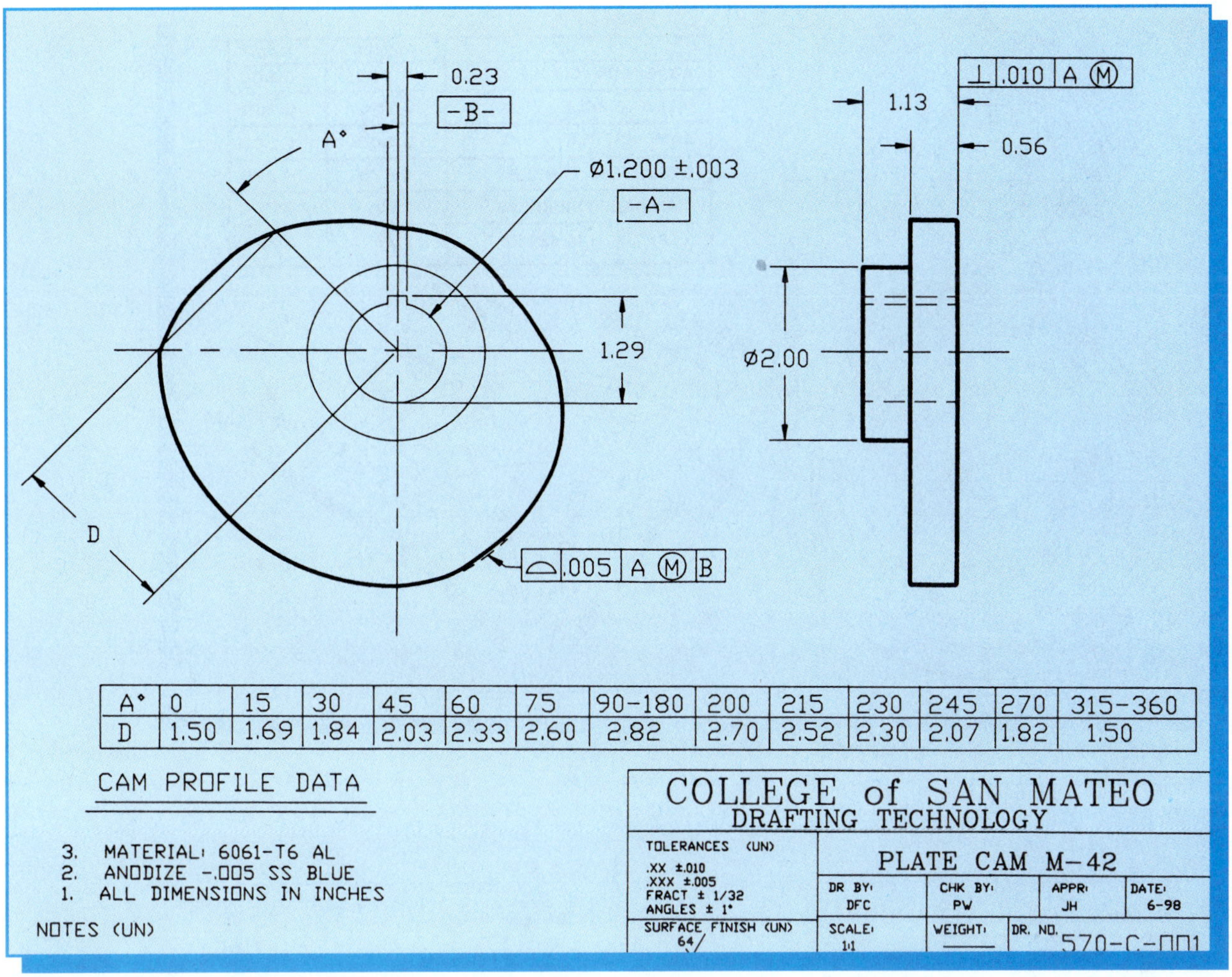

Figure 20-68
Completed CAD-generated plate cam drawing *(Courtesy of the College of San Mateo)*

Drafting Exercises

1. Redraw the pinion and spur gear in **Figure 20-69**.
2. With drawing instruments or a CAD system, draw the spur gears in **Figures 20-70, 20-71,** and **20-72**.
3. With drawing instruments or a CAD system, draw the section of the two miter gears in **Figure 20-73**.
4. With drawing instruments or a CAD system, draw the worm gear in **Figure 20-74**.
5. Draw the cams and their displacement diagrams for **Figures 20-75** through **20-82**.

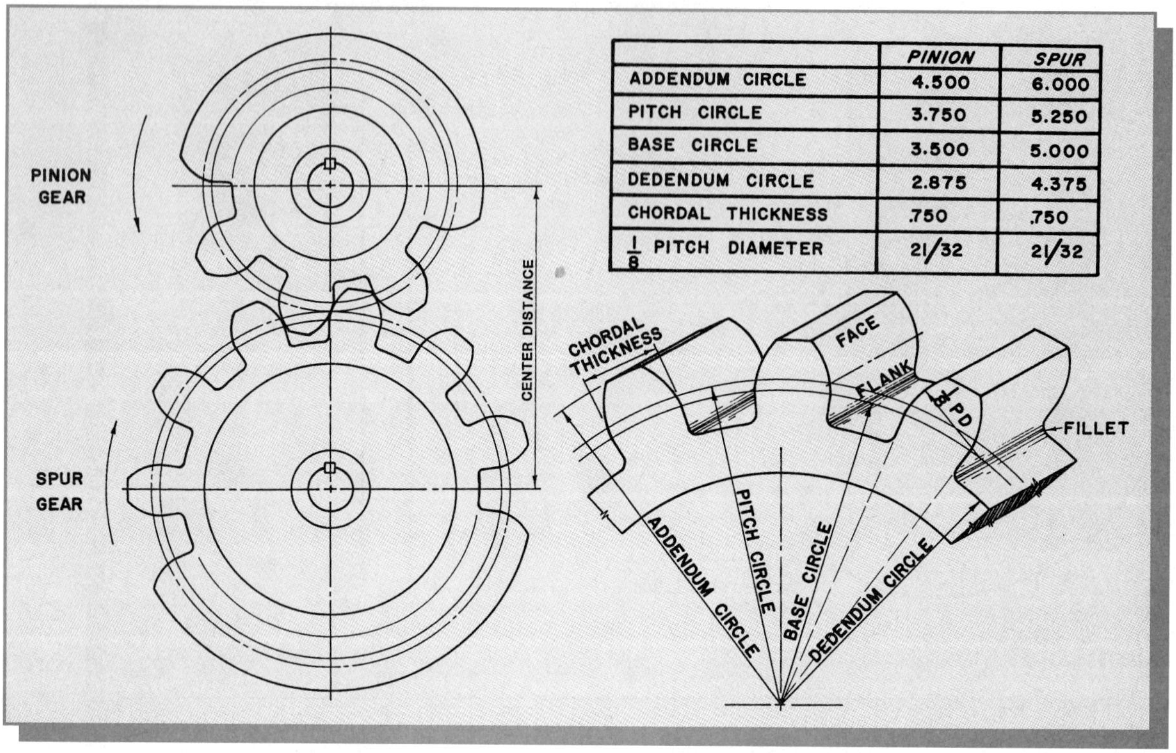

	PINION	SPUR
ADDENDUM CIRCLE	4.500	6.000
PITCH CIRCLE	3.750	5.250
BASE CIRCLE	3.500	5.000
DEDENDUM CIRCLE	2.875	4.375
CHORDAL THICKNESS	.750	.750
1/8 PITCH DIAMETER	21/32	21/32

Figure 20-69
Draw the pinion and spur gear.

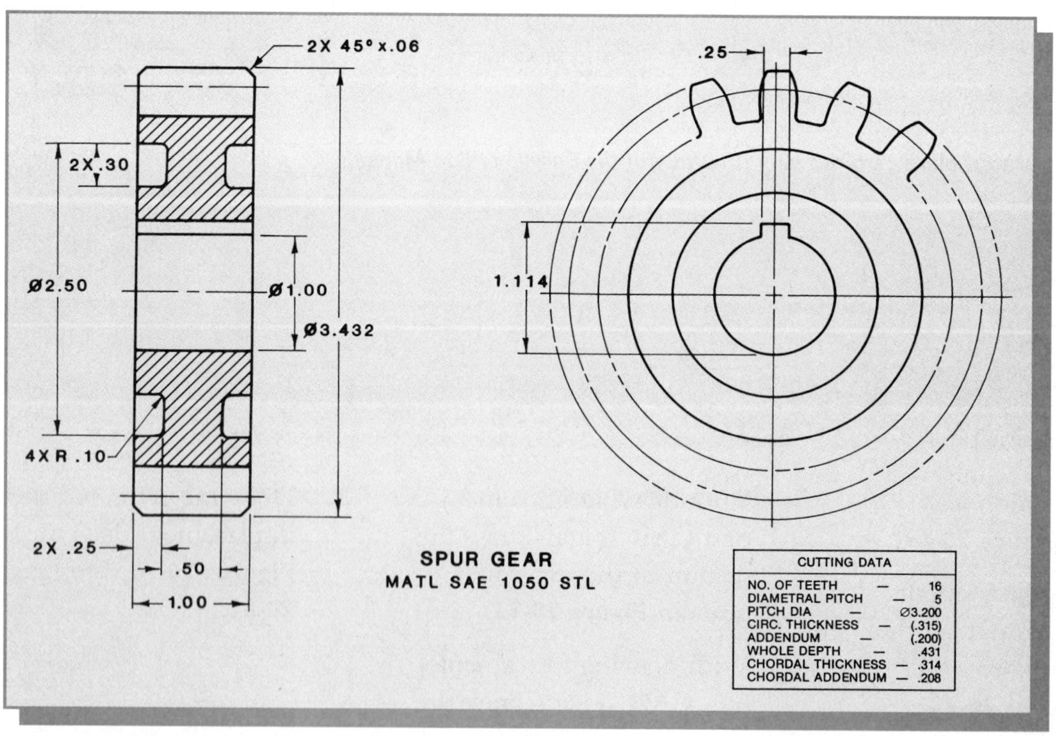

CUTTING DATA	
NO. OF TEETH —	16
DIAMETRAL PITCH —	5
PITCH DIA —	Ø3.200
CIRC. THICKNESS —	(.315)
ADDENDUM —	(.200)
WHOLE DEPTH —	.431
CHORDAL THICKNESS —	.314
CHORDAL ADDENDUM —	.208

Figure 20-70
Draw the spur gear.

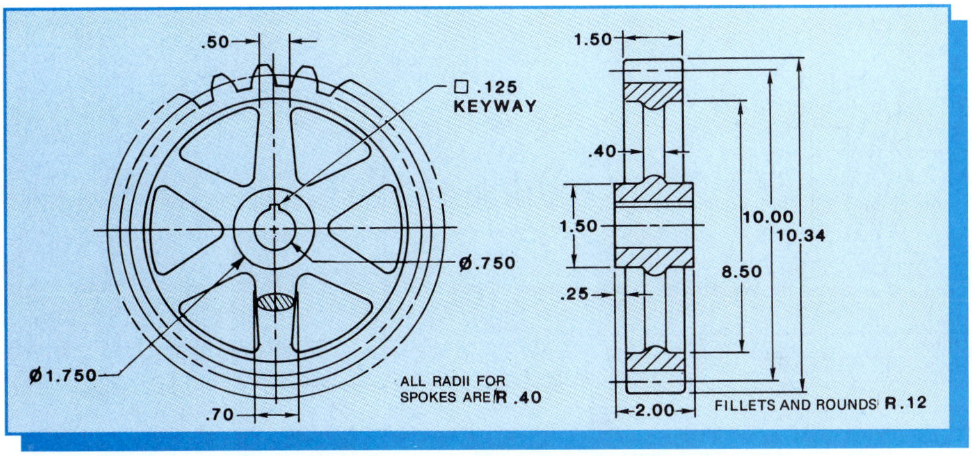

Figure 20-71
Draw the spur gear. Use a gear template for the teeth.

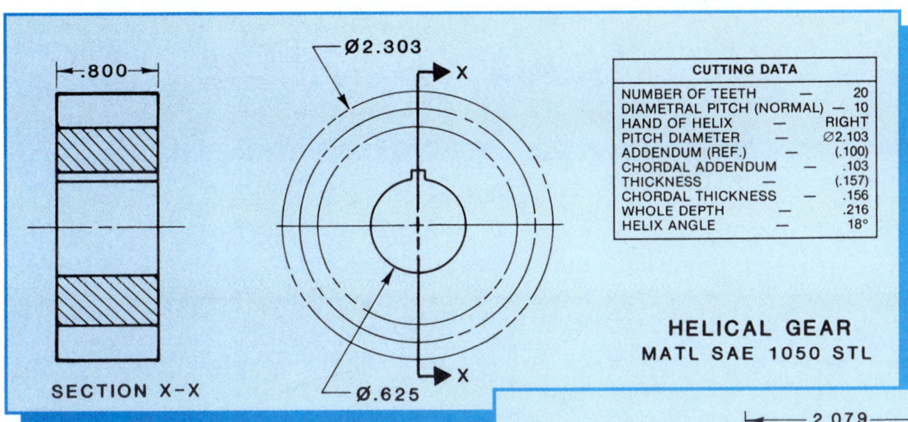

Figure 20-72
Draw the helical gear.

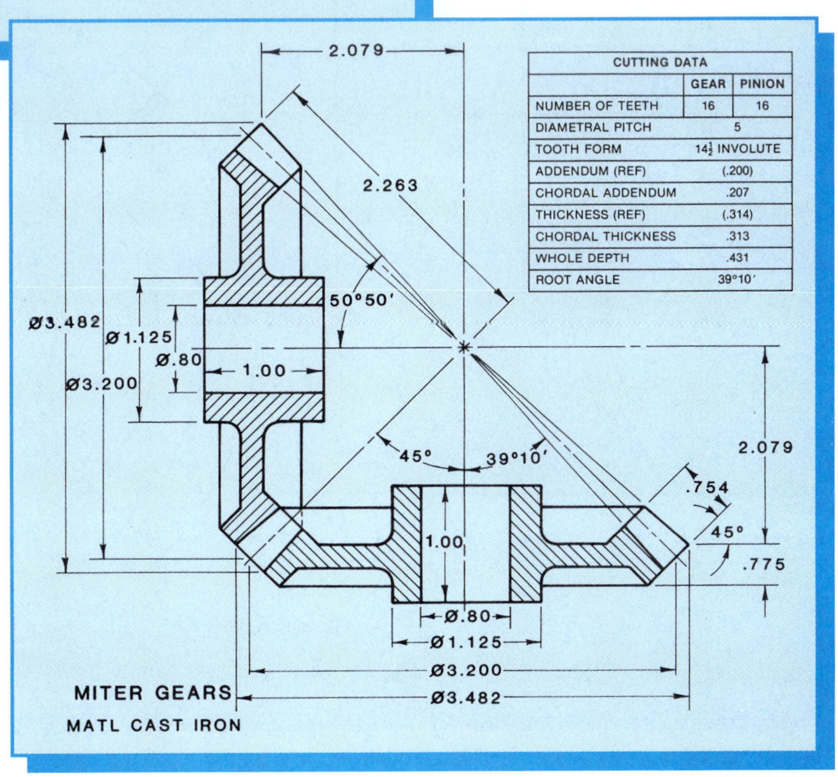

Figure 20-73
Redraw the two miter gears.

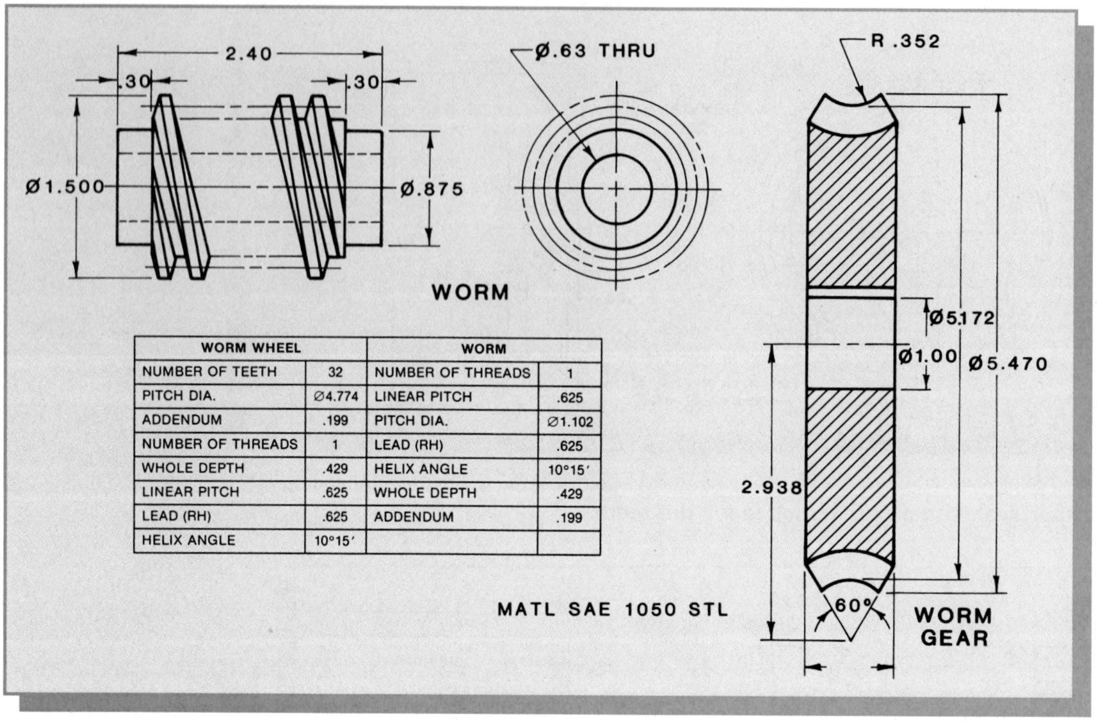

WORM WHEEL		WORM	
NUMBER OF TEETH	32	NUMBER OF THREADS	1
PITCH DIA.	Ø4.774	LINEAR PITCH	.625
ADDENDUM	.199	PITCH DIA.	Ø1.102
NUMBER OF THREADS	1	LEAD (RH)	.625
WHOLE DEPTH	.429	HELIX ANGLE	10°15′
LINEAR PITCH	.625	WHOLE DEPTH	.429
LEAD (RH)	.625	ADDENDUM	.199
HELIX ANGLE	10°15′		

Figure 20-74
Redraw the worm gear.

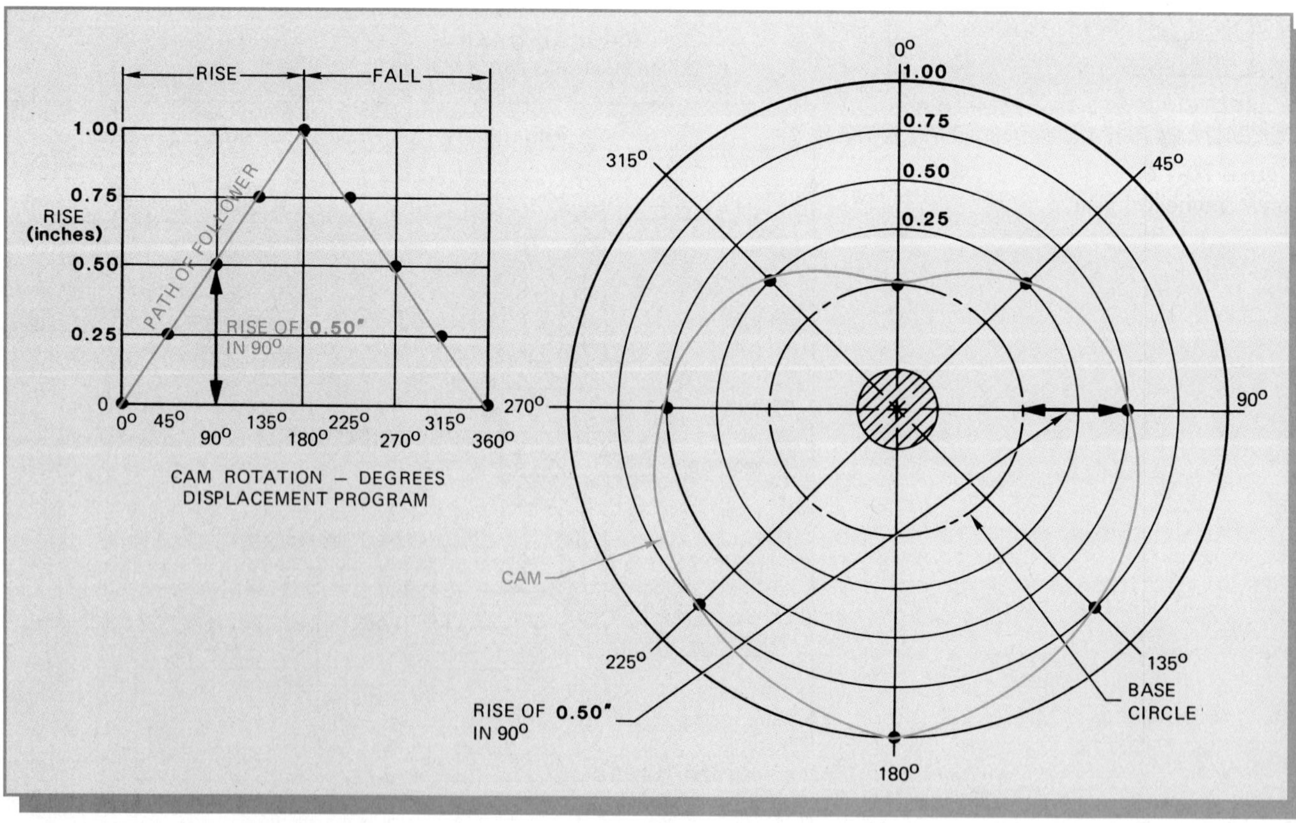

Figure 20-75
Draw the cam and its displacement diagram.

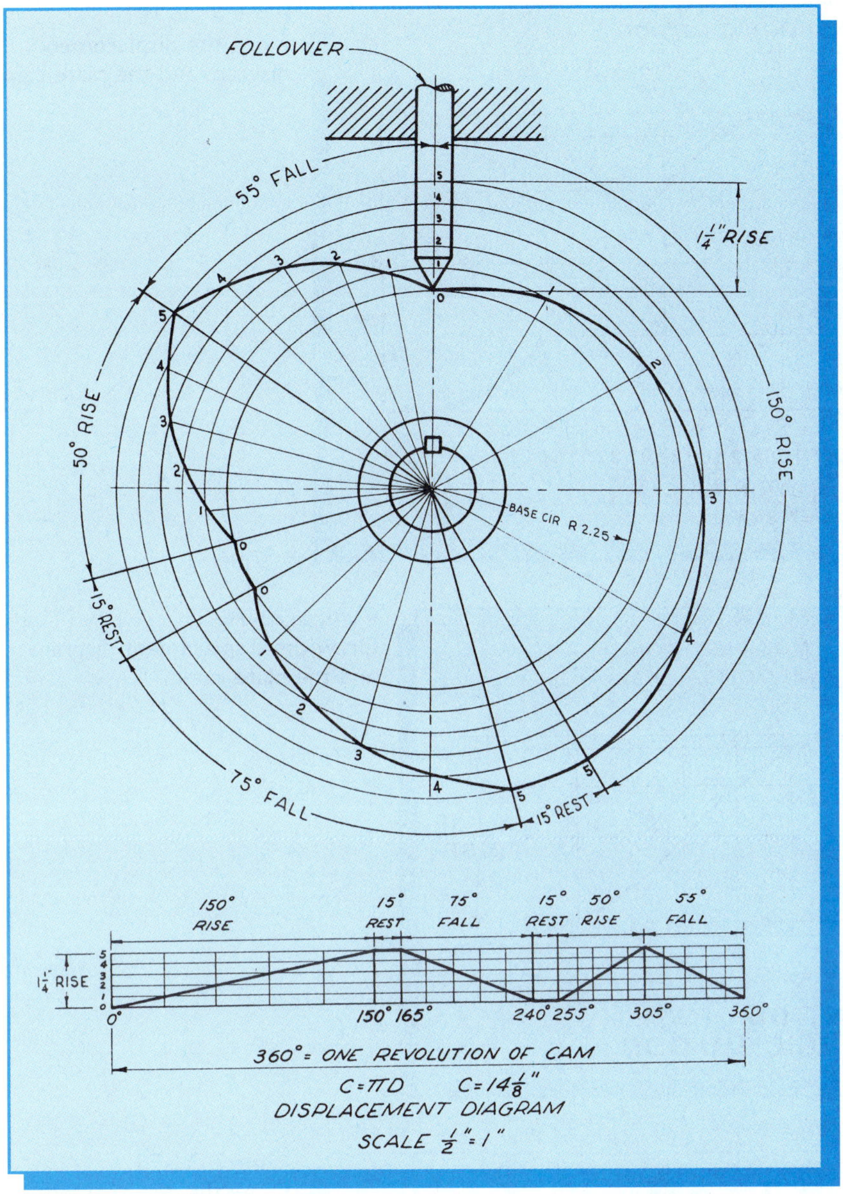

Figure 20-76
Draw the displacement diagram and the face plate.

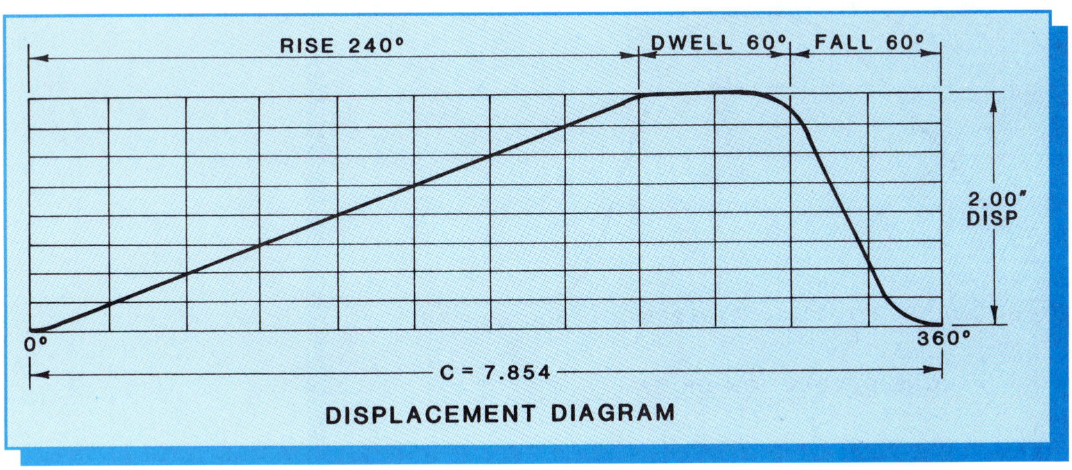

Figure 20-77
Draw the displacement diagram and the plate cam. To find the base circle's diameter, divide pi (π) into the circumference.

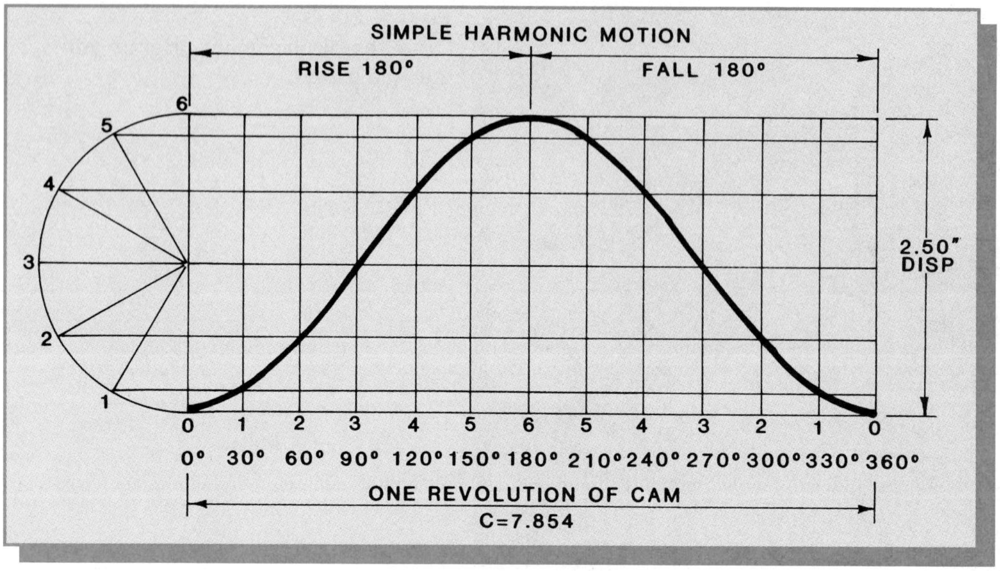

Figure 20-78
Draw the displacement diagram and the plate cam.

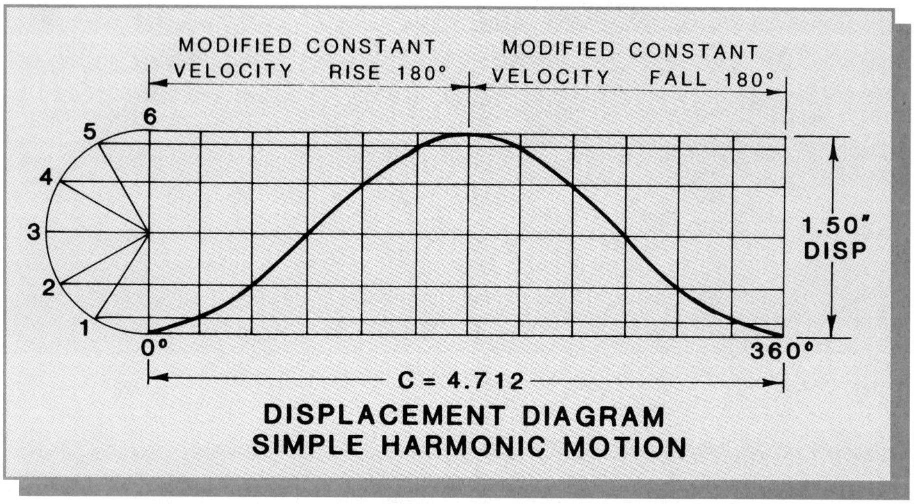

Figure 20-79
Draw the displacement diagram and the plate cam.

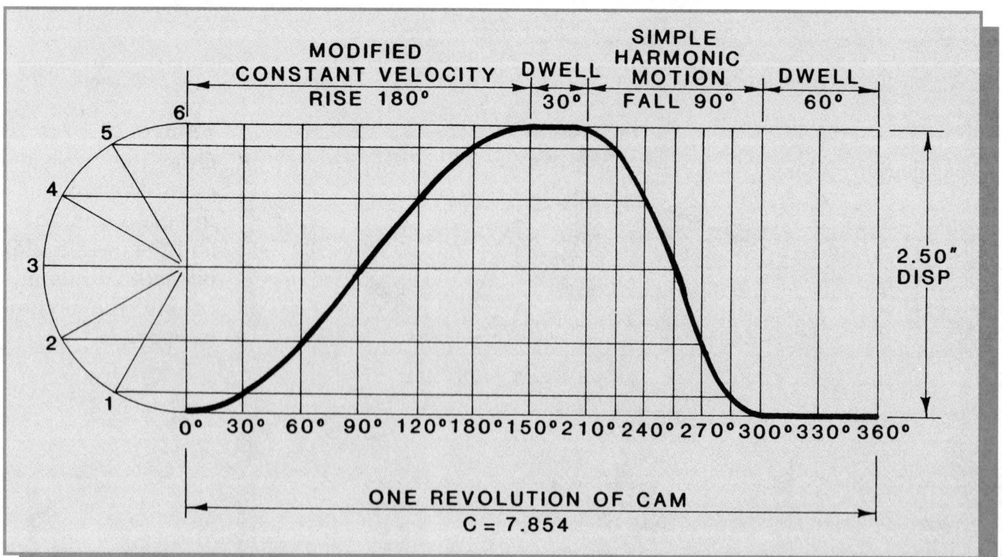

Figure 20-80
Draw the displacement diagram and the plate cam.

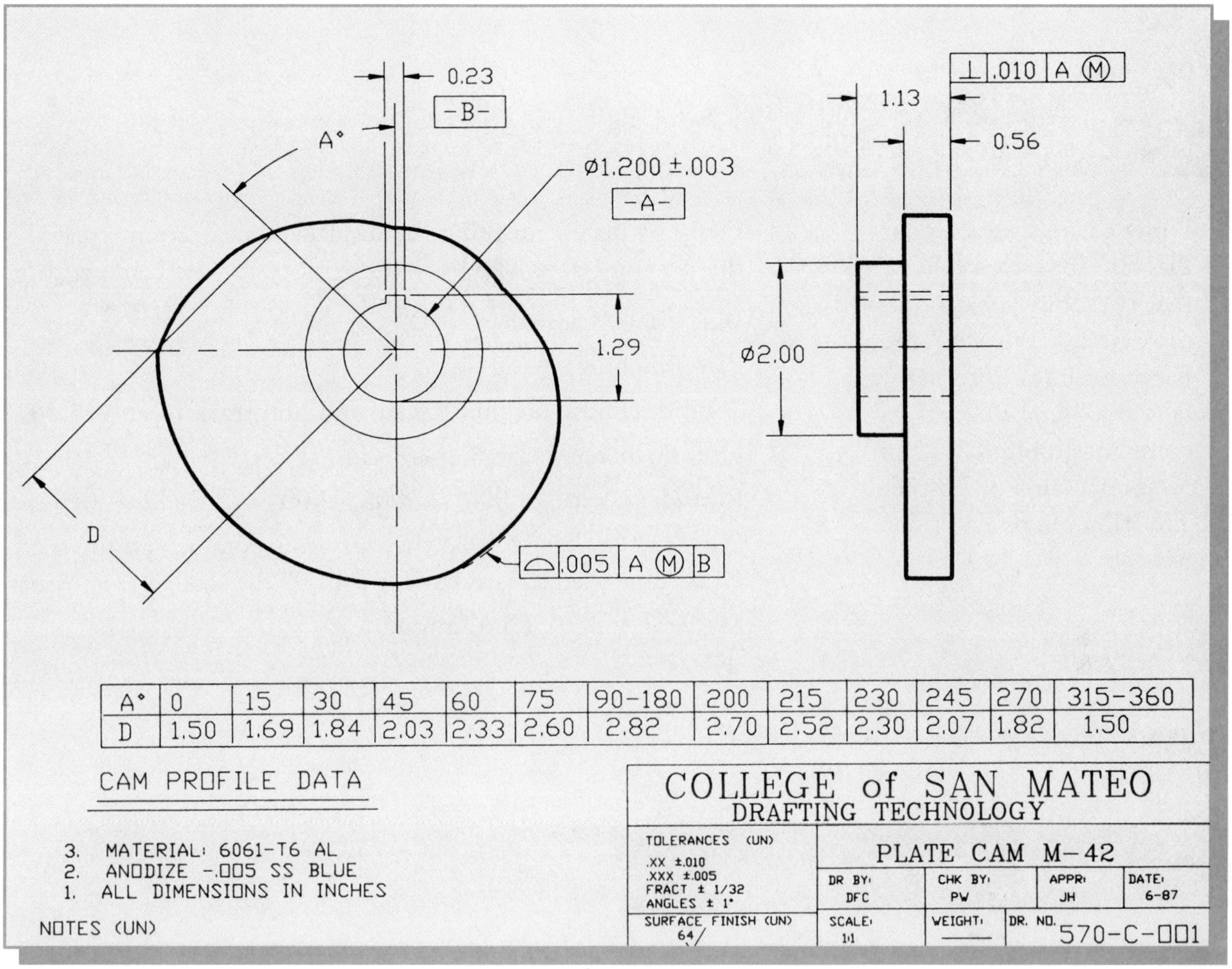

A°	0	15	30	45	60	75	90–180	200	215	230	245	270	315–360
D	1.50	1.69	1.84	2.03	2.33	2.60	2.82	2.70	2.52	2.30	2.07	1.82	1.50

Figure 20-81
Completed CAD-generated plate cam drawing *(Courtesy of the College of San Mateo)*

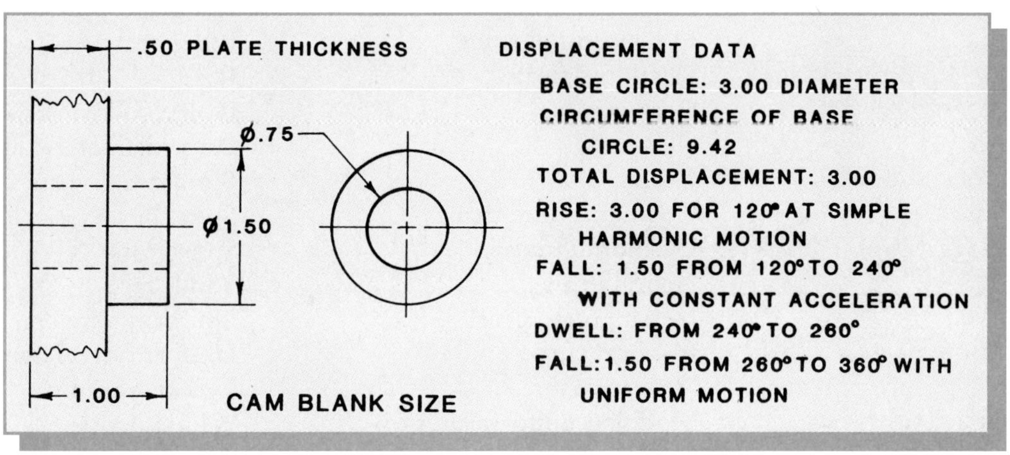

Figure 20-82
Complete the cam's profile drawing from displacement data.

Design Exercises

1. Draw a cam displacement diagram for a cam follower that rises with a harmonic motion 2.00" in 180°, then has a dwell for 30°, then falls 2.00" in 120°, and completes the full 360° of rotation with a 30° dwell. The circumference of the base circle is 12.00".

2. Design a plate cam using the displacement diagram created with the previous data. The basic dimensions for the cam are to be:

   Base circle diameter = 3.82"

   Plate thickness = .50"

   Total thickness, including plate and hub thicknesses = 1.50"

   Hub diameter = 2.50"

   Bore diameter = 1.000" + .005, -.000

   Use a square key to keep the cam from rotating on the shaft. Recommended square key for a 1.00" dia shaft is 3/16" square.

   Material = 6061 T6 Aluminum

Key Terms

Addendum
Addendum circle
Angular gears
Base circle
Belt drives
Bevel gears
Cam
Camshaft
Center distance
Chain and sprocket drives
Chordal thickness
Circular pitch
Clearance
Cycloidal curve
Cycloidal motion
Dedendum
Dedendum circle
Diametral pitch
Displacement diagram
Drum cams
Dwell (rest)
Face cam
Fall
Fillet
Flanged sleeve gears
Flank
Flat-faced followers
Friction wheels
Gear axis
Gear ratio
Gears
Geometric line shapes
Harmonic motion
Helical gears
Herringbone gears
Involute curve
Miter gears
Modified uniform motion
Number of teeth
Parallel shaft gears
Perpendicular shaft gears
Pitch circle
Pitch diameter
Pitch point
Plate cams
Pointed followers
Rack and pinion gears
Ring gears
Rise
Roller followers
Root diameter
Round followers
Spiral bevel gears
Spur gears
Tooth face
Uniform accelerated motion
Uniform decelerated motion
Uniform motion
Whole depth
Working depth
Worm gears

Chapter 21

Piping Drawings

Objectives

The student will be able to:

- recognize pipe drafting symbols
- complete an orthographic single-line piping drawing
- complete an orthographic double-line piping drawing
- complete an isometric single-line piping drawing
- complete an isometric double-line piping drawing
- prepare a piping drawing on a CAD system

Introduction

Pipes are used in a wide range of applications, from small residences to large industrial plants **(Fig. 21-1)**. Pipes carry all types of liquid, gases, and semisolids including water, oil, acid, liquid metal, nitrogen, oxygen, steam, granular materials, and exhausts. Pipes are also used structurally as support columns and for a wide variety of architectural applications, such as handrails and fence components.

Figure 21-1
Industrial plants use many pipes to transport liquids and gases. *(Courtesy of FERRANTI Sciaky, Inc.)*

Pipes are the arteries and veins of plumbing, heating, cooling, drainage, and manufacturing systems. Pipes are made of steel, wrought iron, cast iron, copper, brass, aluminum, plastic, clay, concrete, brass, glass, lead, rubber, wood, and other synthetic material combinations. Drawings that describe the pipe material, size, and location in a piping system are assembly drawings. These use orthographic or pictorial (isometric) projection **(Figs. 21-2** and **21-3)**.

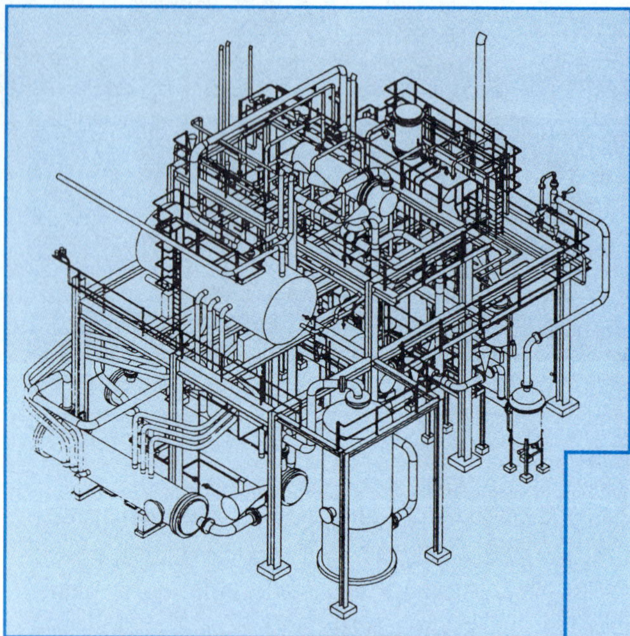

Figure 21-2
Piping drawing using an isometric drawing layout *(Courtesy of Computervision Corporation)*

Figure 21-3
Segment of an industrial plant's piping using an isometric drawing layout *(Courtesy of Computervision Corporation)*

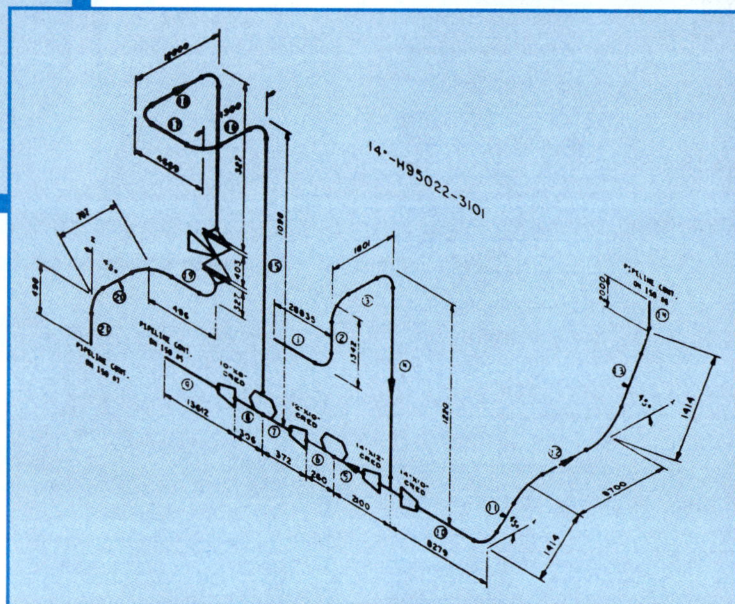

Pipe Connections

Fittings are usually required to join pipes and attach them to other devices, such as pumps and tanks. Pipes are temporarily joined with threaded, flanged, or bell and spigot connections. Others are permanently attached by brazing, soldering, or welding. Some pipes can be bent to change direction.

Threaded Connections

Threaded connections are used primarily on pipes with 2" diameter or less, and are available with straight or tapered threads. Threaded connectors, sometimes called **screwed fittings,** are available in many shapes—elbows **(Fig. 21-4)**, tees, Y bends, reducers, return bends, caps, and crosses. Couplings with inside threads are also used to connect straight pipes or nipples. Nipples are short pipe lengths with external threads on both ends.

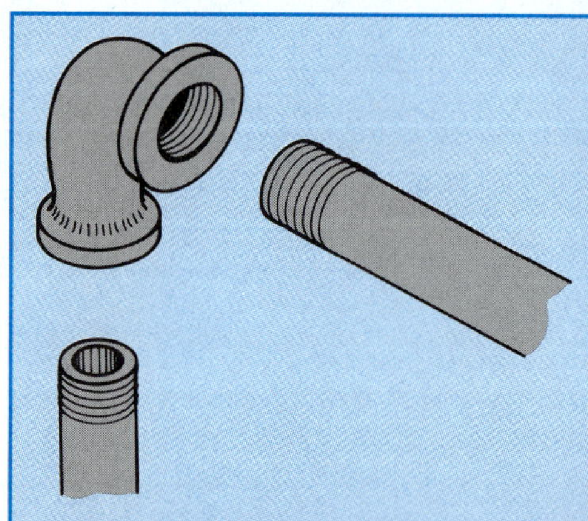

Figure 21-4
Threaded elbow pipe connection

Flange Connections

Flange connections are designed to enable a pipe joint to be disassembled by removing a series of bolts on the face of the flange. Flanges are manufactured in the form of elbows **(Fig. 21-5)**, return bends, Y bends, tees, taper reducers, and crosses.

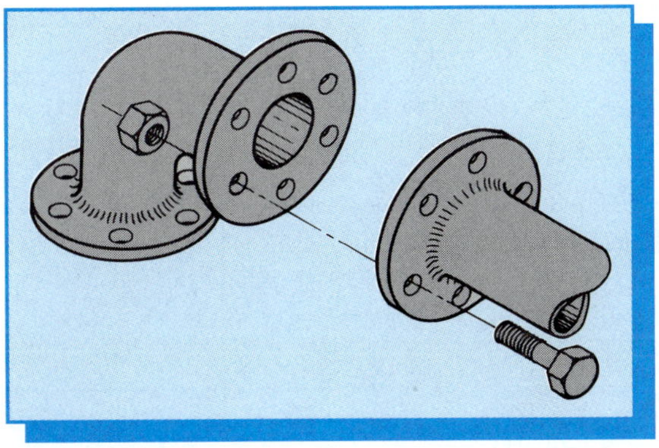

Figure 21-5
Bolted flange pipe connection

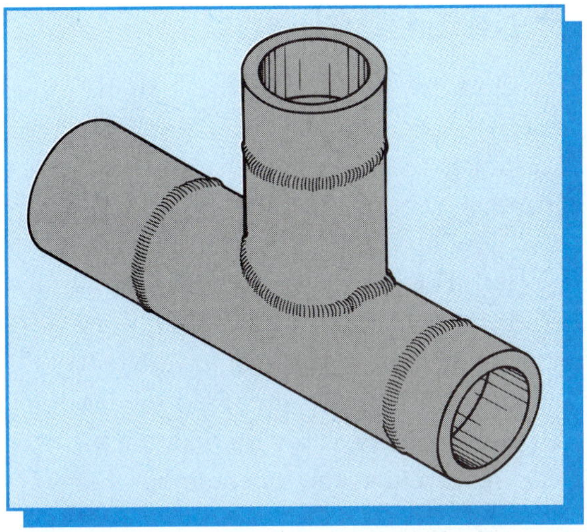

Figure 21-6
Welded pipe connections

Welded Connections

Since **welded connections** are virtually leakproof, welding is used to join pipes containing hot or volatile materials. Welded joints are also more economical to construct. Gas and arc welding methods are most commonly used to join pipe with butt, lap, tee, edge, and corner joints. **Figure 21-6** shows a welded tee shape. See Chapter 19 for welding joint details.

In the process of brazing and soldering, the base metal does not melt. Consequently, brazing and soldering is restricted to joining small copper or brass pipe and tubing. Pipe joints are soldered with the use of connectors **(Fig. 21-7)**. Connectors must be specified to fit tightly over the pipe to aid the sweat soldering process. Remember, solder has a low boiling point, so extremely hot materials, especially under high pressure, must be avoided.

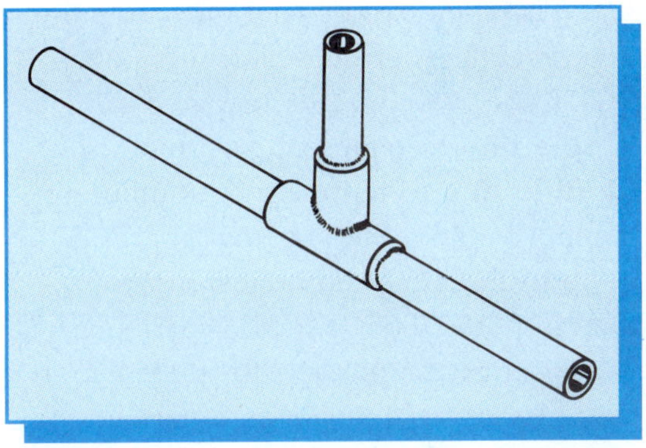

Figure 21-7
Soldered copper pipe connections

Bell and Spigot Connections

Bell and spigot connections are so named because one pipe end is shaped like a bell and inserted into a pipe with a spigot-like shape. These joints are fitted tightly with lead and oakum to make them waterproof **(Fig. 21-8)**.

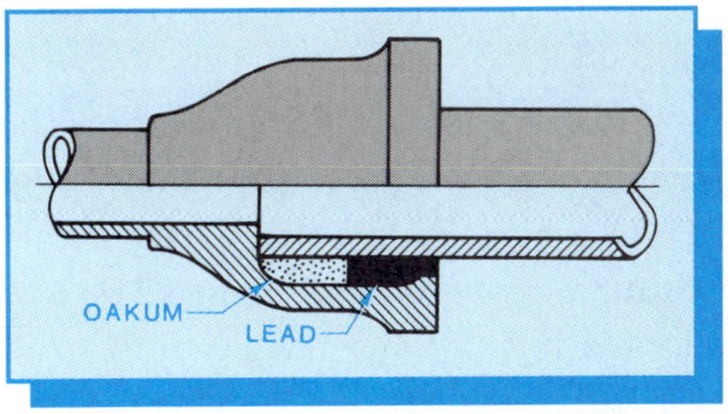

Figure 21-8
Bell and spigot pipe connection

Cementing

Many materials such as plastic, clay, concrete, and glass cannot be welded, soldered, or threaded effectively. Joining pipes made of these materials requires **cementing (Fig. 21-9)**.

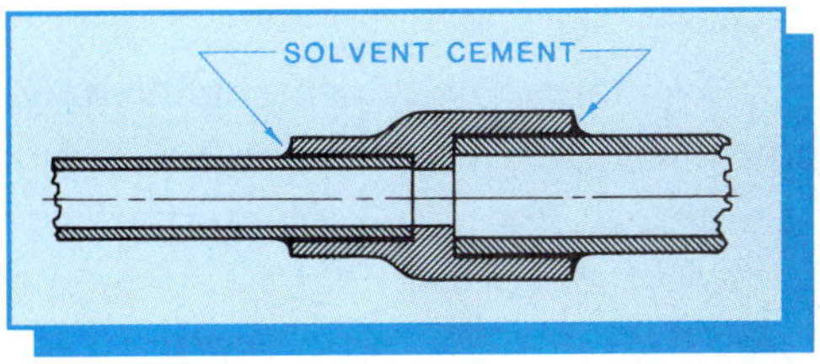

Figure 21-9
Cemented connection for plastic pipes

Pipe Standards

Pipes are specified by the type of material, sizes, and wall thickness. Pipe sizes relate to either inside or outside diameters. The **nominal pipe size** for steel pipe up to 12" is the inside diameter (ID). For pipes over 12", the nominal size is the outside diameter (OD). Pipe wall thicknesses are standardized by schedules of pipe weights. Pipe schedules range from 10 to 160 with wall thickness increasing as the schedule number increases **(Fig. 21-10)**.

STANDARDIZED PIPE SIZES—SCHEDULE 20		
Nominal Size (inches)	Outside Diameter (inches)	Inside Diameter (inches)
1/8	.405	.269
1/4	.540	.364
3/8	.675	.493
1/2	.840	.622
3/4	1.050	.824
1	1.315	1.049
1 1/4	1.660	1.380
1 1/2	1.900	1.610
2	2.375	2.067
2 1/2	2.875	2.469
3	3.500	3.068
3 1/2	4.000	3.548
4	4.500	4.026
5	5.563	5.047
6	6.625	6.065
8	8.625	7.981
10	10.750	10.020
12	12.750	11.938
14	14.000	13.126
16	16.000	15.000
18	18.000	16.876
20	20.000	18.814
24	24.000	22.626

Figure 21-10
Standardized pipe diameters for Schedule 20 pipes

Piping Drawings

There are two general types of piping drawings—orthographic and pictorial. Either type may be prepared using a **single line** to represent each pipe, or a **double line** to represent the outline of a pipe. Double-line drawings appear more realistic but require extensive drafting time. They are used sparingly, except for large pipes or for presentation and instructional purposes. Single-line drawings are used the majority of the time. In single-line drawings, a single thick line is drawn at the location of each pipe centerline, regardless of the pipe size.

When orthographic projection is used, there are usually three basic views. **Figure 21-11** is a single-line orthographic drawing of a simple piping layout. **Figure 21-12** depicts a double-line piping drawing of **Figure 21-11**. Orthographic piping drawings may involve only two views. In architectural work, plumbing systems are depicted in top and front views, and are labeled *plan view* and *elevation view,* respectively.

Orthographic drawings effectively show the layout of piping systems when pipes align on one or two planes. When pipes lie on many different planes and angles, orthographic

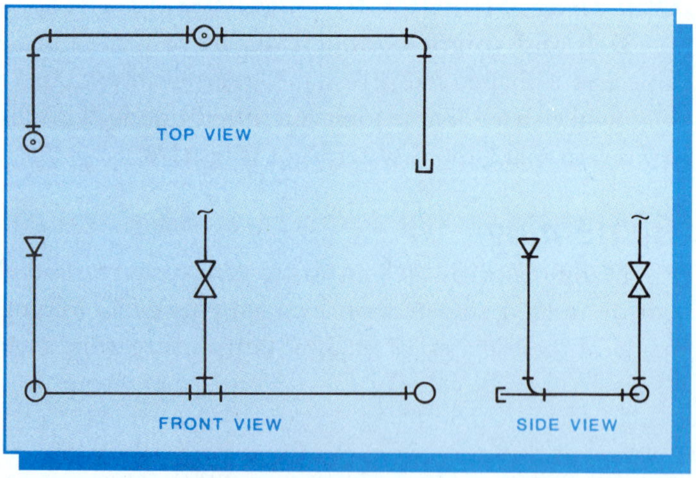

Figure 21-11
Single-line orthographic piping drawing

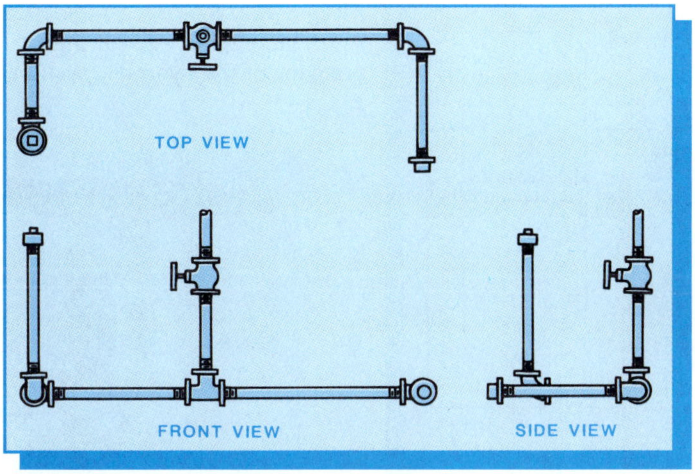

Figure 21-12
Double-line orthographic piping drawing

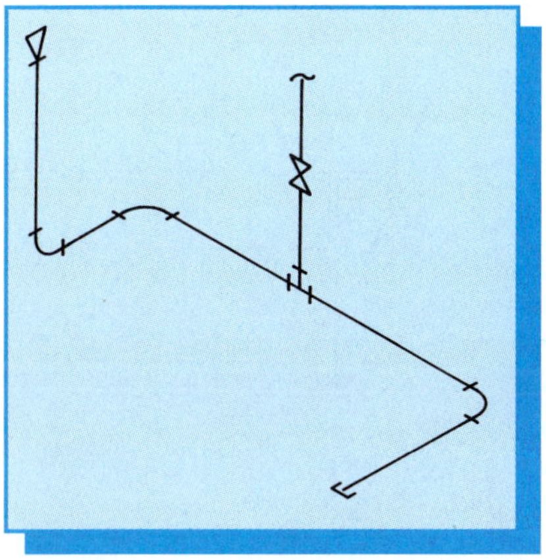

Figure 21-13
Single-line isometric piping drawing

drawings become cluttered and complex. For this reason, pictorial drawings (isometric or oblique) are used to eliminate the confusion created by overlapping pipes, fittings, and fixtures. **Figure 21-13** shows a single-line isometric piping drawing, and **Figure 21-14** shows a double-line isometric drawing of the same layout. Isometric piping drawings are especially effective for describing complex systems that involve a variety of pipe configurations, fittings, valves, fixtures, and devices **(Fig. 21-15)**.

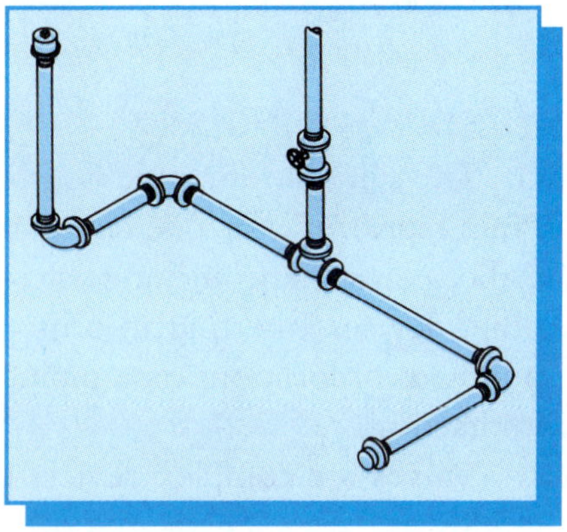

Figure 21-14
Double-line isometric piping drawing

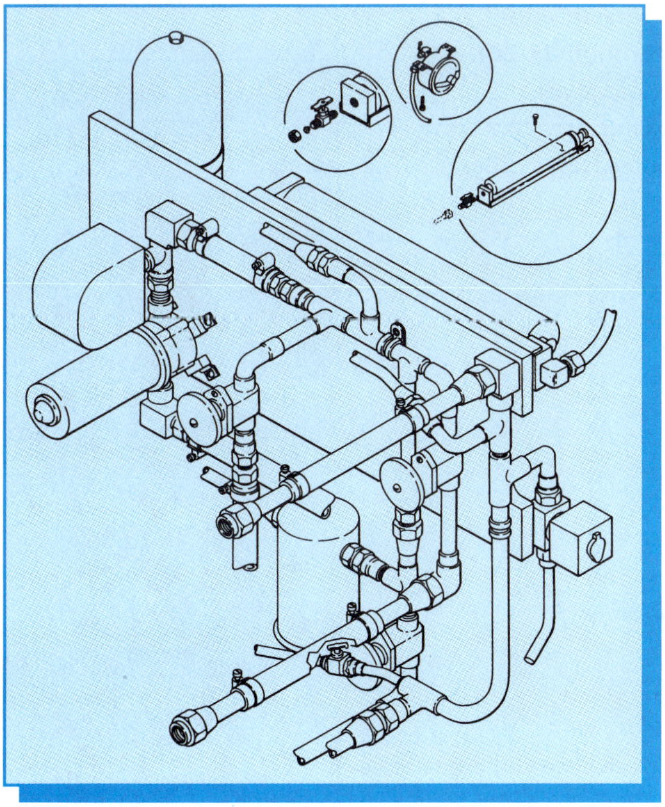

If a single projection plane is involved, an orthographic drawing **(Fig. 21-16)** can be drawn to a more accurate scale. Piping systems are often broken into small segments that fall on one plane. Dimensioned orthographic drawings are prepared for these segments. Isometric drawings are used only to show the general assembly of the entire system or subsystem. Note that the orthographic drawing shown in **Figure 21-16** represents the back of the hydraulic system shown in **Figure 21-15**.

Figure 21-15
Double-line isometric piping drawing of a segment of a hydraulic system

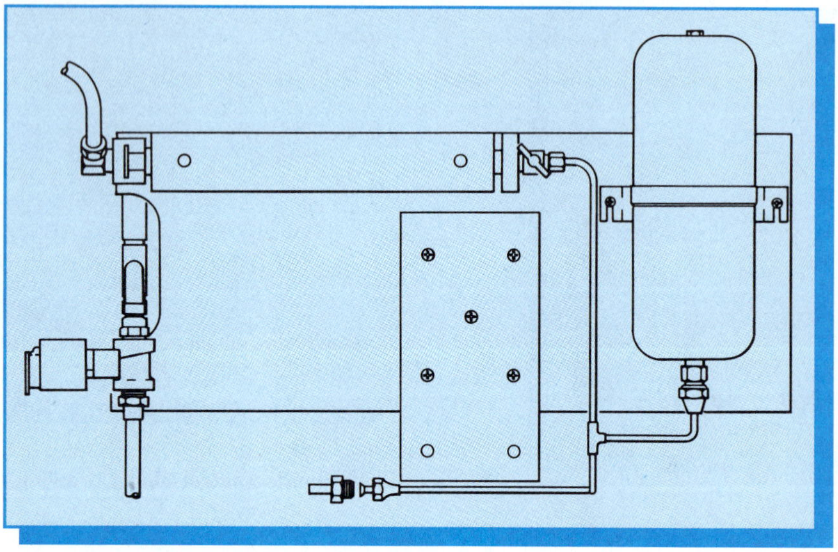

Figure 21-16
An orthographic view of the backside of a segment of the hydraulic system in Figure 21-15

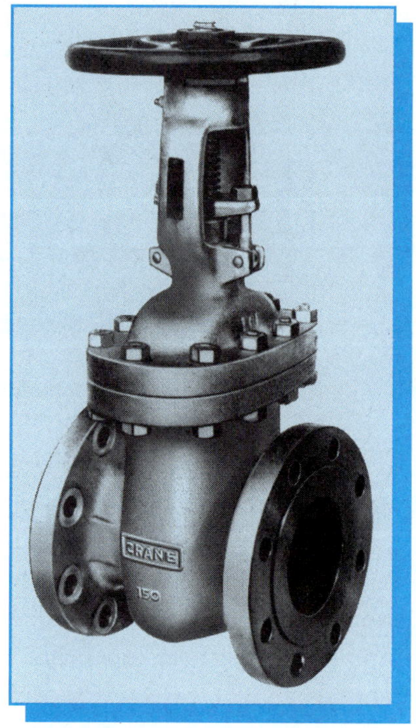

Figure 21-17
Gate valve *(Courtesy of Crane Co.)*

Piping Symbols

The symbol for a pipe is either a single or a double solid line. Pipes represent only one component in a complete piping system. Other components include valves, fittings, instruments, flanges, and other devices. Inserting an orthographic view of these components at hundreds of locations on a piping drawing is time-consuming and impractical.

Valves are complex devices that control the flow of pressurized liquids and gases. There are dozens of valve types, manufactured in hundreds of sizes and types of materials **(Figs. 21-17** through **21-20)**. This makes the preparation of orthographic drawings of valves extremely involved. Templates simply outline the drawing of valves and other components **(Fig. 21-21)**. Nevertheless, most valves, fittings,

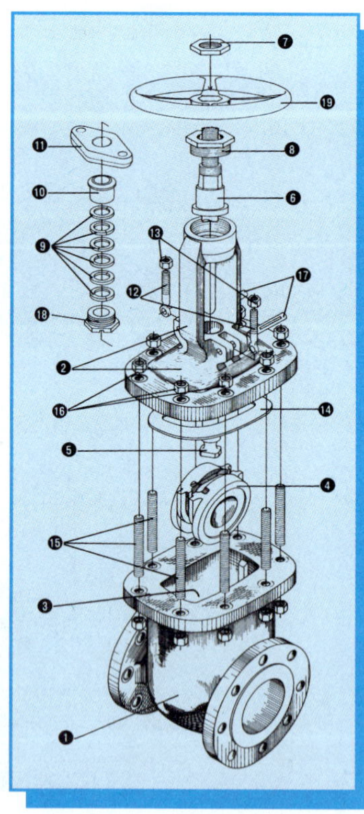

Figure 21-18
Exploded view of a gate valve
(Courtesy of Crane Co.)

Figure 21-19
Ball valve *(Courtesy of Crane Co.)*

Figure 21-21
Piping templates help to speed piping working drawings.

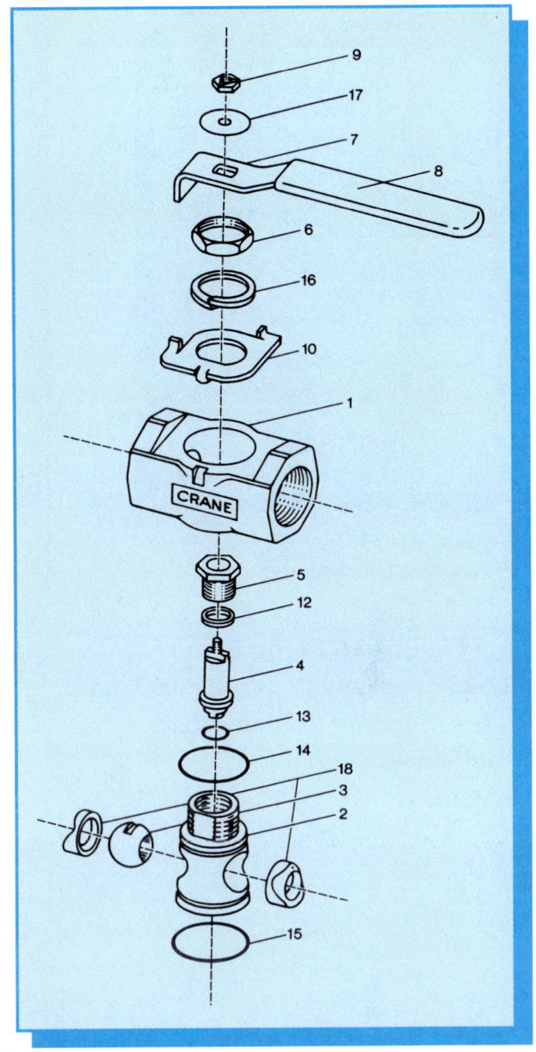

Figure 21-20
Exploded view of a ball valve *(Courtesy of Crane Co.)*

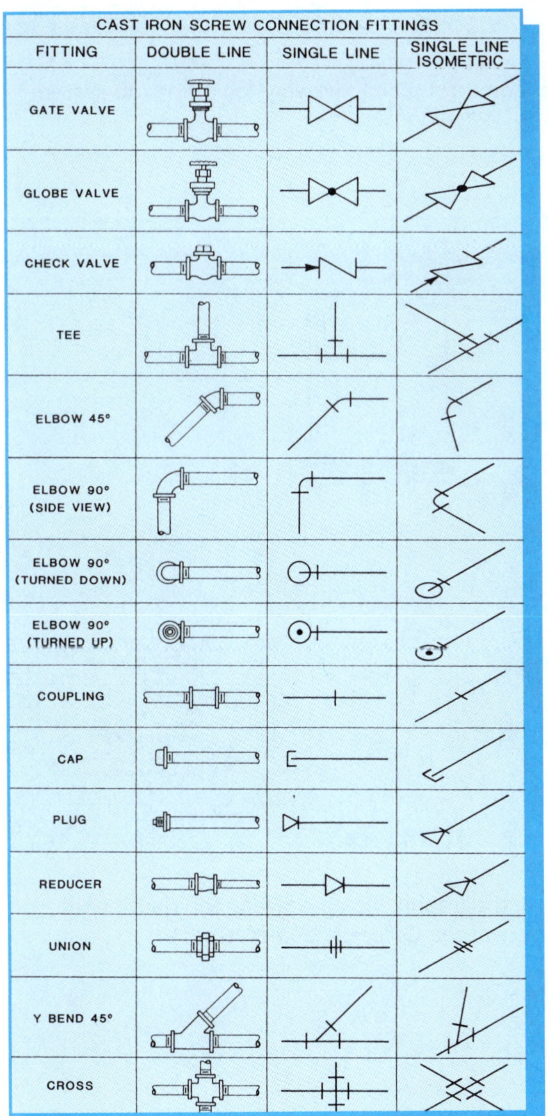

flanges, instruments, fixtures, special lines, and connection methods are represented by simplified symbols on piping drawings. Piping symbols are shown in **Figure 21-22** in orthographic single- and double-line views, and in the isometric single-line view. The various types of pipe connection symbols must be apparent on the piping drawings to insure correct assembly. **Figure 21-23** shows the application of pipe connection symbols.

Figure 21-22
Screw pipe fitting symbols

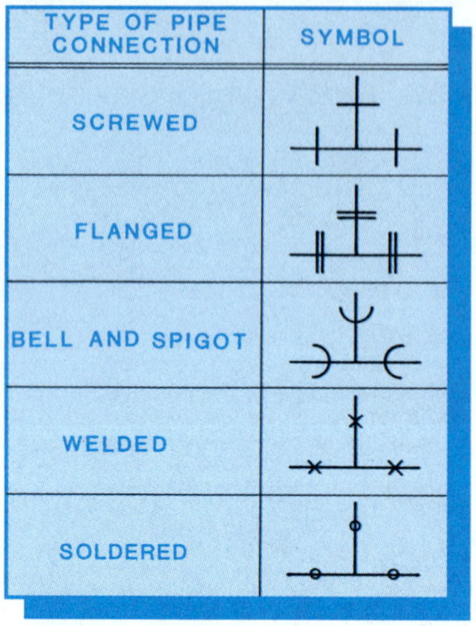

Figure 21-23
Pipe connection symbols

Figures 21-24 through **21-28** are examples of various types of piping diagrams that were drawn with a CAD system.

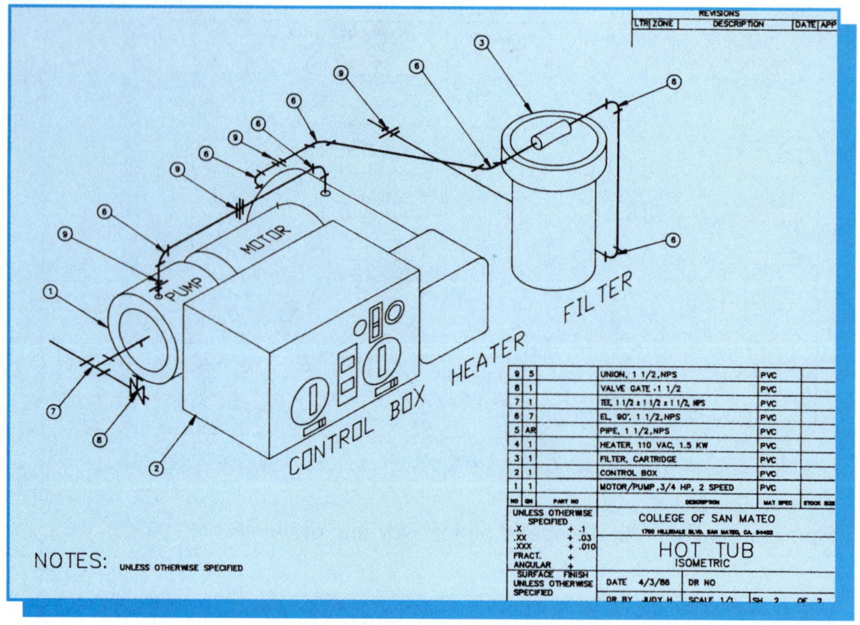

Figure 21-24
CAD-generated, single-line isometric piping drawing for a hot tub system *(Courtesy of the College of San Mateo)*

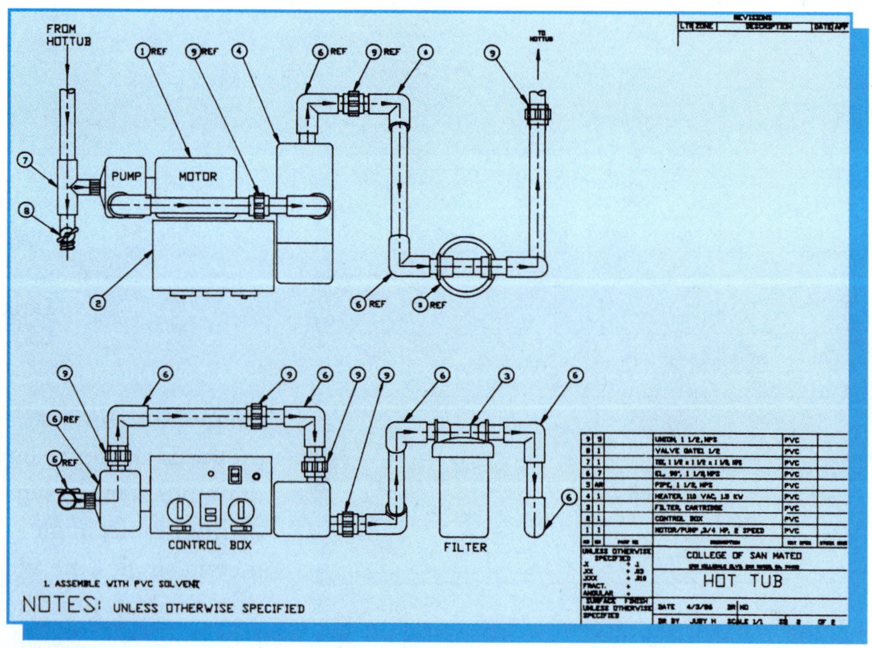

Figure 21-25
CAD-generated, double-line orthographic piping drawing for the hot tub system in Figure 21-24 *(Courtesy of the College of San Mateo)*

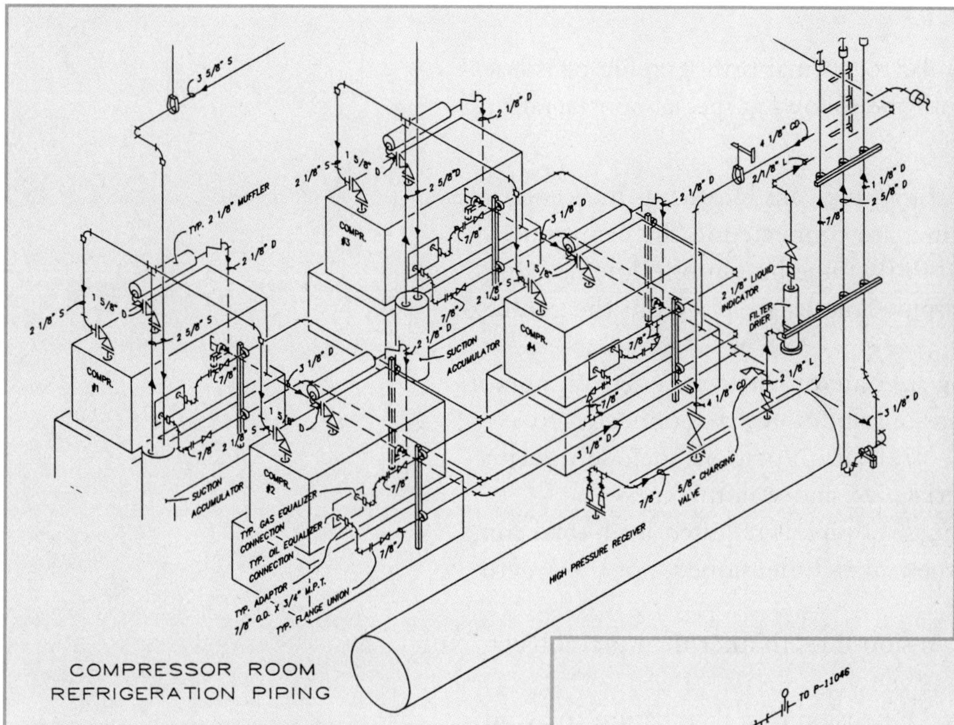

Figure 21-26
CAD-generated, single-line isometric piping drawing for a segment of a refrigeration system

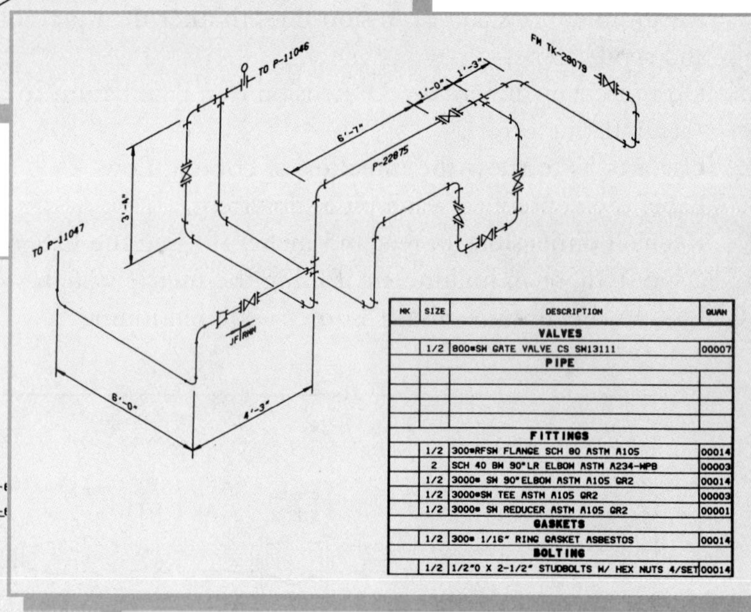

Figure 21-27
CAD-generated, single-line isometric piping drawing with materials list

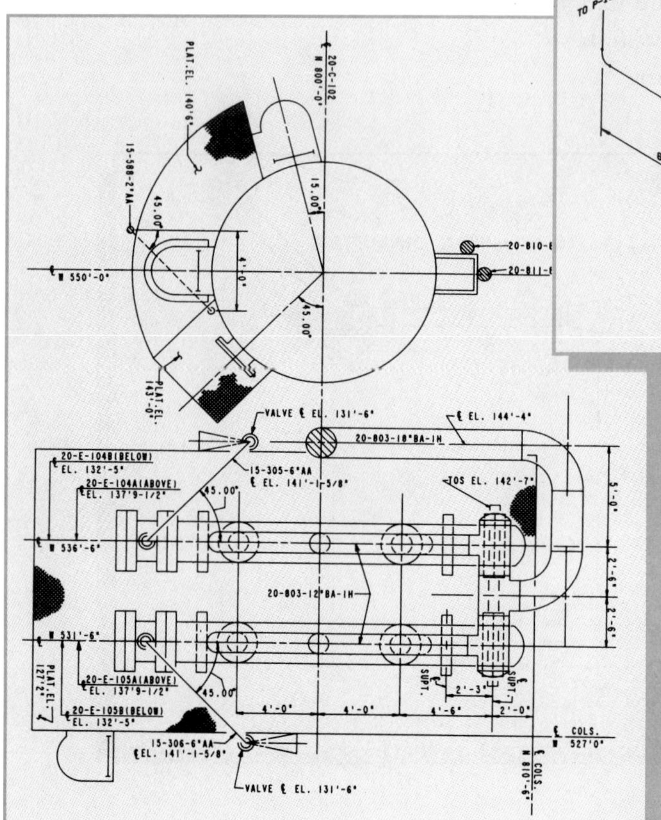

Figure 21-28
CAD-generated, double-line orthographic piping drawing for a segment of a steam system

Dimensioning

Piping dimensions are similar to normal orthographic or isometric dimensioning standards with the following special considerations **(Fig. 21-29)**:

1. Dimensions perpendicular to a pipe are located to the centerline of the pipe. Dimension lines are connected to the centerline in double-line drawings and to the pipe line in one-line drawings.
2. Dimensions parallel with pipe lengths are placed to the center of the pipe or flange.
3. Pipe sizes and pipe fitting sizes are provided in a general note if all pipes are equal. Dimensions and notes are connected to a pipe with a leader if pipe sizes vary. Notations include information on the type of material, size, and wall thickness.
4. When more than one length of pipe is required for a total run, only the total length of the run is dimensioned, not the length of the various pieces.
5. Keep dimensions and extension lines thinner than pipe object and symbol lines.
6. Do not chain-dimension. Dimension to a base datum to avoid accumulating errors.
7. Use arrows to show the direction of content flow.
8. Label the contents of each pipe subsystem.
9. Keep all dimensions in feet and inches if using the U.S. customary system, or in millimeters if using the metric system.
10. Use unilateral numerals for isometric dimensioning.

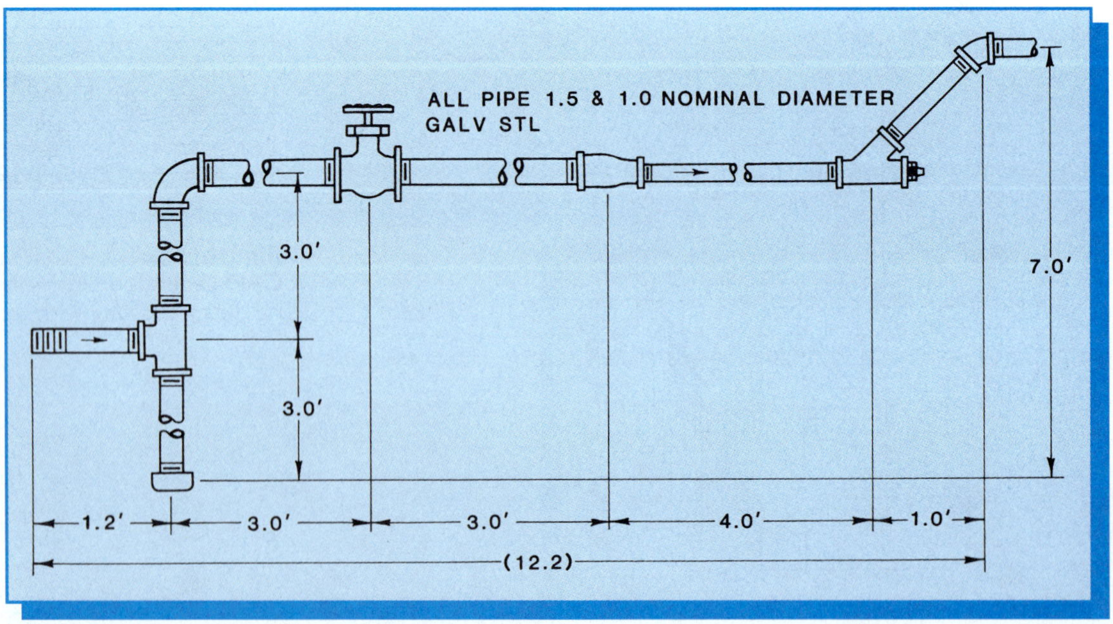

Figure 21-29
Example of a dimensioned pipe drawing

AutoCAD LT Exercises

Identical symbols appear in many different locations on a piping drawing. Repetition of these symbols on the same drawing is time-consuming when drawing with either instruments or templates. Since piping symbols are standardized, they are drawn only once when preparing a piping drawing on a CAD system.

There are several methods of inserting standard symbols into a CAD drawing. The following are the four most common methods:

1. Draw the symbol once, and then use the **Copy** and **Rotate** commands to repeat the symbol on the drawing.
2. Use a graphics tablet overlay that contains piping symbols **(Fig. 21-30)**. The overlay may be a commercial variety or may be created by the CAD operator.
3. Draw the symbol once and create a Block that can be inserted.
4. Use a commercial custom pull-down menu.

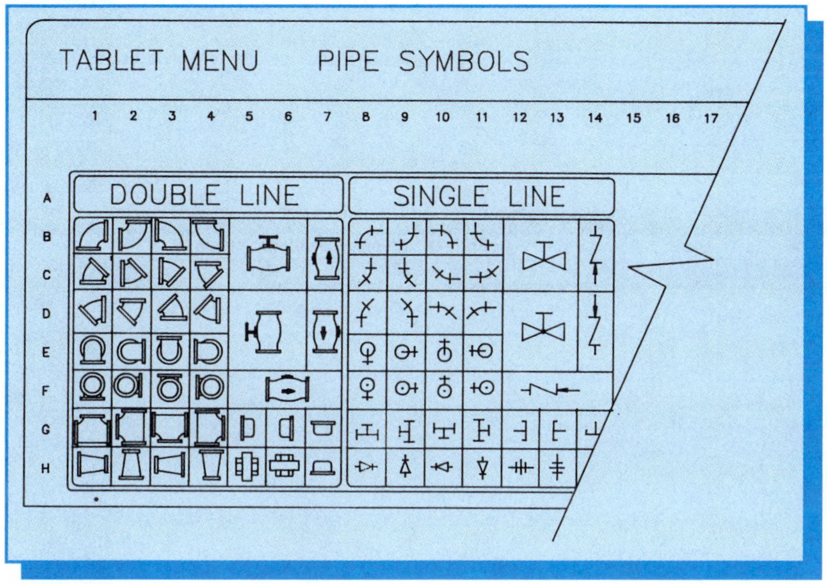

Figure 21-30
Piping symbols on the graphics tablet are stored in the CAD system's library.

Figures 21-31 through **21-35** show five steps in creating a double-line piping drawing, using any one of the systems for inserting standard piping symbols.

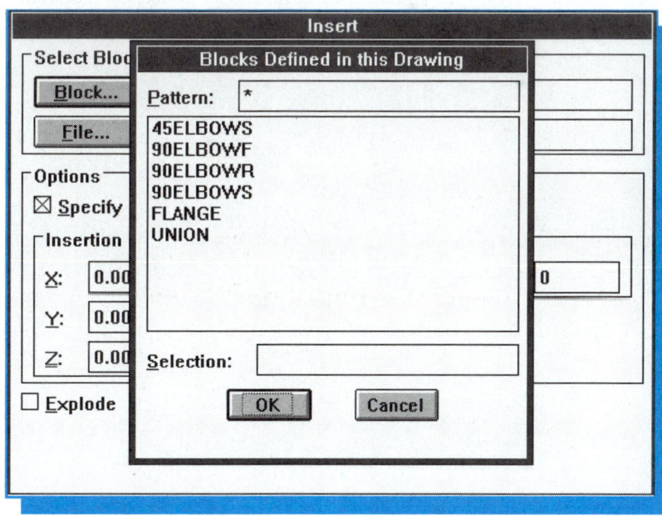

Figure 21-31
Pipe symbols created as Blocks

1. Load the CAD piping program that includes the necessary symbols or create the necessary Blocks. To create shapes, see **Figure 21-22** for symbol shapes. **Figure 21-31** shows the blocks created for this example.
2. Draw the center lines of the piping system to scale **(Fig. 21-32)**.
3. Insert the piping symbols. **(Fig. 21-33)**.

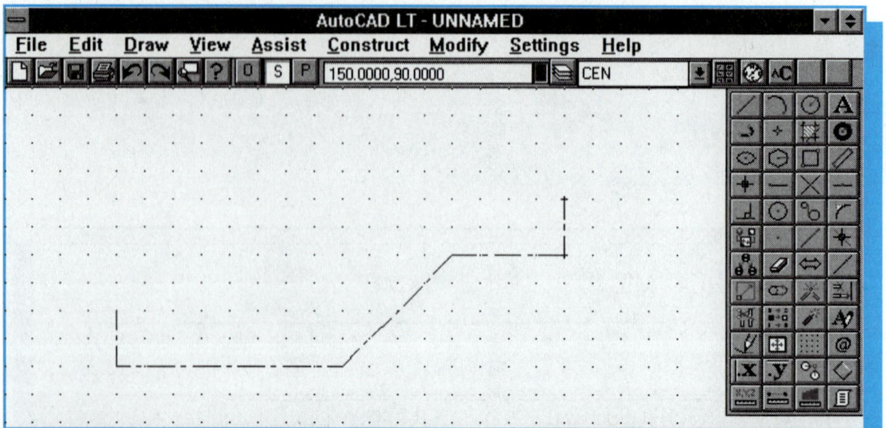

Figure 21-32
Lay out the centerlines of the pipes with Line command.

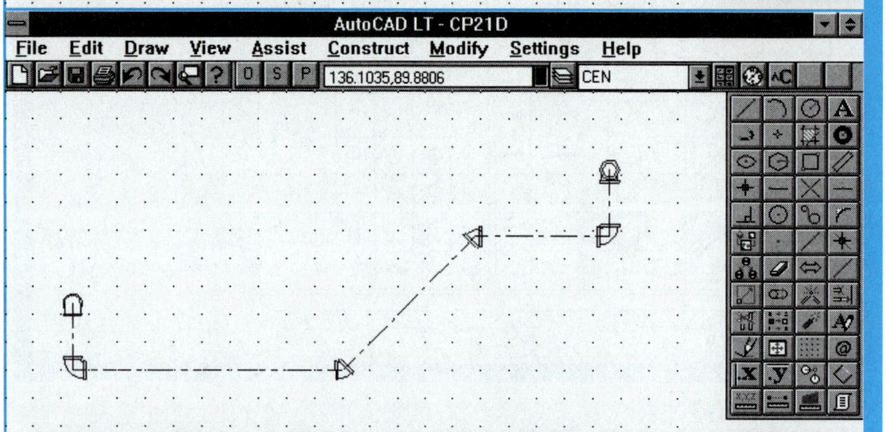

Figure 21-33
Use the Insert command to place the Blocks.

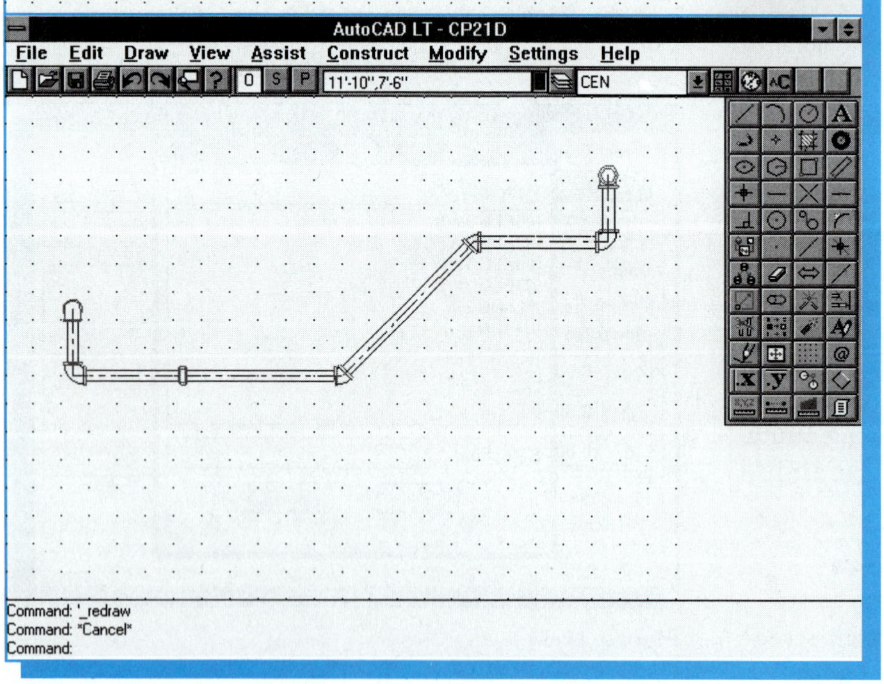

Figure 21-34
Draw lines for pipes with Line command or Double Line command.

4. Draw the pipes using a **Line** or **Double Line** command **(Fig. 21-34)**.
5. Add dimensions to complete the drawing **(Fig. 21-35)**.

Examples of a CAD-generated single-line isometric drawing and a double-line orthographic drawing of the same system (a hot tub heating and filtering system) are shown in **Figures 21-24** and **21-25**. **Figures 21-26** through **21-32** are further examples of computer-generated isometric and orthographic piping working drawings.

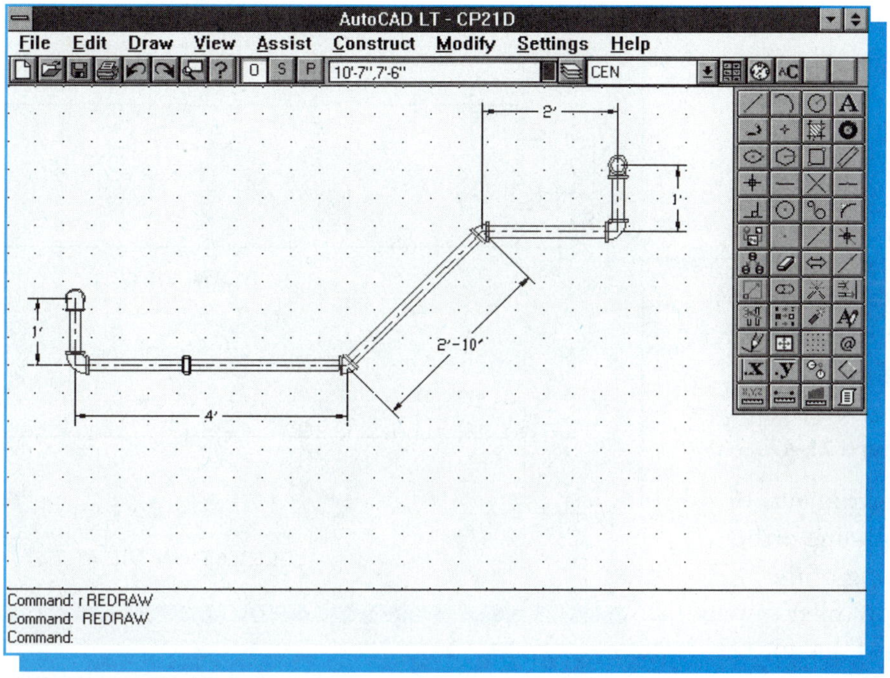

Figure 21-35
Add dimensions.

Drafting Exercises

1. Draw with instruments or CAD the single-line orthographic drawing of the piping system shown in **Figure 21-36**.

2. Draw with instruments or CAD a double-line orthographic drawing of the piping system in **Figure 21-36**.

3. Draw with instruments or CAD a single-line isometric piping drawing of the piping system in **Figure 21-36**.

4. Draw with instruments or CAD a double-line isometric piping drawing of the piping system in **Figure 21-36**.

5. Make the following ECOs to **Figure 21-37**:

a. Change the horizontal pipe diameter to .75".

b. Lengthen pipe #4 to 5" and cap.

c. Correct pipe #7 to 26" in the materials list and drawing.

d. Correct pipe #6 (2) to 38" in the materials list.

e. Shorten pipe #8 (top only) to 28".

f. Update all changes on the parts list.

g. List the ECOs on the drawing.

6. With instruments or CAD, draw an orthographic single- and double-line piping drawing of the sketched piping layout in **Figure 21-38**.

7. With a CAD system, draw the 90° flanged pipe elbow in **Figure 21-39**.

8. Design a single-line isometric pipe drawing plan for a water heater and dryer hookup with a .5" gas pipeline for **Figure 21-40**.

9. Draw with instruments or CAD a single-line isometric drawing of the hot tub filter system in **Figure 21-41** and/or the hot water tank system in **Figure 21-42**.

10. Draw with instruments or CAD a double-line orthographic drawing of the swimming pool filter system in **Figure 21-41** and/or the hot water tank system in **Figure 21-42**.

11. Use the symbols in **Figure 21-43** to make a CAD symbol library or individual blocks.

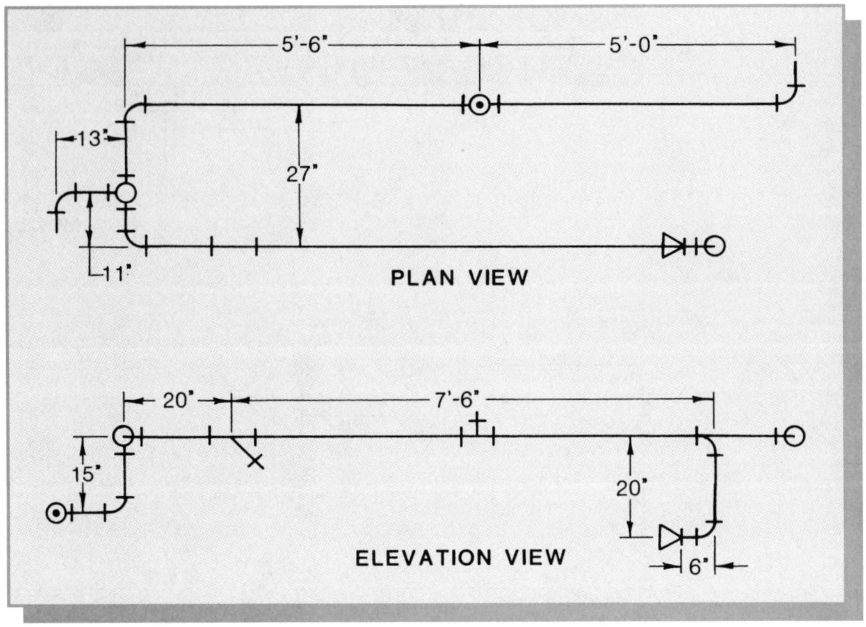

Figure 21-36
Draw the following, using .5" pipe and fittings throughout.

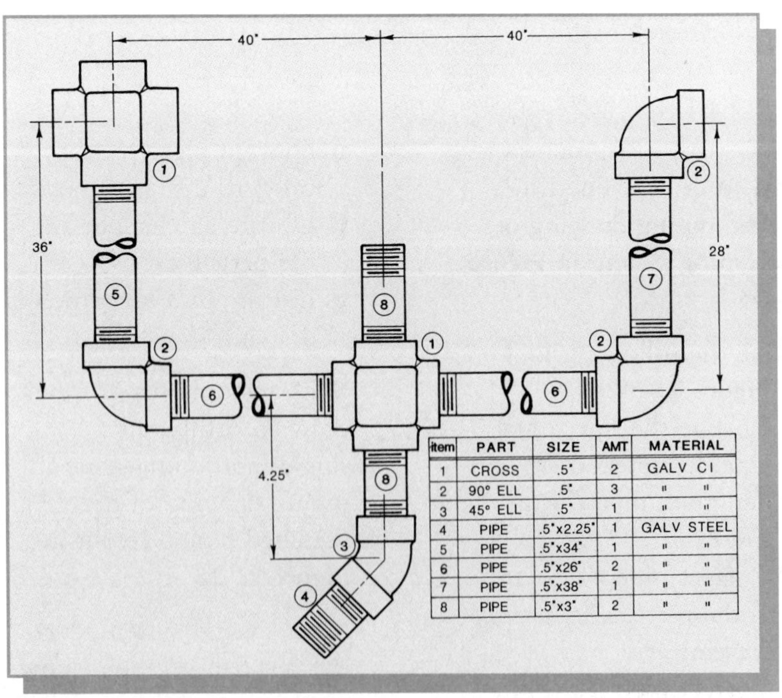

item	PART	SIZE	AMT	MATERIAL
1	CROSS	.5"	2	GALV CI
2	90° ELL	.5"	3	" "
3	45° ELL	.5"	1	" "
4	PIPE	.5"x2.25"	1	GALV STEEL
5	PIPE	.5"x34"	1	" "
6	PIPE	.5"x26"	2	" "
7	PIPE	.5"x38"	1	" "
8	PIPE	.5"x3"	2	" "

Figure 21-37
Draw the double-line orthographic piping drawing. Make the ECOs listed in Exercise 5.

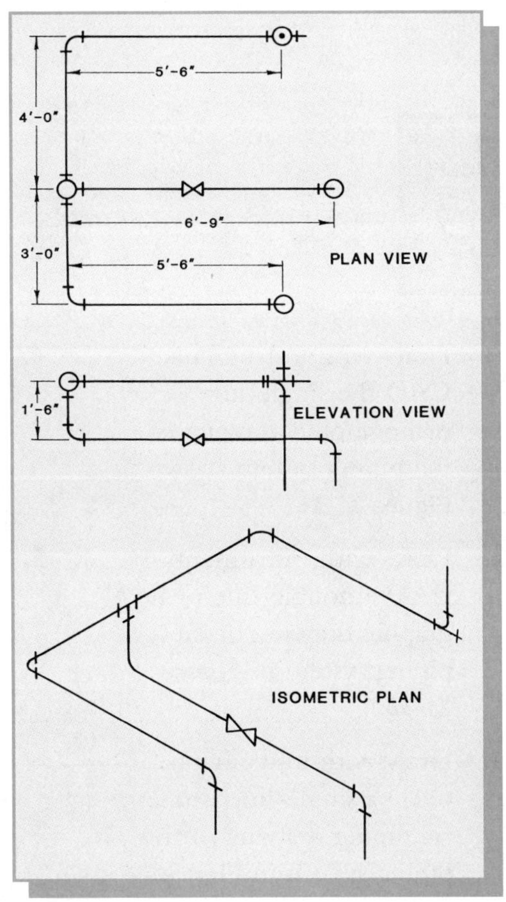

Figure 21-38
Draw the single-line piping drawings and their double-line counterparts with instruments or a CAD system.

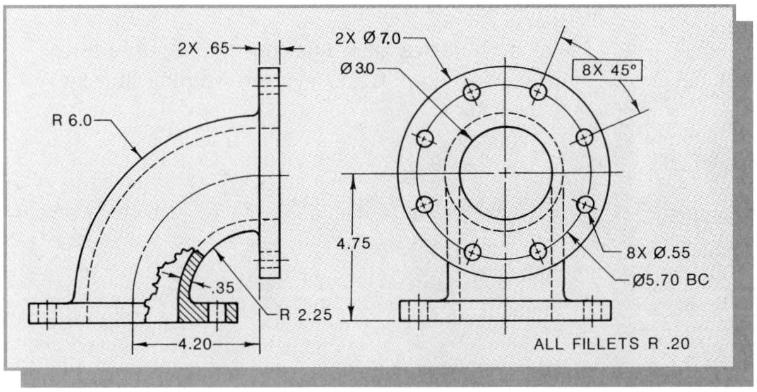

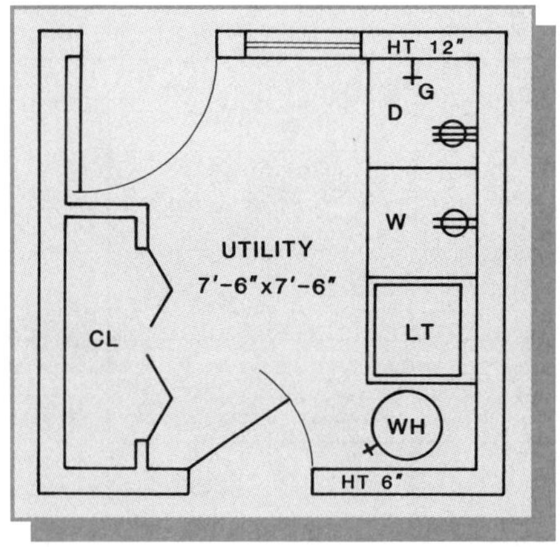

Figure 21-39
Draw the 90° flanged elbow with instruments or a CAD system.

Figure 21-40
Design a single-line isometric piping drawing for the water heater hookup with a .5" gas pipe and dimension.

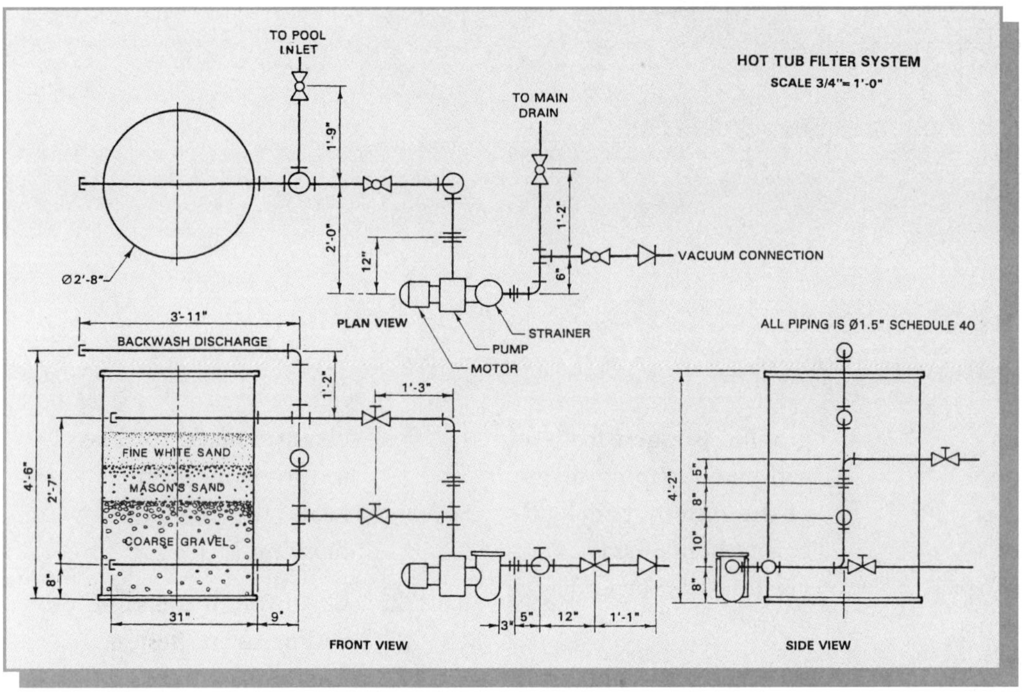

Figure 21-41
Draw a single-line and double-line multiview drawing.

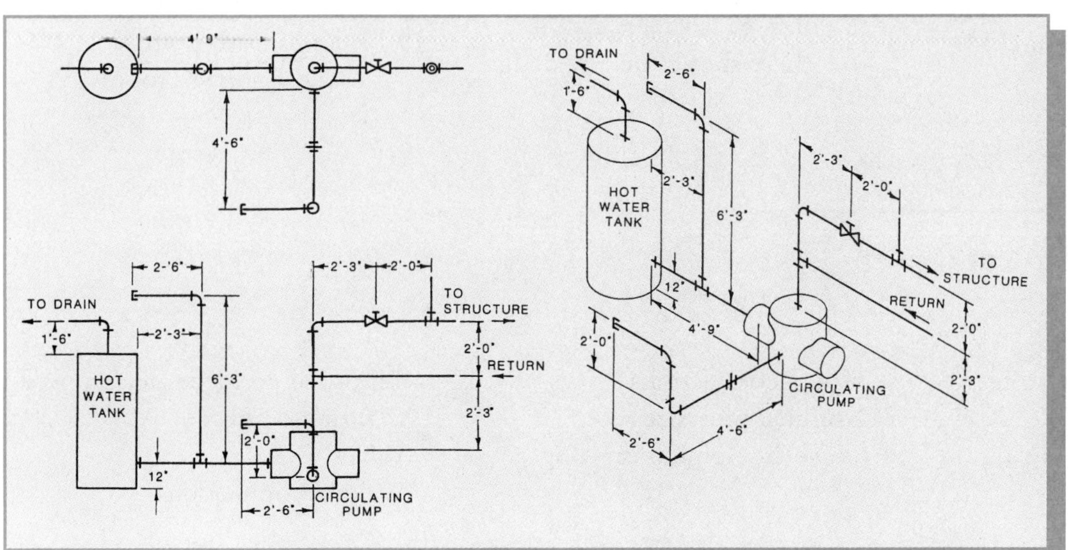

Figure 21-42
Draw a single-line and double-line multiview drawing.

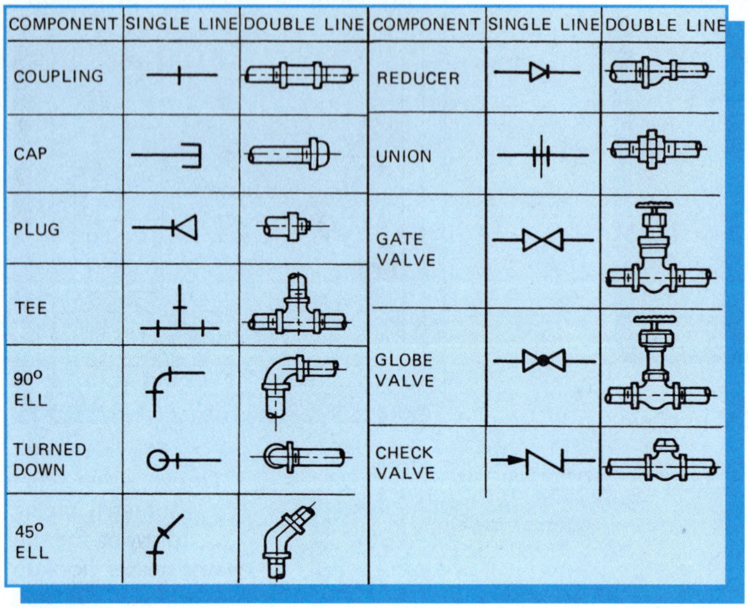

Figure 21-49
Make up a series of single-line and double-line symbols for your CAD system's piping library.

Design Exercises

1. Go to a hot tub or spa showroom and ask to see the motor, pump, heater, and filtering system of their "top-of-the-line" spa. Measure the components and accurately sketch the system, including all pipes and fittings. Draw a single- or double-line orthographic (suggest front and top views) piping diagram of the system. Label all components. Draw a single-line isometric diagram of the same system.

Return to the hot tub store and show the drawings to installers. Ask if there should be any design changes made to the system to simplify their job. Make the changes to your drawings.

2. Go through the same procedure as in Design Exercise 1 for a:

swimming pool
family bathroom
school shower room

Key Terms

Bell and spigot connections
Cementing
Double-line piping drawing
Fittings
Flange connections
Nominal pipe size
Single-line piping drawing
Threaded connections (screwed fittings)
Valves
Welded connections

Chapter 22

Electronics Drafting

Objectives

The student will be able to:

- identify the block diagrams, schematic diagrams, connection diagrams, logic diagrams, and cableform diagram
- differentiate between schematic and simple outline electronic symbols
- with a CAD system or manually, draw a block diagram, schematic diagram, connection diagram, logic diagram, cableform diagram, and a chassis pattern
- name the associations that set electronic standards
- understand how sophisticated CAD systems are used to create electronic drawings

Introduction

The electronics industry is in a virtual explosion of expansion and advancement in research, product development, and manufacturing. Extremely rapid growth has occurred in such high-tech areas as computers, telecommunication, aerospace, automotive, consumer products, and industrial instrumentation **(Figs. 22-1** through **22-4)**. The electronics industry touches almost every facet of our daily lives.

This expanding electronics industry is creating an increasing demand for qualified electronics drafters to produce drawings manually and with computer-aided drafting systems. Before electronics products are produced, drawings are needed to direct the manufacture, assembly, and packaging of all component parts. In addition, drawings provide necessary information for the installation, control, repair, and maintenance of every product. Electronics drafters are a vital communications link in this process.

Figure 22-1
Desktop calculator *(Courtesy of Monroe Systems for Business)*

Figure 22-2
Aerial targets can simulate enemy aircraft, sea-skimming cruise missiles, and antiship missiles, and are used to train U.S. Navy personnel. *(Courtesy of Northrup Corporation)*

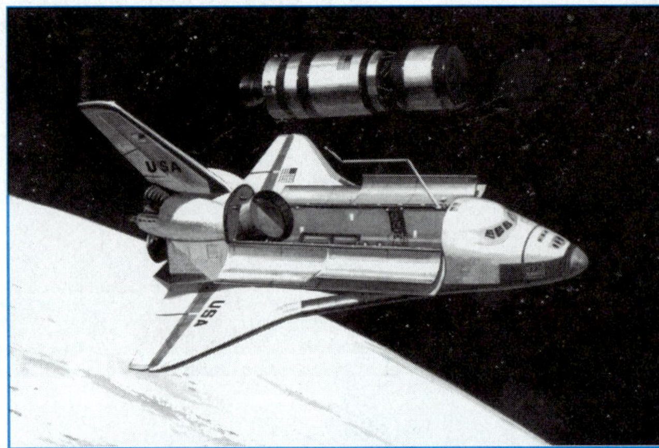

Figure 22-3
Space shuttle *(Courtesy of NASA)*

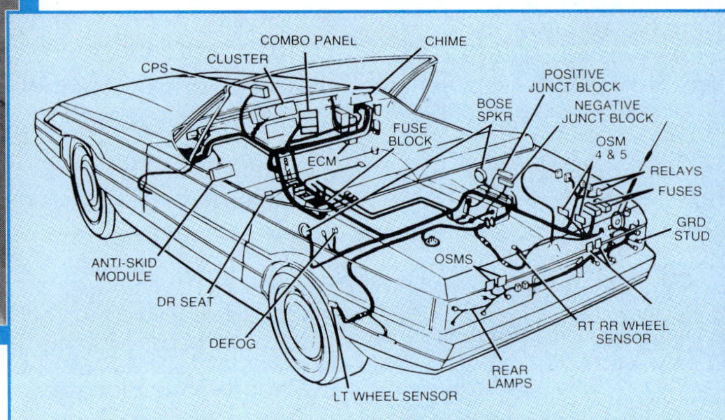

Figure 22-4
Pictorial drawing of an automobile's electrical system

Electronics Industry Standards

Electronics drafting uses a broad combination of governmental and professional association standards in the development of drawings and diagrams. This standardization improves the efficiency and minimizes the potential errors in the interpretation of drawings. Standards cover the uniform selection, formation, symbolism, and reference notations used for all electronics and electrical circuits, equipment, and devices. Standardization for electronics and electrical drawings are prepared by the following organizations:

- **ASME**—American Society of Mechanical Engineers
- **ASA**—American Standards Association
- **IEC**—International Electrotechnical Commission
- **EIA**—Electronics Industries Association
- **IEEE**—Institute of Electrical and Electronics Engineers
- **MIL STD**—Military Standards

Electronics Drawing Symbols

Graphic symbols (electronics shorthand) play a major role in preparing **electronics drawings.** The use of symbols is a much faster drafting method than using orthographic views, pictorials, or notations. Because of the complexity of electronic circuits and devices, the use of graphic symbols is also the most accurate method of preparing electronics and electrical drawings. The most universally accepted set of electronics symbols is the ASME Y32.2 standard symbology, as shown in **Figures 22-19** and **22-20**, page 529. In using electronic symbols to represent electronic circuit devices and components, use the following guidelines:

1. Layout circuits should read from left to right. Complex circuits should read from upper-left to lower-right.
2. Symbols can be drawn at any size, but should be approximately the same size and proportional to the size of the drawing.
3. Symbol directions have no significance and component symbols may be rotated to any direction on the drawing.
4. Avoid line (circuit path) crossovers wherever possible.
5. Keep the length of circuit lines and distance between components to a minimum.
6. Different and discrete parts of a diagram need not adhere to the same scale.
7. Although angular lines have no significance, avoid wherever possible.
8. If additional data is needed to describe a symbol, standard abbreviations should be used next to the symbol.
9. Lay out circuit paths with symbols placed in logical order as the circuit functions, from power source to loads.
10. Use vertical or horizontal lines wherever possible. Avoid using angular lines.
11. If lines are interrupted or progress to another circuit or drawing, identify the lines and indicate where they are connected on the original and receiving drawing.

Most electronic drawings are prepared with pencil lines. However, since many drawings are used in technical manuals, ink drawings are used to produce high-quality lines for this type for reproduction.

Types of Electronics Drawings

There are many types of electronics drawings. Most start as sketches or written instructions from an electronics engineer or designer. The electronics drafter interprets these instructions and graphically transforms them into working drawings. Each type of electronic drawing serves a specific purpose in describing the design, construction, fabrication, maintenance, or repair of a product. It is the responsibility of the electronics drafter to select the best type or combination of drawings to represent each design.

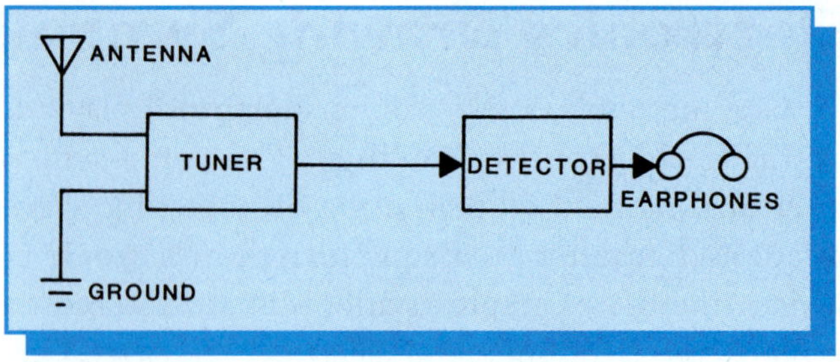

Figure 22-5
Block diagram of a crystal set

Block Diagrams

A **block diagram** is a general presentation drawing used to simplify and show the interrelationship of the various parts of an entire electronics system. Block diagrams are often used in the first planning and design stages because they are a fast and uncomplicated method of recording general design ideas. Circuit details are omitted from block diagrams.

Blocks in the shape of rectangles, squares, or triangles are joined by a single line with arrows. Arrows show the direction of current flow. The direction of block diagram design progresses from left to right, and the function of each block is lettered within the block. **Figure 22-5** shows a simple block diagram for a crystal radio. Examples of a block diagram for a CAD system and a radio are shown in **Figures 22-6** and **22-7**.

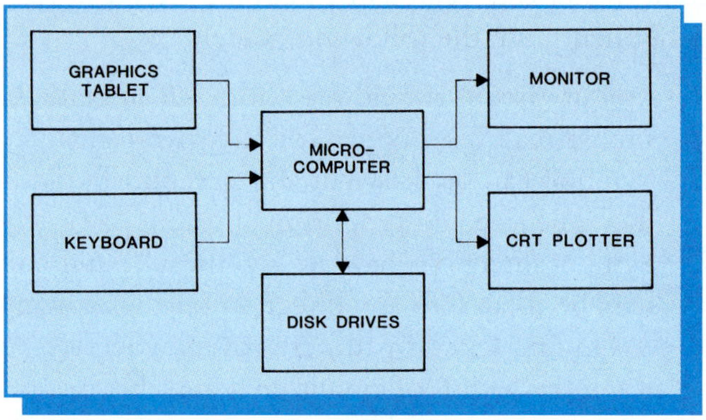

Figure 22-6
Block diagram of a microcomputer CAD system

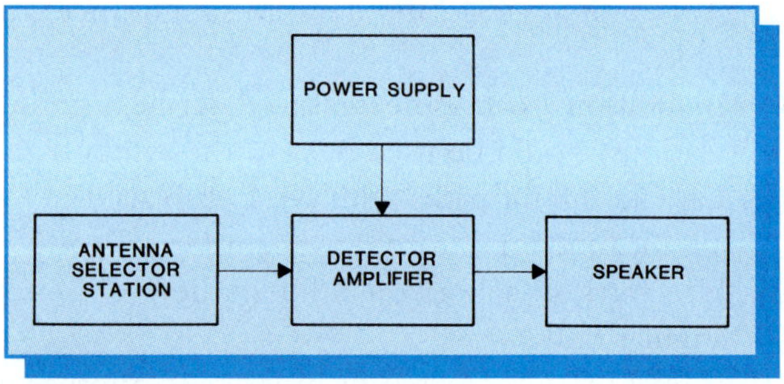

Figure 22-7
Block diagram of a radio

Schematic Diagrams

Schematic diagrams are the most frequently used drawings in the electronics industry. They show the connection and function of each circuit in a simplified form **(Fig. 22-8)**. Nevertheless, **schematic diagrams** show more detail than block diagrams and serve as the master reference for the preparation of production drawings, parts lists, and component specifications. They contain graphic symbols, connections, and noted functions to describe each specific circuit arrangement. Schematic diagrams also show the sequence in which components are connected to complete a circuit. Each component is represented by its own special schematic symbol, and is drawn to show its relationship to other components **(Fig. 22-9)**. Schematic diagrams are sometimes called **elementary diagrams.** The following are general rules for drawing schematic diagrams:

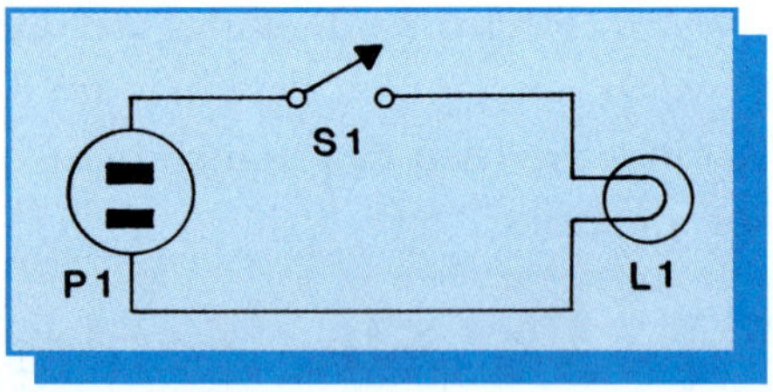

Figure 22-8
Schematic diagram for a lamp and its power supply

1. Current flows from left to right, or upper-left to lower-right.
2. Allow space between symbols and connectors to avoid crowding.

3. Reference callouts are positioned in the same location, next to all symbols.
4. Keep the crossing of connector lines to a minimum.
5. If scale is a factor, use a grid overlay to avoid excessive dimensioning.

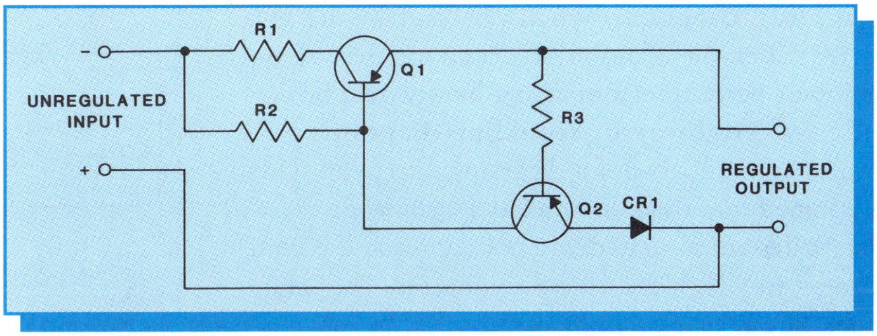

Figure 22-9
Schematic for a regulated power supply

Connection Diagrams

Connection or **wiring diagrams** are electronic assembly drawings that contain details needed to make or trace all electronic connections.

This type of diagram shows precisely where components are located in a circuit and how wires are connected. The components are drawn pictorially or as single orthographic views. Connection diagrams are supplementary drawings used with a schematic diagram for installing and repairing electronic equipment. Types of connection diagrams include point-to-point, highway or trunkline, baseline or airline, and cableform diagrams.

Point-to-Point Diagrams. Point-to-point diagrams show components, drawn as simple outlines, with connecting lines and terminals. These drawings are used by technicians for the assembly of electrical systems. An example of a point-to-point diagram is the motorcycle wiring diagram shown in **Figure 22-10**. Automotive manuals frequently contain point-to-point diagrams for troubleshooting purposes.

Figure 22-10
Point-to-point connection diagram for a dirt bike

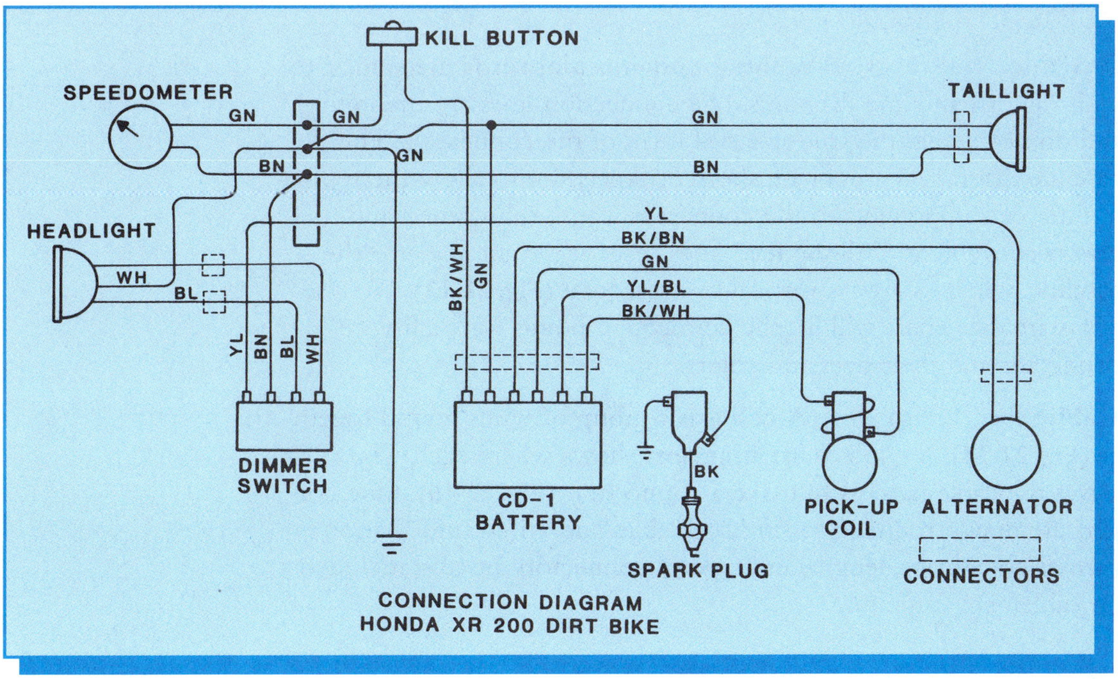

Highway Diagrams. Highway diagrams are preferred when the addition of a large number of point-to-point interconnecting lines would be excessive. **Highway or trunkline diagrams** are similar to point-to-point diagrams, except all wires are joined to a thick line called a *highway* or *trunkline*. Wires are separated where they reach the component to which the wire is connected. The long, thick line is called a **path** or highway, and shorter lines are called **feeders (Fig. 22-11)**. To show where connections are made, wire data is labeled on feeder lines. **Figure 22-12** shows a highway diagram for a meter box with the appropriate wire data. This wire data includes the wire destination (hookup), wire color, and wire size. EIA color code standards are as follows:

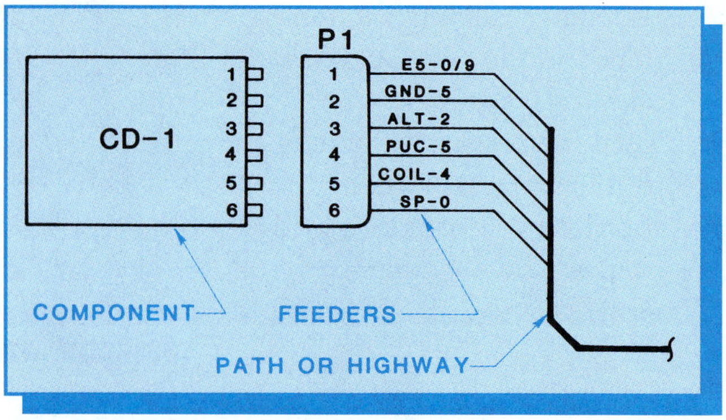

Figure 22-11
Highway diagram's feeders and paths

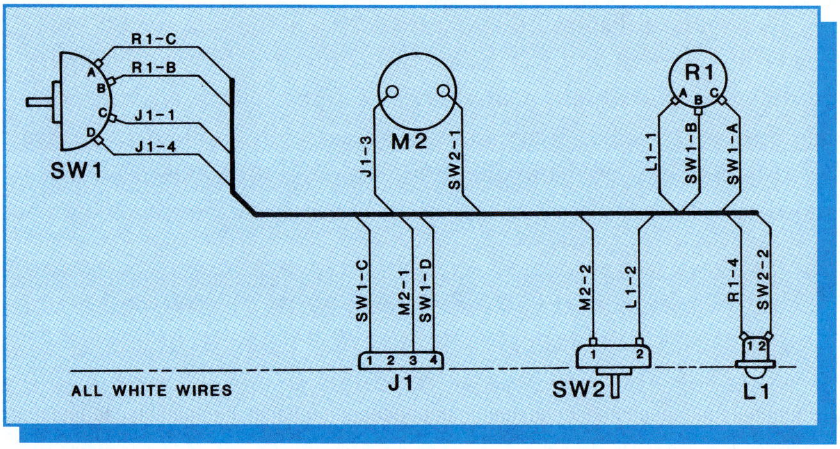

Figure 22-12
Highway diagram for a meter box with wire destinations

Wire Number	Color
0	Black
1	Brown
2	Red
3	Orange
4	Yellow
5	Green
6	Blue
7	Violet
8	Grey
9	White

Baseline Diagrams. **Baseline** or **airline diagrams** are similar to highway or trunkline diagrams. The connection lines are theoretical and do not depict the true size and form of the connectors. The baseline diagram is a drawing shortcut procedure to allow the repositioning of components so the connectors can be shortened and drawn conveniently. All the feed lines meet at a right angle to the baseline, which is always vertical or horizontal **(Fig. 22-13)**. The use of a wire data chart will list the connection points for each wire connector and their sizes and colors.

Cableform Diagrams. A **cable** is a group of wires bound together **(Fig. 22-14)**. A **cableform diagram** shows where each wire leaves the cable harness to make a connection **(Fig. 22-15)**. Line corners are drawn rounded to simulate cables and wires. Information such as color, gage, length, and type of connectors are also included on cableform diagrams.

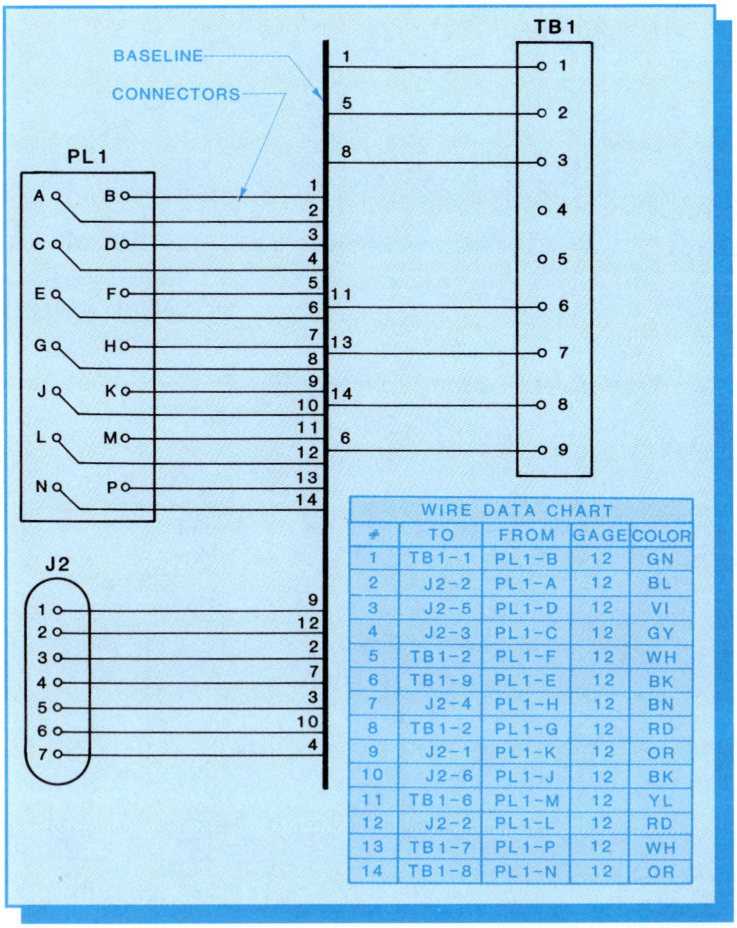

WIRE DATA CHART				
#	TO	FROM	GAGE	COLOR
1	TB1-1	PL1-B	12	GN
2	J2-2	PL1-A	12	BL
3	J2-5	PL1-D	12	VI
4	J2-3	PL1-C	12	GY
5	TB1-2	PL1-F	12	WH
6	TB1-9	PL1-E	12	BK
7	J2-4	PL1-H	12	BN
8	TB1-2	PL1-G	12	RD
9	J2-1	PL1-K	12	OR
10	J2-6	PL1-J	12	BK
11	TB1-6	PL1-M	12	YL
12	J2-2	PL1-L	12	RD
13	TB1-7	PL1-P	12	WH
14	TB1-8	PL1-N	12	OR

Figure 22-13
A baseline diagram with a wire data chart

Figure 22-14
Cable (bundled wires) with jack, ties, and terminal lugs

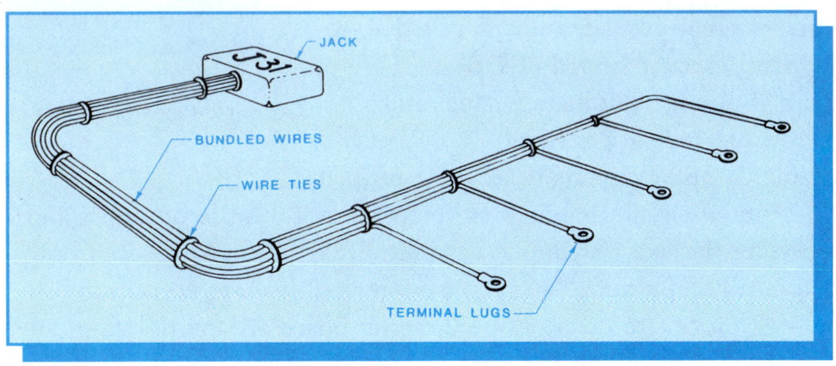

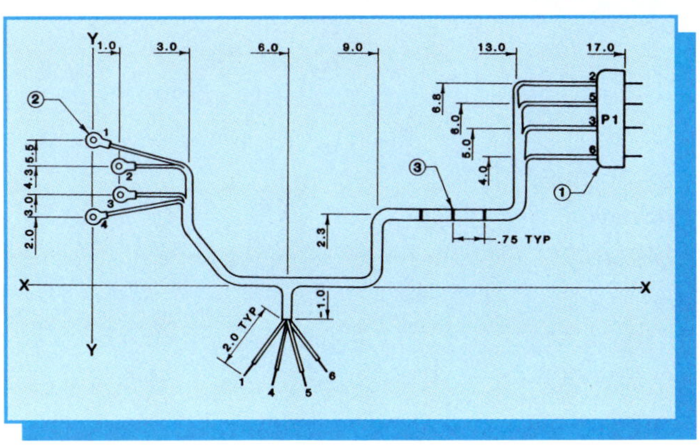

Figure 22-15
Cable form diagram

Logic Diagrams

Logic diagrams are used to design the microcircuitry used in many contemporary devices such as digital watches, calculators, and computers. A logic diagram is a specialized type of block diagram showing circuitry flow details. Blocks of specific shapes are used to represent different logic functions **(Fig. 22-16)**.

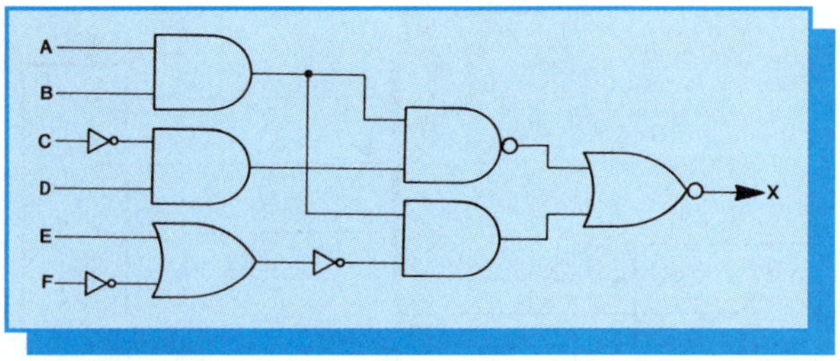

Figure 22-16
Simple logic diagram

Chassis Drawings

Chassis drawings are pattern or development drawings of cabinets that house electronic circuits and equipment. Chassis drawings include overall dimensions and stretchout views of a chassis. They also include the location and size of all holes and other openings needed to mount components and provide openings for wires **(Figs. 22-17a** and **22-17b)**.

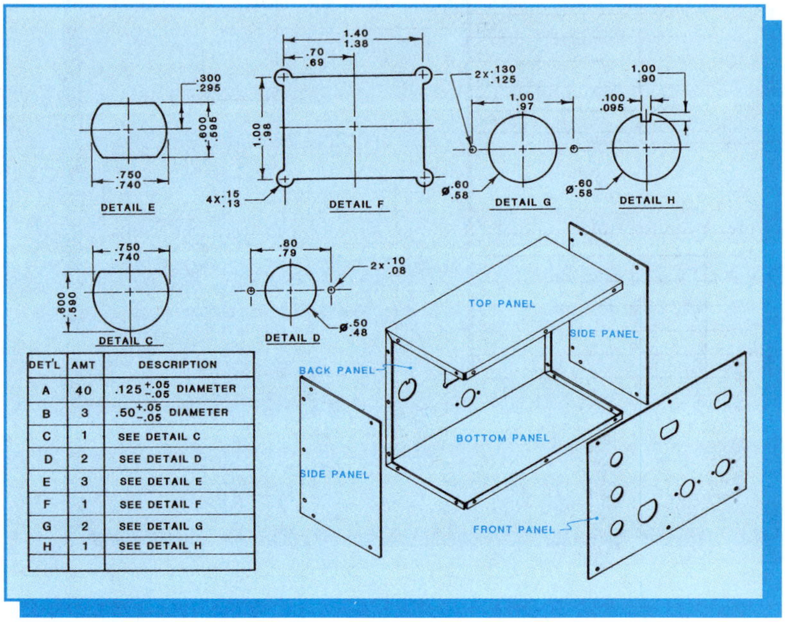

DET'L	AMT	DESCRIPTION
A	40	$.125^{+.05}_{-.05}$ DIAMETER
B	3	$.50^{+.05}_{-.05}$ DIAMETER
C	1	SEE DETAIL C
D	2	SEE DETAIL D
E	3	SEE DETAIL E
F	1	SEE DETAIL F
G	2	SEE DETAIL G
H	1	SEE DETAIL H

Figure 22-17a
Cutout dimensions for a chassis pattern

Printed Circuits

Printed circuits are contemporary replacements for wired circuits. One component is electrically connected to another through printed copper conductors on an insulated board, called a **printed circuit board (PCB).** Printed circuits eliminate wiring faults and errors, reduce production costs, and enable complex circuits to be miniaturized.

Printed circuit drawings are prepared at full scale or larger, and represent the exact layout of a circuit. After the drawings are photographically reproduced and reduced in size, they are used in the etching process to create printed circuit boards. Many drawings are necessary to design and create the documentation necessary to produce a printed circuit board **(Figs. 22-18a** and **22-18e)**.

Standard Symbols Charts

There are two types of standard symbols for electronics components on connecting diagrams—**schematic symbols** and **simple outline symbols (pictorial) (Fig. 22-19)**. Different symbols are used to represent logic diagrams **(Fig. 22-20)**.

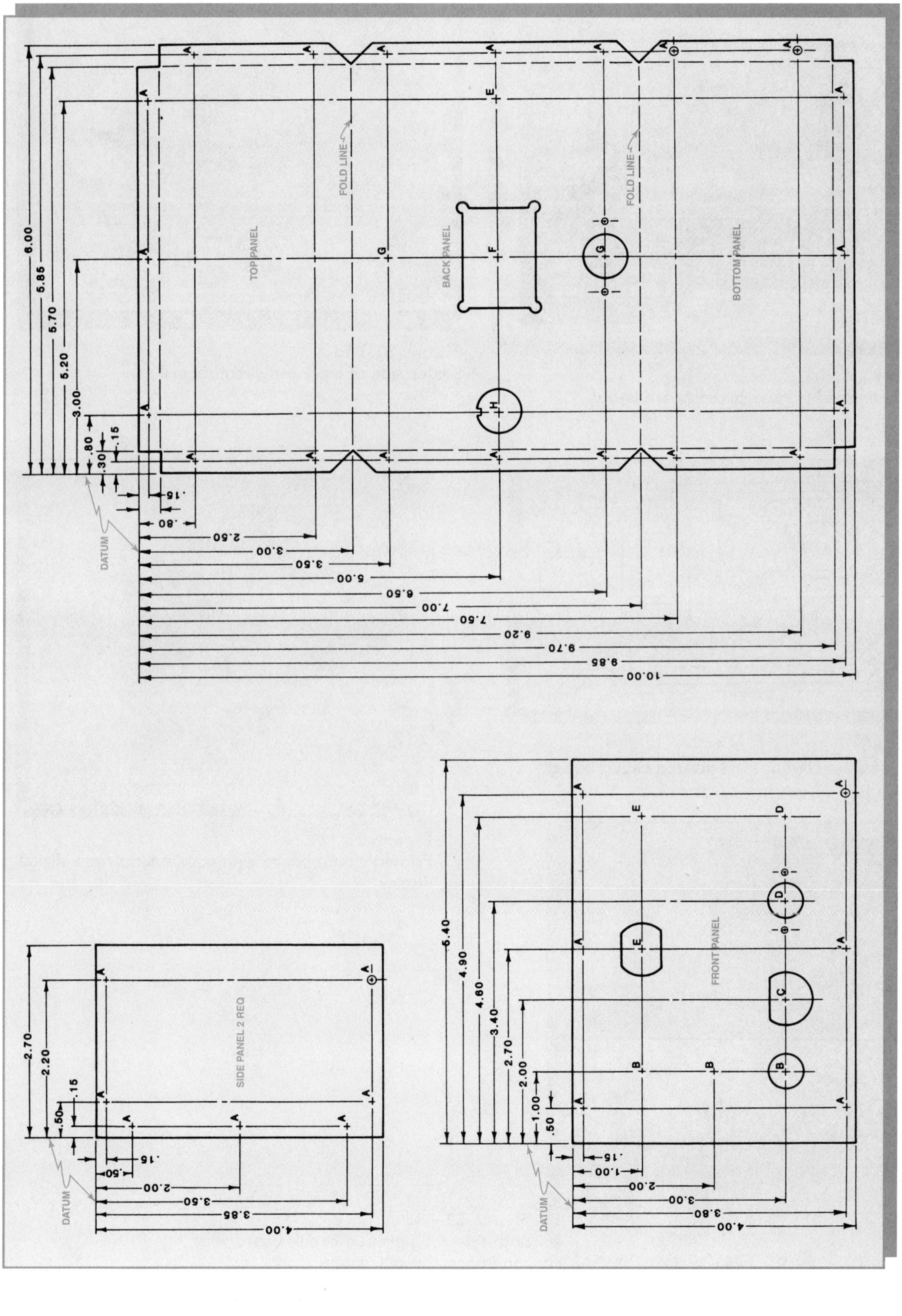

Figure 22-17b
A chassis pattern's working drawing using rectangular coordinate dimensioning

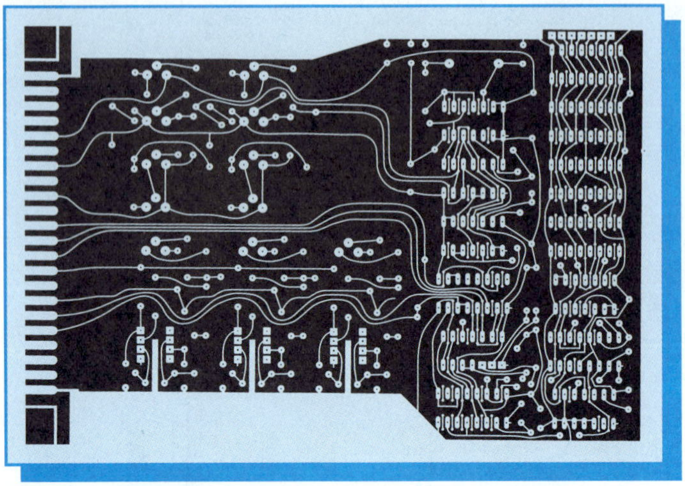

Figure 22-18a
Example of a finished printed circuit board layout

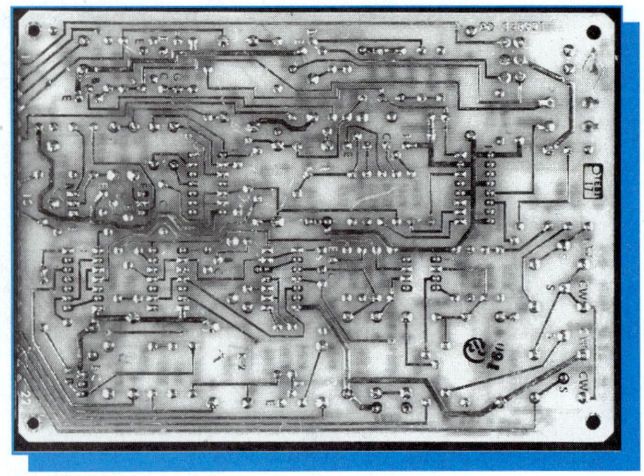

Figure 22-18d
Solder side of a printed circuit board

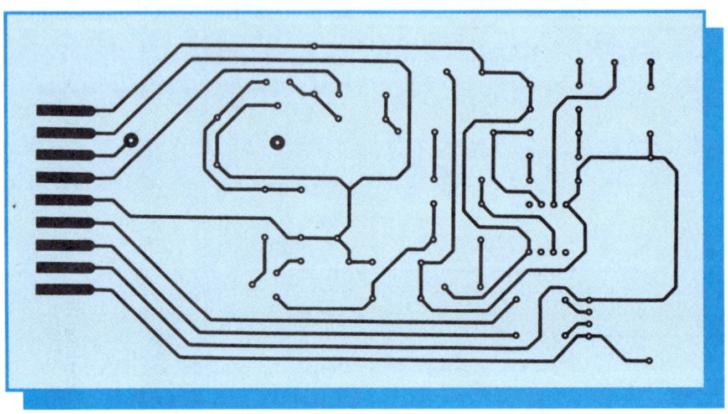

Figure 22-18b
Example of a tape-up layout for a printed circuit board

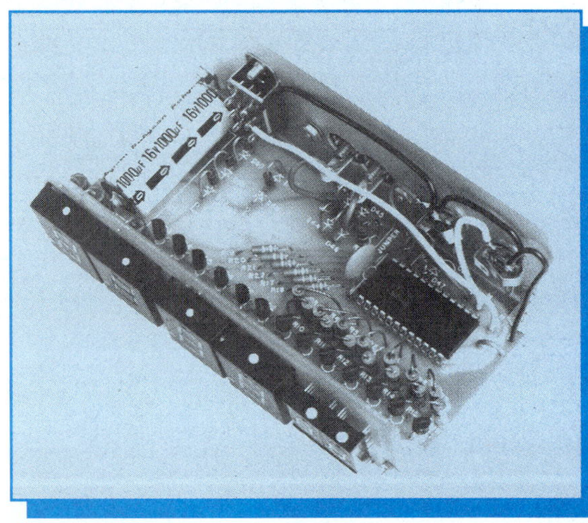

Figure 22-18e
Printed circuit board with components for a digital display

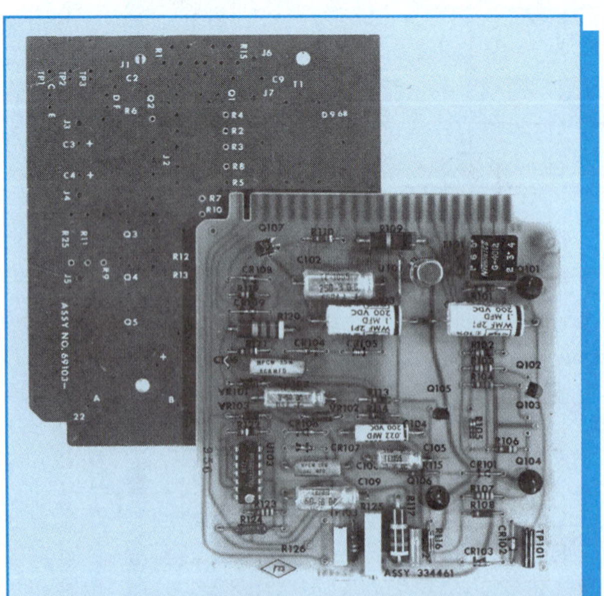

Figure 22-18c
Both sides of a printed circuit board with components installed

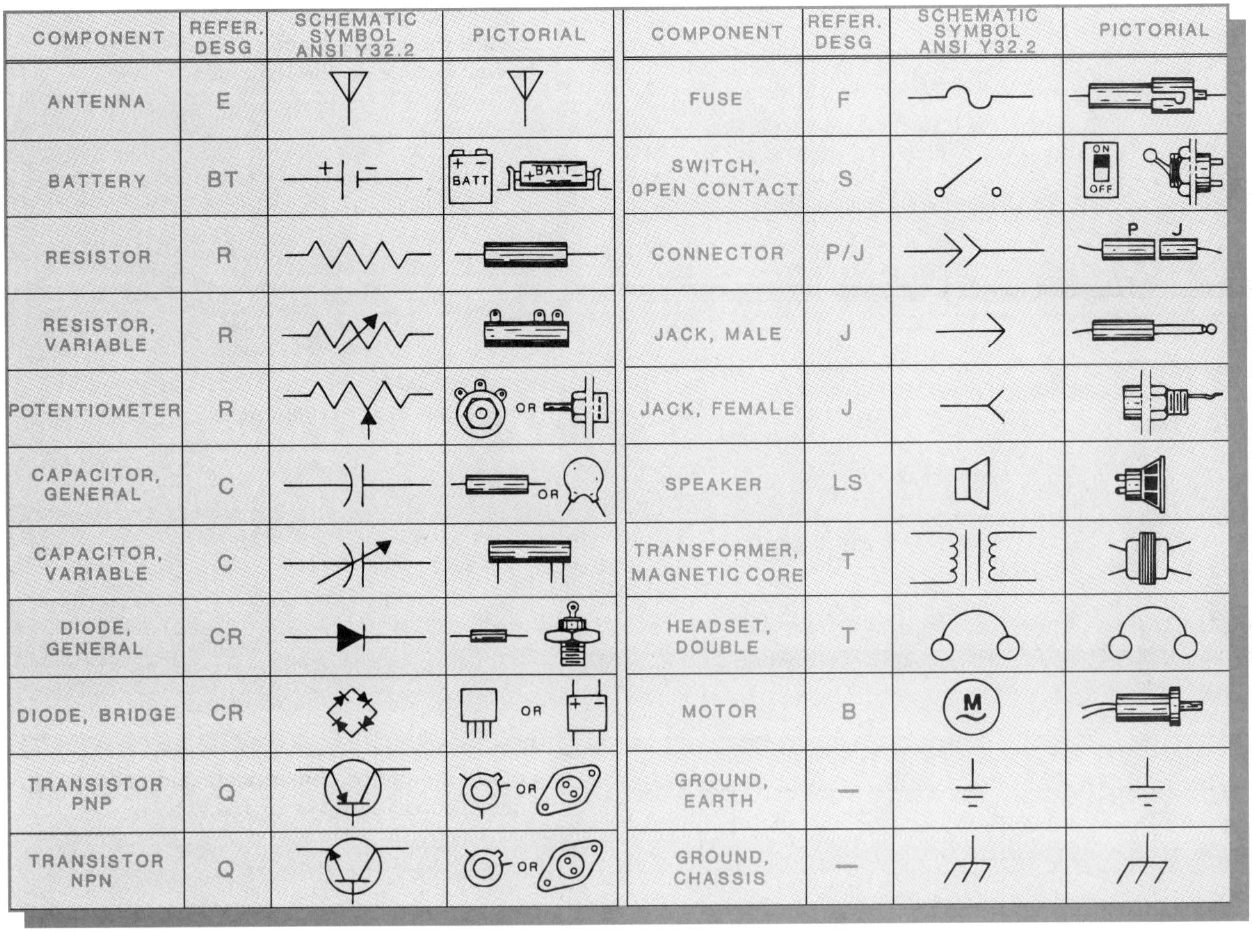

Figure 22-19
Schematic diagram symbols with their pictorial representations

Drawing Schematic Diagrams

Schematic diagrams contain schematic symbols, connecting lines, **reference designations** (part numbers) and **value designations.** Procedures for drawing a simple schematic diagram are shown in **Figures 22-21a** through **22-21d** and include the following steps:

Step 1. Place power input to the left and power output to the right **(Fig. 22-21a)**.

Step 2. Place major components from left to right across the diagram **(Fig. 22-21b)**.

Step 3. Arrange minor components to eliminate crossed lines and keep the flow correct (**(Fig. 22-21c)**.

Step 4. Place the reference designations and value designations **(Fig. 22-21d)**.

Electronics drawing templates greatly speed the layout and drawing of schematic diagrams **(Fig. 22-22)**.

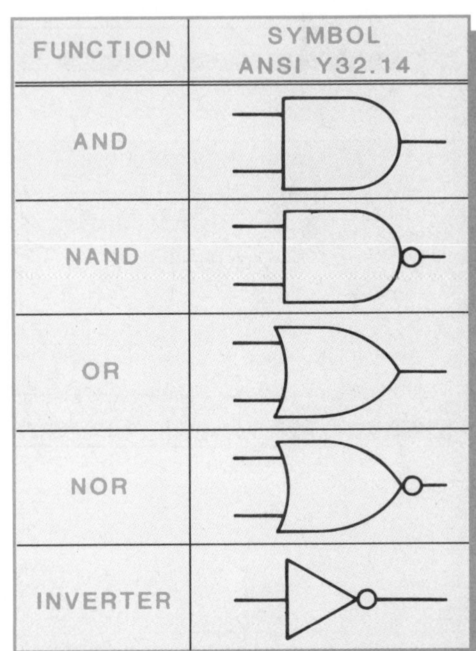

Figure 22-20
Logic diagram symbols

Figure 22-21a
Locate the power input on the left side and power output on the right side.

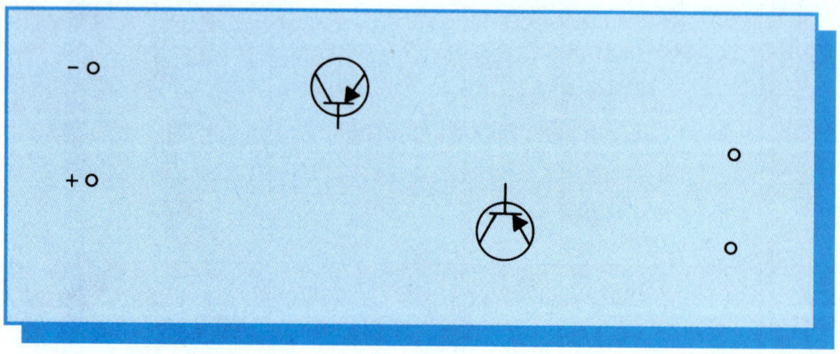

Figure 22-21b
Locate the major components.

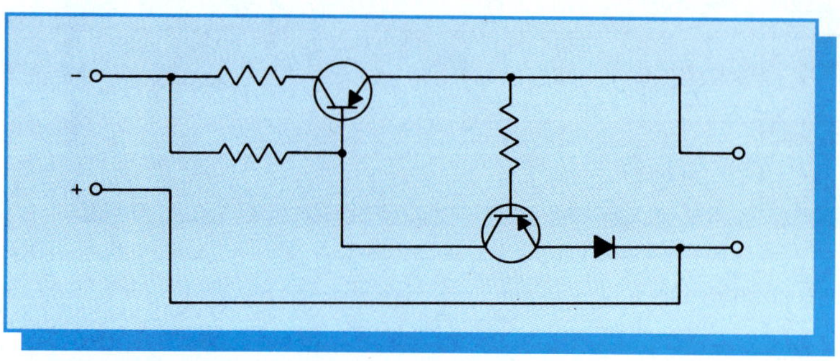

Figure 22-21c
Locate the minor components and connecting lines (also called wires or traces).

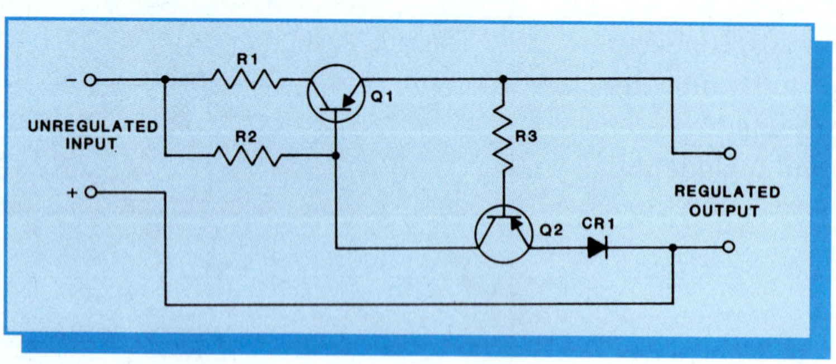

Figure 22-21d
Place reference and value designations to complete the schematic diagram.

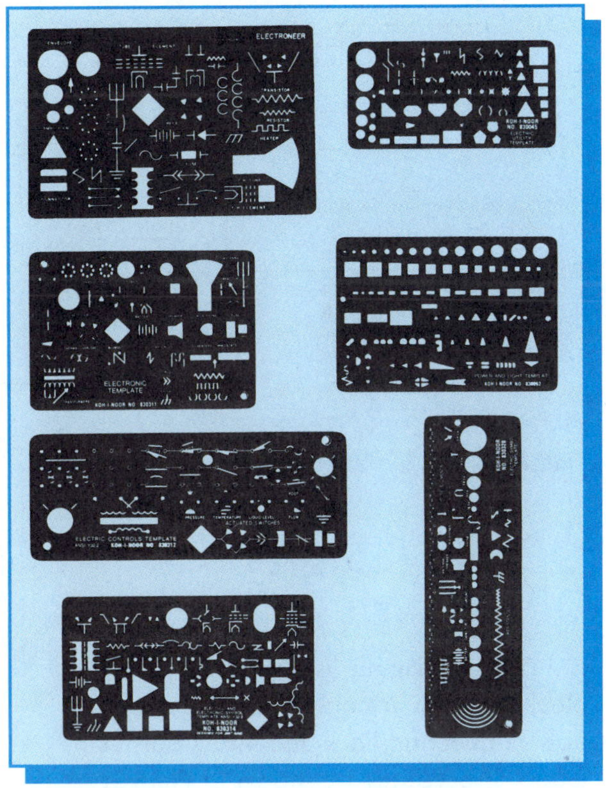

Figure 22-22
Examples of templates used by electronics drafters *(Courtesy of Koh-I-Noor)*

Drawing Connection Diagrams

In a connection diagram, the placement of components (simple outlines) is determined by their relative location on the chassis, rather than by electron flow. Lines that connect components represent wires. It is common to draw the wires with corner radii and color labels. Procedures for drawing a point-to-point connection diagram for a charging system tester are shown in **Figures 22-23a** through **22-23d**, and include the following steps:

Step 1. Draw the outline of the chassis **(Fig. 22-23a)**.

Step 2. Place the components. Arrange them as they should be located in the chassis **(Fig. 22-23b)**.

Step 3. Draw the connecting wires **(Fig. 22-23c)**.

Step 4. Label the reference designations on the components and label wire colors (**(Fig. 22-23d)**.

Figure 22-23a
Draw a chassis outline.

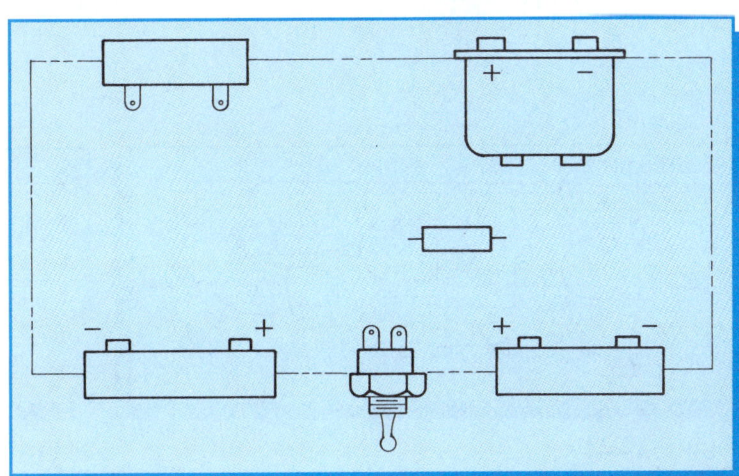

Figure 22-23b
Draw the components in the chassis as per location.

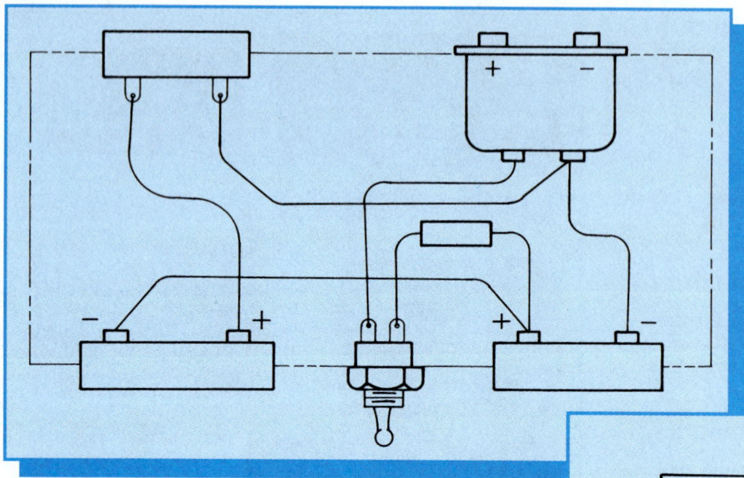

Figure 22-23c
Draw the connecting wires.

Figure 22-23d
Complete the connection diagram with labels and designations.

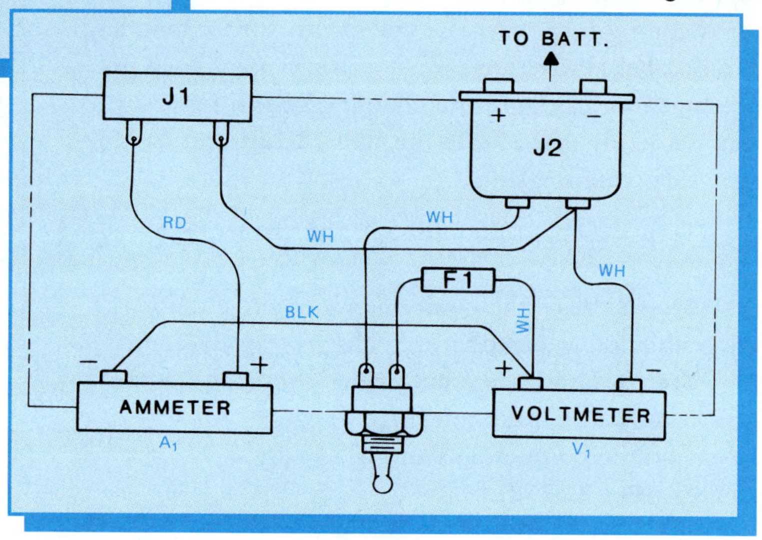

Drawing Logic Diagrams

A logic diagram is a special type of block diagram. It consists of logic symbols, connecting lines, and input/output letters. Procedures are shown in **Figures 22-24a** through **22-24c**, and include the following steps:

Step 1. Arrange logic symbols so they are evenly spaced and the system flows smoothly from left to right **(Fig. 22-24a)**.

Step 2. Place the connecting lines, with significant line contrast between component symbols and connecting lines. Connecting lines are thinner than the symbol lines **(Fig. 22-24b)**.

Step 3. Place input/output letters on the diagram **(Fig. 22-24c)**.

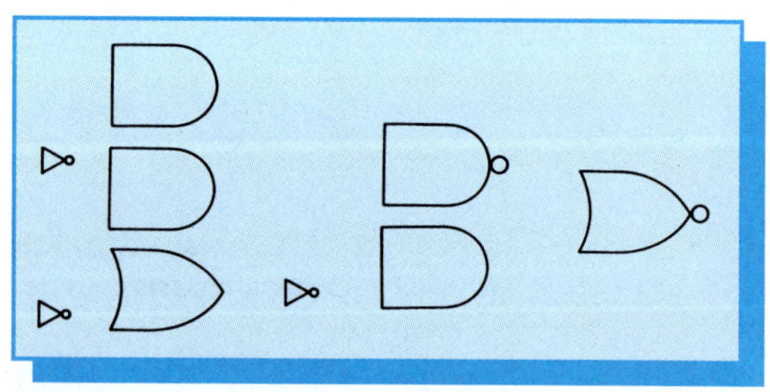

Figure 22-24a
Lay out and draw the logic symbols.

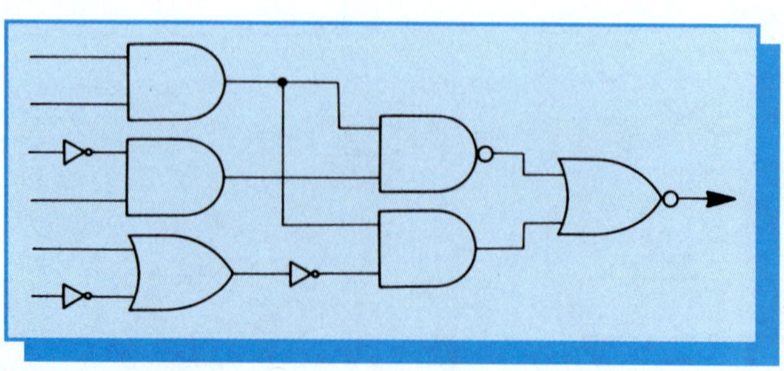

Figure 22-24b
Draw the connecting lines.

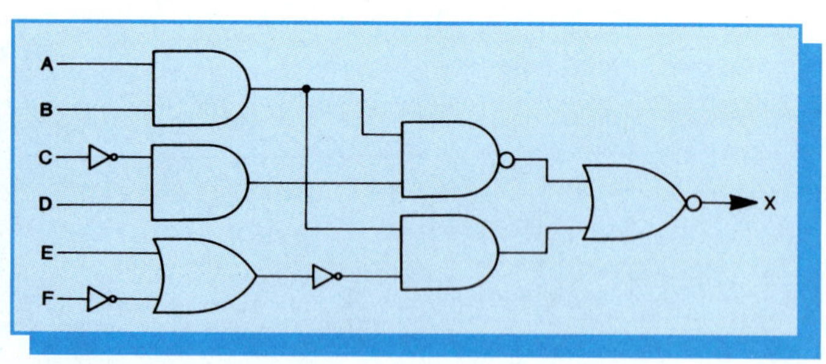

Figure 22-24c
Place input and output letters on the drawing.

AutoCAD LT Exercises

Electronics drafting was one of the first drafting fields to use the computer. This is primarily because early designers of CAD software were familiar with electronics and electronic diagrams. In addition, electronics diagrams contained many standard symbols and repeated details, and did not require complex dimensions or accurate geometric databases. Computer-aided electronics drafting has understandably experienced more growth than any other field of drafting.

CAD system software, now available for all types of computers, includes extensive design capabilities, especially in printed circuit board design. As a CAD operator develops schematic or logic diagrams on a computer, circuit information is stored in the computer's database. This information is processed by printed circuit board software to locate the components on the board and to automatically complete the circuit routing.

Operators may use standard editing commands, such as Move, Copy, and Rotate, to modify designs as desired. The computer then checks the circuit to verify if it matches the schematic diagram. If not, the operator may ask the computer to review the board to match the schematic, or may revise the schematic to match the board.

Some of the major features of sophisticated electronics drawing CAD software are:

- automatic placement of components
- automatic routing of circuit paths
- use of multiple colored layers
- use of multiple PCB trace widths and pad sizes
- automatic circuit and continuity checking capabilities
- creation of artwork for the manufacture of the PCB
- parts list printouts
- creation of assembly drawings

On a simpler level, electronics drawings can easily be created with any CAD system. Since electronic symbols are standardized, they can be drawn and made into blocks or tablet menu icons. Prepared electronics symbols may be purchased as software, saving the time needed to make your own symbols. Grids on a monitor also aid in the placement of components and text. **Figures 22-25** through **22-28** illustrate the following sequence of creating schematic diagrams using standard CAD software:

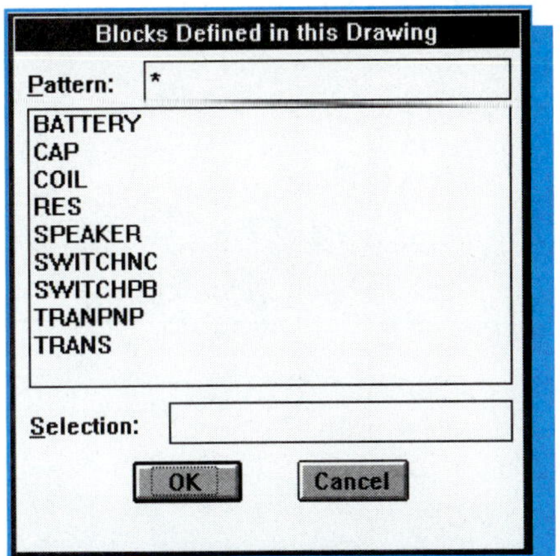

Figure 22-25
Using Blocks to create electronic symbols

1. Create and draw standardized symbols **(Fig. 22-25)** and save them as blocks. See **Figure 22-19** for symbols.

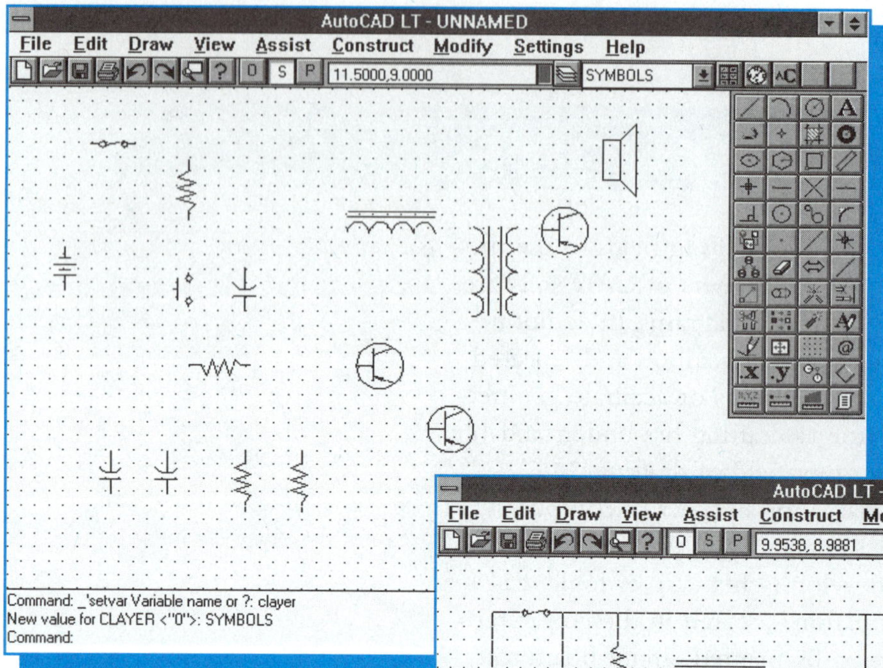

Figure 22-26
Insert and position the symbols.

Figure 22-27
Place the connectors. Reposition the symbols if necessary.

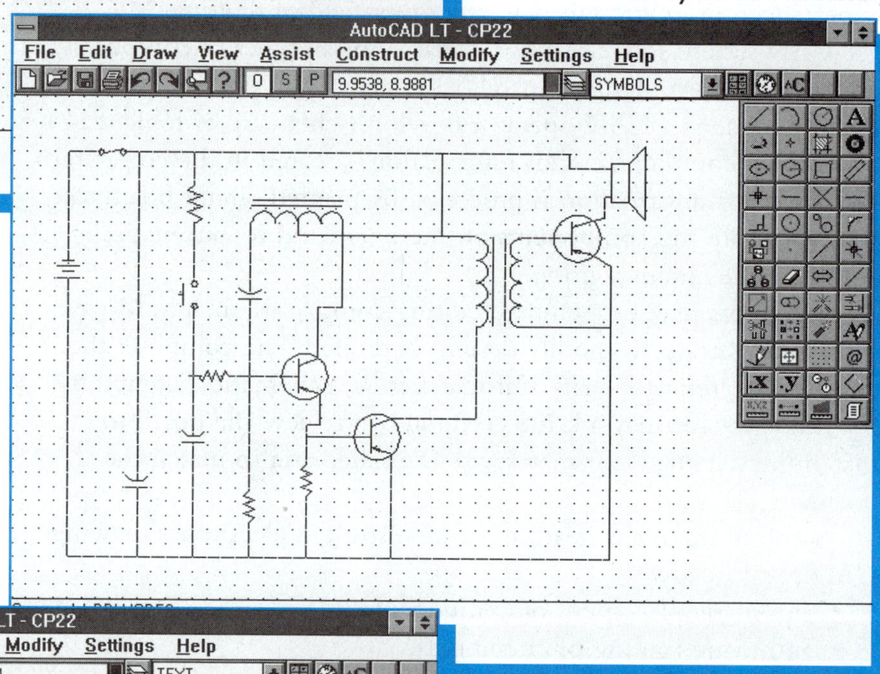

2. Set **grid** to an appropriate size and turn on the Grid to aid in the placement of symbols.
3. Insert the symbols **(Fig. 22-26)**. This placement need not be perfect at the start because it is easy to move symbols at any time with a CAD system to conform with the design.
4. Place the connectors (wires) and move the symbols as necessary **(Fig. 22-27)**.
5. Use the **Text** command to label the diagram **(Fig. 22-28)**.

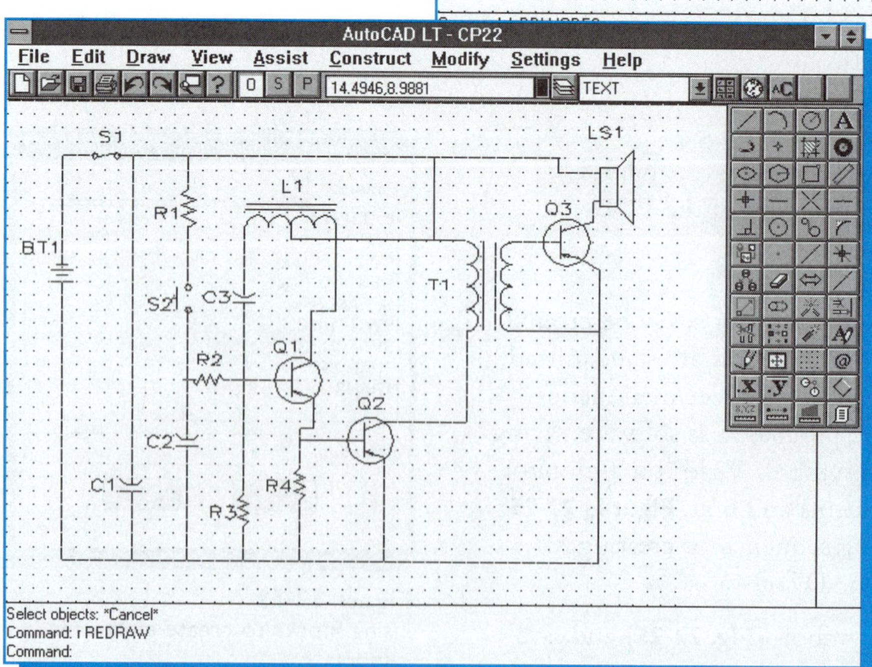

Figure 22-28
Add text to complete diagram.

Drafting Exercises

1. Draw the block diagram for the amplifier system in **Figure 22-29** with drafting instruments or a CAD system.

2. Draw the electronic symbols in **Figure 22-30** with drafting instruments or a CAD system.

3. Draw the electronic schematics in **Figures 22-31** through **22-45** with drafting instruments or a CAD system.

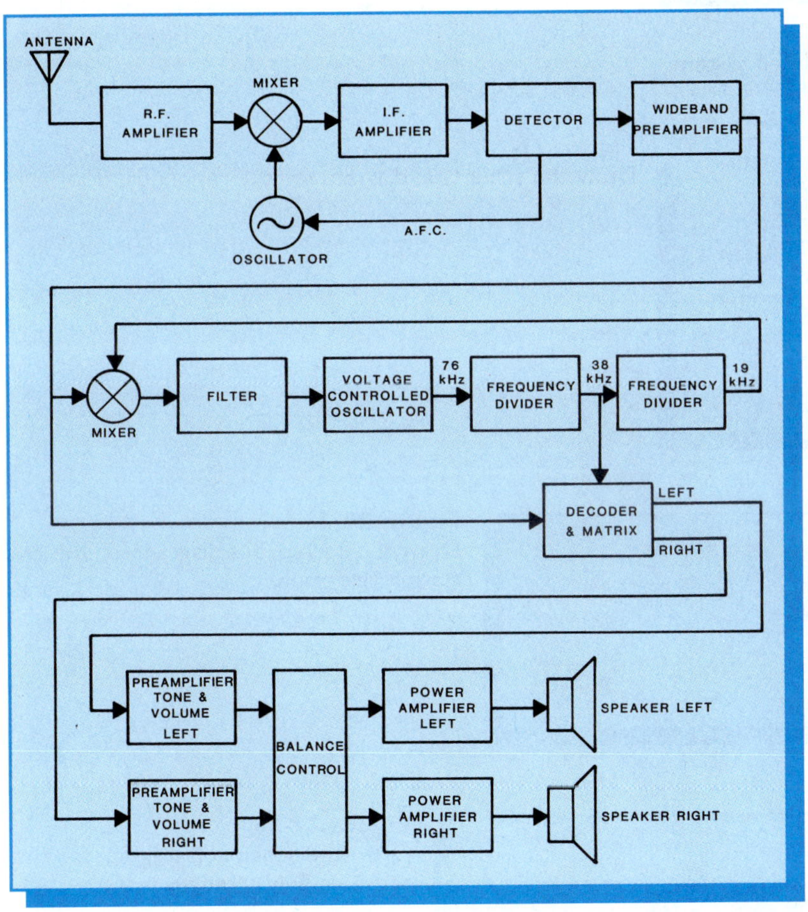

Figure 22-29
Draw the block diagram with drafting instruments or a CAD system.

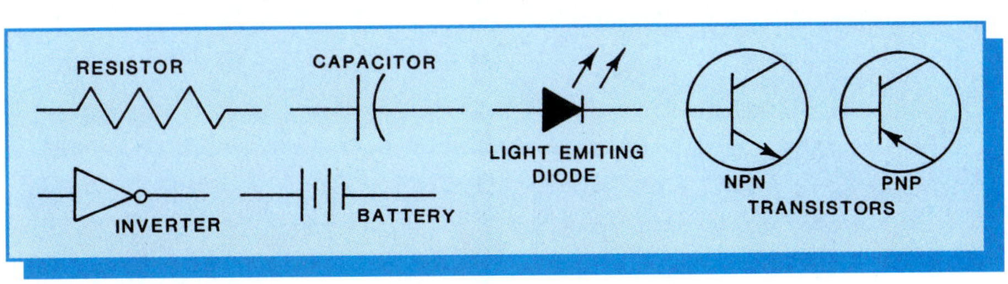

Figure 22-30
Practice drawing the electronics symbols.

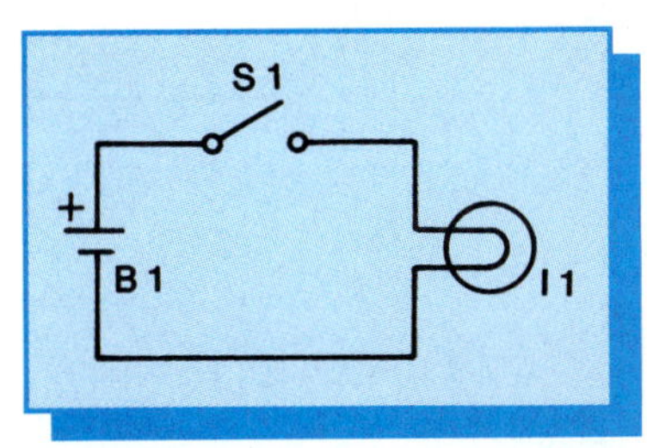

Figure 22-31
Draw the schematic diagram for the flashlight.

Figure 22-32
Draw the schematic for the battery-operated bell system.

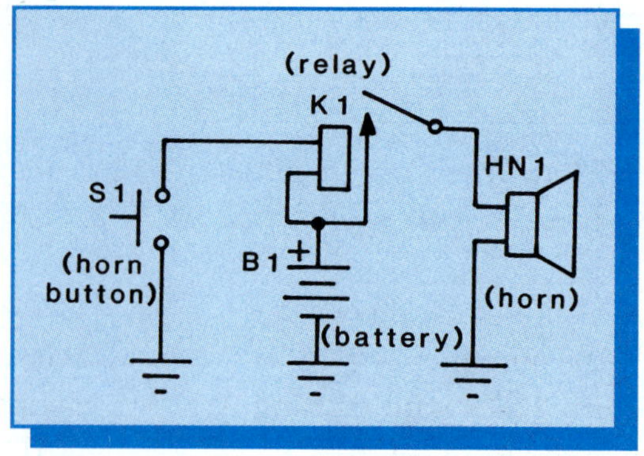

Figure 22-33
Draw the schematic diagram for the automobile horn.

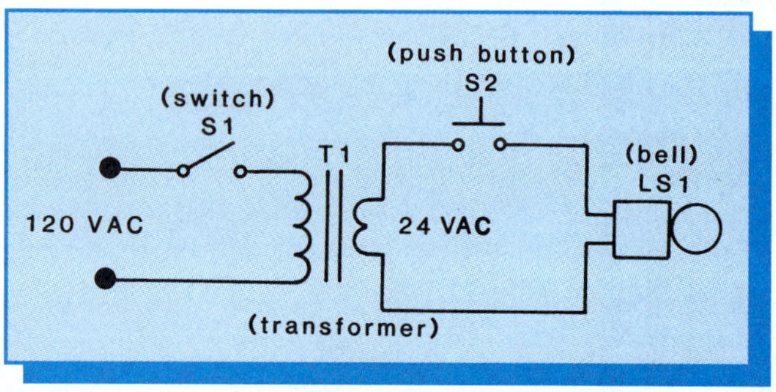

Figure 22-34
Draw the schematic diagram for the residential doorbell.

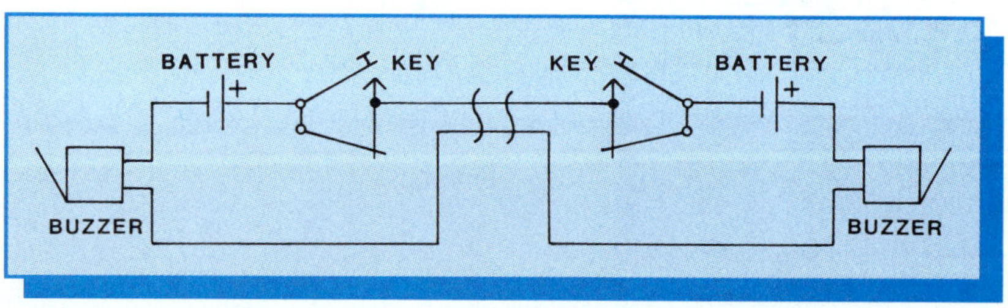

Figure 22-35
Draw the schematics diagram for the two-station telegraph.

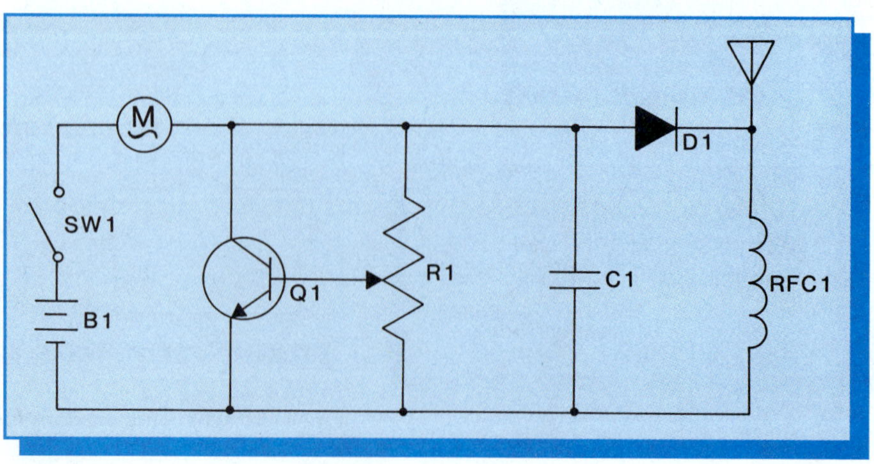

Figure 22-36
Draw the schematic diagram for the sensitive field-strength meter.

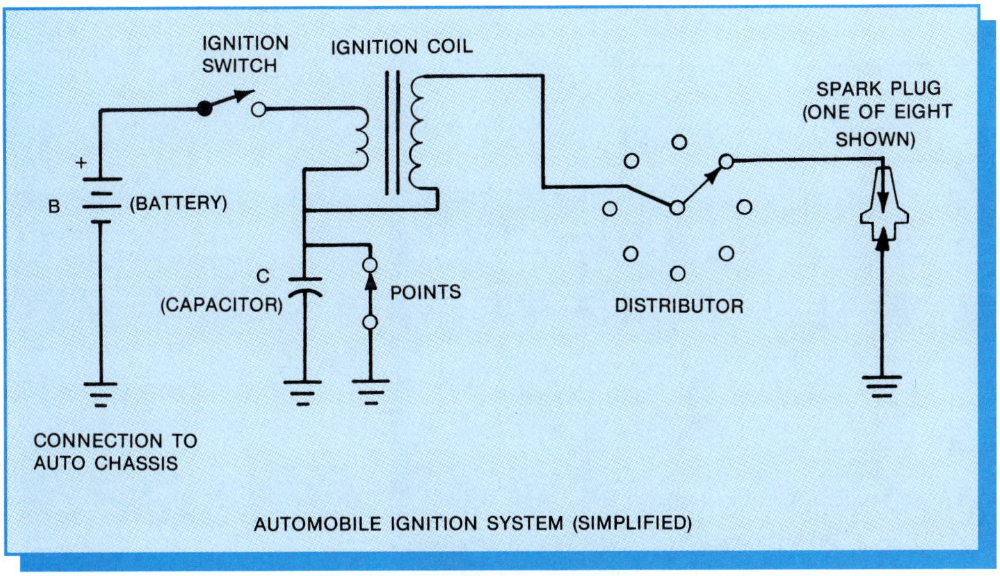

Figure 22-37
Draw the schematic diagram for the automobile ignition system.

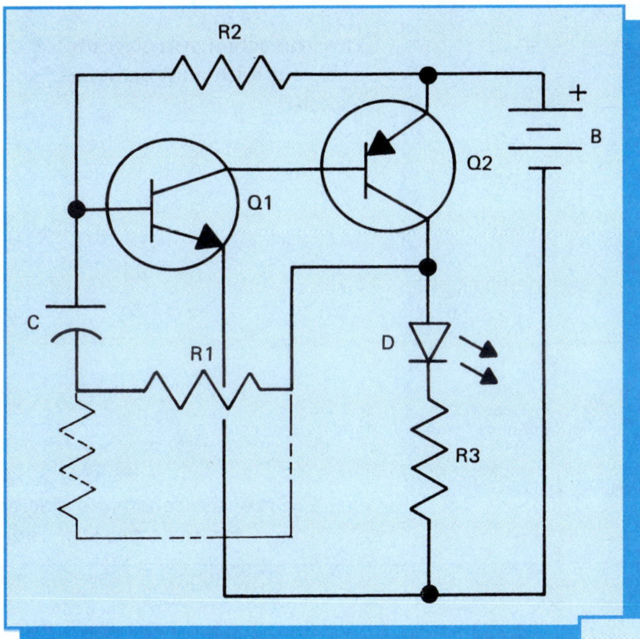

Figure 22-38
Draw the schematic diagram for the flasher unit. Reposition resistor one as shown.

Figure 22-39
Draw the schematic diagram for the code practice oscillator.

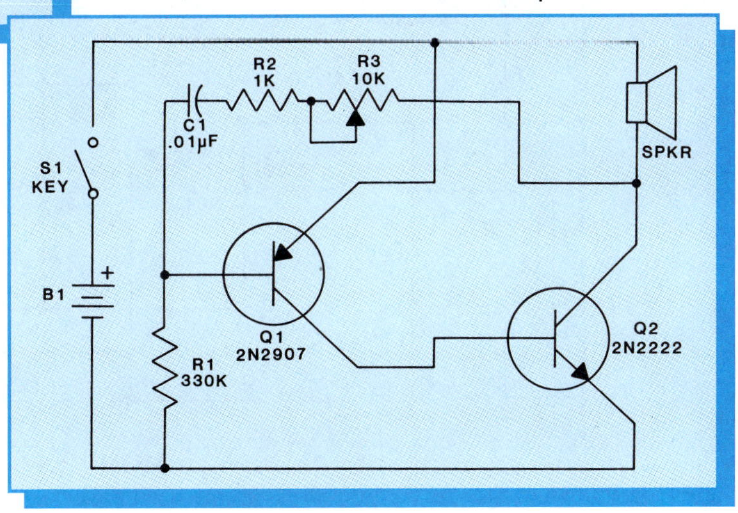

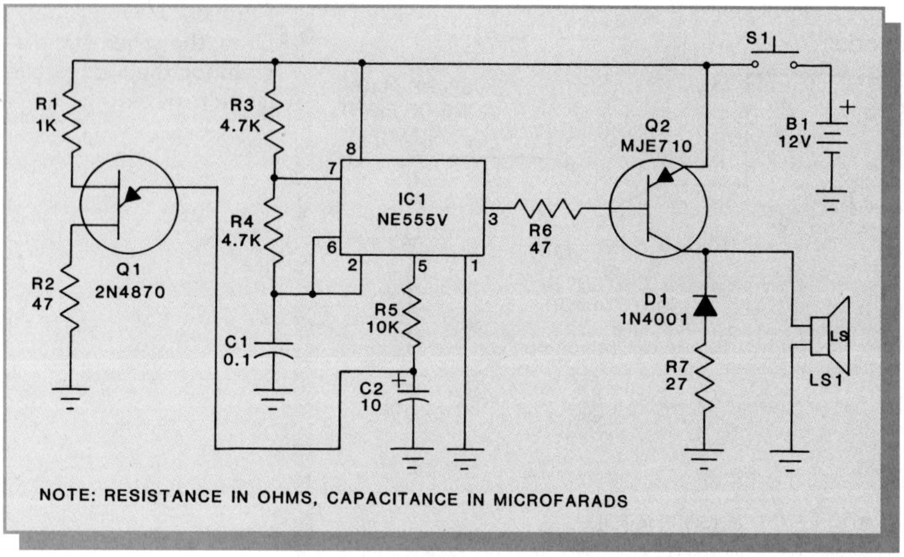

Figure 22-40
Draw the schematic diagram for the electronic siren.

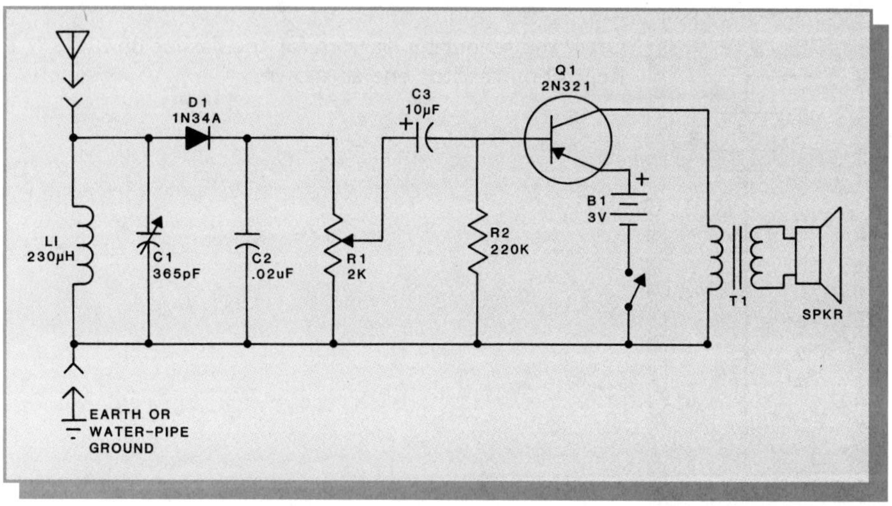

Figure 22-41
Draw the schematic diagram for the crystal set with audio amplifier and speaker.

Figure 22-42
Draw the schematic diagram for the transistor radio.

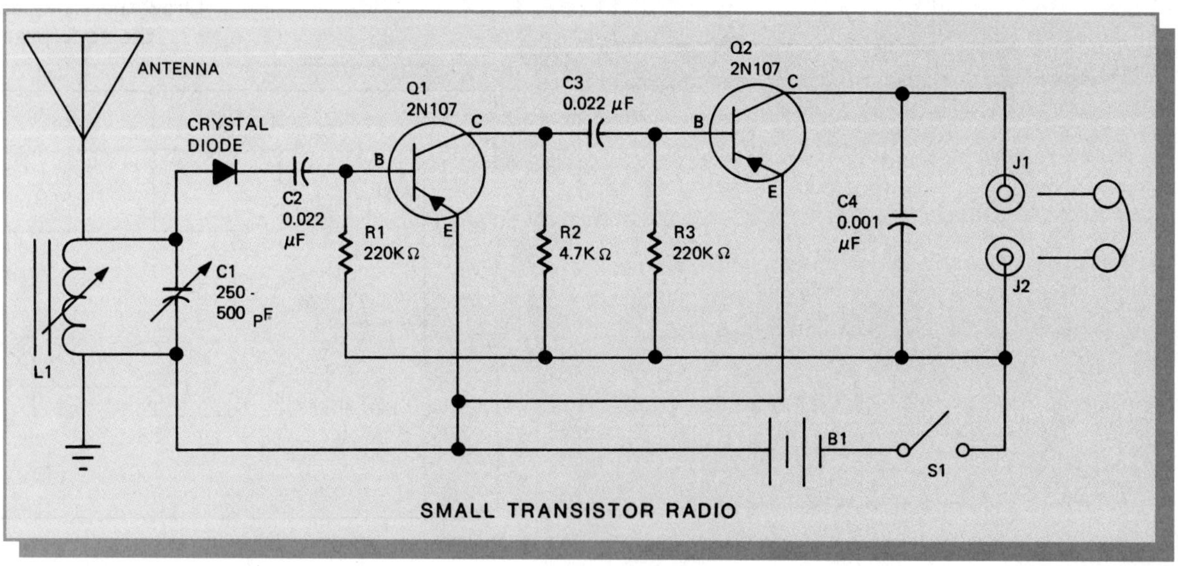

Design Exercises

1. Design a case for the electronic siren in **Figure 22-40**.
2. Design a case for the crystal radio set in **Figure 22-41**.
3. Design a case for the transistor radio in **Figure 22-42**.
4. Design a flashlight that can be attached to your body or clothes so as to leave both hands free. Draw a working drawing for the flashlight and a schematic diagram.

Key Terms

Airline diagrams (baseline diagrams)
ASME (American Society of Mechanical Engineers)
ASA (American Standards Association)
Block diagrams
Cable
Cableform diagrams
Chassis drawings
Connection diagrams (wiring diagrams)
Electronics drawings
EIA (Electronics Industries Association)
Feeder
Highway diagrams (trunkline diagrams)
IEEE (Institute of Electrical and Electronics Engineers)
IEC (International Electrotechnical Commission)
Logic diagrams
MIL STD (Military standards)
Path
Point-to-point diagrams
Printed circuit board (PCB)
Reference designations
Schematic diagrams (elementary diagrams)
Schematic symbols
Simple outline symbols (pictorial)
Value designations

Chapter 23

Tool Design Drafting

Objectives

The student will be able to:

- read jig and fixture working drawings
- design a simple jig
- design a simple fixture
- complete a working drawing for a simple jig
- complete a working drawing for a simple fixture

Introduction

To compete successfully in today's marketplace, industrial products must be accurately and economically produced. This requires effective tool engineering and design. **Tool engineering** is the branch of mechanical engineering responsible for the design and development of tools and processes needed in the mass production of products. This includes the planning, designing, and detail drawing of all tools and devices used in the manufacturing process.

Tool designers must possess a good working knowledge of all machine tool operations. Since they must communicate ideas and describe tools in great detail, a thorough understanding of working drawings and tolerancing is also mandatory. Tool engineers, tool designers, and tool design drafters are all involved at different levels of the machine tool operations and have the following responsibilities:

1. Control production costs by designing tools for rapid production.
2. Design special tools for the manufacture of complete parts.
3. Design tools that aid in the manufacture of the highest quality product within the established limits.
4. Select the proper materials for the manufacture of products.
5. Design tools that require minimum maintenance.
6. Simplify manufacturing and assembly processes.
7. Design tools that are safe under all manufacturing conditions.
8. Design tools that are fault-free and reduce the potential for human error.
9. Design tools that produce accurate interchangeable parts within acceptable tolerances.

To help fulfill these responsibilities, many devices are used to securely hold, guide, move, and position workpieces through a wide variety of manufacturing processes. Most are classified as either jigs or fixtures.

Jigs and Fixtures

Jigs are used in the machining and assembly of products **(Fig. 23-1)**. A **jig** is a device designed to position, hold, and support a workpiece in a fixed position during a machining operation. In addition, a jig holds or guides cutting tools through prescribed paths. A jig may be fastened to a worktable or can be handheld. **Figure 23-2** shows a jig used to hold a workpiece during a drill press operation.

Jigs used in product assembly are designed to hold separate components in fixed positions while final connections are completed. This may involve the use of all types of fasteners, weldments, or adhesives.

Fixtures, like jigs, hold and support workpieces. However, **fixtures** are fixed to a worktable or a machine surface to hold a workpiece rigidly. Fixtures do not hold or guide cutting tools. They are generally larger than jigs and are used in milling, grinding, and honing operations. **Figure 23-3** shows a typical numerically controlled machining center fixture.

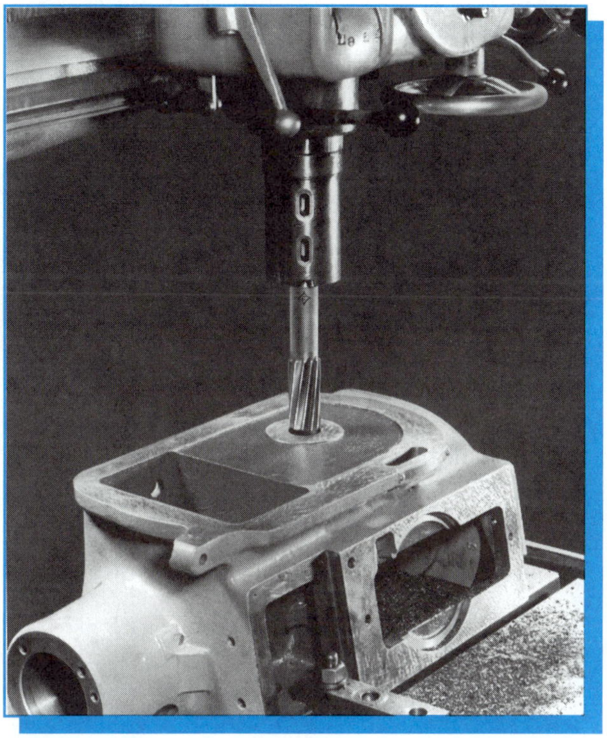

Figure 23-1
Tool design use in industry *(Courtesy of Cleveland Twist Drill Company)*

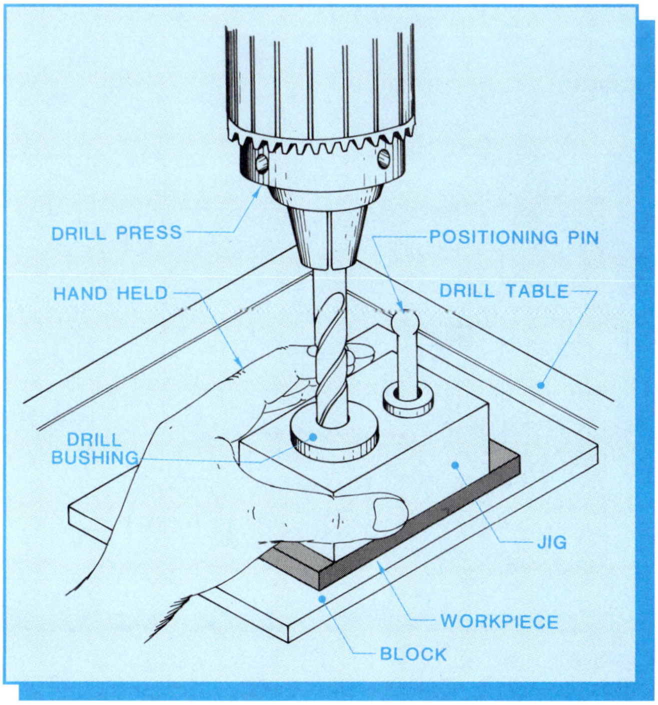

Figure 23-2
Handheld drill jig

Figure 23-3
A typical numerically controlled machining center fixture *(Courtesy of Cincinnati Milacron Inc.)*

Standard Parts

Most jigs and fixtures use standard components. Only when a tool design problem cannot be solved with a standard component will a tool designer use a unique type, size, and configuration of materials. Standard parts in the design and fabrication of jigs and fixtures include locators, feet and rest buttons, clamps, drill bushings, pins, screws, V-blocks, plates, and aluminum or steel shapes. When cast shapes are used, all burrs and sharp edges are removed to protect machine operators.

Locators

Locators are devices, such as pins, dowels, buttons, and pads, used to position a workpiece on a machine surface or worktable. **Figure 23-4** shows the design and size options for standard spherical radius locator buttons. This type of locator is normally positioned with the workpiece touching the tip of the button **(Fig. 23-5)**. This locator also can be used with the workpiece to touch the flattened side of the locator **(Fig. 23-6)**. Note the location and direction of forces in relation to the locators in **Figure 23-6**.

If a workpiece contains holes, locators should be positioned to align with hole centers. **Figure 23-7a** shows the alignment of bullet nose locators with the holes in a workpiece. **Figure 23-7a** also contains standard dimensions for bullet nose locators. **Figure 23-7b** shows standard diamond pin locator dimensions.

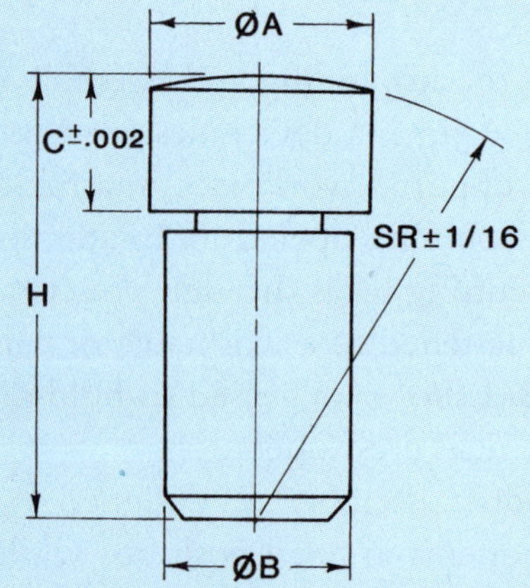

STANDARD SIZES FOR SPHERICAL LOCATORS				
ϕA	ϕB	C	H	SR
5/16	.1890 .1885	1/8	3/8	1/2
3/8	.2515 .2510	3/16	1/2	5/8
1/2	.3765 .3760	3/16	9/16	3/4
1/2	.3765 .3760	1/4	5/8	3/4
1/2	.3765 .3760	3/8	3/4	3/4

Figure 23-4
Standard spherical locators

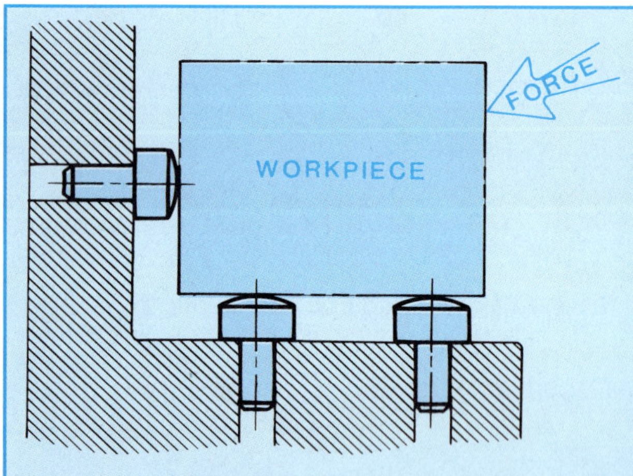

Figure 23-5
Positioning workpiece against ends of locators

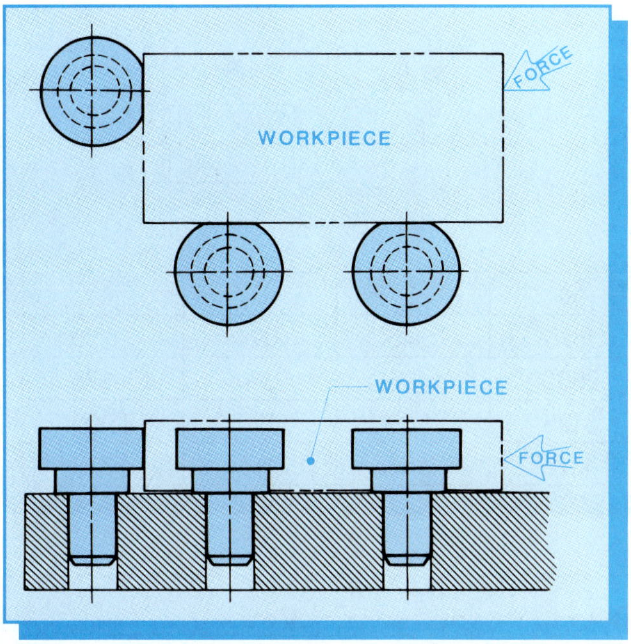

Figure 23-6
Positioning workpiece against flattened sides of locators

Figure 23-7a
Bullet nose locators

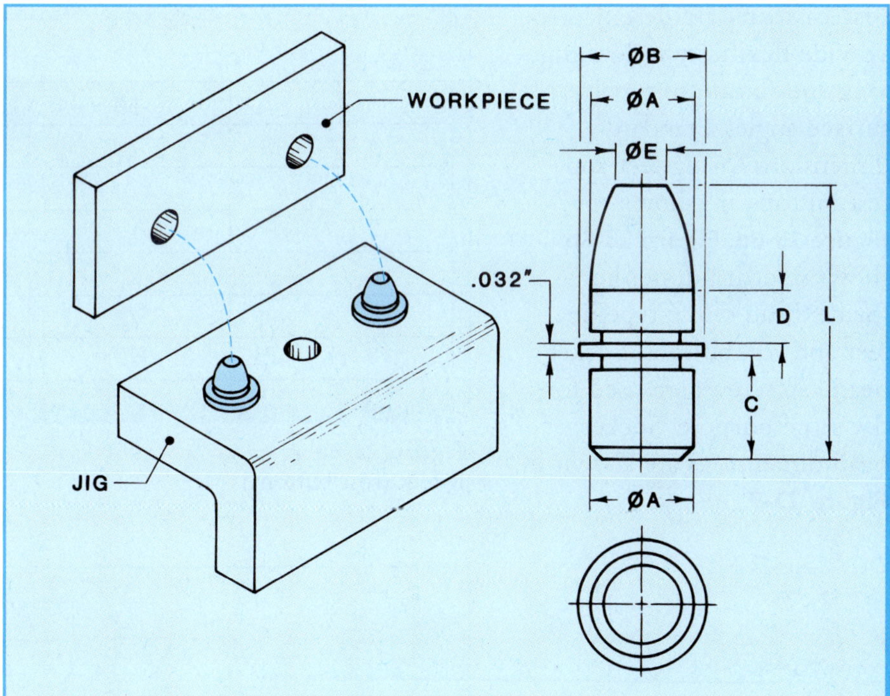

ϕA	ϕB	C	D	L	ϕE	HOLE'S ϕ
.2495	.280	.50	.125	1.000	.125	.2500
.3120	.350	.50	.125	1.000	.156	.3125
.3745	.400	.50	.125	1.000	.188	.3750
.4995	.532	.50	.125	1.000	.250	.5000

Figure 23-7b
Diamond pin locators

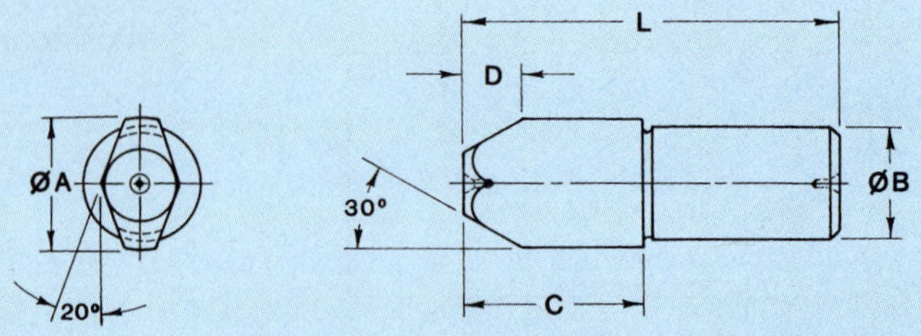

ϕA	ϕB	C	D	L
.2495	.1875	.345	.094	.750
.3120	.2500	.406	.094	.875
.3745	.3125	.500	.125	1.000
.4995	.4375	.656	.156	1.375

Feet and Rest Buttons

Feet are used to elevate jigs or fixtures from a machine surface. **Feet and rest buttons** are designed to provide flexibility in leveling a jig and fixture to precise surface angles. Standard dimensions for jig feet and rest buttons are shown in **Figure 23-8a**. **Figure 23-8b** shows standard dimensions for hexhead screw type jig feet and rest buttons. Socket head cap screws are used for the same purpose. Socket head dimensions are shown in **Figure 23-9**.

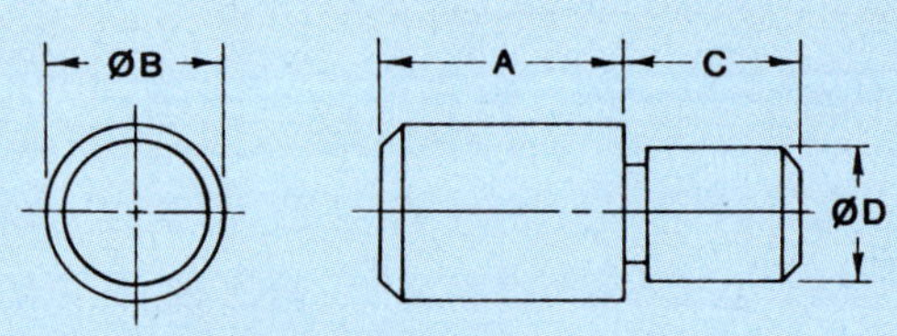

A	ϕB	C	ϕD
1/4"	5/8"	1/2"	.376
3/8"	5/8"	1/2"	.376
1/2"	5/8"	1/2"	.376
5/8"	5/8"	1/2"	.376
3/4"	5/8"	1/2"	.376
1"	5/8"	1/2"	.376

Figure 23-8a
Jig feet (rest buttons)

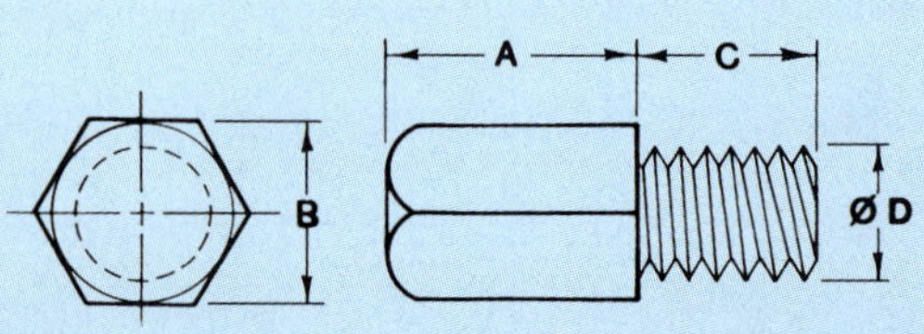

A	B	C	ϕD
1/2"	1/2"	1/2"	3/8"–24
3/4"	1/2"	1/2"	3/8"–24
1"	1/2"	1/2"	3/8"–24

Figure 23-8b
Hexhead screw-type jig feet (rest buttons)

Figure 23-9
Socket head cap screws

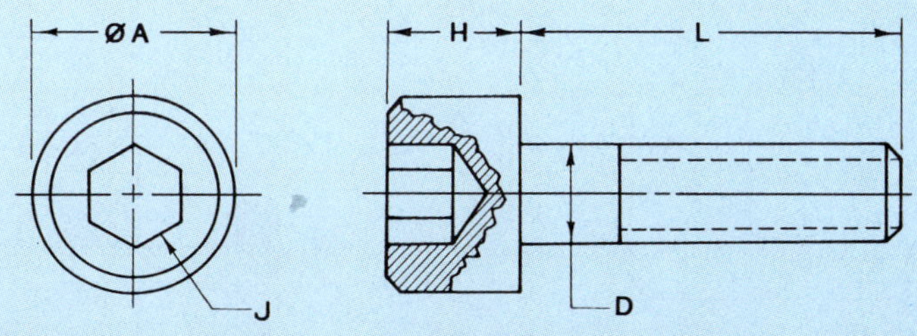

NOMINAL SIZE	BODY DIAMETER		HEAD DIAMETER		HEAD HEIGHT		HEX SOCKET SIZE	
	MAX	MIN	MAX	MIN	MAX	MIN	NOMINAL	
	D		A		H		J	
$\frac{1}{4}$"	0.2500	0.2435	0.375	0.365	0.250	0.244	$\frac{3}{16}$"	0.188
$\frac{5}{16}$"	0.3125	0.3053	0.469	0.457	0.312	0.306	$\frac{1}{4}$"	0.250
$\frac{3}{8}$"	0.3750	0.3678	0.562	0.550	0.375	0.368	$\frac{5}{16}$"	0.312
$\frac{1}{2}$"	0.5000	0.4919	0.750	0.735	0.500	0.492	$\frac{3}{8}$"	0.375
$\frac{5}{8}$"	0.6250	0.6163	0.938	0.921	0.625	0.616	$\frac{1}{2}$"	0.500

Clamps

Clamps are used to hold workpieces firmly against locators and rest buttons without distorting the workpiece. There are many types of clamps for jig and fixture design. The three major categories of clamps are **toggle head clamps, cam clamps,** and **screw clamps.**

Toggle clamps, such as the push-pull toggle clamp shown in **Figure 23-10**, are used where maximum surface contact with the workpiece is required. Cam clamps **(Fig. 23-11)** are used where ease of changing the workpieces is needed to speed up production. Screw clamps **(Fig. 23-12)** reduce the effect of vibration on jigs and workpieces. Using screw clamps often slows down production unless swing-out features are used for loading and unloading workpieces.

Figure 23-10
Push-pull toggle clamp

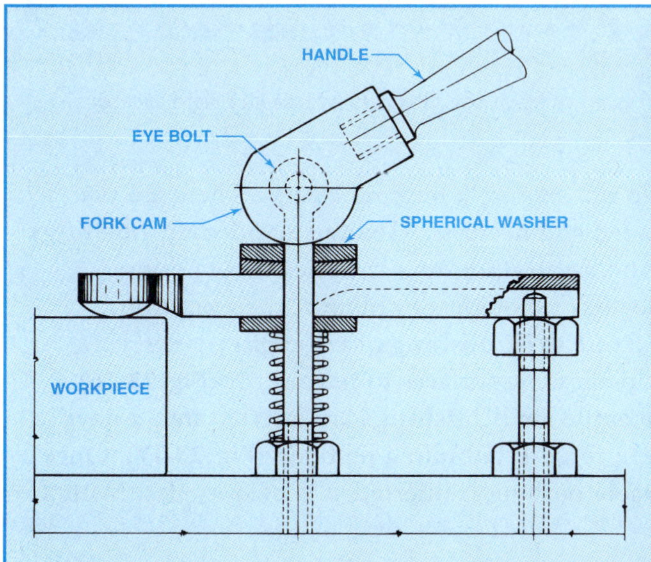

Figure 23-11
Cam clamp assembly

Appropriate application of clamps allows clamping from multiple directions **(Fig. 23-13)**. Generally, cutting actions should be directed not towards a clamp, but towards a rigid locator. **Figure 23-14** also shows an example of a cutting motion directed toward rigid locators and away from a screw clamp.

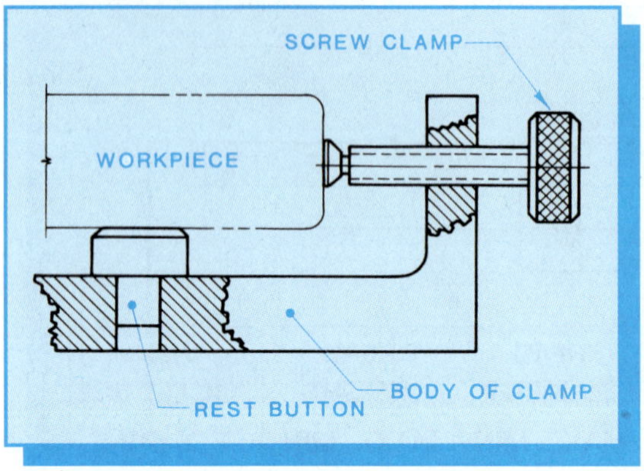

Figure 23-12
Knurled head swivel pad screw clamp

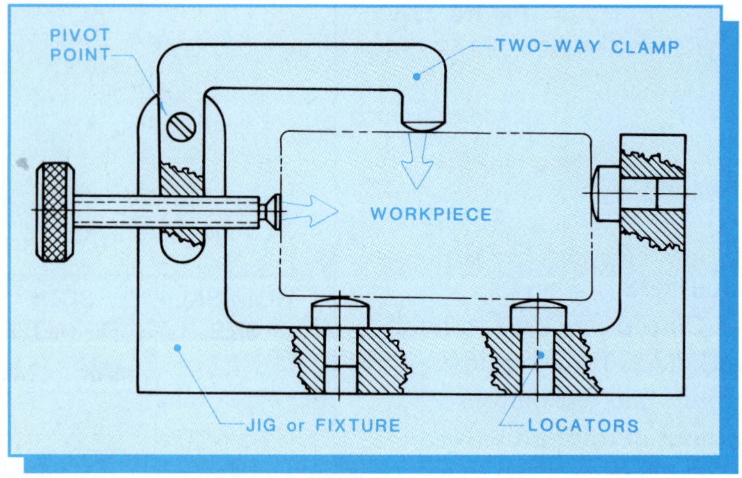

Figure 23-13
Multiple swing-out screw clamp. Note the clamping action is towards the locators.

Drill Bushings

Drill bushings align, hold, and guide drill bits to an exact location, and at a specific angle for drilling **(Fig. 23-15)**. Drill bushings are made of a harder steel than the drill bit. This prevents the drill from damaging the bushings. Drill bushing sizes correspond to standard drill sizes.

Bushings may be either fixed or replaceable. **Fixed drill bushings** are force-fitted into jig bodies. Fixed drill bushings are relatively inexpensive, but are difficult or impossible to replace. They are used for short production runs that do not require a bushing change. There are two kinds—headed and headless. **Headed fixed drill bushings** can withstand greater pressures than headless bushings because they have shoulders to support their position. **Headless fixed drill bushings** can be placed closer together and allow jig surfaces to remain flat **(Fig. 23-16)**.

Replaceable drill bushings are inserted into a liner that has been force-fitted into a jig body **(Fig. 23-17)**. Once the replaceable bushing is inserted, it is held in place with a

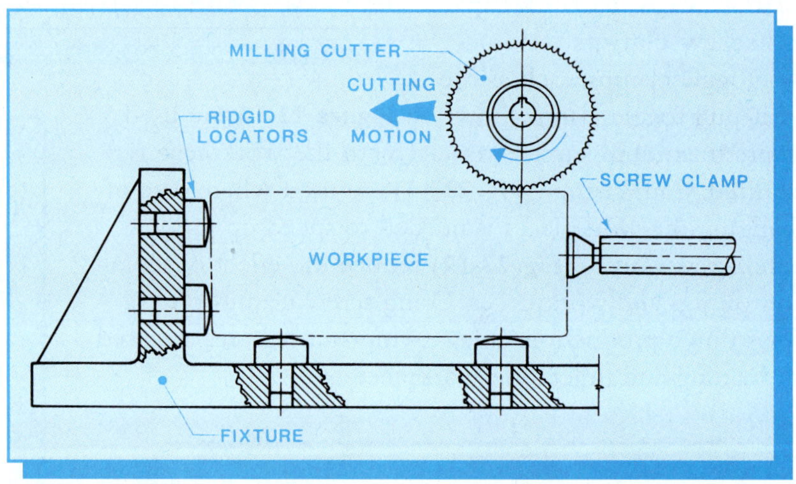

Figure 23-14
Cutting motion is away from the clamp (towards the rigid locators).

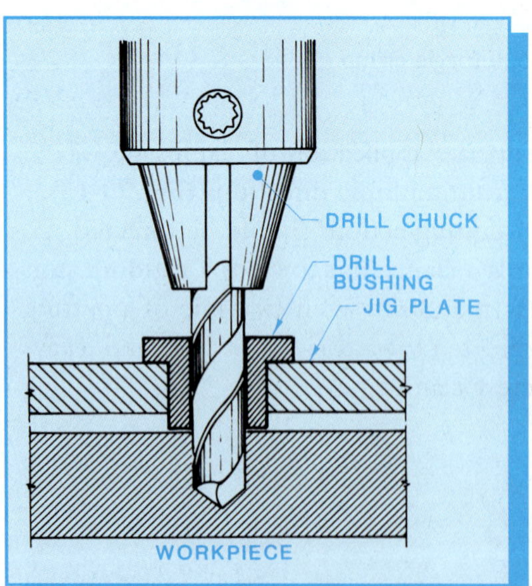

Figure 23-15
Drill bushing guides drills to exact location on workpiece.

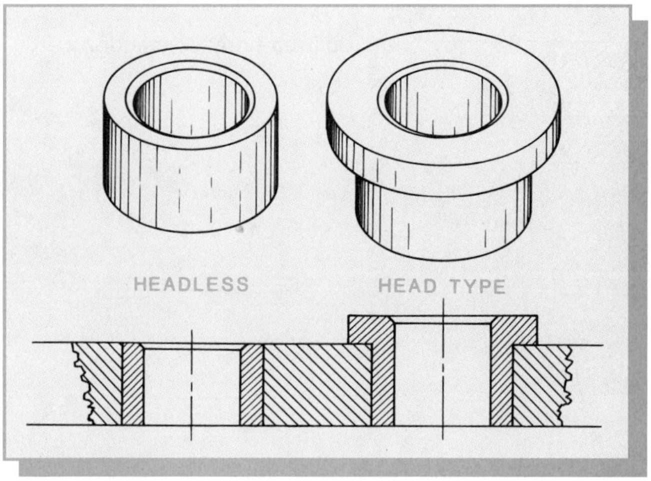

Figure 23-16
Fixed drill bushings (press fit)

lock screw. These bushings can be easily replaced by removing the lock screw.

Bodies

Jig and fixture bodies are made in three general forms—**weld bodies, cast bodies,** and **built-up bodies. Figure 23-18** shows the same jig body produced by these three methods.

The most commonly used form is the built-up body. When attaching plates using the built-up body method, dowel pins and socket head cap screws are used **(Fig. 23-19)**. Dowel pins are used to align the plates in an exact position without slippage, and the cap screws apply pressure to hold the plates rigid. When this standard fastening technique is used, both dowels and camp screws are flush with the plates.

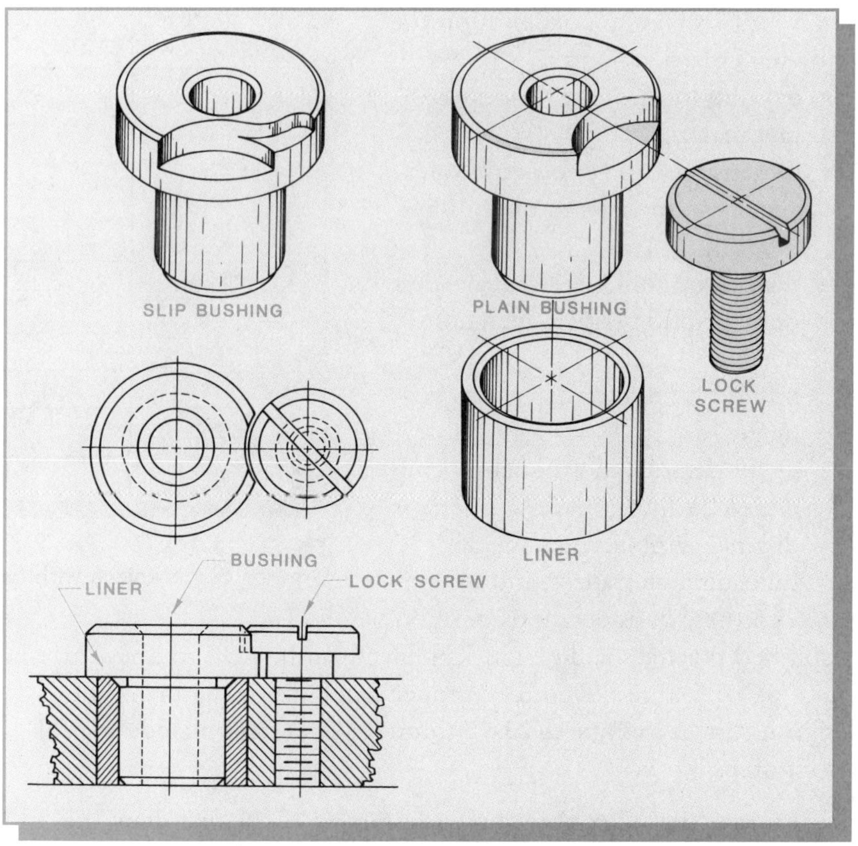

Figure 23-17
Renewable drill bushings

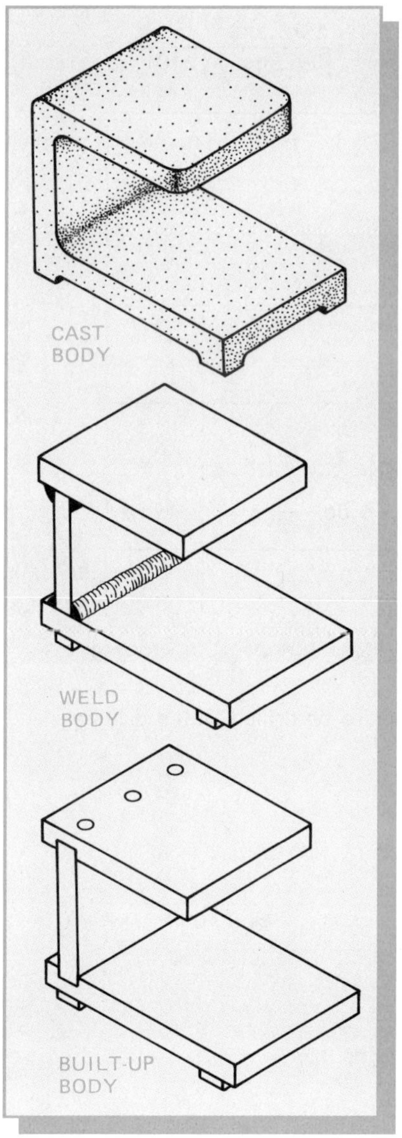

Figure 23-18
Three common forms used to produce bodies for jigs and fixtures

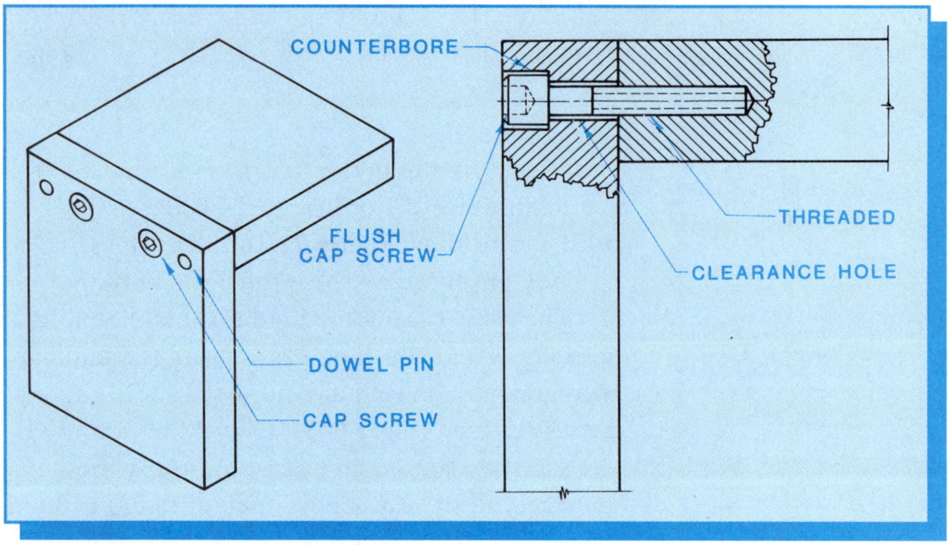

Figure 23-19
Built-up body connections

Jig and Fixture Design

Designing the most appropriate jig for a specific operation depends on the following considerations:

- existing form of the workpiece's raw material
- dimensional and geometric tolerance requirements in the finished product
- type of machine operator needed
- number of repetitive operations needed
- number of duplicate parts to be made
- safety considerations, such as chip clearance, lubricant flow, operator distance, visibility, loading and unloading clearances, and weights.

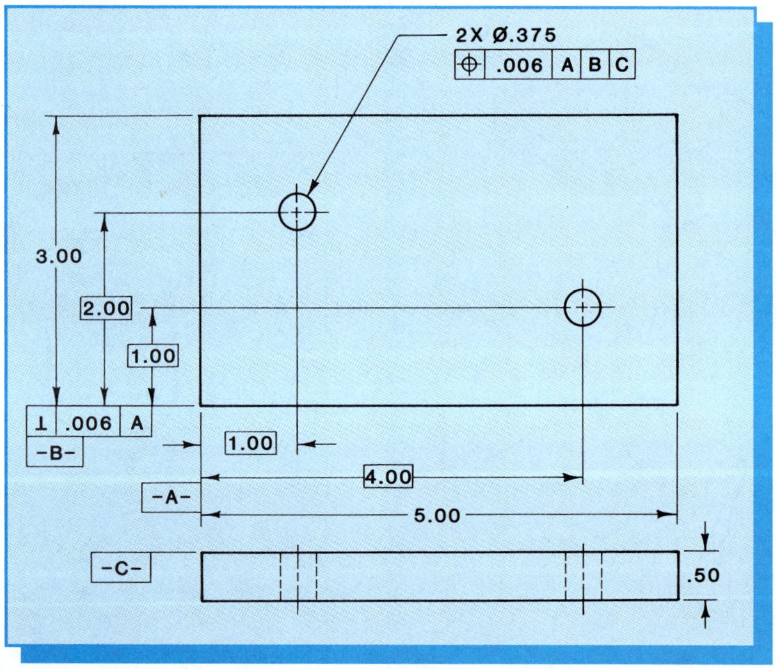

Figure 23-20
Drawing of workpiece with two holes to be drilled with a drill jig

With these considerations in mind, established practices and design sequences can be used to arrive at a sound jig or fixture design. The proper sequence for jig or fixture design is shown in **Figures 23-20** through **23-27**, and include the following steps:

1. To draw the workpiece shown in **Figure 23-20**, first draw the outline with red phantom lines **(Fig. 23-21)**.
2. Since the placement of the locators is critical, study the workpiece drawing carefully before drawing the locators on the workpiece datums **(Fig. 23-22)**.

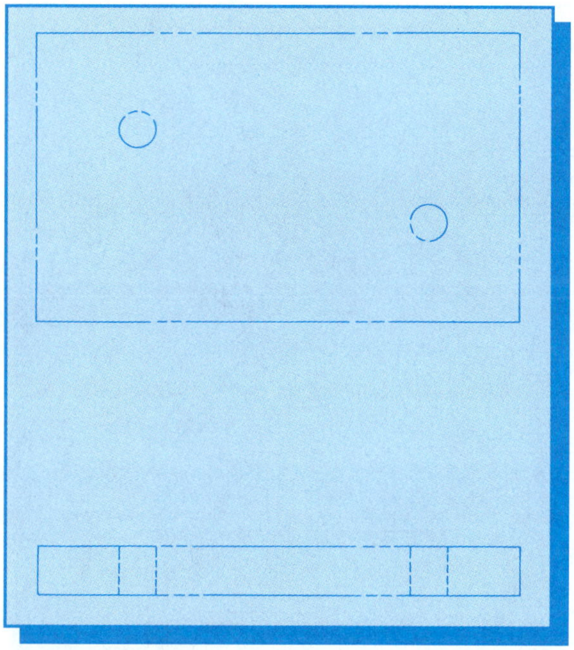

Figure 23-21
Draw an outline of the workpiece with a colored pencil (red preferred).

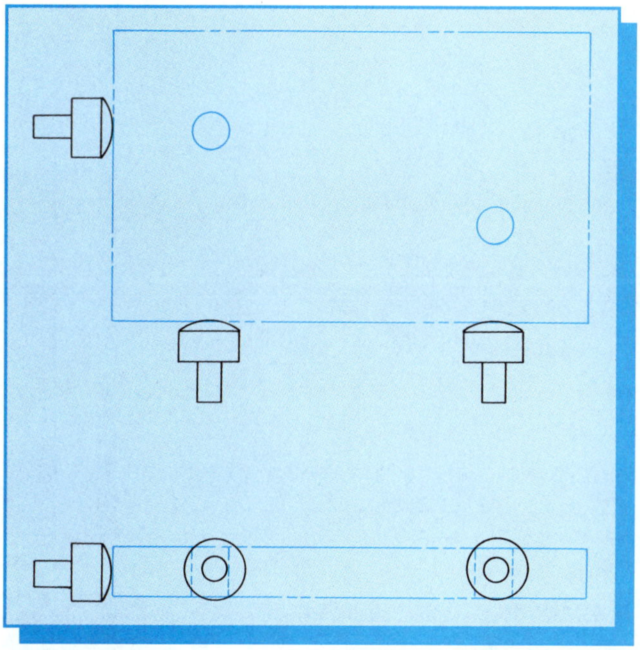

Figure 23-22
Add the locators.

3. Draw the rest buttons so they support the workpiece while the machining operation (drilling) is performed **(Fig. 23-23)**.
4. Draw the bushings so they are placed above the holes to be drilled (fixed head type here). The distance from the bottom of the bushing to the top of the workpiece should equal one-half the drill's diameter, in this case, .188" **(Fig. 23-24)**.
5. Select and draw all clamps so they force the workpiece against the locators and hold it firmly. In this example, a double-acting clamp **(Fig. 23-25)** forces the workpiece against the locators to the left and bottom.
6. Design the body of the jig **(Fig. 23-26)**. The jig body holds all jig components together, and simultaneously allows the workpiece to be inserted and removed easily.
7. Draw the jig feet in position. In this case, socket head cap screws are used. Finish the design by adding fasteners to hold the body plates together and with clamps in place **(Fig. 23-27)**.

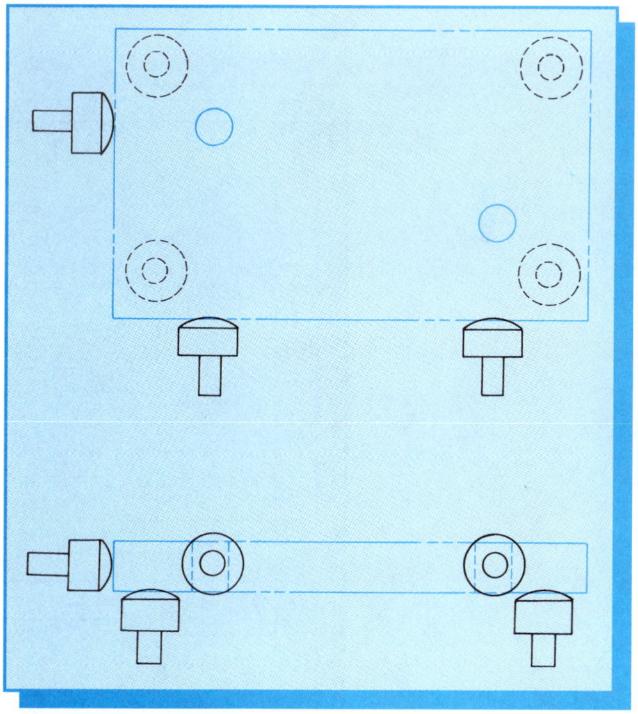

Figure 23-23
Add the rest buttons.

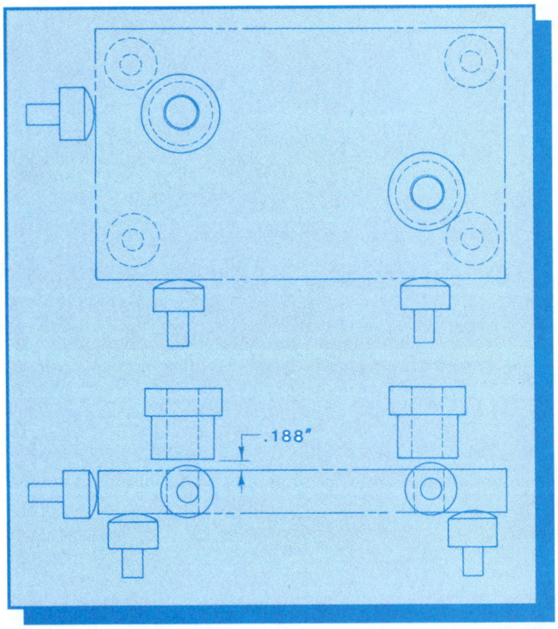

Figure 23-24
Position the bushings .188" above the workpiece.

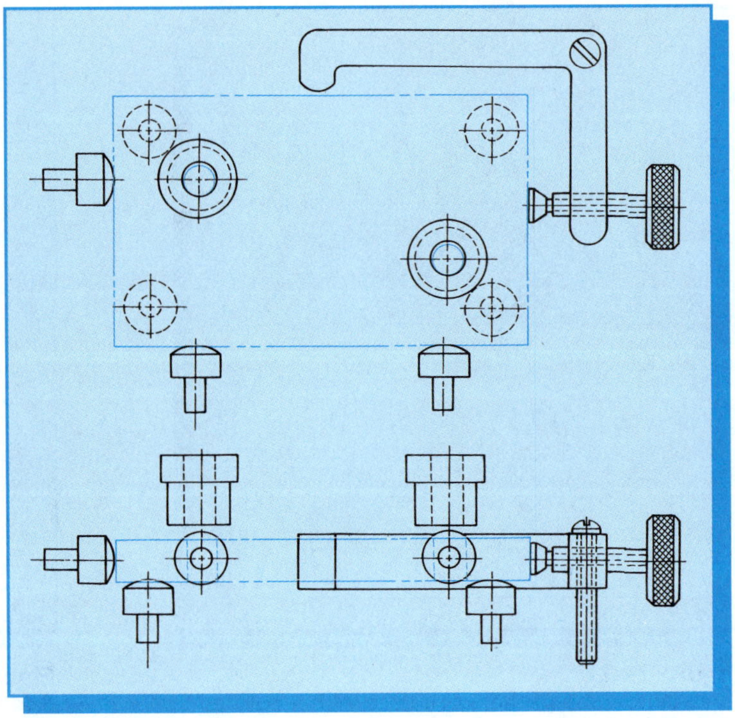

Figure 23-25
Locate the clamp's position

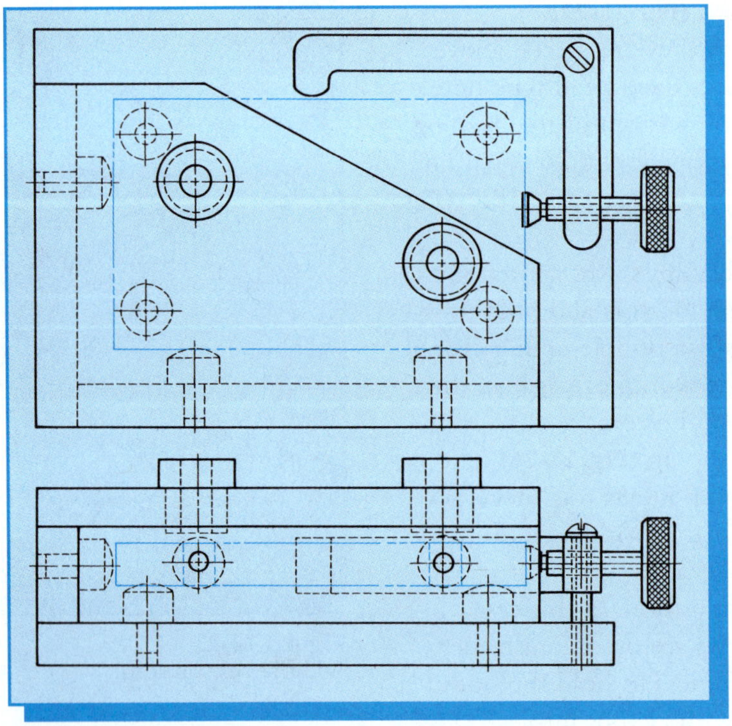

Figure 23-26
Design the body to support all holding components.

Figure 23-27
Add fasteners and feet to the jig body.

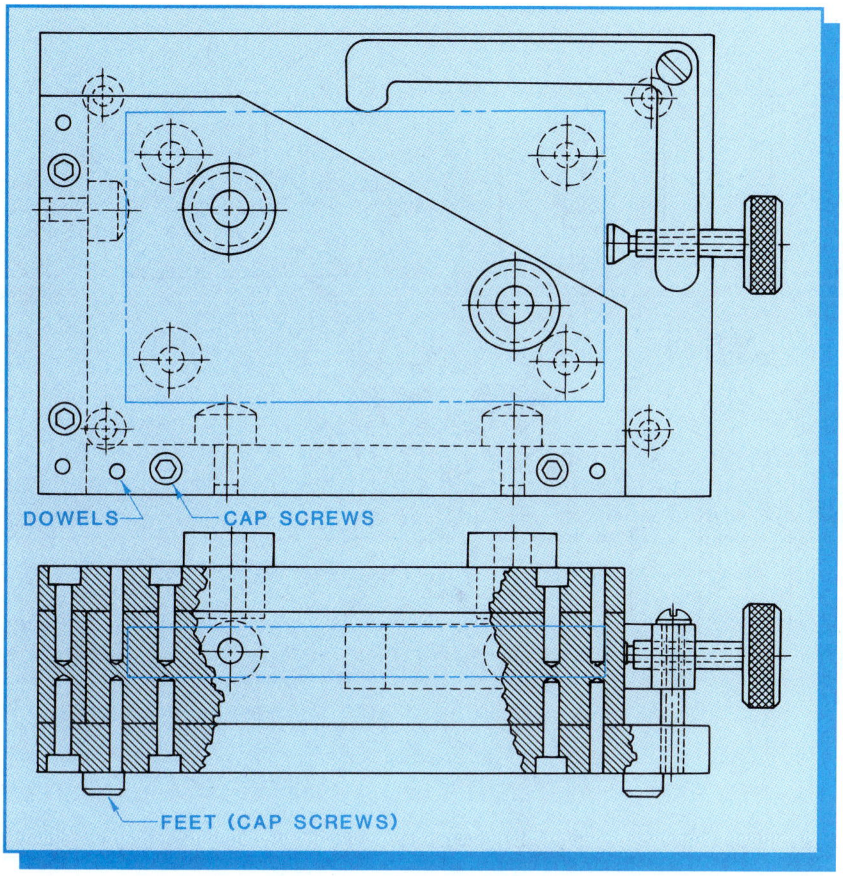

8. Dimension the jig assembly. Jig and fixture dimensioning is completed using the dimensioning and tolerancing practices outlined in Chapters 9 and 10. Since most jigs and fixtures are made of many standard parts, few dimensions are needed because it is not necessary to specify standard parts sizes. Some size dimensions are necessary for the body, but the majority of dimensions on a jig or fixture drawing are location dimensions.

AutoCAD LT Exercises

Because jigs and fixtures are fabricated from standard parts, designing them on a CAD system is simple and fast. Designers need to understand design procedures, and know how to select and position standard locators, rest buttons, bushings, clamps, feet, and bodies.

Drafters must also be able to modify standard parts. This modification may be extensive or simple, such as shortening a locator, trimming a bushing, grinding a rest button, or modifying a clamp. Editing on a CAD system makes it easy to modify standard components and to move parts to their proper position on a drawing. One company that will supply the CAD operator with predrawn symbols is Carr Lane **(Fig. 23-28)**.

In designing jig and fixtures, a CAD-system operator first creates, draws, and saves a library of the standard component drawings. This process is time-consuming at first, but enables a CAD system to

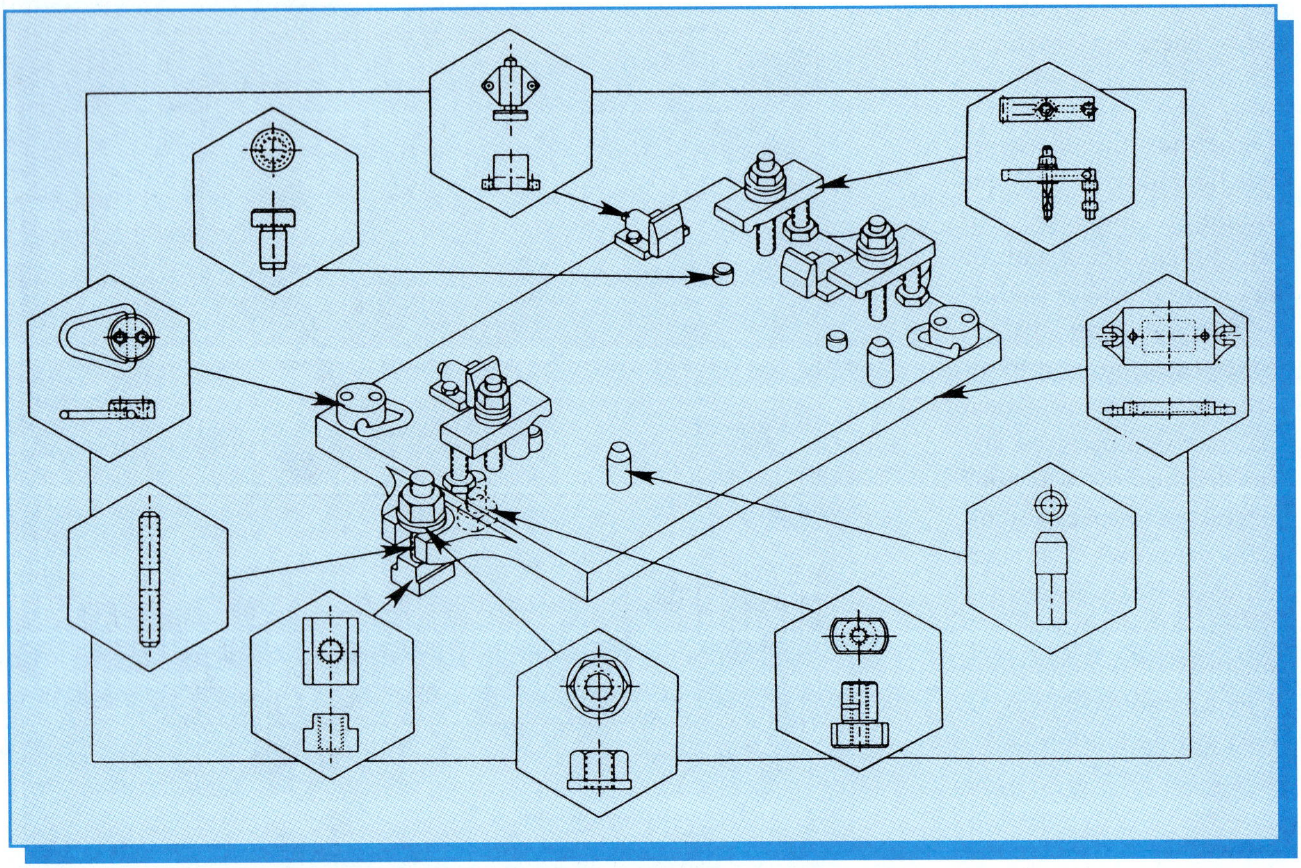

Figure 23-28
Fixture components. *(Courtesy Carr Lane Manufacturing Co.)*

function efficiently and rapidly later. The top and side views are drawn and saved as standard components. **Figure 23-29** is an example of a drill bushing drawing that has been drawn and stored in the library. This bushing is a standard headed press fit bushing. In this case, the CAD-system operator has assigned a number (TD-B1) to this bushing. Each subsequent standard component is drawn and saved in this manner.

Figures 23-30 and **23-31** show how a CAD system is used to modify a standard part.

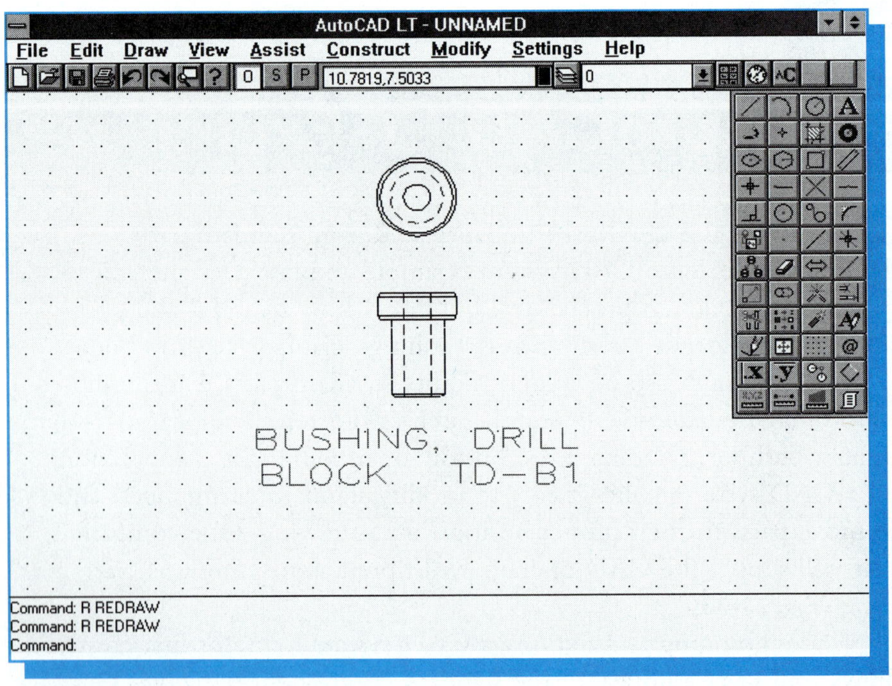

Figure 23-29
A library of jig and fixture components is a necessity. Components may be drawn and saved, or purchased as software.

In these examples, the part to be modified is a drill bushing TD-B1.

1. The operator inserts the block number TD-B1 and erases the unnecessary top view **(Fig. 23-30)**.
2. The modifications are made on the drawing **(Fig. 23-31)**. Note that the bushing has been cut on a 45° angle. The only dimensions on the modified part are the angle information. That is because the bushing is a standard part and its size dimensions are known; the only dimensions needed are the ones describing the modification.

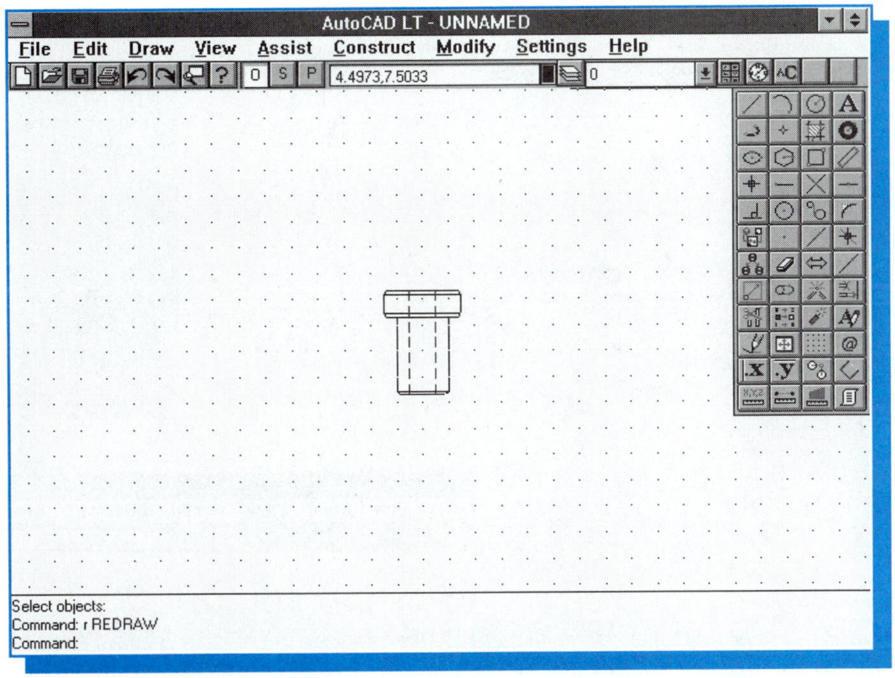

Figure 23-30
Drill bushing TD-B1 is inserted from the library. The block is exploded and the top view erased.

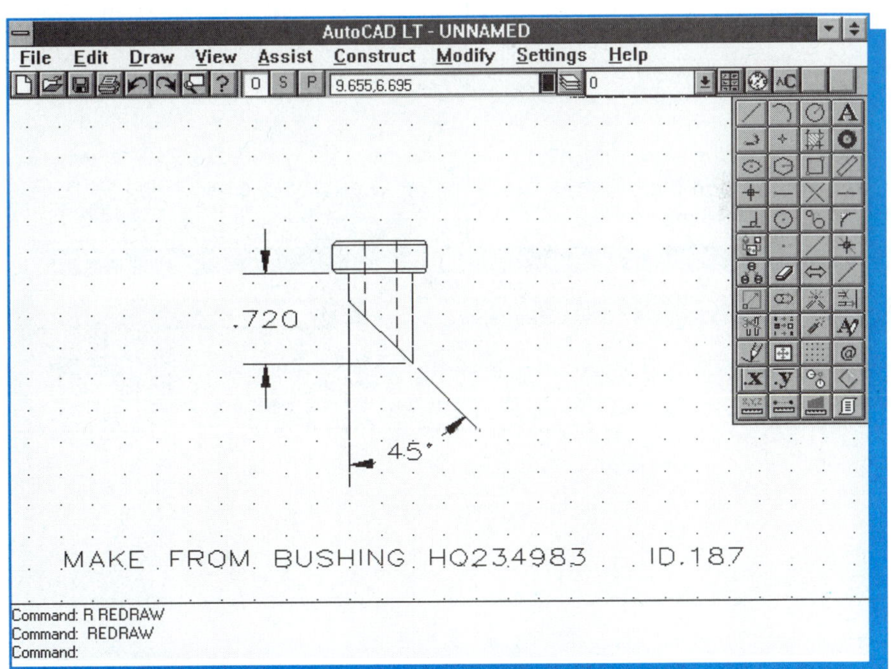

Figure 23-31
Modify the bushing to fit the specific machining requirements.

Figures 23-32 through **23-37** illustrate how to create a jig design using a CAD system. The workpiece to be drilled is shown in **Figure 23-32.** The hole to be drilled is on a slanted surface. If drilled without a jig to support the drill, the drill bit would bend and slide on the inclined surface. A modified drill bushing is needed to solve this problem. The following steps are used in designing a jig using a CAD system:

1. Draw the part with phantom lines using a red layer **(Fig. 23-32)**. In this case, the operator has called the layer BODY.
2. Insert standard locators and rest buttons **(Fig. 23-33)**. Once inserted, they may be moved to any location to support the part.

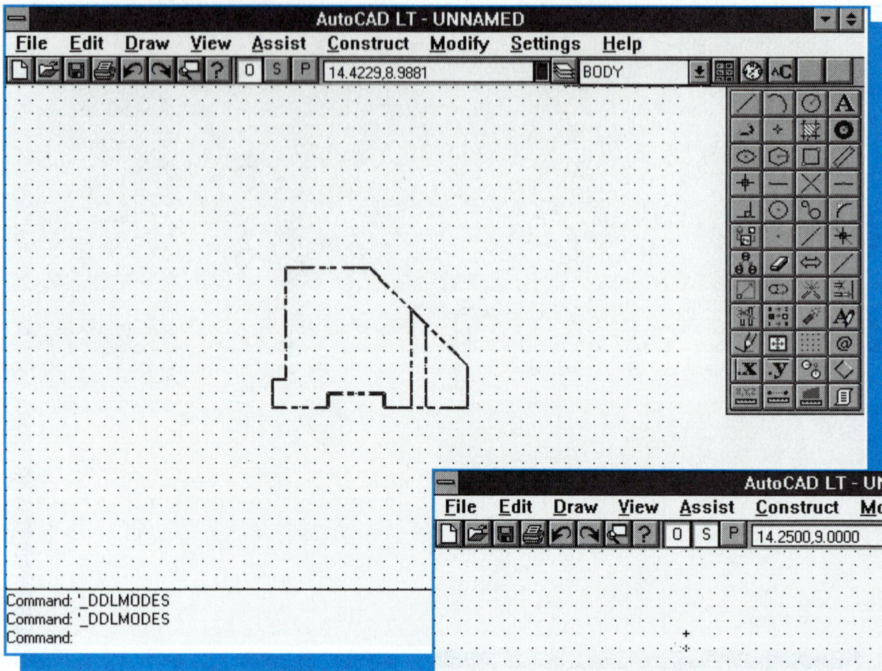

Figure 23-32
Outline the part in color (usually red).

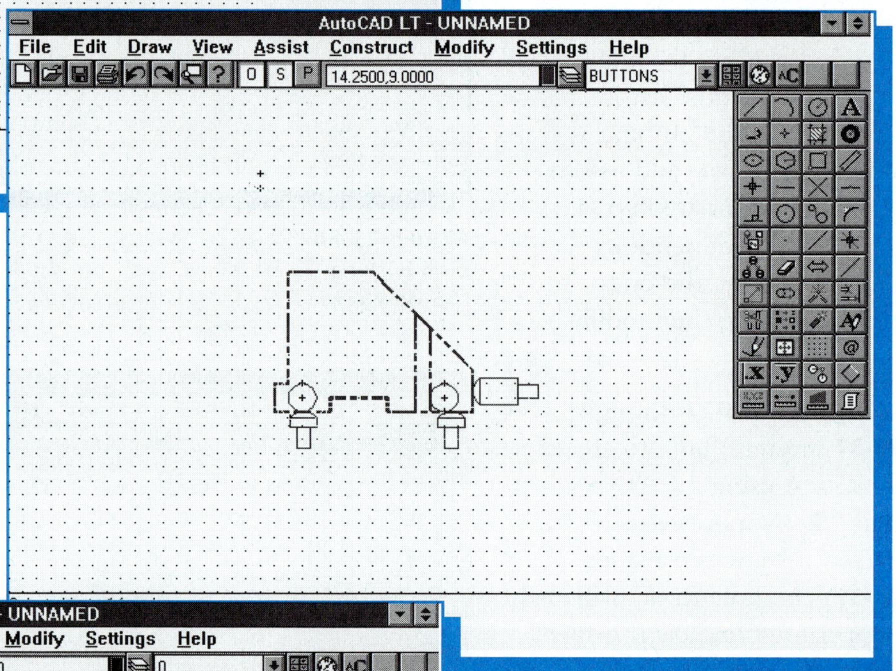

Figure 23-33
Position the locators and rest buttons.

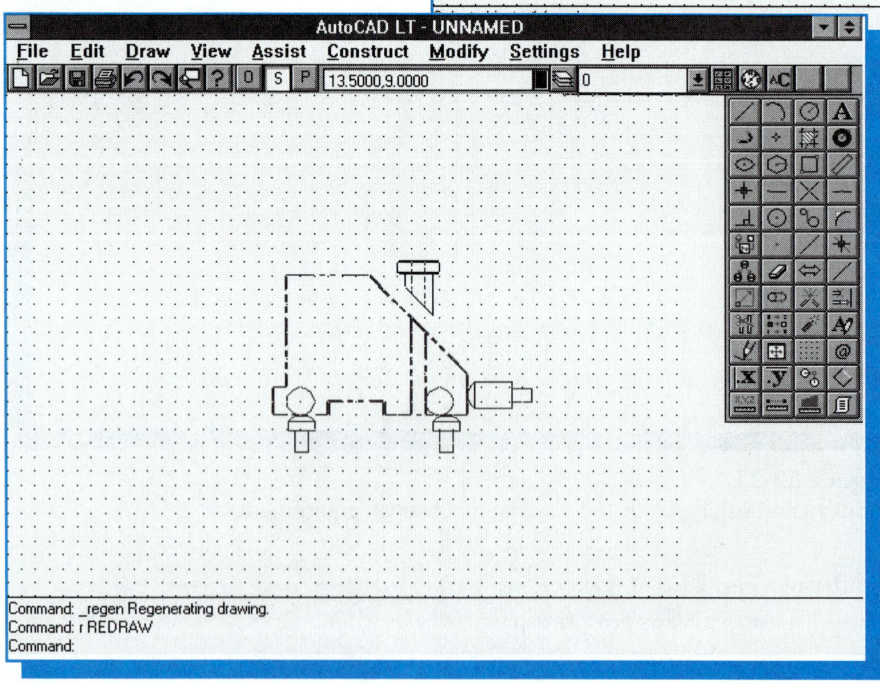

Figure 23-34
Insert and position the modified drill bushing.

3. Insert the modified drill bushing directly above the hole to be drilled **(Fig. 23-34)**.
4. Insert the two thumbscrew clamps **(Fig. 23-35)**.
5. Design a body to hold the standard jig components, using normal drawing commands **(Fig. 23-36)**. If necessary, any standard part may be easily relocated using the Move command.

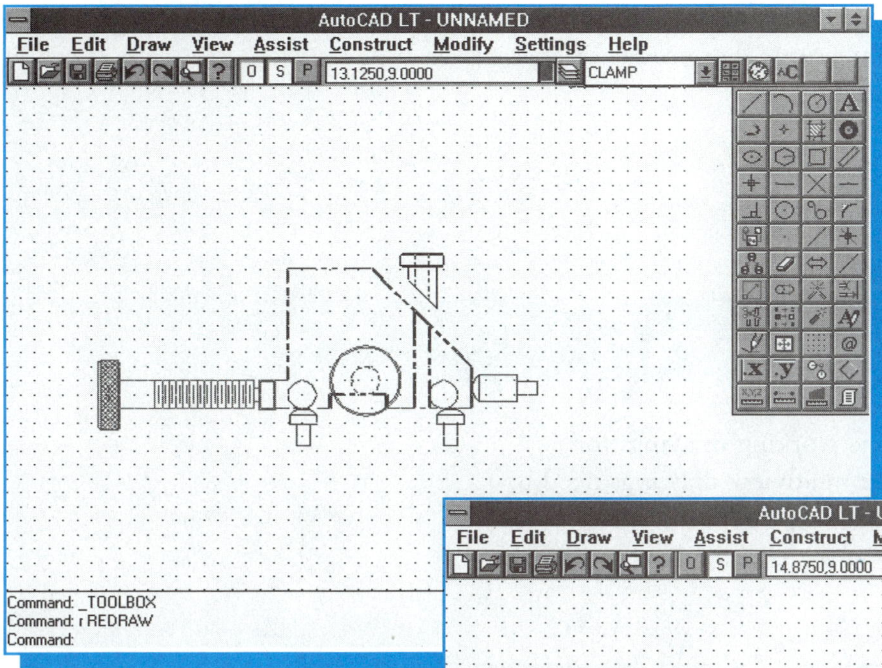

Figure 23-35
Position the clamps.

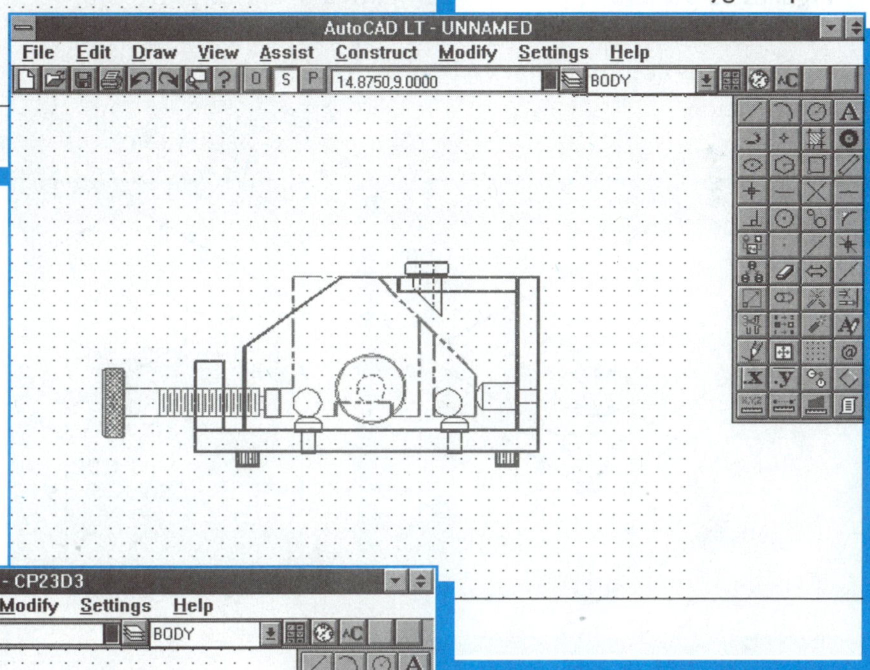

Figure 23-36
Design a body to support the standard jig components.

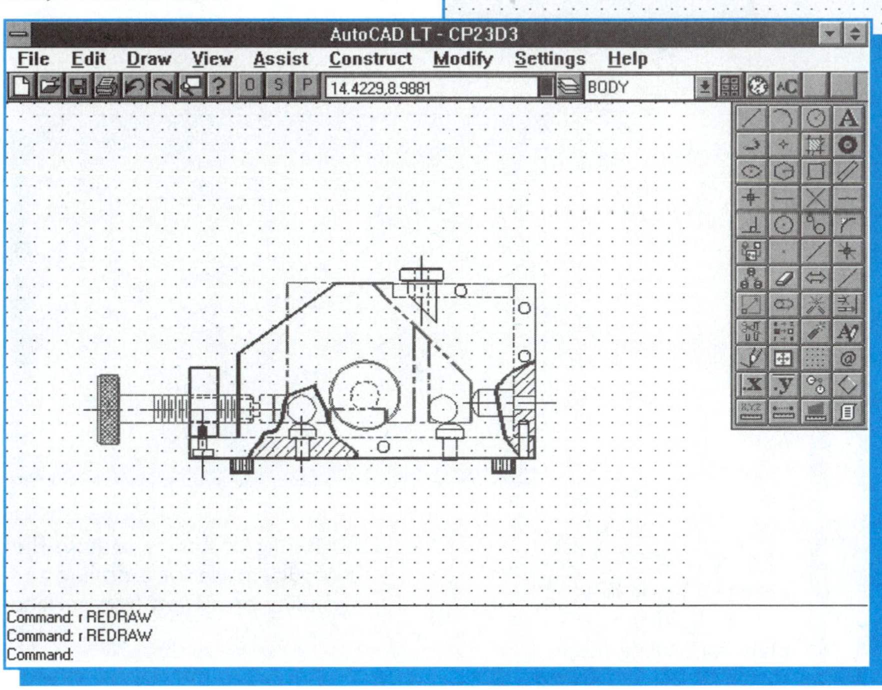

Figure 23-37
Complete the working drawing by adding fasteners, feet, and dimensions.

6. Using CAD Draw and Modify commands, complete the front view of the jig **(Fig. 23-37)**. Hidden lines are changed and broken-out sections are added for clarity. This step also includes the addition of fasteners that hold the body plates.

7. To complete a jig working drawing, a top view, the removed sections, details, and tolerance dimensions are necessary. These tasks are covered in Chapters 8 through 11.

Drafting Exercises

1. Draw the multiview and isometric working drawings for **Figures 23-38** through **23-45**. The multiview drawings need to be created before designing the tooling. The isometric drawing of the part is commonly placed on the tooling drawing to aid in visualization.

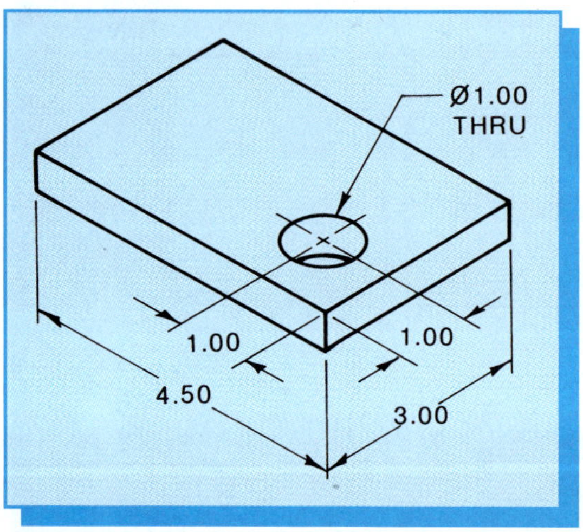

Figure 23-38
Design a jig or fixture to hold the plate for drilling.

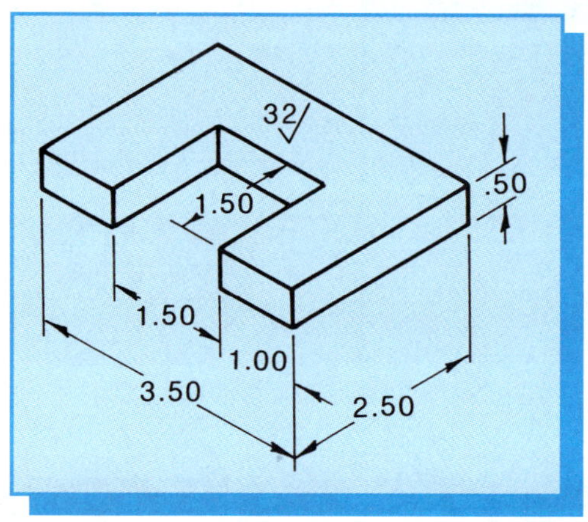

Figure 23-40
Design a jig or fixture to hold the U-plate for surface grinding (top surface only).

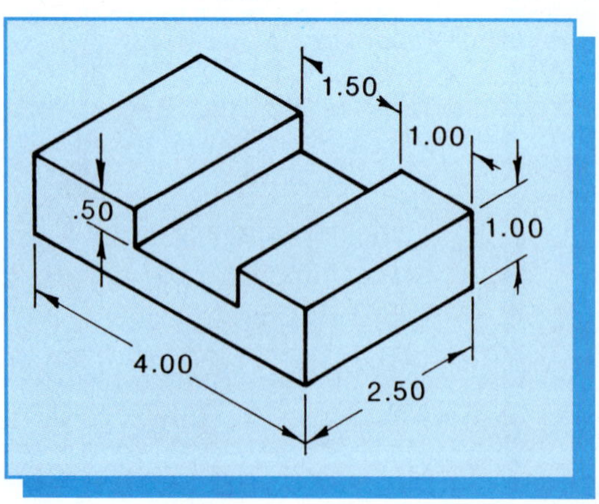

Figure 23-39
Design a jig or fixture to hold the plate for milling the slot.

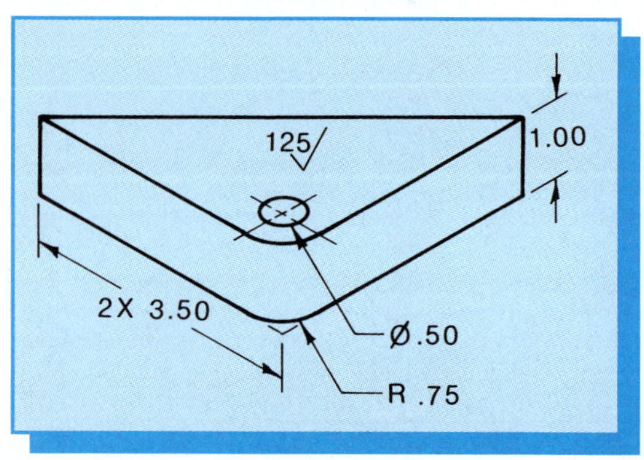

Figure 23-41
Design a jig or fixture to hold the triangular plate for a drilling and grinding operation (top surface only).

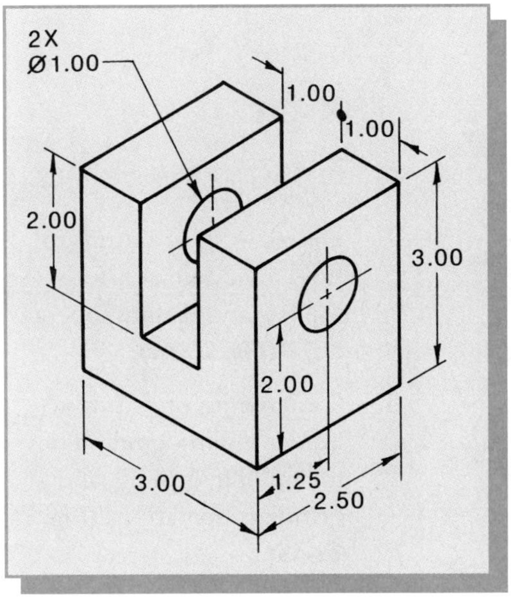

Figure 23-42
Design a jig or fixture to hold the channel block for a milling (U shape) and drilling operation.

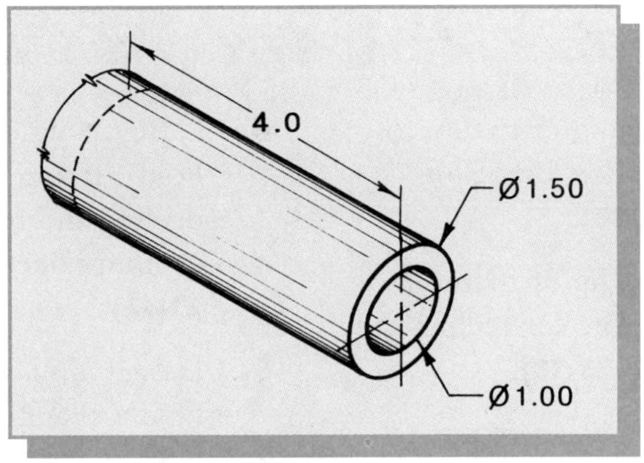

Figure 23-43
Design a jig or fixture to hold the pipe for 4" cutoffs.

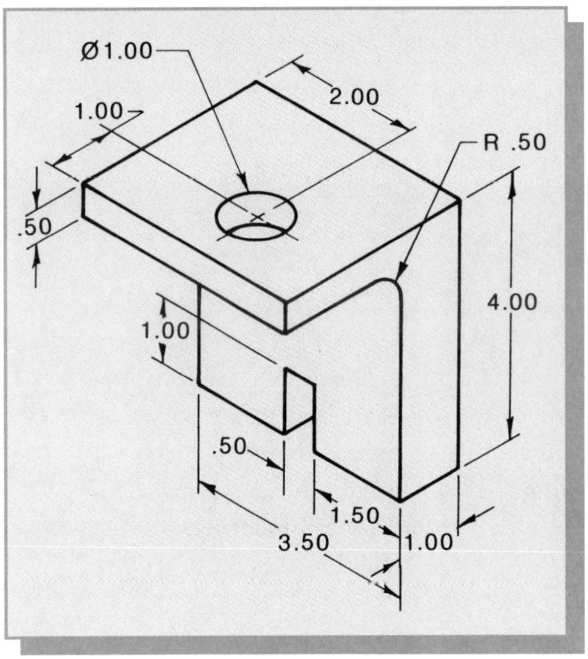

Figure 23-44
Design a jig or fixture to hold the guide block for a slot milling and drilling operation.

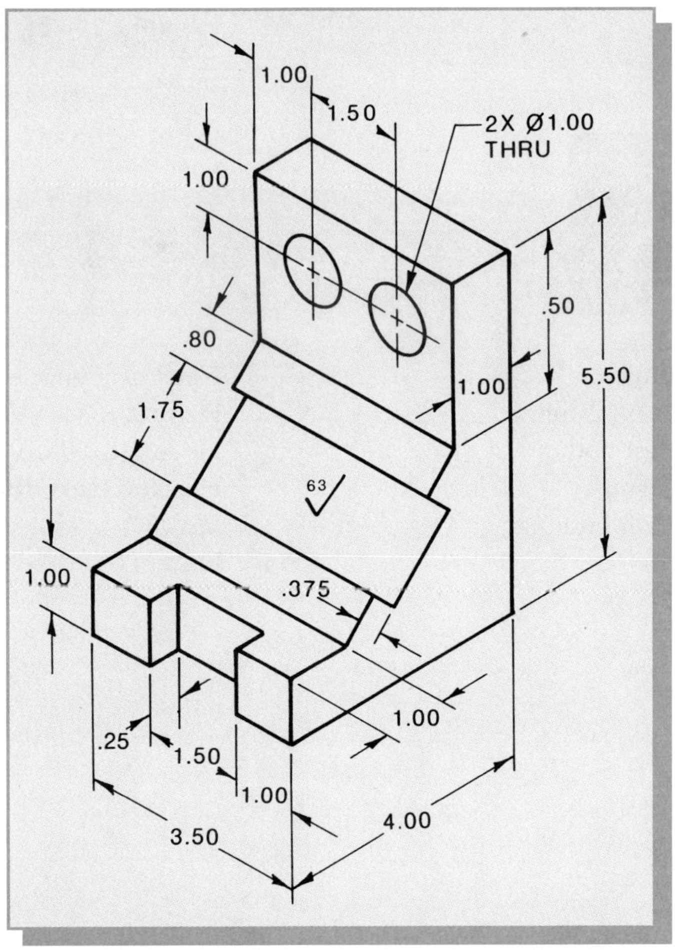

Figure 23-45
Design a jig or fixture to hold the milling guide block for milling two slots (1.50" and 1.75"), drilling operation for two holes, and a grinding operation for the 1.75" slot.

Design Exercises

1. Design a jig or fixture to hold a plate for drilling **(Fig. 23-38)**.
2. Design a jig or fixture to hold a plate for milling the slot **(Fig. 23-39)**.
3. Design a jig or fixture to hold a U-plate for surface grinding of the top only **(Fig. 23-40)**.
4. Design a jig or fixture to hold a triangular plate for drilling and top surface grinding operations **(Fig. 23-41)**.
5. Design a jig or fixture to hold a channel block for milling and drilling operations **(Fig. 23-42)**.
6. Design a jig or fixture to hold a pipe for 4" cutoffs **(Fig. 23-43)**.
7. Design a jig or fixture to hold a guide block for milling and drilling operations **(Fig. 23-44)**.
8. Design a jig or fixture to hold a milling guide block for milling, drilling, and grinding operations **(Fig. 23-45)**.

Key Terms

Built-up bodies
Cam clamps
Cast bodies
Clamps
Drill bushings
Feet
Fixed drill bushings
Fixtures
Headed fixed drill bushings
Headless fixed drill bushings
Jigs
Locators
Replaceable drill bushings
Rest buttons
Screw clamps
Toggle head clamps
Tool engineering
Weld bodies

Chapter 24

Architectural Drafting

Objectives

The student will be able to:

- read architectural drawings
- prepare an exterior two-point perspective drawing
- draw a plot plan
- draw a floor plan
- draw a foundation plan
- draw four exterior elevations
- draw an interior elevation
- draw a sectional construction detail
- design a small structure to meet specific objectives
- draw a set of architectural drawings with manual techniques and a CAD system

The design of any architectural structure follows similar procedures. Only the complexity of the design and drafting process and the levels of structural engineering change as the size of a structure increases. In designing a structure, whether a small residence or a high-rise building, the functions of the structure must first be determined and carefully organized. Only then can steps be taken to create a functional, aesthetically pleasing, and structurally sound design. Because of size and familiarity, residential applications are used in this chapter. However, the principles and practices apply to the design of any structure, regardless of size or function.

Gathering Information

Structures are built primarily for people. Detailed information concerning the goals, habits, and tastes of all building inhabitants must be matched with the conditions of the site, building code limitations, and available funds or financing. The first step in designing a residence **(Fig. 24-1)** is to gather the following information:

1. Number of occupants, present and future, and their age and sex.
2. Price range and monthly mortgage limits.
3. Site conditions that affect the type and orientation of structures **(Fig. 24-2)**. This includes a study of potential views, solar orientation, prevailing winds, slope angles, vegetation, lot shape, street access, and soil conditions.

   The **plat plan (Fig. 24-3)** is a map of a large number of properties and surrounding streets. The following data is included:
   - property-line dimensions
   - building setbacks
   - compass orientation
   - property lines with compass orientation
   - easements
   - street widths
   - street names
   - block numbers
   - lot numbers
   - scale

The **site plan (Fig. 24-4)** is a drawing describing the physical elements of a single piece of property. The following data is included:

- property corner elevations
- contour lines with elevation heights
- compass orientation
- property-line dimensions
- property lines with compass orientation
- landscaping
- legal descriptions
- waterways
- recontoured lines (dotted)
- utility lines

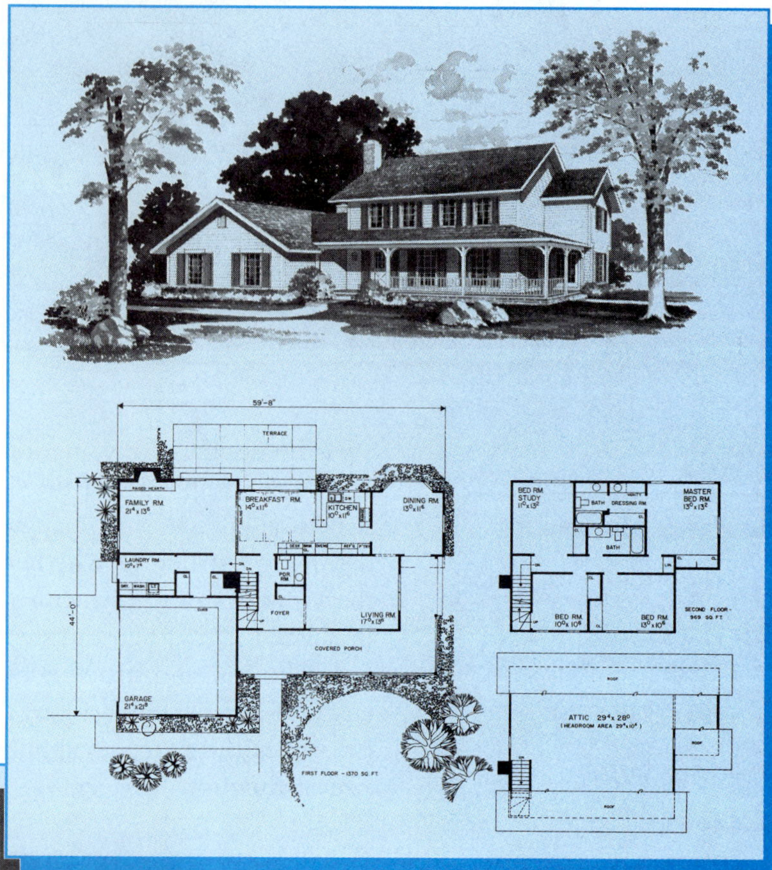

Figure 24-1
Architectural floor-plan drawings for a family residence
(Courtesy of Home Planners, Inc.)

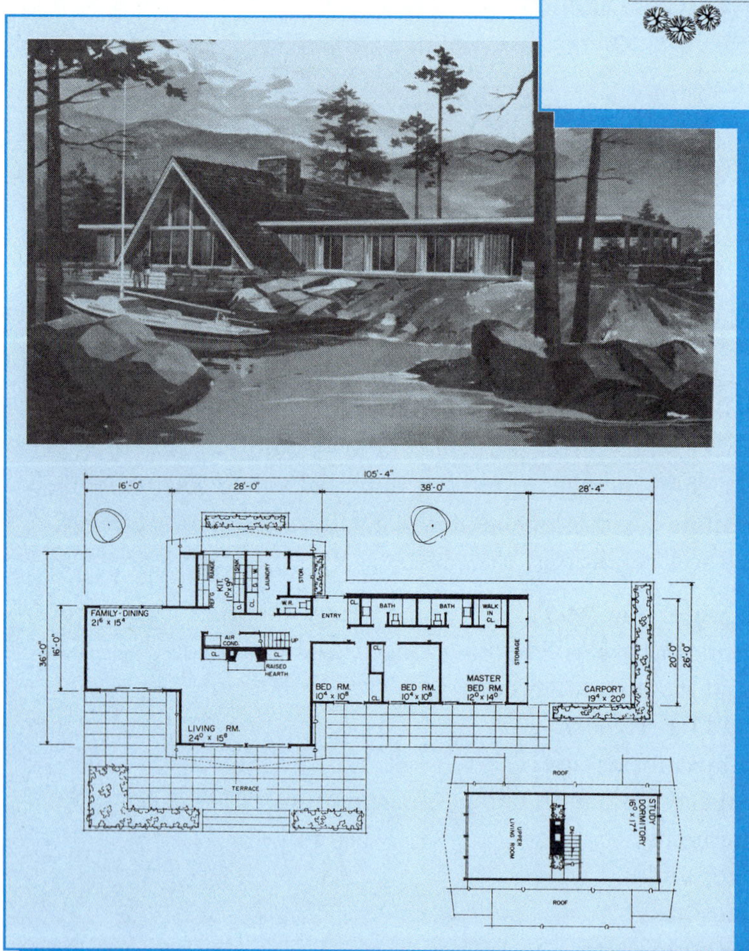

Figure 24-2
The conditions of the building site will affect the design of the house.
(Courtesy of Home Planners, Inc.)

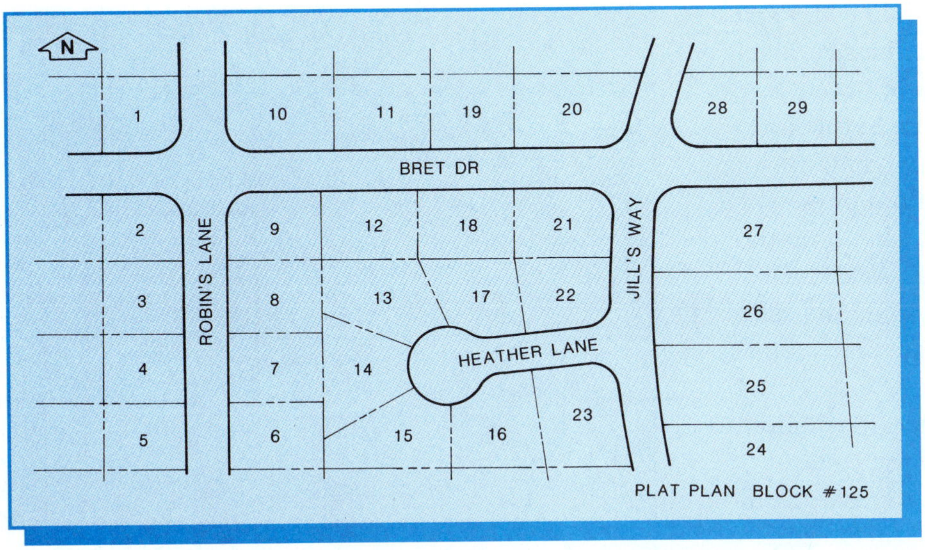

Figure 24-3
Example of a simplified plat plan

The **plot plan (Fig. 24-5)** is a drawing of the structure oriented on the property. The following data is included:

- structure setbacks from property lines
- outline of house and roof
- outline of auxiliary buildings
- property-line dimensions
- property lines with compass orientation
- property corner elevations
- walks, patios
- landscaping
- legal description

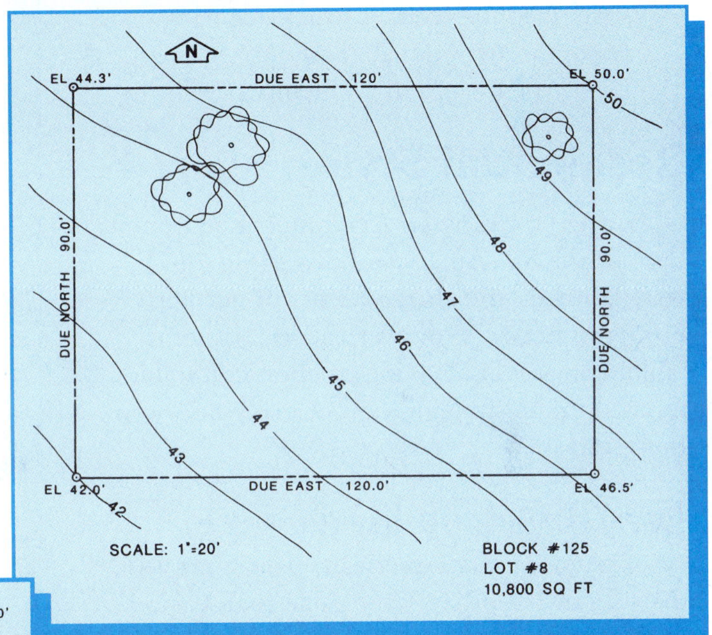

Figure 24-4
Example of a simplified site plan

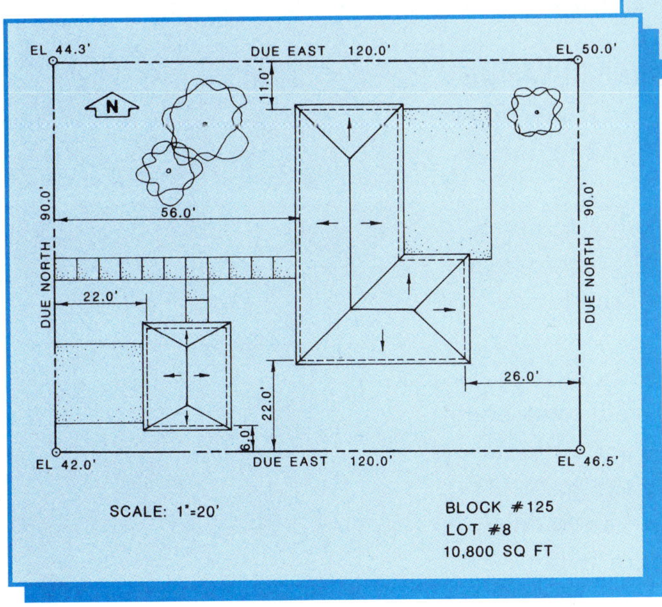

Figure 24-5
Example of a simplified plot plan

4. Local zoning laws and building codes that affect and control minimum lot size, maximum building height, areas of structures, structural types, and setbacks. **Setback** requirements dictate the minimum distance a structure can be located from property lines. The zoning code setbacks set up the boundaries for the buildable area. These dimensions are not shown. Only the actual setbacks for the structures are drawn on the plot plan **(Fig. 24-6)**.
5. Architectural style preference. In addition to personal preference, some local codes restrict or require the use of specific architectural styles and materials. Architectural styles fall into several broad categories: European (traditional), Early American, Later American, and Contemporary.

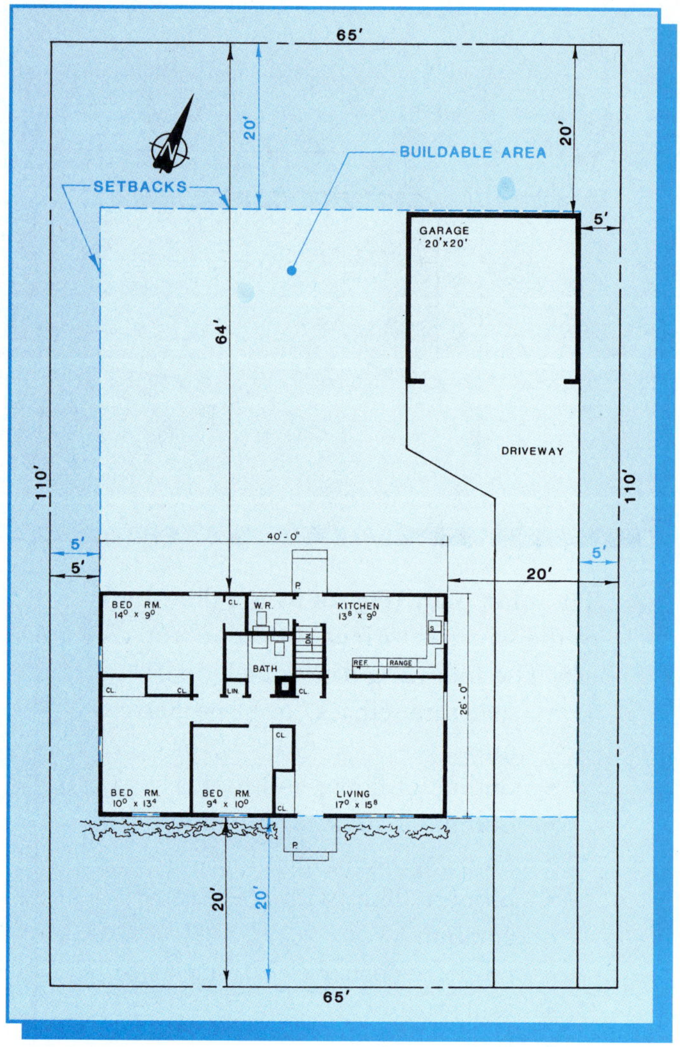

Figure 24-6
Example of typical setback minimums for a structure on a building site

Residential Styles

Since most early settlers immigrated from Europe, the first American houses **(Fig. 24-7)** were based on the characteristics of northern European houses. However, the lack of some building materials and construction constraints resulted in the simplification of many northern European styles.

Early American Style

One of the first American adaptations was the one-and-one-half story, gable roof, Cape Cod style **(Fig. 24-8)**. Cold New England winters contributed to the addition of shutters, large fireplaces, and small window areas. Another early American adaptation was the Dutch colonial **(Fig. 24-9)**, which included dormers and a gambrel (double pitch) roof. Other early American styles added individual and shed dormers to create more second story space **(Fig. 24-10)**. The early American saltbox offset the roof angles to create a full second story on the front of the house **(Fig. 24-11)**.

A full second level, which cantilevered (overhung) the first, was added to create the garrison style **(Fig. 24-12)**. The garrison was built mainly in the north, and therefore had small windows and little outdoor living space. By contrast, southern colonial homes **(Fig. 24-13)** also included a full second floor but expanded the size of windows and added verandas (large porches) for southern outdoor living.

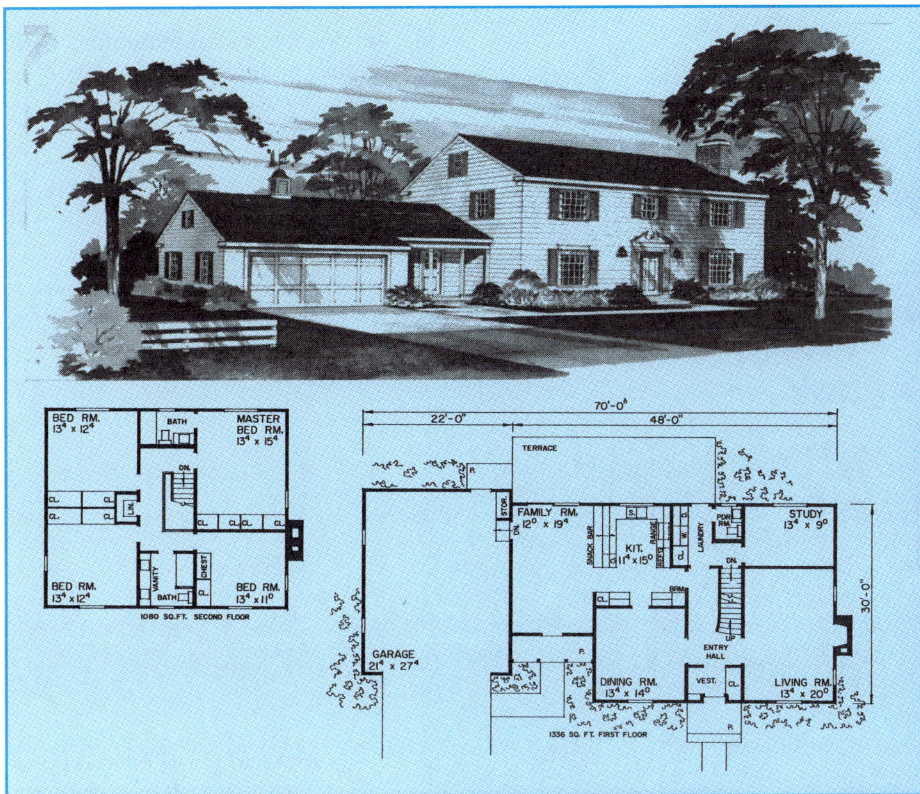

Figure 24-7
An example of contemporary early American colonial architecture *(Courtesy of Home Planners, Inc.)*

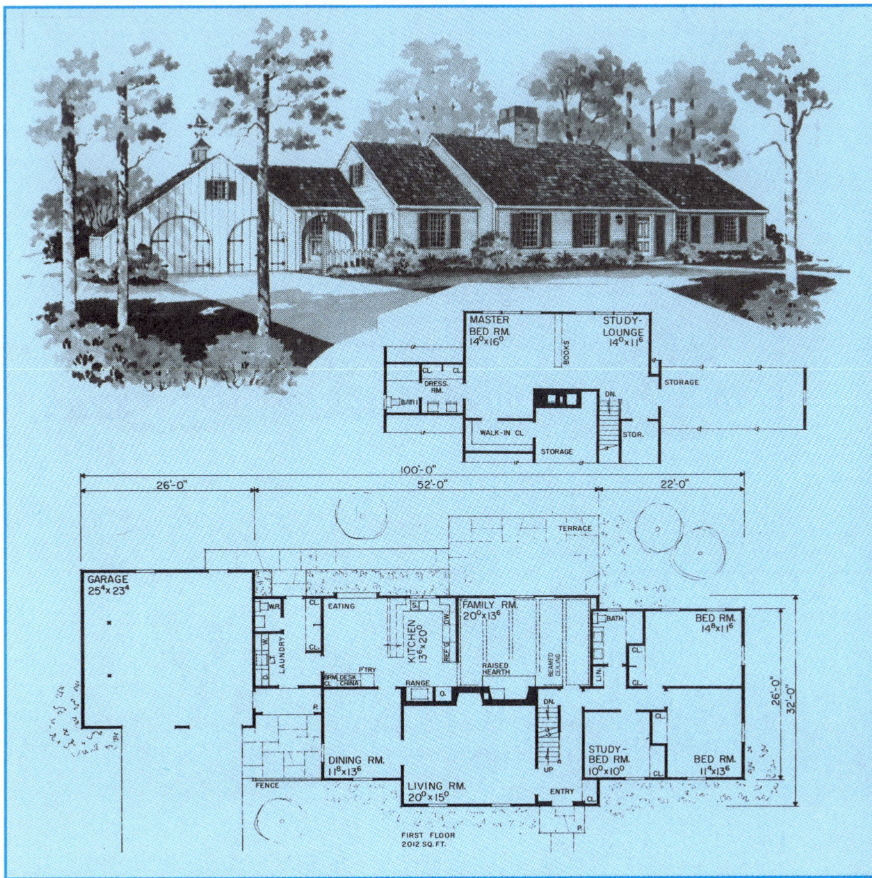

Figure 24-8
An example of contemporary Cape Cod architecture *(Courtesy of Home Planners, Inc.)*

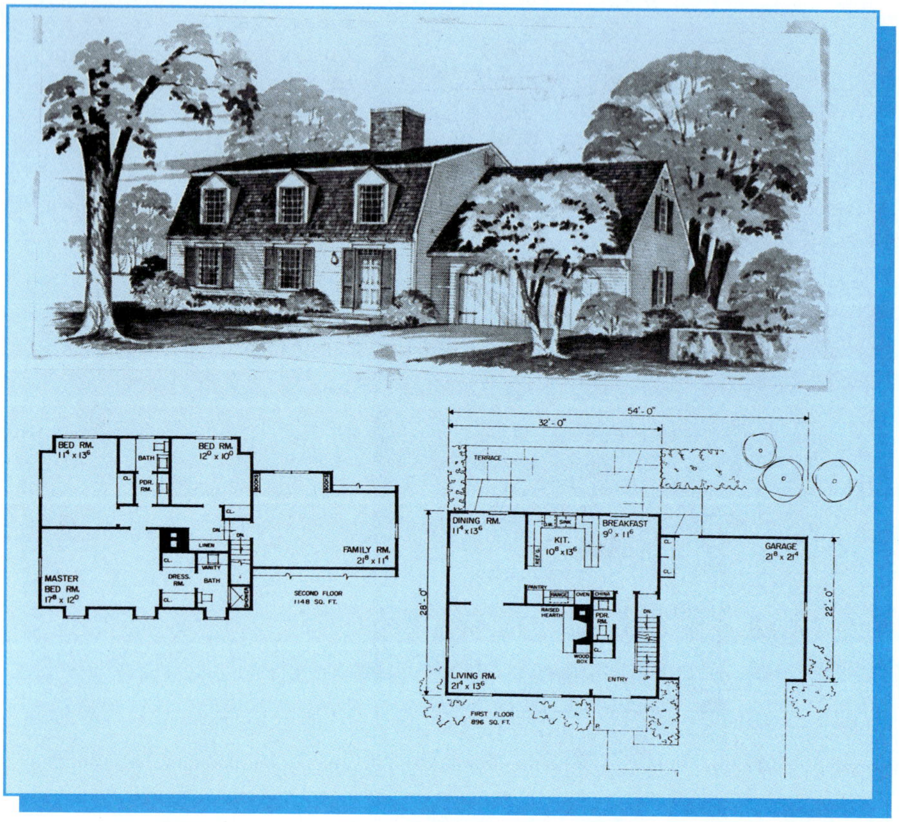

Figure 24-9
An example of contemporary Dutch colonial architecture *(Courtesy of Home Planners, Inc.)*

Figure 24-10
An example of contemporary early American architecture with gable and shed roof dormers *(Courtesy of Home Planners, Inc.)*

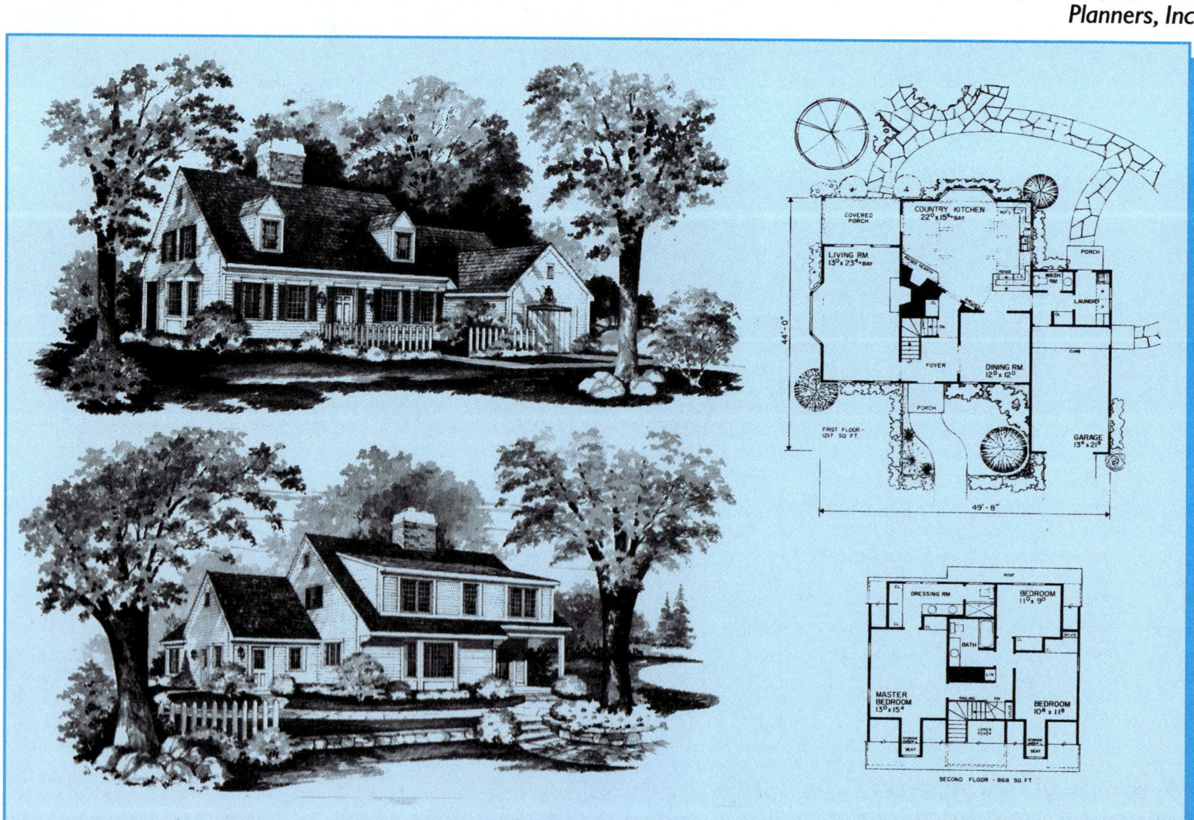

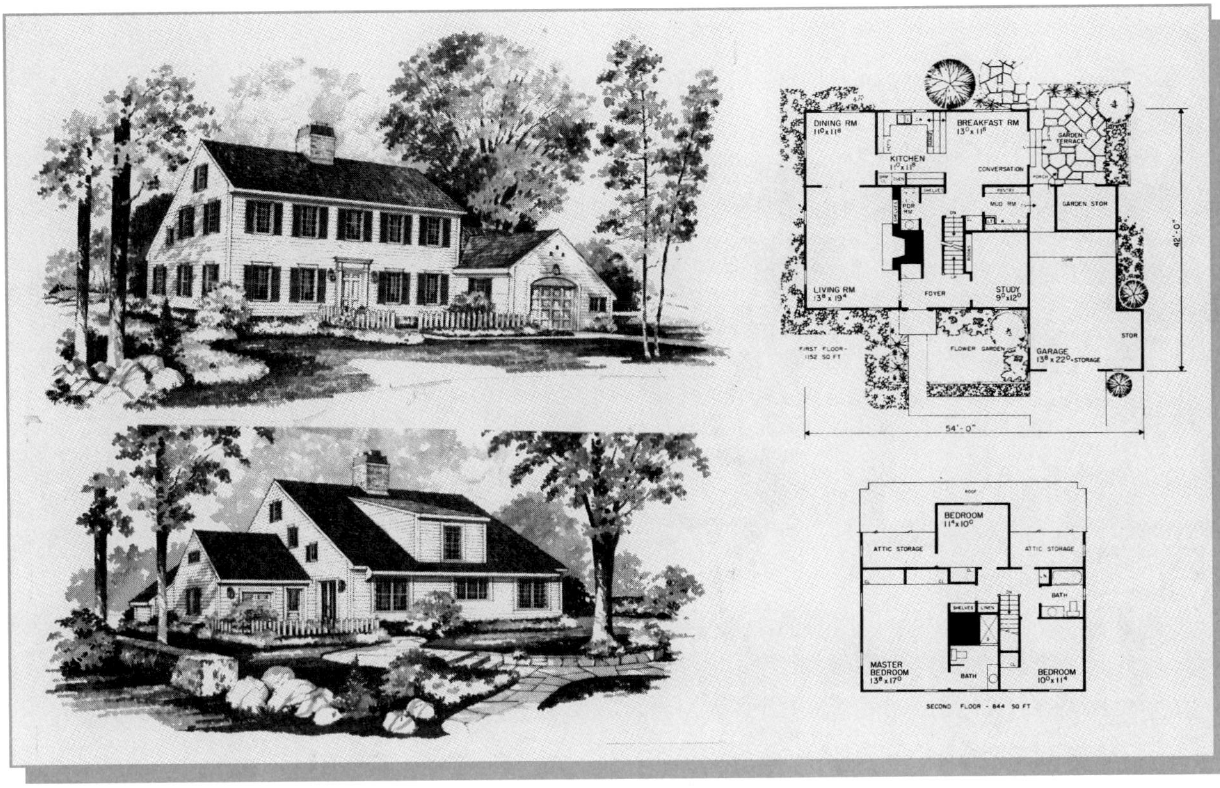

Figure 24-11
An example of contemporary early American saltbox architecture *(Courtesy of Home Planners, Inc.)*

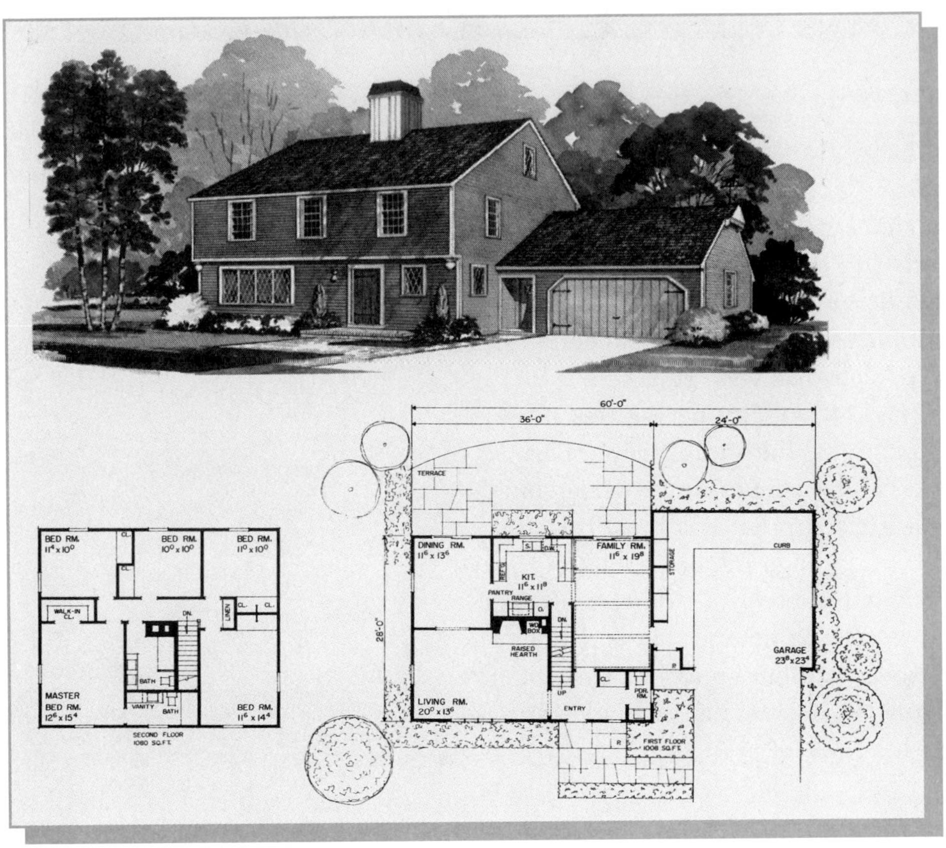

Figure 24-12
An example of contemporary early American garrison architecture *(Courtesy of Home Planners, Inc.)*

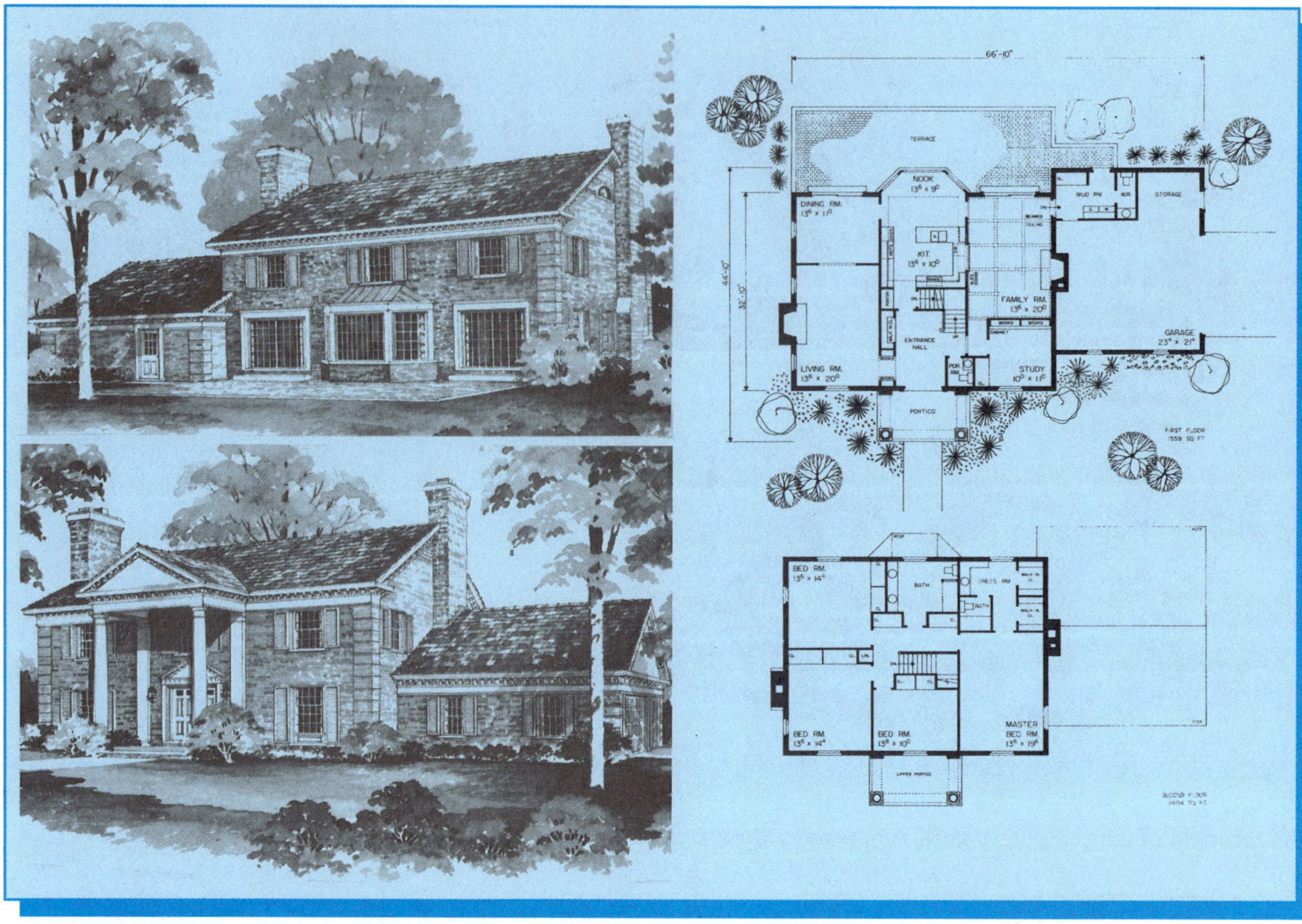

Figure 24-13
An example of contemporary southern colonial architecture *(Courtesy of Home Planners, Inc.)*

Later American Styles

After the colonial period, architectural styles were influenced by climatic conditions and the southern European styles as part of the classical revival movement. Styles developed during this period were strongly affected by French, Spanish, and classical Roman and Greek architecture. They include federal, Victorian, western ranch, Monterey, and French mansard **(Fig. 24-14)**. This period was also marked by the return to some traditional European styling. This became possible as building materials became more abundant, building skills improved, and affluence grew. A popular style of this period was the English Tudor, characterized by high pitched gable roofs, diamond-leaded windows, and brick and stucco half-timber walls **(Fig. 24-15)**. Style changes often lead to eclectic designs. These designs combined the features of many styles without regard for the principles and elements of design. **Figure 24-16** shows an example of an eclectic architectural style.

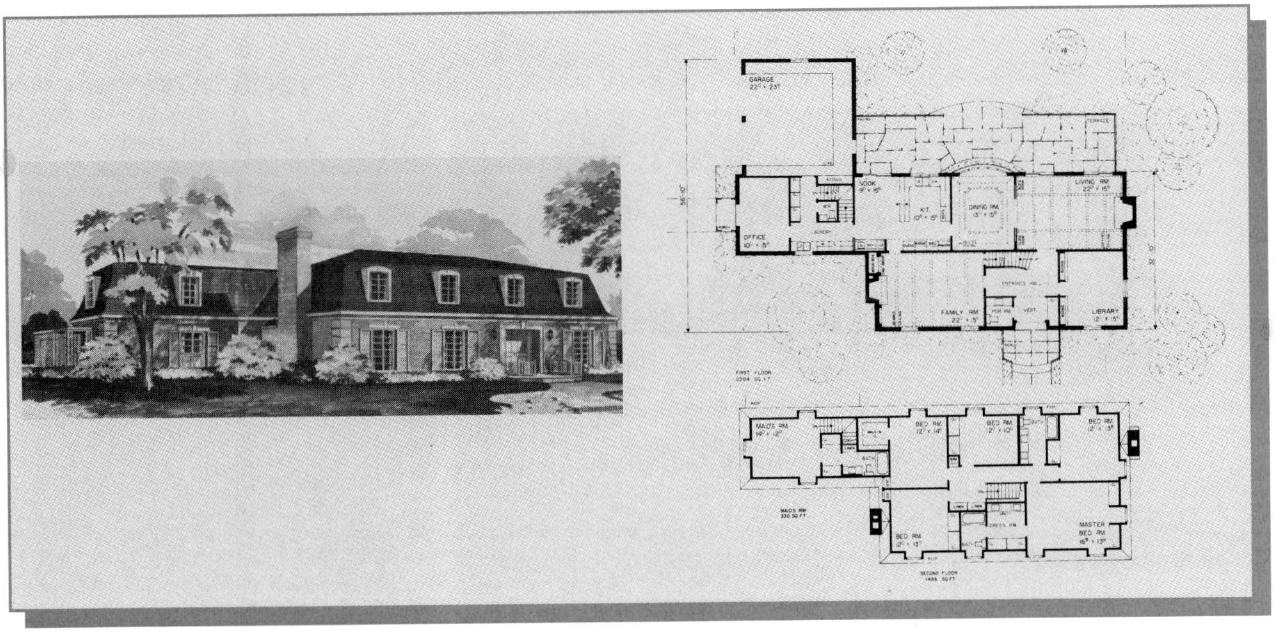

Figure 24-14
An example of contemporary French mansard architecture *(Courtesy of Home Planners, Inc.)*

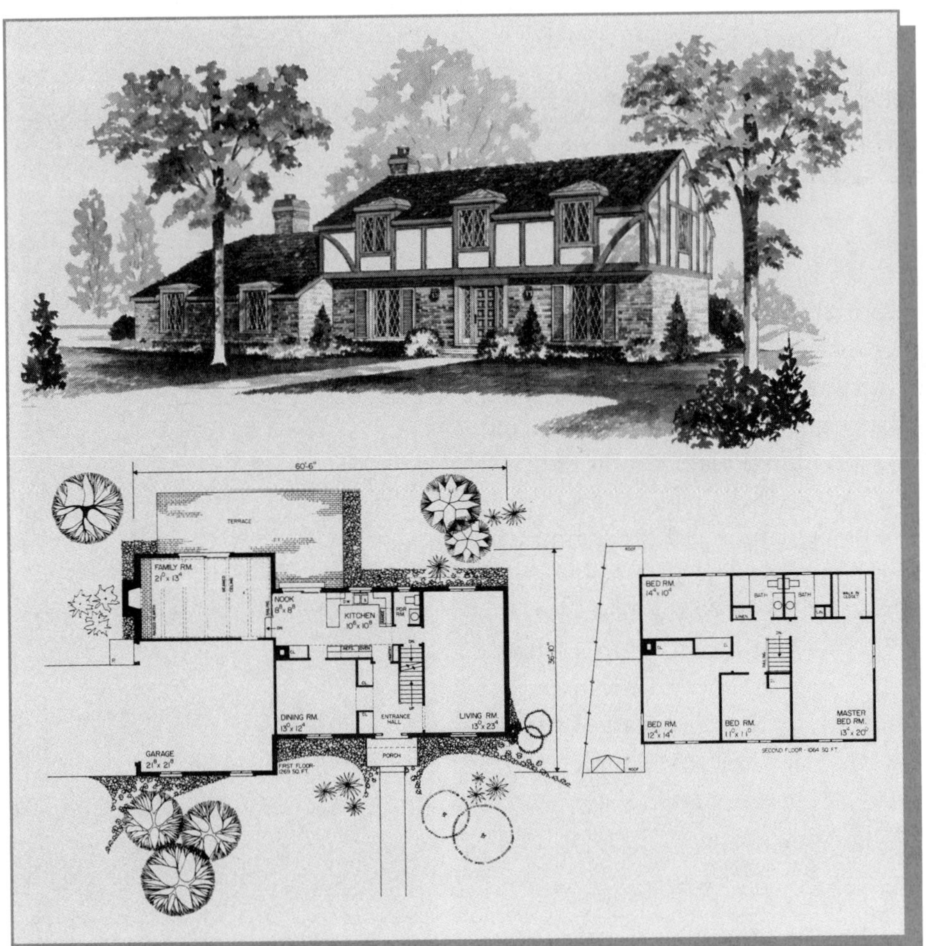

Figure 24-15
An example of contemporary English Tudor architecture *(Courtesy of Home Planners, Inc.)*

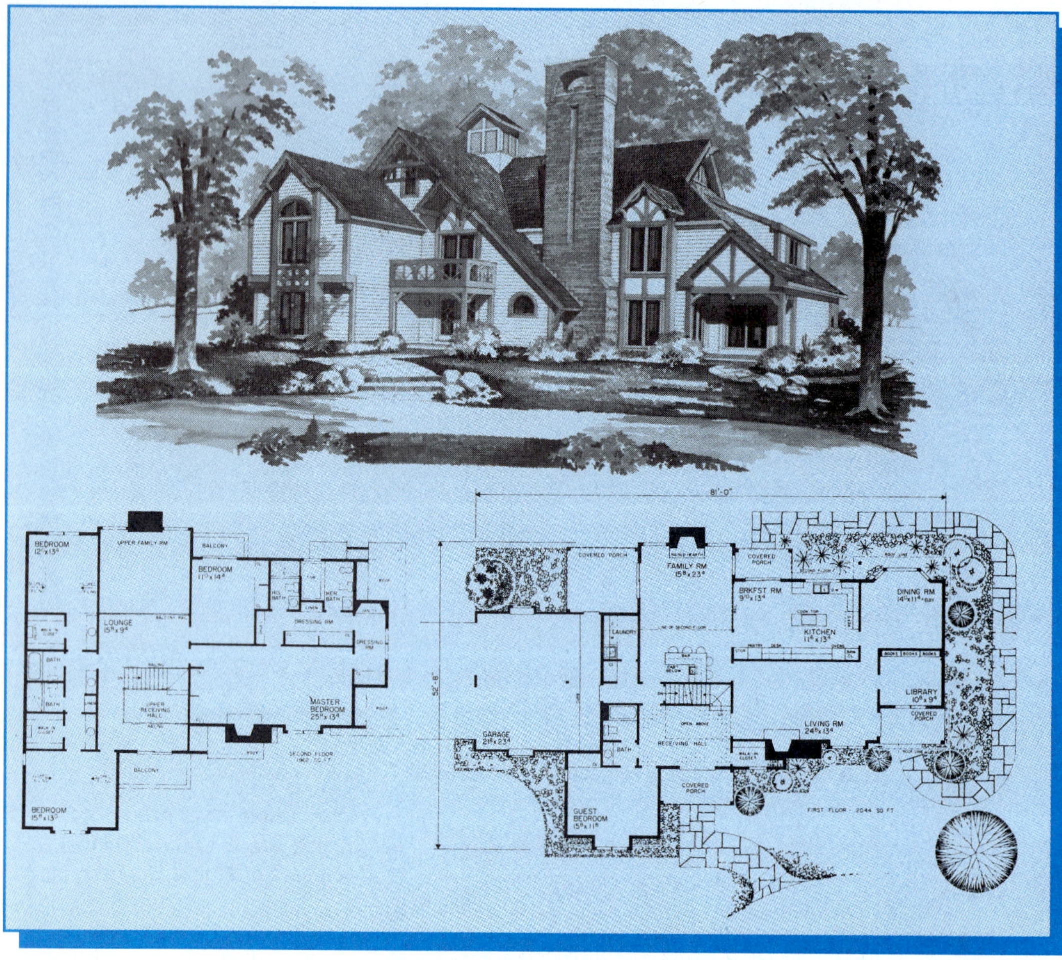

Figure 24-16 An example of mixed architectural styles called *eclectic* *(Courtesy of Home Planners, Inc.)*

Contemporary Styles

As stronger, lighter, and more portable building materials were developed, and as construction methods became more sophisticated, architectural styles were affected. Designs became more functional, broader distances were spanned, larger glass areas manufactured and supported, greater heights reached, and difficult site conditions mastered.

These developments led to the use of larger and more unusual shapes. Circles, spheres, pyramids, and triangles became viable building shapes, breaking the rectangular restrictions of the past. For example, the A-frame structure **(Fig. 24-17)** is based on an ancient building type. However, not until the development of laminated arches and beams did the design become practical for large structures. Concurrently, contemporary styles also became simpler **(Fig. 24-18)**.

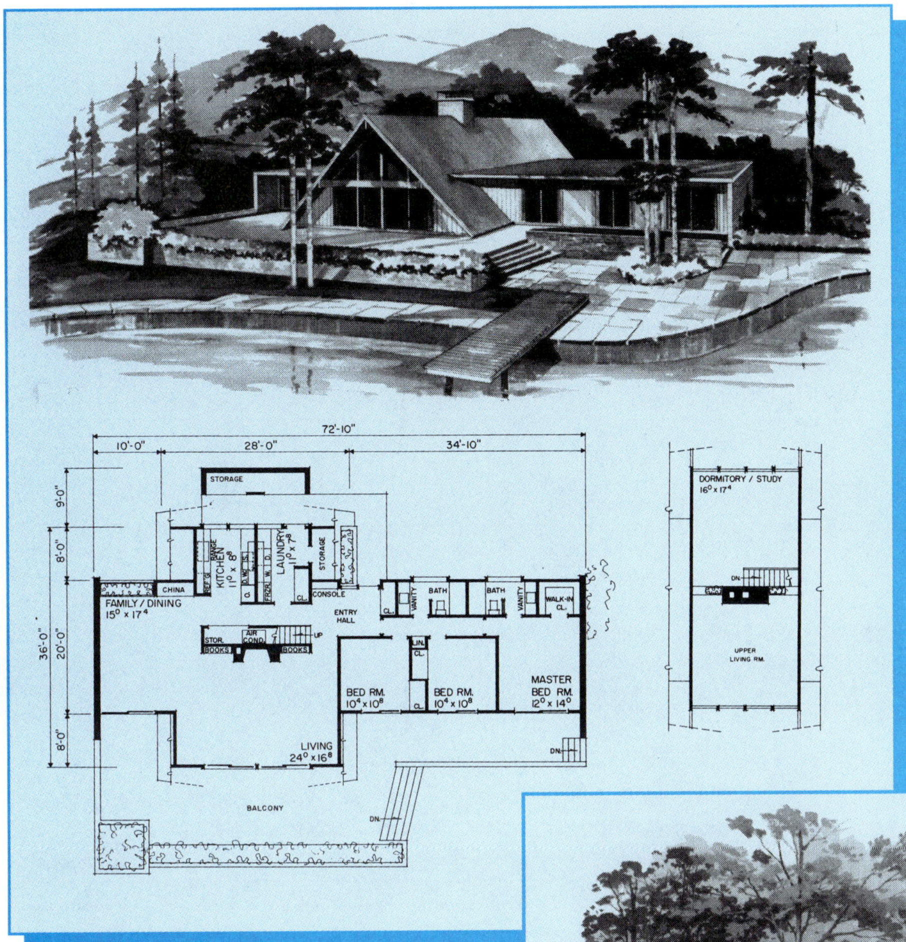

Figure 24-17
An example of an A-frame architectural style *(Courtesy of Home Planners, Inc.)*

Figure 24-18
An example of contemporary style architecture *(Courtesy of Home Planners, Inc.)*

Design Considerations

In the design process, many architectural factors must be considered and compared to the personal goals and requirements of the occupants. The following considerations provide guidelines for the formulation and development of a basic architectural design.

Area Design

For design purposes, each structure is divided into functional areas. In an educational facility, these include classroom areas, administration area, athletic activity areas, technology labs, science labs, food service areas, assembly areas, lavatory facilities, and traffic areas. In an office or industrial building, the areas are quite different.

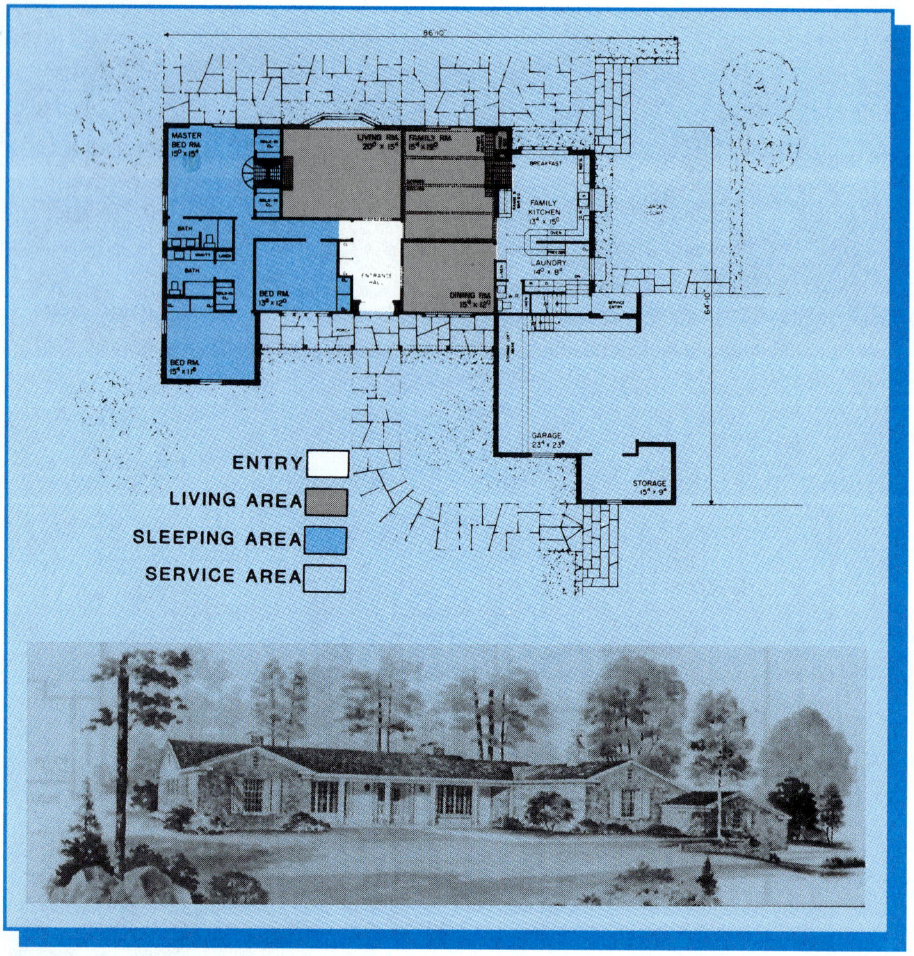

Figure 24-19
Architectural designers must carefully plan the relationship of the entry, living, sleeping, and service areas. *(Courtesy of Home Planners, Inc.)*

In a residence, the design areas include traffic (entry) areas, living area, sleeping area, and service area. The traffic area includes all entrances, foyers, and halls. The living area includes the living room, family room or recreation room, den, study, lavatory, and dining room. Sleeping areas include bedrooms, dressing rooms, baths, and other rooms requiring a quiet surrounding. The service area includes kitchen, utility room, work areas, bath, and garage. A typical breakdown of these areas in a residence is shown in **Figure 24-19**.

Room Size

Room sizes are determined by the function of each room and the total amount of space available. Furniture, appliances, fixtures, equipment, storage, door and window areas, and traffic requirements must be considered in selecting room sizes. **Figure 24-20** shows typical room sizes for small, average, and large rooms for a residence.

Figure 24-20
A general guide to room sizes for primary residences

A GENERAL GUIDE TO ROOM SIZES			
BASIC ROOMS	TYPICAL ROOM SIZES (FEET)		
	SMALL	AVERAGE	LARGE
LIVING ROOM	12 x 18	16 x 20	22 x 28
DINING ROOM	10 x 12	12 x 15	15 x 18
KITCHEN	5 x 10	10 x 16	12 x 20
UTILITY ROOM	6 x 7	6 x 10	8 x 12
BEDROOM	10 x 10	12 x 12	14 x 16
BATHROOM	5 x 7	7 x 9	9 x 12
HALLS	3′ WIDE	3′–6″ WIDE	3′–9″ WIDE
GARAGE	10 x 20	20 x 20	22 x 25
STORAGE WALL	6″ DEEP	12″ DEEP	18″ DEEP
DEN	8 x 10	10 x 12	12 x 16
FAMILY ROOM	12 x 15	15 x 18	15 x 22
WARDROBE CLOSET	2 x 4	2 x 8	2 x 15
ONE ROD WALK-IN CLOSET	4 x 3	4 x 6	4 x 8
TWO ROD WALK-IN CLOSET	6 x 4	6 x 6	6 x 8
PORCH	6 x 8	8 x 12	12 x 20
ENTRY	6 x 6	8 x 10	8 x 15

Traffic Flow

Traffic patterns are critical to functional residential design. External traffic must be adequately planned around the main (front) entry, service entry, and special purpose entries **(Fig. 24-21)**. External traffic also includes the effective design of driveways and walkways. Internal traffic planning includes determining the size and location of foyers, halls, stairs, and circulation through each room.

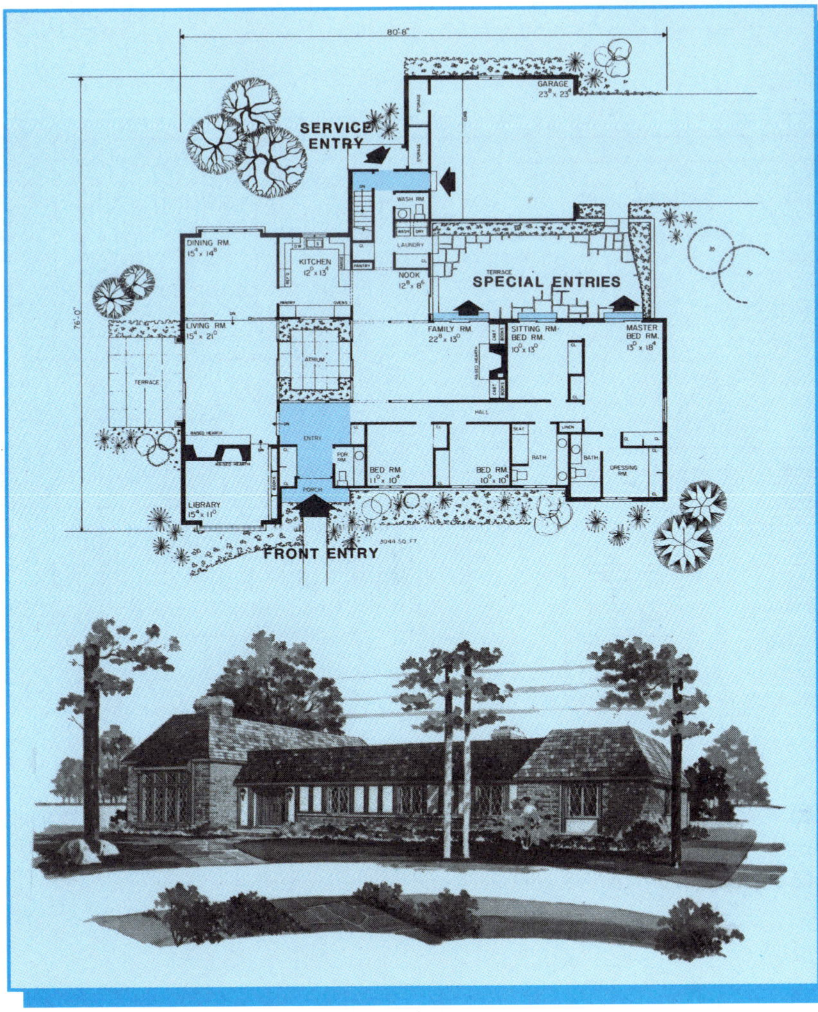

Figure 24-21
Well-placed entrances and exits help make an efficient traffic pattern. *(Courtesy of Home Planners, Inc.)*

Lighting

Architectural lighting is grouped into four types—natural, general, specific (task), and decorative. Natural lighting is all lighting that our sun provides. General lighting provides diffused lighting throughout an entire room. Local (task) lighting provides light for a specific task such as reading, drawing, or cooking. Decorative lighting helps create moods by highlighting architectural features. The lighting in each room must include adequate outlets and switches at each entrance **(Fig. 24-22)**.

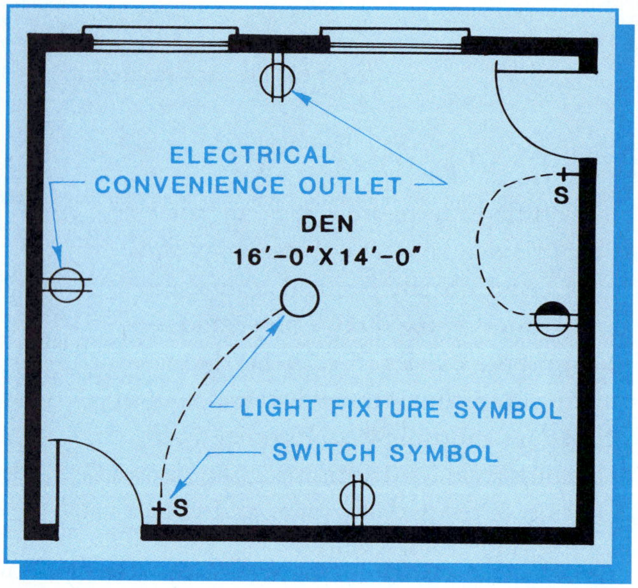

Figure 24-22
Light switches must be accessible at the opening of each door and conform to local building codes for their placement.

Storage

Storage areas are an integral part of every functional design. Adequate storage is planned into a residence through the design of closets, built-in storage facilities, and furniture and general storage areas, such as attics, basements, and garages. **Figure 24-23** shows standard sizes for residential storage facilities.

Doors

Doors are of three basic types—swinging, folding, and sliding. Swinging doors are the most popular since they move with the person opening the door. To conserve wall space, swinging doors should swing into rooms toward an adjacent wall, with the hinge as close to the wall as possible. All types of doors are available in a wide range of panels, materials, and styles **(Fig. 24-24)**. **Figure 24-24** shows the elevation and floor plan symbols for common door types.

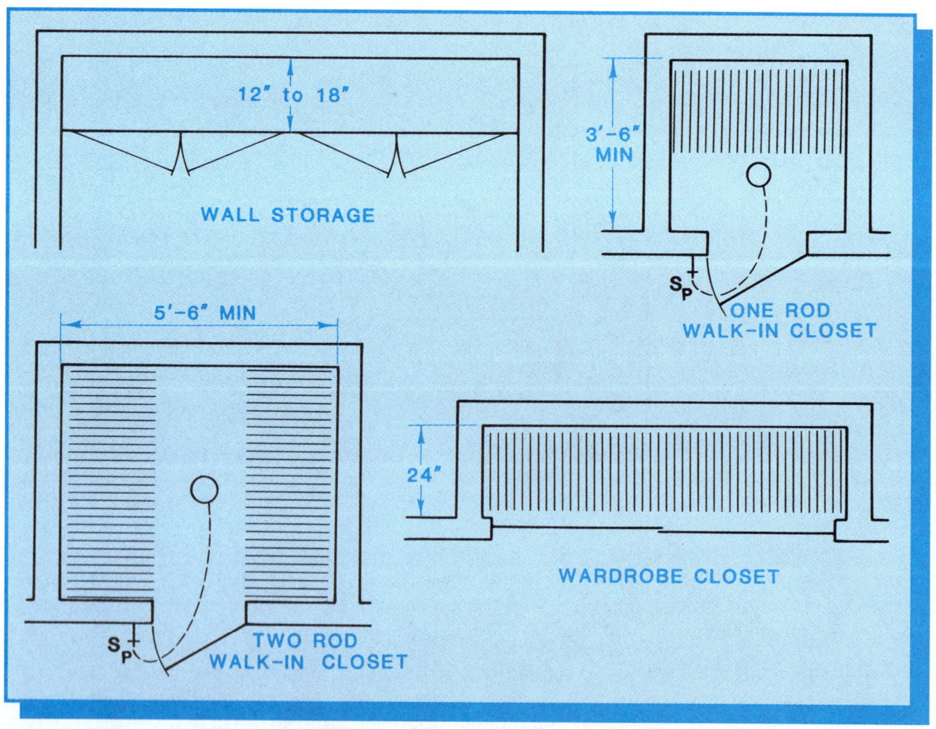

Figure 24-23
Storage and closet areas must be adequate and easily accessible.

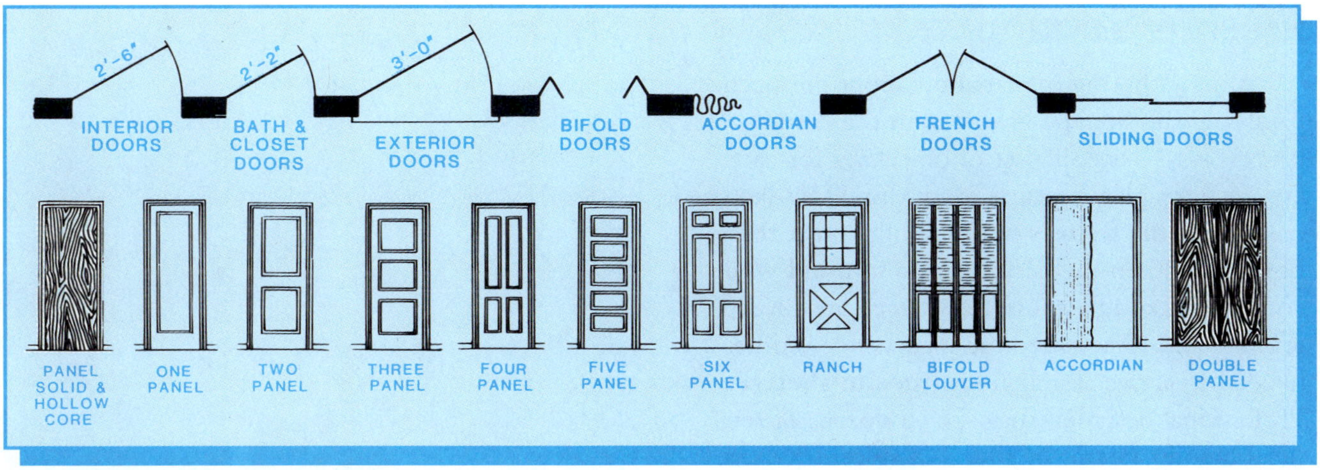

Figure 24-24
Door styles and their symbols

Windows

Windows admit light and air, and provide a view. The effectiveness of a window depends on its size, location, compass orientation, roof overhang, and the amount of exterior and interior reflected light. **Figure 24-25** shows elevation and floor plan symbols used to represent common window types.

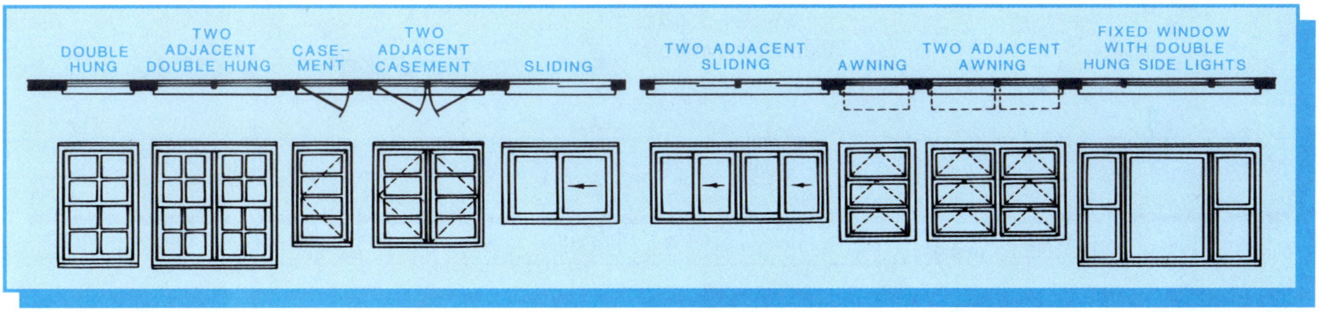

Figure 24-25
Window styles and their symbols

Heating, Ventilation, and Air Conditioning (HVAC)

Comfort levels must be designed into each structure. This involves the maintenance of a desired temperature, humidity, and the movement of clean air. To accomplish this, **heating, ventilation, and air conditioning (HVAC)**, and humidifying systems are used in combination with insulation and ventilation control through vents and windows.

Design Sequence

After gathering information about the occupants and the site, the next step in the design process is to develop a set of objectives for each area. A floor plan for each room should be sketched containing the features that will help meet these objectives. Keep the personal data in mind while completing **rough sketches** of each area. Keep refining and resketching the plan as many times as needed to arrive at a final **composite sketch.** Professional designers may sketch dozens of revisions before arriving at a final composite sketch.

The architectural symbols shown in **Figures 24-26** through **24-32** should be used in sketching the final plan, and in the development of the final working drawings.

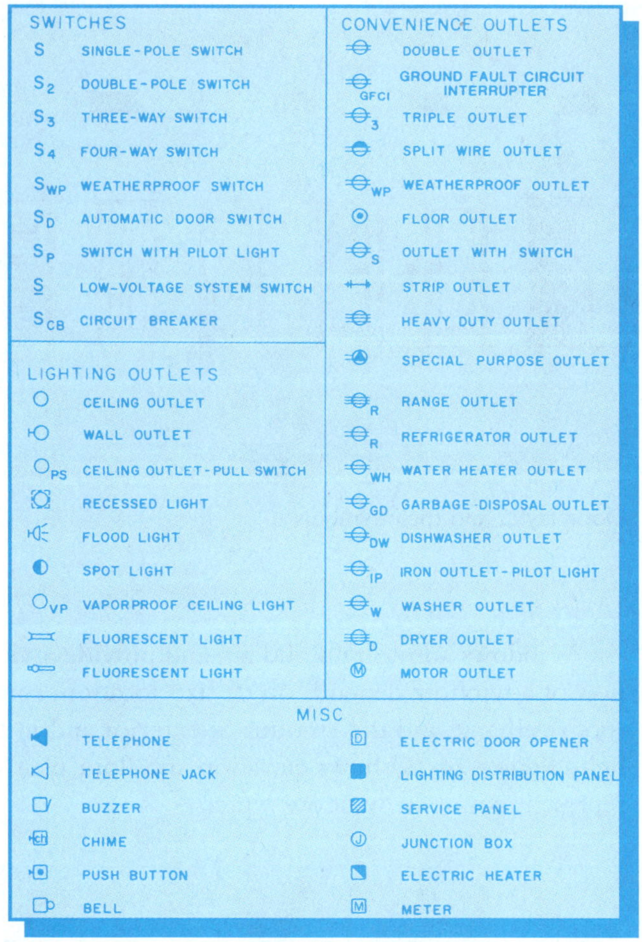

Figure 24-26
Electrical symbols

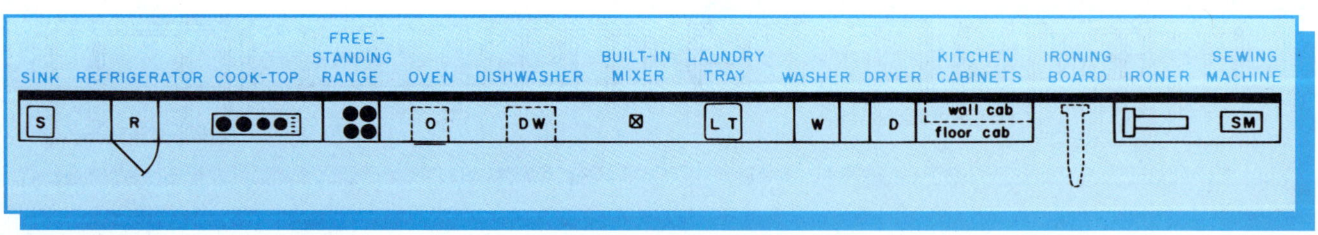

Figure 24-27
Kitchen and utility room symbols

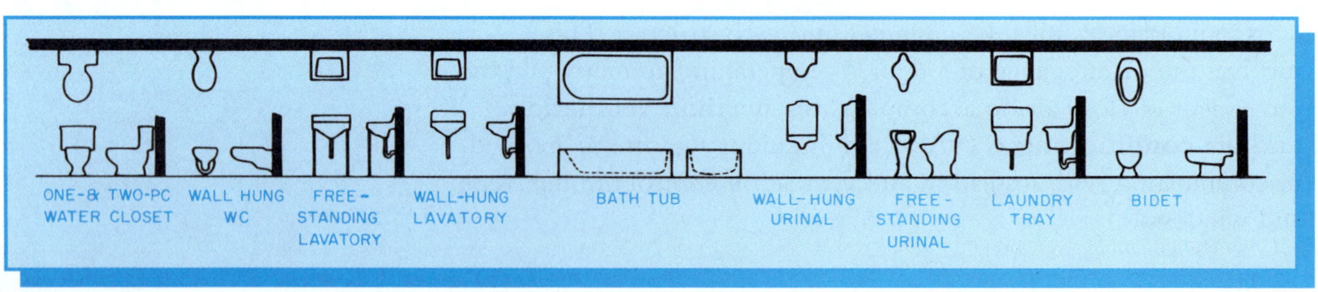

Figure 24-28
Bathroom fixture floor plan and elevation symbols

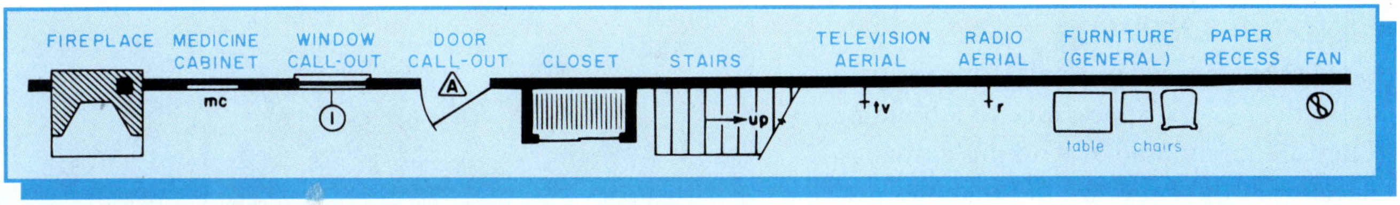

Figure 24-29
Built-in component symbols

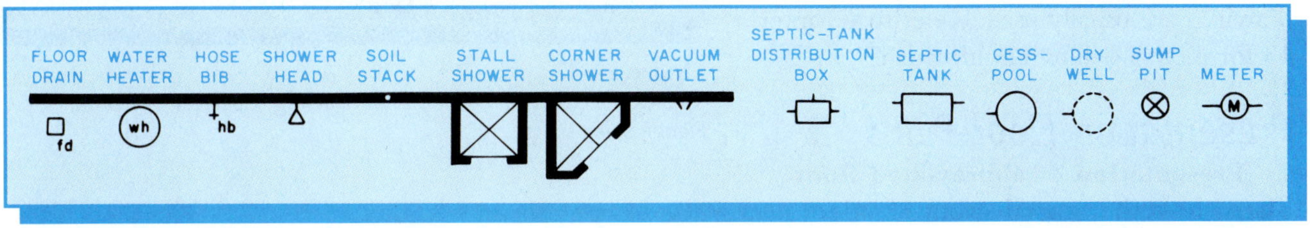

Figure 24-30
Wastewater facility symbols

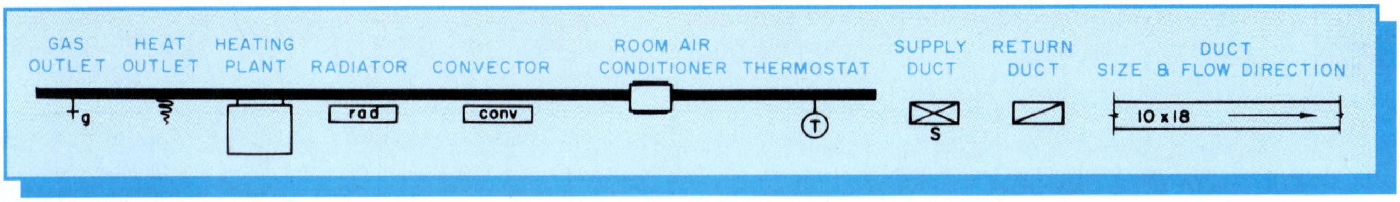

Figure 24-31
HVAC (heating, ventilation, and air conditioning) symbols

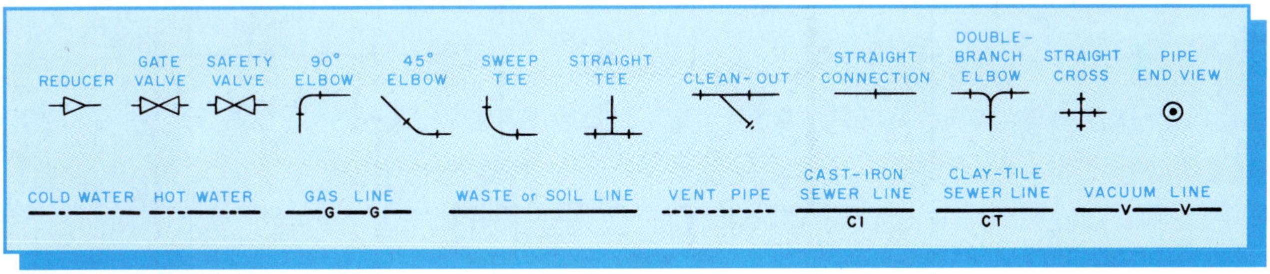

Figure 24-32
Plumbing line symbols

Working Drawings

The final composite sketch is used as a guide in preparing a set of working drawings. A full set of architectural working drawings includes pictorials, floor plans, elevations, and construction detail drawings. A set of working drawings also includes supporting documents such as schedules, and specifications, and may include budgets and contracts.

Pictorial Drawing

Architectural pictorial drawings show the general appearance of a building. They are rarely used as working drawings. Because of the large size of buildings, perspective drawings **(Fig. 24-33)** are used for most architectural pictorials. Where structural details are shown pictorially, isometric drawings are usually used. Refer to Chapter 15 for pictorial drawing information.

Figure 24-33
Two-point exterior perspective drawing *(Courtesy of Home Planners, Inc.)*

Presentation Floor Plans

Presentation or **abbreviated floor plans** show the general layout of a floor plan. They do not contain extensive detailing. These plans are not used as working drawings but are prepared primarily for sales and general reference purposes. **Figure 24-34** shows a typical presentation floor plan. Note the lack of detail dimensions and the use of abbreviated symbols.

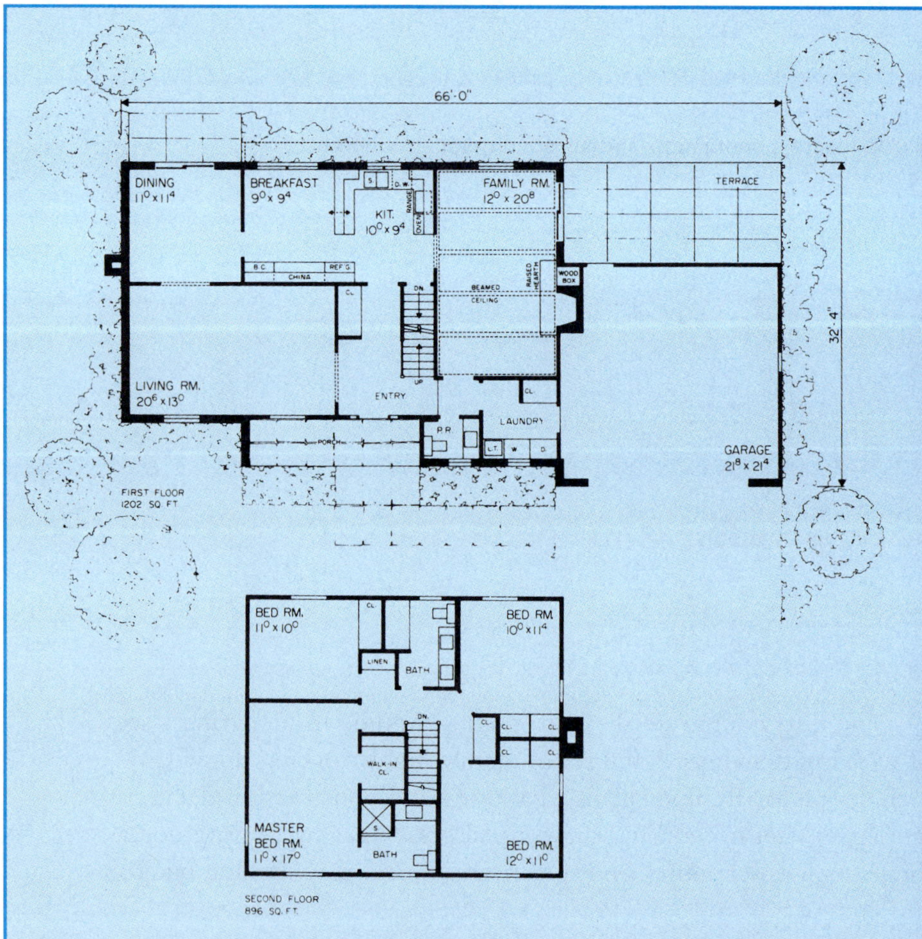

Figure 24-34
Presentation floor plan *(Courtesy of Home Planners, Inc.)*

Plot Plans, Survey Plans, and Site Plans

Before finalizing a floor plan, an outline of the plan must be located on a layout of the lot for which it was designed. A plan of a lot that includes a dimensioned building outline, setback dimensions, driveways, walks, and street locations is a **plot plan (Fig. 24-35)**.

When site contours and utility line locations are added, the plan is known as a **survey plan.** A plan that includes all of these features plus landscaping details is known as a **site plan**.

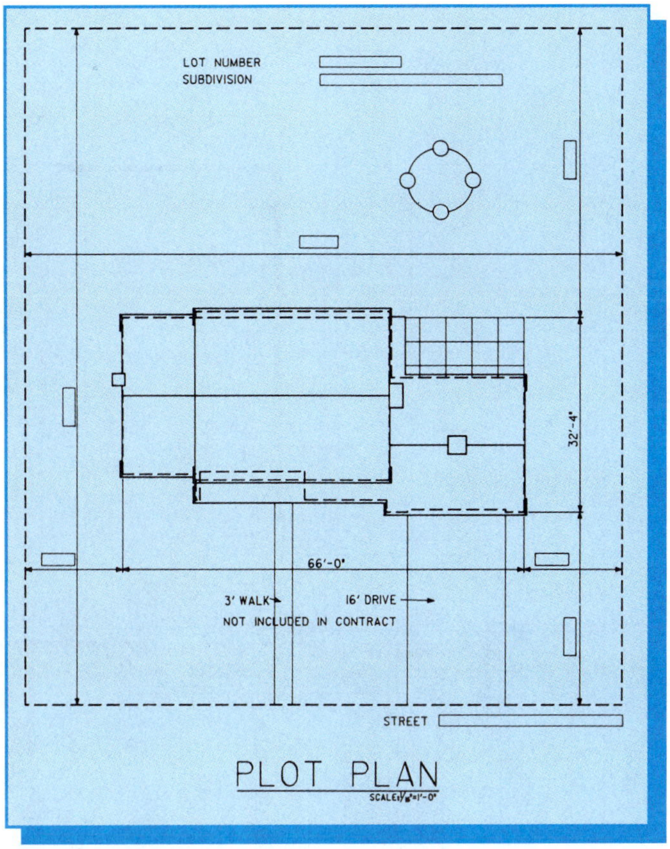

Figure 24-35
Plot plan *(Courtesy of Home Planners, Inc.)*

Floor Plan

Floor plans are horizontal sections of a building projected from a cutting plane line placed approximately 4 ft. above the floor level. Working drawing floor plans contain symbols (see **Figs. 24-26** through **24-32)** that show the top view and position of fixtures, stairs, walls, windows, doors, electrical outlets and switches, plumbing, and HVAC equipment. Floor plans are usually drawn at a scale of $^{1}/_{4}$" = 1'-0", or $^{1}/_{8}$" = 1'-0", depending on the size of the building.

In drawing floor plans such as the one in **Figure 24-36**, first draw the outline of all interior and exterior walls. Next, add the door, window, stair, and fireplace symbols. Finish the plan by adding the fixtures, dimensions, and labeling.

Second-floor plans are drawn on tracing paper placed over the first-floor plan. The position of outside walls, chimneys, and stairs must align with their first-floor counterparts. **Figure 24-37** shows a second-floor plan that can be used as a working drawing. Successive-level floor plans and foundation plans are prepared in the same manner.

Foundation plans are floor plans that show the outline of footings and sections through foundation walls. In addition, a foundation plan shows the location and size of all structural support members such as columns and beams.

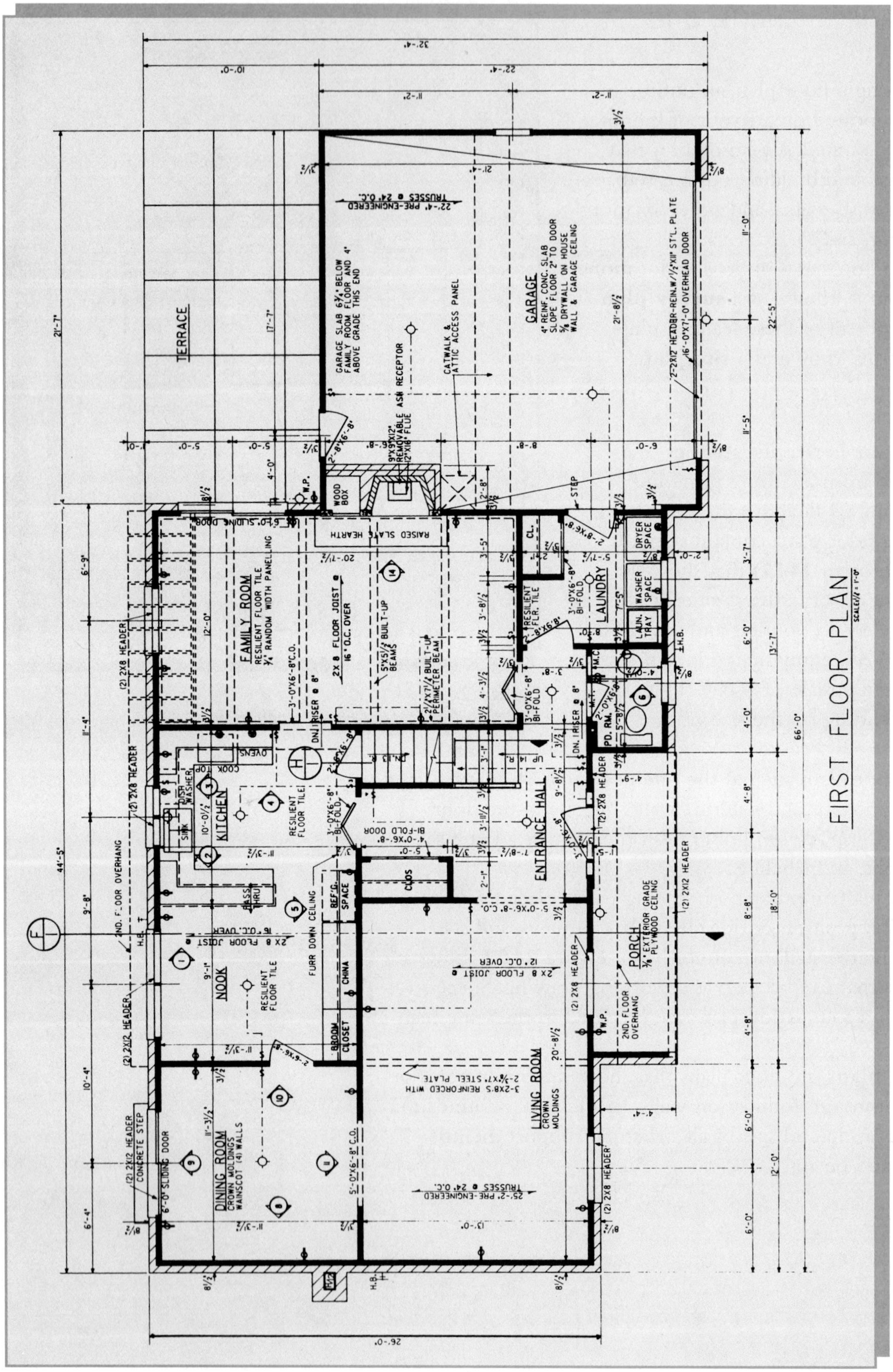

Figure 24-36
Floor plan (*Courtesy of Home Planners, Inc.*)

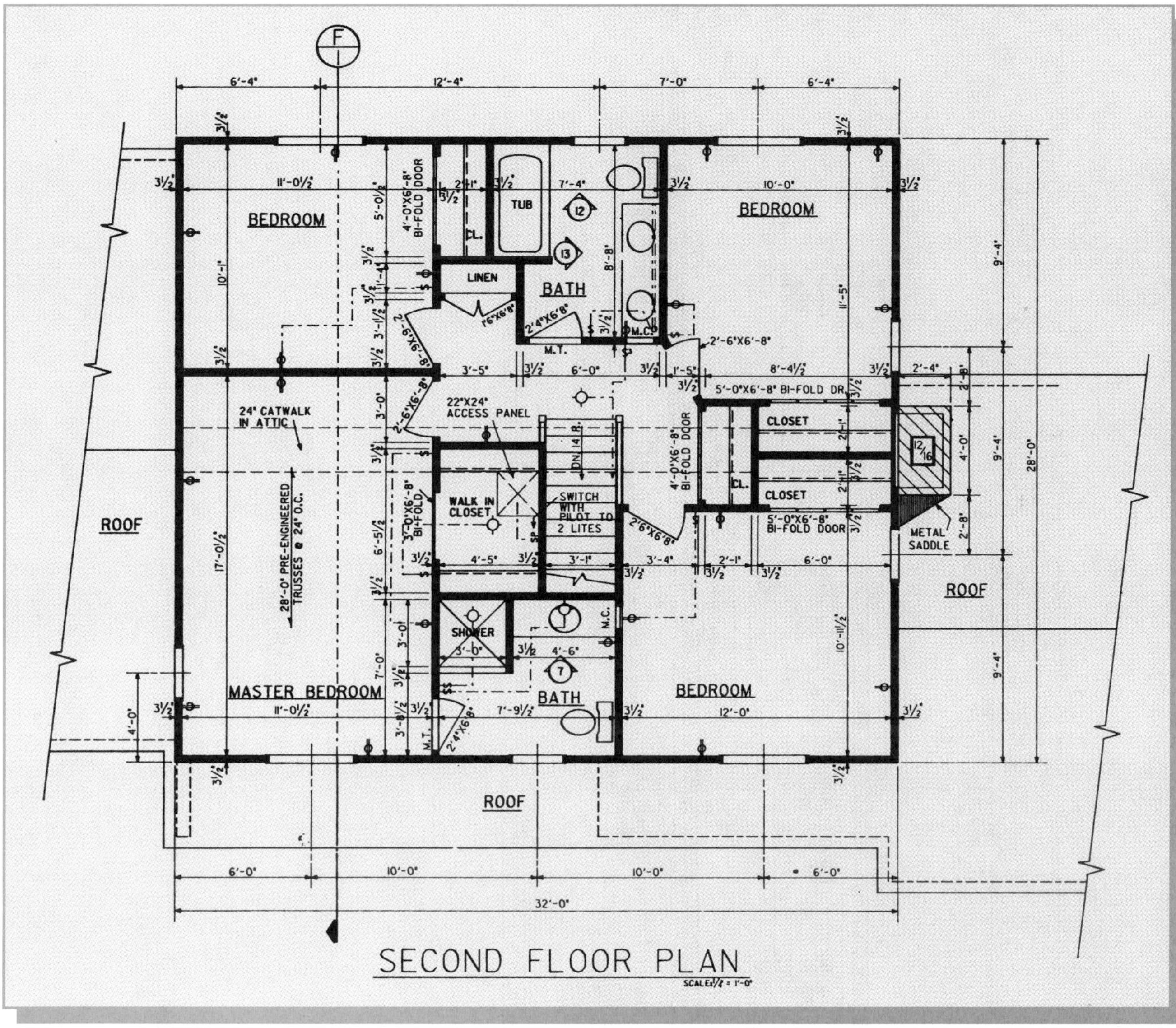

Figure 24-37
Second floor plan *(Courtesy of Home Planners, Inc.)*

Elevations

The orthographic projection of the front, rear, and side views of a building are known as **elevations.** A rectangular building has four sides, or elevations—front, rear, left, and right. These are labeled by compass direction—north, south, east, and west. Horizontal distances on each elevation are projected from related points on each exterior floor plan wall. This is similar to the way orthographic front, rear, and side views are projected from a top view of an object.

Vertical distances are measured and drawn from an established **datum plane,** such as a grade line or first-floor line. **Figures 24-38** through **24-41** show the elevation drawings projected from the floor plans shown in **Figures 24-36** and **24-37**. Note that elevation symbols resemble the appearance of building materials and architectural features more closely than do floor-plan symbols.

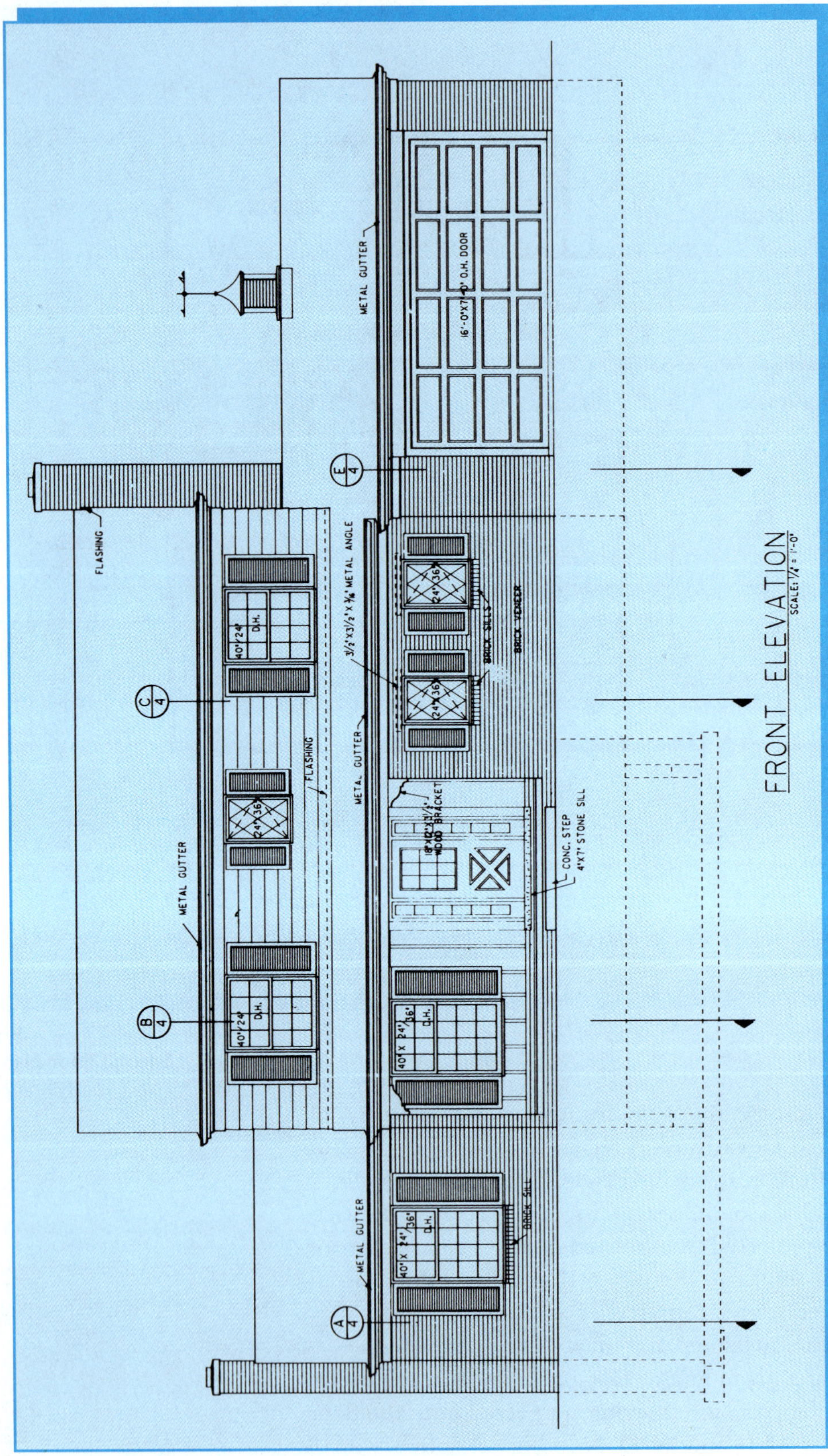

Figure 24-38
Front elevation *(Courtesy of Home Planners, Inc.)*

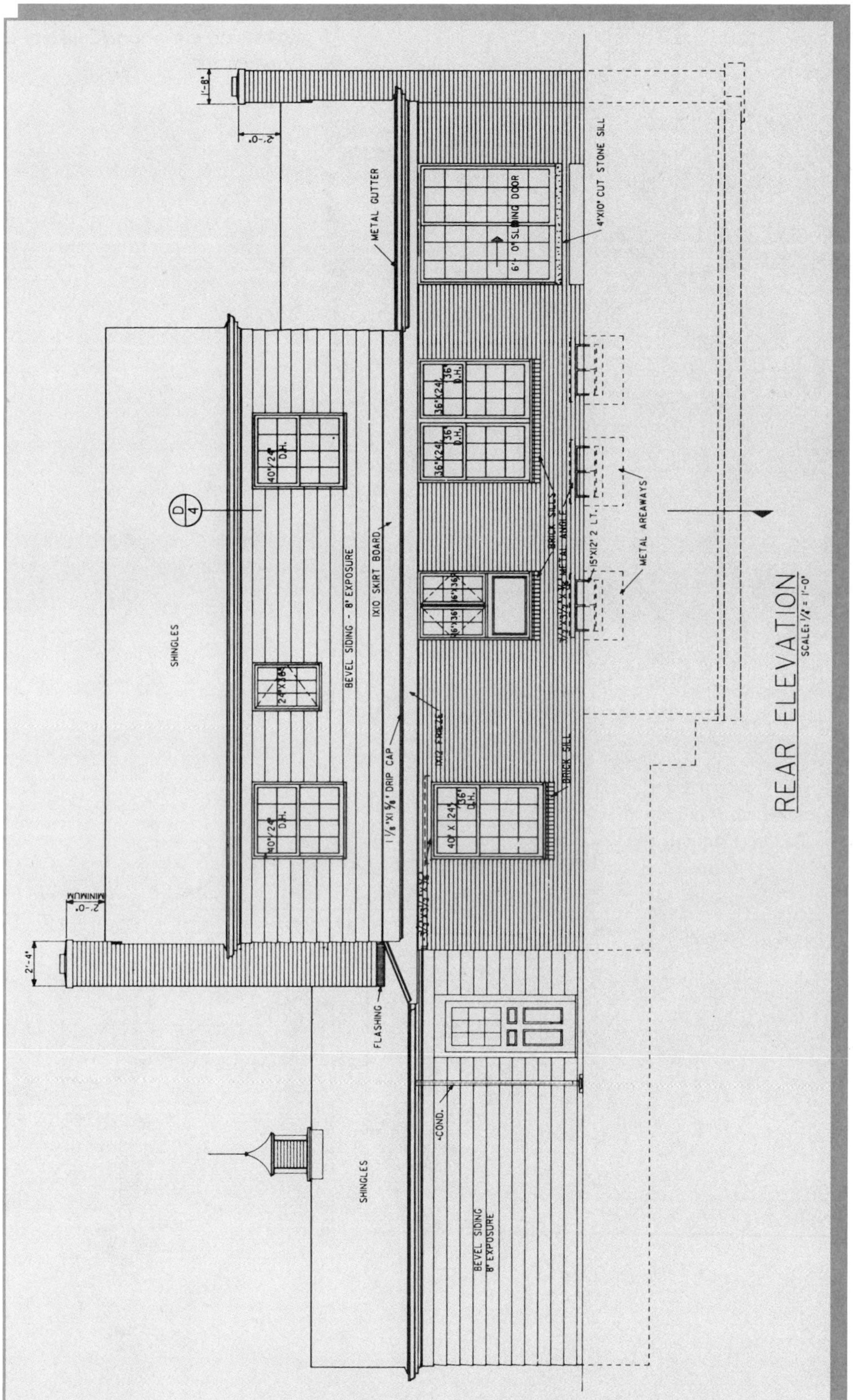

Figure 24-39
Rear elevation *(Courtesy of Home Planners, Inc.)*

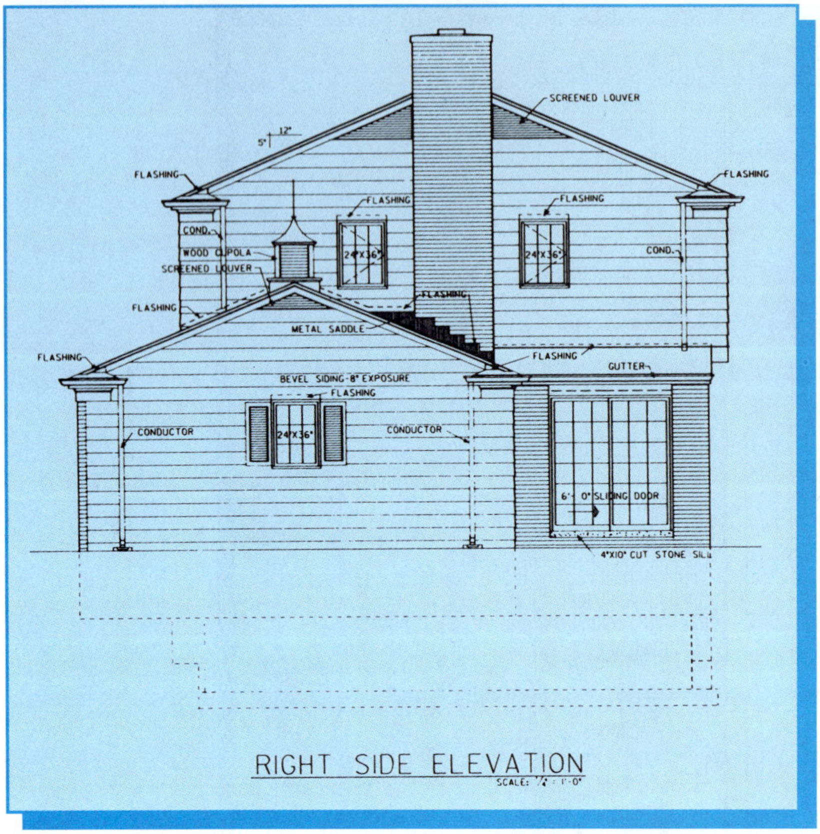

Figure 24-40
Right-side elevation *(Courtesy of Home Planners, Inc.)*

Figure 24-41
Left-side elevation *(Courtesy of Home Planners, Inc.)*

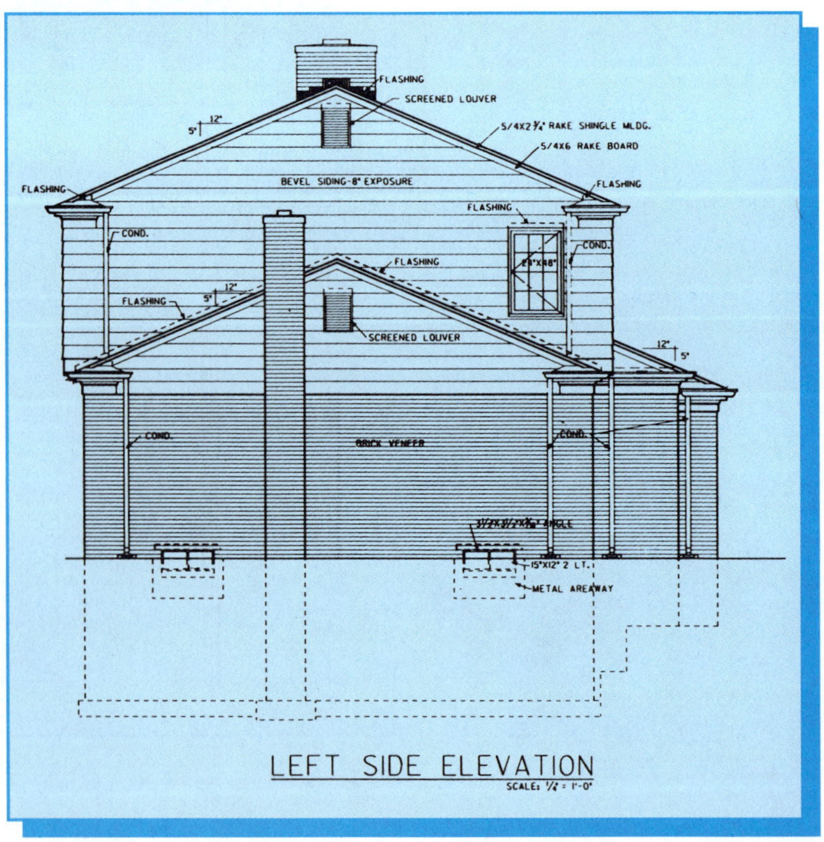

Interior Elevations

An orthographic view of an interior wall is an **interior elevation.** Interior elevations show the materials, windows, doors, openings, fixtures, and built-in details to be constructed or installed on a wall. Interior elevations **(Figs. 24-42** and **24-43)** are identified by room title, compass direction, or detail reference number. For example, an elevation drawing could be labeled kitchen or west wall elevation. The drawing also could be identified with a number or letter corresponding to an index number and arrow on the floor plan. Interior elevations are most commonly prepared for bath and kitchen walls, and walls that contain intricate details such as built-in bookcases, paneling, music centers, or fireplace walls **(Fig. 24-44)**. If interior elevations are drawn to the same scale as the floor plan, horizontal distances can be projected from the floor-plan wall.

Sectional Drawings

Architectural sectional drawings include detail sections and full sections. Detail sections are either **plan (horizontal) sections** or **elevation (vertical) sections.** Detail sections are drawn to a large scale and show the details of a small area of a structure. Detail sections **(Figs. 24-45** through **24-47)** are designed to show the exact size and relationship of construction members and building components.

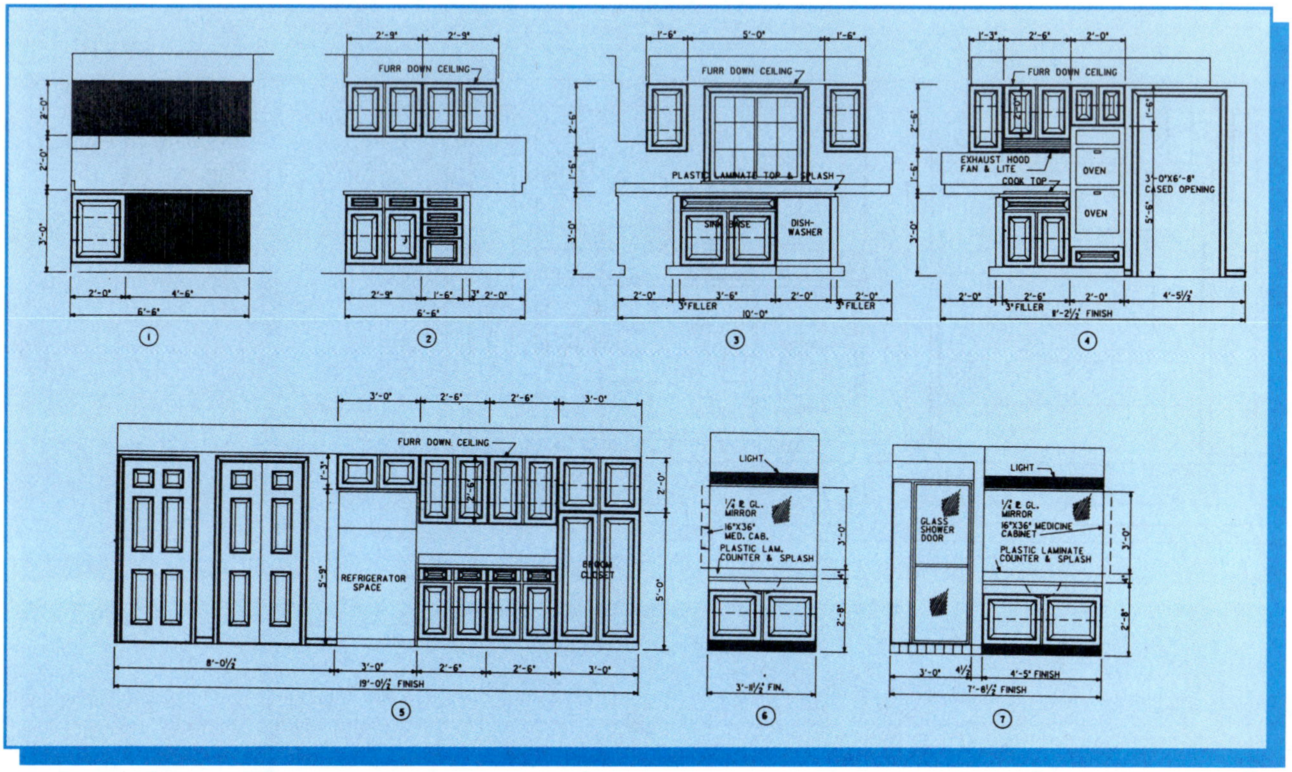

Figure 24-42
Interior elevations *(Courtesy of Home Planners, Inc.)*

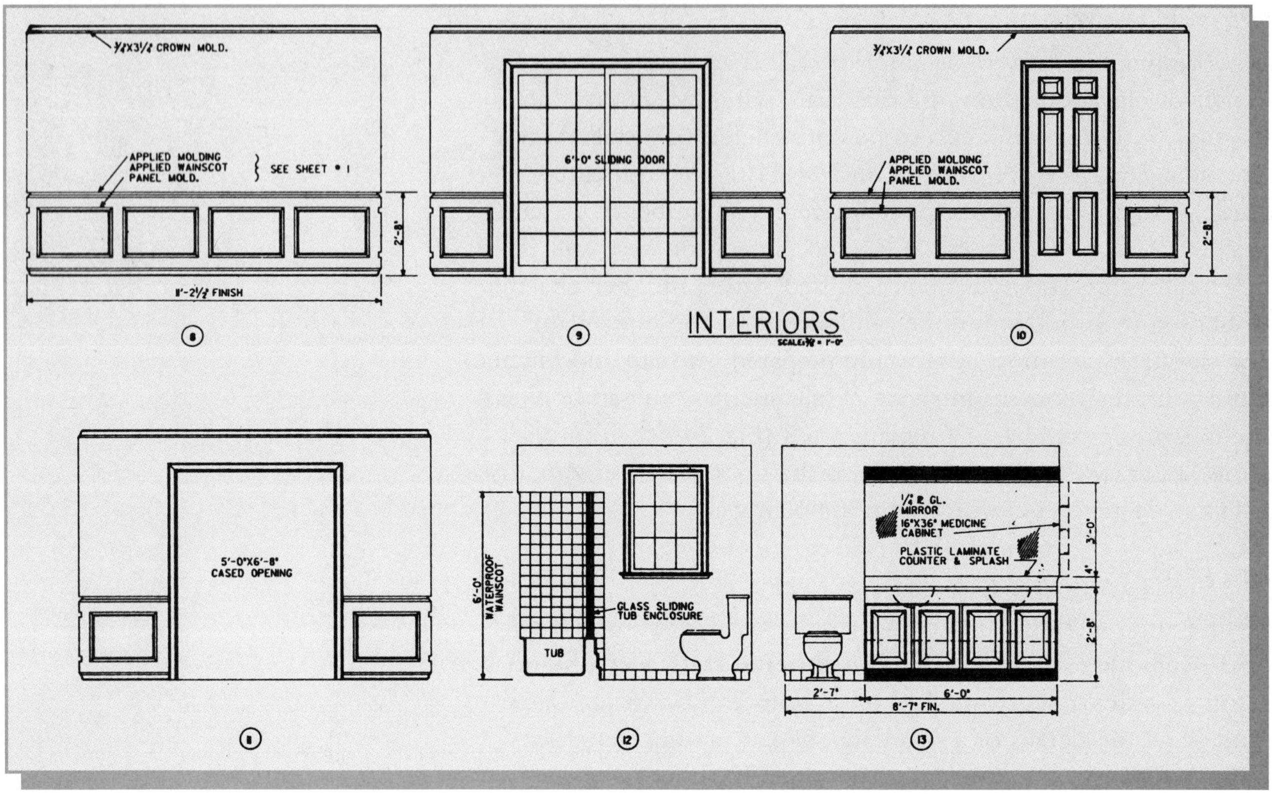

Figure 24-43
Interior elevations *(Courtesy of Home Planners, Inc.)*

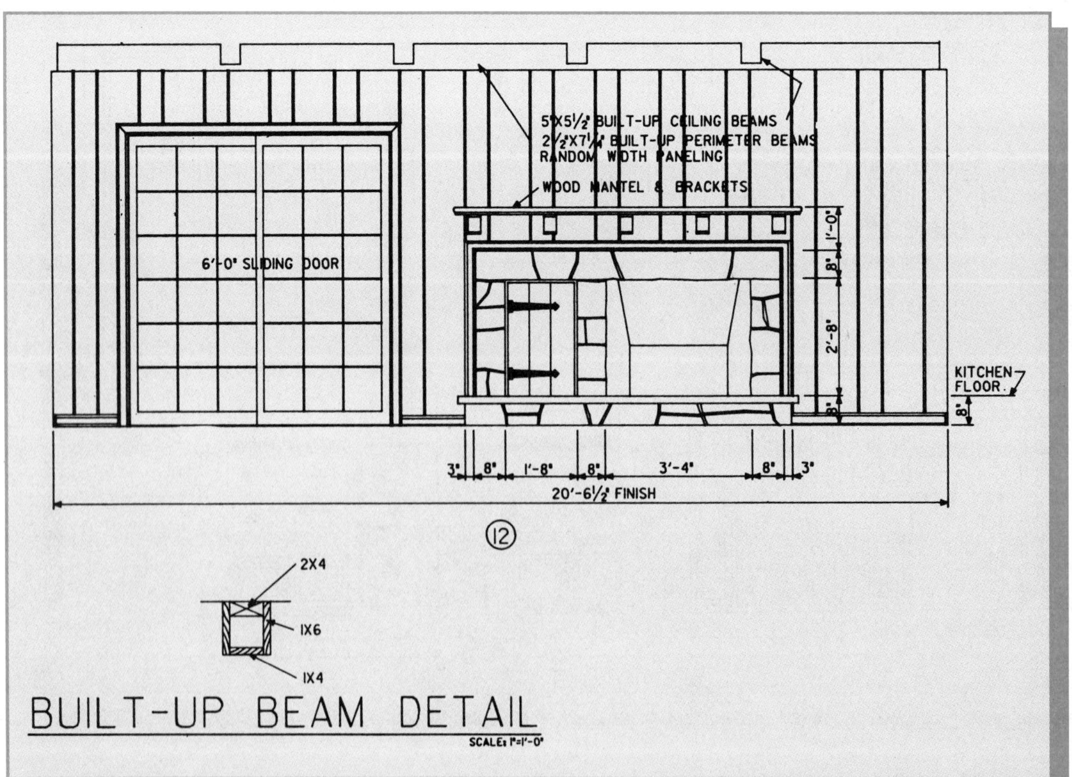

Figure 24-44
Interior elevation of the fireplace wall *(Courtesy of Home Planners, Inc.)*

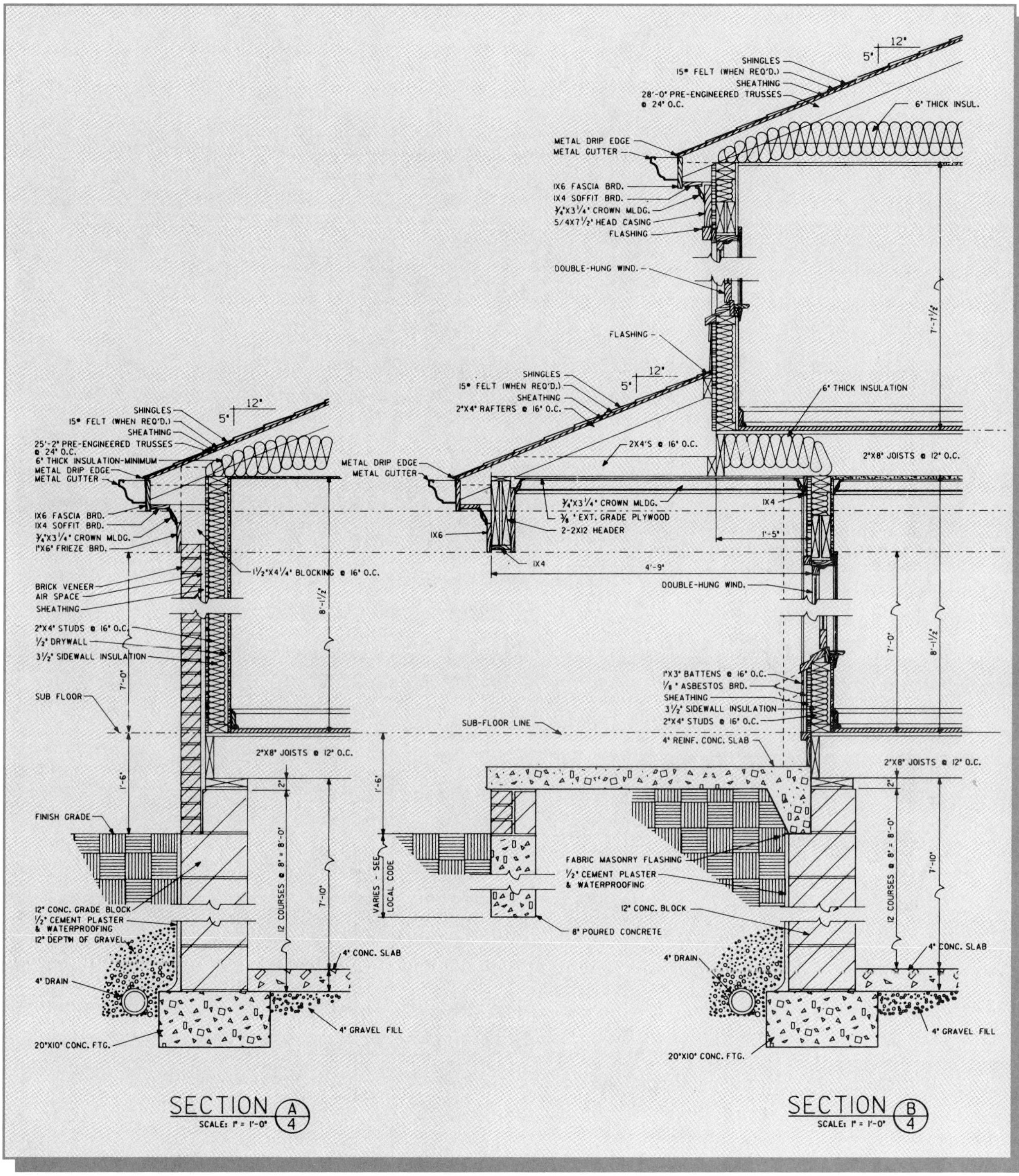

Figure 24-45
Detailed house sections A and B *(Courtesy of Home Planners, Inc.)*

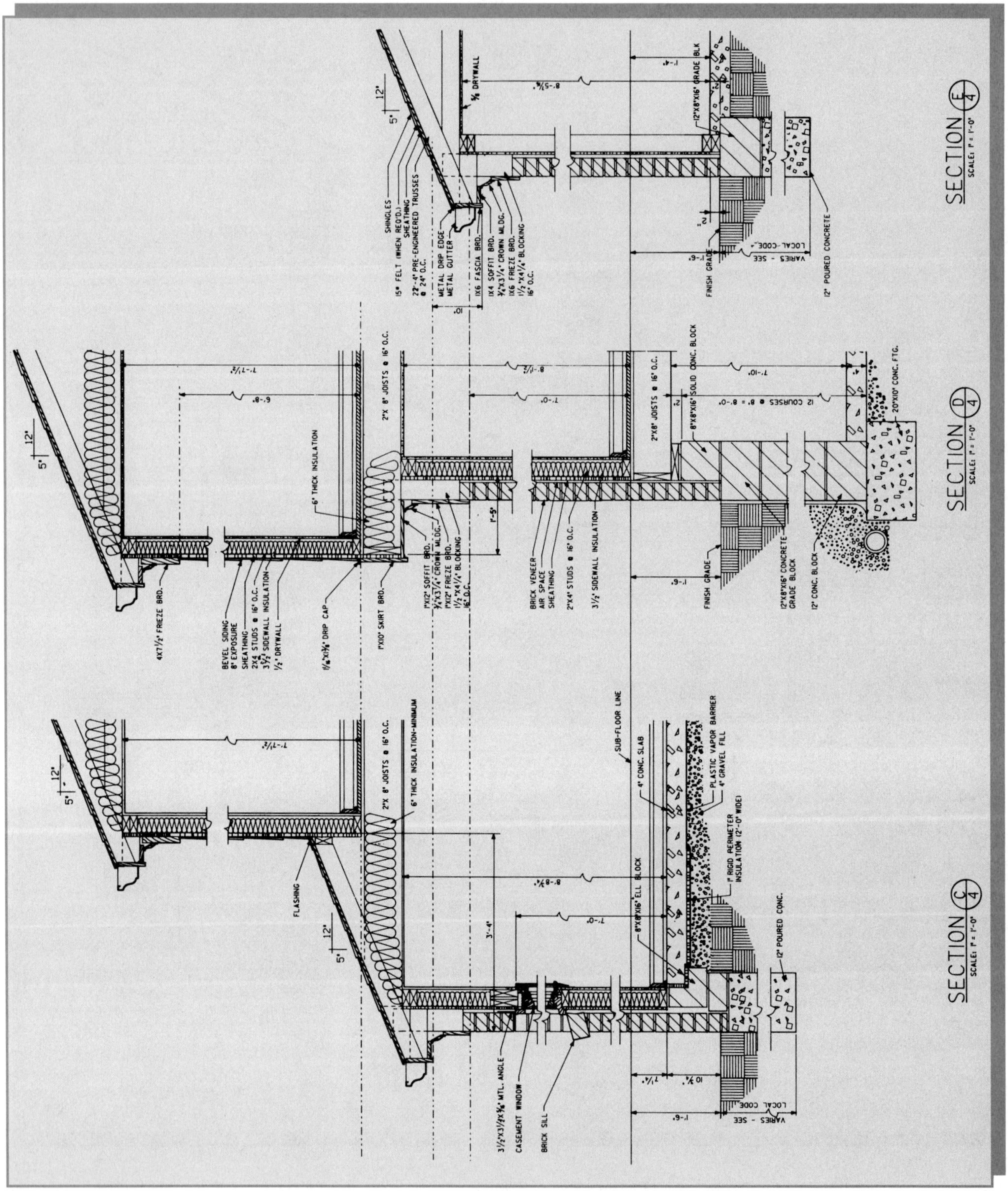

Figure 24-46
Construction details C, D, and E *(Courtesy of Home Planners, Inc.)*

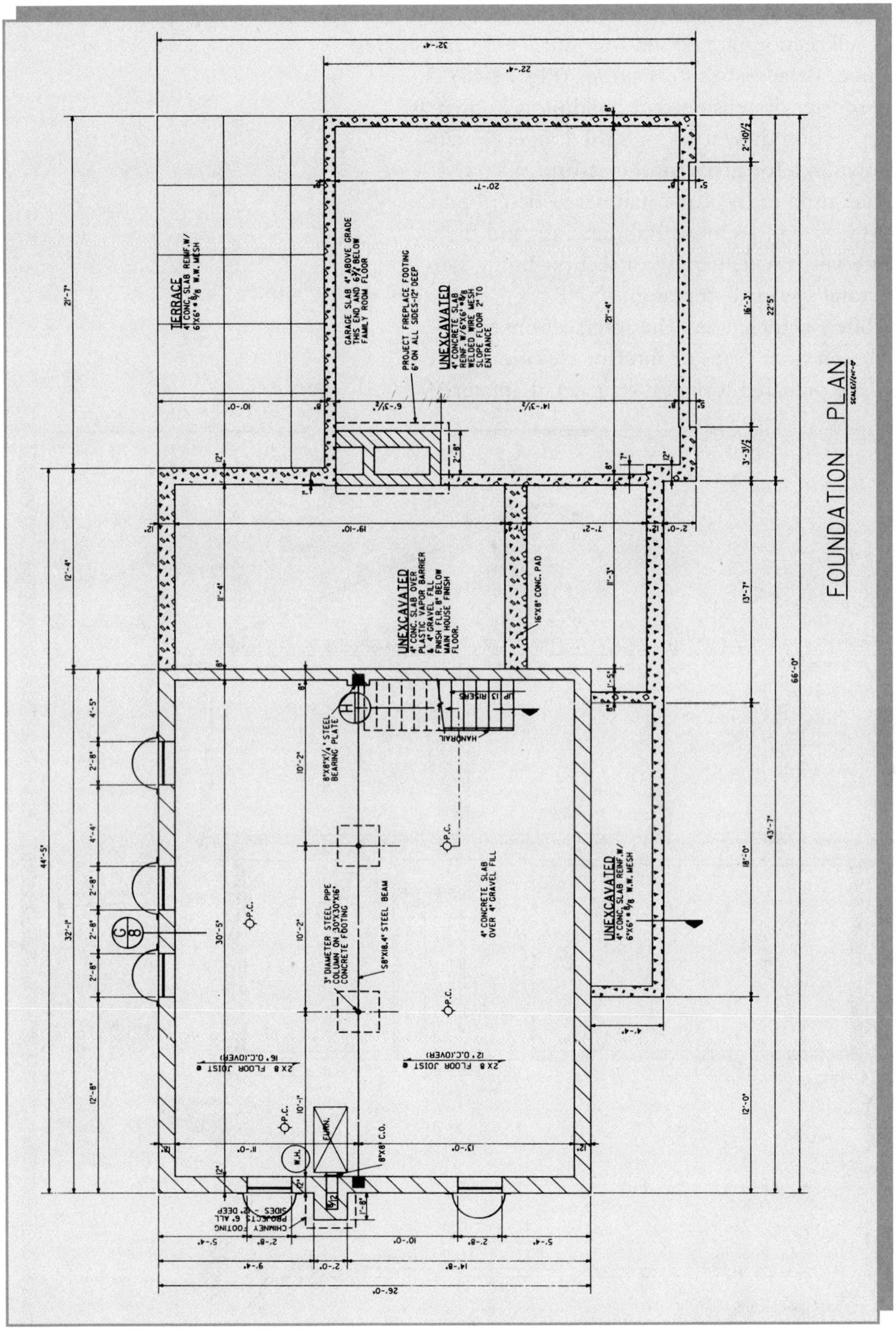

Figure 24-47
Foundation plan (*Courtesy of Home Planners, Inc.*)

Full sections are designed to show the type of construction used for the entire structure. Full sections are drawn to a small scale and do not reveal small member details and intersections **(Fig. 24-48)**. A full section through the shorter dimension of a building is known as a **transverse section.** A section drawn through the longer dimension of a building is known as a **longitudinal section.**

Cutting plane lines are used in the same manner as described in Chapter 11. Because of the size of architectural drawings and the diversity of building materials, a wide variety of section-lining symbols are used in architectural sectional drawings.

Stair assemblies are often complex and the intersections with floors difficult to describe on floor plans or interior elevation drawings. Sectional drawings are prepared to reveal stair details, material

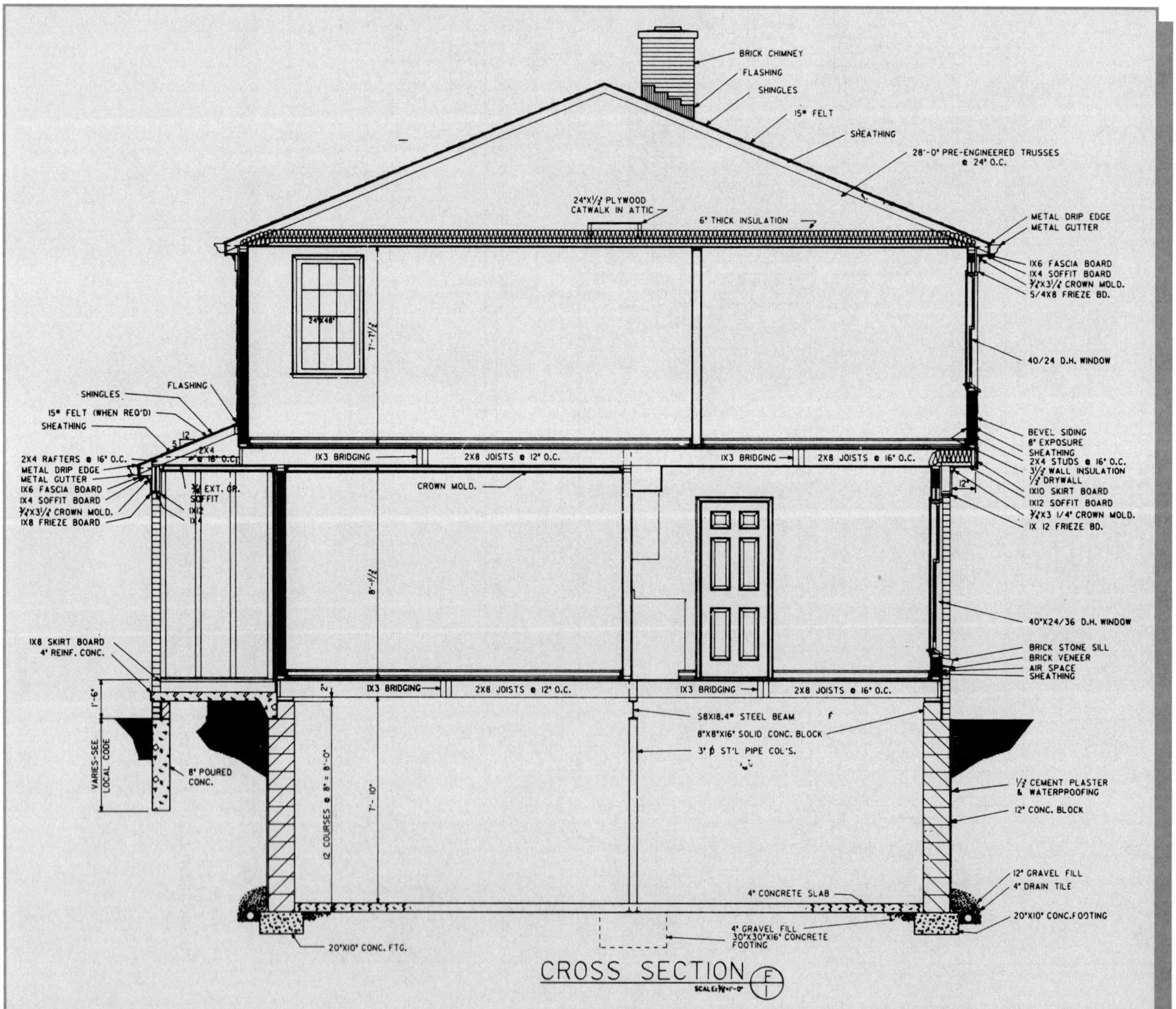

Figure 24-48
Full cross section through house *(Courtesy of Home Planners, Inc.)*

sizes, and dimensional relationships with floors and walls. **Figure 24-49** shows a typical stair detail drawing that covers three floor levels.

The internal structure of a fireplace and chimney system can only be described on a sectional drawing. Fireplace sections **(Fig. 24-50)** describe the size and position of all masonry materials. They also show the methods of intersection with slabs, floors, walls, and roofs.

When a designer wants to specify an exact size and shape of a detail, a full-size sectional drawing may be used. For example, the details shown in **Figure 24-51** are prepared so that the builder knows to use these specific trim shapes. Otherwise, a builder is free to use any molding with these overall dimensions.

Support Documents

In addition to the drawings prepared to describe a building, many other documents are needed to insure a building is constructed exactly as designed. These include schedules,

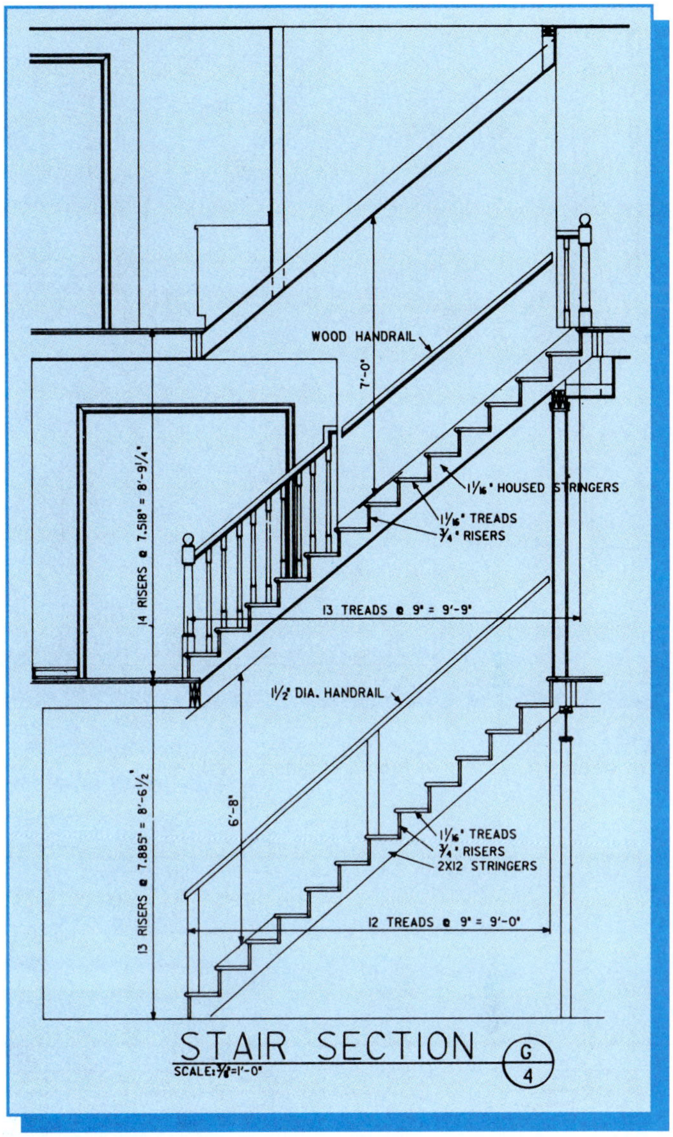

Figure 24-49
Detail of stair section *(Courtesy of Home Planners, Inc.)*

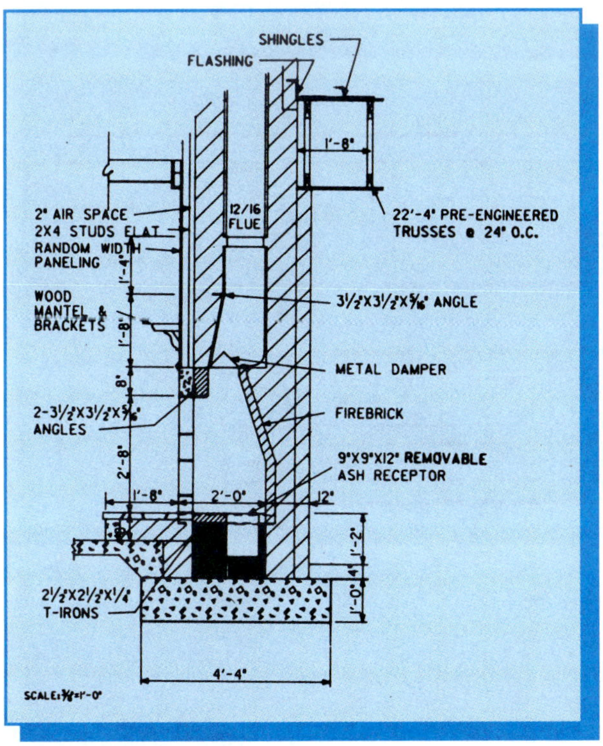

Figure 24-50
Fireplace section *(Courtesy of Home Planners, Inc.)*

specifications, budgets, contracts, and bids. Explanations relating to the set of plans or structural limitations **(Fig. 24-52)** may also be included to emphasize a design concern.

Schedules

Schedules are listings of materials and components used in the construction of a building. Schedules contain information in tabular form that is too extensive or detailed to include on a drawing. This includes such data as detail dimensions, materials, colors, style numbers, and manufacturers.

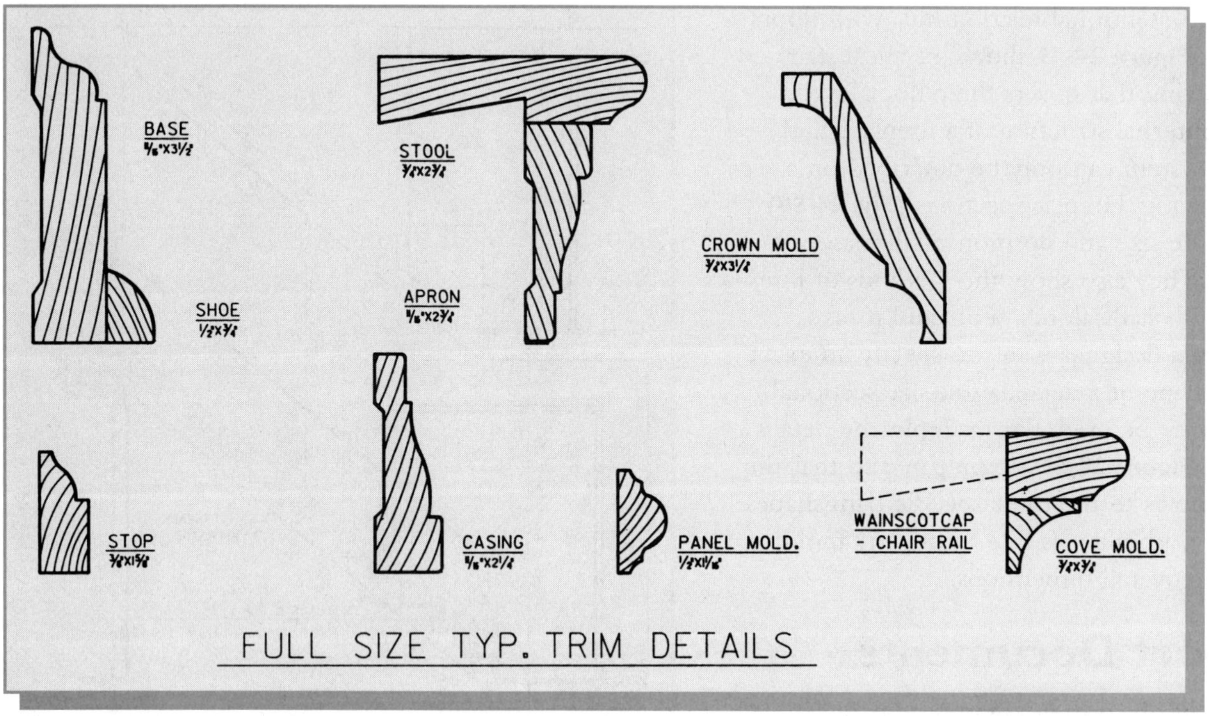

Figure 24-51
Trim details *(Courtesy of Home Planners, Inc.)*

FIRST FLOOR LIVING AREA: 27,385 CU. FT.
1,202 SQ.FT.
SECOND FLOOR LIVING AREA: 896 SQ.FT.
TOTAL: 2,098: SQ.FT.

SCHEDULE OF ABBREVIATIONS

BM	BEAM	D. H.	DOUBLE HUNG	M. T.	MARBLE THRESHOLD	REINF	REINFORCED
BF. D.	BIFOLD DOOR	EXT. GR. PLY.	EXTERIOR GRADE PLYWOOD	M. C.	MEDICINE CABINET	R. F. T.	RESILIENT FLOOR TILE
C. O.	CASED OPENING	F. D.	FLOOR DRAIN	MLDG	MOULDING	SLDG	SLIDING
CER. T.	CERAMIC TILE	GL	GLASS	N. I. C.	NOT IN CONTRACT	S. L. W. B.	STRUCTURAL LAMINATED WOOD BEAM
CL	CLOSET	HDR	HEADER	O. C.	ON CENTER	T & G	TONGUE & GROOVE
COL	COLUMN	H. B.	HOSE BIBB	P. L. C. & S.	PLASTIC LAMINATED COUNTER & SPLASH	W. C.	WATER CLOSET
CONC	CONCRETE	L. T.	LAUNDRY TRAY	P. S.	PULL SWITCH	W. H.	WATER HEATER
C. S. D. S.	CUT STONE DOOR SILL	LAV	LAVATORY	R	RADIUS/RANGE/RISER	W. W. R. M.	WELDED WIRE REINFORCING MESH
DIA	DIAMETER	LIN	LINEN	REFG.	REFRIGERATOR	WD	WOOD

DESIGN LOADS		ROOFS			FLOORS		BALCONIES
		WOOD OR ASPHALT SHINGLES	MISSION TILE	BUILT-UP	WOOD, CARPET OR R.F.T.	CERAMIC TILE, SLATE OR STONE	SPACED DECK
DEAD LOAD	(PSF)	15	25	19	15	25	7
LIVE LOAD	(PSF)	30	30	30	40	40	60
TOTAL	(PSF)	45	55	49	55	65	67

Figure 24-52
Floor area, abbreviations, and design loads *(Courtesy of Home Planners, Inc.)*

Schedules are commonly prepared for doors, windows, finishing methods and materials, fixtures, appliances, and furniture.

Specifications

Specifications are lists of materials that describe the size, type, and quantity of the building materials and components to be used in the construction of a building. It also includes a list for all building procedures not covered in the working drawings. Specifications are organized by building-material type as outlined by the **Construction Specifications Institute (CSI).** The CSI has established a format that subdivides building areas into sixteen headings, and subdivides each heading into the various building trades. This results in better construction documentation, which means safer, more efficient, and more profitable construction for builders.

Types of Buildings

This chapter has referred to residential buildings. It should be understood that there are several types of architecturally designed structures. The following is a general outline of these categories:

1. A single-family dwelling is a residence for one family only.
2. Multiple-family dwellings are multiple-living residences, such as a duplex, triplex, or fourplex. The structure is usually two or three stories high.
3. High-rise dwellings are multiple-living residences of four stories or more. The height of the building is limited by the zoning codes and the structural engineering.
4. Industrial plants are structures that house the production operations of our industries. They may range from small buildings to huge structures covering a vast amount of land.
5. Commercial buildings are structures that range from small business offices to huge skyscrapers. These structures house such operations as business offices and wholesale/retail stores.
6. Other types of structures that must be designed and drawn by architects, engineers, and drafters are government buildings, military structures, educational facilities, medical buildings, entertainment centers, transportation centers, bridges, tunnels, and highways.

Ergonomics

To enable people to do their best, the areas where they live, work, and play must be carefully designed to match their physical size, shape, reach, and mobility. This is the science of **ergonomics.** In order to meet certain design requirements, the human figure must be considered in all phases of the design **(Figs. 24-53** and **24-54)**. Careful planning and an awareness of the specific needs of people is essential for good design.

The Physically Impaired

Part of ergonomic planning is to create an environment sensitive to the needs of individuals confined to a wheelchair, or to those who are sight impaired or hearing impaired. Wherever there is a properly designed environment for the handicapped, an international handicap symbol must be posted nearby **(Fig. 24-55)**.

The following factors will affect a design for persons confined to a wheelchair **(Figs. 24-56** through **24-59)**:

1. Eye level is between 43" and 51".
2. Lap level is about 27"; therefore, clearance lavatories must be 30".
3. Maximum reaching height is 54".
4. Maximum slope of ramps is 1:12 (8.3% slope or 4°-46').
5. Minimum width of ramps and halls is 36".
6. Sidewalks at corners should have gradient ramps to the street.
7. Adjacent to the automobile, the loading zone must be 4' x 20'.
8. Minimum width of doors is 32" (36" preferred).
9. Doors should open 90°.
10. Minimum 5' turning area.
11. Water closet (toilet) height is 18".
12. Area for water closet stall is 36" x 60".
13. Area adjacent to bathtub is 36" x 60".

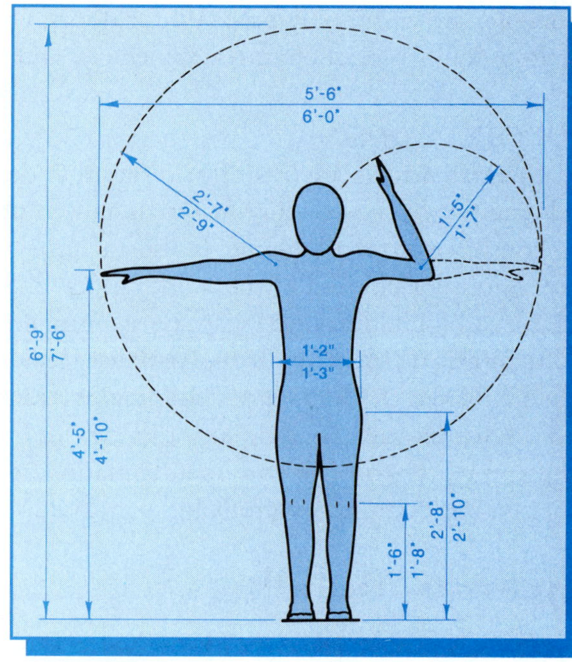

Figure 24-53
Typical human dimensions for women and men

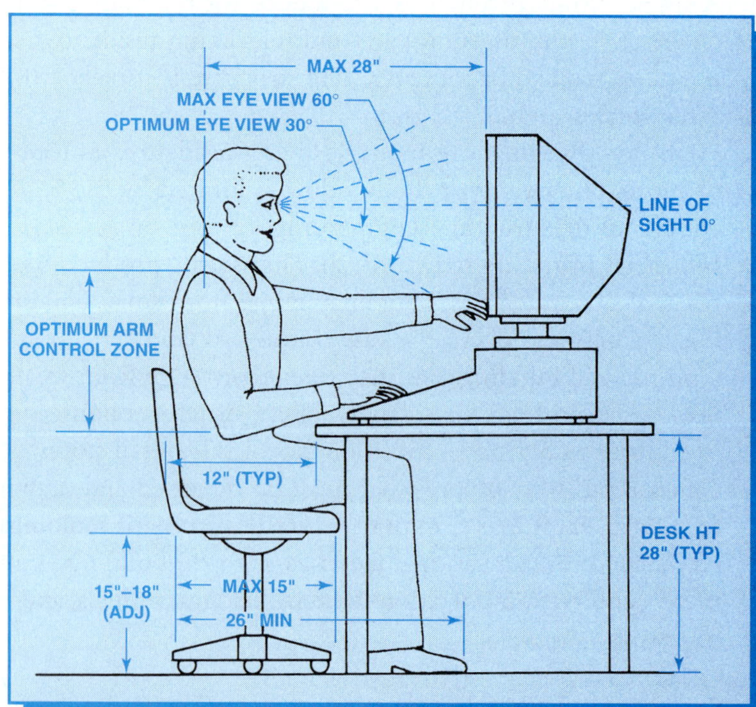

Figure 24-54
Human design data for desk work

Figure 24-55
This sign is posted for public areas that are ergonomically designed for the handicapped.

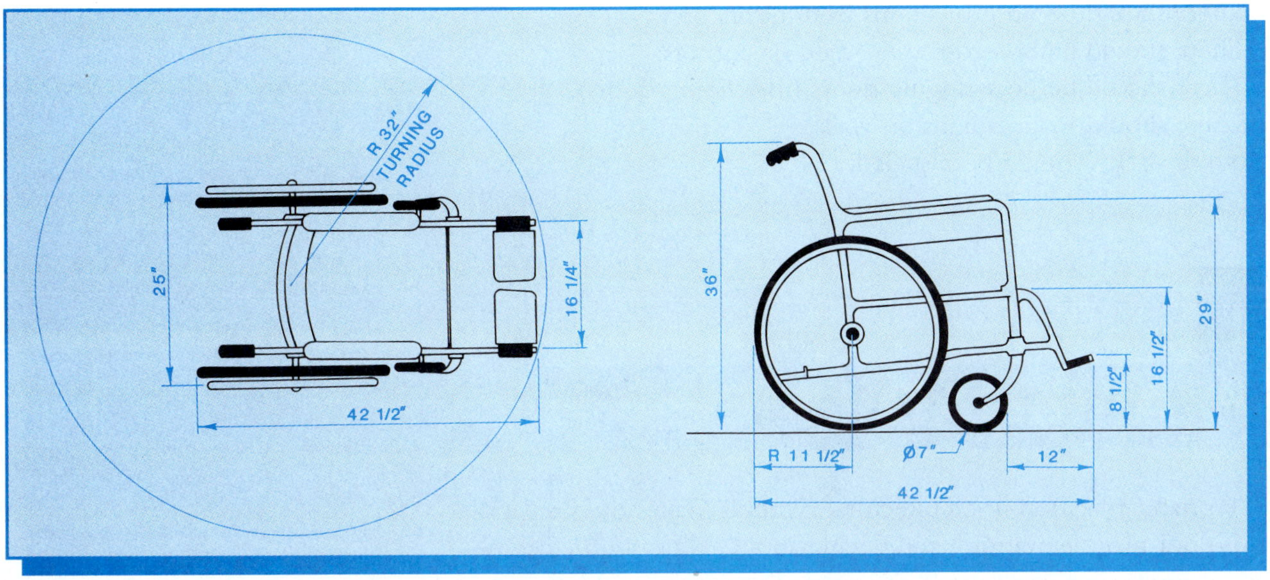

Figure 24-56
Standard size wheel with turning radius and vertical heights needed for clearance

14. Area adjacent to lavatory is 30" x 48".
15. Low mirror installation.

The following items must be included in a design for persons with a hearing impairment:

1. Eye-level signs
2. High-frequency alarms
3. Video warning signals
4. Flashing signal lights
5. Telephone amplifier attachment

The following are factors that will affect a design for persons with a sight impairment:

1. All floors should have nonslip surfaces.
2. Braille signs should be posted.
3. All stairs should have uniform tread widths and riser heights.

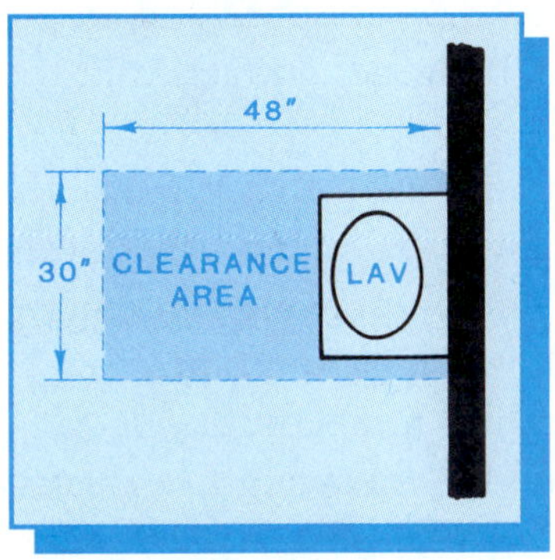

Figure 24-57
The maneuverable area required for a wheelchair under a lavatory

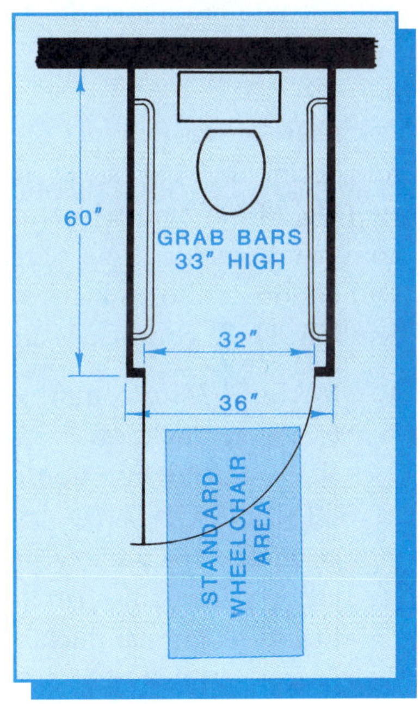

Figure 24-58
The maneuverable area required for a wheelchair next to a bathtub

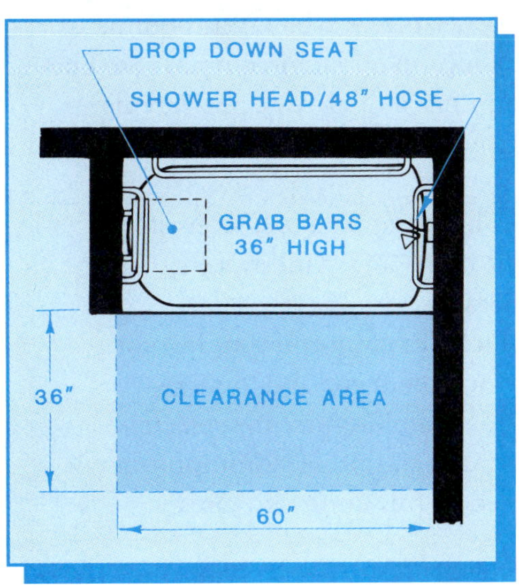

Figure 24-59
The maneuverable area required for a wheelchair in a water closet/toilet stall

4. There should be no projections from floors or walls.
5. There should be no protruding walls or corners.
6. There should be no swinging doors (use sliding doors).
7. There should be guardrails on walls.
8. Noise alarms should be installed.

AutoCAD LT Exercises

The standardization of architectural symbols makes the preparation of floor plans, elevations, and details on a CAD system extremely fast, accurate, and convenient. Specialized drawings, such as electrical, plumbing, and HVAC plans, are also easy to prepare using the Layer command. In addition, computer spreadsheeting and special software programs (third party) greatly improve speed and accuracy in preparing schedules, specifications, budgets, contracts, and bids.

There are many types of architectural drafting software available for specific drawing needs, such as an architectural menu tablet overlay **(Fig. 24-67)**. This tablet contains all common drawing and editing commands, such as Line, Circle, Erase, Move, Copy, Text, and Dimension. It also contains an array of architectural commands and symbols. These commands are divided into the following five areas:

1. WALLS—This command area allows the drafter to draw walls to a scale thickness as easily as drawing a single line. It also allows for the insertion of insulation into walls.
2. DIMENSIONING—In addition to the normal dimensioning commands in the lower-left corner of the tablet, this command area provides automatic dimensions in an architectural style. This includes optional thick marks in place of arrows, measurements in feet and inches, and the placement of dimensions above dimension lines.
3. DOORS and WINDOWS—This command area consists of standard floor-plan symbols for doors and windows that can be inserted into previously drawn walls.
4. TAGS—This command area includes common symbols for removed sections and callouts on architectural drawings.
5. SHAPES—Standard symbols for plumbing fixtures, electrical, and lighting components, furniture, and other commonly used architectural symbols are located in this command area.

Figure 24-67
Architectural menu tablet for a CAD system *(Courtesy of AutoCAD. AutoCAD is a trademark of Auto Desk, Inc.)*

Most of the features of an architectural menu overlay can be accomplished with AutoCAD LT. The following sequence shows how to complete a floor plan using the basic AutoCAD LT program. The same steps can be followed to prepare elevations and structural detail drawings.

1. Set the units you are going to use.
   a. Select **Settings, Units Style**. Pick **Architectural, OK.**
2. Set the Limits of your drawing. In CAD, the drawing is always full size.
   a. Select **Settings, Drawing, Limits**. Press **(EB)**. Type **40'**, **30'**, press **(EB)**.
   Don't forget the foot symbol (') when typing limits.

b. Whenever limits are set, do a Zoom All to get the screen to the full size.
   Type **z** and press **(EB)**. Type **a** and press **(EB)**.

3. Turn on and set snap and grid.
   a. Select **Settings**, **Drawing Aids**.
      Snap **3"**
      Grid **1'**
      Be sure to check the boxes to turn snap and grid on.
      Press **OK**.

4. Draw the outside walls.
   a. Select **Draw**, **Double Line**.
   b. Change the width of the line.
      Type **w**.
      Type **5"** for exterior walls.
   c. Pick absolute point **14'**, **26'**
      Use the coordinate display to find the point.
   d. Use the function key F6 to change the coordinate display from absolute to incremental.
   e. Be sure Snap and Ortho are on and draw a Double Line using these points from absolute starting point of 14', 26'.
   f. Pick points: **4'< 180**
      21'6"<270
      21'6"<0
      21'6"<90
      11'6"<180
      Press **(EB)**.

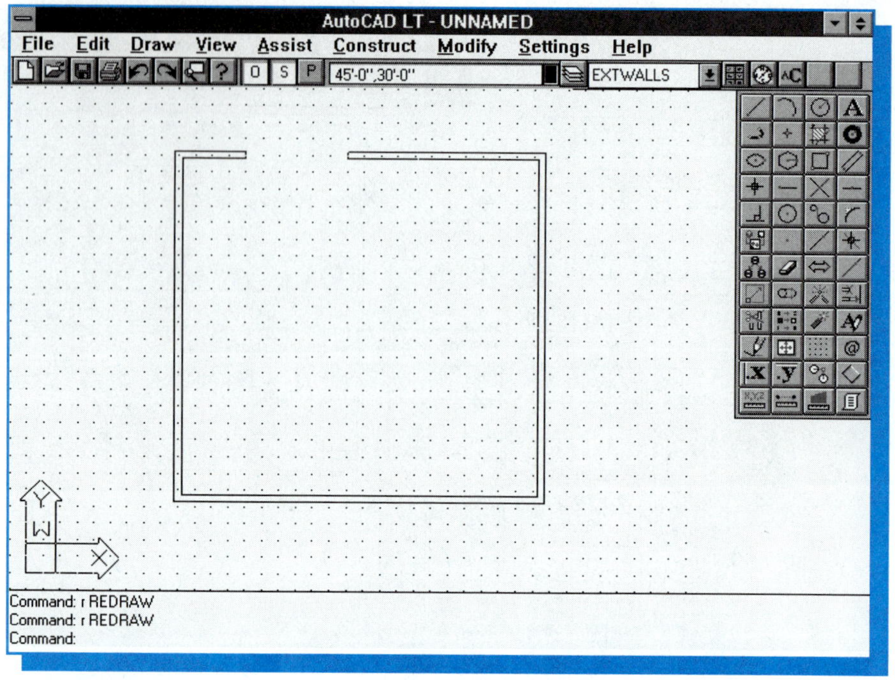

Figure 24-68
The outside walls using the Double Line command

The screen should now look like **Figure 24-68**.

5. Draw the inside walls.
   a. Select **Draw**, **Double Line.**
   b. Change the wall width to **4"**. See above.
   c. Pick points:
      10'3", 15'6"
      10'<0
      7'6"<270
      Press **(EB)**.
      Press **(EB)** again. [This **(EB)** brings back the last command].

d. Pick points:
   24', 25'9"
   5'9"<270
   Press **(EB), (EB)**.
   24'18'
   2'<270
   7'3"<0
   The drawing should now look like **Figure 24-69**.

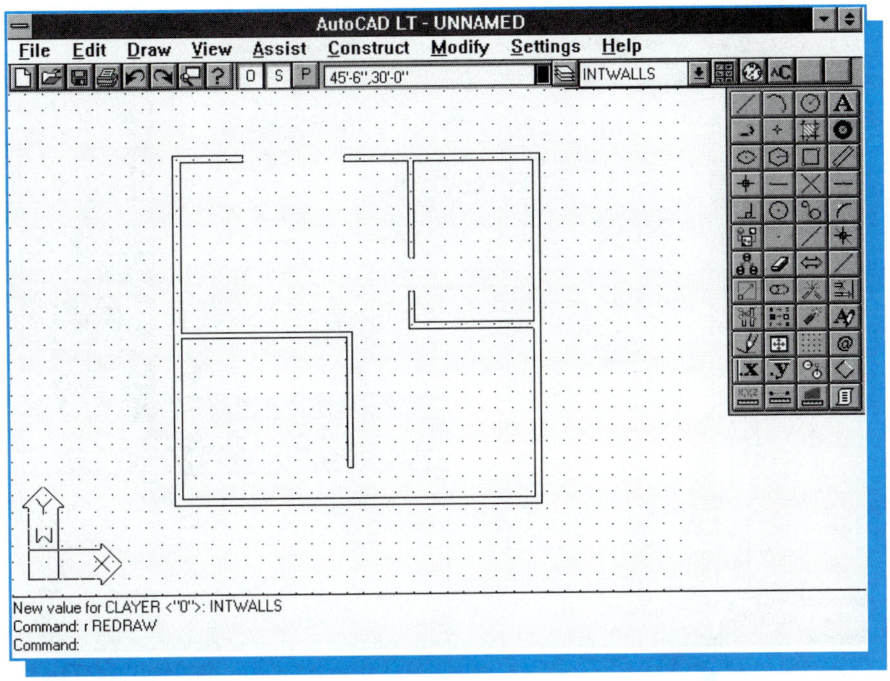

Figure 24-69
Drawing the interior walls using double line

6. Draw the doors and windows.
   Using the Zoom command and the Line and Arc command, draw in the doors and windows. The doors and window symbol may be drawn off to one side and then copied and moved to the proper location on the floor plan. See **Figure 24-24** for door symbols and **Figure 24-25** for window symbols **(Fig. 24-70)**.

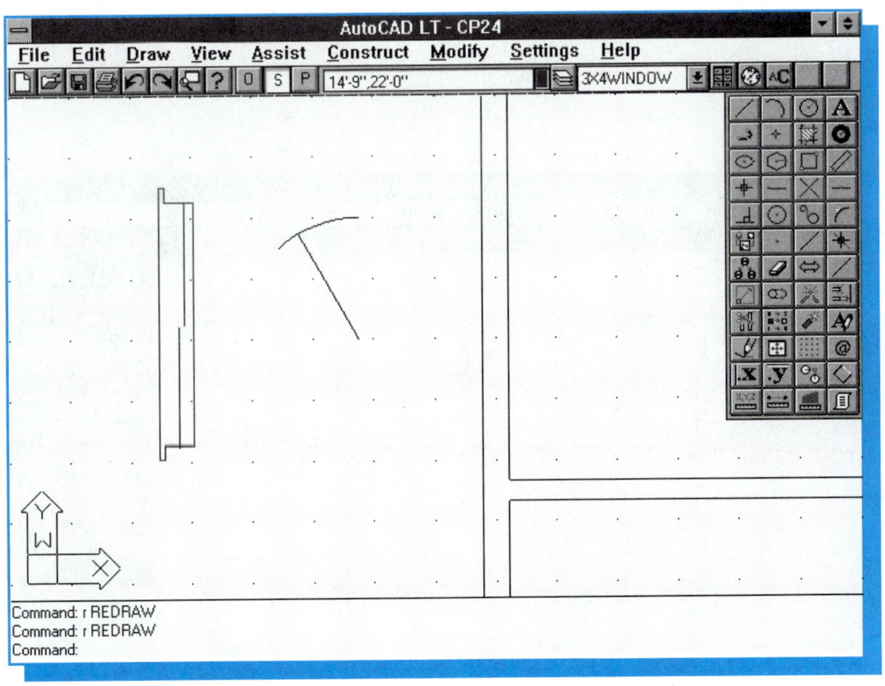

Figure 24-70
Draw the door and window symbols.

7. When you place a window, it will be necessary to trim the wall away.
   a. Copy, move, or rotate the window to its proper location.
   b. Select **Modify**, **Trim.**
   c. The first entities that you select are the cutting edges, in this case the window.
   d. Press **(EB)**.
   e. Now select the lines of the wall that are in-between the cutting lines.
      After selecting the lines to trim select **(EB) (Fig. 24-71)**.
   f. Copy and if necessary rotate the door symbol to its proper location.
   g. The drawing should look like **Figure 24-72**.

tip: *It's a good idea to have a layer for each type of symbol: elec, plumb, text, etc.*

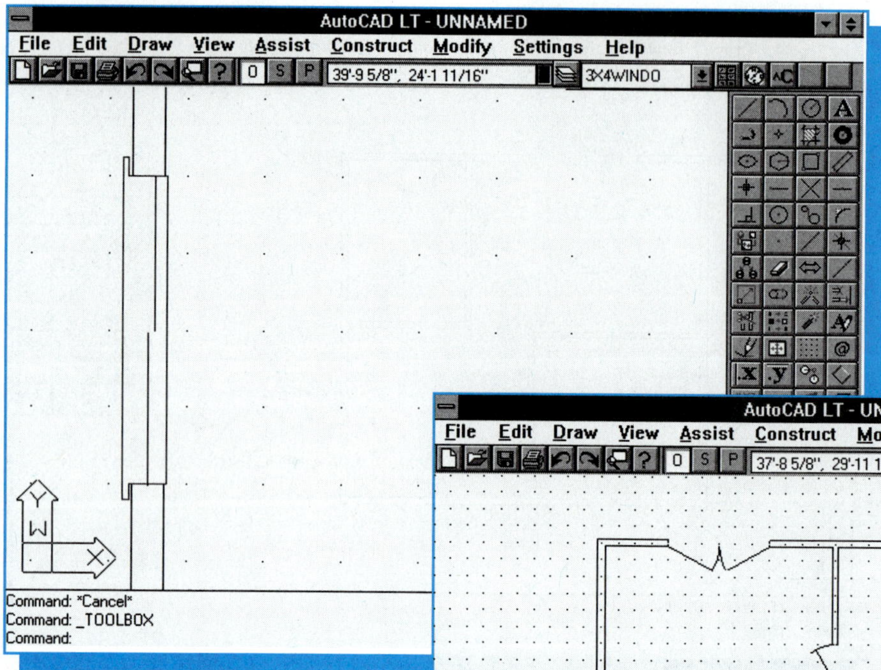

Figure 24-71
Trimming the wall away when inserting a window

Figure 24-72
Floor plan with doors and windows

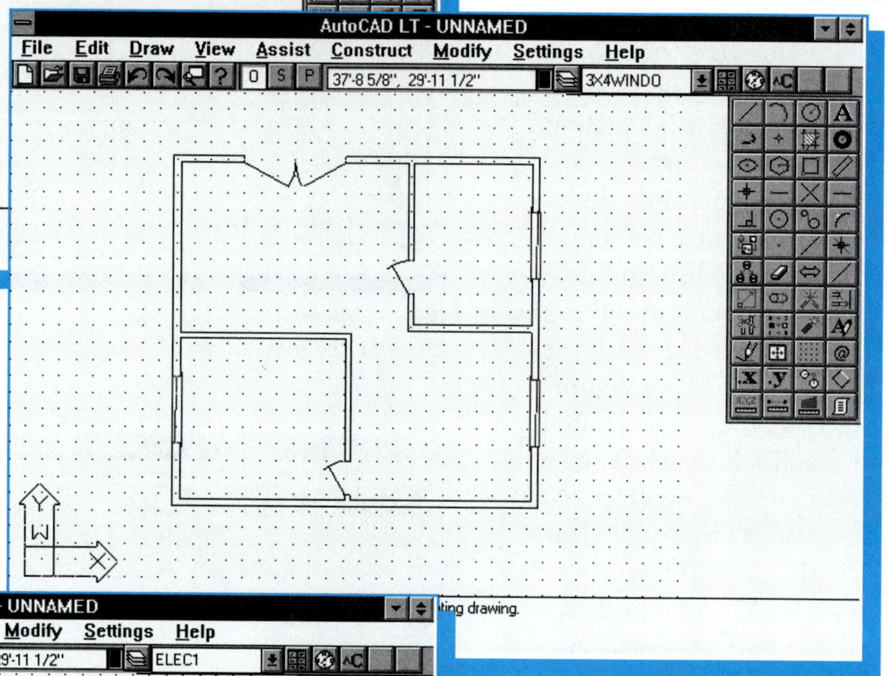

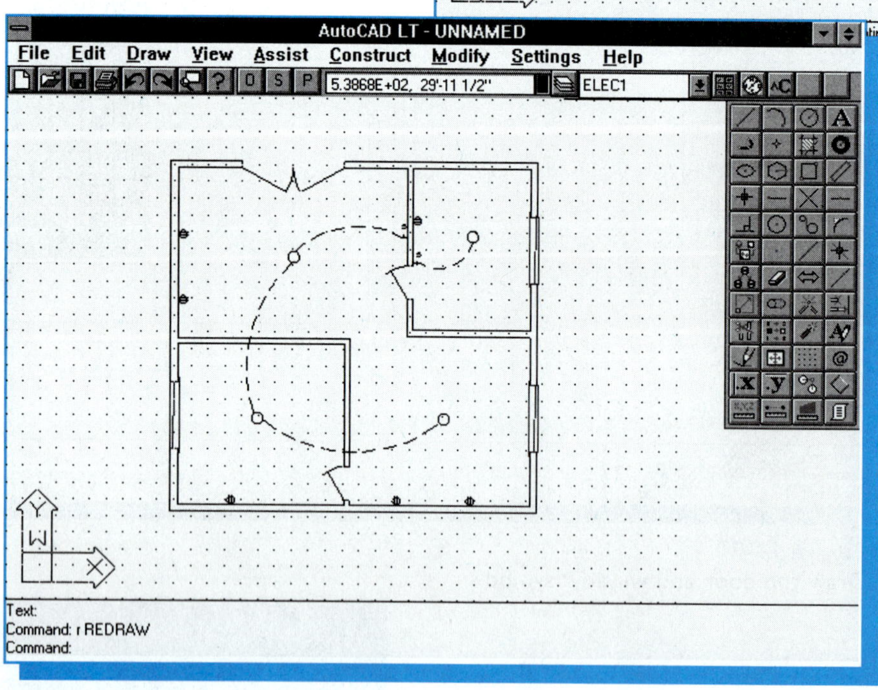

Figure 24-73
Insert electrical symbols.

8. Draw in the electrical symbols as shown in **Figure 24-73**.
9. Label the drawing using the Text command **(Fig. 24-74)**.
10. Using the appropriate dimension command, dimension the floor plan. **(Fig. 24-75)**.

Once an architectural drawing is completed on a CAD system, the drawing can be generated on paper with a plotter or a laser printer **(Figs. 24-76** and **24-77)**.

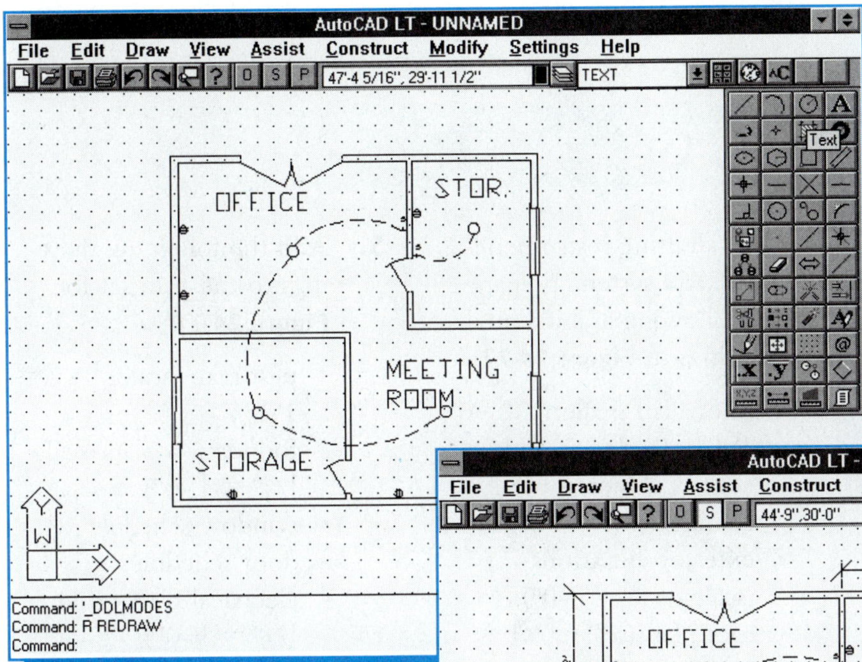

Figure 24-74
Insert text.

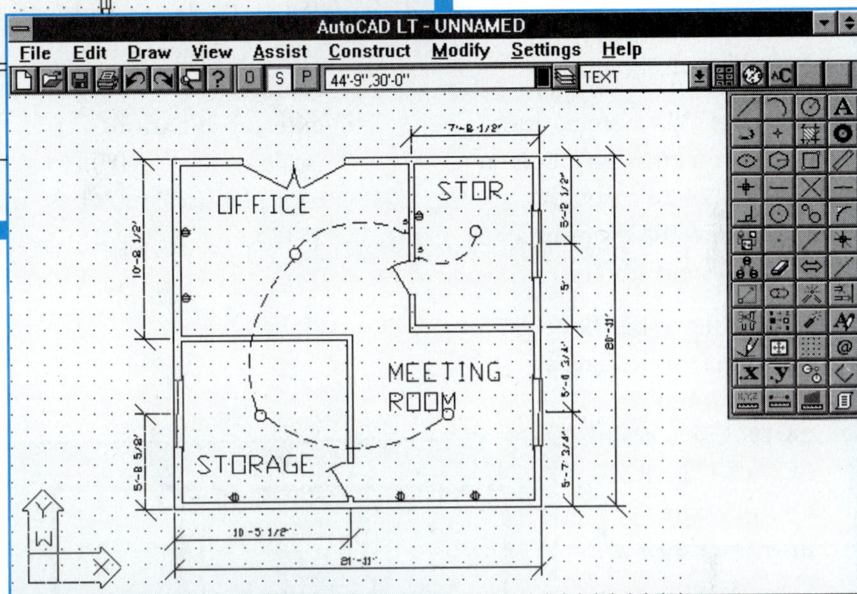

Figure 24-75
Place dimensions.

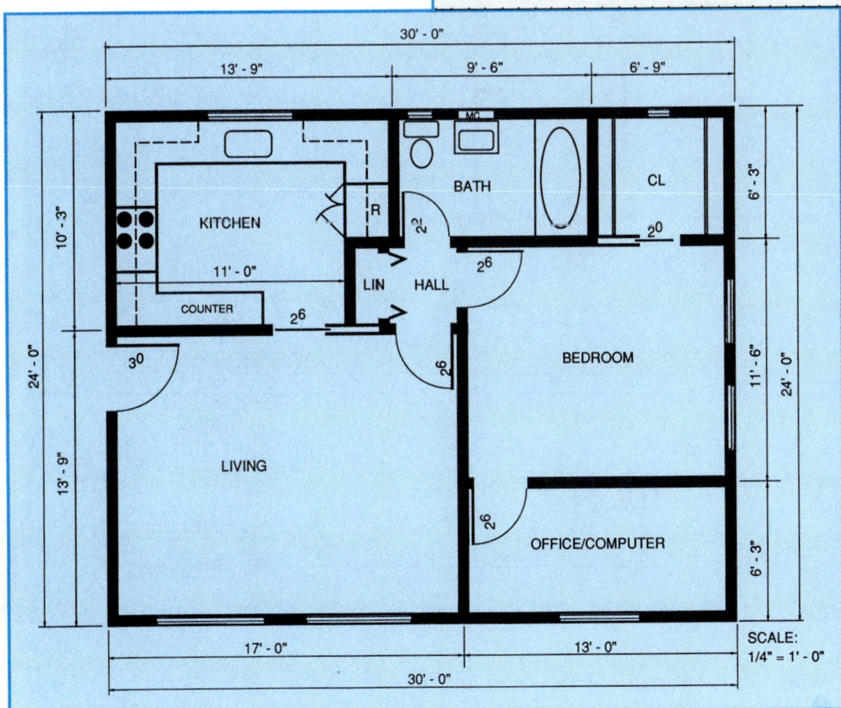

Figure 24-76
Small floor plan designed with a CAD system and printed with a laser printer

Drafting Exercises

1. Redraw the architectural symbols in **Figures 24-26** through **24-32** with drawing instruments. Keep the proportions of each symbol the same, and size them to be easily read on $^{1}/_{4}$" = 1'-0" scaled drawings.

2. With a CAD system, draw the architectural symbols in **Figures 24-26** through **24-32**, and file them in a CAD-system library.

3. Draw one of the floor plans shown in **Figures 24-7** through **24-19**. Use a scale of $^{1}/_{4}$" = 1'-0". Use drawing instruments or a CAD system.

4. With drafting instruments or a CAD system, prepare the following architectural drawings in **Figure 24-77**:

- floor plan (scale: $^{1}/_{4}$" = 1'-0"). Redesign and correct dimensional changes.
- exterior elevations (scale: $^{1}/_{4}$" = 1'-0").
- roof plan (scale: $^{1}/_{8}$" = 1'-0").

5. Add the following drawings to the plan set for **Figure 24-77**:

- plot plan (scale: $^{1}/_{8}$" = 1'-0")
- foundation plan (scale: $^{1}/_{4}$" = 1'-0")
- window schedule
- door schedule
- electrical plan (scale: $^{1}/_{4}$" = 1'-0")

6. Copy the set of working drawings in this chapter to scale as noted.

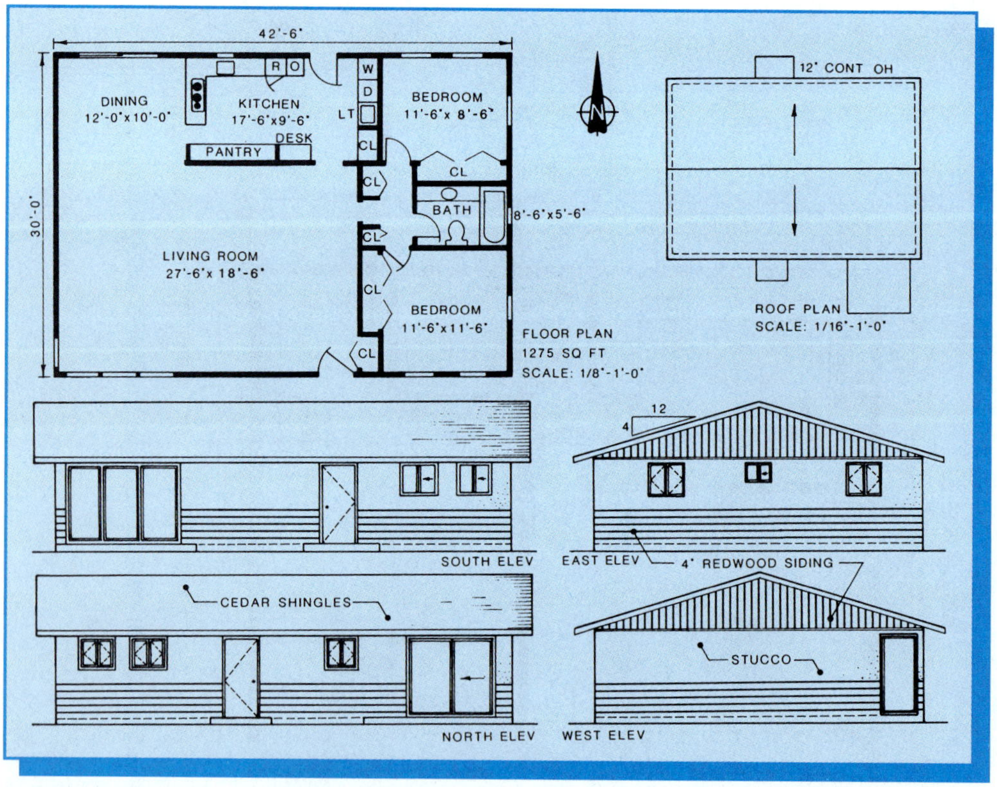

Figure 24-77
Follow the drawing instructions in Exercises 4 and 5.

Design Exercises

1. Remodel the plan in **Figure 24-77** to include the following:

- four bedrooms
- larger living area
- fireplace
- attached two-car garage

2. Design a house to fit the needs of your family. If possible, find an actual building site to use for your design. Draw the floor plan and plot plan.

3. Obtain or design a site plan. Design a house and a set of architectural working drawings for the site.

4. Design a one-bedroom cabin on a CAD system. Draw the floor plan, plot plan, and four elevations with a CAD system.

5. Design a vacation weekend cabin for a family of five. Children include two boys, ages 10 and 13, and one girl, age 7. The site **(Fig. 24-78)** is located adjacent to a large lake suitable for swimming and nonpower boating. The zoning data includes the lot size is 6600 sq. ft. with an allowable building area of 2600 sq. ft. Local codes restrict the structural coverage to a 40 percent maximum. The area of the plan must be 1000 sq. ft. or less. Indicate the square footage on the plan and include two overall dimensions, plus the size of each room. Draw the symbols and location of all fixtures, appliances, windows, and doors. Add the location and symbols for electrical switches, lights, and outlets. Remember to show the compass orientation on the plan as an aid to correct orientation. Be careful to avoid the following:

- Rooms too small
- Miscalculations of square footage
- Offset rooms
- Too much hall space
- Lack of storage or poorly designed storage space
- Disregarding the design criteria and parameters given

6. Redesign the floor plans in **Figures 24-79** and **24-80** for a family of five (two adults, one boy of 9 years, and two girls, 7 and 3 years old).

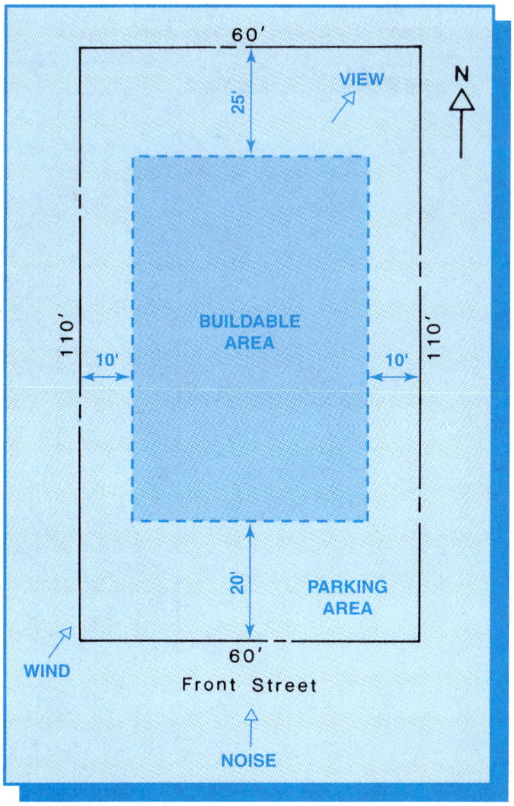

Figure 24-78
Follow the design instructions in Exercise 5.

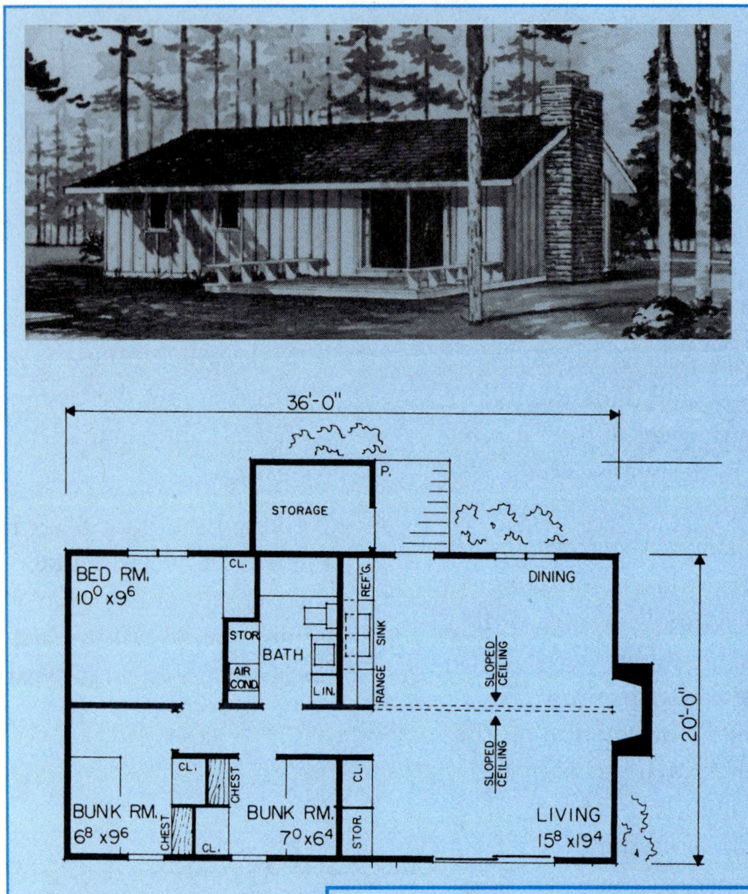

Figure 24-79
Redesign the floor plan for a family of five.

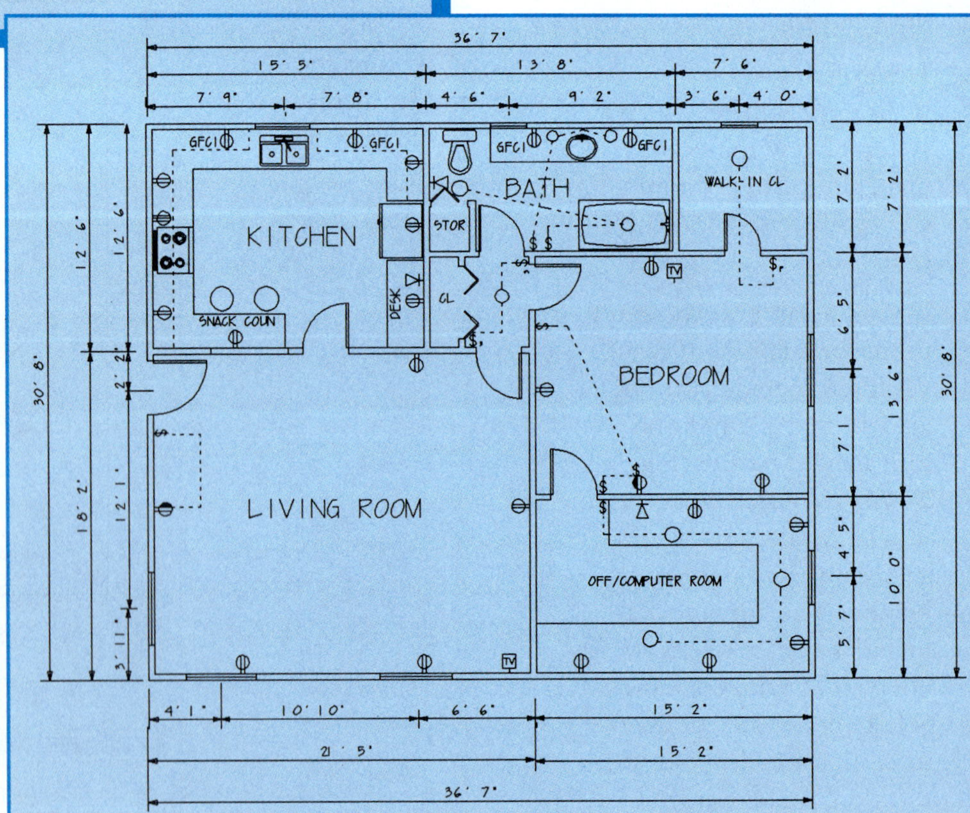

Figure 24-80
Redesign the computer-generated floor plan with a CAD system or drafting instruments. Draw the exterior elevations.

Key Terms

Architectural pictorial drawings
Composite sketch
Construction Specifications Institute (CSI)
Datum plane
Elevation (vertical) sections
Elevations
Ergonomics
Floor plans
Foundations plans
Heating, Ventilation, and Air Conditioning (HVAC)
Interior elevation
Longitudinal section
Plan (horizontal) sections
Plat plan
Plot plan
Presentation (abbreviated) floor plans
Rough sketches
Schedules
Setback
Site plan
Specifications
Survey plan
Transverse section

Appendix A

Tables

Fraction	Decimal Equivalent Customary (in.)	Metric (mm)	Fraction	Decimal Equivalent Customary (in.)	Metric (mm)
1/64	.015625	0.3969	33/64	.515625	13.0969
1/32	.03125	0.7938	17/32	.53125	13.4938
3/64	.046875	1.1906	35/64	.546875	13.8906
1/16	.0625	1.5875	9/16	.5625	14.2875
5/64	.078125	1.9844	37/64	.578125	14.6844
3/32	.09375	2.3813	19/32	.59375	15.0813
7/64	.109375	2.7781	39/64	.609375	15.4781
1/8	.1250	3.1750	5/8	.6250	15.8750
9/64	.140625	3.5719	41/64	.640625	16.2719
5/32	.15625	3.9688	21/32	.65625	16.6688
11/64	.171875	4.3656	43/64	.671875	17.0656
3/16	.1875	4.7625	11/16	.6875	17.4625
13/64	.203125	5.1594	45/64	.703125	17.8594
7/32	.21875	5.5563	23/32	.71875	18.2563
15/64	.234375	5.9531	47/64	.734375	18.6531
1/4	.250	6.3500	3/4	.750	19.0500
17/64	.265625	6.7469	49/64	.765625	19.4469
9/32	.28125	7.1438	25/32	.78125	19.8438
19/64	.296875	7.5406	51/64	.796875	20.2406
5/16	.3125	7.9375	13/16	.8125	20.6375
21/64	.328125	8.3384	53/64	.828125	21.0344
11/32	.34375	8.7313	27/32	.84375	21.4313
23/64	.359375	9.1281	55/64	.859375	21.8281
3/8	.3750	9.5250	7/8	.8750	22.2250
25/64	.390625	9.9219	57/64	.890625	22.6219
13/32	.40625	10.3188	29/32	.90625	23.0188
27/64	.421875	10.7156	59/64	.921875	23.4156
7/16	.4375	11.1125	15/16	.9375	23.8125
29/64	.453125	11.5094	61/64	.953125	24.2094
15/32	.46875	11.9063	31/32	.96875	24.6063
31/64	.484375	12.3031	63/64	.984375	25.0031
1/2	.500	12.7000	1	1.000	25.4000

Table I
Inch/Metric Equivalents
(Reprinted from Drafting for Trades and Industry—Basic Skills *by Nelson, © 1979 Delmar Publishers, Inc.)*

Measures of Length
1 millimeter (mm) = 0.03937 inch
1 centimeter (cm) = 0.39370 inch
1 meter (m) = 39.37008 inches
= 3.2808 feet
= 1.0936 yards
1 kilometer (km) = 0.6214 mile
1 inch = 25.4 millimeters (mm)
= 2.54 centimeters (cm)
1 foot = 304.8 millimeters (mm)
= 0.3048 meter (m)
1 yard = 0.9144 meter (m)
1 mile = 1.609 kilometers (km)
Measures of Area
1 square millimeter = 0.00155 square inch
1 square centimeter = 0.155 square inch
1 square meter = 10.764 square feet
= 1.196 square yards
1 square kilometer = 0.3861 square mile
1 square inch = 645.2 square millimeters
= 6.452 square centimeters
1 square foot = 929 square centimeters
= 0.0929 square meter
1 square yard = 0.836 square meter
1 square mile = 2.5899 square kilometers
Measures of Capacity (Dry)
1 cubic centimeter (cm^3) = 0.061 cubic inch
1 liter = 0.0353 cubic foot
= 61.023 cubic inches
1 cubic meter (m^3) = 35.315 cubic feet
= 1.308 cubic yards
1 cubic inch = 16.38706 cubic centimeters (cm^3)
1 cubic foot = 0.02832 cubic meter (m^3)
= 28.317 liters
1 cubic yard = 0.7646 cubic meter (m^3)
Measures of Capacity (Liquid)
1 liter = 1.0567 U.S. quarts
= 0.2642 U.S. gallon
= 0.2200 Imperial gallon
1 cubic meter (m^3) = 264.2 U.S. gallons
= 219.969 Imperial gallons
1 U.S. quart = 0.946 liter
1 Imperial quart = 1.136 liters
1 U.S. gallon = 3.785 liters
1 Imperial gallon = 4.546 liters
Measures of Weight
1 gram (g) = 15.432 grains
= 0.03215 ounce troy
= 0.03527 ounce avoirdupois
1 kilogram (kg) = 35.274 ounces avoirdupois
= 2.2046 pounds
1000 kilograms (kg) = 1 metric ton (t)
= 1.1023 tons of 2000 pounds
= 0.9842 ton of 2240 pounds
1 ounce avoirdupois = 28.35 grams (g)
1 ounce troy = 31.103 grams (g)
1 pound = 453.6 grams
= 0.4536 kilogram (kg)
1 ton of 2240 pounds = 1016 kilograms (kg)
= 1.016 metric tons
1 grain = 0.0648 gram (g)
1 metric ton = 0.9842 ton of 2240 pounds
= 2204.6 pounds

Table 2
Inch/Metric Conversion

SYMBOL FOR:	ANSI Y14.5	ISO
DIMENSION ORIGIN	⌾→	NONE
FEATURE CONTROL FRAME	⌖ Ø0.5Ⓜ A B C	⌖ Ø0.5Ⓜ A B C
CONICAL TAPER	⌲	⌲
SLOPE	⌳	⌳
COUNTERBORE/SPOTFACE	⌴	NONE
COUNTERSINK	⌵	NONE
DEPTH/DEEP	↧	NONE
SQUARE (SHAPE)	□	□
DIMENSION NOT TO SCALE	15	15
NUMBER OF TIMES/PLACES	8X	8X
ARC LENGTH	⌒105	NONE
RADIUS	R	R
SPHERICAL RADIUS	SR	NONE
SPHERICAL DIAMETER	SØ	NONE

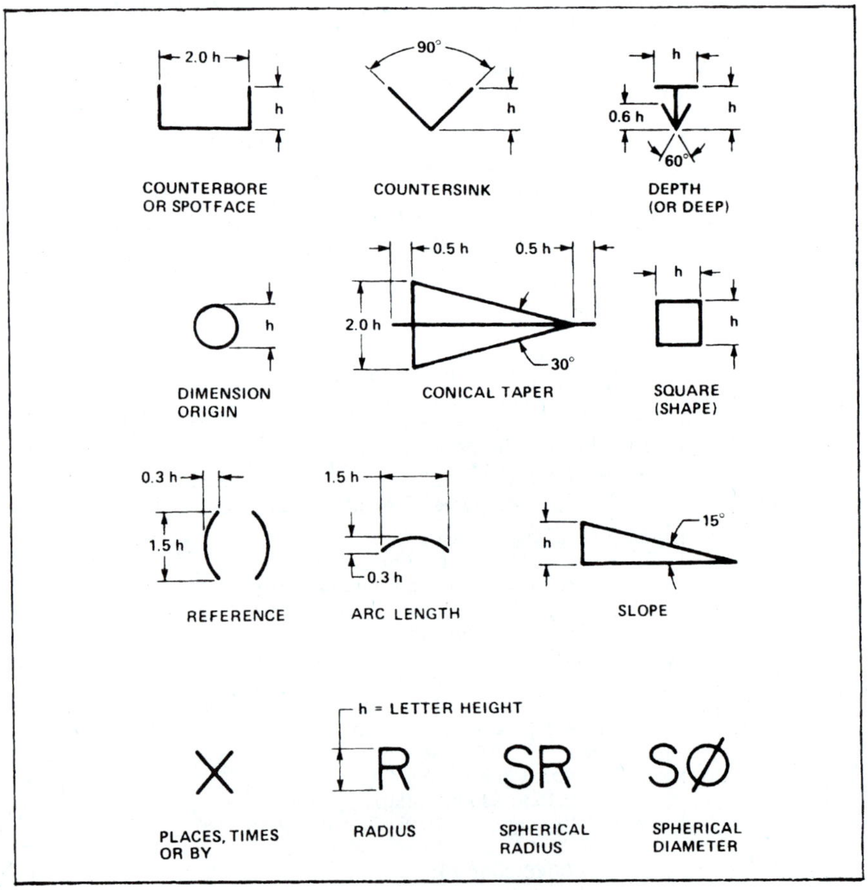

Table 3
Dimensioning Symbols
(Reprinted from The American Society of Mechanical Engineers —ANSI Y14.5M—1982)

Sizes		Basic Major Diameter	THREADS PER INCH											Sizes
			Series with graded pitches			Series with constant pitches								
Primary	Secondary		Coarse UNC	Fine UNF	Extra fine UNEF	4UN	6UN	8UN	12UN	16UN	20UN	28UN	32UN	
0		0.0600	–	80	–	–	–	–	–	–	–	–	–	0
	1	0.0730	64	72	–	–	–	–	–	–	–	–	–	1
2		0.0860	56	64	–	–	–	–	–	–	–	–	–	2
	3	0.0990	48	56	–	–	–	–	–	–	–	–	–	3
4		0.1120	40	48	–	–	–	–	–	–	–	–	–	4
5		0.1250	40	44	–	–	–	–	–	–	–	–	–	5
6		0.1380	32	40	–	–	–	–	–	–	–	–	UNC	6
8		0.1640	32	36	–	–	–	–	–	–	–	–	UNC	8
10		0.1900	24	32	–	–	–	–	–	–	–	–	UNF	10
	12	0.2160	24	28	32	–	–	–	–	–	–	UNF	UNEF	12
1/4		0.2500	20	28	32	–	–	–	–	–	UNC	UNF	UNEF	1/4
5/16		0.3125	18	24	32	–	–	–	–	–	20	28	UNEF	5/16
3/8		0.3750	16	24	32	–	–	–	–	UNC	20	28	UNEF	3/8
7/16		0.4375	14	20	28	–	–	–	–	16	UNF	UNEF	32	7/16
1/2		0.5000	13	20	28	–	–	–	–	16	UNF	UNEF	32	1/2
9/16		0.5625	12	18	24	–	–	–	UNC	16	20	28	32	9/16
5/8		0.6250	11	18	24	–	–	–	12	16	20	28	32	5/8
	11/16	0.6875	–	–	24	–	–	–	12	16	20	28	32	11/16
3/4		0.7500	10	16	20	–	–	–	12	UNF	UNEF	28	32	3/4
	13/16	0.8125	–	–	20	–	–	–	12	16	UNEF	28	32	13/16
7/8		0.8750	9	14	20	–	–	–	12	16	UNEF	28	32	7/8
	15/16	0.9375	–	–	20	–	–	–	12	16	UNEF	28	32	15/16
1		1.0000	8	12	20	–	–	UNC	UNF	16	UNEF	28	32	1
	1 1/16	1.0625	–	–	18	–	–	8	12	16	20	28	–	1 1/16
1 1/8		1.1250	7	12	18	–	–	8	UNF	16	20	28	–	1 1/8
	1 3/16	1.1875	–	–	18	–	–	8	12	16	20	28	–	1 3/16
1 1/4		1.2500	7	12	18	–	–	8	UNF	16	20	28	–	1 1/4
	1 5/16	1.3125	–	–	18	–	–	8	12	16	20	28	–	1 5/16
1 3/8		1.3750	6	12	18	–	UNC	8	UNF	16	20	28	–	1 3/8
	1 7/16	1.4375	–	–	18	–	6	8	12	16	20	28	–	1 7/16
1 1/2		1.5000	6	12	18	–	UNC	8	UNF	16	20	28	–	1 1/2
	1 9/16	1.5625	–	–	18	–	6	8	12	16	20	–	–	1 9/16
1 5/8		1.6250	–	–	18	–	6	8	12	16	20	–	–	1 5/8
	1 11/16	1.6875	–	–	18	–	6	8	12	16	20	–	–	1 11/16
1 3/4		1.7500	5	–	–	–	6	8	12	16	20	–	–	1 3/4
	1 13/16	1.8125	–	–	–	–	6	8	12	16	20	–	–	1 13/16
1 7/8		1.8750	–	–	–	–	6	8	12	16	20	–	–	1 7/8
	1 15/16	1.9375	–	–	–	–	6	8	12	16	20	–	–	1 15/16
2		2.0000	4 1/2	–	–	–	6	8	12	16	20	–	–	2
	2 1/8	2.1250	–	–	–	–	6	8	12	16	20	–	–	2 1/8
2 1/4		2.2500	4 1/2	–	–	–	6	8	12	16	20	–	–	2 1/4
	2 3/8	2.3750	–	–	–	–	6	8	12	16	20	–	–	2 3/8
2 1/2		2.5000	4	–	–	UNC	6	8	12	16	20	–	–	2 1/2
	2 5/8	2.6250	–	–	–	4	6	8	12	16	20	–	–	2 5/8
2 3/4		2.7500	4	–	–	UNC	6	8	12	16	20	–	–	2 3/4
	2 7/8	2.8750	–	–	–	4	6	8	12	16	20	–	–	2 7/8
3		3.0000	4	–	–	UNC	6	8	12	16	20	–	–	3
	3 1/8	3.1250	–	–	–	4	6	8	12	16	–	–	–	3 1/8
3 1/4		3.2500	4	–	–	UNC	6	8	12	16	–	–	–	3 1/4
	3 3/8	3.3750	–	–	–	4	6	8	12	16	–	–	–	3 3/8
3 1/2		3.5000	4	–	–	UNC	6	8	12	16	–	–	–	3 1/2
	3 5/8	3.6250	–	–	–	4	6	8	12	16	–	–	–	3 5/8
3 3/4		3.7500	4	–	–	UNC	6	8	12	16	–	–	–	3 3/4
	3 7/8	3.8750	–	–	–	4	6	8	12	16	–	–	–	3 7/8
4		4.0000	4	–	–	UNC	6	8	12	16	–	–	–	4
	4 1/8	4.1250	–	–	–	4	6	8	12	16	–	–	–	4 1/8
4 1/4		4.2500	–	–	–	4	6	8	12	16	–	–	–	4 1/4
	4 3/8	4.3750	–	–	–	4	6	8	12	16	–	–	–	4 3/8
4 1/2		4.5000	–	–	–	4	6	8	12	16	–	–	–	4 1/2
	4 5/8	4.6250	–	–	–	4	6	8	12	16	–	–	–	4 5/8
4 3/4		4.7500	–	–	–	4	6	8	12	16	–	–	–	4 3/4
	4 7/8	4.8750	–	–	–	4	6	8	12	16	–	–	–	4 7/8
5		5.0000	–	–	–	4	6	8	12	16	–	–	–	5
	5 1/8	5.1250	–	–	–	4	6	8	12	16	–	–	–	5 1/8
5 1/4		5.2500	–	–	–	4	6	8	12	16	–	–	–	5 1/4
	5 3/8	5.3750	–	–	–	4	6	8	12	16	–	–	–	5 3/8
5 1/2		5.5000	–	–	–	4	6	8	12	16	–	–	–	5 1/2
	5 5/8	5.6250	–	–	–	4	6	8	12	16	–	–	–	5 5/8
5 3/4		5.7500	–	–	–	4	6	8	12	16	–	–	–	5 3/4
	5 7/8	5.8750	–	–	–	4	6	8	12	16	–	–	–	5 7/8
6		6.0000	–	–	–	4	6	8	12	16	–	–	–	6

Table 4

Unified Standard Screw Thread Series

(Reprinted from Technical Drawing and Design *by Goestch ad Nelson, © 1986 Delmar Publishers, Inc.)*

AMERICAN NATIONAL STANDARD
SQUARE AND HEX BOLTS AND SCREWS – INCH SERIES

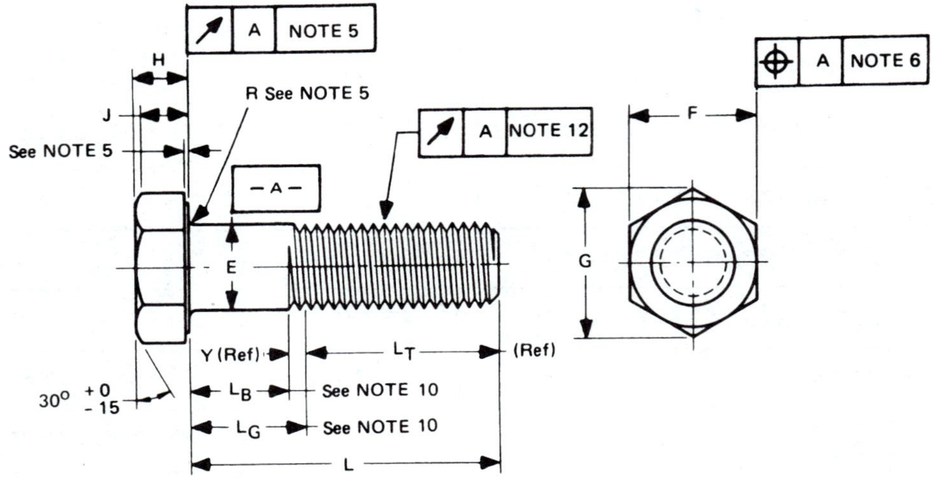

Nominal Size or Basic Product Dia (18)	E Body Dia (8) Max	E Body Dia (8) Min	F Width Across Flats Basic	F Width Across Flats Max	F Width Across Flats Min	G Width Across Corners (4) Max	G Width Across Corners (4) Min	H Height Basic	H Height Max	H Height Min	J Wrenching Height (4) Min	L_T Thread Length For Screw Lengths (10) 6 in. and Shorter Basic	L_T Thread Length For Screw Lengths (10) Over 6 in. Basic	Y Transition Thread Length (10) Max	Runout of Bearing Surface FIM (5) Max
1/4 0.2500	0.2500	0.2450	7/16	0.438	0.428	0.505	0.488	5/32	0.163	0.150	0.106	0.750	1.000	0.250	0.010
5/16 0.3125	0.3125	0.3065	1/2	0.500	0.489	0.577	0.557	13/64	0.211	0.195	0.140	0.875	1.125	0.278	0.011
3/8 0.3750	0.3750	0.3690	9/16	0.562	0.551	0.650	0.628	15/64	0.243	0.226	0.160	1.000	1.250	0.312	0.012
7/16 0.4375	0.4375	0.4305	5/8	0.625	0.612	0.722	0.698	9/32	0.291	0.272	0.195	1.125	1.375	0.357	0.013
1/2 0.5000	0.5000	0.4930	3/4	0.750	0.736	0.866	0.840	5/16	0.323	0.302	0.215	1.250	1.500	0.385	0.014
9/16 0.5625	0.5625	0.5545	13/16	0.812	0.798	0.938	0.910	23/64	0.371	0.348	0.250	1.375	1.625	0.417	0.015
5/8 0.6250	0.6250	0.6170	15/16	0.938	0.922	1.083	1.051	25/64	0.403	0.378	0.269	1.500	1.750	0.455	0.017
3/4 0.7500	0.7500	0.7410	1 1/8	1.125	1.100	1.299	1.254	15/32	0.483	0.455	0.324	1.750	2.000	0.500	0.020
7/8 0.8750	0.8750	0.8660	1 5/16	1.312	1.285	1.516	1.465	35/64	0.563	0.531	0.378	2.000	2.250	0.556	0.023
1 1.0000	1.0000	0.9900	1 1/2	1.500	1.469	1.732	1.675	39/64	0.627	0.591	0.416	2.250	2.500	0.625	0.026
1 1/8 1.1250	1.1250	1.1140	1 11/16	1.688	1.631	1.949	1.859	11/16	0.718	0.658	0.461	2.500	2.750	0.714	0.029
1 1/4 1.2500	1.2500	1.2390	1 7/8	1.875	1.812	2.165	2.066	25/32	0.813	0.749	0.530	2.750	3.000	0.714	0.033
1 3/8 1.3750	1.3750	1.3630	2 1/16	2.062	1.994	2.382	2.273	27/32	0.878	0.810	0.569	3.000	3.250	0.833	0.036
1 1/2 1.5000	1.5000	1.4880	2 1/4	2.230	2.175	2.598	2.480	1 5/16	0.974	0.902	0.640	3.250	3.500	0.833	0.039
1 3/4 1.7500	1.7500	1.7380	2 5/8	2.625	2.538	3.031	2.893	1 3/32	1.134	1.054	0.748	3.750	4.000	1.000	0.046
2 2.0000	2.0000	1.9880	3	3.000	2.900	3.464	3.306	1 7/32	1.263	1.175	0.825	4.250	4.500	1.111	0.052
2 1/4 2.2500	2.2500	2.2380	3 3/8	3.375	3.262	3.897	3.719	1 3/8	1.423	1.327	0.933	4.750	5.000	1.111	0.059
2 1/2 2.5000	2.5000	2.4880	3 3/4	3.750	3.625	4.330	4.133	1 17/32	1.583	1.479	1.042	5.250	5.500	1.250	0.065
2 3/4 2.7500	2.7500	2.7380	4 1/8	4.125	3.988	4.763	4.546	1 11/16	1.744	1.632	1.151	5.750	6.000	1.250	0.072
3 3.0000	3.0000	2.9880	4 1/2	4.500	4.350	5.196	4.959	1 7/8	1.935	1.815	1.290	6.250	6.500	1.250	0.079

Table 5

Dimensions of Hex Cap Screws (Finished Hex Bolts)

(Reprinted from the American Society of Engineers—ANSI B18.2.1—1981)

AMERICAN NATIONAL STANDARD
SQUARE AND HEX BOLTS AND SCREWS – INCH SERIES

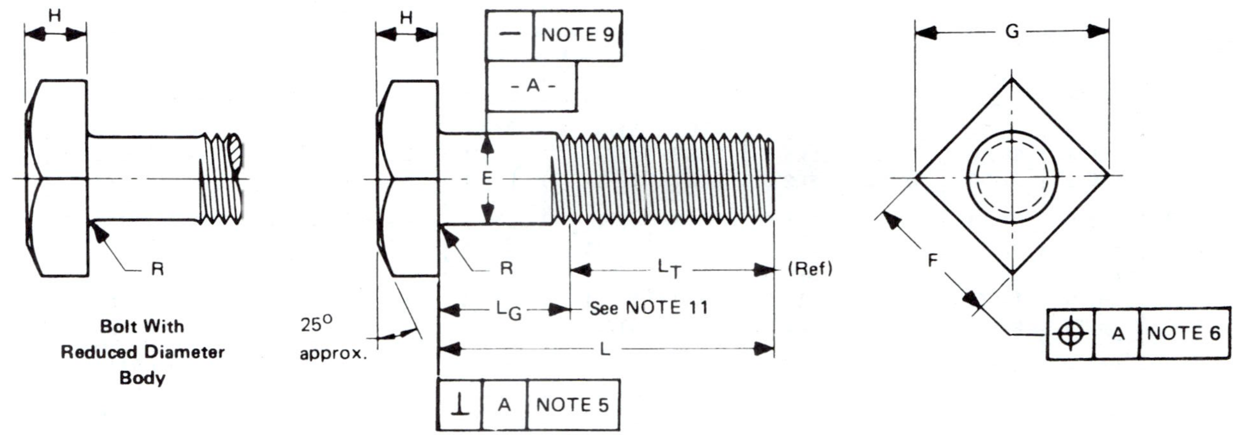

Nominal Size or Basic Product Dia (17)		E	F			G		H			R		L_T	
		Body Dia (7), (14)	Width Across Flats (4)			Wiuth Across Corners		Height			Radius of Fillet		Thread Length For Bolt Lengths (11)	
													6 in. and shorter	over 6 in.
		Max	Basic	Max	Min	Max	Min	Basic	Max	Min	Max	Min	Basic	Basic
1/4	0.2500	0.260	3/8	0.375	0.362	0.530	0.498	11/64	0.188	0.156	0.03	0.01	0.750	1.000
5/16	0.3125	0.324	1/2	0.500	0.484	0.707	0.665	13/64	0.220	0.186	0.03	0.01	0.875	1.125
3/8	0.3750	0.388	9/16	0.562	0.544	0.795	0.747	1/4	0.268	0.232	0.03	0.01	1.000	1.250
7/16	0.4375	0.452	5/8	0.625	0.603	0.884	0.828	19/64	0.316	0.278	0.03	0.01	1.125	1.375
1/2	0.5000	0.515	3/4	0.750	0.725	1.061	0.995	21/64	0.348	0.308	0.03	0.01	1.250	1.500
5/8	0.6250	0.642	15/16	0.938	0.906	1.326	1.244	27/64	0.444	0.400	0.06	0.02	1.500	1.750
3/4	0.7500	0.768	1 1/8	1.125	1.088	1.591	1.494	1/2	0.524	0.476	0.06	0.02	1.750	2.000
7/8	0.8750	0.895	1 5/16	1.312	1.269	1.856	1.742	19/32	0.620	0.568	0.06	0.02	2.000	2.250
1	1.0000	1.022	1 1/2	1.500	1.450	2.121	1.991	21/32	0.684	0.628	0.09	0.03	2.250	2.500
1 1/8	1.1250	1.149	1 11/16	1.688	1.631	2.386	2.239	3/4	0.780	0.720	0.09	0.03	2.500	2.750
1 1/4	1.2500	1.277	1 7/8	1.875	1.812	2.652	2.489	27/32	0.876	0.812	0.09	0.03	2.750	3.000
1 3/8	1.3750	1.404	2 1/16	2.062	1.994	2.917	2.738	29/32	0.940	0.872	0.09	0.03	3.000	3.250
1 1/2	1.5000	1.531	2 1/4	2.250	2.175	3.182	2.986	1	1.036	0.964	0.09	0.03	3.250	3.500

Table 6
Dimensions of Square Bolts
(Reprinted from the American Society of Engineers—ANSI B18.2.1—1981)

AMERICAN NATIONAL STANDARD SQUARE AND HEX NUTS

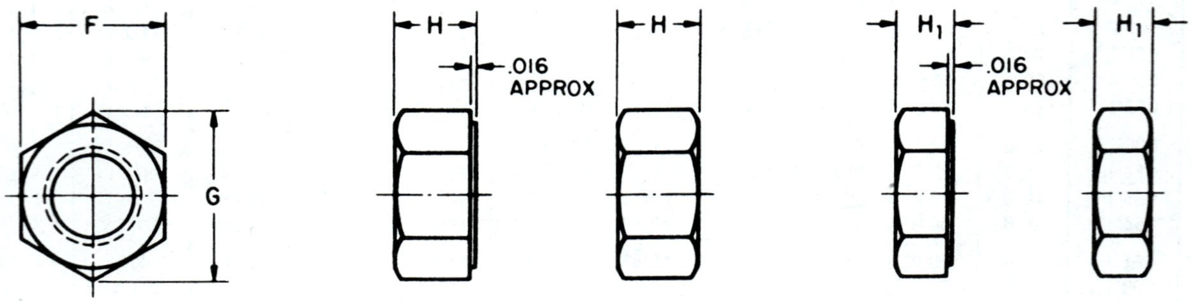

Nominal Size or Basic Major Dia of Thread		F			G		H			H_1			Hex Nuts Specified Proof Load		Jam Nuts
		Width Across Flats			Width Across Corners		Thickness Hex Nuts			Thickness Hex Jam Nuts			Up to 150,000 psi	150,000 psi and Greater	All Strength Levels
		Basic	Max	Min	Max	Min	Basic	Max	Min	Basic	Max	Min	Runout of Bearing Face, FIR Max		
1/4	0.2500	7/16	0.438	0.428	0.505	0.488	7/32	0.226	0.212	5/32	0.163	0.150	0.015	0.010	0.015
5/16	0.3125	1/2	0.500	0.489	0.577	0.557	17/64	0.273	0.258	3/16	0.195	0.180	0.016	0.011	0.016
3/8	0.3750	9/16	0.562	0.551	0.650	0.628	21/64	0.337	0.320	7/32	0.227	0.210	0.017	0.012	0.017
7/16	0.4375	11/16	0.688	0.675	0.794	0.768	3/8	0.385	0.365	1/4	0.260	0.240	0.018	0.013	0.018
1/2	0.5000	3/4	0.750	0.736	0.866	0.840	7/16	0.448	0.427	5/16	0.323	0.302	0.019	0.014	0.019
9/16	0.5625	7/8	0.875	0.861	1.010	0.982	31/64	0.496	0.473	5/16	0.324	0.301	0.020	0.015	0.020
5/8	0.6250	15/16	0.938	0.922	1.083	1.051	35/64	0.559	0.535	3/8	0.387	0.363	0.021	0.016	0.021
3/4	0.7500	1 1/8	1.125	1.088	1.299	1.240	41/64	0.665	0.617	27/64	0.446	0.398	0.023	0.018	0.023
7/8	0.8750	1 5/16	1.312	1.269	1.516	1.447	3/4	0.776	0.724	31/64	0.510	0.458	0.025	0.020	0.025
1	1.0000	1 1/2	1.500	1.450	1.732	1.653	55/64	0.887	0.831	35/64	0.575	0.519	0.027	0.022	0.027
1 1/8	1.1250	1 11/16	1.688	1.631	1.949	1.859	31/32	0.999	0.939	39/64	0.639	0.579	0.030	0.025	0.030
1 1/4	1.2500	1 7/8	1.875	1.812	2.165	2.066	1 1/16	1.094	1.030	23/32	0.751	0.687	0.033	0.028	0.033
1 3/8	1.3750	2 1/16	2.062	1.994	2.382	2.273	1 11/64	1.206	1.138	25/32	0.815	0.747	0.036	0.031	0.036
1 1/2	1.5000	2 1/4	2.250	2.175	2.598	2.480	1 9/32	1.317	1.245	27/32	0.880	0.808	0.039	0.034	0.039
See Notes	9	3			4								2		

Table 7
Dimensions of Hex Nuts and Hex Jam Nuts
(Reprinted from the American Society of Engineers—ANSI B18.2.2—1972)

SLOTTED
FLAT

Type of Head

AMERICAN NATIONAL STANDARD
MACHINE SCREWS AND MACHINE SCREW NUTS

ANSI B18.6.3–1972

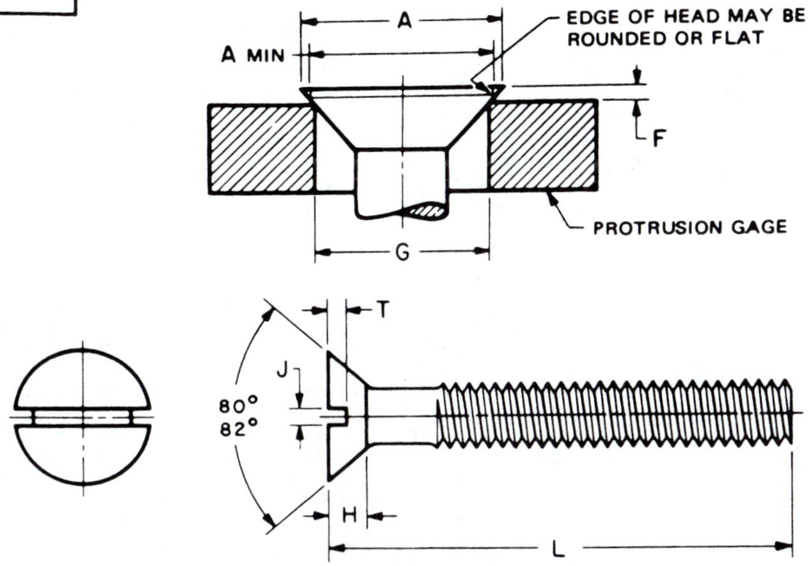

Nominal Size[1] or Basic Screw Diameter	L[2] These Lengths or Shorter are Undercut	A Head Diameter Max, Edge Sharp	A Head Diameter Min, Edge Rounded or Flat	H[3] Head Height Ref	J Slot Width Max	J Slot Width Min	T Slot Depth Max	T Slot Depth Min	F[4] Protrusion Above Gaging Diameter Max	F[4] Protrusion Above Gaging Diameter Min	G[4] Gaging Diameter
0000 0.0210	—	0.043	0.037	0.011	0.008	0.004	0.007	0.003	*	*	*
000 0.0340	—	0.064	0.058	0.016	0.011	0.007	0.009	0.005	*	*	*
00 0.0470	—	0.093	0.085	0.028	0.017	0.010	0.014	0.009	*	*	*
0 0.0600	1/8	0.119	0.099	0.035	0.023	0.016	0.015	0.010	0.026	0.016	0.078
1 0.0730	1/8	0.146	0.123	0.043	0.026	0.019	0.019	0.012	0.028	0.016	0.101
2 0.0860	1/8	0.172	0.147	0.051	0.031	0.023	0.023	0.015	0.029	0.017	0.124
3 0.0990	1/8	0.199	0.171	0.059	0.035	0.027	0.027	0.017	0.031	0.018	0.148
4 0.1120	3/16	0.225	0.195	0.067	0.039	0.031	0.030	0.020	0.032	0.019	0.172
5 0.1250	3/16	0.252	0.220	0.075	0.043	0.035	0.034	0.022	0.034	0.020	0.196
6 0.1380	3/16	0.279	0.244	0.083	0.048	0.039	0.038	0.024	0.036	0.021	0.220
8 0.1640	1/4	0.332	0.292	0.100	0.054	0.045	0.045	0.029	0.039	0.023	0.267
10 0.1900	5/16	0.385	0.340	0.116	0.060	0.050	0.053	0.034	0.042	0.025	0.313
12 0.2160	3/8	0.438	0.389	0.132	0.067	0.056	0.060	0.039	0.045	0.027	0.362
1/4 0.2500	7/16	0.507	0.452	0.153	0.075	0.064	0.070	0.046	0.050	0.029	0.424
5/16 0.3125	1/2	0.635	0.568	0.191	0.084	0.072	0.088	0.058	0.057	0.034	0.539
3/8 0.3750	9/16	0.762	0.685	0.230	0.094	0.081	0.106	0.070	0.065	0.039	0.653
7/16 0.4375	5/8	0.812	0.723	0.223	0.094	0.081	0.103	0.066	0.073	0.044	0.690
1/2 0.5000	3/4	0.875	0.775	0.223	0.106	0.091	0.103	0.065	0.081	0.049	0.739
9/16 0.5625	—	1.000	0.889	0.260	0.118	0.102	0.120	0.077	0.089	0.053	0.851
5/8 0.6250	—	1.125	1.002	0.298	0.133	0.116	0.137	0.088	0.097	0.058	0.962
3/4 0.7500	—	1.375	1.230	0.372	0.149	0.131	0.171	0.111	0.112	0.067	1.186

[1]Where specifying nominal size in decimals, zeros preceding decimal and in the fourth decimal place shall be omitted.
[2]Screws of these lengths and shorter shall have undercut heads as shown in Table 5.
[3]Tabulated values determined from formula for maximum H, Appendix V.
[4]No tolerance for gaging diameter is given. If the gaging diameter of the gage used differs from tabulated value, the protrusion will be affected accordingly and the proper protrusion values must be recalculated using the formulas shown in Appendix I.
*Not practical to gage.

Table 8
Dimensions of Slotted Flat Countersunk Head Machine Screws
(Reprinted from the American Society of Engineers—ANSI B18.6.3—1972)

AMERICAN STANDARD

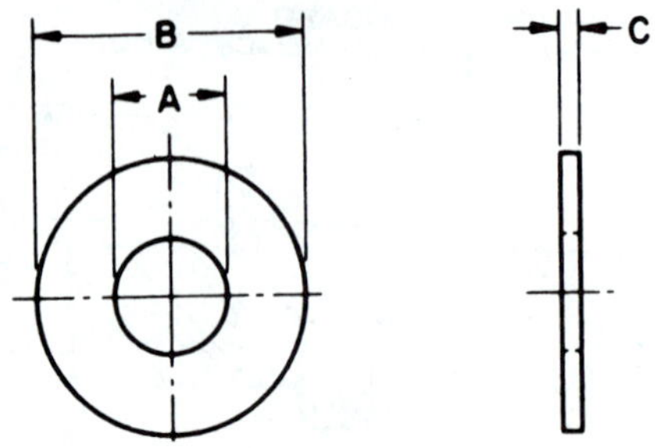

Nominal Washer Size***			Inside Diameter A			Outside Diameter B			Thickness C		
			Basic	Tolerance		Basic	Tolerance		Basic	Max	Min
				Plus	Minus		Plus	Minus			
–	–		0.078	0.000	0.005	0.188	0.000	0.005	0.020	0.025	0.016
–	–		0.094	0.000	0.005	0.250	0.000	0.005	0.020	0.025	0.016
–	–		0.125	0.008	0.005	0.312	0.008	0.005	0.032	0.040	0.025
No. 6	0.138		0.156	0.008	0.005	0.375	0.015	0.005	0.049	0.065	0.036
No. 8	0.164		0.188	0.008	0.005	0.438	0.015	0.005	0.049	0.065	0.036
No. 10	0.190		0.219	0.008	0.005	0.500	0.015	0.005	0.049	0.065	0.036
3/16	0.188		0.250	0.015	0.005	0.562	0.015	0.005	0.049	0.065	0.036
No. 12	0.216		0.250	0.015	0.005	0.562	0.015	0.005	0.065	0.080	0.051
1/4	0.250	N	0.281	0.015	0.005	0.625	0.015	0.005	0.065	0.080	0.051
1/4	0.250	W	0.312	0.015	0.005	0.734*	0.015	0.007	0.065	0.080	0.051
5/16	0.312	N	0.344	0.015	0.005	0.688	0.015	0.007	0.065	0.080	0.051
5/16	0.312	W	0.375	0.015	0.005	0.875	0.030	0.007	0.083	0.104	0.064
3/8	0.375	N	0.406	0.015	0.005	0.812	0.015	0.007	0.065	0.080	0.051
3/8	0.375	W	0.438	0.015	0.005	1.000	0.030	0.007	0.083	0.104	0.064
7/16	0.438	N	0.469	0.015	0.005	0.922	0.015	0.007	0.065	0.080	0.051
7/16	0.438	W	0.500	0.015	0.005	1.250	0.030	0.007	0.083	0.104	0.064
1/2	0.500	N	0.531	0.015	0.005	1.062	0.030	0.007	0.095	0.121	0.074
1/2	0.500	W	0.562	0.015	0.005	1.375	0.030	0.007	0.109	0.132	0.086
9/16	0.562	N	0.594	0.015	0.005	1.156*	0.030	0.007	0.095	0.121	0.074
9/16	0.562	W	0.625	0.015	0.005	1.469*	0.030	0.007	0.109	0.132	0.086
5/8	0.625	N	0.656	0.030	0.007	1.312	0.030	0.007	0.095	0.121	0.074
5/8	0.625	W	0.688	0.030	0.007	1.750	0.030	0.007	0.134	0.160	0.108
3/4	0.750	N	0.812	0.030	0.007	1.469	0.030	0.007	0.134	0.160	0.108
3/4	0.750	W	0.812	0.030	0.007	2.000	0.030	0.007	0.148	0.177	0.122
7/8	0.875	N	0.938	0.030	0.007	1.750	0.030	0.007	0.134	0.160	0.108
7/8	0.875	W	0.938	0.030	0.007	2.250	0.030	0.007	0.165	0.192	0.136
1	1.000	N	1.062	0.030	0.007	2.000	0.030	0.007	0.134	0.160	0.108
1	1.000	W	1.062	0.030	0.007	2.500	0.030	0.007	0.165	0.192	0.136
1 1/8	1.125	N	1.250	0.030	0.007	2.250	0.030	0.007	0.134	0.160	0.108
1 1/8	1.125	W	1.250	0.030	0.007	2.750	0.030	0.007	0.165	0.192	0.136
1 1/4	1.250	N	1.375	0.030	0.007	2.500	0.030	0.007	0.165	0.192	0.136
1 1/4	1.250	W	1.375	0.030	0.007	3.000	0.030	0.007	0.165	0.192	0.136
1 3/8	1.375	N	1.500	0.030	0.007	2.750	0.030	0.007	0.165	0.192	0.136
1 3/8	1.375	W	1.500	0.045	0.010	3.250	0.045	0.010	0.180	0.213	0.153
1 1/2	1.500	N	1.625	0.030	0.007	3.000	0.030	0.007	0.165	0.192	0.136
1 1/2	1.500	W	1.625	0.045	0.010	3.500	0.045	0.010	0.180	0.213	0.153
1 5/8	1.625		1.750	0.045	0.010	3.750	0.045	0.010	0.180	0.213	0.153
1 3/4	1.750		1.875	0.045	0.010	4.000	0.045	0.010	0.180	0.213	0.153
1 7/8	1.875		2.000	0.045	0.010	4.250	0.045	0.010	0.180	0.213	0.153
2	2.000		2.125	0.045	0.010	4.500	0.045	0.010	0.180	0.213	0.153
2 1/4	2.250		2.375	0.045	0.010	4.750	0.045	0.010	0.220	0.248	0.193
2 1/2	2.500		2.625	0.045	0.010	5.000	0.045	0.010	0.238	0.280	0.210
2 3/4	2.750		2.875	0.065	0.010	5.250	0.065	0.010	0.259	0.310	0.228
3	3.000		3.125	0.065	0.010	5.500	0.065	0.010	0.284	0.327	0.249

*The 0.734 in., 1.156 in., and 1.469 in. outside diameters avoid washers which could be used in coin operated devices.

**Preferred sizes are for the most part from series previously designed "Standard Plate" and "SAE." Where common sizes existed in the two series, the SAE size is designated "N" (narrow) and the Standard Plate "W" (wide). These sizes as well as all other sizes for Type A Plain Washers are to be ordered by ID, OD, and thickness dimensions.

***Nominal washer sizes are intended for use with comparable nominal screw or bolt sizes.

Table 9

Dimensions of Preferred Sizes of Type A Plain Washers**

(Reprinted frm The American Society of Mechanical Engineers—ANSI B18.22.1—1965, R[1975])

DECIMAL EQUIVALENTS AND TAP DRILL SIZES
OF WIRE GAGE LETTER AND FRACTIONAL SIZE DRILLS
(TAP DRILL SIZES BASED ON 75% MAXIMUM THREAD)

FRACTIONAL SIZE DRILLS INCHES	WIRE GAGE DRILLS	DECIMAL EQUIVALENT INCHES	TAP SIZES SIZE OF THREAD	TAP SIZES THREADS PER INCH
	80	.0135		
	79	.0145		
1/64	.....	.0156		
	78	.0160		
	77	.0180		
	76	.0200		
	75	.0210		
	74	.0225		
	73	.0240		
	72	.0250		
	71	.0260		
	70	.0280		
	69	.0292		
	68	.0310		
1/32	.....	.0312		
	67	.0320		
	66	.0330		
	65	.0350		
	64	.0360		
	63	.0370		
	62	.0380		
	61	.0390		
	60	.0400		
	59	.0410		
	58	.0420		
	57	.0430		
	56	.0465		
3/64	.....	.0469	0	80
	55	.0520		
	54	.0550		
	53	.0595	1	64
1/16	.....	.0625		72
	52	.0635		
	51	.0670		
	50	.0700	2	56
	49	.0730		64
	48	.0760		
5/64	.....	.0781		
	47	.0785	3	48
	46	.0810		
	45	.0820	3	56
	44	.0860		
	43	.0890	4	40
	42	.0935	4	48
3/32	.....	.0937		
	41	.0960		
	40	.0980		
	39	.0995		
	38	.1015	5	40
	37	.1040	5	44
	36	.1065	6	32
7/64	.....	.1094		
	35	.1100		
	34	.1110		
	33	.1130	6	40
	32	.1160		
	31	.1200		
1/8	.....	.1250		
	30	.1285		
	29	.1360	8	32
	28	.1405		36

FRACTIONAL SIZE DRILLS INCHES	WIRE GAGE DRILLS	DECIMAL EQUIVALENT INCHES	TAP SIZES SIZE OF THREAD	TAP SIZES THREADS PER INCH
9/64	.....	.1406		
	27	.1440		
	26	.1470		
	25	.1495	10	24
	24	.1520		
	23	.1540		
5/32	.....	.1562		
	22	.1570		
	21	.1590	10	32
	20	.1610		
	19	.1660		
	18	.1695		
11/64	.....	.1719		
	17	.1730		
	16	.1770	12	24
	15	.1800		
	14	.1820	12	28
	13	.1850		
3/16	.....	.1875		
	12	.1890		
	11	.1910		
	10	.1935		
	9	.1960		
	8	.1990		
	7	.2010	¼	20
13/64	.....	.2031		
	6	.2040		
	5	.2055		
	4	.2090		
	3	.2130	¼	28
7/32	.....	.2187		
	2	.2210		
	1	.2280		
	A	.2340		
15/64	.....	.2344		
	B	.2380		
	C	.2420		
	D	.2460		
1/4	E	.2500		
	F	.2570	5/16	18
	G	.2610		
17/64	.....	.2656		
	H	.2660		
	I	.2720	5/16	24
	J	.2770		
	K	.2810		
9/32	.....	.2812		
	L	.2900		
	M	.2950		
19/64	.....	.2969		
	N	.3020		
5/16	.....	.3125	⅜	16
	O	.3160		
	P	.3230		
21/64	.....	.3281		
	Q	.3320	⅜	24
	R	.3390		
11/32	.....	.3437		
	S	.3480		
	T	.3580		

FRACTIONAL SIZE DRILLS INCHES	WIRE GAGE DRILLS	DECIMAL EQUIVALENT INCHES	TAP SIZES SIZE OF THREAD	TAP SIZES THREADS PER INCH
23/64	.....	.3594		
	U	.3680	7/16	14
3/8	.....	.3750		
	V	.3770		
	W	.3860		
25/64	.....	.3906	7/16	20
	X	.3970		
	Y	.4040		
13/32	.....	.4062		
	Z	.4130		
27/64	.....	.4219	½	13
7/16	.....	.4375		
29/64	.....	.4531	½	20
15/32	.....	.4687		
31/64	.....	.4844	9/16	12
1/2	.....	.5000		
33/64	.....	.5156	9/16	18
17/32	.....	.5312	⅝	11
35/64	.....	.5469		
9/16	.....	.5625		
37/64	.....	.5781	⅝	18
19/32	.....	.5937		
39/64	.....	.6094		
5/8	.....	.6250		
41/64	.....	.6406		
21/32	.....	.6562	¾	10
43/64	.....	.6719		
11/16	.....	.6875	¾	16
45/64	.....	.7031		
23/32	.....	.7187		
47/64	.....	.7344		
3/4	.....	.7500		
49/64	.....	.7656	⅞	9
25/32	.....	.7812		
51/64	.....	.7969		
13/16	.....	.8125	⅞	14
53/64	.....	.8281		
27/32	.....	.8437		
55/64	.....	.8594		
7/8	.....	.8750	1	8
57/64	.....	.8906		
29/32	.....	.9062		
59/64	.....	.9219		
15/16	.....	.9375	1	12
61/64	.....	.9531		
31/32	.....	.9687		
63/64	.....	.9844	1⅛	7
1	.....	1.0000		

Table 10
Drill and Tap Sizes
Reprinted from Technical Drawing and Design *by Goetsch and Nelson, © 1986 Delmar Publishers, Inc.)*

20 Degree Pressure Angles

Diametral Pitch	Circular Pitch	Circular Thickness	Addend.	Dedend.	Standard Fillet Radius
D.P.	C.P.	C.T.	A	D	
1	3.1416	1.5708	1.0000	1.2500	0.3000
2	1.5708	0.7854	0.5000	0.6250	0.1500
2.5	1.2566	0.6283	0.4000	0.5000	0.1200
3	1.0472	0.5236	0.3333	0.4167	0.1000
3.5	0.8976	0.4488	0.2857	0.3571	0.0857
4	0.7854	0.3927	0.2500	0.3125	0.0750
4.5	0.6981	0.3491	0.2222	0.2778	0.0667
5	0.6283	0.3142	0.2000	0.2500	0.0600
5.5	0.5712	0.2856	0.1818	0.2273	0.0545
6	0.5236	0.2618	0.1667	0.2083	0.0500
6.5	0.4833	0.2417	0.1538	0.1923	0.0462
7	0.4488	0.2244	0.1429	0.1786	0.0429
7.5	0.4189	0.2094	0.1333	0.1667	0.0400
8	0.3927	0.1963	0.1250	0.1563	0.0375
8.5	0.3696	0.1848	0.1176	0.1471	0.0353
9	0.3491	0.1745	0.1111	0.1389	0.0333
9.5	0.3307	0.1653	0.1053	0.1316	0.0316
10	0.3142	0.1571	0.1000	0.1250	0.0300
11	0.2856	0.1428	0.0909	0.1136	0.0273
12	0.2618	0.1309	0.0833	0.1042	0.0250
13	0.2417	0.1208	0.0769	0.0962	0.0231
14	0.2244	0.1122	0.0714	0.0893	0.0214
15	0.2094	0.1047	0.0667	0.0833	0.0200

Table 11

Spur/Pinion Gear Tooth Parts

Reprinted from Technical Drawing and Design *by Goetsch and Nelson, © 1986 Delmar Publishers, Inc.)*

Shaft Diameter	5/16 to 3/8	7/16 to 1/2	9/16 to 3/4	13/16 to 15/16	1 to 1 3/16	1 1/4 to 1 7/16	1 1/2 to 1 3/4	1 13/16 to 2 1/8	2 3/16 to 2 1/2
Key Numbers	204	304 305	404 405 406	505 506 507	606 607 608 609	807 808 809	810 811 812	1011 1012	1211 1212

Table 12

Woodruff Key Sizes for Different Shaft Diameters

Reprinted from Technical Drawing and Design *by Goetsch and Nelson, © 1986 Delmar Publishers, Inc.)*

GAGE NO.	AMERICAN OR BROWN & SHARPE'S A.W.G. OR B. & S.	UNITED STATES STANDARD	MANU-FACTURERS' STANDARD FOR SHEET STEEL	GAGE NO.
0000000	....	.500	....	0000000
000000	.5800	.469	....	000000
00000	.5165	.438	....	00000
0000	.4600	.406	....	0000
000	.4096	.375	....	000
00	.3648	.344	....	00
0	.3249	.312	....	0
1	.2893	.281	....	1
2	.2576	.266	....	2
3	.2294	.250	.2391	3
4	.2043	.234	.2242	4
5	.1819	.219	.2092	5
6	.1620	.203	.1943	6
7	.1443	.188	.1793	7
8	.1285	.172	.1644	8
9	.1144	.156	.1495	9
10	.1019	.141	.1345	10
11	.0907	.125	.1196	11
12	.0808	.109	.1046	12
13	.0720	.0938	.0897	13
14	.0642	.0781	.0747	14
15	.0571	.0703	.0673	15
16	.0508	.0625	.0598	16
17	.0453	.0562	.0538	17
18	.0403	.0500	.0478	18
19	.0359	.0438	.0418	19
20	.0320	.0375	.0359	20
21	.0285	.0344	.0329	21
22	.0253	.0312	.0299	22
23	.0226	.0281	.0269	23
24	.0201	.0250	.0239	24
25	.0179	.0219	.0209	25
26	.0159	.0188	.0179	26
27	.0142	.0172	.0164	27
28	.0126	.0156	.0149	28
29	.0113	.0141	.0135	29
30	.0100	.0125	.0120	30
31	.0089	.0109	.0105	31
32	.0080	.0102	.0097	32
33	.0071	.00938	.0090	33
34	.0063	.00859	.0082	34
35	.0056	.00781	.0075	35
36	.0050	.00703	.0067	36

Table 13
Sheet Metal and Wire Gage Designation
Reprinted from Technical Drawing and Design *by Goetsch and Nelson, © 1986 Delmar Publishers, Inc.)*

TITLE OF STANDARD	
Abbreviations	Y1.1-1972
American National Standard Drafting Practices	
Size and Format	Y14.1-1980
Line Conventions and Lettering	Y14.2M-1979
Multi and Sectional View Drawings	Y14.3-1975(R1980)
Pictorial Drawing	Y14.4-1957
Dimensioning and Tolerancing	Y14.5M-1982
Screw Threads	Y14.6-1978
Screw Threads (Metric Supplement)	Y14.6aM-1981
Gears and Splines	
Spur, Helical, and Racks	Y14.7.1-1971
Bevel and Hyphoid	Y14.7.2-1978
Forgings	Y14.9-1958
Springs	Y14.13M-1981
Electrical and Electronic Diagram	Y14.15-1966(R1973)
Interconnection Diagrams	Y14.15a-1971
Information Sheet	Y14.15b-1973
Fluid Power Diagrams	Y14.17-1966(R1980)
Digital Representation for Communication of Product Definition Data	Y14.26M-1981
Computer-Aided Preparation of Product Definition Data Dictionary of Terms	Y14.26.3-1975
Digital Representation of Physical Object Shapes	Y14 Report
Guideline – User Instructions	Y14 Report No. 2
Guideline – Design Requirements	Y14 Report No. 3
Ground Vehicle Drawing Practices	In Preparation
Chasis Frames	Y14.32.1-1974
Parts Lists, Data Lists, and Index Lists	Y14.34M-1982
Surface Texture Symbols	Y14.36-1978
Illustrations for Publication and Projection	Y15.1M-1979
Time Series Charts	Y15.2M-1979
Process Charts	Y15.3M-1979
Graphic Symbols for:	
Electrical and Electronics Diagrams	Y32.2-1975
Plumbing	Y32.4-1977
Use on Railroad Maps and Profiles	Y32.7-1972(R1979)
Fluid Power Diagrams	Y32.10-1967(R1974)
Process Flow Diagrams in Petroleum and Chemical Industries	Y32.11-1961
Mechanical and Acoustical Element as Used in Schematic Diagrams	Y32.18-1972(R1978)
Pipe Fittings, Valves, and Piping	Z32.2.3-1949(R1953)
Heating, Ventilating, and Air Conditioning	Z32.2.4-1949(R1953)
Heat Power Apparatus	Z32.2.6-1950(R1956)
Letter Symbols for:	
Glossary of Terms Concerning Letter Symbols	Y10.1-1972
Hydraulics	Y10.2-1958
Quantities Used in Mechanics for Solid Bodies	Y10.3-1968
Heat and Thermodynamics	Y10.4-1982
Quantities Used in Electrical Science and Electrical Engineering	Y10.5-1968
Aeronautical Sciences	Y10.7-1954
Structural Analysis	Y10.8-1962
Meteorology	Y10.10-1953(R1973)
Acoustics	Y10.11-1953(R1959)
Chemical Engineering	Y10.12-1955(R1973)
Rocket Propulsion	Y10.14-1959
Petroleum Reservoir Engineering and Electric Logging	Y10.15-1958(R1973)
Shell Theory	Y10.16-1964(R1973)
Guide for Selecting Greek Letters Used as Symbols for Engineering Mathematics	Y10.17-1961(R1973)
Illuminating Engineering	Y10.18-1967(R1977)
Mathematical Signs and Symbols for Use in Physical Sciences and Technology	Y10.20-1975

Table 14

American Society of Mechanical Engineers Standards of Interest to Designers, Architects, and Drafters
Reprinted from Technical Drawing and Design *by Goetsch and Nelson, © 1986 Delmar Publishers, Inc.)*

Appendix B

Abbreviations

Lettered notations and words on working drawings are often abbreviated to save time and space. It is critical that the persons reading the working drawings will be able to interpret the abbreviations correctly. The following are some considerations to understand when using abbreviations:

- Do not use a period after an abbreviation unless the abbreviation is a whole word.
- Use capital letters for (most) abbreviations.
- Singular and plural abbreviations may be the same.
- The same abbreviation may apply to more than one word.
- Many abbreviations are quite close in their spelling.

Using the standardized abbreviations listed here will insure that all abbreviations will be read correctly.

Accessory	ACCESS.
Accumulate	ACCUM
Adaption	ADAPT.
Addition	ADD.
Addendum	ADD.
Alteration	ALT
Alternate	ALT
Alternating current	AC
Altitude	ALT
Aluminum	AL
American Iron and Steel Institute	AISI
American Society for Testing Materials	ASTM
American Society of Mechanical Engineers	ASME
American Wire Gage	AWG
Ampere	AMP
Approved	APPD
Approximate	APPROX
Assembly	ASM
Attach	ATT
Authorize	AUTH
Auxiliary	AUX
Average	AVG
Balance	BAL
Battery	BAT.
Bearing	BRG
Bill of material	B/M
Bracket	BRKT
Brass	BRS
British thermal unit	BTU
Bronze	BRZ
Brown and Sharpe	B&S
Bushing	BUSH.
Cadmium	CAD.
Capacity	CAP.
Carbon steel	CS
Cast iron	CI
Cast steel	CS
Casting	CST
Center	CTR
Centerline	(℄) or CL
Center of gravity	CG
Centigrade	C
Centimeter	CM
Chamfer	CHAM
Change	CHG
Check	CHK
Chief	CH
Chromium	CHR
Circular pitch	CP
Circumference	CIRC
Clockwise	CW
Cold drawn steel	CDS
Cold rolled steel	CRS
Company	CO
Composition	COMP
Compression, Compressor	COMP
Concentric	CONC
Condition	COND
Conductor	COND
Conduit	CDT
Connector	CONN
Continue	CONT
Control	CONT
Copper	CU
Corporation	CORP
Correspond	CORRES
Corrosion resistant steel	CRES
Counterbalance	CBAL
Counterbore	CBORE
Counterclockwise	CCW
Counterdrill	CDRILL
Countersink	CSK

Counterweight	CTWT
Cubic	CU
Cubic centimeter	CC
Cubic feet per minute	CFM
Cubic foot	CU FT.
Cubic inch	CU IN.
Cycle	CY
Dedendum	DED
Degree	(°) or DEG
Department	DEPT
Detail	DET
Develop	DEV
Deviation	DEV
Diagonal	DIAG
Diameter	DIA, θ
Diametral pitch	DP
Dimension	DIM.
Direct current	DC
Distance	DIST
Division	DIV
Drawing	DWG
Each	EA
Eccentric	ECC
Effective	EFF
Electric	ELEC
Elevation	ELEV
Engineer	ENGR
Engineering	ENGRG
Equipment	EQUIP.
Equivalent	EQUIV
Estimate	EST
Etcetera	ETC
Extension	EXT
External	EXT
Extrusion	EXTR
Fahrenheit	F
Feet per minute	FPM
Figure	FIG.
Fillister head	FIL HD
Finish	FIN.
Fitting	FTG
Flat	FL
Flat head	FHD
Flexible	FLEX.
Foot	(') or FT
Foot pounds	FT LB
Forging	FORG
Forward	FWD
Front	FRT
Gage (Gauge)	GA
Gallon	GAL
Galvanize	GALV
Gasket	GSKT
Generator	GEN
Grain	GRN
Gravity	G
Grind	GRD
Harden	HDN
Hardware	HDW
Head	HD
Heat treat	HT TR
Height	HGT
Hexagon	HEX
Horizontal	HORIZ
Hot rolled steel	HRS
Hour	HR
Hydraulic	HYD
Illustration	ILLUS
Inch	(") or IN.
Inch ounce	IN. OZ
Inch pounds	IN. LB
Inclusive	INCL
Information	INFO
Inside diameter	ID
Inside radius	IR
Inspection	INSP
Installation	INSTL
Instrument	INST
Insulation	INSL
Interchangeable	INTCHG
Intermediate	INTER.
Internal	INT
Joint	JT
Kilometer	KM
Knock out	KO
Laboratory	LAB
Left hand	LH
Length	LG
Lock washer	LWASH
Longitude, Longitudinal	LONG.
Lower	LWR
Lubricate	LUB
Machine	MACH
Magnesium	MG
Maintenance	MAINT
Malleable	MAL
Manufacture	MFR
Material	MATL
Maximum	MAX
Maximum material condition	MMC
Mechanical	MECH
Medium	MED
Memorandum	MEMO
Mercury	HG
Mile	MI
Miles per hour	MPH
Millimeter	MM
Minimum	MIN
Minute	(') or MIN
Miscellaneous	MISC
Modify	MOD
Molding	MLDG
Mounting	MTG
National	NATL
National Electrical Mfg. Association	NEMA
National Machine Tool Builders Association	NMTBA
Negative	(-) or NEG
Nickel	NI
No drawing	ND
Nominal	NOM
Nonstandard	NONSTD
Number	NO.
Obsolete	OBS
Of true position	OTP
On Center	OC
Operate	OPER
Opposite	OPP
Optional	OPT
Original	ORIG
Ounce	OZ
Outside diameter	OD
Outside radius	OR.
Oval head	OV HD
Overall	OA
Oxygen	OXY
Package, Packing	PKG
Page	P
Part	PT
Parting line	PL
Patent	PAT.
Pattern	PATT
Perpendicular	PERP
Pint	PT

Pitch	P
Pitch circle	PC
Pitch diameter	PD
Plan view	PV
Point on line	POL
Position	POSN
Positive	(+) or POS
Pound	LB
Pounds per square inch	PSI
Preliminary	PRELIM
Pressure	PRESS
Process	PROC
Product, Production	PROD.
Quality	QUAL
Quantity	QTY
Quart	QT
Quarter	QTR
Rear view	RV
Rectangular	RECT
Reduction	RED.
Reference	REF
Regardless of feature size	RFS
Regular	REG
Reinforce	REINF
Remove	REM
Require	REQ
Required	REQD
Resistor	RES
Reverse	REV
Revision	REV
Revolution	REV
Revolutions per minute	RPM
Right hand	RH
Root diameter	RD
Round	RD
Round head	RD HD
Rubber	RUB.
Screw	SCR
Screw threads	
American National coarse	NC
American National fine	NF
American National extra fine	NEF
American National 8 pitch	8N
American National 12 pitch	12N
American National 16 pitch	16N
American Standard taper pipe	NPT
American Standard straight pipe coupling	NPSC
American Standard taper	NPTF
Unified screw thread coarse	UNC
Unified screw thread fine	UNF
Unified screw thread extra fine	UNEF
Unified screw thread 8 thread	8UN
Unified screw thread 12 thread	12UN
Unified screw thread 16 thread	16UN
Unified screw thread special	UNS
Second	(") or SEC
Section	SECT.
Serial, Series	SER
Sheet	SH
Side view	SV
Sketch	SK
Society of Automotive Engineers	SAE
Special	SPL
Specification	SPEC
Specific gravity	SP GR
Spherical	SPHER
Spot face	SF
Spring	SPR
Square	SQ
Standard	STD
Steel	STL
Support	SUPT
Switch	S W
Symbol	SYM
Symmetrical	SYM
Synthetic	SYN
Tangent	TAN.
Technical	TECH
Teeth	T
Temperature	TEMP
Tensile strength	TS
Terminal	TERM.
Theoretical	THEO
Thickness	THK
Thread	THD
Through	THRU
Tolerance	TOL
Tracer	TCR
Trade mark	TM
Transformer	TRANS
Transmission	TRANS
Transverse	TRANSV
True length	TL
True position	TP
True view	TV
Turnbuckle	TRNBKL
Typical	TYP
Ultimate	ULT
United States	US
United States of America Standards Institute	USASI
United States gage	USG
Universal	UNIV
Unless otherwise specified	UOS
Upper	UPR
Vacuum	VAC
Velocity	V
Versus	VS
Vertical	VERT
Volt	V
Volume	VOL
Water line	WL
Watt	W
Weight	W T
Wood	W D
Wood screw	W D SCR
Wrought iron	WI
Yard	YD
Year	YR

Glossary

A

addendum The radial distance from the pitch circle to the top of a gear tooth.

aligned section Sections which result from the offsetting or revolvement of a part to line up with the cutting plane.

aligned dimensioning The dimensions are always placed in line with the direction of the dimension lines.

allowance Acceptable clearance between mating parts.

alphanumeric keyboard A computer keyboard that allows an operator to input letters and numerals into a computer.

AWG American Wire Gage. A standard numerical system for measuring wire sizes.

angle of thread The included angle formed between the major diameter of a thread and the depth of thread.

arc welding Welding metals with an electric arc.

ASA (American Standards Association)

ASME (American Society of Mechanical Engineers) The organization that coordinates and approves national standards for American industry.

assembly drawing A drawing that shows the relationship of various parts of an item when assembled.

auxiliary view A view of an object other than the normal six orthographic views. Primary auxiliary views are drawings projected directly from any of the six normal orthographic views. A secondary auxiliary view is a drawing projected from a primary auxiliary view.

axis A line around which parts rotate or are concentrically arranged.

axonometric drawing A pictorial drawing which shows the front, side, and top faces of an object in one view. The receding lines in axonometric drawings are always parallel. Isometric, dimetric, and trimetric drawings are types of axonometric drawings.

B

backing weld A weld used to add extra strength to a weld by adding additional welds to the backside of the material being welded.

baseline dimensioning Dimensioning to a common datum point or plane.

basic dimension An exact value used to describe the size, shape, and location of a feature.

basic hole size The standard size of any basic hole cutter.

basic shaft size The standard manufactured size of a shaft.

bell and spigot connections The connection between two pipes. The straight spigot end of one pipe is inserted into the flared-out end of the other pipe. The joint is sealed with caulking or a compression ring.

bevel An inclined surface which is not constructed at a right angle to the adjoining surface.

bilateral tolerance A tolerance which allows variation in both directions from a dimension.

bill of materials A list of materials specified for a project.

block diagram Connected blocks labeled to represent sections of an electronic system.

bolt circle A circular centerline which locates centers of holes positioned about a common center point.

boot A term which means to load a software program into a computer.

bore To enlarge a hole with a hand or machine tool.

bow compass A compass used to draw circles and arcs up to 6" in diameter.

brazing The joining of two pieces of metal with a brass alloy filler which melts about 800°F. This process is similar to soldering.

break line A line used to interrupt a drawing if an object will not fit on a drawing sheet.

brokenout section A section of an object broken away to reveal an interior feature for a sectional drawing.

buffer A device used for temporary electronic storage of information.

building line An imaginary line on a plot beyond which the building cannot extend.

bushing A hollow cylindrical sleeve bearing.

butt weld A welded joint which joins two pieces butted directly at their ends.

byte A single character of computer memory.

C

cabinet oblique drawing A pictorial drawing containing receding lines drawn at half scale, at an angle of 30° or 45° from the horizontal.

cableform diagram A drawing of groups of wires bound together into a cable or harness.

CAD Computer-aided drafting. The use of computers to aid in the preparation of technical drawings.

callout A drawing note connected to a drawing with a leader and arrowhead.

CAM Computer-aided manufacturing. Computer-augmented manufacturing. Computer-automated manufacturing.

cam A machine device used to transmit motion. It may change the speed, direction or type of motion.

cam follower A plunger whose reciprocating motion is produced by contact with the surface of a cam.

cavalier oblique drawing A pictorial drawing containing receding lines drawn to true scale.

center drill A drill used to drill the mounting holes in the end of a workpiece.

centerline A line that defines the common center of symmetrical objects.

chain dimensions Continuous dimensions that connect aligned features of an object without reference to a common datum point or plane.

chain line A line used to indicate a specific area.

chamfer An angle across a corner that eliminates a sharp edge.

character A symbol used to represent data in an organized system.

chordal addendum Distance from the top of a spur gear tooth to the chordal thickness line.

chordal thickness The thickness of a spur gear tooth measured on a chord on the circumference of the pitch diameter.

chord A straight line joining two points on a curve.

circular pitch The length of an arc along the pitch circle between the center of one gear tooth to the center of the next tooth.

circular thickness The length of an arc between the sides of a spur gear tooth measured on the pitch circle.

clearance fit A clearance space between mating parts due to the limits of the dimensions.

collar A flange or ring attached to a shaft or pipe.

command Instructions transmitted to a computer by an operator.

component One part of a multipart object or structure.

concave Surfaces which curve inwardly.

concentric Circles which share a common center.

construction lines Light lines used for the layout of a drawing.

convex Surfaces which curve outwardly.

counterbore To enlarge a hole to allow a screw or bolt head to be recessed below a surface.

coordinates Points of intersection of lines on a grid system.

CNC Computer-numerically controlled.

countersink A recess made to allow the tapered head of a fastener to lie below a surface.

CPU Central processing unit of a computer system.

crest The point of widest diameter of a screw thread which lies on a major diameter plane.

crosshatching Closely spaced parallel lines drawn obliquely to represent sectioned areas.

cursor A marker used to locate points on a video display screen (monitor) while working with a computer program.

cylindrical Having the form or properties of a cylinder.

cutting plane line A line drawn on a view where a cut was made in order to define the location of the imaginary sectional plane.

D

database A stored collection of computer information.

datum A line, point, or plane from which other locations are measured.

datum dimensioning A dimensioning system in which dimensions originate from a common datum.

dedendum The radial distance from the pitch circle to the bottom of a spur gear tooth.

default Action initiated through a computer software program unless an operator specifies other actions.

degree (°) A unit of angular measurement.

design size The size of a part after clearances and tolerances are applied.

detail drawing A dimensioned working drawing of a single part.

development drawing The drawing of a surface of an object unfolded on a flat plane. It is sometimes called a stretchout pattern. Development drawings are used in *pattern drafting.* Parallel line and radial line developments are used to develop patterns for objects that contain flat or single-curved surfaces. *Parallel line development* drawings are used to develop patterns for objects that contain parallel lines. *Radial line development is* used to develop patterns for objects that do not contain parallel lines. These methods cannot be used to develop patterns for warped surfaces. *Triangulation development is* a method of dividing a warped surface and other complex forms that cannot be developed with parallel or radial line development into a series of triangles, and transferring the true size of each triangle to a flat pattern.

diameter The length of a straight line which passes through the center of a circle and terminates at each end on the circumference.

diametral pitch A ratio of the number of teeth on a gear per inch of pitch diameter.

diazo A printing process that produces black, blue, or brown lines on surfaces. The process exposes a drawing with a sensitized material exposed to an ultraviolet light, then processed through an ammonia chamber.

digital circuit Electronic circuits which function through the use of binary numbers.

digital readout Devices that display numbers from the output of calculators and computers.

digitize To locate points and select computer commands with an input device.

digitizer An electronic tablet that serves as a drawing surface for the input of graphic data.

dimension Radial or linear length (width, height, or depth) labeled on a technical drawing.

dimetric drawing Pictorial drawings which contain two axes that form equal angles with the plane of projection.

direct dimensioning To be used when the distance between two points is critical to avoid tolerance buildup.

disk drive A computer input device that can read information on a rotating storage media.

displacement diagram A drawing which shows the path of a cam follower.

dividers A device used to divide and transfer measurements or for scribing arcs on hard surfaces.

documentation Drawings or printed information that contains instructions for assembling, installing, operating, and servicing.

DPI Dots per inch.

drum cam A cylinder with a groove in its surface which guides a follower through the groove to produce reciprocating motion.

drum plotter A cylindrical graphics pen plotter that accepts continuous feed paper.

dual dimensioning The use of the metric and the U.S. customary measuring systems on a drawing.

dump To transfer computer information from one system to another.

E

elbow drafter A manual drafting machine connected to the top of a drawing board.

electronics drawings Working drawings that depict electronic circuits and electronic items.

elements A distinct group of lines, shapes, and text, as defined by a computer operator; also referred to as entities. In development drawings, imaginary lines used to draw the pattern.

elevation drawing A drawing of the exterior or interior walls of a building. It is a perpendicular, or upright, projection from the floor plan to show vertical architectural or design details.

ellipse A closed curve in the form of a symmetrical oval.

engineering change documents Documents used to implement a specific change in a production drawing. Engineering Change Request (ECR), Engineering Change Notice (ECN), and Engineering Change Order (ECO).

engineering drawing A technical drawing in any field of engineering.

enter To insert information into a computer.

ergonomics The study of human space and movement needs.

exploded view A pictorial drawing which shows all parts disassembled, but in relation to each other.

extension lines Lines used to describe the size of an object.

F

face angle The angle between the top of a gear tooth and the gear axis. Also called a bevel angle.

feature An element of a part or object.

feature control frame Contains data for each controlled specification and includes the characteristic symbol, tolerance value, and datum reference letter.

FAO Finish all over.

fastener A device used to hold two or more objects together.

fillet Curved interior intersections between two or more surfaces.

finish Applications of materials to the surface of objects.

fit The relationship between mating parts.

fittings (pipe) The parts of a piping system used for joining pipe lengths together such as bends, elbows, tees, etc.

fixtures In architecture: electrical devices that may be secured to the ceiling or walls; In manufacturing: a tool used to position and clamp an item for a machine operation.

flange A circular rim around a pipe or shaft.

flatbed plotter A pen plotter where the drawing paper is placed on a flat surface and the pen carriage moves over the surface in a manner similar to handheld pens.

flat pattern A development or layout drawing which shows the true shape of an object before bending.

floppy disk A circular, thin, magnetic storage medium for computers available in 8", $5^1/4$", and $3^1/2$" diameters.

floor plan A horizontal, sectioned, working drawing showing all walls, windows, doors, and architectural details with the use of drawing symbols.

flow lines Lines representing the direction of a sequence of operations.

flute A groove on drills, reamers, and taps.

fold line Line of intersection between two planes in a multiview projection, also called a reference line. Representing the position of a fold to be made on flat materials such as paper, sheetmetal, and plastic.

footing An extension at the bottom of a building foundation wall which distributes the load into the ground.

force fit The joining of two mating parts which requires pressure to force the pieces permanently together.

foreshorten To show lines or objects shorter than their true lengths. Foreshortened lines are not perpendicular to the line of sight.

form tolerancing An allowable variation of a feature from the precise form as shown on a drawing.

foundation The basic structure on which a building rests.

free fit Liberal tolerances between mating parts which allow free movement without binding.

fulldivided scale A scale which contains full subdivision lines throughout the entire length of the scale.

full section A sectional drawing based on a cutting plane line which extends completely through an object.

fusion welding A welding process which results in the mixing of molten metals.

G

gage Numerical standard for the thickness of sheet metal and wire.

gas welding A welding process produced from one or more burning gases. A filler metal and pressure may or may not be used.

geometric dimensioning and tolerancing The assignment of tolerances which prescribes allowable variances in shape, angle, straightness, and position of features and parts.

gear A toothed wheel used to transmit power or motion between shafts.

general tolerance note When all tolerances are identical, the tolerance may be listed once in a note.

groove weld A weld which deposits metal in a groove between two members.

graphics tablet An electronic flat surface on which a stylus or mouse is moved to transmit information, digitize points, or pick commands.

grid size (CAD) The size of the grid or series of light dots which appears on the screen to aid the operator. The grid will not print on the final drawing.

H

half-section A sectional drawing based on a cutting plane line that cuts through one quarter of an object. A half-section reveals half of the interior and half of the exterior.

hard copy The output of a computer printed or plotted on paper or film.

hard disk A hard disk is a storage device that may be a peripheral unit or built directly into the CPU. A hard disk holds more data than a microdisk. It stores and retrieves data quickly.

hardware The physical components of a computer system.

harness drawing A drawing showing the layout or pattern for making an electrical harness or cable.

helix An evenly spaced spiral curve around a cylinder.

hems Formed along the edges of sheet metal pieces to seal in the sharp edges.

hexagon A polygon with six equal sides.

hidden lines Dashed lines which represent object lines which fall behind the object's surface.

highway diagram A point-to-point diagram used to simplify electrical interconnecting complex lines, sometimes called a trunkline diagram.

hinged planes The imaginary folded planes used in orthographic and auxiliary projections.

HVAC Heating, ventilation, and air conditioning.

I

IC Integrated circuit. An electronic circuit made on a small chip of silicon, arsenic, or germanium.

IEC International Electrotechnical Commission

IEEE Institute of Electrical and Electronics Engineers

IFI Industrial Fasteners Institute.

inclined Any line or plane located at an angle other than 90° from an orthographic plane.

input The insertion of information into a computer.

interchangeable Parts manufactured with consistent dimensional standards which enable use with any parent object.

interference fit When parts have negative clearances and must be forced together. Used when mating parts are designed to be permanently attached.

involute curve A spiral curve which follows a point on a string as it unwinds from a cylinder or other shapes.

ISO International Organization for Standardization

isometric drawing A pictorial drawing which contains receding planes at 30° from the horizon.

J

jig A device which holds a workpiece and guides the cutting tool.

K

key A device embedded partially into two mating parts to prevent movement.

keyseat A slot which holds a key.

kilo (k) Metric prefix for one thousand (10^3).

L

lay The direction of the dominant surface pattern made by machining operations.

layer Separate overlay drawings using the same base drawing.

lead The lateral distance a screw thread moves in one revolution.

leaders Used to connect a specific area on a drawing to a dimension or notation.

limits Acceptable variances of a dimension, or, on a CAD monitor, the size of the drawing area.

line conventions Standardization of lines used on technical drawings by line weight and style.

line of vision Imaginary lines connecting the item being drawn with the picture plane.

LMC Least material condition. The least amount of material possible in the size of a part. The LMC of an external feature is its lower limit. The LMC of an internal feature is its upper limit.

location dimensions Dimensions which show the exact location of parts of an object.

longitudinal section A full section through the length of a structure.

M

magnetic tape A magnetized strip of plastic film on which information is stored.

mainframe A central processing unit networked to individual satellite terminals (workstations).

major diameter The outside diameter of the threaded portion of a fastener measured from crest to crest.

master An original drawing or intermediate print from which reproductions are produced.

melt-through welds The depth of the penetration of the weld. The depth of the meltthrough is shown on the drawing's weld notation symbol It is used with all types of groove welds.

menu Listings of specialized tasks in a software program.

metric system The international measurement system based on units in multiples of 10.

micro (μ) Metric prefix for one millionth (10^6) microcomputer A computer which uses a microprocessor as a basic CPU element. A microcomputer is a standalone system. It can process information without a link to a mainframe computer.

microfilm Film used to store photographically reduced drawings.

microprocessor A miniaturized electronic circuit device used in personal computers.

MIL STD Military Standards.

milli Metric prefix for one thousandth (10^3).

minor diameter The diameter of the thread measured from root to root, perpendicular to the axis.

MMC Maximum material condition. The most amount of material possible in the size of a part. The MMC of an external feature is its upper limit. The MMC of an internal feature is its lower limit.

modem A device used to transmit data over telephone lines.

monitor The display screen. Monitors may be monochrome (one color, usually green or amber) or may have full-color capabilities.

mouse A graphics input device which is moved across a flat surface to control the movement of a cursor on the screen.

multiview drawings Views of an object projected onto two or more orthographic planes.

mylar Polyester plastic drafting film.

N

National Coarse (NC) American Standard coarse screw thread series.

Nabonal Fine (NF) American Standard fine screw thread series.

NC Numerical control. Control of a machine process with a computer program. Sometimes called computerized numerical control (CNC).

networking When standalone microcomputer systems are connected to a minicomputer or mainframe system in order to use the larger system's data base.

nominal size The designation of the stock size of a standard material before machining the surfaces.

normal surface True size surfaces that are parallel or perpendicular to the plane of projection.

notations Specify materials and any other necessary manufacturing information on a drawing.

O

object line A heavy solid line used on a drawing to represent the outline of an object.

oblique drawing Pictorial drawing in which the front plane is parallel to the plane of projection.

octagon An eight-sided geometric figure with each corner forming a 135-degree angle. Each center angle is 45 degrees.

offset section A sectional drawing created by a cutting plane bent at right angles to features as though they were in the same plane.

open-divided scale A scale where only one major unit is graduated with a full-divided unit. It is adjacent to zero.

optical disks Can store more data than any other storage device. Optical disks are used for permanent storage because laser storage data cannot be changed.

orthographic projection A method of representing three-dimensional objects on a plane having only length and breadth. First angle projections include the front, leftside, and bottom views. Third angle projections include the front, rightside, and top views.

output Processed computer data transmitted to a monitor, hard copy device, storage media, a control device, or another computer.

overall dimensions Dimensions which describe the total depth, height, or width of an object.

oxyacetylene welding The welding process that uses a mixture of oxygen and acetylene for the heat source.

P

parallel Lines which will never meet if extended any distance.

parallel slide A manual drafting system where a horizontal slide is permanently attached to the drawing board. It slides up and down and functions like a T-square.

pattern A drawing or a template prepared to the exact size and shape of a workpiece. Also, a model from which sand molds are made.

pattern development drawing A drawing which shows the outline of all connected surfaces of an object when laid flat.

PC Personal computer

PCB Printed circuit board. The electronic printed circuit assembly including the insulated board.

peripherals The supplemental equipment used in conjunction with, but not as part of, the computer system.

perpendicular A line or plane which is at a right angle to another line or plane.

perspective drawing A pictorial drawing which contains receding lines that converge at vanishing points on the horizon. A perspective drawing can use one, two, or three vanishing points.

pentagon A five-sided geometric figure with each corner forming a 118-degree angle. Each center angle is 72 degrees.

photodrafting The use of photographs for a working drawing.

phantom lines Lines used to show the alternate positions of an object or matching part without interfering with the main drawing.

pictorial drawing A drawing that shows the width, height, and depth of an object in one view.

picture plane Represents the vertical plane upon which the object is viewed.

pin graphics The drafting system that separates segments of working drawings onto separate drawing formats and keeps all the segments aligned together.

pinion gear The smaller gear in a set of mated gears.

piping drawing The working drawings of a pipe system drawn in isometric or with orthographic projecbon. A pipe may be represented by a single line or a double line describing the thickness of both the pipe and the fitting.

pitch A uniform distance from a point on one part to a corresponding point on an adjacent part.

pitch circle An imaginary circle concentric with a gear axis which passes through the thickest point on a gear tooth.

pitch diameter (spur gear) The diameter of a pitch circle.

pixels The series of dots which make up the picture on the monitor.

plane of projection The imaginary plane between the viewer and an object onto which the object is drawn.

plat plan An architectural drawing showing the relationship of the property to the adjacent community.

plot plan An architectural drawing showing the location of all structures on the property.

point-to-point wiring diagram An electronic drawing which shows components, terminals, and connecting wires.

polar coordinate dimensioning The use of both linear and angular dimensions for locating features.

polygon A multisided closed form.

positional tolerancing The allowable variation of a feature from the precise position shown on a drawing.

printed circuit Copper electronic circuits bonded to the surface of an insulated board.

printer A device which converts computer data into printed alphanumeric or graphic images. Printers are usually dot matrrx or laser operated.

prism A solid with base intersections that are parallel polygons and with sides that are parallelograms.

prototype An original functional model of a product constructed prior to mass production.

prompt Instructions on the monitor.

puck A graphics input device which is moved across a graphics tablet to control the movement of the cursor on the monitor.

R

rack A flat bar containing gear teeth which engage with the teeth in a pinion gear.

radius The length of a straight line connecting the center of a circle with a point on the circumference of a circle.

RAM Random-access memory. A temporary computer memory system that includes the data input from the operator, and the software program being used.

rectangular coordinate dimensioning The location of features using linear dimensions from specified XY or XYZ axes.

RFS Regardless of feature size. A condition which requires a tolerance of position regardless of where the feature lies within an object.

relief A groove between surfaces to provide clearance for machining.

rendering Adding embellishments to a drawing to provide a more realistic appearance.

removed section A sectional view removed from the area of the cutting plane and positioned in another location.

resistance welding Welding by using the resistance of metals to the passage of an electric current to produce fusion heat.

revolution drawing A drawing of the repositioned view after the primary orthographic drawing has been rotated. The first view projected from a rotated primary orthographic view is the *primary* revolution. Rotating a primary revolution and projecting a new view is a successive revolution.

revolved section A sectional view that is revolved 90° and perpendicular with the plane of projection.

rise The vertical height of a roof, of a step, or a flight of stairs.

ROM Read-only memory. A permanent computer memory system that allows the computer to read the software programs.

root diameter The diameter of a circle aligned with the gear tooth roots.

root The deepest point or valley of a thread.

root opening The distance between members at the root of a weld joint.

round A curved exterior corner of two surfaces.

run The width of a step or the horizontal distance covered by a flight of stairs. This term is also used to describe the horizontal length of a rafter.

runout The area of intersection of two different shaped surfaces.

S

scale drawing A drawing prepared proportionally smaller or larger than the object it represents.

schematic diagram A drawing of an electronic circuit showing symbols for electronic components and lines representing connecting conductors.

screw thread A helical thread formed on the outside of a cylindrical shaft. Screws with similar combinations of pitches and diameters into several related series: coarse thread (UNC, NC), fine thread (UNF, NF), and extra fine thread (UNEF, NEF).

seam The line of intersection between two materials.

section lines Section lines are used to represent the material through which a cut is made in order to show an interior sectional view.

sectional view A drawing which shows the interior of an object as it would appear if cut in half or quartered.

setback The allowable building distance from the property line of a lot for a building. This distance determines the area of the lot on which a structure can be built. Local zoning regulabons determine these distances.

shaft A cylinder to which rotating machine parts may be attached to transmit power and motion.

SI International System of Units. The metric system of measurement as adopted by the General Conference of Weights and Measures.

site plan An architectural drawing depicting an area or plot of land with defined property lines.

slab foundation A concrete floor and foundation system poured directly on the ground.

slope diagram A drawing that indicates a comparison between the horizontal run and the rise of a roof.

snap interval The distance the cursor will jump (snap) on the display screen.

software Computer programs which contain specific instructions for the functioning of a computer system.

solid modeling The creation of a three-dimensional drawing of the exterior of an object on a CAD system with no hidden features revealed.

solid-state welding The process of heating two metal pieces to their melting point, then holding them together with pressure until the hot atoms for both pieces mix together and cool.

span The distance between structural building supports.

specification A detailed description of components, parts, or materials which includes size, color, manufacturers' number, and sometimes cost.

spline One of a number of keyways cut around a shaft. Flexible strips used for drawing curves.

spot weld A resistance weld operation that joins metal by welding individual spots rather than a conbnuous bead.

spur gear A type of gear with its teeth straight and parallel to an axis shaft.

station point Represents the location of the observer for a perspective drawing.

stretchout A pattern development drawing showing the exact shape of a flat material before forming into a three-dimensional shape.

subassembly An assembled part that is a part of a larger assembly.

surface texture The waviness, roughness, lay, and flaws of a surface.

surface weld The surface buildup of a metal surface using a metal filler and heat.

survey plan An architectural drawing showing the geological aspects of the property.

symbol A graphic form used to represent a standard object, part, material, or component on a drawing.

T

tabular outline dimensioning Dimensions from perpendicular datum planes listed in a table.

tangent A line that intersects a circle or an arc at a point 90° from the radius.

taper A wedge or conical shape that increases or decreases in size at a uniform rate.

taper reamer A reamer which produces tapered holes.

technical illustration A pictorial drawing used to interpret technical working drawings.

template A flat form used as a guide to draw symbols and holes.

thin wall section Used when materials are too thin to add section lining. Thin wall sections remain solid.

thread depth The distance from the crest (major diameter) to the root (minor diameter) measured perpendicular to the axis.

three-dimensional geometric forms Forms possessing height, width, and depth.

tolerance The amount of variation allowed in the dimension of an object. Plus and minus tolerancing dimensions indicate the tolerance range above and below the basic dimension.

torque Rotating or twisting force.

TPI Threads per inch.

track drafter A manual drafting machine that includes a vertical track that slides left and right on a horizontal track. A protractor head is attached to the vertical track.

train Meshed gears in a series.

transitional fit Mating parts that may have either a clearance or interference fit.

transverse section A full section through the width of a structure.

trimetric drawing A pictorial drawing in which the three principal axes are unequally foreshortened.

true length (true size) The condition which results when the line of sight is perpendicular to surfaces or lines.

true position The exact position of a feature.

truncated An object with the apex, vertex, or end removed by an angular plane.

two-dimensional geometric forms Forms possessing height and width.

two-dimensional software 2D CAD programs which can only function on the X (horizontal) and Y (vertical) axes, and not on the Z axis.

typical A term used to describe details or conditions that are identical in many parts of a drawing.

U

undercut To allow an overhanging edge or a cut with inwardly sloping sides.

unidirectional dimensioning All isometric letters, numerals, and arrowheads are positioned vertically. A variation of this is *vertical plane dimensioning,* where all dimensions are drawn in the isometric plane.

unilateral tolerance Tolerance permitted in only one direction from a dimension.

U.S. customary system Units based upon the inch and pound commonly used in the United States.

USMA United States Metric Association.

V

value designation The value of an electronic component expressed as volts, amperes, watts, or ohms.

vanishing point The point at which the receding lines of a perspective drawing meet.

vector A line which has value and direction and defined by its two extremes.

vellum Translucent drafting media.

vernier scale A movable scale attached to a fixed scale which contains subdivisions of the frxed scale.

vertex A point in the intersection of two or more sides.

W

warped surfaces Surfaces that are curved in two directions.

whole depth Full height of a gear tooth which is equal to the sum of the addendum and the dedendum.

wire frame Three-dimensional computer models in which all visible and hidden edges appear as lines.

Woodruff key Crescent-shaped flat key.

working depth Distance a gear tooth extends into a mating space; two times the addendum.

working drawings Drawings that provide all information needed to manufacture or construct a product.

workpiece A piece of material which has machining operations performed on it.

workstation A location where workers perform specific job tasks.

worm The driver of a worm gear that contains at least one complete tooth (endless screw) around the pitch surface.

worm gears Gears with teeth that curve to mesh with and be driven by a worm.

X

X axis The horizontal axis in a rectangular coordinate system.

Y

Y axis The vertical axis in a rectangular coordinate system.

Z

Z axis The axis in a three-dimensional cartesian coordinate system which is perpendicular to the X and Y axis.

zoning The legal restrictions on size, location, and type of structures to be built in a designated area.

Index

A

B

C

D

E

F

G

H

I

J

K

L

M

N

O

P

R

S

T